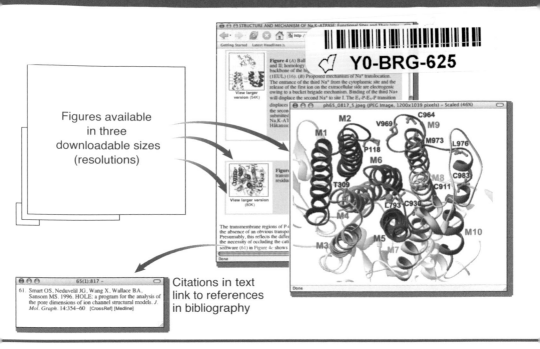

Figures available in three downloadable sizes (resolutions)

Citations in text link to references in bibliography

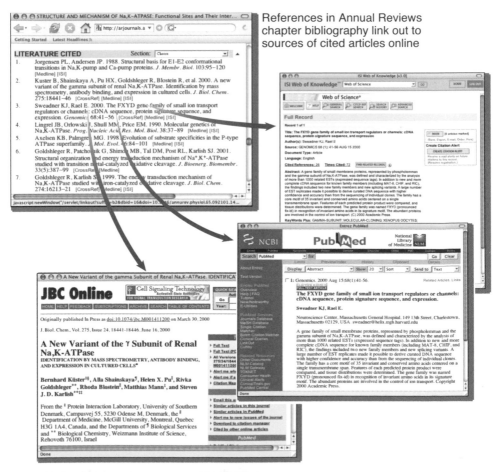

References in Annual Reviews chapter bibliography link out to sources of cited articles online

ANNUAL REVIEW
OF PHYSIOLOGY

ANNUAL REVIEW
OF PHYSIOLOGY

VOLUME 67, 2005

JOSEPH F. HOFFMAN, *Editor*
Yale University School of Medicine

DAVID L. GARBERS, *Associate Editor*
University of Texas, Southwestern Medical Center

www.annualreviews.org   science@annualreviews.org   650-493-4400

ANNUAL REVIEWS
4139 El Camino Way • P.O. Box 10139 • Palo Alto, California 94303-0139

ANNUAL REVIEWS
Palo Alto, California, USA

International Standard Serial Number: 0066-4278
International Standard Book Number: 0-8243-0367-9
Library of Congress Catalog Card Number: 39-15404

TYPESET BY TECHBOOKS, FAIRFAX, VA
PRINTED AND BOUND BY MALLOY INCORPORATED, ANN ARBOR, MI

PREFACE

This volume marks the end of my editorship of the *Annual Reviews of Physiology* (*ARPH*). The future of our series is in good hands because David Garbers now becomes the new editor. I look forward to seeing the expression of David's many talents and his broad understanding of our field in the structuring and composition of forthcoming volumes.

I have learned much during the expanse of my time with the *ARPH*, not only in witnessing the changing scenes of our field but also in the emergence of new applications and attendant insights in the functioning of living cells and organisms. For some time, the prime vectors have necessarily emphasized the application of the more reductive aspects of biological science. It is now evident that there is a transition, as well as a need, to integrate the study of function employing the newly developed tools and discoveries. I have remarked before in this space that there is concern in the physiology community not only about the changing status of departments and names, but also about the state of our science. With each volume of *ARPH*, as they have and will appear, I think it is clear that our science is healthy and shows remarkable promise. Our science has never seen more exciting times nor has it offered as many enticements.

I would like to emphasize that I have greatly benefited from working over the years with the many gifted section editors who have been mainly responsible for the success of our series. Their commitment to quality shows in the excellence of our product. I was first appointed to *ARPH* by Isadore (Izzy) Edelman. Izzy was responsible for sectionalizing *ARPH* and cycling the thematic coverage of the subject within each section. Sadly, Izzy passed away in November, 2004. He will certainly be remembered for he was an outstanding physiologist who made a host of substantive contributions to our field. I then served as Associate Editor under Robert Berne (deceased, October 2001) until I became Editor in 1988. Bob was a leading scientist who provided seminal insights into the action of adenosine in cardiovascular function. Both of these men were pivotal leaders in optimizing *ARPH* for their times.

I also want to stress that I have greatly benefited over the years from the help and advice of a former editor-in-chief of *Annual Reviews*, William Kaufmann. I am equally indebted, as well, to Samuel Gubins, the current President and Editor-in-Chief of *Annual Reviews*, for his wise counsel and support during my tenure. And we all acknowledge the dedication, scholarship, and diplomacy of our Production Editor, Sandra Cooperman, for her invaluable efforts involved with publishing each of our volumes.

This year our Perspectives chapter has been written by Michael Berridge on basic aspects of cell signaling. We also have a special chapter by George Somero and Raul Suarez in memory of Peter Hochachka for his work in comparative physiology. In addition, our Special Topic section, edited by Michael Pusch, is on chloride channels. Note should be taken that in addition to our usual sections, we have a section entitled Ecological, Evolutionary, and Comparative Physiology that formally replaces the previous section called Comparative Physiology. We continue to solicit your comments and suggestions and encourage you to contact us at www.annualreviews.org.

Joseph F. Hoffman
Editor

*Annual Review of Physiology*
Volume 67, 2005

CONTENTS

ERRATA

An online log of corrections to *Annual Review of Physiology* chapters
may be found at http://physiol.annualreviews.org/errata.shtml

Other Reviews of Interest to Physiologists

Annu. Rev. Physiol. 2005. 67:1–21
doi: 10.1146/annurev.physiol.67.040103.152647
Copyright © 2005 by Annual Reviews. All rights reserved
First published online as a Review in Advance on July 16, 2004

UNLOCKING THE SECRETS OF CELL SIGNALING

Michael J. Berridge

*The Babraham Institute, Babraham Research Campus, Babraham, Cambridge CB2 4AT,
United Kingdom; email: michael.berridge@bbsrc.ac.uk*

Key Words calcium, inositol trisphosphate, cyclic AMP, salivary glands

■ **Abstract** My scientific life has been spent trying to understand how cells communicate with each other. This interest in cell signaling began with studies on the control of fluid secretion by an insect salivary gland, and the subsequent quest led to the discovery of inositol trisphosphate (IP_3) and its role in calcium signaling, which effectively divided my scientific career into two distinct parts. The first part was primarily experimental and culminated in the discovery of IP_3, which set the agenda for the second half during which I have enjoyed exploring the many functions of this remarkably versatile signaling system. It has been particularly exciting to find out how this IP_3/Ca^{2+} signaling pathway has been adapted to control processes as diverse as fertilization, proliferation, cell contraction, secretion, and information processing in neuronal cells.

OUT OF AFRICA

My scientific career was shaped by my early life in Africa. I was born in 1938 in a small town called Gatooma (now called Kadoma) in what was then Southern Rhodesia. This country in central Africa has changed its name several times; it became Rhodesia and more recently Zimbabwe. My earliest memories are of Eiffel Flats, a small gold-mining town about five miles from Gatooma. It was the African bush with its rich variety of animals and birds that excited me as a young boy and has continued to fascinate me ever since. This love of nature inspired my initial interest in academic studies. While at school, I spent some of my spare time writing notes and drawing pictures of the animals that surrounded me, and this early passion evolved into a boyhood fantasy of becoming a game warden in one of the National Parks.

At school I was fortunate in being taught biology by Pamela Bates who not only fostered my scholarly interests but also opened my eyes to the fact that there was an academic life after school. None of my family had been to university except for a distant aunt who had a degree in child psychology. Because my parents considered her to be singularly inept at bringing up her own children, they were somewhat dubious about the merits of a university education. However, Miss Bates convinced them that I had the ability to pursue an academic career and I applied for a place at the newly founded University of Rhodesia and Nyasaland in Salisbury to read Zoology and Chemistry.

I approached my university courses with great enthusiasm because I had discovered that the conservation of African wildlife lay in the emerging field of ecology. I had every intention of going on to do research into big game ecology. Indeed, I spent one of my vacations working with Dr. Thane Riney, who was one of the first ecologists to work on African wildlife. One of my tasks was to mark giraffes using a crossbow, which had arrows with paint-filled wax capsules that exploded on contact with the animal thus spreading colored paint over the hide. We were then able to determine their home range by plotting out their locations on subsequent sightings. For a young boy fascinated by wild animals, this was exciting work. However, there was something very imprecise about this embryonic field and my enthusiasm for a career in ecology began to wane. A new influence was at work.

The physiology lectures given by Dr. Eina Bursell began to capture my imagination. I can still remember being particularly fascinated by the beauty of the staircase phenomenon in the heart responding to adrenergic stimulation. Little did I know that many years later I would understand the molecular signaling mechanisms responsible for this intricate process. Bursell's lectures brought out the factual details and precision that were lacking in my initial foray into ecology. After spending my next vacation helping Bursell with his research on water metabolism in tsetse flies, my transformation from a budding big game ecologist to insect physiologist was complete. I had been bitten by the research bug and was fortunate to gain a place in the Department of Zoology at Cambridge University to do a Ph.D. with Sir Vincent Wigglesworth, the father of insect physiology.

THE FIRST CAMBRIDGE PERIOD

My journey out of Africa was somewhat daunting. Within a short period, I was transported from the quiet colonial existence of Rhodesia into the hurly burly of cosmopolitan London and Cambridge. Despite all my reading, nothing could have prepared me for my new life in this ancient seat of learning. Wigglesworth was a fellow of Gonville and Caius College, and he arranged for me to be a member of this college. This was the college of William Harvey, who discovered the circulation of the blood. I felt all the years of academic distinction and tradition crowding in on me, especially as I was given a room in the heart of the college overlooking the Gate of Honor. A feeling of insecurity was exacerbated by the mischievous postdoctoral fellows in the laboratory who took great delight in informing me that Sir Vincent took on three students each year and one of them always fell by the wayside. Imagine my apprehension when I found out that the two other students in my year were Cambridge graduates who had their projects up and running within weeks, while I was still finding my way around the cavernous spaces of the Cambridge Zoology Department. The prediction about the natural wastage of Sir Vincent's students turned out to be correct because one of the students dropped out after a few months making me feel marginally more secure.

Sir Vincent had given me a project to study nitrogen excretion of the African cotton stainer *Dysdercus fasciatus*, which enabled me to use my training in both chemistry and zoology. Once I had settled into my project, I began to enjoy the process of doing research especially when I discovered something quite novel about how these insects excreted nitrogen. The big surprise was that they excreted their excess nitrogen as water-soluble allantoin rather than as the insoluble uric acid favored by most insects. It seems that these sap-sucking insects, which have access to a constant supply of water, have no need to waste energy synthesizing uric acid when they can flush out their nitrogen as allantoin in a plentiful flow of urine (1). I also discovered that this copious flow of urine was maximal at the beginning of each instar when the insects feed avidly and ceases as they prepare for the next moult. Toward the end of my thesis work I began to be interested in how this periodic cycle of fluid secretion by the Malpighian tubules was controlled, and this aspect sparked my interest in cell signaling that has continued ever since. Unfortunately, my time in Cambridge ran out, and I was not able to pursue my project any further because I had to stop doing experiments in order to write up my thesis. However, some of the preliminary work I had done on the hormonal control of excretion formed the basis of a successful postdoctoral application to Professor Dietrich Bodenstein, who was chairman of the Department of Biology at the University of Virginia in Charlottesville.

AN AMERICAN SOJOURN: POSTDOCTORAL STUDIES AT THE UNIVERSITY OF VIRGINIA AND AT CASE WESTERN RESERVE UNIVERSITY

During my time in Cambridge I fully intended to return to Africa, but politics intervened. While I had been studying in England, a vicious civil war had broken out in Rhodesia, and I would certainly have returned to immediate conscription to fight what I believed to be a completely unnecessary bush war. Instead, I decided to emigrate to the United States to begin a new life as a postdoctoral fellow at the University of Virginia in Charlottesville, which is a university town very similar to Cambridge. My intention was to use blowfly Malpighian tubules as a model system to study hormone action. While at Cambridge, I learned how to study Malpighian tubules in vitro from Dr. Arthur Ramsay who had spent his life developing techniques for studying the physiology of these tubular secretory organs (2). The isolated tubule was set up in a drop of saline held under liquid paraffin in a watchmaker's glass dish. A piece of silk was unraveled under liquid paraffin and individual fibers were used as fine ligatures to pull the cut end of the tubule out into the liquid paraffin. The urine that emerged from this cut end was then collected and its volume determined to provide a measure of the rate of fluid secretion. I thus had a simple model system to study hormone action because the effects of adding agents to the bathing medium on the rate of fluid secretion could be easily measured. However, there was a problem in that the tubules did not survive for long in vitro. I spent most of my time in Virginia unsuccessfully trying to develop a culture medium that would promote longer survival.

I continued with this problem of Malpighian tubule survival when I moved to Case Western Reserve University to work first in Michael Locke's laboratory in the Developmental Biology Center and then with Bodil Schmidt-Nielsen in the Department of Biology. As I was making little progress with trying to design media that would support Malpighian tubules, I began to get increasingly desperate, when my luck suddenly changed. One day while dissecting out yet another Malpighian tubule to test out yet another culture medium, I noticed this long clear tube lying alongside the yellow-white Malpighian tubules. Out of curiosity, I dissected it out of the fly and set it up as for the Malpighian tubules and was astonished by the result. This tiny little tubule, considerably smaller than the Malpighian tubules, was secreting at rates 50 to 100 times those I had been recording with the tubules. I soon found out that these tubes were the salivary glands that extend down the length of the fly and secrete a voluminous flow of saliva each time the fly settles down to feed. I calculated that at these rates of secretion the fly would pump itself dry within a short period. This implied that the gland is normally quiescent and is called into action only during feeding. But the glands I had isolated continued to secrete at high rates for long periods, and this could only mean that there was something in my complex medium that was a potent activator of secretion. I soon found out that the stimulant was 5-hydroxytryptamine (5-HT) (3), a contaminant in fetal bovine albumin fraction V that presumably came from the blood platelets during preparation of the albumin fraction. In the absence of 5-HT, the glands were totally quiescent, but upon addition of this agonist they began to secrete rapidly. My foray into the world of cell signaling had taken a giant step forward; I now had a defined agonist capable of switching the gland on and off. My luck was holding because my research was soon to take another major step forward.

An Introduction to the Second Messenger Concept—Cyclic AMP Comes into View

As I made progress developing my model system to study cell signaling, I became aware of the work by Earl Sutherland and Ted Rall on cyclic adenosine monophosphate (cyclic AMP) and their novel second messenger concept (4). I was very excited about the idea that cyclic AMP acted as a second messenger to mediate the action of the first messenger that arrived at the cell surface; in the case of my salivary gland experiments this was 5-HT. Imagine my excitement when I found that the stimulatory effect of 5-HT on fluid secretion was exactly duplicated by the addition of cyclic AMP (3, 5). I had the first steps of an embryonic signaling pathway (Figure 1).

When discussing my latest result with colleagues in the laboratory, I was amazed to find out that Sutherland and Rall had discovered cyclic AMP in the Department of Pharmacology just across the road. I plucked up enough courage to go across to meet them. Sutherland had left the department but Rall was still there and had

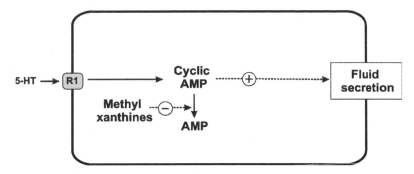

Figure 1   Model 1 (1968). The first model of the signaling pathway used by 5-HT to activate fluid secretion in the insect salivary gland.

become a key player in the field. He was very influential in driving forward the concept of cyclic AMP functioning as a second messenger in cell signaling. The old cliché of being in the right place at the right time was never more apt because my work was helped enormously by frequent pharmacology tutorials from Rall. He patiently taught me about the action of phosphodieterase inhibitors and the concept of synergism, which helped me to establish quickly that cyclic AMP was a key intracellular messenger in the insect salivary gland where it carried out the stimulatory action of 5-HT. Rall took a great interest in this work because it was one of the first examples of cyclic AMP functioning in an invertebrate. My postdoctoral sojourn in America, which for a long time seemed to be drifting rather aimlessly, was ending on a high note. However, it was now time to leave my cozy postdoctoral life to find a more permanent position.

I began looking for a job and went for interviews at Rice University in Houston and various other places. Decision time was fast approaching when a letter arrived from John Treherne in Cambridge offering me a position in a new Unit of Invertebrate Chemistry and Physiology he was setting up in the Zoology department. This position seemed just right for me so in 1969 I crossed back over the Atlantic to begin a second period of research in Cambridge.

RETURN TO THE ZOOLOGY DEPARTMENT IN CAMBRIDGE

This period in Cambridge started off very differently from the first. As I was familiar with the quaint ways of the Zoology department, I soon had my research program in full swing. The work I had done in Cleveland on cyclic AMP provided a robust working hypothesis to guide my new research program. The next obvious step was to determine how cyclic AMP carried out its second messenger role in stimulating fluid secretion. I approached this problem by characterizing the electrophysiological properties of the secretory response in the hope that it might tell me something about how cyclic AMP acted on the ionic mechanisms responsible for fluid secretion. My research student William Prince, who was a talented electrophysiologist, helped to get this project started. We designed a small Perspex perfusion chamber that enabled us to monitor the *trans*-epithelial potential while adding and removing components of the putative signaling pathway (Figure 1) such as 5-HT and cyclic AMP or pharmacological agents such as the methyl xanthines (theophylline and caffeine). When the glands were at rest, the lumen had a slightly positive potential relative to the bathing medium and depolarized rapidly in response to 5-HT. Fully expecting to observe the same response following the addition of cyclic AMP, we were greatly surprised to find exactly the opposite. Instead of depolarizing, the potential was found to hyperpolarize (6). Subsequent experiments revealed that cyclic AMP activated an electrogenic potassium pump, which was the prime mover for fluid secretion. Chloride followed passively, and this transport of KCl created the osmotic gradient to drive the flow of water. Subsequent resistance measurements showed that the passive flux of chloride was facilitated by a 5-HT-dependent increase in chloride conductance that occurred independently of cyclic AMP (7). It seemed that 5-HT was having two actions: one mediated by cyclic AMP to drive potassium transport and a second mechanism facilitating the movement of chloride.

UNCOVERING A ROLE FOR CALCIUM

It soon became apparent that Ca^{2+} might regulate the passive flux of chloride. The notion that Ca^{2+} might function as a second messenger was already well established in muscle cells and had begun to be investigated in secretory cells by Bill Douglas prior to the discovery of cyclic AMP (8). However, the notion that Ca^{2+} might be a more universal intracellular messenger was swept aside by the excitement surrounding cyclic AMP, which was rapidly promoted as a universal second messenger capable of regulating almost every cellular process imaginable. It gradually became apparent, however, that cyclic AMP was not the only messenger operating in cells and Howard Rasmussen at Yale was very much at the forefront of reintroducing the concept that Ca^{2+} could also function as a second messenger (9), an idea that gathered pace in the 1970s. It so happened that in 1971, Rasmussen was on a sabbatical in the Department of Pharmacology in Cambridge. Once again, I was very fortunate to benefit from the advice of an expert in this emerging field of Ca^{2+} signaling. Indeed, we set up a collaboration to study Ca^{2+} signaling in the insect gland, which clearly showed that 5-HT was having a profound effect on Ca^{2+} dynamics. In particular, it suggested that Ca^{2+} was being released from an internal store (10). We subsequently used both electrophysiological and pharmacological techniques to demonstrate that the blowfly salivary gland had two 5-HT receptors operating through separate second messengers (Figure 2) (11–13). The signaling system was becoming a lot more complicated than originally envisaged in that the salivary gland was using two independent systems to regulate secretion (Figure 2). One receptor used cyclic AMP to drive potassium transport, whereas the other employed Ca^{2+} to open chloride channels.

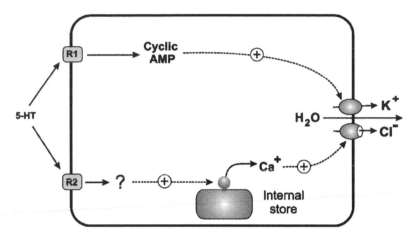

Figure 2   Model 2 (1973). 5-Hydroxytryptamine (5-HT) operates through two signaling pathways. One pathway uses cyclic AMP to activate a potassium pump, whereas the other releases Ca^{2+} from an internal store to activate chloride channels. The big question at the time was how the receptor on the cell surface gained access to the internal store.

Searching for a Ca^{2+}-Mobilizing Second Messenger

The most puzzling aspect of the putative Ca^{2+} messenger system in the blowfly was the source of Ca^{2+}, much of which was derived from an internal store (11). Similar observations were also being made on various mammalian cells so the hunt was on to find the messenger that connected cell surface receptors to the internal store. The question was easy to define but its solution seemed totally intractable. The aspect that frustrated me most was not having any idea of how to approach the problem. Finally, a way forward began to emerge from a somewhat unlikely direction when I became aware of the work by Hokin & Hokin on inositol lipids (14). In 1953, the Hokins discovered that an external agonist stimulated the turnover of phosphatidylinositol (PI) (14). This came to be known as the PI response. It was shown subsequently that the agonist was stimulating the hydrolysis of PI. Although this PI response had been measured in many different cells in response to many different stimulants, its function remained somewhat mysterious. This began to change in 1975 when Bob Michell wrote an extraordinarily insightful and scholarly review where he argued that lipid hydrolysis was responsible for Ca^{2+} signaling (15). A major part of his argument centered on pharmacological observations indicating that certain receptor types used cyclic AMP as a messenger, whereas others seemed to work through Ca^{2+}, and it was the latter group of receptors that also gave a PI response. Because I had shown that the salivary gland was under dual regulation of 5-HT receptors coupled to either cyclic AMP or Ca^{2+} (Figure 2), I was intrigued by Michell's idea that the PI response might be linked to a Ca^{2+}-signaling pathway. My research suddenly lurched from physiology to biochemistry.

Not being a biochemist, I was somewhat daunted by the prospect of starting to work on inositol lipid metabolism, but the transition was greatly helped along its way by John Fain who joined my laboratory as a sabbatical visitor in 1978. Most previous work in this field had used the incorporation of ^{32}P into PI as a way of measuring the PI response. This method necessitated the extraction and separation of the labeled lipids. The small size of the insect gland made this technique well nigh impossible so we turned our attention to using ^3H-inositol. We found that this label was rapidly incorporated into PI and, more significantly, the label could be released from the lipid during agonist stimulation of intact glands and could be collected in the bathing medium. This meant that we could begin to monitor the PI response at regular time intervals in intact glands simply by measuring the efflux of ^3H-inositol (16). The efflux was very small when the glands were at rest, but upon addition of 5-HT there was a dose-dependent increase in the efflux of ^3H-inositol.

In order to relate this hydrolysis of PI to Ca^{2+} signaling, it was necessary to have some way of measuring the latter. At that time, however, there were no established methods for measuring intracellular Ca^{2+}. I began to dabble with Ca^{2+}-sensitive microelectrodes, but these proved to be difficult to prepare and were decidedly unreliable (17). I thus searched for an alternative, and in a separate series of experiments carried out with Herb Lipke, another visitor to the laboratory, I devised a method of measuring Ca^{2+} entry into the insect gland by monitoring the transepithelial flux of ^{45}Ca (18), which proved invaluable as a way to study Ca^{2+} signaling. As I now had methods for monitoring both Ca^{2+} signaling and inositol lipid hydrolysis in intact cells, I could begin to see whether these two processes were related. The first thing we did was to compare their sensitivities to 5-HT. The dose response curve for inositol efflux lay to the left of that for Ca^{2+} flux, which in turn lay to the left of the secretory response. This sensitivity sequence was entirely consistent with Michell's notion (15) that the PI response generated the Ca^{2+} signal responsible for cell stimulation (Figure 3).

Because there had been many false leads regarding the function of the PI response, I felt we needed more direct evidence that it was playing such a central role in the signaling sequence. The fact that inositol was leaking out of the glands (Figure 3), particularly when they were being stimulated, suggested a way of testing the hypothesis more directly. At this stage the hypothesis was based on the assumption that 5-HT was stimulating the hydrolysis of PI to produce diacylglycerol (DAG) and inositol-1-phosphate (IP$_1$), (Figure 3). The latter was then hydrolyzed to inositol (Figure 3), which was then recycled back into PI in order to maintain the membrane level of this precursor lipid. Because our experiments revealed that a proportion of this inositol was being lost from the cell, we reasoned that the cell would begin to run out of it, thus compromising Ca^{2+} signaling owing to a decline in the level of PI. This is exactly what happened. When glands were stimulated for 2 h with a high dose of 5-HT and washed repeatedly to remove the inositol that escaped into the medium (thus reducing the resynthesis of PI), there was a dramatic desensitization of Ca^{2+} signaling (19). What was

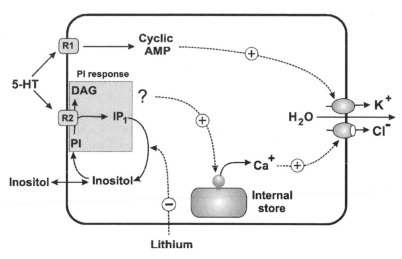

Figure 3 Model 3 (1979). 5-HT has two actions. One is to act through Receptor 1 (R1) to produce cyclic AMP. The other is to stimulate Receptor 2 (R2) to hydrolyze phosphatidylinositol (PI) to produce inositol-1-phosphate (IP$_1$) and diacylglycerol (DAG). Lithium inhibits the enzyme that hydrolyzes IP$_1$ to inositol, which is either incorporated back into PI or lost from the cell. This PI response was proposed to play a role in releasing Ca^{2+} from the internal store through an unknown mechanism.

particularly pleasing was the fact that Ca^{2+} signaling could be rescued by incubating the glands in inositol thus enabling them to reconstitute the level of the PI required for signaling (20). It was this experiment that convinced me that the PI response was directly linked to Ca^{2+} signaling, exactly as proposed by Michell in 1975 (15). But what was the mechanism? The answer finally came through a somewhat circuitous route that began with some studies on lithium (Li$^+$).

LITHIUM PROVIDES AN ENTRÉE TO THE INOSITOL PHOSPHATES

In attempting to find an inroad into the relationship between the PI response and Ca^{2+} signaling, I stopped doing experiments and began an exhaustive search of the large and at times somewhat tedious PI literature in the hope of finding a new way forward. During the course of this reading, I came across some interesting work on Li$^+$ and inositol phosphate metabolism that had been largely ignored (21, 22). Allison and his colleagues found that Li$^+$ was a potent inhibitor of the inositol monophosphatase that hydrolyzed IP$_1$ to inositol (Figure 3). I was immediately struck by the potential significance of this observation because it was this Li$^+$-sensitive step that released the free inositol that we were measuring to monitor PI hydrolysis. This inhibitory action of Li$^+$ has some very unusual properties, which may explain its therapeutic action in controlling manic-depressive illness and may also have a bearing on its teratogenic action of distorting axis formation in early development (23) that I discuss further below. The fact that Li^+ was able to inhibit a putative signaling pathway immediately struck a chord because it reminded me of my earlier studies on cyclic AMP, where the methyl xanthine inhibitors of its metabolism had proved so useful in determining its second messenger function. Therefore, I began to examine whether Li^+ might prove equally valuable for understanding more about the PI response. The inositol efflux method we had used to monitor the PI response offered a simple way of testing the inhibitory effect of Li^+. Because this inositol was derived from the IP_1 produced by hydrolyzing PI, Li^+ should reduce the efflux by inhibiting the hydrolysis of IP_1 to inositol (Figure 3). That is exactly what happened. Li^+ induced a rapid reduction in the efflux of ^3H-inositol from prelabeled glands (24). What came as a surprise, however, was the very large overshoot of inositol efflux that occurred when Li^+ was withdrawn. This large surge in free inositol that flooded into the bathing medium resulted from the inositol monophosphatase suddenly hydrolyzing the IP_1 that had built up behind the Li^+ block. The fact that Li^+ resulted in a large accumulation of inositol phosphates provided an entrée into the next phase of my research on inositol phosphate metabolism.

Inositol Trisphosphate Comes into View

As a physiologist, I had always tried to avoid biochemical techniques that involved destroying the cell. The inositol efflux method described earlier had proved very effective in enabling the PI response to be monitored continuously in the intact cell. However, in order to study the inositol phosphates that were locked up within the cell, there was no escape from breaking the cell open to measure these PI metabolites after separating them out on anion-exchange columns (25). As expected, there was a large accumulation of IP_1 during Li^+ inhibition, but unexpectedly there were two additional peaks running after the IP_1. On the basis of previous observations, these two peaks were likely to be inositol 1,4-bisphosphate (IP_2) and inositol 1,4,5-trisphosphate (IP_3). The whole story was about to get a lot more complex and difficult, especially because there were no commercially available inositol phosphates to use as standards to verify the preliminary identification of the unknown peaks.

Once again I was about to benefit from being in the right place. As far as I can determine, at the time we were doing these studies there were only two sources of inositol phosphates that could be used as standards. One source was in Clint Ballou's laboratory in Berkeley, California and the other was a few miles down the road from me in Rex Dawson's laboratory at the Babraham Institute. Dawson and his colleague Robin Irvine were busily working on various aspects of the PI response and in particular identifying the enzymes that hydrolyzed inositol lipids. Just as Rall and Rasmussen had proved invaluable in guiding me along the cyclic AMP and Ca^{2+} paths respectively, Dawson and Irvine were very patient

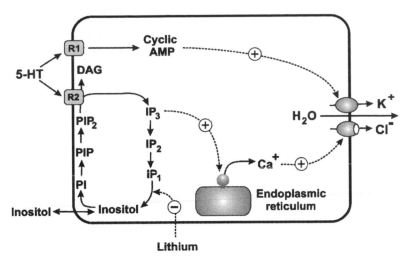

Figure 4   Model 4 (1983). The role of inositol 1,4,5-trisphosphate (IP$_3$) as a Ca^{2+}-mobilizing second messenger. 5-HT acts on either Receptor 1 (R1) to produce cyclic AMP or on Receptor 2 (R2) to hydrolze phosphatidylinositol 4,5-bisphosphate (PIP$_2$) to diacylglycerol (DAG) and IP$_3$. The latter diffuses into the cell to release Ca^{2+} from the endoplasmic reticulum.

in helping me come to grips with the rather arcane world of inositol phosphate biochemistry. More importantly, tucked away in their freezer was a wide range of inositol phosphates that they kindly supplied as standards to verify that the unknown peaks coming off my exchange columns were indeed IP$_2$ and IP$_3$ (25). On the basis of what was known previously, the simplest hypothesis was that IP$_3$ was formed first and was sequentially hydrolyzed to IP$_2$, IP$_1$, and then inositol (Figure 4).

For the sequence shown in Figure 4 to be correct, the substrate used by R2 had to be phosphatidylinositol 4,5-bisphosphate (PIP$_2$) and not the parent lipid PI as originally assumed in the earlier model (Figure 3). Indeed, work being done in other laboratories had already shown that agonists were able to hydrolyze PIP$_2$ (26, 27). In order to obtain further evidence for the IP$_3$ → IP$_2$ → IP$_1$ → inositol sequence, I developed a rapid perfusion system (28) that enabled me to establish the rate at which these inositol phosphates were produced following 5-HT stimulation. These studies on the kinetics of 5-HT-induced inositol phosphate formation revealed that IP$_3$ and IP$_2$ increased first with no apparent latency, whereas IP$_1$ followed by inositol appeared much later. I was excited by this result because it not only confirmed that PIP$_2$ was being hydrolyzed but also revealed that IP$_3$ was being generated quickly. In fact, it was the fastest response to 5-HT I had yet recorded. I already knew that there was a 1–2 s latency before the onset of the electrical change caused by the Ca^{2+}-dependent activation of the chloride channels. I now found the level of IP$_3$ was already elevated by 1 s and, within the limits of my recording system, there appeared to be no latency, which suggested that IP_3 was being generated as soon as 5-HT bound R2. It was this observation that led me to propose that IP_3 might be the diffusible messenger that coupled receptor activation to the mobilization of internal Ca^{2+} (29) (Figure 4).

IP_3-Induced Ca^{2+} Mobilization

It is one thing to suggest a mechanism but quite another to prove it. I already knew from my electrophysiological experiments that Ca^{2+} was coming from an internal store; the problem was to find a way of testing whether IP_3 could release Ca^{2+} from some as yet unknown internal store. The experimental techniques to study the action of intracellular messengers are now relatively easy, but 20 years ago such study was much more difficult. Not only was the supply of IP_3 very limited, but also cellular injection techniques were poorly developed. While grappling with this problem, I attended a workshop in Amsterdam in December 1982 on Biophysical and Biochemical Aspects of Transcellular Transport in Animal Tissues. The title of my lecture was "Phosphatidylinositol Hydrolysis: A General Transducing Mechanism for Calcium-Mobilizing Receptors," and it was the first time that I had publicly put forward the idea that IP_3 might be the long sought after Ca^{2+}-mobilizing second messenger. In the same session, Irene Schulz described a permeabilized pancreatic cell preparation in which she was able to gain access to the internal Ca^{2+} stores. This was exactly what I was looking for to test out the proposed function of IP_3, and she was more than happy to set up a collaborative study. On returning to Cambridge, I contacted Irvine who agreed to prepare the large amounts of IP_3 that were required for these permeabilized cell experiments. A few weeks after Schulz had received the samples, I had an excited phone call to say that IP_3 had mobilized Ca^{2+}. We carried out numerous controls to establish the specificity of IP_3-induced Ca^{2+} mobilization, and we also identified the endoplasmic reticulum as its site of action. The paper summarizing our results was published in *Nature* toward the end of 1983 (30).

The initial evidence that IP_3 was a Ca^{2+}-mobilizing messenger emerged from our work on the permeabilized pancreatic acinar cells, and we were anxious to find out whether it had a similar action in other cell types. Once our paper (30) appeared in *Nature*, we were contacted by several groups anxious to try out this new messenger on their cells. We set up a number of collaborations, and within a short period of time we were able to confirm that IP_3 released Ca^{2+} in many cell types including liver cells (31), Swiss 3T3 cells (32), insulin-secreting cells (33, 34), *Limulus* photoreceptors (35, 36), and leukocytes (37). It seemed clear, therefore, that IP_3 was the long sought after Ca^{2+}-mobilizing second messenger.

It had taken 30 years since the discovery of the PI response by Hokin & Hokin in 1953 (14) to finally find out that one of its functions is to stimulate the release of internal Ca^{2+}. However, that is not its only function. Indeed, a few years earlier Yasutomi Nishizuka had already shown that DAG, the other product of the

PI response (Figure 4), also had a signaling role, i.e., it was a potent activator of a novel protein kinase (38, 39). For those of us involved during that period, one of the really exciting developments was the realization that the PI response was spawning two separate signaling cascades: one activated through DAG to stimulate protein kinase C and the other using IP$_3$ to stimulate the mobilization of Ca^{2+} (39–41):

$$\text{agonist} \rightarrow \text{receptor} \rightarrow \text{PLC} \rightarrow \text{PIP}_2 \text{ hydrolysis} \begin{array}{c} \nearrow \text{IP}_3 \rightarrow \text{Ca}^{2+} \\ \\ \searrow \\ \text{DAG} \rightarrow \text{PKC} \end{array}$$

THE VERSATILITY AND UNIVERSALITY OF Ca^{2+} SIGNALING

The discovery that IP$_3$ was a universal second messenger controlling Ca^{2+} signaling heralded a new phase in my research career as it provided fresh insights into how a variety of cellular processes were regulated. Having spent so much time doing biochemistry, I was anxious to return to physiology, which I did by turning my attention to the spatial and temporal aspects of Ca^{2+} signaling. One of the questions I am asked most frequently is how can this messenger regulate so many different processes, in some instances within the same cell. For example, in smooth muscle cells, Ca^{2+} can regulate both contraction and relaxation. Likewise, in neurons, Ca^{2+} can induce both long-term potentiation (LTP) and long-term depression (LTD) in the same synaptic connection. How is it that Ca^{2+} can separate out these discrete and sometimes opposing signaling functions? We now know that much of this remarkable versatility depends on the way this signal is organized in both time and space (42–45). My interest in these spatiotemporal aspects began with studies on Ca^{2+} oscillations, and when single cell imaging techniques began to appear, attention was also focused on the spatial aspects of Ca^{2+} signaling.

Ca^{2+} Oscillations

Early on in our studies on the electrical responses of the insect salivary gland to 5-HT, we discovered that the transepithelial potential oscillated (6). An interesting feature of these oscillations is that their frequency varies with agonist concentration (6, 46). Because these changes in frequency occurred over the same range of 5-HT concentrations that caused changes in fluid secretion, we proposed that the signaling system operated through frequency modulation (FM) rather than amplitude modulation (AM) (47). These potential oscillations depended on changes of resistance, which were controlled by Ca^{2+}, thus we speculated that they resulted from an underlying oscillation of Ca^{2+} (48). When aequorin was introduced to monitor intracellular Ca^{2+}, particularly striking examples of oscillations were revealed in both liver cells (49) and mammalian oocytes (50–52). Frequency modulation was particularly apparent for the liver cell oscillator. What fascinated me most about such oscillatory activity was its spontaneous nature. In some cases, the level of Ca^{2+} was found to remain constant for up to 60 s and was then interrupted by a spontaneous Ca^{2+} spike that returned back to the resting level for another 60 s before the next spike occurred and so on. In the case of mammalian oocytes undergoing fertilization, these interspike intervals can last for 3–4 min (50–52). The question that intrigued me and still does today is what happens during this interspike interval that results in the appearance of the spontaneous spikes? I began to search for a mechanism.

When we first became aware of such Ca^{2+} oscillations, we had little information about how intracellular Ca^{2+} was regulated. This, however, did not deter Paul Rapp and me from putting forward a model that attempted to explain these oscillations on the basis of feedback loops operating between Ca^{2+} and cyclic AMP (48). However, such models were not particularly plausible because they failed to explain adequately how agonists functioned to increase the intracellular level of Ca^{2+}. This deficiency was resolved once IP_3 had been discovered, and we were then able to develop more refined models to account for both Ca^{2+} oscillations and how frequency might be decoded through protein phosphorylation (53, 54).

An important aspect of these new models was the process of Ca^{2+}-induced Ca^{2+} release (CICR), which provided the positive feedback to account for the rapid regenerative component of each Ca^{2+} transient. The ryanodine receptors of muscle cells were already known to display this process of CICR as first described in 1970 by Makoto Endo (55), and it was soon shown that IP_3 receptors were equally sensitive to Ca^{2+} (56–58). The process of CICR, therefore, applies just as well to IP_3 receptors as it does to ryanodine receptors. Although the IP_3-dependent process of CICR nicely explained the regenerative process responsible for the upstroke of the Ca^{2+} spike, it did not address the critical question of what triggered this spike in the first place. Again, an important clue came from work on muscle cells where a spontaneous release of Ca^{2+} often occurred when the internal store became overloaded with Ca^{2+}. Because ryanodine receptors are sensitive to the level of Ca^{2+} within the lumen of the endoplasmic reticulum, I wondered whether IP_3 receptors might display a similar sensitivity to store loading. Ludwig Missiaen spent his postdoctoral period in Cambridge patiently devising methods of loading up such stores and found that a spontaneous release occurred when the stores were loaded above a critical point (59). It seems that the IP_3 receptors, like their ryanodine receptor counterparts, are sensitive to the level of Ca^{2+} within the lumen of the endoplasmic reticulum. Although this remains a controversial area, I believe that the loading of the internal store plays a critical role in the timing mechanism of Ca^{2+} oscillations because it sets the sensitivity of the IP_3 receptors and thus determines when the next Ca^{2+} spike will initiate (42, 60). The frequency of Ca^{2+} oscillations will thus depend upon how quickly the store can be loaded and this in turn will depend upon the rate at which it enters across the plasma membrane. Such dependence on external Ca^{2+} may explain why oscillations are so sensitive to variations in the external concentration of Ca^{2+}. It may also explain how frequency can be regulated if one assumes that the rate of Ca^{2+} entry is controlled by agonist concentration. The importance of Ca^{2+} entry processes has been established but the nature of these processes remains as one of the major unsolved problems in Ca^{2+} signaling.

Spatial Aspects of Ca^{2+} Signaling

In 1986 I visited Roger Tsien at Berkeley and was shown his new purpose-built, single cell imaging system; immediately I realized the significance of this technology. I obtained the funds to set up such a system in Cambridge and must acknowledge the help that Tsien provided in guiding us into this new technology. Within a short time we had our own system running and began to see how Ca^{2+} signals developed in cells in response to external signals. Some of the first experiments were performed by Tim Cheek who showed that when chromaffin cells were depolarized, the calcium appeared first as a ring confined to the cytoplasm immediately below the plasma membrane (61). Like many other groups, we were also recording regenerative calcium waves that usually initiated in a discrete region of the cell and then propagated through the cytoplasm. A particularly dramatic example of such waves is found in mammalian oocytes following fertilization (51, 52). These waves reflect the spatial organization of the calcium spikes described above. We already knew that a process of CICR caused such spikes, and the next big step was to "see" the elementary events responsible for these calcium waves.

The first indication that one might be able to visualize the activity of the individual building blocks of such calcium signals emerged from work on cardiac cells (62, 63), which revealed the existence of elementary events that were called sparks. These sparks are small bursts of calcium released from a localized group of ryanodine receptors. They occur at the junctional zones and produce the calcium that diffuses onto the myofibrils to activate contraction. Cardiac studies on the microscopic activity of ryanodine receptors established the idea that the global calcium signals we had been studying previously could be broken down into elementary events, and this finding greatly increased our understanding of how calcium signals are constructed (43, 44). By pushing our aging imaging system to its limits, Martin Bootman began to get the first indication that the IP_3/Ca^{2+} signaling system might be similarly organized into elementary events (64). The elementary events produced by IP_3 receptors have been called puffs (65). Bootman then teamed up with Peter Lipp to characterize these puffs and to show how they are recruited to produce global calcium signals in HeLa cells (66, 67). For someone who has been studying covert Ca^{2+} signaling for so long, it was and still is a quasi-religious experience to visualize the generation of Ca^{2+} signals in cells. The spiral waves that pulsate through the cytoplasm of *Xenopus* oocytes are objects of great beauty (68). We have come a long way since our initial studies on the insect salivary gland, which provided the first indication that Ca^{2+} might be oscillating, to the detailed molecular and spatial description we have now. We are still in the midst of exploring the enormous panoply of elementary calcium signaling events that can be manipulated in many different ways to create the versatility that characterizes Ca^{2+} signaling.

THE ROLE OF IP_3/Ca^{2+} SIGNALING IN CELLULAR CONTROL MECHANISMS

Much of my scientific attention at present is focused on trying to understand how the IP_3/Ca^{2+} signaling pathway functions in cellular control mechanisms. One of my first attempts at this was to see whether the new insights we had into the mode of action of Li^+ (24) might help explain its therapeutic action in controlling manic-depressive illness or its teratogenic action during early development. By inhibiting the conversion of inositol phosphate to free inositol with Li^+ (Figure 3), we caused the level of inositol to decline, which led to the inositol depletion hypothesis to explain Li^+'s therapeutic action in manic-depressive illness (23). Some experimental support for this inositol depletion hypothesis has begun to appear (69). The basic idea is that this illness might be caused by overactive inositol lipid signaling, which could be brought back to normal by Li^+ acting to slow down the recycling of inositol. This idea was made all the more plausible because Li^+ exerts its inhibitory action through an unusual uncompetitive mechanism. The Li^+-binding site on the inositol monophosphatase appears only when the enzyme has bound its inositol phosphate substrate. This means that Li^+ functions as a homeostatic drug, i.e., it has little effect if the signaling system is operating normally but begins to exert an ever-increasing inhibitory action as inositol lipid signaling becomes more and more abnormal. I went on to apply this inositol depletion hypothesis to the teratogenic action of Li^+, particularly with regard to its ability to dorsalize *Xenopus* embryos (23). By extending the same idea that Li^+ is most effective when inositol lipid turnover is high, I proposed that dorso-ventral specification might be set up by a gradient of IP_3/Ca^{2+} signaling, with the highest level in the ventral region. In the presence of Li^+ this high level of IP_3/Ca^{2+} signaling in the ventral region would be suppressed down to that seen in the dorsal region, which would account for the dorsalization of the developing embryo (23). There now is growing evidence that IP_3/Ca^{2+} signaling does indeed play such a role in dorso-ventral specification (70, 71).

The fact that Li^+ might be acting to curb inositol lipid signaling raised the real possibility that the IP_3/Ca^{2+} signaling pathway might play a prominent role in the brain. It had long been known that the receptors and enzymes responsible for inositol lipid signaling were strongly expressed in neurons, but their exact function was unknown. With the discovery that IP_3 was a calcium-mobilizing second messenger (30, 40, 41) and that its receptors were located on the endoplasmic reticulum within the spines of Purkinje neurons (72), it became apparent that this signaling pathway had a role in information processing within the brain (73). In order to discern what that function might be, I began to take an interest in the neurobiology of Purkinje neurons and was immediately struck by the fact that the metabotropic receptors responsible for generating IP$_3$ were of central importance in the process of LTD, which is responsible for the motor learning skills that enable us to learn how to ride a bicycle or play the piano. Like many learning mechanisms, LTD occurs when the Purkinje neuron detects the near simultaneous arrival of two synaptic inputs, one coming from the climbing fibers that innervate the base of the dendritic tree and the other arriving through the parallel fibers that innervate the spines. The problem was to identify the molecular coincident detector that enabled the Purkinje cell to associate the information arriving from these two separate inputs. The more I read about the system, the more I realized that the IP$_3$ receptor might be that coincident detector (74). As noted above, opening of the IP$_3$ receptor depends upon the simultaneous presence of both IP$_3$ and Ca^{2+} (56–58), and these are exactly the two messengers being supplied by the two inputs to the Purkinje neurons. The climbing fiber results in depolarization of the dendrites, which then generates a Ca^{2+} signal, whereas the parallel fibers activate the metabotropic receptors to generate IP$_3$. When these two messengers are produced at approximately the same time, there is a large release of Ca^{2+} within the spine, which induces LTD. Although there are contenders for coincident detectors in other neurons, there has been considerable experimental evidence to support the notion that the IP$_3$/Ca^{2+} signaling pathway may be responsible for LTD in Purkinje neurons (75, 76).

What is particularly intriguing about these molecular events is that their temporal characteristics closely match what is known about the timing rule for various forms of motor learning. For example, the most effective learning occurs when the stimulation of the parallel fibers precedes that of the climbing fibers. Likewise, the most effective release of Ca^{2+} by the IP$_3$ receptors occurs when the parallel fibers that generate IP$_3$ are activated 50–200 ms before the climbing fibers that deliver the pulse of Ca^{2+} (75). It is these subtleties that fascinate me as we continue to probe the many functions of the IP$_3$/Ca^{2+} signaling pathway in information processing in the brain.

CONCLUSION

I have been privileged to live through a period when many of the major signaling pathways used for cell communication were being discovered. I was caught up in the excitement of the discovery of cyclic AMP, which introduced the concept of second messengers, and this prepared me for my work on the role of IP$_3$ as a second messenger linking inositol lipid hydrolysis to Ca^{2+} signaling. There has been much excitement along the way. In addition to the real buzz of making new discoveries in the laboratory, I have also enjoyed immensely the intellectual challenge of trying to use all this new information to understand how specific cell functions as divergent as fertilization, cell proliferation, muscle contraction, and synaptic plasticity are controlled.

The *Annual Review of Physiology* is online at http://physiol.annualreviews.org

LITERATURE CITED

1. Berridge MJ. 1965. The physiology of excretion in the cotton stainer *Dysdercus fasciatus* Signoret. III. Nitrogen excretion and ionic regulation. *J. Exp. Biol.* 43:535–52
2. Ramsay JA. 1954. Active transport of water by the Malpighian tubules of the stick insect, *Dixippus morosus* (Orthoptera, Phasmidae). *J. Exp. Biol.* 31:104–13
3. Berridge MJ, Patel NG. 1968. Insect salivary glands: stimulation of fluid secretion by 5-hydroxytryptamine and cyclic AMP. *Science* 162:462–63
4. Sutherland EW, Rall TW. 1958. Fractionation and characterization of a cyclic adenine ribonucleotide formed by tissue particles. *J. Biol. Chem.* 232:1077–91
5. Berridge MJ. 1970. The role of 5-hydroxytryptamine and cyclic 3',5'-adenosine monophosphate in the control of fluid secretion by isolated salivary glands. *J. Exp. Biol.* 53:171–86
6. Berridge MJ, Prince WT. 1972. Transepithelial potential changes during stimulation of isolated salivary glands with 5-hydroxytryptamine and cyclic AMP. *J. Exp. Biol.* 56:139–53
7. Berridge MJ, Lindley BD, Prince WT. 1976. Studies on the mechanism of fluid secretion by isolated salivary glands of *Calliphora. J. Exp. Biol.* 64:311–22
8. Douglas WW, Poisner AM. 1963. The influence of calcium on the secretory response of the submaxillary gland to acetylcholine or to noradrenaline. *J. Physiol.* 165:528–41
9. Rasmussen H. 1970. Cell communication, calcium ion and adenosine monophosphate. *Science* 170:404–12
10. Prince WT, Berridge MJ, Rasmussen H. 1972. Role of calcium and adenosine-3'5'-cyclic monophosphate in controlling fly salivary gland secretion. *Proc. Natl. Acad. Sci. USA* 69:553–57
11. Prince WT, Berridge MJ. 1973. The role

of calcium in the action of 5-hydroxytryptamine and cyclic AMP on salivary glands. *J. Exp. Biol.* 58:367–84
12. Berridge MJ. 1981. Electrophysiological evidence for the existence of separate receptor mechanisms mediating the action of 5-hydroxytryptamine. *Mol. Cell Endocrin.* 23:91–104
13. Berridge MJ, Heslop JP. 1981. Separate 5-hydroxytryptamine receptors on the salivary gland of the blowfly are linked to the generation of either cyclic adenosine 3',5'-mono-phosphate or calcium signals. *Br. J. Pharmacol.* 7:729–38
14. Hokin MR, Hokin LE. 1953. Enzyme secretion and the incorporation of ^{32}P into phospholipids of pancreas slices. *J. Biol. Chem.* 203:967–77
15. Michell RH. 1975. Inositol phospholipids and cell surface receptor function. *Biochim. Biophys. Acta* 415:81–147
16. Fain JN, Berridge MJ. 1979. Relationship between hormonal activation of phosphatidylinositol hydrolysis, fluid secretion and calcium flux in the blowfly salivary gland. *Biochem. J.* 178:45–58
17. Berridge MJ. 1980. Preliminary measurements of intracellular calcium in an insect salivary gland using a calcium-sensitive microelectrode. *Cell Calcium* 1:217–27
18. Berridge MJ, Lipke H. 1979. Changes in calcium transport across *Calliphora* salivary glands induced by 5-hydroxytryptamine and cyclic nucleotides. *J. Exp. Biol.* 78:137–48
19. Berridge MJ, Fain JN. 1979. Inhibition of phosphatidylinositol synthesis and the inactivation of calcium entry after prolonged exposure of the blowfly salivary gland to 5-hydroxytryptamine. *Biochem. J.* 178:56–68
20. Fain JN, Berridge MJ. 1979. Relationship between phosphatidylinositol synthesis and recovery of 5-hydroxytryptamine

responsive Ca^{2+} flux in blowfly salivary gland. *Biochem. J.* 180:655–61

21. Allison JH, Stewart MA. 1971. Reduced brain inositol in lithium-treated rats. *Nature* 233:267–68

22. Allison JH, Blisner ME, Holland WH, Hipps PP, Sherman WR. 1976. Increased brain *myo*-inositol in lithium-treated rats. *Biochem. Biophys. Res. Commun.* 71:664–70

23. Berridge MJ, Downes CP, Hanley MR. 1989. Neural and developmental actions of lithium: a unifying hypothesis. *Cell* 59:411–19

24. Berridge MJ, Downes CP, Hanley MR. 1982. Lithium amplifies agonist-dependent phosphatidylinositol responses in brain and salivary glands. *Biochem. J.* 206:587–95

25. Berridge MJ, Dawson RMC, Downes CP, Heslop JP, Irvine RF. 1983. Changes in the levels of inositol phosphates after agonist-dependent hydrolysis of membrane phosphoinositides. *Biochem. J.* 212:473–82

26. Akhtar RA, Abdel-Latif AA. 1980. Requirement for calcium ions in acetylcholine-stimulated phosphodiesteratic cleavage of phosphatidyl-*myo*-inositol 4,5-bisphosphate in rabbit iris smooth muscle. *Biochem. J.* 192:783–91

27. Michell RH, Kirk CJ, Jones LM, Downes CP, Creba JA. 1981. The stimulation of inositol lipid metabolism that accompanies calcium mobilization in stimulated cells: defined characteristics and unanswered questions. *Philos. Trans. R. Soc. London Ser. B* 296:123–37

28. Berridge MJ, Buchan PB, Heslop JP. 1984. Relationship of polyphosphoinositide metabolism to the hormonal activation of the insect salivary gland to 5-hydroxytryptamine. *Mol. Cell Endocrinol.* 36:37–42

29. Berridge MJ. 1983. Rapid accumulation of inositol trisphosphate reveals that agonists hydrolyse polyphosphoinositides instead of phosphatidylinositol. *Biochem. J.* 212:849–58

30. Streb H, Irvine RF, Berridge MJ, Schulz I. 1983. Release of Ca^{2+} from a nonmitochondrial intracellular store in pancreatic acinar cells by inositol 1,4,5-trisphosphate. *Nature* 306:67–69

31. Burgess GM, Godfrey PP, McKinney JS, Berridge MJ, Irvine RF, Putney JW Jr. 1984. The second messenger linking receptor activation to internal calcium release in liver. *Nature* 309:63–66

32. Berridge MJ, Heslop JP, Irvine RF, Brown KD. 1984. Inositol trisphosphate formation and calcium mobilization in Swiss 3T3 cells in response to platelet-derived growth factor. *Biochem. J.* 222:195–201

33. Prentki M, Biden TJ, Janjic D, Irvine RF, Berridge MJ, Wollheim CB. 1984. Rapid mobilization of Ca^{2+} from rat insulinoma microsomes by inositol 1,4,5-trisphosphate. *Nature* 309:562–64

34. Biden TJ, Prentki M, Irvine RF, Berridge MJ, Wollheim CB. 1984. Inositol 1,4,5-trisphosphate mobilizes intracellular Ca^{2+} from permeabilized insulin-secreting cells. *Biochem. J.* 223:467–73

35. Fein A, Payne R, Corson DW, Berridge MJ, Irvine RF. 1984. Photoreceptor excitation and adaption by inositol 1,4,5-trisphosphate. *Nature* 311:157–60

36. Brown JE, Rubin LJ, Ghalayini AJ, Tarver AP, Irvine RF, et al. 1984. A biochemical and electrophysiological examination of myo-inositol polyphosphate as a putative messenger for excitation in *Limulus* ventral photoreceptor cells. *Nature* 311:160–63

37. Burgess GM, McKinney JS, Irvine RF, Berridge MJ, Hoyle PC, Puntey JW Jr. 1984. Inositol 1,4,5-trisphosphate may be a signal for f-Met-leu-Phe-induced intracellular calcium mobilization in human leukocytes (HL-60 cells). *FEBS Lett.* 176:193–96

38. Takai Y, Kishimoto A, Kikkawa U, Mori T, Nishizuka Y. 1979. Unsaturated diacylglycerol as a possible messenger for the activation of a calcium-activated, phospholipid-dependent protein kinase system. *Biochem. Biophys. Res. Commun.* 91:1218–24

39. Nishizuka Y. 1984. The role of protein kinase C in cell surface signal transduction and tumour promotion. *Nature* 308:693–98

40. Berridge MJ. 1984. Inositol trisphosphate and diacylglycerol as second messengers. *Biochem. J.* 220:345–60

41. Berridge MJ, Irvine RF. 1984. Inositol trisphosphate, a novel second messenger in cellular signal transduction. *Nature* 312:315–21

42. Berridge MJ. 1993. Inositol trisphosphate and calcium signalling. *Nature* 361:315–25

43. Bootman MD, Berridge MJ. 1995. The elemental principles of calcium signalling. *Cell* 83:675–78

44. Berridge MJ. 1997. Elementary and global aspects of calcium signalling. *J. Physiol.* 499:291–306

45. Berridge MJ, Lipp P, Bootman MD. 2000. The versatility and universality of calcium signalling. *Nat. Rev. Mol. Cell Biol.* 1:11–21

46. Galione AG, Berridge MJ. 1988. Pharmacological modulation of oscillations in the blowfly salivary gland. *Biochem. Soc. Trans.* 16:988–90

47. Rapp PE, Berridge MJ. 1981. The control of transepithelial potential oscillations in the salivary gland of *Calliphora erythrocephala*. *J. Exp. Biol.* 93:119–32

48. Rapp PE, Berridge MJ. 1977. Oscillations in calcium-cyclic AMP control loops from the basis of pacemaker activity and other high frequency biological rhythms. *J. Theor. Biol.* 66:497–525

49. Woods NM, Cuthbertson KSR, Cobbold PH. 1986. Repetitive transient rises in cytoplasmic free calcium in hormone-stimulated hepatocytes. *Nature* 319:600–2

50. Cuthbertson KSR, Cobbold PH. 1985. Phorbol ester and sperm activate mouse oocytes by inducing sustained oscillations in cell Ca^{2+}. *Nature* 316:541–42

51. Miyazaki S-I, Hashimoto N, Yoshimoto Y, Kishimoto T, et al. 1986. Temporal and spatial dynamics of the periodic increase in intracellular free calcium at fertilization of golden hamster eggs. *Dev. Biol.* 118:259–67

52. Cheek TR, McGuinness OM, Vincent C, Moreton RB, Berridge MJ, Johnson MH. 1993. Fertilisation and thimerosal stimulate similar calcium spiking patterns in mouse oocytes but by separate mechanisms. *Development* 119:179–89

53. Goldbeter A, Dupont G, Berridge MJ. 1990. Minimal model for signal-induced Ca^{2+} oscillations and for their frequency encoding through protein phosphorylation. *Proc. Natl. Acad. Sci. USA* 87:1461–65

54. Dupont G, Berridge MJ, Goldbeter A. 1991. Signal-induced Ca^{2+} oscillations: properties of a model based on Ca^{2+}-induced Ca^{2+} release. *Cell Calcium* 12:73–85

55. Endo M, Tanaka M, Ogawa Y. 1970. Calcium induced calcium release of calcium from the endoplasmic reticulum of skinned skeletal muscle fibres. *Nature* 228:34–36

56. Iino M. 1990. Biphasic Ca^{2+} dependence of inositol 1,4,5-trisphosphate Ca release in smooth muscle cells of the guinea pig *Taenia caeci*. *J. Gen. Physiol.* 95:1103–22

57. Bezprozvanny I, Watras J, Ehrlich BE. 1991. Bell-shaped calcium response curves of Ins(1,4,5)P_3- and calcium gated channels from endoplasmic reticulum of cerebellum. *Nature* 351:751–54

58. Finch EA, Turner TJ, Goldin SM. 1991. Calcium as a coagonist of inositol 1,4,5-trisphosphate-induced calcium release. *Science* 252:443–46

59. Missiaen L, Taylor CW, Berridge MJ. 1991. Spontaneous calcium release from inositol trisphosphate-sensitive calcium stores. *Nature* 352:241–44

60. Berridge MJ, Dupont G. 1994. Spatial and temporal signalling by calcium. *Curr. Opin. Cell Biol.* 6:267–74

61. Cheek TR, O'Sullivan AJ, Moreton RB, Berridge MJ, Burgoyne RD. 1989. Spatial localization of the stimulus-induced rise in cytosolic Ca^{2+} in bovine adrenal chromaffin cells: distinct nicotinic and muscarinic patterns. *FEBS Lett.* 247:429–34

62. Cheng H, Lederer WJ, Cannell MB. 1993. Calcium sparks: elementary events underlying excitation-contraction coupling in heart muscle. *Science* 262:740–44

63. Lipp P, Niggli E. 1994. Modulation of Ca^{2+} release in cultured neonatal rat cardiac myocytes—insight from subcellular release patterns revealed by confocal microscopy. *Circ. Res.* 74:979–90

64. Bootman MD, Berridge MJ. 1996. Subcellular Ca^{2+} signals underlying waves and graded responses in HeLa cells. *Curr. Biol.* 6:855–65

65. Yao Y, Choi J, Parker I. 1995. Quantal puffs of intracellular Ca^{2+} evoked by inositol trisphosphate in *Xenopus* oocytes. *J. Physiol.* 482:533–53

66. Bootman MD, Niggli E, Berridge MJ, Lipp P. 1997. Imaging the hierarchical Ca^{2+} signalling system in HeLa cells. *J. Physiol.* 499:307–14

67. Bootman MD, Berridge MJ, Lipp P. 1997. Cooking with calcium: the recipes for composing global signals from elementary events. *Cell* 91:367–73

68. Lechleiter JD, Clapham DE. 1992. Molecular mechanisms of intracellular calcium excitability in *X. laevis* oocytes. *Cell* 68:283–94

69. Williams RSB, Cheng L, Mudge AW, Harwood AJ. 2002. A common mechanism of action for three mood-stabilizing drugs. *Nature* 417:292–95

70. Kume S, Muto A, Inoue T, Suga K, Okano H, et al. 1997. Role of inositol 1,4,5-trisphosphate receptor in ventral signalling in *Xenopus* embryos. *Science* 278:1940–43

71. Kühl M, Sheldahl LC, Malbon CC, Moon RT. 2000. Ca^{2+}/calmodulin-dependent protein kinase II is stimulated by Wnt and frizzled homologs and promotes ventral cell fates in *Xenopus*. *J. Biol. Chem.* 275:12701–11

72. Sharp AH, McPherson PJ, Dawson TM, Aoki C, Campbell KP, Synder SH. 1993. Differential immunohistochemical localization of inositol 1,4,5-trisphosphate- and ryanodine-sensitive Ca^{2+} release channels in rat brain. *J. Neurosci.* 13:3051–63

73. Berridge MJ. 1998. Neural calcium signalling. *Neuron* 21:13–26

74. Berridge MJ. 1993. A tale of two messengers. *Nature* 365:388–89

75. Wang SSH, Denk W, Häusser M. 2000. Coincident detection in single dendritic spines mediated by calcium release. *Nat. Neurosci.* 3:1266–73

76. Nishiyama M, Hong K, Mikoshiba K, Poo M-M. 2000. Calcium stores regulate the polarity and input specificity of synaptic modification. *Nature* 408:584–88

Annu. Rev. Physiol. 2005. 67:25–37
doi: 10.1146/annurev.physiol.67.041904.120836
First published online as a Review in Advance on August 20, 2004

PETER HOCHACHKA: Adventures in Biochemical Adaptation

George N. Somero

*Department of Biological Sciences, Hopkins Marine Station, Stanford University,
Pacific Grove, California 93950-3094; email: somero@stanford.edu*

Raul K. Suarez

*Department of Ecology, Evolution, and Marine Biology, University of California,
Santa Barbara, Santa Barbara, California 93106-9610; email: suarez@lifesci.ucsb.edu*

Key Words   adaptation, aerobiosis, anaerobiosis, diving, hydrostatic pressure,
hypoxia, metabolism, scaling, temperature

■ **Abstract**   Peter Hochachka was one of the most creative forces in the field of comparative physiology during the past half-century. His career was truly an exploratory adventure, in both intellectual and geographic senses. His broad comparative studies of metabolism in organisms as diverse as trout, tunas, oysters, squid, turtles, locusts, hummingbirds, seals, and humans revealed the adaptable features of enzymes and metabolic pathways that provide the biochemical bases for diverse lifestyles and environments. In its combined breadth and depth, no other corpus of work better illustrates the principle of "unity in diversity" that marks comparative physiology. Through his publications, his stimulating mentorship, his broad editorial services, and his continuous—and highly infectious—enthusiasm for his field, Peter Hochachka served as one of the most influential leaders in the transformation of comparative physiology.

INTRODUCTION

This essay is an attempt to provide an overview of the extraordinary contributions to physiology made by Peter Hochachka over the course of a career that spanned more than 40 years and was cut short by his death from cancer at age 65 in September 2002. From his first publication in 1959, which dealt with glycogen reserves and resistance to fatigue in rainbow trout (1), to one of his final papers, which focused on the anaerobic metabolism characteristic of the prostate cancer from which he was suffering (2), Peter Hochachka created a picture of metabolic adaptation that stands as one of the great bodies of work in comparative physiology.

Examining Peter's career reveals two different types of important lessons. One stems from the physiology itself, which prospered under his stimulating catalytic guidance. He put intermediary metabolism into a new light, explaining how animals

0066-4278/05/0315-0025$14.00

have finely tuned biochemistry to exploit and adapt to a wide range of environmental conditions. A second lesson relates to how Peter was a great teacher who showed, by example, how one does science best. His importance to the field of comparative physiology can be understood only by taking the broadest possible view of his activities and talents—a perspective that encompasses the impact that Peter's ebullience and charisma, as well as his intellect, had on those around him. He exuded a spirit of excitement and adventure in all facets of his science, and this spirited approach was highly contagious. Peter belonged to the grand tradition of comparative physiologists such as Per Scholander, Laurence Irving, Knut Schmidt-Nielsen, and Kjell Johansen, whose approaches to their field exemplified the benefits and the challenges of studying organisms in their natural settings, as well as in the more controlled confines of the laboratory. The natural diversity available for physiological study might not always be most easily studied in nature—but it usually proved to be a lot more fun to move the laboratory to the animal in its normal surroundings, rather than vice versa.

Peter's approach to comparative physiology rested on two fundamental principles, that of August Krogh—although Claude Bernard probably said it first (3): "For many problems there is an animal on which it can be most conveniently studied."—and, logistical "convenience" aside, that of Kjell Johansen: "If you can study an organism in Cleveland, study something else." He chose his questions and experimental organisms creatively and wisely, and he struck a most effective balance between the field camp and the modern laboratory. The diversity of organisms and methodologies found in Peter's work stands unequalled in comparative physiology. He was a collaborator par excellence who attracted a large and diverse set of research colleagues from around the world. His coauthors number approximately 200 and he mentored over 60 doctoral students and postdoctoral associates. Whereas raw numbers only begin to tell the story, Peter was the leading publisher in his field for three decades, with a total publication list exceeding 400 papers and books.

A major reason for the success of Peter's program and for his ability to spark the interest of others in comparative physiology was his knack for conceptualization. When entering a new field, he was less apt to begin with a short "data" paper—a timid, first-step entry into a new domain—than with a stimulating and provocative concept paper, often published in a major journal such as *Science* (4–7), which cast a subject in a wholly new light. These concept papers frequently struck just the right balance between strongly supported arguments and adventurous flights of imagination. They were typically the sorts of "bold conjectures" that Karl Popper urged scientists to make: concepts that are at once challenging of the conventional wisdom of a field, yet readily subject to rigorous experimental tests that might refute them. That new ideas should play center stage in scientific research was a lesson Peter took to heart very early on in his career (8). Thus by presenting new conceptual frameworks, Peter stimulated others to take a different and a closer look at a wide variety of phenomena that, under his guiding influence, were to become central areas of research in comparative physiology.

Peter's strength in the conceptual realm is illustrated by his ability to appreciate the unity in diversity that characterizes organisms, notably in terms of the metabolic pathways they rely on to provide the energy necessary for such ATP-costly functions as ion transport, locomotion, and protein synthesis. The underlying theme that threads its way through virtually all of his papers is the manner in which the metabolic biochemistry of cells is finely tuned to ensure that appropriate levels of ATP turnover are sustained under a wide range of abiotic and biotic conditions, in organisms as diverse as trout, tunas, oysters, squid, turtles, locusts, hummingbirds, seals, and humans. The unifying principles about metabolic organization and evolution that emerged from Peter's studies of diverse animals represent one of the most creative analyses of metabolism ever developed, one that has deep interest to medical scientists as much as to comparative physiologists. His is a conceptual framework that will continue to support innovative work for decades.

With these generalizations before us, we now elaborate on some of the specific accomplishments Peter made as he moved creatively from one type of study to the next, providing in the process a "view of life" that set the field of comparative physiology on an exciting and productive new course.

Temperature Adaptation of Metabolism

Even in Peter's earliest publications the theme of metabolic adaptation takes center stage. His first half-dozen papers focused on carbohydrate metabolism in fish, and provided some of the initial evidence for the types of intertissue interactions that supported vigorous locomotion in salmonids. Soon, temperature, as well as activity, entered into the picture. In 1962, Hochachka & Hayes (9) published an important paper on the effects of temperature on alternate fates of glucose, work that Peter did for his master's degree at Dalhousie University. This was one of the first investigations to show that metabolism may be qualitatively, as well as quantitatively, different at high and low temperatures. One of us (G. Somero) carried this paper to McMurdo Sound, Antarctica, in 1964. There, during the dark winter months that followed, the insights from this study of trout were applied to cold-adapted Antarctic notothenioid fishes (10). At this same time, Peter was completing his doctoral studies at Duke University, examining the effects of acclimation temperature on expression of lactate dehydrogenase (LDH) isozymes in goldfish (11, 12). These studies, which represented a logical extension of his 1962 work with trout, were focused more on the possible roles of muscle and heart isoforms of LDH in shifting metabolic pathways between aerobic and anaerobic production of ATP than on the effects of temperature on the kinetic properties of the enzymes themselves. However, the data in these initial studies of LDH suggested that temperature adaptation of enzyme kinetic properties might be a fascinating topic to pursue further. The (temporary) redirection of Peter's work toward the study of biochemical adaptation of proteins to temperature, away from a strict focus on intermediary metabolism, coincided with my (Somero) arrival in his laboratory for postdoctoral study a few months after Peter had joined the faculty at the University of British Columbia (UBC), where he remained for all 37 years of his faculty career (13).

At the time that work on enzymatic adaptation to temperature began in Peter's laboratory, little was known about the ways in which orthologous protein variants from organisms adapted to different temperatures differed with respect to thermal characteristics (13). In fact, in the late 1960s, it seems fair to say that we even lacked a sense of what protein traits might be most critical in the context of adaptation to temperature. The few studies published to date on thermal stability and activation energies suggested that adaptation might be prevalent, but the extent to which kinetic properties were finely tuned to the thermal environment in which the species had evolved, and the structural bases of these adaptations in function, were essentially unknown. And, as was so often the case in Peter's career, entry into virgin territory proved to be an exciting and productive adventure. His group, which soon was joined by Thomas Moon and John Baldwin, was the first to demonstrate that a holistic perspective was necessary in the study of protein adaptation. Thus binding ability (as typically indexed by the apparent Michaelis-Menten constant), catalytic power (as indexed by the catalytic rate constant), and structural stability were inextricably linked (13–15). Evolution of one trait could not occur without affecting the other traits. Work in Peter's laboratory thus showed that conservation of appropriate binding ability was balanced against conservation of rate of function and that both traits were, as subsequent studies have shown, linked with the structural stability of the protein. These studies put protein evolution into a new light. They showed that unity of metabolic potential—the ability of organisms from diverse habitats to carry out the same metabolic reactions at their respective habitat temperatures—was based on adaptive diversity in kinetic and structural properties of enzymes. Moreover, in at least some cases, alternate forms of a common type of enzyme could be generated during phenotypic acclimation to different temperatures (16), as Peter's doctoral studies first discovered (11, 12).

Adaptation to Hydrostatic Pressure

The insights that Peter's group attained into the mechanisms of enzymatic adaptation to temperature led to his launching (literally) an expedition to determine whether similar conserved trends—another instance of unity in diversity—characterized adaptation of proteins to elevated hydrostatic pressure. The Alpha Helix—a National Science Foundation-sponsored oceanographic vessel that was the brainchild of one of Peter's heroes, Per Scholander—was a favorite mobile laboratory for Peter, and two expeditions were undertaken to examine adaptations of deep-living animals. Enzymes of fish adapted to deep-sea conditions were found to have responses to pressure different from those of orthologous enzymes of shallow-living forms (17, 18). Although these initial, path-breaking studies of adaptation to pressure were successful in extending the concept of enzymatic adaptation to a new physical variable of the environment, these expeditions did encounter the type of problem that often plagues field study: Capture of large numbers of deep-sea fish proved daunting even to someone with Peter's skills as a field biologist. True to character, however, Peter snatched an additional victory from the jaws of the minor defeat that shortfalls in availability of fish dealt him. He and his collaborators obtained numerous squid and initiated metabolic studies that helped lead him to return the central thrust of his research program to energy metabolism (19).

Facultative Anaerobiosis

Peter's overriding interest in energy metabolism, which remained evident even during the late 1960s and early 1970s when much of his group was studying adaptations to temperature and pressure, returned to the fore with the arrival in his laboratory of several students, notably Tariq Mustafa and Kenneth Storey, who shared this fascination, especially in the context of animals capable of dealing with prolonged anoxia. Over the next three decades and until the end of his life, Peter mentored numerous students and postdoctoral researchers who shared interests in metabolic questions, many of these relevant to limited availability of oxygen. Under Peter's creative guidance, these trainees were perceptive in spotting new theoretical, empirical, and technical developments from diverse fields—ranging from microbiology to biomedicine—and exploiting these developments in a comparative frame of reference.

Some of the earliest metabolic studies to come from the UBC laboratory illustrate the effectiveness of the approach. In the early 1970s, Monod's concept of allostery was still relatively new, and was catalyzing extensive work on metabolic regulation. How enzymes "read" their environment and responded appropriately by modulating their activities was a major topic of biochemical research worldwide. Within this intellectual milieu, research on oyster pyruvate kinase and phosphoenolpyruvate carboxykinase led to the proposal that reciprocal control of these enzymes plays a key role in mediating flux changes at the pyruvate branch-point during aerobic-anaerobic transitions (6, 20–22). Radioisotope studies led to the notion of a new metabolic scheme for the coupling of glucose and aspartate fermentation during anoxia, as well as the detection of an unidentified anaerobic end product (23). It was then discovered that this was a novel compound, alanopine (2,2'-iminodipropionic acid) (24), produced in a reaction catalyzed by alanopine dehydrogenase, an enzyme discovered in the Hochachka laboratory (25).

The studies of facultative anaerobiosis by Peter's group broadened to encompass many taxa. In the late 1970s, radioisotope studies led to the remarkable discovery that goldfish produce ethanol as an anaerobic end product (26). Peter delighted in telling people about the "sophisticated analytical tool" that first suggested the identity of the end-product: His student Eric Shoubridge sniffed the aquarium water the anoxic goldfish were in and remarked, "There's booze in here." Hypoxia-tolerant fish, like hypoxia-tolerant invertebrates, were shown to sustain ATP production under low oxygen conditions, but to achieve this result through pathways of carbon flow and redox balance quite different from those available to most other vertebrates.

Peter's studies of anoxic or hypoxic production of ATP examined not only the sources of energy but also the fine-scale details of acid-base regulation during oxygen-limiting conditions. Anoxia is often associated with acidosis, and a commonly held idea was (and, unfortunately, often still is) that glycolysis produces lactic acid, which then dissociates and yields H^+. Applying acid-base chemistry to the entire glycolytic pathway, Mommsen & Hochachka (5) instead showed that anoxic acidosis (net H^+ production) results from the combined effects of glycolytic ATP synthesis and cellular ATP hydrolysis. This conceptual breakthrough indicated that glycolytic production of ATP is a two-edged sword: It provides energy, but ATP turnover also poses threats owing to acidification.

The rate of ATP turnover during oxygen-limiting conditions thus became an important focus of research on anaerobiosis. Because glycolysis is the central pathway used for ATP synthesis in all anoxia-tolerant animals and the only source of ATP besides creatine phosphate available to vertebrate facultative anaerobes during periods of O_2 limitation, the existing paradigm was that anoxia tolerance in animals results partly from having enzymes with regulatory properties that make them better at increasing glycolytic flux rates (7, 27, 28). The idea, of course, was that the increase in glycolytic ATP production (i.e., the Pasteur effect) allows rates of ATP synthesis to continue to match the rates of ATP hydrolysis while mitochondrial oxidative phosphorylation is inhibited. However, it was eventually realized that instead of glycolytic activation, O_2 limitation results in metabolic depression in anoxia-tolerant animals. Peter and his colleagues published several concept papers that refocused thinking in metabolic biochemistry and pointed researchers in this new direction (4, 29, 30).

Much of the research in Peter's laboratory in the 1980s was directed toward testing hypotheses about metabolic depression or metabolic arrest to discern the quantitative importance of various energy-consuming processes and to elucidate the underlying mechanisms responsible for down-regulating metabolism. During this period, students showed that anoxic turtle hepatocytes in vitro remain in energy balance (31, 32) despite depressed metabolic rates (i.e., rates of ATP synthesis equal rates of hydrolysis, despite decreased rates of ATP turnover). Inhibition of both protein synthesis (33) and degradation (34), as well as reduced Na^+ pump activity (even though membrane potential is constant) (35), were found to account for a significant fraction of the metabolic depression, i.e., to explain why ATP turnover could be dramatically reduced under hypoxia. Similarly, under anoxic conditions, turtle brain slices display depressed rates of metabolism and membrane ion conductance, yet ion gradients are maintained (36–38). Studies using turtles showed that during anoxic submergence, beta-adrenergic stimulation results in hepatic glycogenolysis, which causes hyperglycemia and thus provides high concentrations of glucose for use by anoxic organs (39, 40).

Marine Mammals and Diving

The classic work on vertebrate diving responses by two of Peter's scientific heroes, Scholander and Irving, was laboratory-based studies that involved the forced diving of animals. Research on vertebrate diving underwent a great leap forward as investigators, notably Gerald Kooyman and his colleagues, began to study freely diving animals under natural field conditions. One such group, consisting of an international team led by Warren Zapol of Harvard University, went to the Antarctic to study Weddell seals. Wisely, they invited Peter to participate as their biochemist. Perhaps the most path-breaking of their experiments were those in which seals were fitted with computer-controlled instruments as backpacks and allowed to dive freely under the ice, while physiological parameters were monitored and blood was drawn at timed intervals (41–43). These were landmark studies that would be considered marvels of technological sophistication even today. In later years, collaborative research with other groups involved magnetic resonance imaging (MRI) to study cardiovascular responses and magnetic resonance spectroscopy (MRS) to study high energy phosphate metabolism (44) in northern elephant seals during forced submergence. Along with work done by other eminent investigators in this field, Peter's collaborative work contributed to the development of the complex picture of vertebrate diving responses (45) that now serves as the springboard for further research.

The emergence of evolutionary physiology as a subdiscipline led to the question of whether the mechanisms that make diving mammals so good at what they do are true evolutionary adaptations to the demands of diving. In the strictest sense, this would be true only if such traits enhance diving performance and arise as the outcome of natural selection for this particular capacity. Application of the comparative phylogenetic method to decades of published interspecific data led to the unexpected conclusion (46) that certain diving responses, such as bradycardia, are highly conserved in nondiving as well as in diving mammals and do not appear to have hypertrophied to increase dive duration. In contrast, a positive correlation independent of phylogeny, exists between spleen size and dive duration. This, and the MRI finding that spleen emptying is fast but refilling is slow relative to interdive periods (44), led to the provocative suggestion that hematocrit is always high while divers are at sea, and large spleens might be an adaptation to reduce hematocrit and, thereby, blood kinematic viscosity while the diver is on land!

High Altitude–Adapted Humans

Peter's research concerning adaptations to O_2 limitation led him to investigate human adaptations to hypobaric hypoxia. This project required that he organize research expeditions to Peru to study Quechuas and to the Himalayas to study Sherpas. Much of the rich literature on the subject was of a physiological nature, and research was made difficult by the primitive—and at times dangerous—conditions under which it had to be conducted. At one point, the laboratory in the Andes in which Peter's field work was to be done was blown up by the Shining Path guerillas! Thus despite a deep-seated love of adventure, it seemed prudent in this one case at least to bring the "organism" back to the laboratory to conduct the detailed biochemical studies that were needed to advance the field beyond the physiological work that had previously been done. To this end, Peter adopted the strategy of transporting Quechua and Sherpa volunteers to North America so they could be studied in specialist laboratories using the most modern equipment available. Using positron emission tomography (PET), MRI, and MRS, he and his collaborators showed that the brains of Quechuas metabolize glucose at lower rates than those of lowlanders, indicating maintenance of brain function at lower energetic cost and lower O_2 demand (47, 48). Human hearts use both glucose and fatty acids as oxidative fuels, but generally rely more on the latter. However, using PET, they found that Quechua hearts rely more on glucose than do lowlander hearts (49). Because fatty acid oxidation results in O_2 wasting, i.e., more O_2 is used per unit cardiac work performed compared with glucose oxidation, the greater reliance upon glucose as a premium fuel suggests greater efficiency in both energy production and in O_2 use in the hearts of Quechua compared with those of lowlanders. In Sherpas breathing normoxic or hypoxic air, MRS of calf muscles showed tight coupling between ATP synthesis and hydrolysis rates, despite large changes in work rate (50). The greater reliance on aerobic pathways for energy production and the efficiency of coupling between energy production and energy use under hypoxic conditions was proposed to explain partly why lactate and H^+ do not accumulate to the same extent as in the muscles of lowlanders subjected to these conditions. Peter and colleagues hypothesized that the suite of physiological and biochemical traits found in Quechuas and Sherpas was inherited from prehistoric ancestors selected for survival in colder, higher, and drier habitats more than 50,000 years ago (51).

Metabolism and Exercise

Despite intellectually maturing in a period dominated by Monod's pronouncement that what is true of *Escherichia coli* is true of elephants, Peter had always looked upon biochemical diversity in pathways of energy metabolism as evolved traits (52), relating them to diversity in lifestyles and environments and to how animals run, swim, or fly. Research in the Hochachka laboratory concerning the bioenergetics of exercise covered the full range of levels of biological organization, from enzymes, pathways, and mitochondria, to isolated muscle fibers, perfused hearts, and whole animals. Adherence to both the Krogh and Johansen principles remained evident in all of this work. Thoroughbred racehorses, artificially selected for running, served as subjects for measurements of whole-body metabolite turnover rates during exercise (53). Fishes with discrete red and white skeletal muscles were used to study muscle fuel use and bioenergetics during endurance (54) and sprint (55–57) swimming. Hummingbirds, capable of high wingbeat frequencies and metabolic rates during flight, were used to study metabolic pathway design and the regulation of fuel use in the context of ecology and behavior (58, 59). The role of amino acids in fueling exercise was investigated using salmon in their upstream spawning migration (60). The energetic cost of parenthood was investigated in tree swallows foraging to feed young (61).

Other studies of muscle bioenergetics were comparative in the sense that they took advantage of Nature's experiments and made use of species as a variable. Over the decades, the fishes Peter and his students used for research on muscle metabolism ranged from sluggish Amazon species to high-performance tunas (56) and marlin (62). Studies of bioenergetic scaling in mammalian muscles went from shrews to whales, revealing that oxidative capacities decline, while glycolytic capacities increase with increasing body mass (63, 64).

The wide range of research projects included work on phosphofructokinase, to determine whether its allosteric regulation is sufficient to account for the activation of glycolysis in muscles (65, 66). Studies of muscle physiology and biochemistry also examined fiber-specific (67) and interspecific (68) variation in mitochondrial oxidative properties, the control of respiration in muscles (69), and the effects of exercise training on the control of respiration (70). There was research concerning the purine nucleotide cycle (71) and competition between cytoplasmic dehydrogenases for pyridine nucleotides (72). Peter questioned the idea that the large flux changes that often occur during transitions between rest and exercise could be explained by classical concepts in enzyme regulation (73) and suggested alternative mechanisms (74), providing others with enough research material to last for at least a decade. He challenged popular single-cause models for the allometric scaling of metabolism and, before he died, helped take the first step toward a multiple-cause explanation (75) compatible with what is known in respiratory physiology, biochemistry, and metabolic control theory.

Peter Hochachka's Legacy: A Field Transformed

Areas of science are often born of the marriage between dissimilar fields. Molecular biology, for example, resulted from the polygamous union of genetics, biochemistry, and biophysics. Peter brought biochemistry into comparative physiology, transforming the latter into a truly integrative discipline in which the goal was to understand organismal function at all levels of biological organization. But by asking, throughout his career, how enzymes and pathways are adapted to lifestyles and environments, as well as how they operate in the context of the energetic demands faced by real animals, he also brought the philosophy and approaches of comparative physiology into biochemistry. Articles in any current issue of the *Journal of Experimental Biology, Physiological and Biochemical Zoology* (formerly *Physiological Zoology*), *Comparative Biochemistry and Physiology*, and the *Journal of Comparative Physiology* illustrate how much the field has changed over the past three decades. Peter was a major catalyst in the transformation that united the formerly quite separate disciplines of comparative biochemistry, comparative physiology, ecological physiology, and evolutionary physiology into an intellectual whole. In doing so, he taught us an enormous amount about the nature of life and, of equal importance, he showed us how the right blend of creativity, enthusiasm, and rigor could energize and transform a field and allow it to advance most rapidly.

The *Annual Review of Physiology* is online at http://physiol.annualreviews.org

LITERATURE CITED

1. Miller RB, Sinclair AC, Hochachka PW. 1959. Diet, glycogen reserves and resistance to fatigue in hatchery rainbow trout. *J. Fish. Res. Bd. Can.* 16:321–28

2. Hochachka PW, Rupert JL, Goldenberg L, Gleave M, Kozlowski P. 2002. Going malignant: the hypoxia-cancer connection in the prostate. *BioEssays* 24:749–57

3. Storey KB. 2004. Adventures in oxygen metabolism. *Comp. Biochem. Physiol.* In press

4. Hochachka PW. 1986. Defense strategies against hypoxia and hypothermia. *Science* 212:509–14

5. Hochachka PW, Mommsen TP. 1983. Protons and anaerobiosis. *Science* 219:1391–97

6. Hochachka PW, Mustafa T. 1972. Invertebrate facultative anaerobiosis. *Science* 178:1056–60

7. Hochachka PW, Storey KB. 1975. Metabolic consequences of diving in animals and man. *Science* 187:613–21

8. Hochachka PW. 1995. Ideas unlimited: in appreciation of the 1995 Canada Gold Medal Award for Science and Engineering. *Can. Soc. Zool. Bull.* 26:23–25

9. Hochachka PW, Hayes FR. 1962. The effect of temperature acclimation on pathways of glucose metabolism in the trout. *Can. J. Zool.* 40:261–70

10. Somero GN, Giese AC, Wohlschlag DE. 1968. Cold adaptation of the Antarctic fish *Trematomus bernacchii. Comp. Biochem. Physiol.* 26:223–33

11. Hochachka PW. 1965. Isoenzymes in metabolic adaptation of a poikilotherm: subunit relationships in lactic dehydrogenase of goldfish. *Arch. Biochem. Biophys.* 111:96–103

12. Hochachka PW. 1967. Organization of metabolism during temperature compensation. In *Molecular Mechanisms of Temperature Adaptation*, ed. CL Prosser, pp. 177–

203. Washington, DC: Am. Assoc. Adv. Sci.

13. Somero GN. 2004. Adaptation of enzymes to temperature: search for basic "strategies." *Comp. Biochem. Physiol.* In press

14. Hochachka PW, Somero GN. 1968. The adaptation of enzymes to temperature. *Comp. Biochem. Physiol.* 27:107–18

15. Somero GN, Hochachka PW. 1968. The effect of temperature on catalytic and regulatory properties of pyruvate kinases from the rainbow trout and the Antarctic fish *Trematomus bernacchii. Biochem. J.* 110:395–400

16. Baldwin J, Hochachka PW. 1969. The functional significance of isozymes in thermal acclimation: trout brain acetylcholinesterases. *Biochem. J.* 116:883–87

17. Hochachka PW, Schneider DE, Moon TW. 1971. The adaptation of enzymes to pressure. 2. A comparison of trout liver FDPase with the homologous enzyme from an offshore benthic species. *Am. Zool.* 11:491–502

18. Moon TW, Mustafa T, Hochachka PW. 1971. Effects of hydrostatic pressure on catalysis by epaxial muscle phosphofructokinase from an abyssal fish. *Am. Zool.* 11:467–72

19. Hochachka PW, Moon TW, Mustafa T, Storey KB. 1975. Metabolic sources of power for mantle muscle of a fast swimming squid. *Comp. Biochem. Physiol.* 52B:151–58

20. Mustafa T, Hochachka PW. 1971. Catalytic and regulatory properties of pyruvate kinases in tissues of a marine bivalve. *J. Biol. Chem.* 246:3196–206

21. Mustafa T, Hochachka PW. 1974. Enzymes in facultative anaerobiosis: catalytic properties of PEP carboxykinase of oyster adductor muscle. *Comp. Biochem. Physiol.* 45B:639–55

22. Hochachka PW, Fields JHA, Mustafa T. 1973. Animal life without oxygen. *Am. Zool.* 13:543–55

23. Collicutt J, Hochachka PW. 1977. The anaerobic oyster heart. *J. Comp. Physiol.* 115:147–57

24. Fields JHA, Hochachka PW, Weinstein B. 1980. Alanopine and strombine are novel amino acids produced by a dehydrogenase found in the adductor muscle of the oyster, *Crassostria gigas. Arch. Biochem. Biophys.* 201:110–14

25. Fields JHA, Hochachka PW. 1981. Purification and properties of alanopine dehydrogenase from the adductor muscle of the oyster, *Crassostria gigas. Eur. J. Biochem.* 114:615–21

26. Shoubridge EA, Hochachka PW. 1980. Ethanol: novel end product of vertebrate anaerobic metabolism. *Science* 209:308–9

27. Storey KB, Hochachka PW. 1973. Enzymes of energy metabolism in a vertebrate facultative anaerobe, *Pseudemys scripta.* Turtle heart phosphofructokinase. *J. Biol. Chem.* 249:1417–22

28. Storey KB, Hochachka PW. 1973. Enzymes of energy metabolism in a vertebrate facultative anaerobe, *Pseudemys scripta.* Turtle heart pyruvate kinase. *J. Biol. Chem.* 249:1423–27

29. Hochachka PW, Dunn JF. 1983. Metabolic arrest: the most effective means of protecting tissues against hypoxia. In *Proc. 3rd Banff Int. Hypoxia Symp.*, ed. JR Sutton, CS Houston, NL Jones, pp. 297–309. New York: Liss

30. Hochachka PW. 1986. Metabolic arrest. *Intens. Care Med.* 12:127–33

31. Buck LT, Land SC, Hochachka PW. 1993. Anoxia tolerant hepatocytes: a model system for the study of reversible metabolic suppression. *Am. J. Physiol. Regul. Integr. Comp. Physiol.* 265:R49–56

32. Buck LT, Hochachka PW, Schon A, Gnaiger E. 1993. Microcalorimetric measurement of reversible metabolic suppression induced by anoxia in isolated hepatocytes. *Am. J. Physiol. Regul. Integr. Comp. Physiol.* 265:R1014–19

33. Land SC, Buck LT, Hochachka PW. 1993. Response of protein synthesis to anoxia and recovery in anoxia-tolerant hepatocytes. *Am. J. Physiol. Regul. Integr. Comp. Physiol.* 265:R41–48

34. Land SC, Hochachka PW. 1994. Protein turnover during metabolic arrest in turtle hepatocytes: role and energy dependence of proteolysis. *Am. J. Physiol. Cell Physiol.* 266:C1028–36

35. Buck LT, Hochachka PW. 1993. Suppression of Na^+-K^+-ATPase activity and a constant plasma membrane potential in hepatocytes during anoxia: evidence in support of the channel arrest hypothesis. *Am. J. Physiol. Regul. Integr. Comp. Physiol.* 265:R1020–25

36. Doll CJ, Hochachka PW, Reiner PB. 1991. Effects of anoxia and metabolic arrest on turtle and rat cortical neurons. *Am. J. Physiol. Regul. Integr. Comp. Physiol.* 260:R747–55

37. Doll CJ, Hochachka PW, Reiner PB. 1991. Channel arrest: implications from membrane resistance in turtle neurons. *Am. J. Physiol. Regul. Integr. Comp. Physiol.* 261:R1321–24

38. Doll CJ, Hochachka PW, Land SC. 1994. A microcalorimetric study of turtle cortical slices: insights into brain metabolic depression. *J. Exp. Biol.* 191:141–53

39. Keiver KM, Hochachka PW. 1991. Catecholamine stimulation of hepatic glycogenolysis during anoxia in the turtle *Chrysemys picta. Am. J. Physiol. Regul. Integr. Comp. Physiol.* 261:R1341–45

40. Keiver KM, Hochachka PW. 1992. The role of catecholamines and corticosterone during anoxia and recovery at 5°C in turtles. *Am. J. Physiol. Regul. Integr. Comp. Physiol.* 263:R770–74

41. Guppy M, Hill RD, Schneider RC, Qvist J, Liggins GC, et al. 1986. Microcomputer-assisted metabolic studies of voluntary diving of Weddell seals. *Am. J. Physiol Regul. Integr. Comp. Physiol.* 250:R175–87

42. Qvist J, Hill RD, Schneider RC, Falke KJ, Liggins GC, et al. 1986. Hemoglobin concentrations and blood gas tensions of free-diving Weddell seals. *J. Appl. Physiol.* 61:1560–69

43. Hill RD, Schneider RC, Liggins GC, Guppy M, Hochachka PW, et al. 1987. Heart rate and body temperature during free diving of Weddell seals. *Am. J. Physiol. Regul. Integr. Comp. Physiol.* 253:R344–51

44. Thornton SJ, Spielman DM, Pelc NJ, Block WF, Crocker DE, et al. 2001. Effects of forced diving on the spleen and hepatic sinus in northern elephant seal pups. *Proc. Natl. Acad. Sci. USA* 98:9413–18

45. Hochachka PW, Somero GN. 2002. The diving response and its evolution. In *Biochemical Adaptation. Mechanism and Process in Physiological Evolution.* pp. 158–85. Oxford, UK: Oxford Univ. Press.

46. Mottishaw PD, Thornton SJ, Hochachka PW. 1999. The diving response and its surprising evolutionary path in seals and sea lions. *Am. Zool.* 39:434–50

47. Hochachka PW, Stanley C, Brown WD, Allen PS, Holden JE. 1994. The brain at high altitude: hypometabolism as a defense against chronic hypoxia? *J. Cereb. Blood Flow Metab.* 14:671–79

48. Hochachka PW, Clark CM, Monge C, Stanley C, Brown WD, et al. 1996. Sherpa brain glucose metabolism and defense adaptations against chronic hypoxia. *J. Appl. Physiol.* 81:1355–61

49. Holden JE, Stone C, Brown WD, Nickles RJ, Stanley C, Hochachka PW. 1995. Enhanced cardiac metabolism of plasma glucose in high altitude natives: adaptation against chronic hypoxia. *J. Appl. Physiol.* 79:222–28

50. Allen PS, Matheson GO, Zhu G, Gheorgiu D, Dunlop RS, et al. 1997. Simultaneous ^{31}P magnetic resonance spectroscopy of the soleus and gastrocnemius in sherpas during graded calf muscle exercise and recovery. *Am. J. Physiol. Regul. Integr. Comp. Physiol.* 273:R999–1007

51. Hochachka PW, Gunga HC, Kirsch K. 1998. Our ancestral physiological phenotype: an adaptation for hypoxia tolerance and for endurance performance? *Proc. Natl. Acad. Sci. USA* 95:1915–20

52. Hochachka PW. 1985. Fuels and pathways as designed systems for the support of muscle work. *J. Exp. Biol.* 115:149–64

53. Weber J-M, Parkhouse WS, Hochachka PW, Wheeldon D, Dobson GP. 1987. Lactate kinetics in exercising thoroughbred horses. II. Regulation of turnover rate in plasma. *Am. J. Physiol. Regul. Integr. Comp. Physiol.* 253:R896–903

54. West TG, Arthur PG, Suarez RK, Doll CJ, Hochachka PW. 1992. In vivo utilization of glucose by heart and locomotory muscles of exercising rainbow trout (*Oncorhynchus mykiss*). *J. Exp. Biol.* 177:63–79

55. Driedzic WR, Hochachka PW. 1976. Control of energy metabolism in fish white muscle. *Am. J. Physiol.* 230:579–82

56. Guppy M, Hulbert WC, Hochachka PW. 1979. Metabolic sources of heat and power in tuna muscles. II. Enzyme and metabolite profiles. *J. Exp. Biol.* 82:303–20

57. Dobson GP, Parkhouse WS, Hochachka PW. 1987. Regulation of anaerobic ATP-generating pathways in trout fast-twitch skeletal muscle. *Am. J. Physiol. Regul. Integr. Comp. Physiol.* 253:R186–94

58. Suarez RK, Brown GS, Hochachka PW. 1986. Metabolic sources of energy for hummingbird flight. *Am. J. Physiol. Regul. Integr. Comp. Physiol.* 251:R537–42

59. Suarez RK, Lighton JRB, Moyes CD, Brown GS, Gass CL, Hochachka PW. 1990. Fuel selection in rufous hummingbirds: ecological implications of metabolic biochemistry. *Proc. Natl. Acad. Sci. USA* 87:9207–10

60. Mommsen TP, French CJ, Hochachka PW. 1980. Sites and patterns of protein and amino acid utilization during the spawning migration of salmon. *Can. J. Zool.* 58:1785–99

61. Burness GP, Ydenberg RC, Hochachka PW. 2001. Physiological and biochemical correlates of broodsize and energy expenditure in tree swallows. *J. Exp. Biol.* 204:1491–501

62. Suarez RK, Mallet MD, Daxboek C, Hochachka PW. 1985. Enzymes of energy metabolism and gluconeogenesis in the Pacific blue marlin, *Makaira nigricans. Can. J. Zool.* 64:694–97

63. Emmett B, Hochachka PW. 1981. Scaling of oxidative and glycolytic enzymes in mammals. *Respir. Physiol.* 45:261–72

64. Hochachka PW, Emmett B, Suarez RK. 1988. Limits and constraints in the scaling of oxidative and glycolytic enzymes in homeotherms. *Can. J. Zool.* 66:1128–38

65. Storey KB, Hochachka PW. 1974. Activation of muscle glycolysis—role for creatine phosphate in phosphofructokinase regulation. *FEBS Lett.* 46:337–39

66. Dobson GP, Yamamoto E, Hochachka PW. 1986. Phosphofructokinase control in muscle: nature and reversal of pH-dependent ATP inhibition. *Am. J. Physiol. Regul. Integr. Comp. Physiol.* 250:R71–76

67. Moyes CD, Buck LT, Hochachka PW, Suarez RK. 1989. Oxidative properties of carp red and white muscle. *J. Exp. Biol.* 143:321–31

68. Moyes CD, Suarez RK, Hochachka PW, Ballantyne JS. 1990. A comparison of fuel preferences of mitochondria from vertebrates and invertebrates. *Can. J. Zool.* 68:1337–49

69. Hogan MC, Arthur PG, Bebout DE, Hochachka PW, Wagner PD. 1992. The role of O_2 in regulating tissue respiration. *J. Appl. Physiol.* 73:728–36

70. Burelle Y, Hochachka PW. 2002. Endurance training induced muscle specific changes in mitochondrial function in skinned muscle fibers. *J. Appl. Physiol.* 92:2429–38

71. Mommsen TP, Hochachka PW. 1988. The purine nucleotide cycle in muscle as two temporally separated metabolic units. *Metabolism* 37:552–56

72. Guppy M, Hochachka PW. 1978. Role of dehydrogenase competition in metabolic regulation: the case of lactate and glycerophosphate dehydrogenases. *J. Biol. Chem.* 253:8465–69

73. Hochachka PW, Matheson GO. 1992. Regulating ATP turnover rates over broad dynamic work ranges in skeletal muscles. *J. Appl. Physiol.* 73:1697–703

74. Hochachka PW. 1999. The metabolic implications of intracellular circulation. *Proc. Natl. Acad. Sci. USA* 96:12233–39

75. Darveau C-A, Suarez RK, Andrews RD, Hochachka PW. 2002. Allometric cascade as a unifying principle of body mass effects on metabolism. *Nature* 417:166–70

Annu. Rev. Physiol. 2005. 67:39–67
doi: 10.1146/annurev.physiol.67.040403.114025
Copyright © 2005 by Annual Reviews. All rights reserved
First published online as a Review in Advance on July 21, 2004

Calcium, Thin Filaments, and the Integrative Biology of Cardiac Contractility

Tomoyoshi Kobayashi and R. John Solaro
*Department of Physiology and Biophysics, College of Medicine, University of Illinois
at Chicago, Chicago, Illinois 60612; email: solarorj@uic.edu; tkoba@uic.edu*

Key Words   troponin I, troponin T, troponin C, tropomyosin, actin

■ **Abstract**   Although well known as the location of the mechanism by which the cardiac sarcomere is activated by Ca^{2+} to generate force and shortening, the thin filament is now also recognized as a vital component determining the dynamics of contraction and relaxation. Molecular signaling in the thin filament involves steric, allosteric, and cooperative mechanisms that are modified by protein phosphorylation, sarcomere length and load, the chemical environment, and isoform composition. Approaches employing transgenesis and mutagenesis now permit investigation of these processes at the level of the systems biology of the heart. These studies reveal that the thin filaments are not merely slaves to the levels of Ca^{2+} determined by membrane channels, transporters and exchangers, but are actively involved in beat to beat control of cardiac function by neural and hormonal factors and by the Frank-Starling mechanism.

INTRODUCTION

The objectives of our review are to provide an up-to-date picture of Ca^{2+} and molecular signaling in the thin filaments and to summarize advancements in the translation of this information from the molecular and cellular levels to the level of the integrated biology of the intensity and dynamics of the heart beat. A focal point is the hypothesis that altered thin filament response to Ca^{2+} is an essential element in the control of cardiac dynamics. Publication of the excellent review by Tobacman (1) in *Annual Review of Physiology* on the topic of thin filament regulation was about 10 years ago. Since that time, other excellent and detailed reviews, which track progress in this field, have been published (2–11). However, more recently major new experimental findings have provided a clearer picture of molecular signaling in the thin filament and a better understanding of alterations in this signaling in various physiological and patho-physiological conditions. Moreover, the identification of many new sarcomeric proteins has led to fresh concepts of

0066-4278/05/0315-0039$14.00

what thin filament regulation means. Although space limitations required us to focus here on regions of the thin filament (generally the A-band region) that react with crossbridges, there is an exciting and developing area of research concerned with regions of the thin filament that rarely if ever react with crossbridges and dwell near and within the Z-disk. Recent papers (10, 11) review the significance of direct and indirect interactions of the thin filaments with cytoskeletal and Z-disk proteins in controlling cardiac function, signal transduction, and signal processing.

STRUCTURE AND INTERACTIONS AMONG THIN FILAMENT PROTEINS IN THE A-BAND REGION OF THE SARCOMERE

States of the Thin Filament

Following the original proposal of the steric blocking model of muscle regulation (12–14), structural studies have provided major advances in our perception of the thin filament. Several models for muscle regulation involve either two (on and off) or three (blocked, closed, and open) different states of thin filament (15–17). Our discussions employ the three-state model to describe molecular mechanisms, but other models may also fit the data as discussed in papers by Chalovich (18) and by Squire & Morris (7), which summarize the ambiguities and difficulties in this area of research. Current descriptions of the state of the thin filaments derived from biochemical (16, 19, 20) and structural studies (21, 22) include a blocked state or B-state, reflecting steric block of crossbridges; a closed state (C-state), reflecting weakly bound crossbridges; and an open or strong myosin-binding state (M-state). Figure 1 (see color insert), which is a reconstruction based on electron micrographs, shows the B-, C-, and M-states of the thin filaments as indicated by the position of tropomyosin (Tm). The B-state, which is occupied when the cytoplasmic $[Ca^{2+}]$ is below the threshold for binding to regulatory sites on troponin (Tn), is similar to the original concept that Tm blocks crossbridge reactions with actin by a steric block, although part of the putative myosin-binding site seems to be exposed on actin (21–23). In the B-state, Tm is positioned at the outer domains of actin away from the groove formed by the actin helix. With Ca^{2+} activation, Tm moves closer to the inner domain of actin, nearer the groove. The closed or cocked state reflects a refinement of the original steric hindrance hypothesis that is required to fit biochemical data indicating that some or all crossbridges bind weakly to actin in a non-force-generating but stereo-specific manner. Crossbridges are poised to enter into the force-generating state with the thin filament in the C-state, hence the term cocked (7). Evidence for a Ca^{2+}-dependent movement of Tm on the thin filament has been observed by Förster resonance energy transfer (FRET). Although some studies could not determine Ca^{2+}-induced movement of Tm by FRET measurements (24), others using multi-site lifetime FRET measurements were able to detect Ca^{2+}-dependent Tm movement relative to actin (25). The M-state is associated with the strong binding of force-generating crossbridges that induce further movement of Tm and/or actin subdomains, which is apparently important in spreading activation laterally along the thin filament. Ca^{2+}-dependent crossbridge-induced movement of Tn relative to actin on the thin filament was also observed by FRET measurements (26–28). The functional importance of these movements of Tn and Tm are not fully understood. The discussion concerning their function is couched in two major theories of activation. One is the so-called rate modulation theory originally proposed by Julian (29) and refined by Brenner (30); the other is the classic recruitment-based theory. In the extreme, the rate theory holds that recruitment of crossbridges following Ca^{2+} activation of the thin filaments is relatively unimportant, that crossbridges are all in the C-state, and that Ca^{2+} acts as an allosteric effector increasing the probability that crossbridges enter into force-generating state(s). As discussed here, it appears possible that in heart, a fraction of crossbridges is in each of these states.

Reconstructions of electron micrographs also indicate that Tn may move on the thin filament in order to release the actin-crossbridge reaction from a steric block. To get around the difficulty of visualizing Tn on the thin filament, Lehman et al. (31) used a deletion mutant of Tm, where internal pseudorepeats 2, 3, and 4 were missing. The deletion mutant Tm forms Tm polymers on F-actin, still binds the Tn complex, and demonstrates normal Ca^{2+}-induced movement (32, 33). These modified thin filaments contain seven-fourths-fold more Tn than native thin filaments, which aids in localizing Tn densities and detecting Ca^{2+}-dependent movement of Tn. The three-dimensional images reported by Lehman et al. (31) indicate that in the absence of Ca^{2+}, Tn approached residues 1–4, 23–27, and 47 of actin. This prediction agrees with NMR data (34) and cross-linking studies (35, 36). A different picture of Ca^{2+}-induced structural changes of the thin filament was reported by Narita et al., who used single-particle analysis of cryo-EM images to reconstitute three-dimensional models (37). One structural unit of actin filament (7actin-1Tm-1Tn) was divided into seven segments, I through VII. In the presence of Ca^{2+}, the globular domain of Tn (the T2-TnC-TnI region) was located over the inner domain of actin at segment V. In the absence of Ca^{2+}, it shifted toward the outer domain of actin by $\sim$28 Å. The C-terminal one-third of Tm (segments V–VII) shifted toward the outer domain by $\sim$35 Å, whereas the N-terminal half of Tm shifted less than $\sim$12 Å. In the presence of Ca^{2+}, Tm was located entirely over the inner domain of actin to allow crossbridge attachment. This model suggests that Tn covers at least part of the myosin-binding site on one or two actin molecules among the seven actin molecules in a structural unit.

The Troponin Complex

A major advance in our understanding of the inhibited and active state of the thin filaments came with the determination of the crystal structures of the Ca^{2+} bound form of the core domain of a ternary cardiac troponin (cTn) complex by Takeda et al. (38) (Figure 2, see color insert). They crystallized two different cTn complexes. One had a M_r of 46 kDa and consisted of the full-length Ca^{2+} sensor, cTnC (residues

1–161),* a fragment of the inhibitory protein, cTnI (residues 31–163), missing the first 30 residues and the C-terminal 48 residues, and the C-terminal end of the Tm-binding protein, cTnT (residues 183–288; T2). The other Tn complex had a M_r of 52 kDa owing to inclusion of a larger cTnI fragment (residues 31–210) missing only the N-terminal extension that is unique to the cardiac variant. Thus in both cases, the cardiac-specific N-terminal region of cTnI, the C-terminal region of cTnI, the N-terminal half of TnT (T1), and the C-terminal part of TnT, which may interact with tropomyosin (Tm)-actin, are missing. Two Tn molecules present in an asymmetric unit of each crystal resulted in four different structures (identified as Tn46KA, Tn46KB, Tn52KA and Tn52KB). The overall architecture of each of these four structures is similar, except that TnI in Tn52KB has an extra α-helix spanning residues 164–188. The crystal structure of the core-domain of the cTn complex is divided into two subdomains, the regulatory head and the TnI-TnT arm, which are connected by a flexible linker. The regulatory head is composed of the N-terminal regulatory domain of cTnC and the regulatory or triggering site (downstream of the inhibitory region) of cTnI. The IT arm is made up of the C-domain of cTnC, T2, and two long TnI α-helices, one of which forms a coiled coil with TnT. The D/E linker of cTnC and the inhibitory region of cTnI form a flexible linker that is likely involved in the regulatory mechanism by allowing the two domains to rotate and alter their relative orientation.

Troponin C

TnC ($\sim$18 kDa) is a member of the EF-hand family of Ca^{2+}-binding proteins and consists of two globular domains (39–41). As illustrated in Figure 3 (see color insert), each EF-hand Ca^{2+}-binding site consists of an α-helix-loop-α-helix structural motif (42). The TnC molecule has eight α-helices (designated as A-through H-helix) associated with four Ca^{2+}-binding sites and an extra α-helix at the N terminus (N-helix). Each globular domain contains a pair of Ca^{2+}-binding sites designated sites I through IV from the N terminus. Sites I and II in the N-lobe are generally referred to as Ca^{2+}-specific sites, and sites III and IV in the C-lobe as Ca^{2+} and Mg^{2+} sites. Vertebrate fast skeletal muscle TnC (fsTnC) binds two Ca^{2+} at each domain, whereas in the cardiac/slow skeletal muscle isoform of TnC (cTnC), site I does not bind Ca^{2+} in the physiological range because there are several amino acid replacements in the coordinating sites. Analysis of equilibrium Ca^{2+}-binding data of cTnC in detergent-extracted fiber bundles in ATP-free solutions resolved these two distinct classes of binding sites (43). These sites saturated over the physiological range of Ca^{2+} concentrations with the following binding constants: sites III and IV; $K_{Ca} = 7.4 \times 10^7 \text{ M}^{-1}$, $K_{Mg} = 0.9 \times 10^3 \text{ M}^{-1}$, and site II; $K_{Ca} = 1.2 \times 10^6 \text{ M}^{-1}$, $K_{Mg} = 1.1 \times 10^2 \text{ M}^{-1}$. Following dissociation of rigor complexes by MgATP, there was a fall in K_{Ca} of site II to $\sim$3 $\times$ 10^5 M^{-1}

---

*The residue numbers for cardiac and slow skeletal muscle are from human including the initial Met, and for fast skeletal muscle from rabbit.

with no changes in the affinities for site III and IV. This result, which agreed with earlier data in soluble reconstituted thin filaments reacting with myosin heads (44), indicates that the strong actin crossbridge reaction influences Ca^{2+} binding to the regulatory Ca^{2+}-binding site of cTnC just as Ca^{2+} binding to cTnC influences the actin-crossbridge reaction. Kinetic analysis of exchange of Ca^{2+} with sites III and IV indicated that these sites exchange Ca^{2+} much too slowly for it to occur within a heart beat (43). In contrast, kinetic analysis and site-directed mutagenesis studies indicated that site II is able to exchange Ca^{2+} in the time course of the heart beat and thus must be responsible for its regulation (45–47). However, the precise rate constants occurring with cTnC in its native environment remain unknown. In early studies (48), the kinetics of Ca^{2+} binding to the single regulatory site of cTnC was estimated using the fluorescent probe, IAANS, covalently attached to Cys-35 and Cys-84. These studies gave a dissociation rate constant of ~ 20 s^{-1}. More recent data indicate that this rate constant may reflect a change in structure more than a change in Ca^{2+} binding. When the IAANS label was placed only at Cys-35 as a reporter of Ca^{2+} binding to cTnC in the cTn complex (49, 50), two phases were observed in Ca^{2+} association steps: a fast phase with [Ca^{2+}] dependency and the slower phase without [Ca^{2+}] dependency. Dong et al. (49, 50) concluded that the fast component corresponds to the Ca^{2+}-binding event, whereas the slower component corresponds to the conformational change triggered by Ca^{2+} binding. The results of Hazard et al. (51) are in general agreement with these data. They measured Ca^{2+}-binding kinetics for cTnC using both IAANS-labeled protein at Cys-84 and unlabeled protein. Ca^{2+} dissociation from the unlabeled cTnC was determined using stopped flow and the spectroscopically sensitive chelators, Quin 2 or BAPTA. The off-rate constant for Ca^{2+} binding to site II of cTnC was 700 s^{-1}, whereas the Ca^{2+}-independent off rate determined by IAANS fluorescence was 90 s^{-1}. Together with the results of Dong et al. (49, 50), these data indicate that site II rapidly binds and releases Ca^{2+} and induces relatively slow conformational steps. Because Cys 35 is located in the region surrounding the inactive site I in cTn, the data indicate a more important role for this region in regulation than is generally appreciated. The data of Dong et al. (50) also demonstrated that these steps are sensitive to phosphorylation of cTnI (see below).

Regions surrounding the inactive site I and the D/E linker region, which connects the N- and the C-lobes, may also be of special significance with regard to regulation of thin filament activity by cTnC. The D/E linker region is involved in Ca^{2+}-dependent activation of actomyosin ATPase activity; replacement of acidic amino acid residue(s) in this region with neutral Ala resulted in the loss of activation (52, 53). A direct role of TnT in Ca^{2+} activation is consistent with the observation that the interaction between TnC and TnT is involved in the regulatory function of Tn, at least in fast skeletal muscles (54, 55). In the crystal structure, the C-terminal domain of cTnC interacts with TnT, although interaction sites on TnC are different from those determined by NMR (56). The meaning of these variable results from different approaches should be clarified with the emergence of more cTn structural data in the sarcomeric lattice.

Troponin I

TnI, which was originally named because of its ability to inhibit actin-activated myosin ATPase activity, is a basic protein with a $M_r \sim 21$ kDa for fast skeletal TnI (fsTnI) and slow skeletal TnI (ssTnI) and a $M_r \sim 24$ kDa for cardiac TnI (cTnI). TnI has six distinct functional regions: (*a*) an N-terminal cardiac-specific extension that contains PKA-dependent phosphorylation sites (Ser-23 and Ser-24); (*b*) a region that binds to the C-lobe of TnC; (*c*) a region that binds to C-terminal regions of TnT; (*d*) a basic inhibitory region or first actin-binding region; (*e*) a regulatory or triggering region; and (*f*) a second actin-binding region. A critical function of TnI is to interact with actin and thereby inhibit actomyosin ATPase activity at low cytoplasmic [Ca^{2+}]. TnI is able to inhibit ATPase activity in the absence of Tm and other Tn subunits in vitro. Maximum inhibition occurs at an actin:TnI molar ratio of 1:1. In the presence of Tm, maximum inhibition is observed at a 7:1 ratio of actin:TnI. There is a periodic distribution of TnI along the Tm-actin filaments (57), which indicates that Tm not only amplifies the inhibitory action of TnI but also serves to direct TnI to its optimal thin filament location during sarcomere assembly. A critical actin-binding region on TnI is the inhibitory region (58, 59), whose sequence is highly conserved among muscle types, although Pro-110 in fsTnI and ssTnI is replaced by Thr-143 in cTnI. Thr-143 of cTnI is a major phosphorylation site for protein kinase C (PKC), which may affect velocity of shortening and plays an important role in post-translational modifications of cTn associated with hypertrophic signaling and the transition to heart failure (60). Rarick et al. (61) identified two other actin-binding sites on cTnI that are outside the inhibitory region and contain residues 151–188 of cTnI (residues 120–157 of fsTnI) and 188–210 of cTnI (residues 157–179 of fsTnI). In general agreement with this finding, Ramos (62) reported that residues 166–182 of fsTnI are also involved in actin-Tm binding, and Tripet et al. (63) identified a second actin region in fsTnI containing residues 140–148.

Although not defined in the crystal structure, there are several reports addressing the structure of the TnI inhibitory region. Data from circular dichroism spectroscopy and NMR studies using a binary complex of TnI with TnC indicate that the peptide derived from the fsTnI inhibitory region, residues 96–115, has an extended conformation with a possible kink around Gly-104 (64). A refined model based on further studies employing cross-linking and FRET studies indicates the inhibitory region is a flexible β-hairpin (65). Lindhout & Sykes (66) also determined the solution structure of cTnI inhibitory peptide, residues 129–148 (residues 96–115 of fsTnI), by NMR in a binary complex with the C-domain of cTnC, showing that a segment spanning residues Leu-135 to Lys-140 adopts a helical conformation. Using electron paramagnetic resonance, Brown et al. (67) reported that, among residues 129–145 of cTnI, residues 129–137 adopt an α-helical structure in the Tn ternary complex. Note that in the crystal structure (Figure 1), residues only up to Phe-133 of cTnI form an α-helix, H2(I), with C-cap residue Asp-134. In Tn46KA, the segment (residues 137–144) was not visible. The distance between α carbon atoms of Arg-136 and Arg-145 is 24.4 Å. If the peptide bond is fully extended, the distance between these same atoms of adjacent residues is 32.7 Å. Thus in this case, the conformation of the inhibitory region that was not visible in Tn46KA is not completely extended but rather appears to be a flexible loop. Dong et al. measured the Ca^{2+}-dependent distance change between the inhibitory region and regulatory region of cTnI in a cTn complex, as well as in a cTnI-cTnC complex (68, 69). They observed a large distance increase ($\sim$9 Å) between residues 128 and 152 of cTnI upon Ca^{2+} binding to the regulatory site of cTnC, whereas the distance between the residues in the regulatory site (residues 149 and 166 of cTnI) remained unchanged. Dong et al. concluded that as cTnI is released from actin, residues within the cTnI inhibitory region switch from a β-turn/coil to an extended quasi-α-helical conformation. In contrast, the cTnI regulatory region remains α-helical.

Comparison of the molecular mechanisms by which Ca^{2+} binding to the N-lobe of fsTnC and cTnC triggers contraction reveals unique features in heart versus skeletal muscle. In the case of fsTnC, Ca^{2+} binding to sites I and II induces a conformational change in the N-lobe. Helices B and C move away from helix D, exposing a hydrophobic patch that serves as a new protein-interacting site (70–73). The regulatory site of fsTnI binds to this hydrophobic patch (74–76), and the free energy from this interaction accounts for most of the Ca^{2+}-dependent interaction between fsTnC and fsTnI (77). In studies of isolated cTnC, Ca^{2+} binding to the single regulatory site (site II) cannot open the hydrophobic patch in the N-lobe (78, 79). Opening the hydrophobic patch of cTnC requires an interaction with the regulatory site of cTnI (80, 81). Ultimately, the open structures of the N-lobe of cTnC and fsTnC are indistinguishable (Figure 3). This Ca^{2+}-induced interaction between the N-lobe of TnC and the regulatory region of TnI in turn induces a structural rearrangement among thin filament proteins such that the inhibitory region and the C-terminal part of TnI move away from actin, as determined from fluorescence energy transfer measurements (82–86) in the skeletal system. It is highly likely that a similar mechanism occurs in the cardiac thin filament.

Troponin T

TnT is not only the longest and largest component of Tn but also has the most extensive and diverse interactions with its neighbors on the thin filament. TnT binds to TnI and TnC to form the Tn complex and also binds to Tm and actin. TnT is an elongated protein with a C-terminal region forming part of the globular head of Tn and an N-terminal region forming the tail. Mild chymotryptic digestion yields two soluble fragments, T1 (the tail region) and T2. T1, which was not included in the crystal structure, binds to the Tm-Tm overlap region and confers cooperative myosin S1 binding to the thin filament. As can be seen in the crystal structure (Figure 2), T2 binds to TnC and TnI. T2-TnC-TnI acts as the Ca^{2+} sensor of the thin filament. The M_r of TnT is 31–36 kDa, with 250–300 amino acid residues. Cardiac isoforms are longer than skeletal isoforms at and near the N terminus. In cTnI and cTnC, isoforms arise from separate genes, but in addition, cTnT isoforms arise from alternative splicing of TnT transcripts. In human heart development and in heart failure there are four cTnT isoforms ($cTnT_1$–$cTnT_4$) that are expressed in a variable isoform population (87). The charge differences associated with these shifts in isoform population may contribute to altered myofilament response to Ca^{2+} as reported by Gomes et al. (88).

A role for the tail region of cTnT, T1, in myofilament activation and relaxation is also evident from other studies. The N-terminal region appears to be important in maintaining Tm in the blocked or B-state associated with steric block of the actin-crossbridge reaction. Hinkle et al. (89) reported that $cTnT_{1-153}$ strongly promoted Tm binding to actin and was able to inhibit crossbridge interactions with the thin filament that promote ATPase activity and filament sliding in the motility assay. This region of the cTnT tail also inhibited myosin-S1-ADP binding to the thin filament. Three-dimensional reconstructions from negatively stained images of actin-Tm-$cTnT_{1-153}$ support the idea that the B-state of Tm is associated not only with TnI binding to actin but also with interactions of portions of the TnT tail domains with Tm-actin. The complexity of the functions of the N-terminal tail of TnT is indicated by data showing that deletion of the first 76 amino acids of cTnT results in an inhibition of maximum tension and ATPase rate (90). $cTnT_{77-289}$ bound more tightly to Tm than full-length cTnT. Interestingly, Hinkle et al. (89) reported that whereas deletion of 94 residues from the N terminus of cTnT had no effect on myosin S1 ATPase activity, deletion of 119 residues resulted in a greatly reduced affinity of Tn for actin Tm. Although the fsTnT-T1 region (residues 1–158 of fsTnT) inhibited actomyosin S1 ATPase activity (91), detailed functional mapping of the fsTnT molecule by Oliveira et al. (92) demonstrated that a peptide containing $fsTnT_{77-191}$ was able to activate actomyosin ATPase activity. These results point to the complexity of data interpretation from investigations aimed at functional identification of the various domains of TnT.

In addition to this special role of cTnT in relaxation, a role for cTnT in thin filament activation separate from cTnI has also been proposed. Potter et al. (54) indicated that Ca^{2+} fsTnC not only releases the thin filament from inhibition by fsTnI but also plays a role in promoting the thin filament–myosin interaction via a direct reaction with fsTnT. This hypothesis is based on data generated from experiments with thin filaments reconstituted with a fsTnI N-terminal deletion mutant (fsTnId57), which does not bind to fsTnT. These preparations could still activate myosin ATPase activity, presumably by a direct interaction with fsTnC. Whether this occurs in the cardiac system remains unclear, but there is evidence in general agreement with the studies in fast skeletal preparations. A cardiac deletion mutant ($cTnI_{80-211}$), which is analogous to fsTnId57, also lost its ability to bind to cTnT but retained binding to cTnC, albeit weakly (93). In this case, despite the lack of binding between cTnI and cTnT, Ca^{2+} was able to activate ATPase activity in a reconstituted system to 50% of that obtained with full-length cTnI. The influence of cTnT on Ca^{2+} activation is not restricted to regions that interact with cTnC in the regulatory head identified in the crystal structure. Using metal binding and epitope mapping, Jin & Root (94) reported that the alternatively spliced variable T1 region of chicken fsTnT affects the structure of the T2 region. Data demonstrating that the isoform population of developmental splice variants of the T1 region of cTnT affect the Ca^{2+} sensitivity of tension development of skinned fiber bundles (88) also indicate an effect of the N-terminal tail on the regulatory head of cTn.

Phosphorylation of cTnI and cTnT

Phosphorylation of cTnI and cTnT has been recently reviewed in detail elsewhere (9, 95). Protein kinase A (PKA) specifically phosphorylates Ser-23 and Ser-24 of cTnI. PKA phosphorylates Ser-24 first and, subsequently, much slower phosphorylation occurs at Ser-23 (96, 97). PKC phosphorylates both cTnI and cTnT. Phosphorylation sites on cTnI were previously determined to be Ser-42, Ser-44, and Thr-143 (98). Recently, we found that both beta-PKC and epsilon-PKC phosphorylate Ser-23, Ser-24, and Thr-143 of cTnI in the Tn ternary complex or in the reconstituted thin filament (99). There is also evidence from in vitro studies that Ser 162 of cTnI is a substrate for p21-activated kinase (Pak) (100). However, active Pak in the cellular environment induced dephosphorylation of cTnI via activation of PP2a (101).

Phosphorylation of Ser-23, Ser-24 appears specialized for regulation of sensitivity of the myofilaments to Ca^{2+} and for enhancing crossbridge cycling rate, whereas phosphorylation of the PKC-specific sites appears specialized for depressing crossbridge cycling rate. Data indicate that PKC-dependent phosphorylation of cTnI either enhances (102) or depresses (103) myofilament Ca^{2+} sensitivity. Pak-dependent phosphorylation of Ser-162 increased Ca^{2+} sensitivity of skinned fiber bundles. Residues Thr-194, Ser-198, Thr-203, and Thr-284 are the major PKC-dependent phosphorylation sites in cTnT (104). Among these four phosphorylation sites, Thr-203 appears to be by far the most important for modulation of cTn function (105). Phosphorylation of Thr-203 significantly inhibits tension and Ca^{2+} sensitivity of skinned fiber bundles, indicating the significance of this region of TnT in thin filament function.

Most of the potential phosphorylation sites in cTnI and cTnT were not resolved in the crystal structure. Some were among the segments removed to improve the crystal growth. Others were not seen owing to structural flexibility. In the crystal structure, Ser-42 and Ser-44 of cTnI were visible. Ser-42 is the N-cap residue of a long helix H1(I). Ser-42 forms a side chain backbone hydrogen-bond network with Arg-45, which represents the most common pattern of α-helix capping, i.e., a capping box (Figure 4, see color insert). Phosphorylation of Ser-42 is likely to affect the capping box formation and thus alter cTnI local structure. Ser-44 interacts with Glu-10 of the N-lobe of cTnC, indicating that phosphorylation of Ser-42/Ser-44 might affect the interaction between cTnI and cTnC and thus Ca^{2+} activation. Pseudophosphorylation (replacement of Ser with Glu) of Ser-42/Ser-44 significantly reduces myofilament response to Ca^{2+} (103). Moreover, pseudophosphorylation of these sites, together with pseudophosphorylation of Thr-143, reduces the affinity of cTnI-cTnC interaction (106). Another potential phosphorylation site visible in the crystal structure is Thr-203 of cTnT (Figure 4). This residue also acts as the N-cap of the H1(T2) helix. Modeling of the structural change induced by phosphorylation of Thr-203 predicts an extension of the helix. Compared with controls, skinned fiber bundles containing cTnT-Thr-203-Glu demonstrate a significantly reduced maximum tension and ATPase rate, as well as a reduced Ca^{2+} sensitivity (105).

The domains surrounding Ser-23, Ser-24, and Thr-143 of cTnI were not determined in the crystal structure. Results of NMR structural studies (97, 107) and cross-linking studies (108) support a mechanism in which the dephosphorylated cTnI N-terminal extension interacts with the N-lobe of cTnC. Upon phosphorylation of Ser-23/Ser-24, the N-terminal region dissociates from cTnC, inducing a reduction in Ca^{2+} affinity of the cTn complex and a decrease in the Ca^{2+} sensitivity of myofilament activity. Although the structure of the cardiac-specific region seems flexible, Keane et al. (97) were able to determine the solution structure of the peptide (residues 17–30 of cTnI) derived from this region with phosphorylation at Ser-23 and Ser-24. According to their model structure, this segment adopted a looped conformation with interactions between the side chains of Arg-21 and Arg-22 and phosphate groups on Ser-23 and Ser-24. Thr-143 is located in the middle of the minimum inhibitory region of cTnI. Phosphorylation of this site reduces the affinity for TnC (109) and depresses filament sliding velocity in the motility assay (103).

Tropomyosin

Tm, which consists of 284 amino acids, is a highly extended parallel coiled-coil protein. Structure-function properties of Tm have been reviewed by Perry (4) and by Wolska & Wieczorek (110). The two chains are about 400 Å long and overlap to form continuous strands winding around the actin filaments (Figure 1). The sequence contains a hepta-peptide repeat of hydrophobic amino acid residues, which are essential for coiled-coil formation. The sequence also contains a less characteristic sevenfold periodic repeat that corresponds to seven actin interaction sites. Each periodic repeat has a different functionality (32, 111, 112). Whereas $\alpha\alpha$Tm from cardiac muscle was found in the blocked-state position along the actin filament in the absence of Tn, $\beta\beta$Tm and $\alpha\alpha/\alpha\beta$Tm from skeletal muscle were found in the closed-state position (113). The two isoforms differ by a relatively small number of amino acids, suggesting a low energy barrier between blocked and closed states of the thin filaments. A high-resolution crystal structure of the N-terminal 81 residues of αTm revealed that the coiled coil is staggered at the segment with a high content of Ala, a structural specialization resulting in the bending of the coiled coil (114). Many such "Ala-staggering" segments along the Tm sequence appear to allow Tm to wind on the actin filament. Changing an Ala cluster to a canonical coiled-coil interface with Leu greatly stabilized Tm, but reduced its affinity for actin more than tenfold (115). The structure of the C-terminal 31–34 residues of Tm was also determined by X-ray crystallography (116) and NMR (117). In the structure generated from a crystal formed from a chimera of the C-terminal 31 residues of Tm with GCN-4, the last 22 residues splay and expose a large surface area (116). In the solution structure of the C-terminal 34 residues, residues 253–269 formed a coiled coil, whereas residues 270–279 formed parallel linear helices. This C-terminal uncoiled-coil region may be important for TnT-recognition. Greenfield et al. (118) found that when fragments with residues 251–284 of Tm were titrated with residues 1–33 of Tm, NMR signals from residues 274–284 were perturbed. Further addition of TnT perturbed most of the signals. When the unnatural amino acid hydroxy-Trp was introduced into Tm by Farah & Reinach (119), they observed that Ca^{2+} binding to Tn changed the fluorescence of hydroxy-Trp located in the region outside the Tn-binding site of Tm. Thus Ca^{2+} binding to Tn showed a long-range effect on the Tm molecule.

Variations in the flexibility of Tm may be important for its function in relaxation and activation mechanisms within the thin filament (8, 117, 120–122). Increasing evidence indicates that cooperative activation of the thin filament requires Tm to function as a continuous strand along the thin filament. A model presented by Smith & Geeves (121) considers Tm flexibility in the context of a continuous chain of molecules that includes Tn. They hypothesize that TnI in the relaxed state binds to actin, immobilizes Tm on the thin filament, and impedes the strong actin-crossbridge reaction. Release of TnI from actin by Ca^{2+}-TnC permits movement in Tm that results in crossbridge binding and induces a kink in the Tm-Tn chain. In their concept (121), therefore, the crossbridge and TnI compete for binding sites on actin and induce oppositely directed chain kinks. It is also likely that the tail of TnT may immobilize Tm, as discussed above (89). Smith et al. (122) extended this concept of activation to a model describing the influence of crossbridge binding to cooperatively promote the binding of neighbors on the thin filament.

An alternative model was proposed by Tobacman & Butters (17) on the basis of data from many investigators showing that myosin and Tm strengthen each other's binding to actin. There is a $\sim$10,000-fold strengthening of Tm attachment to actin when crossbridges bind to the thin filament. Implicit in the Smith et al. model is the notion that crossbridge binding to the thin filament should loosen the actin-Tm interaction. A continuous chain of Tm molecules is critical to both models, but the models differ greatly: The Tobacman model invokes specific, tight Tm-actin interactions; the Smith et al. model hypothesizes nonspecific actin-Tm interactions. To account for the observed effects of myosin and Tm on each other, Tobacman proposed that strong myosin-actin attachments alter the actin inner domain so that Tm binds to it tightly (the open or M-state position). Similarly, tropomyosin binding to the actin inner domain alters actin so that myosin binds more tightly on the actin outer domain. These changes in actin permit Tm to shift position on the thin filament cooperatively, removing the steric block to the actin-crossbridge reaction, and positioning Tm at the inner domain of actin, a requirement of strong binding of crossbridges. In the Tobacman model, activation of one longer region of the thin filament is 100 times favored over activation of two shorter regions. In both models the Tm shifts position on actin cooperatively. However, they differ in what mechanism, other than its own rigidity, holds the Tm stiffly in place at the same position along multiple successive actins.

INTEGRATION OF THIN FILAMENT MOLECULAR
SIGNALING AND CONTROL OF CARDIAC FUNCTION

Pressure, Volume, and Time as Descriptors
of Cardiac Function

The dynamics of pressure and volume changes occurring during the heart beat and the relation between volume and pressure during the heart beat (123) provide a framework in which to discuss functional correlates of the molecular interactions at the level of the sarcomere (Figure 5). In the following sections, we discuss molecular reactions involving thin filament proteins in the context of the system's behavior. The two left panels of Figure 5A depict left ventricular pressure (LVP) and volume (V) changes during a beat of a mouse heart under basal conditions and during stimulation with a beta-adrenergic agonist, isoproterenol (ISO). The

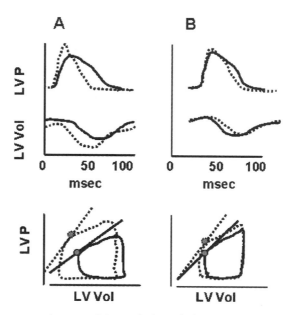

Figure 5 Pressure, volume, and time relations during a beat of the isolated, ejecting mouse heart in control conditions (*solid lines*) and in response to beta-adrenergic stimulation by isoproterenol (*dashed lines*). (*A*) Nontransgenic hearts. (*B*) Transgenic hearts expressing slow skeletal TnI in place of cardiac TnI. The upper panels show the time course of left ventricular pressure (LVP) and volume (LV Vol) and the lower panels show the relation between LV Vol and LVP. Points on the P-Vol loops indicate the end systolic pressure, and lines drawn through these points indicate contractility. Note that the presence of ssTnI signficantly attenuates the enhanced relaxation and contractility associated with adrenergic stimulation. See text for further description.

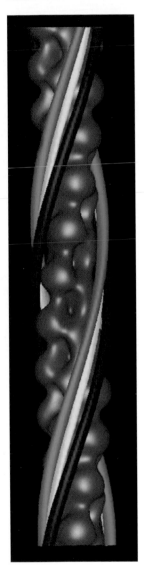

Figure 1 Three-dimensional reconstruction of a surface view of the states of the thin filament under relaxed conditions (Tm, *red*), in the Ca^{2+}-activated condition (Tm, *yellow*), and in rigor (Tm, *green*). Reproduced from Craig & Lehman (22) by permission.

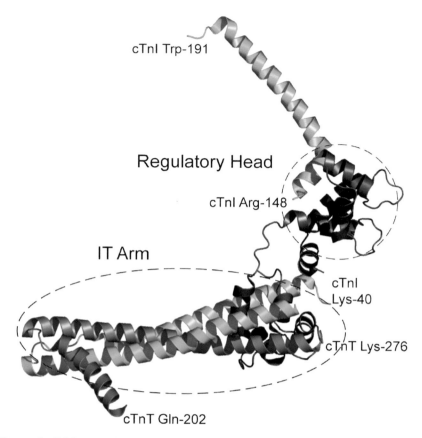

Figure 2   Ribbon model of the crystal structure of the core domain of Tn complex determined by Takeda et al. (38). Shown here are chains D, E, and F of PDB 1J1E. Orange, cTnC; blue, cTnT; green, cTnI.

TnC N-domain closed-form   TnC N-domain open-form

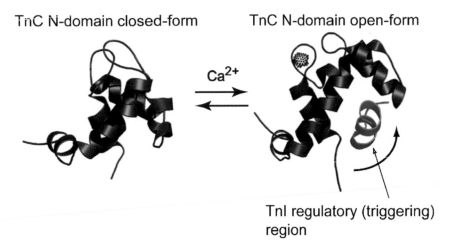

TnI regulatory (triggering)
region

Figure 3 Ca^{2+}-dependent structural change in the N-lobe of TnC. Only the N-terminal domain of TnC and the regulatory (triggering) region of TnI are shown.

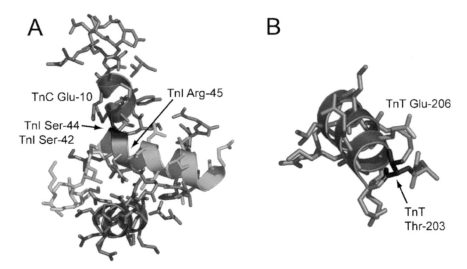

Figure 4 Structural model of phosphorylation sites in cTn. (*A*) cTnI Ser-42 (*yellow*)/Ser-44 (*red*) and nearby residues. Ser-42 forms a capping box with Arg-45 (*magenta*). Whereas Ser-44 interacts with Glu-10 (*blue*) from the N-terminal domain of cTnC in the crystal structure, pseudophosphorylation (Asp-substitution) of Ser-42/Ser-44 perturbs the N-terminal structure of the G-helix (*also shown in blue*) from the C-terminal domain of cTnC (157). (*B*) cTnT Thr-203 (*red*) and nearby residues. Thr-203 forms a capping box with Glu-206 (*pale magenta*) of cTnT.

data are from isolated ejecting mouse hearts beating under conditions of constant venous return and pressure head (afterload) and at constant heart rate (124). The bottom panel of Figure 5A illustrates the P-Vol loop, which is a plot of the relation between LV volume and LVP in a beat. The end systolic pressure (ESP) in Figure 5A is indicated as the level of LVP where ejection from the ventricle stops. Consideration of sarcomeric function in various phases of the P-Vol loop provides a useful framework for relating molecular properties of the thin filaments to cardiac function. Although the relationships between cellular function and ventricular function are complex, the correlate of pressure is sarcomeric tension and the correlate of volume is sarcomere length. The ESP is thus a point at which the sarcomeres are neither shortening nor lengthening. The line drawn through the ESP points in Figure 5 reflects the sarcomere length tension relation in the heart. Points on the line can be generated by systematically altering the venous return (see 123 for further explanation). The increase in ESP with increases in ventricular volume is the essence of Starling's law of the heart.

The response to ISO reported in Figure 5A (dashed lines) identifies important parameters that define the contractile or inotropic state (ability to contract; contractility) and lusitropic state (ability to relax) of the heart. ISO induces an increase in the amplitude of LVP and the rate of rise of pressure. It also induces a decrease in the end systolic volume (ESV) and end diastolic volume (EDV) and an increase in the rate of decay of LVP. Thus there is contractile reserve that can be called upon to meet hemodynamic demands by increasing pressure and the extent of shortening of the sarcomeres during ejection of blood. There is also a relaxation reserve in which enhanced relaxation is called into play. This is especially important in the physiological context where heart rate increases, as during exercise, and thus cycle time of filling and ejection must be reduced. This rotation of the ESP-Vol relation defines a change in contractility; ISO stimulation induced an increase in developed pressure (tension) at a given ventricular volume (~sarcomere length).

Thin Filaments in Diastole

Sarcomeres in the diastolic state operate along the trajectory of the points connecting the ESV to EDV in the P-Vol loops in Figure 5A. It is important that the thin and thick filaments slide past one another during filling in order to maintain the low ventricular pressures that are optimal for filling in light of the low upstream pressures in the atria and large veins of the body and lungs. The nature of the thin filaments in the diastolic state remains unclear. One issue is whether the crossbridges are all sterically blocked during diastole or whether the crossbridges are all weakly bound. Alternatively, it has been proposed that a fraction of crossbridges are blocked (B-state) and a fraction are in the weak binding C-state. Estimates based on kinetics of myosin-S1 binding and fitting to the three state models described above indicate that 50% of the crossbridges are in the B-state (20).

Another issue related to the thin filament in diastole concerns the position of Tm and Tn and the state of actin. Evidence from X-ray structural studies on intact muscle fibers, which formed the basis of the steric blocking model, indicated early on that accessibility of crossbridge-binding sites on actin resulted from a movement of Tm induced by Ca^{2+} binding to Tn (12–14). However, analysis of the X-ray studies focused only on Tm and was interpreted as a movement of Tm over the actin surface in such a way that subdomains 1 and 2 of actin became exposed. Re-evaluation of the data indicates that interpretation of the X-ray diffraction is ambiguous and may, in part, be accounted for by a movement of subdomains of actin and Tn itself (7). This raises questions as to whether the diastolic state requires a particular structure of the actin domain, rather than a static structure waiting for the crossbridge to bind, and whether Tn itself engages in the steric block. Evidence summarized by Squire & Morris (7) supports the concept that both a particular position of the actin subdomains 1 and 2 and the position of Tn may be important determinants of the relaxed state of striated muscle. However, movements of actin subdomains are difficult to resolve in the X-ray data, and data both support (125) and contradict (126) the idea that actin subdomain motion is involved in Tn-Tm regulation of the filaments. Structural studies indicate that, in addition to tethering or latching Tm in a position that inhibits the ATPase rate and force generation, Tn may occupy a position on actin that interferes with both weak and strong binding of crossbridges. The three-dimensional reconstructions from electron micrographs fitted with the atomic model of actin (31) indicate the involvement of residues 1–4, 23–27, and 47 of actin in the interaction with Tn in the relaxed state, as mentioned above. Acidic residues in the N terminus form a likely interface for ionic interactions with basic residues in TnI. However, extensive alterations in charged residues in the N terminus of actin had no effects on the Ca^{2+} sensitivity of thin filaments as determined in the motility assay, or on the inhibitory activity of TnI (127). Tn binding could, however, induce a Ca^{2+}-insensitive change in the structure of the N terminus of actin. This was theorized by Wong et al. (127) to be of possible significance in the effects of Tn-Tm in improving S-1 binding to the thin filament and in the increase in force developed by myosin reacting with regulated thin filaments, as compared with unregulated actin filaments (128). We (129) tested the significance of modifications in actin subdomains 1 and 2 by comparing tension and ATPase rate in heart myofilaments from nontransgenic (NTG) and transgenic (TG-actin) mice in which enteric γ-actin replaced >95% of the cardiac α-actin. The difference between cardiac α-actin and enteric γ-actin is the introduction of an Asp at the N terminus together with Asp9Glu and Asn360Pro substitutions in subdomain 1, and Thr17Cys and Thr89Ser substitutions in actin subdomain 2. Maximum tension and rate of tension redevelopment (k_{tr}) were the same for myofilaments from NTG and TG-actin mice. However, TG-actin myofilaments were less sensitive to Ca^{2+} than the NTG controls, and economy of tension development (unit tension /unit ATP hydrolyzed) was significantly increased. These data indicate that minor amino acid differences in the actin domains reacting with myosin can induce large differences in function when measured in the lattice of myofilament proteins. Studies at the level of the intact heart demonstrated a hypodynamic phenotype in the TG-actin myocardium (130).

In summary, detailed understanding of the interactions among regions of the thin filament proteins that are critical in maintenance of the diastolic state remain to be determined, especially in the lattice of the sarcomere. However, some general conclusions seem reasonable. We think that the diastolic state involves both steric and allosteric effects on regions of actin that react with crossbridges. Relaxation requires an interaction of cTnI with actin and most likely an interaction of TnT with actin-Tm, both of which immobilize Tm, and possibly Tn, at the outer domains of actin. The result is that most of the crossbridges are blocked from reacting with actin, but a substantial number of crossbridges are cocked, in the terminology of Squire & Morris (7), in that they appear to bind weakly with the thin filament. This makes sense for heart muscle contraction in which basal function involves a small fraction of strong crossbridges reacting with the thin filament and in which there is no spatial summation of motor units. Increases in contractility, as indicated by the change in slope of the ESP-V relation, involve recruitment of crossbridges. It is possible that the B-state of the thin filament holds a portion of this pool in reserve and moves to the C-state only with increases in Ca^{2+} delivered to the myofilaments or alterations of the thin filament (discussed below).

Thin Filaments and the Triggering and Sustaining of Systole

At the end of diastole, there is an abrupt transition to systole triggered by Ca^{2+} binding to the thin filament receptor, cTnC. Flow of blood into the left ventricle ceases as pressure rises in the ventricle and the atrio-ventricular valves close. The LV volume does not change until the pressure exceeds that in the aorta, so for a time, pressure rises isovolumically (Figure 5) or, from the perspective of the myocytes, there is relatively little change in cell length. The function of the thin filaments in this transition from the diastolic state early in systole is reflected reasonably well either by studies using release of caged Ca^{2+} to activate sarcomeres from a relaxed state or by studies using a rapid release/restretch protocol to determine the rate constant for crossbridge entry into the force-generating state. As pointed out by Wolff et al. (131), it is not unreasonable to think that k_{tr} may limit the rate and extent of ventricular pressure development during the early isovolumic phase inasmuch as the duration of Ca^{2+} transients is brief compared with tension development. It is noteworthy that submaximal k_{tr} may increase with the increases in sarcomere length associated with increased end diastolic volume (132). This is of some relevance to Starling's law of the heart (see below).

As discussed in detail by Gordon et al. (8), substantial evidence indicates that the mechanism by which Ca^{2+} regulates the transition of crossbridges from weak- to strong-binding states is by affecting the kinetics of thin filament activation. There is little support for an alternative mechanism involving a direct effect of Ca^{2+} on crossbridge kinetics, for example, by Ca^{2+} binding to myosin light chain 2.

However, whether the dynamics of thin filament–related processes can be rate limiting in the transition from weak to strong crossbridge binding remains to be thoroughly tested in heart muscle preparations. Studies in fast skeletal muscle preparations by Brenner & Chalovich (133) indicate that thin filament processes following Ca^{2+} binding to thin filament are in rapid equilibrium and are not rate limiting. They tracked thin filament structural changes employing fluorescent probes attached to TnI in fast skeletal psoas fibers. Changes in TnI fluorescence were not rate limiting in their measurements of force redevelopment after the rapid release/restretch protocol. The data were well-fit assuming a rapid equilibrium among thin filament states and only two (on and off) states of the thin filament. A role for strongly binding crossbridges was minimized by Brenner & Chalovich (133), who argued that the rigor state is short lived in the intact muscle. They also argued that a time-resolved X-ray diffraction study of the position of Tm on the thin filament (134) demonstrated movement well before tension generation and little movement after tension generation. However, as mentioned above, Squire & Morris (7) have brought the interpretation of these early X-ray data into question, especially with regard to motions of actin and Tn. Moreover, when Ca^{2+}-binding properties of fsTnC were modified by mutations or treatment with small molecules that bind with high affinity, k_{tr} at submaximally activating levels of Ca^{2+} is altered. For example, binding of of calmidazolium (CDZ) alters kinetics of Ca^{2+} binding to TnC with no direct effects on crossbridge cycling and also influences the rate of k_{tr} in skinned rabbit psoas muscle fibers (135). These data indicate that modulation of the thin filament Ca^{2+} signaling, which is apparently most well developed in the heart (60), may modulate k_{tr}. Treatment of skinned fiber bundles with PKA induces an increase in rate of force development as determined by measurements of k_{tr} (136) even though the sensitivity of tension and k_{tr} to Ca^{2+} was reduced. A plot of relative tension versus relative k_{tr} revealed the effect of PKA-dependent phosphorylation on k_{tr}. Patel et al. (136) also reported a similar increase in k_{tr} with a switch of Tm isoforms.

Thin Filaments, Relaxation, and the Return to Diastole

In the context of P-Vol loops (Figure 5), left ventricular relaxation is a complex process that begins somewhere around the time of aortic valve closure, which occurs when pressure in the ventricular chamber falls as blood flows into the arterial tree and as activation declines. Relaxation involves Ca^{2+} removal from cTnC, deactivation of the thin filament, dissociation of crossbridges from the thin filament with the transition from strong to weak and blocked states, and, with the loss of strong crossbridge binding, a loss of cooperative activation of the thin filament. It has also been proposed, on the basis of studies with isolated myofibrils (137) relaxed by a rapid drop in Ca^{2+}, that interactions between sarcomeres induce strain-dependent rebinding of inorganic phosphate to myosin, which is a significant factor in rapidly turning off the actin-crossbridge reaction. The rate of relaxation is also dependent on the afterload, i.e., the pressure against which the ventricle is ejecting blood. With increases in afterload, relaxation slows. Although many investigators focus solely on the kinetics of Ca^{2+} removal from the cytoplasm by the sarcoplasmic reticulum as the rate-limiting step in relaxation, there is abundant evidence that this is not the case and that processes in the sarcomere are significant factors in relaxation. Alterations in relaxation kinetics that occur with alterations in the isoform population of myosin heavy chains, which alter crossbridge cycling rates, provide perhaps the best evidence (138). An important point with regard to the thin filaments in relaxation is the demonstration (139, 140) that enhanced relaxation is more closely correlated with enhanced crossbridge cycling than enhanced release of Ca^{2+} from TnC. Therefore, it is important to consider potential ways that altered thin filament function could modify crossbridge cycling. For example, altered thin filament function appears to alter crossbridge cycling by phosphorylation of cTnI at the PKA sites. Direct evidence that cTnI-P affects crossbridge cycling comes from measurements of tension as a function of imposed sinusoidal length oscillations over a broad range of frequencies. The frequency at the minimum tension (dip frequency) is an indirect measure of the forward and back transitions between weakly bound and strong force-generating crossbridges. Studies in both intact (141) and skinned fiber bundles (139) with specific changes in phosphorylation of cTnI at the PKA sites concluded that phosphorylation induces an increase in crossbridge cycling.

Testing the Hypothesis that Altered Thin Filament Response to Ca^{2+} is an Essential Element in the Control of Cardiac Dynamics

The generation of transgenic mouse models has been critical in testing the hypothesis that specific alterations at the level of the thin filament affect the intensity and dynamics of the myocardium. There are many examples of studies involving introduction of mutant thin filament proteins into the cardiac sarcomeres of transgenic mice. In most cases, the mutations mimic those linked to hypertrophic (HCM) and dilated cardiomyopathies (DCM), but in some the mutations involve deletion of phosphorylation sites (102, 142–145) thought to be critical in modulating thin filament response to Ca^{2+}. Compared with controls, hearts of these transgenic mice demonstrate altered dynamics and response to inotropic interventions. However, as expected from the linkage to cardiomyopathies, cardiac-directed expression of familial hypertrophic cardiomyopathy (FHC) or DCM-related mutant thin filament proteins results in other abnormalities such as fibrosis and cellular remodeling, which makes it difficult to interpret the data with respect to the connection between altered thin filament function and altered cardiac function. In the case of mice harboring transgenes that express thin filament proteins lacking phosphorylation sites, it is not clear that the Ala/Ser or Ala/Thr mutations are benign. For example, we have found, both in vitro (146) and in vivo (147), that the mutant cTnI (S42A, S44A) apparently alters the specificity of sites at Ser-22 and Ser-23 of cTnI as substrates for PKA such that they become excellent substrates for PKC. The Ala substitutions also induce a depression in maximum tension (144). Transgenic models in which there is incorporation of naturally occurring isoforms into the thin filament proteins may hold more promise for testing the hypothesis that thin filament response to Ca^{2+} is an essential element in the control of cardiac dynamics. We have investigated two mouse models of this sort. In one we (148) altered the isoform population of Tm, which is predominantly the α-isoform in the adult, by introducing a transgene expressing β-Tm that resulted in a Tm isoform population consisting of about 60% β-Tm. In another model (149, 150), we used transgenesis to completely switch the isoform population of TnI from 100% cTnI to 100% slow skeletal TnI (ssTnI). In both models there was an increase in myofilament response to Ca^{2+} and a depression in the rate of cardiac relaxation in the basal state, which was exacerbated during stimulation with ISO.

A particularly useful model has been one in which the adult form, cTnI, was replaced with ssTnI, which is the isoform expressed in the embryonic and neonatal heart in all species studied to date. Data from investigations of changes in cardiac function with a drop in pH support the hypothesis. Inasmuch as the peak amplitude of the Ca^{2+} transient increases with the fall in tension (151), the most straightforward interpretation of the mechanism for this depression in cardiac activity is a desensitization of the myofilaments to Ca^{2+}. The observation that myofilaments from neonatal dog hearts demonstrate a greatly reduced deactivation by acidic pH indicates a developmental change in the thin filament. With the identification of ssTnI as the embryonic and neonatal isoform in the heart, together with data indicating that isoform switching of TnI may be the most important difference between adult and immature hearts with respect to response to acidosis, it was natural to generate a transgenic (TG) mouse in which ssTnI replaced cTnI. Compared with NTG controls, the Ca^{2+} sensitivity is enhanced in myofilaments from ssTnI-TG mouse hearts, and deactivation by acidic pH is significantly blunted (150). Moreover, papillary muscles from ssTnI-TG hearts generate the same tension when superfused with buffer at physiological pH and buffer at a reduced pH induced by hypercapnia. Under these same acidotic conditions there was the expected significant reduction in force developed by NTG papillary muscle. In view of evidence that expression of ssTnI in the TG mouse heart has very minor effects on cellular Ca^{2+} flux, these results support the hypothesis that altered response to Ca^{2+} may dominate regulation of cardiac activity and dynamics under certain conditions.

The dependence of ventricular systolic pressure on ventricular volume, more commonly known as the Frank-Starling relation, represents another mode of regulation apparently dominated by a length-dependent myofilament response to Ca^{2+}. The relation between ESP and ventricular volume is a measure of Starling's law, which is believed to be rooted in a dependence of cellular tension on sarcomere length (SL). In addition to the increase in maximum tension developed by the myofilaments with increases in SL, there is also an increase in myofilament sensitivity to Ca^{2+}. The mechanism for these changes has been attributed to a decrease in interfilament spacing with increasing SL, but explicit measurements indicate no correlation between lattice spacing and length-dependent activation. More complicated mechanisms involving titin interactions with the thick filament or thin filament may account for length-dependent activation (11). Whatever the case, myofilaments from ssTnI-TG hearts show a significant blunting of length-dependent activation that is not correlated with a change in interfilament spacing

(152). The reduced length dependence of activation predicts a reduction in the steepness of the ESP-ventricular volume relation. This reduction in slope has, in fact, been demonstrated in preliminary studies with ssTG hearts beating in situ during stimulation with ISO. It has also been reported that PKA-dependent phosphorylation of the myofilaments enhances length-dependent activation. Thus under beta-adrenergic stimulation, differences between ssTnI-TG hearts, in which there is no PKA-dependent phosphorylation of TnI, and NTG hearts would be expected to be amplified. These results provide additional and compelling evidence that regulation by alterations in myofilament response to Ca^{2+} is a fundamentally important regulatory device in the heart.

Three recent studies (124, 153, 154) provide strong evidence for a significant role of thin filament proteins as a determinant of contraction and relaxation reserve. These studies also illustrate the importance of evaluating the role of alterations in Ca^{2+} control of thin filament function with the sarcomeres operating under physiological conditions. Investigations by Pena & Wolska (153) of hearts beating in situ (therefore innervated and perfused with blood under physiological loading conditions) support a pivotal role of cTnI phosphorylation as a determinant of relaxation. This study investigated phospholamban knockout (PLBKO) mice expressing either cTnI or ssTnI. PLBKO/cTnI and PLBKO/ssTnI hearts demonstrated similar $-dP/dt$ in the basal state, but during perfusion with ISO, $-dP/dt$ was significantly slower in PLBKO/ssTnI than in PLBKO/cTnI hearts. In another study, Takimoto et al. (154) investigated TG mice expressing cTnI(D22/D23) in which Asp residues replaced Ser in the N-terminal extension unique to cTnI. In vitro studies demonstrated that this substitution produces a pseudophosphorylated cTnI, which mimicked effects of cTnI phosphorylated at the Ser-22 and Ser-23 PKA sites. In the basal state, diastolic and systolic function in hearts of control and cTnI(D22/D23) mice were similar. However, the responses to increases in frequency and afterload were significantly different in these hearts. Rate-dependent increases in contraction and relaxation were enhanced in the cTnI(D22/D23) hearts compared with those in controls. Moreover, prolongation of the time course of relaxation associated with increases in afterload was greatest in the controls. Following beta-adrenergic stimulation, these differences in cardiac function between cTnI(D22/D23) and NTG mice were not evident.

Layland et al. (124) investigated isolated, perfused working hearts from TG mice expressing slow skeletal TnI (ssTnI) in place of cTnI, which lacks the N-terminal extension containing Ser-23, Ser-24 present in cTnI. Replacement of cTnI with ssTnI renders the myofilaments insensitive to effects (reduced sensitivity to Ca^{2+}) of phosphorylation by PKA (155). Perfusion with ISO reduced relaxation time and ESV in control hearts, as illustrated in Figure 5, but not in hearts of ssTnI mice. Contractility, as reflected in the ESV-pressure relation, was also increased by ISO to a greater extent in the controls than in the ssTnI hearts. Isovolumic hearts beating at constant pressure did not demonstrate such significant differences as in the auxotonically loaded, ejecting heart. Similarly, the difference noted above between hearts expressing cTnI(D22/D23) was not present in isometric papillary muscle with increased frequency of stimulation or beta-adrenergic stimulation. A relatively small effect of myofilament phosphorylation had been previously reported in studies comparing tension in isometric papillary muscles or ventricular myocytes isolated from hearts of control mice and mice lacking phospholamban (149, 156). Thus functional effects of phosphorylation of cTnI are fully expressed only when the myofilaments are shortening against a load. This indicates that strain-dependent effects on crossbridge cycling rate may be modified by thin filament proteins.

SUMMARY AND FUTURE DIRECTIONS

Investigation of the control of thin filament function in the heart is at an exciting juncture. The recent data on the role of thin filament alterations in the beating heart have brought fresh understanding to the complexity of control mechanisms at the level of the sarcomere. It is certain that new and exciting lines of investigation will emerge, based on the structural information derived from the crystal structure of the Tn complex. It should not be too long before additional structures emerge showing more detail and different states of the thin filament. The confluence of this information, together with the biochemistry, biophysics, and systems physiology of the heart, will permit a clearer understanding of the role of the thin filament in cardiac function. This information is highly relevant to the diagnosis and treatment of prevalent cardiac diseases genetically linked to thin filament proteins. It is also relevant to a continuing quest for therapeutic agents that modify the sarcomere. A particularly interesting and exciting idea mentioned at the outset is the modern view of the thin filament as it functions in the A-band region and in the I-Z-I region of the sarcomere. Future studies must incorporate the full thin filament in their approaches in order to understand the multiplex functions of the sarcomere apart from shortening and generating force.

ACKNOWLEDGMENTS

The authors thank Dr. Anne Martin for critical reading of the manuscript. The authors also gratefully acknowledge support from the National Institutes of Health and American Heart Association.

The *Annual Review of Physiology* is online at http://physiol.annualreviews.org

LITERATURE CITED

1. Tobacman LS. 1996. Thin filament-mediated regulation of cardiac contraction. *Annu. Rev. Physiol.* 58:447–81
2. Farah CS, Reinach FC. 1995. The troponin complex and regulation of muscle contraction. *FASEB J.* 9:755–67
3. Solaro RJ, Rarick HM. 1998. Troponin and tropomyosin: proteins that switch on and tune in the activity of cardiac myofilaments. *Circ. Res.* 83:471–80
4. Perry SV. 2001. Vertebrate tropomyosin: distribution, properties and function. *J. Muscle Res. Cell Motil.* 22:5–49
5. Perry SV. 1999. Troponin I: inhibitor or facilitator. *Mol. Cell Biochem.* 190:9–32

6. Perry SV. 1998. Troponin T: genetics, properties and function. *J. Muscle Res. Cell Motil.* 19:575–602

7. Squire JM, Morris EP. 1998. A new look at thin filament regulation in vertebrate skeletal muscle. *FASEB J.* 12:761–71

8. Gordon AM, Homsher E, Regnier M. 2000. Regulation of contraction in striated muscle. *Physiol. Rev.* 80:853–924

9. Metzger JM, Westfall MV. 2004. Covalent and noncovalent modification of thin filament action: the essential role of troponin in cardiac muscle regulation. *Circ. Res.* 94:146–58

10. Pyle WG, Solaro RJ. 2004. At the crossroads of myocardial signaling: the role of Z-discs in intracellular signaling and cardiac function. *Circ. Res.* 94:296–305

11. Granzier HL, Labeit S. 2004. The giant protein titin: a major player in myocardial mechanics, signaling, and disease. *Circ. Res.* 94:284–95

12. Haselgrove JC. 1972. X-ray evidence for a conformational change in actin-containing filaments of vertebrate striated muscle. *Cold Spring Harbor Symp. Quant. Biol.* 37:341–52

13. Huxley HE. 1972. Structural changes in actin- and myosin-containing filaments during contraction. *Cold Spring Harbor Symp. Quant. Biol.* 37:361–76

14. Parry DA, Squire JM. 1973. Structural role of tropomyosin in muscle regulation: analysis of the X-ray diffraction patterns from relaxed and contracting muscles. *J. Mol. Biol.* 75:33–55

15. Hill TL, Eisenberg E, Greene L. 1980. Theoretical model for the cooperative equilibrium binding of myosin subfragment 1 to the actin-troponin-tropomyosin complex. *Proc. Natl. Acad. Sci. USA* 77:3186–90

16. McKillop DF, Geeves MA. 1993. Regulation of the interaction between actin and myosin subfragment 1: evidence for three states of the thin filament. *Biophys. J.* 65:693–701

17. Tobacman LS, Butters CA. 2000. A new model of cooperative myosin-thin filament binding. *J. Biol. Chem.* 275:27587–93

18. Chalovich JM. 2002. Regulation of striated muscle contraction: a discussion. *J. Muscle Res. Cell Motil.* 23:353–61

19. Maytum R, Lehrer SS, Geeves MA. 1999. Cooperativity and switching within the three-state model of muscle regulation. *Biochemistry* 38:1102–10

20. Maytum R, Westerdorf B, Jaquet K, Geeves MA. 2003. Differential regulation of the actomyosin interaction by skeletal and cardiac troponin isoforms. *J. Biol. Chem.* 278:6696–701

21. Vibert P, Craig R, Lehman W. 1997. Steric-model for activation of muscle thin filaments. *J. Mol. Biol.* 266:8–14

22. Craig R, Lehman W. 2001. Crossbridge and tropomyosin positions observed in native, interacting thick and thin filaments. *J. Mol. Biol.* 311:1027–36

23. Xu C, Craig R, Tobacman L, Horowitz R, Lehman W. 1999. Tropomyosin positions in regulated thin filaments revealed by cryoelectron microscopy. *Biophys. J.* 77:985–92

24. Miki M, Miura T, Sano K, Kimura H, Kondo H, et al. 1998. Fluorescence resonance energy transfer between points on tropomyosin and actin in skeletal muscle thin filaments: Does tropomyosin move? *J. Biochem.* 123:1104–11

25. Bacchiocchi C, Lehrer SS. 2002. Ca^{2+}-induced movement of tropomyosin in skeletal muscle thin filaments observed by multi-site FRET. *Biophys. J.* 82:1524–36

26. Kobayashi T, Kobayashi M, Collins JH. 2001. Ca^{2+}-dependent, myosin subfragment 1-induced proximity changes between actin and the inhibitory region of troponin I. *Biochim. Biophys. Acta* 1549:148–54

27. Hai H, Sano K, Maeda K, Maeda Y, Miki M. 2002. Ca^{2+}- and S1-induced conformational changes of reconstituted skeletal muscle thin filaments observed

by fluorescence energy transfer spectroscopy: structural evidence for three states of thin filament. *J. Biochem.* 131:407–18

28. Kimura C, Maeda K, Maeda Y, Miki M. 2002. Ca^{2+}- and S1-induced movement of troponin T on reconstituted skeletal muscle thin filaments observed by fluorescence energy transfer spectroscopy. *J. Biochem.* 132:93–102

29. Julian FJ. 1969. Activation in a skeletal muscle contraction model with a modification for insect fibrillar muscle. *Biophys. J.* 9:547–70

30. Brenner B. 1988. Effect of Ca^{2+} on cross-bridge turnover kinetics in skinned single rabbit psoas fibers: implications for regulation of muscle contraction. *Proc. Natl. Acad. Sci. USA* 85:3265–69

31. Lehman W, Rosol M, Tobacman LS, Craig R. 2001. Troponin organization on relaxed and activated thin filaments revealed by electron microscopy and three-dimensional reconstruction. *J. Mol. Biol.* 307:739–44

32. Landis C, Back N, Homsher E, Tobacman LS. 1999. Effects of tropomyosin internal deletions on thin filament function. *J. Biol. Chem.* 274:31279–85

33. Rosol M, Lehman W, Craig R, Landis C, Butters C, Tobacman LS. 2000. Three-dimensional reconstruction of thin filaments containing mutant tropomyosin. *Biophys. J.* 78:908–17

34. Levine BA, Moir AJ, Perry SV. 1988. The interaction of troponin-I with the N-terminal region of actin. *Eur. J. Biochem.* 172:389–97

35. Grabarek Z, Gergely J. 1987. Troponin-I binds to the N-terminal 12-residue segment of actin. *Biophys. J.* 51:A331

36. Luo Y, Leszyk J, Li B, Li Z, Gergely J, Tao T. 2002. Troponin-I interacts with the Met47 region of skeletal muscle actin. Implications for the mechanism of thin filament regulation by calcium. *J. Mol. Biol.* 316:429–34

37. Narita A, Yasunaga T, Ishikawa T, Mayanagi K, Wakabayashi T. 2001. Ca^{2+}-induced switching of troponin and tropomyosin on actin filaments as revealed by electron cryo-microscopy. *J. Mol. Biol.* 308:241–61

38. Takeda S, Yamashita A, Maeda K, Maeda Y. 2003. Structure of the core domain of human cardiac troponin in the Ca^{2+}-saturated form. *Nature* 424:35–41

39. Herzberg O, James MN. 1985. Structure of the calcium regulatory muscle protein troponin-C at 2.8 Å resolution. *Nature* 313:653–59

40. Sundaralingam M, Bergstrom R, Strasburg G, Rao ST, Roychowdhury P, et al. 1985. Molecular structure of troponin C from chicken skeletal muscle at 3-angstrom resolution. *Science* 227:945–48

41. Houdusse A, Love ML, Dominguez R, Grabarek Z, Cohen C. 1997. Structures of four Ca^{2+}-bound troponin C at 2.0 Å resolution: further insights into the Ca^{2+}-switch in the calmodulin superfamily. *Structure* 5:1695–711

42. Kawasaki H, Nakayama S, Kretsinger RH. 1998. Classification and evolution of EF-hand proteins. *Biometals* 11:277–95

43. Pan BS, Solaro RJ. 1987. Calcium-binding properties of troponin C in detergent-skinned heart muscle fibers. *J. Biol. Chem.* 262:7839–49

44. Bremel RD, Weber A. 1972. Cooperation within actin filament in vertebrate skeletal muscle. *Nat. New Biol.* 238:97–101

45. Robertson SP, Johnson JD, Potter JD. 1981. The time-course of Ca^{2+} exchange with calmodulin, troponin, parvalbumin, and myosin in response to transient increases in Ca^{2+}. *Biophys. J.* 34:559–69

46. Johnson JD, Charlton SC, Potter JD. 1979. A fluorescence stopped flow analysis of Ca^{2+} exchange with troponin C. *J. Biol. Chem.* 254:3497–502

47. Putkey JA, Sweeney HL, Campbell ST. 1989. Site-directed mutation of the trigger calcium-binding sites in cardiac troponin C. *J. Biol. Chem.* 264:12370–78

48. Robertson SP, Johnson JD, Holroyde MJ, Kranias EG, Potter JD, Solaro RJ. 1982.

The effect of troponin I phosphorylation on the Ca^{2+}-binding properties of the Ca^{2+}-regulatory site of bovine cardiac troponin. *J. Biol. Chem.* 257:260–63

49. Dong W, Rosenfeld SS, Wang CK, Gordon AM, Cheung HC. 1996. Kinetic studies of calcium binding to the regulatory site of troponin C from cardiac muscle. *J. Biol. Chem.* 271:688–94

50. Dong WJ, Wang CK, Gordon AM, Rosenfeld SS, Cheung HC. 1997. A kinetic model for the binding of Ca^{2+} to the regulatory site of troponin from cardiac muscle. *J. Biol. Chem.* 272:19229–35

51. Hazard AL, Kohout SC, Stricker NL, Putkey JA, Falke JJ. 1998. The kinetic cycle of cardiac troponin C: calcium binding and dissociation at site II trigger slow conformational rearrangements. *Protein Sci.* 7:2451–59

52. Ramakrishnan S, Hitchcock-DeGregori SE. 1996. Structural and functional significance of aspartic acid 89 of the troponin C central helix in Ca^{2+} signaling. *Biochemistry* 35:15515–21

53. Kobayashi T, Zhao X, Wade R, Collins JH. 1999. Involvement of conserved, acidic residues in the N-terminal domain of troponin C in calcium-dependent regulation. *Biochemistry* 38:5386–91

54. Potter JD, Sheng Z, Pan BS, Zhao J. 1995. A direct regulatory role for troponin T and a dual role for troponin C in the Ca^{2+} regulation of muscle contraction. *J. Biol. Chem.* 270:2557–62

55. Bing W, Fraser ID, Marston SB. 1997. Troponin I and troponin T interact with troponin C to produce different Ca^{2+}-dependent effects on actin-tropomyosin filament motility. *Biochem. J.* 327(Part 2):335–40

56. Blumenschein TM, Tripet BP, Hodges RS, Sykes BD. 2001. Mapping the interacting regions between troponins T and C. Binding of TnT and TnI peptides to TnC and NMR mapping of the TnT-binding site on TnC. *J. Biol. Chem.* 276:36606–12

57. Ohtsuki I, Shiraishi F. 2002. Periodic binding of troponin C.I and troponin I to tropomyosin-actin filaments. *J. Biochem.* 131:739–43

58. Syska H, Wilkinson JM, Grand RJ, Perry SV. 1976. The relationship between biological activity and primary structure of troponin I from white skeletal muscle of the rabbit. *Biochem. J.* 153:375–87

59. Talbot JA, Hodges RS. 1981. Synthetic studies on the inhibitory region of rabbit skeletal troponin I. Relationship of amino acid sequence to biological activity. *J. Biol. Chem.* 256:2798–802

60. Solaro R, Wolska B, Arteaga GM, Martin AF, Buttrick PM, De Tombe PP. 2002. Modulation of thin filament activity in long and short term regulation of cardiac function. See Ref. 158, pp. 291–327

61. Rarick HM, Tu XH, Solaro RJ, Martin AF. 1997. The C terminus of cardiac troponin I is essential for full inhibitory activity and Ca^{2+} sensitivity of rat myofibrils. *J. Biol. Chem.* 272:26887–92

62. Ramos CH. 1999. Mapping subdomains in the C-terminal region of troponin I involved in its binding to troponin C and to thin filament. *J. Biol. Chem.* 274:18189–95

63. Tripet B, Van Eyk JE, Hodges RS. 1997. Mapping of a second actin-tropomyosin and a second troponin C binding site within the C terminus of troponin I, and their importance in the Ca^{2+}-dependent regulation of muscle contraction. *J. Mol. Biol.* 271:728–50

64. Hernandez G, Blumenthal DK, Kennedy MA, Unkefer CJ, Trewhella J. 1999. Troponin I inhibitory peptide (96–115) has an extended conformation when bound to skeletal muscle troponin C. *Biochemistry* 38:6911–17

65. Tung CS, Wall ME, Gallagher SC, Trewhella J. 2000. A model of troponin-I in complex with troponin-C using hybrid experimental data: the inhibitory region is a beta-hairpin. *Protein Sci.* 9:1312–26

66. Lindhout DA, Sykes BD. 2003. Structure and dynamics of the C-domain of human cardiac troponin C in complex with the inhibitory region of human cardiac troponin I. *J. Biol. Chem.* 278:27024–34

67. Brown LJ, Sale KL, Hills R, Rouviere C, Song L, et al. 2002. Structure of the inhibitory region of troponin by site directed spin labeling electron paramagnetic resonance. *Proc. Natl. Acad. Sci. USA* 99:12765–70

68. Dong WJ, Xing J, Robinson JM, Cheung HC. 2001. Ca^{2+} induces an extended conformation of the inhibitory region of troponin I in cardiac muscle troponin. *J. Mol. Biol.* 314:51–61

69. Dong WJ, Robinson JM, Stagg S, Xing J, Cheung HC. 2003. Ca^{2+}-induced conformational transition in the inhibitory and regulatory regions of cardiac troponin I. *J. Biol. Chem.* 278:8686–92

70. Gagne SM, Tsuda S, Li MX, Smillie LB, Sykes BD. 1995. Structures of the troponin C regulatory domains in the apo and calcium-saturated states. *Nat. Struct. Biol.* 2:784–89

71. Slupsky CM, Kay CM, Reinach FC, Smillie LB, Sykes BD. 1995. Calcium-induced dimerization of troponin C: mode of interaction and use of trifluoroethanol as a denaturant of quaternary structure. *Biochemistry* 34:7365–75

72. Slupsky CM, Sykes BD. 1995. NMR solution structure of calcium-saturated skeletal muscle troponin C. *Biochemistry* 34:15953–64

73. Strynadka NC, Cherney M, Sielecki AR, Li MX, Smillie LB, James MN. 1997. Structural details of a calcium-induced molecular switch: X-ray crystallographic analysis of the calcium-saturated N-terminal domain of troponin C at 1.75 Å resolution. *J. Mol. Biol.* 273:238–55

74. McKay RT, Tripet BP, Hodges RS, Sykes BD. 1997. Interaction of the second binding region of troponin I with the regulatory domain of skeletal muscle troponin C as determined by NMR spectroscopy. *J. Biol. Chem.* 272:28494–500

75. Luo Y, Leszyk J, Qian Y, Gergely J, Tao T. 1999. Residues 48 and 82 at the N-terminal hydrophobic pocket of rabbit skeletal muscle troponin-C photo-cross-link to Met121 of troponin-I. *Biochemistry* 38:6678–88

76. McKay RT, Tripet BP, Pearlstone JR, Smillie LB, Sykes BD. 1999. Defining the region of troponin-I that binds to troponin-C. *Biochemistry* 38:5478–89

77. Kobayashi T, Zhao X, Wade R, Collins JH. 1999. Ca^{2+}-dependent interaction of the inhibitory region of troponin I with acidic residues in the N-terminal domain of troponin C. *Biochim. Biophys. Acta* 1430:214–21

78. Sia SK, Li MX, Spyracopoulos L, Gagne SM, Liu W, et al. 1997. Structure of cardiac muscle troponin C unexpectedly reveals a closed regulatory domain. *J. Biol. Chem.* 272:18216–21

79. Spyracopoulos L, Li MX, Sia SK, Gagne SM, Chandra M, et al. 1997. Calcium-induced structural transition in the regulatory domain of human cardiac troponin C. *Biochemistry* 36:12138–46

80. Li MX, Spyracopoulos L, Sykes BD. 1999. Binding of cardiac troponin-I147-163 induces a structural opening in human cardiac troponin-C. *Biochemistry* 38:8289–98

81. Dong WJ, Xing J, Villain M, Hellinger M, Robinson JM, et al. 1999. Conformation of the regulatory domain of cardiac muscle troponin C in its complex with cardiac troponin I. *J. Biol. Chem.* 274:31382–90

82. Tao T, Gong BJ, Leavis PC. 1990. Calcium-induced movement of troponin-I relative to actin in skeletal muscle thin filaments. *Science* 247:1339–41

83. Miki M. 1990. Resonance energy transfer between points in a reconstituted skeletal muscle thin filament. A conformational change of the thin filament in response to a change in Ca^{2+} concentration. *Eur. J. Biochem.* 187:155–62

84. Miki M, Kobayashi T, Kimura H, Hagiwara A, Hai H, Maeda Y. 1998. Ca^{2+}-induced distance change between points on actin and troponin in skeletal muscle thin filaments estimated by fluorescence energy transfer spectroscopy. *J. Biochem.* 123:324–31

85. Kobayashi T, Kobayashi M, Gryczynski Z, Lakowicz JR, Collins JH. 2000. Inhibitory region of troponin I: Ca^{2+}-dependent structural and environmental changes in the troponin-tropomyosin complex and in reconstituted thin filaments. *Biochemistry* 39:86–91

86. Li Z, Gergely J, Tao T. 2001. Proximity relationships between residue 117 of rabbit skeletal troponin-I and residues in troponin-C and actin. *Biophys. J.* 81:321–33

87. Anderson PAW. 2002. Thin filament regulation in development. See Ref. 158, pp. 329–77

88. Gomes AV, Guzman G, Zhao J, Potter JD. 2002. Cardiac troponin T isoforms affect the Ca^{2+} sensitivity and inhibition of force development. Insights into the role of troponin T isoforms in the heart. *J. Biol. Chem.* 277:35341–49

89. Hinkle A, Goranson A, Butters CA, Tobacman LS. 1999. Roles for the troponin tail domain in thin filament assembly and regulation. A deletional study of cardiac troponin T. *J. Biol. Chem.* 274:7157–64

90. Chandra M, Montgomery DE, Kim JJ, Solaro RJ. 1999. The N-terminal region of troponin T is essential for the maximal activation of rat cardiac myofilaments. *J. Mol. Cell Cardiol.* 31:867–80

91. Maytum R, Geeves MA, Lehrer SS. 2002. A modulatory role for the troponin T tail domain in thin filament regulation. *J. Biol. Chem.* 277:29774–80

92. Oliveira DM, Nakaie CR, Sousa AD, Farah CS, Reinach FC. 2000. Mapping the domain of troponin T responsible for the activation of actomyosin ATPase activity. Identification of residues involved in binding to actin. *J. Biol. Chem.* 275:27513–19

93. Rarick HM, Tang HP, Guo XD, Martin AF, Solaro RJ. 1999. Interactions at the NH2-terminal interface of cardiac troponin I modulate myofilament activation. *J. Mol. Cell Cardiol.* 31:363–75

94. Jin JP, Root DD. 2000. Modulation of troponin T molecular conformation and flexibility by metal ion binding to the NH2-terminal variable region. *Biochemistry* 39:11702–13

95. Solaro RJ. 2001. Modulation of cardiac myofilament activity by protein phosphorylation. In *Handbook of Physiology: Section 2: The Cardiovascular System*, ed. E Page, H Fozzard, RJ Solaro, pp. 264–300. New York: Oxford Univ. Press

96. Mittmann K, Jaquet K, Heilmeyer LM Jr. 1992. Ordered phosphorylation of a duplicated minimal recognition motif for cAMP-dependent protein kinase present in cardiac troponin I. *FEBS Lett.* 302:133–37

97. Keane NE, Quirk PG, Gao Y, Patchell VB, Perry SV, Levine BA. 1997. The ordered phosphorylation of cardiac troponin I by the cAMP-dependent protein kinase—structural consequences and functional implications. *Eur. J. Biochem.* 248:329–37

98. Noland TA Jr, Raynor RL, Kuo JF. 1989. Identification of sites phosphorylated in bovine cardiac troponin I and troponin T by protein kinase C and comparative substrate activity of synthetic peptides containing the phosphorylation sites. *J. Biol. Chem.* 264:20778–85

99. Kobayashi T, Yang X, Walker L, van Breemen R, Solaro RJ. 2004. Determination of PKC-dependent phosphorylation sites on cardiac troponin I. *Biophys. J.* 86:A395

100. Buscemi N, Foster DB, Neverova I, Van Eyk JE. 2002. p21-activated kinase increases the calcium sensitivity of rat triton-skinned cardiac muscle fiber bundles via a mechanism potentially involving novel phosphorylation of troponin I. *Circ. Res.* 91:509–16

101. Ke YB, Wang L, Pyle WG, De Tombe PP, Solaro RJ. 2004. Intracellular localization and functional effects of P21-activated kinase-1 (Pak1) in cardiac myocytes. *Circ. Res.* 94:194–200

102. Pi Y, Zhang D, Kemnitz KR, Wang H, Walker JW. 2003. Protein kinase C and A sites on troponin I regulate myofilament Ca^{2+} sensitivity and ATPase activity in the mouse myocardium. *J. Physiol.* 552:845–57

103. Burkart EM, Sumandea MP, Kobayashi T, Nili M, Martin AF, et al. 2003. Phosphorylation or glutamic acid substitution at protein kinase C sites on cardiac troponin I differentially depress myofilament tension and shortening velocity. *J. Biol. Chem.* 278:11265–72

104. Jideama NM, Noland TA Jr., Raynor RL, Blobe GC, Fabbro D, et al. 1996. Phosphorylation specificities of protein kinase C isozymes for bovine cardiac troponin I and troponin T and sites within these proteins and regulation of myofilament properties. *J. Biol. Chem.* 271:23277–83

105. Sumandea MP, Pyle WG, Kobayashi T, de Tombe PP, Solaro RJ. 2003. Identification of a functionally critical protein kinase C phosphorylation residue of cardiac troponin T. *J. Biol. Chem.* 278:35135–44

106. Kobayashi T, Dong W, Burkart EM, Cheung HC, Solaro RJ. 2004. Effects of protein kinase C dependent phosphorylation and a familial hypertrophic cardiomyopathy-related mutation of cardiac troponin I on structural transition of troponin C and myofilament activation. *Biochemistry* 43:5996–6004

107. Finley N, Abbott MB, Abusamhadneh E, Gaponenko V, Dong W, et al. 1999. NMR analysis of cardiac troponin C-troponin I complexes: effects of phosphorylation. *FEBS Lett.* 453:107–12

108. Ward DG, Brewer SM, Cornes MP, Trayer IP. 2003. A cross-linking study of the N-terminal extension of human cardiac troponin I. *Biochemistry* 42:10324–32

109. Li MX, Wang X, Lindhout DA, Buscemi N, Van Eyk JE, Sykes BD. 2003. Phosphorylation and mutation of human cardiac troponin I deferentially destabilize the interaction of the functional regions of troponin I with troponin C. *Biochemistry* 42:14460–68

110. Wolska BM, Wieczorek DM. 2003. The role of tropomyosin in the regulation of myocardial contraction and relaxation. *Pflügers Arch.* 446:1–8

111. Landis CA, Bobkova A, Homsher E, Tobacman LS. 1997. The active state of the thin filament is destabilized by an internal deletion in tropomyosin. *J. Biol. Chem.* 272:14051–56

112. Hitchcock-DeGregori SE, Song Y, Greenfield NJ. 2002. Functions of tropomyosin's periodic repeats. *Biochemistry* 41:15036–44

113. Lehman W, Hatch V, Korman V, Rosol M, Thomas L, et al. 2000. Tropomyosin and actin isoforms modulate the localization of tropomyosin strands on actin filaments. *J. Mol. Biol.* 302:593–606

114. Brown JH, Kim KH, Jun G, Greenfield NJ, Dominguez R, et al. 2001. Deciphering the design of the tropomyosin molecule. *Proc. Natl. Acad. Sci. USA* 98:8496–501

115. Singh A, Hitchcock-DeGregori SE. 2003. Local destabilization of the tropomyosin coiled coil gives the molecular flexibility required for actin binding. *Biochemistry* 42:14114–21

116. Li Y, Mui S, Brown JH, Strand J, Reshetnikova L, et al. 2002. The crystal structure of the C-terminal fragment of striated-muscle alpha-tropomyosin reveals a key troponin T recognition site. *Proc. Natl. Acad. Sci. USA* 99:7378–83

117. Greenfield NJ, Swapna GV, Huang Y, Palm T, Graboski S, et al. 2003. The structure of the carboxyl terminus of striated alpha-tropomyosin in solution reveals an unusual parallel arrangement of interacting alpha-helices. *Biochemistry* 42:614–19

118. Greenfield NJ, Palm T, Hitchcock-DeGregori SE. 2002. Structure and interactions of the carboxyl terminus of striated muscle alpha-tropomyosin: it is important to be flexible. *Biophys. J.* 83: 2754–66

119. Farah CS, Reinach FC. 1999. Regulatory properties of recombinant tropomyosins containing 5-hydroxytryptophan: Ca^{2+}-binding to troponin results in a conformational change in a region of tropomyosin outside the troponin binding site. *Biochemistry* 38:10543–51

120. Lehrer SS, Golitsina NL, Geeves MA. 1997. Actin-tropomyosin activation of myosin subfragment 1 ATPase and thin filament cooperativity. The role of tropomyosin flexibility and end-to-end interactions. *Biochemistry* 36:13449–54

121. Smith DA, Geeves MA. 2003. Cooperative regulation of myosin-actin interactions by a continuous flexible chain II: actin-tropomyosin-troponin and regulation by calcium. *Biophys. J.* 84:3168–80

122. Smith DA, Maytum R, Geeves MA. 2003. Cooperative regulation of myosin-actin interactions by a continuous flexible chain I: actin-tropomyosin systems. *Biophys. J.* 84:3155–67

123. Solaro RJ. 1999. Integration of myofilament response to Ca^{2+} with cardiac pump regulation and pump dynamics. *Adv. Physiol. Educ.* 22:S155–63

124. Layland J, Grieve DJ, Cave AC, Sparks E, Solaro RJ, Shah AM. 2004. Essential role of troponin I in the positive inotropic response to isoprenaline in mouse hearts contracting auxotonically. *J. Physiol.* 556.3:835–47

125. Mendelson R, Morris E. 1994. Combining electron microscopy and X-ray crystallography data to study the structure of F-actin and its implications for thin-filament regulation in muscle. *Adv. Exp. Med. Biol.* 358:13–23

126. Gerson JH, Kim E, Muhlrad A, Reisler E. 2001. Tropomyosin-troponin regulation of actin does not involve subdomain 2 motions. *J. Biol. Chem.* 276:18442–49

127. Wong WW, Gerson JH, Rubenstein PA, Reisler E. 2002. Thin filament regulation and ionic interactions between the N-terminal region in actin and troponin. *Biophys. J.* 83:2726–32

128. Homsher E, Lee DM, Morris C, Pavlov D, Tobacman LS. 2000. Regulation of force and unloaded sliding speed in single thin filaments: effects of regulatory proteins and calcium. *J. Physiol.* 524(Part 1):233–43

129. Martin AF, Phillips RM, Kumar A, Crawford K, Abbas Z, et al. 2002. Ca^{2+} activation and tension cost in myofilaments from mouse hearts ectopically expressing enteric gamma-actin. *Am. J. Physiol. Heart Circ. Physiol.* 283:H642–49

130. Kumar A, Crawford K, Close L, Madison M, Lorenz J, et al. 1997. Rescue of cardiac alpha-actin-deficient mice by enteric smooth muscle gamma-actin. *Proc. Natl. Acad. Sci. USA* 94:4406–11

131. Wolff MR, McDonald KS, Moss RL. 1995. Rate of tension development in cardiac muscle varies with level of activator calcium. *Circ. Res.* 76:154–60

132. McDonald KS, Wolff MR, Moss RL. 1997. Sarcomere length dependence of the rate of tension redevelopment and submaximal tension in rat and rabbit skinned skeletal muscle fibres. *J. Physiol.* 501 (Part 3):607–21

133. Brenner B, Chalovich JM. 1999. Kinetics of thin filament activation probed by fluorescence of *N*-((2-(Iodoacetoxy)ethyl)-*N*-methyl)amino-7-nitrobenz-2-oxa-1, 3-diazole-labeled troponin I incorporated into skinned fibers of rabbit psoas muscle: implications for regulation of muscle contraction. *Biophys. J.* 77:2692–708

134. Kress M, Huxley HE, Faruqi AR, Hendrix J. 1986. Structural changes during activation of frog muscle studied by time-resolved X-ray diffraction. *J. Mol. Biol.* 188:325–42

135. Regnier M, Martyn DA, Chase PB. 1996. Calmidazolium alters Ca^{2+} regulation of tension redevelopment rate in skinned skeletal muscle. *Biophys. J.* 71:2786–94

136. Patel JR, Fitzsimons DP, Buck SH, Muthuchamy M, Wieczorek DF, Moss RL. 2001. PKA accelerates rate of force development in murine skinned myocardium expressing alpha- or beta-tropomyosin. *Am. J. Physiol. Heart Circ. Physiol.* 280:H2732–39

137. Stehle R, Kruger M, Pfitzer G. 2002. Force kinetics and individual sarcomere dynamics in cardiac myofibrils after rapid Ca^{2+} changes. *Biophys. J.* 83:2152–61

138. Fitzsimons DP, Patel JR, Moss RL. 1998. Role of myosin heavy chain composition in kinetics of force development and relaxation in rat myocardium. *J. Physiol.* 513(Part 1):171–83

139. Kentish JC, McCloskey DT, Layland J, Palmer S, Leiden JM, et al. 2001. Phosphorylation of troponin I by protein kinase A accelerates relaxation and crossbridge cycle kinetics in mouse ventricular muscle. *Circ. Res.* 88:1059–65

140. Palmer S, Kentish JC. 1998. Roles of Ca^{2+} and crossbridge kinetics in determining the maximum rates of Ca^{2+} activation and relaxation in rat and guinea pig skinned trabeculae. *Circ. Res.* 83:179–86

141. Turnbull L, Hoh JF, Ludowyke RI, Rossmanith GH. 2002. Troponin I phosphorylation enhances crossbridge kinetics during beta-adrenergic stimulation in rat cardiac tissue. *J. Physiol.* 542:911–20

142. James J, Robbins J. 1997. Molecular remodeling of cardiac contractile function. *Am. J. Physiol. Heart Circ. Physiol.* 273:H2105–18

143. Pi YQ, Kemnitz KR, Zhang DH, Kranias EG, Walker JW. 2002. Phosphorylation of troponin I controls cardiac twitch dynamics: evidence from phosphorylation site mutants expressed on a troponin I-null background in mice. *Circ. Res.* 90:649–56

144. Montgomery DE, Wolska BM, Pyle WG, Roman BB, Dowell JC, et al. 2002. alpha-Adrenergic response and myofilament activity in mouse hearts lacking PKC phosphorylation sites on cardiac TnI. *Am. J. Physiol. Heart Circ. Physiol.* 282:H2397–405

145. Pyle WG, Sumandea MP, Solaro RJ, De Tombe PP. 2002. Troponin I serines 43/45 and regulation of cardiac myofilament function. *Am. J. Physiol. Heart Circ. Physiol.* 283:H1215–24

146. Noland TA Jr., Guo X, Raynor RL, Jideama NM, Averyhart-Fullard V, et al. 1995. Cardiac troponin I mutants. Phosphorylation by protein kinases C and A and regulation of Ca^{2+}-stimulated Mg-ATPase of reconstituted actomyosin S-1. *J. Biol. Chem.* 270:25445–54

147. Roman BB, Goldspink PH, Spaite E, Urboniene D, McKinney R, et al. 2004. Inhibition of PKC phosphorylation of cTnI improves cardiac performance in vivo. *Am. J. Physiol. Heart Circ. Physiol.* 286:H2089–95

148. Palmiter KA, Kitada Y, Muthuchamy M, Wieczorek DF, Solaro RJ. 1996. Exchange of beta- for alpha-tropomyosin in hearts of transgenic mice induces changes in thin filament response to Ca^{2+}, strong cross-bridge binding, and protein phosphorylation. *J. Biol. Chem.* 271:11611–14

149. Wolska BM, Arteaga GM, Pena JR, Nowak G, Phillips RM, et al. 2002. Expression of slow skeletal troponin I in hearts of phospholamban knockout mice alters the relaxant effect of beta-adrenergic stimulation. *Circ. Res.* 90:882–88

150. Wolska BM, Vijayan K, Arteaga GM, Konhilas JP, Phillips RM, et al. 2001. Expression of slow skeletal troponin I in adult transgenic mouse heart muscle reduces the force decline observed during acidic conditions. *J. Physiol.* 536:863–70

151. Solaro RJ, Lee JA, Kentish JC, Allen DG. 1988. Effects of acidosis on ventricular

muscle from adult and neonatal rats. *Circ. Res.* 63:779–87

152. Konhilas JP, Irving TC, Wolska BM, Jweied EE, Martin AF, et al. 2003. Troponin I in the murine myocardium: influence on length-dependent activation and interfilament spacing. *J. Physiol.* 547: 951–61

153. Pena JR, Wolska BM. 2004. Troponin I phosphorylation plays an important role in the relaxant effect of beta-adrenergic stimulation in mouse hearts. *Cardiovasc. Res.* 61:756–63

154. Takimoto E, Soergel DG, Janssen PM, Stull LB, Kass DA, Murphy AM. 2004. Frequency- and afterload-dependent cardiac modulation in vivo by troponin I with constitutively active protein kinase A phosphorylation sites. *Circ. Res.* 94:496–504

155. Fentzke RC, Buck SH, Patel JR, Lin H, Wolska BM, et al. 1999. Impaired cardiomyocyte relaxation and diastolic function in transgenic mice expressing slow skeletal troponin I in the heart. *J. Physiol.* 517(Part 1):143–57

156. Li L, Desantiago J, Chu G, Kranias EG, Bers DM. 2000. Phosphorylation of phospholamban and troponin I in beta-adrenergic-induced acceleration of cardiac relaxation. *Am. J. Physiol. Heart Circ. Physiol.* 278:H769–79

157. Finley NL, Ward DG, Trayer IP, Rosevear PR. 2003. Modification of cardiac troponin C upon binding cardiac troponin T and PKC phosphorylated cardiac troponin I. *Biophys. J.* 84:A566

158. Solaro RJ, Moss RL, eds. 2002. *Molecular Control Mechanisms in Striated Muscle Contraction.* Boston: Kluwer

Annu. Rev. Physiol. 2005. 67:69–98
doi: 10.1146/annurev.physiol.67.040403.114521
Copyright © 2005 by Annual Reviews. All rights reserved
First published online as a Review in Advance on July 21, 2004

Intracellular Calcium Release and Cardiac Disease

Xander H.T. Wehrens, Stephan E. Lehnart, and Andrew R. Marks

Department of Physiology and Cellular Biophysics, Center for Molecular Cardiology, Department of Medicine, Columbia University College of Physicians and Surgeons, New York 10032; email: xw80@columbia.edu, sel2004@columbia.edu, arm42@columbia.edu

Key Words calcium channel, heart failure, macromolecular complex, protein kinase A, ryanodine receptor, sudden cardiac death

■ **Abstract** Intracellular calcium release channels are present on sarcoplasmic and endoplasmic reticuli (SR, ER) of all cell types. There are two classes of these channels: ryanodine receptors (RyR) and inositol 1,4,5-trisphosphate receptors (IP$_3$R). RyRs are required for excitation-contraction (EC) coupling in striated (cardiac and skeletal) muscles. RyRs are made up of macromolecular signaling complexes that contain large cytoplasmic domains, which serve as scaffolds for proteins that regulate the function of the channel. These regulatory proteins include calstabin1/calstabin2 (FKBP12/FKBP12.6), a 12/12.6 kDa subunit that stabilizes the closed state of the channel and prevents aberrant calcium leak from the SR. Kinases and phosphatases are targeted to RyR2 channels and modulate RyR2 function in response to extracellular signals. In the classic fight or flight stress response, phosphorylation of RyR channels by protein kinase A reduces the affinity for calstabin and activates the channels leading to increased SR calcium release. In heart failure, a cardiac insult causes a mismatch between blood supply and metabolic demands of organs. The chronically activated fight or flight response leads to leaky channels, altered calcium signaling, and contractile dysfunction and cardiac arrhythmias.

INTRODUCTION

Ryanodine receptor (RyR)/calcium (Ca^{2+}) release channels are large ($>$2.5 MDa) homotetrameric channels that form macromolecular signaling complexes, which control the release of Ca^{2+} from the sarcoplasmic reticulum (SR). They are required for cardiac and skeletal muscle excitation-contraction (EC) coupling. Recent studies have revealed that kinases and phosphatases are targeted directly to the cytoplasmic domain of the RyRs. These enzymes modulate the function of the channel in response to extracellular signals communicated via second

0066-4278/05/0315-0069$14.00

messengers. In failing hearts, abnormal Ca^{2+} release via RyR/intracellular Ca^{2+} release channels contributes to impaired Ca^{2+} cycling. Chronic hyperactivity of the sympathetic nervous system during heart failure results in protein kinase A (PKA) hyperphosphorylation of the cardiac RyR, which leads to dissociation of the stabilizing subunit calstabin2 (the 12.6 kDa FK506-binding protein, FKBP12.6) from the channel macromolecular complex. Calstabin2-depleted channels can leak Ca^{2+} from the SR during diastole (when the channels are supposed to be tightly closed). This diastolic SR Ca^{2+} leak can deplete SR Ca^{2+} stores and contribute to contractile dysfunction, as well as trigger fatal cardiac arrhythmias. Promising new therapeutic agents are currently being developed to inhibit RyR leak in the failing heart by enhancing the affinity of calstabin2 for the receptor.

Whereas the role of intracellular Ca^{2+} release channels in striated muscles is well understood, they are expressed in virtually all cell types and the systems that regulate them are ubiquitous as well. Indeed, the regulation of cytosolic Ca^{2+} concentration modulates numerous cellular processes. In resting nonstimulated cells, cytosolic Ca^{2+} levels are low and rise approximately tenfold upon stimulation. In many systems, the source of this Ca^{2+} rise is the intracellular Ca^{2+} release channels. Thus these channels are required for diverse cellular processes, including transcription, fertilization, synaptic transmission, and EC coupling in skeletal and cardiac muscle (1). Key components of the Ca^{2+} regulatory machinery probably evolved from primordial Ca^{2+} transport systems and then developed into more specialized Ca^{2+} pumps and Ca^{2+} channels on the plasma membrane. Following the evolution of the SR or endoplasmic reticulum (ER) in nonstriated muscle cells into intracellular Ca^{2+} storage organelles, more specialized intracellular Ca^{2+} release channels evolved.

The sarcoplasmic and endoplasmic reticuli play a vital role in Ca^{2+} handling (2), and enable rapid and localized triggered Ca^{2+} release deep inside the cell. The Ca^{2+} release channels on these intracellular organelles have evolved to be distinct from all other known ion channels. First they are much larger (about 10 times larger than voltage-gated ion channels). They are nonselective, high-conductance channels that function as Ca^{2+} release channels by virtue of their localization to organelles that contain very high concentrations of Ca^{2+} (mM concentrations). RyRs and inositol 1,4,5-trisphosphate receptors (IP_3Rs) are similar in structure, both having large N-terminal domains that form scaffolds for channel regulatory proteins in the cytoplasm, and Ca^{2+} channel transmembrane (TM) and pore-forming regions near the C terminus (3, 4). The gene family of Ca^{2+} release channels has three isoforms of RyR and three IP_3Rs in vertebrates (5).

In cardiac myocytes, electrical depolarization of the surface membrane leads to intracellular Ca^{2+} release and contraction of the myofilaments, whereby electrical energy is converted to mechanical energy with Ca^{2+} serving as the second messenger (6). The initial trigger for EC coupling is generated by depolarization of the plasma membrane, which allows Ca^{2+} entry through L-type Ca^{2+} channels located on the transverse (T) tubules. This influx of Ca^{2+} triggers a large intracellular Ca^{2+} release from the SR via RyRs, which elevates cytosolic Ca^{2+} concentrations from

~100 nM during diastole to about 1 μM during systole, and this Ca^{2+} elevation activates cardiac contraction. Defective Ca^{2+} homeostasis plays an important role in the pathogenesis of congestive heart failure (7–12). In this review, we will focus on the alterations in the structure and function of RyRs in the failing heart.

STRUCTURE AND FUNCTION OF RYANODINE RECEPTORS

Ryanodine Receptor Gene Family

Ryanodine receptors were first cloned from mammalian skeletal (RyR1) and cardiac (RyR2) muscle (3, 4, 13). Analysis of the deduced amino acid sequences revealed that these two isoforms are about 66% homologous (3, 4). A third mammalian RyR isoform (RyR3) was cloned from rabbit brain and a mink lung epithelial cell line (14, 15). This isoform was originally thought to be smaller than the others but was later revealed to be the same size, the initial report having been based on only a partial sequence. Similar RyR isoforms have been cloned from nonmammalian vertebrates (e.g., chicken, bullfrog, and blue marlin) (16–18), although only two distinct RyR genes have been identified in these species. The nonmammalian vertebrate RyRα isoforms are homologs of the mammalian skeletal muscle isoform RyR1, whereas the RyRβ isoforms are most closely related to the mammalian RyR3 (16, 18). Although originally described in vertebrates, more recently RyR genes have been identified in invertebrates, including *Caenorhabditis elegans* and *Drosophila melanogaster* (19, 20).

Phylogenetic analysis reveals that RyR and IP$_3$R likely evolved from a common ancestral cation release channel (21). In contrast, intracellular Ca^{2+} release channels have little homology with the large family of voltage-dependent Ca^{2+} channels found in the plasma membrane of all excitable cells (3, 22, 23). It has been predicted that IP$_3$Rs have 6 TMs and RyRs have 4–12 TMs, although definitive structural evidence is currently lacking. Moreover, the overall topology of the intracellular Ca^{2+} release channels is quite different from that of the voltage-gated and other plasma membrane ion channels. Typically, ion channels in the plasma membrane have an α subunit that contains pore-forming sequences most often made up of four sets of six TM domains and a pore-lining sequence located between the fifth and sixth TM segment. The pore-forming sequences of the intracellular Ca^{2+} release channels share some homology with those of the voltage-gated ion channels, as a glycine-valine-aspartic acid (GVD) selectivity filter sequence is present. In contrast, all of the regulatory sequences important for channel function are encoded by a single enormous polypeptide (~565 kDa for the RyRs), which makes these channels distinct from the voltage-gated channels that have multiple nonhomologous subunits. Kinases and phosphatases, which are targeted to both Ca^{2+} release channels and voltage-gated channels, are involved in modulating channel functions. Recently, we have shown that these targeting proteins bind to the channels in several cases via leucine/isoleucine zippers (LIZ)

(24), a feature of the ion channel macromolecular complexes that is shared by intracellular Ca^{2+} release channels (24) and voltage-gated channels (25). The role of LIZs in formation of the IP_3R macromolecular complex (26), as well as in voltage-gated Ca^{2+} channels (27, 28), was recently confirmed.

Ryanodine Receptor Protein Structure

Ryanodine receptors are expressed in a variety of tissues, with high levels found in striated muscles (3, 13, 16, 18, 29–35). RyR1 is in the major Ca^{2+} release channel required for skeletal muscle contraction, although it is also expressed at lower levels in smooth muscle, cerebellum, testis, adrenal gland, and ovaries (3, 13, 16, 30, 31). RyR1 is predominantly expressed in Purkinje cells in the brain, and RyR2 is localized in the somata of neurons (30). RyR2 is abundant in the heart and brain, and at lower levels in the stomach, lung, thymus, adrenal gland, and ovaries (31, 36). RyR3 is expressed in the brain, diaphragm, slow twitch skeletal muscle, as well as abdominal organs (14–16, 31). In nonmammalian vertebrates, RyRα is abundant in skeletal muscle and present in brain, whereas RyRβ is expressed in a variety of tissues, including skeletal and cardiac muscle, lung, stomach, and brain (18). There is evidence that alternative splicing of RyR genes may underlie the tissue-specific expression of certain isoforms, but this does not appear to be a major source of diversity among RyRs (37).

Analysis of the primary amino acid sequence shows that the transmembrane domains are clustered in the C-terminal 10% of the RyR channel (Figure 1A, see color insert) (4, 38). Indeed, truncated RyRs containing only these putative TM domains are capable of forming Ca^{2+}-selective channels when incorporated into planar lipid bilayers (39, 40). Mutational analyses of regions in and around the putative TM domains revealed altered ion selectivity, supporting the concept that this region of the protein contains the TM domains and Ca^{2+}-selective pore of the RyR channel.

The Ryanodine Receptor Macromolecular Complex

The N-terminal domain of the ryanodine receptor serves as a scaffold for proteins that modulate RyR channel function, and the RyR macromolecular complex has been demonstrated using both CHAPS and Triton-X100 solubilized RyR (24, 46, 69, 96).

CALMODULIN   Calmodulin (CaM) was the first protein found to interact with single RyR channels in lipid bilayers (52, 53). The binding site of CaM has been mapped on the linear amino acid sequence using site-directed mutagenesis studies (54, 55), as well as on the three-dimensional surface of RyR2 using cryo-electromicroscopy (48, 56). CaM inhibits the cardiac RyR2 channel activity (57, 58). At present, the functional role of CaM in the modulation of RyR2 during EC coupling is not entirely clear. It appears that apocalmodulin (Ca^{2+}-free calmodulin) is a partial agonist, whereas Ca^{2+} calmodulin is an inhibitor of RyR1 (50).

CALSTABIN2   The Ca^{2+} channel-stabilizing proteins calstabin1 (also known as FKBP12) and calstabin2 (also known as FKBP12.6) associate with RyR1 and RyR2, respectively, such that one calstabin protein is bound to each RyR monomer (59–62). Thus there are four calstabin molecules bound to each RyR1 and RyR2 channel complex. RyR1 and RyR3 channels can bind calstabin1 and calstabin2, although the affinity for calstabin1 seems to be much higher (60, 61, 63). Therefore, *in vivo* RyR1 and RyR3 will have calstabin1 bound to them because of its higher abundance in the cytosol (60, 61). RyR2 channels exhibit a relatively higher affinity for calstabin2 and have predominantly calstabin2 bound to them (61, 64).

Cryo-electronmicroscopy studies of the RyR1 complex show that calstabin1 binds to RyR1 on the outer surface of the cytoplasmic domain (48). Recent observations suggest that the three-dimensional location of calstabin2 on RyR2 is similar to calstabin1 on RyR1 (65). The Val2461 residue on RyR1 (corresponding to Ile2427 in RyR2) is critical for calstabin1 binding (66). The bond formed by Val2461 and Pro2462 (or Ile2427-Pro2428 in RyR2) is analogous to the twisted-amide transition state intermediate of a peptidyl-prolyl bond that calstabin 1 and 2 (FKBP12 and FKBP12.6) bind to with high affinity (55). Mutation of Val2461 to a glycine residue abolishes binding of calstabin1 or calstabin2 to RyR1 (66). Because calstabin binds to RyRs with high affinity, we have reasoned that the target peptidyl-prolyl bond is constrained in the high-energy transition-state intermediate between *cis* and *trans* and that isomerization cannot be completed; otherwise the calstabin would dissociate from the channel, which does not occur. Introduction of increased mobility around the peptidyl-prolyl bond by substituting a smaller amino acid, glycine, for either the valine or isoleucine allows for isomerization to proceed by reducing steric hindrance at that site, and the binding affinity of calstabin to the channel is reduced (66). On the basis of molecular modeling studies, Bultynck et al. recently concluded that the proline in the calstabin-binding region on RyR induces a break in a helix, which imposes a twisted amide transition state on the peptidyl-proline bond and enables calstabin to bind to this domain (67). On the other hand, conflicting reports have been published concerning the amino acids in RyR2 involved in the binding of calstabin2. Masumiya et al. suggested that multiple regions within the N-terminal domain of RyR2 are required for the binding of glutathione S-transferase (GST)-calstabin2 to RyR2 (68), although they did not identify specific residues on RyR2 involved in calstabin2 binding. Therefore, it is likely that multiple RyR domains are involved in forming the calstabin-binding domain.

KINASES AND PHOSPHATASES   The RyR macromolecular complex also includes kinases (PKA and CaMKII). The catalytic subunit of PKA as well as its regulatory subunits (RII) are bound to the targeting protein mAKAP (AKAP6), which in turn is bound to RyR2 (69, 70). The protein phosphatases 1 and 2A (PP1 and PP2A) are bound to RyR2 via their own specific targeting proteins, spinophilin and PR130, respectively (24, 71). Highly conserved LIZ motifs in RyR2 form binding sites for cognate LIZs in the targeting proteins for the kinases and phosphatases (Figure 1*B*)

(24). We and others have recently shown that CaMKII also binds to RyR2, although the binding site has not been identified yet (50, 51).

SORCIN   Sorcin is a ubiquitous 22-kDa Ca^{2+}-binding protein reported to associate with both RyR2 and the L-type Ca^{2+} channel (72, 73). Sorcin may reduce RyR2 open probability, but this effect can be relieved by PKA-dependent phosphorylation of sorcin (45). It is possible that sorcin, similarily to calstabin2, serves as a brake on SR Ca^{2+} release, relievable by PKA-dependent phosphorylation (74, 75). However, the functional role for sorcin in the RyR channel complex is less well defined and requires additional studies.

LUMINAL-BINDING PARTNERS   RyR also binds proteins at the luminal SR surface (e.g., triadin, junctin, and calsequestrin). Junctin (76) and triadrin (77) are presumably involved in anchoring RyR to the SR membrane. Calsequestrin (CSQ) is the main Ca^{2+} binding protein in the SR and provides a high-capacity intracellular Ca^{2+} buffer (78, 79). It has been suggested that Ca^{2+}-dependent conformational changes in CSQ may modulate RyR channel activity (80), although the exact nature of the CSQ-RyR modulation requires further investigations (81, 82).

REGULATION OF RYANODINE RECEPTORS

A variety of cellular mediators may modulate the activity of RyR2 in the heart. These factors include physiologic modulators (e.g., ATP, Ca^{2+}, and Mg^{2+}), direct modifications of the RyR protein (e.g., phosphorylation, dephosphorylation, oxidation), and pharmacologic agents (e.g., ryanodine, caffeine, etc.). For a comprehensive review on RyR modulation see Fill & Copello (83).

Phosphorylation by PKA

The RyR2 protein sequence contains many consensus phosphorylation sites (3). On the basis of phospho-peptide mapping, PKA was shown to phosphorylate Ser2809 on RyR2 (84, 85). These results have been confirmed in several studies using GST-fusion proteins, mutations in the full-length recombinant RyR2 channel, and with a phospho-epitope-specific antibody (50, 69, 86–88).

There have been conflicting reports concerning the effects of exogenously applied PKA on single-channel behavior of ryanodine receptors (Table 1) (69, 86–91). Some studies have shown that phosphorylation by PKA increases the open probability (Po) of RyR2 by increasing the sensitivity of RyR2 to Ca^{2+}-dependent activation (50, 69, 86, 89), whereas other studies demonstrated first an increase in RyR2 Po followed by a slight decrease in the steady-state Po of RyR2 channels (Table 1) (92). In normal cardiomyocytes, PKA phosphorylation of RyR2 does not increase the Ca^{2+} spark frequency under conditions that simulate diastole in the heart when RyR2 is expected to be tightly closed (93). These data are consistent

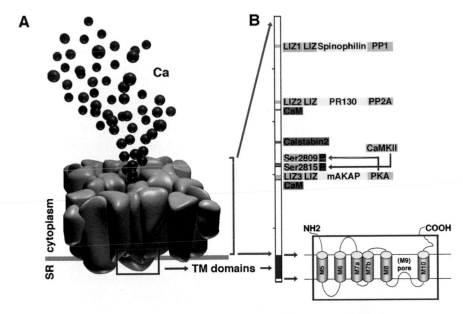

Figure 1 Structural domains in the cardiac ryanodine receptor. (*A*) Three-dimensional artistic impression of RyR2. The blue line represents the SR membrane, the domain within the red box the TM domains. (*B*) Linear amino-acid sequence of RyR2. Indicated are binding domains of protein phosphatases PP1 and PP2, protein kinase A (PKA), the two domains that constitute the calmodulin (CaM)-binding site, as well as the calstabin2 binding site. Also indicated are Ser2809 and Ser2815, which are subject to PKA and Ca^{2+}/calmodulin-dependent protein kinase II (CaMKII) phosphorylation, respectively. The TM domains and pore region are shown in the red box.

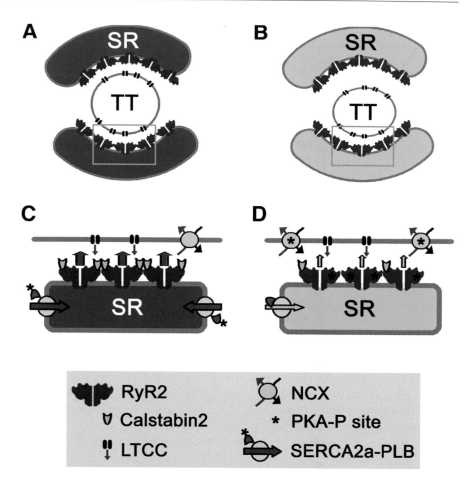

Figure 3 Molecular defects in CRU architecture and EC coupling in heart failure. (A) RyR2 are organized in dense clusters in close proximity of L-type Ca^{2+} channels (LTCC) in the intracellular dyad, which separates the SR Ca^{2+} storage organelle from the plasma membrane invaginations called T tubules (TT). (B) In heart failure, the TTs may become reduced in number or size, resulting in partial uncoupling of RyR2s from LTCCs, which reduces EC coupling gain. (C) Magnification of dyad (*yellow frame in A*). SR Ca^{2+} release via RyR2 channel is activated by CICR via voltage-gated LTCCs. The majority of Ca^{2+} is recycled into the SR via the SERCA2a Ca^{2+} pump shown with PLB dissociated by PKA phosphorylation (*). (D) In heart failure, PKA hyperphosphorylation of RyR2 (*) results in depletion of calstabin2 from the channel complex, which increases SR Ca^{2+} leak and decreases SR Ca^{2+} load (*lighter blue*). Down-regulation of SERCAa expression and function by relative up-regulation of PLB and PLB hypophosphorylation reduces SR Ca^{2+} uptake. Up-regulation of Na^+/Ca^{2+} exchanger (NCX) expression and NCX hyperphosphorylation (*) set the stage for increased depolarizing current (I_{ti}), which may trigger delayed afterdepolarizations (DADs) and arrhythmias.

TABLE 1 Functional effects of PKA phosphorylation of RyR on open probability and calstabin binding

Preparation	Assay	Po	Ca^{2+} sensitivity	Calstabin2	Reference
Canine CSR	[³H]-ryanodine binding	↑	↑	n.d.	(210)
Canine CSR	[³H]-ryanodine binding	↑	n.d.	n.d.	(211)
Canine CSR	[³H]-ryanodine binding	=	n.d.	n.d.	(84)
Rat cardiomyocytes	[³H]-ryanodine binding	↑	n.d.	n.d.	(95)
Rat brain microsomes	[Ca^{2+}] transient	↑	n.d.	n.d.	(212)
Rabbit SkSR	Lipid bilayer	↑	↑	n.d.	(213)
Porcine CSR/SkSR	[³H]-ryanodine binding	↑	n.d.	n.d.	(214)
Canine SkSR	Lipid bilayer	↑	n.d.	n.d.	(215)
Canine CSR	Lipid bilayer	↑	n.d.	n.d.	(89)
Rat trabeculae	[Ca^{2+}] transient	↑	↑	n.d.	(216)
Rabbit SkSR	Ca^{2+} uptake	↑	n.d.	n.d.	(217)
Canine CSR	Lipid bilayer	↑ > ↓	n.d.	n.d.	(92)
Canine CSR	Lipid bilayer	↑	↑	↓	(69)
Rabbit CSR	Lipid bilayer	↑	n.d.	↓	(24)
PKA-TG mouse CSR	Lipid bilayer	↑	n.d.	↓	(166)
Canine CSR	Ca^{2+} leak assay	↑	n.d.	↓	(172)
Canine CSR	Lipid bilayer	↑	n.d.	n.d.	(218)
Mouse myocytes	Ca^{2+} sparks	↑	n.d.	n.d.	(93)
PLB-KO myocytes	Ca^{2+} sparks	=	n.d.	n.d.	(93)
Human CSR	Lipid bilayer	↑	n.d.	↓	(176)
Recombinant human RyR1	Lipid bilayer	↑	n.d.	↓	(219)
Rat SkSR	Lipid bilayer	↑	n.d.	↓	(219)
Recombinant human RyR2	Lipid bilayer	↑	↑	↓	(86)
Recombinant RyR2-S2809D	Lipid bilayer	↑	↑	↓	(86)
β_2-AR TG mouse CSR	Lipid bilayer	↑	n.d.	↓	(176)
Recombinant rabbit RyR2	[³H]-ryanodine binding	=	=	n.d.	(99)
Recombinant RyR2-S2809D	Lipid bilayer	=	n.d.	=	(99)
Rat CSR	Ca^{2+} transient	↑	n.d.	↓	(220)

(*Continued*)

TABLE 1   (*Continued*)

Preparation	Assay	Po	Ca²⁺ sensitivity	Calstabin2	Reference
Recombinant mouse RyR2	Co-IP	n.d.	n.d.	=	(87)
Recombinant human RyR2	Lipid bilayer	↑	↑	↓	(50)
Recombinant RyR2-S2809D	Lipid bilayer	↑	↑	↓	(50)
Rabbit CSR	Lipid bilayer	↑	↑	↓	(50)
Rabbit myocytes	Ca²⁺ transient	↑	n.d.	n.d.	(94)
Mouse CSR	Lipid bilayer	↑	↑	↓	(62)

Legend: Po, open probability; CSR, cardiac sarcoplasmic reticulum microsomes; SkSR, skeletal muscle sarcoplasmic reticulum microscomes; TG, transgenic overexpressing mouse; KO, knockout mouse; n.d., not determined.

with the fact that healthy subjects or animals do not develop SR Ca²⁺ leak and arrhythmias during exercise (62, 86). On the other hand, recent data show that PKA phosphorylation of RyR2 enhances RyR2 activity and increases EC coupling gain during the early phase of EC coupling when only a small number of voltage-gated Ca²⁺ channels are open (Figure 2) (94, 95).

Calstabin2 Binding to RyRs

Of all the proteins that bind to the cytoplasmic/scaffold domain of the RyRs, the functional role for calstabins in the RyR macromolecular complexes is best understood. Calstabin1 was originally identified as KC7, a peptide that copurifies with RyR1 (13, 96). The functional role for calstabin1 in the RyR1 complex was first demonstrated using a heterologous coexpression system in which it was shown that calstabin1 stabilizes the open and closed states of the channel (46); similar findings were reported for calstabin2 in the RyR2 complex (97).

Some groups, including ours, have shown that PKA-phosphorylation of RyR2 can result in the dissociation of calstabin2 from the macromolecular complex (Table 1) (50, 69, 86, 89). In contrast, other groups have shown that calstabin2 may bind to PKA-phosphorylated RyR2 under certain experimental conditions (87, 98, 99). Recent data provide an explanation for these apparent contradictory findings (50). PKA phosphorylation of RyR2 or the mutation Ser2809Asp in RyR2 decreases the affinity of calstabin2 for the channel, which results in calstabin2 release from the RyR2 complex. However, when calstabin2 is present in high concentrations [e.g., when it is overexpressed with RyR2 under nonstoichiometric conditions as in Stange et al. (99) and Xiao et al. (87)], or added in excess amounts [see Xiao et al. (87)], calstabin2 may still be able to bind to PKA-phosphorylated RyR2 or RyR2-Ser2809Asp because it overwhelms the shift in binding affinity induced by PKA phosphorylation of Ser2809 (50).

Calstabin2 concentrations in mammalian hearts are in the range of 200–400 nmol/L. PKA phosphorylation during stress or exercise results in the partial

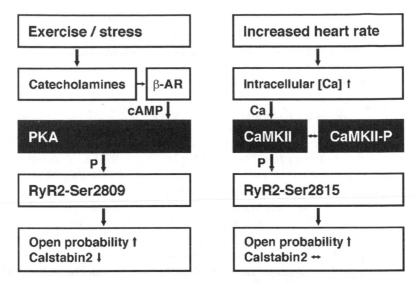

Figure 2 Schematic diagram representing intracellular signaling pathways regulating the RyR2 Po. Exercise and stress cause an increase in plasma catecholamine levels, which activate the β-adrenergic receptors (β-AR) signaling pathway and PKA. PKA phosphorylation of Ser2809 on RyR2 increases the Po and decreases the binding affinity for the channel-stabilizing subunit calstabin2. An increased heart rate raises intracellular calcium (Ca^{2+}) levels, which activate CaMKII. Autophosphorylation of CaMKII (CaMKII-P) keeps the enzyme active. CaMKII phosphorylates Ser2815 on RyR2, which increases the Po without dissociating calstabin2.

dissociation of calstabin2 from RyR2 because it decreases the affinity of calstabin2 binding to RyR2 to about 600 nM (50, 100). The partial release of calstabin2 from the RyR2 channel complex increases the Po of the channel, which results in increased intracellular Ca^{2+} release and augmented cardiac contractility under conditions of increased β-adrenergic signaling (86). Calstabin2 is also critical to normal RyR2 channel operation in the heart under resting conditions (69, 97, 101). Binding of calstabin2 stabilizes the RyR2 channel in the closed state during the resting phase of the heart (diastole) (46, 86). This is important to prevent aberrant diastolic Ca^{2+} release (leaks), which can trigger cardiac arrhythmias (86).

Dissociation of calstabin2 from the RyR2 channel complex by PKA phosphorylation of the channel, or competing it off with the drug rapamycin, results in subconductance states (29, 46, 69, 97, 102). Subconductance states are open events with less than the normal unitary current amplitude. Thus calstabins increase the probability of full conductance openings of RyR channels. Calstabins also functionally couple groups of RyR channels to permit synchronous opening and closing (gating) of arrays of channels (66, 103). RyR channels are present on the terminal cisternae of the SR in a dense array such that the corners of each channel contact its neighbors in a checkerboard type of pattern. This phenomenon, called coupled gating, enables arrays of RyR channels to gate in unison, a process that enhances the efficiency of SR Ca^{2+} release and helps terminate SR Ca^{2+} release as well. Termination of SR Ca^{2+} release is facilitated by coupled gating because when the first channel in the array closes, coupled gating enables all of the channels to close.

Phosphorylation by CaMKII

Based on phospho-peptide mapping, it was proposed that CaMKII phosphorylates the same residue on RyR2 as PKA (Ser2809) (84, 85, 88). A recent study, however, demonstrated that CaMKII phosphorylates Ser2815 (50). Mutating Ser2815 to alanine in full-length recombinant RyR2 channels abolishes CaMKII phosphorylation (50). The identification of the CaMKII phosphorylation site on RyR2, which was also confirmed using a phospho-epitope-specific antibody, has facilitated elucidation of the functional effects of CaMKII phosphorylation of RyR2 (50, 85, 89, 90, 104–106). Single-channel studies of CaMKII-phosphorylated, wild-type RyR2 channels and RyR2-Ser2815Asp mutant channels, which mimic constitutively CaMKII-phosphorylated RyR2, have shown that CaMKII phosphorylation of RyR2 at Ser2815 increases the Po of the channel by augmenting the sensitivity to Ca^{2+}-dependent activation (Figure 2) (50, 105, 106). In contrast to phosphorylation by PKA, CaMKII phosphorylation does not dissociate calstabin2 from the RyR2 channel (50).

Dephosphorylation by Phosphatases

The activity of RyR2 is also regulated by protein phosphatases, which are targeted to the macromolecular channel complex (Figure 1B) (24, 90). Terentyev et al. have reported that protein phosphatase 1 (PP1) may increase RyR2 activity, although this study contradicts several other studies reporting that PP1 decreases RyR2 activity (91, 107–111) and that PKA/ CaMKII phosphorylation increases RyR2 activity (50, 69, 89, 105).

EXCITATION CONTRACTION COUPLING IN THE NORMAL HEART

Release of Intracellular Ca^{2+} During Systole

Contractile force in cardiomyocytes is generated via an increase in cytoplasmic Ca^{2+} concentrations owing to SR Ca^{2+} release via RyR2, which can be measured as Ca^{2+} transients (11, 83). Depolarization of the plasma membrane during the cardiac action potential activates voltage-gated L-type Ca^{2+} channels (LTCC or dihydropyridine receptors) in the sarcolemmal membrane encompassing the T-tubules. Additional Ca^{2+} may enter via the T-type Ca^{2+} channels (TTCC) (112) or the Na^+/Ca^{2+} exchanger (NCX) operating in its reverse mode (113). The ensuing

Ca^{2+} influx then triggers a much greater Ca^{2+} release from the SR via RyR2 via a process called Ca^{2+}-induced Ca^{2+} release (CICR) (6). The tenfold increase in cytoplasmic Ca^{2+} concentrations during this Ca^{2+} transient results in actin-myosin cross-bridge formation that is activated by Ca^{2+} binding to troponin C. This results in displacement of tropomyosin, translocation of the myosin heads along the actin filaments, and contraction of the myocyte.

Calcium Release Units

In order to establish a functional link between Ca^{2+} influx through the plasmalemmal LTCC and the release of SR Ca^{2+} via RyR2, these two channels are organized in structures known as Ca^{2+}-release units (CRUs) (Figure 3A, see color insert). Immunolabeling with specific antibodies and ultrastructural analysis using transmission electron microscopy have established that LTCCs and RyRs are at sites where SR terminal cisternae are apposed to the plasmalemmal T tubules forming intracellular junctions called dyads in cardiac muscle (Figure 3A) (114, 115). CRUs in cardiac muscle contain the two Ca^{2+} channels defined above, the SR docking protein junctophilin (116), the internal Ca^{2+}-binding protein calsequestrin (CSQ) (117), and two proteins that mediate CSQ-RyR interaction (triadin and junctin) (43, 118). Additional proteins associate with and regulate the cytoplasmic domains of RyRs (see above).

In cardiac muscle, there is approximately 1 LTCC for every 5–10 RyR2 channels in the CRU, but the lack of a highly organized pattern suggests that LTCCs are not specifically linked to the RyR2 (114, 119–124). During the cardiac action potential, the LTCC opens to mediate a Ca^{2+} influx that activates RyR2 via the CICR (6, 115, 125). Because there is no direct protein-protein interaction between LTCC and RyR2 in cardiac muscle, the involvement of Ca^{2+} as a diffusible second messenger makes the signal transduction slower than LTCC-RyR1 signaling in skeletal muscle (126, 127).

Removal of Cytoplasmic Ca^{2+} During Diastole

Myocardial relaxation during diastole is initiated by the removal of Ca^{2+} from the cytoplasm. Cytosolic Ca^{2+} is pumped back into the SR by SR Ca^{2+} ATPase (SERCA2a) (128, 129). Activity of this enzyme is inhibited by binding of phospholamban (PLB) (130). In its nonphosphorylated form, PLB inhibits SERCA2a activity (130), whereas phosphorylation of PLB reverses the inhibition. Cytosolic Ca^{2+} can also be extruded from the cardiomyocyte via the sarcolemmal NCX (131).

Modulation of EC Coupling by Kinases and Phosphatases

PHOSPHORYLATION BY PKA   Phosphorylation of the Ca^{2+} handling proteins LTCC (132, 133), RyR2 (50, 69), PLB (134), and NCX by PKA is the downstream event in a signaling cascade that begins with agonist binding to β-adrenergic receptors

(β-AR) on the plasma membrane. This allows for the activation of adenylate cyclase (AC)-mediated by specific G proteins, leading to increased cytosolic levels of cyclic AMP (cAMP), and activation of PKA (135). Because PKA is directly targeted to or present near key intracellular Ca^{2+} cycling proteins, activation of the β-AR signaling pathway increases the amplitude of the Ca^{2+} transient (136), which increases the gain of EC coupling, thereby increasing the amount of Ca^{2+} released by RyR2 per amount of trigger Ca^{2+} entering the cell through L-type Ca^{2+} channels (137). This signaling pathway, also known as the fight-or-flight response, is highly conserved in evolution and allows for rapid enhancement of cardiac contractility during exercise or stress (Figure 2) (11).

PHOSPHORYLATION BY CaMKII   CaMKII plays an important role in the regulation of EC coupling in the heart (Figure 2) (50, 105, 138, 139). Activation of CaMKII at increased heart rates, owing to increased cytoplasmic Ca^{2+} concentrations, mediates increased Ca^{2+} release, leading to enhanced contractile force (e.g., the positive force-frequency relationship). An increase in CaMKII activity results in CaMKII phosphorylation of the LTCC (138, 140), RyR2 (50, 105), and PLB (50, 141–143). The functional effects of CaMKII phosphorylation of these key Ca^{2+} handling proteins include increased Ca^{2+} release from the SR and enhanced reuptake of Ca^{2+} into the SR by SERCA2a (50, 142). More rapid release and reuptake of Ca^{2+} provides more time for diastolic filling of the ventricles at higher heart rates (139).

PHOSPHORYLATION BY PKC-α The Ca^{2+}/phospholipid-dependent protein kinase PKC-α has recently been identified as a fundamental regulator of cardiac contractility. PKCs comprise a family of at least 12 distinct isoforms (144). The conventional PKC isoforms (α, βI, βII, and γ) are activated by Ca^{2+} and lipids, whereas the novel (δ, ε, η, and θ) and atypical (ζ, ι, υ, and λ) PKC isoforms do not require Ca^{2+} for maximal activation. Agonists of angiotensin-II (ATII-R) receptors, α1-adrenergic receptors (α-AR), and endothelin-1 receptors (ET-1R) stimulate PKC via Gq-coupled phospholipase Cβ (PLC-β) (145, 146). PKC-α is the predominant PKC isoenzyme expressed in the heart (147) and can directly phosphorylate protein phosphatase inhibitor-1 (I-1). PKC-α phosphorylation of I-1 augments the activity of the protein phosphatase 1 (PP1) and causes hypophosphorylation of PLB (148). Decreased PLB phosphorylation may result in inhibition of SERCA2a and reduced Ca^{2+} reuptake into the SR.

DEPHOSPHORYLATION BY PP1 AND PP2A   Dephosphorylation of proteins in the EC coupling machinery plays an important role in regulating cardiac contractility (109) and most studies support the role of phosphatases in down-regulating SR Ca^{2+} release and contractile performance (107–110). duBell et al. reported that intracellular dialysis of rat myocytes with either PP1 or PP2A decreased the magnitude of steady-state Ca^{2+} transients without producing measurable changes in SR Ca^{2+} content (108). These data are consistent with lower Po of RyR2 owing to decreased phosphorylation of Ser2809 and Ser2815 (50). However, Terentyev et al. recently reported that PP1 may increase RyR2 Po (149). Therefore, the precise mode of action of PP1 and PP2A on intracellular Ca^{2+} handling remains poorly understood and requires further investigation

PATHOGENESIS OF ALTERED CALCIUM SIGNALING IN HEART FAILURE

Heart failure is a leading cause of morbidity and mortality in the Western world. Any structural or functional cardiac insult that impairs the ability of the heart chambers to fill with or eject blood can result in heart failure, resulting in a syndrome where cardiac performance falls below the needs of the organs it supplies with nutrients. Although these insults usually affect the heart directly, distinct pathologies such as diabetes, hormonal dysregulation, toxins, or medications may contribute to or cause heart failure. Chronically decreased cardiac output activates neurohumoral systems resulting in a *circulus vitiosus* whereby maladaptive changes and elevated levels of catecholamines sustain and exacerbate disease progression via cardiac remodeling. At the level of the cardiomyocyte, chronic activation of cAMP-dependent signaling has direct toxic effects that result in depressed function associated with alterations in EC coupling and changes in intracellular Ca^{2+} metabolism.

In the failing heart, maladaptive changes result in depressed intracellular Ca^{2+} cycling and decreased SR Ca^{2+} concentrations such that any given action potential leading to CICR releases less Ca^{2+} and produces less force during EC coupling. Consistently, the amplitude of the intracellular Ca^{2+} transient ($\Delta[Ca^{2+}]_i$) is reduced and the decay of the Ca^{2+} transient is slowed in cardiomyocytes and heart muscle from heart failure patients and animal models (137, 150–152). The reduced amplitude of the intracellular Ca^{2+} transient results in diminished force production of the failing heart muscle (153, 154). Defective EC coupling, reduced intracellular Ca^{2+} transients, and reduced contractility have been observed in heart failure owing to dilated cardiomyopathy (152, 154), pressure overload (137), myocardial infarction (155), viral myocarditis (156), overexpression of the cytosolic splice variant CaMKIIδ_c (105), and muscle LIM protein knockout (157). Heart failure, when caused by myocardial infarction (155), pacing (158), myocarditis (156), or muscle LIM protein knockout (157), is accompanied by a decrease in the gain of EC coupling ($\Delta[Ca^{2+}]_i/I_{Ca}$).

Heart failure is characterized by a chronic activation of the sympathetic system and maladaptive changes within the β-AR signaling system (159). Downregulation of the βAR-receptor number and function (160), desensitization and uncoupling from stimulatory Gs proteins by receptor kinases (161), antagonistic up-regulation of the inhibitory $G\alpha_i$ subunits (162), down-regulation of adenylyl cyclases (163), and depressed intracellular cAMP synthesis (164) consistently occur in cardiomyocytes from heart failure patients and animal models. Depressed force production during β-adrenergic stimulation of failing cardiomyocytes is related to decreased up-regulation of Ca^{2+} release and EC coupling (165). Although cAMP-dependent signaling is down-regulated in heart failure, it appears that chronically increased sympathetic stimulation sustains PKA phosphorylation of specific intracellular targets such as RyR2, thus exacerbating heart failure or potentially itself causing heart failure (166).

RyR2 Dysfunction and Intracellular Ca^{2+} Leak

In the terminal or dyadic SR, heart failure in patients and animal models causes PKA hyperphosphorylation of RyR2 Ca^{2+} release channels and hyperactive channel function, contributing to SR Ca^{2+} store depletion (Figure 3D) (69, 167). Owing to down-regulation of the β-adrenergic signaling components in heart failure, RyR2 PKA hyperphosphorylation is an unexpected finding; however, a parallel down-regulation of PP1 and PP2A levels in the RyR2 complex in heart failure provides a rationale for significantly increased PKA phosphorylation levels (69, 168). Additionally, despite down-regulation and decreased inotropic response, cAMP-dependent pathways remain functional in heart failure (169). Local control of SR Ca^{2+} release channel activity is mediated by specific targeting of PKA and protein phosphatases to RyR2. These adaptor proteins include mAKAP (PKA), PR130 (PP2A), and sphinophilin (PP1) (24, 26). In heart failure, PKA hyperphosphorylation of other Ca^{2+} handling proteins, such as LTCC and NCX, also occurs presumably via distinct PKA targeting mechanisms (27, 170).

Increased diastolic SR Ca^{2+} leak has been found in heart failure caused by RyR2 PKA hyperphosphorylation (167, 171, 172). Similar to PKA hyperphosphorylation, interventions such as FK506 treatment, which depletes calstabin2 from the RyR2 complex, result in hyperactive channels and intracellular Ca^{2+} leak (101, 173, 174), and interventions that increase calstabin2 binding to RyR2 decrease channel activity or Ca^{2+} leak (62, 86, 175). However, one study reported an intracellular Ca^{2+} leak but no RyR2 PKA hyperphosphorylation in heart failure (98), and another study reported that PKA phosphorylation does not affect the Po or the amount of calstabin2 bound to recombinant RyR2 channels (99). Only recently was it realized that quantification of RyR2 PKA hyperphosphorylation requires phosphatase inhibitors to prevent dephosphorylation and loss of signal (176). Also, physiologic expression levels of calstabin2 are required in recombinant channel experiments because excess calstabin2 expression prevents calstabin2 dissociation from RyR2 following PKA phosphorylation (62, 86, 175).

Blockade of β-adrenergic receptors reduces intracellular cAMP levels and decreases the activity of PKA. The downstream effects of reduced PKA activity owing to β-AR blockade include a reversal of the PKA hyperphosphorylation of RyR2 (168, 172, 177), restoration of the normal stoichiometry of the RyR2 macromolecular complex by increased binding of calstabin2 to RyR2 (69, 168, 172), and normalization of the function of RyR2 channels in failing hearts (168).

Increased NCX expression and function has been consistently demonstrated in heart failure (170, 178). As both NCX and SR Ca^{2+} uptake by SERCA2a

TABLE 2   Ryanodine receptor structure and function in heart failure

Preparation	Assay	PKA phosphorylation	Po	Ca^{2+} sensitivity	Calstabin2	Reference
Human cardiac lysate	[^{3}H]-ryanodine binding	n.d.	=	=	n.d.	(221)
Human CSR	Lipid bilayer	↑	↑	↑	↓	(69)
Canine CSR	Lipid bilayer	↑	↑	↑	↓	(69)
Dog CSR	[^{3}H]-ryanodine binding Ca^{2+} leak MCA fluorescence	n.d.	↑	n.d.	↓	(167)
Dog CSR	[^{3}H]-ryanodine binding Ca^{2+} leak MCA fluorescence	n.d.	n.d.	n.d.	↓	(171)
Mouse CSR	Lipid bilayer	↑	↑	n.d.	↓	(166)
Dog CSR	Lipid bilayer	↑	↑	↑	↓	(177)
Dog CSR	[^{3}H]-ryanodine binding Ca^{2+} leak MCA fluorescence	n.d.	↑	n.d.	↓	(172)
Dog CSR	Lipid bilayer [^{3}H]-ryanodine binding	=	=	=	=	(98)
Rat, mouse CSR	Lipid bilayer	↑	↑	n.d.	↓	(176)
Human CSR	Lipid bilayer	↑	↑	n.d.	↓	(168)
Rabbit cardiomyocytes	Ca^{2+} transient	n.d.	↑	n.d.	n.d.	(222)

compete for intracellular Ca^{2+} extrusion, more Ca^{2+} is extruded from the cell, thus contributing to SR Ca^{2+} store depletion in heart failure (179, 180). In the context of a prolonged intracellular Ca^{2+} transient, Na^+-dependent Ca^{2+} extrusion via NCX compensates for decreased SR Ca^{2+} uptake at the cost of increased membrane depolarization (181, 182). These data indicate that contractile dysfunction in heart failure occurs, at least in part, because of reduced SR Ca^{2+} load, which results from RyR2-dependent Ca^{2+} leak and increased Ca^{2+} extrusion via NCX. Indeed, SR Ca^{2+} load is significantly reduced in failing cardiomyocytes from patients and animal models (154, 183).

Defects in Calcium-Release Units

In animal models of heart failure, the diastolic Ca^{2+} spark frequency is increased and the spark duration prolonged despite reduced SR Ca^{2+} concentrations, which suggests more and longer RyR2 openings (69, 105). Hypertrophied rat cardiomyocytes demonstrate impaired contractility, which is related to impaired coupling between adjacent LTCC and RyR2 molecules as the primary defect of EC coupling (137). A similar defect in LTCC-RyR2 coupling was found in cells derived from failing rat hearts, which points to a common defect that may underlie various forms of cardiac dysfunction (137). Accordingly, altered CICR coupling mechanisms may be involved in increased Ca^{2+} sparks in hypertensive cardiomyopathy (184). In hypertrophy, but not in heart failure, the CICR coupling defect was corrected by increased β-adrenergic stimulation that resulted in a hypercontractile state (137). Impaired EC coupling was proposed to be related to remodeling of the dyad microarchitecture, which might lead to altered spacing between adjacent LTCC and RyR2 molecules, decreasing the ability of I_{Ca} to trigger Ca^{2+} release from the SR in hypertrophied or failing hearts (Figure 3B) (155). Ultrastructural changes in the T tubule system or the plasma membrane side of the junctional complex between LTCCs and RyR2 Ca^{2+} release channels support these findings (185). These changes could contribute to a reduced synchrony of SR Ca^{2+} release and prolonged Ca^{2+}release in failing cardiomyocytes from human hearts (186).

Defective Removal of Cytoplasmic Ca^{2+}

Decreased SR Ca^{2+} uptake is caused by decreased SR Ca^{2+} ATPase (SERCA2a) expression, decreased SR Ca^{2+} uptake rates, and a relative stoichiometric up-regulation of the inhibitory SERCA2a subunit PLB (158, 169, 183). Also, hypophosphorylation of PLB in heart failure is known to decrease SERCA2a function (176, 187), although one study reported PLB hyperphosphorylation in a rabbit infarct heart failure model (188). In addition, expression of distinct and Ca^{2+}-sensitive PKC isoforms is increased in heart failure (189–191) and PKC-α up-regulation may contribute to PLB hypophosphorylation in heart failure (148). The net result of reduced SERCA function would be decreased SR Ca^{2+} concentrations owing to decreased Ca^{2+} uptake and a delayed decay of the intracellular Ca^{2+} transient, leading to increased cytosolic Ca^{2+} concentrations and diastolic dysfunction at higher heart rates (152, 154, 181). Therefore, SERCA2a dysfunction is a generally accepted mechanism contributing to cardiac dysfunction in heart failure.

Accordingly, knockout of PLB, which maximizes SERCA2a function, produces larger intracellular Ca^{2+} transients, which decay faster, and the EC coupling gain is increased (192). However, spontaneous Ca^{2+} sparks are three times more frequent and larger in PLB knockout cells than in wild-type cardiomyocytes because of an increase of the SR Ca^{2+} content due to a higher rate of Ca^{2+} uptake by the SERCA2a in the PLB null cardiomyocytes (192). This can be attributed to the fact that elevated SR Ca^{2+} concentrations directly increase Ca^{2+} spark rates through molecularly unidentified mechanisms that activate RyR2 channels within the SR lumen (93, 193). Transgenic overexpression of the major SR Ca^{2+} storage protein calsequestrin increases SR Ca^{2+} concentrations and release (194). Intracellular Ca^{2+} overload results in spontaneous SR Ca^{2+} release, higher Ca^{2+} sparks rates, and unphysiologic Ca^{2+} waves in cardiomyocytes (193, 195). Intracellular Ca^{2+} overload increases SR Ca^{2+} concentrations, which increases the Po of RyR2 Ca^{2+} release and may cause delayed afterdepolarizations (DADs) (196).

Increasing SR Ca^{2+} uptake by improving SERCA2a function has been suggested to improve systolic and diastolic function through improved Ca^{2+} cycling in the failing myocardium (197–199). However, there is a potential risk of SR Ca^{2+} overload, which may result in DADs and cardiac arrhythmias (200). SR Ca^{2+} uptake may also be increased by decreasing levels of PLB phosphorylation (201) because PLB inhibits SERCA2a function in its unphosphorylated form. However, augmentation of Ca^{2+} cycling by decreasing PLB function may be beneficial only in certain forms of heart failure because two inactivating mutations of PLB result in severe and highly lethal dilated cardiomyopathy in humans (202).

Triggered Arrhythmias in Heart Failure

In the failing heart, aberrant intracellular Ca^{2+} release represents a candidate mechanism for DADs and triggered arrhythmias. Missense mutations in the RyR2 Ca^{2+} release channel result in a gain-of-function defect and intracellular Ca^{2+} leak during β-adrenergic stimulation (86, 203). Mutant RyR2 channels have a decreased calstabin2-binding affinity and accordingly calstabin2 knockout mice develop DADs and exercise-induced arrhythmias in normal hearts (86). In heart failure, additional changes of transmembrane ion transport contribute to electrical instability of the membrane potential, which lowers the threshold to activate DADs (204).

Intracellular Ca^{2+} overload from SR Ca^{2+} leak can activate a transient inward current (I_{ti}) that causes membrane depolarization and results in DADs and triggered arrhythmias (205). It is thought that I_{ti} results from forward mode NCX net Na^+ influx (206) or from a Ca^{2+}-activated Cl^- current ($I_{Ca/Cl}$) (207). In the failing heart, a prominent increase in NCX function contributes significantly more depolarizing current and increases the propensity for arrhythmias triggered by DADs (208, 209). The propensity for triggered arrhythmias may further be increased by activation of β-adrenergic signaling due to up-regulation of SR Ca^{2+} uptake and Ca^{2+} leak in the context of down-regulated repolarizing K^+ currents in heart failure (169).

SUMMARY

Altered intracellular Ca^{2+} handling importantly contributes to impaired contractility in heart failure. Chronic hyperactivity of the β-adrenergic signaling pathway results in PKA-hyperphosphorylation of the cardiac RyR/intracellular Ca^{2+} release channel. This causes the channel-stabilizing protein calstabin2 to dissociate from the RyR2 macromolecular channel complex, which leads to diastolic Ca^{2+} leak from the sarcoplasmic reticulum. One of the few clinically effective classes of drugs, the β-AR blockers, reduces PKA-phosphorylation levels of RyR2, increases calstabin2 binding to RyR2, and may normalize cardiac contractility in failing hearts. Further study of the molecular defects contributing to abnormal Ca^{2+} cycling in the failing heart may lead to a refinement of our understanding of the pathophysiologic and therapeutic importance of these defects.

The *Annual Review of Physiology* is online at http://physiol.annualreviews.org

LITERATURE CITED

1. Bers DM. 2002. Cardiac excitation-contraction coupling. *Nature* 415:198–205

2. Carafoli E. 2002. Calcium signaling: a tale for all seasons. *Proc. Natl. Acad. Sci. USA* 99:1115–22

3. Takeshima H, Nishimura S, Matsumoto T, Ishida H, Kangawa K, et al. 1989. Primary structure and expression from complementary DNA of skeletal muscle ryanodine receptor. *Nature* 339:439–45

4. Otsu K, Willard HF, Khanna VK, Zorzato F, Green NM, MacLennan DH. 1990. Molecular cloning of cDNA encoding the Ca^{2+} release channel (ryanodine receptor) of rabbit cardiac muscle sarcoplasmic reticulum. *J. Biol. Chem.* 265:13472–83

5. Sorrentino V, Barone V, Rossi D. 2000. Intracellular Ca^{2+} release channels in evolution. *Curr. Opin. Genet. Dev.* 10:662–67

6. Fabiato A. 1983. Calcium-induced release of calcium from the cardiac sarcoplasmic reticulum. *Am. J. Physiol. Cell Physiol.* 245:C1–14

7. Braunwald E, Bristow MR. 2000. Congestive heart failure: fifty years of progress. *Circulation* 102:14–23

8. Houser SR, Margulies KB. 2003. Is depressed myocyte contractility centrally involved in heart failure? *Circ. Res.* 92:350–58

9. Lefkowitz RJ, Rockman HA, Koch WJ. 2000. Catecholamines, cardiac beta-adrenergic receptors, and heart failure. *Circulation* 101:1634–37

10. Marks AR. 2000. Cardiac intracellular calcium release channels: role in heart failure. *Circ. Res.* 87:8–11

11. Wehrens XHT, Marks AR. 2003. Altered function and regulation of cardiac ryanodine receptors in cardiac disease. *Trends Biochem. Sci.* 28:671–78

12. Bristow M. 2001. Of phospholamban, mice, and humans with heart failure. *Circulation* 103:787–88

13. Marks AR, Tempst P, Hwang KS, Taubman MB, Inui M, et al. 1989. Molecular cloning and characterization of the ryanodine receptor/junctional channel complex cDNA from skeletal muscle sarcoplasmic reticulum. *Proc. Natl. Acad. Sci. USA* 86:8683–87

14. Giannini G, Clementi E, Ceci R, Marziali G, Sorrentino V. 1992. Expression of a ryanodine receptor-Ca^{2+} channel that is regulated by TGF-β. *Science* 257:91–93

15. Hakamata Y, Nakai J, Takeshima H, Imoto K. 1992. Primary structure and distribution of a novel ryanodine receptor/calcium release channel from rabbit brain. *FEBS Lett.* 312:229–35

16. Ottini L, Marziali G, Conti A, Charlesworth A, Sorrentino V. 1996. Alpha and beta isoforms of ryanodine receptor from chicken skeletal muscle are the homologues of mammalian RyR1 and RyR3. *Biochem. J.* 315:207–16

17. Franck JPC, Morrissette J, Keen JE, Londraville RL, Beamsley M, Block BA. 1998. Cloning and characterization of fiber type-specific ryanodine receptor isoforms in skeletal muscles of fish. *Am. J. Physiol. Cell Physiol.* 275:C401–15

18. Oyamada H, Murayama T, Takagi T, Iino M, Iwabe N, et al. 1994. Primary structure and distribution of ryanodine-binding protein isoforms of the bullfrog skeletal muscle. *J. Biol. Chem.* 269:17206–14

19. Maryon EB, Coronado R, Anderson P. 1996. unc-68 encodes a ryanodine receptor involved in regulating *C. elegans* body-wall muscle contraction. *J. Cell Biol.* 134:885–93

20. Takeshima H, Nishi M, Iwabe N, Miyata T, Hosoya T, et al. 1994. Isolation and characterization of a gene for a ryanodine receptor/calcium release channel in *Drosophila melanogaster. FEBS Lett.* 337:81–87

21. Vazquez-Martinez O, Canedo-Merino R, Diaz-Munoz M, Riesgo-Escovar JR. 2003. Biochemical characterization, distribution and phylogenetic analysis of *Drosophila melanogaster* ryanodine and IP$_3$ receptors, and thapsigargin-sensitive Ca^{2+} ATPase. *J. Cell Sci.* 116:2483–94

22. Takeshima H. 1993. Primary structure and expression from cDNAs of the ryanodine receptor. *Ann. NY Acad. Sci.* 707:165–77

23. Catterall WA. 1995. Structure and function of voltage-gated ion channels. *Annu. Rev. Biochem.* 64:493–531

24. Marx SO, Reiken S, Hisamatsu Y, Gaburjakova M, Gaburjakova J, et al. 2001. Phosphorylation-dependent regulation of ryanodine receptors. A novel role for leucine/isoleucine zippers. *J. Cell Biol.* 153:699–708

25. Marx SO, Kurokawa J, Reiken S, Motoike H, D'Armiento J, et al. 2002. Requirement of a macromolecular signaling complex for beta adrenergic receptor modulation of the KCNQ1-KCNE1 potassium channel. *Science* 295:496–99

26. Tu HP, Tang TS, Wang ZN, Bezprozvanny I. 2004. Association of type 1 inositol 1,4,5-trisphosphate receptor with AKAP9 (Yotiao) and protein kinase A. *J. Biol. Chem.* 279:19375–82

27. Hulme JT, Ahn M, Hauschka SD, Scheuer T, Catterall WA. 2002. A novel leucine zipper targets AKAP15 and cyclic AMP-dependent protein kinase to the C terminus of the skeletal muscle Ca^{2+} channel and modulates its function. *J. Biol. Chem.* 277:4079–87

28. Hulme JT, Lin TW, Westenbroek RE, Scheuer T, Catterall WA. 2003. Beta-adrenergic regulation requires direct anchoring of PKA to cardiac Cav1.2 channels via a leucine zipper interaction with a kinase-anchoring protein 15. *Proc. Natl. Acad. Sci. USA* 100:13093–98

29. Ahern GP, Junankar PR, Dulhunty AF. 1994. Single channel activity of the ryanodine receptor calcium release channel is modulated by FK-506. *FEBS Lett.* 352:369–74

30. Kuwajima G, Futatsugi A, Niinobe M, Nakanishi S, Mikoshiba K. 1992. Two types of ryanodine receptors in mouse brain: skeletal muscle type exclusively in Purkinje cells and cardiac muscle type in various neurons. *Neuron* 9:1133–42

31. Giannini G, Conti A, Mammarella S, Scrobogna M, Sorrentino V. 1995. The ryanodine receptor/calcium channel genes are widely and differentially expressed in murine brain and peripheral tissues. *J. Cell Biol.* 128:893–904

32. Lai FA, Meissner G. 1992. Purification and reconstitution of the ryanodine-sensitive Ca^{2+} release channel complex from muscle sarcoplasmic reticulum. In *Protocols in Molecular Neurobiology*, ed. A Longstaff, P Revest, pp. 287–305. Totowa, NJ: Humana

33. Lai FA, Liu QY, Xu L, el-Hashem A, Kramarcy NR, et al. 1992. Amphibian ryanodine receptor isoforms are related to those of mammalian skeletal or cardiac muscle. *Am. J. Physiol. Cell Physiol.* 263:C365–72

34. Olivares E, Arispe N, Rojas E. 1993. Properties of the ryanodine receptor present in the sarcoplasmic reticulum from lobster skeletal muscle. *Membr. Biochem.* 10:221–35

35. Sutko JL, Airey JA. 1996. Ryanodine receptor Ca^{2+} release channels: does diversity in form equal diversity in function? *Physiol. Rev.* 76:1027–71

36. Nakai J, Imagawa T, Hakamata Y, Shigekawa M, Takeshima H, Numa S. 1990. Primary structure and functional expression from cDNA of the cardiac ryanodine receptor/calcium release channel. *FEBS Lett.* 271:169–77

37. Futatsugi A, Kuwajima G, Mikoshiba K. 1995. Tissue-specific and developmentally regulated alternative splicing in mouse skeletal muscle ryanodine receptor mRNA. *Biochem. J.* 305:373–78

38. Du GG, Sandhu B, Khanna VK, Guo XH, MacLennan DH. 2002. Topology of the Ca^{2+} release channel of skeletal muscle sarcoplasmic reticulum (RyR1). *Proc. Natl. Acad. Sci. USA* 99:16725–30

39. Bhat MB, Hayek SM, Zhao J, Zang W, Takeshima H, et al. 1999. Expression and functional characterization of the cardiac muscle ryanodine receptor Ca^{2+} release channel in Chinese hamster ovary cells. *Biophys. J.* 77:808–16

40. Bhat MB, Zhao J, Takeshima H, Ma J. 1997. Functional calcium release channel formed by the carboxyl-terminal portion of ryanodine receptor. *Biophys. J.* 73:1329–36

41. Coronado R, Morrissette J, Sukhareva M, Vaughan DM. 1994. Structure and function of ryanodine receptors. *Am. J. Physiol. Cell Physiol.* 266:C1485–504

42. Diaz-Munoz M, Hamilton S, Kaetzel M, Hazarika P, Dedman J. 1990. Modulation of Ca^{2+} release channel activity from sarcoplasmic reticulum by annexin V1 (67-kDa Calcimedin). *J. Biol. Chem.* 265:15894–99

43. Jones LR, Zhang L, Sanborn K, Jorgensen AO, Kelley J. 1995. Purification, primary structure, and immunological characterization of the 26-kDa calsequestrin binding protein (junctin) from cardiac junctional sarcoplasmic reticulum. *J. Biol. Chem.* 270:30787–96

44. Knudson CM, Stang KK, Jorgensen AO, Campbell KP. 1993. Biochemical characterization and ultrastructural localization of a major junctional sarcoplasmic reticulum glycoprotein (triadin). *J. Biol. Chem.* 268:12637–45

45. Lokuta AJ, Meyers MB, Sander PR, Fishman GI, Valdivia HH. 1997. Modulation of cardiac ryanodine receptors by sorcin. *J. Biol. Chem.* 272:25333–38

46. Brillantes AB, Ondrias K, Scott A, Kobrinsky E, Ondriasova E, et al. 1994. Stabilization of calcium release channel (ryanodine receptor) function by FK506-binding protein. *Cell* 77:513–23

47. Marks AR. 1996. Immunophilin modulation of calcium channel gating. *Methods: A Companion Methods Enzymol.* 9:177–87

48. Wagenknecht T, Radermacher M, Grassucci R, Berkowitz J, Xin HB, Fleischer S. 1997. Locations of calmodulin and FK506-binding protein on the three-dimensional architecture of the skeletal muscle ryanodine receptor. *J. Biol. Chem.* 272:32463–71

49. Tripathy A, Xu L, Mann G, Meissner G. 1995. Calmodulin activation and inhibition of skeletal muscle Ca^{2+} release channel (ryanodine receptor). *Biophys. J.* 69:106–19

50. Wehrens XHT, Lehnart SE, Reiken SR, Marks AR. 2004. Ca^{2+}/calmodulin-dependent protein kinase II phosphorylation regulates the cardiac ryanodine receptor. *Circ. Res.* 94:E61–70

51. Currie S, Loughrey CM, Craig MA, Smith GL. 2004. Calcium/calmodulin-dependent protein kinase IIδ associates with the ryanodine receptor complex and regulates channel function in rabbit heart. *Biochem. J.* 377:357–66

52. Smith J, Rousseau E, Meissner G. 1989. Calmodulin modulation of single sarcoplasmic reticulum Ca^{2+} channels from cardiac and skeletal muscle. *Circ. Res.* 64:352–59

53. Chu A, Sumbilla C, Inesi G, Jay SD, Campbell KP. 1990. Specific association of calmodulin-dependent protein kinase and related substrates with the junctional sarcoplasmic reticulum of skeletal muscle. *Biochemistry* 29:5899–905

54. Moore CP, Rodney G, Zhang JZ, Santacruz-Toloza L, Strasburg G, Hamilton SL. 1999. Apocalmodulin and Ca^{2+} calmodulin bind to the same region on the skeletal muscle Ca^{2+} release channel. *Biochemistry* 38:8532–37

55. Porter Moore C, Zhang JZ, Hamilton SL. 1999. A role for cysteine 3635 of RYR1 in redox modulation and calmodulin binding. *J. Biol. Chem.* 274:36831–34

56. Samso M, Wagenknecht T. 2002. Apocalmodulin and Ca^{2+}-calmodulin bind to neighboring locations on the ryanodine receptor. *J. Biol. Chem.* 277:1349–53

57. Fruen BR, Bardy JM, Byrem TM, Strasburg GM, Louis CF. 2000. Differential Ca^{2+} sensitivity of skeletal and cardiac muscle ryanodine receptors in the presence of calmodulin. *Am. J. Physiol. Cell Physiol.* 279:C724–33

58. Balshaw DM, Xu L, Yamaguchi N, Pasek DA, Meissner G. 2001. Calmodulin binding and inhibition of cardiac muscle calcium release channel (ryanodine receptor). *J. Biol. Chem.* 276:20144–53

59. Marks AR. 1996. Cellular functions of immunophilins. *Physiol. Rev.* 76:631–49

60. Timerman AP, Ogunbumni E, Freund E, Wiederrecht G, Marks AR, Fleischer S. 1993. The calcium release channel of sarcoplasmic reticulum is modulated by FK-506-binding protein. Dissociation and reconstitution of FKBP-12 to the calcium release channel of skeletal muscle sarcoplasmic reticulum. *J. Biol. Chem.* 268:22992–99

61. Timerman AP, Onoue H, Xin HB, Barg S, Copello J, et al. 1996. Selective binding of FKBP12.6 by the cardiac ryanodine receptor. *J. Biol. Chem.* 271:20385–91

62. Wehrens XHT, Lehnart SE, Reiken SR, Deng SX, Vest JA, et al. 2004. Protection from cardiac arrhythmia through ryanodine receptor-stabilizing protein calstabin2. *Science* 304:292–96

63. Van Acker K, Bultynck G, Rossi D, Sorrentino V, Boens N, et al. 2004. The 12 kDa FK506-binding protein, FKBP12, modulates the Ca^{2+}-flux properties of the type-3 ryanodine receptor. *J. Cell Sci.* 117:1129–37

64. Jeyakumar LH, Ballester L, Cheng DS, McIntyre JO, Chang P, et al. 2001. FKBP binding characteristics of cardiac microsomes from diverse vertebrates. *Biochem. Biophys. Res. Commun.* 281:979–86

65. Sharma MR, Jeyakuma LH, Fleischer S, Wagenknecht T. 2002. Three-dimensional visualisation of FKBP12.6 binding to cardiac ryanodine receptor (RyR2) in open buffer conditions. *Biophys. J.* 82: A644

66. Gaburjakova M, Gaburjakova J, Reiken S, Huang F, Marx SO, et al. 2001. FKBP12 binding modulates ryanodine receptor channel gating. *J. Biol. Chem.* 276:16931–35

67. Bultynck G, Rossi D, Callewaert G, Missiaen L, Sorrentino V, et al. 2001. The conserved sites for the FK506-binding proteins in ryanodine receptors and inositol 1,4,5-trisphosphate receptors are structurally and functionally different. *J. Biol. Chem.* 276:47715–24

68. Masumiya H, Wang RW, Zhang J, Xiao BL, Chen SRW. 2003. Localization of the 12.6-kDa FK506-binding protein (FKBP12.6) binding site to the NH_2-terminal domain of the cardiac Ca^{2+} release channel (ryanodine receptor). *J. Biol. Chem.* 278:3786–92

69. Marx SO, Reiken S, Hisamatsu Y, Jayaraman T, Burkhoff D, et al. 2000. PKA phosphorylation dissociates FKBP12.6 from the calcium release channel (ryanodine receptor): defective regulation in failing hearts. *Cell* 101:365–76

70. Kapiloff MS, Jackson N, Airhart N. 2001. mAKAP and the ryanodine receptor are part of a multi-component signaling

complex on the cardiomyocyte nuclear envelope. *J. Cell Sci.* 114:3167–76

71. Allen P, Ouimet C, Greengard P. 1997. Spinophilin, a novel protein phosphatase 1 binding protein localized to dendritic spines. *Proc. Natl. Acad. Sci. USA* 94:9956–61

72. Meyers MB, Pickel VM, Sheu SS, Sharma VK, Scotto KW, Fishman GI. 1995. Association of sorcin with the cardiac ryanodine receptor. *J. Biol. Chem.* 270:26411–18

73. Meyers MB, Puri TS, Chien AJ, Gao T, Hsu PH, et al. 1998. Sorcin associates with the pore-forming subunit of voltage-dependent L-type Ca^{2+} channels. *J. Biol. Chem.* 273:18930–35

74. Valdivia HH. 1998. Modulation of intracellular Ca^{2+} levels in the heart by sorcin and FKBP12, two accessory proteins of ryanodine receptors. *Trends Pharmacol. Sci.* 19:479–82

75. Marks AR, Marx SO, Reiken S. 2002. Regulation of ryanodine receptors via macromolecular complexes: a novel role for leucine/isoleucine zippers. *Trends Cardiovasc. Med.* 12:166–70

76. Zhang L, Kelley J, Schmeisser G, Kobayashi YM, Jones LR. 1997. Complex formation between junctin, triadin, calsequestrin, and the ryanodine receptor. Proteins of the cardiac junctional sarcoplasmic reticulum membrane. *J. Biol. Chem.* 272:23389–97

77. Flucher BE, Andrews SB, Fleischer S, Marks AR, Caswell A, Powell JA. 1993. Triad formation: organization and function of the sarcoplasmic reticulum calcium release channel and triadin in normal and dysgenic muscle in vitro. *J. Cell Biol.* 123:1161–74

78. Collins J, Tarcsafalvi A, Ikemoto N. 1990. Identification of a region of calsequestrin that binds to the junctional face membrane of sarcoplasmic reticulum. *Biochem. Biophys. Res. Commun.* 167:189–93

79. Viatchenko-Karpinski S, Terentyev D, Gyorke I, Terentyeva R, Volpe P, et al. 2004. Abnormal calcium signaling and sudden cardiac death associated with mutation of calsequestrin. *Circ. Res.* 94:471–77

80. Culligan K, Banville N, Dowling P, Ohlendieck K. 2002. Drastic reduction of calsequestrin-like proteins and impaired calcium binding in dystrophic mdx muscle. *J. Appl. Physiol.* 92:435–45

81. Ohkura M, Furukawa K, Fujimori H, Kuruma A, Kawano S, et al. 1998. Dual regulation of the skeletal muscle ryanodine receptor by triadin and calsequestrin. *Biochemistry* 37:12987–93

82. Szegedi C, Sarkozi S, Herzog A, Jona I, Varsanyi M. 1999. Calsequestrin: more than 'only' a luminal Ca^{2+} buffer inside the sarcoplasmic reticulum. *Biochem. J.* 337:19–22

83. Fill M, Copello JA. 2002. Ryanodine receptor calcium release channels. *Physiol. Rev.* 82:893–922

84. Witcher DR, Kovacs RJ, Schulman H, Cefali DC, Jones LR. 1991. Unique phosphorylation site on the cardiac ryanodine receptor regulates calcium channel activity. *J. Biol. Chem.* 266:11144–52

85. Witcher DR, Strifler BA, Jones LR. 1992. Cardiac-specific phosphorylation site for multifunctional Ca^{2+}/calmodulin-dependent protein kinase is conserved in the brain ryanodine receptor. *J. Biol. Chem.* 267:4963–67

86. Wehrens XHT, Lehnart SE, Huang F, Vest JA, Reiken SR, et al. 2003. FKBP12.6 deficiency and defective calcium release channel (ryanodine receptor) function linked to exercise-induced sudden cardiac death. *Cell* 113:829–40

87. Xiao BL, Sutherland C, Walsh MP, Chen SRW. 2004. Protein kinase a phosphorylation at serine-2808 of the cardiac Ca^{2+}-release channel (ryanodine receptor) does not dissociate 12.6-kDa FK506-binding protein (FKBP12.6). *Circ. Res.* 94:487–95

88. Rodriguez P, Bhogal MS, Colyer J. 2003. Stoichiometric phosphorylation of

cardiac ryanodine receptor on serine 2809 by calmodulin-dependent kinase II and protein kinase A. *J. Biol. Chem.* 278: 38593–600

89. Hain J, Onoue H, Mayrleitner M, Fleischer S, Schindler H. 1995. Phosphorylation modulates the function of the calcium release channel of sarcoplasmic reticulum from cardiac muscle. *J. Biol. Chem.* 270:2074–81

90. Lokuta AJ, Rogers TB, Lederer WJ, Valdivia HH. 1995. Modulation of cardiac ryanodine receptors of swine and rabbit by a phosphorylation-dephosphorylation mechanism. *J. Physiol.* 487:609–22

91. Sonnleitner A, Fleischer S, Schindler H. 1997. Gating of the skeletal calcium release channel by ATP is inhibited by protein phosphatase 1 but not by Mg^{2+}. *Cell Calcium* 21:283–90

92. Valdivia HH, Kaplan JH, Ellis-Davies GC, Lederer WJ. 1995. Rapid adaptation of cardiac ryanodine receptors: modulation by Mg^{2+} and phosphorylation. *Science* 267:1997–2000

93. Li Y, Kranias EG, Mignery GA, Bers DM. 2002. Protein kinase A phosphorylation of the ryanodine receptor does not affect calcium sparks in mouse ventricular myocytes. *Circ. Res.* 90:309–16

94. Ginsburg KS, Bers DM. 2004. Modulation of excitation-contraction coupling by isoproterenol in cardiomyocytes with controlled SR Ca load and ICa trigger. *J. Physiol.* 556:463–80

95. Yoshida A, Takahashi M, Imagawa T, Shigekawa M, Takisawa H, Nakamura T. 1992. Phosphorylation of ryanodine receptors in rat myocytes during beta-adrenergic stimulation. *J. Biochem.* 111:186–90

96. Jayaraman T, Brillantes A-MB, Timerman AP, Erdjument-Bromage H, Fleischer S, et al. 1992. FK506 binding protein associated with the calcium release channel (ryanodine receptor). *J. Biol. Chem.* 267:9474–77

97. Kaftan E, Marks AR, Ehrlich BE. 1996. Effects of rapamycin on ryanodine receptor/Ca^{2+}-release channels from cardiac muscle. *Circ. Res.* 78:990–97

98. Jiang MT, Lokuta AJ, Farrell EF, Wolff MR, Haworth RA, Valdivia HH. 2002. Abnormal Ca^{2+} release, but normal ryanodine receptors, in canine and human heart failure. *Circ. Res.* 91:1015–22

99. Stange M, Xu L, Balshaw D, Yamaguchi N, Meissner G. 2003. Characterization of recombinant skeletal muscle (Ser-2843) and cardiac muscle (Ser-2809) ryanodine receptor phosphorylation mutants. *J. Biol. Chem.* 278:51693–702

100. Marks AR. 2002. Ryanodine receptors, FKBP12, and heart failure. *Front Biosci.* 7:D970–77

101. McCall E, Li L, Satoh H, Shannon TR, Blatter LA, Bers DM. 1996. Effects of FK-506 on contraction and Ca^{2+} transients in rat cardiac myocytes. *Circ. Res.* 79:1110–21

102. Ahern GP, Junankar PR, Dulhunty AF. 1997. Subconductance states in single-channel activity of skeletal muscle ryanodine receptors after removal of FKBP12. *Biophys. J.* 72:146–62

103. Marx SO, Ondrias K, Marks AR. 1998. Coupled gating between individual skeletal muscle Ca^{2+} release channels (ryanodine receptors). *Science* 281:818–21

104. Dulhunty AF, Laver D, Curtis SM, Pace S, Haarmann C, Gallant EM. 2001. Characteristics of irreversible ATP activation suggest that native skeletal ryanodine receptors can be phosphorylated via an endogenous CaMKII. *Biophys. J.* 81:3240–52

105. Maier LS, Zhang T, Chen L, DeSantiago J, Brown JH, Bers DM. 2003. Transgenic CaMKIIdeltaC overexpression uniquely alters cardiac myocyte Ca^{2+} handling: reduced SR Ca^{2+} load and activated SR Ca^{2+} release. *Circ. Res.* 92:904–11

106. Zhang T, Maier LS, Dalton ND, Miyamoto S, Ross J Jr, et al. 2003. The

deltaC isoform of CaMKII is activated in cardiac hypertrophy and induces dilated cardiomyopathy and heart failure. *Circ. Res.* 92:912–19

107. Neumann J, Boknik P, Herzig S, Schmitz W, Scholz H, et al. 1993. Evidence for physiological functions of protein phosphatases in the heart: evaluation with okadaic acid. *Am. J. Physiol. Heart Circ. Physiol.* 265:H257–66

108. duBell WH, Lederer WJ, Rogers TB. 1996. Dynamic modulation of excitation-contraction coupling by protein phosphatases in rat ventricular myocytes. *J. Physiol.* 493:793–800

109. duBell WH, Gigena MS, Guatimosim S, Long X, Lederer WJ, Rogers TB. 2002. Effects of PP1/PP2A inhibitor calyculin A on the E-C coupling cascade in murine ventricular myocytes. *Am. J. Physiol. Heart Circ. Physiol.* 282:H38–48

110. Carr AN, Schmidt AG, Suzuki Y, del Monte F, Sato Y, et al. 2002. Type 1 phosphatase, a negative regulator of cardiac function. *Mol. Cell Biol.* 22:4124–35

111. Santana LF, Chase EG, Votaw VS, Nelson MT, Greven R. 2002. Functional coupling of calcineurin and protein kinase A in mouse ventricular myocytes. *J. Physiol.* 544:57–69

112. Sipido KR, Carmeliet E, Van de Werf F. 1998. T-type Ca^{2+} current as a trigger for Ca^{2+} release from the sarcoplasmic reticulum in guinea-pig ventricular myocytes. *J. Physiol.* 508:439–51

113. Sipido KR, Maes M, Van de Werf F. 1997. Low efficiency of Ca^{2+} entry through the Na^{+}-Ca^{2+} exchanger as trigger for Ca^{2+} release from the sarcoplasmic reticulum. A comparison between L-type Ca2+ current and reverse-mode Na^{+}-Ca^{2+} exchange. *Circ. Res.* 81:1034–44

114. Flucher BE, Franzini-Armstrong C. 1996. Formation of junctions involved in excitation-contraction coupling in skeletal and cardiac muscle. *Proc. Natl. Acad. Sci.USA* 93:8101–6

115. Cannell MB, Soeller C. 1997. Numerical analysis of ryanodine receptor activation by L-type channel activity in the cardiac muscle diad. *Biophys. J.* 73:112–22

116. Takeshima H, Komazaki S, Nishi M, Iino M, Kangawa K. 2000. Junctophilins: a novel family of junctional membrane complex proteins. *Mol. Cell.* 6:11–22

117. Jorgensen AO, Campbell KP. 1984. Evidence for the presence of calsequestrin in two structurally different regions of myocardial sarcoplasmic reticulum. *J. Cell Biol.* 98:1597–602

118. Guo W, Jorgensen AO, Jones LR, Campbell KP. 1996. Biochemical characterization and molecular cloning of cardiac triadin. *J. Biol. Chem.* 271:458–65

119. Carl SL, Felix K, Caswell AH, Brandt NR, Ball WJ Jr, et al. 1995. Immunolocalization of sarcolemmal dihydropyridine receptor and sarcoplasmic reticular triadin and ryanodine receptor in rabbit ventricle and atrium. *J. Cell Biol.* 129:672–82

120. Sun XH, Protasi F, Takahashi M, Takeshima H, Ferguson DG, Franzini-Armstrong C. 1995. Molecular architecture of membranes involved in excitation-contraction coupling of cardiac muscle. *J. Cell Biol.* 129:659–71

121. Protasi F, Sun XH, Franzini-Armstrong C. 1996. Formation and maturation of the calcium release apparatus in developing and adult avian myocardium. *Dev. Biol.* 173:265–78

122. Lai F, Erickson H, Block B, Meissner G. 1988. Evidence for a Ca^{2+} channel within the ryanodine receptor complex from cardiac sarcoplasmic reticulum. *Biochem. Biophys. Res. Commun.* 151:441–49

123. Franzini-Armstrong C. 1996. Functional significance of membrane architecture in skeletal and cardiac muscle. *Soc. Gen. Physiol. Ser.* 51:3–18

124. Sham JS, Cleemann L, Morad M. 1995. Functional coupling of Ca^{2+} channels and ryanodine receptors in cardiac myocytes. *Proc. Natl. Acad. Sci. USA* 92:121–25

125. Bers DM. 1991. *Excitation-Contraction Coupling and Cardiac Contractile Force.* Boston: Kluwer Acad.

126. Ogawa Y, Kurebayashi N, Murayama T. 1999. Ryanodine receptor isoforms in excitation-contraction coupling. *Adv. Biophys.* 36:27–64

127. Bers DM, Stiffel VM. 1993. Ratio of ryanodine to dihydropyridine receptors in cardiac and skeletal muscle and implications for E-C coupling. *Am. J. Physiol. Cell Physiol.* 264:C1587–93

128. Lytton J, Westlin M, Hanley MR. 1991. Thapsigargin inhibits the sarcoplasmic or endoplasmic reticulum Ca-ATPase family of calcium pumps. *J. Biol. Chem.* 266:17067–71

129. Koss KL, Grupp IL, Kranias EG. 1997. The relative phospholamban and SERCA2 ratio: a critical determinant of myocardial contractility. *Basic Res. Cardiol.* 92(Suppl. 1):17–24

130. Jones LR, Simmerman HK, Wilson WW, Gurd FR, Wegener AD. 1985. Purification and characterization of phospholamban from canine cardiac sarcoplasmic reticulum. *J. Biol. Chem.* 260:7721–30

131. Bers DM, Bridge JH. 1989. Relaxation of rabbit ventricular muscle by Na-Ca exchange and sarcoplasmic reticulum calcium pump. Ryanodine and voltage sensitivity. *Circ. Res.* 65:334–42

132. Callewaert G, Cleemann L, Morad M. 1988. Epinephrine enhances Ca current regulated Ca release and Ca reuptake in rat ventricular myocytes. *Proc. Natl. Acad. Sci. USA* 85:2009–13

133. Hussain M, Drago GA, Coyler J, Orchard CH. 1997. Rate-dependent abbreviation of Ca^{2+} transient in rat heart is independent of phospholamban phosphorylation. *Am. J. Physiol. Heart Circ. Physiol.* 273:H695–706

134. Kranias EG, Garvey JL, Srivastava RD, Solaro RJ. 1985. Phosphorylation and functional modifications of sarcoplasmic reticulum and myofibrils in isolated rabbit hearts stimulated with isoprenaline. *Biochem. J.* 226:113–21

135. Rockman HA, Koch WJ, Lefkowitz RJ. 2002. Seven-transmembrane-spanning receptors and heart function. *Nature* 415:206–12

136. Fink MA, Zakhary DR, Mackey JA, Desnoyer RW, Apperson-Hansen C, et al. 2001. AKAP-mediated targeting of protein kinase A regulates contractility in cardiac myocytes. *Circ. Res.* 88:291–97

137. Gomez AM, Valdivia HH, Cheng H, Lederer MR, Santana LF, et al. 1997. Defective excitation-contraction coupling in experimental cardiac hypertrophy and heart failure. *Science* 276:800–6

138. Dzhura I, Wu Y, Colbran RJ, Balser JR, Anderson ME. 2000. Calmodulin kinase determines calcium-dependent facilitation of L-type calcium channels. *Nat. Cell Biol.* 2:173–77

139. DeSantiago J, Maier LS, Bers DM. 2002. Frequency-dependent acceleration of relaxation in the heart depends on CaMKII, but not phospholamban. *J. Mol. Cell Cardiol.* 34:975–84

140. Wu Y, MacMillan LB, McNeill RB, Colbran RJ, Anderson ME. 1999. CaM kinase augments cardiac L-type Ca^{2+} current: a cellular mechanism for long Q-T arrhythmias. *Am. J. Physiol. Heart Circ. Physiol.* 276:H2168–78

141. Napolitano R, Vittone L, Mundina C, Chiappe de Cingolani G, Mattiazzi A. 1992. Phosphorylation of phospholamban in the intact heart. A study on the physiological role of the Ca(2+)-calmodulin-dependent protein kinase system. *J. Mol. Cell Cardiol.* 24:387–96

142. Hagemann D, Kuschel M, Kuramochi T, Zhu W, Cheng H, Xiao RP. 2000. Frequency-encoding Thr17 phospholamban phosphorylation is independent of Ser16 phosphorylation in cardiac myocytes. *J. Biol. Chem.* 275:22532–36

143. Hagemann D, Xiao RP. 2002. Dual site phospholamban phosphorylation and

its physiological relevance in the heart. *Trends Cardiovasc. Med.* 12:51–56

144. Dempsey EC, Newton AC, Mochly-Rosen D, Fields AP, Reyland ME, et al. 2000. Protein kinase C isozymes and the regulation of diverse cell responses. *Am. J. Physiol. Lung Cell Mol. Physiol.* 279:L429–38

145. Wang J, Liu X, Arneja AS, Dhalla NS. 1999. Alterations in protein kinase A and protein kinase C levels in heart failure due to genetic cardiomyopathy. *Can. J. Cardiol.* 15:683–90

146. Sugden PH, Bogoyevitch MA. 1995. Intracellular signalling through protein kinases in the heart. *Cardiovasc. Res.* 30:478–92

147. Pass JM, Zheng YT, Wead WB, Zhang J, Li RCX, et al. 2001. PKCepsilon activation induces dichotomous cardiac phenotypes and modulates PKCepsilon-RACK interactions and RACK expression. *Am. J. Physiol. Heart Circ. Physiol.* 280:H946–55

148. Braz JC, Gregory K, Pathak A, Zhao W, Sahin B, et al. 2004. PKC-alpha regulates cardiac contractility and propensity toward heart failure. *Nat. Med.* 10:248–54

149. Terentyev D, Viatchenko-Karpinski S, Gyorke I, Terentyeva R, Gyorke S. 2003. Protein phosphatases decrease sarcoplasmic reticulum calcium content by stimulating calcium release in cardiac myocytes. *J. Physiol.* 552:109–18

150. Beuckelmann DJ, Nabauer M, Erdmann E. 1992. Intracellular calcium handling in isolated ventricular myocytes from patients with terminal heart failure. *Circulation* 85:1046–55

151. Gwathmey JK, Copelas L, Mackinnon R, Schoen FJ, Feldman MD, et al. 1987. Abnormal intracellular calcium handling in myocardium from patients with end-stage heart failure. *Circ. Res.* 61:70–76

152. Schlotthauer K, Schattmann J, Bers DM, Maier LS, Schutt U, et al. 1998. Frequency-dependent changes in contribution of SR Ca^{2+} to Ca^{2+} transients in failing human myocardium assessed with ryanodine. *J. Mol. Cell Cardiol.* 30:1285–94

153. Hasenfuss G, Mulieri LA, Leavitt BJ, Allen PD, Haeberle JR, Alpert NR. 1992. Alteration of contractile function and excitation-contraction coupling in dilated cardiomyopathy. *Circ. Res.* 70:1225–32

154. Pieske B, Maier LS, Bers DM, Hasenfuss G. 1999. Ca^{2+} handling and sarcoplasmic reticulum Ca^{2+} content in isolated failing and nonfailing human myocardium. *Circ. Res.* 85:38–46

155. Gomez AM, Guatimosim S, Dilly KW, Vassort G, Lederer WJ. 2001. Heart failure after myocardial infarction: altered excitation-contraction coupling. *Circulation* 104:688–93

156. Wessely R, Klingel K, Santana LF, Dalton N, Hongo M, et al. 1998. Transgenic expression of replication-restricted enteroviral genomes in heart muscle induces defective excitation-contraction coupling and dilated cardiomyopathy. *J. Clin. Invest.* 102:1444–53

157. Esposito G, Santana LF, Dilly K, Cruz JD, Mao L, et al. 2000. Cellular and functional defects in a mouse model of heart failure. *Am. J. Physiol. Heart Circ. Physiol.* 279:H3101–12

158. Hobai IA, O'Rourke B. 2001. Decreased sarcoplasmic reticulum calcium content is responsible for defective excitation-contraction coupling in canine heart failure. *Circulation* 103:1577–84

159. Packer M. 1988. Pathophysiological mechanism underlying the adverse effects of calcium channel-blocking drugs in patients with chronic heart failure. *Circulation* 80:IV59–67

160. Bristow MR, Ginsburg R, Minobe W, Cubicciotti RS, Sageman WS, et al. 1982. Decreased catecholamine sensitivity and beta-adrenergic-receptor density in failing human hearts. *N. Engl. J. Med.* 307:205–11

161. Ungerer M, Parruti G, Bohm M, Puzicha M, DeBlasi A, et al. 1994. Expression

of beta-arrestins and beta-adrenergic receptor kinases in the failing human heart. *Circ. Res.* 74:206–13

162. Feldman AM, Cates AE, Veazey WB, Hershberger RE, Bristow MR, et al. 1988. Increase of the 40,000-mol wt pertussis toxin substrate (G protein) in the failing human heart. *J. Clin. Invest.* 82:189–97

163. Ishikawa Y, Sorota S, Kiuchi K, Shannon RP, Komamura K, et al. 1994. Downregulation of adenylylcyclase types V and VI mRNA levels in pacing-induced heart failure in dogs. *J. Clin. Invest.* 93:2224–29

164. Feldman MD, Copelas L, Gwathmey JK, Phillips P, Warren SE, et al. 1987. Deficient production of cyclic AMP: pharmacologic evidence of an important cause of contractile dysfunction in patients with end-stage heart failure. *Circulation* 75:331–39

165. Engelhardt S, Hein L, Dyachenkow V, Kranias EG, Isenberg G, Lohse MJ. 2004. Altered calcium handling is critically involved in the cardiotoxic effects of chronic beta-adrenergic stimulation. *Circulation* 109:1154–60

166. Antos CL, Frey N, Marx SO, Reiken S, Gaburjakova M, et al. 2001. Dilated cardiomyopathy and sudden death resulting from constitutive activation of protein kinase A. *Circ. Res.* 89:997–1004

167. Yano M, Ono K, Ohkusa T, Suetsugu M, Kohno M, et al. 2000. Altered stoichiometry of FKBP12.6 versus ryanodine receptor as a cause of abnormal Ca^{2+} leak through ryanodine receptor in heart failure. *Circulation* 102:2131–36

168. Reiken S, Wehrens XHT, Vest JA, Barbone A, Klotz S, et al. 2003. Beta-blockers restore calcium release channel function and improve cardiac muscle performance in human heart failure. *Circulation* 107:2459–66

169. Pogwizd SM, Schlotthauer K, Li L, Yuan W, Bers DM. 2001. Arrhythmogenesis and contractile dysfunction in heart failure: roles of sodium-calcium exchange, inward rectifier potassium current, and

residual beta-adrenergic responsiveness. *Circ. Res.* 88:1159–67

170. Wei SK, Ruknudin A, Hanlon SU, McCurley JM, Schulze DH, Haigney MC. 2003. Protein kinase A hyperphosphorylation increases basal current but decreases beta-adrenergic responsiveness of the sarcolemmal Na^+-Ca^{2+} exchanger in failing pig myocytes. *Circ. Res.* 92:897–903

171. Ono K, Yano M, Ohkusa T, Kohno M, Hisaoka T, et al. 2000. Altered interaction of FKBP12.6 with ryanodine receptor as a cause of abnormal Ca^{2+} release in heart failure. *Cardiovasc. Res.* 48:323–31

172. Doi M, Yano M, Kobayashi S, Kohno M, Tokuhisa T, et al. 2002. Propranolol prevents the development of heart failure by restoring FKBP12.6-mediated stabilization of ryanodine receptor. *Circulation* 105:1374–79

173. Xiao RP, Valdivia HH, Bogdanov K, Valdivia C, Lakatta EG, Cheng H. 1997. The immunophilin FK506-binding protein modulates Ca^{2+} release channel closure in rat heart. *J. Phys.* 500:343–54

174. Xin HB, Senbonmatsu T, Cheng DS, Wang YX, Copello JA, et al. 2002. Oestrogen protects FKBP12.6 null mice from cardiac hypertrophy. *Nature* 416:334–38

175. Prestle J, Janssen PM, Janssen AP, Zeitz O, Lehnart SE, et al. 2001. Overexpression of FK506-binding protein FKBP12.6 in cardiomyocytes reduces ryanodine receptor-mediated Ca^{2+} leak from the sarcoplasmic reticulum and increases contractility. *Circ. Res.* 88:188–94

176. Reiken S, Gaburjakova M, Guatimosim S, Gomez AM, D'Armiento J, et al. 2003. Protein kinase a phosphorylation of the cardiac calcium release channel (ryanodine receptor) in normal and failing hearts. Role of phosphatases and response to isoproterenol. *J. Biol. Chem.* 278:444–53

177. Reiken S, Gaburjakova M, Gaburjakova J, He KL, Prieto A, et al. 2001. beta-Adrenergic receptor blockers restore

cardiac calcium release channel (ryanodine receptor) structure and function in heart failure. *Circulation* 104:2843–48

178. Studer R, Reinecke H, Bilger J, Eschenhagen T, Bohm M, et al. 1994. Gene expression of the cardiac Na$^+$-Ca^{2+} exchanger in end-stage human heart failure. *Circ. Res.* 75:443–53

179. Pogwizd SM, Qi M, Yuan WL, Samarel AM, Bers DM. 1999. Upregulation of Na$^+$/Ca^{2+} exchanger expression and function in an arrhythmogenic rabbit model of heart failure. *Circ. Res.* 85:1009–19

180. Piacentino V 3rd, Weber CR, Gaughan JP, Margulies KB, Bers DM, Houser SR. 2002. Modulation of contractility in failing human myocytes by reverse-mode Na/Ca exchange. *Ann. NY Acad. Sci.* 976: 466–71

181. Hasenfuss G, Schillinger W, Lehnart SE, Preuss M, Pieske B, et al. 1999. Relationship between Na$^+$-Ca^{2+}-exchanger protein levels and diastolic function of failing human myocardium. *Circulation* 99:641–48

182. Weber CR, Piacentino V 3rd, Margulies KB, Bers DM, Houser SR. 2002. Calcium influx via I(NCX) is favored in failing human ventricular myocytes. *Ann. NY Acad. Sci.* 976:478–79

183. Piacentino V 3rd, Weber CR, Chen XW, Weisser-Thomas J, Margulies KB, et al. 2003. Cellular basis of abnormal calcium transients of failing human ventricular myocytes. *Circ. Res.* 92:651–58

184. Shorofsky SR, Aggarwal R, Corretti M, Baffa JM, Strum JM, et al. 1999. Cellular mechanisms of altered contractility in the hypertrophied heart: big hearts, big sparks. *Circ. Res.* 84:424–34

185. Schaper J, Froede R, Hein S, Buck A, Hashizume H, et al. 1991. Impairment of the myocardial ultrastructure and changes of the cytoskeleton in dilated cardiomyopathy. *Circulation* 83:504–14

186. Louch WE, Bito V, Heinzel FR, Macianskiene R, Vanhaecke J, et al. 2004. Reduced synchrony of Ca^{2+} release with loss of T-tubules-a comparison to Ca^{2+} release in human failing cardiomyocytes. *Cardiovasc. Res.* 62:63–73

187. Schwinger RH, Munch G, Bolck B, Karczewski P, Krause EG, Erdmann E. 1999. Reduced Ca^{2+}-sensitivity of SERCA 2a in failing human myocardium due to reduced serin-16 phospholamban phosphorylation. *J. Mol. Cell Cardiol.* 31:479–91

188. Currie S, Smith GL. 1999. Enhanced phosphorylation of phospholamban and downregulation of sarco/endoplasmic reticulum Ca^{2+} ATPase type 2 (SERCA 2) in cardiac sarcoplasmic reticulum from rabbits with heart failure. *Cardiovasc. Res.* 41:135–46

189. Bowling N, Walsh RA, Song G, Estridge T, Sandusky GE, et al. 1999. Increased protein kinase C activity and expression of Ca^{2+}-sensitive isoforms in the failing human heart. *Circulation* 99:384–91

190. Bayer AL, Heidkamp MC, Patel N, Porter M, Engman S, Samarel AM. 2003. Alterations in protein kinase C isoenzyme expression and autophosphorylation during the progression of pressure overload-induced left ventricular hypertrophy. *Mol. Cell Biochem.* 242:145–52

191. Wang J, Liu X, Sentex E, Takeda N, Dhalla NS. 2003. Increased expression of protein kinase C isoforms in heart failure due to myocardial infarction. *Am. J. Physiol. Heart Circ. Physiol.* 284:H2277–87

192. Santana LF, Kranias EG, Lederer WJ. 1997. Calcium sparks and excitation-contraction coupling in phospholamban-deficient mouse ventricular myocytes. *J. Physiol.* 503:21–29

193. Satoh H, Blatter LA, Bers DM. 1997. Effects of [Ca^{2+}]$_i$, SR Ca^{2+} load, and rest on Ca^{2+} spark frequency in ventricular myocytes. *Am. J. Physiol. Heart Circ. Physiol.* 272:H657–68

194. Jones LR, Suzuki YJ, Wang W, Kobayashi YM, Ramesh V, et al. 1998. Regulation of Ca^{2+} signaling in transgenic mouse cardiac myocytes overexpressing calsequestrin. *J. Clin. Invest.* 101:1385–93

195. Berlin JR, Cannell MB, Lederer WJ. 1989. Cellular origins of the transient inward current in cardiac myocytes. Role of fluctuations and waves of elevated intracellular calcium. *Circ. Res.* 65:115–26

196. Antzelevitch C, Sicouri S. 1994. Clinical relevance of cardiac arrhythmias generated by afterdepolarizations. Role of M cells in the generation of U waves, triggered activity and torsade de pointes. *J. Am. Coll. Cardiol.* 23:259–77

197. Hajjar RJ, Schmidt U, Matsui T, Guerrero JL, Lee KH, et al. 1998. Modulation of ventricular function through gene transfer in vivo. *Proc. Natl. Acad Sci. USA* 95:5251–56

198. Minamisawa S, Hoshijima M, Chu GX, Ward CA, Frank K, et al. 1999. Chronic phospholamban–sarcoplasmic reticulum calcium ATPase interaction is the critical calcium cycling defect in dilated cardiomyopathy. *Cell* 99:313–22

199. Meyer M, Dillmann WH. 1998. Sarcoplasmic reticulum Ca^{2+}-ATPase overexpression by adenovirus mediated gene transfer and in transgenic mice. *Cardiovasc. Res.* 37:360–66

200. Chen Y, Escoubet B, Prunier F, Amour J, Simonides WS, et al. 2004. Constitutive cardiac overexpression of sarcoplasmic/endoplasmic reticulum Ca^{2+}-ATPase delays myocardial failure after myocardial infarction in rats at a cost of increased acute arrhythmias. *Circulation* 109:1898–903

201. Hoshijima M, Ikeda Y, Iwanaga Y, Minamisawa S, Date MO, et al. 2002. Chronic suppression of heart-failure progression by a pseudophosphorylated mutant of phospholamban via in vivo cardiac rAAV gene delivery. *Nat. Med.* 8:864–71

202. Haghighi K, Kolokathis F, Pater L, Lynch RA, Asahi M, et al. 2003. Human phospholamban null results in lethal dilated cardiomyopathy revealing a critical difference between mouse and human. *J. Clin. Invest.* 111:869–76

203. George CH, Higgs GV, Lai FA. 2003. Ryanodine receptor mutations associated with stress-induced ventricular tachycardia mediate increased calcium release in stimulated cardiomyocytes. *Circ. Res.* 93:531–40

204. Volders PG, Kulcsar A, Vos MA, Sipido KR, Wellens HJ, et al. 1997. Similarities between early and delayed afterdepolarizations induced by isoproterenol in canine ventricular myocytes. *Cardiovasc. Res.* 34:348–59

205. Matsuda H, Noma A, Kurachi Y, Irisawa H. 1982. Transient depolarization and spontaneous voltage fluctuations in isolated single cells from guinea pig ventricles. Calcium-mediated membrane potential fluctuations. *Circ. Res.* 51:142–51

206. Giles W, Shimoni Y. 1989. Comparison of sodium-calcium exchanger and transient inward currents in single cells from rabbit ventricle. *J. Physiol.* 417:465–81

207. Collier ML, Levesque PC, Kenyon JL, Hume JR. 1996. Unitary Cl^- channels activated by cytoplasmic Ca^{2+} in canine ventricular myocytes. *Circ. Res.* 78:936–44

208. Schlotthauer K, Bers DM. 2000. Sarcoplasmic reticulum Ca^{2+} release causes myocyte depolarization. Underlying mechanism and threshold for triggered action potentials. *Circ. Res.* 87:774–80

209. Sipido KR, Volders PG, de Groot SH, Verdonck F, Van de Werf F, et al. 2000. Enhanced Ca^{2+} release and Na/Ca exchange activity in hypertrophied canine ventricular myocytes: potential link between contractile adaptation and arrhythmogenesis. *Circulation* 102:2137–44

210. Takasago T, Imagawa T, Shigekawa M. 1989. Phosphorylation of the cardiac ryanodine receptor by cAMP-dependent protein kinase. *J. Biochem.* 106:872–77

211. Takasago T, Imagawa T, Furukawa K, Ogurusu T, Shigekawa M. 1991. Regulation of the cardiac ryanodine receptor by protein kinase-dependent phosphorylation. *J. Biochem.* 109:163–70

212. Yoshida A, Ogura A, Imagawa T, Shigekawa M, Takahashi M. 1992. Cyclic AMP-dependent phosphorylation of the rat brain ryanodine receptor. *J. Neurosci.* 12:1094–100

213. Herrmann-Frank A, Varsanyi M. 1993. Enhancement of Ca^{2+} release channel activity by phosphorylation of the skeletal muscle ryanodine receptor. *FEBS Lett.* 332:237–42

214. Strand MA, Louis CF, Mickelson JR. 1993. Phosphorylation of the porcine skeletal and cardiac muscle sarcoplasmic reticulum ryanodine receptor. *Biochim. Biophys. Acta* 1175:319–26

215. Hain J, Nath S, Mayrleitner M, Fleischer S, Schindler H. 1994. Phosphorylation modulates the function of the calcium release channel of sarcoplasmic reticulum from skeletal muscle. *Biophys. J.* 67:1823–33

216. Patel JR, Coronado R, Moss RL. 1995. Cardiac sarcoplasmic reticulum phosphorylation increases Ca^{2+} release induced by flash photolysis of nitr-5. *Circ. Res.* 77:943–49

217. Mayrleitner M, Chandler R, Schindler H, Fleischer S. 1995. Phosphorylation with protein kinases modulates calcium loading of terminal cisternae of sarcoplasmic

reticulum from skeletal muscle. *Cell Calcium* 18:197–206

218. Uehara A, Yasukochi M, Mejia-Alvarez R, Fill M, Imanaga I. 2002. Gating kinetics and ligand sensitivity modified by phosphorylation of cardiac ryanodine receptors. *Pflügers Arch.* 444:202–12

219. Reiken S, Lacampagne A, Zhou H, Kherani A, Lehnart SE, et al. 2003. PKA phosphorylation activates the calcium release channel (ryanodine receptor) in skeletal muscle: defective regulation in heart failure. *J. Cell Biol.* 160:919–28

220. Ward CW, Reiken S, Marks AR, Marty I, Vassort G, Lacampagne A. 2003. Defects in ryanodine receptor calcium release in skeletal muscle from post-myocardial infarct rats. *FASEB J.* 17:1517–19

221. Schotten U, Schumacher C, Conrads V, Braun V, Schondube F, et al. 1999. Calcium-sensitivity of the SR calcium release channel in failing and nonfailing human myocardium. *Basic Res. Cardiol.* 94:145–51

222. Shannon TR, Pogwizd SM, Bers DM. 2003. Elevated sarcoplasmic reticulum Ca^{2+} leak in intact ventricular myocytes from rabbits in heart failure. *Circ. Res.* 93:592–94

Annu. Rev. Physiol. 2005. 67:99–145
doi: 10.1146/annurev.physiol.67.060603.090918
First published online as a Review in Advance on October 19, 2004

CHEMICAL PHYSIOLOGY OF BLOOD FLOW REGULATION BY RED BLOOD CELLS: The Role of Nitric Oxide and S-Nitrosohemoglobin

David J. Singel
Department of Chemistry and Biochemistry, Montana State University, Bozeman, Montana 59717; email: rchds@montana.edu

Jonathan S. Stamler
Howard Hughes Medical Institute and Departments of Medicine and Biochemistry, Duke University Medical Center, Durham, North Carolina 27710; email: staml001@mc.duke.edu

Key Words hypoxic vasodilation, complexity, S-nitrosothiols

■ **Abstract** Blood flow in the microcirculation is regulated by physiological oxygen (O_2) gradients that are coupled to vasoconstriction or vasodilation, the domain of nitric oxide (NO) bioactivity. The mechanism by which the O_2 content of blood elicits NO signaling to regulate blood flow, however, is a major unanswered question in vascular biology. While the hemoglobin in red blood cells (RBCs) would appear to be an ideal sensor, conventional wisdom about its chemistry with NO poses a problem for understanding how it could elicit vasodilation. Experiments from several laboratories have, nevertheless, very recently established that RBCs provide a novel NO vasodilator activity in which hemoglobin acts as an O_2 sensor and O_2-responsive NO signal transducer, thereby regulating both peripheral and pulmonary vascular tone. This article reviews these studies, together with biochemical studies, that illuminate the complexity and adaptive responsiveness of NO reactions with hemoglobin. Evidence for the pivotal role of S-nitroso (SNO) hemoglobin in mediating this response is discussed. Collectively, the reviewed work sets the stage for a new understanding of RBC-derived relaxing activity in auto-regulation of blood flow and O_2 delivery and of RBC dysfunction in disorders characterized by tissue O_2 deficits, such as sickle cell disease, sepsis, diabetes, and heart failure.

INTRODUCTION

Background

The original identification of endothelium-derived relaxing factor (EDRF) as nitric oxide (NO) was based in part on the ability of hemoglobin (Hb) to inactivate both substances (1, 2). Earlier work had shown that Hb can react rapidly with NO to form nitrate from oxy-Hb or a heme-iron nitrosyl adduct with deoxy-Hb, as summarized in Reactions 1 and 2 (3, 4); neither product exhibits bioactivity characteristic of NO.

$$\text{heme Fe(II)} + \text{NO} \rightarrow \text{heme-Fe(II)NO} \qquad\qquad 1.$$

$$\text{heme Fe(II)O}_2 + \text{NO} \rightarrow \text{heme-Fe(III)} + \text{NO}_3^- \qquad 2.$$

In this light, the vasodilatory bioactivity of NO in blood presented conceptual problems: (*a*) Could this activity coexist with Hb, which can rapidly and efficiently scavenge NO (5–7) and (*b*) would red blood cells (RBCs), through this scavenging chemistry, act as relentless vasoconstrictors (8)?

The first question has been addressed on several complementary levels. Liao and coworkers suggested that owing to the flow of blood, RBCs tend to remain centered in the larger vessels and avoid the walls where NO is produced (9). Lancaster provided a rationale (10), later elaborated by others (11–13), for how the cellular packaging of the Hb retards its reaction with NO. More important, the broader chemistry of NO in biology was shown to include the oxidative formation of S-nitrosothiols (thionitrites), which maintain cardiovascular bioactivity in the presence of Hb, circumventing Reactions 1 and 2 (7).

Thionitrites—including both low-molecular-weight nitrosothiol (SNO) derivatives of cysteine and glutathione, and also S-nitrosylated proteins, such as S-nitrosoalbumin (6, 14)—are among the most potent vasodilatory compounds known. Molar potencies of S-nitrosocysteine and S-nitrosoglutathione in bioassays are equal to or higher than NO (15–19), especially when the comparison is made using the smaller resistance vessels that control blood flow (20). Moreover, S-nitrosothiols appear to be the most abundant compounds to exhibit NO-related bioactivity in the blood and blood vessel walls, existing at basal levels orders of magnitude greater than NO (18). SNOs are unique among the various compounds that derive from NO synthase in that their physiological role in vasoregulation has been demonstrated by strict genetic evidence (21). In particular, SNOs contribute to regulation of vascular resistance under basal conditions and its dysregulation in endotoxic shock (21).

An intriguing idea emerged from consideration of the second question: Do RBCs act as vasoconstrictors? A general principle of physiology holds that cells precisely regulate their primary function. For RBCs this primary function is delivery of oxygen (O_2) to tissues. Vasoconstriction implicated by the NO scavenging chemistry (Reactions 1, 2) would impede blood flow and oppose the primary function. In as much as O_2 delivery is determined primarily by blood flow, rather than by oxy-Hb concentration, this line of thinking implies that, far from having a vasoconstricting effect, RBCs should be capable of dilating blood vessels in the microcirculation to regulate blood flow (8). Furthermore, RBC vasodilation in the pulmonary arteries and arterioles could serve to optimize ventilation-perfusion matching, that is, blood oxygenation, and regulate pulmonary artery pressure (18, 22).

Aspects of Blood Flow Physiology

Blood flow in the microcirculation is principally regulated by physiological O_2 gradients: position-to-position variations in O_2 content, which are immediately reflected in changes in Hb O_2 saturation, are coupled to regulated vasoconstriction or vasodilation (23–27). The overall design matches O_2 delivery with metabolic demand. Thus, decreases in the O_2 content of blood lead to increases in blood flow and vice versa (24, 28). While this regulation of blood flow is exerted through local modulation of arteriolar tone, the mechanism through which graded changes in O_2 content evoke the response is a major unanswered question in biology. Hb would appear to be an ideal O_2 sensor in this regulatory process, particularly since it is the O_2 saturation of blood Hb, not pO_2, that determines blood flow (24, 28) (Figure 1a). In this context, the fixed ideas about Hb's scavenging NO (Reactions 1, 2) presented a conceptual roadblock for understanding how the O_2 signal, detected by Hb, could be transduced to elicit vasodilation.

SNO-Hemoglobin

The resolution of this problem began with the discovery that Hb itself is among the blood proteins that sustain S-nitrosylation. Specific cysteine residues of Hb, conserved in all mammalian and avian species, form S-nitrosothiols both in vivo and in vitro (8). S-nitrosylation of human Hb is linked in vivo to O_2 saturation (Figure 1b) and occurs at Cys-β93 (Figure 2, see color insert). S-nitrosylated-Hb, or SNO-Hb, has further been characterized by mass spectrometry (29) and X-ray crystallography (30, 31). This work provides direct evidence that the scavenging chemistry (Reactions 1, 2) and concomitant loss of NO bioactivity can be avoided to furnish this previously unsuspected product of Hb interactions with NO reagents: SNO-Hb formation is competitive with and/or circumvents the Fe(II)-NO and nitrate-forming reactions in vivo (18, 22, 32–36) (Figure 1b). It has further been demonstrated that the reactivity of these cysteines toward NO reagents is dependent on the quaternary structure of the tetramer (8, 37). SNO-Hb forms preferentially in the oxygenated (or R) structure, whereas conditions favoring T structure, such as low pO_2, favor release of NO groups (8) (Figure 2). The circulating levels of SNO-Hb are thus partly dependent on the O_2 saturation-governing equilibrium between T and R structures, and not on the pO_2 (Figure 1b). Crystal structures (30) and molecular models (30, 37) of SNO-Hb provide a rational, "stereochemical" (38, 39) basis for allosterically regulated dispensing of NO bioactivity; thus, whereas β-cys thiol has no access to solvent in R state (and therefore could not dilate blood vessels), it protrudes into solvent in the deoxygenated (or T) structure (Figure 2). Energy-minimization modeling based on the SNO-Hb crystal structure (30) suggests that the entire SNO moiety is folded back into the protein with no solvent access (30) in R state.

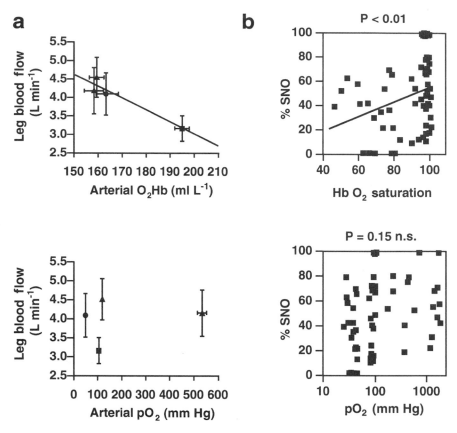

Figure 1 Blood flow and SNO content of Hb are linked to Hb O_2 saturation in humans. Response of limb blood flow (*a*) and Hb S-nitrosylation (*b*) across Hb O_2 saturation (*top*) or pO_2 (*bottom*). Thus blood flow responds to O_2 saturation of Hb and is uncorrelated with pO_2. SNO content of Hb shows a similar behavior, consistent with its role in hypoxic vasodilation [note that many data points overlap at the 100% SNO, 100% sat locus in (*b*)]. In addition to O_2 saturation, other factors including pH, pCO_2, and redox state may influence the O_2-dependent processing of NO by Hb, as discussed below. Panel (*a*) is taken from (24), and (*b*) is from (22), with minor modifications. SNO is presented as a fraction of total NO bound to Hb (%SNO).

Red Blood Cells

NO BIOACTIVITY   In addition to the major pool of cytosolic Hb, which serves in the bulk transport of O_2, a second pool of Hb is localized to the plasma membrane through interaction with the N-terminal cytoplasmic tail of the band 3 protein (anion exchanger 1:AE1). Nitric oxide and related congeners that enter the RBC first encounter the membrane Hb (40): Not only is a major fraction of SNO-Hb directly associated with the membrane, but transfer of NO from β-cysNO of Hb to cysteine thiols of band 3 protein at the RBC membrane was also shown to be necessary and sufficient for robust vasodilation by RBCs under relevant physiological conditions. Moreover, RBCs were shown to actuate a unique, rapid, and graded vasodilator or vasoconstrictor response across a physiological range of pO_2 (22). The primary data that illuminate this response are presented in Figure 3a–d and the mechanistic details are shown in Figure 3e (see color insert). Note that there is some evidence that GSNO may serve as an intermediate in the inport or export from RBCs of NO bioactivity (Figure 3e) (8, 18, 21). The biological activity of SNO-Hb (40, 41) and RBCs (22) is thus seen to exhibit the requisite dependence on Hb O_2 saturation, apparently through the allosteric behavior of Hb. RBCs were shown not only to dilate blood vessels, but to do so in a manner that recapitulated the autoregulation of vessel tone by the physiological O_2 gradient.

Although hypoxic vasodilation by RBCs can be partly blocked by inhibitors of guanylate cyclase (33), cGMP-independent effects have also been reported (42). Vasodilatory effects of RBCs are observed in endothelium-denuded vessels (G. Ahearn, J.R. Pawloski, T.J. McMahon & J.S. Stamler, unpublished results) and are potentiated by the pretreatment of RBCs with NO (33, 40) or SNO (8), consistent with the observation that hypoxic vasodilation in vivo can be entirely endothelium independent (44).

TISSUES, LUNG, AND BRAIN   Whereas arteriolar blood flow in peripheral tissues subserves O_2 delivery (Figure 1a), in the lung it is regulated to optimize O_2 uptake. For example, alveolar hypoxia results in constriction of blood vessels that perfuse alveolar units to preserve V/Q matching. NO counteracts this hypoxic pulmonary vasoconstriction (45), thereby mitigating excessive increases in pulmonary artery pressure and creation of alveolar dead space (46). RBCs entering the lungs contain significant amounts of SNO-Hb (see below), and emerging evidence indicates that dispensing this vasodilator activity may contribute to NO homeostasis; RBC-derived NO bioactivity may thus serve in V/Q matching and maintainance of basal pulmonary arterial tone (Figure 4a). Other studies, by contrast, suggest that by sequestering endothelial NO, RBCs enable hypoxic pulmonary vasoconstriction (47). The extent to which this effect of (infused) RBCs may be an artefact of reduced endogenous SNO-Hb levels is undetermined (Figure 4a). As discussed below, RBCs rapidly lose SNO ex vivo, and RBCs depleted of SNO may accentuate pulmonary hypertension and impair oxygenation. Also of note, RBC-derived SNO can stimulate centers in the brain that control the hypoxic drive to breathe (48), and vasodilation by RBCs within these highly vascularized centers may play a regulatory role (Figure 4b). Thus, although a respiratory cycle for NO is not yet fully understood (49), RBCs may affect essential control mechanisms, not only in peripheral tissue but also in the lungs and brain.

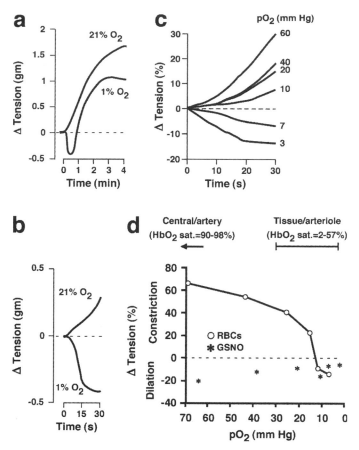

Figure 3 Graded oxygen-dependent vasodilation and vasoconstriction by RBCs, and erythrocytic NO trafficking. (*a*) In organ chamber bioassays, RBCs dilate blood vessels at low pO_2 (1% O_2), which is characteristic of tissues, but are vasoconstrictive in room air. The hypoxic vasodilator response is followed by vasoconstriction in vitro (representing scavenging of endothelial NO), which starts at approximately 1 min following addition of RBCs and therefore has no physiological relevance. (*b*) Responses to RBCs occur over seconds, commensurate with arterial-venous transit times. (*c, d*) The effects of oxygen tension on the activity of RBCs compared with that of the simple endogenous vasodilator S-nitrosoglutathione (GSNO) (*d*). Aortic rings were pre-equilibrated at 2 g and the indicated oxygen pressures. Oxygenated human RBCs were then added at low hematocrit. GSNO (3 nM) evokes a dilatory response independent of oxygen tension. In contrast, RBCs elicit a graded vasodilator response beginning at pO_2 of approximately 60 torr and across the physiological range of hemoglobin O_2 saturations. One should not deduce from panels *c* and *d* that RBCs constrict blood vessels at pO_2 greater than 10 torr, but rather that vasodilatory activity is seen below pO_2 of 60 torr; the data in aortic tissue bioassays cannot be extrapolated to the microcirculation in vivo.

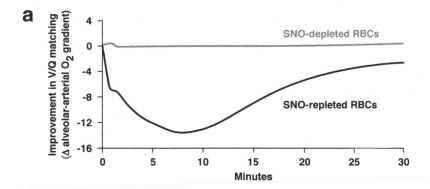

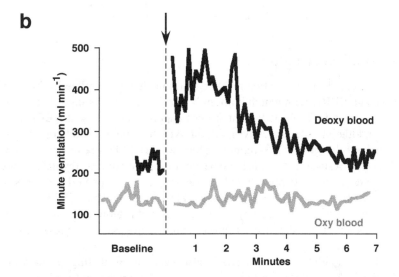

Figure 4   Red blood cell-derived NO bioactivity subserves blood oxygenation in the lungs and mediates central nervous system control of ventilation. (*a*) Infusion of RBCs (50 cc/< 30 s, 30% hematocrit) into the pulmonary artery of an anesthetized pig had little effect if SNO content had been reduced (~ 20% of basal SNO-Hb) by prior deoxygenation and storage, but produced a rapid improvement of ventilation (V)/perfusion (Q) matching (decrease in the alveolar-arterial oxygen gradient) if the SNO content was reconstituted ex vivo (to within twofold of endogenous SNO-Hb content) by exposure to NO. (*b*) Following pretreatment with glutathione, the low-mass fraction from oxygenated (left heart) or deoxygenated (right heart) blood was microinjected into the brainstem nucleus tractus solitarius of conscious rats (*arrow*). Deoxygenation (but not oxygenation) generated low-mass S-nitrosothiol (S-nitrosoglutathione; identified by mass spectrometry), which rapidly and potently stimulated the respiratory drive, as revealed by increased minute ventilation. Figure taken from Lipton et al. (48).

TABLE 1   Characteristics of physiological flow responses

	SNO-Hb	ATP	EDRF-NO
Dependence			
RBC	✓	✓	—
Endothelium	—	✓	✓
Shear	—	—	✓
O_2	✓	✓	—
NO bioactivity	✓	✓	✓

Oxygen concentration also influences vessel tone through a direct effect on the vessel wall; the mechanism is dependent on the duration of change in the O_2 concentration, and is largely independent of EDRF-NO.

Other Vasodilatory Responses

Hb-mediated vasoregulation must be distinguished from related effects, involving ATP and EDRF, with which it might be confused. The distinguishing features are summarized in Table 1. Recent studies support the view that RBCs also release the endothelium-dependent vasodilator, ATP, to regulate blood flow (50, 51). With sustained changes in O_2 saturation, blood levels of ATP rise or fall within minutes. This type of sustained control is to be contrasted with the Hb-regulated (endothelial NO-independent) response in which vasoregulation can be effected over seconds, commensurate with arterial-venous transit times. Thus Hb and ATP may serve complementary roles, respectively, in acute local and prolonged systemic hypoxia. At even longer timescales, transcription-mediated processes, among others, influence blood flow.

The vasodilator effect of RBCs also needs to be distinguished from that of EDRF. Indeed, inherent to this proposition of Hb-mediated vasodilation by RBCs is the idea that endothelial cells and RBCs play complementary roles in the regulation of blood flow. It had been recently argued that EDRF would overwhelm any vasodilation mediated by RBCs, thus eliminating a role for RBCs in vasodilation (34, 52). This contention, which has engendered much controversy, was based on a fundamental misconception of the relevant physiology. EDRF mediates shear- and hormonally induced vasodilation but has no significant role in hypoxic exercise-induced vasodilation (24). Conversely, RBCs dilate in response to low pO_2 (22, 24, 50) but have no direct role in shear-mediated vasodilation.

The proponents of this latter hypothesis have recently reversed their position. They now concur both that Hb/RBCs, through an Hb-allostery regulated, NO-dependent process mediate hypoxic vasodilation (53), and that SNO-Hb can mediate RBC vasodilation (54). Still in dispute are details of the molecular mechanism by which Hb carrries out this function. Accordingly, with this nascent consensus on the basics of RBC regulation of blood flow, it is an especially attractive time to review the progress in this field, to consolidate core ideas, to identify areas that have been and remain in dispute, and to examine critically the experimental results that underlie these disputes, in order to set the stage for a new understanding of the role and function of RBC-derived relaxing factor activity, and of diseases of RBC vasodilator-dysfunction.

BASICS OF SNO-HEMOGLOBIN-MEDIATED RBC-INDUCED VASODILATION

The core elements of Hb's mediation of RBC-induced vasodilation are (*a*) the sensing of oxygen levels by Hb (influenced by allosteric effectors and iron oxidation/spin state), (*b*) the intermediacy of SNO-Hb in RBC vasodilator activity, and (*c*) the release of NO bioactivity in response to reduced oxygen tension (and/or to changes in allosteric effectors, iron oxidation, and spin state). In this paradigm, SNO-Hb is identified as the active species through which oxygen (and oxidation/spin state)-responsive NO-group transfer occurs (7, 8, 41, 42). Thus, this model requires that SNO-Hb can be formed and turned over in amounts sufficient for regulated dilation of constricted vessels. Given the great potency of SNO-Hb (vasodilatory response detected in vitro at <10 nM) (see below) and the typical values reported for its concentration in blood (generally >10 nM, and typically >0.3 μM), Hb clearly dispenses NO in limited quantities—in contrast to its high-throughput delivery of oxygen. It is, moreover, reasonable that Hb interacts with NO in a fashion that tends to avoid the dead ends of nitrate and the putatively un-recoverable heme-Fe(II)NO. The only real requirement in this context, however, is that the NO budget is balanced: the rate of NO loss cannot exceed the daily NO production, which, from NOS, amounts to ~1.0 mmol/day in a human adult (55). The second requirement is the transduction of the ambient oxygen signal to release NO-bioactivity through reactions of SNO-Hb. We have proposed that this process is connected to Hb allostery—the changes in quaternary structure of Hb associated with changes in oxygen saturation, oxidation, spin state, etc. (56).

These requirements, and their correlates, suggest an intriguing principle of this biochemistry. Hb serves as a sensor and reactor that adaptively modifies the chemistry of its interaction with NO to regulate NO bioactivity, blood flow, and ultimately oxygen delivery. This adaptive chemical response presumably includes dispensing of NO-bioactivity in hypoxic vasodilation, capture of NO in hyperoxic vasoconstriction, and, potentially, trapping and/or elimination of NO under conditions of NO overproduction that characterize, for example, septic shock (57, 58). Moreover, at the very high NO levels used in the early in vitro studies, Hb chemistry must faithfully reflect the predominant production of nitrate and heme-Fe(II)-NO. Hb's NO chemistry is complex.

TABLE 2   Salient features of NO binding to Hb

1. Bi-tropic effector (homotropic and heterotropic interactions)

2. Binding of oxidized and reduced hemes

3. Reactions dependent on both NO concentration and NO-to-Hb ratio (physiological: $\ll 1$ micromolar, NO/Hb $<1:250$)

4. Promotion of either T or R structure

5. Pronounced subunit selectivity in reactions (high NO, α; low NO, β)

6. Reactions coupled to heme/thiol redox

COMPLEXITY

In the familiar example of Hb's oxygen-binding function, the presence of interacting subunits gives rise to a distinctive, sigmoidal binding isotherm that is readily distinguished from the simple behavior shown by monomeric myoglobins (and by dimeric Hbs that form at low concentrations). Representative oxygen-binding isotherms are illustrated in Figure 5. This behavior reflects a suppression of the affinity of the tetramer for the first oxygen molecule bound, as compared with the relatively high affinity for the fourth oxygen bound, and it likewise implies a strong tendency toward all-or-nothing binding (zero or four oxygen ligands), and thus a substantial, but not complete, suppression of species with intermediate numbers of ligands. The oxygen-binding function is, moreover, modulated by so-called allosteric effectors, other ligands including protons and certain anions whose binding is thermodynamically "linked" (59) to and affects oxygen binding.

These characteristics provide some lessons on the interactions of NO with Hb. Molecular properties are adaptive: They are coordinated functions of the concentrations and/or saturation of oxygen and the various allosteric effectors, with implicit coupling of the adaptive responses. This latter characteristic, well-evidenced in Hb's oxygen binding (56), suggests how it conducts its adaptive NO chemistries.

NO introduces additional complexity that requires elaboration of the paradigms used in describing the oxygen-binding function. NO attaches both at the heme site and the β-93cys. It is thus both a homotropic and a heterotropic allosteric effector; it is a bi-tropic effector. In addition to binding at the heme iron, in place of oxygen, and coupling to thiol, it reacts to form higher oxides. In its chemistry with Hb, oxidation-reduction plays a central role. It coordinates to both oxidized and reduced heme irons. In further contrast to most heme ligands, NO expresses substantial subunit non-equivalence in its reactions, which are themselves NO-concentration dependent. These distinctive characteristics of NO important to its interactions with Hb are summarized in Table 2.

The products of interactions of Hb with NO, and with related NO-reagents, is dependent not only on the saturation of NO and of O_2 (and concentrations of allosteric effectors) but also on their subunit disposition and the oxidation state

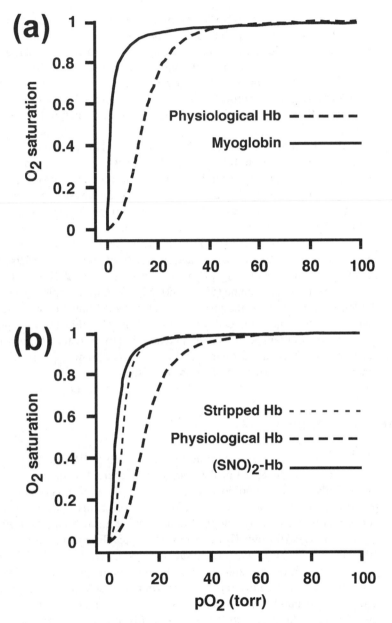

Figure 5 Representative O_2-saturation curves. (*a*) Comparison of monomeric myoglobin and human Hb [pH 7.4 in 100 mM NaCl with 2 mM diphosphoglycerate (DPG)] O_2-saturation curves (56). (*b*) Comparison of different human Hb preparations, with physiological Hb as in Figure 3*a*, stripped Hb without DPG in 7 mM NaCl pH 7.4 (56), and SNO-Hb pH 7.4 phosphate buffered saline (41).

of the accompanying hemes. The landscape of these micropopulations remains to be fully understood, but their differential reactivities are clearly evident. From this perspective, Hb senses ambient levels of oxygen tension, pH, anion levels, etc. It processes this information through structural alterations of the protein that modulate reactive behavior and thus, overall, adaptively modulates NO chemistry (input/output) to yield products that provide the optimal response to the ambient conditions. Salient aspects of these responses are indicated in Figure 6 (see color insert).

In the physiological situation, NO-containing Hb molecules are themselves a micropopulation (1 in 1000–10,000 Hbs possesses an NO), that is, the NO saturations do not vary up to nearly complete saturation, as with oxygen, but stay at levels typically less than 1%. In in vitro studies on the physiological interactions of NO and Hb, the importance of adhering to physiological amounts and proportions of the reagents has been underestimated; changes in these proportions can give rise to stark, nonlinear changes in the distribution of NO micropopulations and reaction products. The disposition and reactivity of NO bound to Hb is a function of many variables, including pH, pCO_2, pO_2, amount of NO, and the ratio of their concentrations to heme (55, 57, 60–65). Overall, complexity emerges from the tetrameric nature of the protein, which provides intersubunit couplings; the heterogeneity of the chemical formulation of the Hb adducts (micropopulations) and their concomitant heterogeneity in reactivity; the allosteric behavior of Hb and concomitant effects on reactivity; and the intricate branched network of coupled kinetic equations underlying this rich chemistry.

CONTROVERSIES

Much of the controversy regarding the role of Hb in RBC-induced vasodilation is traceable to a disregard of the complexity inherent to this system and of the biologically relevant conditions. Conditions used and results obtained in various pertinent studies are summarized in Table 3. Experiments with Hb are typically performed under nonphysiological conditions, e.g., at high pO_2 (typical bioassay is 95% O_2; 700 mm Hg) and low Hb (typical concentration in bioassay is 100 nM–1 μM Hb), whereas tissue pO_2 is much lower (~0.5–3% O_2; 4–20 mm Hg) (25, 66–68) and Hb concentration is much higher (millimolar). NO concentrations in recently reported experiments reach unphysiological levels of many tens to hundreds of micromolar (69–73) (whereas NO is nanomolar in vivo; Table 2). Studies with RBCs have involved lengthy exposures to hundreds of micromolar to high millimolar NO or S-nitrosocysteine (54, 70, 74–77)—conditions that not only obfuscate the relevant chemistry by raising the intracellular iron nitrosyl Hb and SNO concentrations to hundreds of micromolar, but also result in indiscriminate oxidation and nitrosylation of cellular constituents. Such experiments have no relevance to and do not illuminate the physiological situation.

TABLE 3

Author (Ref.)	Amount NO/Hb	NO:Hb ratio	Method	Mode of mixing	Yield SNO-Hb	Absolute amount SNO-Hb
			Reaction of NO with oxy Hb			
Gow[a] (55)	0.2 μM (× 6)/48 μM	1:240	Photolysis	Slow addition	~33%	~0.4 μM
	0.1–0.2 μM (× 8)/25 μM (+ SOD)	1:125	Photolysis/UV	Slow addition	~50%	0.75 μM[d]
Joshi (94)	5 μM/125 μM	1:25	Chem/chemilum[c]	Rapid bolus	~1%	~0.05 μM
	5 μM MAHNO*/125 μM	1:25	Chem/chemilum	Continuous slow release	~1%	~0.05 μM
Herold[a,b] (65)	5 μM/12.5 μM	1:2.5	Saville	Slow addition	~20%	1 μM
	25 μM/12.5 μM	2:1	Saville	Slow addition	~6%	1.5 μM
	50 μM/12.5 μM	4:1	Saville	Slow addition	~3.6%	1.8 μM
	5 μM/25 μM (oxy Hb/met Hb)	1:5	Saville	Slow addition	19%	~1 μM
	50 μM/25 μM (oxy Hb/met Hb)	2:1	Saville	Slow addition	8.6%	4.3 μM
Palmerini (1, 140)	1.5 μM/150 μM	1:100	Electrode	Rapid bolus	High	?
Han (70)	100 μM NO/1.5–5 mM	1:15–1:50	Chem/chemilum	Rapid bolus	1%	0.5–1 μM
	65 μM DEANO (~100 μM)/5 mM	~1:15	Chem/chemilum	Continuous slow release		0[e]
Han (72)	50 μM / 0.1–1 mM	1:2–1:20	Chem/chemilum	Rapid bolus	~1%	~0.5 μM

(Continued)

TABLE 3 (*Continued*)

Author (Ref.)	Amount NO/Hb	NO:Hb ratio	Method	Mode of mixing	Yield SNO-Hb	Absolute amount SNO-Hb
Reaction of NO with deoxy Hb followed by oxygenation						
Gow (64)	2 μM–200 μM/ 200 μM	1:100–1:1	Photolysis Saville	Slow addition Slow addition	~75 → 0%	~1–3 μM (fixed)
Fago (71)	44 μM/175 μM	1:4	?	?	~0–20%	up to 8.8 μM
Herold[a,b] (65)	5 μM/12.5 μM 50 μM/12.5 μM	1:2.5 4:1	Saville Saville	Slow addition Slow addition	16% 1.6%	0.8 μM 1.45 μM
Chen (142)	16 μM/100 μM	1:6	EPR	?	No measurement	No measurement
Reaction of NO with met Hb						
Herold (65)	5 μM/12.5 μM 50 μM/12.5 μM	1:2.5 4:1	Saville Saville	Slow addition Slow addition	42% 13%	2 μM 6.5 μM
Luchsinger (63)	90–660 μM/ 250–450 μM	5:12.5	Saville	—	12–60%	16–210 μM
Palmerini (1, 140)	1.5 μM/150 μM	1:10	Electrode	Continuous	"High"	?

[a]Yields were higher in low phosphate than in high phosphate.
[b]Yields were lower with rapid NO additions than when NO was slowly added.
[c]chem/chemilum = chemical/chemiluminescence.
[d]Numbers were derived from direct nitrosyl Hb measurement and yield of iron nitrosyl Hb by UV-Vis.
[e]Authors subsequently reported low sensitivity of assay. SOD, superoxide dismutase.

NO BIOACTIVE COMPOUNDS IN VIVO:
LEVELS AND ANALYSIS

The distribution of various products obtained from the reaction of NO and related compounds with Hb under physiological conditions, and the amounts of these products (including NO, nitrite, iron nitrosyl, and S-nitrosothiol) in the circulation remain subject to debate. These issues are addressed through both in vitro biochemical studies and analyses of reaction products formed in vivo.

Basal Levels in Vivo

On the central point of nitrosothiol and nitrosyl iron formation and levels, convergence to accurate, reproducible values has been impeded by a fundamental problem. While chemical, electrochemical, and spectroscopic methods have proven adequate for analysis of in vitro chemistry conducted at comfortable analyte concentrations, the low NO:Hb ratios in the physiological situation and the complex chemistry of Hb have made quantitative analysis of in vivo samples very challenging. Among the different assay methods, there are no common sets of practices or rigorous standards that ensure accurate, reproducible measurements of the various protein nitrosyl species and avoid artifacts induced by harsh chemical processing. Recent detailed discussion (35) of these difficulties has led to specific recommendations for improved standards (78) and requirements (78, 143) to help close the gap in quantitative results obtained with existing analytical methods.

The current picture, however, is not altogether bleak. If we think in orders of magnitude, there is a broad consensus for finding protein-nitrosyl levels, in the blood of various mammals examined, in the 10^{-5} to 10^{-7} M range. Kirima et al. (79) reported 1–10 μM Hb-NO derived from L-arginine in the blood of rats at basal conditions. Similar results were reported in one mass spectroscopic study on human blood (80). Other recent electron paramagnetic resonance (EPR) measurements in sheep (81), pigs (82), and humans (83, 84) show similar levels (0.3–3 μM). These EPR measurements are also consistent with our own measurements in rats and humans (0.3–3 μM) in which two different techniques—photolysis-chemiluminesence and a modified Saville assay (8, 22, 37)—have been employed. They are also in keeping with the measurements of James using an electrochemical approach ($\sim$5 μM) (32); of Funai et al. using a modified Saville assay ($\sim$3 μM) (36); of A. Doctor & B. Gaston (personal communication) using a novel copper/cysteine based methodology ($\sim$5 μM); and of Nagababu et al. (85), who detected a pool of reactive species assigned as HbFe(III)NO ($\sim$0.5 μM) using a modified chemiluminescence assay and significantly higher amounts (approximately many micromolar) using an EPR-based technique. Collectively, these results establish the existence of nitrosylated Hb in vivo at levels over 10^2 greater than is required for efficacy in vessel relaxation.

Outside of this range are results of Feelisch and coworkers who measured 70 nM nitrosyl Hb in rat blood (10–100-fold lower than measured by EPR) and, remarkably, zero in human blood (86). These measurements, by the method of Gladwin et al. (74, 87, 88), are thus incompatible with the direct EPR and mass spectroscopic measurements and the photolysis-chemiluminescence and modified Saville and copper/cysteine and electrochemical and chemiluminescence determinations, which agree with the EPR measurements. According to Rifkind (85), Gladwin and colleagues method cannot detect the majority of Hb-NO in human blood, as Hb-NO is readily lost during processing. In our hands, the Gladwin/ Feelisch method is highly sensitive to sample aging, processing, and redox status and thus predisposed to imprecision in quantifying and properly discriminating heme-NO and SNO-Hb. Feelisch and coworkers (89) have reported the lack of recovery of certain test samples in their assay that would lead to severe underestimation of levels in actual samples; they ascribe unique chemistry to heme-NO versus nonheme iron-nitrosyl standards that has no precedent. The method of Gladwin et al. (89a) has also markedly underestimated the levels of plasma SNO-albumin compared with that of mass spectrometry measurements (89b,c), as well as other methods (78). The discrepancy between multiple methods, in particular the direct EPR and mass spectrometry assays, and the method of Gladwin et al. calls for great caution in its application.

EPR measurement of paramagnetic heme-NO species is superior to chemical assays, which typically involve various preparative steps, but the EPR method also has its limitations. First, it is limited to the paramagnetic species, and is thus blind to diamagnetic species including SNO-Hb, nitrite, or low-spin heme-Fe(III)NO species (85, 90). Interconversion of paramagnetic and diamagnetic species during sample preparation can give a misleading picture of in vivo levels. Although the detection limit of EPR may be, with typical instrumentation, as low as $\sim$0.5 μM for the α-subunit 5-coordinate heme Fe(II) NO species that predominates with supraphysiological amounts of NO, the sensitivity is worse for the other species that should predominate under more physiological conditions. The 5-coordinate α-subunit heme-Fe(II)NO Hb species, with its sharp hyperfine structure, is most readily distinguished, whereas it would be comparatively difficult to quantify the spectrum of 6-coordinate β-subunit heme-Fe(II)NO Hb species, which has essentially no resolved hyperfine structure, and no field-domain where it alone would contribute to a composite spectrum in a mixture of species. In addition, EPR (and chemiluminescence), because of multiple correction factors introduced into the measurement, has an absolute accuracy probably no better than $\pm$0.5–1 μM (bridging the range of reported physiological levels, 0.3–3 μM in vivo). The claims of sensitivity of assays to 1 nM and reliance on EPR to establish basal NO-Hb levels are thus open to question (for reviews, see 74, 87, 88).

Altered Levels in Disease States

A final, important line of evidence in this context emanates from studies of the correlation of nitrosylated Hb levels in human subjects in health and in diseased states. Systematic alterations in SNO-Hb and Hb(Fe-NO) levels are reported upon exposure to varied atmospheric oxygen levels (22, 36, 37), and in association with diabetes (84), heart failure (32), pulmonary hypertension (18), sickle cell disease (J.R. Pawloski, D.T. Hess & J.S. Stamler, unpublished results), and septic shock (21). Examination of these relative behaviors provides a means, albeit crude, to mitigate the limitations of assays in determining absolute levels. These studies also raise intriguing ideas concerning the significance of protein-NO function in health and disease.

INTERACTIONS OF NITRIC OXIDE WITH HEMOGLOBIN IN VITRO

SNO-Hb Formation

We have reported the formation of SNO-Hb upon exposure of Hb to thionitrites (8, 37), NO (55, 64), and nitrite (22, 63) in reactions that often involve the intermediacy of iron-nitrosyl Hb (55, 63, 64). SNO-Hb formation through use of both eNOS and iNOS has also been reported (8, 91). As discussed below, these reactions have been confirmed in other laboratories. Details of mechanisms and yields and their dependence on ambient conditions continue to be debated. To date, only James and colleagues have attempted to reproduce our physiological conditions by reaching physiological (submicromolar) levels of NO compounds, and by studying intact RBCs (32, 33).

The first of these reactions presumably involves a simple transnitrosylation process:

$$RSNO + Hb[\beta 93\text{-cys}] \rightarrow RSH + Hb[\beta 93\text{-cys-NO}] \qquad 3.$$

This reaction, where RSNO is S-nitrosocysteine (8), serves as a standard method for preparing SNO-Hb in vitro [although the $(SNO)_2\text{-Hb}[Fe(II)O_2]_4$ molecule produced in this manner is less reactive than is the predominant form of SNO-Hb found in RBCs (22)]. More complicated mechanisms in which RSNO first generates either NO or a heme iron nitrosyl species are not rigorously excluded in this chemistry. Indeed, formation of SNO-Hb is typically accompanied by production of small amounts of met-Hb, and an NO-based mechanism of SNO-Hb formation has been described (92) for the bulky thionitrite GSNO through the intermediate release of NO:

$$SOD[Cu(I)] + H^+ + GSNO \rightarrow GSH + SOD[Cu(II)] + NO^\bullet \qquad 4a.$$

$$Hb[\beta 93\text{-cys}] + NO^\bullet \rightarrow Hb[\beta 93\text{-cysNO}^{\bullet -}] + H^+ \qquad 4b.$$

$$Hb[\beta 93\text{-cysNO}^\bullet]^- + SOD[Cu(II)] \rightarrow Hb[\beta 93\text{-cysNO}] + SOD[Cu(I)]. \qquad 4c.$$

The cys-NO radical suggested in Equation 4b may involve protonated tautomeric forms (K. Houk, personal communication) and may be related to the free radical observed by McMahon et al. (22). In our studies, superoxide dimutase (SOD) (55) increased amounts of nitrosylated Hb. An alternative reaction scheme may be possible. Thus

$$Hb[Fe(II)O_2] \leftrightarrow Hb[Fe(III)O_2^-] \leftrightarrow Hb[Fe(III)] + O_2^- \qquad 4d.$$

$$SOD(Cu(II)) + O_2^- \rightarrow SOD[Cu(I)] + O_2 \qquad 4e.$$

$$SOD[Cu(I)] + H^+ + GSNO \rightarrow GSH + SOD[Cu(II)] + NO^\bullet \qquad 4f.$$

$$Hb[Fe(III)] + NO^\bullet \rightarrow Hb[Fe(II)\beta 93\text{-cysNO}] + H^+ \qquad 4g.$$

$$Hb[Fe(II)]\beta + NO^\bullet \rightarrow Hb[Fe(II)]NO. \qquad 4h.$$

Equation 4g involves heme-iron/NO redox coupling and is discussed further below. These reactions lead to SNO-Hb for R-state Hbs such as oxy-Hb, carbonmonoxy-Hb, nitrosyl-Hb, or met-Hb, but not for T-state Hbs such as deoxy-Hb (8, 30, 31; B.P. Luchsinger & D.J. Singel, unpublished results).

The formation of SNO proteins upon exposure to NO has been recognized since 1992 (6, 7). Electron loss by NO to support the overall chemistry is evidently facile, with numerous electron acceptors (A in Equation 5) (18). In the case of Hb, this chemistry has been reported by numerous laboratories (33, 55, 63, 65, 70–72, 94) with various organic electron acceptors among other electron sinks (95).

$$Hb[\beta 93\text{-cys}] + NO^\bullet + A \rightarrow Hb\,[\beta 93\text{-cysNO}] + A^- + H^+. \qquad 5a.$$

The possible role of O_2 (Equation 5b) or ferriheme (Figure 6) as an acceptor (64, 65) is particularly noteworthy.

$$Hb[Fe(II)NO\beta 93\text{-cys}] + O_2 \rightarrow Hb\,[\beta 93\text{-cysNO}] + O_2^- + H^+ \qquad 5b.$$

S-nitrosylation has also been carried out in Hb crystals by exposure to NO; the electron acceptor was not identified (30).

We recently detailed the competence of the heme-iron of Hb itself as a redox partner in several different reaction scenarios that couple Fe(III)/Fe(II) reduction to formal NO oxidation (63).

$$Hb[Fe(III)\beta 93\text{-cys}] + 2NO \rightarrow Hb[Fe(II)NO\beta 93\text{-cysNO}] + H^+ \qquad 6.$$

$$Hb[Fe(II)NO\beta 93\text{-cys}] + A \rightarrow Hb[Fe(III)NO\beta 93\text{-cys}] + A^- \qquad 7a.$$

$$Hb[Fe(III)NO\beta 93\text{-cys}] \rightarrow Hb[Fe(II)\beta 93\text{-cysNO}] + H^+ \qquad 7b.$$

$$Hb[Fe(II)\beta 93\text{-cys}] + NO_2^- + O_2 \rightarrow Hb[Fe(II)O_2\beta 93\text{-cysNO}] + OH^-. \qquad 8.$$

As with the transnitrosylation reaction (Equation 1), redox-coupled S-nitrosylation of Hb is favored in the R quaternary state. The detailed sequence of bond-breaking, bond-making, and electron transfer in this overall S-nitrosylation chemistry remans to be elucidated. Ford and co-workers, for example, have highlighted the possible intermediacy of N_2O_3, formed alternatively from the reaction of nitrite with Fe(III)NO or the reaction of NO with NO_2, the latter formed through a Fe(III)/nitrite redox couple (95a,b).

The great surprise in this chemistry is that it occurs at all. It continues to be argued that the rapid reactions to form nitrate and heme-Fe(II)NO species (Equations 1, 2) observed in studies at high NO:heme ratios should also predominate in vivo, even though the relevant reactant concentrations are vastly different. If such reactions were indeed to predominate, more NO would be consumed by Hb than is produced by NOS (55). This continuing controversy is reinforced by the ease in obtaining results similar to those obtained in the early studies, if reaction conditions are maladjusted, mass balance is neglected (i.e., not all products are accounted for), behavioral trends are not tested against simple models to illuminate complex behavior, and relevant in vivo studies and conditions are neglected.

The Oxy-Hemoglobin Reaction

In our studies of reactions of NO with Hb (55), we developed a model that delineated the trend, with oxygen saturation, of the product distribution expected under the assumption of simplicity: that the hemes react independently and exclusively to form met-Hb and nitrate from oxy-Hb, and heme Fe(II)-NO from deoxy-Hb (Reactions 1, 2). Under some conditions, this simple model was adhered to precisely, but in others, a marked deviation was observed. This deviation involved discernible, albeit modest, changes (factors of 3–7) in the relative yields of Fe(II)NO and met-Hb. These changes require an oxygen-dependent shift in the relative rates of Reactions 1 and 2. More significantly, the amount of Fe(II)NO and met-Hb produced was considerably less than the amount of NO added: Additional reaction pathways including SNO-Hb formation, presumably via the reactions summarized in Equations 2–8, clearly were occurring at higher O_2 concentrations. We also showed that these SNO-Hb-forming reaction pathways occur in oxygenated RBCs exposed to submicromolar NO. This observation was recently confirmed by James and coworkers (32, 33).

These results were scrutinized in several other studies. Kim-Shapiro and coworkers (73, 76) examined the oxygen-dependence of the relative reaction rates under experimental conditions very different from ours. They checked for excess protein nitrosylation, above the predictions of our simple model, only in the heme-Fe(II)NO product. They did not verify mass balance, did not quantify SNO-formation, and did not perform any studies with physiological amounts of NO with Hb or RBCs. [As noted above, such data do not provide a sound basis for evaluating physiological chemistry, and thus cannot provide any challenge to the role of NO as the "third gas" in the respiratory cycle (22, 76)]. Working under conditions in which precise quantification was hampered by poor signal-to-noise ratios, Kim-Shapiro and coworkers were unable to discern excess heme-Fe(II)NO product, although the systematic deviations between their experimental values and those computed on the basis of our simple model are suggestive of some excess heme Fe(II)NO. Further, they show unanticipated β-heme NO predominance, which is at odds with their perspective of simple kinetics. This group also reprised prior work of Moore & Gibson (96), indicating that the on rate for NO recombination after photolysis in a fully heme-ligated Hb is not distinguishable from the initial on rate of NO addition to deoxy-Hb. The relevance to our studies is unclear. A rate acceleration need not imply an increase in rate constant, but could also derive from an increase in reactant concentration. In any case, their result provides no insight into the production of excess Fe(II)NO, or far more importantly, the SNO-Hb now observed in several studies (see Table 3).

Finally, recent work suggests that oxidized heme can be competitive with oxy-hemes for NO, or more accurately, that chemistry putatively involving Fe(III)NO intermediates is competitive with the reaction of NO with oxygenated hemes (9, 65, 70). Such competitiveness has evolutionary antecedents in the oxygen-detoxification Hb chemistry of *Ascaris suum* (97), but is surprising in human Hb.

Effects of Mixing Methods

Lancaster and coworkers (94) hypothesized that SNO-Hb was formed as a result of bolus addition of NO, which reacts rapidly in locales of high concentration before mixing to homogeneity can occur. In probing this idea, they restricted their studies to fully oxygenated Hb (room air; O_2 saturations of $\sim$99.9%) and to supraphysiological total amounts of NO (5 μM) (94), and compared all-at-once addition of saturated NO solution to slow-release of an equivalent amount in a solution of a NONOate (Table 3). Addition of small aliquots of subsaturated NO solutions might represent an intermediate case between their extremes. Additionally, some evidence supports association of NONOates with proteins, including Hb (G. Ahearn, J.R. Pawloski, T.J. McMahon & J.S. Stamler, unpublished results; B.P. Luchsinger & D.J. Singel, unpublished results; 98); such interactions could interfere with mixing homogeneity and present a high effective molarity of NO for reactions with Hb.

Lancaster and coworkers (94) found no bolus effect on SNO-production, and within experimental error they found the same level of SNO-produced regardless of the method of mixing (see Table 3). Their results are not relevant to the SNO-Hb paradigm. They did appear to obtain considerably less SNO than Gow et al. (55) (and other groups, see Table 3) in both their bolus and slow-release additions. This disparity could easily be explained by a lack of recovery since their analyses rely on the problematic Gladwin assay (74).

In these studies, interactions of NO with deoxy-Hb were ignored despite the abundance of deoxy-Hb in vivo; oxygen-dependent trends in behavior were thus not examined. Similarly, the effect of heme to total-NO ratios on the product distribution was not investigated. Overall, the findings of these investigators, albeit not surprising in view of the selected reactions conditions, are not probative of any key tenet of the model of SNO-Hb function. Their work sheds no light on the complexity—or simplicity—of NO-Hb interactions. Moreover, it has no relevance to physiological situations. In systemic peripheral blood NO is, in fact, undetectable and the pO_2 is much lower than under Lancaster's conditions [tissue pO_2 ~4–20 mm Hg (25, 66–68)]. In the lung capillary, the O_2 saturation is ~99%, but there the NO concentration is four orders of magnitude lower than that employed in Lancaster's experiments (~10 ppb or ~400 picomolar) (99).

More recently, Liao and coworkers (72) again compared bolus additions of saturated NO solutions versus slow release of NO by NONOates, with NO levels corresponding to 50 μM and oxy-Hb concentrations in the range 0.1–1 mM. Predictably, they obtained small SNO-Hb yields, but the reported levels were as much as fivefold greater than those obtained by Lancaster (94). Their most recent study examined, within a limited mechanistic perspective, the possible importance of mixed valence species in SNO-Hb chemistry. Further implications of these species, in particular for intraprotein redox reactions, are discussed below. As with the work of Lancaster and coworkers (94), the connection to physiological conditions is obscure, at best. In neither study is any framework presented through which an extrapolation could be made from the in vitro observations to the physiological situation.

Effects of NO and Heme Concentrations

Such an extrapolation is nontrivial. We have documented that reactions of NO with Hb are critically dependent on both NO concentration and NO:Hb ratio; moreover, the percent yields of the nitrosylated protein are inversely related to NO concentration, with SNO-formation most efficient with nanomolar NO levels (64) (see Table 3). Each aspect of this complex chemical behavior evidenced in our work (55, 64) was recently reproduced in experiments of Herold & Röck (65). Specifically, these authors observed (*a*) production SNO-Hb upon oxygenation of FeNO Hb, (*b*) surprising yields of SNO-Hb relative to met-Hb upon treatment of oxy-Hb with NO, (*c*) potentiation of SNO-Hb forming pathways by met-Hb, (*d*) increased SNO-Hb yields with mixing methods conducive to solution homogeneity, (*e*) increased SNO-Hb production with decreasing NO:Hb ratios, and (*f*) increased SNO-production with decreased phosphate ion concentrations. Indeed, when SNO-Hb yields are viewed as a function of reagent ratio, as illustrated in Figure 7, the quantitative agreement between work from our laboratories and that of Herold & Röck is striking.

These results are difficult to reconcile with those of Lancaster and coworkers (94) and Liao and coworkers (72). They suffice, however, to demonstrate the lack of generality of their results. The trend of increasing SNO-Hb yields with decreasing NO:heme ratios is difficult to reconcile with their work, or with the rationalization of bolus effects by these groups or by Spencer et al. (75).

James et al. (33) recapitulate the trends in behavior observed in vitro by us and by Herold & Röck (65) in experiments on RBCs exposed to physiological NO (240 nM DEA-NO), approximating our results with 200 nM NO (64). The

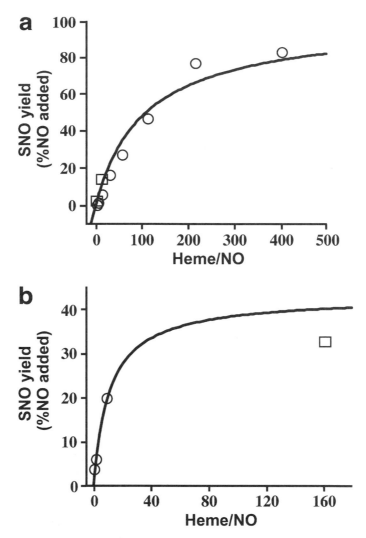

Figure 7 Dependence of SNO-Hb formation on NO:heme ratios in experiments. Comparison of results of Gow et al. (55, 64) with those of Herold & Röck (65). (*a*) SNO-Hb yield upon exposure of deoxy-Hb to NO, followed by oxygenation. Yield is expressed as fraction of added NO. Circles from Gow et al. (64), squares are data from Herold & Röck (65) with best yields (slow mixing, high phosphate). The smooth curve is a best-fit square hyperbola (saturation curve). (*b*) SNO-Hb yield upon exposure of oxygenated-Hb to NO. Circles are data from Herold & Röck (65) with best yields (slow mixing, high phosphate). The square is from Gow et al. (55). The smooth curve is a best-fit square hyperbola (saturation curve). Whereas Herold & Röck (65) called attention to their lower yields compared with those of Gow et al. (55, 64), the data compare agreeably when NO:heme ratios are taken into account.

recent in vivo studies of Mason and coworkers (62) are also noteworthy. Although they introduced sufficient DEA-NO to generate concentrations of NO compounds more characteristic of pathophysiology (e.g., sepsis), they nonetheless observed a striking distribution of products, with heme iron(II) nitrosyl accounting for ~two thirds of the NO released by DEA-NO (or more if less than two NOs are released from each DEA-NO). The fraction of released NO appearing as the heme-nitrosyl product, as predicted by simple competition between the scavenging reactions in Equations 1 and 2, would be a full order of magnitude less. This in vivo result of Mason and coworkers (62) unambiguously underscores the lack of generality of the result of Gladwin et al. in which supraphysiological exposure to NO through inhalation reportedly produced met-Hb as predominant reaction product (87).

Still needed is a complete kinetic/thermodynamic model that would enable prediction of the distribution of products in encounters of Hb with NO under arbitrary reaction conditions. Nevertheless, this biochemistry clearly is complex enough to require caution in interpreting results obtained under different conditions, and the idea that NO-bioactivity survives encounters with Hb can no longer be in doubt.

Effects of Superoxide Dismustase

Recent work from the English group (92) extends the observation by Gow et al. who first reported that SOD (55) increases the yield of Hb nitrosylation (FeNO and SNO) in reactions of oxygenated Hb with NO. In particular, SOD (55, 92) increases the yield of SNO-Hb while decreasing met-Hb accumulation. Most noteworthy is the result, obtained in both laboratories, that under certain physiological conditions, SOD is sufficient to ensure that NO-bioactivity is entirely conserved and channeled to the formation of SNO-Hb, rather than quenched through NO_3^- formation. Romeo et al. (92) provide a specific mechanistic perspective (Equations 4a–c) that encourages further study; the abundance of SOD in the RBCs excites particular interest in this effect.

Novel Intramolecular Biochemistry

HEME-TO-THIOL TRANSFER   Biochemical studies and mutational analyses (β93 cys→ala) (64) support the interconnection between heme- and thiol-nitrosylation in Hb. SNO-Hb forms via heme-to-thiol NO transfer chemistry under conditions that feature physiological amounts (and ratios) of NO and Hb, whereas NO remains bound to the hemes of a β93-cys→ala mutant (64). The amounts that transfer from the heme depend not only on the amounts of NO (64), but also on the NO/Hb ratio, rate of oxygenation, and redox state of the system (63, 100) (see Table 3). Apart from reactions with highly oxidized Hb, SNO levels plateau at ~1 μM. A critical requirement of these reactions is the formal redox activation of the NO group (or alternative one-electron oxidation of the system). For example, the oxidative requirements of this NO-group transfer chemistry are provided by Hb [Fe(III)NO] intermediaries, which can yield SNO-Hb (63), as depicted by Equation 6. Similarly,

Equation 9 accounts for this chemistry, with initial formation of the ferric nitrosyl species prior to SNO formation:

$$Hb[Fe(II)\beta93\text{-cys}] + NO_2^- \rightarrow Hb[Fe(III)NO\beta93\text{-cys}] + OH^-. \qquad 9.$$

More generally, O_2 and redox agents, which can both influence the equilibrium between R and T structures and serve as electron acceptors, promote this NO group transfer chemistry as indicated in Equations 5a,b.

THIOL-TO-HEME TRANSFER   Similar principles apply to the transfer of NO from SNO-Hb to the β-heme. That is, deoxygenation or oxidation of hemes in SNO-Hb decreases β-cysNO stability, thereby promoting NO group release (8, 30, 41, 90). Deoxygenation simultaneously decreases the redox potential of the β hemes, favoring their auto-oxidation (90). Kluger and coworkers (90, 101) described this process as featuring a coupling between heme deoxygenation and β heme/SNO redox that liberates $NO^\bullet$ from the SNO anion radical:

$$Hb[\beta Fe(II)O_2\beta93\text{-cysNO}] \rightarrow Hb[\beta Fe(III)\beta93\text{-cysNO}^{\bullet-}] + O_2 \qquad 10a.$$

$$Hb[\beta Fe(III)\beta93\text{-cysNO}^{\bullet-}] + H^+ \rightarrow Hb[\beta Fe(III)\beta93\text{-cys}^-] + NO^\bullet. \qquad 10b.$$

The released NO is then available for adduct formation with vacated, reduced hemes. Ferric heme accumulation is mitigated by met-Hb reductase (41, 90). This type of chemistry echoes early observations of Rifkind linking heme and exogenous copper redox couplings (102, 103), and provides a specific circuitry for the generic chemistry outlined in Equation 5. This chemistry also should serve as a reminder of the subtleties of β cys-93 thiol modification: such modifications immediately impact not only general oxygenation and heme redox properties, but also this particular internal redox circuit.

Auto-oxidation of SNO-oxy-Hb can lead to analogous chemistry that furnishes a heme iron(II) nitrosyl (63):

$$Hb\{[Fe(II)O_2]_4\beta93\text{-cysNO}\} \rightarrow Hb\{[Fe(II)O_2]_3\beta Fe(III)\beta93\text{-cysNO}\} + O_2 + e^- \qquad 11a.$$

$$Hb\{[Fe(II)O_2]_3\beta Fe(III)\beta93\text{-cysNO}]\} + H^+$$

$$\rightarrow O_2 + Hb\{[Fe(II)NO][[Fe(II)O_2] + [Fe(III)]_2\beta93\text{-cys}\}. \qquad 11b.$$

Dynamical loss of the oxy-ligand on an S-nitrosylated β-subunit, with a coordinated electron-NO transfer, can lead to formation of a β-subunit outfitted with thiolate and Fe(III)NO, as in Equation 11a. Reduction of the latter species by reduction coupled with oxidation of an acceptor, possibly a neighbor heme, furnishes the Fe(II)NO, as in Equation 11b. In this process, β-subunit selectivity emerges both from the proximity of the (β cys-93)-NO and heme on the β-subunit and the redox properties of the hemes. The relative stability of doubly oxidized met-Hb hybrids may enhance the favorability of the process (104, 105).

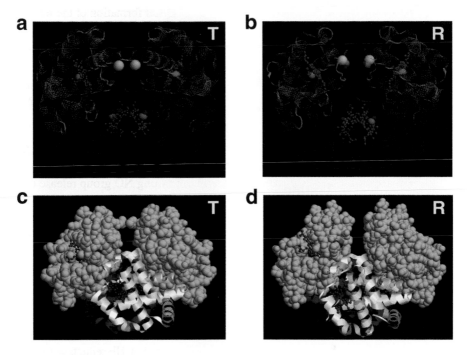

Figure 2 Structures of nitrosylated human hemoglobin in T and R quaternary forms from (30, 31). (*a*, *b*) Strand representations of the globin backbone with β-subunits shown in teal and α-subunits in blue. The hemes are displayed in magenta (ball-and-stick). NO ligands on the hemes, and the β-cys93 sulfur and attached NO group are displayed in space-filling representations with CPK coloration. The T-state nitrosylated protein (*a*) has NO only on the hemes (β-subunit 6-coordinate and α-subunit 5-coordinate), whereas the R-state protein has NO both on heme (all 6-coordinate) and β-cys93. (*c*, *d*). Space-filling representation of the protein β-subunits (*teal*) with ribbon displays of the α-subunits (*blue* and *white*). The heme, β-cys93, and NO groups are represented as in *a* and *b*. In the T structure the β-cys93 sulfur is exposed, whereas in the R structure it is tucked behind its adjoining carbon. In energy-minimized solvated structures (37), the entire SNO-moiety is folded back into the globin and is not solvent accessible.

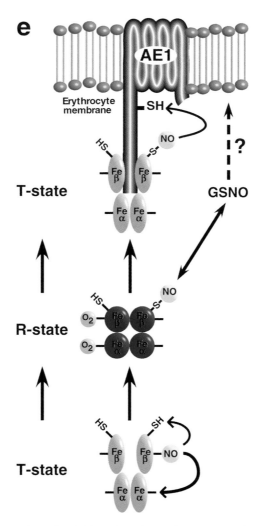

Figure 3e   Generation of NO bioactivity in RBCs occurs preferentially at the membrane. As illustrated, oxygenation (in the lungs) promotes the transition of hemoglobin from T to R state and the transfer of heme-liganded NO to β-cys93; a significant proportion of NO is retained as an iron nitrosyl species. In the vascular periphery, deoxygenation is associated with the transition from R to T state, which is facilitated in a juxta-membrane population of hemoglobin by interaction with the cytoplasmic domain of AE1 (band 3 protein). Concomitantly, NO is transferred from β-cys93 to a cys thiol within AE1 (and the R to T transition may also facilitate a shift in the intraerythrocytic equilibrium between HbSNO and GSNO); a significant proportion of NO is autocaptured to form Fe-NO. Vasodilatory NO bioactivity is conveyed from this membrane-associated compartment. Both intracellular and extracellular GSNO may provide a source of NO groups that is in equilibrium with SNO-Hb and thereby contributes to RBC vasoactivity.

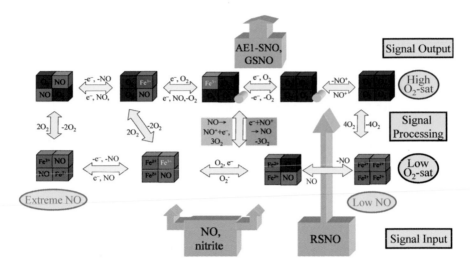

Figure 6   Chemical dynamics of NO interactions with hemoglobin. This perspective envisions Hb as a programmable chemical reactor in which NO chemistry is modulated, as illustrated, by ambient conditions of NO, oxygen, and redox potentials. Allosteric effectors would likewise modulate the chemistry through effects on oxygen saturation, oxidation, and spin states. NO signal input, entailing NO, nitrite, and thionitrites, is processed as directed by ambient conditions to provide appropriate output signals. For hypoxic vasodilation (*low NO, right*), output involves dispensing of vasodilatory activity through formation of S-nitrosylated AE-1 and perhaps GSNO (see Figure 3*e*). In the case of high levels of NO (*left*), for example in sepsis, the adaptive chemistry works to brake NO release (5-coordinate a nitrosyl Hb). The species indicated should be taken as exemplary, but underscore the role of minority species in this chemistry. In the tetrameric hemoglobins, shown as four squares, α chains are upper right and lower left and β chains are lower right and upper left.

Not all of the NO would be expected to be captured by the hemes. Transfer to the heme of Hb simply allows for NO economy in a situation where bioactivity is to be dispensed in a limited manner (41). The functionally important chemistry is transfer from SNO-Hb to other NO-accepting groups that advance the signal transduction (Equation 3, in reverse; Equation 13, below). Indeed, either deoxygenation or heme oxidation was shown to increase RBC bioactivity (8, 32, 40, 42); in addition, both regulated the disposition of NO bound to hemes and thiols in human blood (22, 63). NO transfer to band 3 protein [and/or perhaps ultimately to glutathione (48)] is central to dilating blood vessels (40).

REDOX AND HYBRIDS   The salutary coupling of NO and heme redox/spin states substantially enriches the chemistry of NO as compared with other heme ligands. This coupling is important in understanding how heme Fe(II)NO provides a tappable store of NO bioactivity, rather than a dead-end for NO. Experiments that show an effective loss of NO through the formation of tightly bound 5-coordinate complexes on the α-subunits (61, 106, 107) are often carried out with methods aimed to inhibit oxidative processing. When enabled (55, 106, 107), oxidation of heme (or equivalent redox processes) leads to encounters with NO that produce SNO on (R-state) Hb. The connection between heme redox and NO chemistry again underscores a link to allosteric effectors via their effect on heme redox properties. This coupling also rekindles interest in the intriguing difference in the microstate populations associated with ligand binding (108) versus "hole binding" (i.e., oxidation). Apart from molecules with fully occupied or fully vacant hemes, the former process more strongly suppresses doubly liganded forms, whereas the latter suppresses species with odd numbers of oxidized hemes (104, 105). Analogous chemistry with NO-met hybrids that favors formation of $Hb[Fe(III)]_2[Fe(II)(NO)]_2$ both in vitro and in vivo has also been observed (B.P. Luchsinger & D.J. Singel, unpublished results; 109). Collectively, these reactions are suggestive of facile intramolecular electron-transfer—a chemistry that can be viewed as an emergent property of multimeric Hbs, with fundamental implications for energy landscapes of the NO micropopulation and for Hb reactivity, as suggested in Figure 6.

CRITICISMS   Aspects of this chemistry have been criticized. The transfer of NO from SNO-Hb to heme (8, 22, 90, 101, 110) has recently been suggested ad hoc to be a "nitrite artifact" (75). Isotope-labeling experiments that ostensibly support this contrary view, however, miss the mark. Our results, illustrated in Figure 8, show that over the course of the slow loss of SNO, from various SNO-Hb preparations, a comparable amount of heme-Fe(II)NO is formed (22, 110). With concomitant increases in met-Hb, an overall reaction such as indicated by Equation 11 is possible. The remarkable feature of this chemistry is the preferential β-subunit reactivity. To resolve a point of confusion in the literature, we emphasize that no exogenous nitrite was used in these experiments. Although this reaction is slow in in vitro experiments on neat samples, the reaction is potentiated by physiological levels of thiols and by other reductants (8, 22, 41).

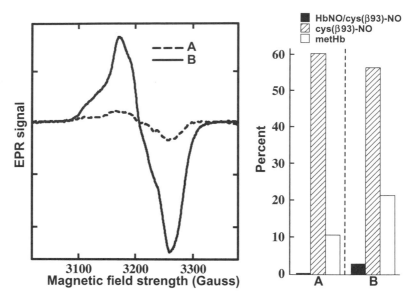

Figure 8   Thiol-to-heme NO-group transfer (22, 110). EPR spectrum of a fresh SNO-oxyHb sample prepared from S-nitrosocysteine (8) (*dashed line A spectrum, left panel*) shows a small signal from heme-Fe(II)NO. Further analysis of the sample shows also a small met-Hb component and 60% S-nitrosylation (*right panel, A*). EPR spectrum of the same sample after aging (*solid line B spectrum, left panel*) shows large heme-Fe(II)NO with predominat β-character. Further analysis shows a correlate decrease in S-nitrosylation and increase in met-Hb content (*right panel, B*).

Nitrite was used to generate heme-Fe(II)NO in other in vitro experiments (22, 63, 110). Again, this reaction can hardly be called artifactual; it represents a specific route to SNO-formation through coupling of heme and NO redox and Hb oxygenation. In these experiments, we demonstrated that the heme-Fe(II)NO species formed from nitrite reaction with deoxygenated Hb was dislodged by oxygenation. This loss is analogous to the oxygenation-induced loss that occurs in samples in which heme Fe(II)NO is formed from reaction of NO with deoxygenated Hb (64, 71). Xu et al. (75) found experimental conditions under which the heme-Fe(II)NO species could not be dislodged by oxygenation, and hastily concluded that our results entailed some "artifact." To clarify this point, Figure 9 illustrates the effect of sample aging on the oxygenation induced loss of heme Fe(II)NO. Immediately after sample preparation, this loss is essentially complete, and the radical signal that accompanies oxygenation is large; some minutes later, however, both the diminution of the iron nitrosyl signal and oxygenation-induced radical signal, as well as the accompanying production of SNO-Hb, are substantially attenuated. At longer intervals (not shown),

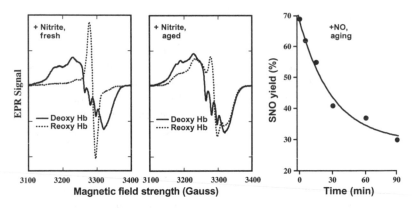

Figure 9   Heme Fe(II)NO displacement accompanying oxygenation (22, 110). Heme Fe(II)NO species revealed by EPR (*left panel, solid line*) of a sample freshly prepared by incubation of deoxygenated Hb with nitrite. Upon oxygenation the spectrum is dramatically altered with the Fe(II)NO signal diasappearing and being replaced by a free-radical spectrum (*dashed line*). After aging, repeated cycling, and/or nitrite exhaustion, the behavior is muted; progressively, oxygenation-induced changes are observed (*middle panel*). All EPR spectra shown in the figure are on the same scale. Analogous experiments have also been conducted with heme Fe(II)NO prepared by NO addition (nitrite-free) at low NO:Hb ratio and establish that the efficiency of SNO-formation on oxygenation is also impaired progressively by sample aging (*right panel*). Aging effects are more rapid in native RBCs (T.J. McMahon & J.S. Stamler, unpublished data).

we obtain the results of Xu et al. (75). Our view is that chemical processes, which occur over the longer time intervals, alter the microscopic composition of the samples and contribute to the alteration in oxygenation-linked reactivity.

   Another aspect of the work of Xu et al. (75) that demands comment is their exposure of RBCs to almost millimolar NO [Fe(II)NO and SNO] levels. Under such extreme conditions, heme-thiol transfer would be expected to be small (55) (Table 3), nowhere near the >100 micromolar values that they errantly impute to our model. Indeed, the imputed values are based on a completely unjustified linear extrapolation of fractional transfer-yields from the quasi-physiological conditions of McMahon et al. (native RBCs), to the extreme conditions of Xu et al. (N.Y. Spencer, personal communication) (RBCs treated with many millimolar NO). Nagababu et al. (85) also dislodge the iron-nitrosyl species derived from nitrite by oxygenation. The results reported by Cosby et al. (53) for nitrite-exposed RBCs also point to this chemistry. Collectively, these results again call attention to condition-dependent chemistry in this complex system.

NITRITE AND HEMOGLOBIN PHYSIOLOGY

It has long been known that nitrite can dilate blood vessels. Nitrite, which is ubiquitous, is by itself $\sim$100–1000 times less potent than NO or S-nitrosothiols and must be converted to either NO or SNO to produce vasodilation. For example, production of nitrite in mitochondria is the first step in the mechanism of nitroglycerin bioactivation (112, 113), and mitochondria support the conversion of nitrite into NO and SNO (113; Z. Chen, M. Foster & J.S. Stamler, unpublished results).

Nitrite Biochemistry

Reutov & Sorokina (114) suggested that the reaction of nitrite with deoxy-Hb, which, as detailed by Doyle and coworkers (107), generates Fe(III) and NO, may represent a source of NO bioactivity. Reutov & Sorokina argued that more NO may be generated through this pathway than though NO synthase, particularly in a low O_2 milieu. Others have recently repeated this idea (53, 85) without addressing its key limitation. Missing from this hypothesis is how the NO could escape the RBC, rather than lodging on vacant Fe(II) hemes that are highly abundant under the low O_2 conditions considered. (Further, most deoxygenated Hb is found in the veins, which is not the region in which blood flow is regulated.) The solution is the (prearteriolar) conversion of FeNO to SNOs, as originally described by McMahon (22) and Luchsinger and coworkers (63). The reaction of deoxy-Hb with limited nitrite leads to production of β FeNO (63), with SNO-Hb formed upon subsequent oxygenation (22, 63) (Equations 5a and 5b, where either O_2 or ferriheme serve as the electron acceptor). Thus Hb/nitrite catalyzes the formation of SNO-Hb through the intermediacy of iron-nitrosyl Hb (nitrite reductase activity). The amounts of SNO-Hb generated correlate directly with RBC bioactivity (G. Ahearn & J.S. Stamler, unpublished results). SNO-Hb formation through interactions of Hb with nitrite—summarized in Equations 8 and 9 and Figure 6, and now observed in several laboratories (22, 53, 63, 85, 115)—overcomes the key limitation of the nitrite reductase hypothesis of Reutov & Sorokina.

An important issue in the utilization of nitrite to generate NO is the possible competition with oxy-Hb reactions to form nitrate and met-Hb (116).

$$4\text{heme-Fe(II)}O_2 + 4NO_2^- + 4H^+ \rightarrow 4\text{heme-Fe(III)} + 4NO_3^- + O_2 + 2H_2O. \quad 12.$$

At physiological nitrite:Hb ratios, however, reaction 12 progresses very slowly compared with the NO-forming, nitrite-reductase reaction, as we have demonstrated (63), owing to lengthening of the cysβ93-dependent lag phase characteristic of the reaction with oxy-Hb (116). The special feature of the nitrite reductase reaction is not the production of an iron-nitrosyl alone but rather its coexistence with a ferric heme within an Hb tetramer, which ultimately, as Hb shifts to its R quaternary state, as discussed above, can support the redox requirements for SNO production. Hybrids of this sort are also likely formed on exposure of partially deoxygenated Hb to NO; under such conditions, reaction with NO inevitably leads to some heme oxidation. Both UV/VIS and unambiguous EPR spectroscopic measurements support this chemistry.

A recent report by Fago et al. challenges a fundamental part of this latter chemistry (71). They report mixing of NO with deoxyHb under conditions where little or no met-Hb is produced, and claim that the observation of met-Hb by Gow & Stamler (64) derives from an error in spectral deconvolution. However, the spectral analysis method discussed by Fago et al. was not used by Gow & Stamler (64), who focused on spectral changes in the 400- and 630-nm regions where confusion over met- and iron-nitrosyl Hbs is least likely. Previous reports on the production of met-Hb, upon exposure of deoxy-Hb to NO, by Hille et al. (106) and by Doyle et al. (107) indicate that yields are dependent on pH, organic phosphates, and mixing methods. Gow & Stamler (64) emphasize the importance of NO concentration and NO:heme ratios on the product distribution. Conditions used by Fago et al. depart significantly from those of Gow & Stamler (64) (see Table 3). There is no basis provided in Fago et al. for asserting that their results are general. As such, they have no obvious bearing on the results of Gow & Stamler and Doyle et al. (107).

Vascular Effects of Nitrite

Results presented in a recent publication ostensibly amplify ideas on the role of nitrite in regulation of blood flow (53). Specifically, the nitrite reductase activity of Hb alone was invoked to explain the response of hypoxic vasodilation (53). Unfortunately, the vascular physiology implicating nitrite (administered by infusion) in hypoxic vasodilation was misinterpreted in this paper. First, the nitrite-mediated increase in blood flow was associated with an increase in venous O_2 saturation, whereas hypoxic vasodilation is the increase in blood flow that accompanies a decrease in venous O_2 saturation (24–26). Second, the flow increment induced by hypoxic exercise was, in fact, unchanged by nitrite. Third, the vasodilation to hypoxic exercise was not blocked by the NO synthase inhibitor L-N^G-monomethyl-arginine (53, 66), which acutely depletes nitrite (117, 118). Thus hypoxic vasodilation is not directly mediated by nitrite. It is improbable that enough nitrite to enable vasodilation can enter the RBC and be converted to NO, or enough NO diffuse out, within the few seconds that blood transits the arteriolar microcirculation. On closer inspection of the in vitro supporting data, nitrite neither potentiated the rate nor the amount of relaxation by RBCs, which would constitute the necessary signatures of NO effects. Further, what appeared as "release" of NO from RBCs is probably the entrainment of NO in an inert gas that was used to sparge RBCs (moreover, the amounts were too low to dilate blood vessels). Hypoxic vasodilation by native RBCs does not require addition of exogenous nitrite (22, 32). Taken together, the observations of Gladwin et al. (53), who observed copious SNO-Hb formation upon infusion of nitrite, are better interpreted as pharmacologic SNO-Hb-mediated RBC vasodilation, not hypoxic vasodilation, although nitrite could also be acting independently of RBCs.

OXYGEN-DEPENDENT SNO DELIVERY

As noted above, the transduction of the ambient oxygen signal to the release of NO-bioactivity through reactions of SNO-Hb is a fundamental tenent of our model of SNO-Hb function. We propose that this process is linked to the changes in quaternary structure of Hb associated with changes in oxygen saturation, i.e., to Hb allostery.

The final steps by which the NO-signal leads to vessel dilation are not yet fully established. In the bioassay, GSH (glutathione) substantially potentiates the response to cell-free SNO-Hb (Figure 10), but is not required for RBC relaxation (Figure 3a–d). Cellular studies demonstrate that transfer of NO groups from β-cysNO to cysteine thiols at the RBC membrane (band 3) is necessary and sufficient for vasodilation by RBCs under representative physiological conditions (40). The means by which S-nitrosylation of band 3 advances vasodilation remain to be elucidated. Clearly, SNO-Hb-mediated bioactivities, analogous to EDRF (119), do not absolutely require free NO (8, 40) or cGMP (42); cGMP involvement has, however, been implicated by James and coworkers (32). There is also evidence that GSH can support RBC-mediated vasodilation. Transgenic mice lacking GSNO reductase, an enzyme that metabolizes GSNO, show increased SNO-Hb levels (21). The addition of GSH to blood under hypoxic conditions, but not normoxic conditions, generates GSNO (48). Collectively, these observations, buoyed by the mechanistic suggestions of English and coworkers (92), suggest an important interplay between GSNO and SNO-Hb. In addition, the chemistry of the oxygen dependencies on the NO reactivities of Hb, although understood in broad strokes, is not yet determined in detail.

Allostery

In a manner reminiscent of Perutz's sterochemical modeling to rationalize ligand-binding, induced allostery, crystal structures (30), and molecular models (30, 37) of SNO-Hb provided a basis for allosterically regulated release of NO bioactivity. The β-cysNO has limited access to solvent in R state (thus hampering release to dilate blood vessels), but it protrudes into solvent in the deoxygenated (or T) structure. McMahon et al. (41) have demonstrated definitively that the transfer of NO groups between SNO-Hb and glutathione is allosterically regulated (Figure 10). They showed that SNO-Hb was stable in air (R structure) in the presence of glutathione, but decomposed upon deoxygenation (Figure 10a), with production of GSNO. Both hypoxia and thiol potentiated SNO-Hb-induced relaxations (Figure 10b), presumably through their effects on the position of the O_2-dependent equilibrium that produces the vasodilator GSNO (Equation 13, where RSH is glutathione). Consistent with this interpretation, the plasma half-life of infused SNO-Hb increases more than twofold in animals breathing pure oxygen (P. Sonveaux, T.J. McMahon, J.S. Stamler & M.W.Dewhirst, unpublished observations).

$$Hb[[Fe(II)O_2]_x\beta93\text{-cysNO}] + RSH$$

$$\leftrightarrow Hb[[Fe(II)O_2]'_x\beta93\text{-cys}] + RSNO + (x\text{-}x')O_2. \qquad 13.$$

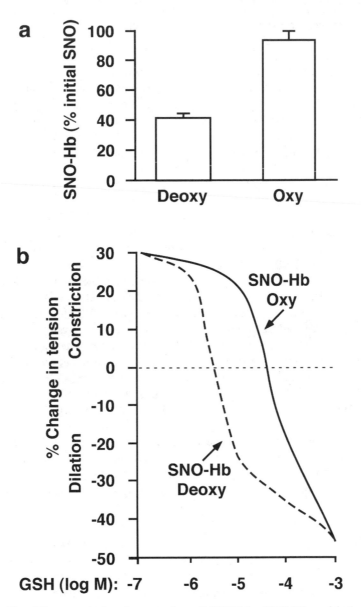

Figure 10 Allostery-regulated generation of GSNO by SNO-Hb and induced vasoactivity. (*a*) SNO-Hb (50 μM) is stable in room air in the presence of glutathione (1 mM) (Oxy) but decays immediately upon deoxygenation (Deoxy) with production of GSNO (not shown). (*b*) Vasorelaxation by SNO-Hb (1 μM) is potentiated by hypoxia (1% O_2; Deoxy versus 21% O_2; Oxy) and glutathione, which favors production of the vasodilatory mediator GSNO (representing a shift in the position of the equilibrium in Equation 13 to the right). Data are derived from McMahon et al. (41).

Under certain experimental conditions, relaxations by SNO-Hb have been observed at both high and low pO_2 (and conversely SNO-Hb vasoconstriction has been seen at both high and low pO_2) (54). Some argue that these data rule out allosteric control of SNO-Hb vasoactivity (54). However, when data are considered over an appropriately broad range of O_2 and thiol concentrations, as illustrated in Figure 10, the flaw in this argument is apparent. The vasoactivity of SNO-Hb at both high and low thiol concentrations is independent of pO_2, as would be expected by a process governed by Equation 13. At intermediate thiol concentrations, however, the O_2 dependence is unmistakenly revealed. In vivo concentrations of SNO-Hb and allosteric effectors, nature of thiol, and rates of reactions will modify the quantitative behavior illustrated in Figure 10. Nevertheless, the data underscore both the operation of allostery and the importance of experimental conditions in probing for it. Crystal structures again corroborate the importance of allostery: Exposure of R-state carbonmonoxyHb crystals to NO leads to iron- and S-nitrosylation, whereas exposure of T-state crystals to NO leads only to iron nitrosylation (30, 31) (Figure 2).

Key evidence in support of the hypoxia-dependent release of NO-bioactivity from RBCs comes from the work of McMahon et al. (22) and, more recently, of James et al. (33). RBCs were shown to actuate a unique, rapid, and graded, vasodilator and vasoconstrictor response across a physiological range of pO_2 (Figure 3). Thus, RBCs were shown not only to dilate blood vessels, but in a manner that recapitulates the autoregulation of vessel tone by the physiological O_2 gradient. This response is uniquely characteristic of SNO-Hb (Figure 10b). Thus, the inference that SNO-Hb relaxations are not allosterically regulated by O_2, drawn from the observation that relaxations by NO itself are also potentiated by hypoxia (120), is ill-conceived: The pO_2 versus activity relationship for NO and SNO-Hb shows little, if any, resemblance (22, 33)—in particular, GSNO relaxations fail to show the same graded responsiveness over the physiological oxygen gradient as does SNO-Hb. The enhanced activity of NO itself under hypoxia does not have an allosteric basis; SNO-Hb relaxations are not mediated by free NO (120, 121).

Linkage

S-nitrosylation of Hb results in enhanced oxygen affinity, as indicated in Figure 5 (41, 122). These measurements were conducted on Hb-$(SNO)_2$, whereas Hb-(SNO) is more likely to predominate in the physiological situation. Enhanced oxygen affinity was seen by some to be inconsistent with the idea that hypoxic vasodilation is regulated by the coupling of NO-bioactivity release to deoxygenation. The argument was, "NO transfer would be limited to regions of extremely low oxygen tension, if this were to occur from deoxygenated hemoglobin" (122). This perspective is fundamentally flawed. At any finite oxygen saturation there is a population of deoxy-Hb subunits. Not only does the number of these molecules increase monotonically with decreasing pO_2, but in view of the cooperativity of oxygen binding, that increase is nearly linear in oxygen saturation. Thus a lowering of oxygen tension that causes a drop in oxygen saturation of the S-nitrosylated Hbs labilizes a proportionate number of S-nitrosothiols. While extremely low oxygen tensions would be required to denitroslyate all nitrosothiols, physiological levels reached in the microcirculation are sufficient for dispensing enough activity through transnitrosylation to modulate vessel tone, particularly in view of the great vasodilatory efficacy of SNO-Hb, in the presence of thiols (Figures 3 and 10). Moreover, the fact that decreased oxygen tension results in destabilization of the SNO-moiety actually implies that S-nitrosylation stabilizes oxygen ligation: The observed behavior, rather than representing a paradox, should be expected. In principle, the thermodynamics could be trumped by kinetic bottlenecks, but such bottlenecks seem not to be significant, provided species such as thiols are present to accept the released NO group (41).

SNO-Hb Stability in RBCs

Gladwin and coworkers (74) published the results of experiments ostensibly demonstrating that SNO-Hb was too unstable in the RBC to exert biological activity, and that the decay of SNO-Hb in the cells was not accelerated by deoxygenation, ruling out allosterically controlled release. However, there is no foundation for asserting that their measured half-life for SNO-Hb ($\sim$10 min) is too short for physiological efficacy inasmuch as A-V transit times are measured in $\sim$seconds. There is also little foundation for asserting that the measured half-life for SNO-Hb under the in vitro conditions employed is relevant to the physiological situation. The cells these authors examined are raised to extreme NO levels by incubation with high millimolar S-nitrosocysteine. The kinetic data presented in the paper are too limited to precisely determine the kinetic order, information required for extrapolation to physiological concentrations (zero to third-order kinetics fit the sparse data equally well). The results of Gladwin et al. are directly refuted by their own subsequent measurements in which no such instability in SNO-Hb content was observed (75). Most significant, untreated RBCs withdrawn from blood into room air likewise show no such instability in their SNO-Hb levels (22).

The treatment of RBCs used by Gladwin and collaborators (74, 75) was originally tried in the Stamler laboratory but abandoned when it was found not only to obscure the relevant chemistry but also to dramatically alter the cellular contents through extensive nitrosylation and oxidation of Hb and all of the protein and peptide contents, including membrane-bound proteins. To the extent that SNO-Hb is activated by deoxygenation for release to nitrosylate band 3 or glutathione then, under Gladwin and colleagues' conditions, this oxygen dependence would be short-circuited owing to the artifactual nitrosylation of band 3 and glutathione. (The fact that Gladwin and colleagues' deoxygenate on a timescale similar to their half-life further masks allosteric release.) Because of these problems, a new method of RBC treatment with NO was introduced that affords NO mass balance among Fe(II)NO and SNO-Hb and maintains the NO concentrations in the physiological realm (40). This method should be the operating standard for NO/RBC studies.

PHYSIOLOGICAL ISSUES: NO DELIVERY
WITH NO GRADIENTS

Observations

In our first report we observed that the amount of SNO-Hb was apparently greater in oxygenated left-ventricular blood than in partially deoxygenated right-ventricular blood, whereas the opposite was true of iron nitrosyl Hb (8). The total amount of NO bound to Hb (i.e., SNO plus FeNO), however, was the same on either side of the heart (within error of the measurement) (8, 37). A correlation between oxygen saturation of Hb (but not pO_2) and amount of SNO-Hb was demonstrated in situ across 80 measurements in human patients (22) (Figure 1). In vitro, oxygenation of the partially deoxygenated human blood samples displaced the NO from the hemes, with accompanying formation of SNO-Hb (22, 40, 64). Much the same chemical behavior, i.e., the displacement of NO from the hemes of venous blood Hb by oxygenation, has recently been reproduced in experiments of Herold & Röck (65), Rifkind et al. (85), and Datta et al. (32).

We interpreted these data as evidence for the oxygenation-induced transfer of some NO from hemes to thiols of Hb within the lung and for some migration of NO back to hemes upon deoxygenation of blood in the microcirculation and veins; the amounts of NO that apparently transferred ranged widely from low nanomolar to micromolar in our studies (22). The adult human placenta serves the function of the fetal lung in utero. Funai et al. have reported analogous gradients in SNO-Hb across the umbilical vein and artery of 19 babies (36). Arterial-venous differences in SNO-Hb/FeNO-Hb have now been measured in humans by five groups (22, 32, 36, 85; A. Doctor & B. Gaston, personal communication), and recently correlated directly with O_2 extraction in both health (22, 32) and disease (32).

Simulation of the oxygen-mediated transfer of NO from heme to thiol of Hb in vitro requires accurate recapitulation of the physiological conditions. In particular, we reported that the efficiency of NO exchange between heme and thiol is an inverse function of both the NO/Hb(FeII) ratio and the amounts of NO present (64), and that the NO-group transfer from heme to thiol shows saturation behavior with a plateau at ~ 1 μM (64). Thus, as heme-bound NO increases, the efficiency of transfer that is coupled to oxygenation decreases such that the transferred NO groups do not exceed micromolar levels. As noted above, these core observations have recently been confirmed in other laboratories (65), and indicate that the regulation of the disposition of the NO group incorporates a brake to limit formation of SNO-Hb when NO exceeds physiological levels. Under physiological conditions, where oxygen delivery is principally regulated by variation in blood flow, the formation of SNO-Hb, which regulates hypoxic vasodilation and hyperoxic vasoconstriction, is enabled (22, 32, 36). In septic shock, however, vasodilation by SNO-RBCs would aggravate hypotension and thus tissue hypoxemia. Hb then effectively circumvents the potential problem of excess SNO-Hb formation by trapping the excess NO on the α hemes (thus lowering the affinity of the other hemes for O_2). This chemistry accomplishes NO "detoxification" while possibly enhancing O_2 delivery, and is exemplified in the $\sim$0.3–1 μM A-V Hb heme-NO gradients observed by Mason and coworkers (62), under conditions ranging from basal NO levels to those characteristic of septic shock. It also consistent with the excess heme-Fe(II)NO over SNO-Hb reported recently in a murine model of sepsis (54) and with inhalation of supraphysiological NO (87).

Nitric Oxide Gradients Contrasted with O_2 Gradients

Notwithstanding these observed A-V NO differences in situ, gradients of NO compounds have been mistakenly analogized to O_2 gradients (52, 88). In the realm of the oxygen-delivery function of Hb, A-V gradients have a particular significance: They measure O_2 delivery to tissues. A-V gradients in vasodilatory NO compounds, however, have no such obvious significance (123). If we assume that NO gradients reflect NO delivery, then the measured gradients, typically in the range of $\sim$0.1–1 μM, are highly problematic because they imply a utilization of NO that vastly exceeds endogenous production ($\sim$1 mmol/70 kg/24 h), and a dose that would lead to lethal hypotension [SNO at 1–10 nM is adequate for regulation of blood flow (22, 41)]. Thus, the argument against SNO delivery by RBCs (88), which is based on measurement of a gradient, is not constrained by any quantitative standards of efficacy. In their study, Gladwin et al. (88) detected A-V gradients of up to 60 nM SNO-Hb, yet, without any rationale, viewed these as insignificant. Datta et al. (32) found micromolar A-V Hb NO gradients that correlate directly with A-V Hb O_2 gradients, both in normal subjects and in patients with heart failure.

Why are NO gradients different from O_2 gradients? Whereas O_2 uptake in the lung and release in tissue are compatible with the simple accounting scheme that balances the gradients and the blood flow with the amount delivered, NO and bioactive products (particularly SNOs) are produced throughout the circulatory system. Salient properties, such as locations, fluxes, etc., of the sources of NO/SNO relevant to their Hb chemistry as well as the sinks of Hb-NO compounds are not fully understood. However, the gradient-determining properties are very different for the O_2 versus the NO systems; the former is designed for wholesale delivery of an electronic sink; the latter, in contrast, is for the regulated dispensing, as a signal, of tiny doses of a high-potency messenger (41). Likewise, the gradient of NO compounds cannot be identified with NO delivery, as is the case for O_2. Total amounts of NO bound to Hb do not change, within the limits of sensitivity of assays, during the A-V cycle (8, 22, 37), a finding in keeping with predictions regarding NO mass balance (41, 55). Thus, the notion that the relative sizes of the A-V gradients, exhibited by various NO-compounds, provides a basis for identifying endogenous vasodilators (52, 53, 88, 123) betrays a misunderstanding of NO biological chemistry in general and the SNO-Hb mechanism in particular.

Dispensing NO

It is well documented (22, 32, 36, 37, 65, 85, 90, 92) that pO_2 regulates the disposition of NO bound in Hb, i.e., the SNO content of blood is proportional to O_2 saturation, whereas heme-NO varies inversely with O_2 saturations. A decline in SNO-Hb is generally accompanied by the capture of NO on β hemes (41, 90), thus creating the appearance of an A-V gradient in SNO-Hb and FeNO [total Hb-NO does not change within limits of detection (8, 22, 37)]. These gradients can be revealed ex vivo provided A-V O_2 differences are carefully maintained during sampling (22, 32). [Gladwin et al. (74) do not detect these gradients, perhaps because they expose both venous and arterial samples to room air.] These gradients may be viewed as informative of steady state situations in vivo. They have no real bearing, however, on the amounts of NO/SNO that transfer per A-V cycle or that are exported from RBCs. Indeed, amounts of NO/SNO delivered by RBCs in vivo should not be expected to be any more accessible to measurement than are the amounts of NO/SNO (EDRF) delivered by endothelial cells in vivo (in the presence of RBCs). Neither should be detectable. Moreover, the gradient-forming transfer of NO from β-cys93 to β hemes (see Thiol-to-Heme Transfer, above) is inevitably accompanied by a second reaction in which NO groups are exchanged with receptor thiols in the RBC membrane (band 3) (40, 121) and in the cytosol (glutathione) (8, 41) to provide (and conserve) NO bioactivity in proportion to degree of hypoxia (22, 41). The pO_2 does not strictly regulate the amount of SNO exported from RBCs (as recently misconstrued) (74), but rather exerts an effect on the position of the equilibrium between SNO-Hb and acceptor thiols (e.g., band 3) that provides bioactivity (Equation 13).

We anticipate therefore that RBCs will have a propensity to deliver NO bioactivity (involving only a small fraction of the total Hb-NO) throughout the low pO_2 circuit, which includes arterioles, capillaries, veins, and pulmonary artery, and thus influence both blood flow systemically, as well as pulmonary arterial oxygenation and pressure (Figure 4a). Conversely, RBC-NO will have a lesser tendency to deliver NO bioactivity in the high pO_2 circuit, including pulmonary veins and large systemic arteries.

PHYSIOLOGICAL VERSUS PATHOPHYSIOLOGICAL NO/HB BIOCHEMISTRY

Principles

The importance of adhering to physiological ratios of the reagents in in vitro studies aimed at elucidating the fundamental chemistry of NO Hb interactions has been chronically underestimated. As noted above, the disposition of NO bound to Hb and reactivity of the molecule is a function of many variables, including the amount of NO, the NO/heme ratio, the details of reagent mixing, and the lag time after NO infusion (55, 57, 60–65). These principles dictate exercise of caution in drawing physiologically relevant inferences from studies conducted with extreme

NO concentrations. Such experiments include the work of Mason and colleagues, who infuse DEA-NO to give concentrations of NO characteristic of septic shock and other pathological conditions (57, 124, 125); Gladwin et al. (74, 75); Xu et al. (75); Huang et al. (73); and Han et al. (70, 72), who add more NO to RBCs (500–10,000 μM) than is found in any pathophysiological situation (74, 75); and Gladwin et al., who introduce inhaled levels of NO at 80 ppm ($\sim$4 μM) in a flow of 40 liters/min—a suprapathological dose (87).

Hemoglobin Structural Effects

As previously shown by Kosaka et al. (57), such high concentrations of NO, in concert with the low pH that accompanies hypoxia at the tissue level, favor production of Hb molecules with 5-coordinate α-subunit heme Fe(II)NO moieties, which are viewed as a signature of the T-state conformation. These T-state molecules switch to R structure upon oxygenation without immediately dislodging NO from the α-chain, but exhibit a low affinity for O_2 on their vacant hemes (61). The lower affinity for O_2 has been suggested as providing a mechanism to improve O_2 delivery (126). This relative lack of affinity for O_2 clearly affects the propensity of bound-NO toward displacement, including β heme-to-thiol NO-group transfer, as [O_2] varies in a physiological cycle. This effect can be understood by recognizing that SNO-Hb is disfavored in T-state molecules (22, 30, 37, 90, 92). Moreover, NO-transfer between 6-coordinate α and β hemes (in R structure), which facilitates production of significant SNO-Hb, is inhibited by the very O_2 molecules that must first bind to β hemes in order to induce the 6-coordinate R structure. Thus Hb may switch the control mechanism by which O_2 delivery is regulated in pathophysiological states. Instead of relying on SNO-Hb vasodilation, which would be counterproductive in shock, nature might exploit NO as an allosteric effector of oxygen delivery (58, 64).

The reactivity of NO and Hb at physiological levels (nanomolar NO, millimolar Hb) is very different. At low NO/heme ratios, selective interactions of NO with β-subunit heme have been observed under a variety of conditions (22, 55, 63, 90, 92), notwithstanding the tendency of the NO groups to reach the α-subunit heme Hb at higher NO levels and/or longer NO incubations (55, 57, 60, 61). In the R state, a partitioning is also established between the 6-coordinate β-subunit heme and thiol (22, 32, 33, 55, 63, 90). The more NO that binds within the T state (and the lower the pH), the greater the tendency to form the relatively inert 5-coordinate α-nitrosyl Hb (more so in rat than in human blood), and β-subunit heme-to-thiol transfer is suppressed. Thus the formation of 5-coordinate α-nitrosyl Hb may be nature's way of limiting SNO-Hb formation (and consequent hypotension) in situations of hypotension characterized by NO plus SNO-Hb excess (54, 58, 64). In the interactions of NO with Hb, the amount of a given product does not respond linearly to the NO concentration; the relative product distribution is not constant. Rather, Hb is a model of complexity whose subunit inequivalences and allosteric states enable a wide-ranging responsiveness to ambient conditions.

OUTLOOK: DISORDERS OF RBC-NO PROCESSING

The integrated vascular response that mediates autoregulation of blood flow in pulmonary and systemic vessels to support O_2 uptake and delivery on physiological timescales has a counterpart in SNO-Hb RBC vasodilation, distinguished from EDRF (shear-induced vasodilation) (35, 49, 58) and pharmacological nitrite vasodilation (53). Inasmuch as this physiological response is paramount to the respiratory cycle, we anticipate that defects in RBC vasodilation might underlie diseases of heart, lung, blood, and potentially other organs.

Impaired release of NO by RBCs (8, 22, 32, 42) may have direct implications for hematological abnormalities such as the vaso-occlusive crises of sickle cell disease, the hypertension of thalessemia, and the thrombotic diathesis of paroxysmal nocturnal hemoglobinuria. More broadly, it may shed light on unexplained and underappreciated medical problems identified with RBCs. Thus, RBC counts, even within the normal range, are directly correlated with blood pressure (127), and RBCs may actively contribute to clot formation (128). These observations may bear directly on the findings that RBC transfusions (128–131, 144), erythropoietin (132) (which raises Hb levels), and Hb-based blood substitutes (133) that are devoid of NO bioactivity have each been associated with increased cardiovascular morbidity and mortality. Indeed, stored blood, which over typical storage intervals likely loses vasodilator SNO, has been shown to raise pulmonary and systemic pressures, induce ventilation-perfusion mismatching, and create mesenteric ischemia (134, 134a). Both erythropoietin (135, 136) and Hb-based blood substitutes (137, 138) induce similar cardiovascular dysfunction, evidently by creating an NO/Hb imbalance. The new understanding that RBCs precisely regulate their function through a controlled NO/Hb balance, thus raises the idea that dysfunction of RBCs (through NO/Hb imbalance or compromised NO processing) may contribute broadly to the pathogenesis of thrombotic, hematologic, and ischemic disorders and perhaps other organ dysfunctions (100). Consistent with this possibility are the initial reports of altered RBC-SNO processing in subjects with diabetes (84), heart failure (32), sickle cell disease (J.R. Pawloski, D.T. Hess & J.S. Stamler, unpublished results), pulmonary hypertension (18), and sepsis (21). For example, patients with pulmonary arterial hypertension (18) and diabetes show impaired RBC relaxations (18), the former associated with SNO-Hb deficiency and the latter with deficient SNO release, as glycosylation of Hb, presumably by promoting the R structure, shifts the equilibrium in Equation 13 to favor reactants (18, 84, 139). Both diabetic (84, 139) and septic (21) patients exhibit excessive SNO-Hb levels, but the latter show increased relaxations (18, 54), whereas the former, because of this shift, show decreased relaxations (18). More generally, impaired vasodilation by RBCs is suggested to originate at multiple levels, ranging from impairments in oxygenation, to Hb itself (allostery), to alterations in interactions of Hb with the membrane band 3 and the RBC NO export function. If true, a new perspective on the alterations in blood flow and tissue oxygenation that characterize many vascular, respiratory, hematologic, and ischemic diseases may emerge from attention to RBC dysfunctions. Reconstitution of RBC-NO activity may also provide a novel therapeutic strategy to treat NO deficiency and/or hypoxemia-associated states (100).

ACKNOWLEDGMENTS

The authors thank Irwin Fridovich for helpful comments and discussions. D.J.S. (MCB00981228) and J.S.S. (5P01-HL42444) were supported by the NSF and NIH, respectively.

**The *Annual Review of Physiology* is online at
http://physiol.annualreviews.org**

LITERATURE CITED

1. Ignarro LJ, Buga GM, Wood KS, Byrns RE, Chaudhuri G. 1987. Endothelium-derived relaxing factor produced and released from artery and vein is nitric oxide. *Proc. Natl. Acad. Sci. USA* 84:9265–69

2. Palmer RM, Ferrige AG, Moncada S. 1987. Nitric oxide release accounts for the biological activity of endothelium-derived relaxing factor. *Nature* 327:524–26

3. Doyle MP, Hoekstra JW. 1981. Oxidation of nitrogen oxides by bound dioxygen in hemoproteins. *J. Inorg. Biochem.* 14:351–58

4. Gibson QH, Roughton FJW. 1957. The kinetics and equilibria of the reactions of nitric oxide with sheep haemoglobin. *J. Physiol.* 136:507–26

5. Lancaster JR Jr. 1994. Simulation of the diffusion and reaction of endogenously produced nitric oxide. *Proc. Natl. Acad. Sci. USA* 91:8137–41

6. Stamler JS, Simon DI, Osborne JA, Mullins ME, Jaraki O, et al. 1992. S-nitrosylation of proteins with nitric oxide: synthesis and characterization of biologically active compounds. *Proc. Natl. Acad. Sci. USA* 89:444–48

7. Stamler JS, Singel DJ, Loscalzo J. 1992. Biochemistry of nitric oxide and its redox-activated forms. *Science* 258:1898–902

8. Jia L, Bonaventura C, Bonaventura J, Stamler JS. 1996. S-nitroso-haemoglobin: a dynamic activity of blood involved in vascular control. *Nature* 380:221–26

9. Liao JC, Hein TW, Vaughn MW, Huang KT, Kuo L. 1999. Intravascular flow decreases erythrocyte consumption of nitric oxide. *Proc. Natl. Acad. Sci. USA* 96:8757–61

10. Liu X, Samouilov A, Lancaster JR Jr, Zweier JL. 2002. Nitric oxide uptake by erythrocytes is primarily limited by extracellular diffusion not membrane resistance. *J. Biol. Chem.* 277:26194–99

11. Tsoukias NM, Kavdia M, Popel AS. 2004. A theoretical model of nitric oxide transport in arterioles: frequency- vs. amplitude-dependent control of cGMP formation. *Am. J. Physiol. Heart Circ. Physiol.* 286:H1043–56

12. Tsoukias NM, Popel AS. 2002. Erythrocyte consumption of nitric oxide in presence and absence of plasma-based hemoglobin. *Am. J. Physiol. Heart Circ. Physiol.* 282:H2265–77

13. Tsoukias NM, Popel AS. 2003. A model of nitric oxide capillary exchange. *Microcirculation* 10:479–95

14. Gow AJ, Chen Q, Hess DT, Day BJ, Ischiropoulos H, Stamler JS. 2002. Basal and stimulated protein S-nitrosylation in multiple cell types and tissues. *J. Biol. Chem.* 277:9637–40

15. Bates JN, Harrison DG, Myers PR, Minor RL. 1991. EDRF: nitrosylated compound or authentic nitric oxide. *Basic Res. Cardiol.* 86(Suppl. 2):17–26

16. Myers PR, Minor RL Jr, Guerra R Jr, Bates JN, Harrison DG. 1990. Vasorelaxant properties of the endothelium-derived relaxing factor more closely resemble S-nitrosocysteine than nitric oxide. *Nature* 345:161–63

17. Gaston B, Drazen JM, Jansen A, Sugarbaker DA, Loscalzo J, et al. 1994. Relaxation of human bronchial smooth muscle by S-nitrosothiols in vitro. *J. Pharmacol. Exp. Ther.* 268:978–84

18. Foster MW, McMahon TJ, Stamler JS. 2003. S-nitrosylation in health and disease. *Trends Mol. Med.* 9:160–68

19. Stamler JS. 1995. S-nitrosothiols and the bioregulatory actions of nitrogen oxides through reactions with thiol groups. *Curr. Top. Microbiol. Immunol.* 196:19–36

20. Sellke FW, Myers PR, Bates JN, Harrison DG. 1990. Influence of vessel size on the sensitivity of porcine coronary microvessels to nitroglycerin. *Am. J. Physiol. Heart Circ. Physiol.* 258:H515–20

21. Liu L, Yan Y, Zeng M, Zhang J, Hanes MA, et al. 2004. Essential roles of S-nitrosothiols in vascular homeostasis and endotoxic shock. *Cell* 116:617–28

22. McMahon TJ, Moon RE, Luchsinger BP, Carraway MS, Stone AE, et al. 2002. Nitric oxide in the human respiratory cycle. *Nat. Med.* 8:711–17

23. Jackson WF. 1987. Arteriolar oxygen reactivity: Where is the sensor? *Am. J. Physiol. Heart Circ. Physiol.* 253:H1120–26

24. Gonzalez-Alonso J, Richardson RS, Saltin B. 2001. Exercising skeletal muscle blood flow in humans responds to reduction in arterial oxyhaemoglobin, but not to altered free oxygen. *J. Physiol.* 530:331–41

25. Gorczynski RJ, Duling BR. 1978. Role of oxygen in arteriolar functional vasodilation in hamster striated muscle. *Am. J. Physiol. Heart Circ. Physiol.* 235:H505–15

26. Duling BR, Berne RM. 1970. Longitudinal gradients in periarteriolar oxygen tension. A possible mechanism for the participation of oxygen in local regulation of blood flow. *Circ. Res.* 27:669–78

27. Guyton AC, Ross JM, Carrier O Jr, Walker JR. 1964. Evidence for tissue oxygen demand as the major factor causing autoregulation. *Circ. Res.* 15(Suppl.):60–69

28. Roach RC, Koskolou MD, Calbet JA, Saltin B. 1999. Arterial O_2 content and tension in regulation of cardiac output and leg blood flow during exercise in humans. *Am. J. Physiol. Heart Circ. Physiol.* 276:H438–45

29. Ferranti P, Malorni A, Mamone G, Sannolo N, Marino G. 1997. Characterisation of S-nitrosohaemoglobin by mass spectrometry. *FEBS Lett.* 400:19–24

30. Chan NL, Rogers PH, Arnone A. 1998. Crystal structure of the S-nitroso form of liganded human hemoglobin. *Biochemistry* 37:16459–64

31. Chan NL, Kavanaugh JS, Rogers PH, Arnone A. 2004. Crystallographic analysis of the interaction of nitric oxide with quaternary-T human hemoglobin. *Biochemistry* 43:118–32

32. Datta B, Tufnell-Barrett T, Bleasdale RA, Jones CJ, Beeton I, et al. 2004. Red blood cell nitric oxide as an endocrine vasoregulator: a potential role in congestive heart failure. *Circulation* 109:1339–42

33. James PE, Lang D, Tufnell-Barret T, Milsom AB, Frenneaux MP. 2004. Vasorelaxation by red blood cells and impairment in diabetes: reduced nitric oxide and oxygen delivery by glycated hemoglobin. *Circ. Res.* 94:976–83

34. Giustarini D, Milzani A, Colombo R, Dalle-Donne I, Rossi R. 2004. Nitric oxide, S-nitrosothiols and hemoglobin: Is

methodology the key? *Trends Pharmacol. Sci.* 25:311–16

35. Stamler JS, Hess DT, Singel DJ. 2003. Reply to "NO adducts in mammalian red blood cells: to much or too little?" *Nat. Med.* 9:481–82

36. Funai EF, Davidson A, Seligman SP, Finlay TH. 1997. S-nitrosohemoglobin in the fetal circulation may represent a cycle for blood pressure regulation. *Biochem. Biophys. Res. Commun.* 239:875–77

37. Stamler JS, Jia L, Eu JP, McMahon TJ, Demchenko IT, et al. 1997. Blood flow regulation by S-nitrosohemoglobin in the physiological oxygen gradient. *Science* 276:2034–37

38. Perutz MF. 1970. Stereochemistry of cooperative effects in haemoglobin. *Nature* 228:726–39

39. Perutz MF, Wilkinson AJ, Paoli M, Dodson GG. 1998. The stereochemical mechanism of the cooperative effects in hemoglobin revisited. *Annu. Rev. Biophys. Biomol. Struct.* 27:1–34

40. Pawloski JR, Hess DT, Stamler JS. 2001. Export by red blood cells of nitric oxide bioactivity. *Nature* 409:622–26

41. McMahon TJ, Stone AE, Bonaventura J, Singel DJ, Stamler JS. 2000. Functional coupling of oxygen binding and vasoactivity in S-nitrosohemoglobin. *J. Biol. Chem.* 275:16738–45

42. Pawloski JR, Swaminathan RV, Stamler JS. 1998. Cell-free and erythrocytic S-nitrosohemoglobin inhibits human platelet aggregation. *Circulation* 97: 263–67

43. Deleted in press

44. Saltin B, Radegran G, Koskolou MD, Roach RC. 1998. Skeletal muscle blood flow in humans and its regulation during exercise. *Acta Physiol. Scand.* 162:421–36

45. Blitzer ML, Loh E, Roddy MA, Stamler JS, Creager MA. 1996. Endothelium-derived nitric oxide regulates systemic and pulmonary vascular resistance during acute hypoxia in humans. *J. Am. Coll. Cardiol.* 28:591–96

46. Ichinose F, Roberts JD Jr, Zapol WM. 2004. Inhaled nitric oxide: a selective pulmonary vasodilator: current uses and therapeutic potential. *Circulation* 109:3106–11

47. Deem S, Swenson ER, Alberts MK, Hedges RG, Bishop MJ. 1998. Red-blood-cell augmentation of hypoxic pulmonary vasoconstriction: hematocrit dependence and the importance of nitric oxide. *Am. J. Respir. Crit. Care Med.* 157:1181–86

48. Lipton AJ, Johnson MA, Macdonald T, Lieberman MW, Gozal D, Gaston B. 2001. S-nitrosothiols signal the ventilatory response to hypoxia. *Nature* 413: 171–74

49. Singel DJ, Stamler J. 2004. Blood traffic control. *Nature* 430:297

50. Gonzalez-Alonso J, Olsen DB, Saltin B. 2002. Erythrocyte and the regulation of human skeletal muscle blood flow and oxygen delivery: role of circulating ATP. *Circ. Res.* 91:1046–55

51. Jagger JE, Bateman RM, Ellsworth ML, Ellis CG. 2001. Role of erythrocyte in regulating local O_2 delivery mediated by hemoglobin oxygenation. *Am. J. Physiol. Heart Circ. Physiol.* 280:H2833–39

52. Gladwin MT, Lancaster JR Jr, Freeman BA, Schechter AN. 2003. Nitric oxide's reactions with hemoglobin: a view through the SNO-storm. *Nat. Med.* 9:496–500

53. Cosby K, Partovi KS, Crawford JH, Patel RP, Reiter CD, et al. 2003. Nitrite reduction to nitric oxide by deoxyhemoglobin vasodilates the human circulation. *Nat. Med.* 9:1498–505

54. Crawford JH, Chacko BK, Pruitt HM, Piknova B, Hogg N, Patel RP. 2004. Transduction of NO-bioactivity by the red blood cell in sepsis: novel mechanisms of vasodilation during acute inflammatory disease. *Blood* 104:1375–82

55. Gow AJ, Luchsinger BP, Pawloski JR, Singel DJ, Stamler JS. 1999. The oxyhemoglobin reaction of nitric oxide. *Proc. Natl. Acad. Sci. USA* 96:9027–32

56. Imai K. 1982. *Allosteric Effects in Haemoglobin.* Cambridge, UK: Cambridge Univ. Press

57. Kosaka H, Sawai Y, Sakaguchi H, Kumura E, Harada N, et al. 1994. ESR spectral transition by arteriovenous cycle in nitric oxide hemoglobin of cytokine-treated rats. *Am. J. Physiol. Cell Physiol.* 266:C1400–5

58. McMahon TJ, Stamler JS. 1999. Concerted nitric oxide/oxygen delivery by hemoglobin. *Methods Enzymol.* 301:99–114

59. Di Cera E. 1995. *Thermodynamic Theory of Site-Specific Binding Processes in Biological Macromolecules.* Cambridge, UK: Cambridge Univ. Press

60. Taketa F, Antholine WE, Chen JY. 1978. Chain nonequivalence in binding of nitric oxide to hemoglobin. *J. Biol. Chem.* 253:5448–51

61. Yonetani T, Tsuneshige A, Zhou Y, Chen X. 1998. Electron paramagnetic resonance and oxygen binding studies of alpha-nitrosyl hemoglobin. A novel oxygen carrier having no-assisted allosteric functions. *J. Biol. Chem.* 273:20323–33

62. Jaszewski AR, Fann YC, Chen YR, Sato K, Corbett J, Mason RP. 2003. EPR spectroscopy studies on the structural transition of nitrosyl hemoglobin in the arterial-venous cycle of DEANO-treated rats as it relates to the proposed nitrosyl hemoglobin/nitrosothiol hemoglobin exchange. *Free Radic Biol. Med.* 35:444–51

63. Luchsinger BP, Rich EN, Gow AJ, Williams EM, Stamler JS, Singel DJ. 2003. Routes to S-nitroso-hemoglobin formation with heme redox and preferential reactivity in the beta subunits. *Proc. Natl. Acad. Sci. USA* 100:461–66

64. Gow AJ, Stamler JS. 1998. Reactions between nitric oxide and haemoglobin under physiological conditions. *Nature* 391:169–73

65. Herold S, Röck G. 2003. Reactions of deoxy-, oxy-, and methemoglobin with nitrogen monoxide. Mechanistic studies of the S-nitrosothiol formation under different mixing conditions. *J. Biol. Chem.* 278:6623–34

66. Honig CR, Gayeski TE. 1993. Resistance to O_2 diffusion in anemic red muscle: roles of flux density to cell PO_2. *Am. J. Physiol. Heart Circ. Physiol.* 265:H868–75

67. Whalen WJ, Nair P. 1967. Intracellular pO_2 and its regulation in resting skeletal muscle of the guinea pig. *Circ. Res.* 21:251–61

68. Coburn R, Mayers L, Luomanmaki K. 1967. An "indicator" method of estimating intracellular oxygen tension in resting muscle. *Fed. Proc.* 26:334 (Abstr.)

69. Huang Z, Ucer KB, Murphy T, Williams RT, King SB, Kim-Shapiro DB. 2002. Kinetics of nitric oxide binding to R-state hemoglobin. *Biochem. Biophys. Res. Commun.* 292:812–18

70. Han TH, Hyduke DR, Vaughn MW, Fukuto JM, Liao JC. 2002. Nitric oxide reaction with red blood cells and hemoglobin under heterogeneous conditions. *Proc. Natl. Acad. Sci. USA* 99:7763–68

71. Fago A, Crumbliss AL, Peterson J, Pearce LL, Bonaventura C. 2003. The case of the missing NO-hemoglobin: spectral changes suggestive of heme redox reactions reflect changes in NO-heme geometry. *Proc. Natl. Acad. Sci. USA* 100:12087–92

72. Han TH, Fukuto JM, Liao JC. 2004. Reductive nitrosylation and S-nitrosation of hemoglobin in inhomogeneous nitric oxide solutions. *Nitric Oxide* 10:74–82

73. Huang Z, Louderback JG, Goyal M, Azizi F, King SB, Kim-Shapiro DB. 2001. Nitric oxide binding to oxygenated hemoglobin under physiological

conditions. *Biochim. Biophys. Acta* 1568:252–60

74. Gladwin MT, Wang X, Reiter CD, Yang BK, Vivas EX, et al. 2002. S-Nitrosohemoglobin is unstable in the reductive erythrocyte environment and lacks O_2/NO-linked allosteric function. *J. Biol. Chem.* 277:27818–28

75. Xu X, Cho M, Spencer NY, Patel N, Huang Z, et al. 2003. Measurements of nitric oxide on the heme iron and beta-93 thiol of human hemoglobin during cycles of oxygenation and deoxygenation. *Proc. Natl. Acad. Sci. USA* 100:11303–8

76. Kim-Shapiro DB. 2004. Hemoglobin-nitric oxide cooperativity: Is NO the third respiratory ligand? *Free Radic. Biol. Med.* 36:402–12

77. Rossi R, Milzani A, Dalle-Donne I, Giannerini F, Giustarini D, et al. 2001. Different metabolizing ability of thiol reactants in human and rat blood: biochemical and pharmacological implications. *J. Biol. Chem.* 276:7004–10

78. Stamler JS. 2004. S-nitrosothiols in the blood: roles, amounts, and methods of analysis. *Circ. Res.* 94:414–17

79. Kirima K, Tsuchiya K, Sei H, Hasegawa T, Shikishima M, et al. 2003. Evaluation of systemic blood NO dynamics by EPR spectroscopy: HbNO as an endogenous index of NO. *Am. J. Physiol. Heart Circ. Physiol.* 285:H589–96

80. Freeman G, Dyer RL, Juhos LT, St John GA, Anbar M. 1978. Identification of nitric oxide (NO) in human blood. *Arch. Environ. Health* 33:19–23

81. Takahashi Y, Kobayashi H, Tanaka N, Sato T, Takizawa N, Tomita T. 1998. Nitrosyl hemoglobin in blood of normoxic and hypoxic sheep during nitric oxide inhalation. *Am. J. Physiol. Heart Circ. Physiol.* 274:H349–57

82. Aldini G, Orioli M, Maffei Facino R, Giovanna Clement M, Albertini M, et al. 2004. Nitrosylhemoglobin formation after infusion of NO solutions: ESR stud-ies in pigs. *Biochem. Biophys. Res. Commun.* 318:405–14

83. Roccatello D, Mengozzi G, Alfieri V, Pignone E, Menegatti E, et al. 1997. Early increase in blood nitric oxide, detected by electron paramagnetic resonance as nitrosylhaemoglobin, in haemodialysis. *Nephrol. Dial Transplant.* 12:292–97

84. Milsom AB, Jones CJ, Goodfellow J, Frenneaux MP, Peters JR, James PE. 2002. Abnormal metabolic fate of nitric oxide in Type I diabetes mellitus. *Diabetologia* 45:1515–22

85. Nagababu E, Ramasamy S, Abernethy DR, Rifkind JM. 2003. Active nitric oxide produced in the red cell under hypoxic conditions by deoxyhemoglobin-mediated nitrite reduction. *J. Biol. Chem.* 278:46349–56

86. Rassaf T, Bryan NS, Maloney RE, Specian V, Kelm M, et al. 2003. NO adducts in mammalian red blood cells: too much or too little? *Nat. Med.* 9:481–82; author reply 2–3

87. Gladwin MT, Ognibene FP, Pannell LK, Nichols JS, Pease-Fye ME, et al. 2000. Relative role of heme nitrosylation and beta-cysteine 93 nitrosation in the transport and metabolism of nitric oxide by hemoglobin in the human circulation. *Proc. Natl. Acad. Sci. USA* 97:9943–48

88. Gladwin MT, Shelhamer JH, Schechter AN, Pease-Fye ME, Waclawiw MA, et al. 2000. Role of circulating nitrite and S-nitrosohemoglobin in the regulation of regional blood flow in humans. *Proc. Natl. Acad. Sci. USA* 97:11482–87

89. Feelisch M, Rassaf T, Mnaimneh S, Singh N, Bryan NS, et al. 2002. Concomitant S-, N-, and heme-nitros(yl)ation in biological tissues and fluids: implications for the fate of NO in vivo. *FASEB J.* 16:1775—85

89a. Wang X, Tanus-Santos JE, Reiter CD, Dejam A, Shiva S, et al. 2004. Biological activity of nitric oxide in the plasmatic compartment. 2004. *Proc. Natl. Acad. Sci. USA* 101:11477–82

89b. Tsikas D, Sandmann J, Gutzki F-M, Stichtenoth DO, Frolich JC. 1999. Measurement of S-nitrosalbumin by gas chromatography-mass spectrometry. II Quantitative determination of S-nitrosoalbumin in human plasma using S-[^{15}N]nitrosoalbumin as internal standard. *J. Chromatogr. B* 726:13–24

89c. Tsikas D, Sandmann J, Frolich JC. 2002. Measurement of S-nitrosalbumin by gas chromatography-mass spectrometry. III Quantitative determination of S-nitrosoalbumin in human plasma after specific conversion of the S-nitroso group to nitrite by cysteine and Cu^{2+} via intermediate formation of S-nitroscysteine and nitric oxide. *J. Chromatogr. B* 772:335–46

90. Pezacki JP, Ship NJ, Kluger R. 2001. Release of nitric oxide from S-nitrosohemoglobin. Electron transfer as a response to deoxygenation. *J. Am. Chem. Soc.* 123:4615–16

91. Mamone G, Sannolo N, Malorni A, Ferranti P. 1999. In vitro formation of S-nitrosohemoglobin in red cells by inducible nitric oxide synthase. *FEBS Lett.* 462:241–45

92. Romeo AA, Capobianco JA, English AM. 2003. Superoxide dismutase targets NO from GSNO to Cysbeta93 of oxyhemoglobin in concentrated but not dilute solutions of the protein. *J. Am. Chem. Soc.* 125:14370–78

93. Deleted in proof

94. Joshi MS, Ferguson TB Jr, Han TH, Hyduke DR, Liao JC, et al. 2002. Nitric oxide is consumed, rather than conserved, by reaction with oxyhemoglobin under physiological conditions. *Proc. Natl. Acad. Sci. USA* 99:10341–46

95. Gow AJ, Buerk DG, Ischiropoulos H. 1997. A novel reaction mechanism for the formation of S-nitrosothiol in vivo. *J. Biol. Chem.* 272:2841—45

95a. Fernandez BO, Ford PC. 2003. Nitrite catalyzes ferriheme protein reductive nitrosylation. *J. Am. Chem. Soc.* 125:10510–11

95b. Fernandez BO, Lorkovic IM, Ford PC. 2004. Mechanisms of ferriheme reduction by nitric oxide: nitrite and general base catalysis. *Inorg. Chem.* 43:5393–402

96. Moore EG, Gibson QH. 1976. Cooperativity in the dissociation of nitric oxide from hemoglobin. *J. Biol. Chem.* 251:2788–94

97. Minning DM, Gow AJ, Bonaventura J, Braun R, Dewhirst M, et al. 1999. Ascaris haemoglobin is a nitric oxide-activated 'deoxygenase.' *Nature* 401:497–502

98. Sun J, Xu L, Eu JP, Stamler JS, Meissner G. 2003. Nitric oxide, NOC-12, and S-nitrosoglutathione modulate the skeletal muscle calcium release channel/ryanodine receptor by different mechanisms. An allosteric function for O_2 in S-nitrosylation of the channel. *J. Biol. Chem.* 278:8184–89

99. Massaro AF, Gaston B, Kita D, Fanta C, Stamler JS, Drazen JM. 1995. Expired nitric oxide levels during treatment of acute asthma. *Am. J. Respir. Crit. Care Med.* 152:800–3

100. Pawloski JR, Stamler JS. 2002. Nitric oxide in RBCs. *Transfusion* 42:1603–9

101. Ship NJ, Pezacki JP, Kluger R. 2003. Rates of release of nitric oxide from HbSNO and internal electron transfer. *Bioorg. Chem.* 31:3–10

102. Rifkind JM, Lauer LD, Chiang SC, Li NC. 1976. Copper and the oxidation of hemoglobin: a comparison of horse and human hemoglobins. *Biochemistry* 15:5337–43

103. Manoharan PT, Alston K, Rifkind JM. 1989. Interaction of copper(II) with hemoglobins in the unliganded conformation. *Biochemistry* 28:7148–53

104. Tomoda A, Tsuji A, Yoneyama Y. 1981. Involvement of superoxide anion in the

reaction mechanism of haemoglobin oxidation by nitrite. *Biochem. J.* 193:169–79

105. Tomoda A, Yoneyama Y, Tsuji A. 1981. Changes in intermediate haemoglobins during autoxidation of haemoglobin. *Biochem. J.* 195:485–92

106. Hille R, Olson JS, Palmer G. 1979. Spectral transitions of nitrosyl hemes during ligand binding to hemoglobin. *J. Biol. Chem.* 254:12110–20

107. Doyle MP, Pickering RA, DeWeert TM, Hoekstra JW, Pater D. 1981. Kinetics and mechanism of the oxidation of human deoxyhemoglobin by nitrites. *J. Biol. Chem.* 256:12393–98

108. Perrella M, Di Cera E. 1999. CO ligation intermediates and the mechanism of hemoglobin cooperativity. *J. Biol. Chem.* 274:2605–8

109. Kruszyna R, Kruszyna H, Smith RP, Thron CD, Wilcox DE. 1987. Nitrite conversion to nitric oxide in red cells and its stabilization as a nitrosylated valency hybrid of hemoglobin. *J. Pharmacol. Exp. Ther.* 241:307–13

110. Luchsinger BP. 2003. *Chemical interaction of nitric oxide and human hemoglobin.* PhD thesis. Montana State Univ., Bozeman. 119 pp.

111. Deleted in proof

112. Chen Z, Zhang J, Stamler JS. 2002. Identification of the enzymatic mechanism of nitroglycerin bioactivation. *Proc. Natl. Acad. Sci. USA* 99:8306–11

113. Sydow K, Daiber A, Oelze M, Chen Z, August M, et al. 2004. Nitroglycerin treatment inhibits mitochondrial aldehyde dehydrogenase and increases mitochondrial reactive oxygen species: central role of mitochondria in nitrate tolerance. *J. Clin. Invest.* 113:482–89

114. Reutov VP, Sorokina EG. 1998. NO-synthase and nitrite-reductase components of nitric oxide cycle. *Biochemistry (Mosc)* 63:874–84

115. Fernandez BO, Ford PC. 2003. Nitrite catalyzes ferriheme protein reductive nitrosylation. *J. Am. Chem. Soc.* 125:10510–11

116. Spagnuolo C, Rinelli P, Coletta M, Chiancone E, Ascoli F. 1987. Oxidation reaction of human oxyhemoglobin with nitrite: a reexamination. *Biochim. Biophys. Acta* 911:59–65

117. Lauer T, Preik M, Rassaf T, Strauer BE, Deussen A, et al. 2001. Plasma nitrite rather than nitrate reflects regional endothelial nitric oxide synthase activity but lacks intrinsic vasodilator action. *Proc. Natl. Acad. Sci. USA* 98:12814–19

118. Kleinbongard P, Dejam A, Lauer T, Rassaf T, Schindler A, et al. 2003. Plasma nitrite reflects constitutive nitric oxide synthase activity in mammals. *Free Radic. Biol. Med.* 35:790–96

119. Bolotina VM, Najibi S, Palacino JJ, Pagano PJ, Cohen RA. 1994. Nitric oxide directly activates calcium-dependent potassium channels in vascular smooth muscle. *Nature* 368:850–53

120. Crawford JH, White CR, Patel RP. 2003. Vasoactivity of S-nitrosohemoglobin: role of oxygen, heme, and NO oxidation states. *Blood* 101:4408–15

121. McMahon TJ, Pawloski JR, Hess DT, Piantadosi CA, Luchsinger BP, et al. 2003. S-nitrosohemoglobin is distinguished from other nitrosovasodilators by unique oxygen-dependent responses that support an allosteric mechanism of action. *Blood* 102:410–11; author reply 2–3

122. Patel RP, Hogg N, Spencer NY, Kalyanaraman B, Matalon S, Darley-Usmar VM. 1999. Biochemical characterization of human S-nitrosohemoglobin. Effects on oxygen binding and transnitrosylation. *J. Biol. Chem.* 274:15487–92

123. Stamler JS. 2003. Hemoglobin and nitric oxide. *N. Engl. J. Med.* 349:402–5; author reply 405

124. Hall DM, Buettner GR, Matthes RD, Gisolfi CV. 1994. Hyperthermia stimulates nitric oxide formation: electron paramagnetic resonance detection of

NO-heme in blood. *J. Appl. Physiol.* 77:548–53

125. Kagan VE, Day BW, Elsayed NM, Gorbunov NV. 1996. Dynamics of haemoglobin. *Nature* 383:30–31

126. Kosaka H, Seiyama A. 1997. Elevation of oxygen release by nitroglycerin without an increase in blood flow in the hepatic microcirculation. *Nat. Med.* 3:456–59

127. Cirillo M, Laurenzi M, Trevisan M, Stamler J. 1992. Hematocrit, blood pressure, and hypertension. The Gubbio Population Study. *Hypertension* 20:319–26

128. Andrews DA, Low PS. 1999. Role of red blood cells in thrombosis. *Curr. Opin. Hematol.* 6:76–82

129. Hebert PC, Wells G, Blajchman MA, Marshall J, Martin C, et al. 1999. A multicenter, randomized, controlled clinical trial of transfusion requirements in critical care. Transfusion Requirements in Critical Care Investigators, Canadian Critical Care Trials Group. *N. Engl. J. Med.* 340:409–17

130. Wu WC, Rathore SS, Wang Y, Radford MJ, Krumholz HM. 2001. Blood transfusion in elderly patients with acute myocardial infarction. *N. Engl. J. Med.* 345:1230–36

131. Vincent JL, Baron JF, Reinhart K, Gattinoni L, Thijs L, et al. 2002. Anemia and blood transfusion in critically ill patients. *J. Am./Med. Assoc.* 288:1499–507

132. Besarab A, Bolton WK, Browne JK, Egrie JC, Nissenson AR, et al. 1998. The effects of normal as compared with low hematocrit values in patients with cardiac disease who are receiving hemodialysis and epoetin. *N. Engl. J. Med.* 339:584–90

133. Saxena R, Wijnhoud AD, Carton H, Hacke W, Kaste M, et al. 1999. Controlled safety study of a hemoglobin-based oxygen carrier, DCLHb, in acute ischemic stroke. *Stroke* 30:993–96

134. Bone RC, Marik PE, Sibbald WJ. 1993. Effect of stored-blood transfusion on

oxygen delivery in patients with sepsis. *J. Am. Med. Assoc.* 269:3024–29

134a. Simchon S, Jan KM, Clien C. 1987. Influence of reduced red cell deformability on regional blood flow. *Am. J. Physiol. Heart Circ. Physiol.* 253:H898–903

135. Ruschitzka FT, Wenger RH, Stallmach T, Quaschning T, deWit C, et al. 2000. Nitric oxide prevents cardiovascular disease and determines survival in polyglobulic mice overexpressing erythropoietin. *Proc. Natl. Acad. Sci. USA* 97:11609–13

136. Casadevall M, Pique JM, Cirera I, Goldin E, Elizalde I, et al. 1996. Increased blood hemoglobin attenuates splanchnic vasodilation in portal-hypertensive rats by nitric oxide inactivation. *Gastroenterology* 110:1156–65

137. Schubert A, O'Hara JF Jr, Przybelski RJ, Tetzlaff JE, Marks KE, et al. 2002. Effect of diaspirin crosslinked hemoglobin (DCLHb HemAssist) during high blood loss surgery on selected indices of organ function. *Artif. Cells Blood Substit. Immobil. Biotechnol.* 30:259–83

138. Alayash AI. 1999. Hemoglobin-based blood substitutes: oxygen carriers, pressor agents, or oxidants? *Nat. Biotechnol.* 17:545–49

139. Padron J, Peiro C, Cercas E, Llergo JL, Sanchez-Ferrer CF. 2000. Enhancement of S-nitrosylation in glycosylated hemoglobin. *Biochem. Biophys. Res. Commun.* 271:217–21

140. Palmerini CA, Saccardi C, Arienti G, Palombari R. 2002. Formation of nitrosothiols from gaseous nitric oxide at pH 7.4. *J. Biochem. Mol. Toxicol.* 16:135–39

141. Palmerini CA, Arienti G, Palombari R. 2004. Electrochemical assay for determining nitrosyl derivatives of human hemglobin: nitrosylhemoglobin and S-nitrosylhemoglobin. *Anal. Biochem.* 330:306–10

142. Chen B, Zhou Y. 1999. Coordinate properties of nitric oxide in hemoglobin

solution containing a minimal amount of nitric oxide. *Tsin. Sci. Tech.* 4:1–6

143. Foster MW, Pawloski JP, Singel DS, Stamler JS, 2004. Role of Circulating S-nitrosothiols in control of blood pressure. *Hypertension.* In press

144. Rao SV, Jollis JG, Harrington RA, Granger CB, Newby LK, et al. 2004. Relationship of blood transfusion and clinical outcomes in patients with acute coronary syndromes. *J. Am. Med. Assoc.* 292:1555–62

Annu. Rev. Physiol. 2005. 67:147–73
doi: 10.1146/annurev.physiol.67.040403.130716
First published online as a Review in Advance on October 25, 2004

RNAi as an Experimental and Therapeutic Tool to Study and Regulate Physiological and Disease Processes

Christopher P. Dillon, Peter Sandy, Alessio Nencioni, Stephan Kissler, Douglas A. Rubinson, and Luk Van Parijs

Center for Cancer Research and Department of Biology, Massachusetts Institute of Technology, Cambridge, Massachusetts 02139; email: cpdillon@mit.edu, psandy@mit.edu, A.nencioni@gmx.net, skissler@mit.edu, drubinso@mit.edu, lukvp@mit.edu

Key Words microRNA, siRNA, shRNA, vector, therapy

■ **Abstract** Over the past four years RNA interference (RNAi) has exploded onto the research scene as a new approach to manipulate gene expression in mammalian systems. More recently, RNAi has garnered much interest as a potential therapeutic strategy. In this review, we briefly summarize the current understanding of RNAi biology and examine how RNAi has been used to study the genetic basis of physiological and disease processes in mammalian systems. We also explore some of the new developments in the use of RNAi for disease therapy and highlight the key challenges that currently limit its application in the laboratory, as well as in the clinical setting.

INTRODUCTION

Mammalian tissue culture and animal models are indispensable tools to study the genetic basis of human physiology and disease. Systematic manipulation of the genetic background by overexpression, deletion, or mutation of genes is the principal method for understanding complex biological processes. Indeed, transgenic, knockout and knockin mice are often the best available in vivo models for human disorders. However, generating these genetically engineered animals requires a significant amount of time, money, and effort. Furthermore, creating a more complex genetic environment with simultaneous gain- and loss-of-function mutations of multiple genes, as is often seen in human diseases, is frequently beyond the reach of these technologies and model systems.

The discovery that long double-stranded RNA molecules (dsRNA) can induce sequence-specific silencing of gene expression in primitive organisms, such as *Caenorhabditis elegans* and *Drosophila melanogaster*, revealed a previously unknown mechanism of gene regulation that is highly conserved throughout

0066-4278/05/0315-0147$14.00

multicellular organisms (1). This process is called RNA interference (RNAi), and is also known as posttranscriptional gene silencing (PTGS) in plants (2). Initial efforts to apply RNAi in mammalian cells were hindered by a potent cellular response to long dsRNAs initiated by the dsRNA-dependent protein kinase (PKR) (3, 4). This stress response has evolved as a defense against viruses and functions to block viral reproduction by halting protein translation and triggering apoptosis of infected cells (5). The breakthrough discovery that dsRNAs of 21 nucleotides in length (termed small interfering RNAs; siRNAs) could trigger sequence-specific gene silencing without inducing the PKR response (6) has opened up revolutionary new approaches to manipulate gene function in mammalian systems.

In the laboratory, RNAi has proven to be a simple, cheap, and powerful tool to generate cells, tissues, or even animals with reduced expression of specific genes. RNAi has been used to interrogate the function of candidate genes and, more recently, following the creation of random and directed siRNA libraries, has permitted phenotype-driven, forward genetic analysis of normal physiological and disease processes. The development of stable and inducible expression vectors driving the expression of short hairpin RNAs (shRNAs) has further expanded the application of RNAi both in tissue culture and in animal models. Furthermore, a number of successful animal trials indicate that RNAi might ultimately become a potent therapeutic approach for the treatment of various human diseases. While the groundwork for using RNAi to manipulate gene expression in animal systems has been laid down, lingering questions about the specificity of the technique and the possibility of inducing adverse cellular responses remain. Future work in this rapidly evolving field will likely address these issues and extend the applications of RNAi both in the laboratory and in the clinic.

THE BIOLOGY OF RNAi

RNAi as a Novel Mechanism that Regulates Development, Normal Physiology, and Disease

The initial description of RNAi resulted from the finding that the introduction of exogenous long dsRNAs in *C. elegans* caused sequence-specific loss of expression of mRNAs (7). Since then RNAi has been observed in most eukaryotes, with the notable exception of *Saccharomyces cerevisiae* (8). The evolutionary conservation of this process is thought to reflect the importance of a class of short noncoding RNAs, termed microRNAs (miRNAs). These RNAs were initially discovered in a screen for genes required for larval development in *C. elegans* (9). More recently, several groups have identified hundreds of miRNAs in species ranging from *C. elegans* to humans through experimental and computational strategies (10–13). A recent study has shown that overexpression of miRNAs alters the development of immune cells in mice (14), indicating that miRNAs are also likely to be critical for normal development and tissue physiology in mammals. Intriguingly, misexpression of miRNAs has been reported in a number of cancers, suggesting that miRNAs may contribute to disease processes as well (15, 16). To date, little is known about the targets of most miRNAs (17). In large part this is because these RNAs show imperfect homology with the mRNAs that they regulate.

Overview of the Biochemistry of RNAi

Significant strides have been made in our understanding of the biochemical mechanisms by which endogenous miRNAs silence gene function. Initially, miRNAs are transcribed as single-stranded precursors up to 2 kb in length that exhibit significant secondary structure owing to the presence of stretches of bases that can undergo extensive base pairing followed by stretches that adopt loop structures (18, 19). Importantly, the pairing regions, or stems, present in miRNAs often contain a small number of mismatched bases that create "bubbles" in the miRNA structure (10–12). Together with loop structures, these are important for the recognition of miRNAs by cellular enzymes and for the ability of these RNAs to silence genes.

Primary miRNA transcripts are identified and processed in the nucleus by the RNase III enzyme, Drosha, into approximately 70-nt-long precursors, known as pre-miRNAs (20). These are exported to the cytoplasm, where they are cleaved by a second RNAse III enzyme, Dicer. Dicer converts miRNAs into double-stranded 21- to 23-nt-long mature miRNAs (21). Mature miRNAs associate with an enzymatic machine known as the RNA-induced silencing complex (RISC) (22). The composition of the RISC is not completely defined but includes Argonaute family proteins (23–28). The RISC unwinds miRNAs and associates stably with the (antisense) strand that is complementary to target mRNA (29). The complete enzymatic machinery required to process miRNAs appears to be expressed in most eukaryotic cells and is essential because mice and zebrafish that lack *dicer1* fail to complete development (30, 31).

Depending on the degree of homology between a miRNA and its target mRNA, miRNA-RISC complexes inhibit gene function by two distinct pathways (17). Most miRNAs pair imperfectly with their targets and silence gene expression by translational repression (32–34). This RNAi mechanism appears to operate most efficiently when multiple miRNA-binding sites are present in the 3'UTR of the target mRNAs (35, 36). In some cases, miRNAs exhibit perfect sequence identity with the target mRNA and inhibit gene function by triggering mRNA degradation (22). As discussed below, this appears to be the dominant mechanism by which synthetic siRNAs and plasmid-expressed shRNAs silence gene expression.

Co-Opting the Endogenous RNAi Machinery for Experimental and Therapeutic Purposes

Much of the success of RNAi as a research and potential therapeutic tool is due to the fact that the enzymatic machinery required to process miRNAs is ubiquitously expressed and can be co-opted by exogenous RNAs to direct sequence-specific gene silencing. In plants and primitive eukaryotic organisms, the sequential activity of Drosha and Dicer converts viral, transgene-encoded, and synthetic dsRNAs into siRNAs that trigger gene silencing (37). To avoid stress responses to long dsR-

NAs, gene silencing in mammalian cells is typically induced experimentally by small exogenous RNAs that enter the RNAi pathway further downstream. Synthetic siRNAs are designed to mimic Dicer products so that they can directly associate with the RISC and target homologous mRNAs for degradation (6). Plasmid-expressed shRNAs are thought to enter the RNAi pathway because they are recognized and cleaved by Dicer into products that bind the RISC (38–40). A recent study suggests that it may even be possible to trigger RNAi in mammalian cells by providing a long dsRNA substrate for Drosha, as long as the dsRNA is confined to the nucleus by removing its 5′ cap structure (41).

RNAi AS AN EXPERIMENTAL TOOL

Discovery and Design of siRNAs for Gene Silencing in Mammalian Systems

Pioneering work by Tuschl and colleagues demonstrated that chemically synthesized short 21-mer dsRNAs (siRNAs) were able to silence genes in a sequence-specific manner when introduced into mammalian cells (6). This exciting observation was quickly reproduced by other groups, and subsequent work rapidly defined the basic structure of siRNAs that were able to induce gene silencing, including the need for a 19-nt RNA duplex with 2-nt overhang on the 3′ ends. Most groups substituted (2′-deoxy)thymidine nucleotides for the overhangs, which did not affect silencing but reduced the cost of the oligonucleotides.

As the use of siRNAs to silence gene expression became more widespread, it became apparent that not all sequences within an mRNA could act as targets for RNAi and that therefore a set of rules was necessary to optimize siRNAs. This contrasts to the situation in primitive organisms in which expression of dsRNAs of 500 base pairs or more typically results in very efficient gene silencing, irrespective of the sequence of the target mRNA (7). Initially the selection of siRNA sequences was largely determined using a limited set of empirical guidelines (6, 42). These included factors such as the GC-content and the region of the mRNA targeted (6, 42). On the basis of these guidelines, most investigators were able to achieve efficient gene silencing with about one quarter of all siRNAs.

An important recent development has been the definition of more effective rational rules for siRNA design. The first breakthrough came from analysis of the biochemistry of RNAi and, in particular, the mechanisms by which siRNAs associate with the RISC complex. These studies revealed that only the antisense strand of the siRNA is incorporated into this enzymatic machinery and that miRNAs and effective siRNAs exhibited decreased stability of the 5′ end of the antisense strand (43, 44). The relevance of this finding to enhancing siRNA efficiency was further underscored by demonstrating that base pair mismatches introduced at the 5′ end of siRNAs improved gene silencing (43, 44). A further step was taken by investigators at Dharmacon, who evaluated a number of characteristics of 180 siRNAs directed against two genes (45). Based on this systematic analysis, a more comprehensive set of criteria was established that significantly increases the likelihood of identifying functional siRNAs (45). It should be noted, however, that highly effective siRNA sequences can be found that do not adhere to these criteria and that some siRNAs that adhere to the criteria do not function well. Therefore, it is likely that future work will continue to define the structure of effective siRNAs.

Stable Induction of RNAi in Mammalian Cells Through Expression of shRNAs

In some organisms, notably *C. elegans*, RNA-dependent RNA polymerases exist that are able to amplify siRNAs and even pass them on through the germ line (46, 47). As a consequence, introduction of dsRNA triggers long-lived, stable gene silencing in this organism (46). These polymerases do not exist in mammalian cells, and consequently, gene silencing induced by siRNAs is limited by the number of RNA molecules introduced into a cell. The number of siRNAs decreases with time by dilution as cells divide and probably also as a consequence of degradation by cellular enzymes. Accordingly, in many cell culture systems, gene silencing is seen for only a few days after siRNAs are administered. On the other hand, siRNA-induced gene silencing has been observed for weeks in slowly proliferating or non-dividing cell types such as macrophages and hepatocytes (48, 49).

The solution to obtaining more universal stable gene silencing through RNAi has been to develop expression systems that stably produce siRNAs. This was first accomplished by several groups (38–40) who used the promoter of either the U6 or H1 splice factor to express a short hairpin RNA (shRNA) that can be processed by Dicer to produce siRNAs. The shRNAs had duplex stems that varied in length from 19 to 29 nt connected by a short loop sequence. The sequence of the loop was critical for effective target silencing; however, several different sequences were shown to be functional. The U6 and H1 promoters recruit RNA polymerase III, a specialized polymerase that is responsible for generating most of the cell's small RNAs (e.g., splicing RNAs and tRNAs) (50). The advantages of using these promoters to create shRNAs are that their transcription initiation site and termination site are well defined and highly conserved and that these promoters are highly active in most, if not all, mammalian cells (50). Importantly, pol III promoter shRNA expression cassettes have been found to be very flexible and have been introduced in a number of different expression systems, including virus-based and those used to create transgenic animals. Several pol III promoters have been extensively characterized and co-opted to express shRNAs. Whether a particular promoter functions significantly better than others in a particular system remains controversial (51–53), and evidence exists that inclusion of an enhancer element may improve their activity (54).

In contrast to promoters for small RNAs, promoters for most cellular genes recruit RNA polymerase II and usually generate transcripts that initiate in a less conserved manner and require a long termination poly (A) sequence. With a few exceptions (41, 55), both features have largely precluded the use of pol II promoters to generate shRNAs. This is not the case for endogenous miRNAs, which are expressed from pol II promoters in a developmentally regulated or tissue-specific manner (18, 19). Furthermore, miRNAs and, possibly, also slightly longer shRNAs may enter the RNAi pathway more efficiently than typical 19-mer shRNAs, potentially leading to more potent gene silencing (14, 40, 56). For these reasons, new stable expression systems for siRNAs that are based on miRNA structures have been developed by a number of groups (14, 56). Expression of a long nuclear-restricted dsRNA in cells and mice has also been shown to silence gene expression effectively (41), suggesting that further improvements and modifications will be made to stable siRNA delivery systems as our understanding of the RNAi machinery improves.

Regulated and Tissue-Specific Gene Silencing by RNAi

Whereas constitutive expression of siRNAs in cells, tissues, and animals has provided important insights into biological processes, the ability to control gene silencing more tightly is likely to significantly extend the applications of experimental RNAi. For this reason, a significant focus has been on the creation of regulated and tissue-specific siRNA delivery systems. To date, most of these systems are based on engineered pol III promoters that are controlled by small molecules or the Cre recombinase. A number of groups have demonstrated that gene silencing can be initiated or inhibited by Cre-driven recombination of modified pol III promoters in cells and in mice and by introduction of DNA elements that bind tetracycline- or ecdysone-regulated transcriptional activators or repressors (see Table 1). Further refinements are likely to result in more effective and flexible RNAi systems to interfere with gene function in mammals. An alternative approach to obtain regulated expression of siRNAs might be to use pol II promoters. With the exception

TABLE 1   In vivo gene silencing in mammals

Silencing	Method	Reference
siRNA or shRNA	Hydrodynamic shock	(48, 69, 70, 146)
	Cationic liposomes/ complexes	(71, 72, 81, 150)
	Peptide conjugation	(82)
	Electroporation	(73–76)
	Adenovirus	(92–94)
Stable	Retrovirus	(58, 86)
	Lentivirus	(84, 85, 87–90)
	Adeno-associated virus	(55, 91, 98, 99)
	Transgenic	(100–102, 104)
Inducible	Tetracycline	(162–167)
	Cre	(168–171)
	Ecdysone	(172)

of the constitutive cytomegalovirus (CMV) promoter (55), this approach has not yet been shown to be successful, although it is likely that by using miRNA-based structures to deliver siRNAs that these promoters will be used to trigger RNAi in a regulated manner in the future.

Off-Target Effects and Interferon Responses: Possible Limitations to the Use of RNAi

A key feature of siRNAs is that they inhibit genes in a highly specific manner. Indeed, the initial description of these RNAs demonstrated that alteration of a single base pair was sufficient to disrupt gene silencing (6). For many subsequent studies with siRNAs, this specificity was shown to hold true (57, 58). A number of more recent studies have suggested that there are situations where mismatches between the siRNA and target sequence can be tolerated (59). This observation has raised the concern that siRNAs may have effects on genes that are not considered targets, so-called off-target effects. This concern has been addressed by a number of groups that have examined genome-wide changes in gene expression following the introduction of siRNAs. Some of these studies found that a number of genes unrelated to the target are changed in expression, mostly by a factor of twofold. These off-target effects have been correlated with the concentration of siRNAs (60), as well as similarities between the off-target transcripts and the 5' ends of siRNAs (61, 62). It seems plausible on the basis of recent work that the decreased off-target mRNA levels are the consequence of siRNAs adopting miRNA-like properties, resulting in slightly decreased levels of mRNAs, possibly through alterations in mRNA stability (61). It remains to be determined whether translational inhibition is seen on these off-target mRNAs.

Currently it is not possible to predict whether a particular mammalian siRNA will induce off-target effects. It is widely assumed that the most informative parameter will be the degree of homology between siRNAs and other gene products, and therefore most experimental siRNAs are designed to have no known perfect matches with mRNAs other than the intended target (62a). It is interesting that off-target effects are not observed when dsRNAs are used in primitive organisms. This may be because the dominant species of siRNAs generated from these dsRNAs are selected by Drosha and Dicer and other components of the endogenous RNAi machinery, which might have a proofreading activity that guards endogenous genes from silencing. Thus it is possible that mammalian siRNAs' generated from dsRNAs' precursors through the action of Drosha and Dicer may be less prone to induce off-target effects.

A second major concern among researchers using RNAi in mammals is the possibility that introducing exogenous dsRNAs may trigger an antiviral interferon response mediated by the PKR. Indeed, many early attempts at silencing gene expression using dsRNAs strategies analogous to those developed for primitive organisms failed because they triggered the production of interferon, nonspecific gene silencing, and apoptosis in mammalian cells (63). Early work by Tuschl and colleagues suggested that dsRNAs that were less than 30 bases in length were able to silence gene expression in a specific manner, while eluding the molecular machinery responsible for triggering the interferon response (6). This finding has been corroborated by the successful use of siRNAs as reagents to interfere specifically with gene function in a wider variety of different mammalian systems. However, a number of recent studies suggest that even short dsRNAs can trigger the expression of some of the target genes of the interferon response and, in some cases, can induce the cellular changes associated with this process (64–66). As of now, it is not clear how often siRNAs and shRNAs trigger the interferon pathway and which conditions favor this response to these RNAs. Chemical features of dsRNAs, as well as their expression levels and delivery routes, may determine whether they become visible to the interferon response machinery (64–66).

Similar to the interferon response, evidence exists that siRNAs and shRNAs can activate dendritic cells and other cells of the immune system through a much more specific and restricted class of receptors, the Toll-like receptors (TLRs), that can recognize foreign nucleic acids including dsRNAs (67, 68). While the consequence of this remains to be determined, these findings do raise the possibility that RNAi reagents may trigger adverse immune responses in vivo.

USE OF RNAi TO STUDY NORMAL TISSUE PHYSIOLOGY AND DISEASE IN ANIMAL MODELS

In Vivo Delivery of siRNAs to Induce RNAi

By allowing efficient and cheap silencing of gene expression, RNAi promises to provide a significant boost to research of the genetic basis of normal tissue physiology, as well as disease processes in animal models. For this reason, many groups have worked on developing strategies to deliver siRNAs or shRNAs to cells and tissues of experimental animals (Table 1). Early efforts focused on direct administration of synthetic siRNAs, and three major delivery methods have been shown to be successful. The first of these, intravenous injection of siRNAs in a large volume (1 ml) of saline solution, works by creating a back-flow in the venal system that forces the siRNA solution into several organs (mainly the liver, but also kidneys and lung with lesser efficiency) (69, 70).

Gene silencing has also been achieved in vivo by injecting smaller volumes of siRNAs that are packaged in cationic liposomes. When siRNAs are administered intravenously using this strategy, silencing is primarily seen in highly perfused tissues, such as the lung, liver, and spleen (71). Local delivery of siRNAs has been shown to be successful in the central nervous system (72). Finally, gene silencing has also been achieved by electroporation of siRNA duplexes directly into target tissues and organs, including muscle, retina, and the brain (73–76).

Although successful, it is likely that these strategies to silence genes are limited by the stability of siRNAs molecules in vivo and the efficiency with which they are taken up by target cells and tissues. Much effort has been directed to increasing the half-life of the siRNAs by modifying the chemistry of the RNAs used (77–80).

A number of groups have also used plasmid-based shRNAs, instead of siRNAs, to obtain relatively long-lived gene silencing in vivo (81). A number of approaches have also been shown to improve cell and tissue delivery of siRNAs and shRNAs, including conjugating RNAs to membrane-permeant peptides and by incorporating specific binding reagents such as monoclonal antibodies into liposomes used to encapsulate siRNAs (81, 82).

Use of Viral Vectors to Induce RNAi in Primary Cells, Tissues and Experimental Animals

To obtain efficient and long-lived gene silencing using RNAi in cells and tissues, many groups have developed a variety of viral vectors to deliver siRNAs both in vitro and in vivo (Table 1). Retrovirus-based vectors that permit stable introduction of genetic material into cycling cells (83) have been engineered to express shRNAs and to trigger RNAi in transformed cells, as well as in primary cells (58, 84–91). Because they infect and are expressed in certain adult stem cells, notably hematopoietic stem cells, retrovirus-based vectors have also been used to create "knockdown" tissues in mice (86).

Even more wide-ranging applications of RNAi have been reported using recombinant lentiviral vectors (Table 1), because these permit infection of noncycling and postmitotic cells such as neurons (84, 85, 87, 91). Lentiviral RNAi vectors have even been used to generate transgenic knockdown animals by infecting embryonic stem cells or single-cell embryos (87, 89). These animals display expected loss-of-function phenotypes and transmit the RNAi vector to their offspring, suggesting that this technique represents an efficient, low-cost alternative to knockout technologies to study normal tissue physiology and disease processes in a variety of experimental animal systems (87, 89).

Highly effective siRNA delivery systems have also been created that are based on adenoviruses and adenovirus-associated viruses (AAV) (Table 1). Adenoviruses can infect a wide range of cells and have been shown to silence gene expression in vivo (55, 92–94). However, they do not integrate into the genome and tend to induce strong immune responses, which may limit their use in some circumstances. In contrast, AAV does not cause disease in humans (95) and can integrate into the genome of infected cells. Unlike retroviruses and lentiviruses, AAV tends to integrate at a defined location in the genome, thus minimizing the chance of a mutagenic effect of the integrated virus (96, 97). Effective gene silencing mediated by AAV-based vectors has been demonstrated following systemic or tissue-specific injection of viral particles (55, 98, 99).

Creation of Transgenic Animal Models Using RNAi

In addition to the use of lentiviral vectors (87, 89), more traditional transgenesis strategies have been used to successfully create loss-of-function models to study gene function in rodents using RNAi, thus providing another strategy by which RNAi might provide an alternative to creating gene knockout animals

(41, 100–102). Inheritable RNAi transgenesis has been achieved both through expression of shRNAs and long dsRNAs whose expression is restricted to the nucleus or oocyte (41, 100–102). On the basis of this small number of pioneering studies, it appears that RNAi is effective at silencing gene expression in many, if not all, tissues (100). In some instances, attempts to create RNAi transgenic animals by injection of plasmids encoding shRNAs into single cell embryos have been unsuccessful, whereas injection of DNA into blastocysts has succeeded (103). This may reflect a toxic effect of overexpression of siRNAs during early development, possibly due to competition with miRNA pathways. How significant an impediment this might be for the generation of RNAi transgenic animals remains to be determined.

Whereas RNAi in transgenic animals has been shown to recapitulate some loss-of-function phenotypes established in knockout animals, there is mounting evidence that the RNAi phenotype will often appear more variegated than the knockout phenotype (87, 104). This is probably due to the fact that RNAi does not abrogate gene expression but rather reduces it to varying levels. Although this may in some cases limit the use of RNAi in vivo, it is also likely to provide important new insights into the genetic basis of normal tissue physiology, disease processes, and therapeutic strategies by demonstrating the effects of altering gene expression to varying degrees. In particular, RNAi may prove especially important in the creation of animal models of human diseases in which susceptibility and resistance are encoded by alleles that show relative, rather than absolute, differences in expression levels (86).

GENETIC SCREENS USING RNAi

Most researchers use RNAi as a simple reverse genetics tool to understand the function of one or a few genes. On the other hand, RNAi represents an ideal strategy to decipher the role of hundreds or thousands of genes simultaneously in screens for specific phenotypic changes. These forward genetic approaches may help to gain insight into complex physiological and pathological processes.

RNAi Screens in Lower Organisms

The discovery of RNAi and sequencing of the genome of popular model organisms such as *C. elegans* and *D. melanogaster* provided the impetus for functional genetic screens (Table 2). The first such screens were performed in *C. elegans* and focused on easily detectable phenotypic changes, such as viability and sterility, and identified the biological role of a few hundred genes located on chromosomes I and III (105, 106). These studies were significantly facilitated by the finding that RNAi can be induced simply by feeding this worm with bacteria overexpressing dsRNA molecules of interest (107, 108). More recently, genome-wide RNAi screens targeting ∼90% of the predicted transcripts have led to the functional annotation of an additional ∼2000 genes (109, 110). Importantly, these and other RNAi-based screens have identified the role of genes that have conserved

TABLE 2   Genetic screens using RNAi

Model organism/ system	Delivery method/ type of RNAi molecule	Number of targeted genes	Phenotypic assay/endpoint	References
C. elegans	Feeding with bacteria overexpressing dsRNA library	~17,000 (~90% of known or predicted transcripts)	Viability, sterility, embryogenesis, fat metabolism, genomic stability, mitochondrial function, life span	(105, 106, 109, 111, 112, 114, 173, 174)
D. melanogaster	Soaking cultured cells in dsRNA-containing medium, dsRNAs are typically synthesized by in vitro transcription	~20,000 (~90% of known or predicted transcripts)	Viability, growth, cell morphology, various signaling pathways, innate immune response	(116–119, 123, 124)
Cultured mammalian cells	Transfected siRNA library (chemically synthesized)	510 (1 siRNA per gene)	TRAIL-induced apoptosis	(125)
	Plasmid shRNA library (transfected)	50 (4 shRNAs per gene)	NF-κB signaling pathway	(126)
	Transfected siRNA library (expressed from PCR products, synthesized in vivo)	~8000 (2 siRNAs per gene)	NF-κB signaling pathway	(129)
	Retroviral shRNA library	~8000 (3 shRNAs per gene)	p53-dependent proliferation arrest	(127)
	Retroviral shRNA library	~5000 (3–9 shRNAs per gene)	26S proteasome function	(128)

orthologs in mice and humans and are involved in processes as diverse as genomic stability, fat metabolism, mitochondrial function, and embryogenesis (111–114). They have also provided evidence that RNAi pathways are under genetic control in this organism (115), a discovery that may have significant impact on the use of RNAi both in basic research and in clinical applications.

RNAi can also be induced with ease in cells derived from fruit flies by adding dsRNAs to the culture medium (3). RNAi-based genetic screens have been performed to systematically explore important signal transduction cascades, such as the Wnt-, Hh-, PI3K- and MAPK-pathways, and more broadly to examine the function of most known kinases and phosphatases (116–118). Other high throughput screens identified genes involved in the regulation of cell shape, cytokinesis, and phagocytosis, as well as in heart development and innate immunity (119–123). Recently, Boutros and colleagues targeted almost all of the predicted *Drosophila* mRNAs to analyze their possible roles in cell growth and viability (124).

RNAi Screens in Mammalian Systems

Soon after the basic requirements to create successful RNAi reagents for mammalian systems were established, multiple groups started to assemble si/shRNA libraries that targeted increasingly larger numbers of genes. These have been successfully applied to study a variety of cellular processes. Initial studies examined the function of a relatively small set of genes in specific signaling or metabolic pathways (125, 126).

Very recently, much larger RNAi libraries attempting genome-wide coverage have been created (Table 2). Generating such libraries and delivering them into mammalian cells has presented real challenges that have been solved in part by using strategies borrowed from technical breakthroughs in the fields of gene sequencing and chemical genetics. Recent reports from Berns et al. and Paddison et al. document the results of the first such large-scale RNAi-based genetic screens in mammalian cells using retrovirus-based vectors to express shRNAs (127, 128). A notable detail of the strategy used by these workers is the inclusion of short unique DNA sequences or the use of the ~60-nt-long shRNA coding sequences as "molecular bar codes" in RNAi vectors. Both groups showed encouraging results that these sequences could be detected using high-density oligonucleotide arrays and thus could circumvent the need to identify shRNAs introduced into cells by sequencing. Zheng and colleagues have also performed a large-scale RNAi-based screen in tissue culture cells using a PCR-based approach to generate siRNAs in vivo (129).

The large-scale genetic screens have successfully identified a number of novel genetic elements of the interrogated signaling or metabolic pathways. On the other hand, certain known components of these pathways were not identified even if the respective targeting constructs were present in the applied RNAi libraries. This discrepancy may be explained by the fact that sometimes not all pathway components can be assessed by the same phenotypic screen. Moreover, despite careful design and use of multiple targeting sequences, some genes may not be effectively silenced to produce the assayed phenotype. However, functional validation of each si/shRNA molecule represents an even bigger challenge than the construction of the library. The recently reported high-throughput methods are based on overexpression of the target gene in fusion with a reporter construct (typically green fluorescent protein; GFP), which require in-frame cloning of the coding sequences of the targeted genes and data handling of hundreds of gigabytes of visual images (130, 131). Therefore, these methods are laborious and expensive and do not allow for the selection of si/shRNA molecules that target untranslated regions of the gene. Currently, other strategies are under development that may surpass the need of in-frame cloning and would also make possible the effective selection of dsRNAs targeting noncoding sequences (P. Sandy, A. Ventura, & T. Jacks, manuscript in preparation).

RNAi should also provide an efficient approach to systematically investigate the genetic basis of normal and disease physiology in animal models. Although RNAi screens performed in vitro have identified a number of physiologically or pathologically relevant genes, these models can often be biased by specific genetic background and/or environmental conditions. Furthermore, many complex physiological or disease processes have no in vitro correlates, such as organ development or metastasis. These limitations will likely be overcome by performing RNAi screens in vivo once appropriate strategies have been developed.

RNAi IN HUMAN THERAPY

RNAi as a Therapeutic Strategy

The potency and flexibility exhibited by RNAi in experimental systems has stimulated efforts to use RNAi-based reagents in the clinic as "molecular targeting" therapeutics to shut down disease-associated genes in humans (132–134). In theory, therapeutic RNAi should be able to alter the expression and function of genes in a wide range of disease settings. Indeed, RNAi has been used to silence the expression of exogenous disease-causing genes, such as those of pathogens, as well as endogenous genes that play an essential role in the disease process (132) (Table 3). The high degree of specificity of RNAi has even been able to distinguish between alleles of genes exhibiting spontaneous or inherited polymorphisms and alternative splicing events that underlie the development of cancer and other diseases (58, 135) (Table 3).

These findings suggest that the expression of disease-associated genes could be inhibited by RNAi-based reagents. These reagents have been shown to be effective as potential therapeutics in a variety of tissue culture and animal preclinical model systems, including those for cancerous disorders, microbial infections, autoimmune and inflammatory disease, and neurological disorders.

Different siRNAs have recently been described that effectively silence cancer-related genes. These include mutated Ras, Bcr-Abl, and vascular endothelial growth factor (VEGF); the focal adhesion kinase (FAK); Bcl-2; MDR-1; human papillomavirus (HPV) E6 and E7 proteins; CDK-2; MDM-2; PKC-α and β; and TGF-β1 (53, 58, 136–144). Studies in the mouse have shown that injection of siRNAs alone or in combination with anticancer drugs is able to promote apoptosis and reduce

TABLE 3   Exogenous and endogenous disease-associated genes successfully targeted by RNAi

Type of disease	Target	Reference
Viral diseases	HIV (viral genes-genome)	(98, 151, 152, 175–184)
	HIV (cellular receptors/enzymes)	(184–189)
	HBV	(190–192)
	HCV	(69, 147, 148, 193–195)
	HDV	(196)
	Cytomegalovirus (CMV)	(197)

(Continued)

TABLE 3   *(Continued)*

Type of disease	Target	Reference
	Influenza virus	(145)
	Rhinovirus	(198)
	SARS coronavirus	(149)
	Prions	(156)
	Gamma herpes virus	(199)
Autoimmune/inflammatory disorders	TNF-α	(71)
	Fas/CD95/Apo1	(48)
	Caspase-8	(154)
Neurological diseases	Mutated SOD (amyotrophic lateral sclerosis)	(135, 200)
	BACE1 (Alzheimer's disease)	(155)
	SCCMS (myastenic disorders)	(201)
	Polyglutamine proteins	(55)
Cancer/malignant hyperproliferative disorders	Bax	(202)
	CXCR4	(163)
	Focal adhesion kinase (FAK)	(140)
	EphA2	(203)
	Matrix metalloproteinase	(204)
	AML1/MTG8	(205)
	BCR-Abl	(136, 137)
	BRAF(V599E)	(206)
	Brk	(207)
	Epstein-Barr virus (EBV)	(208)
	EGFR	(209, 210)
	Fatty acid synthase (FASE)	(211)
	HPV E6	(143)
	Livin/ML-IAP/KIAP	(212)
	MDR	(142)
	BCL-2	(138, 213)
	CDK-2	(138)
	MDM-2	(138)
	PKC-α	(138)
	TGF-β	(138)
	H-Ras	(138)
	K-Ras	(58)
	VEGF	(53, 138)
	PLK1	(214)
	Telomerase	(215)
	S100A10	(216)
	STAT3	(217)
	NPM-ALK	(218)

tumor burden (139, 140). This provides proof of principle that RNAi for cancer-related genes is feasible in vivo and may prove to be helpful in cancer therapy.

The possibility of using siRNAs to combat infections, especially by viruses, has also been extensively explored recently. Many genes from important human viral pathogens, including HIV, HBV, HCV, influenza virus, and SARS cornavirus, have been shown to be targets for RNAi (69, 134, 145, 146). Inhibiting these viral genes has been shown to interfere with viral replication in vitro (145, 147–149) and in mouse models of viral infection (69, 146, 150). Because many viruses, notably HIV, exhibit high mutation rates (151, 152), RNAi-based therapeutic strategies have also been explored that target host genes that are required for viral entry into cells (134, 153) or that contribute to the pathogenic sequelae of virus infection (154). Similar approaches have also been shown to be successful at modulating inflammatory gene expression in experimental models of immune-mediated diseases (48, 71).

RNAi-based therapeutics can also selectively target mutant forms of genes that underlie the development of neurodegenerative disorders. Inhibition by RNAi of genes encoding proteins involved in polyglutamine-induced neurological disorders (spinocerebellar ataxia type 1 and Huntington's disease) (55, 99), Alzheimer's disease (155), amyotrophic lateral sclerosis (135), and prion-based diseases (156) has been shown and may represent a promising therapeutic strategy.

Delivery Routes for RNAi-Based Therapeutics

Perhaps the most significant barrier for RNAi-based therapy is the efficient and effective delivery of RNAi reagents in patients. As discussed above, a number of strategies have been developed that allow siRNAs and shRNAs to be delivered effectively in animals. Hydrodynamic delivery of siRNAs that involves the intravascular injection of large fluid volumes in order to locally increase intravascular pressure (48, 146) might be adapted for local administration of siRNAs by arterial or venous catheterism in organs, such as liver, kidney, heart or lungs, but cannot be performed for systemic treatment. In the mouse, effective silencing of genes in tumor tissues has been reported following intravenous, intraperitoneal, and subcutaneous injections of siRNAs, suggesting that effective delivery of RNAi reagents may be achieved by different parenteral routes (139).

A number of carrier systems and chemical modifications have been explored to enhance the efficiency and specificity of RNAi-based therapeutics. The use of cationic lipids has been shown to significantly enhance gene silencing (71, 157–159). More sophisticated strategies, in which receptor-specific monoclonal antibodies or other targeting proteins are incorporated into pegylated immunoliposomes (PILs), have been shown to direct gene silencing in a number of tissues, including the brain (81).

Virus-based RNAi delivery systems have also been shown to achieve effective gene silencing in vivo. Systemic or tissue-directed injection of adenoviruses encoding shRNAs has been shown to be effective at inhibiting gene expression in the liver, as well as the central nervous system (55, 91). Retroviruses for RNAi could potentially be applied for ex vivo cellular manipulations, including those of dendritic cells for the modulation of immune responses (160). However, the use of these vectors may be associated with a risk of insertional mutagenesis and should be carefully evaluated (161) Other virus-based systems for RNAi, such as those based on AAVs, are also being considered as delivery vehicles for therapeutic RNAi (see above).

SUMMARY AND PROSPECTS

In the past few years, RNAi has come to prominence as a novel and essential biological process, as well as a powerful experimental tool and a potential therapeutic strategy. New discoveries in the field of RNAi biochemistry, coupled with technological breakthroughs, have permitted the creation of effective RNAi reagents that can be used to study normal tissue physiology and disease processes in a range of settings, including experimental animals. By further exploring the biology of RNAi and improving delivery and evaluation technologies for RNAi reagents, these strategies will become more effective and more generally available. Now that the first phase I clinical studies of RNAi are on the horizon, several questions related to the safety and efficacy of using RNAi as a therapeutic strategy must be addressed. Ongoing and future preclinical studies in animal models will hopefully help optimize RNAi therapeutics for applications in humans.

ACKNOWLEDGMENTS

RNAi is a rapidly developing field of research, and as such, we apologize to all the authors whose work could not be cited because of production deadlines and space limitations. C.P.D. is a Howard Hughes Predoctoral Fellow. D.A.R. is supported by a fellowship from the Ludwig foundation. S.K. is supported by a Systems Biology Knowledge Integration Communities grant from the Cambridge MIT initiative (CMI).

**The *Annual Review of Physiology* is online at
http://physiol.annualreviews.org**

LITERATURE CITED

1. Sharp PA. 1999. RNAi and double-strand RNA. *Genes Dev.* 13:139–41
2. Baulcombe DC. 1999. Fast forward genetics based on virus-induced gene silencing. *Curr. Opin. Plant Biol.* 2:109–13
3. Caplen NJ, Fleenor J, Fire A, Morgan RA. 2000. dsRNA-mediated gene silencing in cultured *Drosophila* cells: a tissue culture model for the analysis of RNA interference. *Gene.* 252:95–105
4. Ui-Tei K, Zenno S, Miyata Y, Saigo K. 2000. Sensitive assay of RNA interference in *Drosophila* and Chinese hamster cultured cells using firefly luciferase gene as target. *FEBS. Lett.* 479:79–82

5. Stark GR, Kerr IM, Williams BR, Silverman RH, Schreiber RD. 1998. How cells respond to interferons. *Annu. Rev. Biochem.* 67:227–64

6. Elbashir SM, Harborth J, Lendeckel W, Yalcin A, Weber K, Tuschl T. 2001. Duplexes of 21-nucleotide RNAs mediate RNA interference in cultured mammalian cells. *Nature* 411:494–98

7. Fire A, Xu S, Montgomery MK, Kostas SA, Driver SE, Mello CC. 1998. Potent and specific genetic interference by double-stranded RNA in *Caenorhabditis elegans*. *Nature* 391:806–11

8. Hutvagner G, Zamore PD. 2002. RNAi: nature abhors a double-strand. *Curr. Opin. Genet. Dev.* 12:225–32

9. Lee RC, Feinbaum RL, Ambros V. 1993. The *C. elegans* heterochronic gene *lin-4* encodes small RNAs with antisense complementarity to *lin-14*. *Cell* 75:843–54

10. Lagos-Quintana M, Rauhut R, Lendeckel W, Tuschl T. 2001. Identification of novel genes coding for small expressed RNAs. *Science* 294:853–58

11. Lau NC, Lim LP, Weinstein EG, Bartel DP. 2001. An abundant class of tiny RNAs with probable regulatory roles in *Caenorhabditis elegans*. *Science* 294:858–62

12. Lee RC, Ambros V. 2001. An extensive class of small RNAs in *Caenorhabditis elegans*. *Science* 294:862–64

13. Lim LP, Glasner ME, Yekta S, Burge CB, Bartel DP. 2003. Vertebrate microRNA genes. *Science* 299:1540

14. Chen CZ, Li L, Lodish HF, Bartel DP. 2004. MicroRNAs modulate hematopoietic lineage differentiation. *Science* 303:83–86

15. Calin GA, Dumitru CD, Shimizu M, Bichi R, Zupo S, et al. 2002. Frequent deletions and down-regulation of microRNA genes miR15 and miR16 at 13q14 in chronic lymphocytic leukemia. *Proc. Natl. Acad. Sci. USA* 99:15524–29

16. Calin GA, Sevignani C, Dumitru CD, Hyslop T, Noch E, et al. 2004. Human microRNA genes are frequently located at fragile sites and genomic regions involved in cancers. *Proc. Natl. Acad. Sci. USA* 101:2999–3004

17. Bartel DP. 2004. MicroRNAs: genomics, biogenesis, mechanism, and function. *Cell* 116:281–97

18. Lagos-Quintana M, Rauhut R, Yalcin A, Meyer J, Lendeckel W, Tuschl T. 2002. Identification of tissue-specific micro RNAs from mouse. *Curr. Biol.* 12:735–39

19. Lee Y, Jeon K, Lee JT, Kim S, Kim VN. 2002. MicroRNA maturation: stepwise processing and subcellular localization. *EMBO J.* 21:4663–70

20. Lee Y, Ahn C, Han J, Choi H, Kim J, et al. 2003. The nuclear RNase III Drosha initiates microRNA processing. *Nature* 425:415–19

21. Bernstein E, Caudy AA, Hammond SM, Hannon GJ. 2001. Role for a bidentate ribonuclease in the initiation step of RNA interference. *Nature* 409:363–66

22. Hutvagner G, Zamore PD. 2002. A microRNA in a multiple-turnover RNAi enzyme complex. *Science* 297:2056–60

23. Williams RW, Rubin GM. 2002. ARGONAUTE1 is required for efficient RNA interference in *Drosophila* embryos. *Proc. Natl. Acad. Sci. USA* 99:6889–94

24. Hammond SM, Boettcher S, Caudy AA, Kobayashi R, Hannon GJ. 2001. Argonaute2, a link between genetic and biochemical analyses of RNAi. *Science* 293:1146–50

25. Sasaki T, Shiohama A, Minoshima S, Shimizu N. 2003. Identification of eight members of the Argonaute family in the human genome small star, filled. *Genomics* 82:323–30

26. Tabara H, Sarkissian M, Kelly WG, Fleenor J, Grishok A, et al. 1999. The *rde-1 gene*, RNA interference, and transposon silencing in *C. elegans*. *Cell* 99:123–32

27. Morel JB, Godon C, Mourrain P, Beclin C, Boutet S, et al. 2002. Fertile hypomorphic ARGONAUTE (ago1) mutants

impaired in post-transcriptional gene silencing and virus resistance. *Plant Cell* 14:629–39

28. Mourelatos Z, Dostie J, Paushkin S, Sharma A, Charroux B, et al. 2002. miRNPs: a novel class of ribonucleoproteins containing numerous microRNAs. *Genes Dev.* 16:720–28

29. Martinez J, Patkaniowska A, Urlaub H, Luhrmann R, Tuschl T. 2002. Single-stranded antisense siRNAs guide target RNA cleavage in RNAi. *Cell* 110:563–74

30. Bernstein E, Kim SY, Carmell MA, Murchison EP, Alcorn H, et al. 2003. Dicer is essential for mouse development. *Nat. Genet.* 35:215–17

31. Wienholds E, Koudijs MJ, van Eeden FJ, Cuppen E, Plasterk RH. 2003. The microRNA-producing enzyme Dicer1 is essential for zebrafish development. *Nat. Genet.* 35:217–18

32. Grishok A, Pasquinelli AE, Conte D, Li N, Parrish S, et al. 2001. Genes and mechanisms related to RNA interference regulate expression of the small temporal RNAs that control *C. elegans* developmental timing. *Cell* 106:23–34

33. Hutvagner G, McLachlan J, Pasquinelli AE, Balint E, Tuschl T, Zamore PD. 2001. A cellular function for the RNA-interference enzyme Dicer in the maturation of the let-7 small temporal RNA. *Science* 293:834–38

34. Ketting RF, Fischer SE, Bernstein E, Sijen T, Hannon GJ, Plasterk RH. 2001. Dicer functions in RNA interference and in synthesis of small RNA involved in developmental timing in *C. elegans. Genes Dev.* 15:2654–59

35. Bartel DP, Chen CZ. 2004. Micromanagers of gene expression: the potentially widespread influence of metazoan microRNAs. *Nat. Rev. Genet.* 5:396–400

36. Doench JG, Petersen CP, Sharp PA. 2003. siRNAs can function as miRNAs. *Genes Dev.* 17:438–42

37. Dykxhoorn DM, Novina CD, Sharp PA. 2003. Killing the messenger: short RNAs that silence gene expression. *Nat. Rev. Mol. Cell. Biol.* 4:457–67

38. Brummelkamp TR, Bernards R, Agami R. 2002. A system for stable expression of short interfering RNAs in mammalian cells. *Science* 296:550–53

39. Sui G, Soohoo C, Affar el B, Gay F, Shi Y, Forrester WC. 2002. A DNA vector-based RNAi technology to suppress gene expression in mammalian cells. *Proc. Natl. Acad. Sci. USA* 99:5515–20

40. Paddison PJ, Caudy AA, Hannon GJ. 2002. Stable suppression of gene expression by RNAi in mammalian cells. *Proc. Natl. Acad. Sci. USA* 99:1443–48

41. Shinagawa T, Ishii S. 2003. Generation of Ski-knockdown mice by expressing a long double-strand RNA from an RNA polymerase II promoter. *Genes Dev.* 17:1340–45

42. Tuschl T, Zamore PD, Lehmann R, Bartel DP, Sharp PA. 1999. Targeted mRNA degradation by double-stranded RNA in vitro. *Genes Dev.* 13:3191–97

43. Schwarz DS, Hutvagner G, Du T, Xu Z, Aronin N, Zamore PD. 2003. Asymmetry in the assembly of the RNAi enzyme complex. *Cell* 115:199–208

44. Khvorova A, Reynolds A, Jayasena SD. 2003. Functional siRNAs and miRNAs exhibit strand bias. *Cell* 115:209–16

45. Reynolds A, Leake D, Boese Q, Scaringe S, Marshall WS, Khvorova A. 2004. Rational siRNA design for RNA interference. *Nat. Biotechnol.* 22:326–30

46. Sijen T, Fleenor J, Simmer F, Thijssen KL, Parrish S, et al. 2001. On the role of RNA amplification in dsRNA-triggered gene silencing. *Cell* 107:465–76

47. Smardon A, Spoerke JM, Stacey SC, Klein ME, Mackin N, Maine EM. 2000. EGO-1 is related to RNA-directed RNA polymerase and functions in germ-line development and RNA interference in *C. elegans. Curr. Biol.* 10:169–78

48. Song E, Lee SK, Wang J, Ince N, Ouyang N, et al. 2003. RNA interference targeting

Fas protects mice from fulminant hepatitis. *Nat. Med.* 9:347–51

49. Song E, Lee SK, Dykxhoorn DM, Novina C, Zhang D, et al. 2003. Sustained small interfering RNA-mediated human immunodeficiency virus type 1 inhibition in primary macrophages. *J. Virol.* 77:7174–81

50. Paule MR, White RJ. 2000. Survey and summary: transcription by RNA polymerases I and III. *Nucleic Acids Res.* 28:1283–98

51. Boden D, Pusch O, Lee F, Tucker L, Shank PR, Ramratnam B. 2003. Promoter choice affects the potency of HIV-1 specific RNA interference. *Nucleic Acids Res.* 31:5033–38

52. Kawasaki H, Taira K. 2003. Short hairpin type of dsRNAs that are controlled by tRNA(Val) promoter significantly induce RNAi-mediated gene silencing in the cytoplasm of human cells. *Nucleic Acids Res.* 31:700–7

53. Zhang L, Yang N, Mohamed-Hadley A, Rubin SC, Coukos G. 2003. Vector-based RNAi, a novel tool for isoform-specific knock-down of VEGF and anti-angiogenesis gene therapy of cancer. *Biochem Biophys. Res. Commun.* 303:1169–78

54. Xia XG, Zhou H, Ding H, Affar el B, Shi Y, Xu Z. 2003. An enhanced U6 promoter for synthesis of short hairpin RNA. *Nucleic Acids Res.* 31:e100

55. Xia H, Mao Q, Paulson HL, Davidson BL. 2002. siRNA-mediated gene silencing in vitro and in vivo. *Nat. Biotechnol.* 20:1006–10

56. Zeng Y, Wagner EJ, Cullen BR. 2002. Both natural and designed micro RNAs can inhibit the expression of cognate mRNAs when expressed in human cells. *Mol. Cell.* 9:1327–33

57. McManus MT, Haines BB, Dillon CP, Whitehurst CE, van Parijs L, et al. 2002. Small interfering RNA-mediated gene silencing in T lymphocytes. *J. Immunol.* 169:5754–60

58. Brummelkamp TR, Bernards R, Agami R. 2002. Stable suppression of tumorigenicity by virus-mediated RNA interference. *Cancer Cell* 2:243–47

59. Pusch O, Boden D, Silbermann R, Lee F, Tucker L, Ramratnam B. 2003. Nucleotide sequence homology requirements of HIV-1-specific short hairpin RNA. *Nucleic Acids Res.* 31:6444–49

60. Persengiev SP, Zhu X, Green MR. 2004. Nonspecific, concentration-dependent stimulation and repression of mammalian gene expression by small interfering RNAs (siRNAs). *RNA* 10:12–18

61. Jackson AL, Bartz SR, Schelter J, Kobayashi SV, Burchard J, et al. 2003. Expression profiling reveals off-target gene regulation by RNAi. *Nat. Biotechnol.* 21:635–37

62. Saxena S, Jonsson ZO, Dutta A. 2003. Small RNAs with imperfect match to endogenous mRNA repress translation. Implications for off-target activity of small inhibitory RNA in mammalian cells. *J. Biol. Chem.* 278:44312–19

62a. Mittal V. 2004. Improving the efficiency of RNA interference in mammals. *Nat. Rev. Genet.* 5:355–65

63. Zamore PD. 2001. RNA interference: listening to the sound of silence. *Nat. Struct. Biol.* 8:746–50

64. Bridge AJ, Pebernard S, Ducraux A, Nicoulaz AL, Iggo R. 2003. Induction of an interferon response by RNAi vectors in mammalian cells. *Nat. Genet.* 34:263–64

65. Kim DH, Longo M, Han Y, Lundberg P, Cantin E, Rossi JJ. 2004. Interferon induction by siRNAs and ssRNAs synthesized by phage polymerase. *Nat. Biotechnol.* 22:321–25

66. Sledz CA, Holko M, de Veer MJ, Silverman RH, Williams BR. 2003. Activation of the interferon system by short-interfering RNAs. *Nat. Cell Biol.* 5:834–39

67. Alexopoulou L, Holt AC, Medzhitov R, Flavell RA. 2001. Recognition of

double-stranded RNA and activation of NF-kappaB by Toll-like receptor 3. *Nature* 413:732–38

68. Kariko K, Bhuyan P, Capodici J, Weissman D. 2004. Small interfering RNAs mediate sequence-independent gene suppression and induce immune activation by signaling through Toll-like receptor 3. *J. Immunol.* 172:6545–49

69. McCaffrey AP, Meuse L, Pham TT, Conklin DS, Hannon GJ, Kay MA. 2002. RNA interference in adult mice. *Nature* 418:38–39

70. Lewis DL, Hagstrom JE, Loomis AG, Wolff JA, Herweijer H. 2002. Efficient delivery of siRNA for inhibition of gene expression in postnatal mice. *Nat. Genet.* 32:107–8

71. Sorensen DR, Leirdal M, Sioud M. 2003. Gene silencing by systemic delivery of synthetic siRNAs in adult mice. *J. Mol. Biol.* 327:761–66

72. Baker-Herman TL, Fuller DD, Bavis RW, Zabka AG, Golder FJ, et al. 2004. BDNF is necessary and sufficient for spinal respiratory plasticity following intermittent hypoxia. *Nat. Neurosci.* 7:48–55

73. Kishida T, Asada H, Gojo S, Ohashi S, Shin-Ya M, et al. 2004. Sequence-specific gene silencing in murine muscle induced by electroporation-mediated transfer of short interfering RNA. *J. Gene Med.* 6:105–10

74. Konishi Y, Stegmuller J, Matsuda T, Bonni S, Bonni A. 2004. Cdh1-APC controls axonal growth and patterning in the mammalian brain. *Science* 303:1026–30

75. Kong XC, Barzaghi P, Ruegg MA. 2004. Inhibition of synapse assembly in mammalian muscle in vivo by RNA interference. *EMBO Rep.* 5:183–88

76. Matsuda T, Cepko CL. 2004. Electroporation and RNA interference in the rodent retina in vivo and in vitro. *Proc. Natl. Acad. Sci. USA* 101:16–22

77. Braasch DA, Jensen S, Liu Y, Kaur K, Arar K, et al. 2003. RNA interference in mammalian cells by chemically-modified RNA. *Biochemistry* 42:7967–75

78. Braasch DA, Paroo Z, Constantinescu A, Ren G, Oz OK, et al. 2004. Biodistribution of phosphodiester and phosphorothioate siRNA. *Bioorg. Med. Chem. Lett.* 14:1139–43

79. Czauderna F, Fechtner M, Dames S, Aygun H, Klippel A, et al. 2003. Structural variations and stabilising modifications of synthetic siRNAs in mammalian cells. *Nucleic Acids Res.* 31:2705–16

80. Chiu YL, Rana TM. 2003. siRNA function in RNAi: a chemical modification analysis. *RNA* 9:1034–48

81. Zhang Y, Boado RJ, Pardridge WM. 2003. In vivo knockdown of gene expression in brain cancer with intravenous RNAi in adult rats. *J. Gene Med.* 5:1039–45

82. Muratovska A, Eccles MR. 2004. Conjugate for efficient delivery of short interfering RNA (siRNA) into mammalian cells. *FEBS Lett.* 558:63–68

83. Lois C, Refaeli Y, Qin XF, Van Parijs L. 2001. Retroviruses as tools to study the immune system. *Curr. Opin. Immunol.* 13:496–504

84. Abbas-Terki T, Blanco-Bose W, Deglon N, Pralong W, Aebischer P. 2002. Lentiviral-mediated RNA interference. *Hum. Gene Ther.* 13:2197–201

85. Dirac AM, Bernards R. 2003. Reversal of senescence in mouse fibroblasts through lentiviral suppression of p53. *J. Biol. Chem.* 278:11731–34

86. Hemann MT, Fridman JS, Zilfou JT, Hernando E, Paddison PJ, et al. 2003. An epiallelic series of p53 hypomorphs created by stable RNAi produces distinct tumor phenotypes in vivo. *Nat. Genet.* 33:396–400

87. Rubinson DA, Dillon CP, Kwiatkowski AV, Sievers C, Yang L, et al. 2003. A lentivirus-based system to functionally silence genes in primary mammalian cells, stem cells and transgenic mice by RNA interference. *Nat. Genet.* 33:401–6

88. Stewart SA, Dykxhoorn DM, Palliser D, Mizuno H, Yu EY, et al. 2003. Lentivirus-delivered stable gene silencing by RNAi in primary cells. *RNA* 9:493–501

89. Tiscornia G, Singer O, Ikawa M, Verma IM. 2003. A general method for gene knockdown in mice by using lentiviral vectors expressing small interfering RNA. *Proc. Natl. Acad. Sci. USA* 100:1844–48

90. Qin XF, An DS, Chen IS, Baltimore D. 2003. Inhibiting HIV-1 infection in human T cells by lentiviral-mediated delivery of small interfering RNA against CCR5. *Proc. Natl. Acad. Sci. USA* 100:183–88

91. Hommel JD, Sears RM, Georgescu D, Simmons DL, DiLeone RJ. 2003. Local gene knockdown in the brain using viral-mediated RNA interference. *Nat. Med.* 9:1539–44

92. Arts GJ, Langemeijer E, Tissingh R, Ma L, Pavliska H, et al. 2003. Adenoviral vectors expressing siRNAs for discovery and validation of gene function. *Genome Res.* 13:2325–32

93. Shen C, Buck AK, Liu X, Winkler M, Reske SN. 2003. Gene silencing by adenovirus-delivered siRNA. *FEBS Lett.* 539:111–14

94. Zhao LJ, Jian H, Zhu H. 2003. Specific gene inhibition by adenovirus-mediated expression of small interfering RNA. *Gene* 316:137–41

95. Hildinger M, Auricchio A. 2004. Advances in AAV-mediated gene transfer for the treatment of inherited disorders. *Eur. J. Hum. Genet.* 12:263–71

96. Kay MA, Nakai H. 2003. Looking into the safety of AAV vectors. *Nature* 424:251

97. Thomas CE, Ehrhardt A, Kay MA. 2003. Progress and problems with the use of viral vectors for gene therapy. *Nat. Rev. Genet.* 4:346–58

98. Boden D, Pusch O, Lee F, Tucker L, Ramratnam B. 2004. Efficient gene transfer of HIV-1-specific short hairpin RNA into human lymphocytic cells using recombinant adeno-associated virus vectors. *Mol. Ther.* 9:396–402

99. Xia H, Mao Q, Eliason SL, Harper SQ, Martins IH, et al. 2004. RNAi suppresses polyglutamine-induced neurodegeneration in a model of spinocerebellar ataxia. *Nat. Med.* 10:816–20

100. Hasuwa H, Kaseda K, Einarsdottir T, Okabe M. 2002. Small interfering RNA and gene silencing in transgenic mice and rats. *FEBS Lett.* 532:227–30

101. Stein P, Svoboda P, Schultz RM. 2003. Transgenic RNAi in mouse oocytes: a simple and fast approach to study gene function. *Dev. Biol.* 256:187–93

102. Fedoriw AM, Stein P, Svoboda P, Schultz RM, Bartolomei MS. 2004. Transgenic RNAi reveals essential function for CTCF in H19 gene imprinting. *Science* 303:238–40

103. Carmell MA, Zhang L, Conklin DS, Hannon GJ, Rosenquist TA. 2003. Germline transmission of RNAi in mice. *Nat. Struct. Biol.* 10:91–92

104. Kunath T, Gish G, Lickert H, Jones N, Pawson T, Rossant J. 2003. Transgenic RNA interference in ES cell-derived embryos recapitulates a genetic null phenotype. *Nat. Biotechnol.* 21:559–61

105. Gonczy P, Echeverri C, Oegema K, Coulson A, Jones SJ, et al. 2000. Functional genomic analysis of cell division in *C. elegans* using RNAi of genes on chromosome III. *Nature* 408:331–36

106. Fraser AG, Kamath RS, Zipperlen P, Martinez-Campos M, Sohrmann M, Ahringer J. 2000. Functional genomic analysis of *C. elegans* chromosome I by systematic RNA interference. *Nature* 408:325–30

107. Timmons L, Court DL, Fire A. 2001. Ingestion of bacterially expressed dsRNAs can produce specific and potent genetic interference in *Caenorhabditis elegans*. *Gene* 263:103–12

108. Timmons L, Fire A. 1998. Specific interference by ingested dsRNA. *Nature* 395:854

109. Kamath RS, Fraser AG, Dong Y, Poulin G, Durbin R, et al. 2003. Systematic

functional analysis of the *Caenorhabditis elegans* genome using RNAi. *Nature* 421:231–37

110. Simmer F, Moorman C, Van Der Linden AM, Kuijk E, Van Den Berghe PV, et al. 2003. Genome-wide RNAi of *C. elegans* using the hypersensitive rrf-3 strain reveals novel gene functions. *PLoS Biol.* 1:E12

111. Lee SS, Lee RY, Fraser AG, Kamath RS, Ahringer J, Ruvkun G. 2003. A systematic RNAi screen identifies a critical role for mitochondria in *C. elegans* longevity. *Nat. Genet.* 33:40–48

112. Ashrafi K, Chang FY, Watts JL, Fraser AG, Kamath RS, et al. 2003. Genome-wide RNAi analysis of *Caenorhabditis elegans* fat regulatory genes. *Nature* 421:268–72

113. Pothof J, van Haaften G, Thijssen K, Kamath RS, Fraser AG, et al. 2003. Identification of genes that protect the *C. elegans* genome against mutations by genome-wide RNAi. *Genes Dev.* 17:443–48

114. Piano F, Schetter AJ, Mangone M, Stein L, Kemphues KJ. 2000. RNAi analysis of genes expressed in the ovary of *Caenorhabditis elegans. Curr. Biol.* 10:1619–22

115. Kennedy S, Wang D, Ruvkun G. 2004. A conserved siRNA-degrading RNase negatively regulates RNA interference in *C. elegans. Nature* 427:645–49

116. Muda M, Worby CA, Simonson-Leff N, Clemens JC, Dixon JE. 2002. Use of double-stranded RNA-mediated interference to determine the substrates of protein tyrosine kinases and phosphatases. *Biochem. J.* 366:73–77

117. Clemens JC, Worby CA, Simonson-Leff N, Muda M, Maehama T, et al. 2000. Use of double-stranded RNA interference in *Drosophila* cell lines to dissect signal transduction pathways. *Proc. Natl. Acad. Sci. USA* 97:6499–503

118. Lum L, Yao S, Mozer B, Rovescalli A, Von Kessler D, et al. 2003. Identification of Hedgehog pathway components by RNAi in *Drosophila* cultured cells. *Science* 299:2039–45

119. Kiger A, Baum B, Jones S, Jones M, Coulson A, et al. 2003. A functional genomic analysis of cell morphology using RNA interference. *J. Biol.* 2:27

120. Kim YO, Park SJ, Balaban RS, Nirenberg M, Kim Y. 2004. A functional genomic screen for cardiogenic genes using RNA interference in developing *Drosophila* embryos. *Proc. Natl. Acad. Sci. USA* 101:159–64

121. Somma MP, Fasulo B, Cenci G, Cundari E, Gatti M. 2002. Molecular dissection of cytokinesis by RNA interference in *Drosophila* cultured cells. *Mol. Biol. Cell.* 13:2448–60

122. Ramet M, Manfruelli P, Pearson A, Mathey-Prevot B, Ezekowitz RA. 2002. Functional genomic analysis of phagocytosis and identification of a *Drosophila* receptor for *E. coli. Nature* 416:644–48

123. Foley E, O'Farrell PH. 2004. Functional dissection of an innate immune response by a genome-wide RNAi screen. *PLoS Biol.* 2:E203

124. Boutros M, Kiger AA, Armknecht S, Kerr K, Hild M, et al. 2004. Genome-wide RNAi analysis of growth and viability in *Drosophila* cells. *Science* 303:832–35

125. Aza-Blanc P, Cooper CL, Wagner K, Batalov S, Deveraux QL, Cooke MP. 2003. Identification of modulators of TRAIL-induced apoptosis via RNAi-based phenotypic screening. *Mol. Cell.* 12:627–37

126. Brummelkamp TR, Nijman SM, Dirac AM, Bernards R. 2003. Loss of the cylindromatosis tumour suppressor inhibits apoptosis by activating NF-kappaB. *Nature* 424:797–801

127. Berns K, Hijmans EM, Mullenders J, Brummelkamp TR, Velds A, et al. 2004. A large-scale RNAi screen in human cells identifies new components of the p53 pathway. *Nature* 428:431–37

128. Paddison PJ, Silva JM, Conklin DS, Schlabach M, Li M, et al. 2004. A resource

for large-scale RNA-interference-based screens in mammals. *Nature* 428:427–31

129. Zheng L, Liu J, Batalov S, Zhou D, Orth A, et al. 2004. An approach to genomewide screens of expressed small interfering RNAs in mammalian cells. *Proc. Natl. Acad. Sci. USA* 101:135–40

130. Kumar R, Conklin DS, Mittal V. 2003. High-throughput selection of effective RNAi probes for gene silencing. *Genome Res.* 13:2333–40

131. Mousses S, Caplen NJ, Cornelison R, Weaver D, Basik M, et al. 2003. RNAi microarray analysis in cultured mammalian cells. *Genome Res.* 13:2341–47

132. Lieberman J, Song E, Lee SK, Shankar P. 2003. Interfering with disease: opportunities and roadblocks to harnessing RNA interference. *Trends Mol. Med.* 9:397–403

133. Wall NR, Shi Y. 2003. Small RNA: can RNA interference be exploited for therapy? *Lancet* 362:1401–3

134. Kitabwalla M, Ruprecht RM. 2002. RNA interference—a new weapon against HIV and beyond. *N. Engl. J. Med.* 347:1364–67

135. Ding H, Schwarz DS, Keene A, Affar el B, Fenton L, et al. 2003. Selective silencing by RNAi of a dominant allele that causes amyotrophic lateral sclerosis. *Aging Cell* 2:209–17

136. Wohlbold L, van der Kuip H, Miething C, Vornlocher HP, Knabbe C, et al. 2003. Inhibition of *bcr-abl* gene expression by small interfering RNA sensitizes for imatinib mesylate (STI571). *Blood* 102:2236–39

137. Wilda M, Fuchs U, Wossmann W, Borkhardt A. 2002. Killing of leukemic cells with a BCR/ABL fusion gene by RNA interference (RNAi). *Oncogene* 21:5716–24

138. Yin JQ, Gao J, Shao R, Tian WN, Wang J, Wan Y. 2003. siRNA agents inhibit oncogene expression and attenuate human tumor cell growth. *J. Exp. Ther. Oncol.* 3:194–204

139. Filleur S, Courtin A, Ait-Si-Ali S,

Guglielmi J, Merle C, et al. 2003. SiRNA-mediated inhibition of vascular endothelial growth factor severely limits tumor resistance to antiangiogenic thrombospondin-1 and slows tumor vascularization and growth. *Cancer Res.* 63:3919–22

140. Duxbury MS, Ito H, Benoit E, Zinner MJ, Ashley SW, Whang EE. 2003. RNA interference targeting focal adhesion kinase enhances pancreatic adenocarcinoma gemcitabine chemosensitivity. *Biochem. Biophys. Res. Commun.* 311:786–92

141. Lima RT, Martins LM, Guimaraes JE, Sambade C, Vasconcelos MH. 2004. Specific downregulation of bcl-2 and xIAP by RNAi enhances the effects of chemotherapeutic agents in MCF-7 human breast cancer cells. *Cancer Gene Ther.* 11:309–16

142. Nieth C, Priebsch A, Stege A, Lage H. 2003. Modulation of the classical multidrug resistance (MDR) phenotype by RNA interference (RNAi). *FEBS Lett.* 545:144–50

143. Butz K, Ristriani T, Hengstermann A, Denk C, Scheffner M, Hoppe-Seyler F. 2003. siRNA targeting of the viral E6 oncogene efficiently kills human papillomavirus-positive cancer cells. *Oncogene* 22:5938–45

144. Hall AH, Alexander KA. 2003. RNA interference of human papillomavirus type 18 E6 and E7 induces senescence in HeLa cells. *J. Virol.* 77:6066–69

145. Ge Q, McManus MT, Nguyen T, Shen CH, Sharp PA, et al. 2003. RNA interference of influenza virus production by directly targeting mRNA for degradation and indirectly inhibiting all viral RNA transcription. *Proc. Natl. Acad. Sci. USA* 100:2718–23

146. Giladi H, Ketzinel-Gilad M, Rivkin L, Felig Y, Nussbaum O, Galun E. 2003. Small interfering RNA inhibits hepatitis B virus replication in mice. *Mol. Ther.* 8:769–76

147. Yokota T, Sakamoto N, Enomoto N, Tanabe Y, Miyagishi M, et al. 2003.

Inhibition of intracellular hepatitis C virus replication by synthetic and vector-derived small interfering RNAs. *EMBO Rep.* 4:602–8

148. Kronke J, Kittler R, Buchholz F, Windisch MP, Pietschmann T, et al. 2004. Alternative approaches for efficient inhibition of hepatitis C virus RNA replication by small interfering RNAs. *J. Virol.* 78:3436–46

149. Zhang Y, Li T, Fu L, Yu C, Li Y, et al. 2004. Silencing SARS-CoV Spike protein expression in cultured cells by RNA interference. *FEBS Lett.* 560:141–46

150. Ge Q, Filip L, Bai A, Nguyen T, Eisen HN, Chen J. 2004. Inhibition of influenza virus production in virus-infected mice by RNA interference. *Proc. Natl. Acad. Sci. USA* 101:8676–81

151. Das AT, Brummelkamp TR, Westerhout EM, Vink M, Madiredjo M, et al. 2004. Human immunodeficiency virus type 1 escapes from RNA interference-mediated inhibition. *J. Virol.* 78:2601–5

152. Boden D, Pusch O, Lee F, Tucker L, Ramratnam B. 2003. Human immunodeficiency virus type 1 escape from RNA interference. *J. Virol.* 77:11531–35

153. Wang X, Huong SM, Chiu ML, Raab-Traub N, Huang ES. 2003. Epidermal growth factor receptor is a cellular receptor for human cytomegalovirus. *Nature* 424:456–61

154. Zender L, Hutker S, Liedtke C, Tillmann HL, Zender S, et al. 2003. Caspase 8 small interfering RNA prevents acute liver failure in mice. *Proc. Natl. Acad. Sci. USA* 100:7797–802

155. Kao SC, Krichevsky AM, Kosik KS, Tsai LH. 2004. BACE1 suppression by RNA interference in primary cortical neurons. *J. Biol. Chem.* 279:1942–49

156. Tilly G, Chapuis J, Vilette D, Laude H, Vilotte JL. 2003. Efficient and specific down-regulation of prion protein expression by RNAi. *Biochem. Biophys. Res. Commun.* 305:548–51

157. Sioud M, Sorensen DR. 2003. Cationic liposome-mediated delivery of siRNAs in adult mice. *Biochem. Biophys. Res. Commun.* 312:1220–25

158. Bologna JC, Dorn G, Natt F, Weiler J. 2003. Linear polyethylenimine as a tool for comparative studies of antisense and short double-stranded RNA oligonucleotides. *Nucleosides Nucleotides Nucleic Acids* 22:1729–31

159. Chang FH, Lee CH, Chen MT, Kuo CC, Chiang YL, et al. 2004. Surfection: a new platform for transfected cell arrays. *Nucleic Acids Res.* 32:e33

160. Hill JA, Ichim TE, Kusznieruk KP, Li M, Huang X, et al. 2003. Immune modulation by silencing IL-12 production in dendritic cells using small interfering RNA. *J. Immunol.* 171:691–96

161. Hacein-Bey-Abina S, von Kalle C, Schmidt M, Le Deist F, Wulffraat N, et al. 2003. A serious adverse event after successful gene therapy for X-linked severe combined immunodeficiency. *N. Engl. J. Med.* 348:255–56

162. Wang J, Tekle E, Oubrahim H, Mieyal JJ, Stadtman ER, Chock PB. 2003. Stable and controllable RNA interference: investigating the physiological function of glutathionylated actin. *Proc. Natl. Acad. Sci. USA* 100:5103–6

163. Chen Y, Stamatoyannopoulos G, Song CZ. 2003. Down-regulation of CXCR4 by inducible small interfering RNA inhibits breast cancer cell invasion in vitro. *Cancer Res.* 63:4801–4

164. Czauderna F, Santel A, Hinz M, Fechtner M, Durieux B, et al. 2003. Inducible shRNA expression for application in a prostate cancer mouse model. *Nucleic Acids Res.* 31:e127

165. Wiznerowicz M, Trono D. 2003. Conditional suppression of cellular genes: lentivirus vector-mediated drug-inducible RNA interference. *J. Virol.* 77:8957–61

166. Matsukura S, Jones PA, Takai D. 2003. Establishment of conditional vectors for hairpin siRNA knockdowns. *Nucleic Acids Res.* 31:e77

167. van de Wetering M, Oving I, Muncan V, Pon Fong MT, Brantjes H, et al. 2003. Specific inhibition of gene expression using a stably integrated, inducible small-interfering-RNA vector. *EMBO Rep.* 4: 609–15

168. Tiscornia G, Tergaonkar V, Galimi F, Verma IM. 2004. CRE recombinase-inducible RNA interference mediated by lentiviral vectors. *Proc. Natl. Acad. Sci. USA* 101:7347–51

169. Ventura A, Meissner A, Dillon CP, McManus M, Sharp PA, et al. 2004. Cre-lox-regulated conditional RNA interference from transgenes. *Proc. Natl. Acad. Sci. USA* 101:10380–85

170. Kasim V, Miyagishi M, Taira K. 2003. Control of siRNA expression utilizing Cre-loxP recombination system. *Nucleic Acids Res. Suppl.* 3:255–56

171. Fritsch L, Martinez LA, Sekhri R, Naguibneva I, Gerard M, et al. 2004. Conditional gene knock-down by CRE-dependent short interfering RNAs. *EMBO Rep.* 5:178–82

172. Gupta S, Schoer RA, Egan JE, Hannon GJ, Mittal V. 2004. Inducible, reversible, and stable RNA interference in mammalian cells. *Proc. Natl. Acad. Sci. USA* 101:1927–32

173. Kamath RS, Ahringer J. 2003. Genome-wide RNAi screening in *Caenorhabditis elegans. Methods* 30:313–21

174. Piano F, Schetter AJ, Morton DG, Gunsalus KC, Reinke V, et al. 2002. Gene clustering based on RNAi phenotypes of ovary-enriched genes in *C. elegans. Curr. Biol.* 12:1959–64

175. Yamamoto T, Omoto S, Mizuguchi M, Mizukami H, Okuyama H, et al. 2002. Double-stranded nef RNA interferes with human immunodeficiency virus type 1 replication. *Microbiol. Immunol.* 46:809–17

176. Park WS, Miyano-Kurosaki N, Hayafune M, Nakajima E, Matsuzaki T, et al. 2002. Prevention of HIV-1 infection in human peripheral blood mononuclear cells by specific RNA interference. *Nucleic Acids Res.* 30:4830–35

177. Capodici J, Kariko K, Weissman D. 2002. Inhibition of HIV-1 infection by small interfering RNA-mediated RNA interference. *J. Immunol.* 169:5196–201

178. Nishitsuji H, Ikeda T, Miyoshi H, Ohashi T, Kannagi M, Masuda T. 2004. Expression of small hairpin RNA by lentivirus-based vector confers efficient and stable gene-suppression of HIV-1 on human cells including primary non-dividing cells. *Microb. Infect.* 6:76–85

179. Lee NS, Dohjima T, Bauer G, Li H, Li MJ, et al. 2002. Expression of small interfering RNAs targeted against HIV-1 rev transcripts in human cells. *Nat. Biotechnol.* 20:500–5

180. Coburn GA, Cullen BR. 2002. Potent and specific inhibition of human immunodeficiency virus type 1 replication by RNA interference. *J. Virol.* 76:9225–31

181. Jacque JM, Triques K, Stevenson M. 2002. Modulation of HIV-1 replication by RNA interference. *Nature* 418:435–38

182. Park WS, Miyano-Kurosaki N, Nakajima E, Takaku H. 2001. Specific inhibition of HIV-1 gene expression by double-stranded RNA. *Nucleic Acids Res. Suppl.* 1:219–20

183. Park WS, Hayafune M, Miyano-Kurosaki N, Takaku H. 2003. Specific HIV-1 env gene silencing by small interfering RNAs in human peripheral blood mononuclear cells. *Gene Ther.* 10:2046–50

184. Novina CD, Murray MF, Dykxhoorn DM, Beresford PJ, Riess J, et al. 2002. siRNA-directed inhibition of HIV-1 infection. *Nat. Med.* 8:681–86

185. Butticaz C, Ciuffi A, Munoz M, Thomas J, Bridge A, et al. 2003. Protection from HIV-1 infection of primary CD4 T cells by CCR5 silencing is effective for the full spectrum of CCR5 expression. *Antiviral. Ther.* 8:373–77

186. Martinez MA, Gutierrez A, Armand-Ugon M, Blanco J, Parera M, et al. 2002. Suppression of chemokine receptor

expression by RNA interference allows for inhibition of HIV-1 replication. *AIDS* 16:2385–90

187. Anderson J, Banerjea A, Planelles V, Akkina R. 2003. Potent suppression of HIV type 1 infection by a short hairpin anti-CXCR4 siRNA. *AIDS Res. Hum. Retroviruses* 19:699–706

188. Chiu YL, Cao H, Jacque JM, Stevenson M, Rana TM. 2004. Inhibition of human immunodeficiency virus type 1 replication by RNA interference directed against human transcription elongation factor P-TEFb (CDK9/CyclinT1). *J. Virol.* 78:2517–29

189. Anderson J, Banerjea A, Akkina R. 2003. Bispecific short hairpin siRNA constructs targeted to CD4, CXCR4, and CCR5 confer HIV-1 resistance. *Oligonucleotides* 13:303–12

190. Shlomai A, Shaul Y. 2003. Inhibition of hepatitis B virus expression and replication by RNA interference. *Hepatology* 37:764–70

191. McCaffrey AP, Nakai H, Pandey K, Huang Z, Salazar FH, et al. 2003. Inhibition of hepatitis B virus in mice by RNA interference. *Nat. Biotechnol.* 21:639–44

192. Ying C, De Clercq E, Neyts J. 2003. Selective inhibition of hepatitis B virus replication by RNA interference. *Biochem. Biophys. Res. Commun.* 309:482–84

193. Shlomai A, Shaul Y, Kapadia SB, Brideau-Andersen A, Chisari FV, et al. 2003. Inhibition of hepatitis B virus expression and replication by RNA interference. Interference of hepatitis C virus RNA replication by short interfering RNAs. RNA interference-mediated silencing of mutant superoxide dismutase rescues cyclosporin A-induced death in cultured neuroblastoma cells. *Hepatology* 37:764–70

194. Seo MY, Abrignani S, Houghton M, Han JH. 2003. Small interfering RNA-mediated inhibition of hepatitis C virus replication in the human hepatoma cell line Huh-7. *J. Virol.* 77:810–12

195. Sen A, Steele R, Ghosh AK, Basu A, Ray R, Ray RB. 2003. Inhibition of hepatitis C virus protein expression by RNA interference. *Virus Res.* 96:27–35

196. Chang J, Taylor JM. 2003. Susceptibility of human hepatitis delta virus RNAs to small interfering RNA action. *J. Virol.* 77:9728–31

197. Wiebusch L, Truss M, Hagemeier C. 2004. Inhibition of human cytomegalovirus replication by small interfering RNAs. *J. Gen. Virol.* 85:179–84

198. Phipps KM, Martinez A, Lu J, Heinz BA, Zhao G. 2004. Small interfering RNA molecules as potential anti-human rhinovirus agents: in vitro potency, specificity, and mechanism. *Antiviral. Res.* 61:49–55

199. Jia Q, Sun R. 2003. Inhibition of gammaherpesvirus replication by RNA interference. *J. Virol.* 77:3301–6

200. Maxwell MM, Pasinelli P, Kazantsev AG, Brown RH Jr. 2004. RNA interference-mediated silencing of mutant superoxide dismutase rescues cyclosporin A-induced death in cultured neuroblastoma cells. *Proc. Natl. Acad. Sci. USA* 101:3178–83

201. Abdelgany A, Wood M, Beeson D. 2003. Allele-specific silencing of a pathogenic mutant acetylcholine receptor subunit by RNA interference. *Hum. Mol. Genet.* 12:2637–44

202. Grzmil M, Thelen P, Hemmerlein B, Schweyer S, Voigt S, et al. 2003. Bax inhibitor-1 is overexpressed in prostate cancer and its specific down-regulation by RNA interference leads to cell death in human prostate carcinoma cells. *Am. J. Pathol.* 163:543–52

203. Duxbury MS, Ito H, Zinner MJ, Ashley SW, Whang EE. 2004. EphA2: a determinant of malignant cellular behavior and a potential therapeutic target in pancreatic adenocarcinoma. *Oncogene* 23:1448–56

204. Sanceau J, Truchet S, Bauvois B. 2003. Matrix metalloproteinase-9 silencing by

RNA interference triggers the migratory-adhesive switch in Ewing's sarcoma cells. *J. Biol. Chem.* 278:36537–46

205. Heidenreich O, Krauter J, Riehle H, Hadwiger P, John M, et al. 2003. AML1/MTG8 oncogene suppression by small interfering RNAs supports myeloid differentiation of t(8;21)-positive leukemic cells. *Blood* 101:3157–63

206. Hingorani SR, Jacobetz MA, Robertson GP, Herlyn M, Tuveson DA. 2003. Suppression of BRAF(V599E) in human melanoma abrogates transformation. *Cancer Res.* 63:5198–202

207. Harvey AJ, Crompton MR. 2003. Use of RNA interference to validate Brk as a novel therapeutic target in breast cancer: Brk promotes breast carcinoma cell proliferation. *Oncogene* 22:5006–10

208. Li XP, Li G, Peng Y, Kung HF, Lin MC. 2004. Suppression of Epstein-Barr virus-encoded latent membrane protein-1 by RNA interference inhibits the metastatic potential of nasopharyngeal carcinoma cells. *Biochem. Biophys. Res. Commun.* 315:212–18

209. Nagy P, Arndt-Jovin DJ, Jovin TM. 2003. Small interfering RNAs suppress the expression of endogenous and GFP-fused epidermal growth factor receptor (erbB1) and induce apoptosis in erbB1-overexpressing cells. *Exp. Cell Res.* 285: 39–49

210. Zhang M, Zhang X, Bai CX, Chen J, Wei MQ. 2004. Inhibition of epidermal growth factor receptor expression by RNA interference in A549 cells. *Acta Pharmacol. Sin.* 25:61–67

211. De Schrijver E, Brusselmans K, Heyns W, Verhoeven G, Swinnen JV. 2003. RNA interference-mediated silencing of the fatty acid synthase gene attenuates growth and induces morphological changes and apoptosis of LNCaP prostate cancer cells. *Cancer Res.* 63:3799–804

212. Crnkovic-Mertens I, Hoppe-Seyler F, Butz K. 2003. Induction of apoptosis in tumor cells by siRNA-mediated silencing of the livin/ML-IAP/KIAP gene. *Oncogene* 22:8330–36

213. Futami T, Miyagishi M, Seki M, Taira K. 2002. Induction of apoptosis in HeLa cells with siRNA expression vector targeted against bcl-2. *Nucleic Acids Res. Suppl.* 2:251–52

214. Spankuch-Schmitt B, Bereiter-Hahn J, Kaufmann M, Strebhardt K. 2002. Effect of RNA silencing of polo-like kinase-1 (PLK1) on apoptosis and spindle formation in human cancer cells. *J. Natl. Cancer Inst.* 94:1863–77

215. Kosciolek BA, Kalantidis K, Tabler M, Rowley PT. 2003. Inhibition of telomerase activity in human cancer cells by RNA interference. *Mol. Cancer Ther.* 2: 209–16

216. Zhang L, Fogg DK, Waisman DM. 2004. RNA interference-mediated silencing of the S100A10 gene attenuates plasmin generation and invasiveness of Colo 222 colorectal cancer cells. *J. Biol. Chem.* 279:2053–62

217. Konnikova L, Kotecki M, Kruger MM, Cochran BH. 2003. Knockdown of STAT3 expression by RNAi induces apoptosis in astrocytoma cells. *BMC Cancer* 3: 23

218. Ritter U, Damm-Welk C, Fuchs U, Bohle RM, Borkhardt A, Woessmann W. 2003. Design and evaluation of chemically synthesized siRNA targeting the NPM-ALK fusion site in anaplastic large cell lymphoma (ALCL). *Oligonucleotides* 13:365–73

Ecological, Evolutionary, and Comparative Physiology

Martin E. Feder
Section Editor

Since its inception, physiology has been comparative—whether the comparisons were between healthy and diseased states, experimental preparations and their controls, or alternative genotypes, etc. In a sense, all physiologists are comparative physiologists. One major approach to physiology, i.e., comparative physiology, has in particular exploited the insights that arise from comparing organisms with diverse evolutionary heritages, inhabiting diverse natural environments, or both. The major themes of such physiology have long been consistent with Charles Darwin's own short definition of evolution, "descent with modification." Descent implies both the unity of fundamental physiological mechanisms due to inheritance from common ancestors, and the constraints that these mechanisms impose in defining the range of environments that the inheritors can tolerate. Modification signifies the power of evolutionary processes to adjust physiological mechanisms to function in diverse environments and the relation of function to environment (i.e., ecology). Thus the founders of comparative physiology, such as August Krogh, C. Ladd Prosser, and Knut Schmidt-Nielsen, were simultaneously physiologists, ecologists, and evolutionary biologists—and so are their successors. In this sense, evolutionary physiology and ecological physiology have long been components of comparative physiology, and vice versa. Thus the change in the name of this section of the *Annual Review of Physiology* is but long-overdue truth in advertising.

During the past 30 years, however, both intellectual and technical advances have brought evolutionary physiology and ecological physiology to individual prominence and re-energized comparative physiology. Such technical advances have included the development of molecular biological tools applicable to all organisms and their genomes, miniaturized physiological instrumentation that truly permits physiology outside the laboratory, and a robust methodology for accommodating phylogeny in physiological comparisons. Intellectual advances have included the formalization of adaptive strategies and the infusion of explicitly evolutionary paradigms. These developments have found their reflection in the re-naming of cognate programs (e.g., Ecological and Evolutionary Physiology at the National Science Foundation) and sections of journals and societies (e.g., Comparative and Evolutionary Physiology section of the American Physiological Society and its corresponding component of the *American Journal of Physiology*), and here in the *Annual Review of Physiology*. Indeed, this section of the *Annual Review of Physiology* will welcome reviewers that identify with each, some, or all of evolutionary, ecological, and comparative physiology. Importantly, such approaches have much to offer, and none has a monopoly on insight into physiological unity and diversity.

Annu. Rev. Physiol. 2005. 67:177–201
doi: 10.1146/annurev.physiol.67.040403.105027
Copyright © 2005 by Annual Reviews. All rights reserved
First published online as a Review in Advance on August 13, 2004

BIOPHYSICS, PHYSIOLOGICAL ECOLOGY, AND CLIMATE CHANGE: Does Mechanism Matter?

Brian Helmuth,[1] Joel G. Kingsolver,[2] and Emily Carrington[3]

[1]Department of Biological Sciences, University of South Carolina, Columbia, South Carolina 29208, USA; email: helmuth@biol.sc.edu
[2]Department of Biology, CB-3280, University of North Carolina, Chapel Hill, North Carolina 27599-3280; email: jgking@bio.unc.edu
[3]Department of Biological Sciences, University of Rhode Island, Kingston, Rhode Island 02881; email: carrington@uri.edu

Key Words biogeography, field physiology, insects, rocky intertidal zone, stress

■ **Abstract** Recent meta-analyses have shown that the effects of climate change are detectable and significant in their magnitude, but these studies have emphasized the utility of looking for large-scale patterns without necessarily understanding the mechanisms underlying these changes. Using a series of case studies, we explore the potential pitfalls when one fails to incorporate aspects of physiological performance when predicting the consequences of climate change on biotic communities. We argue that by considering the mechanistic details of physiological performance within the context of biophysical ecology (engineering methods of heat, mass and momentum exchange applied to biological systems), such approaches will be better poised to predict where and when the impacts of climate change will most likely occur.

INTRODUCTION

The earth's climate is changing at a rapid rate (1), but uncertainty is considerable about this perturbation's impact on the planet's biota and about how and where we should look for these effects. The task for biologists is to measure and anticipate the magnitude and rapidity of the effects of climate change on natural ecosystems and to explore any potential for mitigation or prevention (2–4).

Recent studies have documented climate-related mortality events (3–12), changes in population abundances (13–15), shifts in species range boundaries (15–19), and phenological shifts in the timing of reproductive and migratory events (18–20). In particular, meta-analyses of species distribution patterns have elucidated climate-driven effects, but these studies have emphasized discovery of large-scale patterns without necessarily understanding the mechanisms underlying them (18, 19, 21–23).

Concomitantly, however, new biochemical and molecular investigations of the effects of body temperature on organismal and subcellular physiology have flourished (24–28). These latter studies have not only better revealed the mechanisms underlying the physiological responses of organisms to thermal stress but have also provided metrics of stress levels under field conditions (24, 26–39). Meta-analyses of pattern in nature and studies conducted at biomolecular scales, however, have tended to operate independently of one another.

Here we advocate for the inclusion of physiological insight when predicting the effects of climate change on populations and ecosystems (40–43). We describe how the application of biophysical ecology techniques (engineering methods used to explore the exchange of heat, mass, and momentum between organisms and their environment) can explicitly link studies at physiological and biomolecular scales with those at the level of populations and ecosystems (20, 35, 40, 41, 43–47). Moreover, we exemplify how ignorance of the physiological response of individual organisms, or populations of organisms, to environmental parameters can sometimes confound predictions of the effects of climate change on organisms (42, 48, 49). We highlight the importance of studying physiological performance within the context of the organism's local microclimate using a series of examples taken from intertidal, shallow-water marine, and terrestrial environments.

First, the relative importance of environmental factors that may limit a species' geographic or local distribution can vary markedly in space and time (50–53). A clear understanding of how multiple climate-related environmental parameters may alternatively limit species distributions is essential for predicting how each parameter will affect populations. Moreover, even when a single variable (such as body temperature) is a causative factor, which aspect of the environmental signal (e.g., average, maximum, or minimum temperature, or time history of the thermal signal) is most relevant to the species' fitness is not always clear (26, 35, 36, 54–57).

Second, just as environmental stressors can vary over a range of scales, the physiological capacity of an organism to resist or recover from environmental stress can vary significantly in space and time. As a result, it is impossible to predict the level of mechanical or physiological risk to which an organism is likely to be subjected without first quantifying both the magnitude of the environmental stressor and the ability of the organism to resist the environmental condition in question. For example, simply recording the force that a crashing wave imparts on an organism has little meaning without first understanding how the ability of the organism to resist that force may vary (58–60). Thus without a detailed mechanistic understanding of environmental safety factors (58–64), it is possible that our current efforts to characterize environmental signals may be insufficient in many cases.

Third, climate change does not produce a simple increase in global average temperature; rather it involves a characteristic set of changes in diurnal, seasonal, and geographic patterns of temperature, precipitation, and other atmospheric conditions. Analysis of ecological responses to climate change during the past century reveals a diversity of responses (or lack of response) that varies widely among populations and species. We argue that understanding the physiological mechanisms that underlie how specific components of climate affect key life cycle stages will be important for predicting the ecological responses of many species to climate change.

Species and Population Responses to Climate Change

Environmental conditions can influence the physiological performance of organisms through three general mechanisms that vary in response time and reversibility (after Reference 65). First, environmental change can lead to genetic adaptation, a change in allele frequencies in a population via natural selection. This process is typically slow (requiring at least a generation) and irreversible during an individual's lifespan. Second, individuals may exhibit phenotypic plasticity or variable expression of genetic traits in response to environmental conditions. Plasticity can be reversible or irreversible and requires relatively brief times (usually less than one generation). For example, multiple phenotypes can arise where environmental conditions induce alternative developmental programs. The differential activation of developmental switches (e.g., arrested development) can lead to discrete phenotypes, whereas variations in the rates and degrees of expression cause phenotypic modulation (65). Another form of phenotypic plasticity is acclimatization, which involves a short-term compensatory modification of physiological function and tolerance in response to environmental change. Third, organisms can rapidly if not acutely adjust physiological state in response to environmental change. Such adjustments require seconds to days and are generally reversible.

Thus climate change may have three different outcomes depending on the physiological response of the species or population in question, as described by Fields et al. (24). First, if environmental changes are sufficiently small, organisms may acclimatize to those conditions. Second, if environmental conditions exceed the ability of some, but not all, of the individuals to cope with environmental change, then natural selection may favor some genotypes already present in the population. Under this scenario, the species range may be unchanged, but allele frequencies may vary (66–68). Third, if conditions are sufficiently severe, all organisms in the population will die or emigrate and the entire species range will shift (69).

Most difficult to uncover is the first of these scenarios, where environmental conditions do not directly affect mortality but instead alter phenotypic expression of genetic traits. Although acclimatization may imply improved organismal performance, such phenotypic plasticity may not necessarily improve fitness. For example, plasticity often involves the re-allocation of resources to one trait at the expense of another (e.g., trade-off between growth and reproduction). Furthermore, phenotypic variation can strongly influence biological interactions within a community, often in complex and counterintuitive ways (70, 71). Phenotypic plasticity is common in a broad range of plant and animal taxa, yet the consequences of trait-mediated interactions (TMIs) on community dynamics have rarely been documented, largely because of experimental constraints (70, 71). Thus the current challenge for biologists is not only to characterize the plastic responses of organisms to environmental change, but also to improve our understanding of how trait plasticity can affect community dynamics (72).

Here we explore these issues using a series of case studies to illustrate the potential pitfalls that can appear when aspects of physiological performance are not incorporated into the predicted consequences of climate change on biotic communities. It is not our intention to detract from the significant insights that have resulted from large-scale meta-analyses and monitoring studies, but rather to argue that by considering the mechanistic details of physiological performance within the context of biophysical ecology, such approaches will be better poised to predict where the impacts of climate change will most likely occur (40, 41, 46, 47, 73). Although our examples focus on invertebrate animals, we also draw from the plant and vertebrate literature. Our goal is to illustrate the role of physiological performance and biophysical ecology and to emphasize the importance of mechanism in predicting pattern in nature.

Few of the issues raised here are new to physiological ecology, although many have renewed urgency in the face of global climate change. However, the combination of new molecular techniques, coupled with biophysical methods of measuring and modeling environmental parameters, urges a reevaluation of the role of physiological mechanism in studies of climate change. The merger of these approaches presents an unprecedented opportunity to address explicit hypotheses regarding the effects of climate change on natural communities under field conditions (26, 39–41, 46).

HOW DO WE MEASURE THE ENVIRONMENT? THE ROLE OF BIOPHYSICS IN CLIMATE CHANGE BIOLOGY

The response of any individual organism to climate change can be viewed as a series of cascading scales, in which an organism's microclimatic conditions are transduced through the animal or plant's morphology to modify its cellular environment, ultimately eliciting a physiological response. Such measurements and models of climate and climate change are feasible only through large-scale approaches such as remote sensing and observatory networks. These technologies are not only incapable of directly providing information at smaller (microclimate) scales relevant to physiological performance, but they also fail to account for the effects of organism phenotype on physiological response. One promising approach for spanning this diverse range of scales lies in the application of biophysics, which accounts for heat, mass, and momentum exchange between organisms and their physical environment (40, 41, 46, 47).

For example, multiple interacting climatic factors, including short- and long-wave radiation, conduction, convection, and evaporation (43, 74–76), determine body temperature. Furthermore, many characteristics of the organism such as its color, morphology, and mass may modify heat exchange. Thus two ectothermic organisms exposed to identical climatic conditions can exhibit markedly different body temperatures (47, 75). Morphological variation in coefficients of drag, lift, acceleration, and mass transfer (77) can similarly affect the exchange of mass (gas and nutrients) and momentum. For these reasons, the use of environmental temperature indices obtainable by satellite or global observatories [such as sea surface temperature (SST) and air temperature (T_a)] as proxies for body temperature (T_b) require explicit validation, even as relative indicators of thermal stress.

In a specific example, most predictions of coral bleaching are based on measurements of SST (3, 11, 12), even though this metric may indicate the temperature of only the uppermost few centimeters of the water column, and temperatures at depth can often change over much smaller spatial and temporal scales (78, 79). Whereas SST anomalies appear to be good predictors of bleaching over large scales (80), linking small-scale variation in coral physiology with these large-scale indicators of environmental stress is challenging. Indeed, spatial heterogeneity in coral bleaching may be related to concomitant variability in water temperature and flow (4, 81), but these patterns may be difficult to detect with satellite imagery without extrapolating such measurements to scales relevant to the organism.

Similarly, although SST is probably a good indicator of body temperature of intertidal organisms during submergence at high tide, it is not a good proxy for T_b during low tide (47, 82) when these organisms are known to be exposed to damaging thermal conditions (26, 83–85). Likewise, at least for some intertidal organisms, T_a is also a poor indicator of T_b (47, 82).

Biophysical methods, however, provide the critical link between large-scale measurements and individual organisms, and thus may enable large-scale prediction of climate change impact. Such methods explicitly model and measure rates of heat, mass, and momentum exchange between organisms and their environment and quantitatively provide an environmental context for controlled physiological studies (40, 41, 43–47, 59, 75, 85). A biophysical approach is a powerful means of predicting how environmental parameters, such as climate or wave height, are translated into physiologically relevant parameters, such as body temperature or force acting on organisms, and even more importantly of describing the temporal and spatial scales over which these processes occur. Such an understanding is a vital link in knowing where and when to anticipate the effects of climate change.

WHEN DOES CLIMATE SET SPECIES RANGE LIMITS? THE ROCKY INTERTIDAL ZONE AS A MODEL SYSTEM

The assumption that climate and, in particular, temperature set the local and geographic distributions of species is implicit to the idea that climate change will significantly alter natural ecosystems. Indeed, strong correlations exist between temperature and the distribution of species and populations over a range of spatial and temporal scales (14, 26, 86–102). However, several other studies spanning a range of ecosystems have found poor correlations between climate and species range boundaries, challenging this assumption (49, 53, 103, 104). Here we exemplify this issue by considering attempts to understand limits to the distribution of intertidal organisms.

The rocky intertidal zone has been a primary natural laboratory for understanding the role of physical factors in setting range limits (105, 106). It is the interface of the terrestrial and marine environments and is one of the most physically harsh environments on earth. Forces acting on algae and animals from crashing waves in the intertidal zone can regularly exceed those generated by hurricane-force winds on land (58, 61, 107–112), and the body temperatures of intertidal organisms can fluctuate by 25°C or more in a matter of hours during aerial exposure at low tide (74, 82, 85, 86, 113–115). Because gas exchange during aerial exposure concomitantly leads to desiccation, many intertidal invertebrates rely partially or completely on anaerobiosis during low tide, which can lead to marked changes in body chemistry that have cascading influences on physiological performance even after resubmergence (31, 116–120). Temperature and desiccation also interact to influence the growth and survival of macroalgae (121–123). Salinity levels can fluctuate dramatically and, particularly for salt marsh plants, edaphic conditions can have extensive impacts on community structure (52, 124, 125).

Because rocky intertidal invertebrates and algae evolved under marine conditions, these species are generally assumed to be exposed to the highest stress levels at the upper (vertical) edges of their intertidal distributions (105); where temperature and desiccation stresses are assumed to be maximal, periods of maximal aerobic metabolism are greatest (116, 126), and feeding is most limited. The lower distributional limits of these organisms, in contrast, are set by biotic factors such as competition (127) and predation (128). The result is often a striking pattern of zonation in which bands of species replace one another at regular intervals along vertical gradients of physiological stress (105).

Two logical extensions of this paradigm have relevance to the expected effects of climate change on rocky intertidal communities. First, if rocky intertidal species are already living at the absolute limits of their physiological tolerances, then any marked change in the physical environment should be reflected by shifts in upper zonation limits. Specifically, in the face of warming, the upper zonation limits of species should shift downward wherever these limits are set by some aspect of climatic stress (2). Second, if some aspect of physiological stress related to climate sets species range boundaries, then, as in other ecosystems, the distribution of warm- and cold-acclimated species should shift at extremes of range boundaries (13, 15, 129, 130). Several lines of evidence, however, suggest that these expectations are too simplistic.

First, how often upper zonation limits actually are set by climatic factors remains unresolved (104, 131), although some evidence suggests a relatively straightforward linkage. Wethey (50) experimentally transplanted barnacles (*Semibalanus*) above the level at which they naturally occur in New England (United States), and showed that high summer temperatures set the upper zonation limit of this species through post-settlement mortality. Chan & Williams (132) showed similarly that heat stress appears to set the upper distributional limit of the barnacle *Tetraclita* on the rocky shores of Hong Kong. Tomanek & Somero (133) recently demonstrated that the production of heat shock proteins by two congeneric species of inter-tidal snails (*Tegula* spp.) appears to match their intertidal distributional limits, and Hofmann & Somero (30) have reported similar patterns for congeneric species of mussels (*Mytilus* spp.) over latitudinal scales. Southward (134) compared the ther-mal tolerances of several species of intertidal invertebrates in the United Kingdom with the climatic conditions to which they were normally exposed. Whereas he found that organisms at the upper limits of their intertidal distributional limits were seldom subjected to lethal thermal limits, his results did suggest that distributional limits were set by sublethal exposures to physiologically damaging temperatures or by the indirect effects of thermal stress on biotic factors such as competition for space.

In contrast, Wolcott (104) found little correspondence between maximum habi-tat temperature and the upper intertidal distribution of the limpet *Acmaea* and suggested instead that the distribution of these animals was set by biotic factors such as competition or food availability. Sanford (56, 57) presented evidence that the upper foraging limit of the sea star *Pisaster* was driven by seawater temper-ature and appeared to be unrelated to terrestrial conditions. Harley & Helmuth (53) compared microscale patterns of body temperature of barnacles and mussels in Puget Sound (United States) and found that upper zonation limits were tightly correlated with temperature at some sites, but at others the correlation was very weak, suggesting that temperature alternated with an additional limiting factor, such as feeding time, in setting upper zonation heights of these species in this region. Similar uncertainty exists when examining patterns over larger latitudinal scales.

If geographic range limits of intertidal species are set by temperature, as are upper zonation limits, then increased warming should bring about a relative in-crease in warm species at the more equatorial ends of species distributional limits (2, 13, 15, 16). Just such a pattern was observed over a 60-year period at a site in central California (13, 15), and Southward et al. (16) reported similar shifts in distributions of intertidal and planktonic species based on 70 years of climatic and faunal records from the western English Channel. Both studies related the observed shifts in abundance to increases in SST. However, as stated above, SST is not always an effective proxy for the body temperature of intertidal organisms (35, 47, 73, 82), and predicted patterns of thermal stress based on SST can vary significantly from those based on body temperature during aerial exposure at low tide (77). Specifically, Helmuth et al. (73) showed for intertidal mussels on the Pacific coast of North America that instead of a monotonic increase in thermal stress with latitude, a series of "hot spots" exist, where the timing of low tide coincides with hot terrestrial conditions and low wave splash (135). As a result, some sites in northern Washington State and Oregon are likely to become more thermally stressful than sites 2000 km to the south in central California (47, 73). Data collected at multiple intertidal sites (47) suggest that the site examined by

Barry et al. (15) and Sagarin et al. (13) in central California may in fact represent one of these hot spots, and thus it may not be possible to extrapolate results from studies conducted at a single site to the entire Pacific coast of the United States. Predictions of the effects of climate change on the geographic distributions of intertidal organisms, which are based on air temperature or SST alone, are therefore at variance with those based on predictions of the body temperature of the animals (47, 73). As described by Sagarin & Gaines (129, 130), one common model of species range distributions assumes that species peak in abundance near the center of their distributions, where environmental conditions are presumably near the species' physiological optimum, and that abundances taper off near the edges of the species distribution. Sagarin & Gaines measured the abundances of 12 species of intertidal invertebrates at 42 intertidal sites along the Pacific coast of North America, from Mexico to Alaska. Of the 12 species examined, only 2 showed evidence of the abundant center distribution (129, 130). Although these patterns have yet to be explicitly matched against predictions of organismal body temperature, these results are consistent with whatever factor is determining patterns of abundance in these intertidal species not increasing monotonically with latitude along the west coast of the United States.

Similarly, understanding patterns of selection in the field may be difficult without first explicitly quantifying the scales at which environmental stressors operate (35, 107, 136). For example, Schmidt & Rand (137, 138) showed for barnacles that high levels of spatial heterogeneity in the thermal environment were sufficient to maintain genetic polymorphism for alleles affected by thermal stress. Recent studies have suggested that small-scale heterogeneity in thermal niches over the scale of centimeters can often exceed those observed over the scale of thousands of kilometers on the west coast of the United States (35, 37, 47, 73). Clearly, even in a model ecosystem (114, 128, 139) understanding the relationship between climate and climate change in driving species distribution patterns is more challenging than previously assumed.

THE IMPORTANCE OF ENVIRONMENTAL SAFETY FACTORS: EFFECTS OF SHIFTING WAVE CLIMATE ON ROCKY SHORE COMMUNITIES

Vertical zonation in intertidal communities can be influenced by a variety of horizontal gradients in environmental conditions such as wave exposure, substrate orientation, salinity, or nutrient supply (99, 139–141). On rocky shore headlands, for example, increased wave splash and run-up expand and shift zones upward (53, 139). Water motion can also influence the strength of biological interactions, often altering the dominance of a species in a particular zone. One well-known example is the competitive dominance of mussels in the mid-intertidal zone of wave-exposed shores, where they find refuge from mobile predators such as crabs, snails, and sea stars (142–144). Predators are typically effective in reducing mussel abundance in calmer habitats, allowing competitively inferior species, such as barnacles or macroalgae, to persist.

Accordingly, large-scale change in these other variables may affect intertidal communities and is foreseeable. Indeed, one predicted consequence of global warming is an increase in the severity of wave conditions in many parts of the world's oceans. For example, wave heights on the west coast of North America are linked to El Niño (ENSO) conditions, which have increased in frequency and severity over the past few decades (145–147). In the North Atlantic, wave heights have increased at the rate of 2% annually since 1950, and the severity of winter storms is associated with the phase of the North Atlantic Oscillation (NAO) (148–150). The intensity and frequency of hurricanes in the region has also increased in recent years (151).

How will rocky shore organisms respond to increased wave severity? Following the mechanistic approach developed by Denny and colleagues (65, 158, 159), increased wave height should generate increased hydrodynamic forces on organisms (lift, drag, etc.) that could potentially exceed an organism's attachment force and cause mechanical failure. Thus a general prediction is that more severe wave conditions should result in increased dislodgment, shifting species distributions to more wave-protected habitats. Alternatively, species capable of altering hydrodynamic properties (by streamlining morphology, reducing size, etc.) or increasing attachment strength could potentially maintain their distribution. Discerning which of these two scenarios is most likely for a given species has proven difficult for a number of reasons. First, the linkage between offshore wave height (the metric climatologists provide) and the hydrodynamic forces generated on rocky shore organisms has proven difficult to characterize. This is in part because waves are subject to local modification by factors such as seafloor topography, wind speed, and direction. Waves breaking on topographically complex rocky shores undergo further modification (funneling, constructive interference, etc.), yielding maximal flows that are spatially heterogeneous and, in some instances, only weakly correlated with offshore conditions (107, 111, 112, 136). Additionally, our understanding is limited of how a given flow generates hydrodynamic forces on organisms in situ. Despite much progress on the hydrodynamic roles of morphology, flexibility, and aggregation (63, 154–160), direct measures of flow forces on wave-swept organisms are scarce (110, 160).

Second, increasing evidence indicates that the attachment strength of a given organism is often not fixed but instead varies temporally and spatially along gradients of hydrodynamic stress. For example, macroalgae can increase holdfast strength in high-energy habitats (161, 162), and mussel attachment at a given site can vary twofold during the course of a year (58, 163). Therefore, predictions of mechanical failure must evaluate whether organismal attachment strength is sufficient in the context of environmentally and temporally relevant flow forces. Johnson & Koehl (63) conducted such an analysis of environmental safety factors for the subtidal kelp *Nereocystis leutkeana*, concluding that kelps compensate for increased flow in high-energy habitats by increasing strength and becoming more streamlined. Similar observations for other large macroalgae (161, 164, 165) suggest that in many cases increased wave climate will affect algal size, morphology, and tissue mechanics rather than survivorship alone. Such trait plasticity may nonetheless have important fitness consequences by altering reproductive output or biotic interactions.

Not all phenotypic plasticity can be considered acclimatory, as is the case with the blue mussel, *Mytilus edulis*. Environmental safety factors for mussels in southern New England vary considerably during the course of the year owing to a mismatch in the phenology of attachment strength and wave climate (58). Mussels are most sensitive to dislodgement during August to November when large storms coincide with weak mussel attachment. Despite the difficulties in precisely characterizing hydrodynamic forces discussed above, measurements of wave disturbances to mussel beds largely validate these mechanistic predictions (E. Carrington, unpublished data). Thus generic changes in overall wave climate may not be relevant to all organisms; rather, only changes in those extreme events that coincide with periods of low safety factors may be ecologically important. Increases in storm activity during periods of high safety factors (e.g., late season Nor'easters in New England) would likely have little effect on survivorship. Note that these insights would not be possible without a mechanistic understanding of the biophysics of mussel attachment in the surf zone and how it varies seasonally.

The physiological response of mussels to environmental conditions defines the window of time where they are susceptible to dislodgment by severe wave conditions. This window of sensitivity is from late summer to fall for mussels in southern New England, but in spring for the same species on the other side of the Atlantic, in the United Kingdom (58, 59, 163, 166). The two populations therefore differ in their sensitivity to wave climate; mussels in New England will respond strongly to shifts in hurricane activity, whereas those in the United Kingdom will be more responsive to changes in storm activity that occur later in the storm season. This biophysical analysis therefore provides insight into which aspects of wave climate (seasonal storm patterns) may be most important for mussels in a given geographic location. Broad generalization beyond these two mussel populations, however, requires an understanding of the proximal environmental causes of plasticity in mussel attachment. It is likely that environmental parameters other than wave action are also involved (167), and synergistic effects are likely. For example, if low temperature is the proximal cue for increased attachment strength, then a gradual increase in water temperature would shift mussels to calmer habitats, even in the absence of a shift in wave climate.

Another potential synergism involves how changes in wave climate may influence biotic interactions, such as the ability of mussels to dominate space on high-energy intertidal shores. Mussel dominance depends not only on mortality but also on growth and recruitment (168, 169). That is, mussel beds can maintain 100% cover (excluding competitors) with substantial mortality rates as long as the vacated space is reoccupied by growth of the survivors or the arrival of new recruits. These latter two processes are largely driven by other aspects of the environment (temperature, food supply, or currents), which in turn may be subject to climatic shifts. If the effect of climate change on wave height is highly seasonal (i.e., just increased hurricane activity in late summer/fall), then mussels in New England are predicted to shift into what are currently lower-energy habitats. But in such habitats mussels may no longer have a refuge from predation because foraging activities are most intense in summer. In this manner, seasonally dependent climatic change (more severe winters, increased hurricane activity) may shift the distributions of some organisms but not others. Although biophysical models provide an essential component in predicting range shifts, whether the organism of interest persists ultimately depends on the strength and direction of biotic interactions in the new community assemblage.

INSECT RESPONSES TO CLIMATE CHANGE: PATTERNS, PROCESSES, AND PREDICTIONS

The quantitative relationship of weather and climate to distribution and abundance of insects has been extensively explored for over half a century (170). Statistical and semi-mechanistic models relating climate to geographic distribution have been developed for many insects, in particular for agricultural pests and disease vectors. As a result, insects have provided some useful case studies for examining and predicting how climate change can alter insect distribution and abundance (171).

Butterflies are a particularly good example because data are abundant on their geographic ranges and the demographic, behavioral, and physiological mechanisms by which weather and climate affect their ecology (172). For example, seasonal aspects of environmental temperature, radiation, and moisture influence (*a*) larval growth and development rates and the number of generations per year; (*b*) overwintering mortality; (*c*) phenological shifts of the life cycle relative to host plants and natural enemies; and (*d*) thermoregulation, flight activity, and realized fecundity of females [see (172) for a detailed review]. This understanding has been used to predict two major qualitative ecological responses of butterflies to climatic warming: shifts in range and abundance at northern and southern range boundaries (173) and phenological shifts in the seasonal life cycle in spring and fall (174).

Several recent analyses have evaluated aspects of these predictions for butterflies in western Europe, where annual mean temperatures have increased by ∼0.8°C during the past century. Parmesan and colleagues (175) found that during the past 30–100 years, 22 (63%) of 35 nonmigratory species shifted their ranges northward; only 2 of 35 species shifted southward. Typically, northern range shifts were the result of northward shifts in the northern boundaries: Only 22% involved northern shifts in both northern and southern boundaries. The extent of boundary shifts was 35–240 km (along a single boundary), similar to the mean shift in climatic isotherms in Europe during the past century (120 km). As a result, the overall response of most of these species to climate warming has been an increase in geographic range to the north.

Roy & Sparks (176) used data from the British Butterfly Monitoring Scheme to explore phenological shifts in adult flight season during the past two decades. They found that 13 (37%) of 35 species had earlier average spring or summer adult appearance dates; none showed later average appearance dates. For species with multiple generations (flight seasons) per year, the duration of the flight season was also greater. Using regression analyses, Roy & Sparks (176) estimated that a 1°C increase in mean annual temperature would advance adult appearance by 2–10 days in most species.

These analyses clearly demonstrate the consequences of recent climate warming for European butterflies and suggest that predicted climate change during the next 50–100 years will cause range and phenological shifts in butterflies in northern temperate regions. However, there are some important limitations to these conclusions for understanding general responses of butterflies to climate change. For example, these analyses for European range shifts excluded many, perhaps the majority, of nonmigratory European butterfly species (migratory species were also excluded). Species were excluded if they were limited by host-plant distribution, were strongly habitat restricted, or had experienced severe habitat loss (175); thus the analyses naturally focused on those species anticipated to respond to climate. Even among this subset of species is considerable heterogeneity in whether response to climate was detected in northern or southern range boundaries or phenological appearance. As explored before, ecological and physiological differences among species likely contribute to this heterogeneity in response to recent climate change.

Several recently developed models predict how anticipated global climate warming in the coming decades will affect distributions and extinctions of assemblages of insects and other organisms (23, 177). These models combine two main elements. First, a model of the climatic niche or envelope of each species is developed. Although the statistical details and assumptions of different methods vary (178), the basic approach is to use information on geographic occurrences of a species and climatic variables to predict the probability of occurrence (or absence) as a function of climatic variables at different geographic locations. The model for the climatic envelope may use information on only mean annual temperature and precipitation (178) or more detailed information about seasonal or mean monthly temperatures and precipitation (179); none of the current models uses biophysical approaches to measure climate spaces that mechanistically connect environmental conditions to body temperature (75, 180). Second, predictions from general circulation models are used to develop climatic scenarios for future climate conditions. Because the grid size of most general circulation models is quite large, interpolation methods are used to predict local geographic conditions. With this approach, recent analyses predict widespread extinctions and/or distribution changes in the butterfly faunas of Mexico and of Australia (23, 177, 179).

There are several important issues to consider in interpreting these predictions regarding future climate change. First, different annual and seasonal aspects of climate are strongly spatially auto-correlated, making it difficult for climatic envelope models to correctly identify which aspects of climate are associated with occurrence. Yet a diversity of studies illustrates how specific seasonal components of climate during key life stages can dominate the responses of most butterfly species to climate change (43, 172, 181, 182). This probably contributes to the heterogeneity of responses of butterflies to recent (past) climate warming (see above).

Importantly, climate change involves much more than simple increases in global mean temperatures. There is general consensus from predictions of global circulation models about the ways in which greenhouse gases will alter climate (20, 183). First, warming will be greater at higher than at lower latitudes, especially in the northern hemisphere. Second, warming will be greater in winter than in summer, especially at high latitudes. Third, warming will be greater for nocturnal (nighttime) than for diurnal (daytime) temperatures in most seasons and geographic regions, leading to a reduction a daily thermal variation. Indeed, nearly all of the climatic warming seen in north temperate regions in the past century is the result of increased nighttime temperatures. Fourth, the mean and perhaps the variability of global precipitation will increase, with greater winter precipitation at higher latitudes. The general pattern is that anticipated climate change will have greater effects on low temperatures than on high temperatures, and will tend to reduce daily, seasonal, and latitudinal variation in temperatures. As a result, predicted climate change will alter the correlations among different seasonal and spatial components of climate. This will alter the accuracy of predictions of responses to climate change based on statistically estimated climatic envelopes. One way to assess the accuracy of currently available predictions for responses to future climate change is to test the success of current climatic envelope models in predicting responses to past climate change or for range changes in introduced or invading species. Samways et al. (49) provide an interesting example of this approach, making use of data on 15 ladybird species that have been intentionally introduced as biocontrol agents in different areas around the world. They determined climatic envelopes for each species in their native range, using a widely used climatic envelope model (CLIMEX), then compared their introduced geographic range with model predictions. The match between predicted and observed ranges varied from 0 to 100%, and only 4 of 15 species ranges were predicted with high accuracy (49).

All these considerations suggest that a more mechanistic approach is needed to quantitatively predict changes in abundance, geographic range, or phenology of insects in response to climate change. To explore this further we consider several more-detailed case studies of recent responses of butterflies, mosquitoes, and other insects to climate change (20).

The White Admiral Butterfly (*Ladoga camilla* L.) experienced a large northern and western extension of its range in southern Britain during the first half of the last century (172). Detailed field studies at the species range boundary indicated two primary causes of this range extension. First, mean daily temperatures in June caused increased growth and developmental rates, thereby reducing larval and pupal mortality owing to predation. In addition, warm and sunny weather during

July increased the time available for flight because these butterflies behaviorally thermoregulate body temperatures to achieve the elevated temperatures required for active flight (43). This increased the number of eggs laid by females. These two seasonal components of climate are most strongly associated with the observed range extension in White Admirals (172). More generally, mean temperatures and insolation in June-July are the best predictors of butterfly population abundance for more than 80% of British butterfly species that have been studied (172).

Crozier (184, 185) used a more experimental approach to explore climatic determinants of the range expansion of a North American species, the Sachem Skipper (*Atalopedes campestris* L.). The Sachem is a generalist species that feeds a variety of grasses as a larva. During the past half century it has extended its northern range by nearly 500 km into the Pacific Northwest region of the United States. Over this time the mean minimum January temperature has increased by 3°C in this area; the boundary shift is correlated with the –4°C mean January minimum isotherm. The Sachem species lacks a physiological diapause stage and overwinters in larval stages. Measurements of supercooling points and minimum lethal temperatures demonstrate that –5 to –6°C is a critical thermal limit for the Sachem: this limit is not influenced by life stage, acclimation, or population of origin (184). In chronic cold stress experiments, mortality is high in diurnally fluctuating temperatures typical of winter conditions at the current range boundary. For example, larval survival over a 2–3-week period was greater than 70% under a 8 to 0°C thermal cycle, less than 10% at 4 to –4°C, and 0% at 0 to –8°C. Reciprocal field transplant experiments of larvae between the current range boundary and ~100 km inside the range showed significantly lower larval survival rates over the winter at the range boundary compared with that inside the range, with mean overwintering survival rates of 1% or less at the range boundary (185). In contrast, there were no significant differences in larval mortality in summer conditions inside and at the range boundary. Field population censuses of adults show that population sizes increase dramatically from spring to fall each year both inside and at the range boundary; in fact, the magnitude of population expansion is greater at the boundary than inside the range.

The picture that emerges from this case study of the Sachem is that summer population expansion alternates with high overwintering mortality at the range boundary. A demographic model that incorporates these climatic effects shows that winter warming was a prerequisite for this butterfly's range expansion (185). Because future climate change is expected to increase winter temperatures more than summer temperatures, especially at higher latitudes, this may be a major determinant of recent and future range extensions in insects that are freeze-intolerant or that lack a physiological diapause state.

It is well known that tolerance to extreme high and low temperatures varies among insect species. Addo-Bediako et al. (186) recently compiled the data available for supercooling points (SCP), lower lethal temperatures (LLT), upper lethal temperatures (ULT), and critical thermal maxima (CTmax) for 250 insect species representing 87 families and 10 orders. SCP and LLT were strongly correlated for freeze-intolerant species but not for freeze-tolerant species. There was no significant relationship of latitude to either ULT or CTmax; most of the variation in ULT and CTmax was at the family or generic level. In contrast, mean LLT and SCP declined with increasing latitude. This pattern is largely because of the large increases in variation in LTT and SCP among species at higher latitudes, which reflects the fact that there are a variety of physiological and behavioral mechanisms, in addition to reduced SCP and LTT, by which insects may weather low winter temperatures at higher latitudes (187, 188). One crucial factor is the phenological timing and predictability of snow cover, which strongly moderates exposure to extreme low temperatures at higher latitudes and altitudes. There are two important implications of these studies for responses to future climate change. First, increasing minimum winter temperatures at higher latitudes owing to climate change may cause northern range extensions for insect species in which SCP and LTT are the major physiological determinants of overwintering mortality. Second, current global climate change models also predict increased winter precipitation at higher latitudes. However, it is unclear how the combination of increased winter temperature and precipitation will affect timing and extent of snowpack, which will determine temperatures of insects in their overwintering sites. This will likely increase the heterogeneity of the responses of different insects to future climate change and may be one cause of the heterogeneity of response of European butterflies to recent climate change (see above).

The seasonal life cycle of most insects is mediated by seasonal cues that initiate and terminate diapause, estivation, quiescence, or other states in which growth, development, and reproductive processes are suspended. For many temperate and higher latitude insects the transitions to and from diapause are mediated by daylength or photoperiod, and there is widespread geographic variation within and between insect species in the critical photoperiod (CP) that initiates or terminates diapause (189). Thus changes in photoperiodic responses may be an important mechanism by which insects respond to climate change. Bradshaw & Holzapfel (190) provide an elegant demonstration of this response for pitcher-plant mosquitoes, *Wyeomyia smithii*. The larvae of pitcher-plant mosquitoes are restricted to the water-filled leaves of purple pitcher plants, which occur in distinct geographic populations over a wide latitudinal gradient in eastern and central North America. The CP for initiating and terminating larval diapause increases with increasing latitude, such that northern populations terminate diapause later in the spring and initiate diapause earlier in the fall than southern populations. Studies of diapause responses in this species during the past three decades demonstrate genetic shifts in the CP initiating diapause during the past 25 years, coincident with increases in mean annual temperatures during this time period (190). This evolutionary shift has occurred primarily in northern populations, where climate change has been greatest, and corresponds to a nine-day advancement in the date of diapause initation in northern populations. This study illustrates that evolutionary changes may be an important component of biological responses to climate change for insects and other short-lived organisms (171, 191).

MECHANISTIC ECOLOGY IN THE FACE OF CLIMATE CHANGE: WHERE DO WE GO FROM HERE?

Physiological ecologists have long recognized the multitude of interactions between environmental factors in driving physiological stress in the field, and large-scale studies of the effects of climate on ecosystems have rightly pointed to the difficulties inherent in beginning with models at the level of the organism (16, 18, 48). Nevertheless, despite the difficulties in this approach, a failure to consider the mechanistic details of environmental heterogeneity, coupled with the physiological response of organisms to these variables, can lead to erroneous predictions regarding how and where we should look for the effects of climate change.

While the application of physiological techniques to questions of biogeography and ecology has a long history, the advent of new field-based techniques, coupled with biophysical/biomechanical approaches, offers an unprecedented opportunity in the form of hypothesis generation and testing. Specifically, through large-scale meta-analyses we are able to detect patterns and then correlate these patterns with environmental variables; indeed, it may be through these types of approaches that we are best able to determine where trouble spots are occurring on regional and global scales (18, 80). However, although these approaches may be good at detecting pattern, their predictive ability in terms of forecasting future changes may be poor. In contrast, biophysical methods may provide a means of generating explicit hypotheses regarding where and when we should or should not expect to observe physiological stress in the field. Previously, testing such predictions would have required the observation of range shifts or large-scale mortality events, but with the application of new field-based physiological techniques we are now able to precisely test these predictions on small temporal and spatial scales (26–28, 35, 37, 39). The combination of molecular physiology and biophysical ecology, coupled with the detection of large-scale patterns of change, thus presents our best chance of predicting what to expect in the face of global climate change in the coming decades.

ACKNOWLEDGMENTS

We are grateful to Sarah Gilman and Ken Sebens for their editorial input to the paper, and to Lisa Wickliffe for her help in locating literature. We also thank Tom Daniel, Patti Halpin, Jerry Hilbish, Gretchen Hofmann, Scott McWilliams, Saran Twombly, and David Wethey for hours of helpful discussions. Many of the ideas presented in this paper were influenced by our participation in various iterations of the Physical Biology summer course at the University of Washington Friday Harbor Laboratories, and we are thankful for all of the instructors, students, and staff who make the course and learning environment there possible. We also wish to acknowledge support from the National Science Foundation (OCE-0082605 to EC; OCE-0323364 to BH; and IBN-0212798 to J.G.K.).

The *Annual Review of Physiology* is online at http://physiol.annualreviews.org

LITERATURE CITED

1. Intergov. Panel Climate Change. 2001. *Climate Change 2001: The Scientific Basis.* http://www.grida.no/climate/ipcc_tar/wg1/figts-22.htm
2. Lubchenco J, Navarrete SA, Tissot BN, Castilla JC. 1993. Possible ecological consequences to global climate change: nearshore benthic biota of Northeastern Pacific coastal ecosystems. In *Earth System Responses to Global Change: Contrasts Between North and South America*, ed. HA Mooney, ER Fuentes, BI Kronberg, pp. 147–66. San Diego: Academic
3. Wilkinson CR. 1996. Global change and coral reefs: impacts on reefs, economies and human cultures. *Global Change Biol.* 2:547–58
4. West JM, Salm RV. 2003. Resistance and resilience to coral bleaching: implications for coral reef conservation and management. *Conserv. Biol.* 17:956–67
5. Bunkley-Williams L, Williams EH Jr. 1990. Global assault on coral reefs. *Nat. Hist.* 4:47–54
6. Sebens KP. 1994. Biodiversity of coral reefs: What are we losing and why? *Am. Zool.* 34:115–33
7. Glynn PW. 1996. Coral reef bleaching: facts, hypotheses and implications. *Global Change Biol.* 2:495–509
8. Hoegh-Guldberg O. 1999. Climate change, coral bleaching and the future of the world's coral reefs. *Mar. Freshw. Res.* 50:839–66
9. Aronson RB, Precht WF, Toscano MA, Koltes KH. 2002. The 1998 bleaching event and its aftermath on a coral reef in Belize. *Mar. Biol.* 141:435–47
10. Pandolfi JM, Bradbury RH, Sala E, Hughes TP, Bjorndal KA, et al. 2003. Global trajectories of the long-term decline of coral reef ecosystems. *Science* 301:955–58

11. Hughes TP, Baird AH, Bellwood DR, Card M, Connolly SR, et al. 2003. Climate change, human impacts, and the resilience of coral reefs. *Science* 301:929–33
12. Brown BE. 1997. Coral bleaching: causes and consequences. *Coral Reefs* 16(Suppl.):S129–38
13. Sagarin RD, Barry JP, Gilman SE, Baxter CH. 1999. Climate related changes in an intertidal community over short and long time scales. *Ecol. Monogr.* 69:465–90
14. Ottersen G, Planque B, Belgrano A, Post E, Reid PC, Stenseth NC. 2001. Ecological effects of the North Atlantic oscillation. *Oecologia* 128:1–14
15. Barry JP, Baxter CH, Sagarin RD, Gilman SE. 1995. Climate-related, long-term faunal changes in a California rocky intertidal community. *Science* 267:672–75
16. Southward AJ, Hawkins SJ, Burrows MT. 1995. Seventy years' observations of changes in distribution and abundance of zooplankton and intertidal organisms in the western English Channel in relation to rising sea temperature. *J. Thermal Biol.* 20:127–55
17. Hawkins SJ, Southward AJ, Genner MJ. 2003. Detection of environmental change in a marine ecosystem—evidence from the western English Channel. *Sci. Total Environ.* 310:245–56
18. Parmesan C, Yohe G. 2003. A globally coherent fingerprint of climate change impacts across natural systems. *Nature* 421:37–42
19. Root TL, Price JT, Hall KR, Schneider SH, Rosenzweigk C, Pounds JA. 2003. Fingerprints of global warming on wild animals and plants. *Nature* 421:57–60
20. Kingsolver JG. 2002. Impacts of global environmental change on animals. In *The Earth System: Biological and Ecological Dimensions of Global Environmental*

Change, ed. HA Mooney, JG Canadell, pp. 56–66. Chichester, UK: Wiley

21. Noble IR. 1993. A model of the responses of ecotones to climate change. *Ecol. Appl.* 3:396–403

22. Root TL, Schneider SH. 1995. Ecology and climate: research strategies and implications. *Science* 269:334–41

23. Thomas CD, Cameron A, Green RE, Bakkenes M, Beaumont LJ, et al. 2004. Extinction risk from climate change. *Nature* 427:145–48

24. Fields PA, Graham JB, Rosenblatt RH, Somero GN. 1993. Effects of expected global climate change on marine faunas. *Trends Ecol. Evol.* 8:361–67

25. Feder ME, Hofmann GE. 1999. Heat-shock proteins, molecular chaperones, and the stress response. *Annu. Rev. Physiol.* 61:243–82

26. Somero GN. 2002. Thermal physiology and vertical zonation of intertidal animals: optima, limits, and costs of living. *Integr. Comp. Biol.* 42:780–89

27. Dahlhoff EP. 2004. Biochemical indicators of stress and metabolism: applications for marine ecological studies. *Annu. Rev. Physiol.* 66:183–207

28. Costa DP, Sinervo B. 2004. Field physiology: physiological insights from animals in nature. *Annu. Rev. Physiol.* 66:209–38

29. Hofmann GE, Somero GN. 1995. Evidence for protein damage at environmental temperature: seasonal changes in levels of ubiquitin conjugates and hsp70 in the intertidal mussel *Mytilus trossulus*. *J. Exp. Biol.* 198:1509–18

30. Hofmann GE, Somero GN. 1996. Interspecific variation in thermal denaturation of proteins in the congeneric mussels *Mytilus trossulus* and *M. galloprovincialis*: evidence from the heat-shock response and protein ubiquitination. *Mar. Biol.* 126:65–75

31. Stillman JH, Somero GN. 1996. Adaptation to temperature stress and aerial exposure in congeneric species in intertidal porcelain crabs (genus *Petrolisthes*): correlation of physiology, biochemistry and morphology with vertical distribution. *J. Exp. Biol.* 199:1845–55

32. Warner ME, Fitt WK, Schmidt GW. 1996. The effects of elevated temperature on the photosynthetic efficiency of zooxanthellae in hospite from four different species of reef coral: a novel approach. *Plant Cell Environ.* 19:291–99

33. Tomanek L, Somero GN. 1999. Evolutionary and acclimation-induced variation in the heat-shock responses of congeneric marine snails (genus *Tegula*) from different thermal habitats: implications for limits of thermotolerance and biogeography. *J. Exp. Biol.* 202:2925–36

34. Dahlhoff EP, Buckley BA, Menge BA. 2001. Feeding and physiology of the rocky intertidal predator *Nucella ostrina* along an environmental gradient. *Ecology* 82:2816–29

35. Helmuth BST, Hofmann GE. 2001. Microhabitats, thermal heterogeneity, and patterns of physiological stress in the rocky intertidal zone. *Biol. Bull.* 201:374–84

36. Buckley BA, Owen M-E, Hofmann GE. 2001. Adjusting the thermostat: the threshold induction temperature for the heat-shock response in intertidal mussels (genus *Mytilus*) changes as a function of thermal history. *J. Exp. Biol.* 204:3571–79

37. Halpin PM, Sorte CJ, Hofmann GE, Menge BA. 2002. Patterns of variation in levels of Hsp70 in natural rocky shore populations from microscales to mesoscales. *Integr. Comp. Biol.* 42:815–24

38. Tomanek L, Sanford E. 2003. Heat-shock protein 70 (Hsp70) as a biochemical stress indicator: an experimental field test in two congeneric intertidal gastropods (Genus: *Tegula*). *Biol. Bull.* 205:276–84

39. Menge BA, Olson AM, Dahlhoff EP. 2002. Environmental stress, bottom-up effects, and community dynamics: integrating molecular-physiological with

ecological approaches. *Integr. Comp. Biol.* 42:892–908

40. Porter WP, Budaraju S, Stewart WE, Ramankutty N. 2000. Calculating climate effects on birds and mammals: impacts on biodiversity, conservation, population parameters, and global community structure. *Am. Zool.* 40:597–630

41. Porter WP, Sabo JL, Tracy CR, Reichman OJ, Ramankutty N. 2002. Physiology on a landscape scale: plant-animal interactions. *Integr. Comp. Biol.* 42:431–53

42. Hodkinson ID. 1999. Species response to global environmental change or why ecophysiological models are important: a reply to Davis et al. *J. Anim. Ecol.* 68:1259–62

43. Kingsolver J. 1989. Weather and the population dynamics of insects: integrating physiological and population ecology. *Physiol. Zool.* 62:314–34

44. Kingsolver J. 1979. Thermal and hydric aspects of environmental heterogeneity in the pitcher plant mosquito. *Ecol. Monogr.* 49:357–76

45. Grant BW, Porter WP. 1992. Modeling global macroclimatic constraints on ectotherm energy budgets. *Am. Zool.* 32:154–78

46. Porter WP, Munger JC, Stewart WE, Budaraju S, Jaeger J. 1994. Endotherm energetics: from a scalable individual-based model to ecological applications. *Aust. J. Zool.* 42:125–62

47. Helmuth B. 2002. How do we measure the environment? Linking intertidal thermal physiology and ecology through biophysics. *Integr. Comp. Biol.* 42:837–45

48. Davis AJ, Lawton JH, Shorrocks B, Jenkinson LS. 1998. Individualistic species responses invalidate simple physiological models of community dynamics under global environmental change. *J. Anim. Ecol.* 67:600–12

49. Samways MJ, Osborn R, Hastings H, Hattingh V. 1999. Global climate change and accuracy of prediction of species' geographical ranges: establishment success of introduced ladybirds (Coccinellidae, *Chilocorus* spp.) worldwide. *J. Biogeogr.* 26:795–812

50. Wethey DS. 1983. Geographic limits and local zonation: the barnacles *Semibalanus* (*Balanus*) and *Chthamalus* in New England. *Biol. Bull.* 165:330–41

51. Bertness MD, Leonard GH, Levine JM, Bruno JF. 1999. Climate-driven interactions among rocky intertidal organisms caught between a rock and a hot place. *Oecologia* 120:446–50

52. Bertness MD, Ewanchuk PJ. 2002. Latitudinal and climate-driven variation in the strength and nature of biological interactions in New England salt marshes. *Oecologia* 132:392–401

53. Harley CDG, Helmuth BST. 2003. Local and regional scale effects of wave exposure, thermal stress, and absolute vs. effective shore level on patterns of intertidal zonation. *Limnol. Oceanogr.* 48:1498–508

54. Newell RC. 1969. Effect of fluctuations in temperature on the metabolism of intertidal invertebrates. *Am. Zool.* 9:293–307

55. Stillman JH. 2003. Acclimation capacity underlies susceptibility to climate change. *Science* 301:65

56. Sanford E. 1999. Regulation of keystone predation by small changes in ocean temperature. *Science* 283:2095–97

57. Sanford E. 2002. Water temperature, predation, and the neglected role of physiological rate effects in rocky intertidal communities. *Integr. Comp. Biol.* 42:881–91

58. Carrington E. 2002. Seasonal variation in the attachment strength of blue mussels: causes and consequences. *Limnol. Oceanogr.* 47:1723–33

59. Carrington E. 2002. The ecomechanics of mussel attachment: from molecules to ecosystems. *Integr. Comp. Biol.* 42:846–52

60. Schneider KR, Wethey OS, Helmuth B, Hilbish TJ. 2005. Implications of

movement behavior on mussel dislodgement: exogenous selection in a *Mytilus* spp. hybrid zone. *Mar. Biol.* In press

61. Denny MW, Daniel TL, Koehl MAR. 1985. Mechanical limits to size in wave-swept organisms. *Ecol. Monogr.* 55:69–102

62. Lowell RB. 1985. Selection for increased safety factors of biological structures as environmental unpredictability increases. *Science* 228:1009–11

63. Johnson AS, Koehl MAR. 1994. Maintenance of dynamic strain similarity and environmental stress factor in different flow habitats: thallus allometry and material properties of a giant kelp. *J. Exp. Biol.* 195:381–410

64. Koehl MAR. 1999. Ecological biomechanics of benthic organisms: life history, mechanical design, and temporal patterns of mechanical stress. *J. Exp. Biol.* 202:3469–76

65. Willmer P, Stone G, Johnston I. 2000. *Environmental Physiology of Animals.* Malden, MA: Blackwell. 666 pp.

66. Kirby RR, Berry RJ, Powers DA. 1997. Variation in mitochondrial DNA in a cline of allele frequencies and shell phenotype in the dog-whelk *Nucella lapillus* (L.). *Biol. J. Linn. Soc.* 62:299–312

67. Hilbish TJ, Koehn RK. 1985. The physiological basis of natural selection at the *Lap* locus. *Evolution* 39:1302–17

68. Hilbish TJ. 1985. Demographic and temporal structure of an allele frequency cline in the mussel *Mytilus edulis. Mar. Biol.* 86:163–71

69. Holt RD. 1990. The microevolutionary consequences of climate change. *Trends Ecol. Evol.* 5:311–15

70. Werner EE, Peacor SD. 2003. A review of trait-mediated indirect interactions in ecological communities. *Ecology* 84:1083–100

71. Callaway RM, Pennings SC, Richards CL. 2003. Phenotypic plasticity and interactions among plants. *Ecology* 84:1115–28

72. Schmitz OJ, Adler FR, Agrawal AA. 2003. Linking individual-scale trait plasticity to community dynamics. *Ecology* 84:1081–82

73. Helmuth B, Harley CDG, Halpin PM, O'Donnell M, Hofmann GE, Blanchette CA. 2002. Climate change and latitudinal patterns of intertidal thermal stress. *Science* 298:1015–17

74. Helmuth BST. 1999. Thermal biology of rocky intertidal mussels: quantifying body temperatures using climatological data. *Ecology* 80:15–34

75. Porter WP, Gates DM. 1969. Thermodynamic equilibria of animals with environment. *Ecol. Monogr.* 39:245–70

76. Kingsolver JG, Moffat RJ. 1982. Thermoregulation and the determinants of heat transfer in *Colias* butterflies. *Oecologia* 53:27–33

77. Denny MW, Wethey DS. 2001. Physical processes that generate patterns in marine communities. In *Marine Community Ecology*, ed. MD Bertness, SD Gaines, ME Hay, pp. 3–37. Sunderland, MA: Sinauer

78. Leichter JJ, Wing SR, Miller SL, Denny MW. 1996. Pulsed delivery of subthermocline water to Conch Reef (Florida Keys) by internal tidal bores. *Limnol. Oceanogr.* 41:1490–501

79. Leichter J, Helmuth B. 2004. High frequency temperature variability on Caribbean coral reefs: What thermal environments do corals actually experience? *Climate Change and Aquatic Systems: Past, Present and Future.* Plymouth, UK: Univ. Plymouth. Abstr.

80. Gleeson MW, Strong AE. 1995. Applying MCSST to coral reef bleaching. *Adv. Space Res.* 16:151–54

81. Nakamura T, van Woesik R. 2001. Waterflow rates and passive diffusion partially explain differential survival of corals during the 1998 bleaching event. *Mar. Ecol. Prog. Ser.* 212:301–4

82. Helmuth BST. 1998. Intertidal mussel microclimates: predicting the body

temperature of a sessile invertebrate. *Ecol. Monogr.* 68:51–74

83. Hofmann GE, Somero GN. 1996. Protein ubiquitination and stress protein synthesis in *Mytilus trossulus* occurs during recovery from tidal emersion. *Mol. Mar. Biol. Biotech.* 5:175–84

84. Hofmann GE. 1999. Ecologically relevant variation in induction and function of heat shock proteins in marine organisms. *Am. Zool.* 39:889–900

85. Bell EC. 1995. Environmental and morphological influences on thallus temperature and desiccation of the intertidal alga *Mastocarpus papillatus* Kützing. *J. Exp. Mar. Biol. Ecol.* 191:29–55

86. Wethey DS. 2002. Biogeography, competition, and microclimate: the barnacle *Chthamalus fragilis* in New England. *Integr. Comp. Biol.* 42:872–80

87. Zinsmeister WJ. 1982. Late Cretaceous-Early Tertiary molluscan biogeography of the southern circum-Pacific. *J. Paleontol.* 56:84–102

88. Brattström H, Johanssen A. 1983. Ecological and regional zoogeography of the marine benthic fauna of Chile. *Sarsia* 68:289–339

89. Ayres MP, Scriber JM. 1994. Local adaptation to regional climates in *Papilio canadensis* (Lepidoptera: Papilionidae). *Ecol. Monogr.* 64:465–82

90. Pyankov VI, Gunin PD, Tsoog S, Black CC. 2000. C4 plants in the vegetation of Mongolia: their natural occurrence and geographical distribution in relation to climate. *Oecologia* 123:15–31

91. Willot SJ. 1997. Thermoregulation in four species of British grasshoppers (Orthoptera: Acrididae). *Funct. Ecol.* 11: 705–13

92. Hugall A, Moritz C, Moussalli A, Stanisic J. 2002. Reconciling paleodistribution models and comparative phylogeography in the Wet Tropics rainforest land snail *Gnarosophia bellendenkerensis* (Brazier 1875). *Proc. Natl. Acad. Sci. USA* 99: 6112–17

93. Wethey DS. 1986. Climate and biogeography: continuous versus catastrophic effects on rocky intertidal communities. *Estud. Oceanol.* 5:19–25

94. Hilbish TJ. 1981. Latitudinal variation in freezing tolerance of *Malampus bidentatus* (Say) (Gastropoda: Pulmonata). *J. Exp. Mar. Biol. Ecol.* 52:283–97

95. Williams GA, Morritt D. 1995. Habitat partitioning and thermal tolerance in a tropical limpet, *Cellana grata. Mar. Ecol. Prog. Ser.* 124:89–103

96. Ottaway JR. 1973. Some effects of temperature, desiccation, and light on the intertidal anemone *Actinia tenebrosa* Farquhar (Cnidaria: Anthozoa). *Aust. J. Mar. Freshw. Res.* 24:103–26

97. Suchanek TH, Geller JB, Kreiser BR, Mitton JB. 1997. Zoogeographic distributions of the sibling species *Mytilus galloprovincialis* and *M. trossulus* (Bivalvia: Mytilidae) and their hybrids in the North Pacific. *Biol. Bull.* 193:187–94

98. Kennedy VS. 1976. Desiccation, higher temperatures and upper intertidal limits of three species of sea mussels (Mollusca: Bivalvia) in New Zealand. *Mar. Biol.* 35:127–37

99. Harley CDG. 2003. Abiotic stress and herbivory interact to set range limits across a two-dimensional stress gradient. *Ecology* 84:1477–88

100. Hutchins LW. 1947. The bases for temperature zonation in geographical distribution. *Ecol. Monogr.* 17:325–35

101. Bertness MD, Schneider DE. 1976. Temperature relations of Puget Sound thaids in reference to their intertidal distribution. *Veliger* 19:47–58

102. Hilbish TJ, Bayne BL, Day A. 1994. Genetics of physiological differentiation within the marine mussel genus *Mytilus*. *Evolution* 48:267–86

103. Simpson RD. 1976. Physical and biotic factors limiting the distribution and abundance of littoral molluscs on Macquarie Island (Sub-Antarctic). *J. Exp. Mar. Biol. Ecol.* 21:11–49

104. Wolcott TG. 1973. Physiological ecology and intertidal zonation in limpets (*Acmaea*): a critical look at "limiting factors". *Biol. Bull.* 145:389–422

105. Connell JH. 1972. Community interactions on marine rocky intertidal shores. *Annu. Rev. Ecol. Syst.* 3:169–92

106. Paine RT. 1994. *Marine Rocky Shores and Community Ecology: An Experimentalist's Perspective.* Oldendorf/Luhe, Ger: Ecol. Inst.

107. Helmuth B, Denny MW. 2003. Predicting wave exposure in the rocky intertidal zone: do bigger waves always lead to larger forces? *Limnol. Oceanogr.* 48: 1338–45

108. Denny MW. 1987. Life in the maelstrom: the biomechanics of wave-swept shores. *Trends Ecol. Evol.* 2:61–66

109. Carrington E. 1990. Drag and dislodgment of an intertidal macroalga: consequences of morphological variation in *Mastocarpus papillatus* Kutzing. *J. Exp. Mar. Biol. Ecol.* 139:185–200

110. Gaylord B. 2000. Biological implications of surf-zone flow complexity. *Limnol. Oceanogr.* 45:174–88

111. Bell EC, Denny MW. 1994. Quantifying "wave exposure": a simple device for recording maximum velocity and results of its use at several field sites. *J. Exp. Mar. Biol. Ecol.* 181:9–29

112. Gaylord B. 1999. Detailing agents of physical disturbance: wave-induced velocities and acceleration on a rocky shore. *J. Exp. Mar. Biol. Ecol.* 239:85–124

113. Tsuchiya M. 1983. Mass mortality in a population of the mussel *Mytilus edulis* L. caused by high temperature on rocky shores. *J. Exp. Mar. Biol. Ecol.* 66:101–11

114. Carefoot T. 1977. *Pacific Seashores: A Guide to Intertidal Ecology.* Seattle/London: Univ. Wash. Press

115. Elvin DW, Gonor JJ. 1979. The thermal regime of an intertidal *Mytilus californianus* Conrad population on the central Oregon coast. *J. Exp. Mar. Biol. Ecol.* 39:265–79

116. McMahon RF. 1988. Respiratory response to periodic emergence in intertidal mollucs. *Am. Zool.* 28:97–114

117. Shick JM, Gnaiger E, Widdows J, Bayne BL, De Zwaan A. 1986. Activity and metabolism in the mussel *Mytilus edulis* L. during intertidal hypoxia and aerobic recovery. *Physiol. Zool.* 59:627–42

118. Widdows J, Shick JM. 1985. Physiological responses of *Mytilus edulis* and *Cardium edule* to aerial exposure. *Mar. Biol.* 85:217–32

119. Bayne BL, Bayne CJ, Carefoot TC, Thompson RJ. 1976. The physiological ecology of *Mytilus californianus* Conrad 2. Adaptation to low oxygen tension and air exposure. *Oecologia* 22:229–50

120. Burnett L. 1997. The challenges of living in hypoxic and hypercapnic aquatic environments. *Am. Zool.* 37:633–40

121. Bell EC. 1993. Photosynthetic response to temperature and desiccation of the intertidal alga *Mastocarpus papillatus*. *Mar. Biol.* 117:337–46

122. Davison IR, Pearson GA. 1996. Stress tolerance in intertidal seaweeds. *J. Phycol.* 36:197–211

123. Hurd CL. 2000. Water motion, marine macroalgal physiology, and production. *J. Phycol.* 36:453–72

124. Bertness MD, Ellison AM. 1987. Determinants of pattern in a New England salt marsh plant community. *Ecol. Monogr.* 57:129–47

125. Hacker SD, Gaines SD. 1997. Some implications of direct positive interactions for community species diversity. *Ecology* 78:1990–2003

126. McMahon R. 1990. Thermal tolerance, evaporative water loss, air-water oxygen consumption and zonation of intertidal Prosobranchs: a new synthesis. *Hydrobiologia* 193:241–60

127. Connell JH. 1961. The influence of interspecific competition and other factors on the distribution of the barnacle

Chthamalus stellatus. Ecology 42:710–23

128. Paine RT. 1974. Intertidal community structure: Experimental studies on the relationship between a dominant competitor and its principal predator. *Oecologia* 15:93–120

129. Sagarin RD, Gaines SD. 2002. The 'abundant centre' distribution: to what extent is it a biogeographical rule? *Ecol. Lett.* 5:137–47

130. Sagarin RD, Gaines SD. 2002. Geographical abundance distributions of coastal invertebrates: using one-dimensional ranges to test biogeographic hypotheses. *J. Biogeogr.* 29:985–97

131. Robles C, Desharnais R. 2002. History and current development of a paradigm of predation in rocky intertidal communities. *Ecology* 83:1521–36

132. Chan BKK, Williams GA. 2003. The impact of physical stress and molluscan grazing on the settlement and recruitment of *Tetraclita* species (Cirripedia: Balanomorpha) on a tropical shore. *J. Exp. Mar. Biol. Ecol.* 284:1–23

133. Tomanek L, Somero GN. 2000. Time course and magnitude of synthesis of heat-shock proteins in congeneric marine snails (Genus *Tegula*) from different tidal heights. *Physiol. Biochem. Zool.* 73:249–56

134. Southward AJ. 1958. Note on the temperature tolerances of some intertidal animals in relation to environmental temperatures and geographical distribution. *J. Mar. Biol. Assoc. UK* 37:49–66

135. Fitzhenry T, Halpin PM, Helmuth B. 2004. Testing the effects of wave exposure, site, and behavior on intertidal mussel body temperatures: applications and limits of temperature logger design. *Mar. Biol.* 145:339–49

136. Denny MW, Helmuth B, Leonard GL, Harley CDG, Hunt L, Nelson E. 2004. Quantifying scale in ecology: lessons from a wave-swept shore. *Ecol. Monogr.* 74:513–32

137. Schmidt PS, Rand DM. 2001. Adaptive maintenance of genetic polymorphism in an intertidal barnacle: habitat- and life-stage-specific survivorship of *Mpi* genotypes. *Evolution* 55:1336–44

138. Schmidt PS, Rand DM. 1999. Intertidal microhabitat and selection at *Mpi*: interlocus contrasts in the northern acorn barnacle, *Semibalanus balanoides. Evolution* 53:135–46

139. Lewis JR. 1964. *The Ecology of Rocky Shores.* London: English Univ. Press

140. Leonard GH, Levine JM, Schmidt PR, Bertness MD. 1998. Flow-driven variation in intertidal community structure in a Maine estuary. *Ecology* 79:1395–411

141. Levine J, Brewer S, Bertness MD. 1998. Nutrient availability and the zonation of marsh plant communities. *J. Ecol.* 86:285–92

142. Lubchenco J, Menge BA. 1978. Community development and persistence in a low rocky intertidal zone. *Ecol. Monogr.* 48:67–94

143. Menge BA. 1978. Predation intensity in a rocky intertidal community: effect of an algal canopy, wave action and desiccation on predator feeding rates. *Oecologia* 34:17–35

144. Menge BA. 1978. Predation intensity in a rocky intertidal community: relation between predator foraging activity and environmental harshness. *Oecologia* 34:1–16

145. Seymour RJ. 1996. Wave climate variability in southern California. *J. Waterw. Port Coast. Ocean Eng.* 122:182–86

146. Trenberth KE. 1993. Northern hemisphere climate change: physical processes and observed changes. In *Earth System Responses to Global Changes: Contrasts Between North and South America*, ed. HA Mooney, ER Fuentes, BI Kronberg, pp. 35–59. New York: Academic

147. Wellington GM, Dunbar RB. 1995. Stable isotope signature of El Nino-Southern oscillation events in eastern tropical Pacific reef corals. *Coral Reefs* 14:5–25

148. Bacon S, Carter DJT. 1991. Wave climate changes in the North Atlantic and North Sea. *Int. J. Climatol.* 11:545–58

149. Hoozemans MJ, Wiersma J. 1992. Is mean wave height in the North Sea increasing? *Hydrogr. J.* 63:13–15

150. Mantua N, Haidvogel D, Kushnir Y, Bond N. 2002. Making the climate connections: bridging scales of space and time in the U.S. GLOBEC program. *Oceanography* 15:75–86

151. Goldenberg SB, Landsea CW, Mestas-Nuñez AM, Gray WM. 2001. The recent increase in Atlantic hurricane activity: causes and implications. *Science* 293:474–79

152. Denny MW. 1995. Predicting physical disturbance: mechanistic approaches to the study of survivorship on wave-swept shores. *Ecol. Monogr.* 65:371–418

153. Gaylord B, Blanchette CA, Denny MW. 1994. Mechanical consequences of size in wave-swept algae. *Ecol. Monogr.* 64:287–313

154. Koehl MAR. 1984. How do benthic organisms withstand moving water? *Am. Zool.* 24:57–70

155. Koehl MAR. 1996. When does morphology matter? *Annu. Rev. Ecol. Syst.* 27:501–42

156. Denny MW. 1988. *Biology and the Mechanics of the Wave-Swept Environment.* Princeton: Princeton Univ. Press. 329 pp.

157. Denny MW, Gaylord B, Helmuth B, Daniel T. 1998. The menace of momentum: Dynamic forces on flexible organisms. *Limnol. Oceanogr.* 43:955–68

158. Gaylord B, Hale BB, Denny MW. 2001. Consequences of transient fluid forces for compliant benthic organisms. *J. Exp. Biol.* 204:1347–60

159. Johnson AS. 2001. Drag, drafting, and mechanical interactions in canopies of the red alga *Chondrus crispus. Biol. Bull.* 201:126–35

160. Denny M, Gaylord B. 2002. The mechanics of wave-swept algae. *J. Exp. Biol.* 205:1355–62

161. Duggins DO, Eckman JE, Siddon CE, Klinger T. 2003. Population, morphometric and biomechanical studies of three understory kelps along a hydrodynamic gradient. *Mar. Ecol. Prog. Ser.* 265:57–76

162. Kawamata S. 2001. Adaptive mechanical tolerance and dislodgement velocity of the kelp *Laminaria japonica* in wave-induced water motion. *Mar. Ecol. Prog. Ser.* 211:89–104

163. Price HA. 1980. Seasonal variation in the strength of byssal attachment of the common mussel *Mytilus edulis* L. *J. Mar. Biol. Assoc. UK* 60:1035–37

164. Shaughnessy FJ, De Wreede RE, Bell EC. 1996. Consequences of morphology and tissue strength to blade survivorship of two closely related Rhodophyta species. *Mar. Ecol. Prog. Ser.* 136:257–66

165. Blanchette CA, Miner BG, Gaines SD. 2002. Geographic variability in form, size and survival of *Egregria menziesii* around Point Conception, California. *Mar. Ecol. Prog. Ser.* 239:69–82

166. Price HA. 1982. An analysis of factors determining seasonal variation in the byssal attachment strength of *Mytilus edulis* L. *J. Mar. Biol. Assoc. UK* 62:147–55

167. Moeser GM. 2004. *Environmental factors influencing thread production and mechanics in* Mytilus edulis. MS thesis. Univ. Rhode Island. 57 pp.

168. Petraitis PS. 1995. The role of growth in maintaining spatial dominance by mussels (*Mytilus edulis*). *Ecology* 76:1337–46

169. Robles CD, Alvarado MA, Desharnais RA. 2001. The shifting balance of littoral predator-prey interaction in regimes of hydrodynamic stress. *Oecologia* 128:142–52

170. Andreawartha HG, Birch LC. 1954. *The Distribution and Abundance of Animals.* Chicago: Univ. Chicago Press

171. Kareiva PM, Kingsolver JG, Huey RB, eds. 1993. *Biotic Interactions and Global Change.* Sunderland, MA: Sinauer

172. Dennis RLH. 1993. *Butterflies and Climate Change*. Manchester, UK: Manchester Univ. Press

173. Parmesan C. 1996. Climate and species' range. *Nature* 382:765–66

174. Harrington R, Woiwod I, Sparks T. 1999. Climate change and trophic interactions. *Trends Ecol. Evol.* 14:146–50

175. Parmesan C, Ryrholm N, Stefanescu C, Hill JK, Thomas CD, et al. 1999. Poleward shifts in geographical ranges of butterfly species associated with regional warming. *Nature* 399:579–83

176. Roy DB, Sparks TH. 2000. Phenology of British butterflies and climate change. *Global Change Biol.* 6:407–16

177. Peterson AT, Ortega-Huerta MA, Bartley J, Sanchez-Cordero V, Soberon J, et al. 2002. Future projections for Mexican faunas under global climate change scenarios. *Nature* 416:626–29

178. Peterson AT, Stockwell DRB, Kluza DA. 2002. Distributional prediction based on ecological niche modeling of primary occurrence data. In *Predicting Species Occurrences: Issues of Accuracy and Scale*, ed. JM Scott, PJ Heglund, ML Morrison, JB Haufler, MB Raphael, et al., pp. 617–24. Washington, DC: Island Press

179. Beaumont LJ. 2002. Potential changes in the distributions of latitudinally restricted Australian butterfly species in response to climate change. *Global Change Biol.* 8:954–71

180. Kingsolver JG. 1983. Thermoregulation and flight in Colias butterflies: elevational patterns and mechanistic limitations. *Ecology* 64:534–45

181. Dempster JP. 1983. The natural control of populations of butterflies and moths. *Biol. Rev.* 58:461–81

182. Pollard E, Yates T. 1993. *Monitoring Butterflies for Ecology and Conservation*. London: Chapman & Hall

183. Houghton J. 1997. *Global Warming: The Complete Briefing*. Cambridge, UK: Cambridge Univ. Press

184. Crozier L. 2003. Winter warming facilitates range expansion: cold tolerance of the butterfly *Atalopedes campestris*. *Oecologia* 135:648–56

185. Crozier L. 2004. Warmer winters drive butterfly range expansion by increasing survivorship. *Ecology* 85:231–41

186. Addo-Bediako A, Chown SL, Gaston KJ. 2000. Thermal tolerance, climatic variability and latitude. *Proc. R. Soc. London Ser. B* 267:739–45

187. Bale JS. 1987. Insect cold hardiness: freezing and supercooling—an ecophysiological perspective. *J. Insect Physiol.* 33:899–908

188. Duman JG, Wu DW, Xu L, Tursman D, Olsen TM. 1991. Adaptations of insects to subzero temperatures. *Q. Rev. Biol.* 66:387–410

189. Tauber MJ, Tauber CA. 1986. *Seasonal Adaptations of Insects*. New York/Oxford: Oxford Univ. Press. 411 pp.

190. Bradshaw WE, Holzapfel CM. 2001. Genetic shift in photoperiodic response correlated with global warming. *Proc. Natl. Acad. Sci. USA* 98:14509–11

191. Etterson J, Shaw RG. 2001. Constraint to adaptive evolution in response to global warming. *Science* 294:151–54

Annu. Rev. Physiol. 2005. 67:203–23
doi: 10.1146/annurev.physiol.67.040403.104223
Copyright © 2005 by Annual Reviews. All rights reserved
First published online as a Review in Advance on October 14, 2004

COMPARATIVE DEVELOPMENTAL PHYSIOLOGY:
An Interdisciplinary Convergence

Warren Burggren
*Department of Biological Sciences, University of North Texas, Denton, Texas 76203;
email: burggren@unt.edu*

Stephen Warburton
*Department of Biology, Northern Arizona University, Flagstaff, Arizona 86011;
email: Stephen.Warburton@nau.edu*

Key Words   evolution, ontogeny, genomics, animal models, fetal programming

■ **Abstract**   Comparative developmental physiology spans genomics to physiological ecology and evolution. Although not a new discipline, comparative developmental physiology's position at the convergence of development, physiology and evolution gives it prominent new significance. The contributions of this discipline may be particularly influential as physiologists expand beyond genomics to a true systems synthesis, integrating molecular through organ function in multiple organ systems. This review considers how developing physiological systems are directed by genes yet respond to environment and how these characteristics both constrain and enable evolution of physiological characters. Experimental approaches and methodologies of comparative developmental physiology include studying event sequences (heterochrony and heterokairy), describing the onset and progression of physiological regulation, exploiting scaling, expanding the list of animal models, using genetic engineering, and capitalizing on new miniaturized technologies for physiological investigation down to the embryonic level. A synthesis of these approaches is likely to generate a more complete understanding of how physiological systems and, indeed, whole animals develop and how populations evolve.

WHAT IS COMPARATIVE DEVELOPMENTAL PHYSIOLOGY?

Comparative developmental physiology (CDP) is, quite simply, an examination of the comparative physiology of developing animals. Borrowing from the title of Schmidt-Nielsen's (1) wonderful book, *How Animals Work,* we can view CDP as "how developing animals work." Similar to one of its parent disciplines, comparative animal physiology, CDP spans investigations ranging from genomics and proteomics to physiological ecology and evolution. CDP is not a new discipline, and we are not attempting here to provide a comprehensive listing of all studies

0066-4278/05/0315-0203$14.00   **203**

falling under the umbrella of this expanding and vibrant discipline. Rather, our goal is to outline the pathway by which CDP has become an interdisciplinary domain, to highlight the current explosion of studies in this area, and to underscore the important contributions that CDP is making and will continue to make toward the ultimate goal of understanding the connection between evolution, development, and physiology.[1]

The roots of CDP go back millennia. Aristotle (384–322 BC) commented on the pulsing red spot in recently laid chicken eggs, an observation oft-repeated, most notably by Vesalius (1514–1564) and Galileo (1564–1642), before the more detailed characterization of bird embryonic heart rate by numerous seventeenth and eighteenth century proto-physiologists. To this day, the 2–3-Hz heart beat (37°C) of the 3 to 4-day old chick embryo holds fascination for all who observe it.

Until fairly recently the questions asked by investigators of CDP were largely descriptive and not unlike those of Aristotle, perhaps reflecting the classic embryological studies that were themselves so highly descriptive in nature. Thus through most of the twentieth century, typical questions might have been, What is the heart rate of larval bullfrog? (2), or, Can neonatal birds thermoregulate? (3). Few and far between were pioneers such as Adolph (4) or Metcalfe (5) who began to ask more sophisticated questions about the regulation and control of physiological systems during the process of animal development.

While answering descriptive physiological questions (and many such important questions still remain unanswered at the organ system/organismal level) is still important, during the last few decades the field of CDP has expanded from these origins to include experimentation and manipulation. Enabled by new, often miniaturized tools for physiological measurements (see below), a fresh generation of mechanism-based questions has emerged. What are the physiological systems for heart rate regulation during development and what controls them? (6–8, 9, 10). How and when do thermoregulatory mechanisms develop in bird embryos? (11–15). Indeed, in recent years the comparative physiological literature has shown an explosion of developmental studies involving experimental manipulation. Currently, development as a crucial "Z axis" (Figure 1) is a seminal theme. This approach elucidates developmental vectors or trajectories that characterize how physiological processes and their control mechanisms change throughout ontogeny (16). Indeed, physiological studies are increasingly looking to developmental perspectives to explain adult physiological traits, to probe phenotypic plasticity in developmental programs, and to resurrect and refine the intersections of evolution, physiology, and development (see below).

The most recent phase in the maturation of CDP has been the rapid expansion of physiological genomics combined with the use of model organisms and genomic tools such as microarrays, genetic engineering, gene knockouts, etc. These interactions are prompting new questions such as, What genes are involved in the

---

[1]We use the acronym CDP here with some reservation, as acronyms tend to contribute toward the creation of intellectual clusters rather than interdisciplinary gradients.

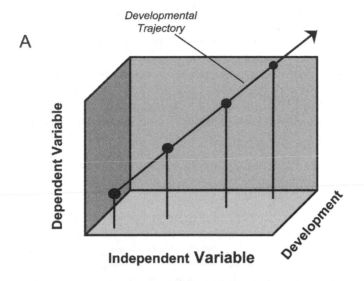

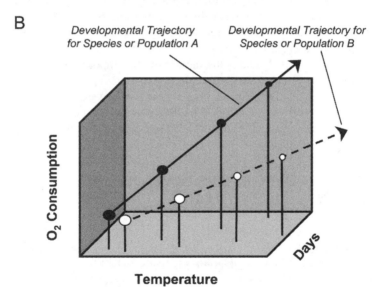

Figure 1   Developmental trajectories. (*A*) Contemporary studies in comparative developmental physiology often examine the interactions between two variables (X and Y as a function of developmental progression, which forms the third variable or Z axis). The result is a distinct and quantifiable developmental trajectory, depicting the nature of interactions between variables (after Reference 16). (*B*) By comparing multiple species or populations, differences in developmental trajectories become apparent. In this example, a differential response to developmental temperature results in two distinct developmental trajectories for oxygen consumption.

formation of embryonic heart chambers? (17), or, What heat shock proteins are induced, and why, during temperature stress in embryos? (18–19). Full realization of the power of genomics requires the ability to relate gene activation first to proteomics and ultimately back to organismal function in both embryos and adults, thereby allowing a comprehensive systems synthesis from gene to adaptive advantage at the population level. Only with this view is the actual interaction of evolution (genotype) and physiology (phenotype) comprehensible.

In summary, the progression in CDP from past to future can be characterized as

Physiological → Physiological → Genome → Systems → Evolution of
description   mechanism   synthesis   characters

What role does physiology play in the postgenomic steps in this progression? Consider the response of Sydney Brenner, Noble laureate (2002) and champion of genetic approaches using *Caenorhabditis elegans* as an experimental model, when asked to comment on systems biology. He replied ". . . .everybody's running around talking about systems biology and integrative biology. It's nothing new. It's called physiology." (20). Indeed, the integrative and synthetic nature of physiology is becoming increasingly apparent as we try to connect genes, proteins, processes, structures, and evolution.

One exciting new trend is the merger of contemporary physiology, genomics, and evolutionary-developmental biology into a previously unrecognized interdisciplinary zone of CDP. From this novel perspective we can better understand how physiological systems develop, how they respond to the environment, and how changes in these systems contribute to the fitness of the animal at each developmental stage.

POSITIONING COMPARATIVE DEVELOPMENTAL PHYSIOLOGY WITHIN BIOLOGY

The disciplines of evolution, physiology, and developmental biology have all helped define the current state of CDP. Their intersection produces fertile interdisciplinary zones where integration is likely to be highly productive (Figure 2). Evolutionary and developmental biology, or "evo-devo," is of escalating significance in elucidating mechanisms linking evolution and development (21–23). Evolutionary physiology, another expanding interdisciplinary zone, is contributing to our understanding of how physiology evolves and how physiology enables and constrains evolution (24–26). The third zone, developmental physiology, helps us understand how structure-function linkages develop from embryonic to adult forms, as well as how physiology plays a permissive role in development. CDP, the nexus of these three zones, is an interdisciplinary crossroads. By explicitly integrating physiology, development, and evolution, CDP provides a unique outlook that enriches our understanding of evolution. This is especially true for

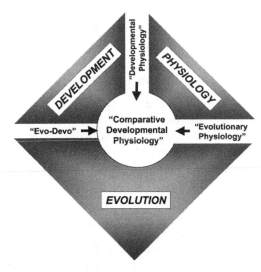

Figure 2 The position of comparative developmental physiology vis-à-vis three major biological disciplines: evolution, physiology, and developmental biology. These disciplines share overlapping zones of integration comprising developmental physiology, evolutionary physiology, and evo-devo. The intersection of these three zones creates a biological crossroads where we are likely to achieve a considerably improved understanding of biology.

understanding the evolution of physiology, which has traditionally received relatively little attention from evolutionary biologists (see 27). Clearly, CDP is playing an increasingly important role in organismal biology. What are the current themes being explored within this burgeoning discipline?

CURRENT THEMES IN COMPARATIVE DEVELOPMENTAL PHYSIOLOGY

Modern studies in CDP are an interdisciplinary fusion of traditional comparative physiology and developmental biology, embryology, and evolutionary biology. Contemporary areas of concentration in CDP are generally (and somewhat arbitrarily) divisible into

- determining how physiological systems apportion regulatory responsibility during their establishment in the developing animal;
- determining how developing physiological systems respond to changing environmental conditions;

- ▪ understanding how differences in physiological capacity of developing animals contribute to differences in fitness and, hence, in the evolution of animals; and
- ▪ understanding developmental constraints on evolution.

Apportioning Regulatory Responsibility

The transformation of a single cell to a multicellular organism requires the transfer of physiological regulation from the cell to shared or autocratic "governance" involving an overarching regulatory system (79). Simply put, any system whose function depends on neural or hormonal integration is limited until the nervous system or appropriate endocrine system has developed functionality. The onset of the regulation of cardiac rhythm is among the most-studied system in this regard (for reviews see 10, 29–33). Early in development, regulation of contractile frequency resides in individual cardiomyocytes. However, as gap junctions begin to connect cardiomyocytes, pacemaker cells come to regulate the cardiac frequency as a whole. Still later, endocrine regulation becomes functional and can modulate pacemakers and then, finally, neural regulation assumes a dominant role. This changing of control modality from local or simple to more remote and/or more complex may also occur in other organ systems. In an amphibian, ventilatory regulation may initially depend upon a simple pacemaker, but later, complex oscillating circuits emerge to dominant the system (34). In the rat gut, modulation of enteric activity via interstitial cells of Cajal is not fully in place at birth and must mature during postnatal life (35). Is the pattern of development from single cell to cellular pacemaker networks indicated in these examples a hallmark of development in all complex animals, where networks cannot exist until there are sufficiently differentiated cells to create such a association? Are these patterns of regulation conserved throughout evolution? This is an important, but largely unexplored sub-genre of CDP.

Response of Developing Systems to Environment

Molecular biologists have emphasized the primacy of genes in the function and development of the organism. Yet, the environment also plays an important role in phenotype determination. Evolutionary biologists use the term phenotypic plasticity for this ability of a single genotype to produce multiple phenotypes. Phenotypic plasticity can be adaptive, maladaptive, or neutral (36–38). The spectrum of phenotypes possible within a given genotype is the reaction norm. Environmental influence on phenotype can appear as both acclimatization and fetal programming. Acclimitization during development has the same definition as when used in reference to adult animals, that is, typically adaptive alteration in characteristics upon exposure to a new, natural environment (e.g., 11, 12, 40, 41). Importantly, acclimitization can reverse or animals can deacclimitize. However, developing animals have the additional capacity (or limitation) permanently to alter adult form and function when perturbing conditions occur during critical windows of development. Examples include blood pressure in animals with reduced glomerular number (42), sound production in midshipman fish (43), olfaction in queenless honeybees (44), and the embryonic determination of glomerular number or coronary geometry (45). Such permanent alterations in the normal developmental trajectory, independent of genotype, are typically termed fetal programming in the medical literature (46, 47), but perhaps should be re-named ontogenetic programming to avoid taxonomic chauvinism (e.g., larva in addition to fetus). The history of studying developmental plasticity is long. Examples include alterations in isoforms of myosin heavy chain expression in carp (48); complex motor patterns in feeding in amphibians, which change during metamorphosis and with diet (49); vascular development during hypoxia in chick embryos, in which vascular lumen diameter and developed ventricular pressure decrease upon incubation in hypoxia (50); and gill surface and hematopoiesis in larval bullfrogs, which hypoxia enhances and hyperoxia inhibits (51, 52). These and numerous other examples have provided much information on developmental plasticity in response to environmental cues.

Regarding mechanisms of evolutionary change, we speculate that in some cases genetic mechanisms of sub-adult organisms have evolved to adjust their host's internal physiological state and so program an appropriate adult morph, and that environmental variation can sometimes override this programming. This suggests that alterations in climate, geography, salinity, etc. may have important effects on speciation. This area is in its infancy but holds exciting possibilities.

Developmental Fitness and Evolutionary Consequences

Comparative physiology has a long-standing relationship with evolutionary biology, and vice versa (Figure 2). Understanding how specific physiological traits allow an organism, population, or species to survive and reproduce better has been a mainstay of comparative physiology almost since its inception. In general, the focus has been on adaptive traits of adult organisms, the rationale being that these are the reproductive individuals. Overlooked is that in many species, the majority of mortality (and thus selection) occurs long before reproductive capacity is realized. Dramatic examples of this abound in both vertebrates and invertebrates that exhibit "r selection," release of thousands to millions of offspring into an unpredictable environment, of which few survive (see 53 for examples and references). Consequently, selection may be stronger on immature forms (embryos, larvae, fetuses, or juveniles), than on adults (see 54), which is a largely neglected area of emphasis for physiological investigation. Potentially, adult forms are at the mercy of juvenile "bottlenecks" in physiological options. Changes in developmental trajectory early in ontogeny, while advantageous then, may yield permanent alterations that are disadvantageous to adults. Suddenly, the notion of tradeoffs emerges, and the importance of life history studies becomes obvious in integrating physiology and evolution. If so, the numerous offspring that some species produce may be an evolutionary necessity to produce those much smaller numbers of offspring that are fit as both juveniles and adults. Perhaps the the adult phenotype is so important that potentially adaptive responses early in ontogeny are "prohibited," and thus sheer numbers are instead required to overcome nonadaptive or maladaptive stages. This connection between embryonic or larval requirements and adult requirements may be direct (i.e., fetal programming) or may be result of correlated selection or genetic correlation (55).

Developmental Constraints on Evolution?

For many years evo-devo, the interdisciplinary zone between development and evolution, languished as a footnote in embryology texts. During the last decade, evo-devo has undergone a renaissance (56, 57). Yet, one aspect that has persisted from evo-devo's embryological roots is the extensive focus on morphological (but rarely physiological) traits to draw evolutionary conclusions (see 27 for earlier commentary on this phenomenon). Presumably, this arises from the relative ease with which morphological alterations are observable both through the fossil record and through experimental genetic manipulations. The consequence of this history of morphological observation is that physiology is rarely mentioned in dicussions of the future directions in evo-devo. Nielsen (58) echoes this morphocentric view in his endnote to *Animal Evolution: Interrelationships of the Living Phyla*, to wit, "Evolutionary developmental biology shows great potential for phylogenetic work, and I hope that there will be close collaboration between morphologists and the 'evo-devo' people." This overly restrictive view apparently is reinforced by the conventional hierarchal view of organismal plasticity, with behavior being most plastic (and thus least useful for evolutionary studies), physiology being less plastic but still inconveniently so, and anatomy being the least plastic, most stable and allegedly the most useful to study (27). Yet, highly inflexible physiological traits are numerous: For example, mammalian blood pressure and ventricular wall tension is predictable and almost invariant regardless of animal size (59), and respiratory frequency and heart rate are highly predictable based on animal size (60). Nonetheless, relatively few attempts have been made to use physiological characters, with evolutionary biologists arguing that physiological invariants (e.g., blood pressure) may be due to evolutionary constraints on anatomy, which are simply mirrored in the physiology of the systems. We would be the first to agree that anatomy puts limits on physiology. However, most physiologists would view the causal relationship between anatomy and physiology as far more tenuous than would anatomists, based on the common observations of large physiological differences enabled by almost undetectable gross anatomical changes. As an extreme example, consider the crab *Scopimera inflata*, where a reduction in the thickness of the chitin covering the meral segments of the walking legs turns these appendages from strictly locomotory organs into respiratory organs (61). Thus relatively trivial anatomical changes in structures can lead to profound reassignment of their physiological function (27, 62)!

The pattern of ontogenetic change in heart rate and metabolism during development in vertebrates is another example of a heritable physiological trait previously overlooked because of the assumption that physiological variation equals physiological plasticity. These studies, which have included birds (63), mammals

(64), and amphibians (65), have revealed that complex changes in heart rate and metabolism during development are highly correlated between sibling groups sharing a common genetic heritage.These findings suggest that even subtle, apparently random, physiological variations may be genetically predetermined.

Given the nongenetic input to developmental programs and the identifiable genetic components of physiology variables, CDP provides an invaluable vantage point from which to examine limitations on evolutionary processes. The study of developmental limitations or constraints on evolution gives us considerable insight into subjects as diverse as life history cycles, heterochrony, and heterokairy. Developmental constraints have been postulated to account for unique traits such as the foramen of Pannizzi in crocodilian reptiles (66). Altering cardiac hemodynamics alters cardiovascular structure (67, 68), and altering cardiovascular structure may alter hemodynamics (28, 68, 69) in the adult. Transgenic or mutant animals certainly provide useful information regarding gross malformations, but the diversity of natural species provides "evolutionary feasibility studies" (27). Studies of the earliest chordate cardiovascular systems or the metabolism of the smallest, fastest living mammals, bats and shrews, provide models of extreme organisms. Investigation of their developmental pathways may reveal constraints on evolutionary options, based not on anatomy or phylogenetic constraints, but on physiological limitations.

The obvious importance to animal survival of physiological function during development should be sufficient to promote the inclusion of physiological characteristics in evolutionary analyses. This may be especially true of comparative studies given the multitude of species available and the relative ease with which developing systems can be perturbed physiologically. Hopefully, as more genetic underpinnings of physiology are elucidated, physiology will be become a full partner in the evo-devo paradigm.

Selected Experimental Approaches and Methodologies of Comparative Developmental Physiology

The field of physiology provides a key point of continuity and connection between cellular/molecular and ecological/evolutionary organization levels. As discussed earlier, CDP occupies a clear crossroads linking evolutionary biology, physiology, and developmental biology. Not surprisingly, the experimental approaches and methodologies employed by comparative developmental physiologists encompass most of contemporary biology. However, here we highlight a few such approaches, and emphasize how the integration of common approaches from different fields can yield distinctive new insights.

Scaling and Development

For more than 50 years comparative physiologists have argued about the significance of scaling parameters that relate physiological functions such as heart rate or metabolic rate to anatomical features such as body weight or surface area.

Ironically, whereas most agree that body size and metabolism display a highly conserved relationship, there is less agreement as to what constitutes the specific nature of that relationship (70, 71). There are highly conserved scaling relationships of individual organ systems across phylogenetic groups as well (59, 72). Across phylogenetic groups, adult organisms conform to well-known and rigid scaling laws, in which adults of a particular body size have a predictable heart rate, metabolic rate, etc. The numerical parameters describing these systems have generated intense interest both for their evolutionary implications of optimality but also as a means to explore the underlying biological mechanisms (73, 74). Moreover, the study of these mechanisms and pathways may contribute to the explanation of the highly conserved scaling relationships. Thus, for example, whereas it may be advantageous to devote a constant percentage of body weight to heart size, we know little of the developmental mechanisms or signaling pathways that result in this conformity. These contentious issues are largely unexplored in the field of physical signals and transduction mechanisms. It may well be that identification of the complex network of developmental cues and cellular responses will answer the questions of scaling parameters with more accuracy than has assessing end results in adult animals. Importantly, advances in this field will reveal which physiological variables are linked and which are independent, if indeed any are. If parameters are invariant, is there a genetic basis (or a genetic constraint or other type of constraint)?

Event Sequences in Physiological Development

One major focus in CDP is the determination of the order of developmental events and whether that order is genetically fixed or plastic (see 9 for examples). That the sequence of key events during development can vary (i.e., heterochrony) is, of course, an old notion (e.g., 75–78). An example of physiological heterochrony is in the differences among vertebrate species in the relative timing and sequence of onset of vagal cardiac control, chemoreflexes, and baroreflexes (29). However, recent papers have argued for the application of heterochrony only to the changes over evolutionary time between species. This argument reserves the term heterokairy for the naturally occurring and experimentally inducible changes in the onset and timing of events within a population between individuals during a single life span (9, 79). For example, the adult metabolic response to hypoxia in the brine shrimp *Artemia* typically occurs simultaneously with segmentation when animals are reared in air-saturated seawater (80). Yet, when reared under conditions of chronic hypoxia, heterokairy is evident because the onset of respiratory regulation in this crustacean now occurs earlier in a sequence of developmental events, i.e., before segmentation.

Even as heterochrony provides insights into physiological evolution, the new conceptual framework of heterokairy provides insights into physiological phenotypic plasticity. Concomitant with these new insights is the requirement for new ways of thinking about development timing and rates, especially when body temperature is a potential variable during development (81).

It is ironic that, despite our earlier protestations about the morphocentric view of evolution, staging in physiological studies is still carried out using anatomically based staging schemes (e.g., Hamilton-Hamburger for chick embryos, Nieuwkoop and Faber for *Xenopus*). Future studies would do well to determine how morphological and physiological plasticity map onto each other during the development of a single animal. That is, are the developmental critical windows of the same width and position for both physiological and anatomical events? Must one consider separate physiological and anatomical heterochronies? If so, what are the ecological and evolutionary implications?

Using Animal Models: Establishing Universal Mechanisms and/or Learning from Diversity

The use of model organisms, historically a mainstay of physiology, has never been more essential to making both pragmatic and conceptual advances in physiology (e.g., 82–86). Numerous discoveries have resulted from focused, persistent investigation of the fruit fly (*Drosophila*), the zebrafish (*Danio rerio*), the chicken (*Gallus gallus*), the nematode worm (*Caenorhabditis elegans*), the mouse (*Mus musculus*), or plants such as *Arabidopsis*. In some instances, however, model organisms have emerged simply because as a species they were the first to be investigated in a particular context—and not because they were best-suited for such investigation. As more and more information was collected, they became wonderful models simply because so much was known about them— a form of self-fulfilling prophecy (87).

When the level of examination is at the molecular or cellular level, cells are cells, and thus the lessons learned from animal models are typically broadly applicable (e.g., the role of Hox genes or the influence of fate mapping on structure/function relationships). However, as molecular and cellular biologists begin to ask broader questions of physiology, ecology, and evolution, some investigators remained focused on the model organisms with which they are familiar. Such models may not be truly representative of a larger taxon, nor represent the full extent of organismal diversity needed to understand ecological and evolutionary relationships (87). Moreover, study of model organisms needs to be informed by an understanding of the conditions in which these species develop in nature. An example of the caution that needs to be exterted when studying model organisms is the use of *C. elegans* and its many relevant mutants to study adaptation and acclimation to hypoxia-induced metabolic-suspended animation. While exciting genomic and proteomic information on hypoxic adaptation is emerging (e.g., 88, 89), often unappreciated is that *C. elegans* evolved in a highly hypoxic soil environment (one also rich in nitric oxide), and that cultures maintained under control laboratory conditions might be more appropriately viewed as continual hyperoxic exposure of this species. Another example is the zebrafish. Many researchers' knowledge of the ecology, life history, and evolution of this important model is summed by the opening sentence of Westerfield's (90) in his widely distributed *The Zebrafish*

Book, "Zebrafish are available at pet stores throughout the world." In an action that flouts their natural thermal evolutionary history, these fish are typically bred and reared at 28.5°C, even though this industry-standard rearing temperature is much closer to the upper rather than lower lethal limit for this fish. Indeed, zebrafish prove more fecund at 25°C, more toward the middle of their thermal range (91).

Both deeply understood model organisms and less well understood but diverse animals deserve study. Consider the study of the ontogeny of cardiovascular regulation in bird embryos. Because bird embryos developed in a self-contained egg, they have long been favored animals for investigating how vertebrate cardiovascular physiological regulation unfolds during development. In this regard, the embryos of the chicken *Gallus gallus* have been the basis for seemingly well-established conclusions. Yet, our recent comparisons of more exotic avian species (such as the emu *Dromiceius novaehollandiae)* with the chick embryo reveal profound differences in the developmental patterns of cardiovascular control between the two species. For example, in developmental patterns reflective of physiological heterochrony, the cardiac vagal tone, chemoreflexive cardiovascular control, and baroreflexes all develop much later in the emu than in the chicken (29). Moreover, these physiological landmarks in emu (chemoreflexes, baroreflexes vagal tone) appear in the exact opposite order in the chicken (vagal tone, baroreflexes, chemoreflexes). Of course, this is only a two-species approach not supported as yet by a more rigorous, cladistic approach, and the question remains as to which of these two species is the more representative of birds (if there is indeed a representative bird). Yet, such data do question the generality of the extensive physiological data available for the chicken but few other birds. Thus, ironically, we investigate an exotic species to calibrate and learn more about a model species! Indeed, the focus on animal models probably slowed our understanding of physiological evolution. We advocate a systematic investigation of other fishes, nematodes, etc. patterned after studies on model organisms, to learn how many of the physiological findings for model animals such as the zebrafish or *C. elegans* are generalizable. Given the importance of Genbank in elucidating phylogenies of genes, consider the impact that could result from a corresponding database of physiological variables from a variety of animals.

Finally, the choice of species for CDP studies depends in part on how early one chooses to look in the overall development process. The earlier the point of investigation, the greater is the interspecific similarity in emerging physiological properties, and the more useful is any given organism as a general physiological model (16). For example, blood pressure, blood flow, and peripheral resistance during the first few days of convective blood flow are similar in the embryos of *Gallus gallus* (92–94), *Xenopus* (95–98), and *Danio rerio* (99, 100). Thus any of these models, when examined early in development, might be equally useful in determining how vertebrate circulations begin their function. Of course, the late bird embryo with a four-chambered heart might tell us little about the larval fish with two-chambered heart. At what time during ontogeny a species ceases to serve as a general developmental model depends upon the system being investigated and the questions being answered, but clearly a heavily comparative approach is effective and warranted at least in understanding early developmental stages of vertebrates.

Technologies for Investigating Comparative Developmental Physiology

Investigations in CDP now span molecules to populations, and, not surprisingly, experimental tools are drawn from all levels. Our intent here is to highlight only a few of the many approaches that have been particularly useful in CDP and to then direct the reader to additional sources on these topics.

MINIATURIZATION   Immature animals are relatively small and embryos sometimes microscopic. Consequently, the relentless drive toward electro-mechanical miniaturization has been a boon to CDP. In some cases, rather astonishing miniaturization of conventional technologies for blood pressure, flow, pH, blood gases, etc. have occurred, allowing unprecedented measurements and insights into physiological function in early development (see 87, 101–103). Perhaps one of the most graphic examples is that of micropressure systems. Using a glass microelectrode with a 2–5 μm diameter tip inserted into a vessel or cardiac chamber, high-frequency response pressure measurements can be made in embryos weighing only milligrams (96, 98–101). Microelectrodes that measure gases and ions have, of course, been available for some time (104, 105).

The emergence of nanotechnology is likely to provide additional experimental tools with unimagined possibilities for CDP. As just one example, "smart dust" is being developed to provide detailed three-dimensional environmental assessments. In this emerging technology, microscopic silicon-based sensors made up of such dust are sprinkled over an environment. The particles are then interrogated with a laser beam (e.g., from an overflying aircraft), and the reflected beam is modified in way that encodes information on variables such as pressure, humidity, temperature, or oxygen levels (106, 107). One can imagine in the near future being able to inject nanotechnology-derived microscopic sensors into near-transparent embryos, and then, using laser interrogation, derive a three-dimensional assessment of internal physico-chemical variables of physiological interest.

IMAGING   Embryos are not only microscopic, but they are often translucent or transparent. Thus there has also been an explosion of approaches using nonintrusive optically based techniques for measurement of physiological variables, techniques collectively termed optophysiology (108). Cardiac output, blood oxygenation, blood flow distribution, and other physiological variables are now commonly measured through such optical techniques (see 101, 103, 108–111). After introduction of various dyes or indicators or even using substances intrinsic to muscle, physiologists now optically derive localized tissue PO_2 (112) and track muscle cell excitation and contraction (13, 113). By keying in on the profound spectral shifts in hemoglobin as it changes oxygenation state, in vivo changes in blood oxygen transport can be determined in real time (108, 111). The advent of multiphoton confocal microscopy (also called nonlinear microscopy) allows in vivo imaging with greater penetration then conventional laser confocal microscopy and with less radical by-product production (114, 115).

GENETIC ENGINEERING   Genetic knockouts in zebrafish, mice, *C. elegans*, and other model species are being widely exploited to gain insight into the assembly of fully functional physiological systems (e.g., 84, 116–122). Indeed, the utility of such approaches has led some to argue that screening studies to "see what is out there" should replace hypothesis-driven research (45)! Yet, the limited number of knockout models available in non-model animals makes it difficult to use a comparative approach to probe the complexity of evolutionary constraints and possibilities. Thus expansions of knockouts beyond the conventional models are to be encouraged for the promise they hold.

One area where CDP can contribute greatly to developmentally directed genomic studies with mutants or knockouts is to expand the scope of some of these studies beyond an analytical approach that seems drawn from traditional toxicology: Do knockout animals die or survive? A more sophisticated question, yielding a more illuminating answer, might be, How well do they survive? or even, What did they die from? Incorporating the techniques and approaches of CDP, namely, quantifying physiological performance and ultimately fitness, into genomic/proteomic studies should prove extremely useful in documenting and understanding the myriads of important but nonlethal effects induced by genetic engineering.

Unanswered Questions in Comparative Developmental Physiology

CDP appears bound for increasing emphasis and significance in evolutionary biology. The specific future of CDP is difficult to predict, but by virtue of this field's position at a biological crossroads (Figure 2), it seems certain to involve enhanced collaboration between physiologists, evolutionists, and developmental biologists. These collaborations will allow us to address key questions (and practical implications), such as

- Do genes or environments make species? (e.g., how much of the variation in phenotype is genetic and how much is environmentally induced?)
- How straightforward is physiological evolution? (e.g., does the evolution from species A to B require more anatomical changes or more physiological changes?)
- Are current, popular animal models most appropriate for advancing developmental physiology? (e.g., should we focus on any one model, or is it important to maintain and explore diversity?)

- Are the basic tenets of developmental physiology overarching across all or most taxa? (e.g., will collaborations between animal and plant biologists provide useful insights to either?)

- What is the role of developmental programming in the ultimate phenotype? (e.g., what are the critical physiological windows, and are they moveable by adaptation or acclimation?)

- How fixed in development are traditional ontogenetic events? (e.g., are developmental landmarks locked in place, or have we just not tried to move them?)

- How interdependent are physiological systems during development? (e.g., when do physiological systems begin to interact and influence each other during the course of development?)

- Does a physiological system have the same function throughout development? (e.g., are there major changes in responsibility of physiological systems as the animal matures and potentially even changes environment?)

- What are the origins of scaling constants? (e.g., how do size and immaturity interrelate in a developmental context?)

- How important is the study of the complete developmental continuum? (e.g., is the aging process a natural extension of development, and can we learn about evolution from its study?)

Transitions in CDP and the fields from which it is formed (Figure 2) will continue. Driven by new forms of collaboration, conceptual advancements both within and outside the field, and by improvements in technology, perhaps the single safe prediction is that additional insights into the critical physiological underpinnings of evolutionary processes will only accelerate during the decade to come.

ACKNOWLEDGMENTS

The authors acknowledge the support of the National Science Foundation and thank their many collaborators, colleagues, and students who have been instrumental in the refinement of the ideas presented in this review.

The *Annual Review of Physiology* is online at
http://physiol.annualreviews.org

LITERATURE CITED

1. Schmidt-Nielsen K. 1972. *How Animals Work.* Cambridge, UK: Cambridge Univ. Press

2. Burggren WW, Doyle ME. 1986. Ontogeny of heart rate regulation in the bull-frog *Rana catesbeiana. Am. J. Physiol. Regul. Integr. Comp. Physiol.* 251:R231–39

3. Wittenberger C, Giurgea R, Coprean D. 1984. Studies on the thermoregulation in

developing chickens. *Arch. Exp. Veterinarmed.* 38:869–74

4. Adolph EF. 1968. *Origins of Physiological Regulations.* New York: Academic

5. Metcalfe J, Bartels H, Moll W. 1967. Gas exchange in the pregnant uterus. *Physiol. Rev.* 47:782–838

6. Chiba Y, Fukuoka S, Niiya A, Akiyama R, Tazawa H. 2004. Development of cholinergic chronotropic control in chick (*Gallus gallus domesticus*) embryos. *Comp. Biochem. Physiol. A* 137:65–73

7. Crossley DA II, Bagatto B, Dzialowski E, Burggren WW. 2003. Maturation of cardiovascular control mechanisms in the embryonic emu (*Dromiceius novaehollandiae*). *J. Exp. Biol.* 206:2703–10

8. Crossley DA II, Burggren WW, Altimiras J. 2003. Cardiovascular regulation during hypoxia in embryos of the domestic chicken *Gallus gallus. Am. J. Physiol. Regul. Integr. Comp. Physiol.* 284:R219–26

9. Spicer JI, Burggren WW. 2003. Development of physiological regulatory systems: altering the timing of crucial events. *Zoology* 106:91–99

10. Tazawa H, Akiyama R, Moriya K. 2002. Development of cardiac rhythms in birds. *Comp. Biochem. Physiol. A* 132:675–89

11. Black JL, Burggren WW. 2004. Acclimation to hypothermic incubation in developing chicken embryos (*Gallus domesticus*): I. Developmental effects and chronic and acute metabolic adjustments. *J. Exp. Biol.* 207:1543–52

12. Black JL, Burggren WW. 2004. Acclimation to hypothermic incubation in developing chicken embryos (*Gallus domesticus*): II. Hematology and blood O_2 transport. *J. Exp. Biol.* 207:1553–61

13. Duchamp C, Rouanet JL, Barre H. 2002. Ontogeny of thermoregulatory mechanisms in king penguin chicks (*Aptenodytes patagonicus*). *Comp. Biochem. Physiol. A* 131:765–73

14. Khandoker AH, Fukazawa K, Dzialowski EM, Burggren WW, Tazawa H. 2004.

Maturation of the homeothermic response of heart rate to altered ambient temperature in developing chick hatchlings (*Gallus gallus domesticus*). *Am. J. Physiol. Regul. Integr. Comp. Physiol.* 286:R129–37

15. Tamura A, Akiyama R, Chiba Y, Moriya K, Dzialowski WM, et al. 2003. Heart rate responses to cooling in emu hatchlings. *Comp. Biochem. Physiol. A* 134:829–38

16. Burggren WW. 1998. Studying physiological development: past, present and future. *Biol. Bull. Nat. Taiwan Normal Univ.* 33:71–84

17. Mooreman AF Christoffels VM. 2003. Cardiac chamber formation: development, genes, and evolution. *Physiol. Rev.* 83:1223–67

18. Leandro NS, Gonzales E, Ferro JA, Ferro MI, Givisiez PE, Macari M. 2004. Expression of heat shock protein in broiler embryo tissues after acute cold or heat stress. *Mol. Reprod. Dev.* 67:172–77

19. Zimmerman JL, Cohill PR. 1991. Heat shock and thermotolerance in plant and animal embryogenesis. *New Biol.* 3:641–50

20. Duncan DE. 2004. Discover dialogue: Sydney Brenner. The man who made worms the workhorses of genetics. *Discover* 25:20–23

21. Arthur W. 2002. The emerging conceptual framework of evolutionary developmental biology. *Nature* 415:757–64

22. Hall BK. 1999. *Evolutionary Developmental Biology.* London: Chapman & Hall

23. Kuratani S, Kuraku S, Murakami Y. 2002. Lamprey as an evo-devo model: from comparative embryology and molecular phylogenetics. *Genesis* 34:175–83

24. Burggren WW. 1991. Does comparative respiratory physiology have a role in evolutionary biology (and vice versa)? In *Physiological Strategies for Gas Exchange and Metabolism*, ed. AJ Woakes, MK Grieshaber, CR Bridges, pp. 1–13. Cambridge, UK: Cambridge Univ. Press

25. Garland T Jr, Carter PA. 1994. Evolutionary physiology. *Annu. Rev. Physiol.* 56:579–621

26. Sibly R, Calow P. 1985. Are patterns of growth adaptive? *J. Theor. Biol.* 125:177–86

27. Burggren WW, Bemis WE. 1990. Studying physiological evolution: paradigms and pitfalls. In *Evolutionary Innovations: Patterns and Processes*, ed. MH Nltecki, pp. 191–228. Oxford, UK: Oxford Univ. Press

28. Burggren WW. 2004. What is the purpose of the embryonic heart? Or how facts can ultimately prevail over dogma. *Physiol. Biochem. Zool.* 77:333–45

29. Burggren WW, Crossley DAII. 2002. Comparative cardiovascular development: improving the conceptual framework. *Comp. Biochem. Physiol. A* 132: 661–74

30. Burggren WW, Keller BB, eds. 1997. *Development of Cardiovascular Systems: Molecules to Organisms*. Cambridge, UK: Cambridge Univ. Press

31. Burggren WW, Warburton S. 1994. Patterns of form and function in developing hearts: contributions from non-mammalian vertebrates. *Cardioscience* 5:183–91

32. Gourdie RG, Kubalak S, Mikawa T. 1999. Conducting the embryonic heart: orchestrating development of specialized cardiac tissues. *Trends Cardiovasc. Med.* (1–2):18–26

33. Phoon CK. 2001. Circulatory physiology in the developing embryo. *Curr. Opin. Pediatr.* 13:456–64

34. Broch L, Morales RD, Sanodval AV, Hedrick MS. 2002. Regulation of the respiratory central pattern generator by chloride-dependent inhibition during development in the bullfrog (*Rana catesbeiana*). *J. Exp. Biol.* 205:1161–69

35. Faussone-Pellegrini MS, Matini P, Stach W. 1996. Differentiation of enteric plexuses and interstitial cells of Cajal in the rat gut during pre- and postnatal life. *Acta Anat.* 155:113–25

36. Behera N, Nanjundiah V. 2004. Phenotypic plasticity can potentiate rapid evolutionary change. *J. Theor. Biol.* 226:177–84

37. LaFiandra EM, Babbitt KJ. 2004. Predator induced phenotypic plasticity in the pinewoods tree from *Hyla femoralis*: necessary cues and the costs of development. *Oecologica* 138:350–59

38. Ashmore GM, Janzen FJ. 2003. Phenotypic variation in smooth softshell turtles (*Apalone mutica*) from eggs incubated in constant versus fluctuating temperatures. *Oecologia* 134:182–88

39. Warburton SJ, Hastings D, Wang T. 1995. Responses to chronic hypoxia in embryonic alligators. *J. Exp. Zool.* 273:44–50

40. O'Steen S, Janzen FJ. 1999. Embryonic temperature affects metabolic compensation and thyroid hormones in hatchling snapping turtles. *Physiol. Biochem. Zool.* 72:520–33

41. Randall DJ, Burggren WW, French K. 2001. *Animal Physiology*. New York: Freeman. 5th ed.

42. Vehaskari VM, Aviles DH, Manning J. 2001. Prenatal programming of adult hypertension in the rat. *Kidney Int.* 59:238–45

43. Sisneros JA, Bass AH. 2003. Seasonal plasticity of peripheral auditory frequency sensitivity. *J. Neurosci.* 23:1049–58

44. Morgan SM, Butz Huryn VM, Downes SR, Mercer AR. 1998. The effects of queenlessness on the maturation of the honey bee olfactory system. *Behav. Brain Res.* 91:115–26

45. Fitzgerald SM, Gan L, Wickman A, Bergstrom G. 2003. Cardiovascular and renal phenotyping of genetically modified mice: a challenge for traditional physiology. *Clin. Exp. Pharmacol. Physiol.* 30:207–16

46. Khan IY, Lakasing L, Poston L, Nicolaides KH. 2003. Fetal programming for

adult disease: where next? *J. Matern. Fetal Neonatal Med.* 13:292–99

47. Sallout B, Walker M. 2003. The fetal origin of adult diseases. *J. Obstet. Gynaecol.* 23:555–60

48. Johnston I, Temple GK. 2002. Thermal plasticity of skeletal muscle phenotype in ectothermic vertebrates and its significance for locomotory behavior. *J. Exp. Biol.* 205:2305–22

49. Wassersug RJ, Yamashita M. 2001. Plasticity and constraints on feeding kinematics in anuran larvae. *Comp. Biochem. Physiol.* 131:183–95

50. Rouwet EV, Tintu AN, Schellings MW, van Bilsen M, Lutgens E, et al. 2002. Hypoxia induces aortic hypertrophic growth, left ventricular dysfunction, and sympathetic hyperinnervation of peripheral arteries in the chick embryo. *Circulation* 105:2791–96

51. Burggren WW, Mwalukoma A. 1983. Respiration during chronic hypoxia and hyperoxia in larval and adult bullfrogs (*Rana catesbeiana*). I. Morphological responses of lungs, skin and gills. *J. Exp. Biol.* 105:191–203

52. Pinder A, Burggren WW. 1983. Respiration during chronic hypoxia and hyperoxia in larval and adult bullfrogs (*Rana catesbeiana*). II. Changes in respiratory properties of whole blood. *J. Exp. Biol.* 105:205–13

53. Burggren WW. 1992. The importance of an ontogenetic perspective in physiological studies: amphibian cardiology as a case study. In *Strategies of Physiological Adaptation, Respiration, Circulation and Metabolism*, ed. R Weber, SC Wood, R Millard, A Hargens, pp. 235–53. New York: Dekker

54. Warburton SJ, Burggren WW, Pelster B, Reiber C Spicer J, eds. 2005. *Comparative Developmental Physiology*. Oxford, UK: Oxford Univ. Press. In press

55. Futuyama DJ. 1998. *Evolutionary Biology*. Sunderland, MA: Sinauer

56. Raff RA. 1996. Th*e Shape of Life: Genes, Development, and the Evolution of Animal Form*. Chicago: Univ. Chicago Press

57. Gerhart J, Kirschner M, Kirschner MW. 1997. *Cells, Embryos, and Evolution: Towards a Cellular and Developmental Understanding of Phenotypic Variation and Evolutionary Adaptability*. Oxford, UK: Blackwell

58. Nielsen C. 2001. *Animal Evolution: Interrelationships of the Living Phyla*. Oxford, UK: Oxford Univ. Press

59. Seymour RS, Blaylock AJ. 2000. The principle of Laplace and scaling of ventricular wall stress and blood pressure in mammals and birds. *Physiol. Biochem. Zool.* 73:389–405

60. Schmidt-Nielsen K. 1984. *Scaling. Why is Animal Size So Important?* Cambridge, UK: Cambridge Univ. Press

61. Maitland DP. 1986. Crabs that breathe air with their legs—*Scopimera* and *Dotilla*. *Nature* 319:493–95

62. Burggren WW. 1992. Respiration and circulation in land crabs: novel variations on the marine design. *Am. Zool.* 32:417–27

63. Burggren WW, Tazawa H, Thompson D. 1994. Intraspecific variability in avian embryonic heart rates: potential genetic and maternal environment influences. *Israel J. Zool.* 40:351–62

64. Bagatto B, Crossley D, Burggren W. 2000. Physiological variability in neonatal armadillo quadruplets: within and between litter differences. *J. Exp. Biol.* 203:1733–40

65. Burggren WW, Crossley D III, Rogowitz G, Thompson D. 2003. Clutch effects explain heart rate variation in embryonic frogs (cave coqui, *Eleutherodactylus cooki*). *Physiol. Biochem. Zool.* 76:672–78

66. Seymour RS, Bennett-Stamper CL, Johnson SD, Carrier DR, Grigg GC, Franklin CE. 2004. Evidence for endothermic ancestors of crocodiles at the stem of archosaur evolution. *Physiol. Biochem. Zool.* In press

67. Reckova M, Rosengarten C, deAlmeida A, Stanley CP, Wessels A, et al. 2003. Hemodynamics is a key epigenetic factor in the development of cardiac conduction system. *Circ. Res.* 93:77–85

68. Ursem NTC, Stekelenburgde Vos S, Wladimiroff JW, Poelmann R, Gittenberger-de Groot AC, et al. 2004. Ventricular diastolic filling characteristics in stage-24 chick embryos after extra-embryonic venous obstruction. *J. Exp. Biol.* 2004 207:1487–90

69. Sedmera D, Pexieder T, Rychterova V, Hu N, Clark EB. 1999. Remodeling of chick embryonic ventricular myoarchitecture under experimentally changed loading conditions. *Anat. Rec.* 254:238–52

70. Aon MA, O'Rourke B, Cortassa S. 2004. The fractal architecture of cytoplasmic organization: scaling, kinetics and emergence in metabolic networks. *Mol. Cell. Biochem.* 256–257:169–84

71. Gunther B, Morgado E. 2003. Dimensional analysis revisited. *Biol. Res.* 36:405–10

72. Li JK-J. 1995. *Comparative Cardiovascular Dynamics of Mammals.* Boca Raton, FL: CRC Press

73. Dawson TH. 2001. Similitude in the cardiovascular system of mammals. *J. Exp. Biol.* 204:395–407

74. Heusner AA. 1988. A theory of similitude may predict a metabolic mass exponent. *Am. J. Physiol. Regul. Integr. Comp. Physiol.* 255:R350–52

75. Gould SJ. 1977. *Ontogeny and Phylogeny.* Cambridge, MA: Harvard University Press

76. Gould SJ. 1992. Heterochrony. In *Keywords in Evolutionary Biology*, ed. E Fox, E Lloyd, pp. 158–67. Cambridge, MA: Harvard University Press

77. Smith KK. 2001. Heterochrony revisited: the evolution of developmental sequences. *Biol. J. Linn. Soc.* 73:169–86

78. Smith KK. 2001. Sequence heterochrony and the evolution of development. *J. Morphol.* 252:82–97

79. Burggren WW. 2005. Complexity during physiological development. See Ref. 54. In press

80. Spicer JI, El-Gamal MM. 1999. Hypoxia accelerates the development of respiratory regulation in brine shrimp—but at a cost. *J. Exp. Biol.* 202:3637–46

81. Rombough P. 2003. Development rate: modeling developmental time and temperature. *Nature* 424:268–69; discussion 270

82. Beck CW, Slack JM. 2001. An amphibian with ambition: a new role for *Xenopus* in the 21st century. *Genome Biol.* 2:R1029

83. Briggs JP. 2002. The zebrafish: a new model organism for integrative physiology. *Am. J. Physiol. Regul. Integr. Comp. Physiol.* 282:R3–9

84. Chen JN, Fishman MC. 1997. Zebrafish tinman homolog demarcates the heart field and initiates myocardial differentiation. *Development* 122:3809–16

85. Feder M. 2005. Sciomics: community/model organism-based and individualistic research strategies for comparative animal developmental physiology. See Ref. 54. In press

86. Reinke V, White KP. 2002. Developmental genomic approaches in model organisms. *Annu. Rev. Genomics Hum. Genet.* 3:153–78

87. Burggren WW. 2000. Developmental physiology, animal models, and the August Krogh principle. *Zoology* 102:148–56

88. Nystul TG, Goldmark JP, Padilla PA, Roth MB. 2003. Suspended animation in *C. elegans* requires the spindle checkpoint. *Science.* 302:1038–41

89. Padilla PA, Nystul TG, Zager RA, Hohgnson AC, Roth MB. 2002. Dephosphorylation of cell cycle-regulated proteins correlates with anoxia-induced suspended animation in *Caenorhabditis elegans.* *Mol. Biol. Cell.* 13:1473–83

90. Westerfield M. *The Zebrafish Book.* 1995. Corvalis, OR: Univ. Oregon Press

91. Bagatto B. 2001. *The developmental physiology of the zebrafish: influence of environment on metabolic and cardiovascular attributes.* PhD thesis. Univ. North Texas, Denton, Texas. 179 pp.

92. Hu N, Clark EB. 1989. Hemodynamics of the stage 12 to stage 29 chick embryo. *Circ. Res.* 65:1665–70

93. Keller BB. 1997. Embryonic cardiovascular function, coupling and maturation: a species view. In *Development of Cardiovascular Systems: Molecules to Organisms.* eds. WW Burggren, BB Keller. New York: Cambridge Univ. Press

94. Wagman AJ, Hu N, Clark EB. 1990. Effect of changes in circulating blood volume on cardiac output and arterial and ventricular blood pressure in the stage 18, 24, and 29 chick embryo. *Circ. Res.* 67:187–92

95. Fritsche R, Burggren W. 1996. Development of cardiovascular responses to hypoxia in larvae of the *Xenopus laevis.* *Am. J. Physiol Regul. Integr. Comp. Physiol.* 271:R912–17

96. Hou P-C L, Burggren WW. 1995. Blood pressures and heart rate during larval development in the anuran amphibian *Xenopus laevis.* *Am. J. Physiol. Regul. Integr. Comp. Physiol.* 269:R1120–25

97. Hou P-C L, Burggren WW. 1995. Cardiac output and peripheral resistance during larval development in the anuran amphibian *Xenopus laevis.* *Am. J. Physiol. Regul. Integr. Comp. Physiol.* 269:R1126–32

98. Warburton SJ, Fritsche R. 2000. Blood pressure control in a larval amphibian *Xenopus laevis.* *J. Exp. Biol.* 203:2047–52

99. Hu N, Sedmera D, Yost HJ, Clark EB. 2000. Structure and function of the developing zebrafish heart. *Anat. Rec.* 260:148–57

100. Pelster B, Burggren WW. 1996. Disruption of hemoglobin oxygen transport does not impact oxygen-dependent physiological processes in developing embryos of zebrafish (*Danio rerio*). *Circ. Res.* 79:358–62

101. Burggren WW, Fritsche R. 1995. Cardiovascular measurements in animals in the milligram body mass range. *Brazil. J. Med. Biol. Res.* 28:1291–305

102. Schwerte T, Fritsche R. 2003. Understanding cardiovascular physiology in zebrafish and *Xenopus* larvae: the use of microtechniques. *Comp. Biochem. Physiol. A.* 135:131–45

103. Schwerte T, Uberbacher D, Pelster B. 2003. Non-invasive imaging of blood cell concentration and blood distribution in zebrafish *Danio rerio* incubated in hypoxic conditions in vivo. *J. Exp. Biol.* 206:1299–307

104. Lee CO. 1988. Measurement of cytosolic calcium: ion selective microelectrodes. *Miner. Electrolyte Metab.* 14:15–21

105. Schneider BH, Hill MR, Prohaska OJ. 1990. Microelectrode probes for biomedical applications. *Am. Biotechnol Lab.* 8:17–18, 20, 22–23

106. Link JR, Sailor MJ. 2003. Smart dust: self-assembling, self-orienting photonic crystals of porous Si. *Proc. Natl. Acad. Sci. USA* 100:10607–10

107. Schmidt KF. 2004. 'Smart dust' is way cool. *US News World Rep.* 136:56–57

108. Colmorgen M, Paul RJ. 1995. Imaging of physiological functions in transparent animals (*Agonus cataphractus, Daphnia magna, Pholcus phalangioides*) by video microscopy and digital image processing. *Comp. Biochem. Physiol.* 111A:583–595

109. Baumer C, Pirow R, Paul RJ. 2002. Circulatory oxygen transport in the water flea *Daphnia magna.* *J. Comp. Physiol. B* 172:275–85

110. Knisley SB, Neuman MR. 2003. Simultaneous electrical and optical mapping in rabbit hearts. *Ann. Biomed. Eng.* 31:32–41

111. Pirow R 2003 The contribution of haemoglobin to oxygen transport in the microcrustacean *Daphnia magna*–a

conceptual approach. *Adv. Exp. Med. Biol.* 510:101–107

112. Koch CJ. 2002. Measurement of absolute oxygen levels in cells and tissues using oxygen sensors and 2-nitroimidazole EF5. *Methods Enzymol.* 352:3–31

113. Delbridge LM, Roos KP. 1997. Optical methods to evaluate the contractile function of unloaded isolated cardiac myocytes. *J. Mol. Cell. Cardiol.* 29:11–25

114. Ragan TM, Huang H, So PT. 2003. In vivo and ex vivo tissue applications of two-photon microscopy. *Methods Enzymol.* 361:481–505

115. Squirrell JM, White JG. 2004. Using multiphoton excitation to explore the murky depths of developing embryos. *Methods Mol. Biol.* 254:113–36

116. Bunz F. 2002. Human cell knockouts. *Curr. Opin. Oncol.* 14:73–78

117. Gingrich JA. 2002. Mutational analysis of the serotonergic system: recent findings using knockout mice. *Curr. Drug Target CNS Neurol. Disord.* 1:449–65

118. Mak TW, Penninger JM, Ohashi PS. 2001. Knockout mice: a paradigm shift in modern immunology. *Nat. Rev. Immunol.* 1:11–19

119. Peachey NS, Ball SL. 2003. Electrophysiological analysis of visual function in mutant mice. *Doc. Ophthalmol.* 107:13–36

120. Sehnert AJ, Stainier DY. 2002. A window to the heart: can zebrafish mutants help us understand heart disease in humans? *Trends Genet.* 18:491–94

121. Stainier DY. 2001. Zebrafish genetics and vertebrate heart formation. *Nat. Rev. Genet.* 2:39–48

122. Takeishi Y, Walsh RA. 2001. Cardiac hypertrophy and failure: lessons learned from genetically engineered mice. *Acta Physiol. Scand.* 173:103–11

Annu. Rev. Physiol. 2005. 67:225–57
doi: 10.1146/annurev.physiol.67.040403.103635
First published online as a Review in Advance on September 22, 2004

Molecular and Evolutionary Basis
of the Cellular Stress Response

Dietmar Kültz

Physiological Genomics Group, Department of Animal Sciences, University of California, Davis, California 95616; email: dkueltz@ucdavis.edu

Key Words   molecular evolution, macromolecular damage, molecular chaperones, redox-regulation, DNA repair

■ **Abstract**   The cellular stress response is a universal mechanism of extraordinary physiological/pathophysiological significance. It represents a defense reaction of cells to damage that environmental forces inflict on macromolecules. Many aspects of the cellular stress response are not stressor specific because cells monitor stress based on macromolecular damage without regard to the type of stress that causes such damage. Cellular mechanisms activated by DNA damage and protein damage are interconnected and share common elements. Other cellular responses directed at re-establishing home-ostasis are stressor specific and often activated in parallel to the cellular stress response. All organisms have stress proteins, and universally conserved stress proteins can be regarded as the minimal stress proteome. Functional analysis of the minimal stress proteome yields information about key aspects of the cellular stress response, includ-ing physiological mechanisms of sensing membrane lipid, protein, and DNA damage; redox sensing and regulation; cell cycle control; macromolecular stabilization/repair; and control of energy metabolism. In addition, cells can quantify stress and activate a death program (apoptosis) when tolerance limits are exceeded.

CELLULAR STRESS: WHAT IS THE THREAT
AND HOW DO CELLS RESPOND?

The study of mechanisms of adaptation to stressful and extreme environments pro-vides the basis for addressing environmental health problems, performing sound toxicological risk assessment, efficiently utilizing bioindication processes to mon-itor global environmental change, and clinically utilizing the inherent healing capacity of the adaptive response to stress. Detailed study of the cellular stress response (CSR) has revealed diverse molecular mechanisms too numerous to be considered comprehensively in this review. Highlighted here are evolutionarily conserved principles of the CSR that are critical for understanding the molecular mechanisms of cellular adaptation to stress.

Classical responses of animals to stress, the "fight-or-flight response" (1) or "general adaptation syndrome" (2), are controlled by hormones at the organismal

0066-4278/05/0315-0225$14.00   **225**

level (3). At the cellular level the CSR is a defense reaction to a strain imposed by environmental force(s) on macromolecules. Such strain commonly results in deformation of/damage to proteins, DNA, or other essential macromolecules (4). The CSR assesses and counteracts stress-induced damage, temporarily increases tolerance of such damage, and/or removes terminally damaged cells by programmed cell death (apoptosis). The capacity of the CSR depends on the proteome expressed in a cell at a particular time and is therefore species- and cell type-dependent.

THE MINIMAL STRESS PROTEOME
OF ALL ORGANISMS

Functional Classification of Stress Proteins
Conserved in All Cells

The CSR is characteristic of all cells. It targets a defined set of cellular functions, including cell cycle control, protein chaperoning and repair, DNA and chromatin stabilization and repair, removal of damaged proteins, and certain aspects of metabolism (4). Proteins involved in key aspects of the CSR are conserved in all organisms. They can be identified experimentally using proteomics approaches such as 2D electrophoresis or 2D chromatography in combination with mass spectrometry analysis. In addition, annotated proteomes of multiple organisms can be compared using bioinformatics approaches that identify evolutionarily conserved stress proteins. Such analysis of human (*Homo sapiens*), yeast (*Saccharomyces cerevisiae*), eubacterial (*Escherichia coli*), and archaeal (*Halobacterium spec.*) proteomes yields circa 300 proteins that are highly conserved in all (4). This protein set corresponds approximately to the size of a minimal gene set and includes tRNA synthetases for all essential amino acids, presumably inherited from the last universal common ancestor (LUCA) (5). Gene ontology and literature analysis of these 300 proteins have revealed 44 proteins with known functions in the CSR (Table 1).

Many more than the 44 proteins in Table 1 participate in the CSR. However, most stress proteins are not ubiquitously conserved in all three superkingdoms and are, therefore, not included in this minimal stress proteome of all organisms. Transcript levels for most universally conserved stress proteins (31 of 44) are up-regulated in response to diverse stresses in yeast (6). However, stress proteins are regulated not only at the mRNA level but also at other levels, e.g., by modulation of protein turnover or by posttranslational modification. Also, high constitutive expression of some conserved stress proteins confers increased cellular stress resistance. Cells with chronic stress exposure constitutively express several stress proteins at very high levels, including Hsp60, Hsp70, peroxiredoxin, and superoxide dismutase in mammalian renal inner medullary cells (7; N. Valkova & D. Kültz, manuscript submitted) and RecA/Rad51 in the extremophile archaeon *Pyrococcus furiosus* (8).

Functionally, the 44 stress proteins cluster into distinct categories that reflect different aspects of the CSR. They include redox-sensitive proteins as well as

TABLE 1   The minimal stress proteome of cellular organisms

Redox regulation	DNA damage sensing/repair	Fatty acid/lipid metabolism
Aldehyde reductase	MutS/MSH	Long-chain fatty acid ABC transporter
Glutathione reductase	MutL/MLH	Multifunctional beta oxidation protein
Thioredoxin	Topoisomerase I/III	Long-chain fatty acid CoA ligase
Peroxiredoxin	RecA/Rad51	
Superoxide dismutase		
MsrA/PMSR	**Molecular chaperones**	**Energy metabolism**
SelB	Petidyl-prolyl isomerase	Citrate synthase (Krebs cycle)
Proline oxidase[a]	DnaJ/HSP40	Ca^{2+}/Mg^{2+}-transporting ATPase[b]
Quinone oxidoreductase[c]	GrpE (HSP70 cofactor)	Ribosomal RNA methyltransferase[d]
NADP-dependent oxidoreductase YMN1[c]	HSP60 chaperonin[d]	Enolase (glycolysis)
Putative oxidoreductase YIM4[c]	DnaK/HSP70	Phosphoglucomutase
Aldehyde dehydrogenase[c]		
Isocitrate dehydrogenase[c]	**Protein degradation**	**Other functions**
Succinate semialdehyde dehydrogenase[c]	FtsH/proteasome-regulatory subunit[d]	Inositol monophosphatase[b]
6 phosphogluconate dehydrogenase[c]	Lon protease/protease La	Nucleoside diphosphate kinase[e]
Glycerol-3-phosphate dehydrogenase[c]	Serine protease	Hypothetical protein YKP1
2-hydroxyacid dehydrogenase[c]	Protease II/prolyl endopetidase	
Hydroxyacylglutathione hydrolase	Aromatic amino acid aminotransferase	
	Aminobutyrate aminotransferase	

[a]Proline oxidase degrades proline to pyrroline 5-carboxylate, hence it is also involved in amino acid degradation.
[b]Signaling functions (Ca^{2+}- and phosphoinositide-mediated).
[c]Many oxidoreductases are also important for energy metabolism.
[d]These proteins are also involved in cell cycle control.
[e]Involved in nucleotide synthesis (possible role in DNA repair).

proteins involved in sensing, repairing, and minimizing macromolecular damage, such as molecular chaperones and DNA repair enzymes. In addition, numerous enzymes (notably oxidoreductases) that are involved in energy metabolism and cellular redox regulation are part of the minimal stress proteome. Some conserved stress proteins also function in cell cycle control (HSP60, FtsH, and ribosomal RNA methyltransferase). Notably, not all aspects of the CSR, in particular signaling-related mechanisms, are based on ubiquitously conserved pathways and proteins. Eukaryotes and prokaryotes differ in the nature of phosphorylation-based signal transduction. Two-component systems based on His/Asp phosphorylation predominate in prokaryotes, whereas more complex eukaryotic signaling cascades are mainly based on Ser/Thr/Tyr phosphorylation. Second, in eukaryotes DNA is packaged into a nucleus, which is absent in prokaryotes, and the degree of packaging is higher because eukaryotic genomes are generally larger. Thus, chromatin organization is more complex and histones and other chromatin proteins have unique roles in eukaryotes. Consequently, eukaryotic mechanisms of transcriptional regulation and cell cycle control are more complex and depend on proteins that differ from those utilized for equivalent functions in bacteria. Exceptions include proteins that constitute the very basic transcription/replication machinery, such as DNA polymerases.

Two Cellular Responses to Environmental Change: Stress Response and Homeostasis Response

In 1974 Tissières and coworkers discovered that heat shock proteins (HSPs) are induced in salivary glands of *Drosophila melanogaster* during heat stress (9). More than a decade later the function of HSPs as molecular chaperones was elucidated. Today, we know that these proteins are induced and activated during many other types of stress as well. They share this responsiveness to diverse stresses with many other proteins, notably most of the proteins included in the conserved minimal stress proteome (6). In addition, diverse stresses activate or induce many more weakly conserved stress proteins (6). The low stressor specificity of stress proteins raises two questions: (*a*) Why are these proteins induced/activated by diverse stresses? (*b*) Where does specificity originate in cell responses to particular environmental perturbations?

Responsiveness to diverse stresses may arise from the most striking and common impact of stress: It deforms and damages macromolecules, mainly membrane lipids, proteins, and/or DNA (4). Some specificity may arise because the types of lesions and damage to proteins, DNA, and membranes vary somewhat depending on the type of stress. Another common feature of diverse stresses is the generation of oxidative stress and change in cellular redox potential (10), referred to as oxidative burst (11, 12). The molecular events that increase reactive oxygen species (ROS) during some types of stress, including exposure of cells to ionizing radiation or highly reactive chemicals, are a direct consequence of the stress. But the molecular basis for oxidative burst is poorly understood in, for example, osmotic stress or heat shock. During many types of stress cellular oxidases such as the plasma membrane NADPH oxidase are very rapidly activated, which may explain increased ROS levels (see below). Different cellular oxidases occur in different compartments (mitochondria, plasma membrane, etc.) and compartment-specific regulation of redox potential may be important for the outcome of the CSR. ROS and cellular redox potential have long been regarded as key regulators of CSR signaling, with ubiquitous roles as second messengers in cells exposed to stress (10).

The molecular basis of stressor-specificity has been a subject of much debate. One way of achieving stressor-specificity with the same set of components (induced/activated stress proteins) is via stressor-specific interactions, posttranslational modifications, and compartmentation of stress proteins resulting from different relative levels of induction within a common set of stress proteins. In addition, every stress also disturbs cellular homeostasis and induces a second type of response distinct from the CSR (Figure 1). In contrast to the transient nature of the CSR, this second type of response, here called the cellular homeostasis response (CHR), is permanent until environmental conditions change again. Its aim is to restore cellular homeostasis with specific regard to the particular environmental variable that has changed. Unlike the CSR, CHR is triggered primarily not by macromolecular damage or oxidative burst but by stressor-specific sensors that monitor changes in particular environmental variables (4). For instance, during osmotic stress the Sln1 and Sho1 membrane proteins function as osmosensors in yeast (13). In mammalian cells a particular transcription factor, the tonicity response element binding protein (TonEBP/NFAT5), activates osmoprotective genes that serve to stably restore cellular ion homeostasis by adjusting the levels of compatible organic osmolytes during osmotic stress (14). Although CSR and CHR signaling pathways are linked and contain common elements, this review focuses only on the former.

Molecular Basis of Cross-Tolerance and Stress-Hardening

Environmental stress tolerance varies widely depending on the species (genome) and on cell type and differentiation state (proteome). The latter is a function of gene-environment interactions during development and of pre-exposure to stress during life history. Stress-hardening (increased tolerance of a stress after preconditioning at low doses of that stress) and cross-tolerance (increased tolerance of one stress after preconditioning by another) are common and significant. For instance, ischemic preconditioning and mild hyperthermia induce HSP70 and decrease reperfusion injury of human muscle and kidney (15). HSP70 induction is also associated with stress-hardening and cross-tolerance to heat and cold stress in the fruit fly *Drosophila melanogaster* (16). Additional stress proteins induce stress-hardening and cross-tolerance of temperature, salinity, ionizing radiation, pH, and chemical stressors in diverse eukaryotic and prokaryotic cells (e.g., 17–20).

The activation and induction of a common set of stress proteins is the molecular basis of both cross-tolerance and stress-hardening. After the initial stress, these

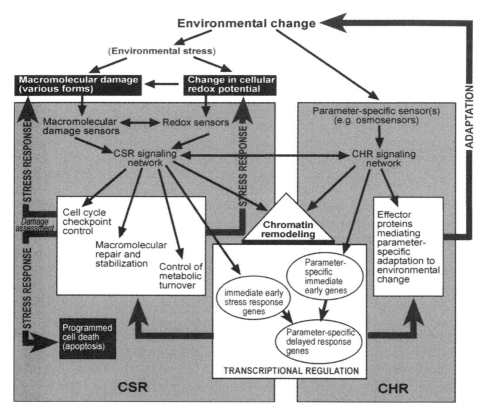

Figure 1   Schematic representation of key aspects of the cellular stress response (CSR) and its interaction with the cellular homeostasis response (CHR). The CSR serves to restore macromolecular integrity and redox potential that are disturbed as a result of stress. In contrast, the CHR serves to restore cellular homeostasis with regard to the particular environmental variable that has changed. Both types of cellular responses to environmental change are interconnected at numerous levels.

proteins remain active/elevated for a period that varies depending on species, cell-type, history of prior stress exposure, gene-environment interactions during development, and stress severity. During this period, activated/elevated stress proteins confer resistance to many different types of stress because of their involvement in general aspects of cellular protection such as protein stabilization, DNA repair, and free radical scavenging. In stark contrast to *Saccharomyces cerevisiae*, the yeast *Candida albicans* does not seem to induce a CSR via changes in gene transcription. Instead, it responds only by activation of the CHR, which correlates with a lack of cross-tolerance (21). This feature of *C. albicans* is exceptional and evolutionarily favored only in extraordinary stable environments (4).

   Differences in constitutive levels of critical stress proteins are also responsible for cell type–specific variation in tolerance thresholds within multicellular organisms. For instance, mammalian renal inner medullary cells tolerate many types of stress much better than do most other mammalian cell types, which correlates with increased constitutive levels of critical stress proteins in these cells (7, 22).

MACROMOLECULAR DAMAGE TRIGGERS THE CELLULAR STRESS RESPONSE

Cellular signal transduction networks commonly encompass three tiers: (a) sensors that perceive a signal; (b) transducers that carry, amplify, and integrate signals; and (c) effectors that adjust cell function corresponding to signals. In much of biology, extracellular signals are perceived by cell membrane receptors, and ligand-receptor interactions are highly specific. In addition, ligands are usually present at very low (nano- or micromolar) concentrations, and the affinity of the corresponding receptors is very high. Both paradigms apply poorly to the CSR. First, specific receptors are inconsistent with the lack of stress specificity in the CSR. Similarly, changes in environmental parameters are usually much more pronounced than minute changes in concentrations of specific ligands. For example, during osmotic stress total osmolyte concentrations can change by several hundreds of millimoles. Stress generally affects all cell compartments, whereas the cell membrane and other boundaries often exclude ligands from certain compartments. Finally, given the nature of stress-induced damage, stress sensors probably monitor the degree of macromolecular integrity in cells rather than an environmental signal per se. This mechanism would provide immediate feedback as to the effectiveness of the CSR once it has been activated. A second quasi-universal property characteristic of cells exposed to stress is an increase in ROS levels, which represents a critical second messenger for CSR signaling networks (23). Hence, sensors of membrane, protein, and DNA damage as well as redox sensors are key regulators of CSR signaling networks.

Lipid Membrane Damage Sensors

The cell membrane is the barrier to (and in direct contact with) the external environment and, therefore, well suited for sensing stress. In addition, secondary, calcium-mediated changes in properties of the mitochondrial membrane (mainly membrane potential and permeability) are important because they affect oxidative phosphorylation and redox potential directly, and thus may contribute to increases in ROS during stress (24, 25). Membrane and lipid damage occurs in all major groups of organisms in response to diverse stresses (e.g., 26–28). The extent of membrane damage and cellular tolerance limits during stress depend on lipid composition, fatty acid saturation, and membrane fluidity of the cell membrane (29, 30). Furthermore, the heat or salinity inducibility of a reporter gene that is driven by the CSR promoter element STRE (stress response element) is inversely correlated to the amount of unsaturated fatty acids in yeast, suggesting that induction of the STRE pathway depends on membrane lipid composition (31). These authors also suggest that stress cross-tolerance may be (at least in part) a lipid-mediated phenomenon. Thus, the three universally conserved stress proteins involved in long-chain fatty acid (LCFA) metabolism and transport (Table 1) may contribute to changes in membrane lipid composition in response to stress. Moreover, the LCFA transporter mediates movement of LCFAs into peroxisomes, where they are metabolized by LCFA CoA ligase. This enzyme, present in multiple isoforms in many organisms, has been implicated in the metabolism of xenobiotics and reactive compounds generated during stress (32). In addition, fatty acyl-CoA esters produced by LCFA CoA ligase are emerging as physiological regulators of cell function, including transcriptional regulation (32).

Membrane damage from physical effects of environmental stress on cells is associated with altered membrane tension or stretch, permeability changes, lipid rearrangement, membrane protein rearrangement, changes in transmembrane potential, and formation of lipid peroxides and lipid adducts. Membrane lipid peroxidation, a common form of damage in response to stress, results from lipid auto-oxidation or catalysis by lipoxygenase (LOX) or the cytochrome P-450 system to yield highly reactive lipid peroxidation products (33, 34). Such products include isoprostanes from arachidonic, eicosapentaenoic, and docosahexaenoic acids; oxysterols from unesterified and esterified cholesterol; various other fatty acid hydroperoxides; and a wide spectrum of aldehydes (35, 36). Such membrane damage represents potential upstream signals for CSR signaling networks, and multiple mechanisms for translating nonspecific membrane damage into activation of CSR signaling pathways have been proposed (Figure 2).

First, nonspecific clustering of growth factor receptor tyrosine kinases and cytokine receptors during osmotic and UV-radiation stress can activate these receptors and the JNK cascade in mammalian cells (37). Activation of such cell surface receptors has other potential consequences, including the activation of PI-3-kinase, which catalyzes conversion of PIP_2 to PIP_3 (38). This activates the small GTP-binding protein Rac1, which, in turn, stimulates NADPH oxidase (38, 39). The NADP-dependent oxidoreductase contained in the minimal stress proteome may function in NADPH oxidase mode under such conditions. NADPH oxidase produces H_2O_2 (hydrogen peroxide), and, therefore, stress-stimulated nonspecific clustering of cell surface receptors provides a possible avenue for oxidative burst and activation of H_2O_2-induced signaling mechanisms during stress (Figure 2). A second potential avenue for oxidative burst and H_2O_2 generation during stress originates from lipid peroxidation (40). Lipid peroxidation products activate multiple signaling pathways, including MAP kinase pathways and the transcription factor AP-1 (activator protein 1), possibly via generation of H_2O_2 as a signaling intermediate (36, 40, 41). Third, processing of integral membrane proteins resulting in liberation of active signaling molecules is common and of extraordinary biological importance. For instance, phospholipase A_2 (PLA$_2$) activity depends on lipid packing density and membrane integrity and is elevated during stress (42). This enzyme catalyzes the hydrolysis of membrane glycerophospholipids, resulting in release

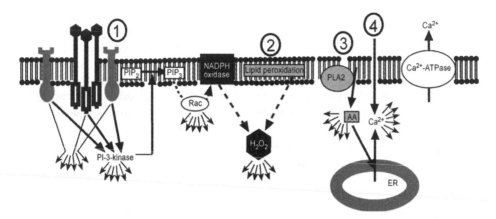

Figure 2 Potential stress sensing mechanisms that are based on lipid membrane damage/rearrangements. (1) Nonspecific clustering of growth factor and cytokine receptors due to membrane rearrangements leads to receptor activation. (2) Activation of NADPH oxidase resulting from receptor activation (1) and lipid auto-oxidation generate oxygen radicals that are converted to the second messenger hydrogen peroxide. (3) Changes in membrane tension or lipid rearrangement result in activation of phospholipase A_2, which leads to liberation of arachidonic acid from membranes. (4) Changes in membrane permeability lead to calcium influx into the cytosol. Multiple arrows emanating from several elements in the figure illustrate possibilities for further signal amplification. Please refer to the text for a discussion of these mechanisms.

of arachidonic acid (AA), an important signaling molecule in cells (43). Another example is the intramembrane proteolysis, liberation, and activation of the yeast transcription factor SPT23, a relative of mammalian NF-κB (nuclear factor kappa B). Because the proteasome-dependent processing of SPT23 is regulated by fatty acid pools, SPT23 may function in sensing membrane composition or fluidity (44). Fourth, changes in membrane permeability and the activity of mechanosensitive ion channels during stress promote calcium influx into the cytosol, an important signal for the CSR (45, 46). A Ca^{2+}-transporting ATPase is part of the minimal stress proteome and may be required to restore cytosolic calcium levels after the initial stress signal has been perceived. Although these mechanisms have not been extensively tested for their universal applicability to a broad spectrum of cells and stresses, they represent potential sensors of membrane lipid damage (Figure 2).

DNA Damage Sensors

Much work during the past decade has focused on DNA damage sensors, and consequently, we now know more about mechanisms of DNA damage sensing than those of membrane lipid damage sensors. Nonetheless, it is still nearly impossible to distinguish primary sensors from secondary transducers of DNA damage.

The problem lies in complex circuits of feedback regulation of proteins involved in sensing DNA damage. For example, many candidate sensor proteins are part of multiprotein complexes and, when activated, they become targets of further modification by their own substrates. Most studies on DNA damage sensors have focused on responses of cells to damage induced by ionizing radiation or highly reactive chemicals. However, recent work has demonstrated that during other types of stress, including osmotic stress and heat shock, DNA damage occurs and key mechanisms involved in eukaryotic DNA damage sensing, transduction, and repair are activated (47–49). These findings, in combination with extensive prior knowledge about ubiquitous effects of many types of stress on protein folding and stability, led to the hypothesis that the CSR represents a universal reaction to macromolecular damage (4).

Damage to DNA occurs in myriads of ways, ranging from common, i.e., certain base modifications such as 8-oxoguanine (8-oxo-7,8-dihydroguanine) formation, to more stressor specific, e.g., pyrimidine dimer formation during UV irradiation. However, despite numerous types of DNA adducts and base modifications, DNA damage can be grouped into a few major types, including DNA double-strand breaks (dsb), DNA nucleotide adduct formation and base modification, DNA base-pairing mismatches, and DNA single-strand breaks (ssb). Accordingly, the major classes of DNA repair are DNA dsb repair by homologous recombination (HR) or nonhomologous end-joining (NHEJ), nucleotide excision repair (NER), and nucleotide mismatch repair (MMR). DNA damage sensors probably recognize common intermediates of major types of DNA damage. Candidate intermediates are DNA ssb that occur during all types of DNA damage (50) and recognition motifs that are common to different base mismatches and modifications (51).

Much of the cellular machinery involved in DNA damage sensing is highly conserved in eukaryotes and prokaryotes but differs considerably between these two major forms of life. Nevertheless, some components of these complex networks are highly conserved in all three superkingdoms (Table 1), including MutS/MSH, MutL/MLH, RecA/RadA/Rad51, Top I/III, Mre11/Rad32, Rad50, and MutT/MTH. The latter two proteins show a lower degree of homology between prokaryotes and eukaryotes and do not meet the criteria for inclusion in the minimal stress proteome outlined above. However, they occur in all three superkingdoms (52) and are critical components of DNA damage sensing and signaling (see below). Mutations in all of these proteins cause defects in CSR signaling networks, resulting in diminished genomic integrity.

Bacterial MutS and eukaryotic MSH proteins recognize and bind to distortions produced by mismatches in DNA base pairing (53). In eukaryotes, the MMR proteins MSH2, MSH6, and MLH1 are part of the large BRCA1-associated genome surveillance supercomplex (BASC). BASC is important for recognizing and repairing base mismatches and in sensing other types of DNA damage in mammalian cells (54). After binding to sites of DNA damage, MutS/MSH proteins recruit MutL (in bacteria) or MLH (in eukaryotes) to those sites and initiate assembly of the MMR complex. The MutS gene in *E. coli* is induced by stress although it is not considered part of the bacterial SOS response, the adaptive activation of stress proteins by the genetic regulator RecA (55). Thus MutS/MSH and MutL/MLH proteins are involved in sensing DNA base mismatches in all organisms, and this sensory capacity is increased by up-regulation of MutS/MSH during stress.

Another important DNA damage sensor in *E. coli* is the single-stranded (ss) DNA binding protein RecA, a recombinase also involved in DNA repair. The RecA protein represents a central part of the bacterial SOS response. RecA functions as a derepressor of LexA (which is a repressor of SOS genes) via its coprotease activity that degrades and inactivates LexA (56). It has been proposed that RecA is activated by recognizing and binding to ssDNA at sites of DNA base modifications or adducts and that this stimulates its coprotease activity (57). The functional homolog of RecA in archaea is the RadA protein, which increases when cells are exposed to stress (58). The eukaryotic homolog of RecA is Rad51, which is the key protein for homologous recombination-mediated repair of DNA dsb. Rad51 catalyzes the central step of homologous recombination, the DNA strand exchange reaction (59). Rad51 also binds ssDNA in eukaryotes (60) and Rad51 knockout mice are not viable, a finding illustrating that this protein is essential (61). Rad51 interacts with other proteins potentially involved in sensing DNA damage, including BRCA1, which is part of the same BASC supercomplex that also includes MSH and MLH proteins (see above; 62, 63).

Mre11 and Rad50 are also part of the mammalian BASC supercomplex (54). These proteins actually form a smaller complex with Nbs1 called the Mre11-Rad50-Nbs1 (MRN) complex, which interacts as a unit with other components of the BASC supercomplex. The MRN complex is required for both homologous recombination and nonhomologous end-joining (NHEJ) and is recruited to DNA dsb by another putative DNA damage sensor called MDC1 (mediator of DNA damage checkpoint 1) (64, 65). In *S. cerevisiae*, MRN complexes are targeted to sites of DNA dsb by direct association of Xrs2 (the yeast ortholog of mammalian Nbs1) with free DNA ends (66). In addition, mammalian Nbs1 regulates the kinetics and magnitude of ATM (ataxia telangiectasia mutated) serine-1981 autophosphorylation (67), and the MRN complex also stimulates ATM kinase activity (68). Thus, the MRN complex represents a critical element of rapid ATM activation resulting from autophosphorylation of serine-1981 and dimer dissociation in response to perturbation of chromatin structure (69). Consequently, it has been proposed that ATM is positioned downstream of the MRN complex and represents a secondary messenger rather than a primary DNA damage sensor (65, 70). ATM phosphorylates some of its own activators, including Nbs1 (70) and histone H2AX (71), both of which are involved in recruitment of the MRN complex to sites of DNA damage (72). Like ATM and its yeast ortholog Tel1, other PIKKs (PI-3-K like kinases) also seem to be early transducers of DNA damage signals rather than primary damage sensors. For instance, the catalytic subunit of mammalian DNA-PK (DNA-dependent protein kinase) is recruited to sites of DNA dsb by two other proteins, Ku70 and Ku80, which seem to be the sensors required for initiation of the NHEJ repair complex (65). Likewise, mammalian ATR (ATM-related) and its yeast ortholog Mec1 are recruited to sites of DNA damage by association with the putative DNA damage sensor protein ATRIP (ATR interacting protein) in mammals and its ortholog Lcd1 in yeast, which results in activation of DNA damage-dependent cell cycle checkpoints (65, 73, 74).

The BASC supercomplex thus emerges as an important sensor of multiple types of DNA damage. The interaction of many highly evolutionarily conserved stress proteins, including MutS/MSH, MutL/MLH, RecA/Rad51, Mre11, and Rad50, with this DNA damage sensory supercomplex suggests that key aspects of DNA damage sensing mechanisms are highly conserved in all organisms. Additional support for the universal conservation of key aspects of genome integrity surveillance mechanisms comes from two other highly conserved proteins: MutT/MTH and topoisomerase. MutT/MTH is part of the nucleotide excision repair pathway that removes oxidized nucleotide precursors and prevents their incorporation into DNA during replication. Thus MutT/MTH is important for preventing replication-dependent oxidative DNA damage during stress-induced oxidative burst. The most stable and deleterious base modification caused by ROS is formation of 8-oxoguanine (8-oxo-7,8-dihydroguanine, 8-oxoG). 8-OxoG is produced not only in nucleotide pools of cells but also in DNA, where it mispairs with adenine and thus damages DNA. MutT/MTH protects cells from the mutagenic effects of 8-oxoG by degrading 8-oxo-dGTP to 8-oxo-dGMP (75).

Another potential DNA damage sensor that is part of the minimal stress proteome is topoisomerase. Cells have various isoforms of this enzyme that participate in different aspects of DNA metabolism. All topoisomerases alter DNA topology by introducing transient ssb into DNA during replication and NER. Because DNA ssb might represent a common intermediate recognized by DNA damage sensors (50), topoisomerase may be a critical element of DNA damage sensing. Indeed, topoisomerase I is involved in NER during the bacterial SOS response to stress-induced DNA damage (76, 77), and the homologous mammalian topoisomerase III is a sensor for the S phase DNA damage checkpoint (78). Additional highly conserved proteins involved in various aspects of DNA repair are candidate DNA damage sensors, notably the RecQ family of helicases that includes the Werner and Bloom syndrome helicases (79). New insights into how highly conserved stress proteins function during stress should further our understanding of DNA damage sensing mechanisms.

Protein Damage Sensors

Protein damage in cells exposed to stress occurs mainly as oxidative or structural (unfolding) damage. Some damaged proteins are repaired by enzymes that reverse oxidative damage or assist in protein refolding. But not all damaged proteins are repaired: Many terminally damaged proteins are removed by proteolytic degradation and regenerated by de novo synthesis. Thus, three processes are mainly responsible for removing protein damage: (*a*) repair of oxidative damage, (*b*) refolding of structurally damaged proteins, and (*c*) proteolysis. Methionine sulfoxide reductase

(MsrA/PMSR) and other conserved redox-regulatory stress proteins contribute to the recognition and repair of oxidative protein damage. The function of these proteins is discussed in the section Redox Sensors and Redox Regulation, below. In addition, five proteins of the minimal stress proteome are molecular chaperones. Because the key function of molecular chaperones pertains to protein maintenance and refolding, their role in the CSR is discussed under Maintenance of Macromolecular Integrity, below. Finally, six highly conserved stress proteins are involved in proteolysis (Table 1). These proteolytic enzymes help cells to monitor protein damage via mechanisms that are best illustrated using FtsH and Lon proteases as examples.

FtsH and Lon are regulatory protease subunits that are critical for removing damaged and abnormal proteins during stress and for controlling levels of key regulatory proteins with short half-lives. They are induced in response to many types of stress, including such unusual conditions as wine toxicity (80). FtsH and Lon both function as molecular chaperones by promoting the insertion of proteins into membranes and supporting the disassembly or oligomerization of protein complexes. FtsH is involved in stress resistance, membrane functions, cell cycle control, gene expression, translocation of secreted proteins, and degradation of some unstable and selected membrane and cytosolic proteins (81–83). Lon also contributes to the regulation of several important cellular functions, including stress resistance, cell division, cell morphology, proteolytic degradation of certain regulatory and abnormal proteins, and DNA maintenance (84–86). However, FtsH is the only membrane-integrated ATP-dependent protease that is universally conserved in all organisms (87). Moreover, in contrast to Lon, for which functionally redundant proteases such as Clp (caseinolytic) protease can substitute, FtsH is essential. But all of these proteases are part of protein complexes consisting of multiple subunits that have proteolytic core domains, regulatory domains with ATPase activity, and molecular chaperone domains (88).

FtsH lacks robust unfoldase activity and can only degrade proteins that are already damaged and partially unfolded. Moreover, FtsH uses the folding state of its protein substrates as a criterion for degradation (89). This feature delineates FtsH as a sensor of stresses that lead to protein unfolding. This property of FtsH further raises the possibility that its substrates are important components of CSR signaling. Indeed, FtsH displays high selectivity in protein degradation. It recognizes key signaling proteins by binding to specific motifs. Important FtsH substrates in bacteria include SecY (90), bacterial cell division protein FtsZ (91), and the heat shock transcription factor RpoH/σ^{32} (92, 93). FtsH is also involved in regulating the activity of σ^{54}-dependent promoters in bacteria (94), although it is not required for all σ^{54}-dependent promoters (95).

Lon also represents a potential sensor of protein damage because, like FtsH, it can recognize specific motifs in key signaling proteins (96–98). The recognition of such motifs depends on how these proteins are folded, which is affected by stress. For degradation of proteins containing specific Lon protease recognition motifs, no tagging (e.g., by ubiquitination) is required for protease activity. However, for recognition of additional, less specific substrates, Lon protease cooperates with molecular chaperones that are also part of the minimal stress proteome, including DnaK/HSP70 (99) and DnaJ/HSP40 (98). A specific substrate of Lon protease in the bacterium *Streptomyces coelicolor* is the negative autoregulator HspR, which represses important CSR genes encoding DnaK/HSP70, Lon protease, and Clp protease when bound to free DnaK/HSP70. During heat stress, the levels of free DnaK/HSP70 decline as a result of increased binding to unfolded proteins, and HspR-mediated repression of DnaK/HSP70, Lon protease, and Clp protease is lifted leading to induction of those genes (100). Such feedback mechanisms illustrate the sensory role of cellular proteases during stress (101).

In eukaryotes, FtsH and Lon are mainly located in mitochondria and plant chloroplasts (102). Consequently, eukaryotic FtsH and Lon are involved in protein degradation in those cellular organelles. In addition to its recognition of protein substrates, Lon also binds to DNA (103) and shares sequence similarity with RecA, a putative DNA damage sensor discussed above (104). It may, therefore, also be involved in sensing damage to mitochondrial and chloroplast DNA, but this potential aspect of Lon function has received little attention. Alternatively, Lon might regulate mitochondrial DNA replication and/or transcription via degradation of DNA binding proteins (103). Recently, a specific isozyme of Lon has been detected in rat liver peroxisomes by an experimental proteomics approach (105). Moreover, FtsH is highly homologous to at least three ATPase regulatory subunits of the eukaryotic 26S proteasome, which is present in the cytosol and nucleus (93, 106). Therefore, functional homologs of Lon and FtsH occur in most compartments of eukaryotic cells.

The 26S proteasome is the only multisubunit ATP-hydrolyzing proteolytic complex in the cytosol and nucleus of eukaryotic cells (107). It is enormously large, consisting of about 50 subunits with a combined molecular mass of 2.4 MDa. It preferentially degrades proteins that are tagged by ubiquitination but also has the capacity to degrade some damaged proteins that lack a ubiquitin tag (108). Certain functional features of the 26S proteasome resemble those of Lon and FtsH. For instance, DnaJ/HSP40 and DnaK/HSP70 molecular chaperones cooperate with the 26S proteasome in protein degradation (107). In addition, recent work indicates that many important cell cycle regulators are targeted selectively for ubiquitination and subsequent degradation by the 26S proteasome (109).

The activation of protein degradation by the 26S proteasome has been studied in detail in plants, yeast, and mammalian cells, and many additional sensors that recognize protein damage/unfolding in the endoplasmic reticulum, cytosol, or plasma membrane have been identified. Such sensors, including BiP (110), ATF6 (111), IRE1 (112), SCF complexes targeting F-box proteins (113), and the COP signalosome (114), are exquisitely important for proteasome-dependent protein degradation. However, they are not as highly evolutionarily conserved as Lon and FtsH and are not considered here. With the exception of mechanisms related to 26S proteasome function, in particular the unfolded protein response (UPR) in the endoplasmic reticulum, protein damage sensors as potential upstream regulators of CSR signaling networks have received less attention than have DNA damage sensing mechanisms. Protein damage sensors need further study and may well be key to understanding the diseases of misfolded protein accumulation, including Alzheimer's, Parkinson's, and Huntington's diseases, prion encephalopathies, cystic fibrosis, myeloma, and some cancers.

Redox Sensors and Redox Regulation

The CSR is intricately associated with free radical formation and changes in cellular redox state. Virtually every gene implicated in response to stress is also affected by changes in cellular redox state or free radical levels (115). Thus, alteration of cellular redox potential is a major trigger of the CSR. Curiously, life originated in an unstable and stressful archaic environment characterized by high ion and free radical density, high and fluctuating temperatures, and large pH gradients. While these extreme environmental conditions may have promoted the free radical reactions that could have led to the origin of life (116, 117), these highly reactive conditions are incompatible with cellular functions relying on homeostasis. This conundrum may have represented a major evolutionary driving force for selection of genes encoding redox-regulatory, free radical scavenging proteins in the last universal common ancestor (LUCA). Such genes have probably aided the transition from anaerobic to aerobic life by providing a means for minimizing oxygen toxicity in developing aerobic mechanisms (118).

All cells have free radical scavenging systems to minimize and repair oxidative damage, including compounds such as ascorbate, glutathione, thioredoxin, and various antioxidant enzymes. In addition, reactive oxygen and nitrogen species (ROS, RNS) are employed as second messengers that carry signals about alterations of cellular redox potential to activate the CSR and other physiological processes such as differentiation, aging, senescence, and pathogen defense (23, 119). In general, two types of free radical-mediated effects can be distinguished: (*a*) direct effects on signaling proteins and (*b*) indirect alteration of signaling pathways by specialized redox-sensitive proteins. Examples of direct alteration of signaling proteins include eukaryotic MAPKs (mitogen-activated protein kinases) and the transcription factors AP-1 and NF-κB (115, 120, 121). Indirect alteration of stress-responsive signaling pathways involves several redox-regulatory proteins that are universally conserved in all cellular life forms (Table 1).

Many oxidoreductases present in the minimal stress proteome are dehydrogenases. Some are elements of basic metabolic pathways, including glycolysis, pentose phosphate cycle, and the Krebs (citrate) cycle, and thus are essential even in the absence of stress. However, these dehydrogenases also influence cellular redox potential and oxidative damage repair by generating reducing equivalents for antioxidant enzymes that depend on NADPH as a cofactor, including thioredoxin reductase, glutathione reductase, and aldehyde reductase. Aldehyde dehydrogenase and aldehyde reductase are important for detoxification of aldehydes, which are common toxic intermediary metabolites during oxidative stress. ROS that are generated during stress are neutralized by the action of antioxidant proteins, many of which are part of the minimal stress proteome.

Superoxide dismutase (SOD) converts superoxide radicals to H_2O_2, which is then further converted to water by peroxidases, including peroxiredoxin. Peroxiredoxin belongs to a family of antioxidative proteins that currently comprises six members in mammals (122). These enzymes are distributed in the cytosol, mitochondria, peroxisomes, and plasma membrane and have peroxidase activity that utilizes thioredoxin and/or glutathione as the electron donor. Peroxiredoxins also modulate cell proliferation, differentiation, and gene expression, probably in similar ways as thioredoxin. SOD and peroxidases such as peroxiredoxin must be coregulated during stress because an imbalance in the ratio of SOD and peroxidases in the presence of heavy metal ions causes conversion of H_2O_2 into noxious hydroxyl radicals via Fenton chemistry. Accumulation of hydroxyl radicals is highly deleterious to cells because they are very effective in causing damage to macromolecules such as DNA, protein, and lipids (123).

Oxidative damage to proteins occurs in multiple forms, most commonly cysteine oxidation (leading to formation of disulfide bonds) or methionine oxidation. The glutathione and thioredoxin systems repair such forms of oxidative protein damage. Glutathione (gamma-glutamyl-cysteinyl-glycine, GSH) is the most abundant low-molecular-weight thiol that is synthesized de novo in animal cells. In its reduced/oxidized forms (GSH/GSSG), it represents the major redox couple in animal cells. The main pathways for GSH metabolism are reduction of hydroperoxides by glutathione peroxidases and peroxiredoxins leading to generation of glutathione disulfide (GSSG). Glutathione-S-transferase catalyzes the conjugation of glutathione. Glutathione reductase catalyzes the NADPH-dependent reduction of GSSG to GSH. Because GSH is a universal free radical scavenger in cells, glutathione reductase is critical during stress, when levels of ROS increase. In contrast to GSSG, which is recycled to GSH by glutathione reductase, glutathione conjugates are excreted from cells (124).

Thioredoxin is a 12-kDa protein in which redox-active dithiol in the active site Cys-Gly-Pro-Cys constitutes a major thiol reducing system (125). The enzymes involved in repairing oxidative cysteine damage via the thioredoxin system are thioredoxin peroxidase and thioredoxin reductase (126). The function of thioredoxin reductase for repairing oxidative cysteine and methionine damage is to recycle oxidized thioredoxin back to its reduced state by using electrons from NADPH. In addition to its overall antioxidant properties, thioredoxin restores transcriptional activity of AP-1, NF-κB, p53, and PEBP2 (23, 128). It also interacts directly with other key signaling molecules such as ASK1 (apoptosis signal regulating kinase 1) (126). Thus, thioredoxin plays multiple roles in cellular processes like proliferation and apoptosis.

Peptide methionine sulfoxide reductase (MsrA) and thioredoxin reductase repair oxidative methionine damage (127, 128). Because of the benefits MsrA confers on oxidatively damaged proteins, it is an important repair enzyme during stress (129). In rats, MsrA was found in all tissues examined but was particularly abundant in tissues routinely exposed to severe stress, such as the renal medulla and retinal epithelium (130). The recent identification of the small heat shock protein HSP21 as a physiological MsrA substrate suggests that heat shock protein activity is protected by MsrA during stress (131).

The seleno-cysteine-specific translation elongation factor SelB is also present in all three lines of organisms (Table 1). SelB, which is homologous to EF-Tu but has a unique C terminus (132), belongs to an ancient subfamily of GTPases (133). SelB is required for the synthesis of seleno-cysteine proteins such as glutathione peroxidases, thioredoxin reductases, and SelR, which has peptide methionine sulfoxide reductase (PMSR) catalytic activity similar to MsrA (134). Most of the other selenoproteins are also key enzymes functioning in antioxidant defense. These enzymes have seleno-cysteine in their active site, which increases their functionality because of the presence of more fully ionized seleno-cysteine compared with the thiol group of cysteine at physiological pH (135). The UGA stop codon encodes seleno-cysteine in archaea, eubacteria, and eukaryotes. Pyrrolysine is the other amino acid encoded by a stop codon (UAG) (136). SelB alters the translational machinery by recognizing a specific motif in mRNAs coding for seleno-cysteine proteins. In addition to SelB, seleno-cysteine tRNA (tRNA-SeC) is universally required for seleno-cysteine protein synthesis. Although SelB recognizes mRNAs encoding seleno-cysteine proteins by similar mechanisms and requires tRNA-SeC in all cases, the cofactors utilized by eubacteria differ from those in archaea and eukaryotes (137). In both cases, however, the incorporation of seleno-cysteine into protein requires several gene products in addition to SelB and tRNA-SeC and is based on the interaction of a C-terminal domain of SelB with a SECIS (seleno-cysteine insertion sequence) element present in mRNAs encoding seleno-cysteine proteins. This example illustrates yet again that large protein complexes functioning in the CSR are organized around universally conserved stress proteins. [For details of eubacterial and archaeal/eukaryotic mechanisms of seleno-cysteine protein synthesis, see (137, 138).]

The CSR utilizes ROS and RNS generated during stress as intracellular messengers. Thus, increased concentrations of free radicals are beneficial for cellular stress sensing and signaling while damaging to cellular macromolecules. The basis and relative importance of these contrasting roles of free radicals during stress merit further investigation.

KEY FUNCTIONS OF THE CELLULAR
STRESS RESPONSE

Stress triggers diverse cellular mechanisms of macromolecular damage that are consequential. The focus here has been on outlining mechanisms that are organized around proteins belonging to the minimal stress proteome and represent early events associated with sensing and transducing common signals generated by stress. These conserved sensory mechanisms activate a very elaborate cellular stress signaling network that involves different proteins in prokaryotes and eukaryotes. Prokaryotic stress response signaling mechanisms involve σ^{32}, σ^{54}, and σ^S transcription factors, but they also rely to a great extent on two-component signal transduction. Two-component systems often consist of a sensor kinase that autophosphorylates on histidine when certain variables in the environment change. Upon autophosphorylation, the sensor is activated as an aspartate kinase that phosphorylates a second protein, the response regulator, on aspartate (139). In some cases, a third protein serves as a mediator of phosphate transfer. The response regulator often functions as a DNA binding protein that modulates the expression of genes with adaptive value during stress.

In eukaryotes, CSR signaling networks are extraordinarily complex and involve numerous proteins. In particular, transcriptional regulation in eukaryotes is much more complex than in prokaryotes. It depends on stability, sequence-specific binding, and nuclear transport of numerous transcription factors and their regulators (140–142). In addition, transcription programs are controlled by chromatin rearrangements and many posttranslational histone modifications (143, 144). Thus for preventative or therapeutic purposes, identification of key elements of such networks that represent efficient targets for manipulating the stress tolerance of cells is critical. One method to do so is by using a comparative approach or, in other words, identifying proteins involved in cellular stress response signaling in many taxa. Examples of such proteins are MAPKs (145), 14-3-3 (146), Bcl-2 (147), ATM and ATR kinases (148), and insulin receptor-like tyrosine kinases (149). These are key regulators of the CSR in eukaryotes and can be regarded as hubs around which other signaling mechanisms are organized.

Recent cDNA microarray experiments have shown that the genome of *S. cerevisiae* is divided into genes preferentially targeted by the SAGA (Spt-Ada-GCN5-acetyltransferase) transcriptional complex ($\sim$10% of the genome) and genes preferentially targeted by the TFIID transcriptional complex ($\sim$90% of the genome) (150). Many SAGA-regulated genes are stress inducible, whereas most TFIID-regulated genes have housekeeping functions. This bimodal transcriptional regulation is indicative of a distinct stress-induced transcription program that mediates global and coordinated activation of yeast stress response genes (150). Similarly, the σ^S (RpoS) subunit of RNA polymerase, whose levels are controlled by proteolysis, is a master regulator of the general stress response in *E. coli* and other bacteria (151). Whether such global stress-induced transcriptional regulation is also utilized by multicellular organisms is presently unclear. Nonetheless, other global changes, such as chromatin organization, posttranslational histone modifications, and rearrangements of chromatin remodeling complexes occur in mammalian cells exposed to stress (152). [For a more detailed exploration of intracellular signaling mechanisms in response to stress, see (153–155).]

Growth Control and Cell Cycle Checkpoints

One universal effect of stress on cells is the impairment of growth and proliferation. Growth arrest represents an adaptive and integrated part of the CSR. It allows for preservation of energy and reducing equivalents and redirects the utilization of these important metabolites toward macromolecular stabilization and repair. In addition, proliferating cells that actively undergo DNA replication and mitosis are more prone to suffer stress-induced damage to macromolecules than are cells in a resting state. In bacteria, the ability to resist stress is greater in stationary phase than in exponential phase, during which cells are rapidly dividing. Thus, rapidly dividing, metabolically active bacteria will experience growth arrest when exposed to stress (156). Similarly, eukaryotic cells also undergo growth arrest when exposed to stress (157). For these reasons, the activation of cell cycle checkpoints is a key aspect of the CSR. Cell cycle checkpoints monitor macromolecular integrity and the successful completion of cellular processes prior to initiating the next phase in the cell cycle (158).

Under extreme stress conditions many bacteria and fungi form stress-resistant spores. Sporulation can be regarded as the ultimate form of growth control and cell cycle regulation during stress. Growth arrest and the onset of sporulation in bacteria involve many proteins. When bacterial DNA replication is interrupted by stress, a component of the SOS response, the inhibitory factor SfiA, is induced and leads to transient inhibition of cell division (159). In addition, the universal stress protein UspA, which belongs to a family of proteins that is conserved in bacteria and many invertebrate eukaryotes, accumulates at very high levels in growth-arrested bacteria (160). The σ^S (RpoS)-driven transcription of stress response genes (see above) promotes growth arrest and counteracts proliferative activities that are primarily directed by σ^{70}. Counteraction of σ^{70} is mediated by an increase in the *E. coli* alarmone guanosine tetraphosphate (ppGpp), which shifts the equilibrium between σ^S and σ^{70} in favor of σ^S. The resulting change in relative competitiveness of these two subunits of the RNA polymerase complex leads to suppression of growth during stress (161).

The extraordinary significance of eukaryotic cell cycle checkpoints for proliferative disorders such as cancer has attracted much attention. Cell cycle regulatory proteins that control such checkpoints maintain the fidelity of DNA replication, repair, and cell division in normal as well as stressed cells. Checkpoints are built into every major transition in the cell cycle, including G1/S, intra-S phase, G2/M, mitotic spindle assembly, and cytokinesis. In mammalian cells such cell cycle checkpoints are controlled by a large number of proteins, of which ATM and ATR kinases, p53, GADD45 proteins, 14–3-3σ, CDC25, CDC2/cyclin B, p21, retinoblastoma protein (pRB), Chk1, Chk2, Polo kinases, and BRCA1 are key (157, 162–165).

One well-known mechanism of G2/M checkpoint induction is based on ATM and ATR kinase activation of Chk1 kinase, which phosphorylates the cell cycle phosphatase CDC25. Phosphorylation leads to binding of 14-3-3 protein on CDC25 and its subsequent sequestration in the cytosol. Cytosolic sequestration prevents CDC25-mediated dephosphorylation of CDC2, which is necessary for the promotion of mitosis by the CDC2/cyclin B complex (166). The p53 protein is involved in the G2/M checkpoint by inhibition of CDC2 via its transcriptional targets GADD45, p21, and 14–3-3σ (167). GADD45 levels increase during stress, not only because of transcriptional regulation but also as a result of posttranscriptional mRNA stabilization (48, 157, 168). However, the mechanism by which GADD45 proteins induce G2 arrest is still elusive. A striking feature of eukaryotic cell cycle checkpoints as well as DNA damage repair pathways is the central role of ATM and ATR kinases (169). These kinases are critical intermediates between DNA damage sensors and effector protein complexes that control key features of the CSR, including cell cycle progression, DNA repair, and apoptosis.

Maintenance of Macromolecular Integrity

A hallmark of the CSR representing one of its first identified features is the induction of heat shock proteins, many of which function as molecular chaperones (9, 170, 171). In combination with the DNA repair machinery, molecular chaperones provide a rapid and direct mechanism of cellular defense against stress-induced damage. Chaperone proteins are required to recognize unfolded proteins and either target them for removal, deter their aggregation, or assist in their refolding into the native, functional state (172, 173). Five molecular chaperones, DnaK/HSP70, DnaJ/HSP40, GrpE, HSP60, and petidyl-prolyl isomerase (cylophilin), are part of the minimal stress proteome. These proteins illustrate the extraordinarily strong evolutionary conservation of this cellular function and the importance of molecular chaperones during stress. They are extensively utilized as bioindicators of environmental stress in many different types of organisms, and their study has extended well beyond laboratory-based analysis into the realm of field-based ecological physiology (174).

Many functional aspects of these five molecular chaperones are known in great detail. In archaea and eubacteria, the molecular chaperones DnaK/HSP70, DnaJ/HSP40, and GrpE are all transcribed from the same locus (175). Some species of archaea have apparently lost this locus, which is surprising and contrasts with the ubiquitous occurrence of these genes in eubacteria and eukarya with no known exception (175). The chaperone activity and induction of DnaK/HSP70, DnaJ/HSP40, and GrpE during stress are well established. Although these molecular chaperones are universally induced during stress, the mechanisms of induction seem to be more diverse. For example, as outlined above in *Streptomyces coelicolor* DnaK/HSP70 operon induction is mediated at the transcriptional level by the HspR repressor, which is degraded by proteolysis during stress (100). In *E. coli*, however, transcriptional induction of the DnaK/HSP70 operon is positively controlled by the RpoH/σ^{32} subunit of RNA polymerase (92, 176).

GrpE and DnaJ/HSP40 function as co-chaperone and nucleotide exchange factors for DnaK/HSP70. They control access of unfolded proteins to the substrate-binding domain of DnaK/HSP70 (177). GrpE is expressed in prokaryotes and eukaryotic mitochondria and plant chloroplasts but it is not present in eukaryotic cytosol, where a GrpE-like function is provided by the BAG1 protein (177). In higher eukaryotes the HSP70 family consists of numerous isoforms, some of which are stress inducible whereas others are constitutively expressed (178). Induction of mammalian HSPs is mediated by heat shock elements in the promoter of their genes. These elements are binding sites for heat shock factors such as mammalian HSF1 that activate transcription of HSPs (179).

Cyclophilins, another universally conserved group of stress proteins, are well known as receptors of the immunosuppressive drug cyclosporin A. They are induced by many types of stress and have molecular chaperone activity (180). The molecular chaperone function of cyclophilins is mediated by their enzymatic peptidyl-prolyl isomerase (PPIase) activity, which has also been suggested to play a regulatory role for transcription and cellular differentiation (181, 182). Moreover, eukaryotic cyclophilin D is a mitochondrial matrix protein and an integral part of the mitochondrial permeability transition pore complex, which is intricately involved in the control of apoptosis (183).

Molecular chaperones have diverse impacts by interacting with proteins involved in other aspects of the CSR. For example, molecular chaperones protect processes for maintaining genomic integrity such as NER, which repairs oxidative nucleotide damage during stress. In a specific case, the UvrA protein in *E. coli* is stabilized and protected from heat inactivation by the DnaK/HSP70, DnaJ/HSP40, GrpE chaperone machinery (184). A related role in NER has recently been attributed to the eukaryotic 26S proteasome. This proteolytic supercomplex interacts with multiple NER proteins, including XPB, Rad4, and Rad23. The latter two proteins form a complex that binds to pyrimidine dimers generated during UV-radiation stress. The 26S proteasome may act as a molecular chaperone to promote disassembly of this NER complex (185). These examples suggest that molecular chaperones participate in NER by targeted protection of DNA repair proteins or disassembly of NER complexes after DNA repair is completed.

These examples also demonstrate that processes of protein and DNA maintenance/repair are co–regulated and share common elements during stress. [More detailed information concerning DNA repair modes is summarized in recent reviews that focus on mechanisms of NER (186–188), MMR (189, 190), and DNA dsb repair (191, 192).] The mechanisms by which molecular chaperones and the DNA repair machinery maintain protein and genomic integrity during stress are at the center of the CSR and remain a captivating subject of investigation.

Energy Metabolism

Another key aspect of the CSR is the modulation of major pathways of energy metabolism, which may be closely linked to the oxidative burst in cells exposed to stress. Minimal stress proteome enzymes such as glycerol-3-phosphate dehydrogenase (G3PDH), 6 phosphogluconate dehydrogenase (6PGDH), enolase, citrate synthase, and isocitrate dehydrogenase (IDH) contribute strongly to the control of key pathways of energy metabolism, including glycolysis, pentose phosphate pathway, and the Krebs (citrate) cycle. Induction of these enzymes during stress may be necessary for generating reducing equivalents (NADH, NADPH) that are needed for cellular antioxidant systems. For example, IDH is strongly elevated in macrophages exposed to pathogen-induced stress. Elevated IDH levels, in turn, lead to increased NADPH production and cellular protection from oxidative damage caused by RNS and ROS (193). Another enzyme in this category, succinate semialdehyde dehydrogenase (SSADH), is also necessary for alleviating oxidative stress. This was demonstrated by constitutively elevated levels of reactive oxygen species in SSADH ($-/-$) mice (194).

Another potential reason for inducing such metabolic pathways lies in the energetic requirements of protein degradation, protein chaperoning, and DNA repair. Many steps in these adaptive processes depend on the hydrolysis of ATP, including the activity of proteolytic complexes (e.g., FtsH, Lon, 26S proteasome), chaperones (e.g., DnaK/HSP70, DnaJ/HSP40, GrpE), and DNA damage sensing/repair complexes (e.g., the BASC supercomplex). Thus, the induction of key enzymes of energy metabolism may provide the reducing and energy equivalents needed for stress-related cell functions. Furthermore, growth arrest results in redirection of NADPH/NADH and ATP utilization from proliferative processes to macromolecular stabilization and repair. Both processes, induction of energy metabolism and growth arrest, are closely coordinated with increased demands for reducing and energy equivalents during stress.

Some energy metabolism enzymes may have additional stress-related functions. For example, 6PGDH has been implicated in cell cycle control during osmotic stress in plants (195) and in increasing glutathione levels through stimulation of the pentose phosphate pathway during oxidative stress in mammalian cells (196).

Apoptosis

A universal response of severely stressed cells is to undergo cell death, but in two alternative ways: necrosis and apoptosis (programmed cell death, cell suicide program). Apoptosis is a common response of metazoan cells when stress exceeds cellular tolerance limits. It is also an important regulatory mechanism during development of multicellular organisms (197). Although mechanisms of programmed cell death that are similar to apoptosis have recently been identified in plants and bacteria (198, 199), apoptosis is best understood in metazoans.

Mammals have two major apoptotic pathways, intrinsic and extrinsic. The intrinsic apoptotic pathway depends on release of cytochrome c (Cyt c) and other apoptogenic factors from mitochondria. Once released, Cyt c binds to APAF-1 (apoptosis protease activating factor 1) and recruits procaspase 9, which is then processed by the apoptosome into active caspase 9. Caspase 9 triggers the caspase pathway by activation of caspase 3, with the final outcome being activation of caspase-activated DNAse (CAD) that digests chromosomal DNA (200). The extrinsic apoptotic pathway is triggered by cell surface receptors (e.g., TNFSF6) when activated by specific ligands. Upon activation, these receptors bind cytoplasmic adapter molecules such as FADD that, in turn, activate procaspases and start the caspase cascade (201).

Both apoptotic pathways are subject to extensive regulation by a complex array of pro- and antiapoptotic signals. The PI-3-K (phosphatidyl inositol 3 kinase)/AKT (V-Akt murine thymoma viral oncogene homolog) pathway is one of the critical pathways that generally suppresses apoptosis (202). Activated AKT kinase phosphorylates several important targets, including the proapoptotic BCL-2 (B cell CLL/lymphoma 2) family member BAD (BCL-2 antagonist of death), forkhead transcription factors, and GADD45. These proteins are bound by 14-3-3 in their phosphorylated form, which results in cytosolic sequestration and inactivation similar to CDC25 inactivation discussed above (202). Molecular chaperones, including HSP60, HSP27, and HSP70, also have antiapoptotic effects. This property of major molecular chaperones is mainly the result of their binding to apoptosis regulating proteins such as AKT (200). The NF-κB pathway also has generally antiapoptotic effects. NF-κB is a transcription factor that induces antiapoptotic members of the BCL-2 family, including BCL-2 and BCL-xL (203). However, in certain cell types, NF-κB also induces proapoptotic genes, including BCL-xS and p53 (203). The p53 pathway promotes apoptosis mostly via p53 transactivation of proapoptotic members of the BCL-2 family, PTEN phosphatase (an inhibitor of the AKT pathway), and GADD45 (201). Like p53 itself, some of its targets such as GADD45 proteins modulate apoptosis (204) as well as cell cycle checkpoints (205). These and other proteins with similar roles in multiple stress response pathways likely are key to how cells encountering stress decide between induction of apoptosis versus cell cycle delay and repair. A mechanism of quantitative macromolecular damage assessment may contribute significantly to this important decision. However, how cells measure the amount of damage during stress and how they recognize whether such damage exceeds their tolerance limits is unclear. These are important unanswered questions that will drive future research on the cellular stress response.

CONCLUSIONS AND PERSPECTIVE

The CSR is a very complex mechanism that ensures survival of healthy (fit) cells and removal of damaged (unfit) cells during stressful environmental conditions. I have selectively summarized major physiological functions of the CSR based on the recent identification of a set of stress proteins that are highly conserved in all organisms. The minimal stress proteome provides an excellent starting point for obtaining insight into key functions of the CSR. This review also analyzes the common nature of stimuli that induce the CSR and discusses common mechanisms by which such stimuli are sensed and integrated into stress response networks by cells. These commonalities create a conceptual framework for further exploration and identification of key elements of the CSR.

ACKNOWLEDGMENTS

This work was supported by grants from the NIH/NIDDK (DK59470) and the NSF (MCB 0244569). I thank George Somero and Martin Feder for their insightful comments that helped improve the manuscript.

The *Annual Review of Physiology* is online at http://physiol.annualreviews.org

LITERATURE CITED

1. Cannon WB. 1929. *Bodily Changes in Pain, Hunger, Fear and Rage.* New York: Appleton-Century-Crofts
2. Selye H. 1936. A syndrome produced by diverse nocuous agents. *Nature* 138:32
3. Charmandari E, Tsigos C, Chrousos G. 2005. Endocrinology of the stress response. *Annu. Rev. Physiol.* 67:259–84
4. Kültz D. 2003. Evolution of the cellular stress proteome: from monophyletic origin to ubiquitous function. *J. Exp. Biol.* 206:3119–24
5. Koonin EV. 2000. How many genes can make a cell: the minimal-gene-set concept. *Annu. Rev. Genomics Hum. Genet.* 1:99–116
6. Gasch AP, Spellman PT, Kao CM, Carmel-Harel O, Eisen MB, et al. 2000. Genomic expression programs in the response of yeast cells to environmental changes. *Mol. Biol. Cell* 11:4241–57
7. Beck FX, Grunbein R, Lugmayr K, Neuhofer W. 2000. Heat shock proteins and the cellular response to osmotic stress. *Cell Physiol. Biochem.* 10:303–6
8. DiRuggiero J, Brown JR, Bogert AP, Robb FT. 1999. DNA repair systems in archaea: mementos from the last universal common ancestor? *J. Mol. Evol.* 49:474–84
9. Tissières A, Mitchell HK, Tracy UM. 1974. Protein synthesis in salivary glands of *Drosophila melanogaster*: relation to chromosome puffs. *J. Mol. Biol.* 84:389–98
10. Pastori GM, Foyer CH. 2002. Common components, networks, and pathways of cross-tolerance to stress: the central role of "redox" and abscisic acid-mediated controls. *Plant Physiol.* 129:460–68
11. Reth M. 2002. Hydrogen peroxide as second messenger in lymphocyte activation. *Nat. Immunol.* 3:1129–34
12. Bolwell GP. 1996. The origin of the oxidative burst in plants. *Biochem. Soc. Trans.* 24:438–42
13. Reiser V, Raitt DC, Saito H. 2003. Yeast osmosensor Sln1 and plant cytokinin receptor Cre1 respond to changes in turgor pressure. *J. Cell Biol.* 161:1035–40
14. Miyakawa H, Woo SK, Dahl SC, Handler JS, Kwon HM. 1999. Tonicity-responsive enhancer binding protein, a rel-like protein that stimulates transcription in response to hypertonicity. *Proc. Natl. Acad. Sci. USA* 96:2538–42
15. Lepore DA, Knight KR, Anderson RL, Morrison WA. 2001. Role of priming stresses and Hsp70 in protection from ischemia-reperfusion injury in cardiac and skeletal muscle. *Cell Stress Chaperones* 6:93–96
16. Sejerkilde M, Sorensen JG, Loeschcke V. 2003. Effects of cold- and heat-hardening on thermal resistance in *Drosophila melanogaster*. *J. Insect Physiol.* 49:719–26
17. Alsbury S, Papageorgiou K, Latchman DS. 2004. Heat shock proteins can protect aged human and rodent cells from different stressful stimuli. *Mech. Ageing Dev.* 125:201–9
18. Koga T, Sakamoto F, Yamoto A, Takumi K. 1999. Acid adaptation induces cross-protection against some environmental stresses in *Vibrio parahaemolyticus*. *J. Gen. Appl. Microbiol.* 45:155–61
19. Mary P, Sautour M, Chihib NE, Tierny Y, Hornez JP. 2003. Tolerance and starvation induced cross-protection against different stresses in *Aeromonas hydrophila*. *Int. J. Food Microbiol.* 87:121–30
20. Alexieva V, Sergiev I, Mapelli S, Karanov E. 2001. The effect of drought and ultraviolet radiation on growth and stress markers in pea and wheat. *Plant Cell Environ.* 24:1337–44

21. Enjalbert B, Nantel A, Whiteway M. 2003. Stress-induced gene expression in *Candida albicans*: absence of a general stress response. *Mol. Biol. Cell* 14:1460–67

22. Santos BC, Pullman JM, Chevaile A, Welch WJ, Gullans SR. 2003. Chronic hyperosmolarity mediates constitutive expression of molecular chaperones and resistance to injury. *Am J. Physiol.* 284: F564–74

23. Mikkelsen RB, Wardman P. 2003. Biological chemistry of reactive oxygen and nitrogen and radiation-induced signal transduction mechanisms. *Oncogene* 22:5734–54

24. Fleury C, Mignotte B, Vayssiere JL. 2002. Mitochondrial reactive oxygen species in cell death signaling. *Biochimie* 84:131–41

25. Lee I, Bender E, Arnold S, Kadenbach B. 2001. New control of mitochondrial membrane potential and ROS formation: a hypothesis. *Biol. Chem.* 382:1629–36

26. Parasassi T, Sapora O, Giusti AM, Destasio G, Ravagnan G. 1991. Alterations in erythrocyte membrane lipids induced by low doses of ionizing radiation as revealed by 1,6-diphenyl-1,3,5-hexatriene fluorescence lifetime. *Int. J. Radiat. Biol.* 59:59–69

27. Bandurska H. 1998. Implication of ABA and proline on cell membrane injury of water deficit stressed barley seedlings. *Acta Physiol. Plant.* 20:375–81

28. Zeng F, An Y, Zhang HT, Zhang MF. 1999. The effects of La(III) on the peroxidation of membrane lipids in wheat seedling leaves under osmotic stress. *Biol. Trace Element Res.* 69:141–50

29. Steels EL, Learmonth RP, Watson K. 1994. Stress tolerance and membrane lipid unsaturation in *Saccharomyces cerevisiae* grown aerobically or anaerobically. *Microbiology* 140:569–76

30. Swan TM, Watson K. 1999. Stress tolerance in a yeast lipid mutant: mem-brane lipids influence tolerance to heat and ethanol independently of heat shock proteins and trehalose. *Can. J. Microbiol.* 45:472–79

31. Chatterjee MT, Khalawan SA, Curran BP. 2000. Cellular lipid composition influences stress activation of the yeast general stress response element (STRE). *Microbiology* 146:877–84

32. Knights KM, Drogemuller CJ. 2000. Xenobiotic-CoA ligases: kinetic and molecular characterization. *Curr. Drug Metab.* 1:49–66

33. Nigam S, Schewe T. 2000. Phospholipase A$_2$s and lipid peroxidation. *Biochim. Biophys. Acta* 1488:167–81

34. Sevanian A, Ursini F. 2000. Lipid peroxidation in membranes and low-density lipoproteins: similarities and differences. *Free Radic. Biol. Med.* 29:306–11

35. Spiteller G. 2003. Are lipid peroxidation processes induced by changes in the cell wall structure and how are these processes connected with diseases? *Med. Hypoth.* 60:69–83

36. Leonarduzzi G, Arkan MC, Basaga H, Chiarpotto E, Sevanian A, Poli G. 2000. Lipid oxidation products in cell signaling. *Free Radic. Biol. Med.* 28:1370–78

37. Rosette C, Karin M. 1996. Ultraviolet light and osmotic stress: activation of the JNK cascade through multiple growth factor and cytokine receptors. *Science* 274:1194–97

38. Rhee SG, Bae YS, Lee SR, Kwon J. 2000. Hydrogen peroxide: a key messenger that modulates protein phosphorylation through cysteine oxidation. *Sci. STKE* 2000:E1–11

39. Rhee SG, Chang TS, Bae YS, Lee SR, Kang SW. 2003. Cellular regulation by hydrogen peroxide. *J. Am. Soc. Nephrol.* 14:S211–15

40. Spiteller G. 2001. Lipid peroxidation in aging and age-dependent diseases. *Exp. Gerontol.* 36:1425–57

41. Marathe GK, Harrison KA, Murphy RC, Prescott SM, Zimmerman GA,

McIntyre TM. 2000. Bioactive phospholipid oxidation products. *Free Radic. Biol. Med.* 28:1762–70

42. Lehtonen JY, Kinnunen PK. 1995. Phospholipase A₂ as a mechanosensor. *Biophys. J.* 68:1888–94

43. Kudo I, Murakami M. 2002. Phospholipase A₂ enzymes. *Prostaglandins Other Lipid. Mediat.* 68–69:3–58

44. Hoppe T, Matuschewski K, Rape M, Schlenker S, Ulrich HD, Jentsch S. 2000. Activation of a membrane-bound transcription factor by regulated ubiquitin/proteasome-dependent processing. *Cell* 102:577–86

45. Zou H, Lifshitz LM, Tuft RA, Fogarty KE, Singer JJ. 2002. Visualization of Ca²⁺ entry through single stretch-activated cation channels. *Proc. Natl. Acad. Sci. USA* 99:6404–9

46. Kass GE, Orrenius S. 1999. Calcium signaling and cytotoxicity. *Environ. Health Perspect.* 107(Suppl. 1):25–35

47. Kültz D, Chakravarty D. 2001. Hyperosmolality in the form of elevated NaCl but not urea causes DNA damage in murine kidney cells. *Proc. Natl. Acad. Sci. USA* 98:1999–2004

48. Kültz D, Madhany S, Burg MB. 1998. Hyperosmolality causes growth arrest of murine kidney cells. Induction of GADD45 and GADD153 by osmosensing via stress-activated protein kinase 2. *J. Biol. Chem.* 273:13645–51

49. Seno JD, Dynlacht JR. 2004. Intracellular redistribution and modification of proteins of the Mre11/Rad50/Nbs1 DNA repair complex following irradiation and heat-shock. *J. Cell. Physiol.* 199:157–70

50. Zhou BB, Elledge SJ. 2000. The DNA damage response: putting checkpoints in perspective. *Nature* 408:433–39

51. Natrajan G, Lamers MH, Enzlin JH, Winterwerp HH, Perrakis A, Sixma TK. 2003. Structures of *Escherichia coli* DNA mismatch repair enzyme MutS in complex with different mismatches: a common recognition mode for diverse substrates. *Nucleic Acids Res.* 31:4814–21

52. Eisen JA, Hanawalt PC. 1999. A phylogenomic study of DNA repair genes, proteins, and processes. *Mutat. Res.* 435:171–213

53. Sixma TK. 2001. DNA mismatch repair: MutS structures bound to mismatches. *Curr. Opin. Struct. Biol.* 11:47–52

54. Wang Y, Cortez D, Yazdi P, Neff N, Elledge SJ, Qin J. 2000. BASC, a super complex of BRCA1-associated proteins involved in the recognition and repair of aberrant DNA structures. *Genes Dev.* 14:927–39

55. Khil PP, Camerini-Otero RD. 2002. Over 1000 genes are involved in the DNA damage response of *Escherichia coli*. *Mol. Microbiol.* 44:89–105

56. Beaber JW, Hochhut B, Waldor MK. 2004. SOS response promotes horizontal dissemination of antibiotic resistance genes. *Nature* 427:72–74

57. Kitagawa J, Yamamoto K, Iba H. 2001. Computational analysis of SOS response in ultraviolet-irradiated *Escherichia coli*. *Genome Inform.* 12:280–81

58. Reich CI, McNeil LK, Brace JL, Brucker JK, Olsen GJ. 2001. Archaeal RecA homologues: different response to DNA-damaging agents in mesophilic and thermophilic archaea. *Extremophiles* 5:265–75

59. Aguilera A. 2001. Double-strand break repair: are Rad51/RecA-DNA joints barriers to DNA replication? *Trends Genet.* 17:318–21

60. Gasior SL, Olivares H, Ear U, Hari DM, Weichselbaum R, Bishop DK. 2001. Assembly of RecA-like recombinases: distinct roles for mediator proteins in mitosis and meiosis. *Proc. Natl. Acad. Sci. USA* 98:8411–18

61. Tsuzuki T, Fujii Y, Sakumi K, Tominaga Y, Nakao K, et al. 1996. Targeted disruption of the Rad51 gene leads to lethality in embryonic mice. *Proc. Natl. Acad. Sci. USA* 93:6236–40

62. Chen JJ, Silver D, Cantor S, Livingston DM, Scully R. 1999. BRCA1, BRCA2, and Rad51 operate in a common DNA damage response pathway. *Cancer Res.* 59:1752s–56s

63. Scully R, Chen J, Plug A, Xiao Y, Weaver D, et al. 1997. Association of BRCA1 with Rad51 in mitotic and meiotic cells. *Cell* 88:265–75

64. Goldberg M, Stucki M, Falck J, D'Amours D, Rahman D, et al. 2003. MDC1 is required for the intra-S-phase DNA damage checkpoint. *Nature* 421:952–56

65. Bradbury JM, Jackson SP. 2003. The complex matter of DNA double-strand break detection. *Biochem. Soc. Trans.* 31:40–44

66. Trujillo KM, Roh DH, Chen L, Van Komen S, Tomkinson A, Sung P. 2003. Yeast Xrs2 binds DNA and helps target Rad50 and Mre11 to DNA ends. *J. Biol. Chem.* 278:48957–64

67. Horejsi Z, Falck J, Bakkenist CJ, Kastan MB, Lukas J, Bartek J. 2004. Distinct functional domains of Nbs1 modulate the timing and magnitude of ATM activation after low doses of ionizing radiation. *Oncogene* 23:3122–27

68. Lee JH, Paull TT. 2004. Direct activation of the ATM protein kinase by the Mre11/Rad50/Nbs1 complex. *Science* 304:93–96

69. Bakkenist CJ, Kastan MB. 2003. DNA damage activates ATM through intermolecular autophosphorylation and dimer dissociation. *Nature* 421:499–506

70. Petrini JHJ, Stracker TH. 2003. The cellular response to DNA double-strand breaks: defining the sensors and mediators. *Trends Cell Biol.* 13:458–62

71. Stiff T, O'Driscoll M, Rief N, Iwabuchi K, Lobrich M, Jeggo PA. 2004. ATM and DNA-PK function redundantly to phosphorylate H2AX after exposure to ionizing radiation. *Cancer Res.* 64:2390–96

72. Stewart GS, Wang B, Bignell CR, Taylor AM, Elledge SJ. 2003. MDC1 is a mediator of the mammalian DNA damage checkpoint. *Nature* 421:961–66

73. Rouse J, Jackson SP. 2002. Lcd1p recruits Mec1p to DNA lesions in vitro and in vivo. *Mol. Cell* 9:857–69

74. Unsal-Kacmaz K, Sancar A. 2004. Quaternary structure of ATR and effects of ATRIP and replication protein A on its DNA binding and kinase activities. *Mol. Cell. Biol.* 24:1292–300

75. Sekiguchi M, Tsuzuki T. 2002. Oxidative nucleotide damage: consequences and prevention. *Oncogene* 21:8895–904

76. Mao Y, Muller MT. 2003. Down modulation of topoisomerase I affects DNA repair efficiency. *DNA Repair* 2:1115–26

77. Kovalsky OI, Grossman L, Ahn B. 1996. The topodynamics of incision of UV-irradiated covalently closed DNA by the *Escherichia coli* Uvr(A)BC endonuclease. *J. Biol. Chem.* 271:33236–41

78. Chakraverty RK, Kearsey JM, Oakley TJ, Grenon M, de La Torre Ruiz MA, et al. 2001. Topoisomerase III acts upstream of Rad53p in the S-phase DNA damage checkpoint. *Mol. Cell. Biol.* 21:7150–62

79. Cobb JA, Bjergbaek L, Gasser SM. 2002. RecQ helicases: at the heart of genetic stability. *FEBS Lett.* 529:43–48

80. Bourdineaud JP, Nehme B, Tesse S, Lonvaud-Funel A. 2003. The ftsH gene of the wine bacterium *Oenococcus oeni* is involved in protection against environmental stress. *Appl. Environ. Microbiol.* 69:2512–20

81. Fischer B, Rummel G, Aldridge P, Jenal U. 2002. The FtsH protease is involved in development, stress response and heat shock control in *Caulobacter crescentus*. *Mol. Microbiol.* 44:461–78

82. Adam Z. 2000. Chloroplast proteases: Possible regulators of gene expression? *Biochimie* 82:647–54

83. Akiyama Y, Yoshihisa T, Ito K. 1995. FtsH, a membrane-bound ATPase, forms a complex in the cytoplasmic membrane of *Escherichia coli*. *J. Biol. Chem.* 270:23485–90

84. Fu GK, Smith MJ, Markovitz DM. 1997. Bacterial protease Lon is a site-specific DNA-binding protein. *J. Biol. Chem.* 272:534–38

85. Ebel W, Skinner MM, Dierksen KP, Scott JM, Trempy JE. 1999. A conserved domain in *Escherichia coli* Lon protease is involved in substrate discriminator activity. *J. Bacteriol.* 181:2236–43

86. Smith CK, Baker TA, Sauer RT. 1999. Lon and Clp family proteases and chaperones share homologous substrate-recognition domains. *Proc. Natl. Acad. Sci. USA* 96:6678–82

87. Karata K, Inagawa T, Wilkinson AJ, Tatsuta T, Ogura T. 1999. Dissecting the role of a conserved motif (the second region of homology) in the AAA family of ATPases: site-directed mutagenesis of the ATP-dependent protease FtsH. *J. Biol. Chem.* 274:26225–32

88. Hlavacek O, Vachova L. 2002. ATP-dependent proteinases in bacteria. *Folia Microbiol.* 47:203–12

89. Herman C, Prakash S, Lu CZ, Matouschek A, Gross CA. 2003. Lack of a robust unfoldase activity confers a unique level of substrate specificity to the universal AAA protease FtsH. *Mol. Cell* 11:659–69

90. Akiyama Y, Kihara A, Tokuda H, Ito K. 1996. FtsH (HflB) is an ATP-dependent protease selectively acting on SecY and some other membrane proteins. *J. Biol. Chem.* 271:31196–201

91. Anilkumar G, Srinivasan R, Anand SP, Ajitkumar P. 2001. Bacterial cell division protein FtsZ is a specific substrate for the AAA family protease FtsH. *Microbiology-UK* 147:516–17

92. Bertani D, Oppenheim AB, Narberhaus F. 2001. An internal region of the RpoH heat shock transcription factor is critical for rapid degradation by the FtsH protease. *FEBS Lett.* 493:17–20

93. Tomoyasu T, Gamer J, Bukau B, Kanemori M, Mori H, et al. 1995. *Escherichia coli* FtsH is a membrane-bound ATP-dependent protease which de-grades the heat shock transcription factor sigma[32]. *EMBO J.* 14:2551–60

94. Carmona M, de Lorenzo V. 1999. Involvement of the FtsH (HflB) protease in the activity of sigma[54] promoters. *Mol. Microbiol.* 31:261–70

95. Sze CC, Bernardo LMD, Shingler V. 2002. Integration of global regulation of two aromatic-responsive sigma[54]-dependent systems: a common phenotype by different mechanisms. *J. Bacteriol.* 184:760–70

96. Nishii W, Maruyama T, Matsuoka R, Muramatsu T, Takahashi K. 2002. The unique sites in SulA protein preferentially cleaved by ATP-dependent Lon protease from *Escherichia coli. Eur. J. Biochem.* 269:451–57

97. Schmidt R, Decatur AL, Rather PN, Moran CP, Losick R. 1994. *Bacillus subtilis* Lon protease prevents inappropriate transcription of genes under the control of the sporulation transcription factor sigma[G]. *J. Bacteriol.* 176:6528–37

98. Jubete Y, Maurizi MR, Gottesman S. 1996. Role of the heat shock protein DnaJ in the Lon-dependent degradation of naturally unstable proteins. *J. Biol. Chem.* 71:30798–803

99. Savel'ev AS, Novikova LA, Kovaleva IE, Luzikov VN, Neupert W, Langer T. 1998. ATP-dependent proteolysis in mitochondria: m-AAA protease and PIM1 protease exert overlapping substrate specificities and cooperate with the mtHsp70 system. *J. Biol. Chem.* 273:20596–602

100. Bucca G, Brassington AME, Hotchkiss G, Mersinias V, Smith CP. 2003. Negative feedback regulation of dnaK, clpB and lon expression by the DnaK chaperone machine in *Streptomyces coelicolor*, identified by transcriptome and in vivo DnaK-depletion analysis. *Mol. Microbiol.* 50:153–66

101. Jenal U, Hengge-Aronis R. 2003. Regulation by proteolysis in bacterial cells. *Curr. Opin. Microbiol.* 6:163–72

102. Adam Z, Adamska I, Nakabayashi K, Ostersetzer O, Haussuhl K, et al. 2001. Chloroplast and mitochondrial proteases in *Arabidopsis*: a proposed nomenclature. *Plant Physiol.* 125:1912–18

103. Fu GK, Markovitz DM. 1998. The human Lon protease binds to mitochondrial promoters in a single-stranded, site-specific, strand-specific manner. *Biochemistry* 37:1905–9

104. Beam CE, Saveson CJ, Lovett ST. 2002. Role for RadA/sms in recombination intermediate processing in *Escherichia coli*. *J. Bacteriol.* 184:6836–44

105. Kikuchi M, Hatano N, Yokota S, Shimozawa N, Imanaka T, Taniguchi H. 2004. Proteomic analysis of rat liver peroxisomes: presence of peroxisome-specific isozyme of Lon protease. *J. Biol. Chem.* 279:421–28

106. Schnall R, Mannhaupt G, Stucka R, Tauer R, Ehnle S, et al. 1994. Identification of a set of yeast genes coding for a novel family of putative ATPases with high similarity to constituents of the 26S protease complex. *Yeast* 10:1141–55

107. Goldberg AL. 2003. Protein degradation and protection against misfolded or damaged proteins. *Nature* 426:895–99

108. Orlowski M, Wilk S. 2003. Ubiquitin-independent proteolytic functions of the proteasome. *Arch. Biochem. Biophys.* 415:1–5

109. Yew PR. 2001. Ubiquitin-mediated proteolysis of vertebrate G1- and S-phase regulators. *J Cell. Physiol.* 187:1–10

110. Gulow K, Bienert D, Haas IG. 2002. BiP is feed-back regulated by control of protein translation efficiency. *J. Cell Sci.* 115:2443–52

111. Hong M, Luo S, Baumeister P, Huang JM, Gogia RK, et al. RK, 2004. Underglycosylation of ATF6 as a novel sensing mechanism for activation of the unfolded protein response. *J. Biol. Chem.* 279:11354–63

112. Welihinda AA, Tirasophon W, Kaufman RJ. 1999. The cellular response to protein misfolding in the endoplasmic reticulum. *Gene Expr.* 7:293–300

113. Craig KL, Tyers M. 1999. The F-box: a new motif for ubiquitin dependent proteolysis in cell cycle regulation and signal transduction. *Prog. Biophys. Mol. Biol.* 72:299–328

114. Wei N, Deng XW. 2003. The COP9 signalosome. *Annu. Rev. Cell Dev. Biol.* 19:261–86

115. Adler V, Yin ZM, Tew KD, Ronai Z. 1999. Role of redox potential and reactive oxygen species in stress signaling. *Oncogene* 18:6104–11

116. Martell EA. 1992. Radionuclide-induced evolution of DNA and the origin of life. *J. Mol. Evol.* 35:346–55

117. Martin W, Russell MJ. 2003. On the origins of cells: a hypothesis for the evolutionary transitions from abiotic geochemistry to chemoautotrophic prokaryotes, and from prokaryotes to nucleated cells. *Philos. Trans. R. Soc. London Ser. B* 358:59–83

118. Bilinski T. 1991. Oxygen toxicity and microbial evolution. *Biosystems* 24:305–12

119. Finkel T. 2003. Oxidant signals and oxidative stress. *Curr. Opin. Cell Biol.* 15:247–54

120. Toone WM, Morgan BA, Jones N. 2001. Redox control of AP-1-like factors in yeast and beyond. *Oncogene* 20:2336–46

121. Wang TL, Zhang X, Li JJ. 2002. The role of NF-kappa B in the regulation of cell stress responses. *Int. Immunopharmacol.* 2:1509–20

122. Fujii J, Ikeda Y. 2002. Advances in our understanding of peroxiredoxin, a multifunctional, mammalian redox protein. *Redox Rep.* 7:123–30

123. Dehaan JB, Cristiano F, Iannello RC, Kola I. 1995. Cu/Zn-superoxide dismutase and glutathione-peroxidase during aging. *Biochem. Mol. Biol. Int.* 35:1281–97

124. Dickinson DA, Forman HJ. 2002. Glutathione in defense and signaling: lessons

from a small thiol. *Ann. NY Acad. Sci.* 973:488–504

125. Arner ESJ, Holmgren A. 2000. Physiological functions of thioredoxin and thioredoxin reductase. *Eur. J. Biochem.* 267:6102–9

126. Yamawaki H, Haendeler J, Berk BC. 2003. Thioredoxin: a key regulator of cardiovascular homeostasis. *Circ. Res.* 93:1029–33

127. Stadtman ER. 2004. Cyclic oxidation and reduction of methionine residues of proteins in antioxidant defense and cellular regulation. *Arch. Biochem. Biophys.* 423:2–5

128. Nishiyama A, Masutani H, Nakamura H, Nishinaka Y, Yodoi J. 2001. Redox regulation by thioredoxin and thioredoxin-binding proteins. *IUBMB Life* 52:29–33

129. Brot N, Weissbach H. 2000. Peptide methionine sulfoxide reductase: biochemistry and physiological role. *Biopolymers* 55:288–96

130. Moskovitz J, Jenkins NA, Gilbert DJ, Copeland NG, Jursky F, et al. 1996. Chromosomal localization of the mammalian peptide-methionine sulfoxide reductase gene and its differential expression in various tissues. *Proc. Natl. Acad. Sci. USA* 93:3205–8

131. Gustavsson N, Kokke BP, Harndahl U, Silow M, Bechtold U, et al. 2002. A peptide methionine sulfoxide reductase highly expressed in photosynthetic tissue in *Arabidopsis thaliana* can protect the chaperone-like activity of a chloroplast-localized small heat shock protein. *Plant J.* 29:545–53

132. Kromayer M, Wilting R, Tormay P, Böck A. 1996. Domain structure of the prokaryotic selenocysteine-specific elongation factor SelB. *J. Mol. Biol.* 262:413–20

133. Leipe DD, Wolf YI, Koonin EV, Aravind L. 2002. Classification and evolution of P-loop GTPases and related ATPases. *J. Mol. Biol.* 317:41–72

134. Kryukov GV, Kumar RA, Koc A, Sun ZH,

Gladyshev VN. 2002. Selenoprotein R is a zinc-containing stereo-specific methionine sulfoxide reductase. *Proc. Natl. Acad. Sci. USA* 99:4245–50

135. Brown KM, Arthur JR. 2001. Selenium, selenoproteins and human health: a review. *Public Health Nutr.* 4:593–99

136. Namy O, Rousset JP, Napthine S, Brierley I. 2004. Reprogrammed genetic decoding in cellular gene expression. *Mol. Cell* 13:157–68

137. Fagegaltier D, Carbon P, Krol A. 2001. Distinctive features in the SelB family of elongation factors for selenoprotein synthesis. A glimpse of an evolutionary complexified translation apparatus. *Biofactors* 14:5–10

138. Driscoll DM, Copeland PR. 2003. Mechanism and regulation of selenoprotein synthesis. *Annu. Rev. Nutr.* 23:17–40

139. Hoch JA, Silhavy TJ. 1995. *Two-Component Signal Transduction.* Washington, DC: Am. Soc. Microbiol. 488 pp.

140. Gill G. 2003. Post-translational modification by the small ubiquitin-related modifier SUMO has big effects on transcription factor activity. *Curr. Opin. Genet. Dev.* 13:108–13

141. Markstein M, Levine M. 2002. Decoding cis-regulatory DNAs in the *Drosophila* genome. *Curr. Opin. Genet. Dev.* 12:601–6

142. Cyert MS. 2001. Regulation of nuclear localization during signaling. *J. Biol. Chem.* 276:20805–8

143. Fischle W, Wang Y, Allis CD. 2003. Histone and chromatin cross-talk. *Curr. Opin. Cell Biol.* 15:172–83

144. Berger SL. 2002. Histone modifications in transcriptional regulation. *Curr. Opin. Genet. Dev.* 12:142–48

145. Kültz D. 1998. Phylogenetic and functional classification of mitogen- and stress-activated protein kinases. *J. Mol. Evol.* 46:571–88

146. Fu H, Subramanian RR, Masters SC. 2000. 14-3-3 proteins: structure, function,

and regulation. *Annu. Rev. Pharmacol. Toxicol.* 40:617–47

147. Wiens M, Diehl-Seifert B, Müller WE. 2001. Sponge Bcl-2 homologous protein (BHP2-GC) confers distinct stress resistance to human HEK-293 cells. *Cell Death Differ.* 8:887–98

148. Craven RJ, Greenwell PW, Dominska M, Petes TD. 2002. Regulation of genome stability by Tel1 and Mec1, yeast homologs of the mammalian ATM and ATR genes. *Genetics* 161:493–507

149. Skorokhod A, Gamulin V, Gundacker D, Kavsan V, Müller IM, Müller WE. 1999. Origin of insulin receptor-like tyrosine kinases in marine sponges. *Biol. Bull.* 197:198–206

150. Huisinga KL, Pugh BF. 2004. A genome-wide housekeeping role for TFIID and a highly regulated stress-related role for SAGA in *Saccharomyces cerevisiae*. *Mol. Cell* 13:573–85

151. Hengge-Aronis R. 2002. Signal transduction and regulatory mechanisms involved in control of the sigma^S (RpoS) subunit of RNA polymerase. *Microbiol. Mol. Biol. Rev.* 66:373–95

152. Rahman I. 2003. Oxidative stress, chromatin remodeling and gene transcription in inflammation and chronic lung diseases. *J. Biochem. Mol. Biol.* 36:95–109

153. Kültz D, Burg MB. 1998. Intracellular signaling in response to osmotic stress. *Contrib. Nephrol.* 123:94–109

154. Amundson SA, Bittner M, Fornace AJ Jr. 2003. Functional genomics as a window on radiation stress signaling. *Oncogene* 22:5828–33

155. Chinnusamy V, Schumaker K, Zhu JK. 2004. Molecular genetic perspectives on cross-talk and specificity in abiotic stress signalling in plants. *J. Exp. Bot.* 55:225–36

156. Aldsworth TG, Sharman RL, Dodd CE. 1999. Bacterial suicide through stress. *Cell Mol. Life Sci.* 56:378–83

157. Smith ML, Fornace AJ. 1996. Mammalian DNA damage-inducible genes as-sociated with growth arrest and apoptosis. *Mutat. Res. Rev. Genet. Toxicol.* 340:109–24

158. Hartwell LH, Weinert TA. 1989. Checkpoints: controls that ensure the order of cell cycle events. *Science* 246:629–34

159. Autret S, Levine A, Holland IB, Seror SJ. 1997. Cell cycle checkpoints in bacteria. *Biochimie* 79:549–54

160. Kvint K, Nachin L, Diez A, Nystrom T. 2003. The bacterial universal stress protein: function and regulation. *Curr. Opin. Microbiol.* 6:140–45

161. Nystrom T. 2003. Conditional senescence in bacteria: death of the immortals. *Mol. Microbiol.* 48:17–23

162. Dai W, Huang X, Ruan Q. 2003. Polo-like kinases in cell cycle checkpoint control. *Front. Biosci.* 8:d1128–33

163. Bartek J, Lukas J. 2001. Mammalian G1- and S-phase checkpoints in response to DNA damage. *Curr. Opin. Cell Biol.* 13:738–47

164. Hartwell LH, Kastan MB. 1994. Cell cycle control and cancer. *Science* 266:1821–28

165. Shiloh Y. 2003. ATM and related protein kinases: safeguarding genome integrity. *Nat. Rev. Cancer* 3:155–68

166. Pietenpol JA, Stewart ZA. 2002. Cell cycle checkpoint signaling: cell cycle arrest versus apoptosis. *Toxicology* 181:475–81

167. Taylor WR, Stark GR. 2001. Regulation of the G2/M transition by p53. *Oncogene* 20:1803–15

168. Chakravarty D, Cai Q, Ferraris JD, Michea L, Burg MB, Kültz D. 2002. Three GADD45 isoforms contribute to hypertonic stress phenotype of murine renal inner medullary cells. *Am. J. Physiol.* 283:F1020–29

169. Nyberg KA, Michelson RJ, Putnam CW, Weinert TA. 2002. Toward maintaining the genome: DNA damage and replication checkpoints. *Annu. Rev. Genet.* 36:617–56

170. Lindquist S. 1986. The heat-shock response. *Annu. Rev. Biochem.* 55:1151–91

171. Morimoto RI. 1998. Regulation of the heat shock transcriptional response: cross talk between a family of heat shock factors, molecular chaperones, and negative regulators. *Genes Dev.* 12:3788–96

172. Ellis RJ, Hartl FU. 1999. Principles of protein folding in the cellular environment. *Curr. Opin. Struct. Biol.* 9:102–10

173. Gething MJ, Sambrook J. 1992. Protein folding in the cell. *Nature* 355:33–45

174. Feder ME, Hofmann GE. 1999. Heat shock proteins, molecular chaperones, and the stress response: evolutionary and ecological physiology. *Annu. Rev. Physiol.* 61:243–82

175. Macario AJ, Malz M, Conway DM. 2004. Evolution of assisted protein folding: the distribution of the main chaperoning systems within the phylogenetic domain archaea. *Front. Biosci.* 9:1318–32

176. Arsene F, Tomoyasu T, Bukau B. 2000. The heat shock response of *Escherichia coli. Int. J. Food Microbiol.* 55:3–9

177. Harrison C. 2003. GrpE, a nucleotide exchange factor for DnaK. *Cell Stress Chaperones* 8:218–24

178. Tavaria M, Gabriele T, Kola I, Anderson RL. 1996. A hitchhiker's guide to the human Hsp70 family. *Cell Stress Chaperones* 1:23–28

179. Christians ES, Yan LJ, Benjamin IJ. 2002. Heat shock factor 1 and heat shock proteins: critical partners in protection against acute cell injury. *Crit. Care Med.* 30:S43–50

180. Bukrinsky MI. 2002. Cyclophilins: unexpected messengers in intercellular communications. *Trends Immunol.* 23:323–25

181. Andreeva L, Heads R, Green CJ. 1999. Cyclophilins and their possible role in the stress response. *Int. J. Exp. Pathol.* 80:305–15

182. Gothel SF, Marahiel MA. 1999. Peptidylprolyl *cis-trans* isomerases, a superfamily of ubiquitous folding catalysts. *Cell Mol. Life Sci.* 55:423–36

183. Waldmeier PC, Zimmermann K, Qian T, Tintelnot-Blomley M, Lemasters JJ. 2003. Cyclophilin D as a drug target. *Curr. Med. Chem.* 10:1485–506

184. Zou Y, Crowley DJ, Van Houten B. 1998. Involvement of molecular chaperonins in nucleotide excision repair: DnaK leads to increased thermal stability of UvrA, catalytic UvrB loading, enhanced repair, and increased UV resistance. *J. Biol. Chem.* 273:12887–92

185. Sweder K, Madura K. 2002. Regulation of repair by the 26S proteasome. *J. Biomed. Biotechnol.* 2:94–105

186. Hanawalt PC. 2001. Controlling the efficiency of excision repair. *Mutat. Res. DNA Repair* 485:3–13

187. van Hoffen A, Balajee AS, van Zeeland AA, Mullenders LH. 2003. Nucleotide excision repair and its interplay with transcription. *Toxicology* 193:79–90

188. Izumi T, Wiederhold LR, Roy G, Roy R, Jaiswal A, et al. 2003. Mammalian DNA base excision repair proteins: their interactions and role in repair of oxidative DNA damage. *Toxicology* 193:43–65

189. Fedier A, Fink D. 2004. Mutations in DNA mismatch repair genes: implications for DNA damage signaling and drug sensitivity. *Int. J. Oncol.* 24:1039–47

190. Schofield MJ, Hsieh P. 2003. DNA mismatch repair: molecular mechanisms and biological function. *Annu. Rev. Microbiol.* 57:579–608

191. Valerie K, Povirk LF. 2003. Regulation and mechanisms of mammalian double-strand break repair. *Oncogene* 22:5792–812

192. van den Bosch M, Lohman PH, Pastink A. 2002. DNA double-strand break repair by homologous recombination. *Biol. Chem.* 383:873–92

193. Maeng O, Kim YC, Shin HJ, Lee JO, Huh TL, et al. 2004. Cytosolic NADP⁺-dependent isocitrate dehydrogenase protects macrophages from LPS-induced nitric oxide and reactive oxygen species. *Biochem. Biophys. Res. Commun.* 317:558–64

194. Gupta M, Hogema BM, Grompe M, Bottiglieri TG, Concas A, et al. 2003. Murine succinate semialdehyde dehydrogenase deficiency. *Ann. Neurol.* 54(Suppl. 6):S81–90

195. Huang J, Zhang H, Wang J, Yang J. 2003. Molecular cloning and characterization of rice 6-phosphogluconate dehydrogenase gene that is up-regulated by salt stress. *Mol. Biol. Rep.* 30:223–27

196. Puskas F, Gergely P Jr, Banki K, Perl A. 2000. Stimulation of the pentose phosphate pathway and glutathione levels by dehydroascorbate, the oxidized form of vitamin C. *FASEB J.* 14:1352–61

197. Kuriyama H, Fukuda H. 2002. Developmental programmed cell death in plants. *Curr. Opin. Plant Biol.* 5:568–73

198. Zhivotovsky B. 2002. From the nematode and mammals back to the pine tree: on the diversity and evolution of programmed cell death. *Cell Death Differ.* 9:867–69

199. Cairns J. 2002. A DNA damage checkpoint in *Escherichia coli*. *DNA Repair* 1:699–701

200. Takayama S, Reed JC, Homma S. 2003. Heat-shock proteins as regulators of apoptosis. *Oncogene* 22:9041–47

201. Fridman JS, Lowe SW. 2003. Control of apoptosis by p53. *Oncogene* 22:9030–40

202. Franke TF, Hornik CP, Segev L, Shostak GA, Sugimoto C. 2003. PI3K/Akt and apoptosis: size matters. *Oncogene* 22: 8983–98

203. Kucharczak J, Simmons MJ, Fan Y, Gelinas C. 2003. To be, or not to be: NF-kappaB is the answer: role of Rel/NF-kappaB in the regulation of apoptosis. *Oncogene* 22:8961–82

204. Mak SK, Kültz D. 2004. GADD45 proteins modulate apoptosis in renal inner medullary cells exposed to hyperosmotic stress. *J. Biol. Chem.* 279:39075–84

205. Sheikh MS, Hollander MC, Fornance AJ Jr. 2000. Role of Gadd45 in apoptosis. *Biochem. Pharmacol.* 59:43–45

Annu. Rev. Physiol. 2005. 67:259–84
doi: 10.1146/annurev.physiol.67.040403.120816

ENDOCRINOLOGY OF THE STRESS RESPONSE[1]

Evangelia Charmandari, Constantine Tsigos, and George Chrousos

*Pediatric and Reproductive Endocrinology Branch, National Institute of Child Health
and Human Development, National Institutes of Health, Bethesda, Maryland 20892,
and Hellenic National Diabetes Center, Athens, 10675, Greece;
email: charmane@mail.nih.gov; chrousosG@aol.com*

Key Words   stress system, endocrinology of stress, stress-related disorders

■ **Abstract**   The stress response is subserved by the stress system, which is located both in the central nervous system and the periphery. The principal effectors
of the stress system include corticotropin-releasing hormone (CRH); arginine vasopressin; the proopiomelanocortin-derived peptides α-melanocyte-stimulating hormone and β-endorphin, the glucocorticoids; and the catecholamines norepinephrine
and epinephrine. Appropriate responsiveness of the stress system to stressors is a crucial prerequisite for a sense of well-being, adequate performance of tasks, and positive
social interactions. By contrast, inappropriate responsiveness of the stress system may
impair growth and development and may account for a number of endocrine, metabolic,
autoimmune, and psychiatric disorders. The development and severity of these conditions primarily depend on the genetic vulnerability of the individual, the exposure
to adverse environmental factors, and the timing of the stressful events, given that
prenatal life, infancy, childhood, and adolescence are critical periods characterized by
increased vulnerability to stressors.

INTRODUCTION

Life exists through maintenance of a complex dynamic equilibrium, termed homeostasis, that is constantly challenged by intrinsic or extrinsic, real or perceived,
adverse forces, the stressors (1, 2). Stress is defined as a state of threatened or perceived as threatened homeostasis. The human body and mind react to stress by activating a complex repertoire of physiologic and behavioral central nervous system
and peripheral adaptive responses, which, if inadequate or excessive and/or prolonged, may affect personality development and behavior, and may have adverse
consequences on physiologic functions, such as growth, metabolism, circulation,
reproduction, and the inflammatory/immune response (1, 2). The state of chronic

---

[1]The U.S. Government has the right to retain a nonexclusive, royalty-free license in and to
any copyright covering this paper.

dyshomeostasis due to inadequate or excessive/prolonged adaptive responses, in which the individual survives but suffers adverse consequences, has been called allostasis.

The present review focuses on the neuroendocrinology of the stress response and the effects of stress on the major endocrine axes. It also provides a brief overview of the altered regulation of the adaptive response in various physiologic and pathologic states that may influence the growth and development of an individual and may define vulnerability of the individual to endocrine, psychiatric, cardiovascular, neoplastic, or immunologic disorders.

ENDOCRINOLOGY OF THE STRESS RESPONSE

Neuroendocrine Effectors of the Stress Response

The stress response is subserved by the stress system, which has both central nervous system (CNS) and peripheral components (1–3). The central components of the stress system are located in the hypothalamus and the brainstem, and include (*a*) the parvocellular neurons of corticotropin-releasing hormone (CRH); (*b*) the arginine vasopressin (AVP) neurons of the paraventricular nuclei (PVN) of the hypothalamus; (*c*) the CRH neurons of the paragigantocellular and parabranchial nuclei of the medulla and the locus ceruleus (LC); and (*d*) other mostly noradrenergic (NE) cell groups in the medulla and pons (LC/NE system). The peripheral components of the stress system include (*a*) the peripheral limbs of the hypothalamic-pituitary-adrenal (HPA) axis; (*b*) the efferent sympathetic-adrenomedullary system; and (*c*) components of the parasympathetic system (1–3) (Figure 1, see color insert).

The central neurochemical circuitry responsible for activation of the stress system has been studied extensively. There are multiple sites of interaction among the various components of the stress system. Reciprocal reverberatory neural connections exist between the CRH and noradrenergic neurons of the central stress system, with CRH and norepinephrine stimulating each other primarily through CRH type 1 and α_1-noradrenergic receptors, respectively (4–6). Autoregulatory negative feedback loops are also present in both the PVN CRH and brainstem noradrenergic neurons (7, 8), with collateral fibers inhibiting CRH and catecholamine secretion via presynaptic CRH and α_2-noradrenergic receptors, respectively (7–9). Both the CRH and the noradrenergic neurons also receive stimulatory innervation from the serotoninergic and cholinergic systems (10, 11), and inhibitory input from the γ-aminobutyric acid (GABA)-benzodiazepine (BZD) and opioid peptide neuronal systems of the brain (7, 12, 13), as well as from the end-product of the HPA axis, the glucocorticoids (7, 14).

Corticotropin-Releasing Hormone, Arginine Vasopressin, and Catecholaminergic Neurons

CRH, a 41-amino acid peptide, is the principal hypothalamic regulator of the pituitary-adrenal axis. CRH and CRH receptors have been detected in many

TABLE 1 Behavioral and physical adaptation during acute stress

Behavioral adaptation: adaptive redirection of behavior	Physical adaptation: adaptive redirection of energy
Increased arousal and alertness	Oxygen and nutrients directed to the CNS and stressed body site(s)
Increased cognition, vigilance, and focused attention	Altered cardiovascular tone, increased blood pressure and heart rate
Euphoria (or dysphoria)	Increased respiratory rate
Heightened analgesia	Increased gluconeogenesis and lipolysis
Increased temperature	Detoxification from toxic products
Suppression of appetite and feeding behavior	Inhibition of growth and reproduction
Suppression of reproductive axis	Inhibition of digestion-stimulation of colonic motility
Containment of the stress response	Containment of the inflammatory/immune response

Adapted from Chrousos & Gold (2).

extrahypothalamic sites of the brain, including parts of the limbic system, the basal forebrain, and the LC-NE sympathetic system in the brainstem and spinal cord. Intracerebroventricular administration of CRH results in a series of behavioral and peripheral responses, as well as activation of the pituitary-adrenal axis and the sympathetic nervous system (SNS), indicating that CRH has a much broader role in coordinating the stress response than initially recognized (1–3) (Table 1).

Since CRH was first characterized, a growing family of ligands and receptors has evolved. The mammalian family members include CRH, urocortinI (UcnI), UcnII, and UcnIII, along with two receptors, CRHR1 and CRHR2, and a CRH-binding protein. These family members differ in their tissue distribution and pharmacology and play an important role in the regulation of the endocrine and behavioral responses to stress. Although CRH appears to play a stimulatory role in stress responsivity through activation of CRHR1, specific actions of UcnII and UcnIII on CRHR2 may be important for dampening stress sensitivity. UcnI is the only ligand with high affinity for both receptors and its role may be promiscuous (15).

CRH receptors belong to the class B subtype of G protein–coupled receptors (GPCR). CRHR1 and CRHR2 are produced from distinct genes and have several splice variants expressed in various central and peripheral tissues (15). The CRH-R1 subtype is widely distributed in the brain, mainly in the anterior pituitary, the neocortex, and the cerebellum, as well as in the adrenal gland, skin, ovary, and testis. CRH-R2 receptors are expressed mostly in the peripheral vasculature, the skeletal muscles, the gastrointestinal tract, and the heart, but also in subcortical structures of the brain, such as the lateral septum, amygdala, hypothalamus, and brain stem. The diversity in CRH receptor subtype and isoform expression is thought to play a key role in modifying the stress response by implicating locally the actions of different ligands (CRH and CRH-related peptides) and different intracellular second messengers (15).

CRH is a major anorexiogenic peptide, whose secretion is stimulated by neuropeptide Y (NPY). NPY is the most potent known orexiogenic factor, which inhibits the LC-NE sympathetic system simultaneously (16–18). The latter may be of particular relevance to alterations in the activity of the stress system in states of dysregulation of food intake, such as malnutrition, anorexia nervosa, and obesity. Glucocorticoids enhance the expression of hypothalamic NPY, whereas they directly inhibit both the PVN CRH and LC-NE sympathetic systems. Substance P (SP) has actions reciprocal to those of NPY, given that it inhibits the PVN CRH neuron while it activates the LC-NE sympathetic system. SP release is likely to be increased centrally secondary to peripheral activation of somatic afferent fibers and may, therefore, have relevance to changes in the stress system activity induced by chronic inflammatory or painful states (19).

A subset of PVN parvocellular neurons synthesize and secrete both CRH and AVP, while another subset secretes AVP only (2, 20, 21). During stress, the relative proportion of the subset of neurons that secrete both CRH and AVP increases significantly. The terminals of the parvocellular PVN CRH and AVP neurons project to different sites, including the noradrenergic neurons of the brainstem and the hypophyseal portal system in the median eminence. PVN CRH and AVP neurons also send projections to and activate proopiomelanocortin (POMC)-containing neurons in the arcuate nucleus of the hypothalamus, which in turn project to the PVN CRH and AVP neurons, innervate LC-NE sympathetic neurons of the central stress system in the brainstem, and terminate on pain control neurons of the hind brain and spinal cord (2, 20, 21). Thus, activation of the stress system via CRH and catecholamines stimulates the secretion of hypothalamic β-endorphin and other POMC-derived peptides, which reciprocally inhibit the activity of the stress system and result in stress- induced analgesia.

The Hypothalamic-Pituitary-Adrenal Axis

CRH is the principal hypothalamic regulator of the pituitary-adrenal axis, which stimulates the secretion of adrenocorticotropin hormone (ACTH) from the anterior pituitary. AVP, although a potent synergistic factor of CRH, has very little ACTH secretagogue activity on its own (22, 23). A positive reciprocal interaction between CRH and AVP also exists at the level of hypothalamus, with each neuropeptide stimulating the secretion of the other. In nonstressful situations, both CRH and AVP are secreted in the portal system in a circadian, pulsatile, and highly concordant fashion (24–27). The amplitude of the CRH and AVP pulses increases early in the morning, resulting in increases primarily in the amplitude of the pulsatile ACTH and cortisol secretion. Diurnal variations in the pulsatile secretion of ACTH and cortisol are often perturbed by changes in lighting, feeding schedules, and activity, as well as following stress.

During acute stress, there is an increase in the amplitude and synchronization of the PVN CRH and AVP pulsatile release into the hypophyseal portal system. AVP of magnocellular neuron origin is also secreted into the hypophyseal portal system via collateral fibers and the systemic circulation via the posterior pituitary (27, 28). In addition, depending on the stressor, other factors, such as angiotensin II, various cytokines, and lipid mediators of inflammation are secreted and act on the hypothalamic, pituitary, and/or adrenal components of the HPA axis and potentiate its activity.

The adrenal cortex is the main target of ACTH, which regulates glucocorticoid and adrenal androgen secretion by the zona fasciculata and reticularis, respectively, and participates in the control of aldosterone secretion by the zona glomerulosa. Other hormones, cytokines, and neuronal information from the autonomic nerves of the adrenal cortex may also participate in the regulation of cortisol secretion (27, 29–31).

Glucocorticoids are the final effectors of the HPA axis. These hormones are pleiotropic, and exert their effects through their ubiquitously distributed intracellular receptors (32–34). In the absence of ligand, the nonactivated glucocorticoid receptor (GR) resides primarily in the cytoplasm of cells as part of a large multiprotein complex consisting of the receptor polypeptide, two molecules of hsp90, and several other proteins (34). Upon hormone binding, the receptor dissociates from hsp90 and other proteins and translocates into the nucleus, where it binds as homodimer to glucocorticoid-response elements (GREs) located in the promoter region of target genes, and regulates the expression of glucocorticoid-responsive genes positively or negatively, depending on the GRE sequence and promoter context. The receptor can also modulate gene expression independently of GRE-binding, by physically interacting with other transcription factors, such as activating protein-1 (AP-1) and nuclear factor-κB (NF-κB) (34).

Glucocorticoids play an important role in the regulation of basal activity of the HPA axis, as well as in the termination of the stress response by acting at extrahypothalamic centers, the hypothalamus, and the pituitary gland. The negative feedback of glucocorticoids on the secretion of CRH and ACTH serves to limit the duration of the total tissue exposure of the organism to glucocorticoids, thus minimizing the catabolic, lipogenic, antireproductive, and immunosuppressive effects of these hormones. A dual-receptor system exists for glucocorticoids in the CNS, which includes the glucocorticoid receptor type I or mineralocorticoid receptor that responds to low concentrations of glucocorticoids, and the classic glucocorticoid receptor type II that responds to both basal and stress concentrations of glucocorticoids. The negative feedback control of the CRH and ACTH secretion is mediated through type II glucocorticoid receptors (1–3).

The LC-NE, Sympathetic, Adrenomedullary, and Parasympathetic Systems

The autonomic nervous system (ANS) responds rapidly to stressors and controls a wide range of functions. Cardiovascular, respiratory, gastrointestinal, renal, endocrine, and other systems are regulated by the SNS and/or the parasympathetic system. In general, the parasympathetic system can both assist and antagonize sympathetic functions by withdrawing or increasing its activity, respectively (3).

Sympathetic innervation of peripheral organs is derived from the efferent preganglionic fibers, whose cell bodies lie in the intermediolateral column of the spinal cord. These nerves synapse in the bilateral chain of sympathetic ganglia with postganglionic sympathetic neurons, which innervate widely the smooth muscle of the vasculature, the heart, skeletal muscles, kidney, gut, fat, and many other organs. The preganglionic neurons are primarily cholinergic, whereas the postganglionic neurons are mostly noradrenergic. The sympathetic system of the adrenal medulla also provides all of circulating epinephrine and some of the norepinephrine.

In addition to the classic neurotransmitters acetylcholine and norepinephrine, both sympathetic and parasympathetic subdivisions of the autonomic nervous system include several subpopulations of target-selective and neurochemically coded neurons that express a variety of neuropeptides and, in some cases, adenosine triphosphate (ATP), nitric oxide, or lipid mediators of inflammation (3). Thus CRH, NPY, somatostatin, and galanin are found in postganglionic noradrenergic vasoconstrictive neurons, whereas vasoactive intestinal peptide (VIP), SP, and calcitonin gene-related peptide are found in cholinergic neurons. Transmission in sympathetic ganglia is also modulated by neuropeptides released from preganglionic fibers and short interneurons, and by primary afferent nerve collaterals.

Adaptive Responses to Stress

The stress system receives and integrates a diversity of cognitive, emotional, neurosensory, and peripheral somatic signals that arrive through distinct pathways. Activation of the stress system leads to behavioral and physical changes that are remarkably consistent in their qualitative presentation and are collectively defined as the stress syndrome (Table 1). These changes are normally adaptive and time limited and improve the chances of the individual for survival.

Behavioral adaptation includes increased arousal, alertness, and vigilance; improved cognition; focused attention; euphoria; enhanced analgesia; elevations in core temperature; and inhibition of vegetative functions, such as appetite, feeding, and reproduction. A concomitant physical adaptation also occurs mainly to promote an adaptive redirection of energy. Oxygen and nutrients are shunted to the CNS and the stressed body sites, where they are most needed. Increases in cardiovascular tone, respiratory rate, and intermediate metabolism (gluconeogenesis, lipolysis) work in concert with the above alterations to promote availability of vital substrates. Detoxification functions are activated to rid the organism of unnecessary metabolic products from the stress-related changes in metabolism, whereas digestive function, growth, reproduction, and immunity are inhibited (3, 35).

During stress, the organism also activates restraining forces that prevent an overresponse from both the central and peripheral components of the stress system.

These forces are essential for successful adaptation. If they are excessive or fail to contain the various elements of the stress response in a timely manner, the adaptive changes may become chronically deficient or excessive, respectively, and may contribute to the development of pathology. Thus the restraining forces may participate in the development of allostasis. Stress is often of a magnitude and nature that allow the subjective perception of control by the individual. In such cases, stress can be pleasant and rewarding, or at least not damaging. On the other hand, stress of a nature, magnitude, or duration that is beyond the adaptive resources of an individual may be associated with a perception of loss of control, dysphoria, and chronic adverse behavioral and physical consequences (1, 3, 35). Frequently allostasis and sense of loss of control go hand-in-hand, with the latter serving as a useful index of the former.

Stress System Interactions with Other CNS Components

In addition to setting the level of arousal and influencing the vital signs, the stress system interacts with three other major CNS components: the mesocorticolimbic dopaminergic or reward system, the amygdala-hippocampus complex, and the hypothalamic arcuate nucleus POMC neuronal system. All three CNS components are activated during stress and, in turn, influence the activity of the stress system. In addition, the stress system interacts with thermoregulatory and appetite-satiety centers of the CNS, as well as the growth, thyroid, and reproductive axes and the immune system (1, 3).

MESOCORTICOLIMBIC SYSTEM   Both the mesocortical and mesolimbic components of the dopaminergic system are innervated by PVN CRH neurons and the LC-NE system and are activated during stress (36, 37). The mesocortical system consists of dopaminergic neurons of the ventral tegmentum, which send projections to the prefrontal cortex, and is involved in anticipatory phenomena and cognitive functions. The mesolimbic system also consists of dopaminergic neurons of the ventral tegmentum, which innervate the nucleus accumbens, and plays a principal role in motivational/reinforcement/reward phenomena and in the formation of the central dopaminergic reward system. Therefore, euphoria or dysphoria is likely to be mediated by the mesocorticolimbic system, which is also considered to be the target of several substances of abuse, such as cocaine. Interestingly, activation of the prefrontal cortex, which is part of the mesocortical dopaminergic system, is associated with inhibition of the stress system (38).

AMYGDALA-HIPPOCAMPUS COMPLEX   The amygdala-hippocampus complex is activated during stress primarily by ascending catecholaminergic neurons originating in the brainstem, and by the end-product of the HPA axis, glucocorticoids, but also by inner emotional stressors, such as fear, which is generated in the amygdala (39). Activation of the amygdala is important for retrieval and emotional analysis of relevant information for any given stressor. The amygdala can directly stimulate both central components of the stress system and the mesocorticolimbic dopaminergic system in response to emotional stressors. The hippocampus exerts tonic and stimulated inhibitory effects on the activity of the amygdala, PVN CRH, and LC-NE-sympathetic system.

POMC NEURONAL SYSTEM   LC-NE-noradrenergic and the CRH/AVP-producing neurons reciprocally innervate and are innervated by opioid peptide (POMC)-producing neurons of the arcuate nucleus of the hypothalamus (7, 40). Activation of the stress system stimulates hypothalamic POMC-derived peptides, such as α-melanocyte-stimulating hormone (α-MSH) and β-endorphin, which reciprocally inhibit the activity of both of the central components of the stress system, produce analgesia through projections to the hind brain and spinal cord, where they inhibit ascending pain stimuli.

TEMPERATURE REGULATION   Activation of the LC-NE and PVN/CRH systems increases the core temperature. Intracerebroventricular administration of norepinephrine and CRH results in elevations in core temperature, probably through prostanoid-mediated actions on the septal and hypothalamic temperature-regulating centers (41, 42). CRH has also been shown to partly mediate the pyrogenic effects of the inflammatory cytokines, tumor necrosis factor-α (TNF-α), interleukin (IL)-1, and IL-6 (3).

APPETITE REGULATION   Stress is also involved in the regulation of appetite by influencing the appetite-satiety centers in the hypothalamus. Acute elevations in CRH concentrations cause anorexia. On the other hand, fasting-stimulated increases in NPY enhance CRH secretion (43), while they concomitantly inhibit the LC-NE-sympathetic system and activate the parasympathetic system, thereby facilitating digestion and storage of nutrients (44). Leptin, a satiety-stimulating polypeptide secreted by the white adipose tissue, is a potent inhibitor of hypothalamic NPY and a stimulant of a subset of arcuate nucleus POMC neurons that secrete α-MSH, another potent anorexiogen that exerts its effects primarily through specific melanocortin receptors type 4 (45, 46).

EFFECTS OF CHRONIC HYPERACTIVATION OF THE STRESS SYSTEM

In general, the stress response is meant to be of short or limited duration. The time-limited nature of this process renders its accompanying antigrowth, antireproductive, catabolic, and immunosuppressive effects temporarily beneficial and/or of no adverse consequences to the individual. However, chronic activation of the stress system may lead to a number of disorders that are the result of increased and/or prolonged secretion of CRH and/or glucocorticoids (Table 2).

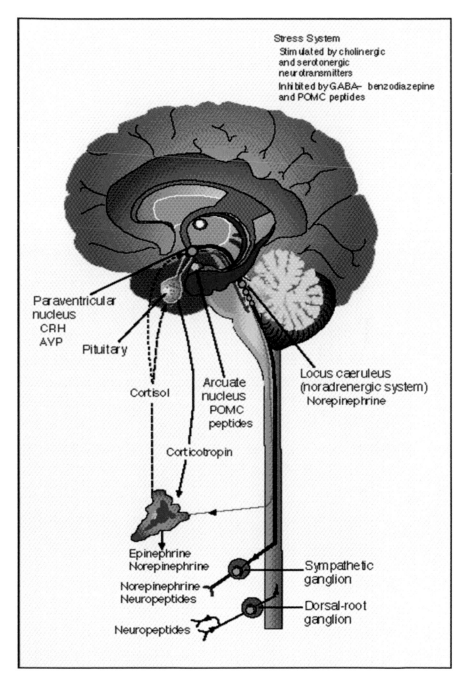

Figure 1 Schematic representation of the central and peripheral components of the stress system, their functional interrelations, and their relation to other central nervous system components involved in the stress response. Adapted from Chrousos (80).

TABLE 2   States associated with altered hypothalamic-pituitary-adrenal (HPA) axis activity and altered regulation or dysregulation of behavioral and/or peripheral adaptation

Increased HPA axis activity	Decreased HPA axis activity
Chronic stress	Adrenal insufficiency
Melancholic depression	Atypical/seasonal depression
Anorexia nervosa	Chronic fatigue syndrome
Malnutrition	Fibromyalgia
Obsessive-compulsive disorder	Hypothyroidism
Panic disorder	Nicotine withdrawal
Excessive exercise (obligate athleticism)	Discontinuation of glucocorticoid therapy
Chronic active alcoholism	After Cushing syndrome cure
Alcohol and narcotic withdrawal	Premenstrual tension syndrome
Diabetes mellitus	Postpartum period
Truncal obesity (Metabolic syndrome X)	After chronic stress
Childhood sexual abuse	Rheumatoid arthritis
Psychosocial short stature	Menopause
Attachment disorder of infancy	
'Functional' gastrointestinal disease	
Hyperthyroidism	
Cushing syndrome	
Pregnancy (last trimester)	

Adapted from Chrousos & Gold (2).

Growth and Development

During stress, the growth axis is inhibited at many levels (Figure 2a). Prolonged activation of the HPA axis leads to suppression of growth hormone (GH) secretion and glucocorticoid-induced inhibition of the effects of insulin-like growth factor I (IGF-I) and other growth factors on target tissues (47–49). Children with Cushing's syndrome have delayed or arrested growth and achieve a final adult height that is on average 7.5–8.0 cm below their predicted height (49). The molecular mechanisms by which glucocorticoids suppress growth are complex and involve both transcriptional and translational mechanisms that ultimately influence GH action (50, 51).

In addition to the direct effects of glucocorticoids, CRH-induced increases in somatostatin secretion, and therefore inhibition of GH secretion, have been implicated as a potential mechanism of chronic stress-related suppression of GH secretion. However, acute elevations of serum GH concentrations may occur at the onset of the stress response or following acute administration of glucocorticoids, most likely due to stimulation of the GH gene by glucocorticoids through GREs in the promoter region of the gene (52).

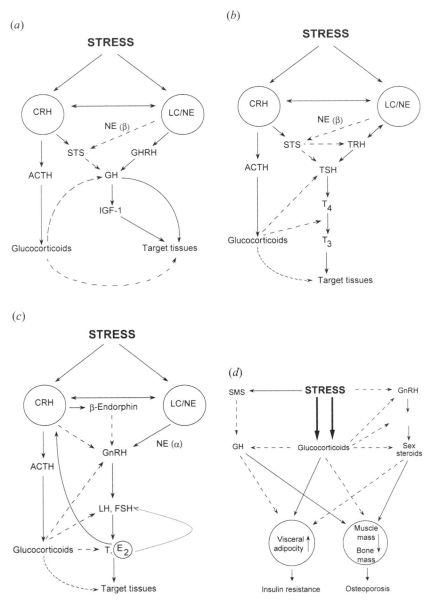

Figure 2   Schematic representation of the interactions between the stress system and (*a*) the GH/IGF-I axis, (*b*) the thyroid axis, (*c*) the hypothalamic-pituitary-gonadal axis, and (*d*) metabolic functions. Adapted from Chrousos & Gold (2).

In several stress-related mood disorders with a hyperactive HPA axis, such as anxiety or melancholic depression, GH and/or IGF-I concentrations are significantly decreased, and the GH response to intravenously administered glucocorticoids is blunted. Compared with healthy control subjects, patients with panic disorder have diminished GH response to intravenously administered clonidine, whereas children with anxiety disorders may have short stature (53, 54). Furthermore, nervous pointer dogs, an animal model of anxiety with both panic and phobic components, have low IGF-I concentrations and deceleration in growth velocity compared with normal animals. The tissue resistance to GH and/or IGF-I of chronically stressed animals can be restored following hypophysectomy or adrenalectomy, a fact that further underlines the importance of glucocorticoids in chronic stress-induced growth suppression (55).

Psychosocial short stature is characterized by severely compromised height in children owing to emotional deprivation and/or physical/psychologic abuse and represents another example of the detrimental effects of a chronically hyperactive stress system on growth. These children display a significant decrease in GH secretion, which is fully restored within a few days following separation of the child from the adverse environment (56, 57). In addition to low GH secretion, they have impaired thyroid function, biochemical findings reminiscent of those of the euthyroid sick syndrome, and a variety of emotional, behavioral and/or psychiatric manifestations.

The inhibited child syndrome usually involves a hyperactive or hyperreactive amygdala, which generates excessive and prolonged fear and anxiety, an activated stress system, which results in the corresponding peripheral physiologic responses, a tachyphylactic or labile mesocorticolimbic dopaminergic system, which generates dysphoria, and/or a hypoactive hippocampus unable to inhibit/limit the activity of the stress system and amygdala (58) (Figure 3). These alterations in the interrelation of the above systems increase the vulnerability of the individual to conditions characterized by a chronically hyperactive or hyperreactive stress, such as chronic anxiety, melancholic depression, eating disorders, substance and alcohol abuse, personality and conduct disorders, as well as psychosomatic conditions, such as chronic fatigue syndrome. Other consequences of hyperactive stress system include delayed growth and puberty, manifestations of the metabolic syndrome, such as visceral obesity, insulin resistance, hypertension, dyslipidemia, cardiovascular disease, and osteoporosis.

Thyroid Function

Thyroid function is also inhibited during stress (Figure 2b). Activation of the HPA axis is associated with decreased production of thyroid-stimulating hormone (TSH), as well as inhibition of peripheral conversion of the relatively inactive thyroxine to the biologically active triiodothyronine (59). These alterations may be due to the increased concentrations of CRH-induced somatostatin and glucocorticoids. Somatostatin suppresses both TRH and TSH, whereas glucocorticoids inhibit the

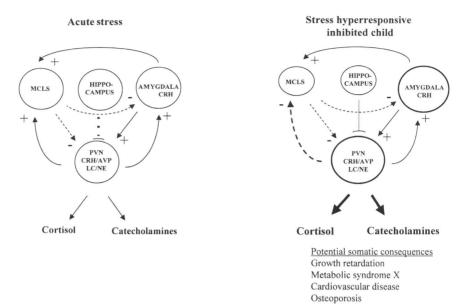

Figure 3   Central neurocircuitry in the stress-hyperresponsive/inhibited child leading to a hyperactive stress system compared with the central neurocircuitry of the normal stress response. The hyperfunctioning amygdala, hypofunctioning hippocampus, and/or hypofunctioning mesocorticolimbic dopaminergic system could predispose an individual to anxiety, melancholic depression, and their somatic consequences. Solid lines represent activation; dashed lines indicate inhibition. Adapted from Chrousos & Gold (58).

activity of the enzyme 5-deiodinase, which converts thyroxine to triiodothyronine. During inflammatory stress, the inflammatory cytokines, such as TNF-α, IL-1, and IL-6, also activate CRH secretion and inhibit 5-deiodinase activity (3, 35).

Reproduction

The reproductive axis is inhibited at all levels by various components of the HPA axis (Figure 2c). CRH suppresses the secretion of gonadotropin-releasing hormone (GnRH) either directly or indirectly, by stimulating the arcuate POMC peptide-secreting neurons (60, 61). Glucocorticoids also exert an inhibitory effect on the GnRH neuron, the pituitary gonadotroph, and the gonads, and render target tissues of gonadal steroids resistant to these hormones (60–63). During inflammatory stress, the elevated concentrations of cytokines also result in suppression of reproductive function via inhibition of both GnRH pulsatile secretion from the hypothalamus and ovarian/testicular steroidogenesis. These effects are exerted both directly and indirectly, by activating hypothalamic neural circuits that secrete CRH and POMC-derived peptides and by increasing the circulating concentrations of glucocorticoids (64).

Suppression of gonadal function secondary to chronic activation of the HPA axis has been demonstrated in highly trained runners of both sexes and ballet dancers (65, 66). These subjects display elevated concentrations of serum cortisol and plasma ACTH in the evening, increased 24-hour urinary-free cortisol excretion, and diminished ACTH responses to exogenous CRH administration. Males have low LH and testosterone concentrations and females have amenorrhea. Interestingly, obligate athletes develop withdrawal symptoms and signs following discontinuation of their exercise routine, which may reflect withdrawal from the daily exercise-induced elevation of opioid peptides and stimulation of the mesocorticolimbic system.

The interaction between CRH and the hypothalamic-pituitary-gonadal axis is bidirectional, given that estrogen increases CRH gene expression via estrogen-response elements in the promoter region of the CRH gene (67). Therefore, the CRH gene is an important target of gonadal steroids and a potential mediator of sex-related differences in the stress-response and the activity of the HPA axis.

Metabolism

In addition to their direct catabolic effects, glucocorticoids also antagonize the actions of GH and sex steroids on fat tissue catabolism (lipolysis) and muscle and bone anabolism (Figure 2d) (3). Chronic activation of the stress system is associated with increased visceral adiposity, decreased lean body (bone and muscle) mass, and suppressed osteoblastic activity, a phenotype observed in patients with Cushing's syndrome, some patients with melancholic depression, and patients with the metabolic syndrome (visceral obesity, insulin resistance, dyslipidemia, hypertension, hypercoagulation, atherosclerotic cardiovascular disease, sleep apnea), many of whom display increased HPA axis activity and demonstrate similar clinical and biochemical manifestations (68–72). The association between chronic stress, hypercortisolism and metabolic syndrome-related manifestations has also been reported in cynomolgus monkeys (70, 71).

Because increased gluconeogenesis is a cardinal feature of the stress response and glucocorticoids induce insulin resistance, activation of the HPA axis may also contribute to the poor control of diabetic patients with emotional stress or concurrent inflammatory or other diseases. Mild, chronic activation of the HPA axis has been demonstrated in type I diabetic patients under moderate or poor glycemic control, and in type II diabetic patients who had developed diabetic neuropathy (71, 73). Over time, progressive glucocorticoid-induced visceral adiposity causes further insulin resistance and deterioration of the glycemic control. Therefore, chronic activation of the stress system in patients with diabetes mellitus may result in a vicious cycle of hyperglycemia, hyperlipidemia, and progressively increasing insulin resistance and insulin requirements.

Low turnover osteoporosis is almost invariably seen in association with hypercortisolism and GH deficiency, and represents another example of the adverse effects of elevated cortisol concentrations and decreased GH/IGF-I concentrations on osteoblastic activity. The stress-induced hypogonadism and the reduced concentrations of sex steroids may further contribute to the development of osteoporosis. Increased prevalence of osteoporosis has been demonstrated in young women with depression or a previous history of depression (74).

Gastrointestinal Function

PVN CRH induces inhibition of gastric acid secretion and emptying, whereas it stimulates colonic motor function (75, 76). These effects are mediated by inhibition of the vagus nerve, which leads to selective inhibition of gastric motility, and by stimulation of the LC-NE-regulated sacral parasympathetic system, which results in selective stimulation of colonic motility. Therefore, CRH may be implicated in mediating the gastric stasis observed following surgery or during an inflammatory process, when central IL-1 concentrations are elevated (77). CRH may also play a role in the stress-induced colonic hypermotility of patients with the irritable bowel syndrome. Colonic contraction and pain in these patients may activate LC-NE-sympathetic neurons, forming a vicious cycle that may account for the chronicity of the condition.

CRH hypersecretion may also be a link between chronic gastrointestinal pain and a history of abuse. A high incidence of physically and sexually abused women has been reported in patients with chronic gastrointestinal pain. Sexually abused women may suffer from chronic activation of the HPA axis (78), and increased CRH secretion may produce colonic pain via activation of the sacral parasympathetic system (79).

Immune Function

Activation of the HPA axis has profound inhibitory effects on the immune/inflammatory response, given that virtually all the components of the immune response are inhibited by glucocorticoids (80, 81). At the cellular level, the main anti-inflammatory and immunosuppressive effects of glucocorticoids include alterations in leukocyte traffic and function, decreases in production of cytokines and mediators of inflammation, and inhibition of their action on target tissues by the latter. These effects are exerted both at the resting, basal state and during inflammatory stress, when the circulating concentrations of glucocorticoids are elevated. A circadian activity of several immune factors has been demonstrated in reverse-phase synchrony with that of plasma glucocorticoid concentrations (82).

During stress, the activated ANS also exerts systemic effects on immune organs by inducing the secretion of IL-6 in the systemic circulation (83). Despite its inherent inflammatory activity, IL-6 plays a major role in the overall control of inflammation by stimulating glucocorticoid secretion (84, 85) and by suppressing the secretion of TNF-α and IL-1. Furthermore, catecholamines inhibit IL-12 and stimulate IL-10 secretion via β-adrenergic receptors, thereby causing suppression of innate and cellular immunity, and stimulation of humoral immunity (86).

The combined effects of glucocorticoids and catecholamines on the monocyte/ macrophage and dendritic cells are to inhibit innate immunity and T helper-1-related cytokines, such as interferon-γ and IL-12, and to stimulate T helper-2-related cytokines, such as IL-10 (87). This suggests that stress-related immuno-suppression refers mostly to innate and cellular immunity, facilitating diseases related to deficiency of these immune responses, such as common cold, tuberculosis, and certain tumors (87).

Psychiatric Disorders

The syndrome of adult melancholic depression represents a typical example of dysregulation of the generalized stress response, leading to chronic dysphoric hyperarousal, activation of the HPA axis and the LC-NE/SNS, and relative im-munosuppression (88, 89). Patients suffering from the condition have hypersecre-tion of CRH, as evidenced by the elevated 24-hour urinary cortisol excretion, the decreased ACTH responses to exogenous CRH administration, and the elevated concentrations of CRH in the cerebrospinal fluid (CSF). They also have elevated concentrations of norepinephrine in the CSF, which remain elevated even during sleep (90), and a marked increase in the number of PVN CRH neurons on autopsy.

Childhood sexual abuse is associated with an increased incidence of adult psy-chopathology, as well as abnormalities in the HPA function. Sexually abused girls have a greater incidence of suicidal ideation, suicide attempts, and dysthymia com-pared with controls (91). In addition, they excrete significantly higher amounts of catecholamines and their metabolites, and display lower basal and CRH-stimulated ACTH concentrations compared with controls. However, the total and free basal and CRH-stimulated serum cortisol concentrations and 24-h urinary-free cortisol concentrations in these subjects are similar to those in controls. These findings reflect pituitary hyporesponsiveness to CRH, which may be corrected for by the presence of intact glucocorticoid feedback regulatory mechanisms (91, 92).

A spectrum of other conditions may also be associated with increased and prolonged activation of the HPA axis. These include anorexia nervosa (93), mal-nutrition (94), obsessive-compulsive disorder, panic anxiety (95), excessive ex-ercise (65, 66), chronic active alcoholism (96), alcohol and narcotic withdrawal (97), diabetes mellitus types I and II (71, 73), visceral obesity (70), and perhaps, hyperthyroidism.

Both anorexia nervosa and malnutrition are characterized by a marked decrease in circulating leptin concentration and an increase in CSF NPY concentration, which could provide an explanation as to why the HPA axis in these subjects is activated in the presence of a profoundly hypoactive LC-NE-sympathetic system (43–46). Glucocorticoids, on the other hand, may produce the hyperphagia and obesity observed in patients with Cushing's syndrome and many rodent models of obesity, such as the Zucker rat, by stimulating NPY and by inhibiting the PVN CRH and the LC-NE sympathetic systems. Glucocorticoids have also been associated with leptin resistance (98). Zucker rats are leptin receptor–deficient with concurrent hypercorticosteronism and decreased LC-NE-sympathetic system activity (99).

EFFECTS OF CHRONIC HYPOACTIVATION
OF THE STRESS SYSTEM

Hypoactivation of the stress system is characterized by chronically reduced secretion of CRH and norepinephrine, and may result in hypoarousal states (Table 2). For example, patients with atypical or seasonal depression and the chronic fatigue syndrome demonstrate chronic hypoactivity of the HPA axis in the depressive (winter) state of the former, and in the period of fatigue of the latter (100). Similarly, patients with fibromyalgia often complain about fatigue and have been shown to have decreased 24-h urinary-free cortisol excretion (101). Hypothyroid patients have clear evidence of CRH hyposecretion, and they often present with depression of the atypical type. Withdrawal from smoking has also been associated with time-limited decreased cortisol and catecholamine secretion, which is associated with fatigue, irritability, and weight gain (102). Decreased CRH secretion in the early period of nicotine abstinence could explain the hyperphagia, decreased metabolic rate, and weight gain frequently observed in these patients. In Cushing's syndrome, the clinical manifestations of atypical depression, hyperphagia, weight gain, fatigue, and anergia are consistent with the suppression of CRH by the elevated cortisol concentrations. The period after cure of hypercortisolism, the postpartum period, and periods after cessation of chronic stress are also associated with suppressed PVN CRH secretion and decreased HPA axis activity (1–3, 62, 103). Chronic hypoactivation of the HPA axis and/or the LC-NE-sympathetic system owing to decreases in the activity of the opioid-peptide system responsible for stress-induced analgesia may also account for the lower pain threshold for visceral sensation reported in patients with functional gastrointestinal disorders.

Hyper- or Hypoactivation of the Stress System
and Immune Function

In theory, an exaggerated HPA axis response to inflammatory stimuli would be expected to mimic the stress or hypercortisolemic state and lead to increased susceptibility of the individual to certain infectious agents or tumors but enhanced resistance to autoimmune inflammatory disease. By contrast, a suboptimal HPA axis response to inflammatory stimuli would be expected to reproduce the glucocorticoid-deficient state and lead to relative resistance to infections and neoplastic diseases but increased susceptibility to autoimmune inflammatory disease (80, 87). These findings have been observed in an interesting pair of near-histocompatible, highly inbred rat strains, the Fischer and Lewis rats, both of which were genetically selected out of Sprague-Dawley rats, for their resistance or susceptibility, respectively, to inflammatory disease (104).

Patients with depression or anxiety have been shown to be more vulnerable to tuberculosis, both in terms of prevalence and severity of the disease (105). Similarly, stress has been associated with increased vulnerability to the common cold virus. A compromised innate and T helper-1 driven immunity may predispose an individual to these conditions. Furthermore, patients with rheumatoid arthritis, a T-helper-1 driven inflammatory disease, have a mild form of central hypocortisolism, as indicated by the normal 24-h cortisol excretion despite the major inflammatory stress, and diminished HPA axis responses to surgical stress (106). Therefore, dysregulation of the HPA axis may play a critical role in the development and/or perpetuation of T helper-1-type of autoimmune disease. The same theoretical concept may explain the high incidence of T helper-1 autoimmune diseases, such as rheumatoid arthritis and multiple sclerosis, observed following cure of hypercortisolism, in the postpartum period and in patients with adrenal insufficiency, who do not receive adequate replacement therapy (87, 107, 108).

GENETICS, DEVELOPMENT, ENVIRONMENT, AND THE STRESS RESPONSE

Appropriate responsiveness of the stress system to stressors is a crucial prerequisite for a sense of well-being, adequate performance of tasks, and positive social interactions. Improper responsiveness has been associated with inadequacies in these functions and increased vulnerability to one or more of the stress-related states. Vulnerability may be the result of genetic, developmental, and environmental factors, and may be considered as the endpoint of converging influences. Depending on the genetic background of the individual and his/her exposure to adverse stimuli in prenatal and/or postnatal life (developmental influences), one might fail to cope with life stressors and may develop any of the above-described states in any combination and any degree of severity (58).

The stress response of an individual is determined by multiple factors, many of which are inherited (1, 3, 109, 110). Genetic polymorphisms, such as those of CRH, AVP, and their receptors and/or regulators, are expected to account for the observed variability in the function of the stress system. This genetic vulnerability is polygenic and allows expression of the clinical phenotype in the presence of environmental triggers. There is a complex genetic background continuum in our population that ranges from extreme resilience to extreme vulnerability to these stress-related comorbid states. Stressors of gradually decreasing intensity may be sufficient to result in the development of these conditions in an individual, whose genetic vulnerability places him on the vulnerable side of the continuum (Figure 4a).

The dose-response relation between the potency of a stressor and the responsiveness of the stress system is represented by a sigmoidal curve, which is expected to differ from individual to individual. One individual's dose-response curve might be shifted to the left of that of an average reactive individual, whereas another individual's dose-response curve might be shifted to the right. The former denotes an excessive reaction, whereas the latter a defective one. Similarly, the dose-response relation between an individual's sense of well-being or performance

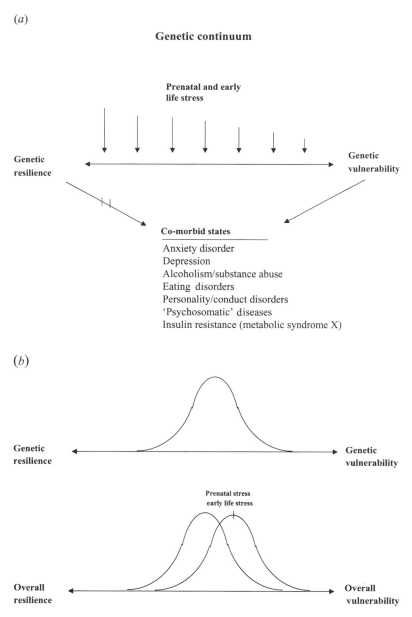

Figure 4 (*a*) Schematic representation of the genetic continuum that defines an individual's genetically determined vulnerability/resilience to stressors. The vertical arrows indicate the magnitude of environmental stressors necessary to result in disease. (*b*) Early environmental stressors may have a permanent effect on the ability of the individual to respond to stress effectively, thus altering the constitutional vulnerability/resilience of an individual to stressors. Adapted from Chrousos (58a).

ability and the activity of the stress system is represented by an inverted U-shaped curve that covers the range of the activity of the latter. Shifts to either the left or the right of this range would result in hypoarousal or hyperarousal, respectively, and a suboptimal sense of well-being or diminished performance (58). Developmental influences, when propitious, may shift an individual toward a more resilient response to stress, whereas, when negative, may have the opposite effect (Figure 4*b*). Therefore, a supportive or an adverse environment may alter the course of one or more of the above stress-related states, indicating that genetics and development define vulnerability, whereas environment may determine the triggering and/or severity of a disease.

The prenatal life, infancy, childhood, and adolescence are periods of increased plasticity for the stress system and, therefore, are particularly sensitive to stressors. Excessive or sustained activation of the stress system during these critical periods may have profound effects on its function (1, 3, 111, 112). These environmental triggers or stressors may have not a transient, but rather a permanent effect on the organism, reminiscent of the organizational effects of several hormones exerted on certain target tissues, which last long after cessation of the exposure to these hormones. Also, sufficiently strong or prolonged stressors may have permanent effects on the organism even if they occur later in life, such as in the adult post-traumatic stress disorders.

These effects of early environment on the development of the HPA axis responses to stress reflect a naturally occurring plasticity whereby factors such as maternal care are able to program rudimentary, biologic responses to threatening stimuli. Developmental programming of CNS responses to stress in early life is likely to be of adaptive value to the adult. Such programming would afford an appropriate HPA response and would minimize the need for a long period of adaptation in adult life.

CONCLUSIONS

The stress system coordinates the adaptive response of the organism to stressors and plays an important role in maintenance of basal and stress-related homeostasis. Activation of the stress system leads to behavioral and peripheral changes that improve the ability of the organism to adapt and increase its chances for survival. Inadequate and/or prolonged response to stressors may impair growth and development and may result in a variety of endocrine, metabolic, autoimmune, and psychiatric disorders. The development and severity of these conditions primarily depend on genetic, developmental, and environmental factors. CRH antagonists may be useful in states characterized by chronic hyperactivity of the stress system, such as melancholic depression and chronic anxiety, whereas CRH agonists may be useful in conditions characterized by chronic hypoactivity of the stress system, such as atypical depression, postpartum depression, and the fibromyalgia/chronic fatigue syndromes (113–116).

The *Annual Review of Physiology* is online at
http://physiol.annualreviews.org

LITERATURE CITED

1. Habib KE, Gold PW, Chrousos GP. 2001. Neuroendocrinology of stress. *Endocrinol. Metab. Clin. North Am.* 30:695–728

2. Chrousos GP, Gold PW. 1992. The concepts of stress and stress system disorders. Overview of physical and behavioral homeostasis. *JAMA* 267:1244–52

3. Chrousos GP. 2002. Organization and integration of the endocrine system. In *Pediatric Endocrinology*, ed. M Sperling, pp. 1-14. Philadelphia: Saunders

4. Calogero AE, Gallucci WT, Chrousos GP, Gold PW. 1988. Catecholamine effects upon rat hypothalamic corticotropin-releasing hormone secretion in vitro. *J. Clin. Invest.* 82:839–46

5. Valentino RJ, Foote SL, Aston-Jones G. 1983. Corticotropin-releasing factor activates noradrenergic neurons of the locus coeruleus. *Brain Res.* 270:363–67

6. Kiss A, Aguilera G. 1992. Participation of alpha 1-adrenergic receptors in the secretion of hypothalamic corticotropin-releasing hormone during stress. *Neuroendocrinology* 56:153–60

7. Calogero AE, Gallucci WT, Gold PW, Chrousos GP. 1988. Multiple feedback regulatory loops upon rat hypothalamic corticotropin-releasing hormone secretion. Potential clinical implications. *J. Clin. Invest.* 82:767–74

8. Silverman AJ, Hou-Yu A, Chen WP. 1989. Corticotropin-releasing factor synapses within the paraventricular nucleus of the hypothalamus. *Neuroendocrinology* 49:291–99

9. Aghajanian GK, VanderMaelen CP. 1982. Alpha 2-adrenoceptor-mediated hyperpolarization of locus coeruleus neurons: intracellular studies in vivo. *Science* 215: 1394–96

10. Calogero AE, Bagdy G, Szemeredi K, Tartaglia ME, Gold PW, Chrousos GP. 1990. Mechanisms of serotonin receptor agonist-induced activation of the hypothalamic-pituitary-adrenal axis in the rat. *Endocrinology* 126:1888–94

11. Fuller RW. 1996. Serotonin receptors involved in regulation of pituitary-adrenocortical function in rats. *Behav. Brain Res.* 73:215–19

12. Calogero AE, Gallucci WT, Chrousos GP, Gold PW. 1988. Interaction between GABAergic neurotransmission and rat hypothalamic corticotropin-releasing hormone secretion in vitro. *Brain Res.* 463:28–36

13. Overton JM, Fisher LA. 1989. Modulation of central nervous system actions of corticotropin-releasing factor by dynorphin-related peptides. *Brain Res.* 488(1–2):233–40

14. Keller-Wood ME, Dallman MF. 1984. Corticosteroid inhibition of ACTH secretion. *Endocr. Rev.* 5:1–24

15. Bale TL, Vale WW. 2004. CRF and CRF receptors: role in stress responsivity and other behaviors. *Annu. Rev. Pharmacol. Toxicol.* 44:525–57

16. Egawa M, Yoshimatsu H, Bray GA. 1991. Neuropeptide Y suppresses sympathetic activity to interscapular brown adipose tissue in rats. *Am. J. Physiol. Regul. Integr. Comp. Physiol.* 260:R328–34

17. Oellerich WF, Schwartz DD, Malik KU. 1994. Neuropeptide Y inhibits adrenergic transmitter release in cultured rat superior cervical ganglion cells by restricting the availability of calcium through a pertussis toxin-sensitive mechanism. *Neuroscience* 60:495–502

18. White BD, Dean RG, Edwards GL, Martin RJ. 1994. Type II corticosteroid receptor

stimulation increases NPY gene expression in basomedial hypothalamus of rats. *Am. J. Physiol. Regul. Integr. Comp. Physiol.* 266:R1523–29

19. Larsen PJ, Jessop D, Patel H, Lightman SL, Chowdrey HS. 1993. Substance P inhibits the release of anterior pituitary adrenocorticotrophin via a central mechanism involving corticotrophin-releasing factor-containing neurons in the hypothalamic paraventricular nucleus. *J. Neuroendocrinol.* 5:99–105

20. Chrousos GP. 1992. Regulation and dysregulation of the hypothalamic-pituitary-adrenal axis: the corticotropin releasing hormone perspective. *Endocrinol. Metab. Clin. North Am.* 21:833–58

21. Tsigos C, Chrousos GP. 1994. Physiology of the hypothalamic-pituitary-adrenal axis in health and dysregulation in psychiatric and autoimmune disorders. *Endocrinol. Metab. Clin. North Am.* 23:451–66

22. Gillies GE, Linton EA, Lowry PJ. 1982. Corticotropin releasing activity of the new CRF is potentiated several times by vasopressin. *Nature* 299:355–57

23. Abou-Samra AB, Harwood JP, Catt KJ, Aguilera G. 1987. Mechanisms of action of CRF and other regulators of ACTH release in pituitary corticotrophs. *Ann. NY Acad. Sci.* 512:67–84

24. Horrocks PM, Jones AF, Ratcliffe WA, Holder G, White A, et al. 1990. Patterns of ACTH and cortisol pulsatility over twenty-four hours in normal males and females. *Clin. Endocrinol. (Oxf).* 32(1):127–34

25. Iranmanesh A, Lizarralde G, Short D, Veldhuis JD. 1990. Intensive venous sampling paradigms disclose high frequency adrenocorticotropin release episodes in normal men. *J. Clin. Endocrinol. Metab.* 71(5):1276–83

26. Veldhuis JD, Iranmanesh A, Johnson ML, Lizarralde G. 1990. Amplitude, but not frequency, modulation of adrenocorticotropin secretory bursts gives rise to the nyctohemeral rhythm of the corticotropic axis in man. *J. Clin. Endocrinol. Metab.* 71:452–63

27. Calogero AE, Norton JA, Sheppard BC, Listwak SJ, Cromack DT, et al. 1992. Pulsatile activation of the hypothalamic-pituitary-adrenal axis during major surgery. *Metabolism* 41:839–45

28. Holmes MC, Antoni FA, Aguilera G, Catt KJ. 1986. Magnocellular axons in passage through the median eminence release vasopressin. *Nature* 319:326–29

29. Andreis PG, Neri G, Mazzocchi G, Musajo F, Nussdorfer GG. 1992. Direct secretagogue effect of corticotropin-releasing factor on the rat adrenal cortex: the involvement of the zona medullaris. *Endocrinology* 131:69–72

30. Ottenweller JE, Meier AH. 1982. Adrenal innervation may be an extrapituitary mechanism able to regulate adrenocortical rhythmicity in rats. *Endocrinology* 111:1334–38

31. Bornstein SR, Chrousos GP. 1999. Clinical review 104: adrenocorticotropin (ACTH)- and non-ACTH-mediated regulation of the adrenal cortex: neural and immune inputs. *J. Clin. Endocrinol. Metab.* 84:1729–36

32. Munck A, Guyre PM, Holbrook NJ. 1984. Physiological functions of glucocorticoids in stress and their relation to pharmacological actions. *Endocr. Rev.* 5:25–44

33. Kino T, Chrousos GP. 2001. Glucocorticoid and mineralocorticoid resistance/hypersensitivity syndromes. *J. Endocrinol.* 169:437–45

34. Bamberger CM, Schulte HM, Chrousos GP. 1996. Molecular determinants of glucocorticoid receptor function and tissue sensitivity to glucocorticoids. *Endocr. Rev.* 17:245–61

35. Chrousos GP. 1997. The neuroendocrinology of stress: Its relation to the hormonal milieu, growth and development. *Growth Genet. Horm.* 13:1–8

36. Roth RH, Tam SY, Ida Y, Yang JX, Deutch AY. 1988. Stress and the meso-corticolimbic dopamine systems. *Ann. NY Acad. Sci.* 537:138–47

37. Imperato A, Puglisi-Allegra S, Casolini P, Angelucci L. 1991. Changes in brain dopamine and acetylcholine release during and following stress are independent of the pituitary-adrenocortical axis. *Brain Res.* 538:111–17

38. Diorio D, Viau V, Meaney MJ. 1993. The role of the medial prefrontal cortex (cingulate gyrus) in the regulation of hypothalamic-pituitary-adrenal responses to stress. *J. Neurosci.* 13:3839–47

39. Gray TS. 1989. Amygdala: role in autonomic and neuroendocrine responses to stress. In *Stress, Neuropeptides and Systemic Disease*, ed. JA McCubbin, PG Kaufman, CB Nemeroff, p. 37. New York: Academic

40. Nikolarakis KE, Almeida OF, Herz A. 1986. Stimulation of hypothalamic beta-endorphin and dynorphin release by corticotropin-releasing factor in vitro. *Brain Res.* 399:152–55

41. Diamant M, de Wied D. 1991. Autonomic and behavioral effects of centrally administered corticotropin-releasing factor in rats. *Endocrinology* 129:446–54

42. Mora F, Lee TF, Myers RD. 1983. Involvement of alpha- and beta-adreno-receptors in the central action of norepinephrine on temperature, metabolism, heart and respiratory rates of the conscious primate. *Brain Res. Bull.* 11:613–16

43. Liu JP, Clarke IJ, Funder JW, Engler D. 1994. Studies of the secretion of corticotropin-releasing factor and arginine vasopressin into the hypophyseal-portal circulation of the conscious sheep. II. The central noradrenergic and neuropeptide Y pathways cause immediate and prolonged hypothalamic-pituitary-adrenal activation. Potential involvement in the pseudo-Cushing's syndrome of endogenous depression and anorexia nervosa. *J. Clin. Invest.* 93:1439–50

44. Egawa M, Yoshimatsu H, Bray GA. 1991. Neuropeptide Y suppresses sympathetic activity to interscapular brown adipose tissue in rats. *Am. J. Physiol. Regul. Integr. Comp. Physiol.* 260:R328–34

45. Rahmouni K, Haynes WG. 2001. Leptin signaling pathways in the central nervous system: interactions between neuropeptide Y and melanocortins. *BioEssays* 23:1095–99

46. Raposinho PD, Pierroz DD, Broqua P, White RB, Pedrazzini T, Aubert ML. 2001. Chronic administration of neuropeptide Y into the lateral ventricle of C57BL/6J male mice produces an obesity syndrome including hyperphagia, hyperleptinemia, insulin resistance, and hypogonadism. *Mol. Cell. Endocrinol.* 185:195–204

47. Burguera B, Muruais C, Penalva A, Dieguez C, Casanueva FF. 1990. Dual and selective actions of glucocorticoids upon basal and stimulated growth hormone release in man. *Neuroendocrinology* 51:51–58

48. Magiakou MA, Mastorakos G, Gomez MT, Rose SR, Chrousos GP. 1994. Suppressed spontaneous and stimulated growth hormone secretion in patients with Cushing's disease before and after surgical cure. *J. Clin. Endocrinol. Metab.* 78(1):131–37

49. Magiakou MA, Mastorakos G, Chrousos GP. 1994. Final stature in patients with endogenous Cushing's syndrome. *J. Clin. Endocrinol. Metab.* 79:1082–85

50. Bamberger CM, Schulte HM, Chrousos GP. 1996. Molecular determinants of glucocorticoid receptor function and tissue sensitivity to glucocorticoids. *Endocr. Rev.* 17:245–61

51. Vottero A, Kimchi-Sarfaty C, Kratzsch J, Chrousos GP, Hochberg Z. 2003. Transcriptional and translational regulation of the splicing isoforms of the growth hormone receptor by glucocorticoids. *Horm. Metab. Res.* 35:7–12

52. Raza J, Massoud AF, Hindmarsh PC, Robinson IC, Brook CG. 1998. Direct effects of corticotrophin-releasing hormone on stimulated growth hormone secretion. *Clin. Endocrinol.* 48:217–22

53. Uhde TW, Tancer ME, Rubinow DR, Roscow DB, Boulenger JP, et al. 1992. Evidence for hypothalamo-growth hormone dysfunction in panic disorder: profile of growth hormone (GH) responses to clonidine, yohimbine, caffeine, glucose, GRF and TRH in panic disorder patients versus healthy volunteers. *Neuropsychopharmacology* 6:101–18

54. Abelson JL, Glitz D, Cameron OG, Lee MA, Bronzo M, Curtis GC. 1991. Blunted growth hormone response to clonidine in patients with generalized anxiety disorder. *Arch. Gen. Psychiatry* 48:157–62

55. Rodgers BD, Lau AO, Nicoll CS. 1994. Hypophysectomy or adrenalectomy of rats with insulin-dependent diabetes mellitus partially restores their responsiveness to growth hormone. *Proc. Soc. Exp. Biol. Med.* 207:220–26

56. Skuse D, Albanese A, Stanhope R, Gilmour J, Voss L. 1996. A new stress-related syndrome of growth failure and hyperphagia in children, associated with reversibility of growth-hormone insufficiency. *Lancet* 348:353–58

57. Albanese A, Hamill G, Jones J, Skuse D, Matthews DR, Stanhope R. 1994. Reversibility of physiological growth hormone secretion in children with psychosocial dwarfism. *Clin. Endocrinol.* 40:687–92

58. Chrousos GP, Gold PW. 1999. The inhibited child syndrome. In *Origins, Biological Mechanisms and Clinical Outcomes*, ed. LA Schmidt, J Schulkin, pp. 193–200. New York: Oxford Univ. Press

58a. Chrousos GP. 1997. The future of pediatric and adolescent endocrinology. *Ann. NY Acad. Sci.* 816:4–8

59. Benker G, Raida M, Olbricht T, Wagner R, Reinhardt W, Reinwein D. 1990. TSH secretion in Cushing's syndrome: relation to glucocorticoid excess, diabetes, goitre, and the 'sick euthyroid syndrome'. *Clin. Endocrinol.* 33:777–86

60. Rivier C, Rivier J, Vale W. 1986. Stress-induced inhibition of reproductive functions: role of endogenous corticotropin-releasing factor. *Science* 231:607–9

61. Vamvakopoulos NC, Chrousos GP. 1994. Hormonal regulation of human corticotropin-releasing hormone gene expression: implications for the stress response and immune/inflammatory reaction. *Endocr. Rev.* 15:409–20

62. Chrousos GP, Torpy DJ, Gold PW. 1998. Interactions between the hypothalamic-pituitary-adrenal axis and the female reproductive system: clinical implications. *Ann. Intern. Med.* 129:229–40

63. MacAdams MR, White RH, Chipps BE. 1986. Reduction of serum testosterone levels during chronic glucocorticoid therapy. *Ann. Intern. Med.* 104:648–51

64. Pau KY, Spies HG. 1997. Neuroendocrine signals in the regulation of gonadotropin-releasing hormone secretion. *Chin. J. Physiol.* 40:181–96

65. Luger A, Deuster PA, Kyle SB, Gallucci WT, Montgomery LC, et al. 1987. Acute hypothalamic-pituitary-adrenal responses to the stress of treadmill exercise. Physiologic adaptations to physical training. *N. Engl. J. Med.* 316:1309–15

66. MacConnie SE, Barkan A, Lampman RM, Schork MA, Beitins IZ. 1986. Decreased hypothalamic gonadotropin-releasing hormone secretion in male marathon runners. *N. Engl. J. Med.* 315:411–17

67. Vamvakopoulos NC, Chrousos GP. 1993. Evidence of direct estrogenic regulation of human corticotropin-releasing hormone gene expression. Potential implications for the sexual dimophism of the stress response and immune/inflammatory reaction. *J. Clin. Invest.* 92:1896–902

68. Gold PW, Loriaux DL, Roy A, Kling MA, Calabrese JR, et al. 1986. Responses

to corticotropin-releasing hormone in the hypercortisolism of depression and Cushing's disease. Pathophysiologic and diagnostic implications. *N. Engl. J. Med.* 314:1329–35

69. Pasquali R, Cantobelli S, Casimirri F, Capelli M, Bortoluzzi L, et al. 1993. The hypothalamic-pituitary-adrenal axis in obese women with different patterns of body fat distribution. *J. Clin. Endocrinol. Metab.* 77:341–46

70. Chrousos GP. 2000. The role of stress and the hypothalamic-pituitary-adrenal axis in the pathogenesis of the metabolic syndrome: neuro-endocrine and target tissue-related causes. *Int. J. Obes. Relat. Metab. Disord.* 249(Suppl.)2:S50–55

71. Roy MS, Roy A, Gallucci WT, Collier B, Young K, et al. 1993. The ovine corticotropin-releasing hormone-stimulation test in type I diabetic patients and controls: suggestion of mild chronic hypercortisolism. *Metabolism* 42:696–700

72. Gold PW, Chrousos GP. 2002. Organization of the stress system and its dysregulation in melancholic and atypical depression: high vs low CRH/NE states. *Mol. Psychiatry* 7:254–75

73. Tsigos C, Young RJ, White A. 1993. Diabetic neuropathy is associated with increased activity of the hypothalamic-pituitary-adrenal axis. *J. Clin. Endocrinol. Metab.* 76:554–58

74. Michelson D, Stratakis C, Hill L, Reynolds J, Galliven E. 1996. Bone mineral density in women with depression. *N. Engl. J. Med.* 335:1176–81

75. Tache Y, Monnikes H, Bonaz B, Rivier J. 1993. Role of CRF in stress-related alterations of gastric and colonic motor function. *Ann. NY Acad. Sci.* 697:233–43

76. Habib KE, Weld KP, Rice KC, Pushkas J, Champoux M, et al. 2000. Oral administration of a corticotropin-releasing hormone receptor antagonist significantly attenuates behavioral, neuroendocrine, and autonomic responses to stress in pri-

mates. *Proc. Natl. Acad. Sci. USA* 97: 6079–84

77. Suto G, Kiraly A, Tache Y. 1994. Interleukin 1 beta inhibits gastric emptying in rats: mediation through prostaglandin and corticotropin-releasing factor. *Gastroenterology* 106:1568–75

78. Heim C, Newport DJ, Heit S, Graham YP, Wilcox M, et al. 2000. Pituitary-adrenal and autonomic responses to stress in women after sexual and physical abuse in childhood. *JAMA* 284:592–97

79. Drossman DA, Leserman J, Nachman G, Li ZM, Gluck H, et al. 1990. Sexual and physical abuse in women with functional or organic gastrointestinal disorders. *Ann. Intern. Med.* 113:828–33

80. Chrousos GP. 1995. The hypothalamic-pituitary-adrenal axis and immune-mediated inflammation. *N. Engl. J. Med.* 332:1351–62

81. Boumpas DT, Chrousos GP, Wilder RL, Cupps TR, Balow JE. 1993. Glucocorticoid therapy for immune-mediated diseases: basic and clinical correlates. *Ann. Intern. Med.* 119:1198–208

82. DeRijk R, Michelson D, Karp B, Petrides J, Galliven E, et al. 1997. Exercise and circadian rhythm-induced variations in plasma cortisol differentially regulate interleukin-1 beta (IL-1 beta), IL-6, and tumor necrosis factor-alpha (TNF alpha) production in humans: high sensitivity of TNF alpha and resistance of IL-6. *J. Clin. Endocrinol. Metab.* 82: 2182–91

83. Van Gool J, van Vugt H, Helle M, Aarden LA. 1990. The relation among stress, adrenalin, interleukin 6 and acute phase proteins in the rat. *Clin. Immunol. Immunopathol.* 57:200–10

84. Mastorakos G, Weber JS, Magiakou MA, Gunn H, Chrousos GP. 1994. Hypothalamic-pituitary-adrenal axis activation and stimulation of systemic vasopressin secretion by recombinant interleukin-6 in humans: potential implications for the syndrome of inappropriate

vasopressin secretion. *J. Clin. Endocrinol. Metab.* 79:934–39

85. Mastorakos G, Chrousos GP, Weber JS. 1993. Recombinant interleukin-6 activates the hypothalamic-pituitary-adrenal axis in humans. *J. Clin. Endocrinol. Metab.* 77:1690–94

86. Elenkov IJ, Papanicolaou DA, Wilder RL, Chrousos GP. 1996. Modulatory effects of glucocorticoids and catecholamines on human interleukin-12 and interleukin-10 production: clinical implications. *Proc. Assoc. Am. Phys.* 108: 374–81

87. Elenkov IJ, Chrousos GP. 1999. Stress hormones, Th1/Th2 patterns, pro/antiinflammatory cytokines and susceptibility to disease. *Trends Endocrinol. Metab.* 10:359–68

88. Gold PW, Goodwin FK, Chrousos GP. 1988. Clinical and biochemical manifestations of depression. Relation to the neurobiology of stress (1). *N. Engl. J. Med.* 319:348–53

89. Gold PW, Goodwin FK, Chrousos GP. 1988. Clinical and biochemical manifestations of depression. Relation to the neurobiology of stress (2). *N. Engl. J. Med.* 319:413–20

90. Wong ML, Kling MA, Munson PJ, Listwak S, Licinio J, et al. 2000. Pronounced and sustained central hypernoradrenergic function in major depression with melancholic features: relation to hypercortisolism and corticotropin-releasing hormone. *Proc. Natl. Acad. Sci. USA* 97: 325–30

91. De Bellis MD, Chrousos GP, Dorn LD, Burke L, Helmers K, et al. 1994. Hypothalamic-pituitary-adrenal axis dysregulation in sexually abused girls. *J. Clin. Endocrinol. Metab.* 78:249–55

92. De Bellis MD, Lefter L, Trickett PA, Putnam FW Jr. 1994. Urinary catecholamine excretion in sexually abused girls. *J. Am. Acad. Child Adolesc. Psychiatry* 33:320–27

93. Kaye WH, Gwirtsman HE, George DT,

Ebert MH, Jimerson DC, et al. 1987. Elevated cerebrospinal fluid levels of immunoreactive corticotropin-releasing hormone in anorexia nervosa: relation to state of nutrition, adrenal function, and intensity of depression. *J. Clin. Endocrinol. Metab.* 64:203–8

94. Malozowski S, Muzzo S, Burrows R, Leiva L, Loriaux L, et al. 1990. The hypothalamic-pituitary-adrenal axis in infantile malnutrition. *Clin. Endocrinol.* 32:461–65

95. Gold PW, Pigott TA, Kling MA, Kalogeras K, Chrousos GP. 1988. Basic and clinical studies with corticotropin-releasing hormone. Implications for a possible role in panic disorder. *Psychiatr. Clin. North Am.* 11:327–34

96. Wand GS, Dobs AS. 1991. Alterations in the hypothalamic-pituitary-adrenal axis in actively drinking alcoholics. *J. Clin. Endocrinol. Metab.* 72:1290–95

97. Von Bardeleben U, Heuser I, Holsboer F. 1989. Human CRH stimulation response during acute withdrawal and after medium-term abstention from alcohol abuse. *Psychoneuroendocrinology* 14:441–49

98. Jeanrenaud B, Rohner-Jeanrenaud F. 2000. CNS-periphery relationships and body weight homeostasis: influence of the glucocorticoid status. *Int. J. Obes. Relat. Metab. Disord.* (Suppl.)2:S74–76

99. Pacak K, McCarty R, Palkovits M, Cizza G, Kopin IJ, et al. 1995. Decreased central and peripheral catecholaminergic activation in obese Zucker rats. *Endocrinology* 136:4360–67

100. Ehlert U, Gaab J, Heinrichs M. 2001. Psychoneuroendocrinological contributions to the etiology of depression, posttraumatic stress disorder, and stress-related bodily disorders: the role of the hypothalamus-pituitary-adrenal axis. *Biol. Psychol.* 57:141–52

101. Demitrack MA, Crofford LJ. 1998. Evidence for and pathophysiologic implications of hypothalamic-pituitary-adrenal

axis dysregulation in fibromyalgia and chronic fatigue syndrome. *Ann. NY Acad. Sci.* 840:684–97

102. Puddey IB, Vandongen R, Beilin LJ, English D. 1984. Haemodynamic and neuroendocrine consequences of stopping smoking—a controlled study. *Clin. Exp. Pharmacol. Physiol.* 11:423–26

103. Gomez MT, Magiakou MA, Mastorakos G, Chrousos GP. 1993 The pituitary corticotroph is not the rate-limiting step in the postoperative recovery of the hypothalamic-pituitary-adrenal axis in patients with Cushing syndrome. *J. Clin. Endocrinol. Metab.* 77:173–77

104. Sternberg EM, Hill JM, Chrousos GP, Kamilaris T, Listwak SJ, et al. 1989. Inflammatory mediator-induced hypothalamic-pituitary-adrenal axis activation is defective in streptococcal cell wall arthritis-susceptible Lewis rats. *Proc. Natl. Acad. Sci. USA* 86:2374–78

105. Petrich J, Holmes TH. 1977. Life change and onset of illness. *Med. Clin. North Am.* 61:825–38

106. Chikanza IC, Petrou P, Kingsley G, Chrousos G, Panayi GS. 1992. Defective hypothalamic response to immune and inflammatory stimuli in patients with rheumatoid arthritis. *Arthritis Rheum.* 35:1281–88

107. Magiakou MA, Mastorakos G, Rabin D, Dubbert B, Gold PW, Chrousos GP. 1996. Hypothalamic corticotropin-releasing hormone suppression during the postpartum period: implications for the increase in psychiatric manifestations at this time. *J. Clin. Endocrinol. Metab.* 81:1912–17

108. Elenkov IJ, Wilder RL, Bakalov VK, Link AA, Dimitrov MA, et al. 2001. IL-12, TNF-alpha, and hormonal changes during late pregnancy and early postpartum: implications for autoimmune disease activ-

ity. *J. Clin. Endocrinol. Metab.* 86:4933–38

109. Plomin R, Owen MJ, McGuffin P. 1994. The genetic basis of complex human behaviors. *Science* 264:1733–39

110. Bouchard TJ Jr. 1994. Genes, environment, and personality. *Science* 264:1700–1

111. Suomi SJ. 1991. Early stress and adult emotional reactivity in rhesus monkeys. *Ciba Found Symp.* 156:171–83; discussion 183–88

112. Goland RS, Jozak S, Warren WB, Conwell IM, Stark RI, Tropper PJ. 1993. Elevated levels of umbilical cord plasma corticotropin-releasing hormone in growth-retarded fetuses. *J. Clin. Endocrinol. Metab.* 77:1174–79

113. Webster EL, Lewis DB, Torpy DJ, Zachman EK, Rice KC, Chrousos GP. 1996. In vivo and in vitro characterization of antalarmin, a nonpeptide corticotropin-releasing hormone (CRH) receptor antagonist: suppression of pituitary ACTH release and peripheral inflammation. *Endocrinology* 137:5747–50

114. Bornstein SR, Webster EL, Torpy DJ, Richman SJ, Mitsiades N, et al. 1998. Chronic effects of a nonpeptide corticotropin-releasing hormone type I receptor antagonist on pituitary-adrenal function, body weight, and metabolic regulation. *Endocrinology* 139:1546–55

115. Habib KE, Weld KP, Rice KC, Pushkas J, Champoux M, et al. 2000. Oral administration of a corticotropin-releasing hormone receptor antagonist significantly attenuates behavioral, neuroendocrine, and autonomic responses to stress in primates. *Proc. Natl. Acad. Sci. USA* 97:6079–84

116. Grammatopoulos DK, Chrousos GP. 2002. Functional characteristics of CRH receptors and potential clinical applications of CRH-receptor antagonists. *Trends Endocrinol. Metab.* 13:436–44

Annu. Rev. Physiol. 2005. 67:285–308
doi: 10.1146/annurev.physiol.67.040403.115914
First published online as a Review in Advance on September 28, 2004

LESSONS IN ESTROGEN BIOLOGY FROM KNOCKOUT AND TRANSGENIC ANIMALS

Sylvia C. Hewitt, Joshua C. Harrell,[1] and Kenneth S. Korach

Receptor Biology Section, Laboratory of Reproductive and Developmental Toxicology, National Institute of Environmental Health Sciences, Research Triangle Park, North Carolina 277009
[1]Current address: University of Colorado Health Sciences Center, Denver, Colorado 80262; email: curtiss@niehs.nih.gov; Joshua.Harrell@UCHSC.edu; korach@niehs.nih.gov

Key Words estrogen receptor, mammary gland, uterus, estrogen mechanisms

■ **Abstract** Tremendous progress has been made in elucidating numerous critical aspects of estrogen signaling. New tools and techniques have enabled detailed molecular analysis of components that direct estrogen responses. At the other end of the spectrum, generation of a multiplicity of transgenic animals has allowed analysis of the physiological roles of the estrogen-signaling components in biologically relevant models. Here, we review the ever-increasing body of knowledge in the field of estrogen biology, especially as applied to the female reproductive processes.

THE ESTROGEN RECEPTOR MOLECULE: STRUCTURES LEAD TO MECHANISMS

Estrogens are essential hormones for successful reproduction in mammals. Although produced locally by the ovary, estrogens circulate systemically and exert selective effects on target tissues. This is mediated by the presence of estrogen receptors (ER), estrogen-regulated transcription factors. Estrogen receptors are members of a family of nuclear transcription factors including receptors for sex steroids, thyroid hormone, vitamin D, retinoids, as well as many orphan receptors, for which no ligands have been identified (1). A second ER gene was cloned from prostate tissue in 1996 (2), and thus there are two ER molecules: the originally described ERα and now the ERβ. By comparing the ER sequences, it is apparent that both share a general domain structure common to ligand-modulated nuclear transcription factors (3). The current understanding of ER mechanisms of action can be summarized with reference to the overall structure of the receptors (Figure 1). The functions of some regions of the ER molecules have been defined using deletion and mutation as well as structural analysis, as described by Nettles & Greene in this volume (3a). The best-characterized functions

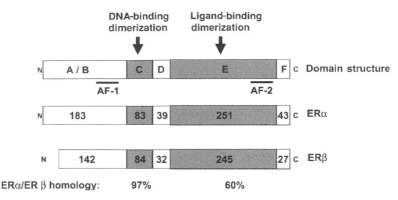

Figure 1 Comparison of domain structures of ERα and ERβ. The estrogen receptors are members of the nuclear receptor superfamily and share a domain structure, which is depicted schematically. The ERs have six domains, A–F, and the number of amino acids in these domains, as well as the functions associated with these domains, are indicated for each form of ER. AF-1 and AF-2 refer to regions that mediate the transcriptional activation functions of the ERs. The degree of homology between ERα and ERβ in the (C) and estrogen domains are indicated below these domains. The ERβ domains are derived from SeqWeb GAP analysis of mouse ERα amino acids 1–599 compared with mouse ERβ 1–530. Reproduced from (26).

include a zinc-finger-containing domain (C domain, Figure 1), which binds with high affinity and specificity to estrogen response elements (EREs) in target genes, and a ligand-binding domain (domain E), which binds estrogen as well as other estrogenic ligands. The consensus ERE is a 13-base pair inverted repeat sequence (GGTCAnnnTGACC); however, the majority of ERE sequences contain one or more variations from the consensus. In vitro DNA-binding studies have indicated that the ER binds as a dimer (4), with one ER molecule contacting each 5-base pair-inverted repeat (5). Although DNA binding is a dimerization stimulus, sequences in the ligand-binding domain are also involved in dimerization (4), and crystallized truncated ER containing only the ligand-binding domain is clearly shown to be a dimer in the presence of an agonist ligand (6).

The AF-1 region in the amino terminus and the AF-2 region within the ligand-binding domain are involved in ligand-independent and ligand-dependent transcriptional activation, respectively, as deletion or mutations of these regions result in a diminished ability to regulate estrogen-responsive genes (7, 8). The mechanism by which transcription is mediated by the ER is thought to occur via interaction of AF-1 and AF-2 with the transcriptional machinery, a general term referring to the complex of molecules that assembles and ultimately results in synthesis of mRNA (9). Much is now known about RNA polymerase II, and the enzymes and factors that orchestrate transcription as discussed in several recent reviews (9–11). Ligand-activated ERs bind to target genes and recruit coregulators and associated chromatin remodeling complexes. Transcriptional coregulators mediate the interaction between ERs and the transcriptional machinery, and many coregulators have been isolated that interact with ER in a ligand-dependent manner. These include members of the steroid receptor complex (SRC)/p160 family or the thyroid receptor-associated protein (TRAP220) complex. Recent reports suggest that p160 and TRAP220 complexes are cyclically exchanged on and off estrogen-responsive genes in a process of repeated chromatin remodeling/transcriptional initiation and transcriptional reinitiation/maintenance, respectively. These mechanisms are more fully discussed elsewhere (11–13). The best-characterized coregulators interact with the AF-2 of the ER, although molecules that interact with the N terminus or AF-1 and AF-2 have also been described. Notably, the coactivator, p68, which interacts with the AF-1 region of ERα, has RNA helicase activity and associates with SRA, a RNA molecule with coactivator activity (14–16).

In the simplest models of estrogen action, estrogen binds to the receptor, which interacts with ERE DNA sequences in target genes. The estrogen-ER complex then recruits the transcriptional comodulators and consequently regulates transcription of target genes. However, numerous variations in this mechanism have been described. Many estrogen-responsive genes lack the canonical ERE sequence and interact with estrogen receptors via a tethering mechanism with a combination of estrogen receptor and SP1 or AP1 transcription factors (17–19). In addition, several mechanisms that account for very rapid nongenomic or nongenotropic estrogen responses indicate that estrogen receptors, either distinct or identical to the nuclear ERs, interact with and activate signal cascades at the cell membrane (20, 21). Finally, several activators of the growth factor receptor pathways, including IGF-1 and EGF, can result in activation of ER-mediated transcription in a ligand-independent manner (20). The mechanistic details of estrogen receptor–directed transcription are more fully described and discussed elsewhere.

USE OF MOUSE MODELS TO STUDY ESTROGEN BIOLOGY

The biological effects of estrogens occur predominantly in female reproductive tissues such as the reproductive tract and mammary glands. However, estrogen has also been shown to play roles in male tissues, as well as in nonreproductive tissues including the central nervous system, the skeletal system, and the cardiovascular system. Historically, surgical and/or pharmacological manipulation has been used to explore the roles of estrogen in reproduction and physiology. The development of knockout or transgenic mice with disruptions, mutation, or overexpression of molecules related to reproduction and hormone action has increased our understanding of their relative roles in developmental and biological processes in the mouse. For example ERα and ERβ knockout mice (αERKO and βERKO) exhibit overt phenotypes related to the essential roles of these receptors in certain tissues and biological responses. There are now hundreds of examples of gene disruptions in mice that result in reproductive phenotypes, as discussed and summarized in recent reviews (22, 23). Here we provide an overview of phenotypes observed in some of these mouse models, particularly as they apply to estrogen biology and female reproduction. We also discuss applications of these models to study hormone responses or pathologic conditions.

Estrogen Biology 101: ERKO and PRKO Models

The description of the losses of function and pathologies in the ERKO model is an ongoing and basic lesson in estrogen biology studies. Disruption of ERα (24) resulted in infertility of both males and females and has been described in detail elsewhere (25, 26), but it is summarized in Table 1. In females, estrogen and progesterone are essential primarily in three biological processes: pubertal development, regulation of estrous cycles, and establishment and maintenance of pregnancy and lactation. At puberty, increasing estrogen directs the maturation of mammary tissue. At birth, the mammary tissue consists of an epithelial rudiment embedded in stromal tissue, and at puberty increased ovarian steroids induce outgrowth of the epithelial ducts until they reach the margins of the stroma (27). The importance of estrogen to this development is illustrated by the lack of mammary duct outgrowth in αERKO mice. In contrast, mice that lack ERβ (βERKO) or progesterone receptors (PR) (PRKO) develop full epithelial ductal structures during puberty (28).

The second biological process for which estrogen and progesterone are needed is regulation of the estrous cycle of the mouse. Normally, adult female mice undergo an estrous cycle every 4–5 days, under the control of gonadotropin-releasing hormone (GnRH) pulses from the hypothalamic region of the brain, which regulate the release of the gonadotropins, luteinizing hormone (LH), and follicle-stimulating hormone (FSH), from the pituitary gonadotroph cells. The gonadotropins then direct the maturation of ovarian follicles, which produce estradiol and progesterone prior to and following ovulation, respectively (29). The interaction of these organs in the regulation of reproduction is termed the hypothalamic-pituitary-gonadal (HPG) axis. Estradiol produced by the ovary in response to the gonadotropin signals is also an essential regulator of the HPG axis, through both ERα and ERβ. Estradiol, for example, down-regulates LHβ gene transcription. αERKO females lack this estradiol-mediated negative feedback on LH and thus have chronically elevated LH, demonstrating the ERα dependence in this process (25). The αERKO females do not undergo estrous cycling and thus do not ovulate. When exogenous gonadotropins are administered to young αERKO females before the LH rises in an attempt to supply a corrected HPG environment (a course of treatment called superovulation), ovulation can be induced (30), indicating that the αERKO ovary responds to LH and FSH and that these processes do not depend on ERα per se, but successful ovulation requires appropriate regulation of HPG components, which does require the ERα.

The estrogen and progesterone produced by the ovary in each estrous cycle prepare the uterus for implantation of embryos (31). The increased weight of the

TABLE 1 Phenotypes leading to ERKO infertility[a]

Tissue	α	β	αβ
Mammary	Immature-ductal rudiment	Normal structure and lactation	Immature-ductal rudiment
Fertility	Both sexes are infertile	Fertile males Subfertile females: infrequent pregnancies, small litter sizes	Both sexes are infertile
Pituitary	LH production is elevated, low prolactin	Normal	Elevated LH production
Ovary	Estrogen and testosterone elevated. Follicles don't mature; hemorrhagic cystic follicles begin developing at puberty as a result of chronic elevated LH. Reduced ovulations in superovulation trial, "trapped follicle" phenotype after superovulation	Reduced number of corpora lutea, inefficient ovulation in superovulation trial, "trapped follicle" phenotype at superovutation. Normal gonadotropins and steroids	Progressive degeneration of germ cells, dramatic loss of granulosa cells; appearance of Sertoli-like cells; elevated LH, elevated estrogen and testosterone
Uterus: estrogen responsiveness	Immature. Insensitive to estrogen, no epithelial proliferation or induction of estrogn-responsive genes	Normal responses to estrogen	Insensitive to estrogen-like αERKO
Uterus: progesterone responsiveness	PR present, progesterone-responsive genes induced, decidualization is estrogen independent	Nd	Nd
Uterus: implantation	No implantation	Not tested, but pregnancies occur and are carried to term, inferring implantation competence	Nd
Testes	Progressive fluid retention and dilation of seminiferous tubules, eventual loss of sperm	Normal	Progressive fluid retention and dilation of seminiferous tubules, eventual loss of sperm

[a]Adapted from (26).
ND: not determined.

TABLE 2

Hours after acute E dosing

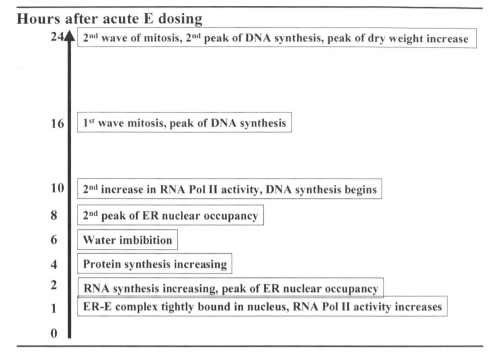

24	2nd wave of mitosis, 2nd peak of DNA synthesis, peak of dry weight increase
16	1st wave mitosis, peak of DNA synthesis
10	2nd increase in RNA Pol II activity, DNA synthesis begins
8	2nd peak of ER nuclear occupancy
6	Water imbibition
4	Protein synthesis increasing
2	RNA synthesis increasing, peak of ER nuclear occupancy
1	ER-E complex tightly bound in nucleus, RNA Pol II activity increases
0	

ovariectomized mouse uterus following a 3-day dosing of estrogenic compounds has long served as a measure of estrogen sensitivity, and the αERKO lacks this response (24). Additionally, following an acute dose of estradiol, the ovariectomized mouse uterus undergoes a series of biochemical and biological changes that are summarized in Table 2. The αERKO lacks these responses, indicating the requirement for ERα to mediate them. The uterine epithelial cells in the αERKO do not proliferate in response to estrogen, as measured by ^{3}H thymidine uptake or BrdU incorporation, nor is there an increase in uterine weight in this 24-h time period, illustrating the lack of water imbibition and hyperemia. Additionally, the infiltration of the uterine tissue by eosinophils following estrogen treatment does not occur in the αERKO (L.L. Hayes, A.-J. Lambert, C.R. Schmidt & H.H. Harris, personal communication). Uterine gene transcription in response to estrogen is robust, but as might be expected, the transcriptional response to estrogen in the αERKO is minimal (33, 34). Post-ovulatory progesterone is important in the regulation of the uterine proliferative response to estrogen. Rising progesterone shifts the proliferative response from the epithelial cells to the uterine stroma (35). Uteri of mice that lack the progesterone receptor (PRKO) develop hyperplasia in response to estrogen, indicating that PR is needed to regulate and temper the proliferative response to estrogen (36). Two PR isoforms are present in uterine tissue: PRA and PRB. Antiproliferative regulation of the estrogen response is recovered in a

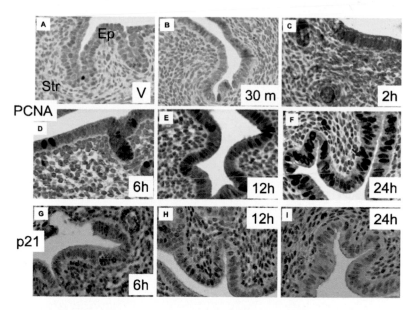

Figure 3 Uterine cross sections from mice treated with an acute dose of estrogen illustrate the synchronous proliferative response between 12 and 24 h. (*A–F*): Proliferating cell nuclear antigen (PCNA) expression in uterine epithelial cells. (*G–I*): p21 is increased in the nucleus of epithelial cells 12 h following estrogen treatment. (*Panel A*): vehicle treated. The stromal (Str) and epithelial (Ep) cells are indicated. (*Panel B*): 30 min estrogen. (*Panel C*): 2 h estrogen. (*Panel D, G*): 6 h estrogen. (*Panel E, H*): 12 h estrogen. Arrow highlights positive p21 epithelial cell in panel H. (*Panel F, I*): 24 h estrogen. Arrow highlights positive PCNA stain in epithelial cells in panel F.

Figure 4 Microarray analysis: genomic pattern mirrors biphasic biological response. Dendogram shows genomic response following estrogen treatment. Each horizontal row represents a comparison of a sample pair (vehicle treated compared with various times of estrogen treatment; 0.5–24 h). Each vertical line represents a single gene. Green indicates repression compared with vehicle; red indicates up-regulated genes. The groups of genes or clusters that are outlined in dark blue are characteristic of late responses; the clusters outlined in white are early genes, whereas others outlined in light blue occur throughout the 24 h. Adapted from (34).

WT   **αERKO**

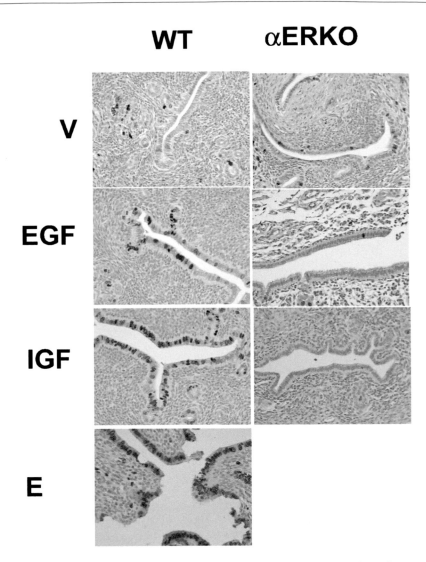

Figure 6 Ki67 expression in response to growth factors is ERα-dependent. Cross sections of uteri from WT or αERKO mice treated with vehicle (V) or for 24 h with EGF, IGF, or estrogen were analyzed for the expression of Ki67 antigen. The increase in Ki67 antigen, indicative of active, proliferative cells, following EGF or IGF treatment, occurs only in the ERα-containing tissue, thus indicating a cross talk mechanism.

PRB-selective knockout (PRBKO, has only PRA), indicating PRA is the isoform that mediates the antiproliferative effect (36). The diminished estrogen sensitivity of mice lacking the steroid receptor coactivator 1 (Src1) indicates uterine response to estrogen also depends on transcriptional coactivator molecules as well (37).

The third biological process that requires estrogen and progesterone is the establishment and maintenance of pregnancy. The αERKO is anovulatory (see Table 1); thus to determine whether the αERKO uterus is capable of implanting embryos, mice were treated with a hormonal regimen mimicking early pregnancy. Although similarly treated normal wild-type (WT) females could implant donor embryos, the αERKOs could not (38). Additionally, in response to implanting embryos, the uterine tissue normally undergoes a massive increase in size as implantation sites are formed in a process called decidualization. The αERKO uterus can be induced to decidualize with an artificial regimen mimicking early pregnancy (39). Together, these experiments indicate that although the uterus does not require ERα to decidualize, it must have ERα for implantation to occur. [Progesterone's role in implantation and decidualization is illustrated by studies in which the PRKO and PRAKO lack decidualization and implantation (36), indicating the essential role of PRA in both of these processes.]

During pregnancy, the progesterone level remains elevated, which not only maintains the pregnancy but also induces development of mammary gland structures necessary for lactation. In a nonpregnant mouse, during each estrous cycle, the mammary tissue is relatively quiescent and might be described as having the appearance of bare tree branches that occupy the extent of the underlying mammary stromal tissue (28). The quiescence of the mammary tissue is illustrated using microarray analysis to examine the global gene responses of the mammary tissue to acute estrogen. When the uterine and mammary gland gene regulation patterns are compared, it is apparent that fewer gene changes occur in the mammary gland, as eight times more uterine than mammary gland genes show significant changes 24 h after estrogen injection (S.C. Hewitt, unpublished data). In addition, 75% of the gene changes that occur in the mammary gland represent decreases in gene expression levels (S.C. Hewitt, unpublished data); in contrast, only 20% of the gene changes in the uterus reflect decreases (34). In response to pregnancy levels of progesterone and prolactin, extensive mammary epithelial proliferation occurs that greatly increases the complexity of the ducts by increasing side branches (which might be described as twigs on a tree branch) (28). Additionally, lobuloalveolar structures develop at the ends of the ducts, which have the appearance of buds on the tree branches and which will begin to fill with milk late in pregnancy. PRKO mice are infertile and thus never become pregnant, and although the ductal tree structure develops at puberty, treatment with pregnancy levels of progesterone does not induce the side-branching or lobuloalveolar development (28, 36). The PRAKO recovers this response, indicating the PRB is sufficient to mediate pregnancy-associated mammary development. The αERKO never develops a full underlying ductal tree structure. However, exogenous progesterone does increase the complexity of the rudimentary ductal structure (40). It has been reported that

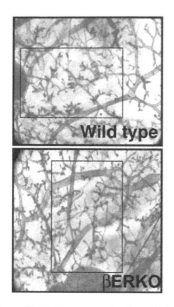
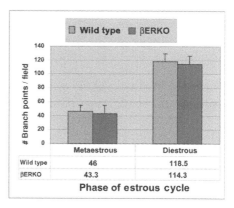

Figure 2   Mammary gland epithelial branching correlates with estrous stage. Comparable areas of wild-type (WT) or βERKO whole mounts (the rectangle shows the field) were analyzed by counting branch points. Each data point was associated with the estrous stage of the mouse, as determined by vaginal smear on the day of necropsy.

the βERKO ductal structures lack complexity (41); however, when correlated with the stage of the estrous cycle, we observed no difference between βERKO and WT structures (Figure 2). Therefore, the reported decreased complexity is a consequence of defects in βERKO estrous cycle leading to fewer individuals progressing to diestrous, the stage at which the greatest mammary complexity occurs. Overall, it is apparent that ERα is essential for pubertal ductal elongation, whereas PR is dispensable for this aspect of mammary gland development, but PRB is essential for pregnancy-associated alveolar formation and mammary gland development.

Estrogen Biology in the Ovary: A Complex Lesson

Study of the biological effects of estrogen on ovarian function is complicated by the level of estrogen synthesis inherent to the tissue. The ovary is made up of developing follicles that contain germ cells at different stages of maturity in a stromal/interstitial tissue (42). Within maturing follicles, the oocyte is surrounded by layers of granulosa cells that are rich in ERβ and are the primary source of estrogen biosynthesis. ERα, in contrast, is localized primarily to thecal and interstitial tissue components of the ovary (25). The ERKO models have proven valuable in that they provide an opportunity to study specific aspects of estrogen signaling within the ovary.

The chronically elevated LH in the αERKO leads to the development of hemorrhagic cysts in the αERKO ovary, an observation also reported in other knockout and transgenic mice that have elevated LH. These examples include mice overexpressing LH (43), FSH (44) or human chorionic gonadotropin (45) (HCG; an LH receptor agonist) transgenes, as well as inhibin α (46) and FSH receptor knockout mice (47), and those in long-term antiestrogen treatment (48, 49). GnRH antagonist treatment of prepubertal αERKO females prevents the rise in LH and the consequent formation of cysts (30). Thus the cysts are characteristic of elevated LH and occur in both the presence and absence of ERα. This is an important facet when drawing conclusions based on observed defects in knockout or transgenic mouse studies. Many homeostatic physiological systems, such as the HPG axis, are made up of interacting elements, and perturbation of components in one tissue (LHβ regulation in the pituitary) may result in physiological defects that are not a direct consequence of gene disruption in another tissue (ovarian cysts). Therefore, it is essential to be cautious in interpreting observations from knockout mice and to consider indirect effects on physiology. Newer models are now being developed to counteract the defect secondary to HPG problems of the ERKOs and allow study of estrogen biology and roles of ERs intrinsic to the ovary.

Lessons in Estrogen Biology: Estrogen Is Important to Male Fertility

The infertility exhibited by αERKO males was unexpected because estrogen was thought to have significant roles only in female reproduction. The αERKO males are infertile, in part, owing to the progressive degeneration of the testicular tissue and eventual loss of sperm because of fluid retention and dilation of the seminiferous tubules (25). αERKO females and males also fail to exhibit successful mating behaviors (25). The successful transmission of the ERα null trait by heterozygous breeders indicates the defect is not intrinsic to the ERα null sperm, but ERα is needed in somatic cells of the male reproductive tract for proper maturation and activation of sperm (50, 51).

Role of ERβ in Estrogen Biology as Revealed by the βERKO Mouse

Because the αERKO still retains ERβ function, the study of the βERKO highlights the contribution of ERβ to estrogen biology. The βERKO mice exhibit less profound phenotypes than the αERKO. The males retain full fertility but in some reported cases develop prostate hyperplasia with aging (52). The βERKO females are subfertile (25), and considering the profound ERβ expression in follicular granulosa cells, it is not surprising that the underlying cause of βERKO subfertility is inefficient ovulatory response. The ability of βERKO females to carry pregnancies to term and nurse their offspring indicates adequate uterine and mammary gland function, but it does not rule out more subtle effects on mammary or uterine responses. Microarray analysis of βERKO mice indicates a transcriptional response to estrogen that is comparable to the WT in the uterine tissue (34). These findings are summarized in Table 1. Overall, the observations in the βERKO females indicate an important role for ERβ in achieving optimal fertility, yet successful pregnancies can occur, which indicates that ERβ is not required. Interestingly, in continuous mating studies, some βERKO females exhibited a normal frequency of pregnancy, with reduced litter sizes; some females had a reduced frequency of pregnancy, with reduced litter sizes; and a third group never became pregnant, suggesting individual differences in sensitivity to loss of ERβ (25, 53). One might hypothesize that all the processes required for successful ovulation are present in the βERKO; however, in some cases there is a defect in fully initiating the response. Indeed, superovulation studies resulted in a very low yield of oocytes, although pathology of the ovary showed many fully developed follicles seemingly ready to ovulate but apparently trapped on the verge of ovulation (54). Several other knockout models exhibit a similar defect, including cyclooxygenase 2 (55), PR (36), cyclin D2 (56), and RIP140 (57), and studies are underway to examine the regulation of these and other genes in superovulated βERKOs.

Male and female knockout mice that lack both ERα and ERβ ($\alpha\beta$ERKO) are infertile, and the underlying causes seem to reflect the previously observed defects in the αERKO (25), indicating the crucial contribution of ERα in mediating reproductive biological events. However, a unique phenotype was observed in the ovary, where a progressive loss of oocytes occurs, and an apparent transdifferentiation of granulosa cells into cells having the appearance of and expressing markers characteristic of Sertoli cells, which are normally found in seminiferous tubules (53, 58, 59). Such a unique ovarian phenotype suggests that maintenance of the proper differentiation state of granulosa cells requires the combined activity of both the ERα and ERβ in ovarian tissue.

Comparing ER-Null to Estrogen-Free Environment: ArKO

The enzyme responsible for synthesis of estradiol is a P450 enzyme, Cyp 19 also called aromatase, which converts testosterone to estradiol. The aromatase knockout (ArKO) mouse retains ERα and ERβ but does not synthesize any estrogen, and many of its observed phenotypes are similar to that of the αERKO, including immature mammary glands and uteri. Additionally, the ovaries progressively develop hemorrhagic cysts like those found in the αERKO mice, again consistent with a secondary effect as a consequence of elevated gonadotropins. Males are infertile owing to progressive loss of spermatids but do not develop the fluid distension seen in the αERKO, suggesting ligand-independent activity of the ERα in the ArKO might prevent the occurrence of the defect. ArKO males also exhibit deficiencies in sexual behaviors. In particular, the mounting response is impaired in both the ArKO and $\alpha\beta$ERKO males but is observed in the αERKO and βERKO males, although intromissions and ejaculations are deficient in the αERKO (60, 61). This indicates that mounting is dependent upon estrogen ligand synthesis but that either ERα or ERβ is sufficient to mediate the mounting response (62). Estrogen replacement in ArKO females results in recovery of uterine weight (62, 63).

Biology Selective to the AP-1 Tethered Mode of ER Signaling: NERKI

A mouse has been engineered to express an ERα with mutations that selectively eliminate classical ERE-mediated signaling while retaining responses mediated via tethering through AP-1 sequences. Females heterozygous for this nonclassical ER knock-in (NERKI) are infertile; they are anovulatory and administering exogenous gonadotropins results in few ovulations (64). It is interesting that superovulation induced hemorrhagic cysts in the NERKI ovaries, yet prior to dosing, the LH levels are in the normal range, suggesting the presence of the nonclassical ER mutant somehow results in increased sensitivity to the gonadatropin, which causes a pathology characteristic of LH overstimulation; superovulation in WT and αERKO mice does not induce hemorrhagic cysts. The NERKI uteri progressively form enlarged hyperplastic endometrial glands, despite normal ovarian steroid levels, suggesting dis-regulated responsiveness of the tissue. The NERKI mammary glands have full ductal development but have decreased complexity, likely a result of anovulatory progesterone levels. The infertility of the heterozygotes has prevented generation of a mouse with two copies of the mutated ER. It is interesting that replacement of one copy of the WT ER results in such a pronounced phenotype, as mice heterozygous for the ERα null allele are fertile, indicating one copy of ERα is sufficient for reproduction. It seems that the combination of one copy of WT ERα and one copy of the nonclassical signaling ER mutation results in a unique perturbation of estrogen physiology. Future analysis of hemizygous NERKI mice produced by crossing the NERKI and the αERKO should indicate the biological consequences of exclusive expression of this nonclassical ER mutant in the animal. Similarly, we have generated a knock-in mouse that expresses ERα with a mutation in the ligand-binding domain of the receptor, which allows it to retain binding to estradiol, but prevents transcriptional activity (AF2ER). In our case, the mice expressing one copy of this mutant ER are fertile; however, embryos homozygous for this mutant ER die before implantation, indicating that unlike the ER-null mutant, the AF2ER mutant disrupts estrogen signaling at a critical point in embryogenesis (65).

APPLICATION OF MOUSE MODELS TO ESTROGEN MECHANISM STUDIES

As described briefly above, acute treatment of ovariectomized mice with estrogen has long served as a key experimental model in which to study the biochemical mechanisms underlying uterine responses. The events that occur following estrogen administration have been divided into those that occur early, within the first hours following estrogen elevation, and subsequent responses that follow up to 24 h later. Thus this acute and rapid response of the uterus has been described as biphasic (66, 67). Early events include nuclear ER occupancy, transcription of early-phase genes such as c-*fos*, fluid uptake (termed water imbibition), hyperemia, and infiltration of immune system cells such as macrophages and eosinophils into the uterine tissue (68, 69). Later phase responses include the transcription of late-phase genes such as *lactoferrin*, increase in uterine wet weight, further accumulation of immune system cells, the development of the epithelial layer into columnar secretory epithelial cells, and subsequent mitosis, which occurs principally in the epithelial layer (70). Coordinated increased uterine DNA synthesis and mitosis are reported to begin 12–24 h following estrogen treatment of ovariectomized mice, indicative of synchronized entry into S phase (70). This is illustrated by the increased proliferating cell nuclear antigen (PCNA) detected in the epithelial cells (Figure 3, see color insert). Microarray analysis of the global gene expression pattern in response to acute estrogen identified clusters of genes characteristic of these early and later responses, some of which overlap but some are distinct to the early or late time points of response (34) (Figure 4, see color insert).

Significantly, estrogen decreased the expression of many genes (Figure 4), yet most mechanisms of ER-mediated gene regulation consider only increases in transcription. Our microarray data indicate ERα also mediates transcriptional repression because (*a*) antiestrogen treatment inhibited gene increases and decreases and (*b*) neither increased nor decreased gene levels were apparent in the αERKO microarray analysis (34). We have identified numerous endogenous uterine genes appropriate for investigation of the mechanisms involved in gene repression by estrogen.

Uterine epithelial cells not only must proliferate but must do so at the proper time as the estrous cycle progresses in preparation for implantation of embryos. Thus it is not surprising that the microarray analysis revealed regulation of several cell cycle modulators, including *p21, Cycin G1, cdc2, and Cyclin E1* (34), (Figure 5). Synchronous/coordinated regulation of the components modulating entry into S phase is one mechanism by which the acute exposure to estradiol might orchestrate the ordered biological response. Estrogen regulation of some cell cycle modulators in the uterus has been previously reported. For example, estrogen treatment induces nuclear relocalization of cyclin D1 protein and an increase in expression of cyclin A and E proteins in the uterine epithelium (35). In some cell types, cyclin D1 is an estrogen-responsive gene; however, in the uterus a minimal increase in transcript occurs (34). p21 inhibits progression into S phase, and its RNA and protein levels are maximally induced and localized to the epithelial cell nuclei 12 h following estradiol treatment (34) (Figures 3 and 5), just prior to the peak of entry into S phase, suggesting that the increased p21 may prevent S phase progression of the epithelial cells until the proper time, allowing coordinated proliferation of the epithelial cells. Thus the properly timed increase in *p21* may act as a gate, coordinating appropriate S phase progression.

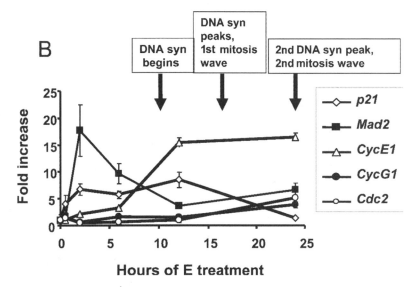

Figure 5 Coordinated regulation of cell cycle regulators by estrogen. (*A*) Table of values from microarray data. Fold increase versus vehicle control. (*B*) Transcripts for *p21*, *Mad2*, *Cyclins G1*, and *E1*, and *cdc2* were assayed following an acute dose of estrogen by RT-real time PCR. The observed times of biological events reflecting S and M phases are indicated. Adapted from (34).

Because the biphasic genomic pattern clearly mirrors the observed biological response, we considered whether the early gene changes were modulating the late genomic responses or whether the later responses depended on continuous ER-mediated activity. To test this, we injected the ER antagonist ICI 182–780 2 h after estrogen treatment to block any further ER activity subsequent to the initial early gene changes. Our preliminary results indicate that in some cases the late gene changes were blocked, suggesting that ER activity throughout the time course is necessary, whereas some late responses still occur, indicating they are most likely secondary responses mediated by early phase regulated genes (S.C. Hewitt, unpublished data).

Using the ERKO to Study ER-Growth Factor Cross Talk

Growth factors, including EGF and IGF-1, are present in uterine tissue, as are their respective receptors. The increases in these growth factors and activation of growth factor–signaling pathways by estrogen indicate roles in uterine biology. Additionally, activators of growth factor (GF) receptor pathways, including IGF-1 and EGF, can result in ER-mediated transcription (20). These cross talk mechanisms were demonstrated in vivo by showing an increase in uterine weight and proliferation of the uterine epithelial cells in ovariectomized mice following EGF or IGF-1 treatment, as illustrated by the Ki67-positive epithelial cells in Figure 6 (see color insert) (71–73). The lack of these responses in similarly tested αERKO mice indicated that the ERα is downstream of the growth factor receptor signaling in this response and that ERα is required (Figure 5) (72–74). The mechanism has been studied in vitro as well, using reporter gene assays, which similarly have shown the requirement for ERα for growth factor receptor activators to regulate estrogen-regulated reporter genes (75, 76). The approaches have been combined, and we showed that IGF-1 treatment increases expression of an estrogen-responsive luciferase reporter gene in transgenic mice (73), providing the first evidence in vivo of ligand-independent ER activation and support for the cross talk hypothesis.

Considering the model of growth factor receptor-ER cross talk, the uterine genomic response of WT and αERKO to EGF or IGF-1 would be expected to exhibit two patterns or clusters of genes with the following response profiles: The first would include genes regulated in response to estrogen or growth factor receptor signaling, representing the cross-talk response; because these responses require ERα, they would be lost in the αERKO. The second pattern would include genes that were directly regulated by growth factor receptor pathways. These genes would depend only on growth factor receptor pathways and therefore would be similarly regulated in the αERKO. Surprisingly, the global genomic response in the uterus did not fit these expected patterns (S.C. Hewitt, submitted manuscript). Clusters of genes that were regulated similarly by either estrogen or growth factors in the WT samples were observed, as described by the cross-talk mechanism. However, these genes retained growth factor responsiveness in the absence of ERα in the αERKO samples. ERα was required for estrogen regulation of these genes, as they were insensitive to estrogen in the αERKO. These responses to growth factors were not inhibited by antiestrogen (ICI 182,780). This seems to indicate that for these genes, growth factors can bypass the requirement for ERα, suggesting the responses are secondary to the increase of IGF by estrogen. However, these responses are occurring as early as 2 h subsequent to estrogen injection in the WT, prior to the peak of IGF induction. Clusters of genes that were regulated only by estrogen in the WT were also apparent. Additionally, genes regulated primarily by growth factors were seen in both WT and αERKOs. Although studies using model reporter genes and αERKO mice have previously demonstrated growth factor-mediated ERα responsiveness through a cross-talk model (72–75, 77), the global response of endogenous uterine genes to estrogen and growth factor appears to encompass greater complexity than this model describes.

Implantation-Associated Signals

In addition to responses initiated by estrogen or growth factors in the uterus, we have also examined responses to a stimulus intended to mimic early pregnancy. The uterine environment encounters a preovulatory estrogen surge, followed by increasing postovulatory progesterone that, as discussed above, prepares the uterus for embryo implantation. Experimental manipulation can mimic this process and elicit the responses of early pregnancy. For the uterus to decidualize, it is necessary to administer a stimulus, in our case infusion of inert oil into the uterus, which mimics the physical apposition of the embryo against the uterine wall. This stimulus must be administered during a specific time of responsiveness that reflects a window of receptivity to implantation (31). We became interested in the nature of the signals elicited by the oil infusion because historically the response to this stimulus had been shown to require prior priming with estrogen for decidualization to progress (78). However, in the αERKO uterus, we observed a relief of this estrogen requirement for decidualization. Use of ICI 182,780, an ER antagonist, also demonstrated that estrogen priming and ER signaling were required in the WT but not αERKO for decidualization responses (39). We examined some novel signals that were initiated and noted activation of phospho-STAT3, a transcriptional activator that is a target for cytokine signal pathways, in WT as well as in αERKO uteri following the oil infusion (38). Thus we evaluated the expression of leukemia-inhibitory factor (LIF), a cytokine increased by estrogen in the uterus and shown to be required for embryo implantation and decidualization (79, 80). We confirmed that estrogen increases uterine *Lif* transcripts, but not in the αERKO. Soon after infusion of oil into the uterus, we also observed an increase in Lif transcript and, unlike the estrogen-induced increase, this response also occurs in the αERKO (38), indicating an ERα-independent pathway. We have subsequently observed a similar regulation pattern in several additional genes, including *Connexin 26 (Cx26)* (81), *c-Fos, cell division cycle 2 homolog A (Cdc2a)*, and *cyclophilin (Cyc)* (S.C. Hewitt, unpublished observations). Therefore, it appears that some estrogen-responsive uterine genes are also regulated by the oil infusion and the latter mode of regulation is ERα independent, allowing uterine response without a need for estrogen priming. ER has been reported to interact with Stat3 (82), which suggests that activation of Stat3 might convey signals to ER-mediated responses.

Model: Converging Signals

Our studies examining uterine responses to acute estrogen, growth factors or the oil infusion stimulus mimicking early pregnancy indicate that divergent signal initiators converge at the level of gene regulation (see model, Figure 7). For example, estrogen initiates nuclear ERα recruitment of transcriptional coregulators (CoAc) and interaction with estrogen-responsive genes directly (ERE) or through tethering (AP-1/SP1), leading to ER-dependent responses. Some studies also indicate gene responses are initiated by estrogen activation of membrane-associated signals (nongenotropic signaling). Additionally, growth factors activate

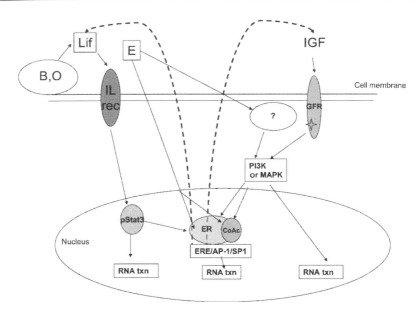

Figure 7   Model: Through divergent signals, genomic convergence. Estrogen (E) binds nuclear ERα, thus recruiting coregulators (CoAc) and modulating gene transcription by directly interacting with ERE DNA sequences or through tethered interaction with AP-1 or SP-1 transcription factors. Initiation of nongenotropic signals at the cell membrane is also depicted where estrogen activates signaling pathways such as the MAPK or PI3 kinase pathways. ERα, AR, or another estrogen-binding molecule may be involved in mediating this response (represented by ?). Growth factors also activate MAPK or PI3 kinase pathways by interacting with and activating their membrane receptors (GFR). Lif, blastocysts (B), or inert oil infusion (O) initiate cytokine-receptor signaling and activate pSTAT3-targeted transcription. These signals converge at the level of genomic modulation; this results in similar genomic responses to estrogen and growth factors or oil infusion either by direct activation of nuclear ER mediated transcription, or by other transcriptional mediators. *Igf-1* is increased by estrogen and further activates the growth factor receptor–mediated pathways, whereas *Lif* is increased by estrogen or oil infusion and activates cytokine signals.

their receptors' intracellular signals such as PI3 kinase and/or MAP kinase, resulting in direct regulation of growth factor–specific genes. Nongenotropic estrogen signaling and growth factor signaling converge as similar intracellular signaling is initiated by either factor and may further converge with nuclear ERα signaling by conveying this activity to the nuclear ERα and associated transcriptional modulators. The initiation of signal by estrogen is sensitive to ICI, as it utilizes ERα. When growth factors directly initiate the signal, ERα is bypassed, thus the initiation is not blocked by ICI and also occurs in the αERKO. In addition, IGF-1, but not EGF or TGFα, is induced by estrogen and can then initiate growth factor receptor–mediated responses. Finally, initiation of a cytokine-like pathway by oil infusion, which results in activation of phospho-Stat3-mediated transcription, also converges with ER transcriptional responses.

STUDY OF ROLES OF ER IN BREAST CANCER MODELS

Knockout models can be applied to studies of pathological conditions such as cancer. As there is a strong correlation between exposure to estrogenic compounds and carcinogenesis and tumor growth in the mammary gland (83, 84), ERα has been a target of chemotherapeutic and chemopreventative therapies in breast cancer. Because ERα is involved in mammary development and its expression correlated with breast cancer growth, we have studied the effect of removing ERα from mice overexpressing mammary-tumor-inducing oncogenes in the mammary gland.

Expression of the MMTV-Wnt-1 transgene, a diffusible factor that signals through frizzled receptors, leads to mammary hyperplasia and tumors. Similarly, transgenic mouse lines that overexpress *erbB2* oncogene (also called neu), an epidermal growth factor receptor (EGFR)-like protein (85) reported to be overexpressed in 20–30% of human breast tumors (86), or a constitutively active mutant neu targeted to the mammary epithelium, exhibit an increased incidence of mammary tumors (87) compared with that of nontransgenic littermates. We further investigated the role of ERα in these transgenic models of mammary carcinogenesis by breeding the transgenes onto the αERKO and comparing tumor incidence and onset rates. In both cases, tumors occurred in both WT and αERKO; however, tumor onset was significantly delayed in the αERKO/MMTV-*Wnt 1* (88) and αERKO/MMTV-*neu* (Table 3) (89) mice in contrast to that in their WT transgenic counterparts. Therefore, functional ERα is not obligatory to MMTV-*Wnt-1* or MMTV-*neu* induced mammary tumors but contributes to the rate of tumor progression.

The effect of removing ERα signaling on mammary tumor induction has also been studied in transgenic mice that express large T antigen in developing mammary duct cells. However, when this transgene was introduced into the αERKO females, it was not expressed, presumably because there was no ductal development in the αERKO, and hence tumors did not occur (90).

Because αERKO mice are anovulatory and have persistent preovulatory progesterone levels, and because the mammary tissue contains only a rudimentary epithelial structure, several experiments were undertaken to determine whether increasing the progesterone level and/or the amount of mammary epithelium of αERKO/*neu* mice would accelerate tumor onset. This is of interest especially in light of the association between increased mammary duct tissue density and increased breast cancer risk in woman (91, 92). When the progesterone level in the αERKO/*neu* mice was increased, either by treatment with progesterone-release pellets or by pituitary xenografts, which secrete prolactin and result in stimulation

TABLE 3 Tumor onset age of neu mice[a]

Group	Onset age	p versus WT/neu
WT (ERα+/+)/neu	50% at 51 weeks	
αERKO (ERα−/−)/neu	50% at 105 weeks	<0.0001
WT/neu ovex	50% at 48 weeks	0.648
WT/neu multiparous	50% at 32 weeks	<0.0001
Pre-pubertal ovex WT/neu	50% at 62 weeks	.038
Pre-pubertal ovex+prog WT/neu	40% at 52 weeks	0.262
WT/neu+P	50% at 37 weeks	<0.0001
WT/neu ovex+P	50% at 40 weeks	0.88 versus WT/neu+P
αERKO/neu+P	50% at 70 weeks	0.0235 versus αERKO/neu without P
WT/neu +pit	50% at 37 weeks	<0.0001
αERKO+pit	50% at 43 weeks	<0.0001 versus αERKO non transplanted, 0.0941 versus WT/neu non transplanted

[a]Age at 50% onset was calculated from the Kaplan-Meier plots, and p values compared with WT/neu or indicated set were determined Adapted from (89).

of ovaian progesterone production, the mammary tumor onset rate equaled or exceeded that of unmanipulated WT/*neu* mice, despite a low content of epithelial tissue in αERKO/*neu* relative to WT/*neu* mice (89) (Table 3). Similarly, when WT/*neu* mice were exposed to elevated progesterone during pregnancy, by direct treatment with progesterone-releasing pellets or by stimulation of the ovaries with pituitary grafts, the onset rate was accelerated (89) (Table 3). In contrast, *Wnt-1* females did not display accelerated tumor onset following pregnancies, although prepubertal ovariectomy did cause some delay in onset (88). The ability of progesterone to accelerate tumor onset in WT/*neu* and αERKO/*neu* mice indicates that the underlying phenotypes of the αERKO (i.e., the lack of postovulatory progesterone, low volume of mammary epithelial cells in which to express transgene) contribute significantly to delayed tumor onset in the αERKO/*neu*. The more rapid onset following progesterone elevation may reflect progesterone's role in pregnancy-associated ductal proliferation and lobuloalveolar development (93) as a possible component to tumor progression.

PRKO mice have been utilized in a mammary tumorgenesis model in which a combination of dimethylbenzanthracene (DMBA) and a pituitary xenograft induce mammary tumors (28). Lack of PR reduced tumor incidence from 60% to 15%, indicating that tumors can occur in the absence of PR but suggesting a role for PR in susceptibility, which is attributed to the loss of epithelial proliferation in the PRKO mammary gland.

Studies in mouse mammary tissues indicate that the PR and ERα are present in a nonuniform pattern, selectively localized in nonproliferating cells, which suggests a paracrine mechanism of progesterone- or estrogen-stimulated mammary proliferation (28, 94, 95). Thus for ductal cells to be proliferative, they may contain neither PR nor ERα, but must respond to proliferative signals from PR- or Erα-positive cells, and it is the abundance of these proliferative cells that may be important indicators of cancer susceptibility, although their numbers may be increased by estrogen or progesterone exposures. If ERα does play such a role in human breast cancer, one might predict that ERα antagonists may delay tumor onset by limiting the population of proliferative cells; however, the identification of drug compounds targeted directly to the proliferative cells themselves, which abrogate the function of signaling molecules involved in breast tumorigenesis, might prevent rather than merely delay tumor formation effectively.

EPILOGUE: LESSONS LEARNED?

In this review we have summarized some of the important and interesting lessons learned utilizing engineered mice as experimental models to increase our understanding of estrogen biology at many levels. We have noted how perturbing homeostasis can reveal mechanisms of many components important in reproduction, both directly related to estrogen signaling and secondary to homeostatic disruptions. We discussed the application of the knockout models to pathologic states to study the role of estrogen signaling in cancer. The uterine model was utilized to examine mechanisms and converging pathways leading to estrogen-dependent and estrogen-independent gene regulation. Much has been learned, yet with every lesson, more questions continue to arise, and considering the techniques and technologies now available, the future promises to advance our understanding in this fascinating and important field of study.

The *Annual Review of Physiology* is online at http://physiol.annualreviews.org

LITERATURE CITED

1. Mangelsdorf DJ, Thummel C, Beato M, Herrlich P, Schutz G, et al. 1995. The nuclear receptor superfamily: the second decade. *Cell* 83:835–39
2. Kuiper GG, Enmark E, Pelto-Huikko M, Nilsson S, Gustafsson JA. 1996. Cloning of a novel receptor expressed in rat prostate and ovary. *Proc. Natl. Acad. Sci. USA* 93: 5925–30
3. Tsai MJ, O'Malley BW. 1994. Molecular

mechanisms of action of steroid/thyroid receptor superfamily members. *Annu. Rev. Biochem.* 63:451–86
3a. Nettles KW, Greene GL. 2005. Ligand control of coregulator recruitment to nuclear receptors. *Annu. Rev. Physiol.* 67:309–33
4. Glass CK. 1994. Differential recognition of target genes by nuclear receptor monomers, dimers, and heterodimers. *Endocr. Rev.* 15: 391–407

5. Klinge CM. 2001. Estrogen receptor interaction with estrogen response elements. *Nucleic Acids Res.* 29:2905–19

6. Pike ACW, Brzozowski AM, Hubbard RE. 2000. A structural biologist's view of the oestrogen receptor. *J. Steroid Biochem. Mol. Biol.* 74:261–68

7. Metzger D, Ali S, Bornert JM, Chambon P. 1995. Characterization of the amino-terminal transcriptional activation function of the human estrogen receptor in animal and yeast cells. *J. Biol. Chem.* 270:9535–42

8. Parker MG. 1995. Structure and function of estrogen receptors. *Vitam. Horm.* 51:267–87

9. Edwards DP. 2000. The role of coactivators and corepressors in the biology and mechanism of action of steroid hormone receptors. *J. Mammary Gland Biol. Neoplasia* 5:307–24

10. McKenna NJ, Lanz RB, O'Malley BW. 1999. Nuclear receptor coregulators: cellular and molecular biology. *Endocr. Rev.* 20:321–44

11. Smith CL, O'Malley BW. 2004. Coregulator function: a key to understanding tissue specificity of selective receptor modulators. *Endocr. Rev.* 25:45–71

12. Glass CK, Rosenfeld MG. 2000. The coregulator exchange in transcriptional functions of nuclear receptors. *Genes Dev.* 14:121–41

13. Hermanson O, Glass CK, Rosenfeld MG. 2002. Nuclear receptor coregulators: multiple modes of modification. *Trends Endocrinol. Metab.* 13:55–60

14. Lanz RB, McKenna NJ, Onate SA, Albrecht U, Wong J, et al. 1999. A steroid receptor coactivator, SRA, functions as an RNA and is present in an SRC-1 complex. *Cell* 97:17–27

15. Endoh H, Maruyama K, Masuhiro Y, Kobayashi Y, Goto M, et al. 1999. Purification and identification of p68 RNA helicase acting as a transcriptional coactivator specific for the activation function 1 of human estrogen receptor alpha. *Mol. Cell Biol.* 19:5363–72

16. Watanabe M, Yanagisawa J, Kitagawa H, Takeyama K, Ogawa S, et al. 2001. A subfamily of RNA-binding DEAD-box proteins acts as an estrogen receptor alpha coactivator through the N-terminal activation domain (AF-1) with an RNA coactivator, SRA. *EMBO J.* 20:1341–52

17. Kushner PJ, Agard DA, Greene GL, Scanlan TS, Shiau AK, et al. 2000. Estrogen receptor pathways to AP-1. *J. Steroid Biochem. Mol. Biol.* 74:311–17

18. Safe S. 2001. Transcriptional activation of genes by 17 beta-estradiol through estrogen receptor-Sp1 interactions. *Vit. Horm.* 62:231–52

19. Jakacka M, Ito M, Weiss J, Chien PY, Gehm BD, Jameson JL. 2001. Estrogen receptor binding to DNA is not required for its activity through the nonclassical AP1 pathway. *J. Biol. Chem.* 276:13615–21

20. Coleman KM, Smith CL. 2001. Intracellular signaling pathways: nongenomic actions of estrogens and ligand-independent activation of estrogen receptors. *Front. Biosci.* 6:D1379–91

21. Cato AC, Nestl A, Mink S. 2002. Rapid actions of steroid receptors in cellular signaling pathways. *Sci STKE* 2002: RE9

22. Matzuk MM, Lamb DJ. 2002. Genetic dissection of mammalian fertility pathways. *Nat. Cell Biol.* 4(Suppl.):s41–49

23. Burns KH, Matzuk MM. 2002. Minireview: genetic models for the study of gonadotropin actions. *Endocrinology* 143: 2823–35

24. Lubahn DB, Moyer JS, Golding TS, Couse JF, Korach KS, Smithies O. 1993. Alteration of reproductive function but not prenatal sexual development after insertional disruption of the mouse estrogen receptor gene. *Proc. Natl. Acad. Sci. USA* 90: 11162–66

25. Couse JF, Korach KS. 1999. Estrogen receptor null mice: What have we learned and where will they lead us? *Endocr. Rev.* 20:358–417

26. Hewitt SC, Korach KS. 2002. Estrogen receptors: structure, mechanisms and

function. *Rev. Endocr. Metab. Disord.* 3: 193–200

27. Visvader JE, Lindeman GJ. 2003. Transcriptional regulators in mammary gland development and cancer. *Int. J. Biochem. Cell Biol.* 35:1034–51

28. Ismail PM, Amato P, Soyal SM, DeMayo FJ, Conneely OM, et al. 2003. Progesterone involvement in breast development and tumorigenesis—as revealed by progesterone receptor "knockout" and "knockin" mouse models. *Steroids* 68:779–87

29. Gharib SD, Wierman ME, Shupnik MA, Chin WW. 1990. Molecular biology of the pituitary gonadotropins. *Endocr. Rev.* 11:177–99

30. Couse JF, Bunch DO, Lindzey J, Schomberg DW, Korach KS. 1999. Prevention of the polycystic ovarian phenotype and characterization of ovulatory capacity in the estrogen receptor-alpha knockout mouse. *Endocrinology* 140:5855–65

31. Paria BC, Song H, Dey SK. 2001. Implantation: molecular basis of embryo-uterine dialogue. *Int. J. Dev. Biol.* 45:597–605

32. Deleted in proof

33. Couse JF, Curtis SW, Washburn TF, Lindzey J, Golding TS, et al. 1995. Analysis of transcription and estrogen insensitivity in the female mouse after targeted disruption of the estrogen receptor gene. *Mol. Endocrinol.* 9:1441–54

34. Hewitt SC, Deroo BJ, Hansen K, Collins J, Grissom S, et al. 2003. Estrogen receptor-dependent genomic responses in the uterus mirror the biphasic physiological response to estrogen. *Mol. Endocrinol.* 17:2070–83

35. Tong W, Pollard JW. 1999. Progesterone inhibits estrogen-induced cyclin D1 and cdk4 nuclear translocation, cyclin E- and cyclin A-cdk2 kinase activation, and cell proliferation in uterine epithelial cells in mice. *Mol. Cell Biol.* 19:2251–64

36. Conneely OM, Mulac-Jericevic B, Lydon JP. 2003. Progesterone-dependent regulation of female reproductive activity by two distinct progesterone receptor isoforms. *Steroids* 68:771–78

37. Xu J, Qiu Y, DeMayo FJ, Tsai SY, Tsai MJ, O'Malley BW. 1998. Partial hormone resistance in mice with disruption of the steroid receptor coactivator-1 (SRC-1) gene. *Science* 279:1922–25

38. Hewitt SC, Goulding EH, Eddy EM, Korach KS. 2002. Studies using the estrogen receptor alpha knockout uterus demonstrate that implantation but not decidualization-associated signaling is estrogen dependent. *Biol. Reprod.* 67:1268–77

39. Curtis SW, Clark J, Myers P, Korach KS. 1999. Disruption of estrogen signaling does not prevent progesterone action in the estrogen receptor or knockout mouse uterus. *Proc. Natl. Acad. Sci. USA* 96:3646–51

40. Bocchinfuso WP, Lindzey JK, Hewitt SC, Clark JA, Myers PH, et al. 2000. Induction of mammary gland development in estrogen receptor-alpha knockout mice. *Endocrinology* 141:2982–94

41. Forster C, Makela S, Warri A, Kietz S, Becker D, et al. 2002. Involvement of estrogen receptor beta in terminal differentiation of mammary gland epithelium. *Proc. Natl. Acad. Sci. USA* 99:15578–83

42. Greenwald GS, Roy SK. 1994. Follicular development and control. In *The Physiology of Reproduction*, ed. E Knobil, J Neill, pp. 629–724. New York: Raven

43. Risma KA, Clay CM, Nett TM, Wagner T, Yun J, Nilson JH. 1995. Targeted overexpression of luteinizing hormone in transgenic mice leads to infertility, polycystic ovaries, and ovarian tumors. *Proc. Natl. Acad. Sci. USA* 92:1322–26

44. Kumar TR, Palapattu G, Wang P, Woodruff TK, Boime I, et al. 1999. Transgenic models to study gonadotropin function: the role of follicle-stimulating hormone in gonadal growth and tumorigenesis. *Mol. Endocrinol.* 13:851–65

45. Matzuk MM, DeMayo FJ, Hadsell LA, Kumar TR. 2003. Overexpression of human chorionic gonadotropin causes multiple reproductive defects in transgenic mice. *Biol. Reprod.* 69:338–46

46. Matzuk MM, Finegold MJ, Su JG, Hsueh AJ, Bradley A. 1992. Alpha-inhibin is a tumour-suppressor gene with gonadal specificity in mice. *Nature* 360:313–19

47. Abel MH, Huhtaniemi I, Pakarinen P, Kumar TR, Charlton HM. 2003. Age-related uterine and ovarian hypertrophy in FSH receptor knockout and FSHbeta subunit knockout mice. *Reproduction* 125:165–73

48. Dukes M, Chester R, Yarwood L, Wakeling AE. 1994. Effects of a non-steroidal pure antioestrogen, ZM 189,154, on oestrogen target organs of the rat including bones. *J. Endocrinol.* 141:335–41

49. Sourla A, Luo S, Labrie C, Belanger A, Labrie F. 1997. Morphological changes induced by 6-month treatment of intact and ovariectomized mice with tamoxifen and the pure antiestrogen EM-800. *Endocrinology* 138:5605–17

50. Mahato D, Goulding EH, Korach KS, Eddy EM. 2000. Spermatogenic cells do not require estrogen receptor alpha for development or function. *Endocrinology* 141:1273–6

51. Mahato D, Goulding EH, Korach KS, Eddy EM. 2001. Estrogen receptor-alpha is required by the supporting somatic cells for spermatogenesis. *Mol. Cell. Endocrinol.* 178:57–63

52. Weihua Z, Makela S, Andersson LC, Salmi S, Saji S, et al. 2001. A role for estrogen receptor beta in the regulation of growth of the ventral prostate. *Proc. Natl. Acad. Sci. USA* 98:6330–35

53. Dupont S, Krust A, Gansmuller A, Dierich A, Chambon P, Mark M. 2000. Effect of single and compound knockouts of estrogen receptors alpha (ERalpha) and beta (ERbeta) on mouse reproductive phenotypes. *Development* 127:4277–91

54. Krege JH, Hodgin JB, Couse JF, Enmark E, Warner M, et al. 1998. Generation and reproductive phenotypes of mice lacking estrogen receptor beta. *Proc. Natl. Acad. Sci. USA* 95:15677–82

55. Lim H, Paria BC, Das SK, Dinchuk JE, Langenbach R, et al. 1997. Multiple reproductive failures in cyclooxygenase 2-deficient mice. *Cell* 91:197–208

56. Sicinski P, Donaher JL, Geng Y, Parker SB, Gardner H, et al. 1996. Cyclin D2 is an FSH-responsive gene involved in gonadal cell proliferation and oncogenesis. *Nature* 384:470–74

57. White R, Leonardsson G, Rosewell I, Jacobs MA, Milligan S, Parker M. 2000. The nuclear receptor co-repressor Nrip1 (RIP140) is essential for female fertility. *Nat. Med.* 6:1368–74

58. Couse JF, Hewitt SC, Bunch DO, Sar M, Walker VR, et al. 1999. Postnatal sex reversal of the ovaries in mice lacking estrogen receptors alpha and beta. *Science* 286:2328–31

59. Dupont S, Dennefeld C, Krust A, Chambon P, Mark M. 2003. Expression of Sox9 in granulosa cells lacking the estrogen receptors, ERalpha and ERbeta. *Dev. Dyn.* 226:103–6

60. Ogawa S, Chester AE, Hewitt SC, Walker VR, Gustafsson JA, et al. 2000. Abolition of male sexual behaviors in mice lacking estrogen receptors alpha and beta (alpha beta ERKO). *Proc. Natl. Acad. Sci. USA* 97:14737–41

61. Vasudevan N, Ogawa S, Pfaff D. 2002. Estrogen and thyroid hormone receptor interactions: physiological flexibility by molecular specificity. *Physiol. Rev.* 82:923–44

62. Simpson ER, Clyne C, Rubin G, Boon WC, Robertson K, et al. 2002. Aromatase—a brief overview. *Annu. Rev. Physiol.* 64:93–127

63. Toda K, Takeda K, Okada T, Akira S, Saibara T, et al. 2001. Targeted disruption of the aromatase P450 gene (Cyp19) in mice and their ovarian and uterine responses to 17beta-oestradiol. *J. Endocrinol.* 170:99–111

64. Jakacka M, Ito M, Martinson F, Ishikawa T, Lee EJ, Jameson JL. 2002. An estrogen receptor (ER)alpha deoxyribonucleic acid-binding domain knock-in mutation provides evidence for nonclassical ER

pathway signaling in vivo. *Mol. Endocrinol.* 16:2188–201

65. Swope DL, Castranio T, Koonce L, Mishna Y, Korach KS. 2003. *An estrogen receptor alpha AF-2 domain knock-in mutation results in early embryonic lethality in mice OR 34–6.* Presented at Endocrine Soc. 85th Annu. Meet. Philadelphia, PA

66. Clark JH, Peck EJ Jr. 1979. *Female Sex Steroids.* Berlin/Heidelberg/New York: Springer-Verlag. 245 pp.

67. Katzenellenbogen BS, Bhakoo HS, Ferguson ER, Lan NC, Tatee T, et al. 1979. Estrogen and antiestrogen action in reproductive tissues and tumors. *Recent Prog. Horm. Res.* 35:259–300

68. Perez MC, Furth EE, Matzumura PD, Lyttle CR. 1996. Role of eosinophils in uterine responses to estrogen. *Biol. Reprod.* 54:249–54

69. Griffith JS, Jensen SM, Lunceford JK, Kahn MW, Zheng Y, et al. 1997. Evidence for the genetic control of estradiol-regulated responses. Implications for variation in normal and pathological hormone-dependent phenotypes. *Am. J. Pathol.* 150:2223–30

70. Pollard JW, Pacey J, Cheng SV, Jordan EG. 1987. Estrogens and cell death in murine uterine luminal epithelium. *Cell Tissue Res.* 249:533–40

71. Nelson KG, Takahashi T, Bossert NL, Walmer DK, McLachlan JA. 1991. Epidermal growth factor replaces estrogen in the stimulation of female genital-tract growth and differentiation. *Proc. Natl. Acad. Sci. USA* 88:21–25

72. Curtis SW, Washburn T, Sewall C, DiAugustine R, Lindzey J, et al. 1996. Physiological coupling of growth factor and steroid receptor signaling pathways: estrogen receptor knockout mice lack estrogen-like response to epidermal growth factor. *Proc. Natl. Acad. Sci. USA* 93:12626–30

73. Klotz DM, Hewitt SC, Ciana P, Raviscioni M, Lindzey JK, et al. 2002. Requirement of estrogen receptor-alpha in insulin-like growth factor-1 (IGF-1)-induced uterine responses and in vivo evidence for IGF-1/estrogen receptor cross-talk. *J. Biol. Chem.* 277:8531–37

74. Klotz DM, Hewitt SC, Korach KS, DiAugustine RP. 2000. Activation of a uterine insulin-like growth factor I signaling pathway by clinical and environmental estrogens: requirement of estrogen receptor-alpha. *Endocrinology* 141:3430–39

75. Ignar-Trowbridge DM, Teng CT, Ross KA, Parker MG, Korach KS, McLachlan JA. 1993. Peptide growth factors elicit estrogen receptor-dependent transcriptional activation of an estrogen-responsive element. *Mol. Endocrinol.* 7:992–98

76. Ignar-Trowbridge DM, Pimentel M, Parker MG, McLachlan JA, Korach KS. 1996. Peptide growth factor cross-talk with the estrogen receptor requires the A/B domain and occurs independently of protein kinase C or estradiol. *Endocrinology* 137:1735–44

77. Ignar-Trowbridge DM, Nelson KG, Bidwell MC, Curtis SW, Washburn TF, et al. 1992. Coupling of dual signaling pathways: epidermal growth factor action involves the estrogen receptor. *Proc. Natl. Acad. Sci. USA* 89:4658–62

78. Finn CA. 1965. Oestrogen and the decidual cell reaction of implantation in mice. *J. Endocrinol.* 32:223–29

79. Cheng JG, Chen JR, Hernandez L, Alvord WG, Stewart CL. 2001. Dual control of LIF expression and LIF receptor function regulate Stat3 activation at the onset of uterine receptivity and embryo implantation. *Proc. Natl. Acad. Sci. USA* 98:8680–85

80. Robb L, Dimitriadis E, Li R, Salamonsen LA. 2002. Leukemia inhibitory factor and interleukin-11: cytokines with key roles in implantation. *J. Reprod. Immunol.* 57:129–41

81. Grummer R, Hewitt SW, Traub O, Korach KS, Winterhager E. 2004. Different regulatory pathways of endometrial connexin expression: pre-implantation hormonal-mediated pathway versus embryo implantation-initiated pathway. *Biol. Reprod.* 71:273–81

82. Yamamoto T, Matsuda T, Junicho A, Kishi H, Saatcioglu F, Muraguchi A. 2000. Cross-talk between signal transducer and activator of transcription 3 and estrogen receptor signaling. *FEBS Lett.* 486:143–48

83. Key TJ, Pike MC. 1988. The role of oestrogens and progestagens in the epidemiology and prevention of breast cancer. *Eur. J. Cancer Clin. Oncol.* 24:29–43

84. Pike MC, Spicer DV, Dahmoush L, Press MF. 1993. Estrogens, progestogens, normal breast cell proliferation, and breast cancer risk. *Epidemiol. Rev.* 15:17–35

85. Bargmann CI, Hung MC, Weinberg RA. 1986. The neu oncogene encodes an epidermal growth factor receptor-related protein. *Nature* 319:226–30

86. Slamon DJ, Godolphin W, Jones LA, Holt JA, Wong SG, et al. 1989. Studies of the HER-2/neu proto-oncogene in human breast and ovarian cancer. *Science* 244:707–12

87. Cardiff RD, Muller WJ. 1993. Transgenic mouse models of mammary tumorigenesis. *Cancer Surv.* 16:97–113

88. Bocchinfuso WP, Hively WP, Couse JF, Varmus HE, Korach KS. 1999. A mouse mammary tumor virus Wnt-1 transgene induces mammary gland hyperplasia and tumorigenesis in mice lacking estrogen receptor-alpha. *Cancer Res.* 59:1869–76

89. Hewitt SC, Bocchinfuso WP, Zhai J, Harrell C, Koonce L, et al. 2002. Lack of ductal development in the absence of functional estrogen receptor alpha delays mammary tumor formation induced by transgenic expression of ErbB2/neu. *Cancer Res.* 62:2798–805

90. Yoshidome K, Shibata MA, Couldrey C, Korach KS, Green JE. 2000. Estrogen promotes mammary tumor development in C3(1)/SV40 large T-antigen transgenic mice: paradoxical loss of estrogen receptor alpha expression during tumor progression. *Cancer Res.* 60:6901–10

91. Byrne C, Schairer C, Brinton LA, Wolfe J, Parekh N, et al. 2001. Effects of mammographic density and benign breast disease on breast cancer risk (United States). *Cancer Causes Control* 12:103–10

92. Byrne C, Schairer C, Wolfe J, Parekh N, Salane M, et al. 1995. Mammographic features and breast cancer risk: effects with time, age, and menopause status. *J. Natl. Cancer Inst.* 87:1622–29

93. Lydon JP, Sivaraman L, Conneely OM. 2000. A reappraisal of progesterone action in the mammary gland. *J. Mammary Gland Biol. Neoplasia* 5:325–38

94. Shim WS, DiRenzo J, DeCaprio JA, Santen RJ, Brown M, Jeng MH. 1999. Segregation of steroid receptor coactivator-1 from steroid receptors in mammary epithelium. *Proc. Natl. Acad. Sci. USA* 96:208–13

95. Soyal S, Ismail PM, Li J, Mulac-Jericevic B, Conneely OM, Lydon JP. 2002. Progesterone's role in mammary gland development and tumorigenesis as disclosed by experimental mouse genetics. *Breast Cancer Res.* 4:191–96

Annu. Rev. Physiol. 2005. 67:309–33
doi: 10.1146/annurev.physiol.66.032802.154710
Copyright © 2005 by Annual Reviews. All rights reserved
First published online as a Review in Advance on October 14, 2004

LIGAND CONTROL OF COREGULATOR RECRUITMENT TO NUCLEAR RECEPTORS

Kendall W. Nettles and Geoffrey L. Greene

*The University of Chicago, The Ben May Institute for Cancer Research, Chicago,
Illinois 60637; email: knettles@uchicago.edu; ggreene@uchicago.edu*

Key Words   coactivator, corepressor, steroid hormone, structure

■ **Abstract**   Nuclear receptors modulate transcription through ligand-mediated recruitment of transcriptional coregulator proteins. The structural connection between ligand and coregulator is mediated by a molecular switch, made up of the most carboxy-terminal helix in the ligand-binding domain, helix 12. The dynamics of this switch are thought to underlie ligand specificity of nuclear receptor signaling, but the details of this control mechanism have remained elusive. This review highlights recent structural work on how the ligand controls this molecular switch and the modulation of this signaling pathway by receptor subtype and dimer partner.

INTRODUCTION

The nuclear receptor (NR) gene family represents a class of 48 known transcription factors that modulate gene expression in response to lipophilic ligands (1, 2). Nuclear receptors are broadly implicated in normal physiological development and metabolism and represent therapeutic targets for a wide range of human diseases, including cancer, endocrine and metabolic disorders, and heart disease (3–5). All NRs have a DNA-binding domain and a carboxy-terminal ligand-binding domain (LBD) with conserved tertiary structures. Some NRs also contain an amino-terminal activation function (AF1), which can activate transcription in a ligand-independent fashion and is divergent among NR family members.

The NR LBD contains a second activation function (AF2) that maps to a surface-exposed hydrophobic pocket, providing a docking site for coregulatory proteins (6). Coactivators, including Mediator and members of the p160 family (SRCs1–3), bind to the AF2 surface via the amino acid motif LxxLL, in which the leucine residues dock into the hydrophobic cleft (7, 8). Binding specificity is added by oppositely charged amino acids at either end of the NR hydrophobic cleft that form a charge clamp with the LxxLL peptide backbone (Supplemental Movie 1; see Supplemental Material link on Annual Reviews home page: http://www.annualreviews.org). The mechanism of transcriptional activation by NRs is via recruitment of these coactivators, which mediate chromatin remodeling and also recruit the basal transcription apparatus (Figure 1A, see color insert).

The most carboxy-terminal helix of the LBD, helix 12, acts as a molecular switch (9), forming one side of the AF2 surface by docking against helices 3 and 11 in the presence of agonist ligands (Figure 2*A*, see color insert) (10). The remainder of the AF2 surface is formed by helices 3–5 for the estrogen receptors (ERα and ERβ), which is equivalent to helices 3–4 in the retinoid X receptor (RXR) family. Antagonists such as tamoxifen obstruct AF2 through a bulky side chain that protrudes into the AF2 surface (Figure 2*B*), displacing helix 12 and preventing coactivator recruitment to the LBD (8, 11). Helix 12 is relocated into the hydrophobic cleft with some antagonists, an effect mediated by sequences in helix 12 homologous to the LxxLL motif. Thus helix 12 provides allosteric control of transcription mediated by its dynamic localization.

The AF2 surface regulates the specificity of NR signaling by differential recruitment of cofactors to the various NR subtypes, such as ERα and ERβ (12–14). NR subtypes have distinct tissue distributions and phenotypes (15–17), making the development of subtype-specific ligands of widespread importance.

Dimerization provides another level of control of NR function (18). Whereas the steroid receptors generally function as homodimers, the so-called type II NRs act as obligate heterodimers with RXR. Permissive dimers, such as RXR with the peroxisome proliferator activator receptor (PPAR), respond to ligands for both receptors (19, 20). In contrast, nonpermissive heterodimers, such as the thyroid receptor (TR), vitamin D receptor (VDR), or retinoic acid receptor (RAR), are generally not activated by RXR ligands and inactivate the RXR AF2 (11, 21). However, RXR ligands are still able to synergize with both PPAR and RAR ligands (20, 22). Unexpectedly, the RXR antagonist LG754 is able to activate heterodimers of PPAR or RAR via stimulation of the partner AF2 (23, 24). This signaling across the heterodimer interface provides an exquisite degree of combinatorial control of gene expression.

Helix 12 is the structural link between ligand and coactivator, but the details of this communication and its differential regulation by receptor, receptor subtype, and dimer partner have remained elusive. A number of recent structural and functional studies are beginning to shed light on the details of these signaling pathways, facilitating an improved understanding of how NR ligands control cofactor recruitment.

SELECTIVE NUCLEAR RECEPTOR MODULATORS

Ligands with mixed agonist/antagonist characteristics have been developed for a number of NRs (reviewed in 25). The prototypical example, tamoxifen, blocks estrogenic activity in the breast (26) while acting as an estrogen or partial estrogen in the bone and uterus (27, 28). This selective modulation derives from a complete blockade of AF2 activity, but a tissue-specific activation of AF1 (Figure 1) (29–31). Second generation selective nuclear receptor modulators (SNRMs), such as raloxifene, maintain the desired therapeutic profile of antagonist activity in the breast and agonist activity in the bone, but do not elicit the uterotrophic effects seen with tamoxifen (32). An understanding of the molecular basis for this tissue selectivity is critical to the development of therapeutics with the desired agonist/antagonist profile.

The mixed agonist/antagonist properties of SNRMs are associated with differential recruitment of coactivators versus corepressors and the tissue-selective expression profiles of these coregulators (25, 33–35). For example, Brown and colleagues (36) showed that in Ishikawa endometrial cancer cells, tamoxifen induced recruitment of p160 coactivator(s) to the c*Myc* gene, whereas raloxifene promoted NCOR recruitment to the same nonclassical ER response element sequence. These and other data (34, 37) suggest that corepressor recruitment to the ligand-binding domain silences the receptor, including AF1.

Nuclear receptor antagonists preclude formation of the AF2 agonist conformation by directly blocking the docking of helix 12 against helices 3 and 11, thereby preventing formation of the coactivator-binding pocket. The crystal structures of several SNRM-NR complexes, including tamoxifen- and raloxifene-ERα LBDs, show that helix 12 relocates into the coactivator-binding cleft (11, 38, 39), blocking coactivator recruitment. These molecules contain an extended side chain that protrudes from the ligand-binding pocket and occupies the same space as the agonist conformation of helix 12 (Figure 2*C*). The question, then, is how to explain corepressor recruitment.

Converging evidence demonstrates that the corepressor-binding site overlaps partially with the coactivator-binding site (40–42). The key difference lies in the longer helical sequence of the corepressor CoRNR box interaction motif (43). The removal of helix 12 from the agonist conformation reveals a long hydrophobic groove that accommodates the extended corepressor-binding sequence (Figure 2*A,C*). The ability of SNRMs to relocate helix 12 into this hydrophobic cleft suggests that helix 12 in fact competes with corepressors for binding. This hypothesis has been confirmed by the large increase in corepressor binding that results from deletion or mutation of helix 12 (34, 44–47).

Crystal structures for unliganded PPAR and RXR show alternate localizations for helix 12, neither of which is consistent with the docking of corepressor in the hydrophobic AF2 cleft. For apo-PPAR, helix 12 is either in the agonist conformation or in the hydrophobic cleft. In contrast, for RXR, helix 12 is extended and contacts the AF2 surface in a neighboring molecule. This conformation appears to contribute to the unique auto-inhibited tetramer found with unliganded RXR and thus seems unlikely to be found in other NRs that do not form tetramers.

The crystal structure of the PPARα LBD bound to a peptide derived from the SMRT corepressor confirms that helix 12 must be relocated out of the AF2 pocket to allow corepressor binding (48). This structure demonstrates that helix 12 relocates to a hydrophobic shelf on helix 3 (Figure 2*C*), adjacent to the AF2 pocket. This hydrophobic shelf is evident in a number of structures, including RAR, TR, LXR, and VDR (Supplemental Figure 1; see Supplemental Material link on Annual Reviews home page: http://www.annualreviews.org). With FXR, a second coactivator peptide was demonstrated to bind to this site [Supplemental Figure 1; (49)]. The interaction of both helix 12 and coactivator with the helix 3 shelf suggests that this region may function as a physiologically significant protein docking site (50).

SNRM Agonist Conformation

While we originally described the localization of helix 12 in the hydrophobic cleft as the antagonist conformation (51), more recent analyses suggest that this is in fact the SNRM agonist conformation, allowing AF1 activity by blocking core-pressor recruitment to the LBD. In this model, full antagonist activity requires relocation of helix 12 out of the cleft. The tissue selectivity of SNRM agonist activity is then a function of the cell-type-specific complement of coactivators and corepressors and the affinity of the coactivators for AF1. This hypothesis regarding the SNRM agonist conformation is supported by a number of observations: (a) The full antagonist ICI bound to ERβ demonstrated no structure for helix 12 (52); (b) deletion of helix 12 blocks SNRM agonist activity; and (c) corepressor binding requires removal of helix 12 from the hydrophobic cleft. This model makes specific predictions about the differences between SNRMs with respect to the dynamics of helix 12.

Structure Activity Relationships for SNRMs

Current data support a model whereby SNRMs regulate corepressor recruitment through control of the dynamic localization of helix 12. This model suggests several structural features that may underlie the differential agonist activity observed for tamoxifen versus raloxifene, for example. In particular, tamoxifen may stabilize helix 12 in the hydrophobic cleft more than raloxifene, consistent with the more limited corepressor recruitment by tamoxifen (34, 36). This hypothesis is supported by the corresponding crystal structures, which demonstrate increased thermal mobility of helix 12 for raloxifene (higher B factors) and the lack of an ordered structure for the h11-h12 loop (51), indicative of a less stable docking of helix 12 within the cleft (Figure 3A, see color insert). The longer side chain of raloxifene directly contacts helix 12, which may account for its destabilizing effect.

The direct contact between raloxifene and helix 12 raises the intriguing possibility that the SNRM side chain can also directly contact corepressor. For example, for the ICI-ERβ LBD complex (52), a lengthy side chain on the ligand extends into the AF2 surface and would likely interfere with corepressor binding (Figure 3B). However, this longer side chain also effectively competes for H12, which is unstructured in this complex. A comparison of the raloxifene-ERα LBD structure with the SMRT-PPARα structure suggests that raloxifene may contact corepressor in a stabilizing fashion (Figure 3C). In fact, ERα in which helix 12 has been deleted demonstrates a much greater affinity for GST-NCOR in the presence of raloxifene than with tamoxifen or ICI-182,780 (K. Nettles & G. Greene, unpublished data), consistent with this observation. Thus SNRMs may directly contact helix12, and possibly corepressor, in ways that are stabilizing or destabilizing.

For ERα, the interaction of the SNRM side chain with a charged helix 3 residue, Asp351, also contributes to selectivity through an unknown mechanism. The positive charge on the raloxifene piperidine nitrogen is critical to its antagonistic behavior, as substitution of a carbon for this nitrogen generates agonist activity similar to tamoxifen (53). A mechanism for this effect is suggested by the PPAR corepressor structure. ERα Asp351 lies in the center of the helix 3 shelf (Figure 2D), generating a polar surface that would not provide a hydrophobic docking site for helix 12. Thus raloxifene, by masking the charge on Asp351, may stabilize h12 binding to helix 3, as seen with PPARα and SMRT. Wheras tamoxifen also contains a positively charged dimethylamino group, it makes a weaker electrostatic interaction with Asp351 because of its positioning (3.4 Å versus 2.7 Å from the COOH), and thus does not mask the Asp351 charge as fully as raloxifene. Mutations at this position in helix 3 have profound effects on SNRM agonist activity, as discussed below in the section on Mutation Analysis.

Understanding SNRM selectivity in terms of the differential dynamics of helix 12 is an internally consistent model, but currently no published data directly relate helix 12 mobility to SNRM behavior. Another model proposes that ligand-mediated receptor proteolysis may contribute to differences among SNRMs. However, there is no direct correlation between activity and NR degradation. For example, full agonists and antagonists promote degradation of ER, whereas SNRMs such as tamoxifen have a stabilizing effect (54). Observed differences in helix 12 conformations in the various SNRM structures could also derive from the packing of the molecules in the crystal lattice, rather than real differences in solution. Fluorescent labeling of helix 12, first applied to PPAR (55), provides a means to measure helix 12 dynamics and could be used to test this model.

INDIRECT ANTAGONISM

A number of full NR antagonists exist that do not have the protypical side chain found in SNRMs. Flutamide is a potent AR antagonist, as is progesterone acting upon the mineralicorticoid receptor, although both of these ligands are similar in size to AR/MR agonists. A structural basis for this phenomenon was suggested by the crystal structure of ERβ LBD bound to the R,R enantiomer of 5,11-cis-diethyl-5,6,11,12-tetrahydrochrysene-2,8-diol (THC), which demonstrated that antagonist activity derived from the ligand-induced mis-positioning of helix 11 (56). With THC, the ligand indirectly relocates helix 12 via the positioning of helix 11.

For all NRs, helix 11 contacts both the ligand and helix 12 in the agonist conformation and is thus positioned to transmit structural information between the ligand and coactivator. When bound to ERα, THC induces the agonist conformation of the receptor. In contrast, when bound to ERβ, a lateral shift in the positioning of the ligand pushes helix 11 into the space that would be occupied by the agonist conformation of helix 12 (Figure 4A, see color insert), and thus obstructs the agonist conformation indirectly via the positioning of helix 11 (57). This mechanism contrasts with SNRMs that have an extended side chain that interferes directly with the helix 12 agonist conformation.

The importance of this phenomenon to other NRs is illustrated by the inverse agonist activity of DES on ERRγ (58). While DES acts as a full agonist on ER, it antagonizes the constitutive activity of apo-ERRγ by altering the agonist conformation of helix 11. Because the ligand-binding pocket is smaller than with ER, DES shifts ERR Phe435 in H11 so as to obstruct helix 12. The phenomenon of indirect antagonism identifies a novel approach to structure-based drug design, in which the orientation of the ligand with respect to helix 11 can determine transcriptional responses. This idea is explored further in the next section.

THE STRUCTURAL BASIS OF PARTIAL AGONIST ACTIVITY

It has been a long-standing mystery that high-affinity ligands can induce low levels of transcription, presumably by eliciting a conformation that is suboptimal for coactivator recruitment. One mechanism for this phenomenon is apparent in the NRs for which the ligand directly contacts and stabilizes helix 12 in the agonist conformation. For PPAR, an acidic head group found in full agonist ligands participates in a hydrogen bond network with both helix 11 and helix 12 (59, 60). The structure of the PPAR partial agonist, GW0072, reveals a loss of this stabilizing interaction (61). This suggests that the dynamics of helix 12 are differentially regulated by partial versus full agonist ligands.

The stabilization of helix 12 in the agonist conformation is also controlled by its interactions with helix 11. For the published high-affinity partial agonist structures, including ligands for PPAR, ER, and RXR, helix 11 is shifted away from helix 12 (Figure 4A), leading to loss of stabilizing interactions (57). Therefore, the ligand can control the dynamics of helix 12 through both direct interactions and indirectly through helix 11, as demonstrated for a full thyroid receptor agonist (GC-24) that distorts helices 3 and 11 into suboptimal positions but has additional stabilizing contacts with helix 12 that allow full activity (62). Thus it is the balance of these interactions that determines the relationship between ligand and helix 12 conformations.

SIGNALING ACROSS THE DIMER INTERFACE

For NRs that dimerize with RXR, heterodimers differ in ligand communication across the dimer interface. The permissive heterodimers, including PPAR and LXR, demonstrate activation by either liganded monomer partner and show a synergy between ligands (63, 64). Conditional heterodimers, such as RAR, demonstrate a reduced effect of RXR ligand (65–67), whereas nonpermissive receptors, including VDR and TR, are generally not responsive to RXR ligand (68, 69). A structural

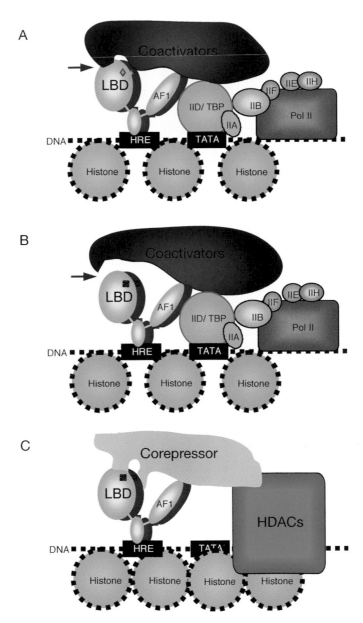

Figure 1 Transcriptional regulation by nuclear receptors. (*A*) Agonist ligands recruit coactivators to the DNA-bound NR via the ligand-binding domain and the amino-terminal AF1. (*B*) Selective nuclear receptor modulator (SNRM) agonist activity is associated with a blockade of coactivator recruitment to the LBD; however, activity in AF1 that is dependent on the tissue-specific presence of coactivators in maintained. (*C*) Full antagonist activity derives from recruitment of corepressors and histone deacety-lase enzymes (HDACs), which precludes AF1 association with coactivators.

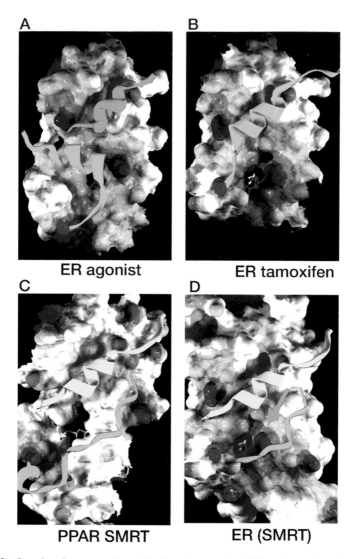

Figure 2 Overlapping coregulator-binding sites on the AF2 surface. A surface rendering of various NR LBDs are colored by electrostatic potential. Helix 12 is depicted as
a blue ribbon, the LxxLL containing peptide is colored green, and the SMRT corepressor peptide is colored yellow. (*A*) ER bound to the agonist DES and an LxxLL peptide
from the Grip protein. (*B*) ER bound to tamoxifen. (*C*) PPAR bound to the antagonist
GW6471 and a peptide from the SMRT corepressor. (*D*) Helix 12 and the SMRT peptide from the PPAR/SMRT structure are superimposed onto the ER LBD. A green arrow
points to Asp351.

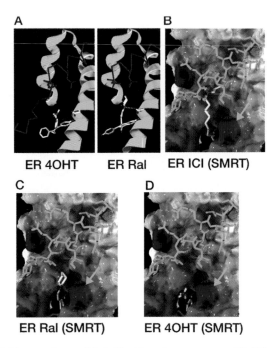

Figure 3 SNRM interactions with helix 12 and corepressor. (*A*) ERα LBD is shown bound to tamoxifen (4OHT) or raloxifene, with helices 3–5 in blue ribbon, and helix 12 in red. Asp 351 is shown forming a hydrogen bond with raloxifene. (*B–D*) The structure of PPAR with the SMRT peptide was superimposed onto the ER LBD over the portions of helices 3–5 that make up the coactivator-binding site. Shown is the surface of ER, colored by electrostatic potential, and the SMRT peptide, colored green. The green arrow points to Asp351.

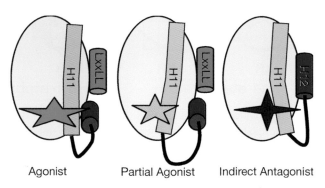

Figure 4 Helix 11 control of Helix 12. A cartoon of the NR LBD illustrates the positioning of helix 11 by the ligand (*stars*). With partial agonists, helix 11 is pulled away from helix 12, destabilizing the agonist conformation. With indirect antagonists, helix 11 is pushed into the space occupied by helix 12 in the agonist conformation, forcing it to relocate.

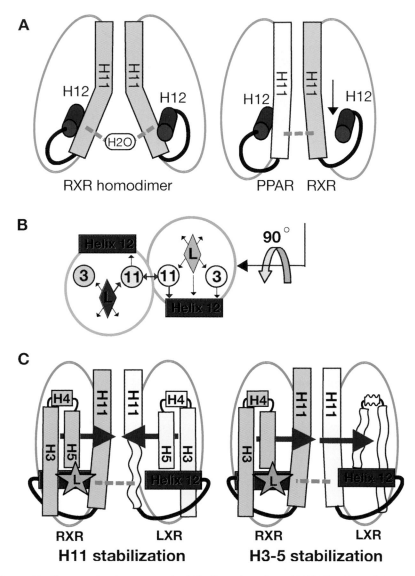

Figure 5 Communication across the NR dimer interface. (*A*) Helix 11 participates in an electrostatic tethering across the dimer interface for several NRs, shown by the green dotted lines. With PPAR heterodimers, RXR helix 11 is pulled toward the dimer interface, resulting in a loss of contacts with helix 12 (*arrow*). (*B*) Helix 11 contacts ligand, helix 12, and dimer partner all in the same plane, suggesting a mechanism for integration of information between these structural elements. (*C*) Two proposed models indicating how the statistically coupled amino acids (SCAs), identified by Mangelsdorf and colleagues (64), mediate allosteric signaling across the dimer interface. The red arrows represent the SCAs, which physically connect and bridge the dimer interface with the hydrophobic core of the AF2 pocket.

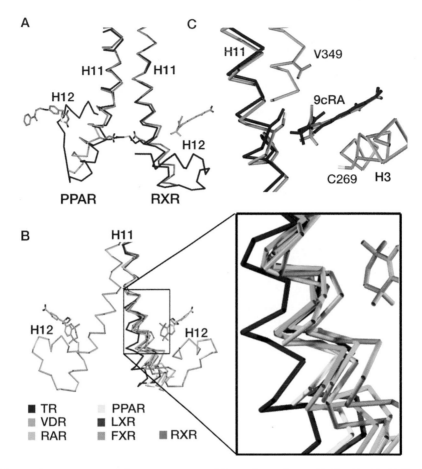

Figure 6 Receptor and dimer partner-specific conformations of helix 11. (*A*) Helix 11 is shown in gray for the RXRα homodimer, and in red for the PPAR/RXR heterodimer. The hydrogen bond across the PPAR/RXR dimer interface at the level of the ligand appears to draw RXR helix 11 closer to the dimer interface and away from both the ligand and helix 12. (*B*) A comparison of 9*cis* retinoic acid bound to RXR in a monomer (*gray*) or in the heterodimer with PPAR (*red*). Two amino acids are shown that directly contact only the ligand in the monomeric structure, using a 4.2 Å cut-off. (*C*) Several NRs were superimposed over the highly conserved N-terminal portion of helix 11, corresponding to residues 413–427 in RXRα.

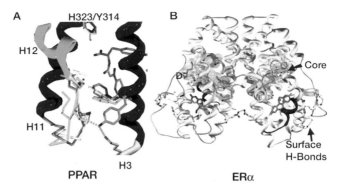

Figure 7 Nuclear receptor subtypes. (*A*) The PPAR subtypes superimposed on PPARγ, showing differences in both the pocket and surface that contribute to the shape of the ligand-binding pocket. (*B*) A dimer of ERα, highlighting residues in the hydrophobic core and surface that are specific to this ER subtype and that control the shape of the ligand-binding pocket.

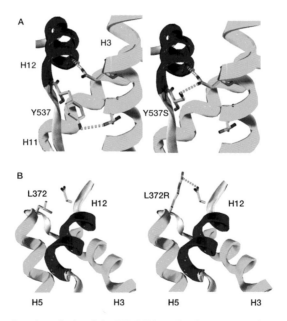

Figure 8 Mutational analysis of the ER LBD. (*A*) The ER LBD bound to the agonist DES is depicted as a ribbon diagram. On the left is shown how Tyr537 interacts with helix 3. The right panel illustrates how the constitutively active mutant Tyr537Ser is positioned to form a hydrogen bond with Asp351, based on molecular modeling. (*B*) The ER LBD bound to the antagonist tamoxifen shows helix 12 in the hydrophobic AF2 cleft. The mutation in helix 5, Leu372Arg, appears to form a hydrogen bond with helix 12 in a molecular model.

model of interdimer signaling needs to account for the following phenomena: (*a*) receptor-specific inactivation or subordination of RXR AF2; (*b*) heterodimer-mediated inhibition of RXR ligand affinity (70); (*c*) the phantom ligand effect whereby an RXR ligand can control the AF2 of the unliganded partner (24, 63); and (*d*) the modulation of these effects by the specific ligand pair (67, 71).

Two distinct structural features in the LBD were recently implicated in control of allosteric communication across the dimer interface. Mangelsdorf and colleagues identified a network of amino acids that are statistically coupled across evolution and specifically implicated in activation of the unliganded dimer partner by RXR (64). These amino acids are physically connected and cluster in the hydrophobic core of the receptor, bridging the dimer interface with the binding pockets for ligand and coactivator (Figure 5*D*, see color insert). A second pathway connects the dimer interface via an electrostatic tethering across helix 11 at the level of the ligand (Figure 4*B*, Figure 5*A*). Mutations at this site block the activation of RAR by RXR ligands (57). Potential structural mechanisms for these findings are discussed below.

Helix 11 and the Dimer Interface

Similar to the effects seen with partial agonists, helix 11 is modulated by the dimer partner, explaining the inactivation of RXR AF2 by the heterodimer partner (57). RXR helix 11 displays distinct conformations in homodimers, or in complex with PPAR, LXR, or RAR (10, 11, 59, 60, 72–80). The dimer partner shifts helix 11 relative to the agonist conformation of helix 12 so that for PPAR, RXR helix 11 resembles the conformation seen with partial agonist ligands (Figure 5*A*), which indicates a significant loss of direct contacts between helices 11 and 12 (57). For RAR, RXR helix 11 is shifted in the opposite direction so that helix 11 clashes with the agonist conformation of helix 12, similar to the indirect antagonism found with THC and ERβ. When heterodimerized with LXR, the backbone of RXR helix 11 is positioned similarly to the homodimer. These structural observations directly correlate with the biological activity of the receptors, as RXR AF2 is silenced by RAR (67, 70), inhibited by PPAR (63, 81), and fully active with LXR (64). Thus heterodimer control of helix 11 position explains how the dimer partner can alter AF2 activity.

The mechanism by which the dimer partner controls helix 11 conformation involves an electrostatic tethering across the dimer interface. An examination of the holo-RXR/PPAR structure reveals a hydrogen bond involving RXR Glu-434 and PPAR Gln-431, which links helix 11 from each molecule across the dimer interface in the plane of the ligand (Figure 5*A*, Figure 6*A*, see color insert). Notably, a charged or polar residue is highly conserved at this position among the RXR heterodimer partners (73), and mutations at this site have profound effects on receptor activity (57; A.I. Shulman & D.J. Mangelsdorf, personal communication).

The nature of the electrostatic tethering across helix 11 varies by receptor. Both ERα and RXR demonstrate an unusual Glu:Glu interaction in the helix 11 tethering that is stabilized by the surrounding amino acids. In both cases, Lys residues in the preceding turn of helix 11 (ERα Lys-520, RXRα Lys-451) form salt bridges with the glutamic acids (ER Glu-523, RXR Glu-454), masking their negative charge and allowing a closer interaction. The Glu:Glu distance varies between 2.6 and 5 Å in the RXR and ER dimers. The close distance in some structures [e.g., PDB code:1H9U (RXRβ), 3ERD (ERα)] suggests that the Glu residues are protonated, implying that the positioning of the lysine significantly increases the pKa of the Glu. In other structures [e.g., PDB code 1MVC, 1MV9 (RXRα)], water mediates bridging of the Glu residues. This structural connection across the dimer interface provides a possible conduit of information for dimer induced changes in the conformation of helix 11 and the resulting effects on helix 12 conformations.

The RAR/RXR heterodimer structure also demonstrates a Glu:Glu distance comparable to that seen with RXR and ER homodimers, suggesting that it may also participate in water-mediated tethering. In contrast, the LXR/RXR structure has the Glu:Glu distance at ∼7 Å, consistent with a lack of tethering for this combination. However mutational analysis of the LXR heterodimer showed that both partners are required for activation by RXR ligand, whereas only the LXR AF2 was needed for the LXR ligand responsivity (64). These data suggest that other pathways are involved in heterodimeric signaling or that the receptor can adopt other conformations not seen in the structure.

In addition to differences in the helix 11 tethering, NRs also differ in the degree of bending and orientation of helix 11 (Figure 6B). A comparison of the RXR homodimer with the PPAR and RAR heterodimer structures shows that the two helices 11 are oriented in concert, so that RXR helix 11 appears to bend with the partner helix 11 (Figure 6A). With LXR, helix 11 is bent more sharply, preventing electrostatic bridging across the dimer interface at the level of the ligand. Thus the receptor-specific bend of helix 11 may also contribute to cross dimer signaling. It is also noteworthy that the statistically coupled amino acids identified by Shulman et al. cluster at the middle of this bend in helix 11 and are thus are positioned to control this structural feature (discussed further below).

Heterodimer Control of Ligand Affinity

Helix 11 is a key determinant of NR ligand affinity and orientation. Dimer-mediated positioning of helix 11 also suggests an explanation for the loss of ligand affinity for RXR in RXR/RAR heterodimers and the variable results obtained with different ligand combinations (19, 66, 71, 82).

The interaction of retinoic acid with RXR demonstrates differential binding in the heterodimer compared with the monomer structure owing to the altered positioning of helix 11. In the PPAR heterodimers, helix 11 is pulled toward the dimer interface and away from the ligand binding pocket, which results in a repositioning of the β-ionone ring and a loss of contact with two amino acids (Figure 6C). Thus dimer-controlled positioning of helix 11 may contribute to RXR subordination by both modulating the conformation of helix 12 and reducing ligand affinity.

The electrostatic tethering of helix 11 suggests a mechanism whereby the ligand in one receptor positions helix 11 and contributes to the positioning of helix 11 in the dimer partner (Figure 5*B*). Helix 11 is positioned not only by the dimer partner but also by the specific ligand. This latter characteristic appears to vary widely by receptor, as some receptors show great variability in the ligand-mediated positioning of helix 11 (e.g., PPAR), whereas others show essentially none (e.g., RAR). Another factor that contributes to the flexibility of helix 11 is the conformation of helix 12, which in the agonist position limits the flexibility of helix 11. This model may explain how ligand affinity can be regulated by both dimer partner and the specific ligand within the dimer partner. Helix 11 contacts ligand, helix 12, and the dimer partner, allowing it to integrate structural information from all three components of NR signaling.

It is also noteworthy that both TR and VDR, which more completely inhibit the binding of RXR ligand, have structural features consistent with this effect. Helix 11 in TR has very little bend (Figure 6*B*) and would be expected to push RXR's helix 11 toward the ligand-binding pocket, thus closing the pocket as seen with the unliganded RXR structures. VDR has a lysine residue at the same position as that of the tethering glutamic acid in helix 11. This longer side chain would also be predicted to hydrogen bond more effectively with the unliganded RXR, resulting in a shift of helix 11 away from the dimer interface.

The Phantom Ligand Effect

The ability of RXR to activate an unliganded dimer partner likely involves dimer-mediated protein folding. The original "mouse-trap" concept of nuclear receptor activation postulated a single switch of conformation upon ligand binding (83). This concept has evolved to the statistical thermodynamic model associated with protein-folding theory (84). In this model, the conformation of receptor is not viewed as simply folded or unfolded, but as an ensemble of conformations. Ligand binding is not a switch from one state to another, but rather as the selection of a subset of the conformations competent to bind ligand and a shift in the population of associated conformations. Nuclear receptor ligands are thus thought to nucleate folding of the ligand and coactivator binding pockets through selection of a subset of conformations. The use of NMR to examine hydrogen-deuterium exchange demonstrates that local unfolding reactions are unevenly distributed within a protein (85), with added flexibility localizing to binding sites, suggesting that this is an important feature of ligand recognition. This hypothesis has been verified for one nuclear receptor, PPARγ, which was shown by NMR to be folded in the hydrophobic core, but not in the ligand-binding pocket in the absence of ligand (86). Ligand binding also protects the LBD from proteolytic digestion (63, 87, 88) and reduces the radius of gyration for various NRs (89), consistent with a key role for ligand in mediating protein folding. This hypothesis was recently verified with hydrogen-deuterium exchange and mass spectrometry of RXR (90), which showed ligand-mediated protection of deuterium exchange throughout the regions of the LBD that contact ligand.

There is also evidence that both the coactivator and dimer interfaces limit the conformational flexibility of receptor LBD. Binding of coactivator can slow the ligand off-rate (91) and limit the conformational flexibility of the ligand (92). Also, similar to the effects of ligand binding, dimerization confers protection from proteolytic digestion, an effect seen with both permissive and nonpermissive receptors (63, 87).

Given that unliganded receptors are conformationally flexible, RXR activation of unliganded dimer partner likely involves stabilization of the partner AF2 surface. The role of helix 11 in this process has been verified for RXR/RAR (57), in which mutation of the tethering Glu in either receptor to Gln or Asn blocks the phantom ligand effect.

RXR may also directly stabilize the AF2 surface via folding of helices 3–5 of the unliganded dimer partner. As mentioned above, Shulman et al. identified a set a 26 amino acids that physically link the dimer interface with the hydrophobic core of the AF2 pocket (64). Mutations in these highly conserved amino acids blocked the ability of RXR to activate the unliganded LXR without effecting activation by LXR ligands. Thus one possible explanation for permissivity is that this pathway elicits folding of the AF2 pocket (helices 3–5) only with permissive heterodimers (Figure 5C). Another possible explanation is that the statistically coupled amino acids act to stabilize the helix 11 portion of the dimer interface but that the specific conformation of helix 11 varies by receptor (Figure 6B).

RECEPTOR SUBTYPE–SPECIFIC SIGNALING

The development of ligands that target a specific subtype or isotype of a protein represents an important therapeutic goal for many classes of receptors. NR subtype–specific signaling occurs through differential regulation of the ligand-binding pocket and helix 12 dynamics. Differences in amino acids that line the ligand-binding pocket account for most of the subtype-selective ligands that have been developed. Full agonist ligands display a higher affinity for one or more subtypes of a receptor d on the basis of a shape that is complementary to the subtype-specific amino acids (79, 93). Another class of ligand acts as an agonist on one subtype but as a full antagonist on the other (94), an effect that depends only on the ligand-binding residues (57, 95).

Differences in amino acids that line the ligand-binding pocket play critical roles in subtype-specific ligand recognition. High-resolution structures of RARγ show that RARγ-selective ligands form an hydrogen bond with Met272, which corresponds to an isoleucine in the other subtypes (75). RARα has a helix 3 Ser232 that corresponds to alanine in the other subtypes and also contributes to differences in the shape of the pocket. As discussed below, this Ser232 also regulates the dynamics of helix 12 and cofactor recruitment because it is poised to contact both ligand and helix 12.

Ligands that are selective for the PPAR subtypes have been designed on the basis of the observed structural differences between subtypes. The L-tyrosine analogue farglitazar binds with higher affinity to PPARγ than to PPARα. The structure of farglitazar with PPARγ suggests that the benzophenone group, which interacts with the PPARγ His323, would clash with the corresponding PPARα Tyr314 (79). The reduction of the benzophenone by three carbons led to the development of GW409544, which binds with high affinity to both PPARγ and PPARα. Several PPARα crystal structures (59) substantiate the importance of Tyr314/His323 in PPAR selectivity (Figure 7A, see color insert). Mutation of this amino acid is sufficient to convert the subtype preferences to a substantial degree, highlighting the importance of a single amino acid as a contributor to differential ligand recognition (79).

Amino acids outside the ligand-binding pocket also make important contributions to subtype selectivity for the ER ligands PPT and THC. For some ligands, including DPN and HPTE, mutagenesis studies demonstrate that subtype selectivity is attributable to the two amino acids that differ in the ligand-binding pockets (57, 93, 95). For other ligands, such as PPT and THC, additional residues are involved. THC acts as an agonist on ERα and antagonist on ERβ. The crystal structures of THC bound to both ER subtypes (56) showed that for ERβ, the ligand was shifted toward helix 11 due to a narrowing of the other end of the pocket. This narrowing of the pocket was associated with amino acid differences in both the hydrophobic core and the surface of the receptor (57).

Regions outside of the pocket also appear to contribute to subtype specificity for PPARs. For PPARδ, there is a narrowing of the pocket near helix 12 that is associated with a Met in helix 11, which is a Leu in the other PPAR receptors (60). However, a conserved Phe in helix 3 is also positioned by a PPARδ-specific conformation of the loop between helices 11 and 12 (Figure 7A). PPARδ has a Tyr in helix 3 that forms a hydrogen bond with the helix 11–12 loop, which stabilizes a specific conformation of both structural elements. There are also amino acid differences in the helix 11–12 loop that contribute to the subtype-specific positioning of the conserved Phe in helix 3.

Subtype-specific differences in the shape of the pocket allow for opposing effects of ligands on different subtypes. For THC, as discussed above, the shift in helix 11 sterically hinders the agonist conformation with ERβ. A similar effect is seen with the mutations in the AR helix 3 that are associated with androgen-insensitivity syndrome. The change of Gly708 to Ala or Val changes the binding mode of AR antagonists, thereby eliminating partial agonist activity (96) and converting them into full antagonists. Differences in the PPAR subtypes allow the same ligand to have differential effects on corepressor recruitment, decreasing affinity for corepressor peptide in PPARγ but stimulating it in the other subtypes (97). These effects appear to be caused by subtype-specific differences in the positioning of the ligand that impinge upon the loop between helix 11 and 12. In PPARγ the interaction is stabilizing, whereas the smaller pockets of the other subtypes force the ligand to remodel this loop in a destabilizing fashion. These studies highlight how the ligand position contributes to cofactor recruitment by altering the dynamics of helix 12.

The localization of helix 12 also contributes to ligand positioning and affinity. As discussed above, THC is differentially positioned by the two ER subtypes, causing helix 11 in ERβ to flex into the space occupied by the agonist conformation of helix 12 (Figure 4). The conversion of THC into an agonist on ERβ required mutations that not only altered the shape of the pocket but also stabilized helix 12 in the agonist conformation (57). By locking helix 12 in the agonist position, these mutations presumably limit the flexibility of helix 11 and reposition the ligand deeper into the pocket. Coactivator binding has also been shown to limit the off-rate of various ligands, again presumably by limiting the dynamics of helix 12 (91). The binding of coactivator to PXR was also demonstrated to directly alter ligand binding (92). In the absence of coactivator, the PXR ligand adopts multiple conformations, which, upon addition of coactivator, resolve into a single, unique position in the crystal structure. The removal of helix 12 from the agonist position allows the ligand access to the solvent, which is thought to be an obligate step in ligand escape (98). Thus the coactivator is an active participant in controlling ligand positioning and dynamics.

NRs demonstrate subtype-specific differences in cofactor recruitment attributed to differential dynamics of helix 12. RAR subtypes differ substantially in the basal recruitment of corepressor and transcriptional repression (99, 100). RARα transcriptional repression and corepressor recruitment was attributed largely to a single amino acid in helix 3 that interacts with the agonist conformation of helix 12. Ser232 in RARα helix 3 is associated with a destabilization of helix 12 and increased corepressor association. Conversion of the corresponding alanine in RARβ or RARγ to any electrophilic residue enhanced corepressor recruitment (99), supporting this interpretation.

Differences in the transactivation profiles of the ER subtypes are also attributable to amino acids that stabilize the agonist conformation (57). ERβ demonstrates reduced transcriptional potency (EC_{50}) with estradiol compared with that of ERα, despite the near identical affinities of both ERs for estradiol. The addition of amino acids that add hydrogen bonds found in ERα increases the potency of estradiol with ERβ, suggesting that the dynamics of helix 12 underlie the discrepancy between affinity and transcriptional potency.

LIGAND-INDEPENDENT ACTIVITY

Recently, the crystal structures of several constitutively active NRs have been solved, revealing the mechanisms by which the active conformations are stabilized in the absence of ligand. With the NGFI-B/Nurr1 family, including the insect ortholog DHR38, the ligand-binding cavity is absent owing to the presence of bulky phenylalanine side chains that completely fill it (101, 102). These NRs are also unusual in that the coactivator-binding cleft is substantially different from other NRs and does not appear to recruit coactivators, relying instead on a strong

AF1. However, deletion of helix 12 still disrupts signaling of the RXR heterodimer, suggesting that activation of RXR AF2 may play a critical role in heterodimeric signaling. As discussed above, the removal of helix 12 from the NR agonist conformation is associated with greater flexibility in helix 11, which provides a potential conduit for structural information across the dimer interface with these receptors. Thus helix 12 of DHR38 may be required for stabilization of the RXR AF2.

Another subset of constitutively active NRs, including ERR, CAR, and LHR-1 (58, 103, 104), which vary widely in the size of the empty cavity, display a ligand-binding cavity but no requirement for ligand to stabilize the active conformation. Although these receptors show high basal activity, they can still be regulated by ligands. ERRγ is antagonized by tamoxifen and the ER agonist diethylstilbesterol (DES) (58). This indirect antagonism (inverse agonism) is similar to that seen with THC and ERβ in that the ligand does not directly relocate helix 12, but instead remodels helix 11 (56, 57). The inactivation of a constitutively active receptor appears to represent a physiologically relevant phenomenon, as CAR activity is suppressed by certain androstane metabolites (105). Both CAR and ERR can also be superactivated: ERR by phytoestrogens (106) and CAR by other agonist ligands (107, 108).

A third class of NRs can bind ligand as a structural factor that does not undergo exchange. For example, HNF-4 was crystallized with a fortuitous fatty acid molecule (109) that could not be exchanged from the receptor (110). In addition, RORα was crystallized with both cholesterol and its sulfate metabolite, leading to the suggestion that it functions as a cholesterol receptor (111, 112). However, it is not clear if physiologically relevant changes in cholesterol are sensed by this receptor or if it is constitutively bound.

These studies highlight the idea that ligands may not always represent the physiological activator for all NRs. Some receptors display ligand-independent activity in certain contexts. For example, in vivo imaging of ER/ERE responses in a transgenic ERE-Luc mouse model suggests that ER functions primarily independently of ligand in several nonreproductive tissues, including bone and brain (113). ER can be rendered ligand independent through phosphorylation of the AF1 region, a process also implicated in the development of tamoxifen resistance in breast cancers (114). Many NRs also actively suppress transcription in the absence of ligand, a critical feature of gene regulation. Thus ligand-independent regulation of transcriptional activity is a critical feature of NR physiology for both NRs that display ligand binding and those that do not.

MUTATIONAL DATA AND HELIX 12 DYNAMICS

There is a large body of experimental and genetic data on mutations in the AF2 surface that can be explained in the context of the overlapping binding sites for helix 12, coactivator, and corepressor. Thus a mutation that increases agonist activity could act by stabilizing helix 12 in an agonist conformation, blocking corepressor recruitment, and/or stimulating coactivator recruitment. The wealth of structural information for the NR LBD suggests a mechanism of action for many of these mutations. A few representative examples are presented to illustrate this principle.

Mutations at the interface of helix 12 and the remainder of the receptor regulate corepressor recruitment, helix 12 localization, and transcriptional activity of NRs. A number of mutations appear to regulate the stability of helix 12 docking in the agonist conformation against helices 3 and 11 (115–118). In a screen of patients with severe insulin resistance, three subjects demonstrated mutations in the PPARγ gene that map to the amino terminus of helix 12 or to helix 3 (116). These mutations were predicted to disrupt the agonist conformation and are associated with a loss of transactivation and coactivator recruitment, and enhanced corepressor recruitment. The role of helix 12 dynamics in this phenomenon was verified by fluorescent labeling of the end of helix 12 (55).

The crystal structures of PPAR bound to rosiglitazone and farglitazar (73) provide a structural explanation for their differential effects on insulin resistance mutations. Whereas rosiglitazone is ineffective in activating these mutant receptors, farglitazar induces a potent transactivation (119). Farglitazar makes more overall contacts with the receptor, consistent with its higher affinity, but also has specific stabilizing contacts with the loop between helices 11 and 12 and with the base of helix 11, which are not found with rosiglitazone. Even at saturating doses of ligand, farglitazar induced a more efficacious recruitment of coactivator with the mutant receptors, consistent with its helix 12 stabilizing interactions (119).

Mutations at the interface of helices 3 and 12 are also associated with constitutive activity (117). For ERα, mutation of Tyr537 to Ser produces ligand-independent activity that can still be blocked by antagonists, suggesting that it is stabilizing the agonist conformation. Tyr537 forms a weak hydrogen bond with helix 3 Asn 348. A serine in this position is postulated to form a hydrogen bond with Aps351 (Figure 8A, see color insert), which already participates in an electrostatic interaction with helix 12, stabilizing the active conformation. This idea is further supported by the observations that the Ser mutation slows both the on and off rate of ligand binding, and provides protection from urea-induced unfolding (120).

Another type of mutation that affects transactivation and cofactor recruitment lies in the hydrophobic cleft. For example, Leu372 lies in the AF2 surface and interacts directly with the GRIP peptide in the agonist-bound ERα structure and with helix 12 in the tamoxifen-bound ERα structure (Figure 8B). The mutation L372R produces a significant increase in transcriptional activity in response to both agonists and SERMs (121). An examination of the ER crystal structures suggests that this effect occurs through distinct mechanisms. In the tamoxifen-bound ER, this mutation is ideally positioned to hydrogen bond with helix 12 (Figure 8B), stabilizing it in the hydrophobic cleft and blocking corepressor recruitment, thereby promoting SERM agonist activity from AF1. In the absence of an ER-corepressor structure, it is not clear if the L372R mutation would directly inhibit corepressor interaction. This hypothesis could be tested by examining corepressor recruitment to the mutant ER in the context of a receptor lacking H12. Huang et al. (121) demonstrated that this mutation increased agonist recruitment of coactivator peptide and blocked SERM recruitment of corepressor motif peptides, consistent with the structural interpretation.

Mutations of Asp351 in helix 3 of ERα appear to have a complex phenotype, as this amino acid hydrogen bonds with helix 12 in the agonist conformation but also interacts directly with the charged side chains of several SERMs (Figure 3). The ER mutation D351Y was originally identified as a naturally occurring mutation that rendered breast cancer cells resistant to tamoxifen treatment in an MCF-7 animal xenograft model (122). The substitution of Tyr at this position significantly reduces corepressor recruitment to ER (123) and allows agonist activity from ER antagonists (53, 124–126). On the basis of published structures, this mutation is seen to both stabilize helix 12 binding to the AF2 surface and destabilize its localization to helix 3, as was shown for PPARα and the SMRT corepressor. Mutation of ERα Asp351 to glycine blocks SERM agonist activity, although this was not directly correlated to increased corepressor binding. This model may also explain the importance of the positive charge in the raloxifene piperidine-containing side chain, which was demonstrated to be crucial for antagonist activity. By interacting directly with Asp 351 and masking the negative charge, the raloxifene side chain may facilitate relocation of helix 12 to the helix 3 shelf and allow corepressor binding.

TRANSCRIPTIONAL POTENCY AND EFFICACY

Potency

Converging evidence suggests a model to explain the structural features that contribute to transcriptional potency and why potency does not necessarily correlate with ligand affinity. Potency describes the concentration of ligand required to obtain a certain signal (e.g., half maximal or EC_{50}). Transcriptional potency is sensitive to the concentrations of receptor, coactivator, DNA, and ligand (91, 127), suggesting that information from all these factors is integrated into a single output, the generation of the mRNA transcript. There are cooperative relationships among the affinities of the various NR cofactors that contribute to the context specificity of transcriptional potency. The DNA response element sequence acts an allosteric ligand, inducing distinct coactivator preferences (36, 128–130). DNA sequence can also control ligand affinity (128). Ligands confer preferences for coactivators and corepressors (12, 34, 97, 131). In fact, transcriptional potency may correlate better with affinity for coactivator than ligand (132). The amount of ligand needed to induce a given number of transcripts is thus determined by the occupancy of response element DNA by receptor, ligand, and coregulators. The assembly of this transcriptional activation complex is a function of both affinity among the various components and their cellular concentrations.

Helix 12 plays a primary role in mediating the structural linkage between ligand and coactivator. For the ER subtypes, ERα has increased transcriptional potency in response to estradiol and increased affinity for the p160 coactivators (13), but has near identical affinity for estradiol when compare with ERβ. The reduced transcriptional potency of ERβ relative to ERα derives from differential dynamics of helix 12 (57). ERα has several structural features, including specific hydrogen bonds, that stabilize the agonist conformation of helix 12. Introduction of these amino acids into ERβ induces an increase in the potency response to estradiol, verifying the critical role for helix 12 (57). The differential dynamics of helix 12 may also account for differences in affinity for coactivator. In this model, ligand-specific effects on transcriptional potency relate to the formation of a high-affinity coactivator-binding site through the stabilization of helix 12.

The coactivator also controls the binding mode and affinity of receptor for ligand (57, 91, 92). The binding of coactivator stabilizes helix 12 in the agonist conformation and thus has direct effects on ligand affinity. Molecular dynamics simulations suggest that movement of helix 12 is an obligate step in ligand escape (98). Coactivator binding has been shown to slow ligand off-rate (91), presumably via stabilization of helix 12 conformation. With receptors such as PPAR, direct hydrogen bonds with helix 12 are an important component of ligand affinity. A stabilization of helix 12 by coactivator shows that the linkage with ligand is bidirectional. Thus increased cellular levels of coactivator may alter transcriptional potency by increasing the occupancy of the receptor by both coactivator and ligand.

Efficacy

Transcriptional efficacy is a measure of the maximum amount of a gene product that can be generated by a saturating dose of a given ligand. Thus it is independent of ligand affinity. Instead, efficacy reflects the ability of a ligand to induce a stable coactivator-binding site. In this model, NRs with antagonist conformations of helix 12 would to bind DNA but would not recruit coactivators, acting instead in a dominant-negative fashion to limit the total number of generated transcripts. Thus a high-affinity ligand can have low efficacy because a subset of the ensemble of conformations is unable to bind coactivator. As discussed above, high-affinity partial agonist ligands (e.g., potent, but with reduced efficacy) display loss of interactions of helix 12 with ligand or helices 3 and 11, thus destabilizing the agonist conformation.

Accumulating evidence suggests that ligand interactions with helix 11 and helix 12 are primary determinants of AF2 stability. Partial agonists such as THC bind ERα with high affinity, but induce a less stable coactivator-binding site through altered localization of helix 11 and destabilization of helix 12. This has been experimentally confirmed by the reduced affinity of coactivator peptide for THC-bound ER compared with that of the estradiol/ER complex (56). Ligand-induced stabilization of helix 12 has also been demonstrated with fluorescent labeling of helix 12 (55). Efficacy thus relates to the ensemble of conformations in which a productive AF2 surface can be induced by a given ligand and by the percentage of receptors with a high-affinity coactivator-binding site.

CONCLUSIONS

Knowledge of the mechanisms through which ligands regulate specific intracellular signals is critical to the development of targeted therapeutics for the nuclear receptor superfamily. Evans and colleagues recognized the importance of helix 12 dynamics as a mediator of hormonal signals nearly a decade ago (9). Recent structural studies offer insight into the regulation of these dynamics, which include positioning of helix 11 by ligand as well as direct interactions between ligand and helix 12 for some NRs. The interaction between ligand, helix 11, and helix 12 explains many aspects of nuclear receptor signaling, including partial agonist and indirect antagonist activity, and RXR subordination by heterodimer partner. As part of the dimer interface, helix 11 integrates signaling across the dimer, contributing to synergistic and permissive ligand activation of NRs. Helix 12 is also regulated directly by antagonist ligands, controlling the relative recruitment of coactivators versus corepressors. Thus many aspects of ligand-specific signaling in the NR superfamily are regulated by the structural connections between ligands and coregulator proteins, which are mediated by helix 12.

The *Annual Review of Physiology* is online at
http://physiol.annualreviews.org

LITERATURE CITED

1. Chambon P. 1996. A decade of molecular biology of retinoic acid receptors. *FASEB J.* 10:940–54
2. Mangelsdorf DJ, Evans RM. 1995. The RXR heterodimers and orphan receptors. *Cell* 83:841–50
3. Kliewer SA, Xu HE, Lambert MH, Willson TM. 2001. Peroxisome proliferator-activated receptors: from genes to physiology. *Recent Prog. Horm. Res.* 56:239–63
4. Chawla A, Repa JJ, Evans RM, Mangelsdorf DJ. 2001. Nuclear receptors and lipid physiology: opening the X-files. *Science* 294:1866–70
5. McDonnell DP, Wijayaratne A, Chang CY, Norris JD. 2002. Elucidation of the molecular mechanism of action of selective estrogen receptor modulators. *Am. J. Cardiol.* 90:35F–43F
6. McKenna NJ, O'Malley BW. 2002. Combinatorial control of gene expression by nuclear receptors and coregulators. *Cell* 108:465–74
7. Darimont BD, Wagner RL, Apriletti JW, Stallcup MR, Kushner PJ, et al. 1998. Structure and specificity of nuclear receptor-coactivator interactions. *Genes Dev.* 12:3343–56
8. Shiau AK, Barstad D, Loria PM, Cheng L, Kushner PJ, et al. 1998. The structural basis of estrogen receptor/coactivator recognition and the antagonism of this interaction by tamoxifen. *Cell* 95:927–37
9. Schulman IG, Juguilon H, Evans RM. 1996. Activation and repression by nuclear hormone receptors: hormone modulates an equilibrium between active and repressive states. *Mol. Cell. Biol.* 16:3807–13
10. Renaud JP, Rochel N, Ruff M, Vivat V, Chambon P, et al. 1995. Crystal structure of the RAR-gamma ligand-binding

domain bound to all-*trans* retinoic acid. *Nature* 378:681–89

11. Bourguet W, Vivat V, Wurtz JM, Chambon P, Gronemeyer H, Moras D. 2000. Crystal structure of a heterodimeric complex of RAR and RXR ligand-binding domains. *Mol. Cell.* 5:289–98

12. Bramlett KS, Wu Y, Burris TP. 2001. Ligands specify coactivator nuclear receptor (NR) box affinity for estrogen receptor subtypes. *Mol. Endocrinol.* 15:909–22

13. Wong CW, Komm B, Cheskis BJ. 2001. Structure-function evaluation of ER alpha and beta interplay with SRC family coactivators. ER selective ligands. *Biochemistry* 40:6756–65

14. Warnmark A, Almlof T, Leers J, Gustafsson JA, Treuter E. 2001. Differential recruitment of the mammalian mediator subunit TRAP220 by estrogen receptors ERalpha and ERbeta. *J. Biol. Chem.* 276:23397–404

15. Matt N, Ghyselinck NB, Wendling O, Chambon P, Mark M. 2003. Retinoic acid-induced developmental defects are mediated by RARbeta/RXR heterodimers in the pharyngeal endoderm. *Development* 130:2083–93

16. Wang YX, Lee CH, Tiep S, Yu RT, Ham J, et al. 2003. Peroxisome-proliferator-activated receptor delta activates fat metabolism to prevent obesity. *Cell* 113:159–70

17. Mueller SO, Korach KS. 2001. Estrogen receptors and endocrine diseases: lessons from estrogen receptor knockout mice. *Curr. Opin. Pharmacol.* 1:613–19

18. Leblanc BP, Stunnenberg HG. 1995. 9-*cis* retinoic acid signaling: changing partners causes some excitement. *Genes Dev.* 9:1811–16

19. DiRenzo J, Soderstrom M, Kurokawa R, Ogliastro MH, Ricote M, et al. 1997. Peroxisome proliferator-activated receptors and retinoic acid receptors differentially control the interactions of retinoid X receptor heterodimers with ligands, coacti-

vators, and corepressors. *Mol. Cell. Biol.* 17:2166–76

20. Kliewer SA, Umesono K, Noonan DJ, Heyman RA, Evans RM. 1992. Convergence of 9-*cis* retinoic acid and peroxisome proliferator signalling pathways through heterodimer formation of their receptors. *Nature* 358:771–74

21. Leo C, Yang X, Liu J, Li H, Chen JD. 2001. Role of retinoid receptor coactivator pockets in cofactor recruitment and transcriptional regulation. *J. Biol. Chem.* 276:23127–34

22. Roy B, Taneja R, Chambon P. 1995. Synergistic activation of retinoic acid (RA)-responsive genes and induction of embryonal carcinoma cell differentiation by an RA receptor alpha (RAR alpha)-, RAR beta-, or RAR gamma-selective ligand in combination with a retinoid X receptor-specific ligand. *Mol. Cell Biol.* 15:6481–87

23. Lala DS, Mukherjee R, Schulman IG, Koch SS, Dardashti LJ, et al. 1996. Activation of specific RXR heterodimers by an antagonist of RXR homodimers. *Nature* 383:450–53

24. Schulman IG, Li C, Schwabe JW, Evans RM. 1997. The phantom ligand effect: allosteric control of transcription by the retinoid X receptor. *Genes Dev.* 11:299–308

25. Smith CL, O'Malley BW. 2004. Coregulator function: a key to understanding tissue specificity of selective receptor modulators. *Endocr. Rev.* 25:45–71

26. Jordan VC. 1992. The strategic use of antiestrogens to control the development and growth of breast cancer. *Cancer* 70:977–82

27. Wolf DM, Jordan VC. 1992. Gynecologic complications associated with long-term adjuvant tamoxifen therapy for breast cancer. *Gynecol. Oncol.* 45:118–28

28. Love RR, Mazess RB, Barden HS, Epstein S, Newcomb PA, et al. 1992. Effects of tamoxifen on bone mineral density in postmenopausal women with

breast cancer. *N. Engl. J. Med.* 326:852–56

29. Berry M, Metzger D, Chambon P. 1990. Role of the two activating domains of the oestrogen receptor in the cell-type and promoter-context dependent agonistic activity of the anti-oestrogen 4-hydroxytamoxifen. *EMBO J.* 9:2811–18

30. McInerney EM, Katzenellenbogen BS. 1996. Different regions in activation function-1 of the human estrogen receptor required for antiestrogen- and estradiol-dependent transcription activation. *J. Biol. Chem.* 271:24172–78

31. Tzukerman MT, Esty A, Santiso-Mere D, Danielian P, Parker MG, et al. 1994. Human estrogen receptor transactivational capacity is determined by both cellular and promoter context and mediated by two functionally distinct intramolecular regions. *Mol. Endocrinol.* 8:21–30

32. Delmas PD, Bjarnason NH, Mitlak BH, Ravoux AC, Shah AS, et al. 1997. Effects of raloxifene on bone mineral density, serum cholesterol concentrations, and uterine endometrium in postmenopausal women. *N. Engl. J. Med.* 337:1641–47

33. Smith CL, Nawaz Z, O'Malley BW. 1997. Coactivator and corepressor regulation of the agonist/antagonist activity of the mixed antiestrogen 4-hydroxytamoxifen. *Mol. Endocrinol.* 11:657–66

34. Webb P, Nguyen P, Kushner PJ. 2003. Differential SERM effects on corepressor binding dictate ERalpha activity in vivo. *J. Biol. Chem.* 278:6912–20

35. Liu Z, Auboeuf D, Wong J, Chen JD, Tsai SY, et al. 2002. Coactivator/corepressor ratios modulate PR-mediated transcription by the selective receptor modulator RU486. *Proc. Natl. Acad. Sci. USA* 99:7940–44

36. Shang Y, Brown M. 2002. Molecular determinants for the tissue specificity of SERMs. *Science* 295:2465–68

37. Lavinsky RM, Jepsen K, Heinzel T, Torchia J, Mullen TM, et al. 1998. Diverse signaling pathways modulate nuclear receptor recruitment of N-CoR and SMRT complexes. *Proc. Natl. Acad. Sci. USA* 95:2920–25

38. Kauppi B, Jakob C, Farnegardh M, Yang J, Ahola H, et al. 2003. The three-dimensional structures of antagonistic and agonistic forms of the glucocorticoid receptor ligand-binding domain: RU-486 induces a transconformation that leads to active antagonism. *J. Biol. Chem.* 278:22748–54

39. Henke BR, Consler TG, Go N, Hale RL, Hohman DR, et al. 2002. A new series of estrogen receptor modulators that display selectivity for estrogen receptor beta. *J. Med. Chem.* 45:5492–505

40. Nagy L, Kao HY, Love JD, Li C, Banayo E, et al. 1999. Mechanism of corepressor binding and release from nuclear hormone receptors. *Genes Dev.* 13:3209–16

41. Marimuthu A, Feng W, Tagami T, Nguyen H, Jameson JL, et al. 2002. TR surfaces and conformations required to bind nuclear receptor corepressor. *Mol. Endocrinol.* 16:271–86

42. Perissi V, Staszewski LM, McInerney EM, Kurokawa R, Krones A, et al. 1999. Molecular determinants of nuclear receptor-corepressor interaction. *Genes Dev.* 13:3198–208

43. Hu X, Lazar MA. 1999. The CoRNR motif controls the recruitment of corepressors by nuclear hormone receptors. *Nature* 402:93–96

44. Zhang J, Hu X, Lazar MA. 1999. A novel role for helix 12 of retinoid X receptor in regulating repression. *Mol. Cell. Biol.* 19:6448–57

45. Jung DJ, Lee SK, Lee JW. 2001. Agonist-dependent repression mediated by mutant estrogen receptor alpha that lacks the activation function 2 core domain. *J. Biol. Chem.* 276:37280–83

46. Yoh SM, Chatterjee VK, Privalsky ML. 1997. Thyroid hormone resistance syndrome manifests as an aberrant interaction between mutant T3 receptors and

transcriptional corepressors. *Mol. Endocrinol.* 11:470–80

47. Gurnell M, Wentworth JM, Agostini M, Adams M, Collingwood TN, et al. 2000. A dominant-negative peroxisome proliferator-activated receptor gamma (PPARgamma) mutant is a constitutive repressor and inhibits PPARgamma-mediated adipogenesis. *J. Biol. Chem.* 275:5754–59

48. Xu HE, Stanley TB, Montana VG, Lambert MH, Shearer BG, et al. 2002. Structural basis for antagonist-mediated recruitment of nuclear co-repressors by PPARalpha. *Nature* 415:813–17

49. Mi LZ, Devarakonda S, Harp JM, Han Q, Pellicciari R, et al. 2003. Structural basis for bile acid binding and activation of the nuclear receptor FXR. *Mol. Cell* 11:1093–100

50. Nettles KW, Greene GL. 2003. Nuclear receptor ligands and cofactor recruitment: is there a coactivator "on deck"? *Mol. Cell.* 11:850–51

51. Brzozowski AM, Pike AC, Dauter Z, Hubbard RE, Bonn T, et al. 1997. Molecular basis of agonism and antagonism in the oestrogen receptor. *Nature* 389:753–58

52. Pike AC, Brzozowski AM, Walton J, Hubbard RE, Thorsell AG, et al. 2001. Structural insights into the mode of action of a pure antiestrogen. *Structure* 9:145–53

53. Levenson AS, Jordan VC. 1998. The key to the antiestrogenic mechanism of raloxifene is amino acid 351 (aspartate) in the estrogen receptor. *Cancer Res.* 58:1872–75

54. Wijayaratne AL, McDonnell DP. 2001. The human estrogen receptor-alpha is a ubiquitinated protein whose stability is affected differentially by agonists, antagonists, and selective estrogen receptor modulators. *J. Biol. Chem.* 276:35684–92

55. Kallenberger BC, Love JD, Chatterjee VK, Schwabe JW. 2003. A dynamic mechanism of nuclear receptor activation and its perturbation in a human disease. *Nat. Struct. Biol.* 10:136–40

56. Shiau AK, Barstad D, Radek JT, Meyers MJ, Nettles KW, et al. 2002. Structural characterization of a subtype-selective ligand reveals a novel mode of estrogen receptor antagonism. *Nat. Struct. Biol.* 9:359–64

57. Nettles KW, Sun J, Radek JT, Sheng S, Rodriguez AL, et al. 2004. Allosteric control of ligand selectivity between estrogen receptors alpha and beta: implications for other nuclear receptors. *Mol. Cell.* 13:317–27

58. Greschik H, Flaig R, Renaud JP, Moras D. 2004. Structural basis for the deactivation of the estrogen-related receptor gamma by diethylstilbestrol or 4-hydroxytamoxifen and determinants of selectivity. *J. Biol. Chem.* 279:33639–46

59. Cronet P, Petersen JF, Folmer R, Blomberg N, Sjoblom K, et al. 2001. Structure of the PPARalpha and -gamma ligand binding domain in complex with AZ 242; ligand selectivity and agonist activation in the PPAR family. *Structure* 9:699–706

60. Xu HE, Lambert MH, Montana VG, Parks DJ, Blanchard SG, et al. 1999. Molecular recognition of fatty acids by peroxisome proliferator-activated receptors. *Mol. Cell* 3:397–403

61. Oberfield JL, Collins JL, Holmes CP, Goreham DM, Cooper JP, et al. 1999. A peroxisome proliferator-activated receptor gamma ligand inhibits adipocyte differentiation. *Proc. Natl. Acad. Sci. USA* 96:6102–6

62. Borngraeber S, Budny MJ, Chiellini G, Cunha-Lima ST, Togashi M, et al. 2003. Ligand selectivity by seeking hydrophobicity in thyroid hormone receptor. *Proc. Natl. Acad. Sci. USA* 100:15358–63

63. Schulman IG, Shao G, Heyman RA. 1998. Transactivation by retinoid X receptor-peroxisome proliferator-activated receptor gamma (PPARgamma) heterodimers: intermolecular synergy requires only the PPARgamma hormone-dependent activation function. *Mol. Cell. Biol.* 18:3483–94

64. Shulman AI, Larson C, Mangelsdorf DJ, Ranganathan R. 2004. Structural determinants of allosteric ligand activation in RXR heterodimers. *Cell* 116:417–29

65. Botling J, Castro DS, Oberg F, Nilsson K, Perlmann T. 1997. Retinoic acid receptor/retinoid X receptor heterodimers can be activated through both subunits providing a basis for synergistic transactivation and cellular differentiation. *J. Biol. Chem.* 272:9443–49

66. Forman BM, Umesono K, Chen J, Evans RM. 1995. Unique response pathways are established by allosteric interactions among nuclear hormone receptors. *Cell* 81:541–50

67. Germain P, Iyer J, Zechel C, Gronemeyer H. 2002. Co-regulator recruitment and the mechanism of retinoic acid receptor synergy. *Nature* 415:187–92

68. MacDonald PN, Dowd DR, Nakajima S, Galligan MA, Reeder MC, et al. 1993. Retinoid X receptors stimulate and 9-*cis* retinoic acid inhibits 1,25-dihydroxyvitamin D_3-activated expression of the rat osteocalcin gene. *Mol. Cell Biol.* 13:5907–17

69. Lemon BD, Freedman LP. 1996. Selective effects of ligands on vitamin D_3 receptor- and retinoid X receptor-mediated gene activation in vivo. *Mol. Cell Biol.* 16:1006–16

70. Westin S, Kurokawa R, Nolte RT, Wisely GB, McInerney EM, et al. 1998. Interactions controlling the assembly of nuclear-receptor heterodimers and co-activators. *Nature* 395:199–202

71. Kersten S, Dawson MI, Lewis BA, Noy N. 1996. Individual subunits of heterodimers comprised of retinoic acid and retinoid X receptors interact with their ligands independently. *Biochemistry* 35:3816–24

72. Egea PF, Mitschler A, Rochel N, Ruff M, Chambon P, Moras D. 2000. Crystal structure of the human RXRalpha ligand-binding domain bound to its natural ligand: 9-*cis* retinoic acid. *EMBO J.* 19:2592–601

73. Gampe RT Jr, Montana VG, Lambert MH, Miller AB, Bledsoe RK, et al. 2000. Asymmetry in the PPARgamma/RXR-alpha crystal structure reveals the molecular basis of heterodimerization among nuclear receptors. *Mol. Cell* 5:545–55

74. Klaholz BP, Mitschler A, Belema M, Zusi C, Moras D. 2000. Enantiomer discrimination illustrated by high-resolution crystal structures of the human nuclear receptor hRARgamma. *Proc. Natl. Acad. Sci. USA* 97:6322–27

75. Klaholz BP, Mitschler A, Moras D. 2000. Structural basis for isotype selectivity of the human retinoic acid nuclear receptor. *J. Mol. Biol.* 302:155–70

76. Klaholz BP, Renaud JP, Mitschler A, Zusi C, Chambon P, et al. 1998. Conformational adaptation of agonists to the human nuclear receptor RAR gamma. *Nat. Struct. Biol.* 5:199–202

77. Love JD, Gooch JT, Benko S, Li C, Nagy L, et al. 2002. The structural basis for the specificity of retinoid-X receptor-selective agonists: new insights into the role of helix H12. *J. Biol. Chem.* 277:11385–91

78. Nolte RT, Wisely GB, Westin S, Cobb JE, Lambert MH, et al. 1998. Ligand binding and co-activator assembly of the peroxisome proliferator-activated receptor-gamma. *Nature* 395:137–43

79. Xu HE, Lambert MH, Montana VG, Plunket KD, Moore LB, et al. 2001. Structural determinants of ligand binding selectivity between the peroxisome proliferator-activated receptors. *Proc. Natl. Acad. Sci. USA* 98:13919–24

80. Svensson S, Ostberg T, Jacobsson M, Norstrom C, Stefansson K, et al. 2003. Crystal structure of the heterodimeric complex of LXRalpha and RXRbeta ligand-binding domains in a fully agonistic conformation. *EMBO J.* 22:4625–33

81. Yang W, Rachez C, Freedman LP. 2000. Discrete roles for peroxisome proliferator-activated receptor gamma

and retinoid X receptor in recruiting nuclear receptor coactivators. *Mol. Cell Biol.* 20:8008–17

82. Minucci S, Leid M, Toyama R, Saint-Jeannet JP, Peterson VJ, et al. 1997. Retinoid X receptor (RXR) within the RXR-retinoic acid receptor heterodimer binds its ligand and enhances retinoid-dependent gene expression. *Mol. Cell. Biol.* 17:644–55

83. Wurtz JM, Bourguet W, Renaud JP, Vivat V, Chambon P, et al. 1996. A canonical structure for the ligand-binding domain of nuclear receptors. *Nat. Struct. Biol.* 3:206

84. Steinmetz AC, Renaud JP, Moras D. 2001. Binding of ligands and activation of transcription by nuclear receptors. *Annu. Rev. Biophys. Biomol. Struct.* 30:329–59

85. Luque I, Leavitt SA, Freire E. 2002. The linkage between protein folding and functional cooperativity: two sides of the same coin? *Annu. Rev. Biophys. Biomol. Struct.* 31:235–56

86. Johnson BA, Wilson EM, Li Y, Moller DE, Smith RG, Zhou G. 2000. Ligand-induced stabilization of PPARgamma monitored by NMR spectroscopy: implications for nuclear receptor activation. *J. Mol. Biol.* 298:187–94

87. Herdick M, Bury Y, Quack M, Uskokovic MR, Polly P, Carlberg C. 2000. Response element and coactivator-mediated conformational change of the vitamin D_3 receptor permits sensitive interaction with agonists. *Mol. Pharmacol.* 57:1206–17

88. Beekman JM, Allan GF, Tsai SY, Tsai MJ, O'Malley BW. 1993. Transcriptional activation by the estrogen receptor requires a conformational change in the ligand binding domain. *Mol. Endocrinol.* 7:1266–74

89. Egea PF, Rochel N, Birck C, Vachette P, Timmins PA, Moras D. 2001. Effects of ligand binding on the association properties and conformation in solution of retinoic acid receptors RXR and RAR. *J. Mol. Biol.* 307:557–76

90. Yan X, Broderick D, Leid ME, Schimerlik MI, Deinzer ML. 2004. Dynamics and ligand-induced solvent accessibility changes in human retinoid X receptor homodimer determined by hydrogen deuterium exchange and mass spectrometry. *Biochemistry* 43:909–17

91. Gee AC, Carlson KE, Martini PG, Katzenellenbogen BS, Katzenellenbogen JA. 1999. Coactivator peptides have a differential stabilizing effect on the binding of estrogens and antiestrogens with the estrogen receptor. *Mol. Endocrinol.* 13:1912–23

92. Watkins RE, Davis-Searles PR, Lambert MH, Redinbo MR. 2003. Coactivator binding promotes the specific interaction between ligand and the pregnane X receptor. *J. Mol. Biol.* 331:815–28

93. Bhat RA, Stauffer B, Unwalla RJ, Xu Z, Harris HA, Komm BS. 2004. Molecular determinants of ER alpha and ER beta involved in selectivity of 16 alpha-iodo-7 beta estradiol. *J. Steroid. Biochem. Mol. Biol.* 88:17–26

94. Gaido KW, Leonard LS, Maness SC, Hall JM, McDonnell DP, et al. 1999. Differential interaction of the methoxychlor metabolite 2,2-bis-(*p*-hydroxyphenyl)-1,1,1-trichloroethane with estrogen receptors alpha and beta. *Endocrinology* 140:5746–53

95. Sun J, Baudry J, Katzenellenbogen JA, Katzenellenbogen BS. 2003. Molecular basis for the subtype discrimination of the estrogen receptor-beta-selective ligand, diarylpropionitrile. *Mol. Endocrinol.* 17:247–58

96. Terouanne B, Nirde P, Rabenoelina F, Bourguet W, Sultan C, Auzou G. 2003. Mutation of the androgen receptor at amino acid 708 (Gly→Ala) abolishes partial agonist activity of steroidal antiandrogens. *Mol. Pharmacol.* 63:791–98

97. Stanley TB, Leesnitzer LM, Montana VG, Galardi CM, Lambert MH, et al. 2003. Subtype specific effects of peroxisome proliferator-activated receptor

ligands on corepressor affinity. *Biochemistry* 42:9278–87

98. Blondel A, Renaud JP, Fischer S, Moras D, Karplus M. 1999. Retinoic acid receptor: a simulation analysis of retinoic acid binding and the resulting conformational changes. *J. Mol. Biol.* 291:101–15

99. Farboud B, Hauksdottir H, Wu Y, Privalsky ML. 2003. Isotype-restricted corepressor recruitment: a constitutively closed helix 12 conformation in retinoic acid receptors beta and gamma interferes with corepressor recruitment and prevents transcriptional repression. *Mol. Cell Biol.* 23:2844–58

100. Hauksdottir H, Farboud B, Privalsky ML. 2003. Retinoic acid receptors beta and gamma do not repress, but instead activate target gene transcription in both the absence and presence of hormone ligand. *Mol. Endocrinol.* 17:373–85

101. Wang Z, Benoit G, Liu J, Prasad S, Aarnisalo P, et al. 2003. Structure and function of Nurr1 identifies a class of ligand-independent nuclear receptors. *Nature* 423:555–60

102. Baker KD, Shewchuk LM, Kozlova T, Makishima M, Hassell A, et al. 2003. The *Drosophila* orphan nuclear receptor DHR38 mediates an atypical ecdysteroid signaling pathway. *Cell* 113:731–42

103. Sablin EP, Krylova IN, Fletterick RJ, Ingraham HA. 2003. Structural basis for ligand-independent activation of the orphan nuclear receptor LRH-1. *Mol. Cell* 11:1575–85

104. Dussault I, Lin M, Hollister K, Fan M, Termini J, et al. 2002. A structural model of the constitutive androstane receptor defines novel interactions that mediate ligand-independent activity. *Mol. Cell Biol.* 22:5270–80

105. Forman BM, Tzameli I, Choi HS, Chen J, Simha D, et al. 1998. Androstane metabolites bind to and deactivate the nuclear receptor CAR-beta. *Nature* 395:612–15

106. Suetsugi M, Su L, Karlsberg K, Yuan YC, Chen S. 2003. Flavone and isoflavone phytoestrogens are agonists of estrogen-related receptors. *Mol. Cancer Res.* 1: 981–91

107. Tzameli I, Pissios P, Schuetz EG, Moore DD. 2000. The xenobiotic compound 1,4-bis[2-(3,5-dichloropyridyloxy)]benzene is an agonist ligand for the nuclear receptor CAR. *Mol. Cell Biol.* 20:2951–58

108. Maglich JM, Parks DJ, Moore LB, Collins JL, Goodwin B, et al. 2003. Identification of a novel human constitutive androstane receptor (CAR) agonist and its use in the identification of CAR target genes. *J. Biol. Chem.* 278:17277–83

109. Dhe-Paganon S, Duda K, Iwamoto M, Chi YI, Shoelson SE. 2002. Crystal structure of the HNF4 alpha ligand binding domain in complex with endogenous fatty acid ligand. *J. Biol. Chem.* 277:37973–76

110. Wisely GB, Miller AB, Davis RG, Thornquest AD Jr, Johnson R, et al. 2002. Hepatocyte nuclear factor 4 is a transcription factor that constitutively binds fatty acids. *Structure* 10:1225–34

111. Kallen J, Schlaeppi JM, Bitsch F, Delhon I, Fournier B. 2004. Crystal structure of the human RORalpha ligand binding domain in complex with cholesterol sulfate at 2.2 Å. *J. Biol. Chem.* 279:14033–38

112. Kallen JA, Schlaeppi JM, Bitsch F, Geisse S, Geiser M, et al. 2002. X-ray structure of the hRORalpha LBD at 1.63 Å: structural and functional data that cholesterol or a cholesterol derivative is the natural ligand of RORalpha. *Structure* 10:1697–707

113. Ciana P, Raviscioni M, Mussi P, Vegeto E, Que I, et al. 2003. In vivo imaging of transcriptionally active estrogen receptors. *Nat. Med.* 9:82–86

114. Schiff R, Massarweh SA, Shou J, Bharwani L, Mohsin SK, Osborne K. 2004. Cross-talk between estrogen receptor and growth factor pathways as a molecular target for overcoming endocrine resistance. *Clin. Cancer Res.* 10:331S–36S

115. Benko S, Love JD, Beladi M, Schwabe JW, Nagy L. 2003. Molecular determinants of the balance between

co-repressor and co-activator recruitment to the retinoic acid receptor. *J. Biol. Chem.* 278:43797–806

116. Barroso I, Gurnell M, Crowley VE, Agostini M, Schwabe JW, et al. 1999. Dominant negative mutations in human PPARgamma associated with severe insulin resistance, diabetes mellitus and hypertension. *Nature* 402:880–83

117. Weis KE, Ekena K, Thomas JA, Lazennec G, Katzenellenbogen BS. 1996. Constitutively active human estrogen receptors containing amino acid substitutions for tyrosine 537 in the receptor protein. *Mol. Endocrinol.* 10:1388–98

118. Collingwood TN, Wagner R, Matthews CH, Clifton-Bligh RJ, Gurnell M, et al. 1998. A role for helix 3 of the TRbeta ligand-binding domain in coactivator recruitment identified by characterization of a third cluster of mutations in resistance to thyroid hormone. *EMBO J.* 17:4760–70

119. Agostini M, Gurnell M, Savage DB, Wood EM, Smith AG, et al. 2004. Tyrosine agonists reverse the molecular defects associated with dominant-negative mutations in human peroxisome proliferator-activated receptor gamma. *Endocrinology* 145:1527–38

120. Carlson KE, Choi I, Gee A, Katzenellenbogen BS, Katzenellenbogen JA. 1997. Altered ligand binding properties and enhanced stability of a constitutively active estrogen receptor: evidence that an open pocket conformation is required for ligand interaction. *Biochemistry* 36:14897–905

121. Huang HJ, Norris JD, McDonnell DP. 2002. Identification of a negative regulatory surface within estrogen receptor alpha provides evidence in support of a role for corepressors in regulating cellular responses to agonists and antagonists. *Mol. Endocrinol.* 16:1778–92

122. Wolf DM, Jordan VC. 1994. The estrogen receptor from a tamoxifen stimulated MCF-7 tumor variant contains a point mutation in the ligand binding domain. *Breast Cancer Res. Treat.* 31:129–38

123. Yamamoto Y, Wada O, Suzawa M, Yogiashi Y, Yano T, et al. 2001. The tamoxifen-responsive estrogen receptor alpha mutant D351Y shows reduced tamoxifen-dependent interaction with corepressor complexes. *J. Biol. Chem.* 276:42684–91

124. Schafer JI, Liu H, Tonetti DA, Jordan VC. 1999. The interaction of raloxifene and the active metabolite of the antiestrogen EM-800 (SC 5705) with the human estrogen receptor. *Cancer Res.* 59:4308–13

125. Levenson AS, Catherino WH, Jordan VC. 1997. Estrogenic activity is increased for an antiestrogen by a natural mutation of the estrogen receptor. *J. Steroid. Biochem. Mol. Biol.* 60:261–68

126. Webb P, Nguyen P, Valentine C, Weatherman RV, Scanlan TS, Kushner PJ. 2000. An antiestrogen-responsive estrogen receptor-alpha mutant (D351Y) shows weak AF-2 activity in the presence of tamoxifen. *J. Biol. Chem.* 275:37552–58

127. Chen S, Sarlis NJ, Simons SS Jr. 2000. Evidence for a common step in three different processes for modulating the kinetic properties of glucocorticoid receptor-induced gene transcription. *J. Biol. Chem.* 275:30106–17

128. Kurokawa R, DiRenzo J, Boehm M, Sugarman J, Gloss B, et al. 1994. Regulation of retinoid signalling by receptor polarity and allosteric control of ligand binding. *Nature* 371:528–31

129. Wood JR, Greene GL, Nardulli AM. 1998. Estrogen response elements function as allosteric modulators of estrogen receptor conformation. *Mol. Cell. Biol.* 18:1927–34

130. Loven MA, Likhite VS, Choi I, Nardulli AM. 2001. Estrogen response elements alter coactivator recruitment through allosteric modulation of estrogen receptor beta conformation. *J. Biol. Chem.* 276:45282–88

131. Routledge EJ, White R, Parker MG, Sumpter JP. 2000. Differential effects of xenoestrogens on coactivator recruitment by estrogen receptor (ER) alpha and ER beta. *J. Biol. Chem.* 275:35986–93

132. Mukherjee R, Sun S, Santomenna L, Miao B, Walton H, et al. 2002. Ligand and coactivator recruitment preferences of peroxisome proliferator activated receptor alpha. *J. Steroid Biochem. Mol. Biol.* 81:217–25

Annu. Rev. Physiol. 2005. 67:335–76
doi: 10.1146/annurev.physiol.67.040403.120151
First published online as a Review in Advance on October 19, 2004

REGULATION OF SIGNAL TRANSDUCTION PATHWAYS BY ESTROGEN AND PROGESTERONE

Dean P. Edwards

University of Colorado Health Sciences Center, Department of Pathology and Program in Molecular Biology, Aurora, Colorado 80045; email: dean.edwards@uchsc.edu

Key Words   steroid receptors, protein kinases, membrane receptors, gene transcription

■ **Abstract**   The female sex steroid hormones 17β-estradiol and progesterone mediate their biological effects on development, differentiation, and maintenance of reproductive tract and other target tissues through gene regulation by nuclear steroid receptors that function as ligand-dependent transcription factors. However, not all effects of 17β-estradiol and progesterone are mediated by direct control of gene expression. These hormones also have rapid stimulatory effects on the activities of a variety of signal transduction molecules and pathways and, in many cases, these effects appear to be initiated from the plasma cell membrane. There is growing evidence that a subpopulation of the conventional nuclear steroid receptor localized at the cell membrane mediates many of the rapid signaling actions of steroid hormones; however, novel membrane receptors unrelated to conventional steroid receptors have also been implicated. This chapter reviews the nature of the receptors that mediate rapid signaling actions of estrogen and progesterone and describes the signaling molecules and pathways involved, the mechanisms by which receptors couple with components of signaling complexes and trigger responses, and the target tissues and cell functions regulated by this mode of steroid hormone action.

INTRODUCTION

Steroid hormones have important regulatory roles in a wide variety of biological processes including reproduction, differentiation, development, cell proliferation, apoptosis, inflammation, metabolism, homeostasis, and brain function (1). Receptors for steroid hormones are members of a superfamily of nuclear receptors that function as ligand- or hormone-dependent transcription factors. Nuclear receptors are modular proteins consisting of a C-terminal or ligand-binding domain (LBD), a centrally located and highly conserved DNA-binding domain (DBD), and an N-terminal domain. Within these domains are at least two transcription activation subdomains or functions (AFs): AF-1 that resides in the N-terminal domain and ligand-dependent AF-2 present within the LBD. The LBD and DBD of nuclear receptors are conserved regions, whereas the N-terminal domain is highly variable with respect to both length and primary sequence but is important for full transcription activity of receptors (2–6). The N-terminal domain also contains multiple Ser/Thr phosphorylation sites that are regulated by various protein kinases. The physiological role of phosphorylation is not well defined but has been suggested to be involved in mediating cross-talk with other signal transduction pathway and modulating activity of AF-1 and interaction with coactivators (7–9).

Direct Control of Gene Transcription by Steroid Hormones

Nuclear receptors are sequence-specific DNA-binding proteins that recognize *cis*-acting hormone response elements (HREs) typically located in the promoter regions of target genes (10–13). As such, a main feature of nuclear receptors is to control expression of specific sets of target genes in response to an activating ligand or other signals. In most vertebrates there are two forms of estrogen and progesterone receptors, ERα and ERβ, and PR-A and PR-B. The two ER subtypes arise from different genes, whereas the PR isoforms are produced from a single gene through alternate use of two promoters that give rise to two different PR mRNAs (14, 15). The precise physiological role of multiple forms of ER and PR is not well understood. The DBD and LBDs of the two ER subtypes are well conserved both at the amino acid sequence level and structurally, the major difference between the two is the N terminus that has little homology (Figure 1A, see color insert). Both ER subtypes recognize similar target DNA sequences and bind and respond similarly to 17β-estradiol, although there are differences in DNA-binding affinity and specificity for pharmacological ligands. Although ERα and ERβ are coexpressed in certain target tissues, they also exhibit different tissue/cell expression patterns and are functionally distinct. ERα is a more potent transcriptional activator than ERβ, and in tissues where both ERs are expressed, ERβ has been suggested to have a role as an attenuator of ERα (16–19).

The A isoform of human PR differs from full-length PR-B by truncation of the N-terminal-most 164 amino acids, thus the two PR proteins have identical sequence in the LBD, DBD, and AF-1 of N-terminal domain (Figure 1B). Despite the fact that PR-A and PR-B have essentially indistinguishable steroid and DNA-binding activities, they have distinct transcription activities. In most contexts, PR-B is a strong activator of gene transcription, whereas PR-A can act as a ligand-dependent *trans*-repressor of PR-B and other steroid receptors including ER. The N-terminal-most 164 residues unique to PR-B harbor a third transcription activation domain, AF-3, which contributes to the stronger transcription activity of PR-B (20–22). PR-A and PR-B are coexpressed in most target tissues; however, the ratio can vary considerably depending on cell type or physiological conditions. Genetic experiments in mice have shown that PR-A has a predominant physiological role in mediating actions of progesterone in the uterus and ovary, whereas PR-B is more important in mammary gland (23, 24).

The basic pathway for how steroid hormones regulate gene transcription has been examined extensively. Upon binding steroid, receptors in target cells become activated through a process that involves dissociation from protein chaperones, conformational change(s), dimerization, and binding to HREs of target genes (1–6). Depending on the cell type or class of receptor, apo-receptors either reside in the cytoplasm and undergo translocation to the nucleus in response to binding ligand or reside largely in the nucleus with ligand promoting a change in intranuclear localization. In either case it is generally accepted that steroid hormone stimulates association of receptor with target gene promoters. DNA-bound receptors alter rates of transcription of target genes through recruitment of coactivator or corepressor proteins. Coactivators themselves are devoid of DNA-binding activity and are recruited through protein-protein interaction with AF-1 or AF-2 interaction surfaces of receptors and function either as enzymatic protein complexes capable of remodeling chromatin or as protein bridging factors to facilitate assembly of the RNA polymerase II initiation complex (4, 5, 25–28).

Rapid Actions of Steroid Hormones Independent of Direct Transcription Regulation

All classes of steroid hormones, vitamin D, and thyroid hormone have been reported to stimulate rapid effects on activities of components of signal transduction pathways, including second messenger production, ion channels, and protein kinase cascades. These actions occur on a time scale of seconds to minutes that is much too fast to involve gene transcription, are not affected by inhibitors of gene transcription and, in some cases, have been reported to occur in the absence of a nucleus, for example, with isolated cell membranes or enucleated cytolplasts (29). In many experimental systems, these rapid effects of steroid hormones are mimicked by cell impermeable steroid-protein conjugates, suggesting a plasma cell membrane initiated event distinguishable from intracellular or nuclear actions of steroids. The rapid action of steroid hormones independent of gene transcription has been commonly termed nongenomic to distinguish it from direct, or genomic, effects on gene expression in the nucleus (see reviews 30–38). One of the earliest reports of rapid nongenomic effects of steroid hormones was by Pietras & Szego in the late 1970s showing that 17β-estradiol induced a rapid stimulation of cAMP production and calcium flux in the endometrium of ovariectomized rats; subsequently, the presence of specific estrogen-binding sites on the cell membranes of endometrial cells was reported (39, 40). This work provided the first evidence of the existence of a receptor at the cell membrane that could be responsible for mediating effects of estrogen.

Major questions in this area of research are the nature of the receptors involved, the precise physiological role of rapid effects of steroid hormones on signaling pathways, and the contribution of rapid nongenomic actions relative to sustained genomic actions to the overall effects of steroid hormones on cell function. Herein we review progress on the actions of the female sex steroids, estrogen and progesterone; other excellent reviews and papers provide comprehensive discussions of the rapid actions of all classes of steroid and thyroid hormones (41–48). Rapid effects of estrogens and progesterone on various signaling pathways have been reported in a wide variety of cell types and tissues. Depending on the nature of the steroid hormone and tissue/cell type-specific response, it appears that different receptors and mechanisms are involved in mediating rapid signaling effects of steroid hormones. Four types of receptors have been proposed: (a) novel transmembrane receptors unrelated to nuclear hormone receptors, (b) transmembrane receptors for neurotransmitters or peptide hormones that are allosterically modulated by steroid hormones, (c) modified forms of the conventional steroid hormone receptor that preferentially localizes to the plasma cell membrane, and (d) a subpopulation of the conventional steroid receptor that associates with sites of signaling complexes in the cytoplasm, or the plasma cell membrane.

Protective Effects of Estrogen in the Cardiovascular System

The cardiovascular system is an important target tissue of estrogen. Both endothelial cells and vascular smooth muscle cells (VSMC), express ERα and ERβ, and estrogen has both acute and long-term effects on protection of blood vessels against injury and atherosclerosis. The long-term protective effects of estrogen are due to the classical actions of ER in mediating direct expression of important vascular cell-specific genes. Acute protective effects of estrogen include maintenance of vessel wall tone by rapidly stimulating vasodilation, inhibiting proliferation of smooth muscle cells and stimulating re-endothelization during vascular remodeling after injury, inhibition of leukocyte adhesion in response to mechanical injury, and cell survival in response to other insults. Acute vasodilation induced by estrogen occurs in seconds or minutes and is the result of ionic fluxes and release of the vasoactive molecule nitric oxide (NO) by endothelial cells to promote relaxation of VSMC by activation of guanylcyclase and cGMP gated-ion channels. NO release also inhibits proliferation of VSMC and adhesion of leukocyte adhesion in response to mechanical injury induced by ischemia or deposition of lipid plaques (49–52). In endothelial cell cultures, estrogen stimulates rapid release of NO in a Ca^{2+}-dependent manner by increasing the activity of eNOS (endothelial cell nitric oxide synthase) without increasing the level of eNOS protein expression. Estrogen activation of eNOS is not inhibited by DRB or cycloheximide and can be mimicked in isolated plasma cell membranes, suggesting this effect of estrogen occurs independently of gene transcription and new protein synthesis (53, 54).

Several lines of compelling evidence indicate that the rapid effect of estrogen on vasodilation dependent on eNOS activation is mediated by a subpopulation of the conventional ER present in the plasma membrane of endothelial cells (53–58). Although the data is somewhat inconclusive to this point, it appears that most of the rapid effects of estrogen on blood vessels are mediated by ERα, whereas only a subset of effects are mediated by ERβ. By immunocytochemistry with antibodies specific for conventional ERα, as well as transfection of endothelial cells with green fluorescent protein (GFP)-tagged ERα, ERα has been visualized in the plasma membrane of endothelial cells along with nuclear and cytoplasmic localization. ERα has also been reported to be present biochemically in isolated plasma membrane preparations of endothelial cells, as well as purified caveolae fractions. Caveolae are specialized lipid subdomains of the plasma membrane enriched in cholesterol and glycosphingolipids, but poor in phospholipids, and are sites of assembly and organization of components of signaling complexes such as Src, ras, MAP kinases, G protein–coupled receptor (GPCRs) and eNOS. ER is detected in both caveolae and noncaveolae membrane fractions, suggesting it is in equilibrium between general and lipid subdomains of the plasma membrane (55). Estrogen is able to rapidly activate eNOS in isolated plasma membranes and caveolae fractions, suggesting that ER is capable of mediating rapid estrogen activation of eNOS in the absence of the cytoplasm and nucleus. Estrogen activation of eNOS is dose dependent over a physiological range of hormone concentration, consistent with the binding affinity of conventional ER, and is blocked by the ER antagonists ICI 182,780 and tamoxifen, suggesting the requirement of conventional ER. In reconstitution cotransfection experiments in ER- and eNOS-negative cells, activation of eNOS was found to be dependent on ER, and using domains of ER, the LBD alone was found to be required and sufficient (53–59). This is consistent with the ER LBD being sufficient for plasma membrane localization and mediating rapid estrogen-induced activation of signaling pathways in other cell types.

Estrogen activation of eNOS and stimulation of NO production in endothelial cells is mediated by the PI3-kinase/Akt signaling pathway (56, 57, 60). Phosphatidylinositol 3-kinase (PI3K) is a lipid kinase composed of a p110 catalytic subunit and p85α regulatory subunit, and produces phosphatidylinositol-3-phosphate second messengers from membrane lipids. ERα has been shown by coimmunoprecipitation and GST-pull down assays to directly associate with the p85α regulatory subunit of PI3K in a hormone agonist-dependent manner, and experiments dissecting components of the PI3K/Akt signaling pathway support the conclusion that this interaction leads to activation of PI3K and downstream activation of eNOS through phosphorylation by the Ser/Thr kinase Akt. The protein surfaces or motifs that mediate interactions between ERα and p85α are not known, nor are the mechanisms by which this interaction leads to activation of PI3K (56, 57, 60). The nonreceptor tyrosine kinase Src has been suggested to be a critical upstream regulator as evidenced by the formation of an ERα/Src/PI3K ternary complex in membranes of EC cells, and by the failure of estrogen to activate eNOS in reconstitution transfection experiments in fibroblasts derived from Src$^{-/-}$ mice (61). As evidence that estrogen activation of eNOS via PI3K/Akt has a role in mediating protective effects of estrogen in vivo, estrogen induced a reduction of leukocyte adhesion to endothelial cells in vessels of the cremaster muscle in mice in an ischemia/reperfusion injury model. This effect of estrogen was reversed by inhibitors of PI3K and eNOS, and when comparing wild type and eNOS$^{-/-}$ mice, little effect of estrogen was observed on leukocyte adhesion in vivo in the absence of eNOS (62).

An important experimental system to decipher whether rapid effects on signaling pathways are mediated by conventional steroid receptors or an unrelated membrane receptor is the use of ER knockout mice. ERα and ERβ knockout mice have been used to study the acute protective effects of estrogen on vascular tissues, but interpretation of the results has been somewhat complicated by the fact the original ERα knockout mouse produced in Chapel Hill (ERKO$_{CH}$) is not a complete ERα null. ERKO$_{CH}$ mice produced by insertion of the neo-cassette into exon 1 express low levels of an N-terminally truncated ERα lacking much of the N-terminal domain AF-1, but containing the DBD and LBD. In transfection experiments, this form of ER is capable of mediating estrogen activation of NO production in blood vessels (63). Although reproductive and uterotropic responses to estrogen are largely abolished in ERKO$_{CH}$ mice, certain estrogen effects on vascular injury response are retained. In wild-type mice, estrogen inhibits medial thickening and proliferation of VSMC and stimulates re-endothelialization in a carotid artery injury model. In ERKO$_{CH}$ mice, this effect was partially retained. Inhibition of vascular remodeling was lost, whereas estrogen continued to inhibit VSMC proliferation (63). These results suggested that ERβ may mediate some of the protective effects of estrogen on vascular injury. However, ERβ knockout mice that are completely null exhibit the same response to estrogen in the carotid artery injury model as do wild-type mice, whereas the ERKO$_{CH}$ and ERβ double knockout mice exhibit a phenotype similar to ERKO$_{CH}$ mice alone (64). Subsequently, a complete ERα knockout mouse was produced in Strasburgh (ERKO$_{ST}$) by deletion of exon 2, and estrogen effects on both vascular remodeling and inhibition of smooth muscle cell proliferation were abrogated in the carotid artery injury model in the ERα null mice (65). These data, taken together, indicate that ERα alone is responsible for mediating the protective effect of estrogen on vascular injury response. However, ERβ knockout mice do show a phenotype for vascular function with respect to estrogen attenuation of smooth muscle cell vasoconstriction (66, 67). Whether the phenotype reflects a rapid effect of estrogen mediated by ERβ modulation of signaling pathways, or a genomic effect, is not known. ERβ in primary endothelial cell cultures is capable of mediating rapid nongenomic estrogen activation of eNOS and, as with ERα, a subpopulation of ERβ was found to be present in isolated caveolae of plasma membranes that is capable of supporting estrogen activation of eNOS (68). The signaling pathway that mediates ERβ activation of eNOS has not been characterized. ERβ was reported to be unable to associate with the p85α subunit of PI3K and to activate the PI3K/Akt signaling pathway, suggesting it may mediate estrogen activation of eNOS through a secondary MAP kinase signaling pathway (60). More studies are needed to define the precise role of two ER subtypes on the protective effects of estrogen on vascular tissues.

MAP kinase pathways have also been implicated in mediating effects of estrogen through membrane ERα on other functions of endothelial cells. In endothelial cell cultures, estrogen rapidly activates the antiapoptotic p38β member of the MAP kinase family while inactivating proapoptotic p38α MAP kinase. Estrogen activation of p38β MAP kinase in turn activates MAPKAP-2 kinase and hsp27 to protect cells from hypoxia-induced apoptosis and to maintain the integrity of the actin cytoskeleton network induced by metabolic stress (69).

Progesterone Effects in the Cardiovascular System

Although much less work has been done on the effects of progesterone on vascular cell function than estrogen, evidence indicates that conventional PR is expressed in endothelial cells and VSMC and mediates rapid effects of progesterone on vascular cell functions (70, 71). Progesterone inhibits proliferation of endothelial cells and VSMC, an effect dependent on conventional PR as shown by blocking with the PR antagonist RU486 and loss of responsiveness in cells isolated from PR knockout (PRKO) mice (72, 73). In endothelial cells, progesterone inhibits proliferation by decreasing expression of cyclins E and A and arresting cells in G1 (72), and was further shown in VSMC to inhibit growth factor activation of MAP kinase and downstream nuclear targets including c-*fos* and c-*myc* (73). To assess the role of PR in repair of vascular injury, ex vivo experiments with isolated intima-induced injured aorta showed that progesterone significantly repressed re-endothelialization of the injured aorta from wild-type mice but not from PRKO mice (74), indicating a requirement of conventional PR for protective effects of progesterone on blood vessels. The extent to which effects of progesterone and PR on vascular function are either genomic or mediated by extranuclear modulation of signaling pathways has not been examined in detail. However, similar to estrogen, progesterone has been shown to have rapid vasoactive effects. Progesterone can act synergistically with estrogen to induce vasodilation, and progesterone alone can relieve agonist-induced vasoconstriction of arteries in vivo and in vitro, an effect that occurs within minutes of administration, and in part is dependent on modulation of L-type calcium channels and intracellular free calcium (75, 76).

Rapid Stimulation of Signal Transduction Components by Estrogen in the CNS

Estrogen has neuroprotective effects and is important for survival of neurons exposed to various injuries including oxidative stress, ischemia, excitotoxicity, and various insults implicated in neurodegenerative diseases. In addition, estrogen is a neurotrophic factor in brain development and differentiation, it can modulate neurotransmitters and synaptic function, and it has a role in transmitting feedback to GNRH neurons in the hypothalamus and to lactotropes in the pituitary to control prolactin production and secretion. The neuroprotective effects of estrogen in several different model systems have been linked to a rapid stimulation of cell signaling molecules including cAMP production, calcium flux, potassium ion channels, and protein kinases of various signal transduction cascades (77–81). Long-term persistent effects of estrogen on primary target gene expression has also been correlated with neuroprotective effects, suggesting that effects of estrogen on neurons are mediated by two pathways: direct regulation of gene expression and rapid extranuclear signaling. Some studies have indicated that rapid effects of estrogen in neurons are mediated by a subpopulation of conventional ER, whereas others have implicated an ER-independent mechanism. Which receptor mechanism is involved may depend on the brain region, type of injury, and whether estrogens are used at pharmacological or physiological concentrations (77–81). Expression of conventional ER has been documented by immunohistochemistry in various regions of the brain, including hippocampus, hypothalamus, preoptic area, amygdala, cerebellum, and cerebral cortex. Furthermore, substantial staining of ER was detected in the cytoplasm and plasma membrane of dendritic spines of neurons, along with nuclear staining (82). ERα and ERβ exhibit distinct spatiotemporal as well as overlapping patterns of expression in the brain. For example, in the cerebral cortex, ERβ expression is persistent during development, whereas ERα expression occurs only transiently during neonatal development (83).

In rodent neo-cortical explants, estrogen stimulates activation of MAP kinase (Erk) by a mechanism that involves phosphorylation and activation of c-*ras*, Src, and raf, and recruitment of the adaptors Shc and Grb2 at the cell membrane. This appears to be mediated by an ERα/ERβ-independent mechanism because the same effect of 17β-estradiol was observed in neo-cortical explants from wild-type and ERα knockout (ERKO) mice and was not blocked by the ER antagonist ICI 182,780. Also, the pharmacology and EC$_{50}$ for responses to different estrogenic compounds did not fit that of ERα or ERβ (83, 84). A putative novel membrane receptor, termed ER-X, of slightly smaller molecular mass (62–63 kDa) than ERα (66 kDa) and with some sequence homology with the LBD of conventional ERα, has been proposed to be involved (79, 85). In cells of neocortex, ER-X is localized primarily in the cytoplasm and membranes of dendritic spines of neurons and is highly enriched in caveolar-like microdomains (CLMs). The origin of ER-X is not known, but it is expressed in ERKO mice at the same level as in wild-type mice, suggesting, along with its molecular size, that it is not equivalent to the N-terminally truncated product of ERα detected in ERKO$_{CH}$ mice. Enriched CLMs of neocortical tissue of ERKO and wild-type mice both exhibit single, high-affinity saturable-binding sites for 17β- and 17α-estradiol, and both estrogens rapidly (30 min) activate Erk1-2 in isolated CLMs, suggesting that ER-X is a binding protein for estrogen that can mediate MAPK activation. That ER-X has a role in acute actions of estrogen in the cerebral cortex is further suggested by the fact that its expression is developmentally regulated and up-regulated in response to injury in an ischemic stroke model (79, 85). Although ER-X is a potentially important protein, there is no evidence that it binds estrogen directly or mediates a signaling response to estrogen, thus meeting the criteria for a receptor. Cloning, expression, and more complete characterization of ER-X will be required to firmly establish its role as novel membrane ER and to understand its relationship with conventional ER, which is also expressed in cortical tissue.

Other ER-independent rapid actions of estrogen have been described in the CNS. Estrogen was reported to bind directly to the β subunit of the Maxi-K ion channel and to stimulate Maxi-K currents. However, the binding affinity of 17β-estradiol for the β subunit of Maxi-K channel is in the low μM range, as is the EC$_{50}$ for estrogen activation, which raises questions about the physiological relevance of this effect of estrogen (86). In hypothalamic neurons, estradiol rapidly activates PKC and PKA by a mechanism dependent on G protein (Gαq)-coupled activation of phospholipase C (87). The fact that a cell impermeable E2-BSA conjugate and a compound structurally related to tamoxifen, STX, which has little affinity for conventional ER, both mimicked the action of free estradiol, was taken as evidence that this effect of estrogen on PKC and PKA is not mediated by the conventional ER (87). In hippocampal CA1 neurons, estrogen rapidly enhanced the amplitude of kainite-induced currents in a manner dependent on GPCR, and similar results were obtained with cells from wild-type and ERKO$_{CH}$ mice, suggesting an ER-independent mechanism (88). In human neuroblastoma cells, SK-N-SH, which lack expression of conventional ERα, an estrogen BSA conjugate induced rapid phosphorylation and activation of Erk and transactivation of a reporter gene controlled by c-*fos*, a nuclear transcription factor target of activated Erk. The ER antagonists, tamoxifen and ICI 182,780, had no effect on estrogen activaton of Erk and c-*fos*, further indicating that this action of estrogen in neuroblastoma cells is not mediated by conventional ER (89).

Other results, however, have suggested a role for conventional ERα or ERβ in mediating rapid signaling actions associated with the neuroprotective effects of estrogen. Estrogen treatment protected primary neuronal cultures against glutamate- and quinoloinic acid–induced excitotoxicity and cell death in a manner dependent on expression of conventional ERα and the ability of estrogen to activate the Src/ras/raf/MAPK signaling pathway (90). Hippocampal-derived immortalized cell lines (HT22), stably transfected with ERα or ERβ, have been used as a model system to study the role of conventional ER in mediating the neuroprotective effect of estrogen. Estrogen stimulated a rapid (15 min) activation of Erk in a manner that was blocked by the ER antagonist ICI 182,780 and the MEK inhibitor PD 98,059, whereas estrogen had no effect in untransfected cells lacking ERα (91). Estrogen also reduced cell death in response to treatment with β amyloid peptide, in a manner dependent on ERα or ERβ, and activated MAPK (91). Thus ERα and ERβ both are capable of mediating estrogen activation of MAPK and the neuroprotective effect of estrogen under certain conditions. The relative importance of the two ER subtypes for the neuroprotective effects of estrogen in vivo is not clear. With in vivo mouse models of ischemia injury of the brain, the protective effect of estrogen was abrogated in ERKO mice, but was preserved in ERβ knockout (BERKO) mice, suggesting ERα is the more important ER subtype (92). However, substantial brain abnormalities have been reported in BERKO mice (93). In these mice abnormal regions of brain development occur at 2-months of life that results in loss of neurons and proliferation of astroglial cells in the limbic system. This becomes progressively worse with age, resulting in significant loss of neuronal bodies throughout the brain (93). This phentotype suggests that ERβ in vivo does play a role in neuronal survival and perhaps in development of degenerative diseases of the CNS.

In pituitary lactotrophs, estrogen stimulates expression and synthesis of the peptide hormone prolactin, but also stimulates a rapid release (a few minutes) of prolactin in a manner dependent on elevated intracellular calcium through release from intracellular stores and influx of Ca^{2+} through voltage-gated Ca^{2+} channels (81, 94–97). That the rapid effect of estrogen on prolactin release is mediated by conventional ER localized in the cell membrane was shown in a subline of GH3/B6 pituitary cells that are enriched for membrane ER. Antibodies to several different epitopes of conventional ERα were shown to react specifically with the surface of nonpermeabilized GH3/B6 cells and were capable of either mimicking estrogen-stimulated release of prolactin or inhibiting release. Also, specific estrogen-binding sites on the surface of G3H/B6 cells were detected with cell-impermeable FITC estrogen conjugates (94–97).

Other Estrogen Target Tissues

ENDOCRINE PANCREAS   The endocrine pancreas is an estrogen-responsive tissue, and 17β-estradiol at physiological concentrations has been shown to act as a modulator of insulin and glucagon secretion by increasing or decreasing the frequency of glucose-induced intracellular [Ca^{2+}] oscillations (98, 99). This effect of estrogen on intracellular calcium signaling is rapid (minutes), involves production of cGMP as a second messenger, activation of protein kinase G (PKG), and phosphorylation and modulation of ATP-dependent potassium channel activity leading to membrane depolarization. Phosphorylation and activation of the cAMP response element–binding (CREB) protein by PKG was also shown to be a rapid response to estradiol in pancreatic islet cells (98, 99). These rapid effects of estrogen on the endocrine pancreas are likely mediated by a novel membrane receptor unrelated to conventional ER, based on the fact that response to estrogen was unaffected by ER antagonists (ICI 182,780 or tamoxifen), was mimicked by a cell-impermeable estrogen conjugated to horseradish peroxidase (EHP), and EHP gave a plasma membrane staining that is distinct from intracellular localization of conventional ERα/ERβ in pancreatic islet cells (100, 101). The EHP cell surface–binding site was characterized as steroid specific for estrogen, but it also binds several catecholamines with a pharmacology similar to γ-adrenergic receptors, suggesting the involvement of a novel membrane protein with characteristics of γ-adrenergic receptors that share binding with estrogens and catecholamines.

MACROPHAGES   Estrogen induces a rapid rise in intracellular free calcium in macrophages through release from intracellular stores as well as influx from extracellular sources. This results in a specific attenuation of serum-stimulated activation of the early immediate response gene c-*fos*, as well as an enchancement of LPS-induced activation of c-*fos*, suggesting a role for estrogen in modulation of macrophage activation (102, 103). This effect of estrogen was mimicked by an E2-BSA conjugate, was not blocked by ER antagonists (ICI 182,780, raloxifine, or tamoxifen), but was sensitive to pertussis toxin, implicating the participation of a GPCR, rather than conventional ER. Use of an estrogen-FITC-BSA conjugate showed a steroid-specific binding site on the surface of intact macrophage cells that, after time of estrogen treatment, was internalized to punctate localization patterns. However, macrophages also express conventional ERα that resides predominantly in the cytoplasm and, to a lesser extent, in the nucleus, but does not colocalize with the estrogen-FITC-BSA surface-binding site. As a further distinction from the estrogen-BSA-FITC binding sites, cytoplasmic ER is not accessible to ER antibodies on the surface of intact cells and does not undergo a sequestration in response to estrogen treatment. These data were taken as evidence for the presence of a novel membrane receptor for estrogen.

BONE   The importance of estrogens in bone homeostasis has been known for some time. Both ERα and ERβ are expressed in osteoblasts and osteoclast cells, and estrogens protect against bone loss by slowing the rate of bone remodeling. Estrogens exert their effect on bone maintenance by stimulating apoptosis of osteoclasts while inhibiting apoptosis of osteoblasts (104). Using mouse osteoblasts and osteoctye cell cultures, the antiapoptotic effect of estrogen was shown to be mediated by rapid activation of the Src/Shc/Erk signaling pathway in a manner dependent on extranuclear actions of the conventional ER. The antiapoptotic effect of estrogen was abrogated by inhibitors of Erk phosphorylation and Src, as well as a dominant-negative Shc and a kinase dead Src. Additionally, fibroblasts derived from Src$^{-/-}$ mice (that also lack Yes and Fyn) and reconstituted to express ER failed to support the antiapoptotic effect of estrogen compared with fibroblast cells derived from Src$^{+/+}$ mice (105). The dependency of the antiapoptotic effect of estrogen in osteoblasts on conventional ER was shown by the blocking effect of the ER antagonist ICI 182,780, and estrogen inhibition of etoposide-induced apoptosis was shown to require ectopic expression of ER in HeLa cells. In HeLa cells transfected to express various domains of ER, and in ER tagged to localize to specific subcellular compartments, the antiapoptotic effect of estrogen was shown to be mediated by an extranuclear action of ER that required only the LBD. Because the LBD cannot transactivate target genes, this indicates that ER possess separable transcription and nontranscription activities. Where conventional ER is located in osteoblast cells to mediate estrogen activation of Src/Shc/Erk pathway and how it couples with signaling complexes are not known. Co-immunoprecipitation of ERα and cSrc was detected in rat osteosarcoma-derived cells, suggesting a physical association of ERα with components of signaling complexes (105).

A synthetic estrogen, 4-estren-3α,17β-diol (estren), has been developed that exhibits effects similar to that of 17β-estradiol in osteoblasts in terms of inhibiting apoptosis by a mechanism that involves activation of the Src/Shc/Erk signaling pathway. This compound has little activity with respect to inducing ER-mediated transcription of target genes, at least for the C3 gene as a representative ER target. When administered to ovariectomized female mice, estren was reported to be at least as effective as 17β-estradiol in preventing ovariectomy-induced apoptosis of osteoblasts in lumbar vertebrae and preserving general bone mineral density. In contrast, estren administration did not stimulate uterine growth of ovariectomized mice and thus did not mimic the uterotropic action of 17β-estradiol in vivo (106).

From these data, it was suggested that estren is capable of dissociating transcription and nontranscription actions of ER (105, 106), and that nontranscription actions of ER are important for mediating protective effects of estrogen in bone, whereas classical transcription actions of ER are more important for proliferative effects of estrogen in reproductive tissues. However, estren has a substantially lower affinity for ER than 17β-estradiol, suggesting that its inability to induce ER target gene activation may be due simply to its poor affinity for ER, while its potent antiapoptotic effect may be mediated by a novel receptor mechanism.

Progesterone and Progesterone Metabolites as Allosteric Modulators of GABA$_A$ and Oxytocin Receptors

Progesterone also has important biological actions in the CNS, including female sexual behavior, release of LHRH from the hypothalamus required for ovulation, and protection against seizures and convulsions. Neuroprotective effects of progesterone are mediated primarily by reduced metabolites of progesterone such as 5α pregnan-3α-ol-20-one (allopregnanolone). Cells in the CNS can synthesize allopregnanolone and other reduced progesterone metabolites de novo or from circulating progesterone, which leads to the concept of "neurosteroids" that act locally in the brain in a paracrine fashion (107, 108). γ-aminobutyric acid type A (GABA$_A$) receptor is a member of the cysteine-*cys*-loop family of membrane transmitter-gated ion channels and is a major inhibitory neurotransmitter in the mammalian CNS. A number of therapeutic drugs act by enhancing interaction of GABA$_A$ with its receptor, resulting in an alteration of Cl$^-$ conductance and neuron excitability. Reduced progesterone metabolites have potent analgesic, sedative, anesthetic, and anticonvulsive activities, mediated by their ability to allosterically enhance interaction of GABA$_A$ with its receptor. The effects of progesterone metabolites on GABA$_A$ receptor occur with low nM concentrations and exhibit a high degree of stereospecificity, which suggests a specific binding site or pocket for progestins. However, a specific progestin-binding site or domain in the GABA$_A$ receptor has not been well defined (107). Although conventional PR is expressed in several areas of the brain, including neo-cortex, hippocampus, amygdala, and limbic system, and is required for female sexual behavior, PR does not appear to be required for the antiseizure and anticonvulsive effects of progesterone (108). These activities of progesterone in vivo were shown to be similar in wild-type and PRKO mice, which suggest the neuroprotective effects of progesterone are mediated primarily by their ability to allosterically modulate the the GABA$_A$ or other neurotransmitter receptors.

Another example where progesterone can allosterically modulate signaling by a known membrane receptor is the oxytocin receptor (OTR). Progesterone has an essential role in maintaining quiescence of the uterus during pregnancy, whereas the posterior pituitary peptide hormone oxytocin has the opposite effect by stimulating contraction of the uterine myometrium. Progesterone decreases the sensitivity of the uterus to oxytocin, and as a mechanism to explain this inhibitory effect of progesterone, several groups have reported that progesterone acts as a negative allosteric modulator of OTR (109–112). The OTR is a GPCR that, in response to binding oxytocin, induces the secretion of prostaglandin F2α mediated by a signal response pathway that involves generation of inositol phosphophate, leading to a rise in intracellular free calcium. Physiological concentrations of progesterone were reported to inhibit this OTR-mediated signaling response to oxytocin in vivo and in vitro (109–111). This effect occurs within minutes, is not blocked by transcription and protein synthesis inhibitors, and can be mimicked by a progesterone-BSA conjugate, suggesting a rapid membrane-initiated effect of progesterone rather than a nuclear action. Progesterone was reported to bind directly to OTR with a high affinity (K_d in the nM range), and to inhibit binding of oxytocin to OTR with an IC_{50} in the nM range. Binding is also highly specific for progestins, does not occur with other steroid hormones, and a species specificity has been observed, e.g., rat OTR preferentially binds progesterone, whereas human OTR preferentially interacts with a progesterone metabolite, 5β-dihydroprogesterone. Uncoupling of OTR from G proteins by the nonhydrolyzable analogue GTPγS reduced the binding of progesterone and the ability of progesterone to inhibit oxytocin binding and signaling, thus implicating GTPase-induced conformational changes of OTR in the actions of progesterone. However, these results on progesterone interaction with OTR have not been consistently reported in the literature. A study with human myometrial membranes failed to detect inhibition of oxytocin binding and OTR-mediated signaling by progesterone or 5β-dihydroprogesterone at any doses tested (113). Another study observed effects only at micromolar concentrations (112). The reason for the discrepancy in results is not known and could be from differences in cell type, species, and experimental conditions. However, the physiological level of serum progesterone during pregnancy is high (500 nM) and can also be exceedingly high in specialized steroidogenic tissues such as the placenta, suggesting that progesterone interaction with OTR even at micromolar concentrations may be physiologically relevant.

Progesterone-Induced Germ Cell Maturation

SPERM CELL ACROSOME REACTION   The most extensively characterized rapid signaling action of steroid hormones, and perhaps the most relevant in terms of a physiological role, is the ability of progesterone to induce germ cell maturation. The acrosome is a secretory granular-like organelle in the sperm head that contains lysosomes and proteases, and the acrosome reaction is a modified exocytotic event that facilitates penetration of the spermatazoa to the ovum and sperm-egg plasma membrane fusion required for fertilization. Progesterone secreted in high concentrations into the cumulus oophorous and follicular fluid surrounding the egg is an initiator of the acrosome reaction, by acting in a rapid nongenomic manner as determined by the usual criteria plus the fact that nuclei of spermatozoa are transcriptionally inactive (38). Progesterone stimulates a rapid influx of extracellular calcium and an efflux of chloride ion that are both essential for induction of acrosome reaction (114). A calcium-dependent increase in cAMP and phosphorylation of several spermatozoa proteins are also involved in mediating progesterone-initiated acrosome reaction, and inhibitor studies have implicated a role for both PKC and cAMP-dependent PKA. PKA and A kinase anchoring proteins (AKAP) have been shown to be expressed in the acrosomal region of sperm, and specific inhibitors of PKA and peptides designed to disrupt PKA-AKAP interactions inhibit progesterone-initiated acrosome reaction (115).

Several groups have reported the existence of high-affinity-specific binding sites for progesterone on the surface of intact spermatozoa, and a strong correlation between affinity constants of different ligands and EC_{50} values for stimulation of calcium influx has been observed, which suggests the biological relevance of these binding sites (116–119). Several candidate membrane receptors have been proposed including a protein(s) related to the LBD of conventional PR, a novel membrane-binding protein for progesterone initially isolated from liver membranes, and the $GABA_A$ receptor/chloride channel complex. Whether full-length conventional PR is involved is inconclusive. Several studies have been unable to detect full-length PR protein or RNA transcript in spermatozoa (116, 118–120), the male PRKO mice are fertile (121), and the steroid specificity and pharmacology of compounds that stimulate the sperm acrosome reaction do not fit the structural activity profile of the conventional PR (116, 118). For example, synthetic progestins such as norethindrone, megestrol, and R5020, which are potent activators of PR-mediated transcription, are poor initiators of the acrosome reaction, whereas the PR antagonist RU486 is only weakly antagonistic of progesterone-initiated acrosome reaction and does not compete for binding of progesterone to the surface of sperm cells. These data suggest that conventional PR is not involved in mediating the effect of progesterone on the sperm acrosome. However, several groups have reported the presence of N-terminally truncated forms of PR (54 and 57 kDa, or 50 and 52 kDA) (depending on the study), or a gene product closely related to PR, as detected by immunoblot with a monoclonal antibody (MAb) (C262) to the C-terminal tail of human PR (118, 122, 123). The C262 MAb recognizes an epitope in the LBD of human PR that is conserved among species of PR, and it competes for binding of progesterone to PR, presumably owing to interaction with amino acid side chains that contact progesterone in the binding pocket of the LBD (124). The 54–57-kDa protein(s) is not detected with antibodies to the N terminus of PR, and ligand blots with horseradish peroxidase conjugated-progesterone have shown that it binds progesterone. As evidence that the 54–57-kDa protein(s) is involved in mediating progesterone-initiated acrosome reaction, the C262 MAb inhibited progesterone-induced calcium influx, chloride ion efflux, and the acrosome reaction, whereas a control unrelated antibody and an antibody to the N-terminal domain of PR had no effect (118, 122, 123). The presence of low copy number PR mRNA in human sperm was detected by RT-PCR, and the mRNA was confirmed by sequencing to encode for a truncated human PR spanning the DBD and LBD (125). Whether this transcript encodes the 50–57-kDa protein reactive with C-terminal C262 antibody has not been determined. A separate study with a different antibody to the C-terminal tail of PR LBD showed a correlation between expression of the putative truncated PR and fertility (126). A significant decrease in immunostaining of the acrosome region with the C-terminal PR antibody was detected in spermatozoa of infertile men, and the ability of spermatozoa to undergo acrosome reaction in vitro correlated with expression of this truncated PR (126). These data collectively suggest that a 50–57-kDa protein(s) detected by antibodies to the C terminus of PR may be an N-terminally truncated membrane form of PR involved in mediating progesterone-initiated acrosome reaction. The origin of the N-terminally truncated PR is not known, nor is it known why it localizes predominantly to the plasma membrane. A C isoform of PR was identified in breast cancer cells that arises through alternate use of an internal ATG codon located within the DBD to generate a PR protein lacking the N-terminal domain, part of the DBD, and extending through the LBD. The C receptor lacks a functional DBD, is inactive as a transcription factor but retains normal hormone-binding activity, and can modulate the transcription activity of PR-A and PR-B (127). Whether PR-C and the 50–57-kDa immunoreactive protein in plasma membranes of sperm are equivalent is not known. A unique PR mRNA has been identified in a testis cDNA library. The message contains an additional novel 5' exon (termed exon S), is lacking the N-terminal domain, and retains exons 4–8 that encode the DBD and LBD. The expression level of PR-S mRNA was higher in spermatozoa than in the uterus (128). However, the S form of PR has been detected only as an RNA transcript; an S form PR protein has not been detected, nor have studies been done to explore a functional role for PR S and its relationship to other N-terminal truncated forms of PR.

A high-affinity progesterone-binding protein unrelated to conventional PR has also been implicated in mediating progesterone-initiated acrosome reaction (117). This protein was isolated and cloned from liver membranes consisting of a 28- and 56-kDa subunit linked by disulfide bonds to form a native $\sim$200-kDa protein (129). Correlative data suggest this protein mediates rapid progesterone effects on calcium release from liver cells. An antibody to the liver progesterone binding protein detects an $\sim$44-kDa protein in sperm plasma membranes and inhibits progesterone-induced calcium flux and progesterone-initiated acrosome reaction in vitro (129). Because the $GABA_A$ receptor/Cl^- ion channel complex is present in spermatozoa and progesterone is known to be an allosteric modulator of brain $GABA_A$ receptor, a role for $GABA_A$ receptor in progesterone-initiated acrosome reaction has been hypothesized (115). Results, however, have been inconclusive. $GABA_A$ alone does not stimulate Ca^{2+} influx and, at best, only weakly induces acrosome reaction in sperm. Additionally, $GABA_A$ does not alter progesterone-induced Ca^{2+} influx, nor does it influence binding of progesterone to the 54–57-kDa protein of sperm membranes. These data suggest that $GABA_A$ receptor is not involved in calcium signaling induced by progesterone. However, the $GABA_A$ Cl^- channel blocker, picrotoxin, partially inhibited progesterone-mediated Cl^- efflux and progesterone-initiated acrosome reaction, which indicates that both the activity of the $GABA_A$ R/Cl^- channel and Cl^- efflux are required for progesterone-initiated acrosome reaction (115). It has been suggested that the 54–57-kDa sperm membrane receptor for progesterone and $GABA_A$ receptor/Cl^- channels are distinct entities that cooperate to mediate influx of calcium and efflux of Cl^- required for progesterone-induction of acrosome reaction.

INDUCTION OF OOCYTE MATURATION   In most animal species, oocytes are arrested in prophase (G2) of meiosis I, and just prior to ovulation they re-enter the meiotic cell cycle and arrest again at metaphase of meiosis II to become competent for fertilization. This process of oocyte maturation has been studied extensively in *Xenopus laevis*, which is an excellent experimental system because isolated *Xenopus* oocytes can be maintained in meiosis I and induced by steroids to progress to meiosis II in vitro (130–132). In M phase arrest, the large nuclear envelope of the *Xenopus* oocyte, termed germinal vesicle, undergoes dissolution, and visual detection of germinal vesicle breakdown (GVBD) is a convenient marker of G2/M progression and oocyte maturation. *Xenopus* oocyte maturation can be induced by progesterone in vitro and presumably reflects progesterone synthesis and secretion by follicular cells surrounding the oocyte in response to gonadotropin. This action of progesterone is mediated by activation of signal transduction pathways that eventually converge upon and activate cyclin B/cdc2 kinase (or maturation-promoting factor; MPF), the key cell cycle kinase that controls G2/M progression (130–133). That this action of progesterone occurs in the absence of gene transcription is based on the usual criteria for nongenomic actions of steroids, plus the fact that some aspects of progesterone-induced maturation occur with enucleated oocytes and are induced more efficiently with extracellular progesterone than with microinjected intracellular progesterone.

The earliest known effect of progesterone on *Xenopus* oocytes is a decrease in intracellular cAMP and PKA owing to inhibition of adenyl cyclase activity. The decrease in cAMP is GTP dependent, suggesting involvement of a GPCR, but the response is insensitive to pertussis toxin and cholera toxin arguing against a receptor coupling with $G\alpha i$ subunits. Progesterone stimulates at least two intracellular protein kinase signaling cascades downstream of cAMP/PKA that converge upon and activate cyclin B/cdc2. One is a MAP kinase pathway consisting of c-*mos* (*Xenopus* equivalent of mammalian c-*raf*), MEK1, MAP kinase (Erk-1/-2), p90Rsk-1/-2, and MyT1. The other signaling cascade consists of the polo-like kinases (Plkk and Plx) that activate the protein phosphatase cdc25C. Cyclin B/cdc2 is a preformed complex that is inactivated constitutively through phosphorylation by the kinase MyT1. Cdc2 can be activated by inhibition of MyT1, achieved through its phosphorylation by p90Rsk in the MAPK pathway, or by dephosphorylation mediated by cdc25C phosphatase in the polo-like kinase pathway (130–133). How the early cAMP/PKA events couple with intracellular kinase signaling cascades has been a long-standing question. Cdc25C was recently shown to be a direct target of PKA. During G2 arrest, elevated PKA phosphorylates cdc25C on a site (ser 287) that inhibits cdc25C phosphatase activity through interaction with and sequestering by 14-3–3. Thus the drop in PKA activity induced by progesterone leads to dephosphorylation of cdc25C, release from 14-3-3, and activation of cdc2 (134). Whether the drop in cAMP/PKA is sufficient for induction of intracellular signaling pathways and oocyte maturation is uncertain. Maintenance of elevated intracellular cAMP by use of cAMP analogs or phosphodiesterase inhibitors blocks oocyte maturation, whereas inhibitors of PKA can induce oocyte maturation in the absence of progesterone (134, 135). These data collectively suggest that progesterone induction of ooctye maturation is a release from cAMP/PKA inhibition mechanism, thought to occur by progesterone acting to relieve a constitutive inhibitory signal mediated by $G\beta\gamma$ subunits (135).

The receptors that mediate progesterone induction of *Xenopus* oocyte maturation have long been sought after, and recent work from several groups indicate the involvement of two receptor systems: a conventional PR localized in the cell membrane and a novel GPCR-related membrane PR (mPR). Two groups independently cloned the *Xenopus* homolog of conventional mammalian PR, termed X-PR (136, 137). The amphibian gene exhibits a high degree of homology with human PR in the DBD (92% identity) and LBD (86% identity) and has little homology in the N-terminal domain. X-PR binds progesterone with a high affinity and stereospecificity expected of conventional PR and, when overexpressed in *Xenopus* or in heterologous mammalian cells, X-PR mediates progesterone-dependent transactivation of a reporter gene controlled by progesterone-response elements (PREs). Several lines of evidence suggest that X-PR is involved in mediating the effects of progesterone on oocyte maturation. Overexpression by microinjection of X-PR mRNA accelerated the rate of progesterone-induced maturation and increased the sensitivity to progesterone, whereas microinjection of antisense oligonucleotides decreased progesterone-induced maturation by 70 to 85% and inhibited progesterone induction of MAPK activation. Moreover, inhibition by antisense oligos was rescued by microinjection and overexpression of either X-PR or human PR. Microinjection of X-PR into enucleated oocytes accelerated the rate of progesterone-induced activation of MAP kinase, and this effect was not altered by treatment with actinomycin D (136, 137). Finally, X-PR localizes predominantly in the cytoplasm of oocytes with approximately 5% of total endogenous X-PR detected in washed plasma membranes, although only minimal nuclear PR was detected (138). Thus it appears that X-PR has a somewhat different intracellular localization than mammalian PR, which is predominantly nuclear. How X-PR in the plasma membrane is mechanistically linked to intracellular signal transduction pathways has not been well defined, but preliminary data suggest an involvement of a physical association between X-PR and PI3K and MAPK in the plasma membrane. Progesterone treatment of oocytes induces a rapid (30 min) activation of PI3K that correlates with physical association of PR with PI3K preceding GVBD. A hormone-dependent association of PR and MAPK was also detected at later times of progesterone treatment subsequent to GVBD when MAPK has become activated, and MAPK was shown to be capable of phosphorylating X-PR (138).

Despite encouraging results with X-PR, the structure activity relationship for steroid induction of oocyte maturation does not fit the stereospecificity for conventional PR, and the EC_{50} (200 nM) for progesterone induction of GVBD is higher than that for progesterone activation (10 nM) of transcription mediated by X-PR (133). Oocyte maturation can be induced efficiently with glucocorticoids, androgens, and mineralocorticoids, and the PR antagonist RU486 is a weak inducer and fails to block progesterone. However, the apparent discrepancy with the actions of RU486 on maturation and X-PR mediated gene expression could be explained by a lower affinity of RU486 for X-PR than mammalian PR. X-PR contains a cysteine residue in place of glycine in the same position of the LBD that greatly reduces affinity of RU486 for chick PR versus human PR (137). That X-PR has a lower affinity for RU486 is suggested by the fact that the concentration of RU486 required to antagonize progestin-induced X-PR-mediated gene transcription is much higher than required to antagonize human PR (137, 139). On the basis of these data, it has been argued that X-PR alone is not sufficient for mediating progesterone induction of oocyte maturation and that other receptors must be involved (133). A conventional androgen receptor (AR) from *Xenopus* oocyte was recently cloned and characterized, and using criteria similar to X-PR studies, an involvement of AR in mediating androgen-induced oocyte maturation was shown (140–142). AR is associated with the plasma membrane, and androgen-induced oocyte maturation is inhibited by the AR antagonist, flutamide, and by knock-down of endogenous AR by RNA interference. Interestingly, after in vivo injection of human chorionic gonadotropin into frogs, androgens (androstenedione and testosterone) were found to be present in the ovary at higher concentrations than progesterone. Characterization of steroidogenic pathways in *Xenopus* ovaries further revealed that the oocyte is capable of metabolizing progesterone to androgens, suggesting that androgens also play a role in promoting oocyte maturation.

In addition to conventional steroid receptors, a novel membrane G protein-coupled-like receptor was recently identified that plays a role in progesterone-induced oocyte maturation of teleosts (143). In sea trout oocytes, the maturation-inducing steroid (MIS) is the progesterone metabolite 20βS (17α, 20β-trihydroxy-4-pregnen-3-one). A novel membrane receptor for 20βS was cloned by screening a sea trout cDNA expression library with a monoclonal antibody raised to a 20βS-binding protein partially purified from sea trout oocyte plasma membranes (144). The gene encodes an ~40,000-kDa protein that bears properties of seven transmembrane GPCRs. By immunochyochemistry and biochemical analysis, membrane PR (mPR) localizes to the surface layer of plasma membrane of oocytes and is not expressed in the cytoplasm or surrounding follicle cells. When expressed as a recombinant protein in bacteria, a solubilized form of mPR directly binds [3H]-progesterone in vitro in a saturable, high affinity (K_d 30 nM) manner, and with a steroid specificity for progesterone and progesterone metabolites. mPR does not bind the synthetic progestin R5020 or the PR antagonist RU486, further distinguishing this protein from conventional PR. When expressed in heterologous mammalian cells (MDA231 breast cancer cells that lack conventional PR), mPR was found predominantly in the cell membrane and was required for a rapid progesterone-induced decrease in intracellular cAMP that was sensitive to pertussis toxin, suggesting coupling with an inhibitory G protein. Ectopically expressed mPR in breast cancer cells also mediated a rapid (5 min) progesterone induction of MAP kinase (Erk-1/-2) (144). A discrepancy was observed in the steroid-binding specificity in that recombinant mPR in vitro bound progesterone only, whereas mPR expressed on the surface of breast cancer cells responded to both progesterone and 20βS. It is possible that recombinant solubilized mPR without its natural cell membrane environment was altered with respect to steroid-binding properties (144). Studies with nonhydrolyzable GTPγ-S confirmed that mPR is coupled to a G protein and activates an inhibitory Gαi subunit (145). That mPR has a role in mediating progesterone-induced oocyte maturation in vivo is supported by several lines of evidence. Its expression is up-regulated in vivo by chorionic gonadotropin, consistent with the timing of up-regulation of 20βS binding sites on the surface of oocytes and acquisition of responsiveness to 20βS. Also, injection of antisense oligonucleotides into zebrafish oocytes inhibited 20βS-induced maturation.

Sea trout mPR is related to a family of 13 vertebrate genes from a variety of different species including humans that fall into three phylogenetic subtypes, α, β, and γ. The α subtype is expressed mostly in reproductive tissues (sea trout mPR is an α subtype), β is expressed exclusively in neural tissue, and γ is expressed in lung, kidney, colon, and adrenal (146). It will be important to determine the role of mPR and family members in mediating progesterone-induced oocyte maturation in additional species (147), and whether these receptors have a role in mediating rapid signaling actions of progesterone in other tissues. It will also be important to determine the relationship, if any, between mPR and conventional PR. In the sea trout, it has been proposed that conventional PR up-regulates mPR gene expression as a priming event required to make the oocyte competent to respond to 20βS. Additionally, mPR and conventional nuclear PR could both be involved in mediating rapid effects of progesterone on signal transduction cascades, with mPR mediating proximal initiating events at the cell surface, and X-PR coupling with intracellular signaling pathways. The progesterone binding site of mPR remains to be characterized. The GPCR-like mPR appears to be the first rigorously characterized membrane receptor for a steroid hormone in animal species that is unrelated to the conventional steroid receptor. The only other well characterized membrane receptor unrelated to the conventional steroid receptor is a receptor for plant brassinosteroids, a transmembrane receptor-like Ser/Thr kinase (RLK) that mediates signal transduction by brassinosteroids across the plasma membrane (148–151). The brassinosteroid receptor gene (BRI1) isolated from *Arabidopsis* has an extracellular domain consisting of 25 leucine-rich repeats required for interaction with brassinosteroids, a transmembrane domain, and an intracellular cytoplasmic kinase domain. Members of the nuclear receptor superfamily are not encoded in the plant genome, suggesting that steroid action at the cell membrane to generate intracellular signaling responses may be an ancient function that has been conserved in evolution.

OVARIAN GRANULOSA CELLS ARE TARGETS OF RAPID PROGESTERONE SIGNALING
Granulosa cells of the ovary surrounding the oocyte synthesize and secrete progesterone, and themselves are targets of progesterone. Conventional PR is expressed in granulosa cells in a regulated fashion by LH during the preovulatory surge, and as shown in studies with PRKO mice and with the antagonist RU486, PR is required for gonadotropin-induced ovulation (121, 152). Progesterone also directly influences granulosa cells prior to the preovulatory period by inhibiting mitosis and apoptosis. This effect of progesterone involves a rapid modulation of intracellular calcium and phosphorylation of Erk-1 and -2, but occurs during a time when conventional full-length PR is not expressed (153). Specific moderate affinity progesterone-binding sites have been characterized on the surface of preovulatory granulosa cells that are not displaced by the PR antagonists RU486 and ZK98299. The C262 MAb to the C-terminal LBD of PR recognizes a 60-kDa protein in the plasma membrane of primary and immortalized granulosa cells, and the immuno-isolated 60-kDA protein was demonstrated to directly bind progesterone in a steroid-specific manner (154–156). As evidence that the 60-kDa protein is involved in mediating these actions of progesterone, the C262 MAb, but not a control antibody, attenuated the antiapoptotic effect of progesterone, and expression of the 60-kDa protein in the ovary was observed to be developmentally regulated. These data suggest that a novel membrane protein related to the LBD of conventional PR mediates antiapoptotic and anti-mitotic effects of progesterone in granulosa cells. The origin and relationship of the 60-kDa protein to conventional PR is not known.

How Do Steroid Receptors Localize to the Cell Membrane?

Conventional steroid receptors in the cell are highly dynamic. They have nuclear localization and nuclear export sequences and continually shuttle between the nucleus and cytoplasm, with hormone addition shifting the equilibrium toward the nuclear compartment (157, 158). Even when associated with specific target genes, receptors in the nucleoplasm are in dynamic equilibrium with target gene loci (159, 160). The majority of steroid receptors in the cell reside in the nucleus in the presence of hormone; only a small fraction of total receptors are localized at or near the cell membrane in either the presence or absence of steroid. Steroid receptors do not contain membrane-targeting signal peptides, hydrophobic *trans*-membrane sequences, or obvious chemical modifications such as fatty acid acylation and glycosylation that could mediate membrane insertion. Thus important questions arise as to how steroid receptors associate with the cell membrane, and what directs the localization of a small fraction of total cellular receptors to the cell membrane. Several studies suggest that ER localization to the cell membrane is facilitated by association with other proteins that themselves translocate to the cell membrane. Candidate proteins reported to fulfill this role are caveolin-1, the adaptor molecule Shc, IGF1 receptor, and Mat1s (a variant form of metastatic tumor antigen 1). Caveolin-1 is a major structural protein of caveolae, and ER and caveolin-1 colocalize in caveloae of endothelial cells and VSMC; the two proteins have been observed by coimmunoprecipitation to physically associate in an estrogen-dependent manner (161). Moreover, overexpression of caveolin-1 in MCF-7 cells stimulated an increased localization of ER in the cell membrane, accompanied by a decrease of ER in the cytoplasm, suggesting caveolin-1 facilitates ER translocation to the cell membrane after binding to ER in the cytoplasm (161). The scaffold domain of caveolin-1 was required for facilitating ER translocation to the cell membrane, and because this domain also interacts with signaling molecules to inhibit their activity, it has been suggested that caveolin-1 dissociation and exchange with ER may be required as an activation signal. The adaptor molecule, Shc, through a coupling of ER with IGF1 receptor, has also been suggested to mediate ER translocation to the cell membrane in an estrogen-dependent manner. An ER/Shc complex was shown to associate in the plasma membrane with IGF1 receptor through Shc binding to phosphorylation sites of the intracellular domain of the IGF1 receptor. Further implicating a role for Shc in facilitating translocation of ER to the cell membrane, estrogen-induced a rapid and transient induction of ER localization to the cell membrane in a Shc and IGF1 receptor-dependent manner, as shown by down-regulation of endogenous Shc or IGFR by RNAi (162).

Mat1 is associated with breast tumor invasion and metastasis and also acts as a corepressor of the nuclear transcriptional activity of ER. MAT1s is a naturally occurring splice variant that, through a frame shift, generates a unique 33 amino acid sequence that contains a nuclear receptor interaction site resembling the LXXLL motif of AF-2 coactivators (27, 163). MAT1s is a cytoplasmic protein and, through physical association, sequesters ER in the cytoplasm, diminishes the transcriptional activity of ER, and enhances the extranuclear actions of ER on signaling pathways such as MAPK. Interestingly, increased MAT1s expression in breast tumors was found to be associated with a predominant cytoplasmic localization of ER, implicating a role for extranuclear signaling by ER in breast cancer progression (163).

Studies designed to examine structural determinants of ER responsible for localization to the cell membrane revealed that the N-terminal A/B domain and DBD are not required and that the LBD alone is sufficient for estrogen-dependent translocation to the cell membrane (105, 164, 165). The LBD alone is also sufficient for mediating many of the described effects of estrogen on signal transduction pathways in different cell types (105, 164, 165). By mutational analysis of ectopically expressed ERα in Chinese hamster ovary (CHO) cells, Ser 522 in the LBD was identified to be important for optimal localization of ER to the cell membrane in response to estrogen treatment and for rapid estrogen-induced activation of Erks (164). Mutation of Ser 522 to an alanine resulted in substantial reduction of estrogen-induced translocation of ER to the cell membrane, association of ER with caveolin-1, activation of Erk, and stimulation of cell cycle progression. The reduced ability of ER S522A to mediate estrogen activation of Erk was not due to an impaired ability to physically associate in vitro with signaling components of the Erk pathway, which suggests the importance of cell membrane localization of ER for rapid effects of estrogen on signal transduction pathways. How Ser 522 contributes to ER membrane localization has not been well defined. The region of the LBD encompassing Ser 522 contains a potential site for lipid modification by palmitoylation. However, no detectable incorporation of [3H] myristic acid or [3H] palmitate was found in full-length ERα (ER66) expressed in CHO cells (164). In contrast, a truncated 46-kDa ERα variant (ER46) expressed in immortalized human endothelial cells was reported to be palmitoylated (but not myristylated), as determined by incorporation of [3H] palmitic acid labeling (166). As evidence that palmitoylation contributes to membrane localization of ER, the palmitoylation inhibitor, tunicamycin, blocked estrogen stimulation of ER46 translocation to the cell membrane, as well as basal levels of membrane ER in the absence of estrogen (166). Comparison of the ability of ER46 and ER66 in immortalized endothelial cells to undergo lipid modification was not reported (166). The failure to detect palmitoylation of ER66 expressed in CHO cells (164) may be from cell-type-specific factors or differences in experimental conditions. Fatty acylation is dynamic and reversible and may simply be difficult to detect, and thus whether ER66 can also undergo lipid modification has not yet been examined exhaustively.

ER46 is generated by alternative splicing, is missing the N-terminal A/B domain, and has been reported to coexist with ER66 in a number of different cell lines including endothelial cells (166), osteoblasts (167, 168), and MCF-7 breast cancer cells (169). Although ER46 is distributed between the cell membrane, cytoplasm, and nuclear fractions, it exhibits a preferential localization in the cell membrane fraction compared with ER66, which is found predominantly in the nucleus. Furthermore, the C terminus of ER46 appears to be accessible to the cell surface as determined by biotinylation reactions and flow cytometry of intact nonpermeabilized cells with C-terminal ER antibodies (166, 169). Several studies suggest that ER46 preferentially mediates estrogen-dependent activation of signaling pathways compared with ER66, which more efficiently mediates activation of gene transcription. For example, in cell transfection experiments, ER46 more efficiently mediated rapid estrogen activation of eNOS in immortalized endothelial cells than did ER66, whereas ER66 more efficiently mediated estrogen transactivation of ERE target genes (166). Additionally, increased expression of ER46 relative to ER66, which occurs when ROS osteoblast cells reach confluence, correlated with activation of PKCα and Src (167). When coexpressed with ER66 in SaOS osteosarcoma cells, ER46 inhibited ER66-mediated transcription (168), and an antibody that recognizes ER46 on the cell surface inhibited estrogen-stimulated growth of MCF-7 cells and rapid estrogen activation of MAPK and Akt (169). One conclusion drawn from these studies is that truncated ER46, through altered folding properties, is a better substrate for lipid modification than ER66 and preferentially mediates rapid signaling actions of estrogen through differential cell compartmentalization.

In contrast to studies on ER, much less work has been done to examine how conventional PR localizes to the plasma membrane. Somewhat analogous to ER studies, truncated proteins related to the LBD of PR have been observed to localize preferentially to the cell membrane of spermatozoa (118, 122, 123) and granulosa cells (154–156). Whether intact PR or modified versions of PR undergo post-translational modifications or interact with other trafficking proteins to facilitate translocation to cell membrane has not been explored.

Subcellular Targeting Dissociates Transcription and Nontranscriptional Actions of Steroid Receptors

Because the conventional steroid receptor is capable of mediating both acute effects on membrane/intracellular signaling pathways and direct effects on nuclear transcription, it is difficult to distinguish between these two functions and to assign a physiological role to each. As one approach to this problem, conventional steroid receptors have been forcedly targeted to one or another subcellular compartment, and effects of steroid hormones were analyzed in these designer cells. When expressed in Cos-1 and HeLa cells, ERα tagged with a strong membrane localization sequence (palmitoylation or membrane localization sequence of GAP-43) and deleted of its nuclear localization sequence (NLS) generated a predominantly membrane-localized ER that was able to mediate rapid estrogen activation of Src/Shc/MAPK (105, 170). As expected, membrane-targeted ER was unable to induce estrogen activation of ERE reporter genes and yet, remarkably, was still able to mediate antiapoptotic effects of estrogen in HeLa cells (105). ER expressed in fibroblast cells lacking a NLS was predominantly cytoplasmic and unable to translocate and activate target genes in the nucleus, yet still mediated rapid estrogen-initiated activation of Src/MAPK, as well as estrogen stimulation of cell proliferation (171). Conversely, attaching a strong NLS from SV40 to ER resulted in a predominantly nuclear-localized ER that retained direct transcriptional activity but had lost its ability to mediate rapid estrogen activation of Src/Shc/MAPK signaling (105).

Although membrane localization of endogenous PR in mammalian cells has not been reported, similar results with that of ER have been obtained by targeting PR to different cell compartments. PR localized predominantly to the cell membrane, by tagging with palmitic acid or to the cytoplasm by deletion of the NLS, mediates rapid progestin-induced activation of Src/MAPK and fails to mediate progestin-induced transactivation of PRE reporter genes. PR tagged with a strong SV40 NLS did not mediate progestin activation of Src/MAPK (172). Results of these subcellular targeting experiments help to distinguish between two different functions of the same receptor in the cell and provide more direct evidence than other experimental approaches that certain cell biological effects of steroid hormones can be mediated by signaling of conventional receptors from the cell membrane in the absence of direct nuclear actions.

How Steroid Receptors Couple with Signaling Pathways and Trigger Responses

To the best of our knowledge, conventional steroid receptors are not themselves protein kinases, nor do they have properties of adaptor or scaffolding components of protein kinase signaling cascades. Thus a significant dilemma is how do steroid receptors that lack obvious properties characteristic of many cell signaling molecules interact with components of signaling pathways, and how do they trigger a signal transduction cascade in response to binding steroid hormones. One of the more extensively characterized extranuclear signaling actions of steroids is the rapid activation of the Src/ras/raf/MAP kinase (Erk) pathway, and studies of the interaction of ER and PR with Src begin to provide some insights into these questions. Src is a key proximal component in the coupling of extracellular-membrane initiated signals with a variety of intracellular signal transduction pathways involved in regulating multiple cell functions including proliferation, differentiation, adhesion, migration, cell-cell interaction, and apoptosis (173, 174). Activation of the Src/MAPK signaling pathway mediated by steroid receptors has been shown to contribute to cell biological effects of estrogens and progesterone in different cell types, including proliferation and survival. Members of the Src family of nonreceptor tyrosine kinases are single-chain polypeptides that contain a unique N-terminal domain, a regulatory region that contains an SH3 and a SH2 domain, a linker region, and a C-terminal catalytic domain (175). Upon activation, the unique N-terminal domain becomes myristoylated and inserts into the cell membrane. SH2 domains recognize and bind tyrosine phosphorylated peptides, whereas SH3 domains recognize short contiguous polyproline PXXPXR (where X is any amino acid and frequently is a P) motifs that form a left-handed helical conformation. Src and closely related Hck are regulated by an auto-inhibitory mechanism that involves intramolecular interactions of a tyrosine phosphorylation site (Y527) in the C-terminal tail with the SH2 domain and the PXXP-like motif in the linker with the SH3 domain. These intramolecular interactions maintain Src/Hck in a closed inactivate conformation. Conversion to an open catalytically active conformation occurs through phosphatases that remove the C-terminal tyrosine phosphorylation to release the SH2 domain interaction, or by competitive displacement with external peptides that recognize SH2 (tyrosine phosphorylated peptides) or SH3 domains (PXXPXR peptide motifs) (173–175).

As determined by coimmunoprecipitation and pull-down assays, ER and PR have been reported to interact with Src in cells in a hormone agonist-dependent manner (105, 165, 176–181). The association of ER with cSrc is mediated primarily by interaction of a phosphorylated tyrosine residue (Y537) in the ER LBD with the SH2 domain of Src (176, 182). However, ER Y537 phosphorylation is not estrogen dependent and in vitro experiments with down-regulated Src/Hck showed that ER interaction with the SH2 domain is not sufficient for activation of Src, which suggests an intermediary or adaptor protein is required for hormone-dependent interaction of ER with Src and for efficient activation of Src (178, 179). Such an adaptor protein, termed MNAR (modulator of nongenomic action of estrogen receptor) has been identified and characterized (178). MNAR is an ~120-kDa protein containing regions of sequence homology with a proline and glutamic acid–rich protein originally designated as PELP1 (183). In cell-free assays, MNAR and ER alone gave little to no activation of Src; however, ER and MNAR together gave a potent synergistic activation of Src that was largely estrogen dependent. An estrogen-stimulated ternary complex between Src/MNAR and ER was detected in cells by coimmunoprecipitation. As evidence that MNAR and ER cooperate to activate Src in intact cells, overexpression of MNAR enhanced estrogen stimulation of Src enzymatic activity and phosphorylation of MAPK in MCF-7 cells, whereas estrogen activation of the Src/MAPK pathway was attenuated by expression of antisense oligonucleotides to MNAR (178). The N-terminal domain of MNAR contains multiple (up to 10) LXXLL motifs similar to those in p160 coactivators that mediate hormone agonist-dependent interaction with AF-2 of nuclear receptors (27), and 3 PXXP motifs that resemble SH3 domain interaction sequences (182). By mutational and functional analysis, MNAR was shown to interact directly with the ER LBD in an estrogen-dependent manner through 2 LXXLL motifs (designated 4 and 5) and with the SH3 domain of Src via its most N-terminal PXXP motif (designated 1). These multiple protein interaction surfaces stabilize the ER/Src complex and facilitate activation of Src via interaction of the PXXP motif with SH3 domain (Figure 2, see color insert). As might be expected, MNAR also interacts in a hormone-agonist-dependent manner through its LXXLL motifs with several other steroid receptors including AR, GR, and PR (178, 182). However, its role in mediating extranuclear signaling actions of these other receptors has not been reported as yet. Thus MNAR appears to provide multiple protein interaction surfaces characteristic of scaffolding proteins in signaling pathways, and may be important in assembly and integration of ER and other steroid receptors with signaling complexes in the cell.

The adaptor protein Shc has also been implicated in rapid estrogen activation of MAP kinase pathways. Shc binds to docking sites on the intracellular domains of several membrane growth factor receptors and, in turn, recruits Grb2/Sos required for coupling with the intracellular MAP kinase pathway. Estrogen rapidly stimulates phosphorylation of Shc and the formation of a Shc/Grb2/Sos complex in MCF-7 breast cancer and other cells (165). By use of a dominant-negative Shc and RNAi, endogenous Shc was shown to be required for estrogen activation of MAPK (162, 165). Activation of Shc by estrogen appears to involve its phosphorylation by upstream Src and an estrogen-dependent physical association of the N-terminal domain of ER with SH2 domain of Shc. The interplay, if any, between ER, MNAR, and Shc is not known, but these data collectively indicate the importance of ER interactions with adaptor proteins in mediating rapid estrogen activation of the MAPK pathway.

Human PR interacts with Src by a different mechanism than ER. The amino terminal domain of human PR contains a short polyproline sequence (PPPPLPPR) located between amino acids 421 to 428 (Figure 1) that mediates a direct interaction with the SH3 domain of Src and a select group of other signaling molecules including other Src family members (Hck, Lyn, Yes) (179). Mutational analysis of PR and competition experiments, with peptides corresponding to the PPPPLPPR sequence in the N-terminal domain of PR, indicate that this polyproline motif is both necessary and sufficient for mediating interaction of PR with the SH3 domain of Src. Point mutations (three proline to alanine mutations) in the polyproline motif abolished interaction of PR with Src, both in cell-free assays and within intact cells, and eliminated the ability of PR to mediate rapid progesterone activation of Src and downstream MAP kinase in several cell types (179). Importantly, these mutations had no effect on other functions of PR including steroid and DNA binding activity, and hormone-dependent gene transcriptional activation (180). Conversely, subtle mutations in the DNA-binding domain and AF-2 that cripple the transcription activity of PR, without affecting steroid binding, had no effect on progesterone activation of Src/MAP pathway. Thus the polyproline sequence of PR represents a structural and functional motif that is separable from other functional domains required for transcriptional activity of PR. PR activates Src by an SH3 domain displacement mechanism dependent on the direct interaction with the SH3 domain mediated by polyproline motif. The K_{act} of closely related downregulated Hck stimulated by PR was in the low nanomolar range, indicating that PR is a potent activator of Src tyrosine kinases equivalent to or greater than other polypeptides that activate Src by an SH3 domain displacement mechanism (Figure 2) (179).

The ability to directly interact with SH3 domains appears to be a unique property of PR. Other steroid receptors and the thyroid hormone receptor either lack the PXXPXR motif or, when tested, did not directly interact with SH3 domains of Src (179). Although AR contains a short polyproline sequence in the N-terminal domain (46) it does not fit well with the consensus SH3 domain interaction motif and does not interact directly with Src in vitro (179). The ability of PR to interact with SH3 domains and to activate Src appears to be an intrinsic property of the receptor, whereas ER and perhaps other steroid receptors require an intermediary adaptor such as MNAR. Although MNAR can interact with the LBD of PR in a hormone-dependent manner via its LXXLL motifs, MNAR did not influence the ability of PR to mediate rapid progesterone activation of Src/MAP kinase when overexpressed in cells (V. Boonyaratanakornkit, B. Cheskis & D.P. Edwards, unpublished data). It has been suggested that adaptors such as MNAR may be required only at very low concentrations of receptors, and this requirement may differ for different classes of steroid receptors (182). Further studies will be needed to determine whether MNAR plays a role in the rapid extranuclear signaling actions of PR under different cellular conditions.

An alternative mechanism for how PR interacts with Src has been suggested to be indirect through ERα. Progesterone was reported to activate the Src/MAPK pathway through a physical interaction of PR-B with unliganded ER, with ER in turn activating Src by a direct interaction with the Src SH2 domain (181). The regions of the receptors involved in mediating PR-ER interactions were mapped to a broad area of the N-terminal domain of PR flanking the polyproline motif and to the LBD of ER (184). On the basis of mutational analysis, this broad region of the N-terminal domain of PR was required for the indirect, ER-dependent mode of progesterone activation of Src/MAPK, whereas the polyproline motif was dispensable. As further evidence of an indirect ER-dependent mechanism, progesterone activation of the Src/MAP pathway in the presence of unliganded ER was reported to be inhibited by PR antagonists as well as by ER antagonists (181). On the basis of these results, it has been proposed that in cells that express both PR and ER, progesterone activation of Src is mediated predominantly through unliganded ER and that direct PR interaction with Src through its intrinsic polyproline motif occurs mainly in the absence of ER. Whether PR-ER interactions are direct or mediated by another protein has not been established, as interactions were detected in cells by yeast two-hybrid assay and by coimmunoprecipitation (181, 184). Because both PR and ER can bind to MNAR in a hormone-dependent manner, it may be possible for both receptors to tether to Src through this or other adaptors, as opposed to direct PR-ER contacts. The existence of a ternary complex consisting of PR associated indirectly with Src through ER has not been shown and is inconsistent with results with purified proteins showing the formation of a ternary ER/PR/Src complex mediated through simultaneous interactions of PR with the SH3 domain and ER with the SH2 domain of Src (178). For reasons that remain to be clarified, the dependency of unliganded ER for efficient progesterone activation of Src/MAPK, and the ability of ER antagonists to block progesterone activation of Src/MAP kinase in the presence of ER, have not been consistent results. Efficient progestin-induced activation of Src dependent on the PXXP motif of PR was observed in different cell types in the absence of ER (179, 180), and ER antagonists had no influence on progesterone activation of Src in the presence of ectopically expressed ER (V. Boonyaratanakornkit & D.P. Edwards, unpublished data). The proposed indirect mechanism of PR acting through ER is also difficult to reconcile in terms of the distinct biological effects of estrogen and progesterone on cell function. One might expect PR to activate Src by a mechanism different from that of ER, otherwise progesterone effects transmitted through ER may mimic the effects of estrogen. Further studies will be needed to define and clarify the mechanisms of the interplay of ER and PR with Src and to determine the role and relative importance of the direct and indirect mechanisms of PR activation of Src/MAP kinase pathways. These studies of ER and PR interaction with Src and adaptor proteins illustrate that conventional steroid receptors do possess motifs for interaction with signaling molecules, which could explain how steroid receptors stimulate a signaling response to hormone. Further studies of ER and PR interaction with components of other signaling complexes and pathways may reveal additional previously unrecognized protein interaction motifs and mechanisms for mediating rapid stimulatory effects of steroid hormones on signal transduction pathways.

Another proposed mechanism for how conventional steroid receptor can stimulate signaling responses at the cell membrane is through cross-talk with GPCRs. Consistent with this notion, rapid estrogen responses dependent on ER in a variety of cell types have been reported to be sensitive to pertussis toxin. Additionally, estrogen stimulation of cellular cAMP production, inositol triphosphate, and calcium in different cell types was determined to be dependent on activation of various G protein subunits. For example, ectopically expressed ERα and ERβ in CHO cells mediates rapid estrogen stimulation of adenyl cyclase activity and formation of inositol triphosphate in a manner dependent on physical association of ER with, and activation of, Gαs and Gαq (185). In endothelial cells, estrogen activation of eNOS is sensitive to pertussis toxin, and there was a report of an estrogen-dependent association of ERα with Gαi blocked by the ER antagonist ICI 182,780 (59). ER localized in the cell membrane bound to estrogen can rapidly activate matrix metalloproteinases (MMPs) by Gαq, Gαi, and Gβγ-dependent mechanisms. Activated MMPs, in turn, are required for estrogen-induced cleavage and release of heparin-binding epidermal growth factor (HB-EGF), making free EGF available to transactivate the EGFR and activate downstream signaling pathways including Erk and PI3K in breast cancer cells, and p38MAPK in endothelial cells (186). A physical association of ER with different G proteins subunits has been detected by coimmunoprecipitation, but whether this represents a direct interaction with G proteins is not known. ER most likely interacts with GPCRs in the cell membrane to influence G protein coupling indirectly.

Integration of Rapid Extranuclear and Nuclear Signaling by Steroid Receptors

Because many cell signaling pathways converge upon and activate nuclear transcription factors by phosphorylation, this suggests that rapid extranuclear activation of signaling pathways by steroid receptors may ultimately affect gene expression patterns in the cell. Several examples of this mode of regulation have been reported for ER. Estrogen induction of the c-*fos* gene occurs through a serum response element (SRE) located in the proximal c-*fos* promoter and is mediated by an extranuclear ER-dependent activation of either the Src/Ras/MAPK pathway or the Src-ras/PI3K pathway (187, 188). Interestingly, the MAPK pathway phosphorylates and stimulates the Elk1 transcription factor to interact with and transactivate the SRE, whereas the PI3K pathway stimulates transactivation of the SRE by a different transcription factor, SRF (serum response factor) (187, 188). Elk-1 and SRF cooperate at the SRE to induce c-*fos* expression and thus can be coordinately regulated by activation of different signaling pathways by estrogen. Estrogen induction of cyclin D1 is also mediated by ER-dependent activation of the Src-PI3K/Akt signaling pathway in the absence of ER interaction with EREs in the promoter of D1 gene (46, 189).

In cardiac myocytes, the protective effects of estrogen are linked to an induction of expression of the early growth responsive gene, *Egr-1*. This gene lacks an ERE, and estrogen induction was shown to be mediated by an ER-dependent activation of MAPK that in turn phosphorylated and activated SRF, resulting in recruitment of SRF to a cluster of SREs located in the *Egr-1* promoter (190). In osteoblast cells, MAP kinase, activated by acute effects of estrogen, translocated to the nucleus and was capable of phosphorylating and activating three transcription factors including Elk-1, C/EBPβ, and CREB. The importance of this signaling was revealed by the finding that dominant-negative Elk-1 reduced the antiapoptotic effect of estrogen in bone cells (191). In HT22 hippocampal cell lines

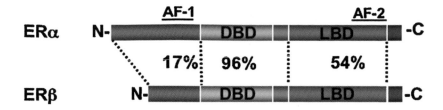

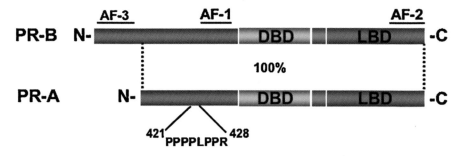

Figure 1   Domain structures of estrogen (ER) and progesterone (PR) receptors. LBD, ligand binding domain; DBD, DNA binding domain; AF, transcription activation domain (AF-1, AF-2, AF-3). The percentages indicate the amino acid identity between domains of ERα and ERβ and between A and B forms of PR. The N-terminal domain of PR contains a polyproline sequence motif that interacts with the SH3 domain of Src.

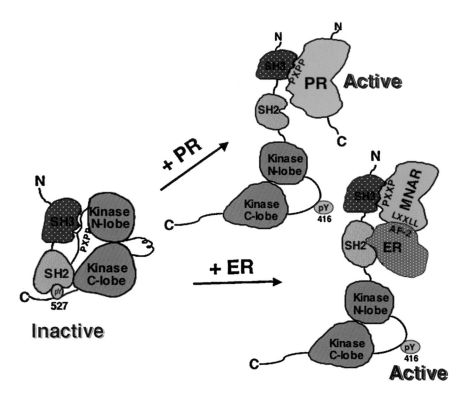

Figure 2   Mechanism of ER and PR activation of Src. PR interacts directly with the SH3 domain of Src through a PXXPXR motif located in the N-terminal domain. This interaction converts Src from an inactive closed conformation to an active open conformation by an SH3 domain displacement mechanism. A direct interaction of ER with the SH2 domain of Src is mediated by a tyrosine phosphorylation site (Y537) in the LBD. However, this interaction is not sufficient to activate Src. Additional interaction surfaces are provided by the adaptor protein MNAR, including an interaction with ER mediated by LXXLL motifs and an interaction with the SH3 domain of Src mediated by a PXXP motif. These additional interaction surfaces provided by MNAR are required for efficient activation of Src, presumably through stabilizing the complex and through interactions with the SH3 domain.

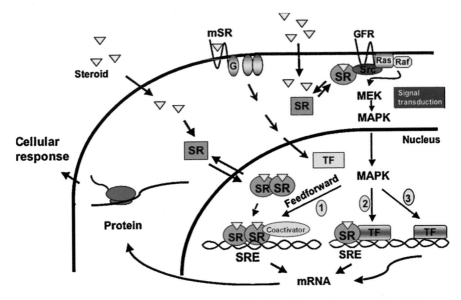

Figure 3 Nuclear transcription and extranuclear signaling pathways regulated by conventional steroid receptors. In the nuclear transcription pathway, steroid hormones activate steroid receptors (SR) by inducing conformational changes that lead to nuclear translocation, dimerization, and binding to steroid response elements (SREs) of target genes. Activated receptor bound to target DNA recruits coactivators that are essential for assembly of a productive transcription complex and for production of new RNA and protein that characterizes the cellular response to the hormone. Subpopulations of steroid receptors (ER and PR) can associate in a hormone-dependent manner with cytoplasmic or cell membrane signaling molecules including the tyrosine kinase Src. This interaction leads to an activation of Src and the downstream Ras, raf, MAP kinase protein phosphorylation cascade. A consequence of steroid-induced activation of MAP kinase is to ultimately influence gene transcription by three potential mechanisms: (1) A feed-forward pathway where activated MAPK increases the direct nuclear transcriptional activity of steroid receptors by phosphorylation of the receptor itself or a receptor interacting coactivator; (2) an activated MAPK phosphorylates and activates other transcription factors (TF) that cooperate with steroid receptors on composite SRE promoters; (3) or a mediated transactivation of genes that lack steroid response elements. Novel membrane receptors (mSR) unrelated to conventional receptors have been identified that mediate rapid steroid-induced activation of signaling pathways.

stably transfected with ERα or ERβ, one consequence of estrogen rapid activation of MAPK is the phosphorylation and activation of CREB (via p90Rsk, a downstream target of MAPK) and transactivation of a CRE (cAMP response element) reporter gene (192). A consequence of progesterone activation of Src/MAPK pathway in breast cancers by direct interaction of PR with the SH3 domain of Src is transactivation of an Elk-1 responsive reporter gene (172). As an attempt to identify endogenous target genes regulated by ER-dependent activation of signaling pathways that converge upon other nuclear transcription factors, a gene microarray experiment with vascular endothelial cells has identified estrogen-upregulated genes at 40 min of treatment that were sensitive to the PI3K inhibitor, LY294002 (193). Although the results of this microarray experiment are difficult to interpret due to the pleiotropic effect of PI3K inhibition, a large number of early estrogen-inducible genes were found to be dependent on active PI3K. These included some genes such as c-*fos* and *Erg-1* that were already known to be regulated indirectly by acute ER activation of PI3K and MAPK signaling pathways. These data collectively suggest that one role of the rapid extranuclear effects of steroid receptors on signaling pathways is to regulate a number of physiologically important genes distinct from those that interact directly with ER in the nucleus.

In pituitary cells, estrogen induction of prolactin gene expression is mediated by a direct interaction of ER with EREs in the promoter of the prolactin gene, but it is also dependent on an acute estrogen-dependent ER-mediated activation of MAP kinase (194). Because the prolactin promoter contains binding sites for other transcription factors whose activities are regulated through phosphorylation by MAPK, it was speculated that estrogen induction requires cooperative interactions between ER (bound to EREs) and other unknown transcription factors that are dependent on MAPK (Figure 3). Alternatively, the dependency of an ERE-regulated target gene on an intact MAPK pathway may be from a feed-forward mechanism whereby MAPK (or other kinases), activated acutely by estrogen, directly phosphorylates and primes the transcriptional activity of ER itself. The estrogen receptor is phosphorylated on multiple serine residues in the N-terminal domain by MAPK, p90Rsk, and other unknown kinases, and these phosphorylation events are important for the intrinsic activity of AF-1 and the full transcription activity of ER (9). Estrogen receptor interacting coactivators essential for transcription activity of ER are also targets of MAPK and other kinases and could be part of such a feed-forward mechanism (Figure 3). Several studies have suggested the existence of such a feed-forward loop. For example, overexpression of the adaptor, MNAR, which facilitates the interaction of ER with Src and the ability of estrogen to acutely activate the Src/MAPK pathway, was observed to enhance ER-mediated transactivation of ERE target genes (178, 182). This enhancement was dependent on activated Src and MAPK and, through use of point mutations, was also shown to require phosphorylation of the N terminus of ER, suggesting that MNAR enhancement of transcription activity is mediated by phosphorylation of the N-terminal domain of ER. This mechanism is distinct from that of classical coactivators, as enhancement of ER transactivation of ERE target genes by ectopically expressed coactivators was not affected by Src or MAPK inhibitors (178, 182). In a separate study, the Src/Erk signaling pathway was shown to potentiate the intrinsic activity of ER through phosphorylation of Ser-118 in the N-terminal domain AF-1, whereas the JNK pathway enhanced transcription activity by modulating ER-interacting coactivators (195). In further support of a priming role, treatment of SK-N-BEC2 neuroblastoma cells with a rapid pulse (20 min) of a cell-impermeable E2-BSA conjugate, which can activate MAPK but not direct ER-mediated transcription, was shown to potentiate ER-mediated transactivation of ERE reporter genes induced by a later administration of free estradiol (196). These data suggest that another physiological role for the acute extranuclear actions of steroid receptors on signaling pathways is to prime or potentiate the later transcription activity of steroid receptors in the nucleus by a feed-forward regulatory loop (Figure 3 see color insert). Such a role implies that the rapid extranuclear and nuclear actions of steroid receptors may be coordinately integrated.

SUMMARY

It has become increasingly evident that conventional ER and PR are dual function proteins capable of acting in the nucleus in their role as direct ligand-dependent transcription factors and outside the nucleus to acutely modulate signal transduction events. Most of our understanding of the detailed biochemistry, structure/function and molecular mechanisms of conventional steroid receptors is in terms of transcriptional signaling in the nucleus. The challenge is to uncover the fundamental properties and mechanisms by which conventional steroid receptors interact with and modulate the activities of extranuclear cell signaling pathways. Because subcellular localization is a key factor, a better understanding of intracellular trafficking of steroid receptors has clearly become of critical importance. A common theme that appears to be emerging from studies is the presence of N-terminally truncated 46-kDa ERα and 50–60-kDa proteins related to the LBD of PR that preferentially localize in the plasma cell membrane and are linked to rapid signaling actions of estrogen and progesterone in different cell types (Table 1). It has been speculated that these variant forms of ER and PR, through distinct protein conformations, may undergo unique post-translation modifications or expose interaction surfaces that promote cell membrane association. Progress has also been made in understanding how conventional steroid receptors can physically interact with components of signaling pathways and trigger an activation response. Studies of ER and PR interaction with Src show how these receptors can interact with and modulate the enzymatic activity of this important protein kinase signaling molecule. A novel adaptor or scaffold protein, termed MNAR, has also been identified that facilitates the assembly of ER and perhaps other steroid receptors with Src signaling complexes in the cell. Future investigations are likely to uncover additional mechanisms by which conventional steroid receptors couple with components of other signaling complexes and activate signal transduction pathways in response to binding hormone.

TABLE 1 Variant forms of ER and PR associated with the cell membrane

Receptor	Cell type	Rapid response	References
62–63-kDa ER-X (related to ERα LBD)	Neo-cortex	E-activation of MAPK	(79, 85)
46-kDa ERα (N-terminal truncation ERα)	Endothelial cells	E-activation of eNOS	(166)
45-kDa ERα (N-terminal truncation ERα)	Osteoblasts	Activation PKCα and Src	(167, 168)
46-kDa ERα (related to ERα LBD)	MCF-7 cells	E-activation MAPK and Akt	(169)
50–52-kDA PR (related to PR LBD)	Spermatozoa	P-initiated acrosome reaction	(123)
54–57-kDa PR (related to PR LBD)	Spermatozoa	P-initiated acrosome reaction	(118, 122)
60-kDa PR (related to PR LBD)	Ovarian granulosa cells	P-induced Ca^{2+} flux and activation of MAPK	(154–156)

Novel membrane receptors unrelated to conventional steroid receptors have also been implicated in mediating specific rapid steroid-induced responses in certain cell types. However, in most studies, potential novel membrane receptors have been described only as steroid binding sites on the surface of cells or as a binding protein, with a pharmacology for steroids and steroid analogs that is distinct from that of conventional steroid receptors. Cloning of the genes and characterization of the proteins are necessary to more firmly establish that these binding entities fulfill the criteria of a membrane receptor capable of generating a downstream signaling response upon binding hormone. The first such gene that fulfills many of the requirements of a novel membrane receptor for progesterone, mPR, has been isolated and characterized as a G protein-like-coupled receptor involved in mediating progesterone-induced oocyte maturation in the sea trout. The identification of mPR gives encouragement that analogous novel membrane receptors for estrogens and other steroid hormones exist and will be discovered in the near future. It will be important to determine the relationship, if any, between mPR and conventional PR, and whether mPR and related family members are involved in mediating any of the rapid progesterone-induced responses described in other animal species and in other cell types.

A major incompletely resolved question is the physiological role of rapid steroid-induced activation of signaling pathways in vivo. Answers will require development of ligands, such as estren in bone studies, that can be used in vivo to separately target rapid extranuclear signaling and nuclear transcriptional signaling pathways. Use of genetically engineered mice will require development of strategies to target steroid receptors exclusively to the cell membrane or nucleus of selected target cells, or to introduce mutations in ER and PR genes that effectively inactivate one or the other functions of receptor (extranuclear rapid signaling versus nuclear transcription). As illustrated in Figure 3, one possible role of rapid steroid-induced activation of cell signaling pathways is to ultimately influence the outcome of steroid-regulated gene expression in the cell through three mechanisms. One way is to regulate specific genes that lack direct HREs through phosphorylation and activatation of other nuclear transcription factors. This mechanism has the potential to expand the gene networks regulated by steroid hormone. Second, complex promoters of direct ER and PR target genes that contain binding sites for other transcription factors may require cooperative interactions between other transcription factors and steroid receptors. Such cooperation may occur through phosphorylation of other transcription factors mediated by protein kinase cascades activated by steroid receptors outside the nucleus. Finally, rapid steroid-induced activation of signaling pathways may have a role in priming the transcriptional activities of ER and PR through a feed-forward mechanism. Protein phosphorylation cascades stimulated by steroid hormones may result in direct phosphorylation of either the receptor itself or receptor-interacting coactivators required for subsequent transcriptional activity of the receptor on direct target genes. Current research in this exciting new area of steroid hormone action is laying the foundation for future studies of the biological roles of rapid steroid-induced signaling response in vivo and how the extranuclear and nuclear pathways of steroid hormone action are integrated to control the overall cell biological responses to estrogens and progesterone.

ACKNOWLEDGMENTS

The author's work described in this chapter was supported in part by National Institutes of Health public health grants DK49030 and CA46938. Jean Sibley (UCHSC) is acknowledged for assistance in preparation of the manuscript.

The *Annual Review of Physiology* is online at
http://physiol.annualreviews.org

LITERATURE CITED

1. Tsai M-J, O'Malley B. 1994. Molecular mechanisms of action of steroid/thyroid receptor superfamily members. *Annu. Rev. Biochem.* 63:451–86

2. Mangelsdorf DJ, Thummel C, Beato M, Herrlich P, Schütz G, et al. 1995. The nuclear receptor superfamily: the second decade. *Cell* 83:835–39

3. Beato M, Klug J. 2000. Steroid hormone receptor: an update. *Hum. Reprod. Update* 6:225–36

4. Edwards DP. 1999. Coregulatory proteins in nuclear hormone receptor action. *Vitam. Horm.* 55:165–218

5. McKenna NJ, O'Malley B. 2002. Combinatorial control of gene expression by nuclear receptors and coregulators. *Cell* 108:465–74

6. Kumar R, Thompson E. 2003. Transactivation functions of the N-terminal domains of nuclear hormone receptors: protein folding and coactivator interactions. *Mol. Endocrinol.* 17:1–10

7. Weigel NL. 1996. Steroid hormone receptors and their regulation by phosphorylation. *Biochem. J.* 319:657–67

8. Lange CA, Shen T, Horwitz KB. 2000. Phosphorylation of human progesterone receptors at serine-294 by mitogen-activated protein kinase signals their degradation by the 26S proteasome. *Proc. Natl. Acad. Sci. USA* 97:1032–37

9. Rochette-Egly C. 2003. Nuclear receptors: integration of multiple signalling pathways through phosphorylation. *Cell. Signal.* 15:355–66

10. Gronemeyer H, Moras D. 1995. How to finger DNA. *Nature* 375:190–92

11. Khorasanizadeh S, Rastinejad F. 2001. Nuclear-receptor interactions on DNA-response elements. *Trends Biochem. Sci.* 26:384–90

12. Freedman LP. 1992. Anatomy of the steroid receptor zinc finger region. *Endocr. Rev.* 13:129–45

13. Glass CK. 1994. Differential recognition of target genes by nuclear receptor monomers, dimers, and heterodimers. *Endocr. Rev.* 15:391–407

14. Kuiper GGJM, Enmark E, Pelto-Huikko M, Nilsson S, Gustafsson J-Å. 1996. Cloning of a novel estrogen receptor expressed in rat prostate and ovary. *Proc. Natl. Acad. Sci. USA* 93:5925–30

15. Kastner P, Krust A, Turcotte B, Stropp U, Tora L, et al. 1990. Two distinct estrogen-regulated promoters generate transcripts encoding the two functionally different human progesterone receptor forms A and B. *EMBO J.* 3:1603–14

16. McInerney EM, Weis KE, Sun J, Mosselman S, Katzenellenbogen BS. 1998. Transcription activation by the human estrogen receptor subtype β (ERβ) studied with ERβ and ERα receptor chimeras. *Endocrinology* 139:4513–22

17. Cowley SM, Parker MG. 1999. A comparison of transcriptional activation by ERα and ERβ. *J. Steroid Biochem. Mol. Biol.* 69:165–75

18. Zhang WH, Saji S, Mäkinen S, Cheng GJ, Jensen EV, et al. 2000. Estrogen receptor (ER) β, a modulator of ERα in the uterus. *Proc. Natl. Acad. Sci. USA* 97:5936–41

19. Hall JM, McDonnell DP. 1999. The estrogen receptor β-isoform (ERβ) of the human estrogen receptor modulates ERα transcriptional activity and is a key regulator of the cellular response to estrogens and antiestrogens. *Endocrinology* 140:5566–78

20. Giangrande PH, McDonnell DP. 1999. The A and B isoforms of the human progesterone receptor: two functionally different transcription factors encoded by a single gene. *Recent Prog. Horm. Res.* 54:291–313

21. Li X, O'Malley BW. 2003. Unfolding the action of progesterone receptors. *J. Biol. Chem.* 278:39261–64

22. Edwards DP. 2004. Progesterone receptor structure/function and crosstalk with cellular signaling pathways. In *Encyclopedia of Hormones*, ed. HL Henry, AW Norman, 3:249–57. San Diego/Oxford: Academic

23. Mulac-Jericevic B, Mullinax RA, DeMayo FJ, Lydon JP, Conneely OM. 2000. Subgroup of reproductive functions of progesterone mediated by progesterone receptor-B isoform. *Science* 289:1751–54

24. Mulac-Jericevic B, Lydon JP, DeMayo FJ, Conneely OM. 2003. Defective mammary gland morphogenesis in mice lacking the progesterone receptor B isoform. *Proc. Natl. Acad. Sci. USA* 100:9744–49

25. Glass C, Rosenfeld M. 2000. The coregulator exchange in transcriptional functions of nuclear receptors. *Genes Dev.* 14:121–41

26. Feng W, Ribeiro RC, Wagner RL, Nguyen H, Apriletti JW, et al. 1998. Hormone-dependent coactivator binding to a hydrophobic cleft on nuclear receptors. *Science* 280:1747–49

27. Heery DM, Kalkhoven E, Hoare S, Parker MG. 1997. A signature motif in transcriptional coactivators mediates binding to nuclear receptors. *Nature* 387:733–36

28. Steinmetz A, Renaud J-P, Moras D. 2001. Binding of ligands and activation of transcription by nuclear receptors. *Annu. Rev. Biophys. Biomol. Struct.* 30:329–59

29. Welshons WV, Cormier EM, Wolf MF, Williams PO Jr, Jordan VC. 1988. Estrogen receptor distribution in enucleated breast cancer cell lines. *Endocrinology* 122:2379–86

30. Cato ACB, Nestl A, Mink S. 2002. Rapid actions of steroid receptors in cellular signaling pathways. *Science STKE* 138:1–11

31. Norman AW, Mizwicki MT, Normal DPG. 2004. Steroid-hormone rapid actions, membrane receptors and a conformational ensemble model. *Nat. Rev.* 3:27–41

32. Valverde MA, Parker MG. 2002. Classical and novel steroid actions: a unified but complex view. *Trends Biochem. Sci.* 27:172–73

33. Cheskis BJ. 2004. Regulation of cell signaling cascades by steroid hormones. *J. Cell. Biochem.* 93:20–27

34. Watson CS, Gametchu B. 1999. Membrane-initiated steroid actions and the proteins that mediate them. *Proc. Soc. Exp. Biol. Med.* 220:9–19

35. Lösel R, Wehling M. 2003. Nongenomic actions of steroid hormones. *Nat. Rev.* 4:46–56

36. Revelli A, Massobrio M, Tesarik J. 1998. Nongenomic actions of steroid hormones in reproductive tissues. *Endocr. Rev.* 19:3–17

37. Greener M. 2003. Steroid action gets a rewrite. *Scientist* Sept. 8:31–32

38. Nemere I, Pietras RJ, Blackmore PF. 2003. Membrane receptors for steroid hormones: signal transduction and physiological significance. *J. Cell. Biochem.* 88:438–45

39. Pietras RJ, Szego CM. 1977. Specific binding sites for oestrogen at the outer surfaces of isolated endometrial cells. *Nature* 265:69–72

40. Pietras RJ, Szego CM. 1980. Partial purification and characterization of oestrogen

receptors in subfractions of hepatocyte plasma membranes. *Biochem. J.* 191:743–60

41. Lösel RM, Falkenstein E, Feuring M, Schultz A, Tillmann H-C, et al. 2003. Nongenomic steroid action: controversies, questions, and answers. *Physiol. Rev.* 83:965–1016

42. Falkenstein E, Tillmann H-C, Christ M, Feuring M, Wehling M. 2000. Multiple actions of steroid hormones—a focus on rapid-nongenomic effects. *Pharmacol. Rev.* 52:513–55

43. Limbourg FP, Liao JK. 2003. Nontranscriptional actions of the glucocorticoid receptor. *J. Mol. Med.* 8:168–74

44. Peterziel H, Mink S, Schonert A, Becker M, Klocker H, Cato ACB. 1999. Rapid signaling by androgen receptor in prostate cancer cells. *Oncogene* 18:6322–29

45. Castoria G, Lombardi M, Barone MV, Bilancio A, Di Domenico M, et al. 2003. Androgen-stimulated DNA synthesis and cytoskeletal changes in fibroblasts by a nontranscriptional receptor action. *J. Cell Biol.* 161:547–56

46. Migliaccio A, Castoria G, Di Domenico M, de Falco A, Bilancio A, et al. 2000. Steroid-induced androgen receptor-oestradiol receptor c-Src complex triggers prostate cancer cell proliferation. *EMBO J.* 19:5406–17

47. Lösel RM, Feuring M, Falkenstein E, Wehling M. 2002. Nongenomic effects of aldosterone: cellular aspects and clinical implications. *Steroids* 67:493–98

48. Zanello LP, Norman AW. 2004. Rapid modulation of osteoblast ion channel responses by $1\alpha,25(OH)_2$-vitamin D_3 requires the presence of a functional vitamin D nuclear receptor. *Proc. Natl. Acad. Sci. USA* 101:1589–94

49. Mendelsohn ME. 2002. Genomic and nongenomic effects of estrogen in the vasculature. *Am. J. Cardiol.* 90:F3-6

50. Haynes MP, Li L, Russell KS, Bender JR. 2002. Rapid vascular cell responses to

estrogen and membrane receptors. *Vasc. Pharmacol.* 38:99–108

51. Ho KJ, Liao JK. 2002. Non-nuclear actions of estrogen: new targets for prevention and treatment of cardiovascular disease. *Mol. Interv.* 2:219–28

52. Michel T, Feron O. 1997. Nitric oxide synthases: which, where, how, and why? *J. Clin. Invest.* 100:2146–52

53. Russell KS, Haynes MP, Sinha D, Clerisme E, Bender JR. 2000. Human vascular endothelial cells contain membrane binding sites for estradiol, which mediate rapid intracellular signaling. *Proc. Natl. Acad. Sci. USA* 97:5930–35

54. Goetz RM, Thatte HS, Prabhakar P, Cho MR, Michel T, Golan DE. 1999. Estradiol induces the calcium-dependent translocation of endothelial nitric oxide synthase. *Proc. Natl. Acad. Sci. USA* 96:2788–93

55. Kim HP, Lee JY, Jeong JK, Bae SW, Lee HK, Jo I. 1999. Nongenomic stimulation of nitric oxide release by estrogen is mediated by estrogen receptor α localized in caveolae. *Biochem. Biophys. Res. Commun.* 263:257–62

56. Simoncini T, Fornari L, Mannella P, Varone G, Caruso A, et al. 2002. Novel non-transcriptional mechanisms for estrogen receptor signaling in the cardiovascular system. Interaction of estrogen receptor α with phosphatidylinositol-3-OH kinase. *Steroids* 67:935–39

57. Simoncini T, Rabkin E, Liao JK. 2003. Molecular basis of cell membrane estrogen receptor interaction with phosphatidylinositol 3-kinase in endothelial cells. *Arterioscler. Thromb. Vasc. Biol.* 23:198–203

58. Chen Z, Yuhanna IS, Galcheva-Gargova Z, Karas RH, Mendelsohn ME, Shaul PW. 1999. Estrogen receptor α mediates the nongenomic activation of endothelial nitric oxide synthase by estrogen. *J. Clin. Invest.* 103:401–6

59. Wyckoff MH, Chambliss KL, Mineo C, Yuhanna IS, Mendelsohn ME, et al.

2001. Plasma membrane estrogen receptors are coupled to endothelial nitric-oxide synthase through $G\alpha_i$. *J. Biol. Chem.* 276:27071–76

60. Simoncini T, Hafezi-Moghadam A, Brazil DP, Ley K, Chin WW, Liao JK. 2000. Interaction of oestrogen receptor with the regulatory subunit of phosphatidylinositol-3-OH kinase. *Nature* 407: 538–41

61. Haynes MP, Li L, Sinha D, Russell KS, Hisamoto K, et al. 2002. Src kinase mediates phosphatidylinositol 3-kinase/Akt-dependent rapid endothelial nitric-oxide synthase activation by estrogen. *J. Biol. Chem.* 278:2118–23

62. Prorock AJ, Hafezi-Moghadam A, Laubach VE, Liao JK, Ley K. 2003. Vascular protection by estrogen in ischemia-reperfusion injury requires endothelial nitric oxide synthase. *Am. J. Physiol. Heart Circ. Physiol.* 284:H133–40

63. Pendaries C, Darblade B, Rochaix P, Krust A, Chambon P, et al. 2002. The AF-1 activation-function of ERα may be dispensable to mediate the effect of estradiol on endothelial NO production in mice. *Proc. Natl. Acad. Sci. USA* 99:2205–10

64. Karas RH, Schulten H, Pare G, Aronovitz MJ, Ohlsson C, et al. 2001. Effects of estrogen on the vascular injury response in estrogen receptor α,β (double) knockout mice. *Circ. Res.* 89:534–39

65. Pare G, Krust A, Karas RH, Dupont S, Aronovitz M, et al. 2002. Estrogen receptor-α mediates the protective effects of estrogen against vascular injury. *Circ. Res.* 90:1087–92

66. Zhu Y, Bian Z, Lu P, Karas RH, Bao L, et al. 2002. Abnormal vascular function and hypertension in mice deficient in estrogen receptor β. *Science* 295:505–8

67. Ábrahám IM, Han S-K, Todman MG, Korach KS, Herbison AE. 2003. Estrogen receptor β mediates rapid estrogen actions on gonadotropin-releasing hormone neurons in vivo. *J. Neurosci.* 23:5771–77

68. Chambliss KL, Yuhanna IS, Anderson RGW, Mendelsohn ME, Shaul PW. 2002. ERβ has nongenomic action in caveolae. *Mol. Endocrinol.* 16:938–46

69. Razandi M, Pedram A, Levin ER. 2000. Estrogen signals to the preservation of endothelial cell form and function. *J. Biol. Chem.* 275:38540–46

70. Morey AK, Pedram A, Razandi M, Prins BA, Hu R-M, et al. 1997. Estrogen and progesterone inhibit vascular smooth muscle proliferation. *Endocrinology* 138:3330–39

71. Simoncini T, Mannella P, Fornari L, Caruso A, Varone G, Genazzani AR. 2003. In vitro effects of progesterone and progestins on vascular cells. *Steroids* 68:831–36

72. Vázquez F, Rodríguez-Manzaneque JC, Lydon JP, Edwards DP, O'Malley BW, Iruela-Arispe ML. 1999. Progesterone regulates proliferation of endothelial cells. *J. Biol. Chem.* 274:2185–92

73. Lee WS, Harder JA, Yoshizumi M, Lee ME, Haber E. 1997. Progesterone inhibits arterial smooth muscle cell proliferation. *Nat. Med.* 3:1005–8

74. Karas RH, van Eickels M, Lydon JP, Roddy S, Kwoun M, et al. 2001. A complex role for the progesterone receptor in the response to vascular injury. *J. Clin. Invest.* 108:611–18

75. Minshall RD, Pavcnik D, Browne DL, Hermsmeyer K. 2002. Nongenomic vasodilator action of progesterone on primate coronary arteries. *J. Appl. Physiol.* 92:701–8

76. Barbagallo M, Dominguez LJ, Licata G, Shan J, Bing L, et al. 2001. Vascular effects of progesterone: role of cellular calcium regulation. *Hypertension* 37:142–47

77. McEwen BS, Alves SE. 1999. Estrogen actions in the central nervous system. *Endocr. Rev.* 20:279–307

78. Moss RL, Gu Q, Wong M. 1997. Estrogen: nontranscriptional signaling pathway. *Recent Prog. Horm. Res.* 52:33–68

79. Toran-Allerand CD. 2004. Minireview: A plethora of estrogen receptors in the brain: where will it end? *Endocrinology* 145:1069–74

80. Wise PM, Dubal DB, Wilson ME, Rau SW, Böttner M. 2001. Minireview: neuroprotective effects of estrogen—new insights into mechanisms of action. *Endocrinology* 142:969–73

81. Kelly MJ, Levin ER. 2001. Rapid actions of plasma membrane estrogen receptors. *Trends Endocrinol. Metab.* 12:152–56

82. Blaustein JD. 1992. Cytoplasmic estrogen receptors in rat brain: immunocytochemical evidence using three antibodies with distinct epitopes. *Endocrinology* 131:1336–42

83. Singh M, Sétáló G Jr, Guan X, Frail DE, Toran-Allerand CD. 2000. Estrogen-induced activation of the mitogen-activated protein kinase cascade in the cerebral cortex of estrogen receptor-α knockout mice. *J. Neurosci.* 20:1694–700

84. Nethrapalli IS, Singh M, Guan XP, Guo QF, Lubahn DB, et al. 2001. Estradiol (E2) elicits Src phosphorylation in the mouse neocortex: the initial event in E2 activation of the MAPK cascade? *Endocrinology* 142:5145–48

85. Toran-Allerand CD, Guan X, MacLusky NJ, Horvath TL, Diano S, et al. 2002. ER-X: a novel, plasma membrane-associated, putative estrogen receptor that is regulated during development and after ischemic brain injury. *J. Neurosci.* 22:8391–401

86. Valverde MA, Rojas P, Amigo J, Cosmelli D, Orio P, et al. 1999. Acute activation of Maxi-K channels (*hSlo*) by estradiol binding to the β subunit. *Science* 285:1929–31

87. Qiu J, Bosch MA, Tobias SC, Grandy DK, Scanlan TS, et al. 2003. Rapid signaling of estrogen in hypothalamic neurons involves a novel G-protein-coupled estrogen receptor that activates protein kinase C. *J. Neurosci.* 23:9529–40

88. Moss RL, Gu Q. 1999. Estrogen: mechanisms for a rapid action in CA1 hippocampal neurons. *Steroids* 64:14–21

89. Watters JJ, Campbell JS, Cunningham MJ, Krebs EG, Dorsa DM. 1997. Rapid membrane effects of steroids in neuroblastoma cells: effects of estrogen on mitogen-activated protein kinase signaling cascade and c-fos immediate early gene transcription. *Endocrinology* 138: 4030–33

90. Singer CA, Figueroa-Masot XA, Batchelor RH, Dorsa DM. 1999. The mitogen-activated protein kinase pathway mediates estrogen neuroprotection after glutamate toxicity in primary cortical neurons. *J. Neurosci.* 19:2455–63

91. Fitzpatrick JL, Mize AL, Wade CB, Harris JA, Shapiro RA, Dorsa DM. 2002. Estrogen-mediated neuroprotection against β-amyloid toxicity requires expression of estrogen receptor α or β and activation of the MAPK pathway. *J. Neurochem.* 82:674–82

92. Dubal DB, Zhu H, Yu J, Rau SW, Shughrue PJ, et al. 2001. Estrogen receptor α, not β, is a critical link in estradiol-mediated protection against brain injury. *Proc. Natl. Acad. Sci. USA* 98:1952–57

93. Wang L, Andersson S, Warner M, Gustafsson J-Å. 2001. Morphological abnormalities in the brains of estrogen receptor β knockout mice. *Proc. Natl. Acad. Sci. USA* 98:2792–96

94. Watson CS, Norfleet AM, Pappas TC, Gametchu B. 1999. Rapid actions of estrogens in GH3/B6 pituitary tumor cells via a plasma membrane version of estrogen receptor-alpha. *Steroids* 64:5–13

95. Watson CS, Campbell CH, Gametchu B. 1999. Membrane oestrogen receptors on rat pituitary tumour cells: immuno-identification and responses to oestradiol and xenoestrogens. *Exp. Physiol.* 84: 1013–22

96. Norfleet AM, Thomas ML, Gametchu B, Watson CS. 1999. Estrogen receptor-α detected on the plasma membrane of aldehyde-fixed GH₃/B6/F10 rat pituitary tumor cells by enzyme-linked immunocytochemistry. *Endocrinology* 140:3805–14

97. Pappas TC, Gametchu B, Watson CS. 1995. Membrane estrogen receptors identified by multiple antibody labeling and impeded-ligand binding. *FASEB J.* 9:404–10

98. Nadal A, Ropero AB, Fuentes E, Soria B. 2001. The plasma membrane estrogen receptor: nuclear or unclear? *Trends Pharmacol. Sci.* 22:597–99

99. Nadal A, Ropero AB, Fuentes E, Soria B, Ripoll C. 2004. Estrogen and xenoestrogen actions on endocrine pancreas: from ion channel modulation to activation of nuclear function. *Steroids* 69:531–36

100. Nadal A, Ropero AB, Laribi O, Maillet M, Fuentes E, Soria B. 2000. Nongenomic actions of estrogens and xenoestrogens by binding at a plasma membrane receptor unrelated to estrogen receptor α and estrogen receptor β. *Proc. Natl. Acad. Sci. USA* 97:11603–8

101. Ropero AB, Soria B, Nadal A. 2002. A nonclassical estrogen membrane receptor triggers rapid differential actions in the endocrine pancreas. *Mol. Endocrinol.* 16:497–505

102. Benten WPM, Stephan C, Lieberherr M, Wunderlich F. 2001. Estradiol signaling via sequestrable surface receptors. *Endocrinology* 142:1669–77

103. Guo Z, Krücken J, Benten WPM, Wunderlich F. 2002. Estradiol-induced nongenomic calcium signaling regulates genotropic signaling in macrophages. *J. Biol. Chem.* 277:7044–50

104. Manolagas SC, Kousteni S, Jilka RL. 2002. Sex steroids and bone. *Recent Prog. Horm. Res.* 57:385–409

105. Kousteni S, Bellido T, Plotkin LI, O'Brien CA, Bodenner DL, et al. 2001. Nongenotropic, sex-nonspecific signaling through the estrogen or androgen receptors: dissociation from transcriptional activity. *Cell* 104:719–30

106. Kousteni S, Chen J-R, Bellido T, Han L, Ali AA, et al. 2002. Reversal of bone loss in mice by nongenotropic signaling of sex steroids. *Science* 298:843–46

107. Lambert JL, Belelli D, Peden DR, Vardy AW, Peters JA. 2003. Neurosteroid modulation of GABA$_A$ receptors. *Prog. Neurobiol.* 71:67–80

108. Reddy DS, Castaneda DC, O'Malley BW, Rogawski MA. 2004. Anticonvulsant activity of progesterone and neurosteroids in progesterone receptor knockout mice. *J. Pharmacol. Exper. Ther.* 310:230–39

109. Grazzini E, Guillon G, Mouillac B, Zingg HH. 1998. Inhibition of oxytocin receptor function by direct binding of progesterone. *Nature* 392:509–12

110. Dunlap KA, Stormshak F. 2004. Nongenomic inhibition of oxytocin binding by progesterone in the ovine uterus. *Biol. Reprod.* 70:65–69

111. Bogacki M, Silvia WJ, Rekawiecki R, Kotwica J. 2002. Direct inhibitory effect of progesterone on oxytocin-induced secretion of prostaglandin F$_{2\alpha}$ from bovine endometrial tissue. *Biol. Reprod.* 67:184–88

112. Burger K, Fahrenholz F, Gimpl G. 1999. Non-genomic effects of progesterone on the signaling function of G protein-coupled receptors. *FEBS Lett.* 464:25–29

113. Astle S, Khan RN, Thornton S. 2003. The effects of a progesterone metabolite, 5β-dihydroprogesterone, on oxytocin receptor binding in human myometrial membranes. *BJOG: Int. J. Obstet. Gynecol.* 110:589–92

114. Meizel S, Turner KO, Nuccitelli R. 1997. Progesterone triggers a wave of increased free calcium during the human sperm acrosome reaction. *Dev. Biol.* 182:67–75

115. Harrison DA, Carr DW, Meizel S. 2000. Involvement of protein kinase A and A kinase anchoring protein in the progesterone-initiated human sperm acrosome reaction. *Biol. Reprod.* 62:811–20

116. Blackmore PF, Fisher JF, Spilman CH, Bleasdale JE. 1996. Unusual steroid specificity of the cell surface progesterone receptor on human sperm. *Mol. Pharmacol.* 49:727–39

117. Falkenstein E, Heck M, Gerdes D, Grube D, Christ M, et al. 1999. Specific progesterone binding to a membrane protein and related nongenomic effects on Ca^{2+}-fluxes in sperm. *Endocrinology* 140:5999–6002

118. Luconi M, Bonaccorsi L, Maggi M, Pecchioli P, Krausz C, et al. 1998. Identification and characterization of functional nongenomic progesterone receptors on human sperm membrane. *J. Clin. Endocrinol. Metab.* 83:877–85

119. Somanath PR, Gandhi KK. 2002. Expression of membrane associated nongenomic progesterone receptor(s) in caprine spermatozoa. *Anim. Reprod. Sci.* 74:195–205

120. Pietrobon EO, De Los Ángeles M, Monclus A, Alberdi AJ, Fornés MW. 2003. Progesterone receptor availability in mouse spermatozoa during epididymal transit and capacitation: ligand blot detection of progesterone-binding protein. *J. Androl.* 24:612–20

121. Lydon JP, DeMayo FJ, Funk CR, Mani SK, Hughes AR, et al. 1995. Mice lacking progesterone receptor exhibit pleiotropic reproductive abnormalities. *Genes Dev.* 9:2266–78

122. Luconi M, Bonaccorsi L, Bini L, Liberatori S, Pallini V, et al. 2002. Characterization of membrane nongenomic receptors for progesterone in human spermatozoa. *Steroids* 67:505–9

123. Sabeur K, Edwards DP, Meizel S. 1996. Human sperm plasma membrane progesterone receptor(s) and the acrosome reaction. *Biol. Reprod.* 54:993–1001

124. Weigel NL, Beck CA, Estes PA, Prendergast P, Altmann M, et al. 1992. Ligands induce conformational changes in the carboxyl-terminus of progesterone receptors which are detected by a site-directed antipeptide monoclonal antibody. *Mol. Endocrinol.* 6:1585–97

125. Sachdeva G, Shah CA, Kholkute SD, Puri CP. 2000. Detection of progesterone receptor transcript in human spermatozoa. *Biol. Reprod.* 62:1610–14

126. Gadkar S, Shah CA, Schdeva G, Samant U, Puri CP. 2002. Progesterone receptor as an indicator of sperm function. *Biol. Reprod.* 67:1327–36

127. Wei LL, Hawkins P, Baker C, Norris B, Sheridan PL, Quinn PG. 1996. An amino-terminal truncated progesterone receptor isoform, PRc, enhances progestin-induced transcriptional activity. *Mol. Endocrinol.* 10:1379–87

128. Hirata S, Shoda T, Kato J, Hoshi K. 2000. The novel isoform of the progesterone receptor cDNA in the human testis and detection of its mRNA in the human uterine endometrium. *Oncology* 59:39–44

129. Buddhikot M, Falkenstein E, Wehling M, Meizel S. 1999. Recognition of a human sperm surface protein involved in the progesterone-initiated acrosome reaction by antisera against an endomembrane progesterone binding protein from porcine liver. *Mol. Cell. Endocrinol.* 158:187–93

130. Ferrell JE Jr. 1999. *Xenopus* oocyte maturation: new lessons from a good egg. *BioEssays* 21:833–42

131. Nebreda AR, Ferby I. 2000. Regulation of the meiotic cell cycle in oocytes. *Curr. Opin. Cell Biol.* 12:666–75

132. Hammes SR. 2003. The further redefining of steroid-mediated signaling. *Proc. Natl. Acad. Sci. USA* 100:2168–70

133. Maller JL. 2001. The elusive progesterone receptor in *Xenopus* oocytes. *Proc. Natl. Acad. Sci. USA* 98:8–10

134. Duckworth BC, Weaver JS, Ruderman JV. 2002. G_2 arrest in *Xenopus* oocytes depends on phosphorylation of cdc25 by protein kinase A. *Proc. Natl. Acad. Sci. USA* 99:16794–99

135. Lutz LB, Kim B, Jahani D, Hammes SR. 2000. G protein $\beta\gamma$ subunits inhibit nongenomic progesterone-induced signaling and maturation in *Xenopus laevis* oocytes. *J. Biol. Chem.* 275:41512–20

136. Tian J, Kim S, Heilig E, Ruderman JV. 2000. Identification of XPR-1, a progesterone receptor required for *Xenopus* oocyte activation. *Proc. Natl. Acad. Sci. USA* 97:14358–63

137. Bayaa M, Booth RA, Sheng Y, Liu XJ. 2000. The classical progesterone receptor mediates *Xenopus* oocyte maturation through a nongenomic mechanism. *Proc. Natl. Acad. Sci. USA* 23:12607–12

138. Bagowski CP, Myers JW, Ferrell JE Jr. 2001. The classical progesterone receptor associates with p42 MAPK and is involved in phosphatidylinositol 3-kinase signaling in *Xenopus* oocytes. *J. Biol. Chem.* 276:37708–14

139. Leonhardt SA, Edwards DP. 2002. Mechanism of action of progesterone antagonists. *Exp. Biol. Med.* 227:969–80

140. Lutz LB, Cole LM, Gupta MK, Kwist KW, Auchus RJ, Hammes SR. 2001. Evidence that androgens are the primary steroids produced by *Xenopus laevis* ovaries and may signal through the classical androgen receptor to promote oocyte maturation. *Proc. Natl. Acad. Sci. USA* 98: 13728–33

141. Lutz LB, Jamnongjit M, Yang W-H, Jahani D, Gill A, Hammes SR. 2003. Selective modulation of genomic and nongenomic androgen responses by androgen receptor ligands. *Mol. Endocrinol.* 17:1106–16

142. Yang W-H, Lutz LB, Hammes SR. 2003. *Xenopus laevis* ovarian CYP17 is a highly potent enzyme expressed exclusively in oocytes. *J. Biol. Chem.* 278:9552–59

143. Thomas P, Zhu Y, Pace M. 2002. Progestin membrane receptors involved in the meiotic maturation of teleost oocytes: a review with some new findings. *Steroids* 67:511–17

144. Zhu Y, Rice CD, Pang Y, Pace M, Thomas P. 2003. Cloning, expression, and characterization of a membrane progestin receptor and evidence it is an intermediary in meiotic maturation of fish oocytes. *Proc. Natl. Acad. Sci. USA* 100:2231–36

145. Thomas P. 2004. Membrane progesterone receptors. *Endocr. Soc. Annu. Meet., Abstr. S62-1*, p. 62, *New Orleans, LA*

146. Zhu Y, Bond J, Thomas P. 2003. Identification, classification, and partial characterization of genes in humans and other vertebrates homologous to a fish membrane progestin receptor. *Proc. Natl. Acad. Sci. USA* 100:2237–42

147. Gill A, Jamnongjit M, Hammes SR. 2004. Androgens promote maturation and signaling in mouse oocytes independent of transcription: a release of inhibition model for mammalian oocyte meiosis. *Mol. Endocrinol.* 18:97–104

148. Li J, Chory J. 1997. A putative leucine-rich repeat receptor kinase involved in brassinosteroid signal transduction. *Cell* 90:929–38

149. Wang Z-Y, Seto H, Fujioka S, Yoshida S, Chory J. 2001. BRI1 is a critical component of plasma-membrane receptor for plant steroids. *Nature* 410:380–83

150. He Z, Wang Z-Y, Li J, Zhu Q, Lamb C, et al. 2000. Perception of brassinosteroids by the extracellular domain of the receptor kinase BRI1. *Science* 288:2360–63

151. Nam KH, Li J. 2002. BRI1/BAK1, a receptor kinase pair mediating brassinosteroid signaling. *Cell* 110:203–12

152. Natraj U, Richards JS. 1993. Hormonal regulation, localization, and functional activity of the progesterone receptor in granulosa cells of rat preovulatory follicles. *Endocrinology* 133:761–69

153. Peluso JJ, Pappalardo A. 1998. Progesterone mediates its anti-mitogenic and anti-apoptotic actions in rat granulosa cells through a progesterone-binding protein with gamma aminobutyric acid A receptor-like features. *Biol. Reprod.* 58:1131–37

154. Peluso JJ, Fernandez G, Pappalardo A, White BA. 2001. Characterization of a putative membrane receptor for progesterone in rat granulosa cells. *Biol. Reprod.* 65:94–101

155. Peluso JJ, Fernandez G, Pappalardo A, White BA. 2002. Membrane-initiated events account for progesterone's ability to regulate intracellular free calcium levels and inhibit rat granulosa cell mitosis. *Biol. Reprod.* 67:379–85

156. Peluso JJ, Bremner T, Fernandez G, Pappalardo A, White BA. 2003. Expression pattern and role of a 60-kilodalton progesterone binding protein in regulating granulosa cell apoptosis: involvement of the mitogen-activated protein kinase cascade. *Biol. Reprod.* 68:122–28

157. DeFranco DB, Madan AP, Tang Y, Chandran UR, Xiao N, Yang J. 1995. Nucleocytoplasmic shuttling of steroid receptors. *Vitam. Horm.* 51:315–38

158. Tyagi RK, Amazit L, Lescop P, Milgrom E, Guiochon-Mantel A. 1998. Mechanisms of progesterone receptor export from nuclei: role of nuclear localization signal, nuclear export signal, and ran guanosine triophosphate. *Mol. Endocrinol.* 12:1684–95

159. Becker M, Baumann C, John S, Walker DA, Vigneron M, et al. 2002. Dynamic behavior of transcription factors on a natural promoter in living cells. *EMBO Rep.* 3:1188–94

160. Stenoien DL, Patel K, Mancini MG, Dutertre M, Smith CL, et al. 2001. FRAP reveals that mobility of oestrogen receptor-α is ligand- and proteasome-dependent. *Nat. Cell Biol.* 3:15–23

161. Razandi M, Oh P, Pedram A, Schnitzer J, Levin ER. 2002. ERs associate with and regulate the production of caveolin: implications for signaling and cellular actions. *Mol. Endocrinol.* 16:100–15

162. Song RX, Barnes CJ, Zhang Z, Bao Y, Kumar R, Santen RJ. 2004. The role of Shc and insulin-like growth factor 1 receptor in mediating the translocation of estrogen receptor α to the plasma membrane. *Proc. Natl. Acad. Sci. USA* 101:2076–81

163. Kumar R, Wang R-A, Mazumdar A, Talukder AH, Mandal M, et al. 2002. A naturally occurring MTA1 variant sequesters oestrogen receptor-α in the cytoplasm. *Nature* 418:654–57

164. Razandi J, Alton G, Pedram A, Ghonshani S, Webb P, Levin ER. 2003. Identification

of a structural determinant necessary for the localization and function of estrogen receptor α at the plasma membrane. *Mol. Cell. Biol.* 23:1633–46

165. Song RX-D, McPherson RA, Adam L, Bao Y, Shupnik M, et al. 2002. Linkage of rapid estrogen action to MAPK activation by ERα-Shc association and Shc pathway activation. *Endocrinology* 16:116–27

166. Li L, Haynes MP, Bender JR. 2003. Plasma membrane localization and function of the estrogen receptor α variant (ER46) in human endothelial cells. *Proc. Natl. Acad. Sci. USA* 100:4807–12

167. Longo M, Brama M, Marino M, Bernardini S, Korach KS, et al. 2004. Interaction of estrogen receptor α with protein kinase C α and c-Src in osteoblasts during differentiation. *Bone* 34:100–11

168. Denger S, Reid G, Ko M, Glouriot G, Parsch D, et al. 2001. ERα gene expression in human primary osteoblasts: evidence for the expression of two receptor proteins. *Mol. Endocrinol.* 15:2064–77

169. Márquez DC, Pietras RJ. 2001. Membrane-associated binding sites for estrogen contribute to growth regulation of human breast cancer cells. *Oncogene* 20:5420–30

170. Zhang Z, Maier B, Santen RJ, Song RX-D. 2002. Membrane association of estrogen receptor α mediates estrogen effect on MAPK activation. *Biochem. Biophys. Res. Commun.* 294:926–33

171. Castoria G, Barone MV, Di Domenico M, Bilancio A, Ametrano D, et al. 1999. Non-transcriptional action of oestradiol and progestin triggers DNA synthesis. *EMBO J.* 18:2500–10

172. Boonyaratanakornkit V, Edwards DP. 2003. Human progesterone receptor A and B isoforms exhibit distinct non-genomic actions on cell signaling pathways. *Endocr. Soc. Annu. Meet., Abstr. P2-119*, p. 335, *Philadelphia, PA*

173. Martin GS. 2001. The hunting of the Src. *Nat. Rev.* 2:467–73

174. Thomas SM, Brugge JS. 1997. Cellular functions regulated by Src family kinases. *Annu. Rev. Cell Dev. Biol.* 13:513–609

175. Xu WQ, Doshi A, Lei M, Eck MJ, Harrison SC. 1999. Crystal structures of c-Src reveal features of its autoinhibitory mechanism. *Mol. Cell* 3:629–38

176. Migliaccio A, Di Domenico M, Castoria G, de Falco A, Bontempo P, et al. 1996. Tyrosine kinase/p21^{ras}/MAP-kinase pathway activation by estradiol-receptor complex in MCF-7 cells. *EMBO J.* 15:1292–1300

177. Castoria G, Migliaccio A, Bilancio A, Di Domenico M, de Falco A, et al. 2001. PI3-kinase in concert with Src promotes the S-phase entry of oestradiol-stimulated MCF-7 cells. *EMBO J.* 20:6050–59

178. Wong C-W, McNally C, Nickbarg E, Komm BS, Cheskis BJ. 2002. Estrogen receptor-interacting protein that modulates its nongenomic activity-crosstalk with Src/Erk phosphorylation cascade. *Proc. Natl. Acad. Sci. USA* 99:14738–88

179. Boonyaratanakornkit V, Scott MP, Ribon V, Sherman L, Anderson SM, et al. 2001. Progesterone receptor contains a proline-rich motif that directly interacts with SH3 domains and activates c-Src family tyrosine kinases. *Mol. Cell* 8:269–80

180. Edwards DP, Wardell SE, Boonyaratanakornkit V. 2003. Progesterone receptor interacting coregulatory proteins and cross talk with cell signaling pathways. *J. Steroid Biochem. Mol. Biol.* 83:173–86

181. Migliaccio A, Piccolo D, Castoria G, Di Domenico M, Bilancio A, et al. 1998. Activation of the Src/p21^{ras}/Erk pathway by progesterone receptor via cross-talk with estrogen receptor. *EMBO J.* 17:2008–18

182. Barletta F, Wong C-W, McNally C, Komm BS, Katzenellenbogen B, Cheskis BJ. 2004. Characterization of the interactions of estrogen receptor and MNAR in the activation of cSrc. *Mol. Endocrinol.* 18:1096–1108

183. Vadlamudi RK, Wang RA, Mazumdar A, Kim Y, Shin J, et al. 2001. Molecular

cloning and characterization of PELP1, a novel human coregulator of estrogen receptor alpha. *J. Biol. Chem.* 276:38272–79

184. Ballaré C, Uhrig M, Bechtold T, Sancho E, Di Domenico M, et al. 2003. Two domains of the progesterone receptor interact with the estrogen receptor and are required for progesterone activation of the c-Src/Erk pathway in mammalian cells. *Mol. Cell. Biol.* 23:1994–2008

185. Razandi M, Pedram A, Greene GL, Levin ER. 1999. Cell membrane and nuclear estrogen receptors (ERs) originate from a single transcript: studies of ERα and ERβ expressed in Chinese hamster ovary cells. *Mol. Endocrinol.* 13:307–19

186. Razandi M, Pedram A, Park ST, Levin ER. 2003. Proximal events in signaling by plasma membrane estrogen receptors. *J. Biol. Chem.* 278:2701–12

187. Duan RQ, Xie W, Li XR, McDougal A, Safe S. 2002. Estrogen regulation of c-*fos* gene expression through phosphatidylinositol-3-kinase-dependent activation of serum response factor in MCF-7 breast cancer cells. *Biochem. Biophys. Res. Commun.* 294:384–94

188. Duan RQ, Xie W, Burghardt RC, Safe S. 2001. Estrogen receptor-mediated activation of the serum response element in MCF-7 cells through MAPK-dependent phosphorylation of Elk-1. *J. Biol. Chem.* 276:11590–98

189. Castro-Rivera E, Samudio I, Safe S. 2001. Estrogen regulation of cyclin D1 gene expression in ZR-75 breast cancer cells involves multiple enhancer elements. *J. Biol. Chem.* 276:30853–61

190. de Jager T, Pelzer T, Müller-Botz S, Imam A, Muck J, Neyses L. 2001. Mechanisms of estrogen receptor action in the myocardium. *J. Biol. Chem.* 276:27873–80

191. Kousteni S, Han L, Chen J-R, Almeida M, Plotkin LI, et al. 2003. Kinase-mediated regulation of common transcription factors accounts for the bone-protective effects of sex steroids. *J. Clin. Invest.* 111:1651–64

192. Wade CB, Dorsa DM. 2003. Estrogen activation of cyclic adenosine 5'-monophosphate response element-mediated transcription requires the extracellularly regulated kinase-mitogen-activated protein kinase pathway. *Endocrinology* 144:832–38

193. Pedram A, Razandi M, Aitkenhead M, Hughes CCW, Levin ER. 2002. Integration of the non-genomic and genomic actions of estrogen. *J. Biol. Chem.* 277:50768–75

194. Watters JJ, Chun T-Y, Kim Y-N, Bertics PJ, Gorski J. 2000. Estrogen modulation of prolactin gene expression requires an intact mitogen-activated protein kinase signal transduction pathway in cultured rat pituitary cells. *Mol. Endocrinol.* 14:1872–81

195. Feng WJ, Webb P, Nguyen P, Liu XH, Li JD, et al. 2001. Potentiation of estrogen receptor activation function 1 (AF-1) by Src/JNK through a serine 118-independent pathway. *Mol. Endocrinol.* 15:32–45

196. Vasudevan N, Kow L-M, Pfaff DW. 2001. Early membrane estrogenic effects required for full expression of slower genomic actions in a nerve cell line. *Proc. Natl. Acad. Sci. USA* 98:12267–71

Annu. Rev. Physiol. 2005. 67:377–409
doi: 10.1146/annurev.physiol.67.031103.153247
First published online as a Review in Advance on September 28, 2004

MECHANISMS OF BICARBONATE SECRETION IN THE PANCREATIC DUCT

Martin C. Steward,[1] Hiroshi Ishiguro,[2] and R. Maynard Case[1]

[1]Faculty of Life Sciences, University of Manchester, Manchester M13 9PT, United Kingdom; email: martin.steward@man.ac.uk; maynard.case@man.ac.uk
[2]Human Nutrition and Internal Medicine, Nagoya University Graduate School of Medicine, Nagoya 464-8601, Japan; email: ishiguro@htc.nagoya-u.ac.jp

Key Words   secretin, CFTR, carbonic anhydrase, intracellular pH, epithelial transport

■ **Abstract**   In many species the pancreatic duct epithelium secretes HCO_3^- ions at a concentration of around 140 mM by a mechanism that is only partially understood. We know that HCO_3^- uptake at the basolateral membrane is achieved by Na^+-HCO_3^- cotransport and also by a H^+-ATPase and Na^+/H^+ exchanger operating together with carbonic anhydrase. At the apical membrane, the secretion of moderate concentrations of HCO_3^- can be explained by the parallel activity of a Cl^-/HCO_3^- exchanger and a Cl^- conductance, either the cystic fibrosis transmembrane conductance regulator (CFTR) or a Ca^{2+}-activated Cl^- channel (CaCC). However, the sustained secretion of HCO_3^- into a HCO_3^--rich luminal fluid cannot be explained by conventional Cl^-/HCO_3^- exchange. HCO_3^- efflux across the apical membrane is an electrogenic process that is facilitated by the depletion of intracellular Cl^-, but it remains to be seen whether it is mediated predominantly by CFTR or by an electrogenic SLC26 anion exchanger.

INTRODUCTION

In many species, including humans, the pancreatic duct epithelium achieves the unique feat of secreting an almost isotonic sodium bicarbonate solution. How it does this has long fascinated physiologists, not only as a problem worthy of solution, but also as a model for bicarbonate transport in other epithelia. In the 10 years or so since the topic was last reviewed, the twin approaches of molecular biology and isolated duct physiology have led to a steady increase in our knowledge of the underlying transport mechanisms. The purpose of this review is to provide a critical account of recent developments and to present an integrated picture of ductal bicarbonate secretion. Those wishing to place this within the wider context of pancreatic physiology are referred to other more general sources (1, 2).

0066-4278/05/0315-0377$14.00

The Classical Model

The year 1988 saw the publication of four innovative studies of duct segments isolated from the rat pancreas. Membrane potential measurements in microperfused ducts revealed the presence of a secretin-stimulated Cl^- conductance and Cl^-/HCO_3^- exchanger at the apical membrane (3) and a K^+ conductance and Na^+/H^+ exchanger at the basolateral membrane (4). Patch-clamp studies defined the properties of a cAMP-activated Cl^- channel at the apical membrane (5), subsequently identified as the cystic fibrosis transmembrane conductance regulator (CFTR), and measurements of intracellular pH (pH_i) confirmed the presence of Na^+/H^+ and Cl^-/HCO_3^- exchangers (6).

Together with previous work on whole preparations of the pancreas (7), these studies led to the model for HCO_3^- secretion shown in Figure 1. In this model, CO_2 diffuses into the duct cell across the basolateral membrane and is hydrated through the action of intracellular carbonic anhydrase (CA). The H^+ ions thus produced are exported across the basolateral membrane by an Na^+/H^+ exchanger driven by the Na^+ gradient that is maintained by the Na^+,K^+-ATPase. The HCO_3^- ions leave

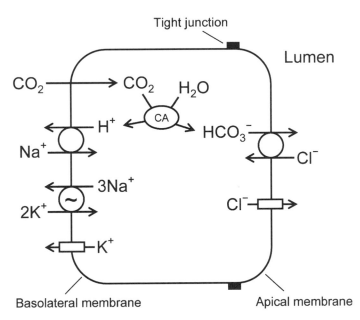

Figure 1 Cellular mechanism proposed for HCO_3^- secretion by pancreatic duct epithelium (3, 5). Intracellular HCO_3^- is derived from CO_2 through the action of carbonic anhydrase (CA). It leaves the cell in exchange for Cl^-, which is supplied to the lumen by a secretin-regulated Cl^- channel. H^+ is extruded at the basolateral membrane by a Na^+/H^+ exchanger and the inward Na^+ gradient is maintained by the Na^+,K^+-ATPase. Basolateral K^+ channels allow the recirculation of K^+ ions brought in by the pump and help to maintain the membrane potential. Na^+ ions enter the secretion via a paracellular pathway through the tight junctions (not shown).

the cell at the apical membrane in exchange for Cl^- ions whose availability in the lumen is determined by secretin-regulated Cl^- channels. Na^+ follows passively via the paracellular pathway, driven by a small transepithelial potential difference, and water follows by osmosis.

Although the involvement of the transporters and channels shown in Figure 1 is not in doubt, this model is unable to account for the ability of the ducts in many species, including humans, to secrete HCO_3^- at concentrations of 140 mM or more. The main challenge of the past few years has been to identify the additional factors that enable such high HCO_3^- concentrations to be achieved.

Technical Developments

Inevitably progress in achieving this goal has been boosted by new experimental approaches. Particularly valuable have been intracellular fluorescent indicators for pH_i, Na^+ (8), and Cl^- (9). Combining microfluorometry with simultaneous microperfusion of the duct lumen (10) has helped to identify and localize transporters to the apical and basolateral membranes. By injecting fluorescent probes for pH and Cl^- into the lumen of an isolated duct (11) it has also been possible to measure net HCO_3^- secretion. Secretory flow rate measurements in isolated ducts (12) have been further refined using fluorescence imaging (13) and video microscopy (14), and concurrent fluorometric measurements of fluid secretion and luminal pH (13) have allowed changes in HCO_3^- secretion to be followed.

Cell lines derived from human ductal adenocarcinomas offer the attractive possibility of Ussing chamber experiments (15, 16) and the chance to evaluate the effects of genetic manipulation. However, most are probably derived from the larger ducts (17) and may therefore have a different phenotype from the intercalated duct cells, which are thought to be the major site of HCO_3^- secretion in human pancreas. In primary culture, bovine duct cells produce confluent monolayers that respond to secretin (18, 19), and canine duct cells have been successfully grown without transformation in long-term culture on filters (20, 21). Recently, immortalized pancreatic cell lines have been obtained from both normal and cystic fibrosis (CF) mice using the ImmortoMouse transgene (22).

Finally, the application of modern molecular biological techniques has led quickly to the identification and characterization of many of the principal acid/base transporters in the ductal epithelium. Studies using heterologous expression systems, such as *Xenopus* oocytes and HEK293 cells, have revealed the properties of individual transporters and channels, and their interactions with each other when coexpressed (23). The development of isoform-specific antibodies has enabled the identification and localization of ductal transporters by immunohistochemistry, particularly in less accessible species such as humans.

Species and Segmental Differences

Although many species secrete almost isotonic $NaHCO_3$ during maximal stimulation, others do not (7). Unfortunately the secretin-stimulated rat pancreas, used in so many pioneering studies, secretes HCO_3^- at a maximal concentration of only around 70 mM (24). Therefore, we switched to the guinea pig pancreas for our experimental studies because, like dog, cat, and human pancreas, it secretes 140 mM HCO_3^- or more (25). Mouse ducts have also been used recently (14, 26, 27) because of the availability of knockout animals. However, this species, like the rat, may be a poor secretor of HCO_3^- (28). Even among the species that can secrete 140 mM HCO_3^- there appear to be significant differences in the ductal transport mechanisms.

There are also variations in the transport mechanism along the ductal tree. Early micropuncture data showed changes in luminal Cl^- and HCO_3^- concentration along the ductal system of the cat pancreas (29), and it is possible that the secretion of high HCO_3^- concentrations involves sequential modification of the juice as it flows along the ducts (30). At least four categories of duct can be distinguished: intercalated ducts (extending into the acini as centroacinar cells), intralobular ducts, interlobular ducts, and main or common ducts. There are clear signs that duct categories may differ in importance between species. For example, in the human pancreas, the key transporters appear to be most highly expressed in the intercalated ducts (31–35), suggesting that these are the main sites of fluid and HCO_3^- secretion. However, in the rat pancreas, both the ultrastructure (36) and immunolocalization data (34, 37–39) suggest that the interlobular ducts may be more important. A further complication is the finding that the interlobular and main ducts in some species may also be sites of HCO_3^- reabsorption or salvage (26).

BASOLATERAL MEMBRANE MECHANISMS

There are two possible sources for the HCO_3^- ions secreted by pancreatic ducts: (a) hydration of intracellular CO_2 by carbonic anhydrase (CA) coupled with basolateral extrusion of H^+ and (b) uptake of HCO_3^- itself by a basolateral anion transporter. The fact that a number of weak acid anions can be secreted in place of HCO_3^- and that the CA inhibitor acetazolamide significantly inhibits secretion seemed initially to indicate that the first mechanism was more likely. However, it is now clear that both mechanisms are present in duct cells and that their relative contributions differ between species.

Na^+/H^+ Exchangers

The idea that Na^+-coupled H^+ extrusion at the basolateral membrane might provide the driving force for HCO_3^- secretion first arose in early studies of the rabbit pancreas (40). Subsequently, Na^+/H^+ exchange was demonstrated directly by the measurement of pH_i in isolated interlobular ducts from rat pancreas (6). The basolateral location of the exchanger was deduced indirectly from electrophysiological studies (4) and was subsequently confirmed by pH_i recovery experiments in microperfused ducts subjected to acid loading (10). Similar results were obtained with isolated interlobular ducts from the guinea pig pancreas (8, 41, 42). Unexpectedly, Na^+/H^+ exchanger activity was also detected at the apical membrane in the main ducts of several species (10, 26, 43), where it probably serves a very different function (see below).

Eight mammalian Na^+/H^+ exchanger isoforms have been identified, and all belong to the SLC9 gene family (44). Immunohistochemistry has confirmed that it is the NHE1 isoform that is expressed at the basolateral membrane of pancreatic duct cells in the mouse (26) and rat (45). There is also immunohistochemical evidence for NHE4 expression in rat ducts and for its colocalization with NHE1 (45). However, attempts to detect NHE4 mRNA in rat pancreas by RT-PCR were unsuccessful (26). Functional identification of the basolateral exchanger as the NHE1 isoform was first demonstrated in microperfused mouse main ducts, where it was shown to be sensitive to the NHE1-specific inhibitor HOE694 (26).

There is no doubt that NHE1 is present at the basolateral membrane of the duct cells in all species and that it participates in the regulation of pH_i. Its contribution to HCO_3^- secretion, on the other hand, is less certain. Amiloride administered to preparations of whole pancreas had no effect at all on secretin-evoked fluid secretion in the rabbit (46) or the pig (47, 48) and only limited effects in the rat (49) and cat (50). There was no evidence of any increase in exchanger activity following stimulation of pig ducts with secretin (51) and, in isolated guinea pig ducts, amiloride only partially inhibited secretin-evoked HCO_3^- uptake across the basolateral membrane (8) and HCO_3^- secretion into the duct lumen (11). Collectively, these results suggest that other basolateral mechanisms contribute significantly to the ductal secretion of HCO_3^-.

H^+-ATPase

An alternative mechanism for H^+ extrusion across the basolateral membrane would be a proton pump (H^+-ATPase). The following observations have led to the proposal that, in the pig pancreas, basolateral H^+-ATPase activity is a major source of HCO_3^- (52): (a) Pig duct cells contain tubulovesicles that are inserted into the basolateral membrane following stimulation with secretin (53), (b) the tubulovesicle interior is acidic (54), and (c) colchicine blocks fusion of the tubulovesicles and also reduces secretin-evoked HCO_3^- secretion by 60% (55). Further support has come from pH_i measurements in pig ducts showing that H^+-ATPase activity is induced by secretin and is capable of restoring pH_i to normal values following standard acid-loading procedures (51). Its sensitivity to bafilomycin A_1 indicates that it is a vesicular-type H^+-ATPase (V-ATPase) (56).

V-ATPase activity has also been demonstrated in microperfused rat ducts following stimulation with secretin or carbachol (10). It is localized predominantly at the basolateral membranes of intra- and interlobular ducts (45). Whether it contributes significantly to HCO_3^- secretion in the rat is unclear because in acid-load experiments, it restores pH_i to only around 6.8 (10), suggesting that it might be relatively inactive in the normal range of pH_i.

Although there is no evidence for V-ATPase activity in unstimulated guinea pig ducts (8), it does become apparent following stimulation with secretin and carbachol, but not when the ducts are pretreated with cytochalasin D and nocodazole to disrupt vesicle translocation (41). However, bafilomycin A_1 does not inhibit secretin-evoked fluid and HCO_3^- secretion in guinea pig ducts (11, 13), suggesting that the V-ATPase does not contribute significantly to HCO_3^- secretion under normal conditions. In contrast, there was a 40% reduction in secretin-evoked HCO_3^- secretion by the pig pancreas treated in vivo with N,N'-dicyclohexylcarbodiimide (DCCD), a general H^+-ATPase inhibitor (57). Furthermore, forskolin-evoked HCO_3^- secretion by human Capan-1 cells appears to be independent of Na^+ and is inhibited by the nonspecific ATPase inhibitor N-ethylmaleimide (15). Thus the contribution of the V-ATPase may prove to be more prominent in some species than others.

Na^+-HCO_3^- Cotransporters

Of all the possible mechanisms for basolateral uptake of HCO_3^- into pancreatic duct cells, the most likely candidate would be a Na^+-HCO_3^- cotransporter. The first evidence for such a mechanism came from studies of the isolated rabbit pancreas in which HCO_3^- transport was observed to be Na^+-dependent and SITS (4-acetamido-4'-isothiocyanatostilbene-2,2'-disulfonic acid)-sensitive but relatively unaffected by CA inhibitors and amiloride (46). The uptake of HCO_3^- was therefore attributed to an Na^+-dependent Cl^-/HCO_3^- exchanger but it now seems likely to be due to an Na^+-HCO_3^- cotransporter.

The involvement of such a transporter in rat ducts was initially excluded on electrophysiological grounds (4). Subsequently, however, Na^+-HCO_3^- cotransport was clearly demonstrated in microperfused rat ducts as a Na^+- and HCO_3^--dependent recovery of pH_i following intracellular acidification (10). The recovery was insensitive to amiloride but reversibly blocked by H_2DIDS (dihydro-4,4'-diisothiocyanatostilbene-2,2'-disulfonic acid) applied to the basolateral membrane. Similar observations followed in the pig (51), where the cotransporter was capable of raising pH_i to higher values than was the V-ATPase, and also in the guinea pig (8, 41, 42).

The electrogenicity of the cotransporter was first recognized in CFPAC-1 cells where manipulation of the membrane potential brought about changes in the rate of recovery of pH_i after an acid load (58). Its electrogenicity was also demonstrated in microperfused guinea pig ducts where a substantial depolarization was observed following application of basolateral H_2DIDS during forskolin stimulation, and a Na^+-dependent hyperpolarization was evoked by raising the HCO_3^- concentration in the bath (59). These data suggested that the electrogenicity of the cotransporter could be an important hyperpolarizing influence during HCO_3^- secretion.

The molecular identity of the Na^+-HCO_3^- cotransporter in the pancreatic duct is now well established. It is one of a group of NBC cotransporters that belongs to the same gene family (SLC4) as the AE anion exchangers (60). The first electrogenic NBC (kNBC1) was cloned from amphibian kidney (61) where its Na^+:HCO_3^- stoichiometry of 1:3 favors HCO_3^- efflux across the basolateral membrane of the proximal tubule cells. The pancreatic splice variant (pNBC1), differing at the N terminus, was cloned from both pancreas (62, 63) and heart (64). Northern blots consistently show strong expression of this variant in pancreas and in other tissues, whereas the kNBC1 variant is largely confined to the kidney.

Evidence for the localization of pNBC1 in both ducts and acini has come from in situ hybridization in mouse pancreas (62). Immunohistochemistry reveals expression of NBC1 at the basolateral membrane of intercalated, intralobular, and interlobular ducts in human (35) and in rat (63). Using variant-specific antibodies, pNBC1 was found to be the dominant variant in rat and human pancreas (34, 65). In rat, it is most strongly expressed at the basolateral membranes of the interlobular and main ducts, with weaker labeling of the apical membrane, whereas in humans the labeling is strongest in the intercalated duct cells and is exclusively basolateral (34). There is also evidence of kNBC1 expression at the apical membrane of some rat interlobular duct cells (34) and colocalization with pNBC1 in the apical membrane of the main duct cells (65). Despite these reports, no signs of functional apical NBC1 activity have been detected in pH_i studies of either rat or guinea pig ducts (10, 42).

The stoichiometry of pNBC1 is of crucial importance when considering its role in pancreatic HCO_3^- secretion. Being electrogenic, the direction in which the transporter operates will depend not only on the Na^+ and HCO_3^- concentration gradients but also on its stoichiometry and the membrane potential. Basolateral uptake of HCO_3^- in pancreatic duct cells would be favored by a 1:2 stoichiometry and accelerated by depolarization. When pNBC1 was expressed in cultured mouse duct cells, measurements of the reversal potential with different Na^+ gradients showed convincingly that the Na^+:HCO_3^- stoichiometry is 1:2 (66). Perhaps surprisingly, the stoichiometry of the two NBC1 variants is not determined by their differing N termini but depends upon the cell type in which they are expressed (67) and can be altered by protein kinase A (PKA)-dependent phosphorylation of a serine residue at the C terminus (68).

What are the relative contributions of pNBC1 and NHE1 during secretin-evoked HCO_3^- secretion in a good HCO_3^- secretor such as the guinea pig pancreas? Our measurements of the fall of pH_i in isolated guinea pig ducts following the application of amiloride and H_2DIDS indicate approximately equal contributions of the two transporters to unstimulated secretion and pH_i regulation (8). But following stimulation with secretin, the activity of pNBC1 increases and is responsible for about 75% of the basolateral HCO_3^- uptake. This conclusion is supported by measurements of fluid and HCO_3^- secretion, which show 56% inhibition by H_2DIDS, 18% by an amiloride analogue, and total inhibition when applied together (13).

The significance of Na^+-HCO_3^- cotransport in rat ducts is less clear. Measurements of pH_i suggest that the recovery from acid loading is not accelerated in the presence of HCO_3^-, nor following stimulation with secretin (69). However, these experiments used intercalated and intralobular ducts that, in the rat, express rather less pNBC1 than do the interlobular ducts, which are probably the main site of fluid secretion (34, 37).

Tenuous indirect evidence for the importance of pNBC1 in human pancreatic secretion comes from a patient with a missense mutation in the common region of pNBC1 and kNBC1 (70). Not only did the patient display proximal renal tubular acidosis, but there was also a raised serum amylase level suggestive of pancreatic duct dysfunction.

Anion Exchangers

A basolateral anion exchanger was discovered unexpectedly in microperfused rat ducts when substitution of Cl^- in the bath led to a HCO_3^--dependent and DIDS-sensitive rise in pH_i (10). Unlike the more familiar apical anion exchanger (see below), the basolateral exchanger in microperfused mouse main ducts is not activated by forskolin (27). In unstimulated guinea pig interlobular ducts, the basolateral exchanger is actually more active than the apical one (42). However, following stimulation with forskolin, measurements of intracellular $[Cl^-]$ reveal a strong activation of the luminal exchanger and probably a decrease in the activity of the basolateral exchanger (9).

The molecular identity of the basolateral exchanger remains uncertain. Of the four AE members of the SLC4 gene family, AE2 is the most likely candidate by analogy with other epithelia (60). AE2 transcripts have certainly been identified in human fetal pancreas by library screening (71), and there is evidence that AE2 is expressed in CFPAC-1 cells, probably at the basolateral membrane (72). However, AE2 could not be detected in rat duct cells by immunohistochemistry even though it was clearly present in the acinar cells (45).

It is difficult to predict the role of a basolateral anion exchanger in ductal HCO_3^- secretion. Assuming that it acts as a neutral Cl^-/HCO_3^- exchanger, it would tend to dissipate the accumulation of intracellular HCO_3^- while favoring basolateral Cl^- uptake, thus developing a larger driving force for Cl^- secretion. Interestingly, this is what appears to happen in rat interlobular ducts stimulated with the GRP-related peptide, bombesin (73). Through an unidentified signaling mechanism, bombesin stimulates a secretion whose inhibitor sensitivity and anion selectivity differs from that evoked by secretin. The possibility that different agonists and signaling pathways can evoke secretions of differing anion composition clearly demands further investigation.

Cation-Chloride Cotransporters

Although early studies indicated that ductal fluid secretion in the rat pancreas was totally dependent on HCO_3^- (12, 74, 75), ducts in some species are capable of secreting a Cl^--rich fluid in the absence of HCO_3^-. This has been observed in interlobular mouse and rat ducts stimulated with forskolin and secretin (14) and is completely blocked by bumetanide, suggesting that it is driven by a Na^+-K^+-$2Cl^-$ cotransporter, probably NKCC1, at the basolateral membrane. Even in the presence of HCO_3^-, Cl^- secretion in the mouse and rat ducts is estimated to contribute as much as 40% of the overall secretion of fluid.

These results contrast markedly with those obtained with guinea pig ducts in which secretion is totally dependent on HCO_3^- (13). Likewise, furosemide and bumetanide have no effect on secretin-evoked secretion by the pig pancreas (47). Nonetheless, there is evidence for NKCC1 activity in human ductal cell lines. Short-circuit current measurements with Capan-1 cells show ~60% inhibition of Cl^- secretion by bumetanide (15), and NKCC1 mRNA has been demonstrated in CFPAC-1 cells by Northern blot (76). Interestingly both the expression and the activity of the cotransporter were doubled in cells transfected with wild-type CFTR. Because these human cell lines are probably derived from larger ducts, it is interesting that main duct cells cultured from bovine pancreas also show a bumetanide-sensitive component in their forskolin-evoked short-circuit current (19).

The K^+-Cl^- cotransporter KCC1, another member of the SLC12 cation-chloride cotransporter superfamily (77), has also been detected in human pancreas (78) and localized to the basolateral membrane in rat intralobular ducts by immunohistochemistry (79). There are no functional data but KCC1 would be expected to mediate coupled electroneutral efflux of K^+ and Cl^- across the basolateral membrane and might be involved in cell volume regulation. It is unlikely to contribute to Cl^- uptake because the steep outward concentration gradient for K^+ will usually exceed the inward Cl^- gradient.

K^+ Channels

Basolateral K^+ channels provide an exit pathway for K^+ brought in by the basolateral Na^+,K^+-ATPase and play a vital role in maintaining the membrane potential, which is a crucial component of the driving force for anion secretion. The substantial basolateral K^+ conductance of pancreatic duct cells was first observed in microelectrode studies of microperfused rat ducts (4). It is more sensitive to Ba^{2+} than to tetraethylammonium (TEA) and accounts for more than 60% of the basolateral conductance in unstimulated ducts (80). The resting K^+ conductance appears to be pH-sensitive but the functional significance of this is not clear.

A transient hyperpolarization, sometimes seen at the onset of stimulation with secretin or VIP (vasoactive intestinal peptide), may indicate a cAMP-evoked increase in the basolateral K^+ conductance (81, 82) but it is usually masked by the depolarizing effect of the increased apical anion conductance. Ca^{2+}-mobilizing agonists such as ATP have more equivocal effects on the K^+ conductance (83, 84).

Basolateral maxi-K^+ channels have been identified in cell-attached and excised patches from rat interlobular ducts (85). As in other cell types, these are Ca^{2+}sensitive and voltage dependent with a linear current-voltage (I/V) relationship in symmetrical K^+ solutions and a conductance of about 200 pS. The maxi-K^+ channels are activated by secretin or cAMP, and the catalytic subunit of PKA appears to act by increasing their sensitivity to Ca^{2+}. There is evidence for a similar channel in cultured dog pancreatic duct cells where ATP activates a charybdotoxin-sensitive, Ca^{2+}-activated K^+ conductance (21). The maxi-K^+ channels are thought unlikely to contribute to the resting K^+ conductance because they have a low open-state probability in unstimulated rat duct cells (85). They are blocked by Ba^{2+} and

TEA (86) and these agents also completely block secretin-evoked fluid secretion in isolated rat ducts (73), so it is likely that the maxi-K^+ channels are actively involved in secretion.

An 82 pS K^+ channel has also been identified in rat duct cells (86). Because it is blocked by Ba^{2+} but not TEA, it might account for the component of resting K^+ conductance that is less sensitive to TEA (4). Nonselective cation channels are seen occasionally, but these appear to be active only at abnormally high cytosolic Ca^{2+} concentrations (87).

Na^+,K^+-ATPase

Apart from the possible contribution of a V-ATPase in the pig pancreas, the primary driving force for ductal HCO_3^- secretion is provided by Na^+,K^+-ATPase at the basolateral membrane. This maintains the inward Na^+ gradient that drives the accumulation of intracellular HCO_3^- and Cl^- by NHE1, pNBC1 and NKCC1, and it maintains the outward K^+ gradient that largely defines the membrane potential. When ouabain is applied to rat interlobular ducts there is a small, rapid depolarization (4), which suggests that the electrogenicity of the pump also makes a minor contribution to the membrane potential.

High levels of pump expression, as assessed by cytochemistry and ^{3}H-ouabain binding, are associated with the basolateral membrane in the interlobular ducts of the rat pancreas (88), whereas autoradiography reveals particularly strong basolateral labeling of the intercalated ducts in the cat pancreas (89), and immunohistochemistry shows a similar pattern in the dog pancreas (90).

Water Channels

As a major site of isotonic fluid secretion, the pancreatic duct epithelium might be expected to express aquaporin (AQP) water channels. Immunohistochemistry has revealed that AQP1 is indeed present at both the apical and basolateral membrane of interlobular ducts in rat pancreas (37, 38). As a functional correlate, both the water permeability of these cells and the process of fluid secretion are significantly reduced by Hg^{2+}, which blocks the AQP1 channels (37).

In the human pancreas, AQP1 expression is greatest in the intercalated duct cells, with decreasing levels of expression farther down the duct system (31). AQP5 is colocalized with AQP1 at the apical membrane of the intercalated ducts and small intralobular ducts, and this distribution pattern maps closely onto that of CFTR (32, 33), suggesting that the smaller ducts are probably the main site of fluid secretion in the human pancreas.

APICAL MEMBRANE MECHANISMS

The secretion of HCO_3^- across the apical membrane could theoretically be achieved by diffusion of CO_2 into the lumen, hydration by CA, and reabsorption of H^+. The apical membrane is certainly permeable to CO_2 (42) and luminal CA activity is also present (see below), but there is no evidence of any mechanism for transporting H^+ from the lumen into the duct cells. It therefore seems more likely that HCO_3^- enters the lumen either by diffusion through an anion channel or via a Cl^-/HCO_3^- exchanger operating in parallel with a Cl^- channel (Figure 1).

CFTR

The presence of an apical Cl^- conductance first became apparent in electrophysiological studies of rat interlobular ducts (3, 5). Small-conductance Cl^- channels were detected in the apical membrane, and their open probability was shown to be raised by secretin and cAMP. From their I/V characteristics, anion selectivity, and pharmacology (91), the channels were quickly identified as CFTR (92). Whole-cell recordings indicated that the channel density is about 4000 per cell, but their low incidence in cell-attached patches suggests that they are not uniformly distributed.

Channels with similar properties were also detected in human duct cells cultured from explants of fetal pancreas (93) and have since been described in mouse (94) and guinea pig duct cells (95), in cultured canine main duct cells (20), in primary cultures of bovine main duct cells (96), and in human Capan-1 cells (97). The expression of CFTR in pancreatic duct cells is further supported by immunohistochemistry, which shows strong labeling at the apical membrane of centroacinar and intercalated duct cells in human pancreas (32, 33) and intralobular and interlobular ducts in rat pancreas (39).

Given the central role of CFTR in HCO_3^- secretion, it is natural to ask whether CFTR itself might provide a pathway for HCO_3^- efflux across the apical membrane. If so, this could account for the Cl^--independent component of HCO_3^- secretion that we have observed in guinea pig ducts (11, 13). A simple calculation shows that the electrochemical gradient for HCO_3^- will be sufficient to achieve luminal concentrations approaching 190 mM if membrane potential is maintained at -60 mV. However, the viability of this proposal also depends upon the HCO_3^- permeability of CFTR. Estimates of the HCO_3^-/Cl^- permeability ratio of CFTR are generally quite low and range from about 0.2 to 0.4 (91, 92, 95, 97–101). However, it is also possible that the HCO_3^-/Cl^- selectivity of CFTR is variable and may be regulated by intracellular signals (102). Perhaps more significantly, recent studies show that CFTR expressed in *Xenopus* oocytes switches to a HCO_3^--permeable state when extracellular Cl^- falls to low concentrations (103), as it does during maximal secretion in the pancreatic duct. Even assuming a fairly modest HCO_3^-/Cl^- permeability ratio (0.4), our calculations suggest that the HCO_3^- permeability of CFTR in the apical membrane of guinea pig duct cells could be sufficient to support the observed rate of HCO_3^- secretion (59).

Corroborative evidence pointing to HCO_3^- secretion via CFTR has been obtained in other epithelia. For example, (*a*) CFTR is thought to provide the apical efflux pathway for both Cl^- and HCO_3^- in Calu-3 airway mucosal gland cells (104); (*b*) HCO_3^- movements across the apical membrane of human nasal epithelium are sensitive to the CFTR blocker DPC and are absent in tissue from CF patients (105); and (*c*) cAMP-stimulated HCO_3^- secretion in rat and rabbit duodenum does not involve any increase in anion exchanger activity and is insensitive to DIDS but is inhibited by the CFTR blocker NPPB (106).

However, there are some problems with this hypothesis. A surprising property of CFTR seen in guinea pig duct cells is that cAMP-activated Cl^- currents are inhibited by high extracellular HCO_3^- concentrations (95). If this also applies to the HCO_3^- permeability of CFTR, then it becomes difficult to see how CFTR could provide a significant route for HCO_3^- secretion. On the other hand, the effects of high extracellular HCO_3^- could reflect a switch to the HCO_3^--permeable form of CFTR induced by the low luminal Cl^- concentration (103). A further puzzling feature of the apical HCO_3^- pathway in guinea pig ducts is that it appears to be insensitive to glibenclamide and NPPB (9, 11), which are normally quite effective blockers of CFTR. This raises the possibility that a channel other than CFTR might be involved in HCO_3^- secretion. One candidate could be the large-conductance g350 channel identified in Capan-1 cells (107), which appears to be equally permeable to Cl^- and HCO_3^- (108).

An intriguing feature of CF is that some CFTR mutants reach the plasma membrane and behave normally as Cl^- channels and yet are associated with pancreatic insufficiency, whereas others are defective as Cl^- channels but support relatively normal pancreatic function (109). Expression studies in HEK293 cells show that CFTR mutants associated with pancreatic sufficiency are still able to activate Cl^-/HCO_3^- exchange, whereas those associated with pancreatic insufficiency are not. The simplest conclusion from this would be that HCO_3^- is secreted by a CFTR-activated anion exchanger rather than via the CFTR channel. However, a 1:1 Cl^-/HCO_3^- exchanger would not be able to secrete HCO_3^- at concentrations as high as 140 mM (see below).

The production of a HCO_3^--rich secretion also demands that Cl^- secretion be minimal. This could be achieved either by a selective decrease in the permeability of CFTR to Cl^- or by a reduction in the electrochemical gradient for Cl^- secretion. Evidence for the latter comes from guinea pig ducts stimulated with forskolin and microperfused with high HCO_3^- solutions to simulate maximal secretion (9). Under these conditions $[Cl^-]_i$ drops to below 10 mM as a result of efflux via CFTR and the lack of basolateral Cl^- uptake. The only Cl^- uptake pathway in guinea pig ducts, which lack a functional NKCC1, is the basolateral anion exchanger, but this does not increase in activity during stimulation and may even be inhibited. The net result is that, in guinea pig duct cells, the secretion may be HCO_3^--rich because of the lack of driving force for Cl^-. In contrast, mouse and rat ducts appear to produce a mixed secretion because they have NKCC1 to accumulate intracellular Cl^- across the basolateral membrane (14), and the basolateral anion exchanger may also contribute to Cl^- uptake (73).

Ca^{2+}-Activated Chloride Channels

Patch-clamp studies of normal and CF mouse ducts have revealed the presence of Ca^{2+}-activated chloride channels (CaCC) in addition to CFTR in the apical membrane. Cyclic AMP activates relatively small CFTR currents in normal mice, whereas ionomycin evokes much larger Ca^{2+}-activated currents in both normal and CF mice (94, 110). Because the presence of CaCCs might explain why these mice show minimal pancreatic pathology (111), there has been considerable interest in the CaCC as a possible therapeutic target in humans (112).

CaCCs have been characterized in the human pancreatic cell line HPAF and in duct cells isolated from normal human pancreas (113). They are activated by intracellular Ca^{2+} over a physiological range (10 nM–1 μM) and show slight outward rectification. They are more permeable to I^- than to Cl^- (unlike CFTR) and are insensitive to DIDS and most other Cl^- channel blockers with the exception of niflumic acid. They are clearly different from the outwardly rectifying Cl^- channel (ORCC), but their molecular identity remains unknown. RT-PCR data from HPAF cells suggest that they are not CLCA-1 or -2 (114), two members of the recently discovered CaCC family.

Channels with quite similar properties have also been described in cultured canine (20) and bovine main duct cells (96). The fact that they are not detected in single-channel recordings is probably because their conductance is very small (115). CaCC current density is much higher in mouse and human compared with that of rat and guinea pig (112) so there is no obvious correlation with these species' ability to secrete HCO_3^-. Nonetheless, a physiological role for CaCC is clearly indicated by the observation that a variety of Ca^{2+}-mobilizing agonists, including acetylcholine (116), luminal ATP (117), and luminal Ca^{2+} (118), evoke ductal secretion. The enhancement of secretin-evoked fluid secretion by ethanol in guinea pig ducts also appears to involve Ca^{2+} mobilization (119). On the other hand both 5-hydroxytryptamine (5-HT) and basolateral ATP raise intracellular Ca^{2+} while inhibiting secretion (117, 120).

CaCCs can be activated by carbachol and ATP in CFPAC-1 cells, which lack functional CFTR (121). They mediate ATP- and angiotensin II-evoked anion secretion (16, 122) which can be enhanced by protein kinase C (PKC) inhibitors (123) but it is not clear whether they can support HCO_3^- secretion. The HCO_3^-/Cl^- permeability ratio for CaCCs is estimated to be about 0.5 in guinea pig duct cells (112) but only 0.1 for endogenous channels in *Xenopus* oocytes (124), so the channels resemble CFTR in having a fairly low permeability to HCO_3^-. The presence of HCO_3^- has no effect on the short-circuit current generated by HPAF monolayers (114), but there is some evidence that CaCCs can support HCO_3^- secretion either via the CaCC channels themselves or via an apical anion exchanger in CFPAC-1 cells (125).

Anion Exchangers

An early suggestion that HCO_3^- secretion might involve Cl^-/HCO_3^- exchange (126) stemmed from the observation that pancreatic HCO_3^- secretion is markedly dependent on Cl^-. Complete replacement of Cl^- with an impermeant anion, while retaining HCO_3^-, inhibits ductal secretion by about 60% in perfused cat pancreas (126), isolated rabbit pancreas (127) and isolated rat ducts (73).

Cl^-/HCO_3^- exchange was first demonstrated in rat interlobular ducts as a DIDS-sensitive rise in pH_i following substitution of Cl^- (6). Its localization at the apical membrane was deduced from the hyperpolarizing effect of luminal SITS in microperfused rat ducts (3). Measurements of pH_i later confirmed the apical expression of the exchanger and, as already mentioned, also revealed the presence of a basolateral exchanger (10).

Consistent with the idea that the apical exchanger is involved in HCO_3^- secretion, stimulation with secretin or forskolin increases its activity in guinea pig interlobular ducts (128) and in mouse main ducts (27). Conversely, substance P inhibits exchanger activity in guinea pig ducts (129), and this may account for the inhibitory effect of substance P on fluid secretion (130, 131).

A further feature of the apical anion exchanger to emerge recently is its interaction with CFTR. The activity of the exchanger is much reduced in mice expressing $\Delta F508$ CFTR, a mutant that fails to reach the apical membrane (27). Probably for the same reason, ATP and trypsin are only able to activate the apical exchanger in CFPAC-1 cells (which also express the $\Delta F508$ mutant) when the cells are transfected with wild-type CFTR (132). This ability of CFTR to activate anion exchangers has subsequently been confirmed in cultured HEK293 cells transfected with either wild-type or mutant CFTRs (23). Interestingly the effectiveness of different CFTR mutants is independent of whether or not they are functional Cl^- channels.

The molecular identity of the apical anion exchanger in pancreatic duct cells has proven to be surprisingly elusive. None of the AE group of SLC4 anion exchangers has yet been firmly identified in duct cells. Furthermore, coexpression studies in HEK293 cells reveal that none of the known AEs is activated by interaction with CFTR (133).

There is a more fundamental problem with the concept of a Cl^-/HCO_3^- exchanger as the primary apical HCO_3^- efflux pathway in pancreatic duct cells. When luminal HCO_3^- and Cl^- concentrations reach the values that they do, for example, in secretin-stimulated guinea pig ducts, a 1:1 Cl^-/HCO_3^- exchanger would reverse and reabsorb HCO_3^- from the lumen. Thus although the apical exchanger could generate the 70 mM HCO_3^- secreted by the rat pancreas (24), it could not continue to secrete HCO_3^- into a luminal fluid containing 140 mM HCO_3^- (134).

According to the model shown in Figure 1, HCO_3^- secretion to the duct lumen should be dependent on luminal Cl^-. We have tested this idea in isolated guinea pig ducts microinjected with weakly buffered solutions containing BCECF-dextran to measure luminal pH (11). In unstimulated ducts, HCO_3^- movement into the lumen is dependent on luminal Cl^- and is inhibited by luminal H_2DIDS, suggesting that the apical anion exchanger is indeed involved in spontaneous HCO_3^- secretion (13). However, secretin, forskolin, and acetylcholine are able to evoke fluid and HCO_3^- secretion in the nominal absence of luminal Cl^-, and this process is not blocked by luminal H_2DIDS (11, 13). Furthermore, the estimated HCO_3^- concentration of the secreted fluid is approximately 140 mM.

Because an apical anion exchanger would be expected to run in reverse when the luminal HCO_3^- concentration is high, one might expect that it would become inhibited under these conditions. There is some evidence for this. Raising the luminal HCO_3^- to 125 mM in microperfused guinea pig ducts has remarkably little effect on pH_i (42), indicating that there is virtually no apical HCO_3^- entry via the exchanger despite the favorable concentration gradients.

SLC26 Exchangers

It now seems probable that the apical membrane anion exchanger in pancreatic duct cells is a member of the recently discovered SLC26 family (135). The ten members of this group encode anion exchangers capable of transporting a wide range of monovalent and divalent anions, including formate and SO_4^{2-}, as well as Cl^-, HCO_3^-, and OH^-. Their stoichiometries vary, some appear to be electrogenic and several show strong interactions with CFTR. The first evidence for their involvement in ductal secretion and for their interaction with CFTR was probably the observation that defective SO_4^{2-} transport in CFPAC-1 cells was restored by transfection with wild-type CFTR (136).

Not all members of the SLC26 family are fully characterized yet, but at least two appear to be expressed in pancreatic ducts, and both mediate Cl^-/HCO_3^- exchange. SLC26A3 (DRA; down-regulated in adenoma) was first identified as a potential tumor suppressor in colon epithelium, but subsequently shown to function as a Cl^-/HCO_3^- exchanger in HEK293 cells (137). Transcripts of SLC26A3 have been found by Northern blot of mouse pancreas RNA, and immunohistochemistry indicates that the protein is expressed at the apical membrane of large mouse ducts (72).

SLC26A6 (PAT1; putative anion transporter-1) was discovered by homology in database searches (138, 139). It is highly expressed in kidney and pancreas, and also in Capan-1 and -2 cell lines, and is localized at the apical membrane of interlobular ducts in human pancreas (138). The mouse ortholog was separately identified as the $Cl^-/formate$ exchanger (CFEX) in renal proximal tubule (140). When expressed in *Xenopus* oocytes, SLC26A6 supports DIDS-sensitive Cl^-/HCO_3^- exchange (141, 142) and shows evidence of electrogenicity (143).

There is good evidence that SLC26A3 and/or SLC26A6 may be the apical membrane Cl^-/HCO_3^- exchanger in CFPAC-1 cells (72). Both are expressed in these cells, and their expression is increased when the cells are transfected with wild-type CFTR, while the expression of AE2 is reduced. Significantly, there is also an increase in DIDS-sensitive $^{36}Cl^-$ uptake by anion exchange across the apical membrane and a decrease at the basolateral membrane. Further evidence comes from coexpression of CFTR and a range of anion exchangers in HEK293 cells and *Xenopus* oocytes (133). Measurements of anion exchange show that CFTR is unable to activate any of the AE family but greatly increases exchange by SLC26A3 and SLC26A6.

The possibility of a physical association with CFTR is strongly indicated by the presence of a PDZ domain at the C terminus of SLC26A6, which is identical to that of CFTR (144). Consequently, SLC26A6, like CFTR, binds through PDZ motifs to the scaffolding proteins NHERF (EBP50) and E3KARP, and it is possible that interactions within a CFTR-NHERF-SLC26A6 complex might account for the observed stimulation of apical anion exchange by CFTR (27). Recent data suggest that the regulatory domain of CFTR interacts with the highly conserved STAS (sulfate transporter and antisigma antagonist) domain of several of the SLC26 transporters, which results in reciprocal activation of both the exchanger and CFTR (145, 146).

Both SLC26 isoforms are electrogenic, but with opposite polarities: SLC26A3 appears to have a $Cl^-:HCO_3^-$ stoichiometry of 2:1 or more, while the ratio for SLC26A6 is 1:2 or less. This has led to the idea that differential expression of the two exchangers along the pancreatic ducts (SLC26A6 in the proximal ducts and SLC26A3 more distally) might help to explain the high HCO_3^- concentration secreted in the final juice (133, 145). This idea will be examined in more detail below.

Na^+/H^+ Exchangers

Juice collected from pancreatic ducts at low flow rates, or after a period of stasis, is well known to be relatively acidic and Cl^--rich (147, 148). It also has a relatively high P_{CO2}, suggesting that there is active acidification of the duct lumen. This has been shown to be the result of Na^+/H^+ exchange in the apical membrane of main duct cells in mouse pancreas (10) and in bovine pancreas (43).

Both NHE2 and NHE3 can be detected in the apical membrane of mouse main ducts by immunohistochemistry (26). As expected, Na^+/H^+ exchange at the apical membrane is much less sensitive to the selective NHE1 inhibitor HOE694 than is Na^+/H^+ exchange at the basolateral membrane. However, there is no loss of apical exchanger activity in NHE2$^{-/-}$ mice and only a 50% loss in NHE3$^{-/-}$ mice. This suggests that only the NHE3 is active in normal animals. The remaining proton extrusion in NHE3$^{-/-}$ mice seems to be mediated by an electroneutral Na^+-HCO_3^- cotransporter (see below).

Apical HCO_3^- reabsorption mediated by these two processes is inhibited by about 50% following stimulation with forskolin (26). Immunoprecipitation studies indicate that this may again be the result of physical interaction with CFTR (149). CFTR certainly enhances the cAMP-induced inhibition of NHE3, and this effect probably involves linkage through a scaffolding protein such as NHERF because it requires the PDZ domain of CFTR. However, HCO_3^- salvage mechanisms such as these are confined to the larger ducts and are absent in the interlobular ducts of rat and guinea pig.

Na^+-HCO_3^- Cotransporters

There is no functional evidence for apical pNBC1 or kNBC1 activity in perfused rat or mouse ducts despite the detection of both variants at the apical membrane of rat interlobular and main ducts by immunolocalization (34, 63, 65). However, the second HCO_3^- salvage mechanism detected in mouse main ducts (see above) may represent the activity of the electroneutral Na^+-HCO_3^- cotransporter NBCn1 or NBC3 (150). This member of the SLC4 family was first identified in human muscle (151) and alternative splicings were subsequently found in various rat tissues (60, 152).

There is clear evidence of NBCn1 expression in mouse pancreas, and it can be coimmunoprecipitated from whole pancreas lysates with CFTR and the scaffolding protein NHERF (150). Its localization in pancreatic ducts has not yet been reported, but it is present in mouse salivary ducts where it serves a similar function in HCO_3^- retrieval (153). Like NHE3, NBCn1 is inhibited by cAMP through its interaction with CFTR when the two are coexpressed in HEK293 cells. This has led to the suggestion that there may be reciprocal control of HCO_3^- secretion and reabsorption through complex interactions between CFTR, SLC26 exchangers, NHE3, and NBCn1 linked by PDZ domains (150).

CARBONIC ANHYDRASE

The enzyme carbonic anhydrase (CA) clearly has an important role in HCO_3^- secretion. Most of the CA activity in the pancreas is located in the ducts (154) and the CA inhibitor acetazolamide has a variable but significant effect on secretin-evoked secretion in human (155–157), rabbit (46), cat (158–160), dog (161, 162), and pig (163). The inhibitory effect of acetazolamide has traditionally been interpreted as evidence that much of the secreted HCO_3^- is derived from the hydration of intracellular CO_2. However, this is not necessarily so because it is now clear that CA activity is also required for the normal function of several HCO_3^- transporters.

There are at least 11 active CA isozymes: Some are intracellular, others are membrane associated and have extracellular activity (164). Messenger RNAs for CAII, -IV, -VI, and -XII have been identified in whole human pancreas by RT-PCR and Northern blot analysis (165). CAII is expressed at a high level in intralobular and interlobular ducts (166, 167). It is present throughout the duct cell cytosol but is also particularly associated with the inner surface of the apical membrane in Capan-1 cells (108), and there may be a specific trafficking pathway from the Golgi apparatus to this location (168). Although this isozyme undoubtedly facilitates the production of intracellular HCO_3^- from CO_2, no pancreatic pathology has been reported for patients with CAII -deficiency syndrome (169) so it may be that HCO_3^- uptake by pNBC1 is sufficient or upregulated in such cases.

The membrane-anchored CAIV isozyme is localized at the apical membrane in Capan-1 cells (108) and also in human pancreatic ducts (170). If HCO_3^- crosses the apical membrane as HCO_3^- rather than CO_2, it is not clear what function CAIV would have in this location. One possibility is that CAIV itself acts as, or forms part of, a HCO_3^- channel. This idea derives from studies of purified CA incorporated into lipid bilayers (171) and from an analysis of red cell volume changes (172). Alternatively, luminal CAIV may simply have a protective role in dissipating acid reflux from the duodenum by converting HCO_3^- to CO_2 (165). CAIX and -XII are membrane-spanning isozymes with extracellular CA activity, but immunohisto-chemistry reveals that they are expressed at the basolateral membranes of the duct cells (173).

Evidence has emerged recently that CA may have other roles in HCO_3^- transport. In the erythrocyte membrane the C terminus of AE1 links the exchanger directly to CAII (174). When AE1 and CAII are coexpressed in HEK293 cells, the activity of AE1 is strongly inhibited by acetazolamide (175). The same seems to be true for AE2 and AE3, which share the same CAII-binding motif, and this might explain why basolateral AE2 activity in mouse salivary acinar cells is markedly inhibited by acetazolamide (176). A similar interaction occurs between the anion exchangers and CAIV on the exterior surface of the cell (177), and this has led to the concept of a HCO_3^- transport "metabolon" in which the close proximity of CAII and CAIV facilitates the anion exchanger process.

What is the significance of this for pancreatic duct cells? The only AE isoform present in these cells seems to be a relatively inactive AE2 expressed at the ba-solateral membrane. The possibility of an interaction between CAII and NBC1, which has a homologous CAII-binding motif, can be excluded because the phos-phorylated NBC1 found at the basolateral membrane of pancreatic duct cells is not inhibited by acetazolamide and appears to be unable to associate with CAII (178). Another possibility is that a direct physical interaction with the membrane-anchored CAIV isozyme is needed for full NBC1 activity (179), but CAIV is ex-pressed at the apical membrane in human pancreatic ducts (170), whereas NBC1 is located at the basolateral membrane.

Although CAIV is expressed in CFPAC-1 cells, which lack functional CFTR, the trafficking of CAIV to the apical membrane is impaired. The fact that it is restored by transfection with wild-type CFTR (180) raises the possibility that there is an interaction of CAIV with CFTR, perhaps as part of the CFTR-NERF-SLC26A6/SLC26A3 complex at the apical membrane. Although SLC26A3 does not have the CAII-binding motif, its activity in HEK293 cells is still inhibited 50% by acetazolamide so it appears to require functional cytosolic CAII activity (181).

CROSSTALK

A crucial property of any secretory epithelium is that the efflux of ions across the apical membrane should be balanced exactly by uptake at the basolateral membrane during sustained secretion. Pancreatic duct cells are remarkable in that the onset of secretin-evoked HCO_3^- secretion is not accompanied by any significant change in pH_i, at least in the guinea pig (8), rat (69), and pig (51). This suggests that the increase in activity of the basolateral NHE, pNBC1, or V-ATPase is not simply because of a fall in pH_i.

Given that several of the ductal transporters are electrogenic, it is possible that membrane potential might be an important link. In rat ducts microperfused with a HCO_3^--free solution, secretin and other agonists cause a 30-mV depolarization from a resting potential of about -60 mV, regardless of whether basolateral HCO_3^- is present (3, 80, 82). In ducts prepared by collagenase digestion and not microperfused, smaller depolarizations of around 20 mV were obtained with secretin and carbachol (182) and also with β-adrenoceptor stimulation (183). This agonist-evoked depolarization can be attributed to Cl^- efflux across the apical membrane through the CFTR or CaCC channels. It is occasionally preceded by a transient hyperpolarization, probably owing to the basolateral K^+ conductance (81). This hyperpolarization is particularly evident following basolateral ATP stimulation (83) but is absent following carbachol, which evokes a similar Ca^{2+} response (184).

The absence of any change in pH_i in guinea pig ducts stimulated with secretin led us to propose that the balance between HCO_3^- efflux and uptake might result from electrical coupling (8). Thus electrogenic HCO_3^- efflux at the apical membrane would lead to a depolarization, as observed in rat ducts, and this would stimulate a compensatory increase in HCO_3^- uptake via the electrogenic pNBC1 at the basolateral membrane. A similar suggestion was made on the basis of experiments with Capan-1 cells (58). However, the membrane potential changes in microperfused guinea pig ducts are complex. With 25-mM HCO_3^- in the lumen and bath, stimulation with secretin or dbcAMP causes a transient hyperpolarization followed by a net depolarization of about 10 mV (59). The transient hyperpolarization results, at least in part, from stimulation of the electrogenic pNBC1 at the basolateral membrane, which actually precedes the depolarization caused by CFTR activation. Therefore, it must be triggered by some other factor, perhaps PKA (185). There is certainly clear evidence for cAMP stimulation of pNBC1 activity in mouse colonic crypt cells (186).

When stimulated guinea pig ducts are microperfused with 125-mM HCO_3^-, simulating the conditions that exist during maximal secretion, the membrane potential shifts back to a more negative value, close to the unstimulated value of -60 mV (59). The hyperpolarizing effect of high luminal HCO_3^- concentrations suggests that the apical HCO_3^- permeability increases relative to Cl^-, perhaps through closure of the CFTR channels by the high luminal HCO_3^- concentration (95) or a change in their selectivity (102, 103).

Although we have no information about cell volume changes in pancreatic ducts, changes in intracellular Cl^- concentration in other epithelia often reflect cell volume changes and are thought to have an important role in activating basolateral transporters (187). There are certainly substantial changes in $[Cl^-]_i$ when guinea pig ducts are stimulated with forskolin. (9). Under resting conditions, with 25-mM HCO_3^- in the lumen and bath, $[Cl^-]_i$ is approximately 30 mM. When the lumen is perfused with 125-mM HCO_3^- and the duct is stimulated with forskolin, simulating maximal secretion of a HCO_3^--rich fluid, $[Cl^-]_i$ drops to less than 10 mM. It remains to be seen whether the fall in intracellular Cl^- or an associated decrease in cell volume is an important trigger for the basolateral transporters. But it is interesting to note that a kinase involved in the regulation of Cl^- transport in the kidney, WNK4, is also expressed in the mouse pancreatic duct (188). WNK4 has powerful inhibitory effects on NKCC1 and SLC26A6 (CFEX) when coexpressed in oocytes. Its upstream regulation is not yet clear, but given its importance in the kidney it might also prove to be a regulator in the pancreatic duct.

REFINEMENT OF THE MODEL

To what extent do these recent discoveries help explain the secretion of 140-mM HCO_3^-? The presence of basolateral pNBC1 and V-ATPase activity does not obviously enhance the driving force for HCO_3^- secretion because pH_i is not significantly greater in species where these transporters are active. The fact that both transporters are electrogenic will help to maintain a large membrane potential but this could favor HCO_3^- and Cl^- secretion equally.

In an updated mathematical model that incorporates the pNBC1 and V-ATPase, raising the CFTR conductance and pNBC1 activity to simulate the effects of secretin generates a fluid containing 67-mM HCO_3^- (30). Interestingly, two thirds of the HCO_3^- is secreted via CFTR (assuming $pHCO_3^-/pCl^- = 0.2$) and only one third by the apical Cl^-/HCO_3^- exchanger. By increasing the activity of the apical exchanger relative to the basolateral exchanger, and by increasing the $pHCO_3^-/pCl^-$ ratio of CFTR to 0.4 (95), the model is capable of secreting no more than 120-mM HCO_3^-. This represents the point at which the apical AE is approaching equilibrium and almost all of the HCO_3^- comes through CFTR (30).

Two-Site Model

As a possible solution to this problem, it has been suggested that ductal secretion of 140-mM HCO_3^- is a two-stage process involving a division of function between proximal and distal ducts (30). Given a small volume of Cl^--rich fluid from the acini, the proximal ducts might produce a large volume of fluid with a HCO_3^- concentration approaching about 120 mM. The proposal is that this relatively high HCO_3^- concentration would inhibit the apical Cl^- permeability in the distal ducts (95) and that they would secrete a relatively small volume of HCO_3^--rich fluid, thus raising the final HCO_3^- concentration. However, if there is a large volume of fluid containing ~40-mM Cl^- arriving at the distal ducts, then they would also have to reabsorb a substantial amount of Cl^-, and it is not clear how this would be achieved.

The discovery that the apical anion exchanger may be SLC26A6 in the proximal ducts and perhaps SLC26A3 in the distal ducts raises the possibility that their asymmetry and electrogenicity might facilitate the production of a HCO_3^--rich secretion (133, 145). Making a few reasonable assumptions it is possible to calculate the maximum luminal HCO_3^- concentrations that could be generated by these exchangers (Figure 2). Taking approximate intracellular values from stimulated guinea pig duct cells (20-mM HCO_3^-, 5-mM Cl^-, and −60 mV) an SLC4 (AE) exchanger with a $HCO_3^-:Cl^-$ stoichiometry of 1:1 could generate a luminal HCO_3^- concentration of up to 128 mM. A 2:1 exchanger, perhaps SLC26A6, could do a little better and approach 136 mM, whereas a 1:2 exchanger, perhaps

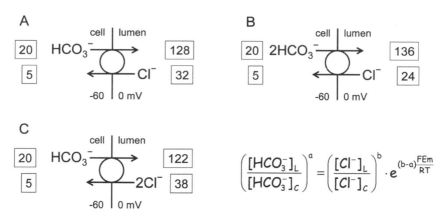

Figure 2 Predicted equilibrium conditions for apical membrane anion exchangers of alternative HCO_3^-:Cl^- stoichiometries: (*A*) 1:1 as for a member of the AE family; (*B*) 2:1 as proposed for SLC26A6 (133, 143); (*C*) 1:2 as proposed for SLC26A3 (133). In each case the intracellular HCO_3^- and Cl^- concentrations are set to 20 mM and 5 mM, respectively, and the membrane potential (E_m) is −60 mV, reflecting measurements made in guinea pig pancreatic duct cells during maximal stimulation and exposure to high luminal HCO_3^- concentrations (9, 59). The luminal concentrations shown are those predicted by the equation at the bottom right for an exchanger with HCO_3^-:Cl^- stoichiometry of a:b. It is assumed that the secretion is isotonic and therefore that the sum of the luminal Cl^- and HCO_3^- concentrations should be 160 mM.

SLC26A3, would do slightly worse and reach only 122 mM. But it should be noted that these are equilibrium values, so the exchangers would run extremely slowly as these concentrations were approached. Above these values, the exchangers would reverse their direction and start to reabsorb HCO_3^- rather than secrete it. Clearly, the proposed stoichiometries of the SLC26 exchangers do not help a great deal. In particular, the expression of SLC26A3 in the distal ducts would certainly not help to raise the luminal HCO_3^- to 140 mM as has been suggested (133, 145).

Single-Site Model

Isolated interlobular ducts from guinea pig pancreas are capable of secreting a HCO_3^--rich fluid (~140 mM) into their own lumen in the nominal absence of Cl^- (13). It may therefore be more appropriate to seek a single-site model for HCO_3^- secretion that does not depend upon sequential modification of the fluid as it flows through the ductal system. This does not preclude a sequential change with time as the duct goes from its resting state to one of maximal stimulation. Indeed, the effects of the rising HCO_3^- and falling Cl^- concentrations on the apical transporters may be vital in achieving maximal HCO_3^- output.

We therefore propose the following sequence of events as a guinea pig duct cell undergoes the transition from its resting state to one of maximal HCO_3^- output (59, 189). Initially, the apical membrane faces a luminal fluid containing a moderate

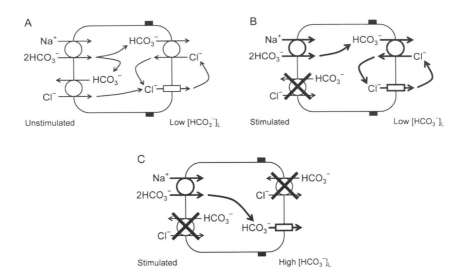

Figure 3   A single-site model to account for the secretion of HCO_3^--rich fluid by pancreatic duct cells. (*A*) Spontaneous secretion by an unstimulated duct cell; (*B*) early stage of secretin-evoked secretion where the luminal Cl^- concentration is still sufficient to support apical HCO_3^- efflux by anion exchange; and (*C*) later stage of stimulated secretion where the luminal HCO_3^- concentration has reached a higher level and HCO_3^- leaves the cell predominantly via an apical channel, possibly CFTR.

amount of Cl^-. Spontaneous secretion (Figure 3*A*) is driven by HCO_3^- uptake via pNBC1 and NHE1 at the basolateral membrane, and there is also Cl^- uptake via the basolateral anion exchanger. Cl^- enters the lumen predominantly via CFTR, and HCO_3^- enters by anion exchange. The result is a spontaneous mixed secretion of Cl^- and HCO_3^-.

Stimulation with secretin (Figure 3*B*) raises the CFTR conductance, stimulates the apical anion exchanger and basolateral pNBC1, and may also inhibit the basolateral anion exchanger. This favors a HCO_3^--rich secretion, which rapidly raises the HCO_3^- concentration in the lumen. The secretion of HCO_3^- will be slightly enhanced if the apical anion exchanger has a 2:1 stoichiometry. As the luminal HCO_3^- rises, several changes occur (Figure 3*C*). The lack of Cl^- in the lumen leads to a depletion of intracellular Cl^-, and it may also switch CFTR to a more HCO_3^--permeable state. The high luminal HCO_3^- concentration may inhibit the luminal anion exchanger, thus preventing its reversal and the consequent reabsorption of HCO_3^-. Alternatively, if it is an SLC26 exchanger with a 2:1 stoichiometry, it may become inactive simply because it is close to equilibrium (Figure 2*B*). Finally, the combination of a declining driving force for Cl^-, combined with a sustained driving force for HCO_3^-, results in a sustained HCO_3^--rich secretion containing very little Cl^-. In this situation virtually all of the HCO_3^- efflux occurs via CFTR.

The main problems with this model are (*a*) the possibility that CFTR is blocked by the high luminal HCO_3^- concentration (95) and (*b*) the observation that the

CFTR mutants associated with pancreatic insufficiency are those that fail to activate the apical anion exchanger, rather than those that lack channel activity (109). However, the unusual properties of the SLC26 exchangers suggest another possibility. If the apical SLC26A6 exchanger operates with a 2:1 stoichiometry and binds HCO_3^- as well as Cl^- (141, 143), then at high luminal HCO_3^- concentrations it might exchange two outwardly directed HCO_3^- ions for one inwardly directed HCO_3^- ion. In this situation SLC26A6 would effectively operate as a HCO_3^- uniport and thus provide an alternative conductive pathway for HCO_3^- efflux across the apical membrane.

CONCLUSIONS

What are the main lessons to be learned from the progress of the past 10 years? First, it is clear that isolated proximal ducts are capable of secreting high HCO_3^- concentrations without any need for secondary modification in the distal ducts. The basolateral transporters NHE1, pNBC1, and V-ATPase can all contribute to HCO_3^- uptake but their relative importance varies between species. The HCO_3^- content of pancreatic juice may be higher in some species simply because they lack Cl^- uptake pathways at the basolateral membrane.

The apical exit pathway for HCO_3^- during maximal secretion is still unclear: A 1:1 Cl^-/HCO_3^- exchanger cannot secrete HCO_3^- into 140-mM HCO_3^-; the electrogenicity of the SLC26 exchangers does not really help unless they operate in HCO_3^- uniport mode; and the CFTR Cl^- channels may be shut down by the high luminal HCO_3^- concentration. CFTR evidently has other functions beyond acting simply as a Cl^- channel: It physically interacts with and activates other transporters and may itself act as a HCO_3^- channel. The role of CA also extends beyond simply catalyzing the hydration of CO_2. CA isozymes appear to be required for optimal activity of other HCO_3^- transporters, several of which seem to form complexes with each other through PDZ interactions. Further investigation of the interactions within these apical transporter complexes should eventually reveal how HCO_3^- crosses the apical membrane.

The *Annual Review of Physiology* is online at http://physiol.annualreviews.org

LITERATURE CITED

1. Go VLW, DiMagno EP, Gardner JD, Lebenthal E, Reber HA, Scheele GA, eds. 1993. *The Pancreas: Biology, Pathobiology, and Disease*. New York: Raven
2. Johnson LR, ed. 1994. *Physiology of the Gastrointestinal Tract*. New York: Raven
3. Novak I, Greger R. 1988. Properties of the luminal membrane of isolated perfused rat

pancreatic ducts: effect of cyclic AMP and blockers of chloride transport. *Pflügers Arch.* 411:546–53
4. Novak I, Greger R. 1988. Electrophysiological study of transport systems in isolated perfused pancreatic ducts: properties of the basolateral membrane. *Pflügers Arch.* 411:58–68

5. Gray MA, Greenwell JR, Argent BE. 1988. Secretin-regulated chloride channels on the apical plasma membrane of pancreatic duct cells. *J. Membr. Biol.* 105: 131–42

6. Stuenkel EL, Machen TE, Williams JA. 1988. pH regulatory mechanisms in rat pancreatic ductal cells. *Am. J. Physiol. Gastrointest. Liver Physiol.* 54:G925–30

7. Case RM, Argent BE. 1993. Pancreatic duct secretion: control and mechanisms of transport. In *The Pancreas: Biology, Pathobiology, and Disease*, ed. VLW Go, EP DiMagno, JD Gardner, E Lebenthal, HA Reber, GA Scheele, pp. 301–50. New York: Raven

8. Ishiguro H, Steward MC, Lindsay ARG, Case RM. 1996. Accumulation of intracellular HCO_3^- by Na^+-HCO_3^- cotransport in interlobular ducts from guinea-pig pancreas. *J. Physiol.* 495:169–78

9. Ishiguro H, Naruse S, Kitagawa M, Mabuchi T, Kondo T, et al. 2002. Chloride transport in microperfused interlobular ducts isolated from guinea-pig pancreas. *J. Physiol.* 539:175–89

10. Zhao H, Star RA, Muallem S. 1994. Membrane localization of H^+ and HCO_3^- transporters in the rat pancreatic duct. *J. Gen. Physiol.* 104:57–85

11. Ishiguro H, Steward MC, Wilson RW, Case RM. 1996. Bicarbonate secretion in interlobular ducts from guinea-pig pancreas. *J. Physiol.* 495:179–91

12. Argent BE, Arkle S, Cullen MJ, Green R. 1986. Morphological, biochemical and secretory studies on rat pancreatic ducts maintained in tissue culture. *Q. J. Exp. Physiol.* 71:633–48

13. Ishiguro H, Naruse S, Steward MC, Kitagawa M, Ko SBH, et al. 1998. Fluid secretion in interlobular ducts isolated from guinea-pig pancreas. *J. Physiol.* 511:407–22

14. Fernández-Salazar MP, Pascua P, Calvo JJ, López MA, Case RM, et al. 2004. Basolateral anion transport mechanisms underlying fluid secretion by mouse, rat and guinea-pig pancreatic ducts. *J. Physiol.* 556:415–28

15. Cheng HS, Leung PY, Chew SBC, Leung PS, Lam SY, et al. 1998. Concurrent and independent HCO_3^- and Cl^- secretion in a human pancreatic duct cell line (CAPAN-1). *J. Membr. Biol.* 164:155–67

16. Chan HC, Cheung WT, Leung PY, Wu LJ, Chew SBC, et al. 1996. Purinergic regulation of anion secretion by cystic fibrosis pancreatic duct cells. *Am. J. Physiol. Cell Physiol.* 271:C469–77

17. Kyriazis AP, Kyriazis AA, Scarpelli DG, Fogh J, Rao MS, Lepera R. 1982. Human pancreatic adenocarcinoma line CAPAN 1 in tissue culture and the nude mouse. *Am. J. Pathol.* 106:250–60

18. Cotton CU, Al-Nakkash L. 1997. Isolation and culture of bovine pancreatic duct epithelial cells. *Am. J. Physiol. Gastrointest. Liver Physiol.* 35:G1328–37

19. Cotton CU. 1998. Ion-transport properties of cultured bovine pancreatic duct epithelial cells. *Pancreas* 17:247–55

20. Nguyen TD, Koh DS, Moody MW, Fox NR, Savard CE, et al. 1997. Characterization of two distinct chloride channels in cultured dog pancreatic duct epithelial cells. *Am. J. Physiol. Gastrointest. Liver Physiol.* 272:G172–80

21. Nguyen TD, Moody MW, Savard CE, Lee SP. 1998. Secretory effects of ATP on nontransformed dog pancreatic duct epithelial cells. *Am. J. Physiol. Gastrointest. Liver Physiol.* 275:G104–13

22. Takacs-Jarrett M, Sweeney WE, Avner ED, Cotton CU. 2001. Generation and phenotype of cell lines derived from CF and non-CF mice that carry the H-2K^b-tsA58 transgene. *Am. J. Physiol. Cell Physiol.* 280:C228–36

23. Lee MG, Wigley WC, Zeng WZ, Noel LE, Marino CR, et al. 1999. Regulation of Cl^-/HCO_3^- exchange by cystic fibrosis transmembrane conductance regulator expressed in NIH 3T3 and HEK 293 cells. *J. Biol. Chem.* 274:3414–21

24. Sewell WA, Young JA. 1975. Secretion of electrolytes by the pancreas of the anaesthetized rat. *J. Physiol.* 252:379–96

25. Padfield PJ, Garner A, Case RM. 1989. Patterns of pancreatic secretion in the anaesthetised guinea pig following stimulation with secretin, cholecystokinin octapeptide, or bombesin. *Pancreas* 4:204–9

26. Lee MG, Ahn W, Choi JY, Luo X, Seo JT, et al. 2000. Na^+-dependent transporters mediate HCO_3^- salvage across the luminal membrane of the main pancreatic duct. *J. Clin. Invest.* 105:1651–58

27. Lee MG, Choi JY, Luo X, Strickland E, Thomas PJ, Muallem S. 1999. Cystic fibrosis transmembrane conductance regulator regulates luminal Cl^-/HCO_3^- exchange in mouse submandibular and pancreatic ducts. *J. Biol. Chem.* 274:14670–77

28. Mangos JA, McSherry NR, Nousia-Arvanitakis S, Irwin K. 1973. Secretion and transductal fluxes of ions in exocrine glands of the mouse. *Am. J. Physiol.* 225:18–24

29. Lightwood R, Reber HA. 1977. Micropuncture study of pancreatic secretion in the cat. *Gastroenterology* 72:61–66

30. Sohma Y, Gray MA, Imai Y, Argent BE. 2000. HCO_3^- transport in a mathematical model of the pancreatic ductal epithelium. *J. Membr. Biol.* 176:77–100

31. Burghardt B, Elkjær ML, Kwon TH, Rácz GZ, Varga G, et al. 2003. Distribution of aquaporin water channels AQP1 and AQP5 in the ductal system of human pancreas. *Gut* 52:1008–16

32. Marino CR, Matovcik LM, Gorelick FS, Cohn JA. 1991. Localization of the cystic fibrosis transmembrane conductance regulator in pancreas. *J. Clin. Invest.* 88:712–16

33. Crawford I, Maloney PC, Zeitlin PL, Guggino WB, Hyde SC, et al. 1991. Immunocytochemical localization of the cystic fibrosis gene product CFTR. *Proc. Natl. Acad. Sci. USA* 88:9262–66

34. Satoh H, Moriyama N, Hara C, Yamada H, Horita S, et al. 2003. Localization of Na^+-HCO_3^- cotransporter (NBC-1) variants in rat and human pancreas. *Am. J. Physiol. Cell Physiol.* 284:C729–37

35. Marino CR, Jeanes V, Boron WF, Schmitt BM. 1999. Expression and distribution of the Na^+-HCO_3^- cotransporter in human pancreas. *Am. J. Physiol. Gastrointest. Liver Physiol.* 277:G487–94

36. Takahashi-Iwanaga H, Yanase H, Orikasa M. 2003. Reexamination of the fine structure of the pancreatic duct system in the rat: a proposal of morphological segmentation. *Biomed. Res.* 24:77–87

37. Ko SBH, Naruse S, Kitagawa M, Ishiguro H, Furuya S, et al. 2002. Aquaporins in rat pancreatic interlobular ducts. *Am. J. Physiol. Gastrointest. Liver Physiol.* 282:G324–31

38. Furuya S, Naruse S, Ko SBH, Ishiguro H, Yoshikawa T, Hayakawa T. 2002. Distribution of aquaporin 1 in the rat pancreatic duct system examined with light- and electron-microscopic immunohistochemistry. *Cell Tissue Res.* 308:75–86

39. Zeng WZ, Lee MG, Yan M, Diaz J, Benjamin I, et al. 1997. Immuno and functional characterization of CFTR in submandibular and pancreatic acinar and duct cells. *Am. J. Physiol. Cell Physiol.* 273:C442–55

40. Swanson CH, Solomon AK. 1975. Micropuncture analysis of the cellular mechanisms of electrolyte secretion by the in vitro rabbit pancreas. *J. Gen. Physiol.* 65:22–45

41. de Ondarza J, Hootman SR. 1997. Confocal microscopic analysis of intracellular pH regulation in isolated guinea pig pancreatic ducts. *Am. J. Physiol.* 272:G124–G34

42. Ishiguro H, Naruse S, Kitagawa M, Suzuki A, Yamamoto A, et al. 2000. CO_2 permeability and bicarbonate transport in microperfused interlobular ducts isolated from guinea-pig pancreas. *J. Physiol.* 528:305–15

43. Marteau C, Silviani V, Ducroc R, Crotte C, Gerolami A. 1995. Evidence for apical Na$^+$/H$^+$ exchanger in bovine main pancreatic duct. *Digest. Dis. Sci.* 40:2336–40
44. Orlowski J, Grinstein S. 2004. Diversity of the mammalian sodium/proton exchanger SLC9 gene family. *Pflügers Arch.* 447:549–65
45. Roussa E, Alper SL, Thevenod F. 2001. Immunolocalization of anion exchanger AE2, Na$^+$/H$^+$ exchangers NHE1 and NHE4, and vacuolar type H$^+$-ATPase in rat pancreas. *J. Histochem. Cytochem.* 49:463–74
46. Kuijpers GAJ, Van Nooy IGP, De Pont JJHHM, Bonting SL. 1984. The mechanism of fluid secretion in the rabbit pancreas studied by means of various inhibitors. *Biochim. Biophys. Acta* 778:324–31
47. Grotmol T, Buanes T, Brors O, Raeder MG. 1986. Lack of effect of amiloride, furosemide, bumetanide and triamterene on pancreatic NaHCO$_3$ secretion in pigs. *Acta Physiol. Scand.* 126:593–600
48. Veel T, Villanger O, Holthe MR, Cragoe EJ, Raeder MG. 1992. Na$^+$-H$^+$ exchange is not important for pancreatic HCO$_3^-$ secretion in the pig. *Acta Physiol. Scand.* 144:239–46
49. Evans LAR, Young JA. 1985. The effect of transport blockers on pancreatic ductal secretion. *Proc. Austral. Physiol. Pharmacol. Soc.* 16:98P (Abstr.)
50. Wizemann V, Schulz I. 1973. Influence of amphotericin, amiloride, ionophores and 2,4-dinitrophenol on the secretion of the isolated cat's pancreas. *Pflügers Arch.* 339:317–38
51. Villanger O, Veel T, Raeder MG. 1995. Secretin causes H$^+$/HCO$_3^-$ secretion from pig pancreatic ductules by vacuolar-type H$^+$-adenosine triphosphatase. *Gastroenterology* 108:850–59
52. Raeder MG. 1992. The origin of and subcellular mechanisms causing pancreatic bicarbonate secretion. *Gastroenterology* 103:1674–84
53. Buanes T, Grotmol T, Landsverk T, Raeder MG. 1987. Ultrastructure of pancreatic duct cells at secretory rest and during secretin-dependent NaHCO$_3$ secretion. *Acta Physiol. Scand.* 131:55–62
54. Veel T, Buanes T, Grotmol T, Ostensen J, Raeder MG. 1991. Secretin dissipates red acridine orange fluorescence from pancreatic duct epithelium. *Acta Physiol. Scand.* 141:221–26
55. Veel T, Buanes T, Engeland E, Raeder MG. 1990. Colchicine inhibits the effects of secretin on pancreatic duct cell tubulovesicles and HCO$_3^-$ secretion in the pig. *Acta Physiol. Scand.* 138:487–95
56. Nishi T, Forgac M. 2002. The vacuolar (H$^+$)-ATPases: nature's most versatile proton pumps. *Nat. Rev. Mol. Cell Biol.* 3:94–103
57. Grotmol T, Buanes T, Raeder MG. 1986. N,N'-dicyclohexylcarbodiimide (DCCD) reduces pancreatic NaHCO$_3$ secretion without changing pancreatic tissue ATP levels. *Acta Physiol. Scand.* 128:547–54
58. Shumaker H, Amlal H, Frizzell R, Ulrich CD, Soleimani M. 1999. CFTR drives Na$^+$-nHCO$_3^-$ cotransport in pancreatic duct cells: a basis for defective HCO$_3^-$ secretion in CF. *Am. J. Physiol. Cell Physiol.* 276:C16–25
59. Ishiguro H, Steward MC, Sohma Y, Kubota T, Kitagawa M, et al. 2002. Membrane potential and bicarbonate secretion in isolated interlobular ducts from guinea-pig pancreas. *J. Gen. Physiol.* 120:617–28
60. Romero MF, Fulton CM, Boron WF. 2004. The SLC4 family of HCO$_3^-$ transporters. *Pflügers Arch.* 447:495–509
61. Romero MF, Hediger MA, Boulpaep EL, Boron WF. 1997. Expression cloning and characterization of a renal electrogenic Na$^+$/HCO$_3^-$ cotransporter. *Nature* 387:409–13
62. Abuladze N, Lee I, Newman D, Hwang J, Boorer K, et al. 1998. Molecular cloning, chromosomal localization, tissue distribution, and functional expression of the

human pancreatic sodium bicarbonate co-transporter. *J. Biol. Chem.* 273:17689–95

63. Thévenod F, Roussa E, Schmitt BM, Romero MF. 1999. Cloning and immunolocalization of a rat pancreatic Na$^+$ bicarbonate cotransporter. *Biochem. Biophys. Res. Commun.* 264:291–98

64. Choi I, Romero MF, Khandoudi N, Bril A, Boron WF. 1999. Cloning and characterization of a human electrogenic Na$^+$-HCO$_3^-$ cotransporter isoform (hhNBC). *Am. J. Physiol. Cell Physiol.* 276:C576–84

65. Roussa E, Nastainczyk W, Thevenod F. 2004. Differential expression of electrogenic NBC1 (SLC4A4) variants in rat kidney and pancreas. *Biochem. Biophys. Res. Commun.* 314:382–89

66. Gross E, Abuladze N, Pushkin A, Kurtz I, Cotton CU. 2001. The stoichiometry of the electrogenic sodium bicarbonate cotransporter pNBC1 in mouse pancreatic duct cells is 2 HCO$_3^-$:1 Na$^+$. *J. Physiol.* 531:375–82

67. Gross E, Hawkins K, Abuladze N, Pushkin A, Cotton CU, et al. 2001. The stoichiometry of the electrogenic sodium bicarbonate cotransporter NBC1 is cell-type dependent. *J. Physiol.* 531:597–603

68. Gross E, Hawkins K, Pushkin A, Sassani P, Dukkipati R, et al. 2001. Phosphorylation of Ser^{982} in the sodium bicarbonate cotransporter kNBC1 shifts the HCO$_3^-$:Na$^+$ stoichiometry from 3:1 to 2:1 in murine proximal tubule cells. *J. Physiol.* 537:659–65

69. Novak I, Christoffersen BC. 2001. Secretin stimulates HCO$_3^-$ and acetate efflux but not Na$^+$/HCO$_3^-$ uptake in rat pancreatic ducts. *Pflügers Arch.* 441:761–71

70. Igarashi T, Inatomi J, Sekine T, Cha SH, Kanai Y, et al. 1999. Mutations in SLC4A4 cause permanent isolated proximal renal tubular acidosis with ocular abnormalities. *Nat. Genet.* 23:264–66

71. Hyde K, Harrison D, Hollingsworth MA, Harris A. 1999. Chloride-bicarbonate exchangers in the human fetal pancreas. *Biochem. Biophys. Res. Commun.* 263:315–21

72. Greeley T, Shumaker H, Wang Z, Schweinfest CW, Soleimani M. 2001. Downregulated in adenoma and putative anion transporter are regulated by CFTR in cultured pancreatic duct cells. *Am. J. Physiol. Gastrointest. Liver Physiol.* 281:G1301–8

73. Ashton N, Argent BE, Green R. 1991. Characteristics of fluid secretion from isolated rat pancreatic ducts stimulated with secretin and bombesin. *J. Physiol.* 435:533–46

74. Kanno T, Yamamoto M. 1977. Differentiation between the calcium-dependent effects of cholecystokinin-pancreozymin and the bicarbonate-dependent effects of secretin in exocrine secretion of the rat pancreas. *J. Physiol.* 264:787–99

75. Petersen OH, Ueda N. 1977. Secretion of fluid and amylase in the perfused rat pancreas. *J. Physiol.* 264:819–35

76. Shumaker H, Soleimani M. 1999. CFTR upregulates the expression of the basolateral Na$^+$-K$^+$-2Cl$^-$ cotransporter in cultured pancreatic duct cells. *Am. J. Physiol. Cell Physiol.* 277:C1100–10

77. Hebert SC, Mount DB, Gamba G. 2004. Molecular physiology of cation-coupled Cl$^-$ cotransport: the SLC12 family. *Pflügers Arch.* 447:580–93

78. Gillen CM, Brill S, Payne JA, Forbush B. 1996. Molecular cloning and functional expression of the K-Cl cotransporter from rabbit, rat, and human—a new member of the cation-chloride cotransporter family. *J. Biol. Chem.* 271:16237–44

79. Roussa E, Shmukler BE, Wilhelm S, Casula S, Stuart-Tilley AK, et al. 2002. Immunolocalization of potassium-chloride cotransporter polypeptides in rat exocrine glands. *Histochem. Cell Biol.* 117:335–44

80. Novak I, Greger R. 1991. Effect of bicarbonate on potassium conductance of isolated perfused rat pancreatic ducts. *Pflügers Arch.* 419:76–83

81. Novak I, Pahl C. 1993. Effect of secretin and inhibitors of $HCO_3^-H^+$ transport on the membrane voltage of rat pancreatic duct cells. *Pflügers Arch.* 425:272–79

82. Pahl C, Novak I. 1993. Effect of vasoactive intestinal peptide, carbachol and other agonists on the membrane voltage of pancreatic duct cells. *Pflügers Arch.* 424:315–20

83. Hug M, Pahl C, Novak I. 1994. Effect of ATP, carbachol and other agonists on intracellular calcium activity and membrane voltage of pancreatic ducts. *Pflügers Arch.* 426:412–18

84. Hede SE, Amstrup J, Christoffersen BC, Novak I. 1999. Purinoceptors evoke different electrophysiological responses in pancreatic ducts—P2Y inhibits K^+ conductance, and P2X stimulates cation conductance. *J. Biol. Chem.* 274:31784–91

85. Gray MA, Greenwell JR, Garton AJ, Argent BE. 1990. Regulation of maxi-K^+ channels on pancreatic duct cells by cyclic AMP-dependent phosphorylation. *J. Membr. Biol.* 115:203–15

86. Argent BE, Gray MA. 1997. Regulation and formation of fluid and electrolyte secretions by pancreatic ductal epithelium. In *Biliary and Pancreatic Ductal Epithelia: Pathobiology and Pathophysiology*, ed. AE Sirica, DS Longnecker, pp. 349–77. New York: Dekker

87. Gray MA, Argent BE. 1990. Nonselective cation channel on pancreatic duct cells. *Biochim. Biophys. Acta* 1029:33–42

88. Madden ME, Sarras MP. 1987. Distribution of Na^+,K^+-ATPase in rat exocrine pancreas as monitored by K^+-NPPase cytochemistry and [^3H]-ouabain binding: a plasma protein found primarily to be ductal cell associated. *J. Histochem. Cytochem.* 35:1365–74

89. Bundgaard M, Moller M, Poulsen JH. 1981. Localization of sodium pump sites in cat pancreas. *J. Physiol.* 313:405–14

90. Smith ZDJ, Caplan MJ, Forbush B, Jamieson JD. 1987. Monoclonal antibody localization of Na^+-K^+-ATPase in the exocrine pancreas and parotid of the dog. *Am. J. Physiol. Gastrointest. Liver Physiol.* 253:G99–109

91. Gray MA, Pollard CE, Harris A, Coleman L, Greenwell JR, Argent BE. 1990. Anion selectivity and block of the small-conductance chloride channel on pancreatic duct cells. *Am. J. Physiol. Cell Physiol.* 259:C752–61

92. Gray MA, Plant S, Argent BE. 1993. cAMP-regulated whole cell chloride currents in pancreatic duct cells. *Am. J. Physiol. Cell Physiol.* 264:C591–602

93. Gray MA, Harris A, Coleman L, Greenwell JR, Argent BE. 1989. Two types of chloride channel on duct cells cultured from human fetal pancreas. *Am. J. Physiol. Cell Physiol.* 257:C240–51

94. Gray MA, Winpenny JP, Porteous DJ, Dorin JR, Argent BE. 1994. CFTR and calcium-activated chloride currents in pancreatic duct cells of a transgenic CF mouse. *Am. J. Physiol. Cell Physiol.* 266:C213–21

95. O'Reilly CM, Winpenny JP, Argent BE, Gray MA. 2000. Cystic fibrosis transmembrane conductance regulator currents in guinea pig pancreatic duct cells: inhibition by bicarbonate ions. *Gastroenterology* 118:1187–96

96. Al-Nakkash L, Cotton CU. 1997. Bovine pancreatic duct cells express cAMP- and Ca^{2+}-activated apical membrane Cl^- conductances. *Am. J. Physiol. Gastrointest. Liver Physiol.* 273:G204–16

97. Becq F, Hollande E, Gola M. 1993. Phosphorylation-regulated low-conductance Cl^- channels in a human pancreatic duct cell line. *Pflügers Arch.* 425:1–8

98. Illek B, Yankaskas JR, Machen TE. 1997. cAMP and genistein stimulate HCO_3^- conductance through CFTR in human airway epithelia. *Am. J. Physiol. Lung Cell Mol. Physiol.* 272:L752–61

99. Illek B, Tam AWK, Fischer H, Machen TE. 1999. Anion selectivity of apical

membrane conductance of Calu 3 human airway epithelium. *Pflügers Arch.* 437:812–22

100. Linsdell P, Tabcharani JA, Rommens JM, Hou YX, Chang XB, et al. 1997. Permeability of wild-type and mutant cystic fibrosis transmembrane conductance regulator chloride channels to polyatomic anions. *J. Gen. Physiol.* 110:355–64

101. Poulsen JH, Fischer H, Illek B, Machen TE. 1994. Bicarbonate conductance and pH regulatory capability of cystic fibrosis transmembrane conductance regulator. *Proc. Natl. Acad. Sci. USA* 91:5340–44

102. Reddy MM, Quinton PM. 2003. Control of dynamic CFTR selectivity by glutamate and ATP in epithelial cells. *Nature* 423:756–60

103. Shcheynikov N, Kim KH, Kim KM, Dorwart MR, Ko SBH, et al. 2004. Dynamic control of cystic fibrosis transmembrane conductance regulator Cl^-/HCO_3^- selectivity by external Cl^-. *J. Biol. Chem.* 279:21857–65

104. Devor DC, Singh AK, Lambert LC, DeLuca A, Frizzell RA, Bridges RJ. 1999. Bicarbonate and chloride secretion in Calu-3 human airway epithelial cells. *J. Gen. Physiol.* 113:743–60

105. Paradiso AM, Coakley RD, Boucher RC. 2003. Polarized distribution of HCO_3^- transport in human normal and cystic fibrosis nasal epithelia. *J. Physiol.* 548:203–18

106. Spiegel S, Phillipper M, Rossmann H, Riederer B, Gregor M, Seidler U. 2003. Independence of apical Cl^-/HCO_3^- exchange and anion conductance in duodenal HCO_3^- secretion. *Am. J. Physiol. Gastrointest. Liver Physiol.* 285:G887–97

107. Becq F, Fanjul M, Mahieu I, Berger Z, Gola M, Hollande E. 1992. Anion channels in a human pancreatic cancer cell line (Capan-1) of ductal origin. *Pflügers Arch.* 420:46–53

108. Mahieu I, Becq F, Wolfensberger T, Gola M, Carter N, Hollande E. 1994. The expression of carbonic anhydrase II and anhydrase IV in the human pancreatic cancer cell line (Capan-1) is associated with bicarbonate ion channels. *Biol. Cell* 81:131–41

109. Choi JY, Muallem D, Kiselyov K, Lee MG, Thomas PJ, Muallem S. 2001. Aberrant CFTR-dependent HCO_3^- transport in mutations associated with cystic fibrosis. *Nature* 410:94–97

110. Winpenny JP, Verdon B, McAlroy HL, Colledge WH, Ratcliff R, et al. 1995. Calcium-activated chloride conductance is not increased in pancreatic duct cells of CF mice. *Pflügers Arch.* 430:26–33

111. Ratcliff R, Evans MJ, Cuthbert AW, MacVinish LJ, Foster D, et al. 1993. Production of a severe cystic fibrosis mutation in mice by gene targeting. *Nat. Genet.* 4:35–41

112. Gray MA, Winpenny JP, Verdon B, O'Reilly CM, Argent BE. 2002. Properties and role of calcium-activated chloride channels in pancreatic duct cells. In *Calcium-Activated Chloride Channels*, ed. CM Fuller, pp. 231–56. San Diego: Academic

113. Winpenny JP, Harris A, Hollingsworth MA, Argent BE, Gray MA. 1998. Calcium-activated chloride conductance in a pancreatic adenocarcinoma cell line of ductal origin (HPAF) and in freshly isolated human pancreatic duct cells. *Pflügers Arch.* 435:796–803

114. Fong P, Argent BE, Guggino WB, Gray MA. 2003. Characterization of vectorial chloride transport pathways in the human pancreatic duct adenocarcinoma cell line HPAF. *Am. J. Physiol. Cell Physiol.* 285:C433–45

115. Ho MWY, Kaetzel MA, Armstrong DL, Shears SB. 2001. Regulation of a human chloride channel. *J. Biol. Chem.* 276:18673–80

116. Ashton N, Evans RL, Elliott AC, Green R, Argent BE. 1993. Regulation of fluid

secretion and intracellular messengers in isolated rat pancreatic ducts by acetylcholine. *J. Physiol.* 471:549–62

117. Ishiguro H, Naruse X, Kitagawa M, Hayakawa T, Case RM, Steward MC. 1999. Luminal ATP stimulates fluid and HCO$_3^-$ secretion in guinea-pig pancreatic duct. *J. Physiol.* 519:551–58

118. Bruce JIE, Yang XS, Ferguson CJ, Elliott AC, Steward MC, et al. 1999. Molecular and functional identification of a Ca^{2+} (polyvalent cation)-sensing receptor in rat pancreas. *J. Biol. Chem.* 274:20561–68

119. Yamamoto A, Ishiguro H, Ko SBH, Suzuki A, Wang Y, et al. 2003. Ethanol induces fluid hypersecretion from guinea-pig pancreatic duct cells. *J. Physiol.* 551:917–26

120. Suzuki A, Naruse S, Kitagawa M, Ishiguro H, Yoshikawa T, et al. 2001. 5-Hydroxytryptamine strongly inhibits fluid secretion in guinea-pig pancreatic duct cells. *J. Clin. Invest.* 108:749–56

121. Warth R, Greger R. 1993. The ion conductances of CFPAC-1 cells. *Cell Physiol. Biochem.* 3:2–16

122. Chan HC, Law SH, Leung PS, Fu LXM, Wong PYD. 1997. Angiotensin II receptor type I-regulated anion secretion in cystic fibrosis pancreatic duct cells. *J. Membr. Biol.* 156:241–49

123. Cheng HS, Wong WS, Chan KT, Wang XF, Wang ZD, Chan HC. 1999. Modulation of Ca^{2+}-dependent anion secretion by protein kinase C in normal and cystic fibrosis pancreatic duct cells. *Biochim. Biophys. Acta* 1418:31–38

124. Qu Z, Hartzell HC. 2000. Anion permeation in Ca^{2+}-activated Cl$^-$ channels. *J. Gen. Physiol.* 116:825–44

125. Zsembery A, Strazzabosco M, Graf J. 2000. Ca^{2+}-activated Cl$^-$ channels can substitute for CFTR in stimulation of pancreatic duct bicarbonate secretion. *FASEB J.* 14:2345–56

126. Case RM, Hotz J, Hutson D, Scratcherd T, Wynne RDA. 1979. Electrolyte secretion by the isolated cat pancreas during replacement of extracellular bicarbonate by organic anions and chloride by inorganic anions. *J. Physiol.* 286:563–76

127. Kuijpers GAJ, Van Nooy IGP, De Pont JJHHM, Bonting SL. 1984. Anion secretion by the isolated rabbit pancreas. *Biochim. Biophys. Acta* 774:269–76

128. Ishiguro H, Lindsay ARG, Steward MC, Case RM. 1995. Secretin stimulation of Cl$^-$-HCO$_3^-$ exchange in interlobular ducts isolated from guinea-pig pancreas. *J. Physiol.* 482:22P (Abstr.)

129. Hegyi P, Gray MA, Argent BE. 2003. Substance P inhibits bicarbonate secretion from guinea pig pancreatic ducts by modulating an anion exchanger. *Am. J. Physiol. Cell Physiol.* 285:C268–76

130. Ashton N, Argent BE, Green R. 1990. Effect of vasoactive intestinal peptide, bombesin and substance P on fluid secretion by isolated rat pancreatic ducts. *J. Physiol.* 427:471–82

131. Evans RL, Ashton N, Elliott AC, Green R, Argent BE. 1996. Interactions between secretin and acetylcholine in the regulation of fluid secretion by isolated rat pancreatic ducts. *J. Physiol.* 496:265–73

132. Namkung W, Lee JA, Ahn WI, Han WS, Kwon SW, et al. 2003. Ca^{2+} activates cystic fibrosis transmembrane conductance regulator- and Cl$^-$-dependent HCO$_3^-$ transport in pancreatic duct cells. *J. Biol. Chem.* 278:200–7

133. Ko SBH, Shcheynikov N, Choi JY, Luo X, Ishibashi K, et al. 2002. A molecular mechanism for aberrant CFTR-dependent HCO$_3^-$ transport in cystic fibrosis. *EMBO J.* 21:5662–72

134. Sohma Y, Gray MA, Imai Y, Argent BE. 1996. A mathematical model of the pancreatic ductal epithelium. *J. Membr. Biol.* 154:53–67

135. Mount DB, Romero MF. 2004. The SLC26 gene family of multifunctional anion exchangers. *Pflügers Arch.* 447:710–21

136. Elgavish A, Meezan E. 1992. Altered sulfate transport via anion exchange in CF-PAC is corrected by retrovirus-mediated CFTR gene transfer. *Am. J. Physiol. Cell Physiol.* 263:C176–86

137. Melvin JE, Park K, Richardson L, Schultheis PJ, Shull GE. 1999. Mouse down-regulated in adenoma (DRA) is an intestinal Cl^-/HCO_3^- exchanger and is up-regulated in colon of mice lacking the NHE3 Na^+/H^+ exchanger. *J. Biol. Chem.* 274:22855–61

138. Lohi H, Kujala M, Kerkela E, Saarialho-Kere U, Kestila M, Kere J. 2000. Mapping of five new putative anion transporter genes in human and characterization of SLC26A6, a candidate gene for pancreatic anion exchanger. *Genomics* 70:102–12

139. Waldegger S, Moschen I, Ramirez A, Smith RJH, Ayadi H, et al. 2001. Cloning and characterization of SLC26A6, a novel member of the solute carrier 26 gene family. *Genomics* 72:43–50

140. Knauf F, Yang C-L, Thomson RB, Mentone SA, Giebisch G, Aronson PS. 2001. Identification of a chloride-formate exchanger expressed on the brush border membrane of renal proximal tubule cells. *Proc. Natl. Acad. Sci. USA* 98:9425–30

141. Jiang Z, Grichtchenko II, Boron WF, Aronson PS. 2002. Specificity of anion exchange mediated by mouse Slc26a6. *J. Biol. Chem.* 277:33963–67

142. Wang ZH, Petrovic S, Mann E, Soleimani M. 2002. Identification of an apical Cl^-/HCO_3^- exchanger in the small intestine. *Am. J. Physiol. Gastrointest. Liver Physiol.* 282:G573–79

143. Xie Q, Welch R, Mercado A, Romero MF, Mount DB. 2002. Molecular characterization of the murine SLC26A6 anion exchanger: functional comparison with SLC26A1. *Am. J. Physiol. Renal Physiol.* 283:F826–38

144. Lohi H, Lamprecht G, Markovich D, Heil A, Kujala M, et al. 2003. Isoforms of SLC26A6 mediate anion transport and have functional PDZ interaction domains.

145. Ko SBH, Zeng WZ, Dorwart MR, Luo X, Kim KH, et al. 2004. Gating of CFTR by the STAS domain of SLC26 transporters. *Nat. Cell Biol.* 6:343–50

146. Gray MA. 2004. Bicarbonate secretion: it takes two to tango. *Nat. Cell Biol.* 6:292–94

147. Gerolami A, Marteau C, Matteo A, Sahel J, Portugal H, et al. 1989. Calcium carbonate saturation in human pancreatic juice: possible role of ductal H^+ secretion. *Gastroenterology* 96:881–84

148. Marteau C, Blanc G, Devaux MA, Portugal H, Gerolami A. 1993. Influence of pancreatic ducts on saturation of juice with calcium carbonate in dogs. *Digest. Dis. Sci.* 38:2090–97

149. Ahn W, Kim KH, Lee JA, Kim JY, Choi JY, et al. 2001. Regulatory interaction between the cystic fibrosis transmembrane conductance regulator and HCO_3^- salvage mechanisms in model systems and the mouse pancreatic duct. *J. Biol. Chem.* 276:17236–43

150. Park M, Ko SBH, Choi JY, Muallem G, Thomas PJ, et al. 2002. The cystic fibrosis transmembrane conductance regulator interacts with and regulates the activity of the HCO_3^- salvage transporter human Na^+-HCO_3^- cotransport isoform 3. *J. Biol. Chem.* 277:50503–9

151. Pushkin A, Abuladze N, Lee I, Newman D, Hwang J, Kurtz I. 1999. Cloning, tissue distribution, genomic organization, and functional characterization of NBC3, a new member of the sodium bicarbonate cotransporter family. *J. Biol. Chem.* 274:16569–75

152. Choi I, Aalkjaer C, Boulpaep EL, Boron WF. 2000. An electroneutral sodium/bicarbonate cotransporter NBCn1 and associated sodium channel. *Nature* 405:571–75

153. Luo X, Choi JY, Ko SBH, Pushkin A, Kurtz I, et al. 2001. HCO_3^- salvage mechanisms in the submandibular gland acinar

and duct cells. *J. Biol. Chem.* 276:9808–16

154. Buanes T, Grotmol T, Landsverk T, Ridderstrale Y, Raeder MG. 1986. Histochemical localization of carbonic anhydrase in the pig's exocrine pancreas. *Acta Physiol. Scand.* 128:437–44

155. Dyck WP, Hightower NC, Janowitz HD. 1972. Effect of acetazolamide on human pancreatic secretion. *Gastroenterology* 62:547–52

156. Dreiling DA, Janowitz HD, Halpern M. 1968. The effect of a carbonic anhydrase inhibitor, Diamox, on human pancreatic secretion. Implications on the mechanism of pancreatic secretion. *Gastroenterology* 54(Suppl.):765–67

157. Anand BS, Goodgame R, Graham DY. 1994. Pancreatic secretion in man: effect of fasting, drugs, pancreatic enzymes, and somatostatin. *Am. J. Gastroent.* 89:267–70

158. Case RM, Harper AA, Scratcherd T. 1969. The secretion of electrolytes and enzymes by the pancreas of the anaesthetized cat. *J. Physiol.* 201:335–48

159. Case RM, Scratcherd T, Wynne RDA. 1970. The origin and secretion of pancreatic juice bicarbonate. *J. Physiol.* 210:1–15

160. Ammar EM, Hutson D, Scratcherd T. 1987. Absence of a relationship between arterial pH and pancreatic bicarbonate secretion in the isolated perfused cat pancreas. *J. Physiol.* 388:495–504

161. Pak BH, Hong SS, Pak HK, Hong SK. 1966. Effects of acetazolamide and acid-base changes on biliary and pancreatic secretion. *Am. J. Physiol.* 210:624–28

162. Banks PA, Sum PT. 1971. Mode of action of acetazolamide on pancreatic exocrine secretion. *Arch. Surg.* 102:505–8

163. Raeder M, Mathisen O. 1982. Abolished relationship between pancreatic HCO_3^- secretion and arterial pH during carbonic anhydrase inhibition. *Acta Physiol. Scand.* 114:97–102

164. Schwartz GJ. 2002. Physiology and molecular biology of renal carbonic anhydrase. *J. Nephrol.* 15(Suppl. 5):S61–74

165. Nishimori I, Fujikawa-Adachi K, Onishi S, Hollingsworth MA. 1999. Carbonic anhydrase in human pancreas: hypotheses for the pathophysiological roles of CA isozymes. *Ann. NY Acad. Sci.* 880:5–16

166. Kumpulainen T, Jalovaara P. 1981. Immunohistochemical localization of carbonic anhydrase isoenzymes in the human pancreas. *Gastroenterology* 80:796–99

167. Parkkila S, Parkkila AK, Juvonen T, Rajaniemi H. 1994. Distribution of the carbonic anhydrase isoenzymes I, II, and VI in the human alimentary tract. *Gut* 35:646–50

168. Alvarez L, Fanjul M, Carter N, Hollande E. 2001. Carbonic anhydrase II associated with plasma membrane in a human pancreatic duct cell line (CAPAN-1). *J. Histochem. Cytochem.* 49:1045–53

169. Sly WS, Hu PY. 1995. Human carbonic anhydrases and carbonic anhydrase deficiencies. *Annu. Rev. Biochem.* 64:375–401

170. Fanjul M, Alvarez L, Salvador C, Gmyr V, Kerr-Conte J, et al. 2004. Evidence for a membrane carbonic anhydrase IV anchored by its C-terminal peptide in normal human pancreatic ductal cells. *Histochem. Cell Biol.* 121:91–99

171. Diaz E, Sandblom JP, Wistrand PJ. 1982. Selectivity properties of channels induced by a reconstituted membrane-bound carbonic anhydrase. *Acta Physiol. Scand.* 116:461–63

172. Widdas WF, Baker GF. 1995. The triphasic volume response of human red blood cells in low ionic strength media: demonstration of a special bicarbonate transport. *Cytobios* 81:135–58

173. Kivela AJ, Parkkila S, Saarnio J, Karttunen TJ, Kivela J, et al. 2000. Expression of transmembrane carbonic anhydrase isoenzymes IX and XII in normal human pancreas and pancreatic tumours. *Histochem. Cell Biol.* 114:197–204

174. Vince JW, Reitheimer RA. 1998. Carbonic anhydrase II binds to the carboxyl terminus of human band 3, the erythrocyte Cl^-/HCO_3^- exchanger. *J. Biol. Chem.* 273:28430–37

175. Sterling D, Reithmeier RAF, Casey JR. 2001. A transport metabolon. Functional interaction of carbonic anhydrase II and chloride/bicarbonate exchangers. *J. Biol. Chem.* 276:47886–94

176. Nguyen H-V, Stuart-Tilley A, Alper SL, Melvin JE. 2004. Cl^-/HCO_3^- exchange is acetazolamide sensitive and activated by a muscarinic receptor-induced $[Ca^{2+}]_i$ increase in salivary acinar cells. *Am. J. Physiol. Gastrointest. Liver Physiol.* 286:G312–20

177. Sterling D, Alvarez BV, Casey JR. 2002. The extracellular component of a transport metabolon. Extracellular loop 4 of the human AE1 Cl^-/HCO_3^- exchanger binds carbonic anhydrase IV. *J. Biol. Chem.* 277:25239–46

178. Gross E, Pushkin A, Abuladze N, Fedotoff O, Kurtz I. 2002. Regulation of the sodium bicarbonate cotransporter kNBC1 function: role of Asp^{986}, Asp^{988} and kNBC1-carbonic anhydrase II binding. *J. Physiol.* 544:679–85

179. Alvarez BV, Loiselle FB, Supuran CT, Schwartz GJ, Casey JR. 2003. Direct extracellular interaction between carbonic anhydrase IV and the human NBC1 sodium/bicarbonate co-transporter. *Biochemistry* 42:12321–29

180. Fanjul M, Salvador C, Alvarez L, Cantet S, Hollande E. 2002. Targeting of carbonic anhydrase IV to plasma membranes is altered in cultured human pancreatic duct cells expressing a mutated (ΔF508) CFTR. *Eur. J. Cell Biol.* 81:437–47

181. Sterling D, Brown NJD, Supuran CT, Casey JR. 2002. The functional and physical relationship between the DRA bicarbonate transporter and carbonic anhydrase II. *Am. J. Physiol. Cell Physiol.* 283:C1522–29

182. Novak I, Hug MJ. 1995. A new preparation of pancreatic ducts for patch-clamp studies. *Cell. Physiol. Biochem.* 5:344–52

183. Novak I. 1998. ß-adrenergic regulation of ion transport in pancreatic ducts: patch-clamp study of isolated rat pancreatic ducts. *Gastroenterology* 115:714–21

184. Hug MJ, Pahl C, Novak I. 1996. Calcium influx pathways in rat pancreatic ducts. *Pflügers Arch.* 432:278–85

185. Gross E, Fedotoff O, Pushkin A, Abuladze N, Newman D, Kurtz I. 2003. Phosphorylation-induced modulation of pNBC1 function: distinct roles for the amino- and carboxy-termini. *J. Physiol.* 549:673–82

186. Bachmann O, Rossmann H, Berger UV, Colledge WH, Ratcliff R, et al. 2003. cAMP-mediated regulation of murine intestinal/pancreatic Na^+/HCO_3^- cotransporter subtype pNBC1. *Am. J. Physiol. Gastrointest. Liver Physiol.* 284:G37–45

187. Robertson MA, Foskett JK. 1994. Na^+ transport pathways in secretory acinar cells: membrane cross talk mediated by $[Cl^-]_i$. *Am. J. Physiol. Cell Physiol.* 267:C146–56

188. Kahle KT, Gimenez I, Hassan H, Wilson FH, Wong RD, et al. 2004. WNK4 regulates apical and basolateral Cl^- flux in extrarenal epithelia. *Proc. Natl. Acad. Sci. USA* 101:2064–69

189. Ishiguro H, Naruse S, San Roman JI, Case RM, Steward MC. 2001. Pancreatic ductal bicarbonate secretion: past, present and future. *JOP J. Pancreas (Online)* 2(4 Suppl.):192–97

Annu. Rev. Physiol. 2005. 67:411–43
doi: 10.1146/annurev.physiol.67.031103.153004
Copyright © 2005 by Annual Reviews. All rights reserved
First published online as a Review in Advance on October 12, 2004

MOLECULAR PHYSIOLOGY OF INTESTINAL Na+/H+ EXCHANGE

Nicholas C. Zachos, Ming Tse, and Mark Donowitz

Departments of Medicine and Physiology and the Hopkins Center for Epithelial Disorders, Johns Hopkins University School of Medicine, Baltimore, Maryland 21205-2195; email: nzachos1@jhmi.edu; mtse@jhmi.edu; mdonowit@jhmi.edu

Key Words   trafficking, brush border, NHE3, NHE2, Na+ absorption

■ **Abstract**   The sodium/hydrogen exchange (NHE) gene family plays an integral role in neutral sodium absorption in the mammalian intestine. The NHE gene family is comprised of nine members that are categorized by cellular localization (i.e., plasma membrane or intracellular). In the gastrointestinal (GI) tract of multiple species, there are resident plasma membrane isoforms including NHE1 (basolateral) and NHE2 (apical), recycling isoforms (NHE3), as well as intracellular isoforms (NHE6, 7, 9). NHE3 recycles between the endosomal compartment and the apical plasma membrane and functions in both locations. NHE3 regulation occurs during normal digestive processes and is often inhibited in diarrheal diseases. The C terminus of NHE3 binds multiple regulatory proteins to form large protein complexes that are involved in regulation of NHE3 trafficking to and from the plasma membrane, turnover number, and protein phosphorylation. NHE1 and NHE2 are not regulated by trafficking. NHE1 interacts with multiple regulatory proteins that affect phosphorylation; however, whether NHE1 exists in large multi-protein complexes is unknown. Although intestinal and colonic sodium absorption appear to involve at least NHE2 and NHE3, future studies are necessary to more accurately define their relative contributions to sodium absorption during human digestion and in pathophysiological conditions.

INTESTINAL WATER/Na ABSORPTION

Every day the gastrointestinal (GI) tract processes ~9 liters of water and ~800 mEq of Na. The majority of the water and Na (~7.5 liters of water and ~650 mEq of Na) is secreted via the GI organs as part of digestion (i.e., the minority of this water, ~1.5 liters, and Na, ~150 mEq, is ingested). A major function of the GI tract is to maintain water/Na homeostasis by absorbing virtually all water and Na to which the GI tract is exposed. The two major organ systems involved in water/Na homeostasis, the intestinal and renal systems, have a similar horizontal organization. The small intestine and renal proximal tubule absorb large amounts of water and Na at low efficiency, whereas the colon and renal distal tubules and collecting ducts absorb less water and Na but with much higher efficiency. The brush border (BB) Na absorptive mechanisms in the jejunum are largely related to end products of digestion, including Na-linked D-glucose/D-galactose and L-amino acid transporters; H^+-linked dipeptide transporters; and to a lesser extent epithelial Na^+/H^+ exchangers (NHE) linked to Cl^-/HCO_3^- exchangers (1, 2). In the ileum, the BB Na^+/H^+ antiporters linked to Cl^-/HCO_3^- exchangers make up the neutral NaCl absorptive process, which accounts for most Na absorption. In the colon there is involvement of BB Na^+/H^+ antiporters linked to Cl^-/HCO_3^- exchangers, as well as to short-chain fatty acid/OH exchangers, whereas in the distal colon, the BB ENaC (epithelial Na channel) plus elevated tight junctional resistance accounts for high efficiency of Na absorption that prevents the body from becoming Na depleted even when on a virtually Na-free diet (1, 3, 4). In this review, we analyze the contribution of Na^+/H^+ exchangers to intestinal absorptive functions.

NEUTRAL NaCl ABSORPTION

Neutral NaCl absorption occurs throughout the mammalian intestine. In humans, there is more Na^+/H^+ exchange than Cl^-/HCO_3^- exchange (almost absent) in the jejunum, with more equivalent amounts of the transporters in the ileum and colon. In the jejunum of other mammalian species (rabbit, rat, mouse), the ratio is closer to 1:1. The rat distal colonic neutral Na absorptive process appears to have three components, with BB Na^+/H^+ exchangers linked to BB short-chain fatty acid/OH exchangers or to BB Cl^-/HCO_3^- exchangers (4).

The intestinal neutral Na absorptive processes account for most intestinal Na absorption in the period between meals and also for the great majority of the increase in ileal Na absorption that occurs post prandially (the contribution to the colonic Na absorptive process has not been documented) (5, 6). This process is both up- and down-regulated as part of digestion, appearing to be inhibited initially during and after eating and then stimulated later in digestion (1, 7). This regulation occurs via the complex eating related intestinal signal transduction with neural/paracrine/endocrine changes in the environment of the intestine. In most diarrheal diseases, there is inhibition of neutral NaCl absorption in the Na absorptive cell, which is accompanied by stimulation of electrogenic Cl secretion from the crypt cells. This accounts for the major GI loss of water and electrolytes in diarrhea (1, 7). Although comprehensive studies of independent regulation of the BB Cl^-/HCO_3^- exchangers have not been reported, it is regulation of the BB Na^+/H^+ exchanger NHE3 that accounts for most of the recognized digestive changes in neutral NaCl absorption, as well as most of the changes in Na absorption that occur in diarrheal diseases.

The nature of the functional cooperation between the BB Na^+/H^+ exchangers and Cl^-/HCO_3^- exchangers in neutral NaCl absorption remains only partially understood. The possibilities include

1. indirect linkage by changes in intracellular pH, with changes in one exchanger generating a pH gradient locally that drives the other (8);

2. direct physical linkage of the NHEs and Cl^-/HCO_3^- exchangers because both NHE3 and SLC26A3 (down-regulated in adenoma; DRA) as well as SLC26A6 (putative anion transporter; PAT1) (the most likely candidates for the BB Cl^-/HCO_3^- exchanger linked to NHE3 activity) bind to brush border PDZ domain-containing proteins (members of the NHERF and PDZK1 families) (9–11);

3. indirect physical linkage because both exchangers appear to be part of large multiprotein complexes, at least partially scaffolded by PDZ proteins, in which multiple regulatory proteins affect the exchangers in coordinated fashion, with regulatory proteins joining or leaving the complexes as part of digestion (12);

4. involvement of cytoskeletal proteins including the ezrin/radixin/moesin (ERM) gene family;

5. changes (potential) in cell volume as affected by one exchanger with subsequent effects of changes in cell volume on the other family of exchangers.

NHE GENE FAMILY

In this review, we highlight the identity, structure, function, regulation, and contributions of the mammalian NHE gene family members to intestinal Na absorption. Other pertinent NHE reviews include (7, 13, 14): In humans, this gene family is called SLC9A (HUGO nomenclature, http://www.gene.ucl.ac.uk/nomenclature). The first NHE was molecularly identified in 1989 by Sardet, Pouyssegur, and colleagues (15). There are 9 identified human SLC9A isoforms (Table 1). Extensive in silico analysis of NHEs from all species suggests that the full complement of mammalian NHEs has been identified (C. Brett, M. Donowitz & R. Rao, in press). The NHE family can be divided into plasma membrane and intracellular, organellar isoforms. The established plasma membrane isoforms include NHE 1–5 (15–21). The plasma membrane isoforms are further divided into those that cycle to and from the recycling endosomes/plasma membrane and include NHE3 (28) and NHE5 (29); and those that permanently reside on the plasma membrane and include NHE1, 2, and 4 (30, 31). The organellar isoforms include NHE6 (in the recycling endosomes) (20, 22, 23) and NHE7 (in the *trans*-Golgi network; TGN) (24) and the related NHE8 (25) and NHE9 (26), although the latter two have not had their intracellular localization determined. In fact NHE8 is thought to be expressed on the plasma membrane and in unidentified intracellular compartments. Of these mammalian NHE paralogues, NHE1–3 and 7 are present in all GI organs, and NHE6, although predicted to be ubiquitous, has been shown to be present only in pancreas and liver (24). NHE4 and 5 do not appear to be present in the intestine. Whether NHE8 and 9 are present in the GI tract is unknown. In this review we emphasize the contributions of NHE3 (SLC9A3) and NHE2 (SLC9A2) to intestinal Na absorption, with some comment about the role of NHE1 (SLC9A1).

Table 1 Cellular and tissue distribution of NHE isoforms in the gastrointestinal tract

NHE	Cellular localization	Jejunum	Ileum	Colon	Stomach	Pancreas	Liver	Gall bladder	References
NHE1	PM (BLM)	+	+	+	+	+	+	+	(15, 16)
NHE2	PM (Ap; BLM)	+	+	+	+	+	+[a]	+	(37, 18)
NHE3	PM (Ap); RE	+	+	+	+	+	+	+	(17, 19)
NHE4	PM (BLM)				+	+	+		(17)
NHE5[b]	PM; SV; RE	−	−	−	−	−	−	−	(20, 21)
NHE6	RE[c]					+	+		(22, 23)
NHE7	TGN	+	+	+	+	+	+	+	(24)
NHE8	PM (Ap)[d]	+	+	+			+		(25)
NHE9	e	+	+	+		+	+		(26)

[a]NHE2 mRNA was expressed in a normal rat cholangiocyte cell line (NRC-1) (27).

[b]NHE5 is expressed only in the mammalian brain and sperm (21).

[c]NHE6 has also been shown to be transiently expressed at the plasma membrane (23).

[d]NHE8 is expressed at the apical membrane in rat renal cortex and in the mouse liver. Although NHE8 appears to be ubiquitous, expression in other GI tissues has yet to be demonstrated (25).

[e]Cellular localization has yet to be determined, although amino acid analysis suggests that NHE9 may be intracellular because it has most homology with NHE6 and NHE7 (26).

Blank spaces indicate tissue expression has not been determined. Abbreviations: PM, plasma membrane; Ap, apical; BLM, basolateral; RE, recycling endosomes; SV, synaptic vesicles; TGN, *trans*-Golgi network.

Emphasis is on intestinal studies with comments about NHEs in other epithelia or nonpolarized cells when relevant areas have not been studied in polarized intestinal cells. On the basis of functional studies an additional colonic NHE called the Cl-dependent isoform has been proposed (32, 33). However, subsequent studies suggested some Cl dependence of multiple NHE isoforms (34) and reported cloning of this Cl-dependent isoform was not confirmed in the human genome, thus implying that it may represent a cloning artifact.

Structually, human NHE1–9 have between 645 and 898 amino acids (aa) and consist of two domains (Figure 1, see color insert) (7, 13): the N-terminal $\sim$500 aa transmembrane transport domain, which carries out exchange of Na^+ for H^+ in a 1:1 electroneutral manner (15, 35–37); and (2) the C-terminal domain, which is involved in most growth factor and protein kinase regulation of the NHE. The N terminus is highly homologous among isoforms (38). There are 12 putative encoded membrane-spanning domains (MSD), the first being a cleaved signal peptide as demonstrated for NHE3 (39) and the yeast NHE homologue, Nhx1 (40). Although not demonstrated for NHE1 or other NHEs, we suggest that all expressed NHEs have a cleaved signal peptide, 11 functional MSDs, and an extracellular N terminus and intracellular C terminus. It is possible, however, that some NHE isoforms have 12 functional MSDs (38). MSDs 4 and 9 contain sequences that are involved in determining sensitivity to amiloride and its analogues. In addition, based on mutagenesis of the yeast NHE homologue, Nhx1, amino acid residues in what appears to be a P-loop (in which residues enter and exit from the same side of the lipid bilayer) between MSD 9 and 10 participate in Na^+ and H^+ transport, as well as in defining differences in cation transport specificity between the plasma membrane and intracellular NHEs. Intracellular NHE isoforms from many species transport K^+ over Na^+, whereas plasma membrane NHEs transport $Na^+ \gg K^+$ (C. Brett & R. Rao, manuscript in preparation). The plasma membrane NHE isoforms have longer C termini than the organellar NHEs. It is tempting to speculate this is because of the more extensive regulation of the former, which requires a greater number of associated proteins. However, functional studies of the organellar isoforms have not been carried out sufficiently to characterize the extent of their regulation physiologically and pathophysiologically.

NHE FUNCTIONS

Plasma membrane NHEs contribute to maintenance of intracellular pH and volume, transcellular absorption of NaCl and $NaHCO_3$, and fluid balance carried out by epithelial cells, especially in the kidney, intestine, gallbladder, and salivary glands, as well as regulation of systemic pH. In addition, both NHE3 (41, 42) and NHE5 (29) function in the intracellular recycling compartment, tending to acidify (Na^+ inside vesicle exchanged for H^+ from cytosol) the early endosomes and perhaps secretory granules, respectively. In the case of NHE3, this function has been shown necessary for albumin absorption in human renal proximal tubule by a plasma membrane receptor/clathrin-dependent uptake process (43). In addition, it now seems that intestinal nutrient absorption is partially under control of the NHE gene family. For instance, PepT1 is an H^+-dipeptide cotransporter that functionally is linked to BB NHE3 function (in Caco-2 cells). This link is postulated to occur by the acid microclimate created at the BB by H^+ exchange for Na^+, which provides the driving force for BB H^+-dipeptide uptake (44).

Functions of the organellar NHEs, although not yet adequately characterized in mammalian cells, are likely to include intracellular Na^+ and K^+ homeostasis, regulation of trafficking to and from the TGN, and the endosomal and lysosmal compartments, and as an H^+ leak pathway to counterbalance acidification by the organellar V-type H-ATPases (C. Brett, M. Donowitz & R. Rao, manuscript in preparation).

GI TRACT LOCALIZATION OF NHE1–3

There is general agreement on the cellular and subcellular localization of NHE1, NHE2, and NHE3 in most GI organs, with only a few points of controversy remaining. NHE1 is uniformly confined to the basolateral membrane throughout the GI tract (Table 1).

Esophagus and Stomach

Rat and rabbit esophagus contain NHE1 with no detectable message for NHE2 or NHE3 (45). Species variation is important in considering which NHE isoforms are present in gastric epithelial cells, although in most species, NHE1, 2, and 4 are expressed (46–48). Rabbit gastric mucous and parietal cells have large amounts of NHE1 and 2 but no NHE3. Rabbit parietal cells also express some NHE4. NHE activity in rabbit parietal cells is $\sim$80% NHE1 and $\sim$20% NHE2, whereas in rabbit gastric mucous cells it is $\sim$65% NHE1 and $\sim$30% NHE2. In rat, NHE3 is present only in parietal cells, whereas NHE1, 2, and 4 are present in all gastric epithelial cells. NHE1 and 4 are on the basolateral membranes of parietal cells; NHE3 is on the apical surface of parietal cells and may remove H^+ between periods of H^+-K^+-ATPase presence on that surface (49).

Duodenum

Both NHE3 and NHE2 are present in human duodenal BB (50). Inhibition of human duodenal NHE2 and NHE3 with amiloride analogues equivalently increased net HCO_3^- secretion (50). This effect also occurred in rat duodenum with NHE3 inhibition, although NHE2 inhibition was without effect. The stimulated HCO_3^- secretion that resulted from NHE3 inhibition occurred with a time delay and required the cystic fibrosis transmembrane regulator (CFTR). The role of CFTR in this process is either as a stimulated BB Cl channel with some HCO_3^- permeability or as a link to a BB Cl^-/HCO_3^- exchanger (11, 51).

Jejunum, Ileum, and Colon

Both NHE2 and NHE3 are present in the apical membrane of villus cells of small intestine and surface cells of the mammalian colon (52–54). The only exception is that NHE2, but not NHE3, is present in the apical membrane of rabbit descending colon surface cells (52). NHE1 is present on basolateral membrane (BLM) of epithelial cells in all intestinal segments, with highest expression in duodenum.

NHE3 appears to be present in larger amounts in the ileum and colon than in the jejunum of several species (52). In human intestine this is also true at the message level (ileum > jejunum > proximal = distal colon) (55). Conversely, at the message level, NHE2 is present in largest amount in the human distal colon > small intestine > proximal colon (55). The ileum/right colon functions as a unit and consistently NHE3 is present more in ileum and right colon, whereas NHE2, at least in humans, is predominantly expressed in the distal colon (56). However, no careful quantitation of the amount of NHE3 or NHE2 protein along the horizontal intestinal axis has been reported.

There is some disagreement as to how far down the villus-crypt axis NHE3 and NHE2 occur. In rabbit ileum, Hoogerwerf et al. (52) demonstrated both NHE2 and NHE3 in the BB of the entire villus of small intestine, in surface cells of colon, and also in the apical membrane of the approximately upper half of the crypt. Given that this distribution encompasses the enterocytes that take part in Na absorption, we suggest that apical NHE2 and NHE3 are responsible and, in fact, define the Na absorptive cells. However, other groups found that message for NHE3 was in the small intestinal and colonic surface but not crypt cells, whereas NHE2 was equally present in surface and crypt cells (53, 55, 56). In rat small intestine, NHE3 protein was reported to be present in only the upper 50% of the villus or in the entire villus, based on different studies, but not in the crypts (57).

Mouse proximal colon had differential distribution of NHE1–3 in surface and crypt cells when studied as units. Message for NHE1 was similar in surface and crypt; NHE3 was much higher in surface than crypt; and NHE2 was higher in crypt than surface. Thus both surface and crypt cells have message for all three NHEs but surface was predominantly NHE3 and crypt NHE2. Mouse distal colon was similar, with NHE2 and NHE3 in BB of surface cells. Also, using functional and immunocytologic methods, mouse distal colon crypt BB Na^+/H^+ exchange activity was identified as NHE2 and not NHE3 (58). These results are supported by a similar finding in the polarized colon cancer cell line HT29-C1 (59). Thus the general pattern throughout the small intestine and colon of several species is the presence of both NHE2 and NHE3 in Na absorptive cells, whereas NHE2 is found predominantly in crypt epithelial cells.

In spite of extensive localization studies in GI tissues, there is insufficient detail to allow determination of NHE1–3 distribution in all cell types. Goblet cells do not appear to contain either NHE2 or NHE3 (52), whereas the distribution in Paneth cells, enteric-nerves and enteroendocrine cells is not known.

Liver and Gallbladder

NHE3 protein is expressed on the apical membranes of mouse and rat cholangio-cytes and in canalicular membranes of hepatocytes; NHE1is present on sinusoidal membranes (60). Unexpectedly, in a normal rat cholangiocyte cell line (NRC-1), NHE1 and NHE2 mRNAs were demonstrated, but NHE3 was not detected (27). NHE3 and NHE2 are present in the apical surface cells of the gallbladder epithe-lial cells (61, 62). In intact gallbladder and gallbladder epithelial cells in primary culture, NHE activity was ~66% NHE2, ~28% NHE3, and ~6% NHE1 (63).

ONTOGENY OF INTESTINAL NHE2 AND NHE3

The ontogenic expression of NHE2 and NHE3 has been reported in rat jejunum (64, 65). Both NHE2 and NHE3 contributed to BB NHE activity at all ages, although total NHE3 activity could not be assessed because of the NHE inhibitors used. In rat jejunum there is a post-suckling (>2 weeks) increase in NHE2 message, protein and contribution to BB NHE activity, which increases further at 6 weeks (post-weaning). However, at all times studied, BB NHE activity from NHE2 was in the minority (maximum ~25%). NHE3 message and protein were increased and sustained at all times post-suckling (>2 weeks). Of interest, jejunal BB NHE activity was four-fold higher at 6 weeks (post-weaning) than at any other period, including adult, which correlated with this time also having the highest NHE2 and NHE3 message and protein levels (65).

SUBCELLULAR LOCALIZATION OF NHE1–3

When overexpressed in AP-1 fibroblasts and PS120 fibroblasts, NHE3 is present in two pools, the plasma membrane and an intracellular, juxtanuclear location. A technique based on study of chimeras of a fluorescent intracellular probe and or-ganellar targeting signals was used to colocalize intracellular NHE3 with markers of the recycling endosome (both transferrin receptor and cellubrevin) but not with the Golgi (41). This supports the view that intracellular NHE3 is predominantly in an endosomal pool. However, others using similar techniques found that cellu-brevin and the transferrin receptor were not in an identical pool (T. Machen, private communication). Not unexpectedly, at least part of the intracellular NHE3 pool colocalizes on sucrose gradient with the early endosomal protein EEA1 (42), sug-gesting its presence in early endosomes; NHE3 also colocalizes with clathrin/AP2, consistent with NHE3's presence in the initiation of the clathrin-dependent endo-cytic pathway (29, 66). NHE2 has not been studied in similar detail, but in fibrob-lasts most occurred on the plasma membrane, with the intracellular NHE2 being in a diffuse vesicular location rather than in the recycling compartment as for NHE3, consistent with NHE2 being resident in the plasma membrane (30).

Quantitation of the amount of plasma membrane versus intracellular NHE indi-cated that only ~15% of NHE3 but ~90% of NHE1 and NHE2 are on the plasma

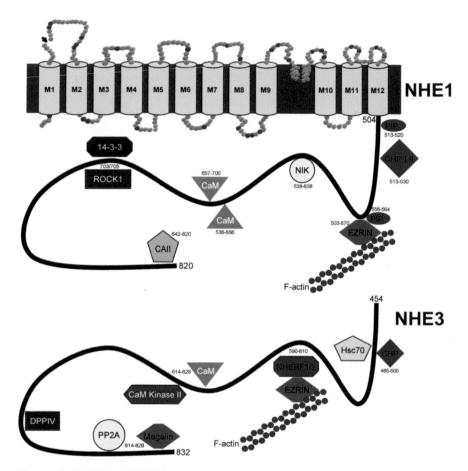

Figure 1 NHE1 and NHE3 topology and C-terminal-binding partners. N-terminal topology based on findings by Zizak et al. (39), suggesting TM1 is a cleaved signal peptide, and Wakabayashi et al. (38), demonstrating location of P-loop between transmembrane domains 9 and 10. Arrow indicates signal peptide cleavage site. Small numbers denote amino acids of NHE1 and NHE3 where binding occurs.

membrane under basal conditions when expressed in PS120 and AP1 (42, 68) cells. A similar predominantly intracellular distribution of NHE3 was found in renal proximal tubule opossum kidney (OK) cells containing endogenous or trans-fected NHE3 (42), although Grinstein described ~50% of NHE3 on the surface in these cells under basal conditions (S. Grinstein, private communication). Not all cells have the majority of NHE3 in an intracellular pool; for instance, although not quantitated to the extent of these cell culture models, rabbit ileal villus cells have ~70% of total NHE3 in a pool that colocalizes with brush border markers (69). In differentiated Caco-2 cells, NHE localization mimics that in other intestinal Na absorptive cells (70, 71). NHE3 and, in some cases, NHE2 is present on the apical membrane, whereas NHE1 is basolateral. In Caco-2 cells, ~80% of total NHE3 is in the brush border and ~20% is in a diffuse supranuclear location (28, 70). The latter is suggested to represent the apical recycling endosome in these cells. The differences in this distribution have important consequences for the regulation of NHEs by endocytosis/exocytosis. If there is a large subapical pool, addition to the plasma membrane is likely to be an effective mechanism to increase NHE activity and a less effective mechanism of down-regulation, and vice versa. No general conclusion can be reached about the differences in the types of cells that express higher versus lower percentage of NHE3 on the plasma membrane. Note that differences in plasma membrane distribution do not appear to be related to cell culture models or to NHE overexpression after transfection.

In the BB, NHE3 is localized to microvilli and to the intervillus clefts (52). Several models have been proposed for the relationship between these NHE3 pools; we favor a model in Na absorptive cells in which NHE3 continually recycles under basal conditions between intervillus clefts in BB and endosomes. Thus in intestinal cells, the consensus for NHE localization is NHE3 and NHE2 on the apical surface, NHE1 on the basolateral surface, with NHE3, in addition, in a supra or juxtanuclear compartment now believed to be the recycling compartment.

ROLE OF NHE2 AND NHE3 IN INTESTINAL Na ABSORPTION

Inhibitor studies have defined the relative roles of NHE2 and NHE3 in intestinal Na absorption. Generally, in the small intestine and colon of most mammalian species both NHE2 and NHE3 contribute to neutral Na absorption. In most cases, NHE3 plays a larger role than NHE2 in NaCl absorption. For example, NHE3 appears to be the major Na^+/H^+ exchanger in rabbit and rat small intestinal NaCl absorption, although NHE2 seems to contribute as well (65, 72). There are exceptions, however. One extreme is dog ileum in which only NHE3 appears involved in Na absorption (5, 6). Using amiloride antagonists in dog ileum, NHE2 did not appear to contribute to intestinal Na absorption under basal and meal-stimulated conditions. In avian ileum (73) studied in vitro, both NHE2 and NHE3 accounted for approximately equal basal-active Na absorption or BB Na^+/H^+ exchange activity.

In colonic studies, both NHE2 and NHE3 contribute to most neutral Na absorption with approximately equal contributions to BB Na^+/H^+ exchange of NHE2 and NHE3 in proximal colon of several species. In rabbit proximal colon, NHE2 and NHE3 contributed approximately equally to Na absorption. There is no NHE3 in rabbit distal colon and Na^+/H^+ exchange accounts for only a small amount of basal Na absorption in this segment. In rat proximal and distal colon, basal BB Na^+/H^+ exchange is ~85% NHE3. In rat distal colon, neutral Na absorption has a HCO_3^--sensitive component in which Na^+/H^+ exchange is linked to BB Cl^-/HCO_3^- exchange (74). The NHE involved is NHE3. In addition, there is a neutral short-chain fatty acid/OH exchanger linked to NHE2 or NHE3 (4). In mouse proximal colon, surface cell NHE activity was 55% NHE3, 20% NHE2, and 20% NHE1. Crypt was 48% NHE2, 40% NHE1, and 12% NHE3 (75). Mouse distal colon surface cells have both NHE2 and NHE3, whereas crypt apical membrane expresses NHE2 but not NHE3. Similarly, in the HT29 cell line, apical membrane Na^+/H^+ exchange activity is entirely NHE2 (59). In avian colon, NHE2 explains most (~85%) BB Na^+/H^+ activity under basal conditions and when aldosterone is elevated by a low Na diet (73). Human colon has not been studied in detail, although BB vesicle studies of cadaveric human colon demonstrated approximately equal contribution of NHE2 and NHE3 (P. Dudeja & K. Ramaswamy, private communication).

The role of BLM NHE1 on intestinal Na absorption has not been evaluated in detail. Of note, in kidney thick ascending limb, inhibiting BLM NHE1 inhibited BB Na^+/H^+ exchange, making functional coupling of intestinal BB and BLM NHEs a possible regulatory mechanism (76).

Thus both NHE2 and NHE3 appear to contribute to intestinal Na absorption, although their relative contributions appear to be dependent on the species, which intestinal segment is studied, and whether basal or regulated transport is involved. It will be important to define in more detail the contributions of NHE2 versus NHE3 in regulated small intestinal and colonic neutral Na absorption in order to develop a picture of the relative contributions of these NHEs to intestinal Na absorption. In determining which small intestinal and colonic models to characterize to understand human disease, deciding which models closely mimic human NHE activity and distribution among NHE isoforms should be considered.

LESSONS FROM NHE1–3 KNOCKOUT MOUSE MODELS FOR INTESTINAL Na ABSORPTION

Intestinal studies of NHE 1–3 knockout (KO) mice have been reported in detail, mostly in studies from Shull et al. (87). In each of these NHE KO genotypes, deletion of NHE isoforms resulted in distinct phenotypes while generally not affecting expression of the other NHEs in the small intestine and colon. Only the NHE3-deficient mouse demonstrated impaired Na absorption in the intestine and displayed a phenotype similar to patients with congenital sodium diarrhea (77–80).

NHE1 is considered the "housekeeping isoform" because it appears to participate in the regulation of intracellular pH and volume. NHE1 KO mice exhibit growth retardation by two weeks of age, ataxia, and seizures (slow-wave epilepsy) (81), as well as altered intracellular pH in pancreatic acinar cells (82). A similar central nervous system phenotype occurred in a spontaneous mutant mouse lacking NHE1 (83). There was also slight gastric atrophy, with normal numbers of parietal cells. The animals died prematurely, 67% before weaning (81). The only reported change in other NHE expression in the NHE1 KO mouse was increased NHE3 expression in the cerebellum (84). Basal intestinal neutral sodium absorption in the NHE1 KO mouse is normal. More detailed evaluation of Na absorptive function in NHE1 KO mice under stimulated and inhibited conditions is warranted, but no abnormality has been identified.

The NHE2 KO mouse does not display any overt disease phenotype, is normal in size, and initially has normal acid secretion (48). However, the long-term viability of parietal cells appears to be decreased by the lack of expression of NHE2 (48). The number of parietal and chief cells of the stomach is reduced in NHE2 KO mice studied from age 10 days to 16 months (i.e., once acid secretion starts in normal mice) (49). As early as day 10, NHE2 KO mice exhibit neutrophil infiltration in the stomach that is subsequently replaced by lymphocytes and plasma cells with advancing age. Therefore, in the stomach of NHE2-deficient mice chronic achlorhydria occurs in the setting of chronic gastritis, and NHE2 appears to serve a protective role against cell damage resulting from acid secretion. Lack of chief cells is attributed to the fact they appear to arise from gastric parietal cells. The role of NHE2 in parietal cell survival is supported by demonstration in isolated gastric epithelial cells exposed to acid that epidermal growth factor treatment reduces damage through an unidentified mechanism that involves NHE2 (85). Although NHE4 and NHE1 are also highly expressed in parietal cells of some species (17), they do not compensate for the loss of NHE2. Of interest, NHE4 KO mice also have loss of parietal cells, suggesting that both NHE2 and NHE4 are required on the basolateral membrane of parietal cells to prevent acid-induced destruction (86). The rate of total Na absorption and that from NHE3 were normal in small intestine of NHE2 KO mice (87). Interestingly, there was double the amount of NHE3 (mRNA and protein) in NHE2 knockout mouse proximal colonic crypts compared with that in wild-type, but this did not appear to affect Na absorption. This was surprising given the recent suggestion of Na absorptive function in colonic crypts and suggests another role for NHE3 in these cells (pH$_i$ control, regulation of endocytosis, etc.). This is the only up-regulation of NHE3 recognized in the intestine of NHE2 KO mice (87).

The results of these studies argue that NHE2 plays little or no role in net Na absorption or fluid absorption in mouse small intestine or colon. What can explain the difference between these results and the pharmacologic studies (see below) that showed contributions of both NHE2 and NHE3 to most intestinal Na absorption including mouse jejunum (although the latter have not been studied in enough detail to accurately define the percentage of BB NHE activity due to NHE2)? There is not a clear answer, although KO mice are chronic models, allowing opportunities for compensatory mechanisms both at the level of transporters and signal transduction. It is possible that the unexpected lack of compensation of NHE2/NHE3 for loss of each other is due to the fact they are differentially regulated or have other important functions. For instance, NHE3 may play a role in regulation of endocytosis or pH control in early endosomes.

The NHE3 KO mouse exhibits modest diarrhea, increased fluid in the small intestine and colon, enlarged diameter of small intestine and colon, disturbed acid-base balance (mild metabolic acidosis), low blood pressure, decreased body fat, plus a high mortality when deprived of oral Na intake (80, 87). NHE3 KO mouse jejunum had residual Na absorption that was ethyl, isopropylamiloride (EIPA) sensitive and inhibited by cAMP. Because cAMP stimulates NHE2, NHE2 probably does not account for this residual Na absorptive process. NHE2/NHE3 double KO mice resembled NHE3 null mice with no change in viability, worsening diarrhea, small intestinal luminal content weight, further increase in plasma aldosterone levels, or sensitivity to Na restriction (81, 88). NHE3 KO mice were able to survive through compensatory mechanisms that include increasing intestinal Na absorption through hypertrophy of the small intestine and colon, as well as increased expression of the apical proteins ENaC and H^+, K^+-ATPase in the colon, which form brush border Na^+, K^+/H^+ exchange. Increased aldosterone production has been shown to be involved in increased expression of ENaC, K^+ channel, and H^+, K^+-ATPase. The increased expression of these ion transporters provides an alternative pathway for Na absorption (change of electroneutral to electrogenic process) in the NHE3-deficient GI tract (89). Although the presence of another NHE isoform, such as NHE8, has not been evaluated, the explanation is not known.

Using cDNA microarray analysis of wild-type and NHE3 KO mouse intestine, a significant number of upregulated genes were found that were known to be responsive to interferon-gamma (INF-γ) (90). Of note, the serum (systemic) INF-γ was not increased whereas intestinal INF-γ was elevated. There was no increase in inflammation in the intestine of the NHE3 KO mouse (90), which removed this as a source of the INF-γ. INF-γ regulates the expression and activity of transporters involved in intestinal absorption and secretion (90–94). The cytokine TNF-α, but not IL-4, IL-2, and IL-6, was increased in the small intestine of these mice (90), whereas IL-1β, but not INF-γ, levels were increased in colon. We speculate that INF-γ might be involved in increasing compensatory Na absorptive mechanisms or regulatory processes to make up for loss of intestinal Na absorption, which depends on NHE3.

There was a residual (~30%) Na-dependent alkalinization present in the jejunum of double NHE2/NHE3 KO mice that was amiloride sensitive.

The NHE3 KO mouse model supports the importance of NHE3 in small intestinal and colonic Na absorption in all mammalian species and in nearly all intestinal segments. This model also provides new insights into compensatory Na absorptive processes and suggests comparisons with whether these processes function at all under normal conditions.

SHORT-TERM REGULATION OF NHE1–3

In the GI tract, electroneutral Na absorption is regulated during the postprandial state as part of the neurohumoral response in digestion. In vivo and in vitro studies have described the regulation of Na absorption in the GI tract using agonists/antagonists that mimic digestion. In general, NHE3 rather than NHE2 or the BB anion exchangers are regulated. Nonetheless, the cellular and molecular mechanisms behind this regulation are not well understood. Given their established important roles in intestinal NaCl absorption, understanding how NHE3 and NHE2 respond to digestive stimuli is important to understand normal digestive physiology. NHE activity in vitro has been determined using techniques that measure either the cellular uptake of ^{22}Na or proton efflux using intracellular pH-sensitive dyes, e.g., 2′,7′-bis (2-carboxyethyl)-5 (6)-carboxyfluorescein (BCECF-AM). Studies initially focused on measuring NHE activity in symmetrical cells such as fibroblasts (7, 13, 14); however, more recent studies have been performed in epithelial cell lines (e.g., Caco-2 and OK cells) (28, 42, 71, 95), as well as in live, intact intestinal tissue. Initial studies suggested that NHE2 and NHE3 were regulated predominantly by changes in the maximal rate of exchange (V_{max}) (7). However, as the ability to measure initial rates of NHE activity improved, regulation is now often seen to involve changes in the H^+ affinity of the allosteric site of the NHE, or pH_i set point in addition to changes in V_{max} (96, 97).

Regulation of NHE1–3

Cyclic AMP, cGMP and elevated intracellular calcium $[Ca^{2+}]_i$, as well as neurohumoral substances and bacterial toxins that cause similar second messenger elevation, inhibit neutral NaCl absorption in ileum and colon and inhibit NHE3 activity. cAMP inhibits intestinal NaCl absorption, including BB Na/H exchange throughout most of the GI tract as well as BB Na/H exchange in the renal proximal tubule. Activation of protein kinase A (PKA) inhibits NHE3 activity in vitro and in vivo (9). cAMP-mediated inhibition of NHE3 in PS120 fibroblasts occurs only in the presence of NHERF1 or NHERF2 (9), which link NHE3 to the actin cytoskeleton via binding ezrin, an association necessary for NHE3 inhibition (100). Furthermore, ezrin also binds the regulatory subunit of PKA II (100–102) and serves as a low-affinity protein kinase A anchoring protein (AKAP) (Table 2).

Although inhibition of NHE3 by cAMP has been well characterized, the role of cGMP continues to remain controversial. In fibroblasts (AP-1 and PS120 cells), cGMP does not affect NHE1–3 (35, 103, 104) but inhibits NHE5 activity (105). In Caco-2/bbe cells, cGMP inhibits NHE3 but does not affect NHE2 (71). Also, nitric oxide (NO) exposure inhibited Caco-2 NHE3 activity via elevated cGMP levels, whereas NHE2 activity was unaffected (106). Elevated $[Ca^{2+}]_i$ acting by protein kinase C (PKC) also inhibits NHE3. In PS120 cells expressing NHERF2, but not NHERF1, increased levels of $[Ca^{2+}]_i$ inhibited NHE3 activity (97). This is associated with increased binding of PKCα and α-actinin-4 to NHERF2, which is

Table 2 Short-term regulation of NHE1, 2, and 3

	NHE1	NHE2	NHE3
Adenosine	—	—	S,I[a,b]
ATP Depletion	I	I	I
Albumin	—	—	S
Angiotensin II	S	—	S
ANP	—	—	I
cAMP	S	S	I
cGMP	—	—	I
Dopamine	I	—	—
EGF	—	—	S
Elevated [Ca^{2+}]$_i$	S	—	I
Endothelin-1	—	—	S
Epinephrine	I	—	S
FBS	S	S	S
FGF	S	S	S
Glucocorticoids	S	—	S
Hyperosmolarity	S	S,I[c]	I
Insulin	—	—	S
LPA	—	—	S
Mineralocorticoids	—	S	—
Nitric oxide	—	—	I
Norepinephrine	—	—	S
PMA	S	S	I
PTH	—	—	I
Okadaic acid	—	S	S
Serotonin	S	—	I
Short chain fatty acid	—	—	S
Somatostatin	I	—	—
Thrombin	S	S	S

[a]S, stimulate; I, inhibit; —, not reported.
[b]Adenosine has been shown to exert concentration-dependent effects on NHE3 activity.
The adenosine analog, N(6)-cyclopentyladenosine, stimulates NHE3 at
concentrations $<10^{-8}$ M and inhibits at concentrations $>10^{-8}$ M (99).
[c]Hyperosmolarity stimulates or inhibits NHE2 depending upon the cell in which it is expressed (13, 37, 98).

constitutively associated with NHE3 (107). In Caco-2/bbe cells, elevating $[Ca^{2+}]_i$ by thapsigargin treatment inhibited NHE2 and NHE3 activity (96). Thus NHE3 is inhibited by multiple signaling pathways in cell culture models that mimic inhibition of neutral NaCl absorption in studies of intact intestine.

Mechanisms of NHE3 Regulation

Two basic regulatory mechanisms are involved in short-term NHE regulation: (*a*) changes in turnover number and (*b*) changes in trafficking. Recent studies have shown that NHE3 is often regulated by changes in its plasma membrane versus intracellular location as a result of changes in the rates of endocytosis and/or exocytosis. NHE3 traffics between the plasma membrane and recycling endosomes under basal and regulated conditions in all cells in which this has been studied.

Table 2 lists short-term stimulatory and inhibitory agonists that regulate NHE1–3. Epidermal growth factor and clonidine stimulate ileal Na absorption by increasing the percentage of total NHE3 in the ileal BB (69). In OK cells, lysophosphatidic acid (LPA) stimulated NHE3 by a NHERF2- dependent process that increased the amount of BB NHE3, by stimulating NHE3 exocytosis (108). Treatment with endothelin-1 also increased surface NHE3 and NHE3 activity (109) through increased exocytosis. In PS120 fibroblasts stably transfected with NHE3, the surface NHE3 amounts were increased by treatment with serum and fibroblast growth factor (FGF), which stimulated exocytosis (110). In contrast, phorbol 12-myristate 12-acetate (PMA) inhibits NHE3 by causing an increase in endocytosis plus an additional change in turnover number in Caco-2 cells (28).

Of the NHE isoforms expressed in the GI tract, only NHE3 has been found to traffic between the plasma membrane and recycling endosomes under basal conditions (28, 41, 111). NHE1 localizes exclusively to the basolateral membrane of epithelial cells and has not been shown to be internalized (112), although it is both stimulated and inhibited (Table 2) (113). NHE2, similar to NHE3, localizes to the apical membrane of epithelial cells and can be stimulated and inhibited, but similar to NHE1 it does not appear to traffic under basal or stimulated conditions (67) (Table 2). Thus NHE2 is regulated by changes in turnover number. NHE3 internalization occurs via clathrin-coated vesicles (61) and involves lipid rafts (66, 69).

Phosphatidylinositol 3-kinase (PI3-K) is involved, at least in regulating exocytic trafficking of NHE3, by what appears to be a lipid raft-dependent process. Approximately 50% of basal NHE3 activity and plasma membrane amount is dependent on PI3-K activity (42, 110). Treatment of PS120 cells stably expressing NHE3 with PI3-K inhibitors, wortmannin, or LY294002, inhibited NHE3 transport activity, decreased the amount of NHE3 on the plasma membrane, and increased the size of the NHE3-containing juxtanuclear compartment (12). Furthermore, PS120/NHE3 cells transfected with a constitutively active form of PI3-K or the downstream Ser/Thr kinase AKT increased NHE3 activity and protein expression at the plasma membrane (114). Similar dependence on PI3-K activity was found for stimulated NHE3 activity/amount in PS120 cells and OK cells treated with basic fibroblast growth factor (110) or epidermal growth factor (12).

NHE Complexes

The C termini of NHE1–3 have been established as the domains necessary for regulation of Na^+/H^+ exchange activity. The C terminus consists of multiple subdomains (7, 115, 116). In studies in which NHE1–3 were progressively truncated to remove nearly the entire C-terminal portion of the protein, each NHE demonstrated severely impaired activity as well as decreased ability to respond to growth factors and protein kinases (7, 115, 117, 118). However, these truncations did not disrupt NHE regulation by serum, hyperosmolarity, and ATP depletion (7, 103, 115, 117), suggesting that these stimuli act on the N-terminal transport domain and/or on the first part of the C-terminal intracellular segment of the protein (119).

Although many regulatory actions are shared among NHE1–3, unique regulation of each NHE has been identified. As shown in Figure 1, the C terminus of NHE1 binds Ca^{2+}/calmodulin (separate domains with high and low affinity) via positively charged clusters of amino acids, which increases transporter activity with elevated $[Ca^{2+}]_i$ and inhibits NHE1 at basal $[Ca^{2+}]_i$ (120–122). In addition, PIP_2 (123), calcineurin B homologous proteins (CHP 1 and 2) (124–126), and ezrin (127, 128) also bind NHE1. Mutation of NHE1 to prevent CHP and PIP_2 association with NHE1 greatly reduces NHE1 activity. Hsp70, 14-3-3 proteins, carbonic anhydrase II, tescalcin (129, 130), and several protein kinases (NIK, ROCK) also bind the C terminus and regulate NHE1 (Figure 1, Table 3) (129–134). NHE1 does not bind NHERF1 or NHERF2, and indirect protein-protein interactions with PDZ proteins have not been identified.

In contrast to the C terminus of NHE1, the organization of the C termini of NHE2 and NHE3 is more complex. The C termini of both NHE2 and NHE3 contain stimulatory and inhibitory domains (Figure 1 for NHE3); (7, 98, 115). Both NHE2 and NHE3 contain subdomains that respond to serum, growth factors, and okadaic acid, as well as amino acid motifs that bind multiple proteins including CaM (NHE1–3 all bind CaM). In addition, NHE3 directly binds NHERF1, NHERF2,

Table 3 Identified associated regulatory proteins that bind to NHE1–3 C terminus

NHE1	Calmodulin (120–122), calcineurin homologous protein (CHP) (124–126), PIP_2 (123); Ste20-like Nck-kinase (NIK) (131), 14-3-3 (132), Hsp70 (133), carbonic anhydrase II (134), ezrin (127), tescalcin (129, 130)
NHE2	Calmodulin (98); c-Src, ras-gap, p85 subunit of PI3-K, α-spectrin, neural-Src (135)
NHE3	Calmodulin (122), CHP (125), NHERF1 (9), NHERF2 (101), megalin (136), dipeptidyl peptidase IV (DPPIV) (137), PP2A (138), PDZK1 (139), Hsc70 (140)

PDZK1, Hsc70, DPPIV, PP2A, megalin, and CaM kinase II (7, 9, 98, 136–141). Although NHE2 and NHE3 appear to interact with some of the same regulatory proteins in similar locations on their C termini, regulation of NHE activity varies between them. For example, PKA and PKC stimulate NHE2 but inhibit NHE3 activity (7, 98, 117). The C termini of NHE1 and NHE2 resemble each other on hydrophobicity analysis more than they resemble the NHE3 C terminus. Thus it is of interest that the regulatory profile of NHE1 resembles that of NHE2 but not NHE3 in terms of nonrecycling and being stimulated by cAMP and PKC. Therefore, simply determining the profile of protein-protein interactions among the NHE isoforms is not sufficient to understand the inherent complex signaling mechanisms behind the regulation of each NHE, and it is also important to consider cell type with emphasis on epithelial cell models.

Recognizing that multiple regulatory proteins bind directly to NHE3, questions arise concerning whether the binding is dynamic, what modifies their binding, and does binding occur all at once or in a competitive manner. NHE3 exists in large protein complexes. Intracellular NHE3 is on complexes on the magnitude of 400kDa. At the plasma membrane, specifically at the apical membrane, NHE3 is in even larger complexes ($\sim$1000 kDa) (12). These complexes change as part of acute regulation. For instance, carbachol exposure increased the size of NHE3 complexes in ileal BB (12). Complexes formed in a cell culture model (PS120 cells) with elevated Ca^{2+} include NHE3, NHERF2, α-actinin-4, and PKCα (12). These complexes are similar to those found in ileal BB after carbachol exposure, which also include NHE3, NHERF2, α-actinin-4, and activated PKCα. This suggests involvement of a conserved regulatory mechanism in the response of NHE3 to elevated Ca^{2+}.

NHE3 and its regulatory binding partners appear to be scaffolded by multi-PDZ-domain-containing proteins that are present in the BB of small intestine and Caco-2 cells. These are members of the brush border PDZ domain gene family, a unique subset among $\sim$250 genes in the human genome that contain PDZ domains. These include NHERF1 and NHERF2 (both have two PDZ domains and a C-terminal ERM binding domain), as well as PDZK1 and IKEPP (both have four PDZ domains). Of note, the large NHE3 complexes seem to involve multiple PDZ domain proteins simultaneously in what may be a network of communicating proteins involving multiple simultaneously interacting PDZ scaffolds. The reason is unclear for having so many similar proteins in the BB, with at least three (NHERF1, NHERF2, PDZK1) shown to interact directly with NHE3. One explanation is that there is some specificity in binding partners that regulates NHE3. For instance, Ca^{2+}-dependent regulation of NHE3 involves NHERF2, which binds α-actinin-4, a protein necessary for aggregation of the NHE3-containing plasma membrane complexes that occur after Ca^{2+} elevation and before NHE3 endocytosis (96). This event is specific to NHERF2. NHERF1 cannot substitute in this Ca^{2+} regulation of NHE3, nor does it bind α-actinin-4.

Thus whereas NHE1–3 are all highly regulated, only NHE3 has been shown to be present in large, dynamic protein complexes. NHE1 has multiple binding partners, but neither NHE1 nor NHE2 have been shown to be present in large complexes that involve multiple proteins simultaneously, and neither one has been shown to bind to PDZ domain proteins. Also, whether the association of binding partners with NHE1 or NHE2 changes as part of acute regulation is unknown. Because NHE3, but not NHE1 and NHE2, traffic from the recycling system to the plasma membrane as part of regulation, it should be considered whether being in these large complexes is needed to allow NHE3 to traffic in this way.

Association With the Actin Cytoskeleton

NHE1 and NHE3 have been shown, to associate with the actin cytoskeleton, and NHE2 has been suggested to do so also (7, 101, 127). Although not studied in detail in epithelial cells, studies in fibroblasts have provided some insights. In fibroblasts, NHE1 links the actin cytoskeleton to the leading edge of lamellipodia through direct interactions with ERM proteins (127). Mutations of NHE1 to disrupt ERM binding resulted in impaired organization of focal adhesions and stress fibers, as well as the shape of the cell, indicating that binding of ERM proteins to NHE1 changes the cytoskeleton (127). This effect of NHE1 was independent of its transport activity (127). In addition, ezrin interaction with PI3-K and Akt were necessary not only for NHE1 function at the plasma membrane but also for cell survival (142). The ability of ERM binding to NHE1 to inhibit endocytosis was increased by stimulating NHE1 activity.

Using an in vitro overlay assay, a C-terminal Pro-rich area of NHE2 was shown to interact with the SH3 domain of α-spectrin (135), a cytoskeletal protein implicated in targeting ENaC to the apical membrane. In LLC-PK$_1$ cells transfected with NHE2, part of the Pro-rich area in the C terminus was necessary for apical localization and when mutated resulted in basolateral membrane NHE2 (130).

In contrast to NHE1, NHE3 links to the actin cytoskeleton at least indirectly via ezrin (100, 101) through interactions with NHERF1, NHERF2, and/or PDZK1 (9, 100, 139), although direct binding to ERM proteins may occur as well. In PS120 cells treated with 8-Br-cAMP, NHE3 forms a multi-protein complex at the plasma membrane that includes NHERF1/NHERF2 and ezrin (100, 101). The formation of this complex is necessary for cAMP-mediated inhibition of NHE3. Rac1, Cdc42, and RhoA, GTPases that have been shown to play integral roles in organizing the actin cytoskeleton, affect NHE3 regulation (143). In AP-1 cells, activation of RhoA, and its downstream effector, p160 Rho-associated kinase 1 (ROCK), were necessary for basal NHE3 activity (144, 145). Some aspects of apical trafficking/targeting retention of NHE3 also were dependent on an intact cytoskeleton. NHE3 and NHERF1 were present on the basolateral membranes in rat proximal tubule cells in which microtubule formation was disrupted by treatment with colchicine (146). Also, *Clostridium difficile* toxin B and inhibition of Rho-kinase, which disrupt the actin cytoskeleton of OK cells, decreased NHE3 expression at the apical plasma membrane (147). Overall results of these studies suggest that the actin cytoskeleton plays an important but complex role in anchoring

NHE3, as well as moving NHE3 in the BB and in NHE3 complex formation, whereas the microtubule network is necessary for proper targeting.

Further insights in cytoskeletal effects on NHE3 regulation in epithelial cells came from a study in OK cells in which FRAP (fluorescence recovery after photobleaching) was used to evaluate the role of the actin cytoskeleton and NHERF1/NHERF2 binding on NHE3 mobility (148). This study showed that NHERF1/NHERF2 binding decreased NHE3 mobility, consistent with a role for these PDZ domain proteins in limiting mobility, perhaps in part by forming NHE3 complexes. When the the PDZ-binding domain of NHE3 was removed, NHE3 still was anchored to the cytoskeleton, and this aspect of cytoskeletal binding was necessary for NHE3 mobility. It was suggested that in microvilli, actin, via polymerization or binding to a myosin motor, perhaps myosin VI, was necessary for movement to the intervillus cleft and/or for endocytosis.

How Are NHE3 Trafficking and Turnover Numbers Regulated?

Given that NHE1–3 bind multiple proteins and, at least, NHE3 is in large complexes, what determines the timing of binding/complex formation as part of short-term regulation? NHE1–3 have been shown to be phosphorylated under basal conditions (7, 149–153), and changes in phosphorylation are the best characterized NHE regulatory mechanism associated with changes in both trafficking and turnover number.

NHE1 PHOSPHORYLATION   NHE1 is phosphorylated at the C terminus (aa 638–815) under basal and stimulated conditions. NHE1 is phosphorylated by the Ser/Thr kinases ROCK, p90^{RSK}, and NIK. Serum-induced stimulation of NHE1 activity in vitro was due to phosphorylation of S^{703} by the ERK-regulated kinase, p90^{RSK} (154). This phosphorylation by p90^{RSK} allowed for direct binding of 14-3-3 proteins, which was necessary for NHE1 ion transport activity (132). Binding of 14-3-3 proteins may protect NHE1 phosphorylation sites from phosphatases (132, 155). Ste20-like Nck-interacting kinase, NIK, binds to NHE1 between aa 538 and 638 and phosphorylates Ser residues C terminal to the NIK-binding site (131) by a process that requires NIK binding to NHE1. Many of the studies concerning NHE1 phosphorylation have been performed in fibroblasts and regulation of NHE1 phosphorylation in epithelial cells remains to be characterized. No studies have been reported of the role of changes in NHE2 phosphorylation as part of its acute regulation.

NHE3 Phosphorylation

NHE3 is phosphorylated under basal conditions, as is NHE1. NHE3 activity is inhibited by acute (10–20 min) activation of PKA and protein kinase G (PKG) through increased phosphorylation of Ser residues (Ser552 and Ser605). However, studies by Kurashima et al. (152) have shown that mutation of Ser605 blocks NHE3 phosphorylation by PKA but reduces inhibition of transport activity by only

~50%. Therefore, in addition to mutating Ser605, Ser634 must also be mutated to abolish PKA-mediated inhibition of NHE3 activity (152). Although PKA does not phosphorylate Ser634 in vitro, other mechanisms (e.g., complex formation) are thought to be responsible for involvement of Ser634 in regulation of NHE3 activity. These results vary from studies performed by Zhao et al. (153) in OK cells. In this study, PKA phosphorylation of Ser552 and Ser605 were both necessary to mediate the effects of PKA on inhibition of NHE3. These changes in phosphorylation were necessary for the initial cAMP-induced changes in NHE3 activity (by stimulation of turnover number), decrease in the surface amount, increased endocytosis, and decreased exocytosis. Similarly, in PS120 cells, cAMP inhibition of NHE3 does not occur under conditions in which changes in phosphorylation do not occur (156). cAMP dependent phosphorylation of NHE3 requires expression of NHERF1 or NHERF2 (156).

Other regulation of NHE3 does not appear to require NHE3 phosphorylation. No changes in NHE3 phosphorylation occurred with FGF stimulation (150). Also, PKC inhibition of NHE3 either was not associated with changes in NHE3 phosphorylation or the changes did not correlate with changes in NHE3 activity (150, 157). The in vivo and in vitro studies thus suggest that phosphorylation of Ser residues may vary among cell types and that other signal transduction events may play a concomitant or dominant role in regulation of NHE3 function. Whereas NHE3 regulation often occurs by changes in NHE3 phosphorylation, there is also NHE3 regulation that does not involve changes in NHE3 phosphorylation, but rather occurs, for example, by some dynamic complex assembly. A future challenge will be to probe the role of phosphorylation in NHE3 complex formation and regulation because it is now clear that although the components involved in NHE3 regulation have been identified (phosphorylation, cytoskeleton, complex formation, etc.), their integration is not understood.

LONG-TERM REGULATION

In addition to the short-term NHE regulation that occurs over minutes to a few hours by growth factors, protein kinases, and changes in osmolarity, these exchangers are also regulated over periods of hours to many days. Digestive aspects of regulation are short-lived and thus are mimicked by the above described short-term NHE regulation. In contrast, the long-term NHE regulation generally mimics intestinal diseases, responses to disease or injury, or disease treatments (e.g., inflammatory bowel diseases).

Many of the responses occur via changes in gene transcription, although details generally are not available. In fact, it is only recently that the promoter regions of NHE2 and NHE3 were partially characterized. The human NHE2 consists of 12 exons separated by 11 introns (158). The rat and human NHE2 promoters lack canonical TATA and CCAAT boxes and are highly GC rich (158). Within both the rat and human NHE2 promoters, conserved transcription factor-binding sites include Sp1, AP-2, CACCC, NF-κB, and Oct-1 (159). There are other *cis* elements that vary between the human and rat promoters, including CdxA and Cdx-2, demonstrating that different mechanisms are involved in regulation of NHE2 expression in different species. The genomic organization of the rat and human NHE3 gene promoters has been characterized. The rat NHE3 promoter contains atypical TATA and CCAAT boxes and putative *cis*-acting elements that include glucocorticoid and thyroid response elements, Ap-1, Ap-2 C/EBP, NF-1, Oct-1, PEA3, and Sp1 transcription-binding sites (159). The human NHE3 promoter contains transcription-binding sites for Sp1, AP-2, and glucocoticoid and thyroid response elements, and these sequences appear at the same positions in both species (160). Whereas both human and rat NHE3 promoters contain TATA boxes and highly rich GC regions, the human NHE3 gene does not possess a CCAAT box, suggesting that NHE3 features are characteristic of both housekeeping and regulatory genes (160). NHE2 and NHE3 gene regulation may occur through the interactions of multiple transcription factors and/or response elements, resulting either in up- or down-regulation of NHE mRNA expression.

RESPONSE TO INCREASED INTESTINAL Na LOAD—SMALL INTESTINAL RESECTION AND DIETARY MANIPULATIONS   The GI tract has a remarkable ability to adapt its capacity to restore digestive and absorptive functions after massive small bowel resection or injury. After recovery from resection of >50% of the small intestine in rats, there was up-regulation of NHE2 and NHE3 mRNA and protein expression in enterocytes. This increase in expression occurred distal to the anastomosis, suggesting that an increased exposure to luminal Na or nutrient-rich chyme might be responsible (57). In addition, an increase in NHE activity occurred in mouse small intestine after bowel resection (161). However, in the mouse, NHE2, but not NHE3, expression and function were up-regulated after 50% small intestinal resection, further suggesting that NHE gene regulation may be species specific (162).

NHE3 mRNA, protein and BB activity were increased in the colon, but not the ileum, of rats fed soluble fiber pectin for 2 days to increase the load of short-chain fatty acids (163). Caco-2/bbe cells treated with short-chain fatty acids for 48 h had a time- and dose-dependent increase in NHE3 protein levels and BB NHE3 activity with no changes in NHE2 protein levels (163). Thus in addition to species variation, NHE2 and NHE3 gene regulation in response to increased Na load to the luminal surface appears to be the result of specific regulatory elements present in the promoters of each of these NHE isoforms

GLUCOCORTICOID AND MINERALOCORTICOID EFFECTS   NHE2 and NHE3 are differently regulated by chronic changes in gluco- and mineralocorticoids, with effects dependent on intestinal segments and species. Yun showed that glucocorticoid treatment for 18–72 h stimulated Na absorption in the rabbit jejunum, ileum, and colon by increasing NHE3 but not NHE1 or NHE2 mRNA and protein expression (164). However, dexamethasone treatment increased NHE3 mRNA expression in the rat ileum and proximal colon but not in the jejunum, distal colon, or whole kidney (165). Oppositely, adrenalectomy lowered NHE3 expression in the rat ileum and proximal colon but not in the jejunum (165). In addition, chronic aldosterone elevation caused by chronic Na and K depletion stimulated NHE3 and NHE2 mRNA and protein in the rat proximal colon but decreased both in the distal colon (166). Not only were the effects of gluco- and mineralocorticoids variable based on tissue specificity but by species as well. In the avian intestine, in response to a low Na diet for several weeks, which elevated aldosterone, BB Na/H exchange in ileum and colon was increased twofold. This stimulation was entirely due to an increase in the amount of NHE2 (73).

Part of long-term dexamethasone stimulation of NHE3 occurred by nontranscriptional mechanisms, and these were mediated by SGK1 (serum- and glucocorticoid-induced protein kinase). SGK1 acted by a PI3-K-dependent mechanism and required NHERF2 to increase the activity of NHE3 (167).

CHRONIC ACID-BASE CHANGES   Chronic metabolic changes in systemic acid-base balance affect the ability of the GI tract to absorb Na (168, 169). Alterations in pH, pCO2, and HCO_3^- affect neutral NaCl absorption in the ileum and colon via effects on the Na^+/H^+ and Cl^-/HCO_3^- exchangers. Metabolic acidosis induced in rats by NH_4Cl feedings resulted in sustained increased NHE2 and NHE3 activities in ileal and colonic BB membranes (170). Both NHE2 and NHE3, but not basolateral NHE1, mRNA and protein expression were increased after 6 days of NH_4Cl feeding (170).

CHRONIC INFLAMMATION   Many studies of NHE regulation have concentrated on chronic up-regulation of NHE activity in the mammalian intestine. Less emphasis has been on long-term down-regulation. Reduced NaCl absorption in the GI tract is usually associated with chronic diseases such as inflammatory bowel diseases (IBD) (171). Patients with IBD have high intestinal levels of cytokines, including INF-γ and TNFα. INF-γ down-regulates NHE3 mRNA and protein expression in the rat intestine and in Caco-2 cells (94). This is of interest given the up-regulation of INF-γ in the NHE3 KO mouse and supports the notion that this up-regulation of INF-γ is not due to a local feedback mechanism but rather is more directly linked to NHE3. Also, the IL-2-deficient mouse, which develops colitis, had inhibited electroneutral NaCl absorption in the right colon and decreased electrogenic Na absorption in the left colon (172). This was associated with reduced mRNA and protein levels of NHE3 (right colon) and ENac (left colon), although how IL-2 deficiency causes these effects is unknown (172). Importantly, the decrease in right-sided Na absorption exceeded the decrease in NHE3 message and protein amount, implicating effects on the function of the remaining NHE3 protein.

INTESTINAL DISEASES ASSOCIATED WITH CONGENITAL ABNORMALITIES IN SLC9As

No identified human intestinal diseases have been found that are due to genetic abnormalities in structure/composition of the SLC9A family. However, in microvillus inclusion disease there is lack of BB Na^+/H^+ exchangers (including NHE3),

along with lack of other apical membrane proteins (173). In congenital Na diarrhea, there are minimally functioning BB Na^+/H^+ exchangers (78, 79). However,
NHE2 and NHE3 were present in the apical membrane of jejunum of the index
case as well as in four other cases, as determined by immunocytochemical analysis
(M. Donowitz, M. Tse & S. Brant, unpublished data). NHE3 in the index case was
sequenced, including the intro/exon splice sites, and was normal. Pedigree analysis
of two affected congenital Na diarrhea families revealed parental consanguinity
and a single common ancestor five generations earlier, and excluded the SLC9A1,
2, 3, and 5 chromosomal loci as potential candidate gene reactions (174). It is
postulated that the functional Na^+/H^+ exchange abnormality is from changes in
associated regulatory proteins, although this has not yet been demonstrated. The
highly regulated nature of NHE3, with no genetic human diseases identified owing
to NHE3 mutations, further suggests that these are diseases of NHE3 regulatory
proteins that remain to be identified.

ACKNOWLEDGMENTS

We thank R. Rao, P. Dudeja, K. Ramaswamy, T. Machen, G. Shull, and S. Grinstein
for information provided in advance of publication. We thank C. Brett for insightful
critique of this review. Supported in part by NIH NIDDK Grants: RO1-DK26523;
RO1-DK32839; PO1-DK44484; R24-DK64388; RO1-CA85428; RO1-CA94012;
T32-DK07632; and the Hopkins Center for Epithelial Disorders.

**The *Annual Review of Physiology* is online at
http://physiol.annualreviews.org**

LITERATURE CITED

1. Donowitz M, Welsh MJ. 1987. Regulation
of mammalian small intestine electrolyte
secretion. In *Physiology of the Gastrointestinal Tract*, ed. LR Johnson, pp. 1351–
88. New York: Raven. 2nd ed.

2. Knickelbein R, Aronson PS, Atherton W,
Dobbins JW. 1983. Sodium and chloride
transport across rabbit ileal brush border. I. Evidence for Na-H exchange. *Am.
J. Physiol. Gastrointest. Liver Physiol.*
245:G504–10

3. Rajendran VM, Binder HJ. 1990. Characterization of Na-H exchange in apical
membrane vesicles of rat colon. *J. Biol.
Chem.* 265:8408–14

4. Krishnan S, Rajendran VM, Binder JJ.
2003. Apical NHE isoforms differentially
regulate butyrate-stimulated Na absorp-

tion in rat distal colon. *Am. J. Physiol. Cell
Physiol.* 285:C1246–54

5. Maher MM, Gontarek JD, Jimenez RE,
Donowitz M, Yeo CJ. 1996. Role of
brush border Na^+/H^+ exchange in canine
ileal absorption. *Dig. Dis. Sci.* 41:651–
59

6. Maher MM, Gontarek JD, Bess RS,
Donowitz M, Yeo CJ. 1997. The Na^+/H^+
exchange isoform NHE3 regulates basal
canine ileal Na^+ absorption in vivo. *Gastroenterology* 112:174–83

7. Donowitz M, Tse CM. 2001. Molecular
physiology of mammalian epithelial Na/H
exchangers NHE2 and NHE3. *Curr. Topics Membr.* 50:437–98

8. Knickelbein R, Aronson PS, Schron CM,
Seifter J, Dobbins JW. 1985. Sodium and

chloride transport across rabbit ileal brush border. II. Evidence for Cl^-/HCO_3^- exchange and mechanism of coupling. *Am. J. Physiol. Gastrointest. Liver Physiol.* 249:G236–45

9. Yun CH, Oh S, Zizak M, Steplock D, Tsao S, et al. 1997. cAMP-mediated inhibition of the epithelial brush border Na^+/H^+ exchanger, NHE3, requires an associated regulatory protein. *Proc. Natl. Acad. Sci. USA* 94:3010–15

10. Lamprecht G, Heil A, Baisch S, Lin-Wu E, Yun CC, et al. 2002. The down-regulated in adenoma (dra) gene product binds to the second PDZ domain of the NHE3 kinase A regulatory protein (E3KARP), potentially linking intestinal Cl^-/HCO_3^- exchange to Na^+/H^+ exchange. *Biochemistry* 41:12336–42

11. Jacob P, Rossmann H, Lamprecht G, Kretz A, Neff C, et al. 2002. Down-regulated in adenoma mediates apical Cl^-/HCO_3^- exchange in rabbit, rat, and human duodenum. *Gastroenterology* 122:709–24

12. Li X, Zhang H, Cheong A, Leu S, Chen Y, et al. 2004. Carbachol regulation of rabbit ileal brush border Na^+/H^+ exchanger 3 (NHE3) occurs through changes in NHE3 trafficking and complex formation and is Src dependent. *J. Physiol.* 556:791–804

13. Orlowski J, Grinstein S. 2004. Diversity of the mammalian sodium/proton exchanger SLC9 gene family. *Pflügers Arch.* 447:549–65

14. Wakabayashi S, Shigekawa M, Pouyssegur J. 1997. Molecular physiology of vertebrate Na^+/H^+ exchangers. *Physiol. Rev.* 77:51–74

15. Sardet C, Franchi A, Pouyssegur J. 1989. Molecular cloning, primary structure, and expression of the human growth factor-activatable Na^+/H^+ antiporter. *Cell* 56:271–80

16. Tse CM, Ma AI, Yang VW, Watson AJ, Levine S, et al. 1991. Molecular cloning and expression of a cDNA encoding the rabbit ileal villus cell basolateral membrane Na^+/H^+ exchanger. *EMBO J.* 10:1957–67

17. Orlowski J, Kandasamy RA, Shull GE. 1992. Molecular cloning of putative members of the Na/H exchanger gene family. cDNA cloning, deduced amino acid sequence, and mRNA tissue expression of the rat Na/H exchanger NHE-1 and two structurally related proteins. *J. Biol. Chem.* 267:9331–39

18. Tse CM, Levine SA, Yun CH, Montrose MH, Little PJ, et al. 1993. Cloning and expression of a rabbit cDNA encoding a serum-activated ethylisopropylamiloride-resistant epithelial Na^+/H^+ exchanger isoform (NHE2). *J. Biol. Chem.* 268:11917–24

19. Tse CM, Brant SR, Walker MS, Pouyssegur J, Donowitz M. 1992. Cloning and sequencing of a rabbit cDNA encoding an intestinal and kidney-specific Na^+/H^+ exchanger isoform (NHE-3). *J. Biol. Chem.* 267:9340–46

20. Klanke CA, Su YR, Callen DF, Wang Z, Meneton P, et al. 1995. Molecular cloning and physical and genetic mapping of a novel human Na^+/H^+ exchanger (NHE5/SLC9A5) to chromosome 16q22.1. *Genomics* 25:615–22

21. Baird NR, Orlowski J, Szabo EZ, Zaun HC, Schultheis PJ, et al. 1999. Molecular cloning, genomic organization, and functional expression of Na^+/H^+ exchanger isoform 5 (NHE5) from human brain. *J. Biol. Chem.* 274:4377–82

22. Numata M, Petrecca K, Lake N, Orlowski J. 1998. Identification of a mitochondrial Na^+/H^+ exchanger. *J. Biol. Chem.* 273:6951–59

23. Brett CL, Wei Y, Donowitz M, Rao R. 2002. Human Na^+/H^+ exchanger isoform 6 is found in recycling endosomes of cells, not in mitochondria. *Am. J. Physiol. Cell Physiol.* 282:C1031–41

24. Numata M, Orlowski J. 2001. Molecular cloning and characterization of a novel $(Na^+, K^+)/H^+$ exchanger localized to

the trans-Golgi network. *J. Biol. Chem.* 276:17387–94

25. Goyal S, Vanden Heuvel G, Aronson PS. 2003. Renal expression of novel Na^+/H^+ exchanger isoform NHE8. *Am. J. Physiol. Renal Physiol.* 284:F467–73

26. de Silva MG, Elliott K, Dahl HH, Fitzpatrick E, Wilcox S, et al. 2003. Disruption of a novel member of a sodium/hydrogen exchanger family and DOCK3 is associated with an attention deficit hyperactivity disorder-like phenotype. *J. Med. Genet.* 40:733–40

27. Spirli C, Granato A, Zsembery K, Anglani F, Okolicsanyi L, et al. 1998. Functional polarity of Na^+/H^+ and Cl^-/HCO_3^- exchangers in a rat cholangiocyte cell line. *Am. J. Physiol. Gastrointest. Liver Physiol.* 275:G1236–45

28. Janecki AJ, Montrose MH, Zimniak P, Zweibaum A, Tse CM, et al. 1998. Subcellular redistribution is involved in acute regulation of the brush border Na^+/H^+ exchanger isoform 3 in human colon adenocarcinoma cell line Caco-2. Protein kinase C-mediated inhibition of the exchanger. *J. Biol. Chem.* 273:8790–98

29. Szaszi K, Paulsen A, Szabo EZ, Numata M, Grinstein S, Orlowski J. 2002. Clathrin-mediated endocytosis and recycling of the neuron-specific Na^+/H^+ exchanger NHE5 isoform. Regulation by phosphatidylinositol 3′-kinase and the actin cytoskeleton. *J. Biol. Chem.* 277:42623–32

30. Cavet ME, Akhter S, Murtazina R, Sanchez de Medina F, Tse CM, Donowitz M. 2001. Half-lives of plasma membrane Na^+/H^+ exchangers. NHE1–3 plasma membrane NHE2 has a rapid rate of degradation. *Am. J. Physiol. Cell Physiol.* 281: C2039–48

31. Pizzonia JH, Biemesderfer D, Abu-Alfa AK, Wu MS, Exner M, et al. 1998. Immunochemical characterization of Na^+/H^+ exchanger isoform NHE4. *Am. J. Physiol. Renal Physiol.* 275:F510–17

32. Rajendran VM, Geibel J, Binder HJ. 1995.

Chloride-dependent Na/H exchange. A novel mechanism of sodium transport in eolonic crypts. *J. Biol. Chem.* 270:11051–54

33. Sangan P, Rajendran VM, Geibel JP, Binder HJ. 2002. Cloning and expression of a chloride-dependent Na^+-H^+ exchanger. *J. Biol. Chem.* 277:9668–75

34. Aharonovitz O, Kapus A, Szaszi K, Coady-Osberg N, Jancelewicz T, et al. 2001. Modulation of Na^+/H^+ exchange activity by Cl^-. *Am. J. Physiol. Cell Physiol.* 281:C133–41

35. Levine SA, Montrose MH, Tse CM, Donowitz M. 1993. Kinetics and regulation of three cloned mammalian Na^+/H^+ exchangers stably expressed in a fibroblast cell line. *J. Biol. Chem.* 268:25527–35

36. Orlowski J. 1993. Heterologous expression and functional properties of amiloride high affinity (NHE-1) and low affinity (NHE-3) isoforms of the rat Na/H exchanger. *J. Biol. Chem.* 268:16369–77

37. Yu FH, Shull GE, Orlowski J. 1993. Functional properties of the rat Na/H exchanger NHE-2 isoform expressed in Na/H exchanger-deficient Chinese hamster ovary cells. *J. Biol. Chem.* 268:25536–41

38. Wakabayashi S, Pang T, Su X, Shigekawa M. 2000. A novel topology model of the human Na^+/H^+ exchanger isoform 1. *J. Biol. Chem.* 275:7942–49

39. Zizak M, Cavet ME, Bayle D, Tse CM, Hallen S, et al. 2000. Na^+/H^+ exchanger NHE3 has 11 membrane spanning domains and a cleaved signal peptide: topology analysis using in vitro transcription/translation. *Biochemistry* 39:8102–12

40. Wells KM, Rao R. 2001. The yeast Na^+/H^+ exchanger Nhx1 is an N-linked glycoprotein. Topological implications. *J. Biol. Chem.* 276:3401–7

41. D'Souza S, Garcia-Cabado A, Yu F, Teter K, Lukacs G, et al. 1998. The epithelial sodium-hydrogen antiporter Na^+/H^+ exchanger 3 accumulates and is functional

in recycling endosomes. *J. Biol. Chem.* 273:2035–43

42. Akhter S, Kovbasnjuk O, Li X, Cavet M, Noel J, et al. 2002. Na⁺/H⁺ exchanger 3 is in large complexes in the center of the apical surface of proximal tubule-derived OK cells. *Am. J. Physiol. Cell Physiol.* 283:C927–40

43. Gekle M, Freudinger R, Mildenberger S. 2001. Inhibition of Na⁺/H⁺ exchanger-3 interferes with apical receptor-mediated endocytosis via vesicle fusion. *J. Physiol.* 531:619–29

44. Thwaites DT, Kennedy DJ, Raldua D, Anderson CM, Mendoza ME, et al. 2002. H/dipeptide absorption across the human intestinal epithelium is controlled indirectly via a functional Na/H exchanger. *Gastroenterology* 122:1322–33

45. Shallat S, Schmidt L, Reaka A, Rao D, Chang EB, et al. 1995. NHE-1 isoform of the Na⁺/H⁺ antiport is expressed in the rat and rabbit esophagus. *Gastroenterology* 109:1421–28

46. Bachmann O, Sonnentag T, Siegel WK, Lamprecht G, Weichert A, et al. 1998. Different acid secretagogues activate different Na⁺/H⁺ exchanger isoforms in rabbit parietal cells. *Am. J. Physiol. Gastrointest. Liver Physiol.* 275:G1085–93

47. Kaneko K, Guth PH, Kaunitz JD. 1992. Na⁺/H⁺ exchange regulates intracellular pH of rat gastric surface cells in vivo. *Pflügers Arch.* 421:322–28

48. Schultheis PJ, Clarke LL, Meneton P, Harline M, Boivin GP, et al. 1998. Targeted disruption of the murine Na⁺/H⁺ exchanger isoform 2 gene causes reduced viability of gastric parietal cells and loss of net acid secretion. *J. Clin. Invest.* 101:1243–53

49. Seidler B, Rossmann H, Murray A, Orlowski J, Tse CM, et al. 1997. Expression of the Na⁺/H⁺ exchanger isoform NHE1–4 mRNA in different epithelial cell types of rat and rabbit gastric mucosa. *Gastroenterology* 110:A285 (Abstr.)

50. Repishti M, Hogan DL, Pratha V, Davy-

dova L, Donowitz M, et al. 2001. Human duodenal mucosal brush border Na⁺/H⁺ exchangers NHE2 and NHE3 alter net bicarbonate movement. *Am. J. Physiol. Gastrointest. Liver Physiol.* 281:G159–63

51. Furukawa O, Bi LC, Guth PH, Engel E, Hirokawa M, Kaunitz JD. 2003. NHE3 inhibition activates duodenal bicarbonate secretion in the rat. *Am. J. Physiol. Gastrointest. Liver Physiol.* 286:G102–9

52. Hoogerwerf WA, Tsao SC, Devuyst O, Levine SA, Yun CH, et al. 1996. NHE2 and NHE3 are human and rabbit intestinal brush-border proteins. *Am. J. Physiol. Gastrointest. Liver Physiol.* 270:G29–41

53. Bookstein C, DePaoli AM, Xie Y, Niu P, Musch MW, et al. 1994. Na⁺/H⁺ exchangers, NHE-1 and NHE-3, of rat intestine. Expression and localization. *J. Clin. Invest.* 93:106–13

54. Bookstein C, Xie Y, Rabenau K, Musch MW, McSwine RL, et al. 1997. Tissue distribution of Na⁺/H⁺ exchanger isoforms NHE2 and NHE4 in rat intestine and kidney. *Am. J. Physiol. Cell Physiol.* 273:C1496–505

55. Dudeja PK, Rao DD, Syed I, Joshi V, Dahdal RY, et al. 1996. Intestinal distribution of human Na⁺/H⁺ exchanger isoforms NHE-1, NHE-2, and NHE-3 mRNA. *Am. J. Physiol. Gastrointest. Liver Physiol.* 271:G483–93

56. Malakooti J, Dahdal RY, Schmidt L, Layden TJ, Dudeja PK, Ramaswamy K. 1999. Molecular cloning, tissue distribution, and functional expression of the human Na⁺/H⁺ exchanger NHE2. *Am. J. Physiol. Gastrointest. Liver Physiol.* 277:G383–90

57. Musch MW, Bookstein C, Rocha F, Lucioni A, Ren H, et al. 2002. Region-specific adaptation of apical Na/H exchangers after extensive proximal small bowel resection. *Am. J. Physiol. Gastrointest. Liver Physiol.* 283:G975–85

58. Chu J, Chu S, Montrose MH. 2002. Apical Na⁺/H⁺ exchange near the base of mouse

colonic crypts. *Am. J. Physiol. Cell Physiol.* 283:C358–72

59. Gonda T, Maouyo D, Rees SE, Montrose MH. 1999. Regulation of intracellular pH gradients by identified Na/H exchanger isoforms and a short-chain fatty acid. *Am. J. Physiol. Gastrointest. Liver Physiol.* 276:G259–70

60. Mennone A, Biemesderfer D, Negoianu D, Yang CL, Abbiati T, et al. 2001. Role of sodium/hydrogen exchanger isoform NHE3 in fluid secretion and absorption in mouse and rat cholangiocytes. *Am. J. Physiol. Gastrointest. Liver Physiol.* 280:G247–54

61. Silviani V, Colombani V, Heyries L, Gerolami A, Cartouzou G, Marteau C. 1996. Role of the NHE3 isoform of the Na+/H+ exchanger in sodium absorption by the rabbit gallbladder. *Pflügers Arch.* 432:791–96

62. Abedin MZ, Giurgiu DI, Abedin ZR, Peck EA, Su X, Smith PR. 2001. Characterization of Na+/H+ exchanger isoform (NHE1, NH32 and NHE3) expression in prairie dog gallbladder. *J. Membr. Biol.* 182:123–34

63. Cremaschi D, Porta C, Botta G, Meyer G. 1992. Nature of the neutral Na+-Cl− coupled entry at the apical membrane of rabbit gallbladder epithelium: IV. Na+/H+, Cl−/HCO3− double exchange, hydrochlorothiazide-sensitive Na+-Cl− symport and Na+-K+-2Cl− cotransport are all involved. *J. Membr. Biol.* 129:221–35

64. Collins JF, Xu H, Kiela PR, Zeng J, Ghishan FK. 1997. Functional and molecular characterization of NHE3 expression during ontogeny in rat jejunal epithelium. *Am. J. Physiol. Cell Physiol.* 273:C1937–46

65. Collins JF, Kiela PR, Xu H, Zeng J, Ghishan FK. 1998. Increased NHE2 expression in rat intestinal epithelium during ontogeny is transcriptionally mediated. *Am. J. Physiol. Cell Physiol.* 275:C1143–50

66. Chow CW, Khurana S, Woodside M, Grinstein S, Orlowski J. 1999. The epithelial Na+/H+ exchanger, NHE3, is internalized through a clathrin-mediated pathway. *J. Biol. Chem.* 274:37551–58

67. Kirchhoff P, Wagner CA, Gaetzschmann F, Radebold K, Geibel JP. 2003. Demonstration of a functional apical sodium hydrogen exchange in isolated rat gastric glands. *Am. J. Physiol. Gastrointest. Liver Physiol.* 285:G1242–48

68. Kurashima K, Szabo EZ, Lukacs G, Orlowski J, Grinstein S. 1998. Endosomal recycling of the Na+/H+ exchanger NHE3 isoform is regulated by the phosphatidylinositol 3-kinase pathway. *J. Biol. Chem.* 273:20828–36

69. Li X, Galli T, Leu S, Wade JB, Weinman EJ, et al. 2001. Na+-H+ exchanger 3 (NHE3) is present in lipid rafts in the rabbit ileal brush border: a role for rafts in trafficking and rapid stimulation of NHE3. *J. Physiol.* 537:537–52

70. Janecki AJ, Montrose MH, Tse CM, de Medina FS, Zweibaum A, Donowitz M. 1999. Development of an endogenous epithelial Na+/H+ exchanger (NHE3) in three clones of caco-2 cells. *Am. J. Physiol. Gastrointest. Liver Physiol.* 277:G292–305

71. McSwine RL, Musch MW, Bookstein C, Xie Y, Rao M, Chang EB. 1998. Regulation of apical membrane Na+/H+ exchangers NHE2 and NHE3 in intestinal epithelial cell line C2/bbe. *Am. J. Physiol. Cell Physiol.* 275:C693–701

72. Wormmeester L, Sanchez de Medina F, Kokke F, Tse CM, Khurana S, et al. 1998. Quantitative contribution of NHE2 and NHE3 to rabbit ileal brush-border Na+/H+ exchange. *Am. J. Physiol. Cell Physiol.* 274:C1261–72

73. Donowitz M, De La Horra C, Calonge ML, Wood IS, Dyer J, et al. 1998. In birds, NHE2 is major brush-border Na+/H+ exchanger in colon and is increased by a low-NaCl diet. *Am. J. Physiol. Renal Physiol.* 274:R1659–69

74. Ikuma M, Kashgarian M, Binder HJ, Rajendran VM. 1999. Differential

regulation of NHE isoforms by sodium depletion in proximal and distal segments of rat colon. *Am. J. Physiol. Gastrointest. Liver Physiol.* 276:G539–49

75. Bachmann O, Riederer B, Rossmann H, Groos S, Schultheis PJ, et al. 2004. The Na^+/H^+ exchanger isoform 2 is the predominant NHE isoform in murine colonic crypts and its lack causes NHE3 upregulation. *Am. J. Physiol. Gastrointest. Liver Physiol.* 287:G125–33

76. Good DW, George T, Watts BA 3rd. 1995. Basolateral membrane Na^+/H^+ exchange enhances HCO_3^- absorption in rat medullary thick ascending limb: evidence for functional coupling between basolateral and apical membrane Na^+/H^+ exchangers. *Proc. Natl. Acad. Sci. USA* 92:12525–29

77. Holmberg C, Perheentupa J. 1985. Congenital Na^+ diarrhea: a new type of secretory diarrhea. *J. Pediatr.* 106:56–61

78. Booth IW, Stange G, Murer H, Fenton TR, Milla PJ. 1985. Defective jejunal brush-border Na^+/H^+ exchange: a cause of congenital secretory diarrhoea. *Lancet* 1:1066–69

79. Keller KM, Wirth S, Baumann W, Sule D, Booth IW. 1990. Defective jejunal brush border membrane sodium/proton exchange in association with lethal familial protracted diarrhoea. *Gut* 31:1156–58

80. Schultheis PJ, Clarke LL, Meneton P, Harline M, Boivin GP, et al. 1998. Targeted disruption of the murine Na^+/H^+ exchanger isoform 2 gene causes reduced viability of gastric parietal cells and loss of net acid secretion. *J. Clin. Invest.* 101:1243–53

81. Bell SM, Schreiner CM, Schultheis PJ, Miller ML, Evans RL, et al. 2001. Targeted disruption of the murine Nhe1 locus induces ataxia, growth retardation, and seizures. *Am. J. Physiol. Cell Physiol.* 276:C788–95

82. Brown DA, Melvin JE, Yule DI. 2003. Critical role for NHE1 in intracellular pH regulation in pancreatic acinar cells. *Am. J. Physiol. Gastrointest. Liver Physiol.* 285:G804–12

83. Cox GA, Lutz CM, Yang CL, Biemesderfer D, Bronson RT, et al. 1997. Sodium/hydrogen exchanger gene defect in slow-wave epilepsy mutant mice. *Cell* 91:139–48

84. Xue J, Douglas RM, Zhou D, Lim JY, Boron WF, Haddad GG. 2003. Expression of Na^+/H^+ and HCO_3^--dependent transporters in Na^+/H^+ exchanger isoform 1 null mutant mouse brain. *Neuroscience* 122:37–46

85. Furukawa O, Matsui H, Suzuki N, Okabe S. 1999. Epidermal growth factor protects rat epithelial cells against acid-induced damage through the activation of Na^+/H^+ exchangers. *J. Pharmacol. Exp. Ther.* 288:620–26

86. Gawenis LR, Miller M, Greeb J, Shull GE. 2004. Impaired gastric acid secretion in NHE4 Na^+/H^+ exchanger knockout mice. *Gastroenterology* 126:A637

87. Gawenis LR, Stien X, Shull GE, Schultheis PJ, Woo AL, et al. 2002. Intestinal NaCl transport in NHE2 and NHE3 knockout mice. *Am. J. Physiol. Gastrointest. Liver Physiol.* 282:G776–84

88. Ledoussal C, Woo AL, Miller ML, Shull GE. 2001. Loss of the NHE2 Na^+/H^+ exchanger has no apparent effect on diarrheal state of NHE3-deficient mice. *Am. J. Physiol. Gastrointest. Liver Physiol.* 281:G1385–96

89. Spicer Z, Clarke LL, Gawenis LR, Shull GE. 2001. Colonic H^+-K^+-ATPase in K^+ conservation and electrogenic Na^+ absorption during Na^+ restriction. *Am. J. Physiol. Gastrointest. Liver Physiol.* 281:G1369–77

90. Woo AL, Gildea LA, Tack LM, Miller ML, Spicer Z, et al. 2002. In vivo evidence for interferon-gamma-mediated homeostatic mechanisms in small intestine of the NHE3 Na^+/H^+ exchanger knockout model of congenital diarrhea. *J. Biol. Chem.* 277:49036–46

91. Yoo D, Lo W, Goodman S, Ali W, Semrad C, Field M. 2000. Interferon-gamma downregulates ion transport in murine small intestine cultured in vitro. *Am. J. Physiol. Gastrointest. Liver Physiol.* 279:G1323–32

92. Sugi K, Musch MW, Field M, Chang EB. 2001. Inhibition of Na$^+$, K$^+$-ATPase by interferon gamma down-regulates intestinal epithelial transport and barrier function. *Gastroenterology* 120:1393–403

93. Colgan SP, Parkos CA, Matthews JB, D'Andrea L, Awtrey CS, et al. 1994. Interferon-gamma induces a cell surface phenotype switch on T84 intestinal epithelial cells. *Am. J. Physiol. Cell Physiol.* 267:C402–10

94. Rocha F, Musch MW, Lishanskiy L, Bookstein C, Sugi K, et al. 2001. IFN-gamma downregulates expression of Na$^+$/H$^+$ exchangers NHE2 and NHE3 in rat intestine and human Caco-2/bbe cells. *Am. J. Physiol. Cell Physiol.* 280:C1224–32

95. Bookstein C, Musch MW, Xie Y, Rao MC, Chang EB. 1999. Regulation of intestinal epithelial brush border Na$^+$/H$^+$ exchanger isoforms, NHE2 and NHE3, in C2bbe cells. *J. Membr. Biol.* 171:87–95

96. Kim JH, Lee-Kwon W, Park JB, Ryu SH, Yun CH, Donowitz M. 2002. Ca^{2+}-dependent inhibition of Na$^+$/H$^+$ exchanger 3 (NHE3) requires an NHE3-E3KARP-alpha-actinin-4 complex for oligomerization and endocytosis. *J. Biol. Chem.* 277:23714–24

97. Cha B, Oh S, Shanmugaratnam J, Donowitz M, Yun CC. 2003. Two histidine residues in the juxta-membrane cytoplasmic domain of Na$^+$/H$^+$ exchanger isoform 3 (NHE3) determine the set point. *J. Membr. Biol.* 191:49–58

98. Nath SK, Kambadur R, Yun CH, Donowitz M, Tse CM. 1999. NHE2 contains subdomains in the COOH terminus for growth factor and protein kinase regulation. *Am. J. Physiol. Cell Physiol.* 276:C873–82

99. Di Sole F, Cerull R, Petzke S, Casavola V, Burckhardt G, Helmle-Kolb C. 2003. Bimodal acute effects of A1 adenosine receptor activation on Na$^+$/H$^+$ exchanger 3 in opossum kidney cells. *J. Am. Soc. Nephrol.* 14:1720–30

100. Lamprecht G, Weinman EJ, Yun CH. 1998. The role of NHERF and E3KARP in the cAMP-mediated inhibition of NHE3. *J. Biol. Chem.* 273:29972–78

101. Yun CH, Lamprecht G, Forster DV, Sidor A. 1998. NHE3 kinase A regulatory protein E3KARP binds the epithelial brush border Na$^+$/H$^+$ exchanger NHE3 and the cytoskeletal protein ezrin. *J. Biol. Chem.* 273:25856–63

102. Dransfield DT, Bradford AJ, Smith J, Martin M, Roy C, et al. 1997. Ezrin is a cyclic AMP-dependent protein kinase anchoring protein. *EMBO J.* 16:35–43

103. Wakabayashi S, Bertrand B, Shigekawa M, Fafournoux P, Pouyssegur J. 1994. Growth factor activation and "H$^+$-sensing" of the Na$^+$/H$^+$ exchanger isoform 1 (NHE1). Evidence for an additional mechanism not requiring direct phosphorylation. *J. Biol. Chem.* 269:5583–88

104. Kandasamy RA, Yu FH, Harris R, Boucher A, Hanrahan JW, Orlowski J. 1995. Plasma membrane Na$^+$/H$^+$ exchanger isoforms (NHE-1, -2, and -3) are differentially responsive to second messenger agonists of the protein kinase A and C pathways. *J. Biol. Chem.* 270:29209–16

105. Attaphitaya S, Nehrke K, Melvin JE. 2001. Acute inhibition of brain-specific Na$^+$/H$^+$ exchanger isoform 5 by protein kinases A and C and cell shrinkage. *Am. J. Physiol. Cell Physiol.* 281:C1146–57

106. Gill RK, Saksena S, Syed IA, Tyagi S, Alrefai WA, et al. 2002. Regulation of NHE3 by nitric oxide in Caco-2 cells. *Am. J. Physiol. Gastrointest. Liver Physiol.* 283:G747–56

107. Lee-Kwon W, Kim JH, Choi JW, Kawano K, Cha B, et al. 2003. Ca^{2+}-dependent inhibition of NHE3 requires PKC alpha

which binds to E3KARP to decrease surface NHE3 containing plasma membrane complexes. *Am. J. Physiol. Cell Physiol.* 285:C1527–36

108. Lee-Kwon W, Kawano K, Choi JW, Kim JH, Donowitz M. 2003. Lysophosphatidic acid stimulates brush border Na^+/H^+ exchanger NHE3 activity by increasing its exocytosis by an NHE3 kinase A regulatory protein-dependent mechanism. *J. Biol. Chem.* 278:16494–501

109. Peng Y, Amemiya M, Yang X, Fan L, Moe OW, et al. 2001. ET(B) receptor activation causes exocytic insertion of NHE3 in OKP cells. *Am. J. Physiol. Renal Physiol.* 280:F34–42

110. Janecki AJ, Janecki M, Akhter S, Donowitz M. 2000. Basic fibroblast growth factor stimulates surface expression and activity of Na^+/H^+ exchanger NHE3 via mechanism involving phosphatidylinositol 3-kinase. *J. Biol. Chem.* 275:8133–42

111. Donowitz M, Janecki A, Akhter S, Cavet ME, Sanchez F, et al. 2000. Short-term regulation of NHE3 by EGF and protein kinase C but not protein kinase A involves vesicle trafficking in epithelial cells and fibroblasts. *Ann. NY Acad. Sci.* 915:30–42

112. Shrode LD, Gan BS, D'Souza SJ, Orlowski J, Grinstein S. 1998. Topological analysis of NHE1, the ubiquitous Na^+/H^+ exchanger using chymotryptic cleavage. *Am. J. Physiol. Cell Physiol.* 75:C431–39

113. Lin CY, Varma MG, Joubel A, Srinivasan M, Lichtarge O, Barber DL. 2003. Conserved motifs in somatostatin, D2-dopamine and α_{2B}-adrenergic receptors for inhibiting the Na/H exchanger, NHE1. *J. Biol. Chem.* 278:15128–35

114. Lee-Kwon W, Johns DC, Cha B, Cavet M, Park J, et al. 2001. Constitutively active phosphatidylinositol 3-kinase and AKT are sufficient to stimulate the epithelial Na^+/H^+ exchanger 3. *J. Biol. Chem.* 276:31296–304

115. Levine SA, Nath SK, Yun CH, Yip JW, Montrose M, et al. 1995. Separate C-terminal domains of the epithelial specific brush border Na^+/H^+ exchanger isoform NHE3 are involved in stimulation and inhibition by protein kinases/growth factors. *J. Biol. Chem.* 270:13716–25

116. Li X, Alvarez B, Casey JR, Reithmeier RA, Fliegel L. 2002. Carbonic anhydrase II binds to and enhances activity of the Na^+/H^+ exchanger. *J. Biol. Chem.* 277:36085–91

117. Cabado AG, Yu FH, Kapus A, Lukacs G, Grinstein S, Orlowski J. 1996. Distinct structural domains confer cAMP sensitivity and ATP dependence to the Na^+/H^+ exchanger NHE3 isoform. *J. Biol. Chem.* 271:3590–99

118. Yun CH, Tse CM, Donowitz M. 1995. Chimeric Na^+/H^+ exchangers: an epithelial membrane-bound N-terminal domain requires an epithelial cytoplasmic C-terminal domain for regulation by protein kinases. *Proc. Natl. Acad. Sci. USA* 92:10723–27

119. Su X, Pang T, Wakabayashi S, Shigekawa M. 2003. Evidence for involvement of the putative first extracellular loop in differential volume sensitivity of the Na^+/H^+ exchangers NHE1 and NHE2. *Biochemistry* 42:1086–94

120. Wakabayashi S, Bertrand B, Ikeda T, Pouyssegur J, Shigekawa M. 1994. Mutation of calmodulin-binding site renders the Na^+/H^+ exchanger (NHE1) highly H^+-sensitive and Ca^{2+} regulation-defective. *J. Biol. Chem.* 269:13710–15

121. Wakabayashi S, Ikeda T, Iwamoto T, Pouyssegur J, Shigekawa M. 1997. Calmodulin-binding autoinhibitory domain controls "pH-sensing" in the Na^+/H^+ exchanger NHE1 through sequence-specific interaction. *Biochemistry* 36:12854–61

122. Bertrand B, Wakabayashi S, Ikeda T, Pouyssegur J, Shigekawa M. 1994. The Na^+/H^+ exchanger isoform 1 (NHE1) is a novel member of the calmodulin-binding proteins. Identification and characterization of calmodulin-binding sites. *J. Biol. Chem.* 269:13703–9

123. Aharonovitz O, Zaun HC, Balla T, York JD, Orlowski J, Grinstein S. 2000. Intracellular pH regulation by Na^+/H^+ exchange requires phosphatidylinositol 4,5-bisphosphate. *J. Cell Biol.* 150:213–24

124. Lin X, Barber DL. 1996. A calcineurin homologous protein inhibits GTPase-stimulated Na-H exchange. *Proc. Natl. Acad. Sci. USA* 93:12631–36

125. Pang T, Su X, Wakabayashi S, Shigekawa M. 2001. Calcineurin homologous protein as an essential cofactor for Na^+/H^+ exchangers. *J. Biol. Chem.* 276:17367–72

126. Pang T, Wakabayashi S, Shigekawa M. 2002. Expression of calcineurin B homologous protein 2 protects serum deprivation-induced cell death by serum-independent activation of Na^+/H^+ exchanger. *J. Biol. Chem.* 277:43771–77

127. Denker SP, Huang DC, Orlowski J, Furthmayr H, Barber DL. 2000. Direct binding of the Na-H exchanger NHE1 to ERM proteins regulates the cortical cytoskeleton and cell shape independently of H^+ translocation. *Mol. Cell.* 6:1425–36

128. Putney LK, Denker SP, Barber DL. 2002. The changing face of the Na^+/H^+ exchanger, NHE1: structure, regulation, and cellular actions. *Annu. Rev. Pharmacol. Toxicol.* 42:527–52

129. Li X, Liu Y, Kay CM, Muller-Esterl W, Fliegel L. 2003. The Na^+/H^+ exchanger cytoplasmic tail: structure, function, and interactions with tescalcin. *Biochemistry* 42:7448–56

130. Mailander J, Muller-Esterl W, Dedio J. 2001. Human homolog of mouse tescalcin associates with Na^+/H^+ exchanger type-1. *FEBS Lett.* 507:331–35

131. Yan W, Nehrke K, Choi J, Barber DL. 2001. The Nck-interacting kinase (NIK) phosphorylates the Na^+-H^+ exchanger NHE1 and regulates NHE1 activation by platelet-derived growth factor. *J. Biol. Chem.* 276:31349–56

132. Lehoux S, Abe Ji, Florian JA, Berk BC. 2001. 14-3-3 Binding to Na^+/H^+ exchanger isoform-1 is associated with serum-dependent activation of Na^+/H^+ exchange. *J. Biol. Chem.* 276:15794–800

133. Silva NL, Haworth RS, Singh D, Fliegel L. 1995. The carboxyl-terminal region of the Na^+/H^+ exchanger interacts with mammalian heat shock protein. *Biochemistry* 34:10412–20

134. Li X, Alvarez B, Casey JR, Reithmeier RA, Fliegel L. 2002. Carbonic anhydrase II binds to and enhances activity of the Na^+/H^+ exchanger. *J. Biol. Chem.* 277:36085–91

135. Chow CW, Woodside M, Demaurex N, Yu FH, Plant P, et al. 1999. Proline-rich motifs of the Na^+/H^+ exchanger 2 isoform. Binding of Src homology domain 3 and role in apical targeting in epithelia. *J. Biol. Chem.* 274:10481–88

136. Biemesderfer D, Nagy T, DeGray B, Aronson PS. 1999. Specific association of megalin and the Na^+/H^+ exchanger isoform NHE3 in the proximal tubule. *J. Biol. Chem.* 274:17518–24

137. Girardi AC, Degray BC, Nagy T, Biemesderfer D, Aronson PS. 2001. Association of Na^+-H^+ exchanger isoform NHE3 and dipeptidyl peptidase IV in the renal proximal tubule. *J. Biol. Chem.* 276:46671–77

138. Quinones H, McLeroy P, Hu MC, Price E, Mumby MC, Moe OW. 2001. Protein phosphatases (PP2A) and the acute regulation of Na^+/H^+ exchanger NHE3 by dopamine (DA): direction interaction of PP2A with NHE3. *J. Am. Soc. Neph.* 12:8A

139. Gisler SM, Pribanic S, Bacic D, Forrer P, Gantenbein A, et al. 2003. PDZK1: I. a major scaffolder in brush borders of proximal tubular cells. *Kidney Int.* 64:1733–45

140. Li X, Zhang H, Cha B, Leu S, Akhter S, Donowitz M. 2004. Heat shock cognate protein 70 (Hsc70) interacts with Na^+/H^+ exchanger 3 (NHE3) and this interaction is necessary for NHE3 activity. *Gastroenterology* 126:A43

141. Donowitz M, Kambadur R, Zizak M, Nath S, Akhter S, Tse M. 1997. NHE3

is a calmodulin binding protein: Calmodulin inhibits NHE3 by binding to the C-terminal 76 amino acids of NHE3. *Gastroenterology* 138:A360

142. Wu KL, Khan S, Lakhe-Reddy S, Jarad G, Mukherjee A, et al. 2004. The NHE1 Na^+/H^+ exchanger recruits ERM proteins to regulate Akt-dependent cell survival. *J. Biol. Chem.* 279:26280–86

143. Kurashima K, D'Souza S, Szaszi K, Ramjeesingh R, Orlowski J, Grinstein S. 1999. The apical Na^+/H^+ exchanger isoform NHE3 is regulated by the actin cytoskeleton. *J. Biol. Chem.* 274:29843–49

144. Szaszi K, Kurashima K, Kapus A, Paulsen A, Kaibuchi K, et al. 2000. RhoA and rho kinase regulate the epithelial Na^+/H^+ exchanger NHE3. Role of myosin light chain phosphorylation. *J. Biol. Chem.* 275:28599–606

145. Szaszi K, Kurashima K, Kaibuchi K, Grinstein S, Orlowski J. 2001. Role of the cytoskeleton in mediating cAMP-dependent protein kinase inhibition of the epithelial Na^+/H^+ exchanger NHE3. *J. Biol. Chem.* 276:40761–68

146. Sabolic I, Herak-Kramberger CM, Ljubojevic M, Biemesderfer D, Brown D. 2002. NHE3 and NHERF are targeted to the basolateral membrane in proximal tubules of colchicines-treated rats. *Kidney Int.* 61:1351–64

147. Hayashi H, Szaszi K, Coady-Osberg N, Furuya W, Bretscher AP, et al. 2004. Inhibition and redistribution of NHE3, the apical Na^+/H^+ exchanger, by clostridium difficile toxin B. *J. Gen. Physiol.* 123:491–504

148. Cha B, Kenworthy A, Murtazina R, Donowitz M. 2004. The lateral mobility of NHE3 on the apical membrane of renal epithelial OK cells is limited by the PDZ domain protein NheRF1/2 but is dependent on an intact actin cytoskeleton as determined by FRAP. *J. Cell Sci.* 117:3353–65

149. Sardet C, Counillon L, Franchi A, Pouyssegur J. 1990. Growth factors induce phosphorylation of the Na^+/H^+ antiporter, glycoprotein of 110 kD. *Science* 247:723–26

150. Yip JW, Ko WH, Viberti G, Huganir RL, Donowitz M, Tse CM. 1997. Regulation of the epithelial brush border Na^+/H^+ exchanger isoform 3 stably expressed in fibroblasts by fibroblast growth factor and phorbol esters is not through changes in phosphorylation of the exchanger. *J. Biol. Chem.* 272:18473–80

151. Moe OW, Amemiya M, Yamaji Y. 1995. Activation of protein kinase A acutely inhibits and phosphorylates Na/H exchanger NHE-3. *J. Clin. Invest.* 96:2187–94

152. Kurashima K, Yu FH, Cabado AG, Szabo EZ, Grinstein S, Orlowski J. 1997. Identification of sites required for down-regulation of Na^+/H^+ exchanger NHE3 activity by cAMP-dependent protein kinase. Phosphorylation-dependent and -independent mechanisms. *J. Biol. Chem.* 272:28672–79

153. Zhao H, Wiederkehr MR, Fan L, Collazo RL, Crowder LA, Moe OW. 1999. Acute inhibition of Na/H exchanger NHE-3 by cAMP. Role of protein kinase A and NHE-3 phosphoserines 552 and 605. *J. Biol. Chem.* 274:3978–87

154. Takahashi E, Abe J, Gallis B, Aebersold R, Spring DJ, et al. 1999. p90(RSK) is a serum-stimulated Na^+/H^+ exchanger isoform-1 kinase. Regulatory phosphorylation of serine 703 of Na^+/H^+ exchanger isoform-1. *J. Biol. Chem.* 274:20206–14

155. Aitken A, Baxter H, Dubois T, Clokie S, Mackie S, et al. 2002. Specificity of 14-3-3 isoform dimer interactions and phosphorylation. *Biochem. Soc. Trans.* 30:351–60

156. Zizak M, Lamprecht G, Steplock D, Tariq N, Shenolikar S, et al. 1999. cAMP-induced phosphorylation and inhibition of Na^+/H^+ exchanger (NHE3) is dependent on the presence but not the phosphorylation of NHERF. *J. Biol. Chem.* 274:24753–58

157. Wiederkehr MR, Zhao H, Moe OW. 1999. Acute regulation of Na/H exchanger NHE3 activity by protein kinase C: role of NHE3 phosphorylation. *Am. J. Physiol. Cell Physiol.* 276:C1205–17

158. Muller YL, Collins JF, Bai L, Xu H, Ghishan FK. 1998. Molecular cloning and characterization of the rat NHE-2 gene promoter. *Biochim. Biophys. Acta* 1442: 314–19

159. Kandasamy RA, Orlowski J. 1996. Genomic organization and glucocorticoid transcriptional activation of the rat Na^+/H^+ exchanger NHE3 gene. *J. Biol. Chem.* 271:10551–59

160. Malakooti J, Memark VC, Dudeja PK, Ramaswamy K. 2002. Molecular cloning and functional analysis of the human Na^+/H^+ exchanger NHE3 promoter. *Am. J. Physiol. Gastrointest. Liver Physiol.* 282:G491–500

161. Sacks AI, Acra SA, Dykes W, Polk DB, Barnard JA, et al. 1993. Intestinal Na^+/H^+ exchanger activity is up-regulated by bowel resection in the weanling rat. *Pediatr. Res.* 33:215–20

162. Falcone RA Jr, Shin CE, Stern LE, Wang Z, Erwin CR, et al. 1999. Differential expression of ileal Na^+/H^+ exchanger isoforms after enterectomy. *J. Surg. Res.* 86:192–97

163. Musch MW, Bookstein C, Xie Y, Sellin JH, Chang EB. 2001. SCFA increase intestinal Na absorption by induction of NHE3 in rat colon and human intestine C2/bbe cells. *Am. J. Physiol. Gastrointest. Liver Physiol.* 280:G687–93

164. Yun CH, Gurubhagavatula S, Levine SA, Montgomery JL, Brant SR, et al. 1993. Glucocorticoid stimulation of ileal Na^+ absorptive cell brush border Na^+/H^+ exchange and association with an increase in message for NHE-3, an epithelial Na^+/H^+ exchanger isoform. *J. Biol. Chem.* 268:206–11

165. Cho JH, Musch MW, DePaoli AM, Bookstein CM, Xie Y, et al. 1994. Glucocorticoids regulate Na^+/H^+ exchange expression and activity in region- and tissue-specific manner. *Am. J. Physiol. Cell Physiol.* 267:C796–803

166. Cho JH, Musch MW, Bookstein CM, McSwine RL, Rabenau K, Chang EB. 1998. Aldosterone stimulates intestinal Na^+ absorption in rats by increasing NHE3 expression of the proximal colon. *Am. J. Physiol. Cell Physiol.* 274:C586–94

167. Yun CC, Chen Y, Lang F. 2002. Glucocorticoid activation of Na^+/H^+ exchanger isoform 3 revisited. The roles of SGK1 and NHERF2. *J. Biol. Chem.* 277:7676–83

168. Charney AN, Dagher PC. 1996. Acid-base effects on colonic electrolyte transport revisited. *Gastroenterology* 111:1358–68

169. Charney AN, Feldman GM. 1984. Systemic acid-base disorders and intestinal electrolyte transport. *Am. J. Physiol. Gastrointest. Liver Physiol.* 247:G1–12

170. Lucioni A, Womack C, Musch MW, Rocha FL, Bookstein C, Chang EB. 2002. Metabolic acidosis in rats increases intestinal NHE2 and NHE3 expression and function. *Am. J. Physiol. Gastrointest. Liver Physiol.* 283:G51–56

171. Sandle GI, Higgs N, Crowe P, Marsh MN, Venkatesan S, Peters TJ. 1990. Cellular basis for defective electrolyte transport in inflamed human colon. *Gastroenterology* 99:97–105

172. Barmeyer C, Harren M, Schmitz H, Heinzel-Pleines U, Mankertz J, et al. 2004. Mechanisms of diarrhea in the interleukin-2-deficient mouse model of colonic inflammation. *Am. J. Physiol. Gastrointest. Liver Physiol.* 286:G244–52

173. Ameen NA, Salas PJ. 2000. Microvillus inclusion disease: a genetic defect affecting apical membrane protein traffic in intestinal epithelium. *Traffic* 1:76–83

174. Muller T, Wijmenga C, Phillips AD, Janecke A, Houwen RH, et al. 2000. Congenital sodium diarrhea is an autosomal recessive disorder of sodium/proton exchange but unrelated to known candidate genes. *Gastroenterology* 119:1506–13

Annu. Rev. Physiol. 2005. 67:445–69
doi: 10.1146/annurev.physiol.67.041703.084745
First published online as a Review in Advance on September 27, 2004

REGULATION OF FLUID AND ELECTROLYTE SECRETION IN SALIVARY GLAND ACINAR CELLS

James E. Melvin,[1] David Yule,[2] Trevor Shuttleworth,[2] and Ted Begenisich[2]

[1]*The Center for Oral Biology in the Aab Institute of Biomedical Sciences and the* [2]*Department of Pharmacology and Physiology, University of Rochester School of Medicine and Dentistry, Rochester, New York, 14642;*
email: james_melvin@urmc.rochester.edu; david_yule@urmc.rochester.edu;
trevor_shuttleworth@urmc.rochester.edu; ted_begenisich@urmc.rochester.edu

Key Words channels, exchangers, cotransporters, Ca^{2+} mobilization, IP_3 receptors

■ **Abstract** The secretion of fluid and electrolytes by salivary gland acinar cells requires the coordinated regulation of multiple water and ion transporter and channel proteins. Notably, all the key transporter and channel proteins in this process appear to be activated, or are up-regulated, by an increase in the intracellular Ca^{2+} concentration ($[Ca^{2+}]_i$). Consequently, salivation occurs in response to agonists that generate an increase in $[Ca^{2+}]_i$. The mechanisms that act to modulate these increases in $[Ca^{2+}]_i$ obviously influence the secretion of salivary fluid. Such modulation may involve effects on mechanisms of both Ca^{2+} release and Ca^{2+} entry and the resulting spatial and temporal aspects of the $[Ca^{2+}]_i$ signal, as well as interactions with other signaling pathways in the cells. The molecular cloning of many of the transporter and regulatory molecules involved in fluid and electrolyte secretion has yielded a better understanding of this process at the cellular level. The subsequent characterization of mice with null mutations in many of these genes has demonstrated the physiological roles of individual proteins. This review focuses on recent developments in determining the molecular identification of the proteins that regulate the fluid secretion process.

INTRODUCTION

Mammals express three major pairs of salivary glands (parotid, submandibular, and sublingual) along with numerous minor glands that are scattered throughout the oral cavity. Glands are composed of two basic cell types, acinar cells, which make up the secretory endpiece, and duct cells, of which there are several varieties (intercalated, striated, etc.). Salivary acinar cells can be either serous or mucous, with the relative distribution of these two cell types in individual glands being species- and gland-specific. The secretion from salivary glands is typically a watery fluid containing electrolytes and a complex mixture of proteins, although

0066-4278/05/0315-0445$14.00

mucous gland secretions are often quite viscous owing to the discharge of large molecular weight mucins. The final composition of saliva varies depending on the origin of the stimulation, i.e., sympathetic versus parasympathetic, and the type of salivary gland. Humans typically secrete more than a liter of saliva each day, the major portion coming from the parotid and submandibular glands. Saliva provides protection and hydration for mucosal structures within the oral cavity, oropharynx, and esophagus. Some of the specialized functions of saliva include the initiation of digestion, antimicrobial defense, and protection from mechanical and chemical insults. Salivation is a highly regulated process, occurring at relatively slow rates between meals, and there is almost none generated during sleep. In contrast, 80–90% of the daily saliva output is produced in response to stimuli such as taste, smell, and masticatory forces during the few hours while eating each day.

The importance of saliva is clearly illustrated in individuals suffering from salivary gland hypofunction. Decreased salivation manifests clinically as oral pain, increased dental caries, and infections by opportunistic microorganisms such as *Candida albicans*. Salivary gland dysfunction is most frequently observed as a consequence of iatrogenic treatments such as medications or irradiation therapy for head and neck cancers, but it is also regularly observed in diseases such as cystic fibrosis, where ductal obstructions are common, or Sjögren's syndrome, an autoimmune disease in which salivary glands are destroyed by lymphocytic infiltration. For a thorough discussion of the histology, function, and disorders of salivary glands see (1–6).

Similar to the fluid secreted by most exocrine organs, e.g., sweat and lacrimal glands (7, 8), saliva formation involves two stages. Micropuncture and microperfusion studies demonstrated that the primary fluid secreted by salivary acinar cells is a plasma-like, isotonic fluid (stage 1). This NaCl-rich solution is subsequently modified upon passage through the ducts (stage 2). Acinar cells are thought to produce all of the fluid in saliva. As the acinar cell secretion passes through the duct system, NaCl is reabsorbed, while K^+ and HCO_3^- are excreted. The hypotonicity of secreted saliva indicates that ducts are relatively impermeant to water. Indeed, depending on the flow rate and the mode of neurotransmitter stimulation, the osmolarity of saliva can be less than 100 mOsm. Acinar cells are also responsible for the secretion of most of the proteins found in saliva ($>85\%$), although duct cells secrete numerous proteins with important biological activities, e.g., nerve growth factor, epidermal growth factor, immunoglobin A, and kallikrein.

FLUID SECRETION MECHANISM

Cl^--Dependent Fluid and Electrolyte Secretion

The salivary gland fluid secretion model has been previously described in depth in several earlier reviews (see 1, 9–11). Therefore, only a brief overview is provided here, as the following sections focus on reviewing recent advances made in determining the molecular identities of the involved transporter mechanisms and in understanding the regulation of fluid secretion.

In currently accepted models, the transepithelial movement of Cl^- is the primary driving force for fluid and electrolyte secretion by salivary gland acinar cells (a typical salivary acinar cell is shown in Figure 1 (see color insert). The Na^+ pump, at the expenditure of ATP, extrudes 3 Na^+ in exchange for 2 extracellular K^+, producing a 10–15-fold inward-directed Na^+ chemical gradient. Various transporters use this gradient to elevate intracellular Cl^- levels. The major Na^+-dependent Cl^- uptake pathway in salivary glands is an electroneutral $Na^+/K^+/2Cl^-$ cotransporter located in the basolateral membrane. Most salivary gland acinar cells also possess a second Cl^- uptake pathway, the paired basolateral Cl^-/HCO_3^- and Na^+/H^+ exchangers. Together, these Na^+-dependent Cl^- uptake mechanisms concentrate intracellular Cl^- greater than five times above its electrochemical gradient (12, 13), a requirement for Cl^- exit via Cl^- channels. Agonist-stimulated secretion is initiated by the opening of K^+ and Cl^- channels located in the basolateral and apical membranes, respectively. Cl^- channels provide the pathway for Cl^- efflux into the acinar lumen. Activation of the K^+ channels is necessary to maintain the electrochemical driving force for Cl^- efflux. As predicted from the secretion model, activation of these channels produces a rapid loss of both intracellular K^+ and Cl^- (12–15). A transepithelial potential difference is created as Cl^- crosses the acinar apical membrane and K^+ enters the interstitial fluid. This lumen-negative, electrical potential difference leads to the passive movement of cations across the tight junctions of acinar cells. Luminal accumulation of ions generates a transepithelial osmotic gradient that drives the movement of water. This process results in the creation of a plasma-like primary secretion, the ionic composition of which reflects the Na^+, Cl^-, and K^+ concentrations of the interstitial fluid bathing the basolateral aspect of the acinar cells. Water movement appears to be mediated by paracellular pathways and transcellular transport via water channels.

HCO_3^--Dependent Fluid and Electrolyte Secretion

Cl^- uptake via $Na^+/K^+/2Cl^-$ cotransporters provides the major driving force for generating fluid secretion. However, it has been known for some time that significant saliva is produced when this bumetanide-sensitive $Na^+/K^+/2Cl^-$ cotransporter is inhibited or absent (~30% of the control rate) (16–18). The residual bumetanide-resistant secretion is HCO_3^- dependent and apparently involves multiple mechanisms (Figure 2, see color insert). As described above, the basolateral membranes of most, but not all (19, 20), salivary acinar cells express paired Cl^-/HCO_3^- and Na^+/H^+ exchanger activities. NaCl uptake in exchange for intracellular HCO_3^- and H^+ mediated by these functionally coupled exchangers provides machinery to drive Cl^--dependent secretion in parallel to the cotransporter. The contribution of the Cl^-/HCO_3^- exchange mechanism to Cl^- movement requires intracellular HCO_3^-. Intracellular carbonic anhydrases generate HCO_3^- by catalyzing the reversible reaction of water and CO_2 to form HCO_3^- and H^+. The protons produced during this process are expelled via Na^+/H^+ exchangers. Numerous observations are in agreement with this model: Salivary glands express carbonic anhydrases (21, 22); carbonic anhydrase inhibitors reduce secretion (16, 23); exchanger activity is dramatically up-regulated during stimulated secretion (24–26); whereas exchanger antagonists inhibit secretion (16, 18, 23).

Similar to Cl^- efflux, HCO_3^- efflux directly drives fluid secretion. Cl^- channels display relatively little discrimination among anions. This nonselectivity and the fact that HCO_3^- is an abundant intracellular anion (second only to Cl^-) indicate that an electrogenic HCO_3^- conductance significantly contributes to fluid secretion. Consistent with channel-mediated HCO_3^- efflux, agonists that generate an increase in $[Ca^{2+}]_i$ induce an acidification in acinar cells that is inhibited by either HCO_3^- depletion or Cl^- channel blockers (25, 27). Several recent studies have detected Na^+/HCO_3^- cotransporter activity in acinar cells as well. Expression of these Na^+/HCO_3^- cotransporters appears to be both gland- and species-specific; these observations suggest a potential role for Na^+/HCO_3^- cotransporters in salivary secretion and/or intracellular pH regulation in at least some salivary glands (28–30).

Furthermore, the intracellular CO_2/HCO_3^- buffering system is important for regulating secretion by maintaining a neutral intracellular pH. The Ca^{2+}-activated Cl^- channel in parotid acinar cells is inhibited by even slightly acidic intracellular pH levels (31). Because of the critical role for this Cl^- pathway, this dependance makes the salivary flow rate similarly sensitive to intracellular pH. In addition, although it has not been directly investigated in salivary tissues, the relevant Ca^{2+}-dependent K^+ channels are also likely inhibited by low internal pH (32, 33). During muscarinic receptor activation, the intracellular pH of acinar cells initially decreases through HCO_3^- efflux (0.2–0.4 unit) via the Ca^{2+}-activated Cl^- channel (25). Consequently, excessive HCO_3^- (as well as Cl^-) efflux via the Ca^{2+}-activated Cl^- channel is prevented, and thus the resulting intracellular pH drop is limited to protect cell homeostasis. Importantly, the agonist-induced increase in intracellular $[Ca^{2+}]$ also stimulates Na^+/H^+ exchangers (24), the increased activity of which reverses the initial acidification and ultimately raises the intracellular pH of acinar cells 0.1–0.3 unit higher than the unstimulated pH. Because the activity of the Ca^{2+}-sensitive K^+ and Cl^- channels increases as the intracellular pH rises, fluid secretion continues even as the cytosolic Ca^{2+} concentration decreases toward the resting level. Thus it is likely that the pH sensitivity of the Ca^{2+}-activated K^+ and Cl^- channels is important in sustaining fluid secretion during prolonged stimulation.

Molecular Identities of Ion and Water Transport Mechanisms in Acinar Cells

Through a combination of molecular, functional, and genetic studies, the identities of many of the key ion, water, and signaling molecules have been confirmed (Table 1). As described above, the primary Cl^- uptake pathway is the $Na^+/K^+/2Cl^-$

TABLE 1 Molecular identity of ion and water transport proteins in salivary glands

Transport mechanism	Gene	Localization	Cell type	Functions
Na$^+$/H$^+$ exchanger	NHE1 (*Slc9a1*)	Basolateral	Acinar Duct	pH$_i$ regulation Fluid secretion
Na$^+$/K$^+$/2Cl$^-$ cotransporter	NKCC1 (*Slc2a2*)	Basolateral	Acinar	Cl$^-$ uptake Fluid secretion
Intermediate K$^+$ channel	IK1 or SK4 (*Kcnn4*)	Basolateral ?	Acinar	Fluid secretion? E$_m$?
Maxi-K K$^+$ channel	SLO (*Kcnmal*)	Basolateral ?	Acinar	Fluid secretion? E$_m$?
Aquaporin-5 water channel	AQP-5 (*Aqp5*)	Apical	Acinar	Transepithelial water movement
Cl$^-$/HCO$_3^-$ exchanger	AE2 (*Slc4a2*)	Basolateral	Acinar Duct?	Cl$^-$ uptake Fluid secretion
cAMP-activated Cl$^-$ channel	CFTR (*Cftr*)	Apical	Acinar? Duct	Cl$^-$ secretion? NaCl reabsorption
Inward rectifier Cl$^-$ channel	CLC-2 (*Clcn2*)	?	Acinar Duct?	E$_m$?

cotransporter. The electroneutral cation/chloride-coupled cotransporter gene family (*SLC12A*) consists of nine members (34) including two bumetanide-sensitive Na$^+$/K$^+$/2Cl$^-$ cotransporters (*SLC12A1-2*), a thiazide-sensitive Na$^+$/Cl$^-$ cotransporter (*SLC12A3*), four K$^+$/Cl$^-$ cotransporters (*SLC12A4-7*), and two with unknown substrate specificity (*SLC12A8-9*). Of the two Na$^+$/K$^+$/2Cl$^-$ cotransporters, the ubiquitously expressed NKCC1 isoform (the *SLC12A2* gene product) was predicted to be the isoform active in salivary gland acinar cells. In agreement with this hypothesis, localization studies placed NKCC1 in the basolateral membrane of rat and mouse salivary glands (17, 35). Moreover, the lack of functional NKCC1 in knockout mice resulted in severe impairment of salivation (>60% reduction) in response to in vivo muscarinic stimulation (17). These *Nkcc1*$^{-/-}$ mice are deaf and also display decreased secretion by the cecum, jejunum, and intestine (36).

Immunohistochemical studies have also localized a second member of this family, KCC1 (a widely expressed K$^+$/Cl$^-$ cotransporter isoform important for cell volume and ion homeostasis), to the basolateral membranes of salivary acinar and duct cells (37). The functional significance of this latter observation remains unclear, as there are no data suggesting a major role for this transporter in salivary acinar cell function. Because KCC1 likely decreases the intracellular [Cl$^-$] in several cell types, activity of this transporter in salivary acinar cells would mediate significant K$^+$ and Cl$^-$ efflux, not uptake, and consequently reduce the driving force for fluid secretion. Paired Cl$^-$/HCO$_3^-$ and Na$^+$/H$^+$ exchangers act in parallel with and are located within the same cell as the Na$^+$/K$^+$/2Cl$^-$ cotransporter to mediate Cl$^-$ uptake (38, 39). Members of two distinct gene families encode for anion exchangers (*SLC4A* and *SLC26A*) (40, 41). Molecular and functional data currently suggest that the *AE2* (or *SLC4A2*) isoform is most likely responsible for the basolateral anion exchanger activity in acinar cells (26, 35). *AE2* is one of four genes that encodes for anion exchangers within the *SLC4A* family (*AE1-3* or *SLC4A1-3*, and *AE4* or *SLC4A9*). *AE4* is also expressed in salivary glands but expression appears to be restricted to the basolateral membranes of duct cells in rat and mouse (42). The other members of this gene family are apparently Na$^+$/HCO$_3^-$ cotransporters.

Immunohistochemical localization studies have detected several of these different Na$^+$/HCO$_3^-$ cotransporters in salivary gland acinar and/or duct cells, both in the apical and basolateral membranes (29, 30, 43, 44). Functional studies detect Na$^+$/HCO$_3^-$ cotransporter activity in acinar cells (28–30), but it remains unclear what function such transporters might play in this cell type. HCO$_3^-$ uptake via Na$^+$/HCO$_3^-$ cotransporters might provide transport substrate for the apical anion channel as well as the basolateral Cl$^-$/HCO$_3^-$ exchanger, and thus, in both cases, increase the driving force for fluid secretion. As for the *SLC26A* gene family, no members have been described in salivary cells; however, the large number of isoforms (41) and the ubiquitous expression pattern of several of these exchangers suggest that one or more members may be involved in salivary gland function.

Na$^+$/H$^+$ exchangers make up a gene family with eight known members (*NHE1-8* or *SLC9A1-8*). NHE1-5 and NHE8 (SLC9A1-5 and SLC9A8) are targeted to the plasma membrane (45, 46), whereas NHE6 and NHE7 (SLC9A6-7) are intracellular Na$^+$/H$^+$ exchangers (47, 48). Like the housekeeping Na$^+$/H$^+$ exchanger NHE1, the intracellular Na$^+$/H$^+$ exchangers are widely expressed, whereas the distribution patterns of NHE2-NHE5 are much more restricted. Salivary glands likely express all these Na$^+$/H$^+$ exchangers with the exception of NHE5, where high-level expression is restricted to the brain. NHE1 is expressed in the basolateral membranes of both acinar and duct cells (very high expression in duct relative to acinar cells: see 35, 49–51). In contrast, NHE2 is targeted to the apical membranes of both acinar and duct cell in some glands, whereas NHE3 is found only in the apical membranes of duct cells (49–51). A NHE4 message was detected in both acinar and duct cells, but the functional significance of this observation is yet to be determined (50). Functional analysis of *Nhe1*, *Nhe2*, and *Nhe3* knockout mice indicates that NHE1 is the primary regulator of intracellular pH in acinar cells (24, 25) and, as such, is likely to be important for salivation. Indeed, *Nhe1*$^{-/-}$ mice secreted 30–40% less saliva in vivo than did wild-type littermates (51).

There are at least 10 aquaporin, or water channel, genes (*Aqp0-Aqp9*). Of these, 5 are expressed in salivary glands including *Aqp1*, *Aqp3*, *Aqp4*, *Aqp5*, and *Aqp8*. *Aqp5* was originally cloned from salivary glands (52) and is highly expressed in the apical membranes of salivary acinar cells (53–55). *Aqp1* is expressed in the endothelial cells of salivary glands (35), whereas *Aqp3* is found in the basolateral membrane of acinar and duct cells of some salivary glands (56). *Aqp4* was detected in the basolateral membrane of duct cells (57), although not all studies have

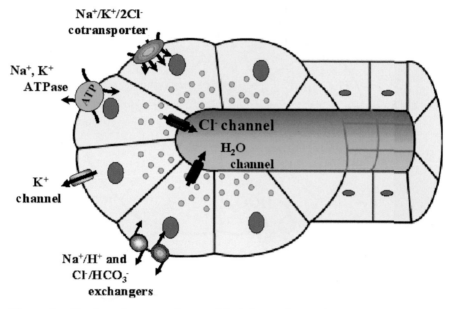

Figure 1 Cl⁻-dependent secretion model. Acinar cell secretion model based on the entry of Cl⁻ across the basolateral membrane mediated by a $Na^+/K^+/2Cl^-$ cotransporter and the paired Na^+/H^+ and Cl^-/HCO_3^- exchangers. Cl⁻ exit across the apical membrane via a Cl⁻ channel. Acinar cells are homogeneous, therefore the different transport elements are spread out for clarity, but all occur in each cell.

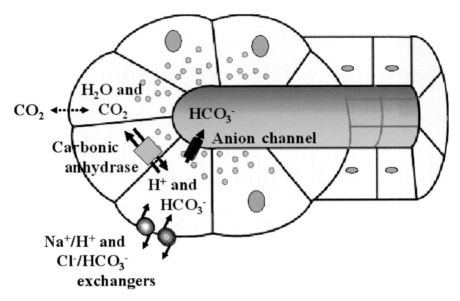

Figure 2 HCO_3^--dependent secretion. In this model intracellular carbonic anhydrases generate HCO_3^- by catalyzing the reversible reaction of water and CO_2 to form HCO_3^- and H^+ to drive acinar cell secretion. HCO_3^- drives secretion via two mechanisms: (a) Cl^- entry occurs in exchange for intracellular HCO_3^- across the basolateral membrane mediated by the paired Na^+/H^+ and Cl^-/HCO_3^- exchangers and Cl^- exits across the apical membrane via a Cl^- channel. (b) HCO_3^- generated by carbonic anhydrases exits across the apical membrane via a Cl^- channel.

current found in most mammalian cells is produced by CLCN3 (77, 78). However, null mutations of the *Clcn3* gene did not eliminate these currents in hepatocytes and pancreatic acinar cells (79), nor in salivary acinar cells (69).

Fluid secretion depends critically on the movement of Cl^- (and HCO_3^-) ions through Ca^{2+}-activated channels in the luminal membrane. Activation of Cl^- channels alone would cause the membrane potential to approach the Nernst potential for these ions and so fluid secretion would cease. Thus other channels must be present to maintain a driving force for Cl^- and fluid secretion. Several types of cation channels have been observed in various types of salivary glands from different animals (reviewed in 1) but the most common is a Ca^{2+}-activated K^+ channel of the maxi-K class (125–250 pS conductance). Channels of this type are expressed in mouse (80) and human (81, 82) parotid acinar cells. The biophysical footprint and pharmacological profile of this channel suggest that it is encoded by the *KCNMA1* (*SLO*) gene (83). Molecular biological evidence in mice points to a parotid-specific splice variant that is highly homologous to human SLO1 (83). The channels in mouse parotid acinar cells appear to be a mix of equal parts homomeric channels with the Slo α subunit and heteromeric channels of Slo and the $\beta 4$ subunit (83).

Cation channels with a lower conductance are often observed in salivary glands—at least some of which do not discriminate among monovalent cations. However, mouse parotid acinar cells robustly express an intermediate (22 pS) conductance, Ca^{2+}-activated K^+-selective channel (83). These channels are encoded by the *KCNN4* gene (previously denoted as *IK1* or *SK4*).

REGULATION OF FLUID SECRETION

Both calcium-mobilizing and cAMP-generating signaling pathways have been linked to salivation. An increase in the intracellular $[Ca^{2+}]$ is the primary fluid secretion signal in salivary acinar cells. In contrast, cAMP signals regulate secretory granule discharge but produce little fluid on their own. Multiple types of receptors linked to Ca^{2+} mobilization are expressed on acinar cells, including muscarinic, α-adrenergic, substance P, and P2y and P2x nucleotide receptor subtypes. It appears, however, that the m3 muscarinic subtype is the key receptor associated with fluid secretion. Knockout of the m1, m2, m4, or m5 muscarinic receptors had little, if any, effect on muscarinic agonist-induced secretion (84). In contrast, disruption of the m3 muscarinic receptor dramatically inhibited secretion (84, 85). The modest effects of knocking out the m1, m4, and m5 muscarinic receptors on fluid secretion potentially suggest nonspecific effects, possibly owing to changes in the response of the vasculature in the salivary glands to stimulation. The cAMP-stimulated discharge of secretory granules by acinar cells is primarily associated with the activation of β-adrenergic receptors; however, other receptors are also coupled to adenylyl cyclase activity through the G protein Gs, e.g., VIP (vasoactive intestinal polypeptide) receptors.

Ca^{2+}-Dependent Fluid Secretion

The initial response to muscarinic receptor activation is primarily the release of intracellular Ca^{2+} stores as discussed below. Fluid secretion proceeds as the rise in $[Ca^{2+}]_i$ activates Ca^{2+}-gated K^+ and Cl^- channels located in the basolateral and apical membranes of acinar cells, respectively. Prolonged salivation, as occurs during the ingestion of a meal, depends on the $[Ca^{2+}]_i$ being maintained by extracellular Ca^{2+} influx. This interpretation of the Ca^{2+} kinetics and dependency of the fluid secretion process arose from several observations. First, removal of external Ca^{2+} rapidly inhibits secretion, and the kinetics of the inhibition mimics the decrease in $[Ca^{2+}]_i$ detected during this maneuver (14). Moreover, the efflux of KCl is mediated by the activation of Ca^{2+}-gated K^+ and Cl^- channels; thus, as predicted from this model, chelation of the rise in intracellular Ca^{2+} prevents KCl efflux (14, 86). Additionally, salivary acinar cells typically react to muscarinic stimulation by a global Ca^{2+} response that results in significant Ca^{2+} loss via plasma membrane Ca^{2+} pumps, thus the need to compensate for the efflux of intracellular Ca^{2+} (see below).

In addition to the Ca^{2+}-gated K^+ and Cl^- channels, the intracellular Ca^{2+} increase during muscarinic stimulation appears to activate other key water and ion transporter proteins. An increase in intracellular Ca^{2+} reportedly up-regulates $Na^+/K^+/2Cl^-$ cotransporter activity $\sim$20-fold relative to the basal flux rate (87). The mechanism for this increased activity is somehow associated with arachidonic acid metabolism, likely a product of the cytochrome P450 pathway of arachidonic acid metabolism. The increase in intracellular Ca^{2+} also appears to up-regulate the Na^+/H^+ and Cl^-/HCO_3^- exchangers (26, 88). Therefore, the Ca^{2+}-induced stimulation of $Na^+/K^+/2Cl^-$ cotransporter activity, as well as the Na^+/H^+ and Cl^-/HCO_3^- exchangers, assures that the intracellular Cl^- concentration remains above its electrochemical gradient. Furthermore, a muscarinic-stimulated rise in $[Ca^{2+}]_i$ acutely regulates the insertion of AQP5 water channels into the plasma membrane (89). This response is short-lived, i.e., the plasma membrane–associated AQP5 content returns to basal levels in $\sim$5 min (90), suggesting that there is short-term increase in the transcellular water permeability. It is interesting to note that this phenomenon correlates with the enhanced secretion rate typically seen during the early stages of secretion (17, 51).

Calcium Mobilization

Acetylcholine (ACh) is the primary neurotransmitter associated with the robust fluid secretion activated during parasympathetic stimulation. Muscarinic receptors, primarily of the m3 subtype (91), are activated on the basolateral surface of acinar cells following neural release of acetylcholine. M3 receptors are coupled to G proteins (G_q and G_{11}), which subsequently activate phospholipase $C\beta$ ($PLC\beta$). $PLC\beta$ cleaves phosphatidylinositol 1,4, bisphosphate (PIP_2) to produce diacylglycerol (DAG) and the soluble signaling molecule inositol 1,4,5, trisphosphate ($InsP_3$). This elevation of $InsP_3$ is the key event in initiating an increase in $[Ca^{2+}]_i$, as $InsP_3$ binds to and subsequently activates a cation-selective ion channel, termed the $InsP_3$ receptor ($InsP_3R$), which is localized in the endoplasmic reticulum (ER) (92). $InsP_3$ receptors are encoded by three genes, resulting in three distinct proteins, the type-1 $InsP_3R$ ($InsP_3R$-1), type-2 ($InsP_3R$-2), and type-3 ($InsP_3R$-3) (93). Differences in the affinity for $InsP_3$ of the $InsP_3R$ types and their regulation of activity is considered a major determinant in defining the characteristics of intracellular Ca^{2+} signaling in a particular cell type (94). Several studies indicate that all three forms of $InsP_3R$ are expressed in both acinar and ductal tissue isolated from submandibular and parotid glands (95–98). At least in rat parotid acinar tissue, $InsP_3R$-2 appears to be the most abundant family member expressed, accounting for approximately 90% of the complement of receptors (97). Substantial expression of $InsP_3R$-3 was also detected in this study with a minor $\sim 1\%$ contribution from $InsP_3R$-1 (97). In common with acinar cells from other exocrine tissue, such as the pancreas, the vast majority of all $InsP_3R$ types is expressed immediately below the apical plasma membrane of the cell in an area bounded by the actin cytoskeleton "terminal-web" (98, 99). The localization of $InsP_3R$, similar to that of pancreatic acinar cells, appears to explain the initiation of Ca^{2+} signaling in the apical region of the cell (96, 98, 100, 101). However, in contrast to pancreatic acinar cells, which are capable of exhibiting localized apical Ca^{2+} signals (101), salivary acinar cells typically react to muscarinic stimulation by displaying rapid global Ca^{2+} signals (96), as shown in Figure 3 (see color insert). The differing kinetics of Ca^{2+} signals described in parotid versus pancreatic acinar cells may in some part be explained by the observation that parotid acinar cells express approximately fourfold higher density of $InsP_3R$ than pancreatic acinar cells (6). In addition, it appears that a major contribution to the apical restriction of Ca^{2+} signals in pancreatic acinar cells is accomplished by the localization of peri-granular mitochondria (102, 103), which, by taking up Ca^{2+}, serve as a "fire wall" to prevent regenerative Ca^{2+} signals from propagating into the basal region of the cell. In contrast, peri-granular mitochondria are absent in parotid acinar cells (104). Instead the predominant localization of mitochondria in parotid tissue is to a region surrounding the nucleus.

The expression of ryanodine receptors (RyR) in salivary acinar cells has also been suggested both functionally (97, 98, 105–107) and physically (97, 98, 105). RyR are intracellular Ca^{2+} release channels that belong to a family related to $InsP_3R$. RyR receptors are gated by elevations in cytoplasmic Ca^{2+} in a process termed calcium-induced calcium release (CICR) and contribute to the mechanisms underpinning regenerative Ca^{2+} waves in a variety of cells (94). PCR performed on rat parotid acinar samples indicates that RyR-1, the skeletal isoform of the protein, is the major isoform expressed. Although the precise role of RyR in the genesis of Ca^{2+} signals in salivary acinar cells is far from well understood, their localization predominately to the basal pole of the cell (97) suggests that they play a role in the globalization of the Ca^{2+} signal. It is tempting to speculate that whereas the invariably global Ca^{2+} signals in salivary acinar cells are necessary for activating spatially separated ionic conductances, the relative paucity of fluid secretion by pancreatic acinar cells reflects the apically localized nature of the Ca^{2+} signals in this cell type.

The clearance of Ca^{2+} from the cytoplasm following the removal of the secretagogue is accomplished by the activity of Ca^{2+}-ATPases localized to both the ER (SERCA-type pump) and plasma membrane (PMCA-type pump) (108, 109). Interestingly, specific isoforms of each of these pumps have been reported to be selectively localized to the apical region of salivary acinar cells (109). The specific localization of these pumps likely contributes to establishing microdomains of Ca^{2+} in the neighborhood of Ca^{2+}-activated ionic conductances. Indeed, local clearance of Ca^{2+} by SERCA pumps has been shown to contribute substantially to the deactivation of chloride currents in parotid acinar cells (96).

Calcium Entry

As noted above, a sustained secretion of salivary fluid requires the influx of Ca^{2+} from the extracellular medium. Despite this importance, and in marked contrast to the $InsP_3$-induced Ca^{2+} release discussed above, the nature and regulation of the Ca^{2+} entry pathways involved in maintaining fluid secretion remain far from clear. In nonexcitable cells in general, such entry would typically involve small-conductance channels activated by agonists acting at receptors coupled to the activation of phospholipase C (PLC). The most thoroughly studied mode of such agonist-activated Ca^{2+} entry is capacitative or store-operated Ca^{2+} entry, which is activated following the emptying of agonist-sensitive intracellular Ca^{2+} stores (110, 111). The channels responsible for this entry are generically described as store-operated Ca^{2+} channels (SOC channels), and it is clear that Ca^{2+} entering via these channels plays a key role in determining the amplitude of the sustained elevated cytosolic Ca^{2+} signals seen at high agonist concentrations, as well as in refilling of the Ca^{2+} stores on termination of the signal. With respect to salivary gland acinar cells, such store-operated Ca^{2+} entry has been fairly extensively described. Indeed, many of the studies establishing and defining this particular mode of Ca^{2+} entry were originally performed on these cells (110, 112). However, these studies have almost exclusively relied on fluorescence measurements of changes in cytosolic Ca^{2+} (or a surrogate such as Mn^{2+} or Ba^{2+}). This technique has several potential pitfalls as it is not always easy to dissect effects on the entry pathway itself from effects on other components of the overall Ca^{2+} regulatory system of the cell, such as SERCA and plasma membrane Ca^{2+} pumps, mitochondria, membrane potential, etc. Perhaps more importantly, although firmly establishing the presence of a store-operated mode of Ca^{2+} entry in salivary gland cells, such studies are less informative regarding the specific nature of the conductances involved. This can be definitively done only when the activity of the Ca^{2+}-permeable conductances are measured directly, which typically requires application of patch-clamp techniques. To date, direct measurement of SOC channel activity using such techniques has been performed on only the HSG salivary gland cell line (113). In these cells, the SOC channels display a rather poor selectivity for Ca^{2+} (114) and, therefore, differ from store-operated currents seen in many other nonexcitable cells including the so-called CRAC channels—the extensively studied, archetypal, highly

Ca^{2+}-selective store-operated channels (115, 116). Potentially, this could be important because the predominant effect of any nonselective cation conductance would be to depolarize the cell, which would tend to reduce Ca^{2+} influx rather than increase it. Whether the same properties are shown by the endogenous SOC channels in native salivary acinar cells remains to be determined.

As to the molecular identity of this conductance, evidence suggests that TRPC1, a member of the TRPC family of proteins originally cloned on the basis of the *trp* mutant in *Drosophila*, may form at least part of this channel. TRPC1 is endogenously present in rat submandibular gland cells, and adenovirus-mediated overexpression of human TRPC1 in this gland resulted in increases in the sustained elevated $[Ca^{2+}]_i$ signal following treatment with either the SERCA pump inhibitor, thapsigargin, or carbachol (100 μM) (117). Importantly, this overexpression of TRPC1 also resulted in a markedly enhanced pilocarpine-stimulated salivary secretion flow. Again, studies in the HSG cell line have generated more detailed information, but it remains to be seen how far this applies to the endogenous SOC channels of acinar cells.

Regulation of the activity of the store-operated channels in salivary gland cells, as in other cell types, is only poorly understood. Reports have implicated a mechanism involving nitric oxide and/or cGMP. Data from submandibular acinar cells suggest that a Ca^{2+}-sensitive NOS is activated by the release of Ca^{2+} from intracellular stores, leading to increases in nitric oxide. This results in elevated cGMP levels that act as the activator of store-operated Ca^{2+} entry (118). In contrast, data from mouse parotid acinar cells suggest that nitric oxide itself increases store-operated Ca^{2+} entry in an indirect, and cGMP-independent fashion by enhancing the release of Ca^{2+} from internal stores (119). Intriguingly, the same group subsequently described a nitric oxide-dependent inhibition of store-operated Ca^{2+} entry in the same cells and have argued that the positive versus negative effects of nitric oxide on store-operated entry may depend on its concentration.

However, an important caveat to all the studies described above is that they all involved the activation of Ca^{2+} entry by processes expected to maximally deplete intracellular stores. Whereas this will obviously provide a powerful signal for the activation of SOC channels, contributions from other Ca^{2+} permeable conductances activated by more subtle stimulation would not be observed under these conditions (120). In this context, recent evidence from a variety of different cell types has shown that at lower, presumably more physiologically relevant, agonist concentrations the influx of Ca^{2+} occurs via a distinct pathway involving a novel arachidonic acid-activated, highly Ca^{2+}-selective channel (ARC channel) (121, 122). Entry via these channels influences the unique spatial and temporal characteristics of the Ca^{2+} signals seen at these lower agonist concentrations (123). Given that these signals are widely considered to be critical for the regulation of a majority of cellular responses, including fluid secretion, the activity and regulation of these ARC channels would seem to be particularly important. Recently, it has been shown that conductances showing all the critical characteristics of ARC channels are present in both mouse and human parotid acinar cells (122). Consistent with this, addition of low concentrations of exogenous arachidonic acid (4–8 μM) to mouse parotid acinar cells induces an enhanced entry of Ca^{2+} that is independent of the release of Ca^{2+} from intracellular stores (i.e., noncapacitative). It should be noted that Watson et al. (124) have described an arachidonic acid-dependent Ca^{2+} entry pathway in mouse parotid acinar cells that is clearly different from that involving the ARC channels. The pathway they describe apparently results from a depletion of ryanodine-sensitive stores and is dependent on the generation of nitric oxide. However, the very high concentrations of exogenous arachidonic acid used in their study (45 μM) are known to induce a variety of nonspecific effects on membrane fluidity and integrity, making interpretation of these data problematic. As for Ca^{2+} entry via the ARC channels, recent evidence indicates that it plays a key specific role in the control of Ca^{2+} signals activated at low agonist concentrations, whereas store-operated entry predominates at high concentrations (O. Mignen & T. Shuttleworth, manuscript in preparation). This would appear to be consistent with the phenomenon of reciprocal-regulation of these two coexisting conductances described in other cell types (120). Given the critical role of Ca^{2+} entry in the generation of a sustained secretory response from the parotid, the precise characterization of these conductances and their respective contribution, relative to the coexisting SOC channels, to overall Ca^{2+} entry under various conditions of stimulation will be critical to understanding the control of secretory activity in the gland.

Cross-Talk Between Signaling Pathways

Although the activation of Ca^{2+}-dependent ion channels is the primary mechanism underlying fluid secretion from the parotid gland (125), secretion can be significantly enhanced when both Ca^{2+} and cAMP signaling systems are activated concurrently (126, 127). Despite extensive studies demonstrating that the elevation of $[Ca^{2+}]_i$ is critically important for initiating fluid secretion, a considerable body of evidence indicates that this is not the complete story for the overall control of secretion in salivary glands. Several in vivo studies have indicated a key role in the primary neural control of salivary fluid secretion for various neuropeptides such as vasoactive intestinal peptide (VIP) and calcitonin gene-related peptide (CGRP), as well as β-adrenergic agonists, whose actions are predominately mediated by an elevation in intracellular cAMP (128–131). Interestingly, evidence indicates that these agents have little or no effect on fluid secretion by themselves but have potent synergistic effects on the fluid secretion induced by submaximal stimulation with ACh (126, 127). VIP also markedly increases blood flow to the glands (127), but there is clear evidence for effects on fluid secretion that are independent of any vasomotor actions.

The generality of this response has been demonstrated by numerous reports in a variety of animal models including rodents, ferrets, and cats. These studies show essentially the same potentiating effects on parotid fluid secretion induced by Ca^{2+}-mobilizing agonists, such as substance P or ACh, or by other agents that

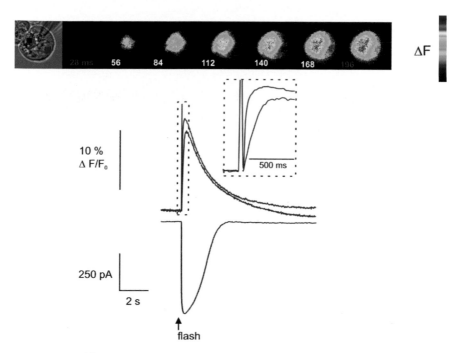

Figure 3 Ca^{2+} mobilization. The cell, shown in brightfield, was loaded via the patch pipette with 3 μM NPE-caged 1,4,5-InsP$_3$ and 75 μM OGB-2. The pseudocolored panel shows images at the indicated times, post-photolysis of the NPE-caged 1,4,5-InsP$_3$. Uncaging of the InsP$_3$ induces an increase in $[Ca^{2+}]_i$ that is initiated in the extreme luminal portion of the cell after a short latency (~1 image frame). This increase in $[Ca^{2+}]_i$ becomes global within 200 ms after photolysis. The upper traces depict the changes in $[Ca^{2+}]_i$ initiated by photolysis. The blue trace was taken from a region of interest (ROI) within the luminal region and the red trace was from a ROI in the basal pole of the cell. The inset represents the area on an expanded time scale to illustrate the very short delay between the initiation of the apical and basal signal. The lower trace shows the I_{CaCl} activated by the increase in $[Ca^{2+}]_i$, and illustrates the high temporal coincidence of these signals.

observed this expression pattern (56). *Aqp8* is strongly expressed in myoepithelial cells (58). Of these water channels, AQP5 appears to be the only aquaporin to play a major role in salivation. A null mutation in the *Aqp5* gene dramatically inhibited in vivo saliva production by about 60% (55, 59). Consistent with these observations, mice lacking *Aqp5* also displayed decreased acinar cell water permeability of a magnitude comparable to the loss of secretion (59). In contrast, disruption of the *Aqp1*, *Aqp3*, or *Aqp4* genes had no significant effects on secretion (55, 60). It is not clear whether AQP5 is associated with some forms of salivary gland dysfunction. Subjects with Sjögren's syndrome suffer from a severe deficit in fluid production by salivary glands and other exocrine organs. It was found in salivary acinar cells from individuals with Sjögren's syndrome that AQP5 targets to the basolateral membrane, but it distributes to the apical membrane in patients with other salivary gland hypofunction disorders, as well as in control subjects (61). These observations suggest that correct AQP5 targeting may be sensitive to the disease state of the gland and that its localization may be critical to its function in salivation. It has been conjectured that in addition to its water channel function, AQP5 acts as an osmosensor that provides a signal to regulate the tonicity of saliva and the permeability of acinar cell tight junctions (62). This interesting model might explain the increased tonicity of the saliva produced by AQP5-deficient mice (55); however, there is currently no direct evidence to support this hypothesis.

At least five types of Cl^- channels have been identified in the plasma membrane of salivary gland acinar cells. This classification scheme is based on the electrophysiological fingerprints of each current, including, e.g., gating mechanism, activation kinetics, and pharmacology. The activation properties for each of these Cl^- channels are unique, such that they can be functionally isolated during patch-clamp recordings. Channels that are dependent for activation on an increase in intracellular Ca^{2+}, an increase in intracellular cAMP, exposure to extracellular ATP, hyperpolarization of the plasma membrane, or cell swelling have been described (63–65). As discussed below, functional studies indicate that the Ca^{2+}-gated Cl^- channel is the major apical Cl^- efflux pathway. Activation of the Ca^{2+}-dependent Cl^- channel is both time and voltage dependent, and the current-voltage relation is strongly outwardly rectifying (66). The molecular identity of the acinar Ca^{2+}-activated Cl^- current has not been determined, although members of three gene families have been implicated as Ca^{2+}-activated Cl^- channels, i.e., *CLCN3*, *CLCA*, and *Bestrophin*. Human CLCN3 was cloned from the T84 colon tumor cell line and, when expressed in the mammalian epithelial cell line tsA, produces CaMKII-dependent Cl^- currents (67). However, these currents look quite different from the native CaMKII-dependent Cl^- currents in T84 cells (68). Expressed CLCN3 currents displayed nearly linear current-voltage relations and no time dependence; furthermore, they did not require Ca^{2+}, as chelation with BAPTA did not inhibit the currents activated by the catalytic subunit of CaMKII (67). Unlike these currents and those in T84 cells, the Ca^{2+}-activated Cl^- currents in rat salivary acinar cells are insensitive to CaMKII inhibitors (68). Moreover, it has been unequivocally demonstrated that Clcn3 is not the Ca^{2+}-activated Cl^-

channel, at least in mouse parotid acinar cells, because knockout of the *Clcn3* gene had no effect on either muscarinic-induced fluid secretion or on the magnitude or activation kinetics of the Ca^{2+}-activated Cl^- currents (69).

The first member of another putative Ca^{2+}-activated Cl^- channel gene family, bCLCA1, was cloned from cow (70). At least four different members (*CLCA1-4*) have subsequently been identified (71). Because of the high sequence homology within this gene family, it has been difficult to precisely determine the tissue expression patterns of the various CLCA isoforms. With this caveat in mind, it appears that mCLCA1 is highly expressed in secretory organs including salivary glands (C. Ovitt & J.E. Melvin, unpublished observation). However, controversy remains as to the functions of the members of this gene family. It is clear that expression of these genes leads to Ca^{2+}-modulated Cl^- currents (72), but the currents do not recapitulate the functional properties of the native currents in the tissues from which these channels were cloned. Interestingly, other reports suggest that CLCAs act as tumor suppressor and/or signaling molecules (73). More recently, an additional Ca^{2+}-gated Cl^- channel gene family has been identified, the so-called *Bestrophins*, of which there are currently four identified members, *BEST1–BEST4*. Mutations in *BEST1* result in Best's disease, an autosomal-recessive progressive form of vitelliform macular dystrophy (74). It is not know whether members of this gene family are expressed in salivary glands. Expression of mBest2 generates Ca^{2+}-gated Cl^- currents that are sensitive to physiological intracellular Ca^{2+} concentrations ($K_d = 230$ nM) (75). However, unlike acinar cell Ca^{2+}-gated Cl^- currents, heterologously expressed mBest2 currents are neither voltage nor time dependent, and the current-voltage relation is linear. It is tempting to speculate that either a CLCA or a Bestrophin is an integral part of the Ca^{2+}-gated Cl^- channel in salivary acinar cells. However, the failure of heterologous expression systems to produce currents like those in native acinar cells raises concern. If a CLCA or a Bestrophin is the Ca^{2+}-gated Cl^- channel in acinar cells, these results suggest that an auxiliary subunit is required to faithfully recapitulate native currents.

At least four other types of Cl^- channels are also expressed in salivary acinar cells, but the functional roles of these channels remain unknown. These channels include those activated by exposure to extracellular ATP, by cell swelling, by an increase in intracellular cAMP, or through hyperpolarization of the plasma membrane. Molecular identification of the two latter channels has been recently made. *CFTR* and *CLCN2* encode for the cAMP-dependent (64) and the hyperpolarization-activated (76) Cl^- channels, respectively. Mutations in functional domains of the *CFTR* gene result in cystic fibrosis, but this disease apparently has little or no effect on the salivary flow rate. However, there are reports of increased blockage of mucin-secreting salivary glands, suggesting that there may be changes in the properties and/or composition of the saliva in affected glands. Targeted disruption of the *Clcn2* gene led to degeneration of the retina and testis, but the loss of this hyperpolarization-activated channel in acinar cells did not affect secretion (76). The identities of the cell swelling-dependent and ATP-activated Cl^- channels remain unclear. Considerable evidence suggests that the cell swelling-dependent raise cAMP, including VIP, CGRP, and isoproterenol, or by forskolin (to directly activate adenylyl cyclase) (126–132). Despite this considerable body of evidence from in vivo studies indicating this marked potentiation of $[Ca^{2+}]_i$-induced fluid secretion by agents that elevate cAMP, and the demonstration of its critical role in the maximal stimulation of fluid secretion by the salivary glands, the cellular basis for this effect remains largely unknown.

Conceptually, enhanced fluid secretion resulting from the interplay between the Ca^{2+} and cAMP signaling systems could occur at multiple levels. Although augmented fluid secretion must ultimately occur by the increased transcellular flux of ions, it potentially could occur at the level of generation of the $[Ca^{2+}]_i$ signal because the key ionic conductances are Ca^{2+} dependent; alternatively, it could occur by directly influencing the activity of the ion channels themselves. Whereas some evidence does exist for the direct modulation of ion channels, there is strong support in the literature for the idea that raising cAMP can modulate the generation of the Ca^{2+} signal per se. For example, isoproterenol treatment of rat parotid acinar cells at concentrations that alone do not elicit Ca^{2+} changes significantly potentiates Ca^{2+} signals stimulated by muscarinic agonists (133). Similarly, direct activation of adenylyl cyclase by forskolin, or direct activation of protein kinase A (PKA) using cell permeable analogs of cAMP, markedly enhances Ca^{2+} signals through PLCβ-coupled agonists (106). Specifically, the initial increase in $[Ca^{2+}]_i$ is enhanced, subthreshold stimulation is transformed to a measurable $[Ca^{2+}]_i$ increase, and an oscillatory increase can be converted to a sustained $[Ca^{2+}]_i$ increase, all consistent with PKA-activation resulting in a left shift in sensitivity to stimulation by Ca^{2+}-mobilizing agonists (106). The signal transduction machinery activated by muscarinic agonists is a rich source of potential sites for modulation by PKA. Indeed, in other cell systems, PKA has been shown to regulate $InsP_3$ levels, $InsP_3R$ sensitivity, RyR sensitivity, Ca^{2+} influx, and Ca^{2+} clearance (for review see 134). Although cAMP has been reported to enhance $InsP_3$ levels in submandibular acinar cells (135), no measurable effect of cAMP on $InsP_3$ levels could be measured in two independent studies in parotid acinar cells (106, 133). Similarly, no effect on Ca^{2+} influx or SERCA pump activity was observed upon activating PKA (108). Consistent with the marked enhancement of the initial rise of the Ca^{2+} signal, the major effect appears to be enhanced $InsP_3$-induced Ca^{2+} release as a result of phosphorylation of $InsP_3R$ (106). Moreover, this effect was attributed to the $InsP_3R$-2 in particular, because this isoform has the greatest sensitivity to $InsP_3$, is the most abundant in mouse parotid, and was shown to be specifically phosphorylated (106). Although reports in other exocrine tissues suggest that phosphorylation of $InsP_3R$-3 results in attenuated Ca^{2+} release, potentiated Ca^{2+} release through the small population of type-I $InsP_3R$ ($InsP_3R$-I) could also contribute to this observation.

Whereas in mouse parotid acinar cells no evidence for the participation of RyR in the potentiation of Ca^{2+} release has been reported, several studies performed in permeabilized rat parotid acinar cells or microsomal vesicles have reported cAMP-induced Ca^{2+} release, without elevation of $InsP_3$. This appears to occur through a mechanism requiring RyR (105). However, the physiological relevance of these observations is not clear because cAMP application resulted in a relatively small release of Ca^{2+} compared with that of $InsP_3$, and this release occurred under conditions where Ca^{2+} was markedly elevated above baseline levels, and thus presumably RyR would already be activated. It should also be noted that several studies in intact cells have reported no effect of cAMP on $[Ca^{2+}]_i$ in the absence of stimulation with agonists, and thus it seems likely that the major effect of raising cAMP on Ca^{2+} signaling events occurs at the level of sensitization of the $InsP_3R$.

In addition to the above effects of cAMP on Ca^{2+} mobilization, potential interactions between Ca^{2+} entry and the cAMP system have been described. However, to date these have focused primarily on the influence of Ca^{2+} entry on cAMP levels. It has been demonstrated that the store-operated entry of Ca^{2+} in mouse parotid acinar cells can increase intracellular cAMP levels, apparently by activating a Ca^{2+}-sensitive type 8 adenylyl cyclase (136). This effect is also seen in a variety of other cell types, where it has been shown to reflect a close spatial association of the adenylyl cyclase with the mouth of the SOC channel pore (137). Significantly, this appears to be a response that is specific to Ca^{2+} entering via the store-operated pathway, because in HEK293 cells it has been shown that entry via the coexisting ARC channels fails to affect adenylyl cyclase activity (138). It will be interesting to determine whether the same applies to parotid cells. More importantly, although Ca^{2+} entry can influence cAMP levels, it is not known whether cAMP (or PKA) has any effect on the entry.

FUTURE DIRECTIONS

Although considerable progress has been made over the past decade in determining the function of the transporter and regulatory molecules involved in salivary gland fluid and electrolyte secretion, we are still some way from a complete understanding. For example, the IK1 and Slo Ca^{2+}-activated K^+ channels will certainly provide the needed driving force for Cl^- (and subsequent fluid) secretion, but why are there two types of channels when only one would appear to be needed? Perhaps redundancy is worth the cost in order to protect an important physiological activity. Or is it possible that there is a specific role for each of these channels? Cook & Young (139) showed that fluid secretion could, in principle, be increased if some of the basolateral cell K^+ conductance were distributed in the luminal membrane. In order to address this issue, it would be useful to determine the special localization of these two types of K^+ channels—perhaps a difficult task considering the low number of expressed Slo channels (see Reference 1). Single-channel patch-clamp studies almost certainly sample the basolateral membrane and show activity of both channel types. Direct patching of the luminal membrane has not been reported, but it has been suggested that zymogen granules in pancreas acini contain K^+ channels, and these may become part of the luminal membrane during secretion (140).

The large number of Cl^- channels, in salivary acinar cells raises similar questions. Are the functions of these channels overlapping or are they distinct?

Ca^{2+}-activated Cl^- channels clearly play a central role in the fluid secretion process, but this issue cannot be resolved until the molecular identity of the channel is determined. At present, it is uncertain whether the CLCA or the Best Ca^{2+}-activated Cl^- channels are involved, or if a new family of Ca^{2+}-activated Cl^- channel proteins will be identified. The recent discovery of an ATP-activated Cl^- channel in mouse parotid acinar cells (65) further complicates the overall picture but suggests an attractive alternative mechanism to explain the synergism often seen when both Ca^{2+} and cAMP signaling systems are activated (126, 127). ATP is released from secretory granules, and if the ATP-activated Cl^- channel is expressed in the apical surface of the acinar cell, then activation of this channel in parallel with the Ca^{2+}-activated Cl^- channels could result in enhanced secretion. This makes sense functionally, as the increased fluid production would be important for washing the proteins discharged from the secretory granules through the ductal system. Many challenges remain, most urgent the molecular identification of this channel and its localization in acini.

It is clear that the agonist-induced increase in the intracellular $[Ca^{2+}]$ up-regulates most, if not all, key downstream ion transporters, but the mechanism for this regulation remains ambiguous in most cases. Does Ca^{2+} directly stimulate each transporter or is a kinase or other regulatory protein required? Moreover, a better understanding of the spatial and temporal aspects of the $[Ca^{2+}]_i$ signal and the specific pathways modulating such signals might shed light on the mechanisms involved in regulating these ion transporters. Further elucidation of the details of these regulatory processes will ultimately lead to improved therapies for salivary gland dysfunction.

ACKNOWLEDGMENTS

This work was supported in part by National Institutes of Health Grant DE13539 (J.E.M.).

NOTE ADDED IN PROOF

Two additional members of the *CLCA* gene family have recently been cloned.

Evans SR, Thoreson WB, Beck CL. 2004. Molecular and functional analyses of two calcium-activated chloride channel family members from mouse eye and intestine. *J. Biol. Chem.* 279:1792–800

The *Annual Review of Physiology* is online at http://physiol.annualreviews.org

LITERATURE CITED

1. Cook DI, Van Lennep EW, Roberts ML, Young JA. 1994. Secretion by the major salivary glands. In *Physiology of the Gastrointestinal Tract*, ed. LR Johnson, pp. 1061–17. New York: Raven. 3rd ed.

2. Melvin JE. 1991. Saliva and dental diseases. *Curr. Opin. Dent.* 1:795–801

3. Melvin JE. 1999. Chloride channels and salivary gland function. *Crit. Rev. Oral Biol. Med.* 10:199–209

4. Melvin JE, Culp DJ. 2004. Salivary Glands, Physiology. In *Encyclopedia of Gastroenterology*, ed. LR Johnson, pp. 318–25. San Diego: Academic

5. Pedersen AM, Bardow A, Jensen SB, Nauntofte B. 2002. Saliva and gastrointestinal functions of taste, mastication, swallowing and digestion. *Oral Diseases* 8:117–29

6. Tabak LA. 1995. In defense of the oral cavity: structure, biosynthesis, and function of salivary mucins. *Annu. Rev. Physiol.* 57:547–64

7. Mircheff AK. 1989. Lacrimal fluid and electrolyte secretion: a review. *Curr. Eye Res.* 8:607–17

8. Quinton PM. 1999. Physiological basis of cystic fibrosis: a historical perspective. *Physiol. Rev.* 79(Suppl. 1):S3–S22

9. Melvin JE, Nehrke K, Arreola J, Begenisich T. 2002. Ca^{2+}-activated Cl^- currents in salivary and lacrimal glands. *Curr. Topics Membr.* 53:203–223

10. Nauntofte B. 1992. Regulation of electrolyte and fluid secretion in salivary acinar cells. *Am. J. Physiol. Gastrointest. Liver Physiol.* 263:G823–37

11. Turner RJ, Sugiya H. 2002. Understanding salivary fluid and protein secretion. *Oral Diseases* 8:3–11

12. Foskett JK. 1990. $[Ca^{2+}]_i$ modulation of Cl^- content controls cell volume in single salivary acinar cells during fluid secretion. *Am. J. Physiol. Cell Physiol.* 259:C998–1004

13. Zeng W, Lee MG, Muallem S. 1997. Membrane-specific regulation of Cl^- channels by purinergic receptors in rat submandibular gland acinar and duct cells. *J. Biol. Chem.* 272:32956–65

14. Melvin JE, Koek L, Zhang GH. 1991. A capacitative Ca^{2+} influx is required for sustained fluid secretion in sublingual mucous acini. *Am. J. Physiol. Gastrointest. Liver Physiol.* 261:G1043–50

15. Nauntofte B, Dissing S. 1988. K^+ transport and membrane potentials in isolated rat parotid acini. *Am. J. Physiol. Cell Physiol.* 255:C508–18

16. Case RM, Hunter M, Novak I, Young JA. 1984. The anionic basis of fluid secretion by the rabbit mandibular salivary gland. *J. Physiol.* 349:619–30

17. Evans RL, Park K, Turner RJ, Watson GE, Nguyen H-V, et al. 2000. Severe impairment of salivation in $Na^+/K^+/2Cl^-$ cotransporter (NKCC1)-deficient mice. *J. Biol. Chem.* 275:26720–26

18. Lau KR, Howorth AJ, Case RM. 1990. The effects of bumetanide, amiloride and Ba^{2+} on fluid and electrolyte secretion in rabbit salivary gland. *J. Physiol.* 425:407–27

19. Paulais M, Valdez IH, Fox PC, Evans RL, Turner RJ. 1996. Ion transport systems in human labial acinar cells. *Am. J. Physiol. Gastrointest. Liver Physiol.* 270:G213–19

20. Zhang GH, Cragoe EJ Jr, Melvin JE. 1992. Regulation of cytoplasmic pH in rat sublingual mucous acini at rest and during muscarinic stimulation. *J. Membr. Biol.* 129:311–21

21. Fujikawa-Adachi K, Nishimori I, Sakamoto S, Morita M, Onishi S, et al. 1999. Identification of carbonic anhydrase IV and VI mRNA expression in human pancreas and salivary glands. *Pancreas* 18:329–35

22. Hennigar RA, Schulte BA, Spicer SS. 1983. Immunolocalization of carbonic anhydrase isozymes in rat and mouse salivary and exorbital lacrimal glands. *Anat. Rec.* 207:605–14

23. Pirani D, Evans LA, Cook DI, Young JA. 1987. Intracellular pH in the rat mandibular salivary gland: the role of Na-H and $Cl-HCO_3$ antiports in secretion. *Pflügers Arch.* 408:178–84

24. Evans RL, Bell SM, Schultheis PJ, Shull GE, Melvin JE. 1999. Targeted disruption of the *Nhe1* gene prevents muscarinic-induced upregulation of Na^+/H^+ exchange in mouse Parotid acinar cells. *J. Biol. Chem.* 274:29025–30

25. Nguyen H-V, Shull GE, Melvin JE. 2000. Muscarinic receptor-induced acidification in sublingual mucous acinar cells. Loss of pH recovery in NHE1-null mice. *J. Physiol.* 523:139–46

26. Nguyen H-V, Stuart-Tilley A, Alper SL, Melvin JE. 2004. Cl^-/HCO_3^- exchange is acetazolamide-sensitive and activated by a muscarinic receptor-induced $[Ca^{2+}]_i$ increase in salivary gland acinar cells. *Am. J. Physiol. Gastrointest. Liver Physiol.* 286:G312–20

27. Melvin JE, Moran A, Turner RJ. 1988. The role of HCO_3^- and Na^+/H^+ exchange in the response of rat parotid acinar cells to muscarinic stimulation. *J. Biol. Chem.* 263:19564–69

28. Kim YB, Yang BH, Piao ZG, Oh SB, Kim JS, Park K. 2003. Expression of Na^+/HCO_3^- cotransporter and its role in pH regulation in mouse parotid acinar cells. *Biochem. Biophys. Res. Commun.* 304:593–98

29. Luo X, Choi JY, Ko SB, Pushkin A, Kurtz I, et al. 2001. HCO_3^- salvage mechanisms in the submandibular gland acinar and duct cells. *J. Biol. Chem.* 276:9808–16

30. Park K, Hurley PT, Roussa E, Cooper GJ, Smith CP, et al. 2002. Expression of a sodium bicarbonate cotransporter in human parotid salivary glands. *Arch. Oral Biol.* 47:1–9

31. Arreola J, Melvin JE, Begenisich T. 1995. Inhibition of Ca^{2+}-dependent Cl^- channels from secretory epithelial cells by low internal pH. *J. Membr. Biol.* 147:95–104

32. Pedersen KA, Jorgensen NK, Jensen BS, Olesen SP. 2000. Inhibition of the human intermediate-conductance, Ca^{2+}-activated K^+ channel by intracellular acidification. *Pflügers Arch.* 440:153–56

33. Klaerke DA. 1995. Purification and characterization of epithelial Ca^{2+}-activated K^+ channels. *Kidney Int.* 48:1047–56

34. Delpire E, Mount DB. 2002. Human and murine phenotypes associated with defects in cation-chloride cotransport. *Annu. Rev. Physiol.* 64:803–43

35. He X, Tse CM, Donowitz M, Alper SL, Gabriel SE, Baum BJ. 1997. Polarized distribution of key membrane transport proteins in the rat submandibular gland. *Pflügers Arch.* 433:260–68

36. Flagella M, Clarke LL, Miller ML, Erway LC, Giannella RA, et al. 1999. Mice lacking the basolateral Na-K-2Cl cotransporter have impaired epithelial chloride secretion and are profoundly deaf. *J. Biol. Chem.* 274:26946–55

37. Roussa E, Shmukler BE, Wilhelm S, Casula S, Stuart-Tilley AK, et al. 2002. Immunolocalization of potassium-chloride cotransporter polypeptides in rat exocrine glands. *Histochem. Cell Biol.* 117:335–44

38. Lee SI, Turner RJ. 1991. Mechanism of secretagogue-induced HCO_3^- and Cl^- loss from rat parotid acini. *Am. J. Physiol. Gastrointest. Liver Physiol.* 261:G111–18

39. Turner RJ, George JN. 1988. $Cl^--HCO_3^-$ exchange is present with $Na^+-K^+-Cl^-$ cotransport in rabbit parotid acinar basolateral membranes. *Am. J. Physiol. Cell Physiol.* 254:C391–96

40. Simon J, Deshmukh G, Couch FJ, Merajver SD, Weber BL, et al. 1996. Chromosomal mapping of the rat Slc4a family of anion exchanger genes, Ae1, Ae2, and Ae3. *Mamm. Genome* 7:380–82

41. Sterling D, Casey JR. 2002. Bicarbonate transport proteins. *Biochem. Cell Biol.* 80:483–97

42. Ko SB, Luo X, Hager H, Rojek A, Choi JY, et al. 2002. AE4 is a DIDS-sensitive Cl^-/HCO_3^- exchanger in the basolateral membrane of the renal CCD and the SMG duct. *Am. J. Physiol. Cell Physiol.* 283:C1206–18

43. Gresz V, Kwon TH, Vorum H, Zelles T, Kurtz I, et al. 2002. Immunolocalization of electroneutral Na^+-HCO cotransporters in human and rat salivary glands. *Am. J. Physiol. Gastrointest. Liver Physiol.* 283:G473–80

44. Roussa E. 2001. H^+ and HCO_3^+ transporters in human salivary ducts. An

immunohistochemical study. *Histochem. J.* 33:337–44

45. Burckhardt G, Di Sole F, Helmle-Kolb C. 2002. The Na⁺/H⁺ exchanger gene family. *J. Nephrol.* 15(Suppl.)5:S3–21

46. Goyal S, Vanden Heuvel G, Aronson PS. 2003. Renal expression of novel Na⁺/H⁺ exchanger isoform NHE8. *Am. J. Physiol. Renal Physiol.* 284:F467–73

47. Miyazaki E, Sakaguchi M, Wakabayashi S, Shigekawa M, Mihara K. 2001. NHE6 protein possesses a signal peptide destined for endoplasmic reticulum membrane and localizes in secretory organelles of the cell. *J. Biol. Chem.* 276:49221–27

48. Numata M, Orlowski J. 2001. Molecular cloning and characterization of a novel Na⁺,K⁺/H⁺ exchanger localized to the *trans*-Golgi network. *J. Biol. Chem.* 276:17387–94

49. Lee MG, Schultheis PJ, Yan M, Shull GE, Bookstein C, et al. 1998. Membrane-limited expression and regulation of Na⁺-H⁺ exchanger isoforms by P2 receptors in the rat submandibular gland duct. *J. Physiol.* 513:341–57

50. Park K, Olschowka JA, Richardson LA, Bookstein C, Chang EB, Melvin JE. 1999. Expression of multiple Na⁺/H⁺ exchanger isoforms in rat parotid acinar and ductal cells. *Am. J. Physiol. Gastrointest. Liver Physiol.* 276:G470–78

51. Park K, Evans RL, Watson GE, Nehrke K, Bell SM, et al. 2001. Defective secretion and NaCl absorption in the parotid glands of mice lacking Na⁺/H⁺ exchange. *J. Biol. Chem.* 276:27042–50

52. Raina S, Preston GM, Guggino WB, Agre P. 1995. Molecular cloning and characterization of an aquaporin cDNA from salivary, lacrimal, and respiratory tissues. *J. Biol. Chem.* 270:1908–12

53. Funaki H, Yamamoto T, Koyama Y, Kondo D, Yaoita E, et al. 1998. Localization and expression of AQP5 in cornea, serous salivary glands, and pulmonary epithelial cells. *Am. J. Physiol. Cell Physiol.* 275:C1151–57

54. Matsuzaki T, Suzuki T, Koyama H, Tanaka S, Takata K. 1999. Aquaporin-5 (AQP5), a water channel protein, in the rat salivary and lacrimal glands: immunolocalization and effect of secretory stimulation. *Cell Tissue Res.* 295:513–21

55. Ma T, Song Y, Gillespie A, Carlson EJ, Epstein CJ, Verkman AS. 1999. Defective secretion of saliva in transgenic mice lacking aquaporin-5 water channels. *J. Biol. Chem.* 274:20071–74

56. Gresz V, Kwon TH, Hurley PT, Varga G, Zelles T, et al. 2001. Identification and localization of aquaporin water channels in human salivary glands. *Am. J. Physiol. Gastrointest. Liver Physiol.* 281:G247–54

57. Frigeri A, Gropper MA, Umenishi F, Kawashima M, Brown D, Verkman AS. 1995. Localization of MIWC and GLIP water channel homologs in neuromuscular, epithelial and glandular tissues *J. Cell Sci.* 108:2993–3002

58. Elkjaer ML, Nejsum LN, Gresz V, Kwon TH, Jensen UB, et al. 2001. Immunolocalization of aquaporin-8 in rat kidney, gastrointestinal tract, testis, and airways. *Am. J. Physiol. Renal Physiol.* 281:F1047–57

59. Krane CM, Melvin JE, Nguyen H-V, Richardson L, Towne JE, et al. 2001. Salivary gland acinar cells from aquaporin 50-deficient mice have decreased membrane permeability and altered cell volume regulation. *J. Biol. Chem.* 276:23413–20

60. Moore M, Ma T, Yang B, Verkman AS. 2000. Tear secretion by lacrimal glands in transgenic mice lacking water channels AQP1, AQP3, AQP4 and AQP5. *Exp. Eye Res.* 70:557–62

61. Steinfeld S, Cogan E, King LS, Agre P, Kiss R, Delporte C. 2001. Abnormal distribution of aquaporin-5 water channel protein in salivary glands from Sjögren's syndrome patients. *Lab. Invest.* 81:143–48

62. Murakami M, Shachar-Hill B, Steward MC, Hill AE. 2001. The paracellular component of water flow in the rat

submandibular salivary gland. *J. Physiol.* 537:899–906

63. Arreola J, Park K, Melvin JE, Begenisich T. 1996. Three distinct chloride channels control anion movements in rat parotid acinar cells. *J. Physiol.* 490:351–62

64. Zeng W, Lee MG, Yan M, Diaz J, Benjamin I, et al. 1997. Immuno and functional characterization of CFTR in submandibular and pancreatic acinar and duct cells. *Am. J. Physiol. Cell Physiol.* 273:C442–55

65. Arreola J, Melvin JE. 2003. ATP-activated chloride conductance in mouse parotid acinar cells. *J. Physiol.* 547:197–208

66. Arreola J, Melvin JE, Begenisich T. 1996. Activation of calcium-dependent chloride channels in rat parotid acinar cells. *J. Gen. Physiol.* 108:35–47

67. Huang P, Liu J, Di A, Robinson NC, Musch MW, et al. 2001. Regulation of human CLC-3 channels by multifunctional Ca^{2+}/calmodulin-dependent protein kinase. *J. Biol. Chem.* 276:20093–100

68. Arreola J, Melvin JE, Begenisich T. 1998. Differences in regulation of Ca^{2+}-activated Cl^- channels in colonic and parotid secretory cells. *Am. J. Physiol. Cell Physiol.* 274:C161–66

69. Arreola J, Begenisich T, Nehrke K, Nguyen H-V, Park K, et al. 2002. Secretion and cell volume regulation by salivary gland acinar cells in mice lacking expression of the *Clcn3* Cl^- channel gene. *J. Physiol.* 545:207–16

70. Cunningham SA, Awayda MS, Bubien JK, Ismailov II, Arrate MP, et al. 1995. Cloning of an epithelial chloride channel from bovine trachea. *J. Biol. Chem.* 270:31016–26

71. Pauli BU, Abdel-Ghany M, Cheng HC, Gruber AD, Archibald HA, Elble RC. 2000. Molecular characteristics and functional diversity of CLCA family members. *Clin. Exp. Pharmacol. Physiol.* 27:901–5

72. Fuller CM, Ji HL, Tousson A, Elble RC, Pauli BU, Benos DJ. 2001. Ca^{2+}-activated Cl^- channels: a newly emerging anion transport family. *Pflügers Arch.* 443:S107–10

73. Abdel-Ghany M, Cheng HC, Elble RC, Pauli BU. 2002. Focal adhesion kinase activated by $beta_4$ integrin ligation to mCLCA1 mediates early metastatic growth. *J. Biol. Chem.* 277:34391–400

74. Sun H, Tsunenari T, Yau KW, Nathans J. 2002. The vitelliform macular dystrophy protein defines a new family of chloride channels. *Proc. Natl. Acad. Sci. USA* 99:4008–13

75. Qu Z, Fischmeister R, Hartzell C. 2004. Mouse bestrophin-2 is a bona fide Cl^- channel: identification of a residue important in anion binding and conduction. *J. Gen. Physiol.* 123:327–40

76. Nehrke K, Arreola J, Nguyen H-V, Begenisich T, Richardson L, et al. 2002. Loss of the inward rectifying Cl^- current in parotid acinar cells from mice lacking expression of the *Clcn2* gene. *J. Biol. Chem.* 277:23604–11

77. Duan D, Zhong J, Hermoso M, Satterwhite CM, Rossow CF, et al. 2001. Functional inhibition of native volume-sensitive outwardly rectifying anion channels in muscle cells and *Xenopus* oocytes by anti-ClC-3 antibody. *J. Physiol.* 531:437–44

78. Wang GX, Hatton WJ, Wang GL, Zhong J, Yamboliev I, et al. 2003. Functional effects of novel anti-ClC-3 antibodies on native volume-sensitive osmolyte and anion channels in cardiac and smooth muscle cells. *Am. J. Physiol. Heart Circ. Physiol.* 285:H1453–63

79. Stobrawa SM, Breiderhoff T, Takamori S, Engel D, Schweizer M, et al. 2001. Disruption of ClC-3, a chloride channel expressed on synaptic vesicles, leads to a loss of the hippocampus. *Neuron* 29:185–96

80. Maruyama Y, Gallacher DV, Petersen OH. 1983. Voltage and Ca^{2+}-activated K^+ channel in baso-lateral acinar cell membranes of mammalian salivary glands. *Nature* 302:827–29

81. Maruyama Y, Nishiyama A, Teshima T. 1986. Two types of cation channels in the basolateral cell membrane of human salivary gland acinar cells. *Jpn. J. Physiol.* 36:219–23

82. Park K, Case RM, Brown PD. 2001. Identification and regulation of K$^+$ and Cl$^-$ channels in human parotid acinar cells. *Archiv. Oral Biol.* 46:801–10

83. Nehrke K, Quinn CC, Begenisich T. 2003. Molecular identification of Ca^{2+}-activated K$^+$ channels in parotid acinar cells. *Am. J. Physiol. Cell Physiol.* 284:C535–46

84. Bymaster FP, Carter PA, Yamada M, Gomeza J, Wess J, et al. 2003. Role of specific muscarinic receptor subtypes in cholinergic parasympathomimetic responses, in vivo phosphoinositide hydrolysis, and pilocarpine-induced seizure activity. *Eur. J. Neurosci.* 17:1403–10

85. Matsui M, Motomura D, Karasawa H, Fujikawa T, Jiang J, et al. 2000. Multiple functional defects in peripheral autonomic organs in mice lacking muscarinic acetylcholine receptor gene for the M3 subtype. *Proc. Natl. Acad. Sci. USA* 97:9579–84

86. Nauntofte B, Poulsen JH. 1986. Effects of Ca^{2+} and furosemide on Cl$^-$ transport and O$_2$ uptake in rat parotid acini. *Am. J. Physiol. Cell Physiol.* 251:C175–85

87. Evans RL, Turner RJ. 1997. Upregulation of Na$^+$-K$^+$-2Cl$^-$ cotransporter activity in rat parotid acinar cells by muscarinic stimulation. *J. Physiol.* 499:351–59

88. Manganel M, Turner RJ. 1990. Agonist-induced activation of Na$^+$/H$^+$ exchange in rat parotid acinar cells is dependent on calcium but not on protein kinase C. *J. Biol. Chem.* 265:4284–89

89. Ishikawa Y, Eguchi T, Skowronski MT, Ishida H. 1998. Acetylcholine acts on M3 muscarinic receptors and induces the translocation of aquaporin5 water channel via cytosolic Ca^{2+} elevation in rat parotid glands. *Biochem. Biophys. Res. Commun.* 245:835–40

90. Ishikawa Y, Skowronski MT, Ishida H. 2000. Persistent increase in the amount of aquaporin-5 in the apical plasma membrane of rat parotid acinar cells induced by a muscarinic agonist SNI-2011. *FEBS Lett.* 477:253–57

91. Watson EL, Abel PW, DiJulio D, Zeng W, Makoid M, et al. 1996. Identification of muscarinic receptor subtypes in mouse parotid gland. *Am. J. Physiol. Cell Physiol.* 271:C905–13

92. Taylor CW, Genazzani AA, Morris SA. 1999. Expression of inositol trisphosphate receptors. *Cell Calcium* 26:237–51

93. Patel S, Joseph SK, Thomas AP. 1999. Molecular properties of inositol 1,4,5-trisphosphate receptors. *Cell Calcium* 25:247–64

94. Berridge MJ, Lipp P, Bootman MD. 2000. The versatility and universality of calcium signalling. *Nat. Rev. Mol. Cell Biol.* 1:11–21

95. Nezu A, Tanimura A, Morita T, Irie K, Yajima T, Tojyo Y. 2002. Evidence that zymogen granules do not function as an intracellular Ca^{2+} store for the generation of the Ca^{2+} signal in rat parotid acinar cells. *Biochem. J.* 363:59–66

96. Giovannucci DR, Bruce JI, Straub SV, Arreola J, Sneyd J, et al. 2002. Cytosolic Ca^{2+} and Ca^{2+}-activated Cl$^-$ current dynamics: insights from two functionally distinct mouse exocrine cells. *J. Physiol.* 540:469–84

97. Zhang X, Wen J, Bidasee KR, Besch HR Jr, Wojcikiewicz RJ, et al. 1999. Ryanodine and inositol trisphosphate receptors are differentially distributed and expressed in rat parotid gland. *Biochem. J.* 340:519–27

98. Lee MG, Xu X, Zeng W, Diaz J, Wojcikiewicz RJ, et al. 1997. Polarized expression of Ca2$^+$ channels in pancreatic and salivary gland cells. Correlation with initiation and propagation of [Ca^{2+}]$_i$ waves. *J. Biol. Chem.* 272:15765–70

99. Yule DI, Ernst SA, Ohnishi H, Wojcikiewicz RJ. 1997. Evidence that

zymogen granules are not a physiologically relevant calcium pool. Defining the distribution of inositol 1,4,5-trisphosphate receptors in pancreatic acinar cells. *J. Biol. Chem.* 272:9093–98

100. Tanimura A, Matsumoto Y, Tojyo Y. 1998. Polarized Ca^{2+} release in saponin-permeabilized parotid acinar cells evoked by flash photolysis of 'caged' inositol 1,4,5-trisphosphate. *Biochem. J.* 332:769–72

101. Thorn P, Lawrie AM, Smith PM, Gallacher DV, Petersen OH. 1993. Local and global cytosolic Ca^{2+} oscillations in exocrine cells evoked by agonists and inositol trisphosphate. *Cell* 74:661–68

102. Tinel H, Cancela JM, Mogami H, Gerasimenko JV, Gerasimenko OV, et al. 1999. Active mitochondria surrounding the pancreatic acinar granule region prevent spreading of inositol trisphosphate-evoked local cytosolic Ca^{2+} signals. *EMBO J.* 18:4999–5008

103. Straub SV, Giovannucci DR, Yule DI. 2000. Calcium wave propagation in pancreatic acinar cells: functional interaction of inositol 1,4,5-trisphosphate receptors, ryanodine receptors, and mitochondria. *J. Gen. Physiol.* 116:547–60

104. Bruce JI, Giovannucci DR, Blinder G, Shuttleworth TJ, Yule DI. 2004. Modulation of $[Ca^{2+}]_i$ signaling dynamics and metabolism by perinuclear mitochondria in mouse parotid acinar cells. *J. Biol. Chem.* 279:12909–17

105. Zhang X, Wen J, Bidasee KR, Besch HR Jr, Rubin RP. 1997. Ryanodine receptor expression is associated with intracellular Ca^{2+} release in rat parotid acinar cells. *Am. J. Physiol. Cell Physiol.* 273:C1306–14

106. Bruce JI, Shuttleworth TJ, Giovannucci DR, Yule DI. 2002. Phosphorylation of inositol 1,4,5-trisphosphate receptors in parotid acinar cells. A mechanism for the synergistic effects of cAMP on Ca^{2+} signaling. *J. Biol. Chem.* 277:1340–48

107. Yao J, Li Q, Chen J, Muallem S. 2004. Subpopulation of store-operated Ca^{2+} channels regulate Ca^{2+}-induced Ca^{2+} release in non-excitable cells. *J. Biol. Chem.* 279:21511–19

108. Bruce JI, Yule DI, Shuttleworth TJ. 2002. Ca^{2+}-dependent protein kinase—a modulation of the plasma membrane Ca^{2+}-ATPase in parotid acinar cells. *J. Biol. Chem.* 277:48172–81

109. Lee MG, Xu X, Zeng W, Diaz J, Kuo TH, et al. 1997. Polarized expression of Ca^{2+} pumps in pancreatic and salivary gland cells. Role in initiation and propagation of $[Ca^{2+}]_i$ waves. *J. Biol. Chem.* 272:15771–76

110. Putney JW Jr. 1986. A model for receptor-regulated calcium entry. *Cell Calcium* 7: 1–12

111. Putney JW Jr. 1990. Capacitative calcium entry revisited. *Cell Calcium* 11:611–24

112. Takemura H, Putney JW Jr. 1989. Capacitative calcium entry in parotid acinar cells. *Biochem. J.* 258:409–12

113. Liu X, O'Connell A, Ambudkar IS. 1998. Ca^{2+}-dependent inactivation of a store-operated Ca^{2+} current in human submandibular gland cells. Role of a staurosporine-sensitive protein kinase and the intracellular Ca^{2+} pump. *J. Biol. Chem.* 273:33295–304

114. Liu XB, Ambudkar IS. 2001. Characteristics of a store-operated calcium-permeable channel—sarcoendoplasmic reticulum calcium pump function controls channel gating. *J. Biol. Chem.* 276: 29891–98

115. Hoth M, Penner R. 1993. Calcium release-activated calcium current in rat mast cells. *J. Physiol.* 465:359–86

116. Zweifach A, Lewis RS. 1993. Mitogen-regulated Ca^{2+} current of T lymphocytes is activated by depletion of intracellular Ca^{2+} stores. *Proc. Natl. Acad. Sci. USA* 90:6295–99

117. Singh BB, Zheng CY, Liu XB, Lockwich T, Liao D, et al. 2001. Trp1-dependent enhancement of salivary gland

fluid secretion: role of store-operated calcium entry. *FASEB J.* 15:NIL96–112

118. Xu X, Zeng W, Diaz J, Lau KS, Gukovskaya AC, et al. 1997. nNOS and Ca^{2+} influx in rat pancreatic acinar and submandibular salivary gland cells. *Cell Calcium* 22:217–28

119. Watson EL, Jacobson KL, Singh JC, Ott SM. 1999. Nitric oxide acts independently of cGMP to modulate capacitative Ca^{2+} entry in mouse parotid acini. *Am. J. Physiol. Cell Physiol.* 277:C262–70

120. Mignen O, Thompson JL, Shuttleworth TJ. 2001. Reciprocal regulation of capacitative and arachidonate-regulated noncapacitative Ca^{2+} entry pathways. *J. Biol. Chem.* 276:35676–83

121. Mignen O, Shuttleworth TJ. 2000. I_{ARC}, a novel arachidonate-regulated, noncapacitative Ca^{2+} entry channel. *J. Biol. Chem.* 275:9114–19

122. Mignen O, Thompson JL, Shuttleworth TJ. 2003. Ca^{2+} selectivity and fatty acid specificity of the noncapacitative, arachidonate-regulated Ca^{2+} (ARC) channels. *J. Biol. Chem.* 278:10174–81

123. Shuttleworth TJ, Mignen O. 2003. Calcium entry and the control of calcium oscillations. *Biochem. Soc. Trans.* 31:916–19

124. Watson EL, Jacobson KL, Singh JC, DiJulio DH. 2004. Arachidonic acid regulates two Ca^{2+} entry pathways via nitric oxide. *Cell. Signal.* 16:157–65

125. Putney JW Jr. 1986. Identification of cellular activation mechanisms associated with salivary secretion. *Annu. Rev. Physiol.* 48:75–88

126. Lundberg JM, Anggard A, Fahrenkrug J. 1982. Complementary role of vasoactive intestinal polypeptide (VIP) and acetylcholine for cat submandibular gland blood flow and secretion. *Acta Physiol. Scand.* 114:329–37

127. Lundberg JM, Anggard A, Fahrenkrug J, Hokfelt T, Mutt V. 1980. Vasoactive intestinal polypeptide in cholinergic neurons of exocrine glands: functional significance of coexisting transmitters for vasodilation and secretion. *Proc. Natl. Acad. Sci. USA* 77:1651–55

128. Ekstrom J, Olgart L. 1986. Complementary action of substance P and vasoactive intestinal peptide on the rat parotid secretion. *Acta Physiol. Scand.* 126:25–31

129. Chernick W, Bobyock E, Bradford P. 1989. 5-Hydroxytryptamine modulation of rat parotid salivary gland secretion. *J. Dent. Res.* 68:59–63

130. Bobyock E, Chernick WS. 1989. Vasoactive intestinal peptide interacts with alpha-adrenergic-, cholinergic-, and substance-P-mediated responses in rat parotid and submandibular glands. *J. Dent. Res.* 68:1489–94

131. Ekstrom J, Ekman R, Hakanson R, Sjogren S, Sundler F. 1988. Calcitonin gene-related peptide in rat salivary glands: neuronal localization, depletion upon nerve stimulation, and effects on salivation in relation to substance P. *Neuroscience* 26:933–49

132. Larsson O, Olgart L. 1989. The enhancement of carbachol-induced salivary secretion by VIP and CGRP in rat parotid gland is mimicked by forskolin. *Acta Physiol. Scand.* 137:231–36

133. Tanimura A, Nezu A, Tojyo Y, Matsumoto Y. 1999. Isoproterenol potentiates alpha-adrenergic and muscarinic receptor-mediated Ca^{2+} response in rat parotid cells. *Am. J. Physiol. Cell Physiol.* 276:C1282–87

134. Bruce JI, Straub SV, Yule DI. 2003. Crosstalk between cAMP and Ca^{2+} signaling in non-excitable cells. *Cell Calcium* 34:431–44

135. Martinez JR, Zhang GH. 1998. Cross-talk in signal transduction pathways of rat submandibular acinar cells. *Eur. J. Morphol.* 36(Suppl.):190–93

136. Watson EL, Jacobson KL, Singh JC, Idzerda R, Ott SM, et al. 2000. The type 8 adenylyl cyclase is critical for

Ca^{2+} stimulation of cAMP accumulation in mouse parotid acini. *J. Biol. Chem.* 275:14691–99

137. Fagen KA, Mahey R, Cooper DMF. 1996. Functional co-localization of transfected Ca^{2+}-stimulable adenyl cyclases with capacitative Ca^{2+} entry sites. *J. Biol. Chem.* 271:12438–44

138. Shuttleworth TJ, Thompson JL. 1999. Discriminating between capacitative and arachidonate-activated Ca^{2+} entry path-

ways in HEK293 cells. *J. Biol. Chem.* 274:31174–78

139. Cook DI, Young JA. 1989. Effect of K^{+} channels in the apical plasma membrane on epithelial secretion based on secondary active Cl^{-} transport. *J. Membr. Biol.* 110:139–46

140. Fuller CM, Eckhard L, Schulz I. 1989. Ionic and osmotic dependence of secretion from permeabilised acini of the rat pancreas. *Pflügers Arch.* 413:385–94

Annu. Rev. Physiol. 2005. 67:471–90
doi: 10.1146/annurev.physiol.67.031103.153530
First published online as a Review in Advance on November 1, 2004

SECRETION AND ABSORPTION BY COLONIC CRYPTS

John P. Geibel

Department of Surgery, Department of Cellular and Molecular Physiology,
Yale University School of Medicine, New Haven, Connecticut 06520;
email: john.geibel@yale.edu

Key Words   colon, electrolyte transport, sodium, chloride, fluid

■ **Abstract**   The intestines play an important role in the absorption and secretion of nutrients. The colon is the final area for recapturing electrolytes and water prior to excretion, and in order to maintain this electrolyte homeostasis, a complex interaction between secretory and absorptive processes is necessary. Until recently it was thought that secretion and absorption were two distinct processes associated with either crypts or surface cells, respectively. Recently it was demonstrated that both the surface and crypt cells can perform secretory and absorptive functions and that, in fact, these functions can be going on simultaneously. This issue is important in the complexities associated with secretory diarrhea and also in attempting to develop treatment strategies for intestinal disorders. Here, we update the model of colonic secretion and absorption, discuss new issues of transporter activation, and identify some important new receptor pathways that are important modulators of the secretory and absorptive functions of the colon.

INTRODUCTION

The intestines are charged with managing the complex issues of nutrient, electrolyte and fluid absorption and the secretion of waste materials, as well as excess electrolytes and fluid. In addition to these complex tasks, a further important issue is the fact that there is a rapid and continuous replenishment of the apical surface, thus making this epithelium unique in that it is constantly turning over and regenerating. In this environment the colon is the final conduit into which bidirectional transport of electrolytes and fluids can occur. Not only has this segment been broken up into a proximal and distal zone, but it has further been defined as surface and crypt to delineate the two main classes of epithelial cells that make up the complex apical surface of the colon. These changes in the cells that make up the surface and the crypt are further associated with the changes in the complexity of the tight junctions, cytoskeletal-associated proteins, and some associated enzymes (15, 95).

   This review focuses on the current views of ion transport along the colon and identifies some common misconceptions in the association of transport within

0066-4278/05/0315-0471$14.00

specific sections of the colon. We attempt to elucidate both the secretory and absorptive characteristics of the epithelium and define what parameters can lead to either secretion or absorption of electrolytes and fluid. The goal is to show that the colon uses a combination of balanced absorption and secretion to obtain the net transport of ions needed by the body.

In this review we challenge the long-standing dogma that surface cells are the reabsorptive component of the colon and the crypt cells are the secretory component. Recent data have shown that absorption and secretion can be handled by both cell types and that the fixed rule of surface versus crypt is no longer correct (55–57, 64). The colon has to withstand extreme changes in electrolyte movement, as illustrated by severe secretory states (infection, diarrhea) (17, 89) and reabsorptive states (cystic fibrosis, CF) (64, 73, 75). In both extremes it is apparent that secretion and absorption have to be tightly regulated so that net fluid and electrolyte transport is maintained.

We also review the identified transporters in native tissue and address the regulation of these proteins in both surface and crypt cells when known. We do not address the issues of mucous formation and secretion in great detail, but rather identify the mechanisms that have been associated with modifying electrolyte transport in this tissue.

BASIC ANATOMY OF COLONIC EPITHELIUM

Cell Types

To date, morphological studies of mammalian colonic epithelium have identified three different cell types: columnar, mucous, and enterochromaffin cells (18). Columnar and goblet cells make up ~95% of the total cells, the remaining ~5% contain enterochromaffin cells. A further subdivision of the columnar cells has been defined on the basis of their proliferative activity, which has been defined as the expression of differentiation markers and their functional properties (48, 91). Using this classification system, we note that the base of the crypt shows the highest proliferative activity and few differentiation markers. The surface cells have a lower tendency to proliferate, show increased levels of differentiation markers in addition to lectins, and are thought to have primarily an absorptive function (47). When the crypt-to-surface axis is viewed, the farther away from the base of the crypt the more differentiated the cells become, with maximal differentiation at the surface. This unique morphometry provides an important means to constantly replace the epithelium because cells migrate toward the surface (11, 27).

Ion Transport: Surface versus Crypt

Until recently a debate raged as to whether crypts and surface cell epithelium were two distinct compartments with specialized functions for absorption and secretion. One view was that the crypt played a role predominantly to clear the mucous that is being secreted from the goblet and columnar epithelial cells

(30, 100). Although this may be true, mucous secretion also takes place in the surface cells. As a result of these and other studies (9, 10, 19, 20, 38, 39, 84, 96), it is no longer reasonable to assume that crypts are purely a secretory epithelium and surface cells a reabsorptive epithelium. Direct evidence shows that surface cells as well as crypts can be involved in secretion (55), with data obtained from vibrating electrodes (26, 55, 57) and patch-clamp studies of Cl^- channels (26). Also, our laboratory has shown with measurements of fluid secretion that the crypt cells are capable of both secretion and absorption (10, 19, 38, 39). In the case of the isolated perfused crypt we found that in the absence of neuronal or other secretagogues the crypt becomes an absorptive epithelium and with the addition of these secretory agents the crypt will robustly secrete fluid (10, 19, 38, 39). When addressing the issue of reabsorption, there is even more controversy. As discussed below, absorption can be electroneutral via Na/H and Cl/HCO_3^- exchange or electrogenic via sodium channels (ENaC). The results of some studies based on the theory that a high osmotic pressure gradient is required in order to absorb water from the lumen of the intestine and to generate fecal material rely on a scheme that traps ions in the crypts rather than in the surface cells (77, 78, 80, 96). These studies have combined the use of fluorescent dextrans with direct measurements of fluid reabsorption from the crypt. In addition to these data, there are also demonstrations of NaCl absorption in both the crypt and the surface cells. We examine the absorptive function of the colon and each of the identified transport proteins that are involved. Included is a discussion of a calcium-sensing receptor (which we recently identified) that can act as a potent modulator of both secretion and absorption. We also examine the known secretory transport proteins in the colon. Owing to the restrictions in length herein, we concentrate on only the major transport proteins and touch briefly on other factors that may modulate the absorption of ions (Figure 1).

ABSORPTION

There are two main characterizations of absorption in the colon: electroneutral and electrogenic. It is noteworthy that definite species differences exist in terms of the fraction of electroneutral and electrogenic absorption and that hormonal balance also plays a key role in the distribution of the proteins. The ratio of transports is further complicated owing to regulation by the hormonal status of the epithelium. In the case of increased levels of aldosterone, the predominate mechanism is amiloride-sensitive Na^+ channels (21, 65). In addition, species-specific differences are illustrated by the fact that the electrogenic absorption is the predominant pathway in the rabbit, and electroneutral absorption is the predominant pathway in the rat (7).

Electroneutral NaCl Absorption

The majority of bulk transport of NaCl is through electroneutral absorption of ions via the luminal Na/H and Cl/HCO_3^- exchangers (Figure 2).

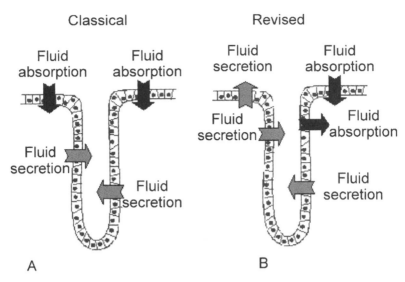

Figure 1 Schematic model of colonic transport. Panel *A* represents the classical model of secretion and absorption in which surface cells reabsorb and crypt cells secrete. Panel *B* shows the new model in which both crypts and surface cells can secrete and reabsorb fluid and electrolytes.

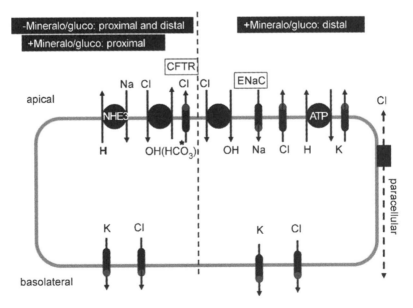

Figure 2 Cell model of the proximal and distal colon. This model summarizes the known transport proteins in the proximal and distal colon in the presence or absence of mineralocorticoid stimulation.

Na/H Exchange

To date, three separate types of Na/H exchangers (NHE) have been detected in the colon. NHE1 is found in the basolateral membrane and appears to be unaffected by Na^+ depletion in the epithelia. NHE2 and NHE3 have been identified on the apical surface, with the predominant type being NHE3 under control conditions (92). The regulation of NHE2 and NHE3 differs in proximal and distal colon, with both being unregulated under Na^+ depletion in proximal colon and becoming attenuated in the distal colon (51, 92). A potential third and unique member of this family of proteins appears to express a Cl^- dependence in rat crypt cells (9, 10, 84). This protein is likely coupled to a Cl^- channel that has some functional properties of the cystic fibrosis transmembrane regulator (CFTR) rather than to a Cl^- anion exchange (10, 84).

It is clear, however, that Na/H exchange occurs in both the surface and crypt cells and is, for the most part, tightly coupled to Cl/HCO_3^- exchange, and in some cases may be associated with CFTR (84). Recently Na/H exchanger regulatory factor (NHERF) was identified in the small intestine but not in cultured colonic carcinoma cells. NHERF has been shown to play a key role in the kidney in the cAMP-mediated inhibition of Na absorption by luminal NHE3 and basolateral Na/HCO_3^- (99, 103).

It is well accepted that NHE1–3 are found in the colon and that all play important roles in the electroneutral absorption of NaCl. Although likely, the roles of NHERF and CFTR need to be further investigated before a definitive fraction of absorption can be linked to either or both of these proteins.

Cl/HCO_3^- Exchange

To date there are at least two identified types of Cl^- exchangers that have been linked to the apical membranes of colonic epithelium. The basolateral membrane expresses a third type of Cl/HCO_3^- (50, 82, 83). The apical exchangers consist of a Cl/HCO_3^- system and a Cl/OH exchanger, which have been associated with an AE1 type protein and DRA (down-regulated in colonic adenomas) (83). DRA has been shown to be up-regulated in mice lacking NHE3 but is not tightly regulated by Na^+ depletion; on the other hand, expression of AE1 is inhibited by aldosterone (76).

In summary, there are several different isoforms of NHE as well as three different isoforms of anion exchangers. Furthermore, there is evidence that the activity of both NHE and Cl/HCO_3^- is influenced by CFTR and that inhibition or mutations in CFTR can have dramatic effects on fluid absorption in the colon.

Electrogenic Absorption

The distal colon has, in addition to the electroneutral pathway for Na^+ absorption, an electrogenic pathway to enhance Na^+ uptake. This pathway has been identified as the epithelial Na^+ channel (ENaC) and is localized to apical membranes of colonic epithelial cells. The ENaC protein is inhibited by amiloride or amiloride analogues (6, 97). Because of the existing electrochemical gradient for Na^+ and the negative membrane voltage, there is a large driving force for luminal Na^+ uptake via a channel. This Na^+ absorption is accompanied by Cl^-, which acts as the negative charge ion. It is thought that Cl^- enters the cell via the apical Cl^- channel, as well as through the paracellular shunt (59, 73). The Na^+ ions are then transported out of the cell via the basolateral Na/K-ATPase, thereby producing net Na^+ absorption. The Cl^- is transported into the blood either by basolateral Cl^- channels or via basolateral Cl/HCO_3^- exchange (41). With enhanced NaCl absorption, one could expect that water would move in parallel either via a transcellular pathway or via paracellular shunts. However, to date, a major role for aquaporins in the colon has yet to be established.

This complex absorption of Na^+ relies on a multilevel feedback mechanism that assures parallel up-regulation of basolateral exit on the ions with enhanced apical uptake. An example is aldosterone, which can act to increase basolateral Na/K ATPase activity as well as apical ENaC (25, 85). Furthermore, recent evidence indicates that CFTR can directly interact with ENaC and modulate Na^+ transport (58, 64).

Electrogenic Na^+ absorption is tightly linked to enhanced ENaC activity, with parallel Cl^- entry and basolateral NaCl exit that results in NaCl absorption. ENaC is also regulated through hormonal (aldosterone) and ion channels (CFTR).

SECRETION

In addition to the absorption of electrolytes, the colon has a further role as a final secretory segment. This secretion runs in parallel and is balanced by the absorption of electrolytes so that homeostasis is maintained; a balance that is critical for preventing secretory diarrhea and the potential fatal loss of electrolytes.

One theory regarding the role of the secretory pathway in healthy tissues is that it aids in the transport of mucous out of the crypt while maintaining hydration of this material. Mucous secretion has been associated with an increase in intracellular cAMP levels, which also enhances fluid secretion (45). In the nonstimulated cells, apical K and Cl^- channels are active and lead to the secretion of KCl, which is balanced by increased Na^+, K, and Cl^- entry via the basolateral NKCC protein (24). The secretion of both KCl and NaCl has been linked to a variety of secretagogues using various intracellular messengers (7, 60, 62–64, 100). Therefore, a key to maintaining the secretion is coordinated apical efflux with basolateral entry so that net cell ion concentration remains constant. A fall in either the basolateral entry or apical exit could lead to large alterations in intracellular ion concentrations and resulting catastrophic effects on cell ionic homeostasis.

NKCC Cotransport

In order to maintain secretion, the basolateral membrane needs to uptake ions at a sufficient rate to ensure that the cellular milieu remains constant and that no net changes in cellular ion concentration occur owing to enhanced apical exit. The Na^+ 2Cl K cotransporter type 1 (NKCC1) has been identified in the basolateral membrane of the colon. This protein is relatively insensitive to the loop diuretic furosemide but is blocked by the more potent drug azosemide (40). At rest, the transporter expresses a relative low affinity for the inhibitors, and it is not until stimulation by Cl^- secretion that the drugs show any hyperpolarizing effect on the basolateral membrane. The detectable change in membrane potential is the result of inhibition of Cl^- uptake. This change in intracellular Cl^- creates a new electrochemical equilibrium potential that is close to that of K (68–70). The importance of this transport process is demonstrated by impaired secretion of Cl^- in NKCC1 knockout mice, thereby confirming the importance of the NKCC1 protein in Cl^- secretion (25). It should be noted that although Cl^- secretion was reduced in these mice, it could be activated in other intestinal segments (jejunum and cecum) by raising intracellular cAMP levels with *Escherichia coli* toxin. This result illustrates that although the NKCC1 protein plays an important role, it is possible under certain conditions to activate additional basolateral uptake mechanisms such as a Cl/HCO_3^- and Na/H exchangers that could contribute to net Cl^- secretion. Under standard conditions, increases in cAMP lead to enhanced expression of NKCC1 via a direct phosphorylation by protein kinase A (PKA) and other protein kinases and phosphatases (43, 44, 79). In the mammalian colon, a parallel activation of NKCC1 and apical Cl^- excretion via CFTR or other Cl^- channels leads to enhanced net secretion. In order to enhance NKCC1 activity, the colonic cell senses the fall in intracellular Cl^- and/or cell volume at the onset of secretion (72) (Figure 3).

Cl^- Channels

CFTR   This channel has been targeted as the primary source for apical Cl^- efflux in the colon due to its high abundance and its associated properties for activation. The CFTR protein, similar to the NKCC1 protein, is activated by PKA and other second messenger pathways. In addition, it has also been shown to be activated by protein kinase C (PKC), calcium/calmodulin-dependent kinase, and a cGMP-dependent kinase (12, 40, 42, 49, 59, 61, 62, 74). It is not surprising that this channel could be up-regulated under conditions that raise cAMP(cholera toxin) or cGMP (E. Choli, personal communication). With enhanced Cl^- channel activity from increased second messenger levels, an enhanced Cl^- entry and channel activation occur via NKCC1, with the net result being enhanced Cl^- secretion.

INTERMEDIATE CONDUCTANCE Cl CHANNELS   Due to inherent technical difficulties with patch-clamping native colonic tissue, numerous studies have been carried out in colonic cell lines. It should be pointed out that many of these studies have been conducted in colonic carcinoma cell lines and that the channels expressed in these lines may not give an accurate presentation of what is occurring in the native tissues. With this in mind, one particular type of chloride channel was found quite frequently in excised patches of rat colonic epithelial cells. The observed channel

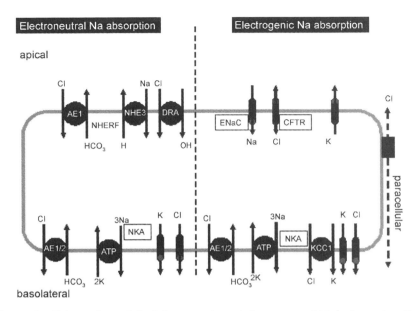

Figure 3 Schematic model of electrogenic and electroneutral Na$^+$ absorption. Left panel shows apical electroneutral Na$^+$ absorption, which relies on NHE3, DRA, and AE1. Right panel shows electrogenic Na$^+$ absorption, which is linked to ENaC, CFTR, and a K channel.

had a single conductance of ~50 pS and typical outward rectification characteristics. Owing to this rectification, the channel was called an ORCC or intermediate conductance outward rectifying chloride channel (ICOR) (7). With high abundance of the ORCC in these cells, it was thought that this could be the pathway for Cl$^-$ ion secretion. Further compelling evidence indicated that the ORCC could be activated by PKA (66). However, additional recent studies have failed to reveal a role for this type of channel in native human mucosa (94, 100, 103, 124) (Figure 4).

CALCIUM-ACTIVATED Cl CHANNELS As the name calcium-activated Cl$^-$ channels (CaCC) implies, they are a class of channels activated by increasing levels of calcium at the intracellular face of the channel. These channels have been found in small intestine and in T84 cells, as well as in human colonic carcinoma cell lines (22, 60, 63, 102).

BASOLATERAL Cl CHANNELS In the rat colonic epithelium evidence has been given for the existence of a Cl$^-$ channel on the basolateral membrane (26). In the turtle colon there is evidence that the basolateral channel is regulated by cholinergic stimulation and also by increased levels of cAMP (24). To date there has been no description of these channels in the human, and the exact role these channels play

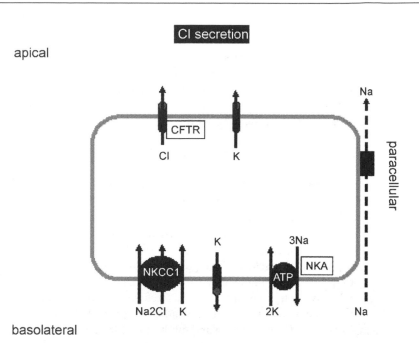

Figure 4   Model of colonic Cl⁻ secretion.

is not clear. Future studies will be required to determine if these channels have a significant role in cell volume regulation or in the maintenance of *trans*-cellular Cl⁻ flux.

Potassium Channels

As is the case with the Cl⁻ channels, reports suggest the presence of both apical and basolateral membrane K channels.

APICAL K CHANNELS   A variety of reports show at least two types of K channels in the colon apical membrane. The initial reports characterized a large conductance ~150–200 pS apical channel (14). Recently it was suggested that the ROMK channel is also present on the apical membrane of the colon (98).

One common characteristic of the apical K channels is that they are up-regulated by aldosterone and glucocorticoids in parallel to the Na⁺ channels (13, 14, 35, 36, 93, 94).

BASOLATERAL K CHANNELS   In order to maintain a hyperpolarizing membrane voltage and the driving force for Cl⁻ secretion and Na⁺ absorption, functional basolateral K channels are required. To date, the basolateral K conductance consists of at least two different K channels that are activated either by cAMP or increases in intracellular Ca^{2+} (71, 74).

Regulated Ion Secretion

As is clear in the above discussion, both secretion and absorption are controlled by a complex interaction of endocrine, paracrine, autocrine, and neuronal stimuli. Of particular interest is the ability of certain compounds (e.g., food additives, bile acids, or bacterial toxins) that reach the lumen of the intestine to release secretagogues, which can dramatically alter the secretory and/or absorptive transporters that control secretion (5, 81). The effects of these various agents on the colon have been shown not only to be membrane specific but also to be segmentally specific (5).

Owing to the complex nature of these various secretagogues and their interrelated roles, we have chosen to give a brief overview of two main modulators of ion secretion that are associated with many of the compounds and bacterial agents listed above: cAMP-dependent secretion and cGMP-dependent secretion.

cAMP-DEPENDENT SECRETION   Changes in cAMP levels can be brought on by a variety of different membrane or cytosolic processes (5), all resulting in increased intracellular concentrations of cAMP. As the level of cAMP increases, activation of net efflux pathways for Cl^- (CFTR) and Cl^- entry (NKCC) occur. This results in a rapid onset of hypersecretion and inhibition of the absorptive components of the colon (103).

cGMP-DEPENDENT SECRETION   It has been postulated that cGMP leads to enhanced Cl^- secretion through a cGMP-regulated protein kinase G type II and that it may furthermore exert additional inhibitory effects on phosphodiesterases leading to an increase in cAMP (23, 54). Phosphodiesterase type 5 was found in human colonic epithelial cells and appears to play an important role in the regulation of intracellular cGMP levels (1, 2, 4).

BICARBONATE SECRETION   In addition to active KCl secretion, the colonic epithelium can also secrete HCO_3^- into the lumen, resulting in a slightly alkaline lumen pH. There has been direct evidence in the guinea pig, rabbit, and rat that bicarbonate is secreted using a variety of pathways including (*a*) electrogenic efflux, (*b*) transport via a luminal Cl/HCO_3^- exchange, and (*c*) SCFA bicarbonate exchange (64). We have recently shown that the predominant pathway for HCO_3^- excretion in the rat is via the crypt through apical Cl^- channels; however, we could not find evidence for an apical Cl/HCO_3^- component (39, 96). Thus there are two possibilities for apical Cl^- channel action: either in electrogenic secretion of HCO_3^- or as a recycling pathway for Cl^- if a Cl/HCO_3^- system were functional on this membrane (Figure 5).

TRANSPORT OF WATER IN THE COLON   One of the major functions of the colonic epithelium is the net reabsorption of ~1.5 liters of water per day (8, 82), and unidirectional transport rates are in all likelihood higher. We see this important

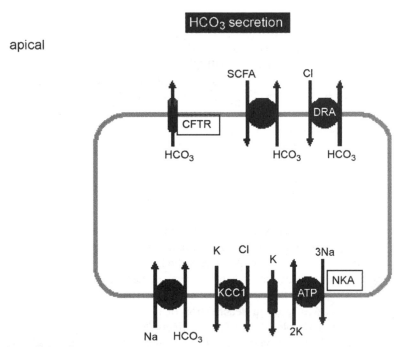

apical

basolateral

Figure 5 Model of colonic HCO_3^- secretion. Note that HCO_3^- can be transported via the apical CFTR Cl^- channel.

effect of the unidirectional transport when disturbances in the transport function occur such as in secretory diarrhea; for example, in extreme cases such as in cholera up to 6 liters h^{-1} can be secreted. When faced with the high osmotic gradient generated by the feces, it was difficult to understand how water could be reabsorbed by the surface cells. With the recent identification that crypts can also actively absorb fluid (10, 19, 20, 96), it now becomes easy to rectify this apparent discrepancy.

Pathological Modifications of Ion Transport

The colon has a variety of associated modulators of ion transport that lead to huge unidirectional shifts. In extreme cases, secretory diarrhea can lead to liters of fluid per hour being secreted. Conversely, in constipation stool can become completely dehydrated so that most of the electrolytes and water have been removed from the lumen of the colon. Because of these complex absorptive secretory issues, the colon must tightly modulate electrolyte movement to prevent huge swings in total electrolyte levels.

SECRETORY DIARRHEA   Uncontrolled modifications in electrolyte transport can be caused by either congenital or acquired modification of ion transport. In the case of congenital defects, we see disease states such as chloride-dependent diarrhea caused by a defect in DRA or sodium diarrhea resulting from a defective Na/H exchanger (49, 86, 88–90). In the case of colonization by pathogenic micro-organisms, there is a similar activation of the secretory components of intestinal transport via cAMP- or cGMP-dependent mechanisms that results in down-regulation of the absorptive components of intestinal transport while causing the hypersecretion of Cl^- (64).

BACTERIAL TOXINS   A variety of bacterial agents can induce secretory and inflammatory diarrhea, including *Shigella flexneria, E. coli, Salmonella typhimurium,* and *Vibrio cholerae.* Each of these agents attacks the colon in slightly different ways; however, all cause an alteration in ion transport and in many cases disruption of the tight junctions, which can lead to a large amount of fluid secretion and activation of an inflammatory response in the intestine (16, 28, 31, 64, 86, 88–90). All these agents activate secretion by changes in the levels of cAMP either directly, as is the case with cholera, or indirectly, as is the case with *E. coli,* in which elevated levels of cGMP result in deactivation of phosphodiesterase, which leads to enhanced cAMP levels. Many of these effects result in hyperactivation of the CFTR protein on the apical membrane. It is of interest that bacterial toxins fail to elicit secretory diarrhea in CF-deficient mice (37) and could explain why patients that are heterozygotes for CF may have a natural protection from secretory diarrhea.

As many of these diarrheas share the common pathway of enhanced cAMP production, it may be possible to target therapies based on lowering cAMP levels in the intestine. We have recently show that the colon contains a calcium-sensing receptor (CaSR) localized to the apical and basolateral membranes. Activation of this receptor results in a suppression of fluid secretion and a change in cAMP levels (19, 20). This receptor is discussed in detail below.

Inflammatory Bowel Disease

The term inflammatory bowel disease describes a wide variety of pathological states that modify the colon and intestinal morphometry. Included in this group are the common conditions of Crohn's disease and ulcerative colitis (32– 34). In both these conditions there is a loss of tight junctions, and there is speculation that the Cl/HCO_3^- system, as well as the DRA and potentially the Na/KATPase, are down-regulated (46). This results in alterations in membrane integrity and in enhanced secretion of electrolytes.

Long-term changes in morphometry as a result of these disease states lead to enhanced risk for tumorigenic activity because the high levels of cAMP can result in dedifferentiation and unregulated proliferation of the mucosa (28, 32, 33, 48, 52, 63, 67).

Novel Transport Modulation

The colon is a complex combination of surface and crypt cells that is able to secrete and reabsorb fluid. It is the balance between these two phases that allows for the maintenance of barrier function and prevents secretory diarrhea. Should either secretion or absorption be stimulated at the expense of the other, the individual will suffer from diarrhea or constipation, respectively. We have recently identified a novel receptor pathway located on both the apical and basolateral membranes of the colon that plays an important role in fine-tuning fluid and electrolyte transport and may be a natural biosensor that helps to maintain a balance between secretion and absorption (19, 20) (Figure 6).

CALCIUM-SENSING RECEPTOR   The CaSR protein and its transcripts have been identified in the intestinal tract in a variety of mammals, birds, and fish (19). We have recently identified this receptor in the colon of the rat on both the apical and basolateral membranes (20). We have shown that modification in the levels of divalent ions on either the apical or basolateral membrane leads to a change in the levels of fluid secretion in the colon and that this process is taking place via modulation in CaSR activity (20). Activation of the CaSR by Ca^{2+}, Gd^{3+}, or antibiotics such as neomycin leads to rapid rises in intracellular Ca^{2+} in both surface and crypt cells (19, 20). Recent evidence shows that the elevation in intracellular Ca^{2+}

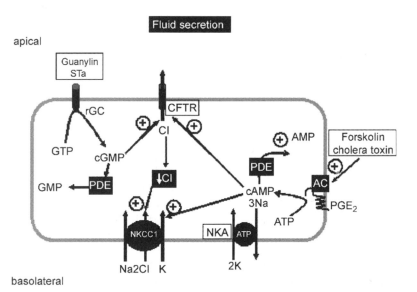

Figure 6 Colonic fluid secretion. Schematic model showing cAMP- and cGMP-linked fluid secretion in the colon. Classic toxin secretagogues STa and cholera toxin are indicated, both of which lead to unregulated fluid and electrolyte secretion.

occurring within a few seconds following stimulation of the receptor is consistent with activation of the phosphatidylinositol/phospholipase C/inositol 1,4,5-trisphosphate (PI-PLC-IP3) pathway by this family of G protein–coupled cell membrane receptors. Stimulation-induced Ca^{2+}-sensing, receptor-mediated increases in intracellular Ca^{2+} can be prevented by pre-treatment of crypt cells with U-73122, a specific inhibitor of PI-PLC. Inhibition by a PLC-specific compound demonstrated that the intracellular Ca^{2+} transients induced by Ca^{2+}-sensing receptor agonists were not the result of altered entry of extracellular Ca^{2+} into colonic epithelial cells but rather from receptor-mediated activation of PI-PLC. We have previously shown that the receptor-mediated increase in intracellular Ca^{2+} in colon comes from the release of Ca^{2+} from thapsigargin-sensitive cell stores (19).

Increasing the levels of calcium on either the apical or basolateral membrane of the intact colon in Ussing chambers or in the isolated perfused crypt leads to a decrease in fluid secretion (19, 20). We further demonstrated, even in the presence of potent secretagogues such as forskolin, that we could cause a reduction in secretion and could, in fact, cause absorption of fluid in the continued presence of potent secretagogues (19, 20). This observation demonstrates the important role the CaSR plays in modulating fluid secretion in the colon. As the colon becomes a secreting epithelium following secretagogue activation of the crypt cells by either enhanced cAMP or cGMP production, the CaSR acts as a brake by increasing PKC activity and thus enhancing the destruction of cAMP or cGMP and suppressing fluid secretion (19, 20) (see Figure 6). This damping of fluid secretion can take place with only very small changes in calcium concentration due to the fact that polyamines as well as amino acids are important allosteric modifiers of the receptor. Enhanced delivery of either of these agents to the receptor left shifts the activation curve and thereby allows for enhanced activation with minute changes in calcium concentration (19).

This receptor could play a major role in determining the levels of fluid in the stool by modulating the levels of electrolyte reabsorption. In fact, ingestion of high levels of calcium (enhanced delivery of calcium to the colon) results in constipation. Using this system, in combination with allosteric modifiers, the body could experience an enhanced absorptive state with slight increases in levels of divalent ions in the blood; the resultant increase in fluid absorption would cause increased levels of electrolyte absorption. Conversely inactivation of the CaSR either by decreased concentration of the allosteric modifiers and/or calcium or by decreased delivery of calcium to the colon would lead to a return to a state of secretion.

In disease states or infectious states, as outlined above, fluid and electrolyte secretion can occur at pronounced levels and, if unchecked, can lead to dehydration and potentially death. By modulating the CaSR through increased delivery of calcium to the receptor, it appears possible that secretion could be stopped. This aspect of the CaSR could serve as an important new therapeutic target to modulate secretion and reabsorption of electrolytes along the colon.

SUMMARY

We have shown that the classical model of secretion and reabsorption along the colon is flawed and that recent studies have now conclusively demonstrated that surface cells as well as crypt cells have dual functionality and can secrete and absorb electrolytes and fluid. This important finding helps us to better understand the complexities of the colonic epithelium and allows us to more clearly explain the colon's ability to secrete massive amounts of fluid and electrolytes under disease states. We have also presented evidence for a calcium-sensing receptor that can serve as a potent modulator of colonic electrolyte transport and can furthermore halt secretion against the face of secretagogue-induced secretion.

These important insights into colonic electrolyte transport will allow the development of new therapies for secretory diarrhea and maintenance of cell and whole-body ionic homeostasis.

The *Annual Review of Physiology* is online at
http://physiol.annualreviews.org

LITERATURE CITED

1. Bakre MM, Ghanekar Y, Visweswariah SS. 2000. Homologous desensitization of the human guanylate cyclase C receptor. Cell-specific regulation of catalytic activity. *Eur. J. Biochem.* 267:179–87

2. Bakre MM, Sopory S, Visweswariah SS. 2000. Expression and regulation of the cGMP-binding, cGMP-specific phosphodiesterase (PDE5) in human colonic epithelial cells: role in the induction of cellular refractoriness to the heat-stable enterotoxin peptide. *J. Cell Biochem.* 77:159–67

3. Deleted in proof

4. Bakre MM, Visweswariah SS. 1997. Dual regulation of heat-stable enterotoxin-mediated cGMP accumulation in T84 cells by receptor desensitization and increased phosphodiesterase activity. *FEBS Lett.* 408:345–49

5. Barrett KE, Keely SJ. 2000. Chloride secretion by the intestinal epithelium: molecular basis and regulatory aspects. *Annu. Rev. Physiol.* 62:535–72

6. Benos DJ, Cunningham S, Baker RR, Beason KB, Oh Y, Smith PR. 1992. Molecular characteristics of amiloride-sensitive sodium channels. *Rev. Physiol. Biochem. Pharmacol.* 120:31–113

7. Binder HJ, Foster ES, Budinger ME, Hayslett JP. 1987. Mechanism of electroneutral sodium chloride absorption in distal colon of the rat. *Gastroenterology* 93:449–55

8. Binder HJ, McGlone F, Sandle GI. 1989. Effects of corticosteroid hormones on the electrophysiology of rat distal colon: implications for Na^+ and K^+ transport. *J. Physiol.* 410:425–41

9. Binder HJ, Rajendran VM, Geibel JP. 2000. Cl-dependent Na-H exchange. A novel colonic crypt transport mechanism. *Ann. NY Acad. Sci.* 915:43–53

10. Binder HJ, Singh SK, Geibel JP, Rajendran VM. 1997. Novel transport properties of colonic crypt cells: fluid absorption and Cl-dependent Na-H exchange. *Comp. Biochem. Physiol. A* 118:265–69

11. Bleich M, Ecke D, Schwartz B, Fraser G, Greger R. 1997. Effects of the carcinogen dimethylhydrazine (DMH) on the function of rat colonic crypts. *Pflügers Arch.* 433:254–59

12. Briel M, Greger R, Kunzelmann K. 1998. Cl⁻ transport by cystic fibrosis transmembrane conductance regulator (CFTR) contributes to the inhibition of epithelial Na⁺ channels (ENaCs) in *Xenopus* oocytes co-expressing CFTR and ENaC. *J. Physiol.* 508:825–36

13. Budinger ME, Foster ES, Hayslett JP, Binder HJ. 1986. Sodium and chloride transport in the large intestine of potassium-loaded rats. *Am. J. Physiol. Gastrointest. Liver Physiol.* 251:G249–52

14. Butterfield I, Warhurst G, Jones MN, Sandle GI. 1997. Characterization of apical potassium channels induced in rat distal colon during potassium adaptation. *J. Physiol.* 501:537–47

15. Cartwright CA, Mamajiwalla S, Skolnick SA, Eckhart W, Burgess DR. 1993. Intestinal crypt cells contain higher levels of cytoskeletal-associated pp60c-src protein tyrosine kinase activity than do differentiated enterocytes. *Oncogene* 8:1033–39

16. Chang EB, Bergenstal RM, Field M. 1985. Diarrhea in streptozocin-treated rats. Loss of adrenergic regulation of intestinal fluid and electrolyte transport. *J. Clin. Invest.* 75:1666–70

17. Chang EB, Brown DR, Field M, Miller RJ. 1984. An antiabsorptive basis for precipitated withdrawal diarrhea in morphine-dependent rats. *J. Pharmacol. Exp. Ther.* 228:364–69

18. Chang EB, Fedorak RN, Field M. 1986. Experimental diabetic diarrhea in rats. Intestinal mucosal denervation hypersensitivity and treatment with clonidine. *Gastroenterology* 91:564–69

19. Cheng SX, Geibel JP, Hebert SC. 2004. Extracellular polyamines regulate fluid secretion in rat colonic crypts via the extracellular calcium-sensing receptor. *Gastroenterology* 126:148–58

20. Cheng SX, Okuda M, Hall AE, Geibel JP, Hebert SC. 2002. Expression of calcium-sensing receptor in rat colonic epithelium: evidence for modulation of fluid secretion. *Am. J. Physiol. Gastrointest. Liver Physiol.* 283:G240–50

21. Clauss W, Schafer H, Horch I, Hornicke H. 1985. Segmental differences in electrical properties and Na-transport of rabbit caecum, proximal and distal colon in vitro. *Pflügers Arch.* 403:278–82

22. Cliff WH, Frizzell RA. 1990. Separate Cl⁻ conductances activated by cAMP and Ca²⁺ in Cl⁻-secreting epithelial cells. *Proc. Natl. Acad. Sci. USA* 87:4956–60

23. Cuthbert AW, Hickman ME, MacVinish LJ, Evans MJ, Colledge WH, et al. 1994. Chloride secretion in response to guanylin in colonic epithelial from normal and transgenic cystic fibrosis mice. *Br. J. Pharmacol.* 112:31–36

24. Dawson DC. 1991. Ion channels and colonic salt transport? *Annu. Rev. Physiol.* 53:321–39

25. Delpire E, Rauchman MI, Beier DR, Hebert SC, Gullans SR. 1994. Molecular cloning and chromosome localization of a putative basolateral Na⁺-K⁺-2Cl⁻ cotransporter from mouse inner medullary collecting duct (mIMCD-3) cells. *J. Biol. Chem.* 269:25677–83

26. Diener M, Rummel W, Mestres P, Lindemann B. 1989. Single chloride channels in colon mucosa and isolated colonic enterocytes of the rat. *J. Membr. Biol.* 108:21–30

27. Ecke D, Bleich M, Greger R. 1996. Crypt base cells show forskolin-induced Cl⁻ secretion but no cation inward conductance. *Pflügers Arch.* 431:427–34

28. Fargeas MJ, Theodorou V, More J, Wal JM, Fioramonti J, Bueno L. 1995. Boosted systemic immune and local responsiveness after intestinal inflammation in orally sensitized guinea pigs. *Gastroenterology* 109:53–62

29. Field M, Rao MC, Chang EB. 1989. Intestinal electrolyte transport and diarrheal disease (2) *N. Engl. J. Med.* 321:879–83

30. Field M, Rao MC, Chang EB. 1989. Intestinal electrolyte transport and diarrheal disease (1) *N. Engl. J. Med.* 321:800–6

31. Field M, Semrad CE. 1993. Toxigenic diarrheas, congenital diarrheas, and cystic fibrosis: disorders of intestinal ion transport. *Annu. Rev. Physiol.* 55:631–55

32. Fiocchi C. 1997. Intestinal inflammation: a complex interplay of immune and nonimmune cell interactions. *Am. J. Physiol. Gastrointest. Liver Physiol.* 273:G769–75

33. Fiocchi C. 1998. Inflammatory bowel disease: etiology and pathogenesis. *Gastroenterology* 115:182–205

34. Fiocchi C. 1999. From immune activation to gut tissue injury: the pieces of the puzzle are coming together. *Gastroenterology* 117:1238–41

35. Foster ES, Budinger ME, Hayslett JP, Binder HJ. 1986. Ion transport in proximal colon of the rat. Sodium depletion stimulates neutral sodium chloride absorption. *J. Clin. Invest.* 77:228–35

36. Foster ES, Jones WJ, Hayslett JP, Binder HJ. 1985. Role of aldosterone and dietary potassium in potassium adaptation in the distal colon of the rat. *Gastroenterology* 88:41–46

37. Gabriel SE, Brigman KN, Koller BH, Boucher RC, Stutts MJ. 1994. Cystic fibrosis heterozygote resistance to cholera toxin in the cystic fibrosis mouse model. *Science* 266:107–9

38. Geibel JP, Rajendran VM, Binder HJ. 2001. Na^+-dependent fluid absorption in intact perfused rat colonic crypts. *Gastroenterology* 120:144–50

39. Geibel JP, Singh S, Rajendran VM, Binder HJ. 2000. HCO_3^- secretion in the rat colonic crypt is closely linked to Cl^- secretion. *Gastroenterology* 118:101–7

40. Greger R. 2000. Role of CFTR in the colon. *Annu. Rev. Physiol.* 62:467–91

41. Greger R, Mall M, Bleich M, Ecke D, Warth R, et al. 1996. Regulation of epithelial ion channels by the cystic fibrosis transmembrane conductance regulator. *J. Mol. Med.* 74:527–34

42. Greger R, Schreiber R, Mall M, Wissner A, Hopf A, et al. 2001. Cystic fibrosis and CFTR. *Pflügers Arch.* 443(Suppl. 1):S3–S7

43. Haas M, Forbush B III. 1998. The Na-K-Cl cotransporters. *J. Bioenerg. Biomembr.* 30:161–72

44. Haas M, Forbush B III. 2000. The Na-K-Cl cotransporter of secretory epithelia. *Annu. Rev. Physiol.* 62:515–34

45. Halm DR, Halm ST. 2000. Secretagogue response of goblet cells and columnar cells in human colonic crypts. *Am. J. Physiol. Cell Physiol.* 278:C212–33

46. Harris J, Shields R. 1970. Absorption and secretion of water and electrolytes by the intact human colon in diffuse untreated proctocolitis. *Gut* 11:27–33

47. Hermiston ML, Gordon JI. 1995. Organization of the crypt-villus axis and evolution of its stem cell hierarchy during intestinal development. *Am. J. Physiol. Gastrointest. Liver Physiol.* 268:G813–22

48. Ho SB. 1992. Cytoskeleton and other differentiation markers in the colon. *J. Cell Biochem. Suppl.* 16G:119–28

49. Hopf A, Schreiber R, Mall M, Greger R, Kunzelmann K. 1999. Cystic fibrosis transmembrane conductance regulator inhibits epithelial Na^+ channels carrying Liddle's syndrome mutations. *J. Biol. Chem.* 274:13894–99

50. Ikuma M, Geibel J, Binder HJ, Rajendran VM. 2003. Characterization of Cl^-HCO_3 exchange in basolateral membrane of rat distal colon. *Am. J. Physiol. Cell Physiol.* 285:C912–21

51. Ikuma M, Kashgarian M, Binder HJ, Rajendran VM. 1999. Differential regulation of NHE isoforms by sodium depletion in proximal and distal segments of rat colon. *Am. J. Physiol. Gastrointest. Liver Physiol.* 276:G539–49

52. Izzo AA, Mascolo N, Capasso F. 1998. Nitric oxide as a modulator of intestinal water and electrolyte transport. *Dig. Dis. Sci.* 43:1605–20

53. Jaffe LF, Nuccitelli R. 1974. An ultrasensitive vibrating probe for measuring

steady extracellular currents. *J. Cell Biol.* 63:614–28

54. Jarchau T, Hausler C, Markert T, Pohler D, Vanderkerckhove J, et al. 1994. Cloning, expression, and in situ localization of rat intestinal cGMP-dependent protein kinase II. *Proc. Natl. Acad. Sci. USA* 91: 9426–30

55. Kockerling A, Fromm M. 1993. Origin of cAMP-dependent Cl⁻ secretion from both crypts and surface epithelia of rat intestine. *Am. J. Physiol. Cell Physiol.* 264:C1294–301

56. Kockerling A, Sorgenfrei D, Fromm M. 1993. Electrogenic Na⁺ absorption of rat distal colon is confined to surface epithelium: a voltage-scanning study. *Am. J. Physiol. Cell Physiol.* 264:C1285–93

57. Konig J, Schreiber R, Mall M, Kunzelmann K. 2002. No evidence for inhibition of ENaC through CFTR-mediated release of ATP. *Biochim. Biophys. Acta* 1565:17–28

58. Kunzelmann K. 2003. ENaC is inhibited by an increase in the intracellular Cl⁻ concentration mediated through activation of Cl⁻ channels. *Pflügers Arch.* 445:504–12

59. Kunzelmann K. 1997. Regulation and amiloride-binding site of epithelial Na⁺ channel. *Kidney Blood Press. Res.* 20: 151–53

60. Kunzelmann K, Grolik M, Kubitz R, Greger R. 1992. cAMP-dependent activation of small-conductance Cl⁻ channels in HT29 colon carcinoma cells. *Pflügers Arch.* 421:230–37

61. Kunzelmann K, Kiser GL, Schreiber R, Riordan JR. 1997. Inhibition of epithelial Na⁺ currents by intracellular domains of the cystic fibrosis transmembrane conductance regulator. *FEBS Lett.* 400:341–44

62. Kunzelmann K, Koslowsky T, Hug T, Gruenert DC, Greger R. 1994. cAMP-dependent activation of ion conductances in bronchial epithelial cells. *Pflügers Arch.* 428:590–96

63. Kunzelmann K, Kubitz R, Grolik M, Warth R, Greger R. 1992. Small-conductance Cl⁻ channels in HT29 cells: activation by Ca²⁺, hypotonic cell swelling and 8-Br-cGMP. *Pflügers Arch.* 421:238–46

64. Kunzelmann K, Mall M. 2002. Electrolyte transport in the mammalian colon: mechanisms and implications for disease. *Physiol. Rev.* 82:245–89

65. Levitan R, Fordtran JS, Burrows BA, Ingelfinger FJ. 1962. Water and salt absorption in the human colon. *J. Clin. Invest.* 41:1754–59

66. Li M, McCann JD, Liedtke CM, Nairn AC, Greengard P, Welsh MJ. 1988. Cyclic AMP-dependent protein kinase opens chloride channels in normal but not cystic fibrosis airway epithelium. *Nature* 331:358–60

67. Lipsky MS, Adelman M. 1993. Chronic diarrhea: evaluation and treatment. *Am. Fam. Physician* 48:1461–66

68. Lohrmann E, Burhoff I, Nitschke RB, Lang HJ, Mania D, et al. 1995. A new class of inhibitors of cAMP-mediated Cl⁻ secretion in rabbit colon, acting by the reduction of cAMP-activated K⁺ conductance. *Pflügers Arch.* 429:517–30

69. Lohrmann E, Greger R. 1993. Isolated perfused rabbit colon crypts: stimulation of Cl⁻ secretion by forskolin. *Pflügers Arch.* 425:373–80

70. Lohrmann E, Greger R. 1995. The effect of secretagogues on ion conductances of in vitro perfused, isolated rabbit colonic crypts. *Pflügers Arch.* 429:494–502

71. Loo DD, Kaunitz JD. 1989. Ca²⁺ and cAMP activate K⁺ channels in the basolateral membrane of crypt cells isolated from rabbit distal colon *J. Membr. Biol.* 110:19–28

72. Lytle C, Forbush B III. 1996. Regulatory phosphorylation of the secretory Na-K-Cl cotransporter: modulation by cytoplasmic Cl. *Am. J. Physiol. Cell Physiol.* 270:C437–48

73. Mall M, Bleich M, Kuehr J, Brandis M, Greger R, Kunzelmann K. 1999. CFTR-mediated inhibition of epithelial

Na$^+$ conductance in human colon is defective in cystic fibrosis. *Am. J. Physiol. Gastrointest. Liver Physiol.* 277:G709–16

74. Mall M, Bleich M, Schurlein M, Kuhr J, Seydewitz HH, et al. 1998. Cholinergic ion secretion in human colon requires coactivation by cAMP. *Am. J. Physiol. Gastrointest. Liver Physiol.* 275:G1274–81

75. Mall M, Kreda SM, Mengos A, Jensen TJ, Hirtz S, et al. 2004. The DeltaF508 mutation results in loss of CFTR function and mature protein in native human colon. *Gastroenterology* 126:32–41

76. Melvin JE, Park K, Richardson L, Schultheis PJ, Shull GE. 1999. Mouse down-regulated in adenoma (DRA) is an intestinal Cl$^-$/HCO$_3^-$ exchanger and is up-regulated in colon of mice lacking the NHE3 Na$^+$/H$^+$ exchanger. *J. Biol. Chem.* 274:22855–61

77. Naftalin RJ, Pedley KC. 1999. Regional crypt function in rat large intestine in relation to fluid absorption and growth of the pericryptal sheath. *J. Physiol.* 514:211–27

78. Naftalin RJ, Zammit PS, Pedley KC. 1999. Regional differences in rat large intestinal crypt function in relation to dehydrating capacity in vivo. *J. Physiol.* 514:201–10

79. Payne JA, Xu JC, Haas M, Lytle CY, Ward D, Forbush B III. 1995. Primary structure, functional expression, and chromosomal localization of the bumetanide-sensitive Na-K-Cl cotransporter in human colon. *J. Biol. Chem.* 270:17977–85

80. Pedley KC, Naftalin RJ. 1993. Evidence from fluorescence microscopy and comparative studies that rat, ovine and bovine colonic crypts are absorptive. *J. Physiol.* 460:525–47

81. Quist RG, Ton-Nu HT, Lillienau J, Hofmann AF, Barrett KE. 1991. Activation of mast cells by bile acids. *Gastroenterology* 101:446–56

82. Rajendran VM, Binder HJ. 2000. Characterization and molecular localization of anion transporters in colonic epithelial cells. *Ann. NY Acad. Sci.* 915:15–29

83. Rajendran VM, Black J, Ardito TA, Sangan P, Alper SL, et al. 2000. Regulation of DRA and AE1 in rat colon by dietary Na depletion. *Am. J. Physiol. Gastrointest. Liver Physiol.* 279:G931–42

84. Rajendran VM, Geibel J, Binder HJ. 1999. Role of Cl channels in Cl-dependent Na/H exchange. *Am. J. Physiol. Gastrointest. Liver Physiol.* 276:G73–78

85. Rajendran VM, Kashgarian M, Binder HJ. 1989. Aldosterone induction of electrogenic sodium transport in the apical membrane vesicles of rat distal colon. *J. Biol. Chem.* 264:18638–44

86. Rao MC, Field M. 1984. Enterotoxins and ion transport. *Biochem. Soc. Trans.* 12:177–80

87. Rao MC, Guandalini S, Laird WJ, Field M. 1979. Effects of heat-stable enterotoxin of *Yersinia enterocolitica* on ion transport and cyclic guanosine 3′,5′-monophosphate metabolism in rabbit ileum. *Infect. Immun.* 26:875–78

88. Rao MC, Guandalini S, Smith PL, Field M. 1980. Mode of action of heat-stable *Escherichia coli* enterotoxin. Tissue and subcellular specificities and role of cyclic GMP. *Biochim. Biophys. Acta* 632:35–46

89. Rao MC, Nash NT, Field M. 1984. Differing effects of cGMP and cAMP on ion transport across flounder intestine. *Am. J. Physiol. Cell Physiol.* 246:C167–71

90. Rao MC, Orellana SA, Field M, Robertson DC, Giannella RA. 1981. Comparison of the biological actions of three purified heat-stable enterotoxins: effects on ion transport and guanylate cyclase activity in rabbit ileum in vitro. *Infect. Immun.* 33:165–70

91. Rijke RP, Plaisier HM, Langendoen NJ. 1979. Epithelial cell kinetics in the descending colon of the rat. *Virchows Arch. B* 30:85–94

92. Robert ME, Singh SK, Ikuma M, Jain D, Ardito T, Binder HJ. 2001. Morphology

of isolated colonic crypts. *Cells Tissues Organs* 168:246–51

93. Sandle GI, Foster ES, Lewis SA, Binder HJ, Hayslett JP. 1985. The electrical basis for enhanced potassium secretion in rat distal colon during dietary potassium loading. *Pflügers Arch.* 403:433–39

94. Sandle GI, Hayslett JP, Binder HJ. 1984. Effect of chronic hyperaldosteronism on the electrophysiology of rat distal colon. *Pflügers Arch.* 401:22–26

95. Saxon ML, Zhao X, Black JD. 1994. Activation of protein kinase C isozymes is associated with post-mitotic events in intestinal epithelial cells in situ. *J. Cell Biol.* 126:747–63

96. Singh SK, Binder HJ, Boron WF, Geibel JP. 1995. Fluid absorption in isolated perfused colonic crypts. *J. Clin. Invest.* 96:2373–79

97. Smith PR, Benos DJ. 1991. Epithelial Na^+ channels. *Annu. Rev. Physiol.* 53:509–30

98. Warth R, Bleich M. 2000. K^+ channels and colonic function. *Rev. Physiol. Biochem. Pharmacol.* 140:1–62

99. Weinman EJ, Steplock D, Donowitz M, Shenolikar S. 2000. NHERF associations with sodium-hydrogen exchanger isoform 3 (NHE3) and ezrin are essential for cAMP-mediated phosphorylation and inhibition of NHE3. *Biochemistry* 39:6123–29

100. Welsh MJ, Smith PL, Fromm M, Frizzell RA. 1982. Crypts are the site of intestinal fluid and electrolyte secretion. *Science* 218:1219–21

101. Deleted in proof

102. Worrell RT, Butt AG, Cliff WH, Frizzell RA. 1989. A volume-sensitive chloride conductance in human colonic cell line. T84. *Am. J. Physiol. Cell Physiol.* 256: C11110–19

103. Yun CH, Oh S, Zizak M, Steplock D, Tsao S, et al. 1997. cAMP-mediated inhibition of the epithelial brush border Na^+/H^+ exchanger, NHE3, requires an associated regulatory protein. *Proc. Natl. Acad. Sci. USA* 94:3010–15

Annu. Rev. Physiol. 2005. 67:491–514
doi: 10.1146/annurev.physiol.67.031103.151256
First published online as a Review in Advance on October 12, 2004

RETINAL PROCESSING NEAR ABSOLUTE THRESHOLD: From Behavior to Mechanism

Greg D. Field, Alapakkam P. Sampath, and Fred Rieke

*Department of Physiology and Biophysics, University of Washington, Seattle,
Washington 98195; email: gfield@salk.edu; apsampat@u.washington.edu;
rieke@u.washington.edu*

Key Words   visual sensitivity, photon detection, signal processing, physical limits,
scotopic vision

■ **Abstract**   Vision at absolute threshold is based on signals produced in a tiny
fraction of the rod photoreceptors. This requires that the rods signal the absorption
of single photons, and that the resulting signals are transmitted across the retina and
encoded in the activity sent from the retina to the brain. Behavioral and ganglion
cell sensitivity has often been interpreted to indicate that these biophysical events
occur noiselessly, i.e., that vision reaches limits to sensitivity imposed by the division of light into discrete photons and occasional photon-like noise events generated
in the rod photoreceptors. We argue that this interpretation is not unique and provide a more conservative view of the constraints behavior and ganglion cell experiments impose on phototransduction and retinal processing. We summarize what is
known about how these constraints are met and identify some of the outstanding open
issues.

INTRODUCTION

Sensory signals are inherently variable. These variations set a fundamental limit
to the performance of any system designed to detect and process these signals.
In the case of vision, sensitivity cannot exceed the limit set by the quantization
of light into discrete photons and the consequent Poisson fluctuations in photon
absorption. Several aspects of dark-adapted visual processing approach this limit.
Rod photoreceptors reliably signal the absorption of single photons (1, 2), and behavioral detection may require absorption of only a few photons (3). This exquisite
sensitivity is crucial for normal night vision. On a moonless night only one rod in
10,000 receives a photon during the integration time of the rod signals [reviewed
in (4)]. Thus visually guided behavior requires the retina to read out and process
single-photon signals carried by a tiny fraction of the rods while rejecting noise
generated in the remaining rods.

Reaching the sensitivity limit imposed by the quantal nature of light requires
that intrinsic noise is small compared with that generated by Poisson fluctuations

0066-4278/05/0315-0491$14.00

491

in photon absorption. Both behavioral (5–7) and physiological (2, 8, 9) studies have identified conditions under which this requirement is not met; under these conditions intrinsic noise limits sensitivity. The intrinsic noise has been associated with photon-like noise events produced by spontaneous activation of rhodopsin in the rod photoreceptors (5, 8, 9). If correct, this idea implies that the retinal readout of single-photon responses in the rod array is efficient and effectively noiseless. However, uncertainty in existing estimates of behavioral sensitivity and of rod noise makes this identification tenuous. Several additional noise sources, including other sources of rod noise, synaptic noise in retina and cortex, and noise in spike generation, could also contribute to limiting behavioral sensitivity. Although it is not clear whether these other noise sources are negligible, behavioral sensitivity does require that they are small.

Behavioral work has motivated studies of how absorbed photons are transduced by rods and how the resulting signals are processed by the retinal circuitry. This work has focused on several questions: (*a*) How do rods detect single photons (1, 2)? (*b*) How are the resulting signals reliably transmitted across retinal synapses (10–12)? (*c*) How are the signals produced by absorption of a few photons coded in the pattern of activity sent from the retina to the brain (8)? Similar problems occur elsewhere in the nervous system. For example, signals in some cortical areas, like those in the rod array, are sparsely coded (13, 14). In such cases, appropriate strategies for computations based on the encoded signals must take the sparseness into account. These general questions can be posed clearly in the retina because the signal and noise properties of the rod inputs can be measured and the stimuli can be precisely controlled.

Here we review what is known about how the mammalian retina works near absolute visual threshold. We begin by discussing evidence from behavioral experiments and ganglion cell recordings that the visual system can encode the absorption of a small number of photons. We then describe the properties of the noise inferred from behavioral and ganglion cell experiments. A clear prediction from behavioral and ganglion cell sensitivity is that rod photoreceptors can detect single photons; we summarize the mechanisms responsible and highlight the rod noise sources that could limit sensitivity. Finally we discuss the retinal circuitry that conveys the rod signals to the ganglion cells, emphasizing how this circuitry processes single-photon responses.

BEHAVIOR

Behavioral experiments provide estimates of the minimum number of photon absorptions that can be detected and the noise that limits the reliability of detection. These experiments are often interpreted to indicate that humans can detect <10 absorbed photons (3, 6, 15), limited by the spontaneous activation of rhodopsin in the rod photoreceptors (5). We argue that these interpretations are not unique. Thus it is currently unclear what process or processes limit behavioral sensitivity.

Consequently, although behavior constrains the physiological mechanisms responsible for dark-adapted visual sensitivity, the extent of these constraints is not clear.

Frequency of Seeing Experiments

Not long after it was established that light is quantized into discrete photons, Lorentz realized that a just-detectable flash delivered about 100 photons to the cornea and hence that the quantization of light might be of relevance for visual sensitivity [reviewed in (16)]. Identifying the minimum number of photons required for seeing from this observation is difficult because of uncertainty in the quantum efficiency, i.e., the fraction of photons at the cornea absorbed by the rods. Estimates of the quantum efficiency based on measurements of the scatter and absorption properties of structures in the eye (the absorptive quantum efficiency) range from 0.1 to 0.3 (17), indicating that 10–30 photons are required for seeing a flash.

The frequency of seeing experiments of Hecht and colleagues (3) and van der Velden (15) estimated the detection threshold and quantum efficiency directly from behavior. Measurement of the fraction of trials in which a flash was seen as a function of the number of photons at the cornea showed a broad transition from flashes that were rarely seen to those frequently seen (e.g., Figure 1A). The threshold and quantum efficiency were estimated on the basis of three assumptions: First, variability in a subject's responses was attributed to the Poisson statistics of photon absorption and the consequent trial-to-trial fluctuations in the number of absorbed photons. Second, only flashes producing at least Θ absorbed photons were seen. Third, the average number of photons contributing to seeing, $\bar{n}$, was related to the average number of photons at the cornea, $\bar{N}$, by an unknown quantum efficiency, $\bar{n} = Q_E \bar{N}$. Thus the probability of seeing a flash delivering an average of $\bar{N}$ photons to the cornea is

$$P_{see} = \sum_{n \geq \Theta}^{\infty} \frac{\exp(-Q_E \bar{N})(Q_E \bar{N})^n}{n!}. \qquad 1.$$

This approach estimates the behavioral threshold, Θ, from the steepness of the transition between flashes that are rarely and almost always seen. If Θ is small, Poisson fluctuations in the number of absorbed photons from one trial to the next will make the transition broad. The transition becomes steeper with increasing Θ.

When the probability of seeing from Equation 1 is plotted against the logarithm of the flash strength, the unknown quantum efficiency shifts the curve along the flash strength axis but does not change its shape. Thus the quantum efficiency is estimated from the shift required to align Equation 1 with the frequency of seeing data (Figure 1A). From this analysis, Hecht and colleagues estimated a threshold of 5–7 photons and a quantum efficiency of ~0.06. This behavioral quantum efficiency is considerably lower than estimates of the absorptive quantum efficiency, a point we return to below. Even with the higher quantum efficiency, the likelihood that an individual rod absorbed >1 photon on any trial is small

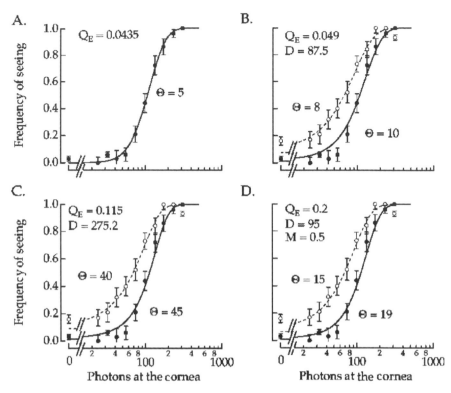

Figure 1 Frequency of seeing data adapted, with permission, from Teich and colleagues (7). (*A*) Data for low false-positive rate fit with Equation 1. Θ is the threshold number of photons, and Q_E is the quantum efficiency. (*B*) Frequency of seeing curves for high (○) and low (●; identical to *A*) false-positive rates from a single subject. Fits from Equation 2 include additive Poisson noise, constrained to be equal for each curve, and different thresholds. (*C*) Data in *B* replotted and fit with Equation 2 with a higher quantum efficiency, higher thresholds, and increased additive Poisson noise. (*D*) Data replotted and fit with Equation 3, which includes multiplicative Poisson noise. The multiplicative noise allows for a higher quantum efficiency, a lower threshold, and less additive Poisson noise.

because the flashes covered an area of the retina containing ~500 rods. Thus these experiments require that individual rod photoreceptors detect single photons.

False Positives and Internal Noise

The frequency of seeing analysis described above assumes that all the noise in the visual system is the result of Poisson fluctuations in photon absorption. If true, visual sensitivity would reach the limit imposed by the division of light into discrete photons. However, observers occasionally report seeing a flash even when none is delivered, and detection threshold depends on the rate of these false-positive responses. Barlow demonstrated this dependence by allowing observers to adopt two criteria, answering either yes or maybe in a frequency of seeing experiment; the maybe responses had a lower threshold and higher false-positive rate than the yes responses (5) (e.g., Figure 1B). Barlow used these results to argue that the false positives were produced by internal noise that occasionally generated the false perception of a flash. This noise, together with Poisson fluctuations in photon absorption, was interpreted as limiting detection performance.

Sakitt (6) did a more complete version of Barlow's yes-maybe experiment by asking subjects to rate the strength of each of a series of dim flashes on a scale from 0 to 6, where 0 corresponds to "did not see anything" and 6 to "very bright light." She constructed frequency of seeing curves and estimated thresholds for ratings of 1 or more, 2 or more, 3 or more, etc. This amounts to measuring thresholds with six different criteria. The false-positive rate and sensitivity decreased as the criterion increased. The experiments of Sakitt and Barlow show that false positives can trade for detection threshold across a wide range of criteria. In this view, different criteria correspond to different signal-to-noise ratios, and observers can choose where to operate on the basis of how many mistakes they are allowed to make. These experiments indicate that a small number of photons, perhaps even a single photon, contributes to detection.

Barlow and Sakitt converted the false-positive rates in their respective experiments into an estimate of the internal noise limiting performance. They assumed this noise could be expressed as an additive dark light complete with Poisson fluctuations. In this case, following Equation 1, the probability of seeing is

$$P_{see} = \sum_{n \geq \Theta}^{\infty} \frac{\exp[-Q_E(\bar{N} + D)][Q_E(\bar{N} + D)]^n}{n!}.$$ 2.

Here D is the additive Poisson noise (the dark light), expressed as an equivalent number of photons at the cornea. In darkness (i.e., $\bar{N} = 0$) the probability of seeing is nonzero because of the dark light. Barlow fit the yes and maybe results with Equation 2, assuming they shared a common amount of dark light but had different thresholds (e.g., Figure 1B). Sakitt similarly fit her data such that only the threshold changed between different criteria. Estimated values of the dark light and difficulties in obtaining a unique estimate are discussed below.

Ambiguities in Behavioral Measurements

The central problem with interpretation of the behavioral experiments summarized above is that the fits to the frequency of seeing curves are not unique. We have described the discrepancy between the behavioral and absorptive quantum efficiencies: Behavioral measurements place the quantum efficiency between 0.03 and 0.06, whereas direct estimates based on losses within the eye range from 0.1 to 0.3. Barlow suggested that this discrepancy could originate because the behavioral quantum efficiency can trade for additive Poisson noise when fitting frequency of seeing data with Equation 2 (5). This is illustrated in Figure 1*C*, which shows fits to the frequency of seeing data from Teich and colleagues (7) with a quantum efficiency of 0.11. Raising the quantum efficiency from 0.05 to 0.11 increased the threshold by a factor of ∼4.5 and the dark light by a factor of ∼3 (Figure 1*B*).

Uncertainty in the quantum efficiency produces a 10-fold range in estimates of threshold and dark light. Expressed as an equivalent rate of photon-like noise events in each rod photoreceptor (using estimates of the spatial and temporal summation of the rod array, the rod density, and an assumed quantum efficiency), the dark light ranges from 0.002 to 0.03 s^{-1} (18). This large range of values makes it impossible to draw a strong association between the dark light and noise originating in the rods.

Figures 1*B*, *C* illustrate two examples in a range of possible fits to the frequency of seeing data. At one end of this continuum, the number of photons required for detection is <10, the dark light is close to estimates of rod noise, but the behavioral quantum efficiency is very low compared with the absorptive quantum efficiency. A possible explanation is that at least half of the photons absorbed by the rods are not processed. Although it sounds counterintuitive, this can be an effective processing strategy (discussed below). At the other end of the continuum, the behavioral and absorptive quantum efficiencies agree, but the threshold number of photons and the amount of dark light are high. These extremes provide qualitatively different views of retinal processing. In the first, the retina efficiently and noiselessly processes the rod signals. In the second, the dark light cannot be explained by rod noise alone and instead post-rod processing must be noisy or inefficient.

Another possible resolution of the discrepancy between behavioral and absorptive quantum efficiencies is that the additive noise model is wrong. Lillywhite (19) showed that multiplicative noise could help resolve this discrepancy. Such noise could arise if several Poisson noise sources operate sequentially, and hence the product of their probability distributions determines the response statistics. Additive noise is still required to explain the false positives. With both additive and multiplicative Poisson noise, the probability of seeing becomes

$$P_{see} = \sum_{n \geq 0}^{\infty} \frac{\exp[-Q_E(\bar{N} + D)][Q_E(\bar{N} + D)]^n}{n!} \sum_{s \geq \Theta}^{\infty} \frac{\exp(-Mn)(Mn)^s}{s!}. \qquad 3.$$

The term inside the first sum is the probability of *n* photon-like events given the quantum efficiency Q_E, the mean number of photons at the cornea $\bar{N}$, and the dark light *D*. The second sum determines the probability that the response *s* is equal to or greater than the threshold Θ given *n* absorbed photons. *M* is the gain from *n* to *s*, such as might be observed if one photon generated two spikes on average in a ganglion cell. Figure 1*D* shows fits for the frequency of seeing data according to Equation 3. A combination of additive and multiplicative noise allows for a quantum efficiency of 0.2 and relatively low thresholds (7, 20).

Summary

Behavioral measurements of dark-adapted sensitivity place several constraints on visual processing: (*a*) Individual rods can detect single photons; (*b*) false positives and sensitivity can trade for one another; and (*c*) signals from a small number of photons, perhaps a single photon, influence perception. A key open issue is whether the retina is efficiently processing the rod signals. Can the dark light be attributed to noise events in the rods? Does the apparent discrepancy between behavioral and absorptive quantum efficiencies result from the discarding of photon responses and/or the presence of multiplicative noise? Alternatively, is the discrepancy a result of incorrectly assuming both the threshold number of photons and the amount of dark light are relatively small? The physiological work described below provides some answers.

RETINAL GANGLION CELLS

The elegance and the curse of the behavioral experiments is that they lump together all factors that could lower sensitivity, including many central factors (17, 21) that are a nuisance if the aim is to constrain retinal processing. An alternative is to focus on the fidelity of signals in the retinal ganglion cells. The most extensive of such studies in mammals are in the anesthetized cat.

Barlow, Levick, and colleagues (8, 22, 23) recorded extracellular spikes from individual ON ganglion cells probably ON-X according to (Reference 9) in darkness and in response to dim light flashes. The spontaneous activity in darkness consisted of bursts of several spikes. The distribution of intervals between bursts was approximately exponential, indicating that the bursts occur largely independently. This behavior is consistent with the idea that the maintained discharge is produced by additive Poisson noise consisting of independent, discrete noise events, each of which produces a burst of $\sim$3 spikes.

Barlow and colleagues also analyzed responses to flashes delivering 5–50 photons at the cornea, using a frequency of seeing analysis similar to that used in Sakitt's behavioral measurements (8) and a receiver operator characteristic analysis (23). They concluded that the performance of dark-adapted ganglion cells could be explained with the following assumptions: (*a*) The quantum efficiency was 0.18, (*b*) each absorbed photon caused the ganglion cell to generate 2–3 extra spikes, and (*c*) detection was limited by Poisson fluctuations in photon absorption and additive Poisson noise arising from discrete noise events (dark light) occurring at a rate of 5–6 s^{-1}.

Mastronarde provided additional evidence for Barlow's observations by recording simultaneously from pairs of cat retinal ganglion cells (9). He found that nearby cells exhibited correlated bursts of 2–3 spikes in darkness and that weak backgrounds increased the rate of these correlated bursts. Cross-correlation functions calculated between the spike trains of two nearby cells had a slow component with a timescale of $\sim$50 ms; this relatively long timescale is consistent with a common input to the cell pair. Pairs of ON cells and pairs of OFF cells were positively correlated, and ON-OFF pairs were anticorrelated. These correlations included ON and OFF center X (β or midget-like) and Y (α or parasol-like) ganglion cells.

On the basis of the similarity of the correlated bursts in darkness and in the presence of a dim background, Mastronarde argued that the slow correlations resulted from photon-like events in the rods, either from spontaneous or light-activated rhodopsin. Assuming a quantum efficiency of 0.12 and an approximate receptive field size, Mastronarde estimated that each photon produced 1.5–2 spikes, similar to the 2–3 spikes per burst estimated from autocorrelation functions computed from the cell's spike trains in darkness. Noise bursts occurred at a rate of 2–6 s^{-1}. A rate of 6 s^{-1} corresponds to ∼0.002 events/rod/s given the rod density (24) and receptive field size (25, 26). These conclusions are in agreement with those of Barlow and colleagues (8).

The results from Mastronarde and Barlow are consistent with the idea that the retina can detect and process single-photon responses. However, as with the behavioral data, this is not a unique interpretation. More conservatively, their results argue for a source of discrete, independent noise events originating in the rods or in the retinal circuitry. This noise likely originates in the AII amacrine cells or earlier (see Retinal Circuitry and Interneurons, below) to explain the anticorrelation between ON and OFF cells. We provide additional arguments below that the identification of the ganglion cell bursts with rod noise is tenuous.

Several studies provide evidence for nonadditive noise originating downstream of the rods. Frishman & Levine (27) recorded from ganglion cells in the presence of a steady dim background light or a modulated light with the same mean intensity. They showed that the statistics of ganglion cell spiking under these two conditions were inconsistent with models with only photoreceptor noise. In addition, Lillywhite (19) and Saleh & Teich (28) argue that multiplicative noise can account for many aspects of the ganglion cell signals previously attributed to additive noise. Multiplicative noise can explain the observation that the variance of the ganglion cell spike count is greater than the mean (28). As with the behavioral data, multiplicative noise may help explain why the inferred quantum efficiency of ∼0.15 is lower than the absorptive quantum efficiency of ∼0.3 in cat (7, 17, 19). If each photon produces more than one spike, however, multiplicative noise will play a lesser role in limiting sensitivity than Poisson fluctuations in photon absorption. Furthermore, multiplicative noise alone cannot explain the ganglion cell's spontaneous spiking activity in darkness.

PHOTOTRANSDUCTION

The first prediction from behavioral and ganglion cell experiments, that rods can detect single photons, was confirmed more than 25 years ago by Baylor and colleagues (1, 2). These studies identified three important functional properties of the rod's single-photon responses (e.g., Figure 2A): (a) The electrical response triggered by photon absorption is highly amplified, (b) the rod maintains low noise in darkness, and (c) individual single-photon responses show low trial-to-trial

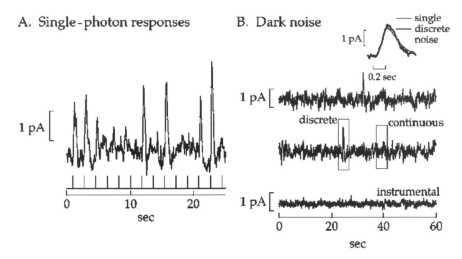

Figure 2   Single-photon responses and dark noise in a primate rod. (*A*) Responses to a repeated dim flash producing an average of 0.5 Rh*. (*B*) Two sections of dark record showing discrete and continuous dark noise. The average of eight discrete noise events is compared with the cell's average single-photon response in the inset. A section of record in saturating light to eliminate the transduction current and isolate instrumental noise is shown below.

variability. Combined efforts in biochemistry, molecular biology, and physiology have since explored the molecular basis of each of these properties (29).

The activity of a single rhodopsin molecule is amplified in several stages [reviewed in (30)]. First, rhodopsin itself catalyzes the activation of hundreds or thousands of copies of the G protein transducin (31, 32). Each transducin activates a single phosphodiesterase (efficiency ~ 0.8), and each phosphodiesterase hydrolyzes many cGMP molecules. This results in the closure of cGMP-gated channels and a 1–2 pA reduction in inward current (2). In total, activation of a single rhodopsin leads to the degradation of 10^5–10^6 cGMP molecules and the failure of $\sim 10^6$ Na$^+$ ions to flow into the outer segment. This scheme applies to all vertebrate rods; however, the fidelity of the single-photon response varies considerably from one species to another. For instance, the signal-to-noise ratio of the single-photon response is ~ 6 in primate rods (2, 33) and ~ 3 in mouse rods (34).

Dark noise in the rod signals comes from spontaneous activation of rhodopsin and phosphodiesterase (2, 35, 36). Spontaneous rhodopsin activation produces discrete noise events indistinguishable from single-photon responses (Figure 2*B*); these occur once every 200–400 s in a monkey rod (2; G.D. Field & F. Rieke, unpublished observation). Thus each of the rod's $\sim 10^8$ rhodopsin molecules activates spontaneously every 500–1000 years on average. Spontaneous phosphodiesterase activation produces continuous fluctuations in the rod current with a magnitude about one fourth the size of the single-photon response (Figure 2*B*).

These current fluctuations have a frequency composition similar to the rod's single-photon response, and thus occasional large continuous noise deviations will mimic single-photon responses (2).

In addition to the low dark noise of the phototransduction cascade, the rod's single-photon responses vary little from one to the next, particularly when compared with the variations expected for signals initiated by a single molecule (33, 37, 38). This low variability poses an interesting molecular design question: How is the activity of a single rhodopsin molecule regulated to avoid the expected statistical variations in its lifetime? Low variability is also required if the visual system is to count photons, as suggested by Sakitt's behavioral results, and may enable the rod responses to encode precisely the time of photon arrival (37).

COMPARISON OF ROD NOISE WITH BEHAVIORAL AND GANGLION CELL SENSITIVITY

Behavioral measurements and recordings from ganglion cells provide an estimate of the intrinsic noise that limits absolute sensitivity. This noise has usually been associated with spontaneous activation of rhodopsin, which produces additive Poisson noise (5, 8, 9). Indeed, the rate of photon-like noise events in monkey rods is in approximate agreement with estimates of the noise limiting human behavior [reviewed in (18)]. Furthermore, the noise limiting the sensitivity of cat ganglion cells consists of discrete bursts of spikes similar to those produced by dim backgrounds (8, 9).

Several issues make the association of the dark light with photon-like noise events in the rods tenuous. First, fits to the frequency of seeing curves depend on several factors (quantum efficiency, additive Poisson noise, and multiplicative Poisson noise) that can trade for one another (Figure 1). This precludes a unique interpretation. Second, in the cat ganglion cell experiments, the number of rods providing input to the recorded cells was not accurately determined, and the properties of noise in cat rods have not been measured. Experiments in toad provide additional evidence that behavior approaches limits set by spontaneous rhodopsin activation. In particular, the temperature dependence of behavioral threshold is correlated with the temperature dependence of the rate of photon-like noise events in the rods (39). However, the effect of temperature on behavior is very different in frogs (40). This difference among closely related species, along with differences in both the signal-to-noise ratio of the single-photon responses and the underlying retinal circuitry in amphibian and mammalian retina, makes it impossible to generalize from toads to humans.

What other sources of noise might limit sensitivity? First, continuous noise in mammalian rods can generate large fluctuations that look like true photon responses (2, 34). Second, noise downstream of the rods, e.g., synaptic noise or noise in spike generation, could contribute to either additive or multiplicative noise. Determining the identity of the noise limiting ganglion cell sensitivity remains a key question in how the retina works at low light levels. If the noise can be entirely attributed to the rods, the retinal readout of the rod array must be efficient and effectively noiseless. This is a strong constraint.

RETINAL CIRCUITRY AND INTERNEURONS

Behavioral, ganglion cell, and rod experiments illustrate that single photons are detected at the initial stage of retinal processing and suggest that single-photon responses reliably traverse the retina. Understanding the retinal mechanisms responsible for this sensitivity and the identity of the noise limiting sensitivity will require tracking the rod's single-photon responses across the retina. Our current understanding of how single-photon responses are processed by retinal interneurons is primitive compared with our knowledge about rod and ganglion cell physiology.

Rod signals are transmitted across the mammalian retina through at least three pathways: the rod bipolar pathway, the rod-cone pathway, and the rod-OFF pathway (Figure 3; also see 41, 42). The rod bipolar pathway, which is a special feature of mammalian retina, has long been considered the primary route for rod signals at light levels near absolute threshold. Only recently has a solid experimental basis for this idea been developed (43–45). In particular, isolation of the pathways through a combination of genetic and pharmacological manipulations indicates that the

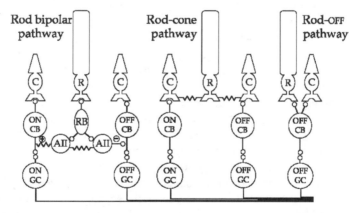

Figure 3 Pathways for rod signals. In mammalian retina, responses generated by rod photoreceptors (R) can reach ganglion cells (GC) by three routes. In the rod bipolar pathway (46–48), rod signals are sent to rod bipolar cells (an ON-type bipolar) (RB), which subsequently send these signals to a network of electrically coupled AII amacrine cells. AII amacrines are electrically coupled to ON-cone bipolar (CB) cells and connected by a glycinergic synapse to OFF-cone bipolar cells. These cone bipolars send signals to ON and OFF ganglion cells. In the rod-cone pathway (24, 49, 50), rods are electrically coupled to cones (C), which relay both ON and OFF signals to ganglion cells through cone bipolars (CB). In the rod-OFF pathway (43, 51, 52), ganglion cells receive rod input through OFF-cone bipolar cells, which make synaptic contacts with rods.

rod bipolar pathway operates at light levels at least 10-fold lower than the others. Thus behavior at light levels near absolute threshold is mediated by the rod bipolar pathway.

Rod signals in the rod bipolar pathway are passed to rod bipolar cells, a type of ON or depolarizing bipolar cell that receives exclusive rod input (46–48). Rod bipolars differ from prototypical bipolar cells because they make few or no direct contacts with ganglion cells. Instead, each rod bipolar contacts several electrically coupled AII amacrine cells (48). AII amacrine cells contact ON-cone bipolar cells through gap junctions, and OFF-cone bipolar cells through sign-inverting glyciner-gic synapses. Cone bipolar cells transmit rod signals from the rod bipolar pathway to ganglion cells. Below, we describe what is known about how each component of this pathway processes single-photon responses and identify some of the outstand-ing open questions. A more detailed account of the anatomy and general function can be found in Bloomfield & Dacheux (42).

The Rod-to-Rod Bipolar Synapse

Work on transmission of single-photon responses from rods to rod bipolar cells has emphasized three issues: (a) the presynaptic mechanisms permitting reli-able transmission of the small voltage changes produced by photon absorption, (b) separation of the rod's single-photon responses from continuous noise, and (c) a speeding of the single-photon response in bipolar cells compared with that in rods.

Behavioral sensitivity requires that rods generate electrical responses to single photons and that some of these responses are transmitted to rod bipolar cells. The presynaptic hyperpolarization produced by absorption of a photon is only 1–2 mV in amplitude (50). The small size of the presynaptic signal makes transmission challenging compared with the situation at central synapses where the voltage changes are 100 times larger. Rods are depolarized in darkness and continuously release glutamate at a specialized ribbon-type synapse (53, 54). Photon absorption hyperpolarizes the rod and produces a reduction or pause in release. Random pauses or slowing generated by statistical fluctuations in release will masquerade as true photon events and thus produce a source of noise potentially limiting visual sensitivity (10). Such a noise source could account for some of the dark light discussed in the Behavior and Ganglion Cell sections. The vesicle release rate and statistics determine the magnitude of this synaptic noise.

Salamander rods release vesicles at a rate of at least 400 s^{-1} at the dark po-tential of -40 mV, as indicated by capacitance measurements (55); these rods, however, have larger active zones and more release sites than mammalian rods. Rao-Mirotznik and colleagues (12) argue that mammalian rods must maintain a dark release rate of at least 80–100 s^{-1} to ensure that random pauses in release produce fewer false photon-like events than spontaneous rhodopsin activation in the rod outer segment. Their argument is based on two assumptions: Vesicle re-lease obeys Poisson statistics, and release is completely abolished for $\sim$100 ms during the single-photon response.

The first assumption of Rao-Mirotznik (12) is supported by work on synapses made by spiking cells. In particular, the release of vesicles at the neuromuscular junction follows Poisson statistics as long as the release probability at each fusion site is low (56); however, these are synapses that can reload between the brief bouts of exocytosis produced by action potentials. The statistics of vesicle release could be different when the synaptic machinery is forced to operate continuously as for rods in darkness; each of the $\sim$40 fusion sites (11) in a mammalian rod spherule would have to release several vesicles per second to support a rate of $100\ s^{-1}$. Variability in vesicle release could be reduced if each vesicle fusion site exhibited a refractory period following release, just as a refractory period can reduce variability in spike generation (57, 58). Indeed, optical tracking of single vesicles in goldfish bipolar terminals (also a ribbon-type synapse) indicates that they are not available for exocytosis for $\sim$0.1 s after arrival at the membrane (59). If present, a similar refractory period at release sites in the rod terminal could reduce the variance in the number of vesicles released, thus reducing the required mean rate.

The second assumption made by Rao-Mirotznik and colleagues (12), that release is completely suppressed by the single-photon response, has not been directly tested. This is because the small size of mammalian rods makes direct measurement of the voltage dependence of transmitter release difficult. This relation has been studied in amphibians by recording simultaneously from rods and postsynaptic cells (60, 61), by monitoring capacitance changes produced by exocytosis (62), and by detecting released glutamate photometrically (63). These studies indicate that the release rate changes exponentially for a 2–5 mV voltage change. A similar voltage dependence in mammalian rods would cause the 1–2 mV single-photon response to suppress the rate of vesicle fusion by at most two thirds. Keeping synaptic noise below rod outer segment noise in this case would require a dark release rate exceeding $400\ s^{-1}$. Direct measurement of the statistics of vesicle release and the consequences of the resulting synaptic noise remains a key experiment.

A second issue at the rod-to-rod bipolar synapse is the separation of single-photon responses from continuous noise. The continuous noise appears relatively innocuous in recordings from single rods (e.g., Figure 2); however, when signals from multiple rods are combined at light levels near absolute threshold, noise generated in all the rods threatens to overwhelm the light responses generated in a few rods. Linear summation of the rod inputs is a poor strategy under these conditions because each rod input contributes to the sum without regard to the likelihood that the rod absorbed a photon. A more effective strategy is to identify rods that likely absorbed a photon. This can be done by passing the rod inputs through a thresholding nonlinearity that retains signals from those rods likely generating single-photon responses and rejects signals from those rods likely generating noise (2, 64) (see Figure 4). Convergence of 20–100 rods onto a rod bipolar makes the rod-to-rod bipolar synapse the final opportunity to implement such a threshold. Indeed, Field & Rieke (34) found that such a nonlinearity exists at the rod-to-rod bipolar synapse. As a consequence, the single-photon responses of mouse rod bipolar cells are much more identifiable than those of the rods (Figure 4).

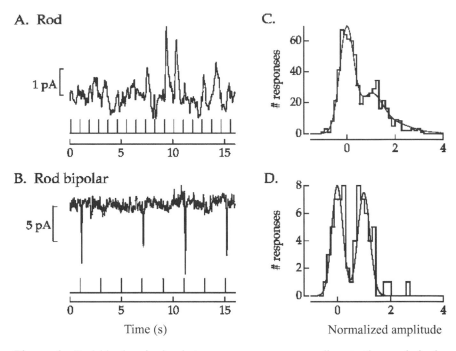

A. Rod

1 pA

0   5   10   15

B. Rod bipolar

5 pA

0   5   10   15

Time (s)

C.

responses

60

40

20

0

0   2   4

D.

responses

8

6

4

2

0

0   2   4

Normalized amplitude

Figure 4   Rod bipolar single-photon responses are more discrete than rod single-photon responses. (*A*) Rod photocurrent and (*B*) rod bipolar current recorded while delivering periodic flashes that generate ∼0.6 activated rhodopsins per rod. Rod currents were measured with a suction electrode. Rod bipolar currents were measured under voltage clamp with the cell held at −60 mV. Histograms of the normalized response amplitudes are plotted in (*C*) rods and (*D*) rod bipolar cells. Each histogram has been fit with a sum of weighted Gaussian functions.

Although convergence dictates the need for a thresholding nonlinearity at the rod-to-rod bipolar synapse, it does not predict where the threshold should be positioned relative to the single-photon response. Instead, the optimal position of the threshold is set by the light level and properties of the rod noise (Figure 5*A,B*). The light level at visual threshold can be thought of as a prior probability of 0.0001 that an individual rod generates a single-photon response. Photon-like noise events increase this probability to ∼0.0007 (for an 0.1 s integration time) because they cannot be distinguished from real responses (these noise events were neglected in Reference 34). Because of this, all signals smaller than the average single-photon response in a mouse rod are more likely to be noise than signal and should be rejected. Thus only relatively large single-photon responses should be retained, i.e., those to the right of the crossing point between the signal and noise distributions in Figure 5*B*. The observed threshold is close to this optimal position (34). Rejection of single-photon responses sounds like a poor processing strategy; however, the

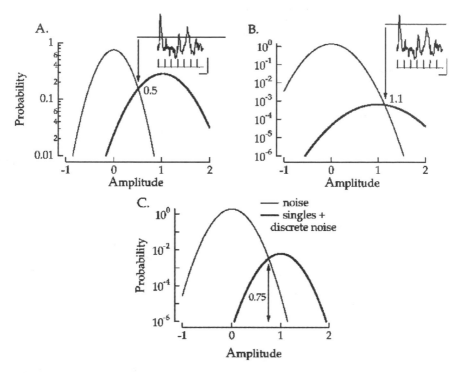

Figure 5 Optimal separation of signal and noise by a thresholding nonlinearity. (*A*) Comparison of the distribution of the continuous dark noise (*thin line*) and the distribution of signals generated by single-photon responses plus the spontaneous activation of rhodopsin (*thick line*) for a rod photoreceptor at a light level of 0.6 Rh*. The crossing point of the distributions (0.5) indicates the position of the optimal threshold. Insets show the position of such a threshold relative to the rod responses. (*B*) Same as (*A*) but at a light level of 0.0001 Rh* (near absolute threshold). The optimal position of the threshold moves to a higher value as the light level is decreased. (*C*) Same comparison as (*B*) for a rod bipolar cell at visual threshold. The separation of signal and noise in rod bipolar cells (see Figure 4) lowers the position of the optimal threshold (0.75).

resulting decrease in noise dramatically improves the fidelity of the rod signals. Rejection of some single-photon responses may contribute to the low quantum efficiencies inferred from behavior and ganglion cell recordings when compared with the absorptive quantum efficiency.

Electroretinograms (ERG) provide additional evidence for a nonlinearity in signal transfer from rods to rod bipolars (65, 66). The inferred position of the threshold from ERG recordings is lower than that found in slice recordings, suggesting that few photon responses are discarded at the rod-to-rod bipolar synapse. The position inferred from ERG measurements implies one of the following scenarios: (*a*) The threshold is not well placed to separate signal and noise; (*b*) the component of the

ERG attributed to the rod bipolar cells does not effectively isolate events occurring at the rod-to-rod bipolar synapse; or (c) existing rod noise measurements (2, 35, 36, 67) do not accurately describe rod noise in vivo. Because the ERG does not measure rod noise, resolving this issue will require measuring signal and noise from single cells under conditions close to those in vivo.

Sampath & Rieke (69) found that the thresholding nonlinearity at the rod-to-rod bipolar synapse was produced by saturation of the G protein cascade in the rod bipolar dendrites. Glutamate from the rods activates metabotropic receptors on the rod bipolar dendrites, which leads to closure of nonselective cation channels through a poorly understood signaling cascade (70). Reduction in glutamate release during the rod light response reduces receptor activity, opens channels, and leads to depolarization of the rod bipolar cell. In darkness, glutamate release in the rod is sufficient to saturate a component of the rod bipolar transduction cascade and cause the bipolar current to be insensitive to small changes in rod voltage and the corresponding changes in glutamate release. Larger changes, such as those produced during the single-photon response, relieve this saturation and produce an electrical response in the rod bipolar cell.

The ability of a threshold to separate single-photon responses from noise requires that signals are not mixed between neighboring rods prior to reaching the rod synaptic terminal. This is clearly not the case in amphibians, where rods display strong electrical coupling (71–74). Gap junctions are also present between rods in some rodent retinas (52) but have not been described in cat or primate. The discreteness of the single-photon responses in the rod bipolar currents and the sensitivity inferred from behavior and ganglion cells suggest that gap junctions between rods are not functional near absolute threshold (24).

A third aspect of signal transfer from rods to bipolar cells is the speeding of the response in the bipolar cells compared with that of the rods (34, 65, 75) (Figure 6). The response speeding indicates a reduced synaptic gain at low temporal frequencies, i.e., a high-pass filtering. Indeed, paired recordings in amphibian retina indicate that presynaptic mechanisms cause synaptic transmission between rods and bipolars to preferentially transmit temporal frequencies near 2 Hz (68). This filtering is well suited for the task of identifying the times of photon arrival on the basis of the rod currents (76, 77), as temporal frequencies below 1 Hz carry little information about photon arrival time and temporal frequencies above 4 Hz are dominated by noise. This qualitative argument can be formalized to predict the kinetics of the bipolar response based solely on the rod signal and noise. In mouse, the kinetics of rod-mediated responses in retinal ganglion cells is only slightly sped compared with that in the rod bipolars (Figure 6); thus the time course of the rod signals is dominated by rod phototransduction and the rod-to-rod bipolar synapse.

The work summarized in this section identifies the rod-to-rod bipolar synapse as a key component of both transmission and processing of the rod responses. Not only must this synapse maintain low noise, but linear and nonlinear mechanisms serve to speed the rod single-photon responses and separate them from continuous noise

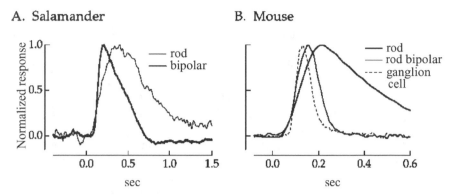

Figure 6 (*A*) Normalized dim flash responses from a salamander rod and bipolar cell recorded simultaneously (68). (*B*) Normalized dim flash responses from a mouse rod, rod bipolar cell, and ON ganglion cell.

produced in rod phototransduction. Processing at the rod-to-rod bipolar synapse appears to be well matched to the signal and noise properties of the rod signals.

The Rod Bipolar-to-AII Amacrine Synapse

Rod bipolar cells make ribbon synapses onto several types of amacrine cells (46), but they have not been observed to contact ganglion cells directly. The connections to the AII and A17 amacrine cells are the best studied. The A17 amacrines (S1/S2 amacrines in rabbit retina) modulate the rod signals by providing inhibitory feedback to rod bipolar synaptic terminals (78–82). Although the AII amacrines are a requisite component of the rod bipolar pathway, little is known about how single-photon responses traverse the rod bipolar-to-AII amacrine synapse. Below, we focus on open questions at the rod bipolar-to-AII amacrine synapse, in particular how the pattern of convergence and divergence affects photon detection.

AII amacrine cells in cat receive converging input from 20–30 rod bipolar cells, each of which gets input from ~15 rods (48). Thus the AII amacrine cell gets direct input from ~300–450 rods. At visual threshold, a small fraction (<0.01) of the converging rod bipolar inputs to an AII amacrine carry a light-driven signal. Thus the AII amacrine, similar to the rod bipolar, faces the problem of convergence of sparse, noisy inputs. Convergence implies that a thresholding nonlinearity such as that operating at the rod-to-rod bipolar synapse could serve to retain selectively rod bipolar single-photon responses while rejecting noise (83). As is the case at the rod-to-rod bipolar synapse, the best position of such a threshold is dictated by the light level and the rod bipolar signal and noise, independent of convergence.

Noise in the rod bipolar inputs consists of fluctuations in the rod bipolar electrical signals (Figure 5*C*) and noise produced by stochastic fluctuations in vesicle fusion at the synapse. These noise sources, particularly the synaptic noise, have not been characterized sufficiently to determine how severely linear summation by the AII amacrine would compromise sensitivity. Figure 5C gives a rough estimate of the distributions of photon-like events and noise in the rod bipolar cell at absolute threshold assuming synaptic noise is negligible. The separation of the signal and noise distributions is greater than that in the rod responses (Figure 5B,C); thus an optimally positioned threshold at the rod bipolar-to-AII synapse (arrow, Figure 5C) would reject a smaller fraction of the single-photon responses than are rejected at the rod-to-rod bipolar synapse. Figure 5C suggests that at absolute visual threshold the smallest 10–15% of the rod bipolar single-photon responses would have a greater probability of being noise than signal and should be discarded. This nonlinear summation would improve the signal-to-noise ratio of the AII signals by a factor of ~20 compared with linear summation. There is at present no evidence for or against a threshold at the rod bipolar-to-AII amacrine synapse.

Divergence causes the signal from a single rod to be represented in several AII amacrine cells. A single rod typically sends outputs to 2 rod bipolar cells, and these 2 rod bipolar cells contact ~5 AII amacrine cells (48). This divergence will correlate light-dependent signals in nearby AIIs. Noise generated at the rod bipolar-to-AII synapse should be independent in different AII amacrines, however, and thus the fidelity of the rod signal may be improved by averaging across AIIs (83, 84). Such averaging is provided by electrical coupling between nearby AII amacrines (85–87), which causes the signals to spread from 1 AII amacrine to ~20 neighbors under dark-adapted conditions (88). AII amacrine cells, unlike rods and rod bipolar cells, can also generate Na^+ action potentials (89). Modeling work suggests that the resulting nonlinear electrical properties may amplify single-photon responses more than noise in a collection of electrically coupled AIIs (83). The role of these action potentials in signal transmission at visual threshold has not been determined experimentally.

The AII-to-Cone Bipolar and Cone Bipolar-to-Ganglion Cell Synapses

Single-photon responses in the AII amacrine cell network are passed to ON- and OFF-cone bipolar cells through distinct mechanisms. ON signals from AII amacrine cells pass through gap junctions to ON-cone bipolar cell terminals. Electrical signaling through these gap junctions preferentially transmits signals from AII amacrines to cone bipolars owing to the large difference (~fivefold) in input impedance of the AII amacrines and the cone bipolar cells (87). Signals from AII amacrines are passed to OFF-cone bipolar cell terminals through inhibitory glycinergic synapses (90). The properties of glycinergic signal transmission from AII amacrines to OFF-cone bipolar cells are not well understood. Connections to ON and OFF bipolars do not appear to be entirely promiscuous because some ON and OFF ganglion cells receive almost exclusive cone input (44, 45, 91), requiring bipolar cells that get little or no rod input (92). This also suggests that some cones do not make functional gap junctions with rods (see Figure 3).

Cone bipolar cells convey rod signals directly to ganglion cells. Because all rod signals are conveyed to ganglion cells through cone bipolar cells, no ganglion cells should receive exclusive rod input. Although the cone bipolar-to-ganglion cell synapse has not been studied at light levels near absolute threshold, a good deal is known about how these synapses operate at cone light levels [reviewed in (93)].

CONCLUSIONS

We have reviewed what is currently known about retinal processing near absolute visual threshold. Behavioral experiments (3, 6, 7) and in vivo retinal ganglion cell recordings (8, 9) predict that rod photoreceptors can detect single photons and indicate a source or soures of additive Poisson noise that can be expressed as an equivalent input light or dark light (2). Some of the dark light clearly originates from spontaneous activation of rhodopsin in the rods. However, attributing all of the dark light to this source causes the behavioral and absorptive quantum efficiencies to differ substantially. Imposing that the quantum efficiencies are similar requires a large increase in the dark light, making it unlikely that rod noise alone limits sensitivity. These possibilities provide two very different views of retinal processing: The first implies the rod signals are efficiently and noiselessly read out by the retinal circuitry, the other implies that a significant amount of additive noise is introduced post-photoreceptor.

The comparison of the noise limiting behavior with noise from spontaneous activation of rhodopsin neglects at least three issues. First, such a comparison fails to consider other sources of noise in the rods, particularly continuous noise (35, 36). Single-photon responses can be separated from continuous noise by a threshold-like nonlinearity at the rod-to-rod bipolar synapse (2, 64). Such a threshold, however, would also reject a substantial fraction of the rod's single-photon responses (34) and thus contribute to limiting sensitivity. This may be one of the reasons that the absorptive and behavioral quantum efficiencies differ. Second, the assumption that the intrinsic noise can be modeled as an additive Poisson source may be wrong. A combination of additive and multiplicative noise (7, 19) can explain behavioral and ganglion cell sensitivity while providing closer agreement of behavioral and absorptive quantum efficiencies. Third, there is more to rod vision than detecting the presence or absence of dim lights. For example, estimating motion relies on extracting temporal information from the rod array. The noise limiting such computations may differ from that limiting detection. Reproducibility of the rod's single-photon responses may permit photon arrival times to be encoded precisely (37), and bandpass filtering at the rod synapse may be a first step in extracting this temporal information (68, 76).

To date, the main success story in our understanding of the retinal basis of photon detection has been work establishing how rod photoreceptors detect single photons. The same sort of quantitative and mechanistic description is lacking in our understanding of how the retina reads out the rod signals. We are beginning to understand how the first synapse in the retina may be optimized for processing retinal signals, but even here significant gaps in our knowledge remain. In particular, we know neither the statistics of vesicle release nor how the rod signals are being sped. Our knowledge of how single-photon responses are processed at subsequent synapses and encoded in the electrical activity of downstream retinal interneurons is even more incomplete. In addition, recent anatomical and physiological work has identified alternative pathways that rod signals can take through the mammalian retina. The function of these additional pathways is not clear, although a possible explanation is that they serve to process the rod signals over a wide range of light levels (41).

Vision is one of several examples in neurophysiology where sensory performance is exquisitely sensitive [reviewed in (94)]. Pheromone detection in insects may approach the single-molecule limit (95). Hearing approaches sensitivity limits set by thermal motion of the auditory hair cell stereocilia (94). Electroreceptors can detect voltage gradients of $\sim 10\,\mathrm{nV/cm}$ (96), much smaller than voltage changes required to open most ion channels. These examples of extreme sensitivity challenge our understanding of sensory transduction, processing, and encoding. Studies of photon detection in the retina provide an opportunity to understand how the underlying biophysical mechanisms meet these challenges.

ACKNOWLEDGMENTS

We thank Horace Barlow, E.J. Chichilnisky, Thuy Doan, Felice Dunn, and Valerie Uzzell for valuable comments on the manuscript, and W. Rowland Taylor and Robert Smith for helpful discussions. Support was provided by the National Institutes of Health through grant EY-11850 (FR) and NRSA EY-14784 (APS). G.D. Field and A.P. Sampath contributed equally to this review.

The *Annual Review of Physiology* is online at
http://physiol.annualreviews.org

LITERATURE CITED

1. Baylor DA, Lamb TD, Yau K-W. 1979. Responses of retinal rods to single photons. *J. Physiol.* 288:613–34
2. Baylor DA, Nunn BJ, Schnapf JL. 1984. The photocurrent, noise and spectral sensitivity of rods of the monkey *Macaca fascicularis*. *J. Physiol.* 357:575–607
3. Hecht S, Shlaer S, Pirenne MH. 1942. Energy, quanta, and vision. *J. Gen. Physiol.* 25:819–40
4. Walraven J, Enroth-Cugell C, Hood DC, Dia ML, Schnapf JL. 1990. The control

of visual sensitivity. In *Visual Perception: The Neurophysiological Foundations*, ed. L Spillmann, SJ Werner, pp. 53–101. San Diego: Academic
5. Barlow HB. 1956. Retinal noise and absolute threshold. *J. Opt. Soc. Am.* 46:634–39
6. Sakitt B. 1972. Counting every quantum. *J. Physiol.* 223:131–50
7. Teich MC, Prucnal PR, Vannucci G, Breton ME, McGill WJ. 1982. Multiplication noise in the human visual system at threshold. 1. Quantum fluctuations and

minimum detectable energy. *J. Opt. Soc. Am.* 72:419–31

8. Barlow HB, Levick WR, Yoon M. 1971. Responses to single quanta of light in retinal ganglion cells of the cat. *Vision Res. Suppl.* 3:87–101

9. Mastronarde DN. 1983. Correlated firing of cat retinal ganglion cells. II. Responses of X- and Y-cells to single quantal events. *J. Neurophysiol.* 49:325–49

10. Falk G, Fatt P. 1972. Physical changes induced by light in the rod outer segment of vertebrates. In *The Handbook of Sensory-Physiology, Volume VII/1*, ed. HJA Dartnall, pp. 200–44. Berlin: Springer-Verlag

11. Rao R, Buchsbaum G, Sterling P. 1994. Rate of quantal transmitter release at the mammalian rod synapse. *Biophys. J.* 67:57–63

12. Rao-Mirotznik R, Buchsbaum G, Sterling P. 1998. Transmitter concentration at a three-dimensional synapse. *J. Neurophysiol.* 80:3163–72

13. Young MP, Yamane S. 1992. Sparse population coding of faces in the inferotemporal cortex. *Science* 256:1327–31

14. Vinje WE, Gallant JL. 2000. Sparse coding and decorrelation in primary visual cortex during natural vision. *Science* 287:1273–76

15. van der Velden HA. 1946. The number of quanta necessary for the perception of light in the human eye. *Opthalmologica* 111:321–31

16. Bouman MA. 1961. History and present status of quantum theory in vision. In *Sensory Communication*, ed. W Rosenblith, pp. 377—401. Cambridge, MA: MIT Press

17. Barlow HB. 1977. Retinal and central factors in human vision limited by noise. In *Vertebrate Photoreception*, ed. HB Barlow, P Fatt, pp. 337—51. New York: Academic

18. Donner K. 1992. Noise and the absolute thresholds of cone and rod vision. *Vision Res.* 32:853–66

19. Lillywhite PG. 1981. Multiplicative intrinsic noise and the limits to visual performance. *Vision Res.* 21:291–96

20. Prucnal PR, Teich MC. 1982. Multiplication noise in the human visual system at threshold. 2. Probit estimation of parameters. *Biol. Cybern.* 43:87–96

21. Hallett PE. 1969. The variations in visual threshold measurement. *J. Physiol.* 202:403–19

22. Barlow HB, Levick WR. 1969. Changes in the maintained discharge with adaptation level in the cat retina. *J. Physiol.* 202:699–718

23. Levick WR, Thibos LN, Cohn TE, Catanzaro D, Barlow HB. 1983. Performance of cat retinal ganglion cells at low light levels. *J. Gen. Physiol.* 82:405–26

24. Smith RG, Freed MA, Sterling P. 1986. Microcircuitry of the dark-adapted cat retina: functional architecture of the rod-cone network. *J. Neurosci.* 6:3505–17

25. Goodchild AK, Ghosh KK, Martin PR. 1996. Comparison of photoreceptor spatial density and ganglion cell morphology in the retina of human, macaque monkey, cat, and the marmoset *Callithrix jacchus. J. Comp. Neurol.* 366:55–75

26. Cleland BG, Levick WR, Wassle H. 1975. Physiological identification of a morphological class of cat retinal ganglion cells. *J. Physiol.* 248:151–71

27. Frishman LJ, Levine MW. 1983. Statistics of the maintained discharge of cat retinal ganglion cells. *J. Physiol.* 339:475–94

28. Saleh BE, Teich MC. 1985. Multiplication and refractoriness in the cat's retinal-ganglion-cell discharge at low light levels. *Biol. Cybern.* 52:101–7

29. Rieke F, Baylor DA. 1998. Single-photon detection by rod cells of the retina. *Rev. Mod. Phys.* 70:1027–36

30. Pugh EN, Lamb TD. 1993. Amplification and kinetics of the activation steps in phototransduction. *Biochim. Biophys. Acta* 1141:111–49

31. Vuong TM, Chabre M, Stryer L. 1984. Millisecond activation of transducin in the cyclic nucleotide cascade of vision. *Nature* 311:659–61

32. Leskov IB, Klenchin VA, Handy JW, Whitlock GG, Govardovskii VI, et al. 2000. The gain of rod phototransduction: reconciliation of biochemical and electrophysiological measurements. *Neuron* 27:525–37

33. Field GD, Rieke F. 2002. Mechanisms regulating variability of the single photon responses of mammalian rod photoreceptors. *Neuron* 35:733–47

34. Field GD, Rieke F. 2002. Nonlinear signal transfer from mouse rods to bipolar cells and implications for visual sensitivity. *Neuron* 34:773–85

35. Baylor DA, Matthews G, Yau KW. 1980. Two components of electrical dark noise in toad retinal rod outer segments. *J. Physiol.* 309:591–21

36. Rieke F, Baylor DA. 1996. Molecular origin of continuous dark noise in rod photoreceptors. *Biophys. J.* 71:2553–72

37. Rieke F, Baylor DA. 1998. Origin of reproducibility in the responses of retinal rods to single photons. *Biophys. J.* 75:1836–57

38. Whitlock GG, Lamb TD. 1999. Variability in the time course of single photon responses from toad rods: termination of rhodopsin's activity. *Neuron* 23:337–51

39. Aho AC, Donner K, Hyden C, Larsen LO, Reuter T. 1988. Low retinal noise in animals with low body temperature allows high visual sensitivity. *Nature* 334:348–50

40. Aho AC, Donner K, Reuter T. 1993. Retinal origins of the temperature effect on absolute visual sensitivity in frogs. *J. Physiol.* 463:501–21

41. Sharpe LT, Stockman A. 1999. Rod pathways: the importance of seeing nothing. *Trends Neurosci.* 22:497–504

42. Bloomfield SA, Dacheux RF. 2001. Rod vision: pathways and processing in the mammalian retina. *Prog. Retin. Eye Res.* 20:351–84

43. Soucy E, Wang Y, Nirenberg S, Nathans J, Meister M. 1998. A novel signaling pathway from rod photoreceptors to ganglion cells in mammalian retina. *Neuron* 21:481–93

44. Deans MR, Volgyi B, Goodenough DA, Bloomfield SA, Paul DL. 2002. Connexin36 is essential for transmission of rod-mediated visual signals in the mammalian retina. *Neuron* 36:703–12

45. Volgyi B, Deans MR, Paul DL, Bloomfield SA. 2002. Role of AII amacrine cell coupling in rod vision: studies using a connexin36 knockout mouse. *Invest. Ophthal. Vis. Sci. Suppl.* S1944 (Abstr.)

46. Kolb H, Nelson R. 1983. Rod pathways in the retina of the cat. *Vision Res.* 23:301–12

47. Dacheux RF, Raviola E. 1986. The rod pathway in the rabbit retina: a depolarizing bipolar and amacrine cell. *J. Neurosci.* 6:331–45

48. Sterling P, Freed MA, Smith RG. 1988. Architecture of rod and cone circuits to the ON-beta ganglion cell. *J. Neurosci.* 8:623–42

49. Nelson R. 1977. Cat cones have rod input: a comparison of the response properties of cones and horizontal cell bodies in the retina of the cat. *J. Comp. Neurol.* 172:109–35

50. Schneeweis DM, Schnapf JL. 1995. Photovoltage of rods and cones in the macaque retina. *Science* 268:1053–56

51. Hack I, Peichl L, Brandstatter JH. 1999. An alternative pathway for rod signals in the rodent retina: rod photoreceptors, cone bipolar cells, and the localization of glutamate receptors. *Proc. Natl. Acad. Sci. USA* 96:14130–35

52. Tsukamoto Y, Morigiwa K, Ueda M, Sterling P. 2001. Microcircuits for night vision in mouse retina. *J. Neurosci.* 21:8616–23

53. Trifonov Y. 1968. Study of synaptic transmission between the photoreceptor and the horizontal cell using electrical stimulation of the retina. *Biofizika* 13:809–17

54. Dowling JE, Ripps H. 1973. Effect of magnesium on horizontal cell activity in the skate retina. *Nature* 242:101–3

55. Rieke F, Schwartz EA. 1996. Asynchronous transmitter release: control of exocytosis and endocytosis at the salamander rod synapse. *J. Physiol.* 493:1–8

56. del Castillo J, Katz B. 1953. Statistical nature of facilitation at a single nerve-muscle junction. *Nature* 171:1016–17

57. de Ruyter van Steveninck RR, Lewen GD, Strong SP, Koberle R, Bialek W. 1997. Reproducibility and variability in neural spike trains. *Science* 275:1805–8

58. Berry MJ, Meister M. 1998. Refractoriness and neural precision. *J. Neurosci.* 18:2200–11

59. Zenisek D, Steyer JA, Almers W. 2000. Transport, capture and exocytosis of single synaptic vesicles at active zones. *Nature* 406:849–54

60. Attwell D, Borges S, Wu SM, Wilson M. 1987. Signal clipping by the rod output synapse. *Nature* 328:522–24

61. Belgum JH, Copenhagen DR. 1988. Synaptic transfer of rod signals to horizontal and bipolar cells in the retina of the toad (*Bufo marinus*). *J. Physiol.* 396:225–45

62. Thoreson WB, Rabl K, Townes-Anderson E, Heidelberger R. 2004. A highly Ca^{2+}-sensitive pool of vesicles contributes to linearity at the rod photoreceptor ribbon synapse. *Neuron* 42:595–605

63. Witkovsky P, Schmitz Y, Akopian A, Krizaj D, Tranchina D. 1997. Gain of rod to horizontal cell synaptic transfer: relation to glutamate release and a dihydropyridine-sensitive calcium current. *J. Neurosci.* 17:7297–306

64. van Rossum MC, Smith RG. 1998. Noise removal at the rod synapse of mammalian retina. *Vis. Neurosci.* 15:809–21

65. Robson JG, Frishman LJ. 1995. Response linearity and kinetics of the cat retina: the bipolar cell component of the dark-adapted electroretinogram. *Vis. Neurosci.* 12:837–50

66. Saszik SM, Robson JG, Frishman LJ. 2002. The scotopic threshold response of the dark-adapted electroretinogram of the mouse. *J. Physiol.* 543:899–916

67. Schneeweis DM, Schnapf JL. 2000. Noise and light adaptation in rods of the macaque monkey. *Vis. Neurosci.* 17:659–66

68. Armstrong-Gold CE, Rieke F. 2003. Band-pass filtering at the rod to second-order cell synapse in salamander (*Ambystoma tigrinum*) retina. *J. Neurosci.* 23:3796–806

69. Sampath AP, Rieke F. 2004. Selective transmission of single photon responses by saturation at the rod-to-rod bipolar synapse. *Neuron* 41:431–43

70. Nawy S. 1999. The metabotropic receptor mGluR6 may signal through G_o, but not phosphodiesterase, in retinal bipolar cells. *J. Neurosci.* 19:2938–44

71. Schwartz EA. 1976. Electrical properties of the rod syncytium in the retina of the turtle. *J. Physiol.* 257:379–406

72. Copenhagen DR, Owen WG. 1976. Coupling between rod photoreceptors in a vertebrate retina. *Nature* 260:57–59

73. Detwiler PB, Hodgkin AL, McNaughton PA. 1980. Temporal and spatial characteristics of the voltage response of rods in the retina of the snapping turtle. *J. Physiol.* 300:213–50

74. Attwell D, Wilson M, Wu SM. 1984. A quantitative analysis of interactions between photoreceptors in the salamander (*Ambystoma*) retina. *J. Physiol.* 352:703–37

75. Schnapf JL, Copenhagen DR. 1982. Differences in the kinetics of rod and cone synaptic transmission. *Nature* 296:862–64

76. Bialek W, Owen WG. 1990. Temporal filtering in retinal bipolar cells. Elements of an optimal computation? *Biophys. J.* 58:1227–33

77. Rieke F, Owen WG, Bialek W. 1991. Optimal filtering in the salamander retina. In *Advances in Neural Information Processing Systems*, ed. D Touretzky, J Moody, 3:377–83. San Mateo, CA: Kaufmann

78. Nelson R, Kolb H. 1985. A17: a broad-field amacrine cell in the rod system of the cat retina. *J. Neurophysiol.* 54:592–614

79. Zhang J, Li W, Trexler EB, Massey SC. 2002. Confocal analysis of reciprocal feedback at rod bipolar terminals in the rabbit retina. *J. Neurosci.* 22:10871–82

80. Hartveit E. 1999. Reciprocal synaptic interactions between rod bipolar cells and

amacrine cells in the rat retina. *J. Neurophysiol.* 81:2923–36

81. Euler T, Masland RH. 2000. Light-evoked responses of bipolar cells in a mammalian retina. *J. Neurophysiol.* 83:1817–29

82. Dong CJ, Hare WA. 2003. Temporal modulation of scotopic visual signals by A17 amacrine cells in mammalian retina in vivo. *J. Neurophysiol.* 89:2159–66

83. Smith RG, Vardi N. 1995. Simulation of the AII amacrine cell of mammalian retina: functional consequences of electrical coupling and regenerative membrane properties. *Vis. Neurosci.* 12:851–60

84. Vardi N, Smith RG. 1996. The AII amacrine network: Coupling can increase correlated activity. *Vision Res.* 36:3743–57

85. Mills SL, Massey SC. 1995. Differential properties of two gap junctional pathways made by AII amacrine cells. *Nature* 377:734–37

86. Vaney DI. 1991. Many diverse types of retinal neurons show tracer coupling when injected with biocytin or neurobiotin. *Neurosci. Lett.* 125:187–90

87. Veruki ML, Hartveit E. 2002. Electrical synapses mediate signal transmission in the rod pathway of the mammalian retina. *J. Neurosci.* 22:10558–66

88. Bloomfield SA, Xin D, Osborne T. 1997. Light-induced modulation of coupling between AII amacrine cells in the rabbit retina. *Vis. Neurosci.* 14:565–76

89. Boos R, Schneider H, Wassle H. 1993. Voltage- and transmitter-gated currents of all-amacrine cells in a slice preparation of the rat retina. *J. Neurosci.* 13:2874–88

90. Haverkamp S, Muller U, Harvey K, Harvey RJ, Betz H, Wassle H. 2003. Diversity of glycine receptors in the mouse retina: localization of the alpha3 subunit. *J. Comp. Neurol.* 465:524–39

91. DeVries SH, Baylor DA. 1995. An alternative pathway for signal flow from rod photoreceptors to ganglion cells in mammalian retina. *Proc. Natl. Acad. Sci. USA* 92:10658–62

92. Pang JJ, Gao F, Wu SM. 2004. Light evoked current responses in rod bipolar cells, cone depolarizing bipolar cells and AII amacrine cells in dark-adapted mouse retina. *J. Physiol.* 558:897–912

93. Masland RH, Raviola E. 2000. Confronting complexity: strategies for understanding the microcircuitry of the retina. *Annu. Rev. Neurosci.* 23:249–84

94. Bialek W. 1987. Physical limits to sensation and perception. *Annu. Rev. Biophys. Biophys. Chem.* 16:455–78

95. Angioy AM, Desogus A, Barbarossa IT, Anderson P, Hansson BS. 2003. Extreme sensitivity in an olfactory system. *Chem. Senses* 28:279–84

96. Kalmijn AJ. 1982. Electric and magnetic field detection in elasmobranch fishes. *Science* 218:916–18

Annu. Rev. Physiol. 2005. 67:515–29
doi: 10.1146/annurev.physiol.67.040403.101353
First published online as a Review in Advance on September 27, 2004

A Physiological View of the Primary Cilium

Helle A. Praetorius[1] and Kenneth R. Spring[2]

[1]The Water and Salt Research Center, Clinical Institute, University of Aarhus,
8200 Aarhus N, Denmark; email: helle.praetorius@ki.au
[2]LKEM, NHLBI, The National Institutes of Health, Bethesda, Maryland, 20892;
email: springk@direcway.com

Key Words   function, Ca^{2+}, mechano-sensation, chemo-sensation, signal transduction, flow, polycystin

■ **Abstract**   The primary cilium, an organelle largely ignored by physiologists, functions both as a mechano-sensor and a chemo-sensor in renal tubular epithelia. This forgotten structure is critically involved in the determination of left-right sidedness during development and is a key factor in the development of polycystic kidney disease, as well as a number of other abnormalities. This review provides an update of our current understanding about the function of primary cilia. Much new information obtained in the past five years has been stimulated, in part, by discoveries of the primary cilium's key role in the genesis of polycystic kidney disease as well as its involvement in determination of left-right axis asymmetry. Here we focus on the various functions of the primary cilium rather than on its role in pathology.

BACKGROUND

The term primary cilium arose from a classification of the cilia of eukaryotic cells into nonmotile primary and motile secondary cilia (1). The primary cilium is a solitary cellular structure known to cytologists for over a hundred years (2), and it has been shown to be present in almost all vertebrate cells (3). The most important exceptions are the bone marrow–derived cells and the intercalated cells of the kidney collecting duct. The function of the primary cilium has been studied in only a few cell types. Much of the recent work that has led to better functional understanding of the primary cilium has been carried out on invertebrates such as *Caenorhabditis elegans* and has been an inspiration for the experiments done on vertebrate cells.

The primary cilium has a so-called 9 + 0 axoneme, which refers to its nine peripherally located microtubule pairs and the absence of the central microtubule pair seen in motile cilia or 9 + 2 cilia (Figure 1). In addition to the central microtubules, the 9 + 2 motile cilia possess radial spokes linked to the microtubules. These cilia are motile because they have a combination of inner and outer arm

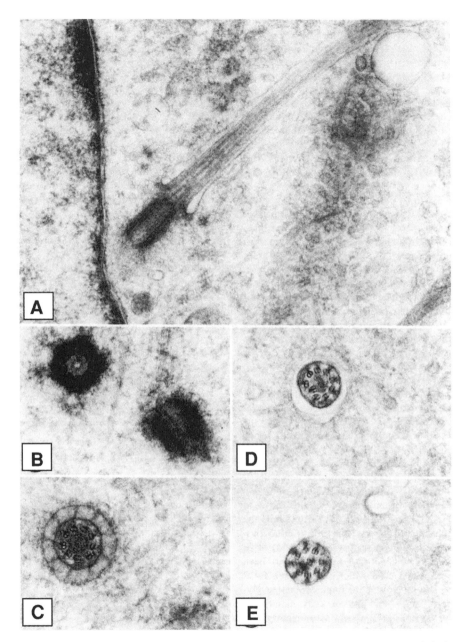

Figure 1   Electron micrographs of a primary cilium emerging from the centriole in PtK₁ cells (*A*). Cross sections through the centriole (*B*) and at various levels of the primary cilium (*C–E*). Reprinted from Jensen et al. with permission (4).

dynein motors and dynein-regulatory proteins with ATPase activity. The primary cilium is generally considered to be nonmotile because it lacks the central microtubule pairs, the radial spokes, and the dynein apparatus. When visualized by video microscopy, the cilium shows only Brownian movements, i.e., small fast movements induced by thermal energy.

Nodal Cilia—Motile Form of the Primary Cilium

An exception to the lack of motility of the 9 + 0 cilium is a specialized primary cilium, the nodal cilium in the blastocyst, which does have radial spokes and dynein arms but lacks a central pair of microtubules. Instead of beating from side to side as does the 9 + 2 cilia, nodal cilia exhibit a characteristic twirling movement that is responsible for the nodal flow crucial for the development of left-right sidedness in the developing mouse embryo (5). The nodal cilium moves forward to an almost upright position and the return stroke is quite close to the membrane. Only the forward movement has any significance for generation of the nodal flow (5). Although nodal flow occurs during only a few hours in the late stage of neural plate development (6), its disruption results in a random distribution between situs inversus and the normal situs solitus. Nodal flow can be disturbed by mutations in either the genes coding for the motor molecules or those coding for the proteins necessary for normal intraflagellar transport (IFT) (see below). Randomization of left-right specification is seen in mice lacking the kinesins KIF3A (7, 8) and KIF3B (5) or lacking the ciliary transport protein, Polaris (9), which results in the absence or stunting of primary cilia. Mutations that result in defective dynein, for example, the *inv* gene that encodes for left-right dynein, also result in nonmotile nodal cilia and situs inversus (10, 11).

The evidence determining that nodal flow is responsible for left-right asymmetry was carried out with live embryos in a flow chamber. Superimposed flow from left to right could reverse left-right asymmetry in wild-type embryos and create normal left-right development in embryos with defective dynein, which did not have nodal flow of their own (6).

Two models have been posited to explain how the nodal flow determines the left-right axis of the embryo. Initially, the nodal cells were suggested to secrete a signal substance. The nodal flow would then increase the concentration of this substance at one pole, thus creating a concentration gradient across the node (5, 7, 8, 10, 11). However, external applied flow at high rates was able to override the nodal flow and determine the later organ position (6). This argues against a soluble factor because the external applied flow would tend to wash out or dilute a soluble factor.

On the basis of the observation that there are two types of cilia in the node, the motile nodal cilia and nonmotile primary cilia (12), another hypothesis was put forward. The primary cilia were proposed to sense the nodal flow produced by the nodal cilia. This hypothesis was supported by the observation that wild-type mouse gastrula have higher intracellular Ca^{2+} concentrations on the left margin of the node, which was not observed in left-right dynein mutants (12). Interestingly, mice with mutations in polycystin 2 have, in addition to the polycystic phenotype, abnormal left-right development (13).

Mutation in the *inv* gene results in situs inversus in the vast majority of the mice offspring. The corresponding protein, inversin, has been localized to the primary cilium in the node (14). These mice have a leftward-directed nodal flow, although it is slower and more turbulent (11). These findings are not easily reconciled with the current theory of nodal flow, and thus the question remains unresolved.

The Primary Cilium and the Centriole

The primary cilium develops from and is continuously anchored to the cell's mother centriole. In some cell types, the primary cilium is located in a membrane invagination under the surface of the cell (15). In other cell types, however, the primary cilium protrudes several microns away from the cell surface. In cultured Kangaroo rat kidney cells, the primary cilia may be as long as 30 μm (16). The primary cilium is surrounded by a membrane that is continuous with the plasma membrane (15). However, it has been shown frequently that cells direct specific proteins to the membrane of the primary cilium by the mechanism of intraflagellar or intraciliary transport (17–20). Of interest in this regard is the fact that the primary cilium, as it emerges from the centriole, comes into close proximity to the Golgi apparatus (21).

The centriole is a diplosomal structure consisting of a basal body tied to a proximal centriole (22) (Figure 2). The complex is structurally stabilized by attachment to the plasma membrane via transitional fibers called the alar shields. The centriole functions as the microtubule-organizing center and the cell's microtubular meshwork radiates from it (23). The internal structure of the centriole and alar shields restricts the size of molecules that can pass through them into or out of the cilium.

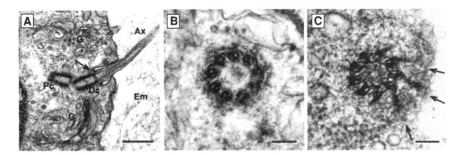

Figure 2   Electron micrograph of diplosomal centrioles and centrosomes of chondrocytes. (*A*) The relationship between the primary cilium (Ax), the distal (Dc), and proximal (Pc) centriole. Bar 500 nm. (*B*) Cross section of the proximal centriole. Bar 100 nm. (*C*) Distal tip of the basal body. Bar 100 nm. Reprinted from Jensen et al. with permission (22).

Unpublished data referred to in Reference 22 indicate that particles less than only 10 kDa can pass on from the cytosol into the primary cilium shaft.

Intraciliary Transport

The cilia and flagella elongate from the centriole by extension at the tip of the cilium (24, 25). Cilia and flagella are unable to synthesize the various ciliary proteins, and thus all the proteins needed for elongation of the primary cilium have to be transported out to the ciliary tip (24, 25). This is accomplished by a transport apparatus known as the intraciliary (ICT) or IFT system. The IFT system was first discovered in green alga, *Chlamydomonas*, by Kozminski et al. (26). The transport system, as visualized by electron microscopy, has been described as macromolecular rafts localized between the flagella membrane and the outer doublet of microtubules (27–28) (Figure 3). The rafts are made up of various components denoted as the IFT particles. About 15–20 polypeptides have been identified and all are necessary for normal intraflagellar transport. These polypeptides are subdivided into complex A and B (29, 30). All the IFT particles tend to accumulate around the basal bodies. This means that at any given time a vast pool of transport particles is available for docking and, therefore, an enormous reserve capacity exists for IFT (31). In *Chlamydomonas*, 14 different IFT particles have been identified (31). A gene of special interest is *IFT88*, an orthologue of the murine *Tg737* (32) and of the corresponding OSM-5 in *C. elegans* (33). The gene products participate in flagella and cilia assembly, and a mutation in *IFT88* results in short flagella in *Chlamydomonas* (32); mutations in *Tg737* result in inability to form primary cilia (9, 32). Mice with the *Tg737* mutation show a phenotype with multiple renal, pancreatic, and hepatic cysts, and have hydrocephalus and cartilage and bone malformations (32). The multiple kidney cysts were the first indication of a relationship between the primary cilium and polycystic kidney disease (see below).

All microtubules of cilia and flagella are oriented with the plus end at the tip of the organelle (34, 35). Because microtubule motors move particles in only one direction, two types are needed to fully account for the ciliary transport. Kinesin-II is responsible for the anterograde movement of the IFT (36) at an approximate

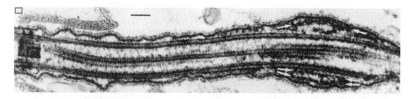

Figure 3   Electron micrograph of flagella from wild-type *Chlamydomonas*. The IFT rafts (*white arrows*) are associated with either the flagella membrane or the microtubules. Bar 100 nm. Reprinted from Witman et al. with permission (28).

speed of 2.5 μm s^{-1} (26). Kinesin-II consists of the two motor subunits, IKF3A and KIF3B, as well as the nonmotor subunit kinesin-associated protein (37). The retrograde motor was found to involve a cytoplasmic dynein 1b that moves IFT particles at a speed of approximately 4 μm s^{-1} (26). Defects in the gene encoding for cytoplasmic dynein light chain in *Chlamydomonas* LC8 resulted in significant shortening of the flagella 50 to 70% of the wild-type length (28). The speed of movement of IFT itself cannot be the rate-limiting factor in the growth of cilia and flagella.

Primary cilia develop rather slowly in kidney epithelia. In Madin-Darby canine kidney (MDCK) cells, the primary cilium starts to appear several days after the cells have reached confluency (20). After the primary cilia have been removed by treatment with chloral hydrate, short cilia are seen 72 h later, and cilia of 3–5 μm of length are first observed after 96 h (38). In contrast, flagella regrowth is a rather fast process. After deflagellation in *Chlamydomonas*, the flagella regenerate within 2 h (39). Even in this situation the speed of movement of IFT itself cannot be the rate-limiting factor in cilia and flagella growth.

Flow Sensing

The function of the primary cilium has been speculated upon for decades but only recently has been addressed. The specialized primary cilia such as those in the retina that connect the rod outer and inner segments have received some attention, but the nearly ubiquitous primary cilium remained a mystery to physiologists until evidence for a sensory function started to emerge around 2000.

In 1997, Bowser and coworkers (40) showed that the primary cilium of cultured renal cells was able to bend, when the cells were superfused at flow rates comparable to those seen in renal tubules (Figure 4). The study was carried out on living kidney epithelial cells grown on flexible support. When the tissue and its support were folded over, it was possible to visualize the primary cilium without any underlying structures. Measurement of the degree of bending of the cilium, modeled as a heavy elastica beam, showed that the strain on the ciliary membrane must be largest on the convex side of the bent cilium.

Although the findings of Schwartz et al. (40) showed that cilia could deform with reasonable flows, a link between the deformation and cell signaling was not established. Any attempt to study the role of the cilium in flow sensing would be clouded by the possibility that the sensors were located on the cell membrane, as suggested in the shear-stress hypothesis in endothelia. We, therefore, devised experimental procedures to test whether bending of the cilium itself was able to produce an intracellular response. MDCK was used as a test system. These cells, derived from dog kidney, have a mixture of principal cells and intercalated cells, quite similar to that of the collecting duct. The principal cells of the MDCK cell line express primary cilia, whereas the intercalated cells do not. Although the length of the primary cilium is limited to 2 to 3 μm in vivo, cultured cells exhibit cilia up to 30 μm (16).

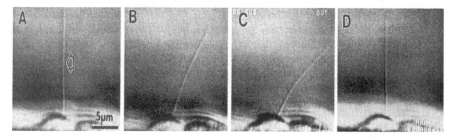

Figure 4 Bending of primary cilia by increasing flow rates. PtK$_1$ cells grown on flexible support are viewed from the side by folding over the support. (*A*) The primary cilium is indicated by the arrow. (*B*) At the onset of flow from left to right, the cilium bends accordingly. (*C*) As the flow rate increases, the bending increases. (*D*) The cilium straightens up after the flow has ceased. Reprinted from Schwartz et al. with permission (40).

Our experimental protocol employed cells whose cilia were ∼5–8 μm long. To avoid disturbing putative cell membrane flow sensors, the cilium was bent by suction from a micropipette positioned ∼4 μm above the cell membrane. We chose to measure intracellular Ca^{2+} because most mechanically sensitive processes alter that parameter. The intracellular Ca^{2+} was monitored with a fluorescent indicator dye. It was soon evident that the MDCK cell primary cilium acts as a flow sensor (38, 41). Bending the primary cilium either by micropipette or by increasing the perfusate flow rate led to a remarkable increase in the intracellular Ca^{2+} concentration (Figure 5). This flow response was absolutely dependent on the presence of a

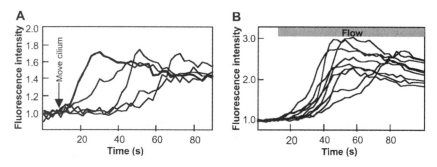

Figure 5 The intracellular Ca^{2+} increase induced by bending of the primary cilium in MDCK cells. (*A*) Bending of a single primary cilium with a micropipette. The pipette was positioned ∼4 μm above the apical membrane and the primary cilium was bent toward the micropipette (*arrow*) by application of negative pressure in the pipette. The MDCK cells were grown to confluency on cover slips and loaded with the fluorescent probe fluo-4. (*B*) Bending primary cilia by flow. The traces represent increments of the intracellular Ca^{2+} concentration after changing the perfusate flow rate from 0 to 8 μl/s. Reprinted from Praetorius & Spring with permission (41).

primary cilium, because immature cells that do not present a primary cilium or cells, from which the cilium was removed by chloral hydrate treatment did not respond to increasing flow rates by increasing their intracellular Ca^{2+} concentration (38, 41).

Data indicating that the primary cilium might act as a flow sensor in the collecting duct were obtained by Liu et al. (42). They developed a mathematical model to explain their observations in intact perfused rabbit collecting ducts and concluded that the primary cilium is the sensor of fluid flow in the intact collecting duct. Thus the heavy elastica model for cilium bending developed by Schwartz et al. (40) was challenged. Liu et al. (42) argued that the primary cilia of the collecting duct are not long enough, according to the heavy elastica model, to bend at the fluid shear present in the collecting duct. Rather, the fluid shear provides a torque that is transmitted via the primary cilium to the cytoskeleton. Newer data from fluid shear on chondrocytes show that cilia of 2–3 μm are able to bend according to the heavy elastica model (22). Resolution of this issue awaits additional functional studies on the intact collecting duct.

The ability of the kidney epithelia to sense changes in luminal fluid flow rate appears to be important for the regulation of the fluid composition. In renal distal tubule, an increase in luminal flow rate is known to induce K^+ secretion and increase the lumen-negative transepithelial potential differences (43). For MDCK cells, bending of the primary cilium was shown to induce a substantial hyperpolarization of the apical membrane as a result of the activation of intermediate Ca^{2+}-dependent K^+ channels (44). Thus the activation of apical Ca^{2+}-dependent K^+ channels might be responsible for the flow-induced K^+ secretion observed in the intact tubule (45). In addition, data obtained from polycystin 1- and 2-deficient cells (see below) suggest that the sensing of fluid flow might be important for maintaining cell polarization.

Details of the Calcium Response

The signal transduction pathway for the cilium-dependent Ca^{2+} response depends on extracellular Ca^{2+} for initiation of the signal. After the cilium has been bent, Ca^{2+} can be omitted from the medium without interrupting further development of the signal. This, in turn, means that only the initial part of the signal transduction event requires Ca^{2+} influx. The Ca^{2+} influx is not sensitive to blockers of voltage-gated Ca^{2+} channels, but is inhibited by Gd^{3+} and amiloride, the somewhat nonselective inhibitors of stretch-activated Ca^{2+} channels. Recent evidence shows that the Ca^{2+} permeant channel involved in flow sensing is polycystin 2 (46). Nauli et al. were able to show that the embryonic kidney epithelial cells in which either polycystin 1 or 2 was modified lost their flow-induced Ca^{2+} response (46). When the cells were rescued by reinserting the gene, flow sensing returned. Polycystin 2 is a member of the transient receptor potential (TRP) channel superfamily (47) and in that nomenclature is denoted as TRPP (48).

Barr et al. (49) were the first to recognize that the analogue of the PKD2 gene product in *C. elegans* was targeted to the primary cilia of that organism. Soon after, polycystin 1 and 2 were immunocytochemically localized to the primary cilium of human and mice kidney epithelial cells (19, 20) and later were shown in embryonic kidney cells (46) and MDCK cells (50). At present, a connection between mechano-sensation and cyst formation as seen in polycystic kidney disease is purely speculative. However, it has been suggested that such an association might exist through the action of inversin. Localization of inversin to primary cilia and mutation in the coding sequence results phenotypically in polycystic kidneys in mice (51). Inversin is thought to interfere with the cell cycle through interaction with the Apc2 subunit of the anaphase-promoting complex (51). Because inversin binds to calmodulin in the absence of Ca^{2+}, it has been speculated that inversin represents the link between the lack of mechanically induced calcium increases and re-entry in cell cycle (51, 52).

Both in MDCK cells and in embryonic kidney cells, the primary Ca^{2+} entry is followed by a global Ca^{2+} increase (41, 46). This increase is thought to be a consequence of Ca^{2+}-induced Ca^{2+} release, but at present there has been no testing to determine whether an intermediate signal transduction molecule is involved. In MDCK cells, the global Ca^{2+} response involves the IP_3-sensitive Ca^{2+} stores, whereas in the embryonic kidney cells, the ryanodine-sensitive stores are activated. In the perfused rabbit collecting ducts (occluded and non-occluded), the flow-induced Ca^{2+} signal is independent of the availability of Ca^{2+} on the apical side of the tubule, but absolutely dependent on Ca^{2+} at the basolateral side (42). Either the flow-induced signal of the perfused tubule is a mix of more processes or the signal flow-sensing signal cascade is substantially different from that in cultured cells.

The flow-induced Ca^{2+} response is not instantaneous, as flow has to be sustained for at least 3 s to get a full-scale Ca^{2+} increase. In MDCK cells, the development of the Ca^{2+} signal is slow and is maximal about 30 s after the start of the stimulus. These findings are consistent with a rather slow signal transduction mechanism, as would be the case if Ca^{2+} had to diffuse down the primary cilium and initiate a global response from a highly localized area at the base of the cilium. Force transmission through the cytoskeleton would not account for the latency found in MDCK cells. The primary cilium's narrow geometry has been suggested to be ideal for initiation of intracellular Ca^{2+} signals (58) because calcium is unlikely to be heavily chelated in the cilium, and very few calcium ions are needed to increase the cilia Ca^{2+} concentration substantially (38). Furthermore, after the initial flow-induced Ca^{2+} response, MDCK cells are relatively refractory, and full amplitude of the Ca^{2+} response is first observed after 25 min. However, these cells are quite responsive to other stimuli. Stimulating the P2Y receptors with 100-μM ATP results in a substantial Ca^{2+} increase within minutes after the flow response (H.A. Praetorius & K.R. Spring, unpublished data). Thus the refractory period is not a consequence of depleted internal stores. One can speculate that the refractory period represents the time it takes to restore normal Ca^{2+} levels in the primary cilium.

Mechano-Sensing in Other Tissues

The primary cilium has been suggested to act as mechano-sensors in osteocytes (54) and chondrocytes (21, 22). It is likely, however, that the primary cilium is needed only for flow sensation in low-pressure systems. In human endothelial cells, laminar shear stress in the range seen in the arterial system surprisingly causes disassembly of primary cilia (58). This may mean that cells use the primary cilia as an amplifier to sense changes in flow rates too low to produce any significant shear stress effect on the plasma membrane.

The ubiquitous distribution of primary cilia in tissues that have no obvious requirement for flow sensing, e.g., nerve, muscle, or endocrine cells, leads us to conclude that this organelle serves functions other than mechano-sensing in these tissues.

The Primary Cilium and Chemo-Sensing

In *C. elegans*, primary cilia are essential for chemo- and osmo-sensation. In contrast to mammalian cells, the adult *C. elegans* hermaphrodite expresses primary cilia in only 60 out of its 302 neurons (59). Mutations that affect the structure of the cilia result in sensory defects such as altered osmotic avoidance, poor mating behavior, and defective chemotaxia (60). It is established that modified primary cilia, such as olfactory cilia and cilia of the sensory neurons, the acousto-vestibular system, and photoreceptors in the central nervous system (CNS), have sensory functions (53). It appears that the majority of the neurons of the rat CNS express solitary primary cilia, as visualized immunohistochemically with an antibody against G11 (53). In immunohistochemical localization studies of the somatostatin receptor (SST_3) and the 5-HT6 receptor in the CNS, these receptors were found on primary cilia, consistent with a sensory function. The SST_3 receptor is found in the primary cilia of the rat and mouse CNS (17, 61) and the 5-HT6 receptor, similarly, in neurons of the rat brain (18). Both receptors are G protein–coupled, indicating that the initial parts of signal-transduction could occur in the primary cilium.

Because the growth of the cilium and intraciliary transport are so closely regulated, it is reasonable to assume that proteins are inserted into the ciliary membrane for only a specific functional reason. Perhaps presenting important hormonal receptors in the cilium instead of having them generally distributed in the plasma membrane increases the likelihood that they will encounter an agonist.

An example of a receptor inserted into the membrane of the primary cilium of MDCK cells (and renal tubules) is $\beta 1$-integrin, a widely distributed transmembrane protein that binds a number of agonists. $\beta 1$-integrin, one of the molecules involved in anchoring of the cells to matrix proteins, is thought to play an important role in sensing of mechanical stress by endothelial cells (62). Twisting of integrins bound to metallic beads by altering a magnetic field produces an intracellular Ca^{2+} signal in bovine capillary endothelial cells (63). Integrin-dependent mechano-sensing is also believed to be linked to the release of cellular nucleotides. Indeed, RGD-peptides bind to $\beta 1$-integrin and are able to inhibit ATP release in *Xenopus* oocytes (64). The subunit associated with $\beta 1$-integrin is an important determinant of the affinity for a particular agonist. In MDCK cells, it is predominantly the α_3 subunit. Although integrins are commonly thought to be expressed only at the basolateral side of epithelia where they interact with matrix proteins, they are also present on the apical membrane of MDCK cells and rat collecting duct (65). β1-integrin has also been shown to be present in the primary cilium of MDCK cells and in the cilium in the rat nephron (66).

The renal primary cilium extends far into the lumen of the tubules and is ideally situated to intercept agonists that bind to β1-integrin. It was recently shown that the presence of β1-integrin on the apical membrane and primary cilium is not necessary for the flow-induced Ca^{2+} response in MDCK cells (66). On the other hand, agonist stimulation of β1-integrin by fibronectin or the lectin from sambucus nigra (which is known to bind to β1-integrin in MDCK cells) results in a dose-dependent increase in the intracellular Ca^{2+} concentration of MDCK cells. The dose-response curve for fibronectin was strongly dependent on the presence of the primary cilium (66). In preparations in which the cells were either too immature to develop a primary cilium or in which the cilium had been removed with chloral hydrate, the dose-response curve was shifted significantly to the right in contrast to that seen in mature MDCK cells with primary cilia. In this connection it should be mentioned that the apical distribution of β1-integrin was quite similar in mature, immature, and chloral hydrate–treated MDCK cells (the amount of protein per cell is substantially the same). These findings indicate that expression of β1-integrin on the primary cilium enables epithelial cells to detect an agonist in much lower concentrations than would occur with the receptor present on the apical membrane alone. This could be a consequence of extending the receptors beyond the unstirred layer and into the lumen of the tubule.

A Role for Primary Cilia in Cartilage and Bone

Expression of integrins on the primary cilia of other cell types might be important for interaction with matrix proteins. It was recently shown that matrix proteins, such as collagen, bind to the primary cilium in chondrocytes (22). When the cilium interacts with collagen, alterations occur in the plasma membrane and in the ciliary microtubules that may be responsible for the acute bending of the primary cilia seen in the presence of matrix proteins (22). In bone tissue, however, it has been speculated that the primary cilium functions as flow sensor. Osteocytes reside in laculae and create a network throughout the bone tissue by extending their processes through canaliculi, which reach other members of the network (54). Stain pulses along the bones induce fluid movement in the canaliculi (55–57). It has been suggested that the osteocyte primary cilia are able to sense the changes in flow rate in the canaliculi and thereby transmit signals for bone remodeling (54).

Primary Cilia in Secretory Cells

Because the cilium is present in almost all cells, its presence alone does not indicate its functional significance. A primary cilium is frequently observed in alpha, beta, and gamma cells of the endocrine pancreas (67) where it may potentially play a significant role in chemo-sensing. Rates and composition of secretion are often modulated in response to environmental or hormonal factors, but the nature of the link between these two events has often not been explored. The unique characteristics of the primary cilium make it ideal for this purpose. The challenge to physiologists is to devise suitable experiments to test this possibility.

Conclusion

Even though evidence of specific functions of the primary cilium has begun to emerge, many of the central theories about its potential functions have not been addressed. Most prominently is an understanding of why the organelle is such a generalized feature among cells. Although it is clear that the primary cilium can fulfill various functions, there does not yet seem to be a common denominator for all tissues.

The *Annual Review of Physiology* is online at http://physiol.annualreviews.org

LITERATURE CITED

1. Williams PL, Warwick P, Dyson M, Bannister LH, eds. 1989. Cells and tissues. In *Grays Anatomy*, pp. 31–33. London: Churchill Livingstone. 37th ed.
2. Zimmermann KW. 1898. Beiträge zur Kenntnis einiger Drüsen und Epithelien. *Arch. Mikr. Anat. Entwicklungsmech.* 52:552–706
3. Wheatley DN. 1995. Primary cilia in normal and pathological tissues. *Pathology* 63:222–38
4. Jensen CG, Jensen LC, Rieder CL. 1979. The occurrence and structure of primary cilia in a subline of *Potorous tridactylus*. *Exp. Cell Res.* 123:444–49
5. Nonaka S, Tanaka Y, Okada Y, Takada S, Harada A, et al. 1998. Randomization of left-right asymmetry due to loss of nodal cilia generating leftward flow of embryonic fluid in mice lacking KIF3B motor protein. *Cell* 84:829–37
6. Nonaka S, Shiratori H, Saijoh Y, Hamada H. 2002. Determination of left-right patterning of the mouse embryo by artificial nodal flow. *Nature* 418:96–99
7. Marszalek JR, Ruiz-Lozano P, Roberts E, Chien KR, Goldstein LS. 1999. Situs inversus and embryonic ciliary morphogen-

esis defects in mouse mutants lacking the KIF3A subunit of kinesin-II. *Proc. Natl. Acad. Sci. USA* 96:5043–48
8. Takeda S, Yonekawa Y, Tanaka Y, Okada Y, Nonaka S, Hirokawa N. 1999. Left-right asymmetry and kinesin superfamily protein KIF3A: new insights in determination of laterality and mesoderm induction by kif3A$^{-/-}$ mice analysis. *J. Cell Biol.* 145:825–36
9. Yoder BK, Tousson A, Millican L, Wu JH, Bugg CE Jr, et al. 2002. Polaris, a protein disrupted in *orpk* mutant mice, is required for assembly of renal cilium. *Am. J. Physiol. Renal Physiol.* 282:F541–52
10. Supp DM, Brueckner M, Kuehn MR, Witte DP, Lowe LA, et al. 1999 Targeted deletion of the ATP binding domain of left-right dynein confirms its role in specifying development of left-right asymmetries. *Development* 126:495–504
11. Okada Y, Nonaka S, Tanaka Y, Saijoh Y, Hamada H, Hirokawa N. 1999. Abnormal nodal flow precedes situs inversus in iv and inv mice. *Mol. Cell* 4:459–68
12. McGrath J, Brueckner M. 2003. Cilia are at the heart of vertebrate left-right asymmetry. *Curr. Opin. Genet. Dev.* 13:385–92

13. Pennekamp P, Karcher C, Fischer A, Schweickert A, Skryabin B, et al. 2002 The ion channel polycystin-2 is required for left-right axis determination in mice. *Curr. Biol.* 12:938–43

14. Watanabe D, Saijoh Y, Nonaka S, Sasaki G, Ikawa Y, et al. 2003 The left-right determinant Inversin is a component of node monocilia and other 9+0 cilia. *Development* 130:1725–34

15. Barnes BG. 1961. Ciliated secretory cells in the pars sistialis of the mouse hypophysis. *J. Ultrastruct. Res.* 5:453–61

16. Wheatley DN, Bowser SS. 2000. Length control of primary cilia: analysis of monociliate and multiciliate PtK1 cells. *Biol. Cell.* 92:573–82

17. Handel M, Schulz S, Stanaius A, Schreff M, Erdtmann-Viourliotis M, et al. 1999. Selective targeting of somatostatin receptor 3 to neuronal cilia. *Neuroscience* 89:909–26

18. Brailov I, Bancilia M, Brisorgueil M, Miquel M, Hamon M, Verge D. 2002. Localization of 5-HT6 receptor at the plasma membrane of neuronal cilia in the rat brain. *Brain Res.* 872:271–75

19. Pazour GJ, San Agustin JT, Follit JA, Rosenbaum JL, Witman GB. 2002. Polycystin-2 localizes to the kidney cilia and the ciliary level is elevated in ORPK mice with polycystic kidney disease. *Curr. Biol.* 12:R378–80

20. Yoder BK, Hou X, Guay-Woodford LM. 2002. The polycystic kidney disease proteins, polycystin-1, polycystin-2, polaris, and cystin, are co-localized in renal cilia. *J. Am. Soc. Nephrol.* 13:2508–16

21. Poole CA, Jensen CG, Snyder JA, Gray CG, Hermanutz VL, Wheatley DN. 1997. Confocal analysis of primary cilia structure and colocalization with the Golgi apparatus in chondrocytes and aortic smooth muscle cells. *Cell Biol. Int.* 21:483–94

22. Jensen CG, Poole CA, McGlashan SR, Marko M, Issa ZI. 2004. Ultrastructural, tomographic and confocal imaging of the chondrocyte primary cilium in situ. *Cell Biol. Int.* 28:101–10

23. Kellogg DR, Field CM, Alberts BM. 1989. Identification of microtubule-associated proteins in the centrosome, spindle, and kinetochore of the early *Drosophila* embryo. *J. Cell Biol.* 109:2977–91

24. Rosenbaum JL, Child FM. 1967. Flagellar regeneration in protozoan flagellates. *J. Cell Biol.* 34:345–64

25. Johnson KA, Rosenbaum JL. 1992. Polarity of flagellar assembly in *Chlamydomonas*. *J. Cell Biol.* 119:1605–11

26. Kozminski KG, Johnson KA, Forscher P, Rosenbaum JL. 1993. A motility in the eukaryotic flagellum unrelated to flagellar beating. *Proc. Natl. Acad. Sci. USA* 90:5519–23

27. Kozminski KG, Beech PL, Rosenbaum JL. 1995. The *Chlamydomonas* kinesin-like protein FLA10 is involved in motility associated with the flagellar membrane. *J. Cell Biol.* 131:1517–27

28. Pazour GJ, Wilkerson CG, Witman GB. 1998. A dynein light chain is essential for the retrograde particle movement of intraflagellar transport (IFT). *J. Cell Biol.* 141:979–92

29. Piperno G, Mead K. 1997. Transport of a novel complex in the cytoplasmic matrix of *Chlamydomonas* flagella. *Proc. Natl. Acad. Sci. USA* 94:4457–62

30. Cole DG, Diener DR, Himelblau AL, Beech PL, Fuster JC, Rosenbaum JL. 1998. *Chlamydomonas* kinesin-II-dependent intraflagellar transport (IFT): IFT particles contain proteins required for ciliary assembly in *Caenorhabditis elegans* sensory neurons. *J. Cell Biol.* 141:993–1008

31. Rosenbaum JL, Witman GB. 2002. Intraflagellar transport. *Nat. Rev. Mol. Cell Biol.* 3:813–25

32. Pazour GJ, Dickert BL, Vucica Y, Seeley ES, Rosenbaum JL, et al. 2000. *Chlamydomonas* IFT88 and its mouse homologue, polycystic kidney disease gene *tg737*, are required for assembly of

cilia and flagella. *J. Cell Biol.* 151:709–18

33. Haycraft CJ, Swoboda P, Taulman PD, Thomas JH, Yoder BK. 2001. The *C. elegans* homolog of the murine cystic kidney disease gene *Tg737* functions in a ciliogenic pathway and is disrupted in *osm-5* mutant worms. *Development* 128:1493–505

34. Allen C, Borisy GG. 1974. Structural polarity and directional growth of microtubules of *Chlamydomonas* flagella. *J. Mol. Biol.* 90:381–402

35. Binder LI, Dentler WL, Rosenbaum JL. 1975. Assembly of chick brain tubulin onto flagellar microtubules from *Chlamydomonas* and sea urchin sperm. *Proc. Natl. Acad. Sci. USA* 72:1122–26

36. Goodson HV, Kang SJ, Endow SA. 1994. Molecular phylogeny of the kinesin family of microtubule motor proteins. *J. Cell Sci.* 107:1875–84

37. Scholey JM. 1996. Kinesin-II, a membrane traffic motor in axons, axonemes, and spindles. *J. Cell Biol.* 133:1–4

38. Praetorius HA, Spring KR. 2003. Removal of the MDCK cell primary cilium abolishes flow sensing. *J. Membr. Biol.* 191:69–76

39. Dentler WL, Adams C. 1992. Flagellar microtubule dynamics in *Chlamydomonas*: cytochalasin D induces periods of microtubule shortening and elongation; and colchicine induces disassembly of the distal, but not proximal, half of the flagellum. *J. Cell Biol.* 117:1289–98

40. Schwartz EA, Leonard ML, Bizios R, Bowser SS. 1997. Analysis and modeling of the primary cilium bending response to fluid shear. *Am. J. Physiol. Renal Physiol.* 272:F132–38

41. Praetorius HA, Spring KR. 2001. Bending the MDCK cell primary cilium increases intracellular calcium. *J. Membr. Biol.* 184:71–79

42. Liu W, Xu S, Woda C, Kim P, Weinbaum S, Satlin LM. 2003. Effect of flow and stretch on the $[Ca^{2+}]_i$ response of principal and intercalated cells in cortical collecting duct. *Am. J. Physiol. Renal Physiol.* 285:F998–1012

43. Malnic G, Berliner RW, Giebisch G. 1989. Flow dependence of K^+ secretion in cortical distal tubules of the rat. *Am. J. Physiol. Renal Physiol.* 256:F932–41

44. Praetorius HA, Frokiaer J, Nielsen S, Spring KR. 2003. Bending the primary cilium opens Ca^{2+}-sensitive intermediate-conductance K^+ channels in MDCK cells. *J. Membr. Biol.* 191:193–200

45. Woda CB, Bragin A, Kleyman TR, Satlin LM. 2001. Flow-dependent K^+ secretion in the cortical collecting duct is mediated by a maxi-K channel. *Am. J. Physiol. Renal Physiol.* 280:F786–93

46. Nauli SM, Alenghat FJ, Luo Y, Williams E, Vassilev P, et al. 2003. Polycystins 1 and 2 mediate mechanosensation in the primary cilium of kidney cells. *Nat. Genet.* 33:129–37

47. Koulen P, Cai Y, Geng L, Maeda Y, Nishimura S, et al. 2002. Polycystin-2 is an intracellular calcium release channel. *Nat. Cell Biol.* 4:191–97

48. Montell C. 2001. Physiology, phylogeny, and functions of the TRP superfamily of cation channels. *Sci. STKE* 90:E1

49. Barr MM, DeModena J, Braun D, Nguyen CQ, Hall DH, Sternberg PW. 2001. The *Caenorhabditis elegans* autosomal dominant polycystic kidney disease gene homologs *lov-1* and *pkd-2* act in the same pathway. *Curr. Biol.* 11:1341–46

50. Scheffers MS, van der Bent P, van de Wal A, van Eendenburg J, Breuning MH, et al. 2004. Altered distribution and co-localization of polycystin-2 with polycystin-1 in MDCK cells after wounding stress. *Exp. Cell Res.* 292:219–30

51. Morgan D, Eley L, Sayer J, Strachan T, Yates LM, Craighead AS, Goodship JA. 2002. Expression analyses and interaction with the anaphase promoting complex protein Apc2 suggest a role for inversin in primary cilia and involvement in

the cell cycle. *Hum. Mol. Genet.* 11:3345–50

52. Nauli SM, Zhou J. 2004. Polycystins and mechanosensation in renal and nodal cilia. *BioEssays* 26:844–56

53. Fuchs JL, Schwark HD. 2004. Neuronal primary cilia: a review. *Cell Biol. Int.* 28:111–18

54. Whitfield JF. 2003. Primary cilium—is it an osteocyte's strain-sensing flowmeter? *J. Cell Biochem.* 89:233–37

55. Knothe Tate ML, Knothe U, Niederer P. 1998. Experimental elucidation of mechanical load-induced fluid flow and its potential role in bone metabolism and functional adaptation. *Am. J. Med. Sci.* 316:189–95

56. Knothe Tate ML, Steck R, Forwood MR, Niederer P. 2000. In vivo demonstration of load-induced fluid flow in the rat tibia and its potential implications for processes associated with functional adaptation. *J. Exp. Biol.* 203:2737–45

57. Smit TH, Burger EH, Huyghe JM. 2002. A case for strain-induced fluid flow as a regulator of BMU-coupling and osteonal alignment. *J. Bone Miner. Res.* 17:2021–29

58. Iomini C, Tejada K, Mo W, Vaananen H, Piperno G. 2004. Primary cilia of human endothelial cells disassemble under laminar shear stress. *J. Cell Biol.* 164:811–17

59. Ware RW, Clark D, Crossland K, Russell RL. 1975. Nerve ring of nematode *Caenorhabditis elegans*—sensory input

and motor output. *J. Comp. Neurol.* 162:71–110

60. Apfeld J, Kenyon C. 1999. Regulation of lifespan by sensory perception in *Caenorhabditis elegans. Nature* 402:804–9

61. Stepanyan Z, Kocharyan A, Pyrski M, Hubschle T, Watson AM, et al. 2003. Leptin-target neurons of the rat hypothalamus express somatostatin receptors. *J. Neuroendocrinol.* 15:822–30

62. Shyy JY, Chien S. 2003. Interactions of mechanotransduction pathways. *Biorheology* 40:47–52

63. Wang N. 1998. Mechanical interactions among cytoskeletal filaments. *Hypertension* 32:162–65

64. Maroto R, Hamill OP, Brefeldin. 2001. A block of integrin-dependent mechanosensitive ATP release from *Xenopus* oocytes reveals a novel mechanism of mechanotransduction. *J. Biol. Chem.* 276:23867–72

65. Praetorius J, Spring KR. 2002. Specific lectins map the distribution of fibronectin and beta 1-integrin on living MDCK cells. *Exp. Cell Res.* 276:52–62

66. Praetorius HA, Praetorius J, Frokiear J, Nielsen S, Spring KR. 2004. Beta1 integrins in the primary cilium of MDCK cells potentiate fibronectin induced Ca^{2+} signaling. *Am. J. Physiol. Renal Physiol.* In press

67. Aughsteen AA. 2001. The ultrastructure of primary cilia in the endocrine and excretory duct cells of the pancreas of mice and rats. *Eur. J. Morphol.* 39:277–83

Annu. Rev. Physiol. 2005. 67:531–55
doi: 10.1146/annurev.physiol.67.031103.154456
First published online as a Review in Advance on September 16, 2004

CELL SURVIVAL IN THE HOSTILE ENVIRONMENT OF THE RENAL MEDULLA

Wolfgang Neuhofer and Franz-X. Beck

*Department of Physiology, University of Munich, D-80336 Munich,
Germany; email: Wolfgang.Neuhofer@med.uni-muenchen.de;
FX.Beck@physiol.med.uni-muenchen.de*

Key Words   hypoxia, osmotic stress, osmolytes, urea, heat shock proteins

■ **Abstract**   The countercurrent system in the medulla of the mammalian kidney provides the basis for the production of urine of widely varying osmolalities, but necessarily entails extreme conditions for medullary cells, i.e., high concentrations of solutes (mainly NaCl and urea) in antidiuresis, massive changes in extracellular solute concentrations during the transitions from antidiuresis to diuresis and vice versa, and low oxygen tension. The strategies used by medullary cells to survive in this hostile milieu include accumulation of organic osmolytes and heat shock proteins, the extensive use of the glycolysis for energy production, and a well-orchestrated network of signaling pathways coordinating medullary circulation and tubular work.

INTRODUCTION

Depending on the hydration status, the mammalian kidney produces urine of widely varying osmolality. The ability to excrete a highly concentrated urine during states of water shortage is linked intimately to the specific transport properties and to the unique architecture of the tubular and vascular structures in the renal medulla, which form countercurrent multiplication (tubular structures) and countercurrent exchange (vascular structures) systems, respectively (1, 2). Although these unique features allow the establishment and maintenance of high medullary interstitial solute concentrations (predominantly NaCl and urea), countercurrent exchange in the vasa recta and the requirement for limited blood flow to and through the medulla necessarily entail unusually low oxygen tensions within this kidney region (3, 4). Hence, in antidiuresis the cells of the renal medulla function in an exceptionally inhospitable environment characterized by high salt and urea concentrations and low oxygen tensions. In addition, during the transitions from antidiuresis to diuresis and vice versa, these cells are challenged by massive changes in extracellular solute concentrations. In the following we describe some of the strategies that have evolved to allow renal medullary cells to survive and function in this hostile milieu.

CELL SURVIVAL IN A CHRONICALLY HYPOXIC ENVIRONMENT

The renal medullary circulation must satisfy the conflicting demands of preserving the cortico-medullary gradients of NaCl and urea while maintaining adequate oxygen and nutrient delivery to the medulla. This is achieved by arrangement of the vasa recta in a hairpin form, allowing countercurrent exchange of low-molecular-weight solutes and water between the ascending (AVR) and descending (DVR) vasa recta, thus minimizing washout of medullary solutes. The low medullary oxygen tension results from the relatively low flow rates within the vasa recta, from oxygen diffusion from the DVR to the AVR, and from oxygen consumption by tubular epithelium, in particular the medullary thick ascending limbs (mTAL) (3, 5–7). The partial pressure of oxygen is of the order of 10 and 20 mm Hg in the inner and outer medulla, respectively, compared with about 50 mm Hg in the cortex (3, 5, 7). Thus medullary hypoxia is an inevitable accompaniment of efficient urinary concentration. The medullary blood flow, if excessive, would disrupt the osmolality gradient; if too low, hypoxic injury of the tubules (mTAL) may occur. The DVR play an important role in the regulation of medullary blood flow because they distribute most or all of the blood flow from juxtamedullary cortex to the renal inner and outer medulla. In contrast to the AVR, the DVR possess pericytes, contractile smooth muscle-like cells that provide the basis for vasomotion. Local or systemic mediators work in concert to regulate medullary perfusion and tubular work, and most have the ability to enhance medullary blood flow and inhibit salt reabsorption (8, 9) (Figure 1).

Factors Influencing Medullary Oxygen Tension

In the following, we briefly summarize some of the most relevant autocrine or paracrine factors modulating oxygen supply and tubular work in the renal medulla. For a more comprehensive overview, the reader is referred to recent reviews (8–10).

PROSTAGLANDINS   Prostaglandin E_2, formed by cyclooxygenases-1 and -2 (COX-1, COX-2) in the renal medulla, both dilates renal medullary vessels (DVR) and reduces solute reabsorption by the mTAL and the collecting duct, thus improving the medullary oxygen balance (11–13). Prostaglandin E_2 inhibits the Na^+/K^+-ATPase in tubular epithelial cells and appears to mediate the vasodilatory actions of peptide hormones such as interleukin-1 and atrial natriuretic peptide (ANP) (12, 14). Whereas COX-1 is expressed constitutively in inner medullary collecting duct (IMCD), the tonicity-inducible COX-2 appears to be produced by interstitial cells of the outer and inner medulla and by IMCD and mTAL cells (15–17). Because COX-2 is induced by dehydration (16, 17) and COX-2 inhibition is associated with damage of medullary interstitial and collecting duct cells (18, 19), COX-2-derived prostanoids may buffer the vasoconstrictor effects of vasopressin or angiotensin II on the DVR during antidiuresis and volume depletion (20) (Figure 1).

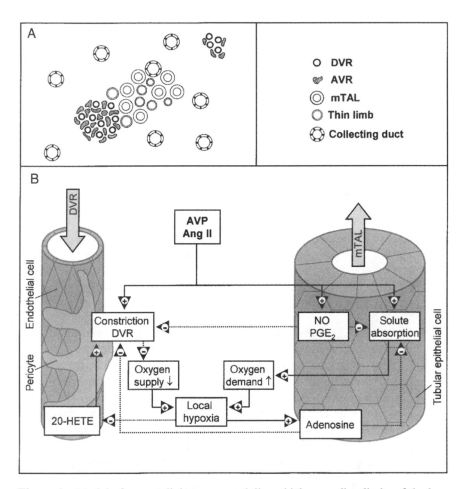

Figure 1   Model of cross-talk between medullary thick ascending limbs of the loop of Henle (mTAL) from long-looped nephrons and descending vasa recta (DVR). (*A*) Cross-section of the inner stripe of the outer medulla demonstrating the spatial tubulo-vascular relationship. (*B*) Schematic summary of tubulo-vascular cross-talk. Angiotensin II (Ang II) and arginine vasopressin (AVP) increase tubular solute re-absorption in the mTAL and reduce blood flow in the DVR by constrictor effects on pericytes, thus reducing oxygen availability in proximity to the mTAL. NO and prostaglandin E_2 (PGE_2), locally released by the mTAL, may buffer the constrictor effects on DVR and reduce tubular solute reabsorption. In addition, at low oxgen tension, formation of 20-hydroxyeicosatetraenoic acid (20-HETE) is reduced, whereas release of adenosine is increased, both resulting in elevated medullary blood flow. Adapted from (2, 8).

In addition, COX-2-derived prostanoids may promote the expression of heat shock proteins (HSPs) and accumulation of low-molecular-weight organic osmoeffectors (organic osmolytes) (18, 19). Another arachidonic acid-derived lipid mediator relevant for the renal medullary perfusion is 20-hydroxyeicosatetraenoic acid (20-HETE), formed by cytochrome P450-dependent pathways (21). 20-HETE is a potent vasoconstrictor and, interestingly, the local oxygen concentration is the rate-limiting step in 20-HETE production, suggesting that reduced formation of 20-HETE increases medullary blood flow in response to a decrease in tissue oxygenation (10, 22).

NITRIC OXIDE (NO)   Locally synthesized NO is a major player in the regulation of medullary blood flow and stimulates natriuresis by reducing Na^+ reabsorption in the collecting duct and the TAL (23–25). Considerable evidence indicates that the vasodilator NO, produced by mTAL cells, counterbalances circulating vasoconstrictors in control of the renal medullary circulation. In agreement, elevations in circulating levels of angiotensin II, norepinephrine, and vasopressin also increase medullary NO production, whereas pharmacological inhibition of NO production potentiates the vasoconstrictor effects of angiotensin II on DVR (26, 27). Studies in knockout mice indicate that NO synthase III is the isoform responsible for NO production by the TAL (28).

ADENOSINE   Adenosine is generated in the renal medulla under hypoxic conditions and constricts cortical vessels via A_1 receptors, reduces glomerular filtration rate (and thus tubular solute load), inhibits salt reabsorption by the mTAL, and increases medullary blood flow, effects likely to improve the medullary O_2 balance (29–32). The effect on medullary perfusion is probably mediated by vasodilatory A_2 receptors on DVR (29, 31). The finding that the mTAL produces adenosine (30) and its closeness to the DVR, which is located in the periphery of the outer medullary vascular bundles, suggests that adenosine released by the mTALs acts in a local feedback mechanism to increase medullary perfusion in response to elevated tubular work (33). Thus similar to prostaglandins and NO, adenosine is a vasodilator in the DVR that also inhibits salt reabsorption and thus oxygen consumption in the mTAL (Figure 1).

OTHER VASODILATORS   Kinins may also play a role in the control of medullary oxygen status because blockade of B_2 kinin receptors selectively reduces medullary blood flow and blunts the natriuretic response to volume expansion without affecting the glomerular filtration rate (34, 35). In contrast, the local application of bradykinin selectively increases medullary blood flow as well as Na^+ and water excretion (9). Several members of the ANP family including ANP itself and urodilatin probably elevate medullary oxygen tension by inhibition of IMCD Na^+ reabsorption and transient increase of medullary blood flow (36, 37).

VASOCONSTRICTORS   The most potent DVR vasoconstrictors are angiotensin II and endothelins, which cause intense focal constrictions along the DVR (9, 10).

Arginine vasopressin (AVP), on the other hand, is a mild vasoconstrictor that reduces inner medullary blood flow (via V_1 receptors) to a greater extent than outer medullary or cortical blood flow. The use of V_1-specific inhibitors has demonstrated that AVP-mediated reduction in inner medullary blood flow contributes to the antidiuretic action of AVP (38, 39). However, as discussed above, any of the vasoconstrictors may induce a negative feedback loop imposed by vasodilatory substances acting either in an autocrine or paracrine manner to prevent hypoxic damage to the renal medulla (Figure 1).

Medullary Hypoxia: Metabolic Implications

Given the relatively low oxygen tension in the inner medulla, it is not surprising that medullary cells have a far higher capacity for anaerobic glycolytic ATP production than proximal tubule cells (40, 41). Lactate production plays an important role in medullary cells in regenerating NAD^+ by reduction of pyruvate to lactate, thus providing the basis for ATP synthesis by anaerobic glycolysis (40, 41). In addition, lactate is subject to countercurrent exchange, and each glucose molecule is converted to two molecules of lactate by anaerobic glycolysis, thus producing net osmols (42). Hence, interstitial lactate concentrations increase along the cortico-medullary axis and may contribute to the high interstitial osmolality in this region (42).

Heat shock proteins (HSPs), some of which are expressed abundantly in the renal medulla, may represent an intrinsic molecular defense system for medullary cells. HSPs are expressed at low levels in unstressed cells and are induced vigorously in cells exposed to various types of cell stress, including hypoxia and hypertonicity, characteristic features of the renal medulla. HSPs are molecular chaperones that recognize and form complexes with incorrectly folded or denatured proteins, which ultimately leads to correct folding, compartmentalization, or degradation (43, 44). In addition, they inhibit apoptosis by interfering with the apoptosis signaling cascade at several levels (45, 46). By these mechanisms, HSPs contribute to cell survival under otherwise lethal conditions.

HSPs that are more highly expressed in the renal medulla than in the cortex include HSP70, osmotic stress protein (OSP) 94 (OSP94), HSP110, HSP27, and the HSP27 homolog αB-crystallin (47–51). For each of these HSPs, experimental evidence supporting a protective role under hypoxic conditions or energy depletion has been obtained in various cell types (52–56).

CELL SURVIVAL IN HIGH EXTRACELLULAR SALT CONCENTRATIONS

With few exceptions, cell membranes are highly water permeable, implying that medullary cells will rapidly equilibrate osmotically with their hypertonic interstitium in antidiuresis (1). In the steady state, medullary cells achieve this without major changes in intracellular ionic strength by accumulating metabolically neutral organic osmolytes (see below).

Acute Adaptation

After prolonged diuresis (a situation associated with a drastic reduction of intracellular organic osmolytes), and the countercurrent system is stimulated by infusion of AVP, and interstitial NaCl concentration at the tip of the papilla doubles within 2 h (57). The increased extracellular NaCl concentration causes cell shrinkage and a nonspecific rise in the concentration of all intracellular solutes, i.e., of intracellular ionic strength (57). The mTAL and IMCD cells respond rapidly to an increase in extracellular tonicity by activating transmembrane transport pathways mediating the uptake of NaCl. This results in osmotically obliged water entry and recovery of cell volume (regulatory volume increase). NaCl uptake is achieved by parallel operation of ADH/cAMP-dependent Na^+/H^+ and Cl^-/HCO_3^- exchangers in the basolateral membranes (58–63). In IMCD cells, an amiloride-sensitive cation channel may contribute to apical Na^+ influx (59, 64). Assuming similar activation of these transporters in the inner medulla in vivo, the increase in intracellular Na^+ concentration in papillary collecting duct cells after an acute rise in interstitial NaCl concentration is quite moderate compared with the substantial increase in intracellular K^+ concentration (57). This may reflect acute activation of the Na^+/K^+-ATPase by the enhanced availability of Na^+. Hence, Na^+ entering the cells would appear to be replaced eventually by K^+ from the interstitial compartment, thus elevating intracellular K^+ concentration. Interestingly, hypertonicity-induced activation of the Na^+/K^+-ATPase is associated with increased oxidative stress in mTAL cells (65).

Chronic Adaptation

Renal medullary cells chronically exposed to elevated extracellular NaCl concentrations activate a variety of protective mechanisms that allows them to survive in this inhospitable environment.

SODIUM/POTASSIUM-ATPase   Prolonged exposure of renal tubular cells to elevated extracellular concentrations of non-permeant solutes leads to enhanced expression of the Na^+/K^+-ATPase α_1, β_1, γ_a, and γ_b subunits and to higher Na^+/K^+-ATPase activity (66–68). Renal medullary cells are able to maintain an intracellular Na^+ concentration not substantially higher than in cortical cells, despite greatly elevated extracellular NaCl concentrations. Thus they are able to maintain transmembrane transport pathways indispensable for cell homeostasis [Na^+/Ca^{2+}, Na^+/H^+ exchangers, sodium/*myo*-inositol cotransporter (SMIT), betaine/GABA transporter (BGT)] and transepithelial electrolyte transport (Na channels). Enhanced expression of the γ subunits is vitally important for IMCD cells exposed to hypertonic stress, and suppression of this response greatly reduces survival of IMCD cells under this condition (68, 69). The up-regulation of the γ subunits, but not of the α_1

subunit, depends on Cl^- entry and activation of c-Jun NH_2-terminal kinase 2 and phosphatidylinositol 3-kinase and contributes to the survival of cells challenged with hypertonic stress (68, 69).

Shrinkage-induced activation of ion transport pathways can restore cell volume but does not, a priori, normalize the elevated intracellular ionic strength. A chronic increase in intracellular ionic strength, however, may have numerous adverse effects on cell function such as DNA damage, growth arrest, inhibition of protein synthesis, and disturbance of physiological transmembrane electrolyte gradients (70, 71).

SURVIVAL DESPITE TONICITY-INDUCED DNA DAMAGE   Exposure of cultured cells to elevated concentrations of NaCl, but less so to membrane-permeant solutes (i.e., glycerol or urea), may cause chromosome aberrations, sister-chromatid exchanges, and DNA single- and double-strand breaks (70, 72). These phenomena are accompanied by growth arrest and increased incidence of apoptosis (73, 74). Although the exact mechanism of hypertonicity-induced increase in DNA damage is not clear, both direct breakage of the phosphodiester backbone and inhibition of DNA damage repair may contribute. Specifically, hypertonic stress impedes the translocation of Ku70 and Ku80 proteins from the cytosol to the nucleus in rat fibroblasts and displaces Mre11, normally a nuclear protein, from the nucleus of mIMCD3 cells and mouse embryonic fibroblasts (75, 76). All these proteins participate in the nuclear repair of DNA double-strand breaks (77). The p53 tumor suppressor plays a decisive role in initiating the cell cycle delay after hypertonic stress, thus possibly providing sufficient time for repair of DNA damage or, otherwise, for initiating apoptosis, which could prevent the propagation of damaged DNA (78, 79). In response to hypertonic stress, the abundance of p53 increases because of reduced proteasome-mediated degradation, and its transcriptional activity rises (79). Interestingly, the p38 kinase is indispensable for hypertonicity-induced p53 activation and cell cycle delay (79, 80). Because hypertonic stress hampers repair of DNA damage and because DNA damage may occur during normal cell replication, it is not surprising that cells with high rates of DNA replication display increased incidence of hypertonicity-induced DNA damage and apoptosis (81). Conversely, the low proliferation rate of inner medullary cells in the kidney in vivo may contribute to the high resistance of these cells to NaCl-induced apoptosis (81). The comparatively high proliferation rates of cells in culture complicate the extrapolation of findings from most of the studies cited to renal medullary cells in vivo.

Gradual accumulation of metabolically neutral organic osmolytes, particularly the trimethylamines [glycerophosphorylcholine (GPC) and betaine] and the polyols (*myo*-inositol and sorbitol) allows K^+ and Cl^- concentrations to normalize, thus restoring normal intracellular ionic strength (82).

GPC   In antidiuresis, GPC content is low in the cortex, intermediate in the outer medulla, and highest in the papilla (83, 84). GPC is liberated from phosphatidylcholine by phospholipase A_1, phospholipase A_2, and lysophospholipase (85, 86). Studies on cultured renal epithelial cells, however, suggest that, following acute hyperosmotic stress, the intracellular GPC concentration rises primarily because of reduced GPC degradation (86, 87). The relevance of low rates of GPC degradation for intracellular GPC content in the intact kidney is highlighted by the much lower activity of GPC-degrading enzymes in papillary than in cortical tissue (83). Both elevated NaCl and urea concentrations cause a rapid decrease in the activity of GPC:choline phosphodiesterase, which catalyzes the degradation of GPC to choline and glycerol 3-phosphate (87). However, whereas the inhibitory effect of high urea concentrations on GPC:choline phosphodiesterase activity is sustained, elevated NaCl concentrations inhibit this enzyme only transiently (87). The lessening of the inhibitory effect by elevated NaCl concentrations is probably through the gradual accumulation of organic osmolytes counteracting the adverse influences of cell shrinkage, elevated intracellular ionic strength, or/and increased intracellular crowding of macromolecules on GPC:choline phosphodiesterase activity. This notion is supported by the observation that in MDCK cells, exposure to hypertonic medium containing millimolar concentrations of betaine and *myo*-inositol leads to increased contents of these compatible osmolytes, attenuation of GPC:choline phosphodiesterase inhibition, and a concomitant decrease in GPC content (88). In the medulla of the concentrating kidney in vivo, however, urea-induced inhibition of GPC degradation may prevail because the inhibitory effect of high urea concentrations predominates. The close relation between medullary urea and GPC contents is consistent with this notion (89, 90). Because prolonged exposure of MDCK cells to hypertonic, NaCl-containing medium increases phospholipase activity (91), enhanced production of GPC may also contribute to intracellular GPC accumulation in the renal medulla. There is as yet no evidence for enhanced GPC uptake from the extracellular compartment.

BETAINE   Betaine, like GPC, is a trimethylamine. The distribution of betaine in the concentrating kidney, however, clearly differs from that of GPC. Compared with GPC, betaine contents rise less steeply between the cortico-medullary boundary and the tip of the papilla, and in the outermost part of the inner medulla, betaine contents may even be lower than in the adjacent outer medulla (84, 90, 92). Intracellular accumulation of betaine is mediated by uptake from the extracellular fluid via BGT-1, a protein of 614 amino acids with sorting signals and basolateral retention motifs in the cytosolic C-terminal domain (93–96). Accordingly, in MDCK cells, BGT-1 expressed after hypertonic stress is inserted primarily into the basolateral membrane (97–99). However, evidence for a preferentially apical localization of BGT-1 has been obtained in cells derived from the TAL (100). BGT-1 combines the uphill transport of betaine with the downhill import of 3 Na^+ and 1 or 2 Cl^- (94, 101).

The production of betaine from choline via betaine aldehyde by the sequential action of choline dehydrogenase and betaine aldehyde dehydrogenase is much higher in the proximal tubule, notably the pars recta of the proximal tubule, than in any other tubule segment (102–104). Evidence suggests that during hypertonic stress, the increased demand of inner medullary cells for betaine may be met, at least partly, by enhanced production by cortical cells (103, 105). It is thus tempting to speculate that betaine synthesized by proximal tubule cells is released into the tubular fluid and/or into the interstitial/vascular compartment and finally taken up by medullary cells.

The required stimulation of betaine uptake in response to hypertonic stress is achieved primarily by enhanced transcription of the BGT-1 gene, as reflected by the elevated abundance of BGT-1 mRNA in the renal medulla (106–109). Transcriptional, tonicity-responsive enhancer-binding protein (TonEBP)/nuclear factor of activated T-lymphocytes 5 (NFAT5)-dependent up-regulation of BGT-1 expression is of prime importance in this context (110). However, adjustment of betaine cotransport activity by protein kinases A and C and modulation of betaine uptake via changes in microtubule-mediated BGT-1 transfer to the plasma membrane may also participate in the control of betaine uptake (111, 112). In addition, evidence indicates a role of the actin-based microfilament network in the regulation of BGT-1 activity (113).

SORBITOL   In cultured renal epithelial cells, the production of this polyol from glucose by aldose reductase (AR) is subject to pronounced osmotic regulation (114, 115). Consistent with these results obtained on cells in culture are observations in the kidney in situ demonstrating that, in the cortex and outer medulla of the concentrating kidney, sorbitol is usually not detectable. However, along with the abundance of both AR mRNA and protein, sorbitol content increases steeply from the outer-inner medullary boundary to the tip of the papilla, thus paralleling the rise in interstitial solute concentrations (84, 92, 109, 116–118). The hypertonicity-induced increase in AR expression, and hence sorbitol production, depends heavily on TonEBP/NFAT5 binding to several TonEs in the 5'-flanking region of the AR gene (110, 119–121). In contrast, conversion of sorbitol to fructose by sorbitol dehydrogenase is regulated only weakly by changes in extracellular tonicity (109, 122, 123).

The gradual increase in the concentrating ability of the mammalian kidney after birth is accompanied by a rise in AR mRNA and AR immunoreactivity in the inner medulla (118, 124). During this developmental period, immature papilla-resident TALs are transformed into mature ascending thin limbs (ATLs). This process involves the differentiation of AR-positive TAL cells into mature ATL cells and apoptotic deletion of AR-negative cells (118). The postnatal rise in medullary tonicity may contribute to initiation of the apoptotic program in those "non-protected" cells (118).

myo-INOSITOL   In the concentrating kidney, myo-inositol content is usually highest in the outer medulla and tends to decline toward the tip of the papilla (57, 84, 116). Hypertonicity-induced accumulation of myo-inositol by renal medullary cells is achieved by enhanced TonEBP/NFAT5-dependent expression of SMIT (110, 121,

125–127). This protein of 718 amino acids belongs to the Na^+/glucose cotransporter family, is thought to contain 14 transmembrane domains, and couples the uptake of 1 *myo*-inositol to the entry of 2 Na^+ (126, 128–130). In renal epithelial cells, Na^+-dependent *myo*-inositol uptake proceeds either predominantly across the basolateral membrane [Madin-Darby canine kidney (MDCK) cells] (98) or across both basolateral and apical membranes (mTAL-derived cells) (100). Northern blot analyses on kidneys from antidiuretic animals demonstrate much higher SMIT mRNA abundances in outer and inner medulla than in cortex (109, 131, 132). Both in outer and inner medulla, SMIT mRNA abundance decreases significantly after induction of diuresis (109, 131).

FREE AMINO ACIDS   In the steady state of osmotic adaptation of renal medullary cells to high interstitial NaCl concentrations in antidiuresis, free amino acids, including taurine, contribute less to the pool of organic osmolytes than either trimethylamines or polyols (57, 89, 133). At first glance, this is somewhat surprising because free amino acids are used widely as intracellular osmolytes by animal cells exposed to hypertonic stress, and uptake of free amino acids by renal epithelial cells via system A and the Na^+- and Cl^--dependent taurine transporter (TauT) is activated by hypertonic stress (134–136). In addition, TauT mRNA abundance is higher in antidiuresis in both the outer medulla and papilla of the rat kidney than it is in diuresis (137). However, TonEBP does not appear to play a major role in regulating the transcription of the TauT gene (121). It has been proposed that uptake of free amino acids precedes the relatively slow intracellular accumulation of trimethylamines and polyols and abates gradually as the intracellular concentrations of the latter rise (135, 136, 138).

CELL SURVIVAL IN THE PRESENCE OF FLUCTUATING EXTRACELLULAR SALT CONCENTRATIONS

Medullary cells must adapt not only to high but also to widely fluctuating extracellular solute concentrations. During the transition from antidiuresis to diuresis, when extracellular tonicity falls rapidly, medullary cells release organic osmolytes quickly, thus avoiding undue cell swelling (116, 139). Inhibition of this adaptive response enhances expression of HSPs in the rat renal medulla, which underscores the importance of transmembrane osmolyte movement for volume regulation of medullary cells (140). The future molecular identification of the transmembrane pathway(s) mediating this fast initial response of medullary cells to declining extracellular solute concentrations will allow a better understanding of osmolyte efflux from medullary cells (141, 142). Adaptation of medullary cells to chronically lower extracellular tonicities (relative to strict antidiuresis) involves the eventual downregulation of tonicity-sensitive genes, such as those encoding BGT-1, AR, SMIT, and HSP70, and reflects the diminished requirement for organic osmolytes and stress proteins (115, 133). In contrast to the fast response to falling extracellular solute concentrations, adaptation of medullary cells to rising extracellular salt and urea concentrations during the transition from diuresis to antidiuresis (see above) takes many hours or even days (89).

CELL SURVIVAL IN HIGH UREA CONCENTRATIONS

The renal concentrating mechanism generates urea concentrations higher than those in any other organ. Because, with few exceptions, cell membranes are relatively permeable to this solute, the intracellular urea concentration is comparably high (133). In concentrations of the order of those reached in the papilla of the concentrating kidney of many mammals, urea is a potent protein-destabilizing agent and may compromise protein structure and function. Such a high urea concentration has been shown repeatedly to reduce the activity of various enzymes, including pyruvate kinase, lactate dehydrogenase, and aldose reductase, enzymes that are essential for papillary cells (143–146). Furthermore, high concentrations of urea induce apoptosis in cultured medullary cells (74, 147, 148). Trimethylamine osmolytes (e.g., betaine, GPC) are believed to ameliorate the deleterious effects of high urea concentrations on protein structure and function, and the loss in enzyme activity in the presence of high urea concentrations is attenuated in the presence of betaine or GPC (82, 146, 149). On the other hand, more recent studies indicate that trimethylamines may even aggravate the negative effects of high urea concentrations on macromolecular structure and function under certain conditions, so that the role and effectiveness of trimethylamine osmolytes in protecting renal medullary cells against high urea concentrations is again under debate (144, 150).

Recently, it has become increasingly clear that specific HSPs, some of which are expressed abundantly in the renal inner medulla, are key factors in the resistance of papillary cells against high urea concentrations. The expression of HSP70 is much higher in the hyperosmotic renal medulla than in the iso-osmotic cortex, and its abundance correlates with the diuretic state (151, 152). In the renal papilla, HSP70 contributes 0.5% of total cellular protein (145, 153). Experiments involving forced over-expression or down-regulation of HSP70 indicate that HSP70 protects medullary cells from the detrimental effects of high urea concentrations by counteracting urea-mediated inhibition of enzymes and protection from urea-induced apoptosis (145, 148, 154). The underlying mechanism preventing loss of enzyme activity is probably related to the chaperoning functions of HSP70. Urea is a potent protein-unfolding agent and HSP70, by binding reversibly to hydrophobic side chains exposed by unfolded or partially unfolded proteins, promotes reestablishment of the correct conformation, thereby preventing incorrect intramolecular interactions, intermolecular aggregation, and irreversible loss of function (44). In addition, HSP70 prevents the execution of the apoptotic pathway by several mechanisms including inhibition of Apaf-1 and cytochrome-c release from mitochondria and subsequent processing of procaspase-9 (45, 46, 155). The relevance of these properties for protection against urea-induced apoptosis in the kidney needs to be addressed further, because exposure of IMCD cells to high urea concentrations is not associated with mitochondrial dysfunction or cytochrome-c release (156). Recent studies have demonstrated that cultured IMCD cells exposed continuously to hypertonic medium, and thus constitutively expressing high amounts of HSP70, are more resistant to heat, H_2O_2, cyclosporine A, and apoptotic inducers than IMCD cells kept in isotonic medium, which further supports a functional role for HSP70 in the renal papilla (157). In addition, targeted disruption of the tonicity-inducible HSP70 gene in mice results in an increased incidence of apoptosis in the renal medulla following osmotic stress (158).

Other stress proteins much more abundant in the renal inner medulla than in the cortex include HSP27 and the structurally similar αB-crystallin, OSP94, and HSP110. Because these chaperones are induced by tonicity stress in both cultured renal medullary cells in vitro and in the renal papilla in vivo after water deprivation (49, 51, 151, 152, 159, 160), it seems likely that they play a significant biological role in the hyperosmotic renal papilla. Experimental evidence for a protective role in the context of hyperosmolality has been obtained only for αB-crystallin, which protects glial cells from acute hypertonic stress (161).

In addition to its destabilizing effects on cellular macromolecules, urea also induces oxidative stress in several cell types, as evidenced by the appearance of 8-oxoguanine lesions and single-strand breaks in genomic DNA after urea exposure (162, 163). Further urea-mediated effects indicative of oxidative stress include decreased cellular content of reduced glutathione (163), increased expression of growth arrest and DNA damage-inducible gene-153 (GADD153) (163), and increased expression of heme oxygenase-1 (164), which protects against oxidative stress (162). Also supporting the view that renal medullary cells may be exposed to oxidative stress in vivo is the observation that heme oxygenase-1, which is primarily an inducible isoform, is expressed constitutively in the renal inner medulla and its abundance increases from cortex to papilla (165, 166). Under physiological conditions, urea dissociates spontaneously into ammonia and cyanate (167). The reactive form of cyanate, isocyanic acid, can carbamylate various proteins (167), which results in altered biological activity (168–171) and produces modified proteins with properties similar to those of molten-globule intermediates in protein folding and unfolding pathways (168–171). In particular, the latter effect may necessitate the constitutive expression of molecular chaperones as present in the renal medulla.

FUNCTIONAL NETWORK OF GENES REGULATED BY HYPERTONICITY

High interstitial NaCl concentrations during antidiuresis not only provide the driving force for osmotic water abstraction along the collecting ducts but also induce a variety of genes that are essential either for the production of concentrated urine or for prevention of damage to medullary cells (Figure 2). TonEBP/NFAT5, a rel-like

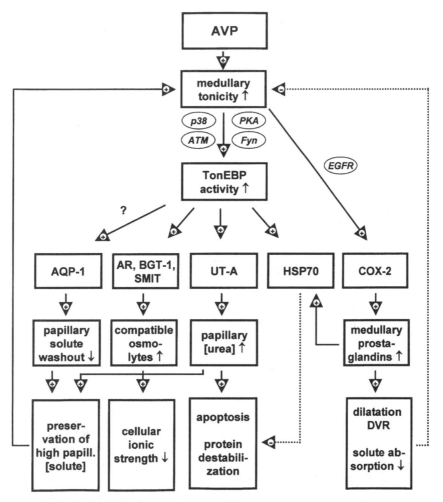

Figure 2 Functional network of medullary genes regulated by tonicity. Hypertonic stress increases the transcriptional activity of the tonicity-responsive, enhancer-binding protein (TonEBPNFAT5), which leads to the enhanced expression of various target genes. Induction of aldose reductase (AR), the betaine/GABA transporter (BGT-1) and the sodium/*myo*-inositol cotransporter (SMIT) entails intracellular accumulation of organic osmolytes, thus protecting the cells from hypertonicity. Increased abundance of urea transporters (UT-A1, UT-A3) is essential for the establishment of high papillary urea concentrations. The 70-kDa heat shock protein (HSP70) protects against the toxic effects of high urea concentrations. Induction of aquaporin-1 (AQP-1) mediates effective H_2O abstraction from the (outer medullary) descending loop of Henle and countercurrent flux of H_2O from the DVR to the AVR, thus preventing papillary solute washout. The osmotic induction of cyclooxygenase-2 (COX-2) appears to be independent of TonEBP. COX-2-derived prostanoids stimulate HSP70 expression and reduce medullary solute absorption.

transcription factor, stimulates transcription of genes involved in the intracellular accumulation of organic osmolytes (BGT-1, SMIT, AR), thus (as noted above) allowing osmotic equilibration of medullary cells with the hypertonic interstitium, while maintaining essentially normal intracellular ionic strength (115, 120, 172, 173). TonEBP/NFAT5 also induces transcription of HSP70, which confers resistance against the potentially lethal effects of high urea concentrations that are present during antidiuresis (110, 148, 174). In addition, TonEBP/NFAT5 stimulates transcription of the papillary UT-A1 and UT-A3 urea transporters, which are essential for generating high interstitial urea concentrations during antidiuresis (175). In the 5′-flanking region of these TonEBP target genes, there are multiple TonEs to which TonEBP binds and stimulates transcription in response to elevated ambient tonicities (Figure 2). Hypertonicity-induced activation of TonEBP entails its dimerization, hyperphosphorylation, nuclear translocation and, at least in cell cultures, enhanced expression (172, 173, 176–179). Evidence suggests the involvement of various kinases (p38 MAPK, Fyn, PKA and ATM) in TonEBP activation (180–182), although the precise signaling pathways are largely unknown. Apart from changes in intracellular ionic strength, additional factors such as changes in cell volume or/and intracellular macromolecular crowding may modulate TonEBP activity (120).

However, other genes, whose regulation is more complex than that of classic TonEBP/NFAT5 target genes, are indispensable for the proper function of the renal medulla during antidiuresis (Figure 2). AQP-1 is expressed in the descending limb of the loop of Henle and in the DVR where it promotes H_2O abstraction and countercurrent H_2O exchange from the DVR to the AVR, thus preventing excessive H_2O entry into the papillary interstitium and dilution of the high interstitial solute concentrations. Transgenic mice lacking AQP-1 water channels displayed a severe urinary concentrating defect (183). Interestingly, the abundance of AQP-1 increases with tonicity both in vivo and in vitro, which is mediated by a hypertonicity-response element different from that described in other TonEBP/NFAT5 target genes (184). Although it is not clear whether TonEBP/NFAT5 drives AQP-1 expression directly in response to hypertonicity, tonicity-mediated up-regulation of AQP-1 is essential for urinary concentration (184). Furthermore, tonicity-induced induction of COX-2 in the renal papilla appears to be independent of TonEBP/NFAT5, as is the case for TauT (121, 185). Nevertheless, COX-2 may play an important role during antidiuresis because its inhibition during water deprivation is associated with damage to papillary interstitial and collecting duct cells (18, 19).

CONSEQUENCES OF IMPAIRED ADAPTATION

The significance of tonicity-induced accumulation of organic osmolytes and up-regulation of HSP expression for the survival of medullary cells and for the integrity of the renal medulla is underscored by studies on TonEBP/NFAT5-null mice, on mice that specifically lack the tonicity-sensitive *hsp70.1* gene, and on rats in which uptake of *myo*-inositol is specifically inhibited by 2-*O*, *C*-methylene-*myo*-inositol (121, 158, 186). These animals display atrophy of the kidney medulla (121), increased incidence of apoptosis of medullary cells, especially following NaCl-loading (121, 158), or even acute renal failure associated with severe injury of outer medullary cells (186).

STUDIES ON CELL CULTURES: COMPLICATION BY SIMPLIFICATION

Most studies on cell cultures cited so far have employed step increases in extracellular solute concentration when the osmolality of the incubation medium is to be elevated. However, recent studies have shown that a gradual rise in extracellular NaCl concentration, which more closely resembles the situation of medullary cells in vivo, is better tolerated by cultured medullary cells and increases mRNA abundance of a number of tonicity-sensitive genes (AR, BGT-1, HSP70, OSP94) more efficiently than an equivalent step increase (187, 188). In addition, high urea or K^+ concentrations, as present in the extracellular spaces of the inner medulla in antidiuresis (89, 189), may modify NaCl-elicited responses of cultured renal epithelial cells substantially (120, 187, 190). Hence, considerable caution must be exercised in extrapolating results on cell adaptation to and survival in elevated extracellular solute concentrations obtained from reduced (i.e., simplified) systems such as cultured cells to medullary cells in the kidney in vivo.

ACKNOWLEDGMENTS

Work in the authors' laboratory was supported by the Deutsche Forschungsgemeinschaft, the Deutsche Nierenstiftung, and the Münchener Medizinische Wochenschrift. We thank Dr. J. Davis for his help.

The *Annual Review of Physiology* is online at http://physiol.annualreviews.org

LITERATURE CITED

1. Sands JM, Layton HE. 2000. Urine concentrating mechanism and its regulation. In *The Kidney: Physiology and Pathophysiology*, ed. DW Seldin, G Giebisch, pp. 1175–216. Philadelphia: Lippincott Williams & Wilkins. 3rd. Ed.
2. Kriz W, Kaissling B. 2000. Structural organization of the mammalian kidney. See Ref. 1, pp. 587–654
3. Baumgartl H, Leichtweiss HP, Lübbers DW, Weiss C, Huland H. 1972. The oxy-gen supply of the dog kidney: measurement of intrarenal pO_2. *Microvasc. Res.* 4:247–57
4. Silva P. 1987. Renal fuel utilization, energy requirements, and function. *Kidney Int.* 32(Suppl. 22):S9–14
5. Brezis M, Heyman SN, Epstein FH. 1994. Determinants of intrarenal oxygenation. II. Hemodynamic effects. *Am. J. Physiol. Renal Physiol.* 267:F1063–68

6. Brezis M, Rosen S. 1995. Hypoxia of the renal medulla—its implications for disease. *N. Engl. J. Med.* 332:647–55

7. Ostensen J, Stokke ES. 1996. Energy requirement of sodium reabsorption in the thick ascending limb of Henle's loop in the dog kidney: effects of bumetanide and ouabain. *Acta Physiol. Scand.* 157:275–81

8. Cowley AW Jr, Mori T, Mattson D, Zou AP. 2003. Role of renal NO production in the regulation of medullary blood flow. *Am. J. Physiol. Regul. Integr. Comp. Physiol.* 284:R1355–69

9. Pallone TL, Zhang Z, Rhinehart K. 2003. Physiology of renal medullary microcirculation. *Am. J. Physiol. Renal Physiol.* 284:F253–66

10. Pallone TL, Silldorff EP. 2001. Pericyte regulation of renal medullary blood flow. *Exp. Nephrol.* 9:165–70

11. Stokes JB, Kokko JP. 1977. Inhibition of sodium transport by prostaglandin E_2 across the isolated, perfused rabbit collecting tubule. *J. Clin. Invest.* 59:1099–104

12. Jabs K, Zeidel ML, Silva P. 1989. Prostaglandin E_2 inhibits Na^+-K^+-ATPase activity in the inner medullary collecting duct. *Am. J. Physiol. Renal Physiol.* 257:F424–30

13. Lear S, Silva P, Kelley VE, Epstein FH. 1990. Prostaglandin E_2 inhibits oxygen consumption in rabbit medullary thick ascending limb. *Am. J. Physiol. Renal Physiol.* 258:F1372–78

14. Zeidel ML. 1993. Hormonal regulation of inner medullary collecting duct sodium transport. *Am. J. Physiol. Renal Physiol.* 265:F159–73

15. Khan KN, Venturini CM, Bunch RT, Brassard JA, Koki AT, et al. 1998. Interspecies differences in renal localization of cyclooxygenase isoforms: implications in nonsteroidal antiinflammatory drug-related nephrotoxicity. *Toxicol. Pathol.* 26:612–20

16. Yang T, Schnermann JB, Briggs JP. 1999. Regulation of cyclooxygenase-2 expression in renal medulla by tonicity in vivo and in vitro. *Am. J. Physiol. Renal Physiol.* 277:F1–9

17. Hao CM, Yull F, Blackwell T, Kömhoff M, Davis LS, et al. 2000. Dehydration activates an NF-κB-driven, COX2-dependent survival mechanism in renal medullary cells. *J. Clin. Invest.* 106:973–82

18. Moeckel GW, Zhang L, Fogo AB, Hao C-M, Pozzi A, et al. 2003. COX2 activity promotes organic osmolyte accumulation and adaptation of renal medullary interstitial cells to hypertonic stress. *J. Biol. Chem.* 278:19352–57

19. Neuhofer W, Holzapfel K, Fraek ML, Ouyang N, Lutz J, et al. 2004. Chronic COX-2 inhibition reduces medullary HSP70 expression and induces papillary apoptosis in dehydrated rats. *Kidney Int.* 65:431–41

20. Qi Z, Hao CM, Langenbach RI, Breyer RM, Redha R, et al. 2002. Opposite effects of cyclooxygenase-1 and -2 activity on the pressor response to angiotensin II. *J. Clin. Invest.* 110:61–69

21. Carroll MA, Sala A, Dunn CE, McGiff JC, Murphy RC. 1991. Structural identification of cytochrome P450-dependent arachidonate metabolites formed by rabbit medullary thick ascending limb cells. *J. Biol. Chem.* 266:12306–12

22. Harder DR, Narayanan J, Birks EK, Liard JF, Imig JD, et al. 1996. Identification of a putative microvascular oxygen sensor. *Circ. Res.* 79:54–61

23. Biondi ML, Dousa T, Vanhoutte P, Romero JC. 1990. Evidences for the existence of endothelium-derived relaxing factor in the renal medulla. *Am. J. Hypertens.* 3:876–78

24. Garcia NH, Pomposiello SI, Garvin JL. 1996. Nitric oxide inhibits ADH-stimulated osmotic water permeability in cortical collecting ducts. *Am. J. Physiol. Renal Physiol.* 270:F206–10

25. Ortiz PA, Garvin JL. 2002. Role of nitric oxide in the regulation of nephron

transport. *Am. J. Physiol. Renal Physiol.* 282:F777–84

26. Brezis M, Heyman SN, Dinour D, Epstein FH, Rosen S. 1991. Role of nitric oxide in renal medullary oxygenation. Studies in isolated and intact kidneys. *J. Clin. Invest.* 88:390–95

27. Alberola AM, Salazar FJ, Nakamura T, Granger JP. 1994. Interaction between angiotensin II and nitric oxide in control of renal hemodynamics in conscious dogs. *Am. J. Physiol. Regul. Integr. Comp. Physiol.* 267:R1472–78

28. Plato CF, Shesely EG, Garvin JL. 2000. eNos mediates L-arginine-induced inhibition of thick ascending limb chloride flux. *Hypertension* 35:319–23

29. Dinour D, Brezis M. 1991. Effects of adenosine on intrarenal oxygenation. *Am. J. Physiol. Renal Physiol.* 261:F787–91

30. Beach RE, Good DW. 1992. Effects of adenosine on ion transport in rat medullary thick ascending limb. *Am. J. Physiol. Renal Physiol.* 263:F482–87

31. Agmon Y, Dinour D, Brezis M. 1993 Disparate effects of adenosine A_1- and A_2-receptor agonists on intrarenal blood flow. *Am. J. Physiol. Renal Physiol.* 265:F802–6

32. Zou AP, Nithipatikom K, Li PL, Cowley AW Jr. 1999. Role of renal medullary adenosine in the control of blood flow and sodium excretion. *Am. J. Physiol. Regul. Integr. Comp. Physiol.* 276:R790–98

33. Silldorff EP, Kreisberg MS, Pallone TL. 1996. Adenosine modulates vasomotor tone in outer medullary descending vasa recta of the rat. *J. Clin. Invest.* 98:18–23

34. Fenoy FJ, Roman RJ. 1992. Effect of kinin receptor antagonists on renal hemodynamic and natriuretic responses to volume expansion. *Am. J. Physiol. Regul. Integr. Comp. Physiol.* 263:R1136–40

35. Mattson DL, Cowley AW. 1993. Kinin actions on renal papillary blood flow and sodium excretion. *Hypertension* 21:961–65

36. Hansell P, Ulfendahl HR. 1986. Atriopeptins and renal cortical and papillary blood flow. *Acta Physiol. Scand.* 127:349–57

37. Brenner BM, Ballermann BJ, Gunning ME, Zeidel ML. 1990. Diverse biological actions of atrial natriuretic peptide. *Physiol. Rev.* 70:665–99

38. Franchini KG, Cowley AW Jr. 1996. Renal cortical and medullary blood flow responses during water restriction: role of vasopressin. *Am. J. Physiol. Regul. Integr. Comp. Physiol.* 270:R1257–64

39. Turner MR, Pallone TL. 1997. Vasopressin constricts outer medullary descending vasa recta isolated from rat kidneys. *Am. J. Physiol. Renal Physiol.* 272:F147–51

40. Lee JB, Peter HM. 1969. Effect of oxygen tension on glucose metabolism in rabbit kidney cortex and medulla. *Am. J. Physiol.* 217:1464–71

41. Bagnasco S, Good D, Balaban R, Burg M. 1985. Lactate production in isolated segments of the rat nephron. *Am. J. Physiol. Renal Physiol.* 248:F522–26

42. Thomas SR. 2000. Inner medullary lactate production and accumulation: a vasa recta model. *Am. J. Physiol. Renal Physiol.* 279:F468–81

43. Beck FX, Neuhofer W, Müller E. 2000. Molecular chaperones in the kidney: distribution, putative roles, and regulation. *Am. J. Physiol. Renal Physiol.* 279:F203–15

44. Fink AL. 1999. Chaperone-mediated protein folding. *Physiol. Rev.* 79:425–49

45. Mosser DD, Caron AW, Bourget L, Meriin AB, Sherman MY, et al. 2000. The chaperone function of hsp70 is required for protection against stress-induced apoptosis. *Mol. Cell. Biol.* 20:7146–59

46. Ravagnan L, Gurbuxani S, Susin SA, Maisse C, Daugas E, et al. 2001. Heat-shock protein 70 antagonizes apoptosis-inducing factor. *Nat. Cell Biol.* 3:839–43

47. Iwaki T, Iwaki A, Liem RK, Goldman JE. 1991. Expression of alpha B-crystallin

in the developing rat kidney. *Kidney Int.* 40:52–56

48. Cowley BD Jr, Muessel MJ, Douglass D, Wilkins W. 1995. In vivo and in vitro osmotic regulation of HSP-70 and prostaglandin synthase gene expression in kidney cells. *Am. J. Physiol. Renal Physiol.* 269:F854–62

49. Müller E, Neuhofer W, Ohno A, Rucker S, Thurau K, et al. 1996. Heat shock proteins HSP25, HSP60, HSP72, HSP73 in isoosmotic cortex and hyperosmotic medulla of rat kidney. *Pflügers Arch.* 431:608–17

50. Kojima R, Randall J, Brenner BM, Gullans SR. 1996. Osmotic stress protein 94 (Osp94). A new member of the HSP110/SSE gene subfamily. *J. Biol. Chem.* 271:12327–32

51. Santos BC, Chevaile A, Kojima R, Gullans SR. 1998. Characterization of the Hsp110/SSE gene family response to hyperosmolality and other stresses. *Am. J. Physiol. Renal Physiol.* 274:F1054–61

52. Turman MA, Kahn DA, Rosenfeld SL, Apple CA, Bates CM. 1997. Characterization of human proximal tubular cells after hypoxic preconditioning: constitutive and hypoxia-induced expression of heat shock proteins HSP70 (A, B, and C), HSC70, and HSP90. *Biochem. Mol. Med.* 60:49–58

53. Suzuki K, Sawa Y, Kaneda Y, Ichikawa H, Shirakura R, et al. 1998. Overexpressed heat shock protein 70 attenuates hypoxic injury in coronary endothelial cells. *J. Mol. Cell. Cardiol.* 30:1129–36

54. Yagita Y, Kitagawa K, Ohtsuki T, Tanaka S, Hori M, et al. 2001. Induction of the HSP110/105 family in the rat hippocampus in cerebral ischemia and ischemic tolerance. *J. Cereb. Blood Flow Metab.* 21:811–19

55. Morrison LE, Hoover HE, Thuerauf DJ, Glembotski CC. 2003. Mimicking phosphorylation of alphaB-crystallin on serine-59 is necessary and sufficient to provide maximal protection of cardiac myocytes from apoptosis. *Circ. Res.* 92:203–11

56. Ruchalski K, Mao H, Singh SK, Wang Y, Mosser DD, et al. 2003. HSP72 inhibits apoptosis-inducing factor release in ATP-depleted renal epithelial cells. *Am. J. Physiol. Cell Physiol.* 285:C1483–93

57. Sone M, Albrecht GJ, Dörge A, Thurau K, Beck FX. 1993. Osmotic adaptation of renal medullary cells during transition from chronic diuresis to antidiuresis. *Am. J. Physiol. Renal Physiol.* 264:F722–29

58. Hebert SC. 1986. Hypertonic cell volume regulation in mouse thick limbs. II. Na^+-H^+ and Cl^--HCO_3^- exchange in basolateral membranes. *Am. J. Physiol. Cell Physiol.* 250:C920–31

59. Sun A, Hebert SC. 1989. Rapid hypertonic cell volume regulation in the perfused inner medullary collecting duct. *Kidney Int.* 36:831–42

60. Alper SL, Stuart-Tilley AK, Biemesderfer D, Shmukler BE, Brown D. 1997. Immunolocalization of AE2 anion exchanger in rat kidney. *Am. J. Physiol. Renal Physiol.* 273:F601–14

61. Paillard M. 1998. H^+ and HCO_3^- transporters in the medullary thick ascending limb of the kidney: molecular mechanisms, function and regulation. *Kidney Int.* 53(Suppl. 65):S36–41

62. Sun AM, Liu Y, Centracchio LJ, Dworkin LD. 1998. Expression of Na^+/H^+ exchanger isoforms in inner segment of inner medullary collecting duct (IMCD₃). *J. Membr. Biol.* 164:293–300

63. Eladari D, Blanchard A, Leviel F, Paillard M, Stuart-Tilley AK, et al. 1998. Functional and molecular characterization of luminal and basolateral Cl^-/HCO_3^- exchangers of rat thick limbs. *Am. J. Physiol. Renal Physiol.* 275:F334–42

64. Light DB, McCann FV, Keller TM, Stanton BA. 1988. Amiloride-sensitive cation channel in apical membrane of inner medullary collecting duct. *Am. J. Physiol. Renal Physiol.* 255:F278–86

65. Mori T, Cowley AW Jr. 2004. Renal oxidative stress in medullary thick ascending limbs produced by elevated NaCl and glucose. *Hypertension* 43:341–46

66. Ohtaka A, Muto S, Nemoto J, Kawakami K, Nagano K, et al. 1996. Hyperosmolality stimulates Na-K-ATPase gene expression in inner medullary collecting duct cells. *Am. J. Physiol. Renal Physiol.* 270:F728–38

67. Capasso JM, Rivard CJ, Berl T. 2001. Long-term adaptation of renal cells to hypertonicity: role of MAP kinases and Na-K-ATPase. *Am. J. Physiol. Renal Physiol.* 280:F768–76

68. Capasso JM, Rivard CJ, Enomoto LM, Berl T. 2003. Adaptation of murine inner medullary collecting duct (IMCD3) cell cultures to hypertonicity. *Ann. NY Acad. Sci.* 986:410–15

69. Capasso JM, Rivard CJ, Enomoto LM, Berl T. 2003. Chloride, not sodium, stimulates expression of the γ subunit of Na/K-ATPase and activates JNK in the response to hypertonicity in mouse IMCD3 cells. *Proc. Natl. Acad. Sci. USA* 100:6428–33

70. Kültz D, Chakravarty D. 2001. Hyperosmolality in the form of elevated NaCl but not urea causes DNA damage in murine kidney cells. *Proc. Natl. Acad. Sci. USA* 98:1999–2004

71. Brigotti M, Petronini PG, Carnicelli D, Alfieri RR, Bonelli MA, et al. 2003. Effects of osmolarity, ions and compatible osmolytes on cell-free protein synthesis. *Biochem. J.* 369:369–74

72. Galloway SM, Deasy DA, Bean CL, Kraynak AR, Armstrong MJ, et al. 1987. Effects of high osmotic strength on chromosome aberrations, sister-chromatid exchanges and DNA strand breaks, and the relation to toxicity. *Mutat. Res.* 189:15–25

73. Kültz D, Madhany S, Burg MB. 1998. Hyperosmolality causes growth arrest of murine kidney cells. Induction of GADD45 and GADD153 by osmosensing via stress-activated protein kinase 2. *J. Biol. Chem.* 273:13645–51

74. Michea L, Ferguson DR, Peters EM, Andrews PM, Kirby MR, et al. 2000. Cell cycle delay and apoptosis are induced by high salt and urea in renal medullary cells. *Am. J. Physiol. Renal Physiol.* 278:F209–18

75. Endoh D, Okui T, Kon Y, Hayashi M. 2001. Hypertonic treatment inhibits radiation-induced nuclear translocation of the Ku proteins G22p1 (Ku70) and Xrcc5 (Ku80) in rat fibroblasts. *Radiat. Res.* 155:320–27

76. Dmitrieva NI, Bulavin DV, Burg MB. 2003. High NaCl causes Mre11 to leave the nucleus, disrupting DNA damage signaling and repair. *Am. J. Physiol. Renal Physiol.* 285:F266–74

77. Bradbury JM, Jackson SP. 2003. The complex matter of DNA double-strand break detection. *Biochem. Soc. Trans.* 31:40–44

78. Dmitrieva N, Kültz D, Michea L, Ferraris J, Burg M. 2000. Protection of renal inner medullary epithelial cells from apoptosis by hypertonic stress-induced p53 activation. *J. Biol. Chem.* 275:18243–47

79. Kishi H, Nakagawa K, Matsumoto M, Suga M, Ando M, et al. 2001. Osmotic shock induces G_1 arrest through p53 phosphorylation at Ser[33] by activated p38[MAPK] without phosphorylation at Ser[15] and Ser[20]. *J. Biol. Chem.* 276:39115–22

80. Dmitrieva NI, Bulavin DV, Fornace AJ Jr, Burg MB. 2002. Rapid activation of G_2/M checkpoint after hypertonic stress in renal inner medullary epithelial (IME) cells is protective and requires p38 kinase. *Proc. Natl. Acad. Sci. USA* 99:184–89

81. Zhang Z, Cai Q, Michea L, Dmitrieva NI, Andrews P, et al. 2002. Proliferation and osmotic tolerance of renal inner medullary epithelial cells in vivo and in cell culture. *Am. J. Physiol. Renal Physiol.* 283:F302–8

82. Somero GN, Yancey PH. 1997. Osmolytes and cell-volume regulation: physical and evolutionary principles. In

Handbook of Physiology, Section 14: Cell Physiology, ed. JF Hoffman, JD Jamieson, pp. 441–84. New York/Oxford: Oxford Univ. Press

83. Wirthensohn G, Beck FX, Guder WG. 1987. Role and regulation of glycerophosphorylcholine in rat renal papilla. *Pflügers Arch.* 409:411–15

84. Yancey PH, Burg MB. 1989. Distribution of major organic osmolytes in rabbit kidneys in diuresis and antidiuresis. *Am. J. Physiol. Renal Physiol.* 257:F602–7

85. Zablocki K, Miller SPF, Garcia-Perez A, Burg MB. 1991. Accumulation of glycerophosphocholine (GPC) by renal cells: osmotic regulation of GPC:choline phosphodiesterase. *Proc. Natl. Acad. Sci. USA* 88:7820–24

86. Bauernschmitt HG, Kinne RKH. 1993. Metabolism of the 'organic osmolyte' glycerophosphorylcholine in isolated rat inner medullary collecting duct cells. I. Pathways for synthesis and degradation. *Biochim. Biophys. Acta* 1148:331–41

87. Kwon ED, Zablocki K, Jung KY, Peters EM, Garcia-Perez A, et al. 1995. Osmoregulation of GPC:choline phosphodiesterase in MDCK cells: different effects of urea and NaCl. *Am. J. Physiol. Cell Physiol.* 269:C35–41

88. Kwon ED, Zablocki K, Peters EM, Jung KY, Garcia-Perez A, et al. 1996. Betaine and inositol reduce MDCK cell glycerophosphocholine by stimulating its degradation. *Am. J. Physiol. Cell Physiol.* 270:C200–7

89. Sone M, Ohno A, Albrecht GJ, Thurau K, Beck FX. 1995. Restoration of urine concentrating ability and accumulation of medullary osmolytes after chronic diuresis. *Am. J. Physiol. Renal Physiol.* 269:F480–90

90. Yancey PH. 1988. Osmotic effectors in kidneys of xeric and mesic rodents: corticomedullary distributions and changes with water availability. *J. Comp. Physiol.* 158:369–80

91. Kwon ED, Jung KY, Edsall LC, Kim HY, Garcia-Perez A, et al. 1995. Osmotic regulation of synthesis of glycerophosphocholine from phosphatidylcholine in MDCK cells. *Am. J. Physiol. Cell Physiol.* 268:C402–12

92. Beck FX, Schmolke M, Guder WG, Dörge A, Thurau K. 1992. Osmolytes in renal medulla during rapid changes in papillary tonicity. *Am. J. Physiol. Renal Physiol.* 262:F849–56

93. Yamauchi A, Uchida S, Kwon HM, Preston AS, Robey RB, et al. 1992. Cloning of a Na^+- and Cl^--dependent betaine transporter that is regulated by hypertonicity. *J. Biol. Chem.* 267:649–52

94. Rasola A, Galietta LJV, Barone V, Romeo G, Bagnasco S. 1995. Molecular cloning and functional characterization of a GABA/betaine transporter from human kidney. *FEBS Lett.* 373:229–33

95. Perego C, Bulbarelli A, Longhi R, Caimi M, Villa A, et al. 1997. Sorting of two polytopic proteins, the γ-aminobutyric acid and betaine transporters, in polarized epithelial cells. *J. Biol. Chem.* 272:6584–92

96. Perego C, Vanoni C, Villa A, Longhi R, Kaech SM, et al. 1999. PDZ-mediated interactions retain the epithelial GABA transporter on the basolateral surface of polarized epithelial cells. *EMBO J.* 18:2384–93

97. Ahn J, Mundigl O, Muth TR, Rudnick G, Caplan MJ. 1996. Polarized expression of GABA transporters in Madin-Darby canine kidney cells and cultured hippocampal neurons. *J. Biol. Chem.* 271:6917–24

98. Yamauchi A, Kwon HM, Uchida S, Preston AS, Handler JS. 1991. *myo*-inositol and betaine transporters regulated by tonicity are basolateral in MDCK cells. *Am. J. Physiol. Renal Physiol.* 261:F197–202

99. Kempson SA, Parikh V, Xi L, Chu S, Montrose MH. 2003. Subcellular redistribution of the renal betaine transporter during hypertonic stress. *Am. J. Physiol. Cell Physiol.* 285:C1091–100

100. Grunewald RW, Oppermann M, Schettler V, Fiedler GM, Jehle PM, et al. 2001. Polarized function of thick ascending limbs of Henle cells in osmoregulation. *Kidney Int.* 60:2290–98

101. Matskevitch I, Wagner CA, Stegen C, Bröer S, Noll B, et al. 1999. Functional characterization of the betaine/γ-aminobutyric acid transporter BGT-1 expressed in *Xenopus* oocytes. *J. Biol. Chem.* 274:16709–16

102. Wirthensohn G, Guder WG. 1982. Studies on renal choline metabolism and phosphatidylcholine synthesis. In *Biochemistry of Kidney Function*, ed. F Morel, pp. 119–28. Amsterdam: Elsevier

103. Grossman EB, Hebert SC. 1989. Renal inner medullary choline dehydrogenase activity: characterization and modulation. *Am. J. Physiol. Renal Physiol.* 256:F107–12

104. Lohr J, Acara M. 1990. Effect of dimethylaminoethanol, an inhibitor of betaine production, on the disposition of choline in the rat kidney. *J. Pharmacol. Exp. Ther.* 252:154–58

105. Moeckel GW, Lien YH. 1997. Distribution of de novo synthesized betaine in rat kidney: role of renal synthesis on medullary betaine accumulation. *Am. J. Physiol. Renal Physiol.* 272:F94–99

106. Uchida S, Yamauchi A, Preston AS, Kwon HM, Handler JS. 1993. Medium tonicity regulates expression of the Na^+- and Cl^--dependent betaine transporter in Madin-Darby canine kidney cells by increasing transcription of the transporter gene. *J. Clin. Invest.* 91:1604–7

107. Kaneko T, Takenaka M, Okabe M, Yoshimura Y, Yamauchi A, et al. 1997. Osmolarity in renal medulla of transgenic mice regulates transcription via 5′-flanking region of canine BGT1 gene. *Am. J. Physiol. Renal Physiol.* 272:F610–16

108. Moeckel GW, Lai LW, Guder WG, Kwon HM, Lien YH. 1997. Kinetics and osmoregulation of Na^+- and Cl^--dependent betaine transporter in rat renal medulla. *Am. J. Physiol. Renal Physiol.* 272:F100–6

109. Burger-Kentischer A, Müller E, Neuhofer W, März J, Thurau K, et al. 1999. Expression of aldose reductase, sorbitol dehydrogenase and Na^+/*myo*-inositol and Na^+/Cl^-/betaine transporter mRNAs in individual cells of the kidney during changes in the diuretic state. *Pflügers Arch.* 437:248–54

110. Na KY, Woo SK, Lee SD, Kwon HM. 2003. Silencing of TonEBP/NFAT5 transcriptional activator by RNA interference. *J. Am. Soc. Nephrol.* 14:283–88

111. Preston AS, Yamauchi A, Kwon HM, Handler JS. 1995. Activators of protein kinase A and of protein kinase C inhibit MDCK cell *myo*-inositol and betaine uptake. *J. Am. Soc. Nephrol.* 6:1559–64

112. Basham JC, Chabrerie A, Kempson SA. 2001. Hypertonic activation of the renal betaine/GABA transporter is microtubule dependent. *Kidney Int.* 59:2182–91

113. Bricker JL, Chu S, Kempson SA. 2003. Disruption of F-actin stimulates hypertonic activation of the BGT1 transporter in MDCK cells. *Am. J. Physiol. Renal Physiol.* 284:F930–37

114. Bagnasco SM, Uchida S, Balaban RS, Kador PF, Burg MB. 1987. Induction of aldose reductase and sorbitol in renal inner medullary cells by elevated extracellular NaCl. *Proc. Natl. Acad. Sci. USA* 84:1718–20

115. Neuhofer W, Bartels H, Fraek ML, Beck FX. 2002. Relationship between intracellular ionic strength and expression of tonicity-responsive genes in rat papillary collecting duct cells. *J. Physiol.* 543:147–53

116. Schmolke M, Bornemann A, Guder WG. 1996. Site-specific regulation of organic osmolytes along the rat nephron. *Am. J. Physiol. Renal Physiol.* 271:F645–52

117. Cowley BD Jr, Ferraris JD, Carper D, Burg MB. 1990. In vivo osmoregulation of aldose reductase mRNA, protein, and

sorbitol in renal medulla. *Am. J. Physiol. Renal Physiol.* 258:F154–61

118. Jung JY, Kim YH, Cha JH, Han KH, Kim MK, et al. 2002. Expression of aldose reductase in developing rat kidney. *Am. J. Physiol. Renal Physiol.* 283:F481–91

119. Ruepp B, Bohren KM, Gabbay KH. 1996. Characterization of the osmotic response element of the human aldose reductase gene promoter. *Proc. Natl. Acad. Sci. USA* 93:8624–29

120. Neuhofer W, Woo SK, Na KY, Grünbein R, Park WK, et al. 2002. Regulation of TonEBP transcriptional activator in MDCK cells following changes in ambient tonicity. *Am. J. Physiol. Cell Physiol.* 283:C1604–11

121. Lopez-Rodriguez C, Antos CL, Shelton JM, Richardson JA, Lin F, et al. 2004. Loss of NFAT5 results in renal atrophy and lack of tonicity-responsive gene expression. *Proc. Natl. Acad. Sci. USA* 101:2392–97

122. Sands JM, Schrader DC. 1990. Coordinated response of renal medullary enzymes regulating net sorbitol production in diuresis and antidiuresis. *J. Am. Soc. Nephrol.* 1:58–65

123. Boulanger Y, Legault P, Tejedor A, Vinay P, Theriault Y. 1988. Biochemical characterization and osmolytes in papillary collecting ducts from pig and dog kidneys. *Can. J. Physiol. Pharmacol.* 66:1282–90

124. Bondy CA, Lightman SL, Lightman SL. 1989. Developmental and physiological regulation of aldose reductase mRNA expression in renal medulla. *Mol. Endocrinol.* 3:1409–16

125. Veis JH, Molitoris BA, Teitelbaum I, Mansour JA, Berl T. 1991. *myo*-inositol uptake by rat cultured inner medullary collecting tubule cells: effect of osmolality. *Am. J. Physiol. Renal Physiol.* 260:F619–25

126. Kwon HM, Yamauchi A, Uchida S, Preston AS, Garcia-Perez A, et al. 1992. Cloning of the cDNA for a Na^+/myo-inositol cotransporter, a hypertonicity

stress protein. *J. Biol. Chem.* 267:6297–301

127. Rim JS, Atta MG, Dahl SC, Berry GT, Handler JS, et al. 1998. Transcription of the sodium/*myo*-inositol cotransporter gene is regulated by multiple tonicity-responsive enhancers spread over 50 kilobase pairs in the 5′-flanking region. *J. Biol. Chem.* 273:20615–21

128. Nakanishi T, Turner RJ, Burg MB. 1989. Osmoregulatory changes in *myo*-inositol transport by renal cells. *Proc. Natl. Acad. Sci. USA* 86:6002–6

129. Hager K, Hazama A, Kwon HM, Loo DD, Handler JS, et al. 1995. Kinetics and specificity of the renal Na^+/myo-inositol cotransporter expressed in *Xenopus* oocytes. *J. Membr. Biol.* 143:103–13

130. Wright EM, Turk E. 2004. The sodium/glucose cotransporter family SLC5. *Pflügers Arch.* 447:510–18

131. Yamauchi A, Nakanishi T, Takamitsu Y, Sugita M, Imai M, et al. 1994. In vivo osmoregulation of Na/*myo*-inositol cotransporter mRNA in rat kidney medulla. *J. Am. Soc. Nephrol.* 5:62–67

132. Wiese TJ, Matsushita K, Lowe WL, Stokes JB, Yorek MA. 1996. Localization and regulation of renal Na^+/myo-inositol cotransporter in diabetic rats. *Kidney Int.* 50:1202–11

133. Beck FX, Burger-Kentischer A, Müller E. 1998. Cellular response to osmotic stress in the renal medulla. *Pflügers Arch.* 436:814–27

134. Uchida S, Nakanishi T, Kwon HM, Preston AS, Handler JS. 1991. Taurine behaves as an osmolyte in Madin-Darby canine kidney cells. Protection by polarized, regulated transport of taurine. *J. Clin. Invest.* 88:656–62

135. Chen JG, Coe M, McAteer JA, Kempson SA. 1996. Hypertonic activation and recovery of system A amino acid transport in renal MDCK cells. *Am. J. Physiol. Renal Physiol.* 270:F419–24

136. Kempson SA. 1998. Differential activation of system A and betaine/GABA

transport in MDCK cell membranes by hypertonic stress. *Biochim. Biophys. Acta* 1372:117–23

137. Bitoun M, Levillain O, Tappaz M. 2001. Gene expression of the taurine transporter and taurine biosynthetic enzymes in rat kidney after antidiuresis and salt loading. *Pflügers Arch.* 442:87–95

138. Law RO. 1991. Alterations in renal inner medullary levels of amino nitrogen during acute water diuresis and hypovolaemic oliguria in rats. *Pflügers Arch.* 418:442–46

139. Beck FX, Sone M, Dörge A, Thurau K. 1992. Effect of loop diuretics on organic osmolytes and cell electrolytes in the renal outer medulla. *Kidney Int.* 42:843–50

140. Ohno A, Müller E, Fraek ML, Rucker S, Beck FX, et al. 1996. Ketoconazole inhibits organic osmolyte efflux and induces heat shock proteins in rat renal medulla. *Kidney Int.* 50(Suppl. 57):S110–18

141. Perlman M, Goldstein L. 1999. Organic osmolyte channels in cell volume regulation in vertebrates. *J. Exp. Zool.* 283:725–33

142. Tomassen SFB, Fekkes D, De Jonge HR, Tilly BC. 2004. Osmotic swelling-provoked release of organic osmolytes in human intestinal epithelial cells. *Am. J. Physiol. Cell Physiol.* 286:C1417–22

143. Hand SC, Somero GN. 1982. Urea and methylamine effects on rabbit muscle phosphofructokinase. Catalytic stability and aggregation state as a function of pH and temperature. *J. Biol. Chem.* 257:734–41

144. Burg MB, Peters EM, Bohren KM, Gabbay KH. 2000. Factors affecting counteraction by methylamines of urea effects on aldose reductase. *Proc. Natl. Acad. Sci. USA* 96:6517–22

145. Neuhofer W, Fraek ML, Beck FX. 2002. Heat shock protein 72, a chaperone abundant in renal papilla, counteracts urea-mediated inhibition of enzymes. *Pflügers Arch.* 445:67–73

146. Yancey PH, Clark ME, Hand SC, Bowlus RD, Somero GN. 1982. Living with water stress: evolution of osmolyte systems. *Science* 217:1214–22

147. Colmont C, Michelet S, Guivarc'h D, Rousselet G. 2001. Urea sensitizes mIMCD3 cells to heat shock-induced apoptosis: protection by NaCl. *Am. J. Physiol. Cell Physiol.* 280:C614–20

148. Neuhofer W, Lugmayr K, Fraek ML, Beck FX. 2001. Regulated overexpression of heat shock protein 72 protects Madin-Darby canine kidney cells from the detrimental effects of high urea concentrations. *J. Am. Soc. Nephrol.* 12:2565–71

149. Burg MB, Kwon ED, Peters EM. 1996. Glycerophosphocholine and betaine counteract the effect of urea on pyruvate kinase. *Kidney Int.* 50(Suppl. 57):S100–4

150. Burg MB, Peters EM. 1997. Urea and methylamines have similar effects on aldose reductase activity. *Am. J. Physiol. Renal Physiol.* 273:F1048–53

151. Medina R, Cantley L, Spokes K, Epstein FH. 1996. Effect of water diuresis and water restriction on expression of HSPs-27, -60 and -70 in rat kidney. *Kidney Int.* 50:1191–94

152. Müller E, Neuhofer W, Burger-Kentischer A, Ohno A, Thurau K, et al. 1998. Effects of long-term changes in medullary osmolality on heat shock proteins HSP25, HSP60, HSP72 and HSP73 in the rat kidney. *Pflügers Arch.* 435:705–12

153. Smoyer WE, Ransom R, Harris RC, Welsh MJ, Lutsch G, et al. 2000. Ischemic acute renal failure induces differential expression of small heat shock proteins. *J. Am. Soc. Nephrol.* 11:211–21

154. Neuhofer W, Müller E, Burger-Kentischer A, Fraek ML, Thurau K, et al. 1999. Inhibition of NaCl-induced heat shock protein 72 expression renders MDCK cells susceptible to high urea concentrations. *Pflügers Arch.* 437:611–16

155. Beere HM, Wolf BB, Cain K, Mosser DD, Mahboubi A, et al. 2000. Heat-shock protein 70 inhibits apoptosis by preventing

recruitment of procaspase-9 to the Apaf-1 apoptosome. *Nat. Cell Biol.* 2:469–75

156. Michea L, Combs C, Andrews P, Dmitrieva N, Burg MB. 2002. Mitochondrial dysfunction is an early event in high-NaCl-induced apoptosis of mIMCD3 cells. *Am. J. Physiol. Renal Physiol.* 282:F981–90

157. Santos BC, Pullman JM, Chevaile A, Welch WJ, Gullans SR. 2003. Chronic hyperosmolarity mediates constitutive expression of molecular chaperones and resistance to injury. *Am. J. Physiol. Renal Physiol.* 284:F564–74

158. Shim EH, Kim JI, Bang ES, Heo JS, Lee JS, et al. 2002. Targeted disruption of *hsp70.1* sensitizes to osmotic stress. *EMBO Rep.* 3:857–61

159. Takenaka M, Imai E, Nagasawa Y, Matsuoka Y, Moriyama T, et al. 2000. Gene expression profiles of the collecting duct in the mouse renal inner medulla. *Kidney Int.* 57:19–24

160. Neuhofer W, Müller E, Burger-Kentischer A, Fraek ML, Thurau K, et al. 1998. Pretreatment with hypertonic NaCl protects MDCK cells against high urea concentrations. *Pflügers Arch.* 435:407–14

161. Kegel KB, Iwaki A, Iwaki T, Goldman JE. 1996. αB-crystallin protects glial cells from hypertonic stress. *Am. J. Physiol. Cell Physiol.* 270:C903–9

162. Zhang Z, Dmitrieva NI, Park JH, Levine RL, Burg MB. 2004. High urea and NaCl carbonylate proteins in renal cells in culture and in vivo, and high urea causes 8-oxoguanine lesions in their DNA. *Proc. Natl. Acad. Sci. USA* 101:9491–96

163. Zhang Z, Yang XY, Cohen DM. 1999. Urea-associated oxidative stress and Gadd153/CHOP induction. *Am. J. Physiol. Renal Physiol.* 276:F786–93

164. Tian W, Bonkovsky HL, Shibahara S, Cohen DM. 2001. Urea and hypertonicity increase expression of heme oxygenase-1 in murine renal medullary cells. *Am. J. Physiol. Renal Physiol.* 281:F983–91

165. Zou AP, Billington H, Su N, Cowley AW

Jr. 2000. Expression and actions of heme oxygenase in the renal medulla of rats. *Hypertension* 35:342–47

166. Agarwal A, Nick HS. 2000. Renal response to tissue injury: lessons from heme oxygenase-1 gene ablation and expression. *J. Am. Soc. Nephrol.* 11:965–73

167. Stark GR, Stein WH, Moore S. 1960. Reaction of the cyanate present in aqueous urea with amino acids and proteins. *J. Biol. Chem.* 235:3177–81

168. Horkko S, Savolainen MJ, Kervinen K, Kesaniemi YA. 1992. Carbamylation-induced alterations in low-density lipoprotein metabolism. *Kidney Int.* 41:1175–81

169. Kraus LM, Elberger AJ, Handorf CR, Pabst MJ, Kraus AP Jr. 1994. Urea-derived cyanate forms ε-amino-carbamoyl-lysine (homocitrulline) in leukocyte proteins in patients with end-stage renal disease on peritoneal dialysis. *J. Lab. Clin. Med.* 123:882–91

170. Prabhakar SS, Zeballos GA, Montoya-Zavala M, Leonard C. 1997. Urea inhibits inducible nitric oxide synthase in macrophage cell line. *Am. J. Physiol. Cell Physiol.* 273:C1882–88

171. Harding JJ. 1991. Post-translational modification of lens proteins in cataract. *Lens Eye Toxic. Res.* 8:245–50

172. Miyakawa H, Woo SK, Dahl SC, Handler JS, Kwon HM. 1999. Tonicity-responsive enhancer binding protein, a Rel-like protein that stimulates transcription in response to hypertonicity. *Proc. Natl. Acad. Sci. USA* 96:2538–42

173. Stroud JC, Lopez-Rodriguez C, Rao A, Chen L. 2002. Structure of the TonEBP-DNA complex reveals DNA encircled by a transcription factor. *Nat. Struct. Biol.* 9:90–94

174. Woo SK, Lee SD, Na KY, Park WK, Kwon HM. 2002. TonEBP/NFAT5 stimulates transcription of HSP70 in response to hypertonicity. *Mol. Cell. Biol.* 22:5753–60

175. Bagnasco SM, Peng T, Nakayama Y, Sands JM. 2000. Differential expression of individual UT-A urea transporter isoforms in rat kidney. *J. Am. Soc. Nephrol.* 11:1980–86

176. Woo SK, Dahl SC, Handler JS, Kwon HM. 2000. Bidirectional regulation of tonicity-responsive enhancer binding protein in response to changes in tonicity. *Am. J. Physiol. Renal Physiol.* 278:F1006–12

177. Dahl SC, Handler JS, Kwon HM. 2001. Hypertonicity-induced phosphorylation and nuclear localization of the transcription factor TonEBP. *Am. J. Physiol. Cell Physiol.* 280:C248–53

178. Lee SD, Woo SK, Kwon HM. 2002. Dimerization is required for phosphorylation and DNA binding of TonEBP/NFAT5. *Biochem. Biophys. Res. Commun.* 294:968–75

179. Cha JH, Woo SK, Han KH, Kim YH, Handler JS, et al. 2001. Hydration status affects nuclear distribution of transcription tonicity responsive enhancer binding protein in rat kidney. *J. Am. Soc. Nephrol.* 12:2221–30

180. Ko BCB, Lam AKM, Kapus A, Fan L, Chung SK, et al. 2002. Fyn and p38 signaling are both required for maximal hypertonic activation of the osmotic response element-binding protein/tonicity-responsive enhancer-binding protein (OREBP/TonEBP). *J. Biol. Chem.* 277: 46085–92

181. Ferraris JD, Persaud P, Williams CK, Chen Y, Burg MB. 2002. cAMP-independent role of PKA in tonicity-induced transactivation of tonicity-responsive enhancer/osmotic response element-binding protein. *Proc. Natl. Acad. Sci. USA* 99: 16800–5

182. Irarrazabal CE, Liu JC, Burg MB, Ferraris JD. 2004. ATM, a DNA damage-inducible kinase, contributes to activation by high NaCl of the transcription factor TonEBP/OREBP. *Proc. Natl. Acad. Sci. USA* 101:8809–14

183. Ma T, Yang B, Gillespie A, Carlson EJ, Epstein CJ, et al. 1998. Severely impaired urinary concentrating ability in transgenic mice lacking aquaporin-1 water channels. *J. Biol. Chem.* 273:4296–99

184. Umenishi F, Schrier RW. 2003. Hypertonicity-induced aquaporin-1 (AQP1) expression is mediated by the activation of MAPK pathways and hypertonicity-responsive element in the AQP1 gene. *J. Biol. Chem.* 278:15765–70

185. Zhao H, Tian W, Tai C, Cohen DM. 2003. Hypertonic induction of COX-2 expression in renal medullary epithelial cells requires transactivation of the EGF receptor. *Am. J. Physiol. Renal Physiol.* 285:F281–88

186. Kitamura H, Yamauchi A, Sugiura T, Matsuoka Y, Horio M, et al. 1998. Inhibition of *myo*-inositol transport causes acute renal failure with selective medullary injury in the rat. *Kidney Int.* 53:146–53

187. Cai Q, Ferraris JD, Burg M. 2004. Greater tolerance of renal medullary cells for a slow increase in osmolality is associated with enhanced expression of HSP70 and other osmoprotective genes. *Am. J. Physiol. Renal Physiol.* 286:F58–67

188. Cai Q, Michea L, Andrews P, Zhang Z, Rocha G, et al. 2002. Rate of increase of osmolality determines osmotic tolerance of mouse inner medullary epithelial cells. *Am. J. Physiol. Renal Physiol.* 283:F792–98

189. Diezi J, Michoud P, Aceves J, Giebisch G. 1973. Micropuncture study of electrolyte transport across papillary collecting duct of the rat. *Am. J. Physiol.* 224:623–34

190. Tian W, Cohen DM. 2001. Urea inhibits hypertonicity-inducible TonEBP expression and action. *Am. J. Physiol. Renal Physiol.* 280:F904–12

Annu. Rev. Physiol. 2005. 67:557–72
doi: 10.1146/annurev.physiol.67.031103.153949
First published online as a Review in Advance on September 28, 2004

Novel Renal Amino Acid Transporters

François Verrey, Zorica Ristic, Elisa Romeo,
Tamara Ramadan, Victoria Makrides, Mital H. Dave,
Carsten A. Wagner, and Simone M.R. Camargo
*University of Zurich, Institute of Physiology, CH-8057 Zurich,
Switzerland; email: verrey@access.unizh.ch*

Key Words proximal kidney tubule, membrane carrier, Hartnup disorder,
epithelial transport, aminoaciduria

■ **Abstract** Reabsorption of amino acids, similar to that of glucose, is a major task
of the proximal kidney tubule. Various amino acids are actively transported across the
luminal brush border membrane into proximal tubule epithelial cells, most of which
by cotransport. An important player is the newly identified cotransporter (symporter)
B^0AT1 (*SLC6A19*), which imports a broad range of neutral amino acids together with
Na^+ across the luminal membrane and which is defective in Hartnup disorder. In
contrast, cationic amino acids and cystine are taken up in exchange for recycled neutral
amino acids by the heterodimeric cystinuria transporter. The basolateral release of
some neutral amino acids into the extracellular space is mediated by unidirectional
efflux transporters, analogous to GLUT2, that have not yet been definitively identified.
Additionally, cationic amino acids and some other neutral amino acids leave the cell
basolaterally via heterodimeric obligatory exchangers.

INTRODUCTION

The absorption of amino acids in the intestine and their reabsorption in the proximal
kidney tubule are evolutionarily old functions, as shown by the high degree of
conservation of many involved transporters throughout vertebrate evolution (1–
3). This review focuses on the reabsorption machinery of amino acids across the
kidney proximal tubule epithelium, the task of which is to prevent the urinary loss
of these essential substrates that are quantitatively filtered in the kidney glomeruli.
Assuming a plasma concentration of 2.5 mM amino acids with a mean molecular
weight of 115 and a glomerular filtration rate of 120 ml/min, the amount of amino
acids recovered per 24 h is ∼50 g.

Including the proximal tubule cells, which additionally express the transcellular
amino acid reabsorbtion machinery, all kidney cells express some amino acid
transporters that are involved mostly in house-keeping functions and appear to
localize to the basolateral membrane of tubular cells. Some kidney cells also

0066-4278/05/0315-0557$14.00

557

require additional uptake of amino acids, which are used as precursors for the synthesis of paracrine and/or endocrine substances such as NO (4).[1]

In addition to amino acid transport, proximal tubule epithelial cells have a role in many other reabsorption and excretion tasks, for example, the excretion of protons. This activity requires that the tubular cells take up L-glutamine, the nontoxic precursor/carrier of secreted ammonium (5–7). Because L-glutamine is progressively cleared from the tubular fluid by the amino acid reabsorption machinery, the cellular uptake of L-glutamine takes place across the basolateral membrane in the later segments of the proximal tubule.

AXIAL HETEROGENEITY ALONG THE PROXIMAL TUBULE

To assume that the amino acid transport machinery is uniform along the proximal tubule is a simplification that does not account for the differential role of the early (beginning of S1) and the late segments (S2, S3) of the proximal tubule. For instance, just after glomerular filtration occurs, the amino acid concentration is similar to that found in plasma, whereas at the end of the proximal tubule it is low. Thus whereas the first epithelial cells of the proximal tubule face a high solute concentration that might necessitate the control or limitation of the reabsorption activity in order to protect the cells, more distal epithelial cells face decreasing substrate concentrations that can more effectively be reabsorbed by transporters of high affinity. Such a sequential organization of Na^+cotransporters with differential affinity for their substrate has been described for glucose, with the low-affinity SGLT2 being expressed in the early segments of the proximal tubule (S1, S2) and the high-affinity cotransporter SGLT1, which is also the intestinal cotransporter, being expressed in the later segments of proximal tubule (S2, S3) (8). Although several cases of differential amino acid transporter expression along the proximal tubule axis are mentioned in this review, there is no detailed knowledge and understanding of this complex axial arrangement. Differences between proximal tubules of superficial and deep nephrons are also not well understood.

---

[1]Abbreviations: 4F2hc (CD98): glycoprotein heavy chain of heterodimeric amino acid transporters that is basolateral in kidney proximal tubule; LAT2: catalytic light chain of heteromeric amino acid transporter associated with 4F2hc, which performs L-type transport (exchange of large neutral amino acids); y^+LAT1: catalytic light chain of heteromeric amino acid transporter associated with 4F2hc, which performs y^+L-type transport; rBAT (NBAT, D2): glycoprotein heavy chain of heteromeric amino acid transporter that is luminal in kidney proximal tubule; $b^{0,+}$AT: catalytic light chain of heteromeric amino acid transporter associated with rBAT, which performs $b^{0,+}$-type transport; TAT1: T-type amino acid transporter-1; ASCT: Na^+-dependent ASC-type amino acid exchangers; EAAT3 (EAAC1): excitatory amino acid transporter 3; PAT1: proton-dependent amino acid transporter.

LIMITATIONS OF THE TECHNIQUES TO DESCRIBE SPECIFIC AMINO ACID TRANSPORT SYSTEMS

The characterization of the amino acid transport systems has expanded since they are now mostly attributed to cloned transporters that can be tested in expression systems. However, the understanding of their nature and their cooperation for the transport of amino acids across epithelial cells has been slowed down by a number of technical difficulties. The first limitation is the fact that different transporters display overlapping substrate selectivity (2, 9, 10). Thus it is difficult to determine quantitatively the contribution of each. Additional difficulties arise from the lack of selective inhibitors and from the intrinsic limits of the techniques. A further level of complexity is given by the fact that several transporters function as more or less obligatory and symmetrical exchangers of amino acids (11). A number of approaches have been used to characterize the transport of amino acids across the luminal and the basolateral membranes of epithelial cells and across epithelia. Three sources of experimental models have been used: (*a*) animals and ex vivo organs/tissues/cells, (*b*) cultured cells derived from epithelia of interest, and (*c*) expression systems for cloned transporters.

a. Amino acid transport has been tested in animals, for instance, by clearance studies, micropuncture and microperfusion experiments using flux measurements, and electrophysiological recordings (12). Ex vivo transport studies were made by in vitro perfusion of tubules and by using membrane vesicles, in particular from purified brush border or basolateral membranes (13–16). A major limitation of all these approaches is that many different transporters are expressed in each preparation and thus the interpretation of the data is often difficult. A powerful possibility is the analysis of the phenotype of organisms with specific gene defects, for instance humans with genetic diseases (i.e., cystinuria, lysinuric protein intolerance, or Hartnup disorder) or more recently of animals (mostly mice) carrying a targeted gene modification (i.e., knockout etc.) (17–19).

b. Studies have been made on kidney epithelial cell cultures (primary cultures and immortalized cell lines). From a technical point of view, the major advantage is the structural simplicity and the large number of identical cells, which facilitates their use for biochemical experiments. The major limitation is generally the dedifferentiation of the cultured cells, whether primary or immortalized. The cell line with the most conserved proximal tubule transport properties available to date is the opossum kidney (OK) cell line (genome sequence not yet available). This cell line has been useful in characterizing the function of several transporters, in particular $b^{0,+}$AT-rBAT, LAT2-4F2hc, and B^0 (20, 21) (Z. Ristic, S.M.R. Camargo, E. Romeo, R. Warth & F. Verrey, unpublished results). A useful possibility is the knock-down of specific transporters using antisense RNA or siRNA that can demonstrate the specific contribution of given transporters (21).

c. With the identification of cDNAs encoding specific amino acid transporters, their expression in heterologous systems has become possible. Major advantages of this approach are that cells not expressing the exogenous proteins are a perfect control and mutant transporters can be tested easily. The *Xenopus* oocytes expression system has been most used (10, 22). In addition, fibroblasts, nonconfluent epithelial cells, and mammalian cells forming epithelia have also been used as recipients for amino acid transporters. In particular, proximal tubule transporters have been expressed in Madin-Darby canine kidney (MDCK) cells, reconstituting part of the transepithelial transport machinery (23, 24).

LUMINAL AMINO ACID TRANSPORTERS OF THE PROXIMAL TUBULE

B^0AT1 (SLC6A19) Corresponds to the Na^+-Amino Acid Cotransport System B^0

An uphill Na^+-dependent transport system for neutral amino acids was first detected in brush-border membrane vesicles of the intestine and proximal kidney tubule approximately 30 years ago; it was called B or B^0 (10, 13, 14, 25, 26). Subsequently, results of in situ microperfusion coupled with electrophysiological measurements on rat kidney proximal tubules supported the notion that neutral amino acids are taken up by a Na^+ cotransport system (15, 16). Kinetic analysis suggested the existence of at least two such rheogenic transport systems, one for L-α-amino acids and another for L-proline and L-glycine.

A familial condition with urinary loss and decreased intestinal absorption of neutral amino acids, called Hartnup disorder, was described in the 1960s (27, 28). This genetic condition was sometimes accompanied by additional symptoms, such as a pellagra–like rash and neurological symptoms (i.e., ataxia), and was later proposed to be the result of a defect in the B^0 transporter (10).

Unlike other transporters that were molecularly identified several years ago, no transporter corresponding to system B^0 was identified before 2004. The Na^+-dependent amino acid transporter ATB^0 (identified in 1996) was rapidly recognized as an ASC-type transporter that performs Na^+-dependent exchange of amino acids and is mostly expressed in sites other than the intestinal and renal brush border membrane (29). Recently, successful identification of a mouse B^0-type transporter was reported by Broer and colleagues (30). This transporter was named B^0AT1 (Slc6a19) and belongs to a cluster of orphan transporters within the family of Na^+- and Cl^--dependent neurotransmitters and amino acid transporters (SLC6). Expressed in *Xenopus* oocytes, the B^0AT1 cDNA induces a Na^+-dependent, Cl^--independent uptake of neutral amino acids with an apparent affinity for L-leucine uptake on the order of magnitude previously measured in other systems (15, 25) (Table 1). We have shown actual brush border localization of this gene product

TABLE 1   Amino acid reabsorption in kidney proximal tubule

Amino acid type	Luminal influx	Basolateral efflux
Neutral amino acids (not L-proline)[a]	Na$^+$-cotransporter (B^0)[b] $\underline{B^0AT1}$[d] and others?	Heterodimeric exchanger[c](L) LAT2-4F2hc
Aromatic neutral amino acids	Na$^+$-cotransporter (B^0) $\underline{B^0AT1}$ and others?	Facilitated[e] diffusion transporter TAT1
Gly	H$^+$-cotransporter[f] (*imino*) PAT1	Heterodimeric exchanger (L) LAT2-4F2hc
L-proline	H$^+$-cotransporter (*imino*) PAT1	(ASC)[g]? ASCT1
L-cystine	Heterodimeric exchanger[h] ($b^{0,+}$) $\underline{b^{0,+}AT}$-rBAT	Reduction to two L-cysteines Heterodimeric exchanger (L) LAT2-4F2hc
Cationic amino acids	Heterodimeric exchanger ($b^{0,+}$) $\underline{b^{0,+}AT}$-rBAT	Heterodimeric exchanger (y^+L) y$^+$LAT1-4F2hc
Anionic amino acids	Na$^+$-dependent cotransporter $\underline{EAAT3}$ (EAAC1)	Partially metabolized to glutamine or/and exchanger ASCT1?

[a]The transported amino acids are indicated by groups without specifying differential apparent affinities.

[b]Corresponding transport systems (name of transport functions/entities before molecular identification) are indicated in parentheses and italic. The logic of the acronyms can be found elsewhere (75).

[c]The potential efflux of L-proline via LAT2-4F2hc and/or TAT1 has not been tested.

[d]The transporters for which the kidney proximal kidney tubule subcellular localization has been confirmed are underlined.

[e]Facilitated diffusion efflux pathways that transport neutral amino acids recycled by parallel functioning obligatory exchangers.

[f]The correspondence of PAT1 to system imino has been proposed for rat and human (not rabbit) because the apparent Na$^+$dependence of their imino systems can be explained by functional coupling with Na$^+$/H$^+$ exchange.

[g]The expression of ASCT1 in the basolateral membrane of kidney proximal tubule is hypothetical.

[h]b$^{0,+}$AT-rBAT also takes large neutral amino acids, but normally as intracellular efflux substrates that are exchanged against extracellular cystine or cationic amino acids.

by immunofluorescence in mouse kidney, and the hypothesis that homozygote mutations of the corresponding gene leads to Hartnup disorder was verified in a large fraction of the tested cases, including in the original Hartnup family (18, 31). The question remains as to whether the cases of Hartnup disorder that do not map to *SLC6A19* are from mutations in other transporters or from defects in another gene product necessary for the correct expression and/or function of B^{0}AT1 (18).

Taken together, B^{0}AT1 appears to be the major neutral amino acid transporter of the small intestine and kidney (Figure 1, see color insert). It is noteworthy, however, that several related gene products, two in human (products of *SLC6A18* and *20*) and three in mice (*Slc6a18* and two genes corresponding to human *SLC6A20*), are expressed in the kidney brush border and, to some extent, in the small intestine as well. Interestingly, whereas the gene product of Slc6a19 (B^{0}AT1) is expressed mainly in the early part of the proximal tubule, the product of Slc6a18 is most abundant in the late proximal segments. In contrast, the gene products derived from the *SLC6A20* homologues appear to be expressed all along the proximal tubule (E. Romeo, Z. Ristic, J. Loffing, M.H. Dare, C.A. Wagner & F. Verrey, unpublished data). It is tempting to hypothesize that these gene products perform B^0-type Na^+-amino acid cotransport similar to that of B^0AT1, but with somewhat different selectivity, apparent affinity, and axial distribution. A major limitation for their characterization has been the lack of reproducible detection of induced amino acid transport upon expression in *Xenopus* oocytes and in other epithelial and nonepithelial expression systems (Z. Ristic & F. Verrey, unpublished results).

The H^+-Cotransporter PAT1 Corresponds to System Imino

Proline and hydroxyproline were recognized in functional experiments to share a renal tubular reabsorptive system with glycine (32, 33). This transport system, referred to as system imino, also transports other substrates such as L-and D-alanine, GABA, MeAIB, and D-serine. In Caco-2 cells (human colon carcinoma-derived cell line), this transport system was shown to function in the absence of Na^+ and to depend on the driving force of protons (34) (Table 1). The fact that the local luminal pH is set by the activity of the Na^+/H^+ exchanger explains much of the controversy concerning the Na^+ dependence of this transport system. However, unlike humans and rats, rabbits and guinea pigs appear to express an additional imino-type transporter that is Na^+ dependent (25).

With identification of PAT1, a molecular substrate for system imino has been determined that fulfills the prediction in terms of substrate selectivity and confirms the observation that its substrates are cotransported with protons (35). PAT1 mRNA is broadly expressed. particularly in small intestine and kidney (Figure 1). The protein PAT1 was shown to localize apically in Caco-2 cells (36). Interestingly, the same transporter is expressed in many other cells, in particular in neurons, mostly intracellularly at the lysosomal membrane, where the proton gradient drives the efflux of amino acids into the cytosol (37). The physiological role of the related transporter PAT2 has not yet been clearly determined (37).

Thus PAT1 appears to be the major imino-type transporter expressed in the brush border membrane of the small intestine and kidney proximal tubule, although its localization in the latter structure has not yet been demonstrated.

The Heterodimeric Exchanger $b^{0,+}AT$-rBAT is the Cystinuria Transporter

Cystinuria, the most common primary inherited aminoaciduria, is characterized by urinary excretion of cationic amino acids and cystine (a dimer of L-cysteine). Its associated pathology is caused by renal cystine lithiasis (38–40). The heteromeric nature of the causative transporter was suggested after the expression cloning of the type II glycoprotein rBAT (single transmembrane segment) in *Xenopus laevis*

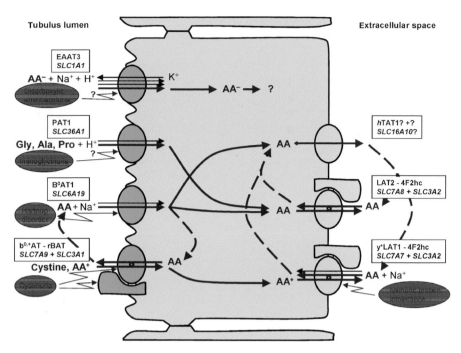

Figure 1 Luminal and basolateral transporters belonging to the amino acid absorption/reabsorption machinery. Names of transporters and of the corresponding genes are indicated in rectangles. Recycling pathways are shown by dotted lines. Types of aminoacidurias, possibly (?) owing to defects of the corresponding transporters, are indicated in ellipses. AA, neutral amino acids; AA$^+$, cationic amino acids; AA$^-$ anionic amino acids.

oocytes, in which it induced a high affinity transport of cationic amino acids and cystine (Table 1). The actual catalytic subunit of this transporter, $b^{0,+}AT$, is apparently expressed constitutionally in *Xenopus* oocytes and requires association with rBAT to reach the cell surface membrane (41). It was finally identified on the basis of its similarity to LAT1, the first catalytic subunit of heterodimeric amino acid transporters to be identified (38, 41–44).

Mutations in the rBAT gene (*SLC3A1*) were detected in most cases of type I cystinuria (absence of subclinical aminoaciduria in heterozygote relatives), and mutations in the $b^{0,+}AT1$ gene (*SLC7A9*) were found in most cases of nontype I cystinuria (subclinical aminoaciduria in heterozygotes) (38, 45–49), confirming the identity and the important role of this transporter (Figure 1).

The catalytic subunit $b^{0,+}AT$ has been shown to be covalently linked to rBAT and to be expressed in the brush border membrane of the initial part of the proximal tubule and in the jejunum and the ileum. Its axial distribution is thus similar to that of basolateral y^+LAT1 and LAT2 in these tubular structures (24, 41, 50, 51). In respect to the axial distribution, it is important to note that although rBAT is additionally highly expressed in the brush border of the last proximal tubule segment S3, its additional function in S3 is not known.

In heterologous expression experiments, it has been demonstrated that the rBAT-$b^{0,+}AT$-induced L-cystine and cationic amino acid uptake is an obligatory exchange against mostly neutral amino acids (41, 52, 53). The direction of the exchange is given by the relatively high extracellular apparent affinity of this transporter for cystine and cationic amino acids relative to the lower one for neutral amino acids, and by the driving forces that are given by the low intracellular cystine (reduction to two L-cysteines), the membrane potential (for cationic substrates), and the outward gradient for neutral amino acids that are recycled into the cell via the B^0-type Na^+ cotransporter (53).

Thus although it functions as an obligatory exchanger, $b^{0,+}AT$-rBAT mediates the apical step of the directional transport of cationic amino acids and of cystine. It can fulfill its exchange function correctly only if sufficient neutral amino acids are available inside the cell. This requires the function of the parallel B^0-type transporter and/or, at least, that of the basolateral influx pathways for neutral amino acids (Bauch et al. 2003).

Luminal Na^+-Dependent Cotransporter EAAT3 (EAAC1)

The high-affinity transporter for anionic amino acids, originally called EAAC1, and now mostly referred to as EAAT3 (*SLC1A1*), is expressed in the intestine and kidney, as well as in the brain (22, 54) (Table 1). In the kidney, it localizes to the proximal tubule brush border membrane with an axial gradient: low amounts in S1 and highest levels in the later segments S2 and S3 (55) (Figure 1). It is still not clear whether there is another low-affinity transporter, as suggested originally by brush border membrane vesicle studies (54). The fact that transport by early proximal tubule is saturated, when the filtered load is only slightly increased, is compatible with the possibility that EAAT3 (which is expressed at a lower level in S1) is the only selective transporter for anionic amino acids of the proximal tubule. Importantly, EAAT3 is the major transporter for L-glutamic acid, which is present only at a low concentration in plasma and can be produced in the lumen of the proximal kidney tubule from L-glutamine by deamidation via phosphate-independent glutaminase (gamma-glutamyltransferase). Once L-glutamic acid has reached the inside of the cell via EAAT3, its fate remains intimately linked to that of L-glutamine. These amino acids can be interconverted and/or metabolized to produce energy, ammonia, and bicarbonate and thus play an important role in the context of pH regulation.

BASOLATERAL AMINO ACID TRANSPORTERS OF THE KIDNEY PROXIMAL TUBULE

Basolateral Efflux of Neutral Amino Acids by Facilitated Diffusion: Role of T-Type Amino Acid Transporter-1

The best characterized basolateral transporters of the proximal tubule amino acid reabsorption machinery function as obligatory exchangers (y$^+$LAT1-4F2hc and LAT2-4F2hc; see below) and thus do not perform net basolateral amino acids export. They can play an important role by extending the export selectivity range to their intracellular efflux substrates (24). However, this export activity requires the recycling of extracellular substrate amino acids. Thus at least one transporter that mediates net amino acid export is required to drive net basolateral efflux (11). This could be a facilitated diffusion pathway (analogous to GLUT2) mediating the net transport (generally efflux) along the transmembrane substrate amino acid concentration gradient. In reabsorbing cells, this concentration is generally higher inside owing to apical amino acid influx and basolateral recycling of efflux substrates through the exchangers.

It is conceivable that TAT1 exerts this crucial function of a facilitated diffusion pathway. The original characterization of this transporter, which belongs to the gene family *SLC16* monocarboxylate transporters (58), indicated that rat and human orthologues correspond to the T-type amino acid transport system, namely that they mediate the uptake of aromatic amino acids, e.g., L-tryptophan, L-tyrosine, L-phenylalanine, and L-DOPA, with apparent affinities in the millimolar range (56, 57). This transporter was also shown to mediate the efflux of L-tryptophan from *Xenopus laevis* oocytes independently of the presence of an extracellular substrate, suggesting it functions as a facilitated efflux diffusion pathway (56) (Table 1, Figure 1). In the same publication, TAT1 was also shown to localize to the basolateral surface of the small intestine enterocytes. We have confirmed that it functions as a facilitated efflux-diffusion pathway for aromatic amino acids upon expression of the mouse orthologue in *Xenopus* oocytes and have shown that it localizes to the basolateral membrane of kidney proximal tubule (T. Ramadan & F. Verrey, unpublished data). Thus this aromatic amino acid transporter potentially controls the rate of efflux of the other neutral and cationic amino acids because they can efflux basolaterally in exchange for aromatic amino acids (see below).

We have tested the possibility that another member of the *SLC16* family, namely XPCT, would participate in this facilitated amino acid efflux diffusion function. Indeed, this transporter shares a high level of amino acid identity with TAT1 (46%) and was shown in rat to localize to kidney, in addition to heart and liver (59). Our preliminary experiments could not substantiate this hypothesis. It is interesting, however, that both TAT1 and XPCT from mouse similarly mediate the uptake of thyroid hormone triiodothyronine (T_3) (T. Ramadan & F. Verrey, unpublished data).

In summary, TAT1, as a facilitated diffusion pathway, might control the rate of efflux of neutral and cationic amino acids from the small intestine and human kidney proximal tubule cells. More experiments are required to test and quantify such a cooperation with exchangers and the possible involvement of other directional transporters.

System L Heterodimeric Exchanger LAT2-4F2hc is Highly Expressed in the Early Kidney Proximal Tubule and Small Intestine Basolateral Membranes

The LAT2-4F2hc transporter, an obligatory exchanger of neutral amino acids, is highly expressed in the small intestine and proximal kidney tubule (60, 61) (Figure 1). Kinetic analysis has shown that it mediates the efflux of all neutral amino acids, including glycine, with a relatively low intracellular apparent affinity that controls its rate of exchange function (Table 1). At the outside, the concentration of preferred influx substrates of LAT2-4F2hc (aromatic and other large neutral amino acids) is sufficient to support near-maximal transport velocity (62). The role of LAT2-4F2hc in transepithelial transport and, specifically, the basolateral efflux of L-cysteine, has been demonstrated in cultured proximal kidney cells (OK cell line) (21). Thus this exchanger participates in the net vectorial transport of some amino acids providing that a parallel unidirectional pathway permits the directional efflux of recycled substrates. Its localization along the small intestine and the kidney proximal tubule supports the notion that it is part of the amino acid reabsorption machinery (24, 50).

System y^+L Heterodimeric Exchanger y^+LAT1-4F2hc is Highly Expressed in the Proximal Kidney Proximal Tubule and Small Intestine Basolateral Membranes

y^+LAT1-4F2hc has been shown to preferentially perform the basolateral efflux of cationic amino acids in exchange for aromatic and other large neutral amino acids together with Na$^+$ (24, 39, 63–65) (Table 1). It has an axial distribution along the kidney proximal tubule and small intestine similar to that of LAT2-4F2hc (24, 50) (Figure 1). Its rate of function also depends on the availability of intracellular substrates, because, owing to its high extracellular apparent affinity for large neutral amino acids, the substrate concentration outside the cell is generally not limiting. In other words, the function of this exchanger is to control the intracellular concentration of its cationic substrates. Its limiting role in cationic amino acid reabsorption is confirmed by the human genetic disease lysinuric protein intolerance, as well as by expression studies performed in MDCK epithelia (24, 39, 40). Lysinuric protein intolerance is a complex disease that maps to *SLC7A7* (y$^+$LAT1) and presents many symptoms including the malabsorption and urinary loss of cationic amino acids; other symptoms might be because y$^+$LAT1-4F2hc lacks function in other cells. These observations demonstrate that the expression of concentrative cationic amino acid transporters such as CAT1 do not compensate for its absence (24).

Metabolism and/or Basolateral Efflux of Anionic Amino Acids

A basolateral export pathway for L-glutamic or L-aspartic acid has not been described in the kidney as yet. However, the apparent lack of reproducible arteriovenous L-glutamic acid concentration difference suggests that such a basolateral efflux should exist (66). From the known anionic amino acid transporters, EAAT1 could be the basolateral uptake transporter in proximal tubule cells (67). Its transport cycles would, however, essentially not permit a reverse transport under physiological conditions, although in ischemia this transporter is responsible for the efflux of glutamate in the brain. As mentioned below, it is conceivable that ASCT exchangers might export the anionic amino acids basolaterally with a low affinity that would be appropriate in regard to their high intracellular concentration. In this context, it is worth recalling that the anionic amino acids can also be converted intracellularly to L-glutamine and L-asparagine, respectively, and thus play an important role in energy and pH metabolism.

Taken together, L-glutamic acid is taken up apically by proximal tubule cells, and the extent to which is it metabolized to glutamine or further to α-keto glutarate or pyruvate depends on metabolic conditions. It appears that the remainder of L-glutamic acid leaves the cell basolaterally via an export pathway that has not yet been identified. It can not be ruled out that some particular mechanism, such as the exocytosis of vesicles containing L-glutamic acid, might contribute to the basolateral efflux of anionic amino acids.

Non-Epithelial Amino Acid Transporters Expressed in Proximal Tubule

ASCT transporters are Na$^+$-dependent obligatory exchangers of amino acids (in particular alanine, serine, cysteine, and threonine) that are structurally related to the EAAT transporters. The mRNA of ASCT2 is expressed in kidney at a level similar to that observed in lung, pancreas, skeletal muscle, and testis, but less so in the proximal tubule (29, 68) (M.H. Dave & F. Verrey, unpublished data). It is not yet clear how to reconcile this observation with the results of an immunofluorescence study performed on rabbit tissues that shows an apical brush border localization for ASCT2 in the small intestine and proximal kidney tubule (69). In addition to the fact that ASCT2 mRNA is expressed in the kidney at a relatively low level similar to that in heart, lung, placenta, and liver, no other data are available on its expression in kidney (70, 71). Thus it is not yet clear to which epithelial cell membrane the ASCTs are localized in proximal kidney tubule. As amino acid exchangers, both might extend the transmembrane amino acid transport selectivity range. This may be important for the basolateral efflux of some amino acids, in particular L-proline, which has been shown to be transported by ASCT1 (72). ASCT2 (or ASCT1?) could also play a role in the basolateral efflux of L-aspartate and/or L-glutamate, both potential low-affinity substrates (73).

The system N (transport of L-glutamine, L-asparagine, and L-histidine) transporter SNAT3, also called SN1 (*SLC38A3*), is expressed in proximal kidney tubule (C. Moret, F. Verrey & C.A. Wagner, unpublished data) and controls the basolateral uptake and eventually also export of glutamine (74). It might thus play an important role in the context of pH regulation.

System y^+ transporters are also expressed on the basolateral membrane of proximal tubule cells and play a house-keeping role of importing cationic amino acids if the intracellular level is too low. In the context of transepithelial transport, this transport system does not appear to play a major role. Similarly, a system A (alanine-preferring transport of amino acids) transporter (*SLC38A2* and or *SLC38A3*) is probably also expressed at the basolateral surface of proximal tubule cells. Such a regulated uptake transporter would be required in cells that do not import a sufficient amount of amino acids from the lumen for their metabolism.

CONCLUSIONS AND PERSPECTIVES

Nearly all the different elements of the transepithelial amino acid reabsorption machinery have been molecularly identified; most recent of these the luminal neutral amino acid- Na^+-cotransporter of B^0-type (B^0AT1, *SLC6A19*). In parallel with this cotransporter, the $b^{0,+}AT$-rBAT exchanger mediates the influx of cationic amino acids and of cystine (in exchange for neutral amino acids). EAAT3 imports apically anionic amino acids together with three Na^+ ions and one proton in exchange for one K^+ ion. Another secondary active luminal transporter is PAT1 (system imino) that takes up, for instance, L-proline or glycine together with a proton. It is still an open issue as to whether the orphan transporters *SLC6A18* and *20* (XT2 and XT3) represent B^0-type transporters that complement the activity of B^0AT1, as suggested by their structural similarity and by their brush border localization. Another uncertainty is the potential presence of the exchanger ASCT2 in the brush border membrane, where it could equilibrate the transport of different neutral amino acids with differential affinities for the B^0-type transporters.

At the level of the basolateral membrane, the efflux of amino acids is somewhat less well understood. This can be explained by the fact that this transport is more difficult to test experimentally than luminal uptake. The best characterized basolateral efflux transporters function as obligatory amino acid exchangers and can thus only play the role of extending the overall amino acid efflux selectivity across the basolateral membrane for specific amino acids. The major players that quantitatively control the efflux of amino acids are probably one or more facilitated diffusion pathways that might be regulated or at least strongly (kinetically) activated by an increase in intracellular amino acid concentration (low apparent K_m). One potentially important facilitated efflux transporter in human kidney proximal tubule, TAT1, does selectively transport aromatic amino acids. These amino acids could in turn be recycled into the cell by the parallel exchangers y$^+$LAT1-4F2*hc* and LAT2-4F2*hc* to drive the efflux of their intracellular substrates (cationic amino acids and neutral amino acids, respectively).

Because the cooperation of these different transporters is required for the overall directional transport of amino acids, it will be interesting to reconstitute such a system in order to test the role of individual elements of this machinery (24).

ACKNOWLEDGMENTS

The laboratory of F. V. is supported by Swiss National Science Foundation grant 31–59141.99/02 and the European FP6 Project Eugindat.

The *Annual Review of Physiology* is online at http://physiol.annualreviews.org

LITERATURE CITED

1. Broer S. 2002. Adaptation of plasma membrane amino acid transport mechanisms to physiological demands. *Pflügers Arch.* 444:457–66

2. Verrey F, Jack DL, Paulsen IT, Saier MH, Pfeiffer R. 1999. New glycoprotein-associated amino acid transporters. *J. Membr. Biol.* 172:181–92

3. Veljkovic E, Stasiuk S, Skelly PJ, Shoemaker CB, Verrey F. 2004. Functional characterization of *Caenorhabditis elegans* heteromeric amino acid transporters. *J. Biol. Chem.* 279:7655–62

4. Zewde T, Wu F, Mattson DL. 2004. Influence of dietary NaCl on L-arginine transport in the renal medulla. *Am. J. Physiol. Regul. Integr. Comp. Physiol.* 286:R89–93

5. Karinch AM, Lin CM, Wolfgang CL,

Pan M, Souba WW. 2002. Regulation of expression of the SN1 transporter during renal adaptation to chronic metabolic acidosis in rats. *Am. J. Physiol. Renal Physiol.* 283:F1011–19

6. Curthoys NP. 2001. Role of mitochondrial glutaminase in rat renal glutamine metabolism. *J. Nutr.* 131:2491S–95S; discussion 6S-7S

7. Nissim I. 1999. Newer aspects of glutamine/glutamate metabolism: the role of acute pH changes. *Am. J. Physiol. Renal Physiol.* 277:F493–97

8. Wright EM, Turk E. 2004. The sodium/glucose cotransport family SLC5. *Pflügers Arch.* 447:510–18

9. Christensen HN. 1990. Role of amino acid transport and countertransport in nutrition

and metabolism. *Physiol. Rev.* 70:43–77

10. Palacin M, Estevez R, Bertran J, Zorzano A. 1998. Molecular biology of mammalian plasma membrane amino acid transporters. *Physiol. Rev.* 78:969–1054

11. Verrey F, Meier C, Rossier G, Kuhn LC. 2000. Glycoprotein-associated amino acid exchangers: broadening the range of transport specificity. *Pflügers Arch.* 440:503–12

12. Silbernagl S. 1988. The renal handling of amino acids and oligopeptides. *Physiol. Rev.* 68:911–1007

13. Sigrist-Nelson K, Murer H, Hopfer U. 1975. Active alanine transport in isolated brush border membranes. *J. Biol. Chem.* 250:5674–80

14. Evers J, Murer H, Kinne R. 1976. Phenylalanine uptake in isolated renal brush border vesicles. *Biochim. Biophys. Acta* 426:598–615

15. Samarzija I, Fromter E. 1982. Electrophysiological analysis of rat renal sugar and amino acid transport. III. Neutral amino acids. *Pflügers Arch.* 393:119–209

16. Fromter E. 1982. Electrophysiological analysis of rat renal sugar and amino acid transport. I. Basic phenomena. *Pflügers Arch.* 393:179–89

17. Palacin M, Bertran J, Zorzano A. 2000. Heteromeric amino acid transporters explain inherited aminoacidurias. *Curr. Opin. Nephrol. Hypertens.* 9:547–53

18. Kleta R, Romeo E, Ristic Z, Ohura T, Stuart C, et al. 2004. Mutations in *SLC6A19* (hB0AT1) are associated with the Hartnup disorder. *Nat. Genet.* 36:999–1002

19. Nicholson B, Sawamura T, Masaki T, MacLeod CL. 1998. Increased Cat3-mediated cationic amino acid transport functionally compensates in Cat1 knockout cell lines. *J. Biol. Chem.* 273:14663–66

20. Mora C, Chillaron J, Calonge MJ, Forgo J, Testar X, et al. 1996. The rBAT gene is responsible for L-cystine uptake via the $b^{0,+}$-like amino acid transport system in a "renal proximal tubular" cell line (OK cells). *J. Biol. Chem.* 271:10569–76

21. Fernandez E, Torrents D, Chillaron J, Martin Del Rio R, Zorzano A, Palacin M. 2003. Basolateral LAT-2 has a major role in the transepithelial flux of L-cystine in the renal proximal tubule cell line OK. *J. Am. Soc. Nephrol.* 14:837–47

22. Kanai Y, Hediger MA. 1992. Primary structure and functional characterization of a high-affinity glutamate transporter. *Nature* 360:467–71

23. Bauch C, Verrey F. 2002. Apical heterodimeric cystine and cationic amino acid transporter expressed in MDCK cells. *Am. J. Physiol. Renal Physiol.* 283:F181–89

24. Bauch C, Forster N, Loffing-Cueni D, Summa V, Verrey F. 2003. Functional cooperation of epithelial heteromeric amino acid transporters expressed in Madin-Darby canine kidney cells. *J. Biol. Chem.* 278:1316–22

25. Stevens BR, Ross HJ, Wright EM. 1982. Multiple transport pathways for neutral amino acids in rabbit jejunal brush border vesicles. *J. Membr. Biol.* 66:213–25

26. Lynch AM, McGivan JD. 1987. Evidence for a single common Na^+-dependent transport system for alanine, glutamine, leucine and phenylalanine in brush-border membrane vesicles from bovine kidney. *Biochim. Biophys. Acta* 899:176–84

27. Baron DN, Dent CE, Harris H, Hart EW, Jepson JB. 1956. Hereditary pellagra-like skin rash with temporary cerebellar ataxia, constant renal amino-aciduria, and other bizarre biochemical features. *Lancet* 268:421–28

28. Scriver CR. 1965. Hartnup disease: a genetic modification of intestinal and renal transport of certain neutral alpha-amino acids. *N. Engl. J. Med.* 273:530–32

29. Kekuda R, Prasad PD, Fei YJ, Torres Zamorano V, Sinha S, et al. 1996. Cloning of the sodium-dependent, broad-scope, neutral amino acid transporter B-0 from a human placental choriocarcinoma cell line. *J. Biol. Chem.* 271:18657–61

30. Broer A, Klingel K, Kowalczuk S, Rasko JE, Cavanaugh J, Broer S. 2004. Molecular

cloning of mouse amino acid transport system B0, a neutral amino acid transporter related to Hartnup disorder. *J. Biol. Chem.* 279:24467–76

31. Seow HF, Broer S, Broer A, Baily CG, Potter SJ, et al. 2004. Hartnup disorder is caused by mutations in the neutral amino acid transporter, *SLC6A19. Nat. Genet.* 36: 1003–7

32. Stevens BR, Kaunitz JD, Wright EM. 1984. Intestinal transport of amino acids and sugars: advances using membrane vesicles. *Annu. Rev. Physiol.* 46:417–33

33. Munck LK, Munck BG. 1992. The rabbit jejunal 'imino carrier' and the ileal 'imino acid carrier' describe the same epithelial function. *Biochim. Biophys. Acta* 1116:91–96

34. Thwaites DT, McEwan GT, Brown CD, Hirst BH, Simmons NL. 1993. Na^+-independent, H^+-coupled transepithelial beta-alanine absorption by human intestinal Caco-2 cell monolayers. *J. Biol. Chem.* 268:18438–41

35. Boll M, Foltz M, Rubio-Aliaga I, Kottra G, Daniel H. 2002. Functional characterization of two novel mammalian electrogenic proton dependent amino acid cotransporters. *J. Biol. Chem.* 277:22966–73

36. Chen Z, Fei YJ, Anderson CM, Wake KA, Miyauchi S, et al. 2003. Structure, function and immunolocalization of a proton-coupled amino acid transporter (hPAT1) in the human intestinal cell line Caco-2. *J. Physiol.* 546:349–61

37. Sagne C, Agulhon C, Ravassard P, Darmon M, Hamon M, et al. 2001. Identification and characterization of a lysosomal transporter for small neutral amino acids. *Proc. Natl. Acad. Sci. USA* 98:7206–11

38. Feliubadalo L, Font M, Purroy J, Rousaud F, Estivill X, et al. 1999. Non-type I cystinuria caused by mutations in *SLC7A9*, encoding a subunit ($b^{0,+}$AT) of rBAT. *Nat. Genet.* 23:52–57

39. Torrents D, Mykkanen J, Pineda M, Feliubadalo L, Estevez R, et al. 1999. Identi-

fication of *SLC7A7*, encoding y^+LAT-1, as the lysinuric protein intolerance gene. *Nat. Genet.* 21:293–96

40. Borsani G, Bassi MT, Sperandeo MP, De Grandi A, Buoninconti A, et al. 1999. *SLC7A7*, encoding a putative permease-related protein, is mutated in patients with lysinuric protein intolerance. *Nat. Genet.* 21:297–301

41. Pfeiffer R, Loffing J, Rossier G, Bauch C, Meier C, et al. 1999. Luminal heterodimeric amino acid transporter defective in cystinuria. *Mol. Biol. Cell* 10:4135–47

42. Mastroberardino L, Spindler B, Pfeiffer R, Skelly PJ, Loffing J, et al. 1998. Amino acid transport by heterodimers of 4F2hc/CD98 and members of a permease family. *Nature* 395:288–91

43. Kanai Y, Segawa H, Miyamoto K, Uchino H, Takeda E, Endou H. 1998. Expression cloning and characterization of a transporter for large neutral amino acids activated by the heavy chain of 4F2 antigen (CD98). *J. Biol. Chem.* 273:23629–32

44. Reig N, Chillaron J, Bartoccioni P, Fernandez E, Bendahan A, et al. 2002. The light subunit of system $b^{0,+}$ is fully functional in the absence of the heavy subunit. *EMBO J.* 21:4906–14

45. Verrey F, Schaerer E, Zoerkler P, Paccolat MP, Geering K, et al. 1987. Regulation by aldosterone of Na^+,K^+-ATPase mRNAs, protein synthesis, and sodium transport in cultured kidney cells. *J. Cell Biol.* 104:1231–37

46. Dello Strologo L, Pras E, Pontesilli C, Beccia E, Ricci-Barbini V, et al. 2002. Comparison between *SLC3A1* and *SLC7A9* cystinuria patients and carriers: a need for a new classification. *J. Am. Soc. Nephrol.* 13:2547–53

47. Leclerc D, Boutros M, Suh D, Wu Q, Palacin M, et al. 2002. *SLC7A9* mutations in all three cystinuria subtypes. *Kidney Int.* 62:1550–59

48. Chairoungdua A, Segawa H, Kim JY, Miyamoto K, Haga H, et al. 1999.

Identification of an amino acid transporter associated with the cystinuria-related type II membrane glycoprotein. *J. Biol. Chem.* 274:28845–48

49. Calonge MJ, Gasparini P, Chillaron J, Chillon M, Gallucci M, et al. 1994. Cystinuria caused by mutations in rBAT, a gene involved in the transport of cystine. *Nat. Genet.* 6:420–25

50. Dave MH, Schulz N, Zecevic M, Wagner CA, Verrey F. 2004. Expression of heteromeric amino acid transporters along the intestine. *J. Physiol.* 588:597–610

51. Fernandez E, Carrascal M, Rousaud F, Abian J, Zorzano A, et al. 2002. rBAT-b$^{0,+}$AT heterodimer is the main apical reabsorption system for cystine in the kidney. *Am. J. Physiol. Renal Physiol.* 283:F540–48

52. Busch AE, Herzer T, Waldegger S, Schmidt F, Palacin M, et al. 1994. Opposite directed currents induced by the transport of dibasic and neutral amino acids in *Xenopus oocytes* expressing the protein rBAT. *J. Biol. Chem.* 269:25581–86

53. Chillaron J, Estevez R, Mora C, Wagner CA, Suessbrich H, et al. 1996. Obligatory amino acid exchange via systems b$^{0,+}$-like and y$^+$L-like. A tertiary active transport mechanism for renal reabsorption of cystine and dibasic amino acids. *J. Biol. Chem.* 271:17761–70

54. Hediger MA. 1999. Glutamate transporters in kidney and brain. *Am. J. Physiol. Renal Physiol.* 277:F487–92

55. Shayakul C, Kanai Y, Lee WS, Brown D, Rothstein JD, Hediger MA. 1997. Localization of the high-affinity glutamate transporter EAAC1 in rat kidney. *Am. J. Physiol. Renal Physiol.* 273:F1023–29

56. Kim DK, Kanai Y, Chairoungdua A, Matsuo H, Cha SH, Endou H. 2001. Expression cloning of a Na$^+$-independent aromatic amino acid transporter with structural similarity to H$^+$/monocarboxylate transporters. *J. Biol. Chem.* 276:17221–28

57. Kim do K, Kanai Y, Matsuo H, Kim JY, Chairoungdua A, et al. 2002. The human

T-type amino acid transporter-1: characterization, gene organization, and chromosomal location. *Genomics* 79:95–103

58. Halestrap AP, Meredith D. 2004. The *SLC16* gene family-from monocarboxylate transporters (MCTs) to aromatic amino acid transporters and beyond. *Pflügers Arch.* 447:619–28

59. Friesema EC, Ganguly S, Abdalla A, Manning Fox JE, Halestrap AP, Visser TJ. 2003. Identification of monocarboxylate transporter 8 as a specific thyroid hormone transporter. *J. Biol. Chem.* 278:40128–35

60. Segawa H, Fukasawa Y, Miyamoto K, Takeda E, Endou H, Kanai Y. 1999. Identification and functional characterization of a Na$^+$-independent neutral amino acid transporter with broad substrate selectivity. *J. Biol. Chem.* 274:19745–51

61. Rossier G, Meier C, Bauch C, Summa V, Sordat B, et al. 1999. LAT2, a new basolateral 4F2hc/CD98-associated amino acid transporter of kidney and intestine. *J. Biol. Chem.* 274:34948–54

62. Meier C, Ristic Z, Klauser S, Verrey F. 2002. Activation of system L heterodimeric amino acid exchangers by intracellular substrates. *EMBO J.* 21:580–89

63. Pfeiffer R, Rossier G, Spindler B, Meier C, Kuhn L, Verrey F. 1999. Amino acid transport of y$^+$L-type by heterodimers of 4F2hc/CD98 and members of the glycoprotein-associated amino acid transporter family. *EMBO J.* 18:49–57

64. Kanai Y, Fukasawa Y, Cha SH, Segawa H, Chairoungdua A, et al. 2000. Transport properties of a system y$^+$L neutral and basic amino acid transporter. Insights into the mechanisms of substrate recognition. *J. Biol. Chem.* 275:20787–93

65. Deves R, Chavez P, Boyd CA. 1992. Identification of a new transport system (y$^+$L) in human erythrocytes that recognizes lysine and leucine with high affinity. *J. Physiol.* 454:491–501

66. van de Poll MC, Soeters PB, Deutz NE, Fearon KC, Dejong CH. 2004. Renal

metabolism of amino acids: its role in interorgan amino acid exchange. *Am. J. Clin. Nutr.* 79:185–97

67. Cheng C, Glover G, Banker G, Amara SG. 2002. A novel sorting motif in the glutamate transporter excitatory amino acid transporter 3 directs its targeting in Madin-Darby canine kidney cells and hippocampal neurons. *J. Neurosci.* 22:10643–52

68. Utsunomiya-Tate N, Endou H, Kanai Y. 1996. Cloning and functional characterization of a system ASC-like Na^+-dependent neutral amino acid transporter. *J. Biol. Chem.* 271:14883–90

69. Avissar NE, Ryan CK, Ganapathy V, Sax HC. 2001. Na^+-dependent neutral amino acid transporter ATB^0 is a rabbit epithelial cell brush-border protein. *Am. J. Physiol. Cell Physiol.* 281:C963–71

70. Arriza JL, Kavanaugh MP, Fairman WA, Wu YN, Murdoch GH, et al. 1993. Cloning and expression of a human neutral amino acid transporter with structural similarity to the glutamate transporter gene family. *J. Biol. Chem.* 268:15329–32

71. Shafqat S, Tamarappoo BK, Kilberg MS, Puranam RS, McNamara JO, et al. 1993. Cloning and expression of a novel Na^+-dependent neutral amino acid transporter structurally related to mammalian Na^+/glutamate cotransporters. *J. Biol. Chem.* 268:15351–55

72. Pinilla-Tenas J, Barber A, Lostao MP. 2003. Transport of proline and hydroxyproline by the neutral amino-acid exchanger ASCT1. *J. Membr. Biol.* 195:27–32

73. Tetsuka K, Takanaga H, Ohtsuki S, Hosoya K, Terasaki T. 2003. The L-isomer-selective transport of aspartic acid is mediated by ASCT2 at the blood-brain barrier. *J. Neurochem.* 87:891–901

74. Gu S, Roderick HL, Camacho P, Jiang JX. 2000. Identification and characterization of an amino acid transporter expressed differentially in liver. *Proc. Natl. Acad. Sci. USA* 97:3230–35

75. Christensen HN, Albritton LM, Kakuda DK, MacLeod CL. 1994. Gene-product designations for amino acid transporters. *J. Exp. Biol.* 196:51–57

Annu. Rev. Physiol. 2005. 67:573–94
doi: 10.1146/annurev.physiol.67.031103.154845
Copyright © 2005 by Annual Reviews. All rights reserved
First published online as a Review in Advance on September 22, 2004

RENAL TUBULE ALBUMIN TRANSPORT

Michael Gekle

*Physiologisches Institut, University of Würzburg, 97070 Würzburg,
Germany; email: michael.gekle@mail.uni-wuerzburg.de*

Key Words   endocytosis, megalin, cubilin, vesicular pH, interstitial nephropathy

■ **Abstract**   Albumin is the most abundant protein in serum and contributes to the
maintenance of oncotic pressure as well as to transport of hydrophobic molecules. Although albumin is a large anionic protein, it is not completely retained by the glomerular
filtration barrier. In order to prevent proteinuria, albumin is reabsorbed along the proximal tubules by receptor-mediated endocytosis, which involves the binding proteins
megalin and cubilin. Endocytosis depends on proper vesicle acidification. Disturbance
of endosomal acidification or loss of the binding proteins leads to tubular proteinuria.
Furthermore, endocytosis is subject to modulation by different signaling systems, such
as protein kinase A (PKA), protein kinase C (PKC), phosphatidylinositol 3-kinase
(PI3-K) and transforming growth factor beta (TGF-β). In addition to being reabsorbed
in the proximal tubule, albumin can also act as a profibrotic and proinflammatory
stimulus, thereby initiating or promoting tubulo-interstitial diseases.

A SHORT INTRODUCTION TO SERUM ALBUMIN

The word albumin derives from *albus*, the Latin word for white (1). It dates back
to the recognition that the portion of an egg (also called albumen), which appears white after coagulation, consists mainly of proteins. Nowadays albumin is
generally regarded to mean serum or plasma albumin, a single protein species.
Quantitatively, albumin is the most important plasma protein because it represents
$\sim$60% of all plasma proteins. The concentration of albumin in human serum is
$\sim$45 g/l or $652 \cdot 10^{-6}$ mol/l, and the molecular mass obtained from physical data is
69,000 kDa (1). According to the physico-chemical data, the following conformation of the albumin molecule in a physiological environment can be derived
(Figure 1): The amino acid chain is arranged in nine loops, stabilized by disulfide
bonds, that form three spherical domains (called I–III, starting from the amino
terminus) with different net charges. The net charges of the three domains are not
identical, ranging from -9, -8 to $+2$, and leading to an overall net charge of -15.
Albumin can be described as an ellipsoid molecule of 15 nm length and 3.8 nm diameter ("stubby cigar" shape) (1). An important physiological function of albumin
is the maintenance of an oncotic pressure difference between plasma and the interstitial space, whereby albumin is involved in the regulation of fluid exchange across

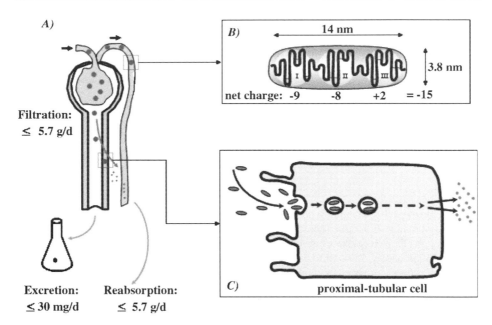

Figure 1   (*A*) Overview of renal handling of serum albumin in healthy subjects. (*B*) Basic data and configuration of serum albumin. (*C*) Reabsorption of filtered albumin occurs via endocytosis in the proximal tubule. Inside the cell (in lysosomes) albumin is degraded to amino acids, which are released across the basolateral membrane.

the capillary walls. Furthermore, albumin serves as a carrier for a variety of substances, such as Ca^{2+}, bilirubin, fatty acids, and drugs. Thus albumin influences the renal elimination kinetics of bound small-molecular substances, because binding to albumin dramatically reduces the filtration rate of these substances in the kidney.

GLOMERULAR FILTRATION OF SERUM ALBUMIN

Analysis of the urine of healthy individuals shows that only a small amount of albumin (<100 mg/d) (Figure 1) is excreted in the final urine (2). Taking into account the plasma concentration of albumin (see above) and the high rate of renal blood flow, renal albumin clearance is low. Nevertheless, the small amount of albumin in final urine shows that albumin can reach the tubular lumen. Furthermore, there are pathophysiological states in which the daily albumin excretion is dramatically increased and results in albuminuria. The best-known form is glomerular proteinuria, which always includes albuminuria. Here, pathological alterations in the glomeruli result in enhanced filtration of proteins.

How does a protein of the size and charge of albumin cross the glomerular filtration barrier? The composition of glomerular ultrafiltrate depends on the permeability properties of the glomerular filter barrier, which is negatively charged and has pores with a theoretical mean diameter of ∼4 nm (3). Thus under physiological conditions, molecules with effective diameters greater than 4 nm are not freely filtered but retained to an increasing extent as the diameter increases, i.e., the fractional filtration (substrate concentration in renal ultrafiltrate/substrate concentration in plasma) decreases from 1 to 0. The model of a single population of log-normal distributed pores may be oversimplified. Two other models have been proposed (4). One model postulates the existence of a population of small restrictive pores in parallel with a population of nonrestrictive pores (shunt). Another model postulates two populations of restrictive pores plus a shunt pathway. In addition, negatively charged macromolecules with effective diameters close to the filter pore diameter can be restricted to a greater extent compared with neutral molecules of comparable size (2, 3). However, the precise contribution of charge selectivity and the leakiness of glomerular capillaries are still under debate (5–7). For albumin, the effective radius is in the range of 7.5 nm (2, 3), resulting in a low-fractional filtration, which is even further impaired, possibly by the negative net charge, and ranges between 0.0005 and 0.0007 (2, 8, 9). Recent calculations indicate that fractional filtration may be even lower (10). Using theses values, we can estimate the filtered amount of serum albumin. As mentioned above, the plasma albumin concentration is ∼45 g/liter. Thus albumin concentration in renal ultrafiltrate is in the range of 22–32 mg/liter. These values are in good agreement with albumin concentrations determined by micropuncture studies (2, 8). Studies using more indirect techniques (11) in order to determine the albumin concentration in renal ultrafiltrate concluded that the concentrations are much higher, which initiated a controversy. However, to date the experimental evidence clearly favors the above mentioned numbers. In healthy adults, the daily glomerular filtration rate is 150–180 liters and therefore the daily filtered load of albumin is in the range of 3300–5760 mg. This corresponds to up to ∼4.3% of total plasma albumin or to ∼10% of the proximal tubular amino acid load.

In the meantime knockout mouse models, such as the CLC-5 or the megalin knockout mouse, have been generated with dramatically reduced renal protein reabsorption (12, 13). In addition, there is a dog model with reduced protein reabsorption (cubilin-deficient dogs) (14, 15). Comparison of albumin excretion in these animals with normal controls has also been used to estimate the amount of filtered albumin, assuming that filtration is not affected and reabsorption abolished. These comparisons indicate that in the absence of reabsorption, albumin excretion increases by a factor of 10 or more. If the data were transferable to humans, they would lead to numbers similar to those described above. These animal models strongly argue against the possibility that albumin concentrations in glomerular ultrafiltrate are in the high mg/liter- or even low g/liter-range, as postulated by some investigators (11). Regardless of the precise filtered amount, it is clear that under physiological conditions less than 1% of filtered albumin appears in final urine. Consequently, it must be reabsorbed efficiently along the renal tubular apparatus.

Receptor-Mediated Endocytosis of Serum Albumin

Albumin appears to be reabsorbed almost evenly in early and late proximal convoluted tubules as well as in the straight tubules, as elegantly shown by a micropuncture study (8). Beyond the proximal tubule there is no significant reabsorption of albumin. Because of its size and the tubule-to-blood concentration ratio, albumin cannot be reabsorbed passively on the paracellular route across the tight junctions. Furthermore, unlike linear peptides, albumin is not cleaved in the tubular lumen and therefore does not cross the apical membrane of proximal tubular cell in the form of free amino acids or dipeptides (2, 8, 16). This is confirmed by enhanced albumin excretion in animals with reduced endocytic activity (12, 17). Thus the only mechanism able to mediate albumin reabsorption is endocytosis (Figure 2).

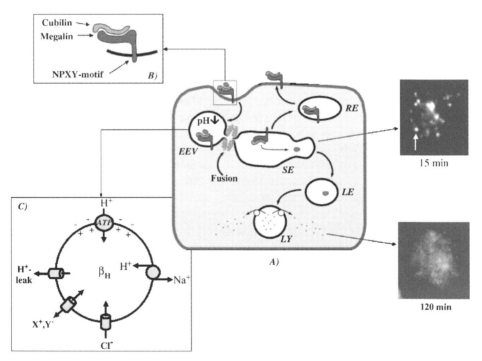

Figure 2   (A) The endocytic pathway of albumin. The two photomicrographs show the endosomal localization of FITC-albumin after 15 min incubation and the diffuse distribution of the degradation products of FITC-albumin after 120 min [experiments were performed in proximal tubule-derived opposum kidney (OK) cells]. EEV, early endocytic vesicle; SE, sorting endosome; LE, late endosome; RE, recycling endosome; LY, lysosome. (B) Scheme of the megalin-cubilin complex at the apical membrane. (C) Model showing the components that contribute to vesicular pH homeostasis. β_H, buffer capacity for protons; X^+, Y^-, unknown ions that may serve as counterions.

During the process of endocytosis, small plasma membrane invaginations are formed at the microvillar base that contain some tubular fluid and thus also molecules dissolved in the fluid. The endocytic invaginations detach from the membrane to form endocytic vesicles that deliver their content to the sorting endosomal (SE) compartment. From the early endosomal compartment part of the material taken up reaches the lysosomal (LY) compartment via late endosomes (LE), whereas another part is brought back to the plasma membrane via recycling endosomes (RE).

Functionally, one can distinguish roughly two basic forms of endocytosis: (*a*) fluid-phase endocytosis and (*b*) adsorptive or receptor-mediated endocytosis (2, 18). During fluid-phase endocytosis the concentration of any given substrate in the endocytic invagination is the same as in the extracellular space, indicating that these substrates are not enriched at the plasma membrane. Thus uptake increases proportionally to the extracellular concentration of the substrate. Classical substrates for this kind of endocytic uptake are dextran or inulin. Inulin is also the prototypical marker for the determination of glomerular filtration rate. Obviously, uptake in the form of fluid-phase endocytosis is a very slow process as can be easily derived from the fact that virtually all inulin filtered in the glomeruli reaches the final urine. Thus fluid-phase endocytosis is quantitatively neglectable. During adsorptive or receptor-mediated endocytosis, substances are concentrated at the cell membrane, and the concentration in the endocytic invagination exceeds the concentration in the extracellular space severalfold. In the case of albumin, we estimated an enrichment at the apical membrane of up to 40-fold, based on comparison with dextran endocytosis (19). This enrichment in the endocytic invaginations makes adsorptive endocytosis far more effective than fluid-phase endocytosis and renders quantitative uptake possible.

Albumin is reabsorbed efficiently along the proximal tubule under physiological conditions. Hence, reabsorption of albumin must be accomplished by adsorptive or receptor-mediated endocytosis. Enrichment of proteins at the plasma membrane can be either from interactions with the negative surface charges of adjacent microvilli (selective constraint model) (2) or from binding to specific sites. Because albumin is an anionic protein, charge interactions cannot explain the quantitative reabsorption of native albumin. This notion is supported by the fact that neutralization of negative surface charges enhances albumin binding, whereas an increase in ionic strength has only a minor effect (M. Gekle, unpublished results).

Furthermore, uptake of albumin is specific in the sense that endocytosis of labeled albumin can be prevented by an excess of unlabeled albumin (19). Functional characterization of albumin binding to the apical membrane of proximal tubule-derived OK cells revealed one binding site with an apparent dissociation constant (K_d) of $100–300 \cdot 10^{-9}$ mol/liter, corresponding to 7–20 mg/liter (20, 21). Binding of labeled albumin can be almost completely prevented by an excess of unlabeled albumin but not by transferrin or lactalbumin. Furthermore, unlabeled albumin also stimulates the dissociation of labeled albumin from the apical membrane (21). The equilibrium-binding affinity is almost the same as the value of apparent affinity for albumin endocytosis in OK cells and in isolated proximal tubules (2, 19, 22). These data show that there is at least one albumin-binding site in the apical membrane of proximal tubular cells with an affinity constant in the range of physiological albumin concentrations in the tubular lumen. Furthermore, these data indicate that under physiological conditions albumin reabsorption is not saturated, especially in later parts of the proximal tubule where luminal albumin concentration is lower. Consequently, an increase in albumin filtration leads to enhanced albumin reabsorption and the protein-load of proximal tubular cells rises during glomerular proteinuria.

The mechanisms underlying receptor-mediated endocytosis can be roughly subdivided into three types with respect to vesicle formation: (*a*) endocytosis via clathrin-coated pits, (*b*) caveolae-mediated endocytosis, and (*c*) clathrin- and caveolae-independent endocytosis by a largely unknown mechanism (18, 23, 24). Clathrin-mediated endocytosis is the best characterized endocytic mechanism and is the predominant pathway for macromolecule uptake along epithelia (16, 23–25). One example for clathrin-mediated endocytosis is the uptake of filtered serum albumin across the apical membrane of renal proximal tubular cells (16, 26, 27). In contrast to endocytosis by clathrin-coated pits, caveolae-mediated endocytosis seems not to be involved in proximal tubular albumin endocytosis (26).

With respect to cargo recruitment during receptor-mediated endocytosis, three basic scenarios can be distinguished on the basis of the class of surface receptors (24). Some receptors are constitutively concentrated in coated pits independently of cargo, such as the transferrin receptor, the low-density lipoprotein receptor (LDLR), or megalin (see below). The cytoplasmic tail of these receptors contains an internalization motif (e.g., YXRF or NPXY) that allows constitutive interaction with the $\mu 2$-subunit of the assembly particle AP2 (part of the clathrin-coat). Consequently, these receptors undergo a constant shuttling between the plasma membrane and the endosomal compartment, like a public bus transportation system. Recently, it was proposed that this process may not be entirely constitutive but may have a certain degree of regulation because components of the AP2 complex can be phosphorylated (18). A second group of receptors undergoes endocytosis only in the presence of a ligand, which enhances the affinity of the receptor for clathrin-coated vesicles. This is comparable to a taxi, only moving when there is a passenger. The best-studied example is probably the EGF-receptor. A third group of receptors, for example, CD4, is retained in the plasma membrane and released only in response to certain intracellular signaling events.

Pharmacological studies have shown that receptor-mediated endocytosis of albumin depends on the integrity of the actin cytoskeleton, because disruption of the cytoskeleton abolishes albumin uptake almost completely (26), which, in part, is from the microvilli stabilizing effect of actin. Furthermore, the apical submembrane actin network, together with myosin motors, seems to help albumin-containing endocytic vesicles to move away from their place of formation in a coordinate fashion. In addition, albumin endocytosis is accelerated by microtubules (26). Disruption of microtubules with nocodazole led to a significantly reduced uptake rate without stopping uptake completely. In summary, the initial phase of vesicle movement from the plasma membrane to the endocytic compartment is essentially dependent on an intact actin cytoskeleton, whereas the latter phase is supported by the interaction with microtubules.

ALBUMIN-BINDING SITES

Knowledge about albumin-binding proteins in renal epithelial cells has been limited for a long time in contrast to data available on albumin-binding proteins in endothelial cells (28). Four endothelial albumin-binding proteins have been described with different preferences for native or modified albumins. One of these albumin-binding proteins, gp60 or albondin, appears responsible for binding and transcytosis of native serum albumin in endothelial cells (28). Apparent affinity of albumin binding to albondin is in the same range as binding to proximal tubular cells ($150 \cdot 10^{-9}$ mol/liter) and is not inhibited by transferrin (28). Cessac-Guillemet et al. attempted to isolate albumin-binding proteins from renal brush-border membranes by affinity chromatography (29). They described two binding proteins with apparent molecular masses of 55 and 31 kDa, similar to the proximal tubular OK cell line (30). Immunohistochemistry shows these binding proteins on microvilli, in endocytic vacuoles, and in lysosomes. No labeling was detected in the inner part of the outer medullary zone. Thus the data are in agreement with micropuncture studies showing that reabsorption takes place only in the proximal tubule. In contrast, functional studies in rat kidney suggest that megalin, a 517-kDa monomeric protein in proximal tubular brush-border (31), is involved in albumin endocytosis (32). Originally, megalin was identified as the primary antigen in Heymann nephritis. Further studies on OK cells, as well as in rat kidney, supported this hypothesis (14, 33, 34). Thus it has been shown that RAP (receptor-associated protein, a classical megalin ligand), as well as antimegalin antibodies, reduces albumin endocytosis (33). Finally, megalin knockout mice show enhanced urinary albumin excretion (14, 34). Megalin, a member of the LDLR family, is a large polyspecific type I cell surface receptor with one transmembrane domain and a relative short cytosolic tail. This cytosolic tail contains a typical cytosolic NPXY motif for clathrin-mediated internalization. The extracellular domain is composed of four clusters of cysteine-rich ligand-binding repeats, of which the second cluster seems to be of special importance for ligand binding (35–37). The binding regions are separated by spacers containing YWTD motifs that are involved in pH-dependent release of ligand (14). Megalin appears to undergo constitutive endocytosis (see above). When ligands are present, they bind to megalin, which delivers its cargo to early and/or late endosomes where a pH-dependent dissociation occurs (35). Thereafter, megalin recycles to the plasma membrane, whereas the ligand may be delivered to lysosomes, as in the case of albumin. This behavior is in agreement with several functional data on albumin endocytosis. Being a polyspecific receptor, megalin binds a variety of ligands [for review see (14)] including vitamin-binding proteins (e.g., vitamin-D-binding protein), carrier proteins (e.g., transthyretin), lipoproteins (e.g., apolipoprotein B), hormones (e.g., parathyroid hormone, PTH), drugs (e.g., aminoglycosides), enzymes (e.g., lipoprotein lipase), immune-related proteins (e.g., light chains), and myoglobin (38). Furthermore, megalin is also expressed in nonrenal tissue such as intestine, choroid plexus, type II pneumocytes, epididymidis, parathyroid gland, thyroid, endometrium, and inner ear.

In addition to acting as a polyspecific-binding protein, megalin can also act as membrane anchor for a peripheral membrane protein, cubilin (16, 39), which is a 460-kDa protein with no apparent transmembrane domain and no GPI-anchor (14). Therefore, cubilin is retained at the apical membrane of proximal tubules by binding to megalin (39) and the two proteins form a scavenger receptor complex (Figure 2). In proximal tubular cells, megalin and cubilin are colocalized in the brush-border, coated pits, endocytic vesicles, and recycling endosomes. Cubilin consists mainly of 27 CUB (complement subcomponents C1r/C1s, EGF-related sea urchin protein, and bone morphogenic protein-1) domains, indicating the potential to bind various ligands (39). The most renown function of cubilin is intestinal reabsorption of intrinsic factor vitamin B_{12} (39), but it also binds to other proteins such as light chains, transferrin, vitamin-D-binding protein, and high-density lipoprotein (14). Proximal tubular cubilin, together with megalin, is involved in albumin endocytosis. Recent studies provide convincing evidence for this model (15, 33). Birn et al. determined that the affinity of albumin for cubilin is ~ 0.6 μmol/liter (15), a value close to the apparent affinity constants for albumin endocytosis (see above). These data suggest that cubilin is the major receptor for renal albumin reabsorption under physiological conditions. Obviously, there is a discrepancy regarding the molecular mass of the supposed binding proteins described by Cessac-Guillemet et al. or Brunskill et al. compared with that of megalin and cubilin. Possibly additional proteins of the proximal tubular apical membrane have the ability to bind albumin, but not all are responsible for physiological reabsorption, depending on their binding affinities. In addition, Western blot experiments showed that anticubilin antibodies also recognize 30-kDa and 60-kDa proteins (15), which are most likely cubilin degradation products. The role of cubilin in proximal tubular protein reabsorption has received further support by the detection of mutations in patients suffering from hereditary megaloblastic anemia I/Imerslund-Gräsbecks disease, which is associated with proteinuria (40).

The binding sites for albumin seem to be subject to long-term regulation in response to protein exposure. It has been shown that there is a dose-dependent decrease in the maximum binding capacity, without changes in binding affinity after exposure to increasing albumin concentrations (20). The underlying mechanisms seem to include enhanced ammoniagenesis, which might affect lysosomal pH and/or act as a stress signal in order to reduce protein uptake. The physiological implication of this long-term downregulation could be the protection of proximal tubular cells from protein overload when glomerular filtration is enhanced.

THE IMPORTANCE OF ENDOSOMAL pH

Once albumin has been taken up by receptor-mediated endocytosis, it is directed to the endosomal compartment and its final destination is the lysosomes. In the lysosomal compartment albumin will be cleaved and the resulting amino acids will exit the cell across the basolateral membrane (Figure 1). Because the albumin-binding proteins megalin and cubilin recycle back to the plasma membrane and are not directed to lysosomes, ligand-receptor dissociation must occur along the endocytic pathway. Subcellular localization of the ligand dissociation process has been investigated in more detail for megalin and its ligand lipoprotein lipase (35). Newly formed megalin-ligand complexes are delivered to early endosomes where dissociation occurs. A common mechanism triggering receptor-ligand dissociation is the drop in pH in the different endocytic compartments (41). Because binding affinity for albumin is pH dependent (21), its dissociation from the binding site is triggered, at least in part, by acidification (pH < 6.5). As long as pH ranges from 7.4 to 6.7, values corresponding to the pH at the beginning and at the end of proximal tubular lumen, binding affinity remains virtually constant. Thus albumin can bind effectively to receptors in the apical membrane of proximal tubular cells and subsequently dissociate as soon as pH has dropped appropriately along the early endocytic compartment. This pathway seems to be the major route for megalin ligands taken up by constitutive endocytosis (35). However, there is at least one interesting exception, namely RAP (35). RAP attaches to megalin in the endoplasmic reticulum during protein synthesis. Because it competes for the binding of other ligands, RAP must dissociate from megalin before it can serve as scavenger receptor. In order to be freed from RAP, which has not detached before reaching the plasma membrane, megalin travels from the plasma membrane to the late endosomal compartment where dissociation occurs in a pH-dependent manner. RAP binding is possibly less pH sensitive and thus requires the lower pH of late endosomes for dissociation.

The functional importance of pH-dependent dissociation for effective albumin reabsorption has been demonstrated experimentally (19, 42). Vesicular alkalinization by either bafilomycin A_1 or millimolar NH_4Cl leads to a dramatic decrease of apparent uptake affinity, as well as maximum uptake rate. At the same time, fluid-phase endocytosis is virtually unaffected, indicating that vesicular alkalinization does not prevent vesicle formation. Furthermore, binding site recycling to the plasma membrane and albumin degradation are impaired. Enhanced renal albumin excretion in states of increased NH_4^+ blood levels may be explained, at least in part, by endosomal alkalinization (2). Of course, these observations do not rule out the possibility that vesicular alkalinization affects endocytosis by additional mechanisms, for example, vesicle trafficking, endosomal fusion events, and coat formation (19, 21, 23, 41–44).

Vesicular acidification is partially accomplished by the vacuole-type H^+-ATPase, which works in parallel with a counterion conductance (Figure 2). The H^+-ATPase not only translocates protons across the membrane but also translocates positive charges counteracting further proton transport. In order to limit the formation of an endosomal-positive membrane potential, either negative charge carriers have to enter vesicles or positive charge carriers have to leave (45). In most cases the counterion conductance seems to consists of Cl^- channels (19, 41, 46, 47), although the presence of cation counter-conductances has also been proposed (48–51). The precise nature of these Cl^- channels is not completely understood, and different studies suggest the involvement of different Cl^- channels in different cell types, such as the cystic fibrosis transmembrane regulator (CFTR), Ca^{2+}-activated Cl^- channels, or CLC-5 Cl^- channels. Furthermore, different counterion conductance might exist in different endosomal subpopulations of one cell type. In proximal tubular cells, CLC-5-type Cl^- channels have been reported to play an important role for counterion conductance in albumin-containing endosomes (52, 53). Furthermore, it has been reported that CLC-5 colocalizes with albumin-containing endosomes but not with dextran-containing endosomes (54), although both populations show bafilomycin A_1-sensitive acidification, which indicates the presence of an H^+-ATPase. Consequently, the conclusion is that at least two different populations of endocytic vesicles exist with respect to counterion conductance and albumin-containing vesicles expressing CLC-5. It is noteworthy that a blocker of CFTR and Ca^{2+}-activated Cl^- channels, NPPB, which does not affect CLC-5 at relevant concentrations, induced alkalinization of dextran-containing endosomes but not of albumin-containing endosomes (55) and exerted virtually no effect on albumin uptake. In addition, loss of CLC-5 reduces proximal tubular protein endocytosis but not fluid-phase endocytosis of dextran (56). The involvement of CLC-5 in albumin endocytosis is also of interest with respect to an inherited kidney disease, Dent's disease, where CLC-5 is mutated (57, 58) and one of the first symptoms observed is proteinuria. The importance of CLC-5 for renal protein reabsorption in vivo has been nicely proven by the generation of CLC-5 knockout mice (12, 13).

The intracellular route of albumin can be summarized as follows: After detachment from the apical membrane, endocytic vesicles deliver the albumin-cubilin-megalin complex to the sorting endosomal compartment. Albumin dissociates from the cubilin-megalin complex because of the low pH. The binding proteins recycle to the apical plasma membrane and albumin itself reaches the lysosomal compartment where it is cleaved.

A Surprising Role for Na^+/H^+-Exchange-3 (NHE-3)

Recently, evidence was presented for the involvement of Na^+/H^+-exchange isoform 3 (NHE-3) in endosomal acidification (46, 59, 60). Na^+/H^+ exchangers are ubiquitous plasma membrane transport proteins involved in cellular pH homeostasis and volume regulation. NHE-3 is the predominant isoform in the apical membrane of proximal tubule cells and cycles between the plasma membrane and the early endosomal compartment, contributing on its way to vesicular acidification (61, 62). Pharmacological and cell biological studies suggest that endosomal

NHE-3, but not plasma membrane NHE-3, plays a role in cubilin-megalin-mediated endocytosis of albumin owing to its contribution to endosomal pH-homeostasis (63, 64). Of course, NHE-3 can acidify endocytic vesicles only in the presence of a sufficient Na^+ gradient. Most likely there is a sufficient Na^+ outward gradient (vesicle-to-cytosol) in early endocytic vesicles that drives NHE-3 in the appropriate direction (Figure 2). The Na^+ concentration in early endocytic vesicles is in the same range as extracellular Na^+ because the ionic composition of early vesicular fluid corresponds to the extracellular milieu. Inhibition of NHE-3 has been shown to disturb early vesicular acidification and to retard albumin endocytosis. Impaired receptor recycling seems to be one consequence, which only partially explains reduced albumin uptake. Another possible mechanism transducing endosomal alkalinization to albumin uptake is vesicle fusion, which is reduced by NHE-3 inhibition (65). In addition, megalin-cubilin trafficking and the formation of carrier vesicles can be affected (23). The importance of vesicular pH in megalin-cubilin trafficking has been confirmed in the CLC-5 knockout mouse (56). Of course, NHE-3 can contribute to only the very early phase of vesicular acidification and cannot compensate for H^+-ATPase. This has been shown indirectly with the CLC-5 knockout model where the H^+-ATPase is functionally impaired and albuminuria occurs (12, 13). The question now was whether NHE-3 inhibition affects endocytosis in vivo and if renal protein handling is affected in NHE-3 knockout mice (66). We tested the effect of pharmacological NHE-3 inhibition, using S3226 and EIPA, on megalin-mediated cytochrome c reabsorption by micropuncture in rat kidney and detected a significant reduction of endocytosis (66a). In addition we could show that NHE-3 knockout mice have tubular proteinuria (66a).

ALBUMIN ENDOCYTOSIS CAN BE MODULATED BY DIFFERENT SIGNALING PATHWAYS

It is well known that endocytosis can be regulated by several mechanisms. However, in the case of albumin reabsorption little is known about regulation. Recent studies have shed some light on the role of different signaling pathways, including heterotrimeric G proteins, protein kinases, PI3-K, and Ca^{2+} in albumin endocytosis. Ca^{2+} does not act as a regulator of albumin endocytosis, but it is a prerequisite for physiological uptake rates (32, 67). Alterations of the cytoplasmic Ca^{2+} concentration had only minor effects on albumin endocytosis. In contrast, complete removal of extracellular Ca^{2+} leads to a dramatic decrease of the apparent uptake affinity (K_m increases from $0.1 \cdot 10^{-6}$ to $2.5 \cdot 10^{-6}$ mol/liter) without affecting the maximal uptake rate, whereas a partial reduction of extracellular Ca^{2+} to $\sim 10^{-6}$ mol/liter is without effect. One possible explanation for the dependence of albumin uptake on the presence of extracellular Ca^{2+} is the Ca^{2+} sensitivity of ligand binding to megalin, especially of cubilin (16). Thus Ca^{2+} is likely a prerequisite for the very first step of the endocytic process, which is binding to the receptor.

Overexpression of a G_i protein subunit in proximal tubule-derived OK cells results in an increase of albumin uptake (68), which is inhibited by pertussis toxin, an inactivator of G_i and G_o proteins. Furthermore, endocytosis in nontransfected cells is reduced by pertussis toxin, whereas binding of albumin is not affected. Activation of G_s proteins by cholera toxin has no effect on albumin endocytosis. These data support a critical role for certain types of heterotrimeric G proteins in the regulation of albumin endocytosis, as was also shown for other endocytic processes. Yet, it is currently not known at which stage G proteins interact with albumin endocytosis. Because albumin binding is not affected, G proteins do not seem to interact with the homeostasis of binding sites but might affect trafficking of internalized ligand. According to the widely accepted model of G protein action, an upstream receptor is linked to a downstream effector via G proteins. Further studies will have to determine whether this is also true for albumin endocytosis regulation and, if so, which is the downstream regulator and which are the physiological upstream activators.

Stimulation of protein kinase A (PKA) via cAMP, forskolin, or by parathyroid hormone (PTH) leads to a decrease of the maximum uptake rate in OK cells without affecting the apparent affinity (26, 67). Kinetic analysis revealed that the effect of PKA is rapid compared with that of protein kinase C (PKC) stimulation (see below) (26). Part of the action of PKA may be attributed to cAMP-induced alkalinization of early endocytic vesicles (see above). In a study using OK cells, NHE-3 deficiency was shown to reduce dramatically cAMP sensitivity of albumin endocytosis and endocytic pH (64). Transfection of NHE-3-deficient cells with human NHE-3 restored cAMP sensitivity. This study indicates that NHE-3 serves as a molecular tool for cAMP-mediated regulation of albumin endocytosis. NHE-3 deficiency results in a redistribution of megalin to intracellular compartments, indicating impaired trafficking (64). Fluid-phase endocytosis was not affected. Of course, phosphorylation of other factors involved in the formation of endocytic invaginations and in budding cannot be excluded (69). Because the formation of intracellular cAMP by the adenylate cyclase is under the negative control of G_i proteins, the downstream effector of G_i protein manipulation (see above) could well be PKA (68). In this case, stimulation of G_i protein would reduce the activity of adenylate cyclase and subsequently the stimulation of PKA. Thus an inhibitory effector for albumin endocytosis would be reduced, thus resulting in an increase of net endocytosis.

In addition to PKA, albumin endocytosis is also modulated by PKC but not by protein kinase G (PKG) (26). Stimulation of PKC led to a reduction of albumin net uptake. In contrast to PKA activation, the effect of PKC activation was characterized by a lag phase of several minutes (26). One explanation for the different time course comes from the sites of action along the endocytic pathway. PKA inhibits internalization of albumin, whereas PKC has a stimulatory effect on re-exocytosis.

Phosphatidylinositol 3-kinase (PI3-K) has been implicated in endocytosis, as well as exocytosis, in a variety of cell types (62, 70–74). Thus PI3-K is also a good candidate for modulation of albumin uptake. Inhibition of PI3-K, using either wortmannin or LY294002, reduced albumin uptake significantly (75). The time course was similar to PKA stimulation and the rate of re-exocytosis was not affected. These data indicate that PI3-K supports internalization of ligand but does not affect re-exocytosis or receptor recycling. Interestingly, NHE-3 is also regulated by PI3-K (62). Inhibition of PI3-K, using either wortmannin or LY294002, reduced the apical NHE-3 activity significantly. Thus it is conceivable that regulation of albumin uptake and NHE-3 by PI3-K are interconnected. Several other pathways potentially involved in regulation of endocytosis, including phospholipase A2 (76), phospholipase D (77), and Rho and Rac (78), have not been investigated with respect to albumin uptake.

Another possibility for regulation of albumin uptake results from interaction of the cytosolic tail of megalin with signaling molecules. There is evidence for an interaction with Disabled protein 2 (79), membrane-associated guanylate kinase with inverted orientation-1 (MAGI-1) (80), and ankyrin-repeat family A protein (ANKRA) (80). In addition, PKA and PKC motifs, as well as SH2 and SH3 recognition sites, have been identified in the cytoplasmic tail of megalin on the basis of sequence comparison (79). Whether these interactions affect endocytosis has to be investigated in detail. Finally, it has been proposed that megalin and NHE-3 form a multimeric complex at the apical membrane of the proximal tubule (81). Because NHE-3 appears to be a regulatory switch for endocytosis, the close interaction with megalin may contribute to specificity of regulation.

PATHOPHYSIOLOGICAL CONSIDERATIONS: THE PROXIMAL TUBULE AS THE SITE OF ALBUMINURIA

In healthy subjects, filtration and reabsorption of albumin are in equilibrium and less than 1% of filtered albumin reaches the final urine. Pathophysiological states with increased renal albumin excretion, albuminuria, can result from two malfunctions: (*a*) An increase of the filtered load can result in glomerular overflow albuminuria when the binding capacity at the brush-border membrane is exceeded (the maximum transport rate of high-affinity reabsorption of albumin in isolated proximal tubules is in the range of 0.05 ng/min per mm tubular length) (2). (*b*) Furthermore, renal excretion of albumin may increase without any change of the tubular load, in the case of a reduced tubular reabsorption capacity. If the reabsorption capacity decreases at a constant filtered load, a growing disequilibrium originates and causes a tubular albuminuria. Thus, in the case of enhanced albumin excretion, one should always consider malfunction of proximal tubular reabsorption.

Impairment of endosomal acidification can affect endocytic albumin reabsorption. Simulation of affinity changes by a simple mathematical model (2, 27) shows the high susceptibility of the system toward small changes of its kinetic parameters: An increase of the affinity constant from $370 \cdot 10^{-9}$ to $1300 \cdot 10^{-9}$ mol/liter would result in an increase of the fractional albumin excretion from values of ~1% to ~60%. Thus it is not surprising that a defect in vesicular counterion conductance of CLC-5, as it occurs in Dent's disease (12, 13, 57, 58), leads to proteinuria. This mechanism is likely also responsible for proteinuria induced by cadmium or cisplatin. Both toxins reduce albumin uptake in proximal tubular cells and lead to endosomal alkalinization (82–84). Finally, proteinuria as a side effect of the COX-inhibitor tenidap has been shown to result from reduced reabsorption along the proximal tubule, probably resulting from endosomal alkalinization (85). Changes in tubular uptake kinetics may also occur by other mechanisms, as in the case of autosomal-dominant polycystic kidney diseases (ADPKD) (86), Imerslund-Gräsbecks disease (40), or the food contaminant ochratoxin A (87). For ADPKD, a loss of the endocytic machinery has been described, although the underlying mechanisms are not known. In patients suffering from Imerslund-Gräsbecks disease, cubilin mutations have been detected. Ochratoxin A reduces the number of albumin-binding sites in the apical membrane and increases the rate of re-exocytosis, thereby causing tubular proteinuria (87). The underlying molecular mechanism in this case has not yet been unveiled but may involve the interaction of the nephrotoxin with different cellular signaling pathways (88, 89). Alterations in proximal tubular signaling by other pathophysiological stimuli may underlie certain forms of proteinuria of unknown origin. Animal studies provided evidence that transforming growth factor-β_1 (TGF-β_1) contributes to proteinuria during hypertension or diabetes (90), possibly by affecting lysosomal activity. Cell culture studies showed that TGF-β_1 inhibits albumin endocytosis by reducing megalin and cubilin expression, as well as by slowing down internalization and degradation. These effects depend on the transcription factors Smad2 and 3 (91).

Tubular protein loss, if massive, leads to reduced intravascular protein concentrations and consequently to reduced oncotic pressure, finally leading to the formation of edema. However, this scenario is not frequent in case of a pure tubular proteinuria. More importantly, one has to consider the proximal tubule as a site of extensive and regulated metabolism and transport. In this respect, transport and metabolism of vitamins and PTH is of especial importance (14). Reabsorption of vitamin-D-binding protein is necessary in order to supply proximal tubular cells with sufficient precursors to synthesize 1,25-OH-calcitriol, and a lack of megalin-cubilin-mediated reabsorption can lead to bone disease (34, 92). On the other hand, reduced PTH-reabsorption will increase tubular PTH concentrations and therefore enhance PTH-signaling (12). Enhanced PTH signaling would increase the rate of 1,25-OH-calcitriol synthesis and thereby potentially counteract the effect of impaired vitamin-D reabsorption. Consequently the effects on bone physiology depend on the delicate balance between these two events (93, 94).

PATHOPHYSIOLOGICAL CONSIDERATIONS: THE PROXIMAL TUBULE AS A TARGET OF ALBUMIN-INDUCED CELL DAMAGE

The interaction of albumin with proximal tubular cells is not restrained to reabsorption. Filtered albumin and some of the substances bound to albumin (such as fatty acids) act as profibrotic and proinflammatory stimuli in proximal tubular cells

(95). Some time ago it was recognized that albuminuria was a major risk factor for the progression of renal diseases (96), contributing to the development of tubulo-interstitial inflammation and fibrosis, although the underlying reasons were not completely understood. The hypothesis then was that albuminuria represented a marker for the extent of glomerular damage, which determined the progression of the disease. However, it is now becoming clear that albumin is not just a marker but a pathogenic factor itself. The interaction of filtered albumin with proximal tubular cells triggers several proinflammatory and profibrotic events within proximal tubular cells (95–99). Although the precise mechanisms leading to protein-induced tubulo-interstitial damage are not completely understood, it has been shown that expression and/or activation of certain signaling molecules, such as NF-κB, AP1, and ERK1/2, play a role (95, 100–103). In addition, the involvement of PKC-activation and reactive oxygen species generation has been suggested (104, 105). Subsequently, proximal tubular cells synthesize various hormones and cytokines (e.g., RANTES, ET-1, MCP-1, IL-6, TNF-α) that transduce the albumin signal to the interstitial tissue, where inflammatory cells are attracted, fibroblasts are stimulated, and blood vessels are constricted. As a consequence, tubulo-interstitial fibrosis develops. In this sense, the proximal tubule can be regarded as transducer of pathophysiological signals. In addition, proximal tubular cells also contribute directly to the formation of extracellular matrix when exposed to elevated albumin concentrations (106). Under these conditions formation and secretion of collagen I, III, and IV are stimulated by NF-κB, PKC, and PKA. These events in concert make proteinuria the most decisive determinant of the progression of renal disease. Recent cell culture studies suggest that proteinuria enhances NHE-3 expression and activity (107, 108). Such a mechanism would support volume expansion and hypertension.

The importance of albumin uptake in proteinuria-induced interstitial nephropathies has not been determined systematically, and it is conceivable that enhanced apical protein-binding suffices in order to elicit the above mentioned effects. However, in recent cell culture studies uptake of albumin was shown to be essential for the pathophysiological effects (106, 109). Moreover, increased extracellular albumin concentrations can even protect cells with low endocytic activity from enhanced collagen formation. Interestingly, cubilin- or megalin-deficient animals seem to show less interstitial inflammatory cell accumulation (110). These data clearly show that endocytosis is necessary for the proinflammatory and profibrotic effects and also help to explain why other tissues do not respond with pathological alterations in the presence of comparable high albumin concentrations in the interstitial space. Of course, these mechanisms also apply to other filtered proteins such as immunoglobulin light chains (111) and advanced glycosylated endproducts (112–114). Recent studies indicate that important pathophysiological agents are small molecular compounds bound to albumin (115–118). Thus albumin also serves as carrier for nephritogenic substances and delivers them to the proximal tubule.

ACKNOWLEDGMENTS

I thank Ruth Freudinger, Sigrid Mildenberger, Birgit Gassner, and Katharina Völker for their excellent technical assistance during the past years. Work from the author's laboratory has been supported by the Deutsche Forschungsgemeinschaft and the Wilhelm-Sander-Stiftung.

The *Annual Review of Physiology* is online at http://physiol.annualreviews.org

LITERATURE CITED

1. Peters T. 1985. Serum albumin. *Adv. Protein Chem.* 37:161–245
2. Maack T, Park CH, Camargo MJF. 1992. Renal filtration, transport and metabolism of proteins. In *The Kidney: Physiology and Pathophysiology*, ed. DW Seldin, G Giebisch, pp. 3005–38. New York: Raven
3. Comper WD, Glasgow EF. 1995. Charge selectivity in kidney ultrafiltration. *Kidney Int.* 47:1242–51
4. D'Amico G, Bazzi C. 2003. Pathophysiology of proteinuria. *Kidney Int.* 63:809–25
5. Guimaraes MA, Nikolovski J, Pratt LM, Greive K, Comper WD. 2003. Anomalous fractional clearance of negatively charged Ficoll relative to uncharged Ficoll. *Am. J. Physiol. Renal Physiol.* 285:F1118–24
6. Russo LM, Bakris GL, Comper WD. 2002. Renal handling of albumin: a critical review of basic concepts and perspective. *Am. J. Kidney Dis.* 39:899–919
7. Haraldsson B, Sorensson J. 2004. Why do we not all have proteinuria? An update of our current understanding of the glomerular barrier. *News Physiol. Sci.* 19:7–10
8. Tojo A, Endou H. 1992. Intrarenal handling of proteins in rats using fractional micropuncture technique. *Am. J. Physiol. Renal Physiol.* 263:F601–6
9. Lund U, Rippe A, Venturoli D, Tenstad O, Grupp A, Rippe B. 2003. Glomerular filtration rate dependence of sieving of albumin and some neutral proteins in rat kidney. *Am. J. Physiol. Renal Physiol.* 284:F1226–34
10. Norden AGW, Lapsley M, Lee PJ, Pusey CD, Scheinman SJ, et al. 2001. Glomerular protein sieving and implications for renal failure in Fanconi syndrome. *Kidney Int.* 60:1885–92
11. Eppel GA, Osicka TM, Pratt LM, Jablonski P, Howden BO, et al. 1999. The return of glomerular-filtered albumin to the rat renal vein. *Kidney Int.* 55:1861–70
12. Piwon A, Günther W, Schwake M, Bösl MR, Jentsch TJ. 2000. CIC-5 Cl⁻-channel disruption impairs endocytosis in a mouse model for Dent's disease. *Nature* 408:369–73
13. Wang SS, Devuyst O, Courtoy PJ, Wang XT, Wang H, et al. 2000. Mice lacking renal chloride channel, CLC-5, are a model for Dent's disease, a nephrolithiasis disorder associated with defective receptor-mediated endocytosis. *Hum. Mol. Genet.* 9:2937–45
14. Christensen EI, Birn H. 2002. Megalin and cubilin: multifunctional endocytic receptors. *Nat. Rev. Mol. Cell Biol.* 3:256–66
15. Birn H, Fyfe JC, Jacobsen C, Mounier F, Verroust PJ, et al. 2000. Cubilin is an albumin binding protein important for renal tubular albumin reabsorption. *J. Clin. Invest.* 105:1353–61
16. Christensen EI, Birn H, Verroust P, Moestrup SK. 1998. Membrane receptors for endocytosis in the renal proximal tubule. *Int. Rev. Cytol.* 180:237–84
17. Willnow TE, Hilpert J, Armstrong SA, Rohlmann A, Hammer RE, et al. 1996.

Defective forebrain development in mice lacking pg30/megalin. *Proc. Natl. Acad. Sci. USA* 93:8460–64

18. Conner SD, Schmid SL. 2003. Regulated portals of entry into the cell. *Nature* 422:37–44

19. Gekle M, Mildenberger S, Freudinger R, Silbernagl S. 1995. Endosomal alkalinization reduces J_{max} and K_m of albumin receptor-mediated endocytosis in OK cells. *Am. J. Physiol. Renal Physiol.* 268:F899–906

20. Gekle M, Mildenberger S, Freudinger R, Silbernagl S. 1998. Long-term protein exposure reduces albumin binding and uptake in proximal tubule-derived opossum kidney cells. *J. Am. Soc. Nephrol.* 9:960–68

21. Gekle M, Mildenberger S, Freudinger R, Silbernagl S. 1996. Functional characterization of albumin binding to the apical membrane of OK cells. *Am. J. Physiol. Renal Physiol.* 271:F286–91

22. Schwegler JS, Heppelmann B, Mildenberger S, Silbernagl S. 1991. Receptor-mediated endocytosis of albumin in cultured opossum kidney cells: a model for proximal tubular protein reabsorption. *Pflügers Arch.* 418:383–92

23. Mukherjee S, Ghosh RN, Maxfield FR. 1997. Endocytosis. *Physiol. Rev.* 77:759–803

24. Schmid SL. 1997. Clathrin-coated vesicle formation and protein sorting: an integrated process. *Annu. Rev. Biochem.* 66:511–48

25. Marshansky V, Bourgoin S, Londono I, Bendayan M, Maranda B, Vinay P. 1997. Receptor-mediated endocytosis in kidney proximal tubules: Recent advances and hypothesis. *Electrophoresis* 18:2661–76

26. Gekle M, Mildenberger S, Freudinger R, Schwerdt G, Silbernagl S. 1997. Albumin endocytosis in OK cells: dependence on actin and microtubules, regulation by protein kinases. *Am. J. Physiol. Renal Physiol.* 272:F668–77

27. Gekle M. 1998. Renal proximal tubular albumin reabsorption: the daily prevention of albuminuria. *News Physiol. Sci.* 13:5–11

28. Schnitzer JE, Oh P. 1994. Albondin-mediated capillary permeability to albumin. *J. Biol. Chem.* 269:6072–82

29. Cessac-Guillemet AL, Mounier F, Borot C, Bakala H, Perichon M, et al. 1996. Characterization and distribution of albumin binding protein in normal rat kidney. *Am. J. Physiol. Renal Physiol.* 271:F101–7

30. Brunskill NJ, Nahorski S, Walls J. 1997. Characteristics of albumin binding to opossum kidney cells and identification of potential receptors. *Pflügers Arch.* 433:497–504

31. Christensen EI, Nielsen S, Moestrup SK, Borre C, Maunsbach AB, et al. 1995. Segmental distribution of the endocytosis receptor gp330 in renal proximal tubules. *Eur. J. Cell Biol.* 66:349–64

32. Cui S, Verroust PJ, Moestrup SK, Christensen EI. 1996. Megalin/gp330 mediates uptake of albumin in renal proximal tubule. *Am. J. Physiol. Renal Physiol.* 271:F900–7

33. Zhai XY, Nielsen H, Birn H, Drumm K, Mildenberger S, et al. 2000. Cubilin- and megalin-mediated uptake of albumin in cultured proximal tubule cells of Opossum kidney. *Kidney Int.* 58:1523–33

34. Nykjaer A, Dragun D, Walther D, Vorum H, Jacobsen C, et al. 1999. An endocytic pathway essential for renal uptake and activation of the steroid 25-(OH) vitamin D3. *Cell* 96:507–15

35. Czekay R-P, Orlando RA, Woodward L, Lundstrom M, Farquhar MG. 1997. Endocytic trafficking of megalin/RAP complexes: dissociation of the complexes in late endosomes. *Mol. Biol. Cell* 8:517–32

36. Orlando RA, Exner M, Czekay R-P, Yamazaki H, Saito A, et al. 1997. Identification of the second cluster of ligand-binding repeats in megalin as a site for

receptor-ligand interactions. *Proc. Natl. Acad. Sci. USA* 94:2368–73

37. Saito A, Pietromonaco S, Kwor-Chieh Loo A, Farquhar MG. 1994. Complete cloning and sequencing of rat gp330/ "megalin," a distinctive member of the low density lipoprotein receptor gene family. *Proc. Natl. Acad. Sci. USA* 91: 9725–29

38. Gburek J, Birn H, Verroust P, Goj B, Jacobsen C, et al. 2003. Renal uptake of myoglobin is mediated by the endocytic receptors megalin and cubilin. *Am. J. Physiol. Renal Physiol.* 285:F451–58

39. Moestrup SK, Kozyraki R, Kristiansen M, Kaysen JH, Rasmussen HH, et al. 1998. The intrinsic factor-vitamin B_{12} receptor and target of teratogenic antibodies is a megalin-binding peripheral membrane protein with homology to developmental proteins. *J. Biol. Chem.* 273:5235–42

40. Aminoff M, Carter JE, Chadwick RB, Johnson C, Grasbeck R, et al. 1999. Mutations in CUBN, encoding the intrinsic factor-vitamin B12 receptor, cubilin, cause hereditary megaloblastic anaemia 1. *Nat. Genet.* 21:309–13

41. Mellman I, Fuchs R, Helenius A. 1986. Acidification of the endocytic and exocytic pathways. *Annu. Rev. Biochem.* 55: 663–700

42. Friis UG, Johansen T. 1996. Dual regulation of the Na^+/H^+-exchange in rat peritoneal mast cells: role of protein kinase C and calcium on pH_i regulation and histamine release. *Br. J. Pharmacol.* 118:1327–34

43. Papkonstanti EA, Emmanouel DS, Gravanis A, Stournaras C. 1996. Na^+/P_i cotransport alters rapidly cytoskeletal protein polymerization dynamics in opossum kidney cells. *Biochem. J.* 315:241–47

44. Storrie B, Desjardins M. 1996. The biogenesis of lysosomes: is it a kiss and run, continuous fusion and fission process? *BioEssays* 18:895–903

45. Rybak SL, Lanni F, Murphy RF. 1997. Theoretical considerations on the role of

membrane potential in the regulation of endosomal pH. *Biophys. J.* 73:674–87

46. Marshansky V, Vinay P. 1996. Proton gradient formation in early endosomes from proximal tubules. *Biochim. Biophys. Acta* 1284:171–80

47. Steinmeyer K, Schwappach B, Bens M, Vandewalle A, Jentsch TJ. 1995. Cloning and functional expression of rat CLC-5, a chloride channel related to kidney disease. *J. Biol. Chem.* 270:31172–77

48. Lukacs GL, Rotstein OD, Grinstein S. 1991. Determinants of the phagosomal pH in macrophages. In situ assessment of vacuolar H^+-ATPase activity, counterion conductance, and H^+ "leak." *J. Biol. Chem.* 266:24540–48

49. Grabe M, Oster G. 2001. Regulation of organelle acidity. *J. Gen. Physiol.* 117:329–44

50. Fuchs R, Male P, Mellman I. 1989. Acidification and ion permeabilities of highly purified rat liver endosomes. *J. Biol. Chem.* 264:2212–20

51. Gaete V, Nunez MT, Glass J. 1991. Cl^-, Na^+, and H^+ fluxes during the acidification of rabbit reticulocyte endocytic vesicles. *J. Bioenerg. Biomembr.* 23:147–60

52. Luyckx VA, Goda FO, Mount DB, Nishio T, Hall A, et al. 1998. Intrarenal and subcellular localization of rat CLC5. *Am. J. Physiol. Renal Physiol.* 275:F761–69

53. Sakamoto H, Sado Y, Naito I, Kwon T-H, Inoue S, et al. 1999. Cellular and subcellular immunolocalization of ClC-5 channel in mouse kidney: colocalization with H^+-ATPase. *Am. J. Physiol. Renal Physiol.* 277:F957–65

54. Devuyst O, Christie PT, Courtoy PJ, Beauwens R, Thakker RV. 1999. Intrarenal and subcellular distribution of the human chloride channel, CLC-5, reveals a pathophysiological basis for Dent's disease. *Hum. Mol. Genet.* 8:247–57

55. Gekle M, Mildenberger S, Freudinger R, Silbernagl S. 1994. pH of endosomes labelled by receptor-mediated and fluid-phase endocytosis and its possible role for

the regulation of endocytotic uptake. In *Studies in Honour of Karl Julius Ullrich. An Australian Symposium*, ed. P Poronnik, DI Cook, JA Young, pp. 45–49. Glebe, NSW, Aust: Wild & Woolley

56. Christensen EI, Devuyst O, Dom G, Nielsen R, Van Der Smissen P, et al. 2003. Loss of chloride channel ClC-5 impairs endocytosis by defective trafficking of megalin and cubilin in kidney proximal tubules. *Proc. Natl. Acad. Sci. USA* 100:8472–77

57. Igarashi T, Günther W, Sekine T, Inatomie J, Shiraga H, et al. 1998. Functional characterization of renal chloride channel, CLCN5, mutations associated with Dent's Japan disease. *Kidney Int.* 54:1850–56

58. Morimoto T, Uchida S, Sakamoto H, Kondo Y, Hanamizu H, et al. 1998. Mutations in CLCN5 chloride channel in Japanese patients with low molecular weight proteinuria. *J. Am. Soc. Nephrol.* 9: 811–18

59. Kapus A, Grinstein S, Wasan S, Kandasamy R, Orlowski J. 1994. Functional characterization of three isoforms of the Na^+/H^+ exchanger stably expressed in Chinese hamster ovary cells. *J. Biol. Chem.* 269:23544–52

60. D'Souza S, Garcia-Cabado A, Yu F, Teter K, Lukacs G, et al. 1998. The epithelial sodium-hydrogen antiporter Na^+/H^+ exchanger 3 accumulates and is functional in recycling endosomes. *J. Biol. Chem.* 273:2035–43

61. Janecki AJ, Montrose MH, Zimniak P, Zweibaum A, Tse CM, et al. 1998. Subcellular redistribution is involved in acute regulation of the brush border Na^+/H^+ exchanger isoform 3 in human colon adenocarcinoma cell line Caco-2. *J. Biol. Chem.* 273:8790–98

62. Kurashima K, Szabó EZ, Lukacs G, Orlowski J. 1998. Endosomal recycling of the Na^+/H^+ exchanger NHE3 isoform is regulated by the phosphatidylinositol 3-kinase pathway. *J. Biol. Chem.* 273:20828–36

63. Gekle M, Drumm K, Mildenberger S, Freudinger R, Gaßner B, Silbernagl S. 1999. Inhibition of Na^+-H^+ exchange impairs receptor-mediated albumin endocytosis in renal proximal tubule-derived epithelial cells from opossum. *J. Physiol.* 520:709–21

64. Gekle M, Drumm K, Serrano OK, Mildenberger S, Freudinger R, et al. 2002. Na^+/H^+-exchange-3 serves as a molecular tool for cAMP-mediated regulation of receptor-mediated endocytosis. *Am. J. Physiol. Renal Physiol.* 283:F549–58

65. Gekle M, Freudinger R, Mildenberger S. 2001. Inhibition of Na^+-H^+ exchanger-3 interferes with apical receptor-mediated endocytosis via vesicle fusion. *J. Physiol.* 531.1:619–29

66. Schultheis PJ, Clarke LL, Meneton P, Miller ML, Soleimani M, et al. 1998. Renal and intestinal absorptive defects in mice lacking the NHE3 Na^+/H^+ exchanger. *Nat. Genet.* 19:282–85

66a. Gekle M, Völker K, Mildenberger S, Freudinger R, Shull GE, Wiemann M. 2004. NHE3 Na^+/H^+ exchanger supports proximal tubular protein reabsorption in vivo. *Am. J. Physiol. Renal Physiol.* 287:F469–73

67. Gekle M, Mildenberger S, Freudinger R, Silbernagl S. 1995. Kinetics of receptor-mediated endocytosis of albumin in cells derived from the proximal tubule of the kidney (Opossum kidney cells): influence of Ca^{2+} and cAMP. *Pflügers Arch.* 430:374–80

68. Brunskill NJ, Cockcroft N, Nahorski S, Walls J. 1996. Albumin endocytosis is regulated by heterotrimeric GTP-binding protein $G\alpha_{i-3}$ in opossum kidney cells. *Am. J. Physiol. Renal Physiol.* 271:F356–64

69. Slepnev VI, Ochoa G-C, Butler MH, Grabs D, De Camilli P. 1998. Role of phosphorylation in regulation of the assembly of endocytic coat complexes. *Science* 281:821–24

70. Jones SM, Howell KE. 1997. Phosphatidylinositol 3-kinase is required for the formation of constitutive transport vesicles from the TGN. *J. Cell Biol.* 139: 339–49

71. Li G, D'Souza-Schorey C, Barbieri MA, Roberts RL, Klippel A, et al. 1995. Evidence for phosphatidylinositol 3-kinase as a regulator of endocytosis via activation of Rab5. *Proc. Natl. Acad. Sci. USA* 92:10207–11

72. Yano H, Nakanishi S, Kimura K, Hanai N, Saitoh Y, et al. 1993. Inhibition of histamine secretion by Wortmannin through the blockade of phosphatidylinositol 3-kinase in RBL-2H3 cells. *J. Biol. Chem.* 268:25846–56

73. De Camilli P, Emr SD, McPherson PS, Novick P. 1996. Phosphoinositides as regulators in membrane traffic. *Science* 271:1533–39

74. Joly M, Kazlauskas A, Corvera S. 1995. Phosphatidylinositol 3-kinase activity is required at a postendocytic step in platelet-derived growth factor receptor trafficking. *J. Biol. Chem.* 270:13225–30

75. Brunskill NJ, Stuart J, Tobin AB, Walls J, Nahorski S. 1998. Receptor-mediated endocytosis of albumin by kidney proximal tubule cells is regulated by phosphatidylinositide 3-kinase. *J. Clin. Invest.* 101:2140–50

76. Mayorga LS, Colombo MI, Lennartz M, Brown EJ, Rahman KH, et al. 1993. Inhibition of endosome fusion by phospholipase A_2 (PlA2) inhibitors points to a role for PLA2 in endocytosis. *Proc. Natl. Acad. Sci. USA* 90:10255–59

77. Jones AT, Clague MJ. 1997. Regulation of early endosome fusion by phospholipase D activity. *Biochem. Biophys. Res. Commun.* 236:285–88

78. Lamaze C, Chuang TH, Terlecky LJ, Bokock GM, Schmid SL. 1996. Regulation of receptor-mediated endocytosis by Rho and Rac. *Nature* 382:177–79

79. Oleinikov AV, Zhao J, Makker SP. 2000. Cytosolic acceptor protein Dab2 is an intracellular ligand of endocytic receptor gp600/megalin. *Biochem. J.* 347:613–21

80. Patrie KM, Drescher AJ, Goyal M, Wiggins RC, Margolis B. 2001. The membrane-associated guanylate kinase protein MAGI-1 binds megalin and is present in glomerular podocytes. *J. Am. Soc. Nephrol.* 12:667–77

81. Biemesderfer D, Nagy T, DeGray B, Aronson PS. 1999. Specific association of megalin and the Na^+/H^+ exchanger isoform NHE3 in the proximal tubule. *J. Biol. Chem.* 274:17518–24

82. Choi JS, Kim KR, Ahn DW, Park YS. 1999. Cadmium inhibits albumin endocytosis in opossum kidney epithelial cells. *Toxicol. Appl. Pharmacol.* 161:146–52

83. Herak-Kramberger CM, Brown D, Sabolic I. 1998. Cadmium inhibits vacuolar H^+-ATPase and endocytosis in rat kidney cortex. *Kidney Int.* 53:1713–26

84. Takano M, Nakanishi N, Kitahara Y, Sasaki Y, Murakami T, Nagai J. 2002. Cisplatin-induced inhibition of receptor-mediated endocytosis of protein in the kidney. *Kidney Int.* 62:1707–17

85. Aleo MD, Wang T, Giebisch G, Sanders MJ, Walsh AH, Lopez-Anaya A. 1996. Model development and analysis of tenidap-induced proteinuria in the rat. *J. Pharmacol. Exp. Ther.* 279:1318–26

86. Obermüller N, Kränzlein B, Blum WF, Gretz N, Witzgall R. 2001. An endocytosis defect as a possible cause of proteinuria in polycystic kidney disease. *Am. J. Physiol. Renal Physiol.* 280:F244–53

87. Gekle M, Mildenberger S, Freudinger R, Silbernagl S. 1994. The mycotoxin ochratoxin A impairs protein uptake in cells derived from the proximal tubule of the kidney (opossum kidney cells). *J. Pharmacol. Exp. Ther.* 271:1–6

88. Benesic A, Mildenberger S, Gekle M. 2000. Nephritogenic ochratoxin A interferes with hormonal signaling of immortalized human kidney epithelial cells. *Pflügers Arch.* 439:278–87

89. Schramek H, Wilflingseder D, Pollack V, Freudinger R, Mildenberger S, Gekle M. 1997. Ochratoxin A-induced stimulation of extracellular signal-regulated kinases 1/2 is associated with Madin-Darby canine kidney-C7 cell dedifferentiation. *J. Pharmacol. Exp. Ther.* 283:1460–68

90. Russo LM, Osicka TM, Bonnet F, Jerums G, Comper WD. 2002. Albuminuria in hypertension is linked to altered lysosomal activity and TGF-β1 expression. *Hypertension* 39:281–86

91. Gekle M, Knaus P, Nielsen R, Mildenberger S, Freudinger R, et al. 2003. TGF-1 reduces megalin/cubilin-mediated endocytosis of albumin in proximal tubule derived OK-cells. *J. Physiol.* 552:471–81

92. Nykjaer A, Fyfe JC, Kozyraki R, Leheste JR, Jacobsen C, et al. 2001. Cubilin dysfunction causes abnormal metabolism of the steroid hormone 25(OH) vitamin D_3. *Proc. Natl. Acad. Sci. USA* 98:13895–900

93. Devuyst O, Guggino WB. 2002. Chloride channels in the kidney: lessons learned from knockout animals. *Am. J. Physiol. Renal Physiol.* 283:F1176–91

94. Gunther W, Piwon N, Jentsch TJ. 2003. The ClC-5 chloride channel knock-out mouse—an animal model for Dent's disease. *Pflugers Arch.* 445:456–62

95. Burton C, Harris KPG. 1996. The role of proteinuria in the progression of chronic renal failure. *Am. J. Kidney Dis.* 27:765–75

96. Jerums G, Panagiotopoulos S, Tsalamandris C, Allen TJ, Gilbert RE, Comper WD. 1997. Why is proteinuria such an important risk factor for progression in clinical trials. *Kidney Int.* 52:S87–92

97. Bruzzi I, Benigni A, Remuzzi G. 1997. Role of increased glomerular protein traffic in the progression of renal failure. *Kidney Int.* 52:S29–31

98. Chen L, Wang Y, Tay Y-C, Harris DCH. 1997. Proteinuria and tubulointerstitial injury. *Kidney Int.* 52:S60–62

99. Zoja C, Donadelli R, Colleoni S, Figliuzzi M, Bonazzola S, et al. 1998. Protein overload stimulates RANTES production by proximal tubular cells depending on NF-κB activation. *Kidney Int.* 53:1608–15

100. Guijarro C, Egido J. 2001. Transcription factor-κB (NF-κB) and renal disease. *Kidney Int.* 59:415–24

101. Dixon R, Brunskill NJ. 1999. Activation of mitogen pathways by albumin in kidney proximal tubule epithelial cells: implications for proteinuric states. *J. Am. Soc. Nephrol.* 10:1487–97

102. Wang Y, Rangan GK, Tay Y-C, Harris DCH. 1999. Induction of monocyte chemoattractant protein-1 by albumin is mediated by nuclear factor κB in proximal tubule cells. *J. Am. Soc. Nephrol.* 10:1204–13

103. Drumm K, Bauer B, Freudinger R, Gekle M. 2003. Albumin induces NF-κB expression in human proximal tubule-derived cells (IHKE-1). *Cell Physiol. Biochem.* 12:187–96

104. Morigi M, Macconi D, Zoja C, Donadelli R, Buelli S, et al. 2002. Protein overload-induced NF-κB activation in proximal tubular cells requires H_2O_2 through a PKC-dependent pathway. *J. Am. Soc. Nephrol.* 13:1179–89

105. Drumm K, Gassner B, Silbernagl S, Gekle M. 2001. Albumin in the mg/l-range activates NF-κB in renal proximal tubule-derived cell lines via tyrosine kinases and protein kinase C. *Eur. J. Med. Res.* 6:247–58

106. Wohlfarth V, Drumm K, Mildenberger S, Freudinger R, Gekle M. 2003. Protein uptake disturbs collagen homeostasis in proximal tubule-derived cells. *Kidney Int. Suppl.* 84:103–9

107. Klisic J, Zhang J, Nief V, Reyes L, Moe OW, Ambuhl PM. 2003. Albumin regulates the Na^+/H^+ exchanger 3 in OKP cells. *J. Am. Soc. Nephrol.* 14:3008–16

108. Lee EM, Pollock CA, Drumm K, Barden JA, Poronnik P. 2003. Effects of pathophysiological concentrations of albumin on NHE3 activity and cell proliferation

in primary cultures of human proximal tubule cells. *Am. J. Physiol. Renal. Physiol.* 285:F748–57

109. Drumm K, Gassner B, Silbernagl S, Gekle M. 2001. Inhibition of Na⁺/H⁺ exchange decreases albumin-induced NF-κB activation in renal proximal tubular cell lines (OK and LLC-PK1 cells). *Eur. J. Med. Res.* 6:422–32

110. Abbate M, Remuzzi G. 2001. Novel mechanism(s) implicated in tubular albumin reabsorption and handling. *Am. J. Kidney Dis.* 38:198–204

111. Sengul S, Zwizinski C, Batuman V. 2003. Role of MAPK pathways in light chain-induced cytokine production in human proximal tubule cells. *Am. J. Physiol. Renal Physiol.* 284:F1245–54

112. Saito A, Nagai R, Horiuchi S, Hama H, Cho K, et al. 2003. Role of megalin in endocytosis of advanced glycation end products: implications for a novel protein binding to both megalin and advanced glycation end products. *J. Am. Soc. Nephrol.* 14:1123–31

113. Bendayan M, Londono I. 1996. Reabsorption of native and glycated albumin by renal proximal tubular epithelial cells. *Am. J. Physiol. Renal Physiol.* 271:F261–68

114. Chen S, Cohen MP, Ziyadeh FN. 2000. Amadori-glycated albumin in diabetic nephropathy: pathophysiologic connections. *Kidney Int.* 58:S40–44

115. Thomas ME, Harris KP, Walls J, Furness PN, Brunskill NJ. 2002. Fatty acids exacerbate tubulointerstitial injury in protein-overload proteinuria. *Am. J. Physiol. Renal Physiol.* 283:F640–47

116. Arici M, Brown J, Williams M, Harris KP, Walls J, Brunskill NJ. 2002. Fatty acids carried on albumin modulate proximal tubular cell fibronectin production: a role for protein kinase C. *Nephrol. Dial. Transplant* 17:1751–57

117. Arici M, Chana R, Lewington A, Brown J, Brunskill NJ. 2003. Stimulation of proximal tubular cell apoptosis by albumin-bound fatty acids mediated by peroxisome proliferator activated receptor-gamma. *J. Am. Soc. Nephrol.* 14:17–27

118. Kamijo A, Kimura K, Sugaya T, Yamanouchi M, Hase H, et al. 2002. Urinary free fatty acids bound to albumin aggravate tubulointerstitial damage. *Kidney Int.* 62:1628–37

Annu. Rev. Physiol. 2005. 67:595–621
doi: 10.1146/annurev.physiol.67.040403.102553
Copyright © 2005 by Annual Reviews. All rights reserved
First published online as a Review in Advance on October 19, 2004

EXOCYTOSIS OF LUNG SURFACTANT: From the Secretory Vesicle to the Air-Liquid Interface

Paul Dietl[1] and Thomas Haller[2]

[1]Department of General Physiology, University of Ulm, Ulm, D 89069, Germany;
[2]Department of Physiology, Medical University of Innsbruck, A-6020 Innsbruck, Austria;
email: paul.dietl@medizen.uni-ulm.de; thomas.haller@uibk.ac.at

Key Words type II pneumocyte, secretion, fusion pore, lysosome, surface tension

■ **Abstract** Exocytosis is fundamental in biology and requires an orchestra of pro-teins and other constituents to fuse a vesicle with the plasma membrane. Although the molecular fusion machinery appears to be well conserved in evolution, the process it-self varies considerably with regard to the diversity of physico-chemical and structural factors that govern the delay between stimulus and fusion, the expansion of the fusion pore, the release of vesicle content, and, finally, its extracellular dispersion. Exocytosis of surfactant is unique in many of these aspects. This review deals with the secretory pathway of pulmonary surfactant from the type II cell to the air-liquid interface, with focus on the distinct mechanisms and regulation of lamellar body (LB) fusion and release. We also discuss the fate of secreted material until it is rearranged into units that finally function to reduce the surface tension in the lung.

INTRODUCTION

Lung tissue exerts a strong retractive force that tends to squeeze out air, which may result in atelectasis. The discovery of surface tension as the major component of this force was made as early as 1929 (1), but it was not until the 1950s and early 1960s that active surface material from the lung was isolated and characterized (2, 3), thereby demonstrating that its deficiency causes IRDS (infant respiratory distress syndrome) (4, 4a) and identifying the lamellar body (LB) as the intracellular storage site of surfactant (5, 6). The first convincing evidence in favor of an exocytotic release mechanism was provided by electron microscopy (EM) studies (7).

Surfactant consists of lipids (mainly phospholipids) and specific proteins. Its biosynthesis and composition have been reviewed elsewhere in detail and thus we give only a brief summary of these mechanisms (7a,b).

The phospholipid composition of LBs isolated from type II pneumocytes is similar to that of whole lung surfactant obtained from broncho-alveolar lavage (BAL) (8, 9). Therefore, essentially all alveolar surfactant phospholipids are se-creted via exocytosis of LBs (10). Whole surfactant additionally contains four specific proteins (SP-A, SP-B, SP-C, and SP-D), which account for about 10% by weight (11). Their respective distributions within LBs and BAL are different: The small hydrophobic SP-B and SP-C are localized within LBs and co-secreted with LB contents. Both proteins are believed to play an important role in squeezing out non-DPPC (dipalmitoyl phosphatidylcholine) components during film formation and compression, which results in a highly DPPC-enriched surface film (12). Hence, their exocytotic pathway appears to match that of lipids. In contrast, the large hydrophilic SP-A and SP-D are secreted largely independently of LB contents. Although SP-A inhibits exocytosis of LBs in vitro (see below), these proteins appear to be mainly involved in pulmonary host defense. As this review deals with exocytosis of LBs, their secretory pathways are not discussed further.

SURFACTANT SECRETION AND TURNOVER

In the alveolar lumen, surfactant lipids and proteins are long-lived materials for which biological half lives of more than 10 to several tens of hours have been found (13–15). These high values result from a slow clearance from the alveoli toward the upper airways, and, additionally, from continuous recycling between alveolar lining fluid (ALF) and type II cells by reuptake and resecretion. Owing to this dynamic behavior, the biological half life yields no specific information about the rate of secretion. Therefore, the turnover time (defined as the time of appearance of an amount of a substance equal to its amount present in the compartment) was introduced by several investigators (15–17). It represents an estimate of the net transfer of a material from one compartment to the other, and results—in the case of a recycling substance—from the difference between two unidirectional fluxes (18). For the secretory path, i.e., the transfer of surfactant material from the LB fraction to the alveolar lumen, turnover times between about 4 and 11 h (18) were calculated. According to the above definition, this is the time needed to replenish the alveolar surfactant pool by ongoing secretion and recycling.

In vitro, the rate of secretion as a unidirectional flux can be assessed in an experimental setting in which the extracellular fluid volume is almost unlimited and/or continuously replaced. This situation applies for lung slices or isolated type II cells grown on an artificial substrate and bathed in an experimental solution. The first isolation techniques for alveolar type II cells evolved in the 1970s and were further improved for a high yield and purity of type II cells (19). The insufficient sensitivity of conventional biochemical assays for the detection of minute amounts of surfactant prompted experimenters to use radiolabeled precursors of disaturated phospholipids. Evidently, the accumulation of radioactive material over time in the cell supernatant is a parameter for the time course of LB exocytosis. Time/activity plots of released radiolabels have been established in response to a multitude of stimuli (20–29) and reveal many important insights into the factors that directly stimulate the type II cell in culture. As a common important result, these studies also demonstrated that the period between cell stimulation and termination of phospholipid secretion is exceedingly long, up to several hours, indicating that the entire exocytotic process is very slow.

STEPS WITHIN THE EXOCYTOTIC PROCESS

The methods described above have detected the endpoint of a complex process. Within the prefusion phase, directed vesicle trafficking, docking, and a series of reactions defined as priming take place to make a vesicle fusion-competent. Fusion competence generally refers to a condition in which a vesicle is able to fuse with the plasma membrane in response to a sudden increase of the cytoplasmic Ca^{2+} concentration ($[Ca^{2+}]_c$). These definitions are mainly derived from studies on excitable cells, and far less is known in epithelial cells, although some basic mechanisms appear to be the same. The post-fusion phase denotes the expansion of the fusion pore and the release and dispersal of the stored substances.

To investigate LB fusion, different approaches have been used. Haller et al. developed a method based on the lipid-staining properties of FM 1–43 to study LB fusion in intact type II cells (30). With this method it is possible to observe how LB content is released and to analyze the dynamics of the fusion pore (see below). In vitro fusion assays were based on LB-liposome (31, 32) and LB-plasma membrane fusion (33). Patch-clamp techniques were introduced to study the change in cell capacitance during secretion (34), and permeabilized cell preparations were used to investigate the roles of soluble intracellular components for surfactant secretion (35).

THE PREFUSION PHASE

Modes of Stimulation

Regulated exocytosis is initiated by a certain stimulus, either chemical or physical. Many agonists have been described that directly activate the type II cell and include β_2-mimetic drugs (20, 23, 25), ATP and other purinergic receptor agonists (24, 36), gastrin-releasing peptide (37), prostaglandin derivatives (38), vasopressin (39), and endothelin-1 (40). Alkalosis activates the isolated perfused lung (29) and type II cells (41). Distension of the lung in situ (42–46) or ex situ (47–50), and of type II cells in culture (28, 51), is a potent stimulus. Secretion is highly temperature dependent, consistent with an active process (52–54). For a comprehensive summary of agonists, second messengers, and other factors reported to affect surfactant secretion, which would greatly exceed the scope of this article, the reader is referred to several excellent reviews (55–58).

Among the various stimuli found under experimental conditions, lung distension is considered to be physiologically most relevant. At least part of this effect is probably mediated by a direct mechanical strain of type II cells (59–61). Adrenergic stimulation is likely to play a role during labor and birth (62), but its relevance during exercise and other conditions is less clear. Cholinergic stimulation is indirect in rats and higher vertebrates (23, 63–65) but important in lower animals (52). The physiological relevance of purinergic stimulation is still obscure. The model of ATP release during strain in terms of an autocrine mechanism could not be confirmed in type II cells grown on silastic membranes (51). Other modes of ATP release in the lung, which could be significant, have not yet been determined.

The Role of Ca^{2+} for LB Fusion

Ca^{2+}-induced fusion is perceived as the defining step of regulated exocytosis, considering the ubiquitous role of Ca^{2+} signaling in secretion (66). Substantial direct and indirect evidence indicates that Ca^{2+} is also the major, if not exclusive (see below), second messenger that triggers the LB fusion event in the type II cell (19, 28, 30, 35, 36, 40, 51, 67–72). In general, Ca^{2+}-induced fusion is highly diverse in terms of the origin and spatio-temporal spreading of the Ca^{2+} signal, the size and intracellular location of the exocytotic vesicles, and the resulting temporal and spatial behavior of the fusion response. Owing to fixed and mobile Ca^{2+} buffers in the cytoplasm, both the duration and amplitude of an elevation of $[Ca^{2+}]_c$ critically depend on the distance from the Ca^{2+} mobilization or entry site (73). For example, Neher calculated that a vesicle at 20 nm distance from a voltage-gated Ca^{2+} channel is subject to a $[Ca^{2+}]_c$ rise to about 100 μM, which lasts for microseconds, whereas a vesicle 200 nm away experiences a $[Ca^{2+}]_c$ rise of 5–10 μM lasting for $\sim$10 ms. Both the duration of the Ca^{2+} signal and the distance from the plasma membrane define the likelihood for a vesicle to fuse. Kasai established a theoretical model in which this probability equals the product of the rate constant of exocytosis (which in turn depends on the proximity between vesicle and plasma membrane and vesicle translocation barriers) and the duration of the Ca^{2+} spike (74). This temporal aspect between Ca^{2+} signaling and fusion is consistent with type II cells, where the agonist-induced fusion response is a function of the integrated $[Ca^{2+}]_c$ over time (69). With regard to Ca^{2+} concentrations, attention must be paid to Ca^{2+} microdomains, which are hardly accessible to analysis with Ca^{2+}-specific dyes because they are themselves Ca^{2+} buffers and report $[Ca^{2+}]_c$ changes with a limited time resolution only. To overcome these problems, caged Ca^{2+} compounds are frequently employed to generate an instantaneous and presumably homogeneous global $[Ca^{2+}]_c$ increase. In the type II cell, this approach and others (ionomycin, etc.) yielded a very high Ca^{2+} sensitivity of the fusion machinery with a threshold $\sim$320 nM $[Ca^{2+}]_c$ (70, 71). This is among the lowest values reported for exocytotic fusion in mammalian cells. In general, this Ca^{2+} sensitivity exhibits a high degree of diversity between cell types and even vesicle populations within a single cell (such as small versus large dense-core vesicles). Ca^{2+}-binding constants between a few and several hundreds of micromolar were reported (66). Owing to the complex spatio-temporal behavior of vesicle fusion noted above, these variations do not necessarily imply differences in the intrinsic Ca^{2+} sensitivities of individual vesicles or the coexistence of both high- and low-affinity Ca^{2+} sensors. Although there is certainly more than one Ca^{2+}-sensitive element involved in the entire exocytotic process, it is still unclear whether a unique or various Ca^{2+} sensors with different affinities trigger the terminal steps of fusion (66).

Intracellular Pathways for Ca^{2+}

Pathways for Ca^{2+} and spatio-temporal aspects of Ca^{2+} signals in type II cells are still poorly described. Adenosine nucleotides, vasopressin, and mastoparan have been reported to cause phosphatidylinositol hydrolysis and generation of inositol 1,4,5-trisphosphate (IP$_3$) (39, 75, 76). Furthermore, agonist-induced mobilization of Ca^{2+} from intracellular stores (68, 72, 76) is consistent with the known action of IP$_3$ to activate a ligand-gated ion channel in the endoplasmic reticulum. This is in accordance with other exocrine cells, in which the endoplasmic reticulum is considered as the major source of Ca^{2+} that triggers secretion (74). In addition to the endoplasmic reticulum, LBs are another storage site for Ca^{2+} with a reported concentration of 6.6 mmol/kg dry weight (77). LBs actively pump Ca^{2+}, H$^+$, and Cl$^-$ ions into their lumen (77–81). Therefore, a role for Ca^{2+} signaling has been suggested but is still questionable. A possible involvement for stimulated LB fusion is conceivable because increased Ca^{2+} concentrations are found in those LBs that are located near the apical membrane (77). Furthermore, LBs in cultured type II cells continuously lose Ca^{2+} (79), a process that is accompanied by a gradual loss of secretory potency. Finally, lysosomes, which share many properties with LBs (see below), are a site of IP$_3$-mediated Ca^{2+} release in other epithelial cells (82). However, Ca^{2+} in LBs may serve functions unrelated to fusion (see below), but related to storage, packaging, and processing of surfactant components (83), or to supply the ALF with Ca^{2+} (84, 85).

Ca^{2+} Entry

Ca^{2+} channels in the plasma membrane of type II cells have attracted little attention. Agonists that mobilize Ca^{2+} from intracellular stores generally activate a Ca^{2+} entry pathway, known as capacitative or store-operated [initially proposed by Putney (86)]. Its presence in type II cells is strongly supported by the observation (M. Frick, unpublished data) that inhibition of the endoplasmic Ca^{2+} ATPase by thapsigargin results in large elevations of [Ca^{2+}]$_c$ when Ca^{2+} is added to the cells, but neither the mode of activation nor the molecular entities of this pathway are yet clearly defined. However, recent data suggest that part of a large family of ion channels, designated TRP, is involved. Thus far a Ca^{2+}-impermeable TRP channel with yet unknown physiological function was recently observed in rat and human type II cells (86a). The significance of capacitative Ca^{2+} entry for sustained LB fusion is supported by the observation that the fusion activity in response to ATP is considerably shortened when Ca^{2+} is removed from the bath (69).

Voltage-gated (L-type) Ca^{2+} channels in type II cells were suggested to mediate Ca^{2+} entry because Ca^{2+} channel blockers inhibit secretion (40, 87, 88). However, the actions of dihydropyridine derivatives in type II cells were found to be nonselective and nonstereospecific, and no voltage-gated currents could be found in patch-clamp experiments (89). Hence, as in other epithelial cells, these channels should be considered as not expressed.

Mechanosensitive ion channels are particularly intriguing candidates for the generation of Ca^{2+} signals (59). Single-cell analysis of $[Ca^{2+}]_c$ during the entire strain/relaxation cycle revealed activation of a Gd^{3+}-insensitive Ca^{2+} entry pathway when the cell surface area was increased by more than 8% (51). The activation of this Ca^{2+} entry pathway, which was not identified in earlier investigations (28), was found to be a prerequisite for LB fusion (51). Ca^{2+} entry, in turn, activates a Ca^{2+}-induced Ca^{2+} release mechanism, both entry and release adding to a strain-induced elevation of $[Ca^{2+}]_c$ (28, 51). This synergism is different from the action of Ca^{2+}-dependent agonists, where intracellular Ca^{2+} release does not depend on the presence of Ca^{2+} in the extracellular space (68, 90). Yet, it is not clear if, or to what extent, the intracellular Ca^{2+} pool involved in Ca^{2+}-induced Ca^{2+} release overlaps with the IP_3-sensitive pool; for example, Ca^{2+} entry could activate a Ca^{2+}-sensitive phospholipase C (PLC) and/or increase the affinity of the IP_3 receptor, as proposed for mechanical indentation of bone cells (91).

Strain-induced Ca^{2+} channels in the plasma membrane are particularly elusive because of the technical difficulties in applying an equibiaxial cell strain and performing a patch-clamp experiment simultaneously. Therefore, functional characteristics (ion selectivity, conductance, etc.) of the strain-induced Ca^{2+} channels have not yet been determined and could be different from mechanosensitive pathways activated by other mechanical stimuli, such as hypotonic swelling (outlined in detail in Reference 92).

Mechanotransduction in the alveolus is certainly complex and not restricted to the type II cells. In an elegant study applying single-cell imaging to intact lungs, Ashino et al. found inflation-induced Ca^{2+} signals to originate in type I cells and to spread onto type II cells through gap junctions (93). Recently the same group demonstrated that epithelial Ca^{2+} signals and LB fusions are also inducible by an increase in pulmonary capillary pressure (94). The gap junction blocker heptanol abolished this effect, which was interpreted by these authors as further evidence that type I cells are the prime mechanosensors in the alveolar units.

Irrespective of the origin of Ca^{2+} signals in the lung, which is not entirely clear, pulmonary mechanotransduction bears many additional open questions: Cell-cell contacts, for example, are per se important prerequisites for the generation of strain-induced Ca^{2+} entry because type II cells lacking these contacts in vitro are not responsive to even high amounts of strain (51). This suggests that cell-cell contacts determine the tensegrity structure and hence the sensitivity to strain by influencing the microfilament network, which propagates a certain deformation to a mechanosensitive element (95). Hence, alveolar remodeling in the course of epithelial injury/repair could severely alter the mechanosensitive properties of alveolar cells and the secretory response to lung inflation. Whereas strain-induced LB fusion appears to be exclusively related to the elevation of $[Ca^{2+}]_c$, both in vitro and in vivo (28, 51, 93), other responses, such as differentiation and proliferation, are likely mediated via additional pathways (60, 96). It is reasonable to assume that a single mechanosensor does not exist but that the combination of various mechanosensitive elements determines the final outcome of a certain mechanical stimulus. Finally, we propose that there may be a considerable overlap, if not a common mechanism, between mechanically and agonist-induced LB fusion (see below).

Structural Aspects and Possible Ca^{2+} Sensors Involved in LB Fusion

LBs are large organelles that contain lysosomal enzymes (97) and late endosomal marker proteins (98). They also have an acidic pH < 6.1, established by a vacuolar H^+-ATPase as the prime active component driving H^+ (and also Ca^{2+}) uptake (81). According to these features, they can be classified as lysosome-related organelles (83).

It is becoming increasingly evident that lysosomes can undergo regulated secretion in various cells (99). These secretory lysosomes share many features with LBs. They may also be as large as >1 μm, and they slowly fuse with the plasma membrane upon a moderate elevation of $[Ca^{2+}]_c$ to low micromolar levels (99). The molecular machinery and Ca^{2+} sensors for LB exocytosis are, as for all secretory lysosomes, poorly understood (99). Pharmacological (100, 101) and ultrastructural (102) evidence suggests the involvement of calmodulin. Liu et al. (35) found, in permeabilized type II cells, that several annexins increased secretion in the submicromolar $[Ca^{2+}]_c$ range. Fusion between liposomes and LBs and surfactant secretion were mediated by synexin in the presence of Ca^{2+} (67, 103). Recently, evidence for the involvement of the SNARE proteins SNAP 23, syntaxin1, and alpha-SNAP was presented (104, 105). Synaptobrevin, syntaxin, and Snap-25 were identified by Western blot analysis (106). The role of synaptotagmin, a major candidate for Ca^{2+}-triggered fusion and putative Ca^{2+} sensor in other secretory lysosomes (99), as well as the effects of clostridial toxins, has not yet been investigated in type II cells. As for all exocytotic systems, a definite role of the SNARE proteins has not been established.

A Mechanistic Model for LB Fusion

Upon stimulation, the first LB fusion events follow with a considerable delay (usually several seconds) after the $[Ca^{2+}]_c$ peak. Other vesicles then fuse sequentially over a time period of minutes up to half an hour (69). This behavior may reflect different distances of LBs to the plasma membrane, corresponding to their poorly organized, scattered distribution within type II cells in vitro and in vivo. Considering the transient nature of Ca^{2+} signals, it is evident that early LB fusions take place when $[Ca^{2+}]_c$ is still high, whereas late fusions occur when $[Ca^{2+}]_c$ has already declined to resting levels. As a consequence, $[Ca^{2+}]_c$ at the time of individual fusion events is highly variable, but, as noted above, the overall cellular fusion response is significantly correlated with the integrated $[Ca^{2+}]_c$ over time (69).

This time dependency suggests that an elevated $[Ca^{2+}]_c$ maintains a condition in which LB fusion is facilitated. This condition most likely involves facilitated LB transfer to and docking with the plasma membrane. In addition to translocation of LBs by contractile elements, such a facilitated movement could be mediated by the Ca^{2+}-dependent removal of a physical barrier, in particular the cortical filamentous actin as demonstrated in chromaffin cells (107). Several lines of evidence are in favor of such a mechanism in type II cells.

Rose et al. (108) found that a toxin-induced decay of cellular F-actin with a concomitant increase in monomeric actin stimulated basal secretion. Tsilibary & Williams (109) reported that LBs in the process of secretion are surrounded by actin-like material and that the F-actin-disrupting agent cytochalasin D caused a considerable change in the shape of type II cells. They suggested a contractile mechanism by which LBs are moved toward the plasma membrane. Furthermore, the microtubule-disrupting agent colchicine inhibits inflation- and agonist-induced secretion (110, 111). Cytoskeletal involvement during constitutive and regulated secretion is also suggested by proteolysis and reorganization of the cytoskeleton by calpain (112–114).

The notion that LB fusion with the plasma membrane is brought about by pulling/pushing forces along with the disassembly of a dominant cortical fusion clamp prompts the question whether all types of LB fusion can be explained by a common mechanical mechanism. This corresponds to a hypothesis put forward by Muallem et al. (115) that actin filament disassembly is a sufficient final trigger for exocytosis in nonexcitable cells. Indeed, many EM images demonstrate bulging of the plasma membrane by LBs in the process of secretion (116, 117). Interestingly, ATP-induced Ca^{2+} spikes coinciding with single LB fusion events were observed, and these Ca^{2+} spikes depended on the presence of bath Ca^{2+} (116). A yet untested hypothesis proposes that local stress of the plasma membrane, caused by the protrusion of an LB where the actin clamp is released, activates a Ca^{2+} entry pathway, which completes fusion and accelerates the expansion of the fusion pore (see below).

Ca^{2+}-Independent Pathways

In addition to Ca^{2+}, several other signaling pathways were reported in the control of surfactant secretion. The most powerful stimulus is treatment with phorbol esters (27), which activates protein kinase C (PKC) but does not cause $[Ca^{2+}]_c$ to increase. Exocytosis without a Ca^{2+} signal is common to many epithelial cells (118). Nevertheless, the action of phorbol ester is not Ca^{2+} independent because chelation of intracellular Ca^{2+} completely blocks fusion elicited by PMA or flash photolysis of caged diacylglycerol (DAG) (69, 71). A direct action of PKC in vesicle fusion, bypassing Ca^{2+}, has not been described. In pituitary gonadotropes, PKC was suggested to sensitize the exocytotic machinery for Ca^{2+} (119), an action that we also proposed for type II cells (69). The target proteins by which PKC exerts its effect on LB exocytosis are not known.

The significance of PKC activation for LB exocytosis under physiological conditions is based on circumstantial evidence. Liu et al. (120) found that cyclic

strain causes activation of phospholipases C and D (PLC, PLD) and a translocation of PKC from cytosol to membrane. However, these effects were related to
an increase of cell proliferation, not secretion. The activation of PKC and its subtypes by purinergic stimulation has been intensively investigated (58). However,
as noted above, purinergic compounds also stimulate intracellular Ca^{2+} release,
and it is difficult to dissect the overall secretory response into single components.
A role of PKC was inferred by the use of PKC antagonists (27, 121). Chander
et al. (21) reported that downregulation of PKC by pretreatment with phorbol ester
causes a greatly blunted secretory response to ATP (40% of control). However,
this procedure also downregulates the Ca^{2+} signal (101). It is likely, therefore, that
in addition to its possible role as a Ca^{2+} sensitizer (see above), PKC modulates
purinergic Ca^{2+} signals. This idea is supported by the observation (G. Siber, unpublished data) that the ATP-induced elevation of $[Ca^{2+}]_c$ is often biphasic, with
an initial peak immediately after its addition, and a second one following ~ 10 min
later. This temporal pattern nearly coincides with that of DAG formation (75).

Less potent than purinergic stimulation is that via the β_2-adrenergic receptor.
β_2-mimetic drugs lead to a production of cAMP (20, 25). Isolated rat type II
cells respond to β_2-mimetic drugs, xanthine derivatives, and cell-permeable cAMP
but not cGMP analogues with an increased secretion (20, 23, 25). In addition,
terbutaline caused a small and transient elevation of $[Ca^{2+}]_c$ (72). At present, it
is unknown if the cAMP-dependent stimulation of secretion is mediated by this
elevation of $[Ca^{2+}]_c$ exclusively or if, in addition, cAMP potentiates the actions
of Ca^{2+}. As in the case of PKC, a direct and Ca^{2+}-independent effect of protein
kinase A on exocytotic vesicle fusion has not yet been determined.

THE POST-FUSION PHASE

Fusion Pore Expansion

The earliest event in secretion is the formation of a fusion pore, an aqueous channel between LB lumen and the extracellular space, which results in a stepwise
increase of the type II cell capacitance (34). In general, initial fusion pores are
unstable structures, eventually causing a capacitance flicker before they fully open
or close again (122). However, this has not yet been shown in type II cells. Following formation, the fate of the pore depends on factors related to the composition
of vesicle contents and on those related to the cell and its structural components.
Hydrophilic secretory products in dense core vesicles, for example, are bound to
a proteoglycan matrix, which swells when exposed to the surrounding medium,
thereby aiding the expansion of the fusion pore and liberating the bound material.
Alvarez de Toledo et al. (123) proposed that this release mechanism is advantageous for large vesicles where simple diffusion would take a "prohibitively long
time."

The situation is quite different in LBs. As described below, their lipid content does not favor rapid water uptake and vesicle swelling. This assumption is

consistent with morphological (124) and functional evidence that fusion pores in type II cells are stable, long-lasting structures (125). If swelling of LB contents is not a major determinant, what are the other factors governing the expansion of these pores to permit full release? Fusion pore diameters have been quantified by the analysis of diffusion rates of the fluorescent dye FM 1–43. These studies showed that fusion pores expand discontinuously (at randomly changing rates) and slowly (up to hours) (116). Any elevation of $[Ca^{2+}]_c$ accelerates this process, as does mechanical strain of the type II cell (116, 126). In addition, fusion pores are also mechanical barriers for surfactant release (126). This resistance to mechanical forces suggests that a rigid structure surrounds the pore and counteracts the tendency of the membrane curvature at the pore to increase its radius and to flatten out. Naturally, Ca^{2+}-responsive cytoskeletal elements are the most likely candidates, although it is not yet clear if the effect of strain is a direct one (a mechanical transduction through cytoskeletal elements) or a consequence of Ca^{2+} entry (see above). SP-A appears to inhibit fusion pore expansion, as it inhibits release without affecting secretagogue-induced LB fusion (127). This clearly emphasizes the need to distinguish between pre- and post-fusion events in the mode of action of potential modulators of secretion. As the molecular composition of fusion pores is not clear, mechanistic views on post-fusion actions remain purely speculative.

Transient Fusion

Transient fusion denotes a condition where a fusion pore is transiently formed but recloses again. In other cell types, this was shown to be associated with partial release of vesicle contents (123). Observations of selective dye uptake into a restricted number of LBs suggest that transient fusions may also take place in type II cells (122). It is questionable if transient LB fusion would be accompanied by partial surfactant release. It could, however, result in a significant release of small and unbound LB constituents into the ALF (see below). Whether this might contribute to the Ca^{2+} and pH homeostasis of the hypophase remains speculative.

On the other hand, transient fusions could possibly allow substances from the hypophase to gain "backward" access into the fused LB (via the fusion pore) and thus be taken up into the type II cell again. Such a new mechanism in cell biology could operate in parallel to the endocytotic pathway and contribute to the recycling of spent surfactant.

LB-LB Fusion and Compound Exocytosis

Fusion between multivesicular bodies and LBs is frequently observed (128). Similarly, two or more lamellated inclusions can be surrounded by a common limiting membrane, suggesting fusion between these organelles even at late stages of vesicle maturation. Although it seems clear that interorganellar fusion contributes to the biogenesis of LBs (83), much less is known about the physiological importance of LB-LB fusion for compound exocytosis. This defines a condition where intracellular vesicles fuse with each other and share only one fusion pore at the plasma membrane. Compound exocytosis is a mechanism to focus the release of considerable quantities of substrates to a target region of the cell (129), which otherwise lacks sufficient area to support many fused vesicles side by side. The existence of compound exocytosis in type II cells was clearly demonstrated by simultaneous patch-clamp and fluorescence microscopy experiments (34) and is also suggested by EM studies (116). It is not known if this homotypic fusion is governed by the same factors that regulate LB-plasma membrane fusion. It also remains to be determined if compound exocytosis is a necessary process to allocate a sufficient amount of surfactant to the hypophase under conditions of increased demand. Furthermore, compound exocytosis is likely associated with a delayed and sustained mode of release.

From Fused LBs into the ALF

Fusion pore formation marks the transition of surfactant from intracellular storage to its further destiny in the alveolar space. It is characterized by the separation of the membranes, which previously were connected to each other during a state called hemifusion, and which ultimately expose the vesicle lumen to the extracellular fluid. Any further consideration of the surfactant's life cycle is therefore based on the particular assumptions made for the physical and chemical conditions prevailing outside the fused vesicles.

ALF

The most accepted view of the type II cell's microenvironment is that of an aqueous layer, ranging in thickness from several nanometers up to >1 μm (84), with a tendency to be thickest in the septal corners, the preferred location of the type II cells. Owing to the difficulties encountered in analyzing such a thin and dynamic layer, its volume and extent are not exactly known, and its continuity over the alveolar surface is disputed (130). Undoubtedly, however, these cells have to release surfactant into an exceedingly small volume, estimated to be 0.1–0.5 ml/kg for humans. Thus the hypophase would still qualify as a bulk liquid phase for small molecules but not with respect to the dimensions of LB contents. This microenvironment is certainly unique among exocytotic cells and could critically affect the mode of release, depending on such localized variations in ALF-thickness or its effective lateral extension (131). Also, a high concentration of hydrophobic material in close proximity to the fusion pore could have profound effects. Indeed, ultrastructural studies of the ALF provide us with the view of a physically heterogeneous layer, differing in thickness as well as morphological appearance. Because it is difficult to mimic such an extracellular matrix, most studies on secretion have been carried out with a bulk water phase, ignoring the cell-specific microenvironment, in particular an air-liquid interface in close proximity to the cells. Similarly, the problem of surfactant secretion without exposure to an aqueous phase (into a "dry alveolus") has been addressed (131) but not tested so far. Only a few attempts have been made to include an air-liquid interface (132) or phospholipids (133) in cell culture systems.

Whatever the local physical condition in the ALF, the bulk of it is reported to contain 1.6 mM of Ca^{2+} (85) and to have a pH of about 6.9 (134). As noted above, this composition is similar to that of intact LBs. However, an accurate determination even of this bulk composition is not trivial to perform, and many factors, such as osmolarity, are not precisely known. The aqueous phase of LBs, in turn, is even more inaccessible to analysis, but a water content of ~15% has been inferred from NMR studies (135). The water is located between the lipid bilayers and might be immobilized by hydration to the phospholipid head groups. Yet, to classify LBs as dry (136) may not be accurate, as they are able to accumulate hydrophilic tracers in considerable quantities (30). The amount of water taken up by LB material in fused vesicles is difficult to predict owing to unknown osmotic gradients. In contrast, water uptake is likely to occur when LB content is completely freed from the limiting membrane, as the released material appears in a swollen state. In contrast, the pH gradient should quickly dissipate once the lumen of LBs and the ALF are conjunct. Indeed, polar tracers trapped within prefused LBs disappear after pore formation, suggesting an equilibration of aqueous LB content with that of the ALF. This applies for all solutes for which a gradient may exist (e.g., Na^+) (77). However, in the type II cells, equilibration still proceeds at a much slower pace (30) than that in exocytotic vesicles, which lack such an immobile content. Thus, the impact of fluid and electrolyte exchange on LB content immediately after vesicle fusion is not known.

Through the Pore

After pore formation, surfactant starts to extrude into the extracellular space. This process has been captured by numerous ultrastructural studies using transmission or, less frequently, scanning electron microscopy (125). As a general feature, the limiting membrane of LBs is in continuity with the plasma membrane in an omega-shape, and this seems to be preserved even when vesicle content has been extruded completely (137). By morphometric analysis, Kliewer et al. (124) observed that the fusion pores (elliptical $0.2 \times 0.4 \, \mu m$) are about half the diameter of LBs. They were the first to propose that the restriction of these pores imposes a deformation of LB material during release, which could be important for the transformation into functional units. This is eventually the case because shear stress on membrane sheets has been postulated to promote tubular myelin formation (138). Furthermore, handling of surfactant preparations was found to have profound effects on surfactant subtype conversion (139). To our knowledge, essentially all EM studies show that during the process of extrusion LB material has a constriction at the site of the fusion pore and a less densely packed structure protruding into the extracellular space (116), which is sometimes interpreted as self-decomposition (140). Williams also noted (141) that the innermost lamellae appear to be extruded first, whereas the more peripheral layers are left behind, severely affecting LB structure. In line with this, real-time observations by laser scanning microscopy demonstrated occasional but profound changes in the structure of extruded LB material in living cells. Long tubes ($\sim 100 \, \mu m$) resembling those of a retractable telescope have been reported (142, 143). These and the observations by EM are intriguing as surfactant seems to be squeezed through the pores (132). However, we do not know the forces that could drive this process, and it remains to be determined whether structural changes at this early stage of extracellular surfactant metabolism are of functional significance. Despite the restriction of the fusion pore, it is likely that nonsoluble LB content is extruded as a single unit and not by partial release as some investigators have claimed (140), because it is difficult to envision the nature of forces responsible for the separation of this compact material into smaller units during release.

In the ALF

After release, LB material appears in manifold structures: lamellar body–like particles (LBPs, see below), tubular myelin (TM), a lipid-protein film, small vesicles (SVs) and several others (18, 144). These structures are in a metabolic sequence, firmly established by many investigations (145, 146), and represent different steps of conversion of a parent material, undoubtedly the LBPs. Their structure is generally similar to that of LB contents within the cells (147). Multiple layers ($\sim$3–40) of tightly packed membranes, each $\sim$8 nm thick, are arranged to form spherical particles (141). In many cases they also have a homogeneous core that additionally may contain a few unilamellated vesicles (141). These structures are reminiscent of the proposed origin of LBs from multivesicular bodies (137, 83). From cytochemical studies, it was demonstrated that lysosomal enzymes are also present in secreted LBPs. In particular, both LBs and LBPs contain an acid phosphatase reactive core (141). In addition to these criteria, evidence for a proximate state of LBPs in the course of conversion stems from functional studies. Before the onset of air breathing in fetal animals, surfactant is mainly present in its most compact form (predominantly LBPs), but with the commencement of air breathing, it is increasingly found in smaller, less dense aggregates (148). This is consistent with in vivo studies demonstrating a sequential turnover of radiolabels from intracellular LBs via LBPs/TM to light fractions and in vitro cycling experiments (see below) where a decline of LBPs is counterbalanced by an increase in the smaller, less surface active fractions (145); for the terminology used to characterize surfactant subtypes, see Reference 149. Finally, LBPs exhibit a high surface activity similar to that of LBs isolated from homogenized lungs (146, 150, 151). From all these observations, it is justified to define LBPs as the proximate and immediate product released by the type II cells and as the direct precursor of many, if not all, other surfactant forms.

So far, LBPs are released into the ALF, but what is their further fate? Certainly, the transit of material from LBPs into the alveolar air-liquid interface is not instantaneous (135). The turnover time of LBPs in the lung has been estimated to be of the order of $\sim$0.7 h (145). Therefore, the important implication is that LPBs are indeed stable for that period of time. However, it is this apparent stability that bears many unresolved questions. For example, some investigators found that even intact LBs isolated from rat lung homogenates are highly unstable and rapidly disorganize into extended lamellar whorls (~30 min) (152). In contrast, the majority of other experimental data support the opposite view: LBPs in the process of release are not subject to a rapid dispersal, and when released into a physiological saline, they preserve their size and typical morphology for many days (143). This is indirectly supported by subtype fractionation studies carried out with protracted centrifugation times (e.g., 60 h). These measurements would simply not be meaningful if LBPs progressively dissolve during the experiments. Likewise, amniotic fluids contain LBPs in considerable quantities ($>5 \times 10^7$/ml in humans), which is incompatible with a short life time of these particles. Finally, a study using laser tweezers (153) reported an amazingly high resistance of LBPs to mechanical forces. These particles could be stretched reversibly into long tethers, recoiling to a spherical granule after each stretch, and no spontaneous disorganization could be observed—a further clue that other mechanisms must exist to convert LBPs in vivo.

Cycling

The situation in the alveolus could differ from experimental settings with respect to physical, chemical, or biological conditions. Obviously, the most particular situation in the alveolus is an air-liquid interface subject to periodic changes during the ventilation cycle. To imitate this situation, Gross & Narine developed a protocol they called cycling: Plastic tubes, half filled with experimental solution, were rotated end-over-end to create cyclic changes in the surface area by a factor of ~9, 80 times per min. Since its introduction, this technique has been widely used to study the mechanisms of particle conversions in vitro (145). By exposing nascent surfactant, harvested from pretreated intact lungs, to a 2-h cycling protocol, Gross & Schultz found a progressive conversion of LBPs to TM and SVs (154). These subtypes were separated by virtue of their specific buoyant densities in linear sucrose gradients. Interestingly, subtype conversion proceeded from one distinct subtype into the other, with no intermediate forms. Further examinations confirmed that the subtypes produced by cycling were identical in many aspects to those found in BAL, suggesting that this procedure indeed mimics the physical situation in vivo. Moreover, particle conversion was related to the surface area available, the number of cycles, and the relative changes in surface area during each cycle, but it occurred neither in static suspensions nor in suspension subject to stirring only (155). In line with Hall et al. (156), Veldhuizen et al. (157) assumed adsorption of surfactant at a changing air-liquid interface to be the most important step for conversion. They argued that factors known to increase the rate of adsorption (e.g., high amplitude surface change, 37°C) also increased conversion, whereas factors known to decrease adsorption (addition of proteins) also decreased conversion. Recently, a microscopic technique has been introduced to study LBP interactions at an air-liquid interface directly (143). This study revealed that contact of LBPs with the interface is essential for particle conversion and occurs instantaneously (<1 s), as soon as LBPs hit the interface. Furthermore, this process stopped when the interface was occupied by previously transformed material or when surface tension was <40 mN/m. Thus, it was concluded that the surface tension is the driving force for LBP disintegration and deposition of transformed material in the interface. This conclusion is consistent with that of Hall et al. (156), who emphasized the importance of interfacial effects on both particle adsorption and conversion. However, as Veldhuizen pointed out (157), conversion is a more complex process that cannot depend on adsorption exclusively. For example, surface preoccupancy by proteins affects this process, although conflicting results were reported [decreased (156, 157) versus increased (157, 158) conversion rates], leaving space for an additional mechanism.

Convertase

Conversion of LBPs by cycling has been demonstrated in surfactant preparations from a variety of species, e.g., sheep, rabbit, and bovine. Common to all these studies is the finding of a strict dependency of conversion on cycling, which supports in vivo observations that small tidal volume ventilation results in a lower rate of aggregate conversion (159, 160). However, a further level of complexity was introduced when it was found that subtype conversion is temperature dependent (145) and can be blocked when serine protease inhibitors (in particular α_1-antitrypsin) are added to murine preparations (154). The conclusion was straightforward, as was the name given for this putative enzyme, convertase (144). At first glance, this hypothesis seems appealing because it provides a new level of regulation for extracellular surfactant metabolism and plausible because it was known previously that LBs contain cytochemically demonstrable enzymes. Indeed, the convertase could be a product of the type II cells, co-secreted with lipids into the ALF (161, 162). However, as Williams pointed out earlier (141), a co-secretion of enzymes with their substrates would necessitate that their activity be blocked within the vesicles and reactivated once the material is released. In principle, this could be caused by a change in a specific ionic condition after exocytosis. However, this simple view is complicated by several facts: The convertase does not require Ca^{2+} and has a broad pH range (155). Thus there is no apparent reason for activation by release. Instead, convertase has the highly unusual requirement of a cyclically compressed and expanded air-liquid interface in order to express activity (155). In addition, the substrates of convertase and their specific mode of interaction with the enzyme are still obscure (155). Furthermore, convertase was found to be ineffective in converting LBPs directly, but was able to act on structures that are already substantially transformed (TM). This is supported by the finding that enzyme inhibitors attenuate degradation of compact subtypes, but without impairing the rate of film formation, the minimum surface tension, and surface tension area loops (163). Likewise, reconstituted surfactants (PL + SP-A + SP-B) are surface active but require the purified enzyme to show significant particle conversion on cycling (164). In contrast, other reports challenged the significance of this concept: Particle conversion is seen in many preparations despite a lack of convertase (157, 164), and Hall et al. (156) provided evidence that conversion is the result of interfacial phenomena rather than enzymatic processes. Nevertheless, a role for convertase, probably in the generation of SVs from a preexisting film, is suggested, but thus far no enzymatic process of physiological relevance is known to degrade LBPs directly.

Tubular Myelin

TM was considered an intermediate step between LBPs and the surface monolayer (151). This remarkable structure, first described by Policard (165), is a semicrystalline orientation of phospholipid membranes into a regular three-dimensional lattice. Its tubular nature was first established by Weibel et al. (166), giving rise to the now accepted nomenclature for this organized complex, although its fine structure (crossed versus corrugated bilayers) is still not clear. An intriguing argument that TM directly originates from LBPs stems from morphological observations. Membranes of LBPs, in particular the outermost, were found to be contiguous with TM in many preparations (141, 167), suggesting that LBPs, usually several in concert (168), unspool, and their membranes become stabilized in this lattice structure. Williams (141) further observed that extracellular LBPs contain structurally specialized areas (mini-lattices) with some similarities to TM. Others found parallel ribs on the lamellar fracture planes on freeze-fractured LBs (169). Both findings led to the speculation either that TM could be preassembled within LBs and simply expand when freed from the limiting membrane or that LBs contain prearranged lipoprotein complexes (parallel ribs) that serve as a template to organize TM in the extracellular fluid (141). To date, however, both speculations still lack direct experimental proof. Gil & Reiss (151), who first isolated TM from homogenized lungs, provided the first hint on its formation. When EDTA was added to their isolation media, no TM was found, suggesting that Ca^{2+} is required for its formation. Two subsequent reports, both using intact LBs, underscored this effect. Paul et al. (150) showed that LBs are highly surface active in the presence of Ca^{2+}, and Sanders et al. (152) demonstrated that incubation of LBs with Ca^{2+} leads to TM formation, obviously without the need of other factors. At first glance, such an effect is not surprising, as divalent cations neutralize the surface charge of polar head groups, leading to such diverse effects as wound sealing or membrane merger in various biological systems. However, Ca^{2+} might not be the only factor. SP-A aggregates surfactant lipids in a Ca^{2+}-dependent fashion (170), induces formation of multilamellar structures (171), is associated with the corners of intersecting TM bilayers (172), and increases the rate of adsorption of isolated LBs (173). In fact, the particles seen on transition membranes and TM (141, 174) might indeed represent ongoing SP-A interaction with uncoiling LBP membranes. Therefore, in stabilizing this nanostructure (175), Ca^{2+}/SP-A could act in driving the unfolding of LBPs into a planar lattice, and the completion of membrane transition into tubular structures could then be brought about by SP-B by promoting further lipid mixing and/or lipid bilayer fusions (176). This final assembly of phospholipids, which contain proteins not present in LBPs (151), could, in fact, serve as the immediate storage form of functional surfactant, as evidenced by its capability of forming surface films that rapidly adsorb and greatly decrease surface tension (146, 151). Thus TM has the morphological appearance, the chemical composition (86% saturated lecithins), and the physical properties to be considered a metabolic product of LBPs and a precursor of the surface film (151). Several EM studies even show TM figures that are closely associated with the alveolar-lining layer (159, 177), and models exist to explain how a three-dimensional structure could feed and stabilize a monomolecular film (178).

However, there are several conflicting aspects arguing against a strict product (surface film) -precursor (TM) relationship. A high surface activity of TM is not confirmed by all investigators (53, 140). Organic solvent extracts of lavaged lung surfactant exhibit rapid adsorption, equivalent to that of whole surfactant, as do a variety of synthetic phospholipid/protein mixtures, but owing to the lack of SP-A, neither of these preparations forms TM (179). Similarly, rapid film formation is not abolished after TM structure is destroyed by an excess of Ca^{2+} (180). Transgenic models of SP-A null mutant mice have almost no TM despite normal lung functions under resting as well as exercise-induced stress conditions (181). Moreover, their surfactant was able to restore lung volume when applied to surfactant-deficient preterm rabbits. Therefore, it seems that TM is associated with, but not entirely required for, an optimized surfactant function in vivo. In light of the known lipid polymorphism common to apolar/polar phase systems, it is likely that TM represents a reversible (180) and energetically favored dispersion of amphiphilic matter, but serves no specific function other than being an extracellular reservoir (151). With regard to its relative low abundance in natural surfactants (159, 180), this function could be questioned as well. Alternatively, however, TM could be a breakdown product desorbed from an already existing film. Much less attention has been paid to this possibility, although some evidence supports this view. For example, TM could be found in type II cell cultures only when the extracellular fluid surface was exposed to air (132). In this regard our knowledge of TM function has not much advanced since 1987, when Wright & Clements noted that the data were too inconclusive to allow a clear statement (18).

In conclusion, the molecular details of LBP conversion are still unclear. The main obstacle to understanding these transitions in more detail is the fact that LBPs are highly organized pleomorphic structures, combined with the technical challenge to analyze LBP interactions with an air-liquid interface directly. Experiments relying on bulk measurements only (e.g., cycling) are not suited to unravel this complexity, in particular to define the forces acting on LBPs and to dynamically resolve bilayer transitions. Some recent advances, including scanning force microscopy of small submerged bubbles (182) and a fluorescent-inverted interface using nascent LBPs (143), are examples of potentially promising techniques.

SUMMARY

The delivery of surfactant from its intracellular storage to its final destination is unique in many aspects. Most notably, the secretory material is a highly organized lipid structure that has to pass an aqueous environment before rearrangement at the air-exposed surface. Therefore, a sequence of steps, each under distinct control, follows the initial formation of the exocytotic fusion pore, including a period before LB material permeates the expanding pore, its transformation within the ALF, and its integration into the air-liquid interface. The regulated way by which these steps finally reduce surface tension makes the post-fusion phase of surfactant exocytosis particularly important. In contrast to other exocytotic systems, a tight coupling between vesicle fusion with the plasma membrane and the final function of the secretory product is not apparent. Rather, it appears that the intracellular fusion machinery is adapted to a regulated but sluggish supply of stored material. In this regard, LBs share many aspects with secretory lysosomes in other cells, where secretion is at the interface between constitutive and regulated exocytosis.

The *Annual Review of Physiology* is online at
http://physiol.annualreviews.org

LITERATURE CITED

1. Von Neergaard K. 1929. Neue Auffassungen über einen Grundbegriff der Atemmechanik. *Z. Ges. Expt. Med.* 66:373–94

2. Clements JA. 1997. Lung surfactant: a personal perspective. *Annu. Rev. Physiol.* 59:1–21

3. Pattle RE. 1965. Surface lining of lung alveoli. *Physiol. Rev.* 45:48–79

4. Avery ME, Mead J. 1959. Surface properties in relation to atelectasis and hyaline membrane disease. *Am. J. Dis. Child.* 97:517–23

4a. Avery ME, Taeusch HW, Floros J. 1986. Surfactant replacement. *N. Engl. J. Med.* 315:825–26

5. Buckingham S, Avery ME. 1962. Time of appearance of lung surfactant in the foetal mouse. *Nature* 193:688–89

6. Campiche M. 1960. Les inclusions lamellaires des cellules alveolaires dans le pneumon du raton. *J. Ultrastruct. Res.* 3:302–12

7. Ryan US, Ryan JW, Smith DS. 1975. Alveolar type II cells: studies on the mode of release of lamellar bodies. *Tissue Cell* 7:587–99

7a. Van Golde LMG, Batenburg JJ, Robertson B. 1988. The pulmonary surfactant system: biochemical aspects and functional significance. *Physiol. Rev.* 68:374–455

7b. Rooney SA, Young S, Mendelson CR. 1994. Molecular and cellular processing of surfactant. *FASEB J.* 8:957–67

8. Veldhuizen R, Nag K, Orgeig S, Possmayer F. 1998. The role of lipids in pulmonary surfactant. *Biochim. Biophys. Acta* 1408:90–108

9. Kikkawa Y, Smith F. 1983. Cellular and biochemical aspects of pulmonary surfactant in health and disease. *Lab. Invest.* 49:122–39

10. Askin FB, Kuhn C. 1971. The cellular origin of pulmonary surfactant. *Lab. Invest.* 25:260–68

11. Veldhuizen EJ, Haagsman HP. 2000. Role of pulmonary surfactant components in surface film formation and dynamics. *Biochim. Biophys. Acta* 1467:255–70

12. Schurch S, Green FH, Bachofen H. 1998. Formation and structure of surface films: captive bubble surfactometry. *Biochim. Biophys. Acta* 1408:180–202

13. Jobe A. 1977. The labeling and biological half-life of phosphatidylcholine in subcellular fractions of rabbit lung. *Biochim. Biophys. Acta* 489:440–53

14. King RJ, Martin H, Mitts D, Holmstrom FM. 1977. Metabolism of the apoproteins in pulmonary surfactant. *J. Appl. Physiol.* 42:483–91

15. Young SL, Kremers SA, Apple JS, Crapo JD, Brumley GW. 1981. Rat lung surfactant kinetics biochemical and morphometric correlation. *J. Appl. Physiol.* 51:248–53

16. Baritussio AG, Magoon MW, Goerke J, Clements JA. 1981. Precursor-product relationship between rabbit type II cell lamellar bodies and alveolar surface-active material. Surfactant turnover time. *Biochim. Biophys. Acta* 666:382–93

17. Jacobs H, Jobe A, Ikegami M, Jones S. 1982. Surfactant phosphatidylcholine source, fluxes, and turnover times in 3-day-old, 10-day-old, and adult rabbits. *J. Biol. Chem.* 257:1805–10

18. Wright JR, Clements JA. 1987. Metabolism and turnover of lung surfactant. *Am. Rev. Respir. Dis.* 136:426–44

19. Dobbs LG, Gonzalez R, Williams MC. 1986. An improved method for isolating type II cells in high yield and purity. *Am. Rev. Respir. Dis.* 134:141–45

20. Brown LA, Longmore WJ. 1981. Adrenergic and cholinergic regulation of lung surfactant secretion in the isolated perfused rat lung and in the alveolar type II cell in culture. *J. Biol. Chem.* 256:66–72

21. Chander A, Sen N, Wu AM, Spitzer AR. 1995. Protein kinase C in ATP regulation of lung surfactant secretion in type II cells. *Am. J. Physiol. Lung Cell Mol. Physiol.* 268:L108–16

22. Delahunty TJ, Johnston JM. 1976. The effect of colchicine and vinblastine on the release of pulmonary surface active material. *J. Lipid Res.* 17:112–16

23. Dobbs LG, Mason RJ. 1979. Pulmonary alveolar type II cells isolated from rats. Release of phosphatidylcholine in response to beta-adrenergic stimulation. *J. Clin. Invest.* 63:378–87

24. Gilfillan AM, Rooney SA. 1987. Purinoceptor agonists stimulate phosphatidylcholine secretion in primary cultures of adult rat type II pneumocytes. *Biochim. Biophys. Acta* 917:18–23

25. Mettler NR, Gray ME, Schuffman S, LeQuire VS. 1981. beta-Adrenergic induced synthesis and secretion of phosphatidylcholine by isolated pulmonary alveolar type II cells. *Lab. Invest.* 45:575–86

26. Rice WR, Osterhoudt KC, Whitsett JA. 1984. Effect of cytochalasins on surfactant release from alveolar type II cells. *Biochim. Biophys. Acta* 805:12–18

27. Sano K, Voelker DR, Mason RJ. 1985. Involvement of protein kinase C in pulmonary surfactant secretion from alveolar type II cells. *J. Biol. Chem.* 260:12725–29

28. Wirtz HR, Dobbs LG. 1990. Calcium mobilization and exocytosis after one mechanical stretch of lung epithelial cells. *Science* 250:1266–69

29. Chander A. 1989. Regulation of lung surfactant secretion by intracellular pH. *Am. J. Physiol. Lung Cell Mol. Physiol.* 257:L354–60

30. Haller T, Ortmayr J, Friedrich F, Volkl H, Dietl P. 1998. Dynamics of surfactant release in alveolar type II cells. *Proc. Natl. Acad. Sci. USA* 95:1579–84

31. Liu L, Fisher AB, Zimmerman UJ. 1995. Lung annexin II promotes fusion of isolated lamellar bodies with liposomes. *Biochim. Biophys. Acta* 1259:166–72

32. Liu L, Tao JQ, Li HL, Zimmerman UJ. 1997. Inhibition of lung surfactant secretion from alveolar type II cells and annexin II tetramer-mediated membrane fusion by phenothiazines. *Arch. Biochem. Biophys.* 342:322–28

33. Chattopadhyay S, Sun P, Wang P, Abonyo B, Cross NL, Liu L. 2003. Fusion of lamellar body with plasma membrane is driven by the dual action of annexin II tetramer and arachidonic acid. *J. Biol. Chem.* 278:39675–83

34. Mair N, Haller T, Dietl P. 1999. Exocytosis in alveolar type II cells revealed

by cell capacitance and fluorescence measurements. *Am. J. Physiol. Lung Cell Mol. Physiol.* 276:L376–82

35. Liu L, Wang M, Fisher AB, Zimmerman UJ. 1996. Involvement of annexin II in exocytosis of lamellar bodies from alveolar epithelial type II cells. *Am. J. Physiol. Lung Cell Mol. Physiol.* 270:L668–76

36. Rice WR, Singleton FM. 1987. P2Y-purinoceptor regulation of surfactant secretion from rat isolated alveolar type II cells is associated with mobilization of intracellular calcium. *Br. J. Pharmacol.* 91:833–38

37. Asokananthan N, Cake MH. 1996. Stimulation of surfactant lipid secretion from fetal type II pneumocytes by gastrin-releasing peptide. *Am. J. Physiol. Lung Cell Mol. Physiol.* 270:L331–37

38. Marino PA, Rooney SA. 1980. Surfactant secretion in a newborn rabbit lung slice model. *Biochim. Biophys. Acta* 620:509–19

39. Brown LA, Chen M. 1990. Vasopressin signal transduction in rat type II pneumocytes. *Am. J. Physiol. Lung Cell Mol. Physiol.* 258:L301–7

40. Sen N, Grunstein MM, Chander A. 1994. Stimulation of lung surfactant secretion by endothelin-1 from rat alveolar type II cells. *Am. J. Physiol. Lung Cell Mol. Physiol.* 266:L255–62

41. Gerboth GD, Effros RM, Roman RJ, Jacobs ER. 1993. pH-induced calcium transients in type II alveolar epithelial cells. *Am. J. Physiol. Lung Cell Mol. Physiol.* 264:L448–57

42. Massaro GD, Massaro D. 1983. Morphologic evidence that large inflations of the lung stimulate secretion of surfactant. *Am. Rev. Respir. Dis.* 127:235–36

43. McClenahan JB, Urtnowski A. 1967. Effect of ventilation on surfactant, and its turnover rate. *J. Appl. Physiol.* 23:215–20

44. Nicholas TE, Power JH, Barr HA. 1982. Surfactant homeostasis in the rat lung during swimming exercise. *J. Appl. Physiol.* 53:1521–28

45. Nicholas TE, Power JH, Barr HA. 1982. The pulmonary consequences of a deep breath. *Respir. Physiol.* 49:315–24

46. Oyarzun MJ, Clements JA. 1978. Control of lung surfactant by ventilation, adrenergic mediators, and prostaglandins in the rabbit. *Am. Rev. Respir. Dis.* 117:879–91

47. Faridy EE. 1976. Effect of distension on release of surfactant in excised dogs' lungs. *Respir. Physiol.* 27:99–114

48. Hildebran JN, Goerke J, Clements JA. 1981. Surfactant release in excised rat lung is stimulated by air inflation. *J. Appl. Physiol.* 51:905–10

49. Nicholas TE, Barr HA. 1981. Control of release of surfactant phospholipids in the isolated perfused rat lung. *J. Appl. Physiol.* 51:90–98

50. Nicholas TE, Barr HA. 1983. The release of surfactant in rat lung by brief periods of hyperventilation. *Respir. Physiol.* 52:69–83

51. Frick M, Bertocchi C, Jennings P, Haller T, Mair N, et al. 2004. Ca^{2+} entry is essential for cell strain-induced lamellar body fusion in isolated rat type II pneumocytes. *Am. J. Physiol. Lung Cell Mol. Physiol.* 286:L210–20

52. Daniels CB, Orgeig S. 2001. The comparative biology of pulmonary surfactant: past, present and future. *Comp. Biochem. Physiol. A* 129:9–36

53. Massaro D, Clerch L, Massaro GD. 1981. Surfactant aggregation in rat lungs: influence of temperature and ventilation. *J. Appl. Physiol.* 51:646–53

54. Suwabe A, Mason RJ, Smith D, Firestone JA, Browning MD, Voelker DR. 1992. Pulmonary surfactant secretion is regulated by the physical state of extracellular phosphatidylcholine. *J. Biol. Chem.* 267:19884–90

55. Chander A, Fisher AB. 1990. Regulation of lung surfactant secretion. *Am. J. Physiol. Lung Cell Mol. Physiol.* 258:L241–53

56. Mason RJ, Voelker DR. 1998. Regulatory mechanisms of surfactant secretion. *Biochim. Biophys. Acta* 1408:226–40

57. Wright JR, Dobbs LG. 1991. Regulation of pulmonary surfactant secretion and clearance. *Annu. Rev. Physiol.* 53:395–414

58. Rooney SA. 2001. Regulation of surfactant secretion. *Comp. Biochem. Physiol. A.* 129:233–43

59. Dietl P, Frick M, Mair N, Bertocchi C, Haller T. 2004. The pulmonary consequences of a deep breath revisited. *Biol. Neonate* 85:299–304

60. Edwards YS. 2001. Stretch stimulation: its effects on alveolar type II cell function in the lung. *Comp. Biochem. Physiol. A.* 129:245–60

61. Wirtz H, Schmidt M. 1992. Ventilation and secretion of pulmonary surfactant. *Clin. Invest.* 70:3–13

62. Marino PA, Rooney SA. 1981. The effect of labor on surfactant secretion in newborn rabbit lung slices. *Biochim. Biophys. Acta* 664:389–96

63. Abdellatif MM, Hollingsworth M. 1980. Effect of oxotremorine and epinephrine on lung surfactant secretion in neonatal rabbits. *Pediatr. Res.* 14:916–20

64. Corbet AJ, Flax P, Rudolph AJ. 1977. Role of autonomic nervous system controlling surface tension in fetal rabbit lungs. *J. Appl. Physiol.* 43:1039–45

65. Massaro D, Clerch L, Massaro GD. 1982. Surfactant secretion: evidence that cholinergic stimulation of secretion is indirect. *Am. J. Physiol. Cell Physiol.* 243:C39–45

66. Burgoyne RD, Morgan A. 2002. Secretory granule exocytosis. *Physiol. Rev.* 83:581–632

67. Chander A, Wu RD. 1991. In vitro fusion of lung lamellar bodies and plasma membrane is augmented by lung synexin. *Biochim. Biophys. Acta* 1086:157–66

68. Dorn CC, Rice WR, Singleton FM. 1989. Calcium mobilization and response recovery following P2-purinoceptor stimulation of rat isolated alveolar type II cells. *Br. J. Pharmacol.* 97:163–70

69. Frick M, Eschertzhuber S, Haller T, Mair N, Dietl P. 2001. Secretion in alveolar type II cells at the interface of constitutive and regulated exocytosis. *Am. J. Respir. Cell Mol. Biol.* 25:306–15

70. Haller T, Auktor K, Frick M, Mair N, Dietl P. 1999. Threshold calcium levels for lamellar body exocytosis in type II pneumocytes. *Am. J. Physiol. Lung Cell Mol. Physiol.* 277:L893–900

71. Pian MS, Dobbs LG, Duzgunes N. 1988. Positive correlation between cytosolic free calcium and surfactant secretion in cultured rat alveolar type II cells. *Biochim. Biophys. Acta* 960:43–53

72. Sano K, Voelker DR, Mason RJ. 1987. Effect of secretagogues on cytoplasmic free calcium in alveolar type II epithelial cells. *Am. J. Physiol. Cell Physiol.* 253:C679–86

73. Neher E. 1998. Vesicle pools and Ca^{2+} microdomains: new tools for understanding their roles in neurotransmitter release. *Neuron* 20:389–99

74. Kasai H. 1999. Comparative biology of Ca^{2+}-dependent exocytosis: implications of kinetic diversity for secretory function. *Trends Neurosci.* 22:88–93

75. Griese M, Gobran LI, Rooney SA. 1991. ATP-stimulated inositol phospholipid metabolism and surfactant secretion in rat type II pneumocytes. *Am. J. Physiol. Lung Cell Mol. Physiol.* 260:L586–93

76. Rice WR, Dorn CC, Singleton FM. 1990. P2-purinoceptor regulation of surfactant phosphatidylcholine secretion. Relative roles of calcium and protein kinase C. *Biochem. J.* 266:407–13

77. Eckenhoff RG, Somlyo AP. 1988. Rat lung type II cell and lamellar body: elemental composition in situ. *Am. J. Physiol. Cell Physiol.* 254:C614–20

78. Chander A, Johnson RG, Reicherter J, Fisher AB. 1986. Lung lamellar bodies maintain an acidic internal pH. *J. Biol. Chem.* 261:6126–31

79. Eckenhoff RG, Rannels SR, Fisher AB. 1991. Secretory granule calcium loss after isolation of rat alveolar type II cells.

Am. J. Physiol. Lung Cell Mol. Physiol. 260:L129–35

80. Wadsworth SJ, Chander A. 2000. H^+- and K^+-dependence of Ca^{2+} uptake in lung lamellar bodies. *J. Membr. Biol.* 174:41–51

81. Wadsworth SJ, Spitzer AR, Chander A. 1997. Ionic regulation of proton chemical (pH) and electrical gradients in lung lamellar bodies. *Am. J. Physiol. Lung Cell Mol. Physiol.* 273:L427–36

82. Haller T, Dietl P, Deetjen P, Volkl H. 1996. The lysosomal compartment as intracellular calcium store in MDCK cells: a possible involvement in $InsP_3$-mediated Ca^{2+} release. *Cell. Calcium* 19:157–65

83. Weaver TE, Na CL, Stahlman M. 2002. Biogenesis of lamellar bodies, lysosome-related organelles involved in storage and secretion of pulmonary surfactant. *Semin. Cell Dev. Biol.* 13:263–70

84. Bastacky J, Lee CY, Goerke J, Koushafar H, Yager D, et al. 1995. Alveolar lining layer is thin and continuous: low-temperature scanning electron microscopy of rat lung. *J. Appl. Physiol.* 79:1615–28

85. Nielson DW. 1986. Electrolyte composition of pulmonary alveolar subphase in anesthetized rabbits. *J. Appl. Physiol.* 60:972–79

86. Putney JW Jr. 1986. A model for receptor-regulated calcium entry. *Cell. Calcium* 7:1–12

86a. Mair N, Frick M, Bertocchi C, Haller T, Amberger A, et al. 2004. Inhibition by cytoplasmic nucleotides of a new cation channel in freshly isolated human and rat type II pneumocytes. *Am. J. Physiol. Lung Cell Mol. Physiol.* 287:1284–92

87. Sen N, Wu AM, Spitzer AR, Chander A. 1997. Activation of protein kinase C by 1,4-dihydro-2,6-dimethyl-5-nitro-4-[2-(trifluoromethyl)-phenyl]-3-pyridine carboxylic acid methyl ester (Bay K 8644), a calcium channel agonist, in alveolar type II cells. *Biochem. Pharmacol.* 53:1307–13

88. Warburton D, Parton L, Buckley S, Cosico L. 1989. Verapamil: a novel probe of surfactant secretion from rat type II pneumocytes. *J. Appl. Physiol.* 66:1304–8

89. Frick M, Siber G, Haller T, Mair N, Dietl P. 2001. Inhibition of ATP-induced surfactant exocytosis by dihydropyridine (DHP) derivatives: a non-stereospecific, photoactivated effect and independent of L-type Ca^{2+} channels. *Biochem. Pharmacol.* 61:1161–67

90. Dobbs LG, Gonzalez RF, Marinari LA, Mescher EJ, Hawgood S. 1986. The role of calcium in the secretion of surfactant by rat alveolar type II cells. *Biochim. Biophys. Acta* 877:305–13

91. Charras GT, Horton MA. 2002. Single cell mechanotransduction and its modulation analyzed by atomic force microscope indentation. *Biophys. J.* 82:2970–81

92. Hamill OP, Martinac B. 2001. Molecular basis of mechanotransduction in living cells. *Physiol. Rev.* 81:685–740

93. Ashino Y, Ying X, Dobbs LG, Bhattacharya J. 2000. $[Ca^{2+}]_i$ oscillations regulate type II cell exocytosis in the pulmonary alveolus. *Am. J. Physiol. Lung Cell Mol. Physiol.* 279:L5–13

94. Wang PM, Fujita E, Bhattacharya J. 2002. Vascular regulation of type II cell exocytosis. *Am. J. Physiol. Lung Cell Mol. Physiol.* 282:L912–16

95. Goldmann WH. 2002. Mechanical aspects of cell shape regulation and signaling. *Cell Biol. Int.* 26:313–17

96. Vlahakis NE, Hubmayr RD. 2003. Response of alveolar cells to mechanical stress. *Curr. Opin. Crit. Care* 9:2–8

97. Hook GE, Gilmore LB. 1982. Hydrolases of pulmonary lysosomes and lamellar bodies. *J. Biol. Chem.* 257:9211–20

98. Wasano K, Hirakawa Y. 1994. Lamellar bodies of rat alveolar type 2 cells have late endosomal marker proteins on their limiting membranes. *Histochemistry* 102:329–35

99. Andrews NW. 2000. Regulated secretion of conventional lysosomes. *Trends Cell Biol.* 10:316–21

100. Griese M, Gobran LI, Rooney SA. 1993. Signal-transduction mechanisms of ATP-stimulated phosphatidylcholine secretion in rat type II pneumocytes: interactions between ATP and other surfactant secretagogues. *Biochim. Biophys. Acta* 1167:85–93

101. Voyno-Yasenetskaya TA, Dobbs LG, Williams MC. 1991. Regulation of ATP-dependent surfactant secretion and activation of second-messenger systems in alveolar type II cells. *Am. J. Physiol.* 261(4 Suppl.):105–9

102. Hill DJ, Wright TC Jr, Andrews ML, Karnovsky MJ. 1984. Localization of calmodulin in differentiating pulmonary type II epithelial cells. *Lab. Invest.* 51:297–306

103. Chander A, Sen N, Spitzer AR. 2001. Synexin and GTP increase surfactant secretion in permeabilized alveolar type II cells. *Am. J. Physiol. Lung Cell Mol. Physiol.* 280:L991–98

104. Abonyo BO, Wang P, Narasaraju TA, Rowan WH III, McMillan DH, et al. 2003. Characterization of alpha-soluble *N*-ethylmaleimide-sensitive fusion attachment protein in alveolar type II cells: implications in lung surfactant secretion. *Am. J. Respir. Cell Mol. Biol.* 29:273–82

105. Abonyo BO, Gou D, Wang P, Narasaraju T, Wang Z, Liu L. 2004. Syntaxin 2 and SNAP-23 are required for regulated surfactant secretion. *Biochemistry* 43:3499–506

106. Zimmerman UJ, Malek SK, Liu L, Li HL. 1999. Proteolysis of synaptobrevin, syntaxin, and SNAP-25 in alveolar epithelial type II cells. *IUBMB. Life* 48:453–58

107. Vitale ML, Rodriguez DC, Tchakarov L, Trifaro JM. 1991. Cortical filamentous actin disassembly and scinderin redistribution during chromaffin cell stimulation precede exocytosis, a phenomenon not exhibited by gelsolin. *J. Cell Biol.* 113:1057–67

108. Rose F, Kurth-Landwehr C, Sibelius U, Reuner KH, Aktories K, et al. 1999. Role of actin depolymerization in the surfactant secretory response of alveolar epithelial type II cells. *Am. J. Respir. Crit. Care Med.* 159:206–12

109. Tsilibary EC, Williams MC. 1983. Actin in peripheral rat lung: S1 labeling and structural changes induced by cytochalasin. *J. Histochem. Cytochem.* 31:1289–97

110. Brown LA, Pasquale SM, Longmore WJ. 1985. Role of microtubules in surfactant secretion. *J. Appl. Physiol.* 58:1866–73

111. Corbet AJ, Voelker RM, Murphy FM, Owens ML. 1988. Inhibition by colchicine of phospholipid secretion induced by lung distension. *J. Appl. Physiol.* 65:1710–15

112. Li HL, Feinstein SI, Liu L, Zimmerman UJ. 1998. An antisense oligodeoxyribonucleotide to m-calpain mRNA inhibits secretion from alveolar epithelial type II cells. *Cell Signal.* 10:137–42

113. Zimmerman UJ, Speicher DW, Fisher AB. 1992. Secretagogue-induced proteolysis of lung spectrin in alveolar epithelial type II cells. *Biochim. Biophys. Acta* 1137:127–34

114. Zimmerman UJ, Wang M, Liu L. 1995. Inhibition of secretion from isolated rat alveolar epithelial type II cells by the cell permeant calpain inhibitor II (*N*-acetyl-leucyl-leucyl- methioninal). *Cell. Calcium* 18:1–8

115. Muallem S, Kwiatkowska K, Xu X, Yin HL. 1995. Actin filament disassembly is a sufficient final trigger for exocytosis in nonexcitable cells. *J. Cell Biol.* 128:589–98

116. Haller T, Dietl P, Pfaller K, Frick M, Mair N, et al. 2001. Fusion pore expansion is a slow, discontinuous, and Ca^{2+}-dependent process regulating secretion from alveolar type II cells. *J. Cell Biol.* 155:279–89

117. Risco C, Romero C, Bosch MA, Pinto dS. 1994. Type II pneumocytes revisited: intracellular membranous systems, surface characteristics, and lamellar body secretion. *Lab. Invest.* 70:407–17

118. Hille B, Billiard J, Babcock DF, Nguyen T, Koh DS. 1999. Stimulation of exocytosis without a calcium signal. *J. Physiol.* 520:23–31

119. Zhu H, Hille B, Xu T. 2002. Sensitization of regulated exocytosis by protein kinase C. *Proc. Natl. Acad. Sci. USA* 99:17055–59

120. Liu M, Xu J, Liu J, Kraw ME, Tanswell AK, Post M. 1995. Mechanical strain-enhanced fetal lung cell proliferation is mediated by phospholipase C and D and protein kinase C. *Am. J. Physiol. Lung Cell Mol. Physiol.* 268:L729–38

121. Rooney SA, Gobran LI. 1997. Influence of the protein kinase C inhibitor 3-[1-[3-(amidinothio) propyl]-1H-indoyl-3-yl]-3-(1-methyl-1H-indoyl-3-yl) maleimide methane sulfonate (Ro-318220) on surfactant secretion in type II pneumocytes. *Biochem. Pharmacol.* 53:597–601

122. Dietl P, Haller T. 2000. Persistent fusion pores but transient fusion in alveolar type II cells. *Cell Biol. Int.* 24:803–7

123. Alvarez de Toledo G, Fernandez-Chacon R, Fernandez JM. 1993. Release of secretory products during transient vesicle fusion. *Nature* 363:554–58

124. Kliewer M, Fram EK, Brody AR, Young SL. 1985. Secretion of surfactant by rat alveolar type II cells: morphometric analysis and three-dimensional reconstruction. *Exp. Lung Res.* 9:351–61

125. Haller T, Pfaller K, Dietl P. 2001. The conception of fusion pores as rate-limiting structures for surfactant secretion. *Comp. Biochem. Physiol. A* 129:227–31

126. Singer W, Frick M, Haller T, Bernet S, Ritsch-Marte M, Dietl P. 2003. Mechanical forces impeding exocytotic surfactant release revealed by optical tweezers. *Biophys. J.* 84:1344–51

127. Bates SR, Tao JQ, Notarfrancesco K, De-Bolt K, Shuman H, Fisher AB. 2003. Effect of surfactant protein-A on granular pneumocyte surfactant secretion in vitro. *Am. J. Physiol. Lung Cell Mol. Physiol.* 285:L1055–65

128. Stahlman MT, Gray MP, Falconieri MW, Whitsett JA, Weaver TE. 2000. Lamellar body formation in normal and surfactant protein B-deficient fetal mice. *Lab. Invest.* 80:395–403

129. Scepek S, Lindau M. 1993. Focal exocytosis by eosinophils—compound exocytosis and cumulative fusion. *EMBO J.* 12:1811–17

130. Hills BA. 1999. An alternative view of the role(s) of surfactant and the alveolar model. *J. Appl. Physiol.* 87:1567–83

131. Notter RH, Finkelstein JN. 1984. Pulmonary surfactant: an interdisciplinary approach. *J. Appl. Physiol.* 57:1613–24

132. Dobbs LG, Pian MS, Maglio M, Dumars S, Allen L. 1997. Maintenance of the differentiated type II cell phenotype by culture with an apical air surface. *Am. J. Physiol. Lung Cell Mol. Physiol.* 273:L347–54

133. Dobbs LG, Wright JR, Hawgood S, Gonzalez R, Venstrom K, Nellenbogen J. 1987. Pulmonary surfactant and its components inhibit secretion of phosphatidylcholine from cultured rat alveolar type II cells. *Proc. Natl. Acad. Sci. USA* 84:1010–14

134. Nielson DW, Goerke J, Clements JA. 1981. Alveolar subphase pH in the lungs of anesthetized rabbits. *Proc. Natl. Acad. Sci. USA* 78:7119–23

135. Grathwohl C, Newman GE, Phizackerley PJ, Town MH. 1979. Structural studies on lamellated osmiophilic bodies isolated from pig lung. ^{31}P NMR results and water content. *Biochim. Biophys. Acta* 552:509–18

136. Morley CJ, Banhham AD, Johnson P, Thorburn GD, Jenkin G. 1978. Physical and physiological properties of dry lung surfactant. *Nature* 271:162–63

137. Kuhn C III. 1968. Cytochemistry of pulmonary alveolar epithelial cells. *Am. J. Pathol.* 53:809–33

138. Sanderson RJ, Vatter AE. 1977. A mode of formation of tubular myelin from lamellar bodies in the lung. *J. Cell Biol.* 74:1027–31

139. Putman E, Creuwels LA, van Golde LM, Haagsman HP. 1996. Surface properties, morphology and protein composition of pulmonary surfactant subtypes. *Biochem. J.* 320:599–605

140. Sato S, Kishikawa T. 2001. Ultrastructural study of the alveolar lining and the bronchial mucus layer by block staining with oolong tea extract: the role of various surfactant materials. *Med. Electron Microsc.* 34:142–51

141. Williams MC. 1977. Conversion of lamellar body membranes into tubular myelin in alveoli of fetal rat lungs. *J. Cell Biol.* 72:260–77

142. Dietl P, Haller T, Mair N, Frick M. 2001. Mechanisms of surfactant exocytosis in alveolar type II cells in vitro and in vivo. *News Physiol. Sci.* 16:239–43

143. Haller T, Dietl P, Stockner H, Frick M, Mair N, et al. 2004. Tracing surfactant transformation from cellular release to insertion into an air-liquid interface. *Am. J. Physiol. Lung Cell Mol. Physiol.* 286:L1009–15

144. Gross NJ. 1995. Extracellular metabolism of pulmonary surfactant: the role of a new serine protease. *Annu. Rev. Physiol.* 57:135–50

145. Gross NJ, Narine KR. 1989. Surfactant subtypes of mice: metabolic relationships and conversion in vitro. *J. Appl. Physiol.* 67:414–21

146. Magoon MW, Wright JR, Baritussio A, Williams MC, Goerke J, et al. 1983. Subfractionation of lung surfactant. Implications for metabolism and surface activity. *Biochim. Biophys. Acta* 750:18–31

147. Stratton CJ. 1984. Morphology of surfactant producing cells and of the alveolar lining layer. In *Pulmonary Surfactant*, ed. B Robertson, LM van Golde, JJ Batenburg, pp. 67–118. Amsterdam: Elsevier

148. Spain CL, Silbajoris R, Young SL. 1987. Alterations of surfactant pools in fetal and newborn rat lungs. *Pediatr. Res.* 21:5–9

149. Gross NJ, Kellam M, Young J, Krishnasamy S, Dhand R. 2000. Separation of alveolar surfactant into subtypes. A comparison of methods. *Am. J. Respir. Crit. Care Med.* 162:617–22

150. Paul GW, Hassett RJ, Reiss OK. 1977. Formation of lung surfactant films from intact lamellar bodies. *Proc. Natl. Acad. Sci. USA* 74:3617–20

151. Gil J, Reiss OK. 1973. Isolation and characterization of lamellar bodies and tubular myelin from rat lung homogenates. *J. Cell Biol.* 58:152–71

152. Sanders RL, Hassett RJ, Vatter AE. 1980. Isolation of lung lamellar bodies and their conversion to tubular myelin figures in vitro. *Anat. Rec.* 198:485–501

153. Singer W, Frick M, Haller T, Bernet S, Ritsch-Marte M, Dietl P. 2003. Mechanical forces impeding exocytotic surfactant release revealed by optical tweezers. *Biophys. J.* 84:1344–51

154. Gross NJ, Schultz RM. 1990. Serine proteinase requirement for the extracellular metabolism of pulmonary surfactant. *Biochim. Biophys. Acta* 1044:222–30

155. Gross NJ, Schultz RM. 1992. Requirements for extracellular metabolism of pulmonary surfactant: tentative identification of serine protease. *Am. J. Physiol. Lung Cell Mol. Physiol.* 262:L446–53

156. Hall SB, Hyde RW, Kahn MC. 1997. Stabilization of lung surfactant particles against conversion by a cycling interface. *Am. J. Physiol. Lung Cell Mol. Physiol.* 272:L335–43

157. Veldhuizen RA, Yao LJ, Lewis JF. 1999. An examination of the different variables affecting surfactant aggregate conversion in vitro. *Exp. Lung Res.* 25:127–41

158. Ueda T, Ikegami M, Jobe A. 1994. Surfactant subtypes. In vitro conversion, in vivo

function, and effects of serum proteins. *Am. J. Respir. Crit. Care Med.* 149:1254–59

159. Savov J, Silbajoris R, Young SL. 1999. Mechanical ventilation of rat lung: effect on surfactant forms. *Am. J. Physiol. Lung Cell Mol. Physiol.* 277:L320–26

160. Veldhuizen RA, Marcou J, Yao LJ, McCaig L, Ito Y, Lewis JF. 1996. Alveolar surfactant aggregate conversion in ventilated normal and injured rabbits. *Am. J. Physiol. Lung Cell Mol. Physiol.* 270:L152–58

161. Clark H, Allen L, Collins E, Barr F, Dobbs L, et al. 1999. Localization of a candidate surfactant convertase to type II cells, macrophages, and surfactant subfractions. *Am. J. Physiol. Lung Cell Mol. Physiol.* 276:L452–58

162. Oulton M, Edwards E, Handa K. 1999. Convertase activity in alveolar surfactant and lamellar bodies in fetal, newborn, and adult rabbits. *J. Appl. Physiol.* 86:71–77

163. Gross NJ, Veldhuizen R, Possmayer F, Dhand R. 1997. Surfactant convertase action is not essential for surfactant film formation. *Am. J. Physiol. Lung Cell Mol. Physiol.* 273:L907–12

164. Dhand R, Sharma VK, Teng AL, Krishnasamy S, Gross NJ. 1998. Protein-lipid interactions and enzyme requirements for light subtype generation on cycling reconstituted surfactant. *Biochem. Biophys. Res. Commun.* 244:712–19

165. Policard A, Collet AJ, Pregermain S. 1957. Etude au microscope electronique des figures myeliniques dans les processus inflammatoires. *Bull. Micr. Appl.* 7:49–53

166. Weibel ER, Kistler GS, Toendury G. 1966. A stereologic electron microscopy study of "tubular myelin figures" in alveolar fluids of rat lungs. *Z. Zellforsch. Mikrosk. Anat.* 69:418–27

167. Leeson TS, Leeson CR. 1966. Osmiophilic lamellated bodies and associated material in lung alveolar spaces. *J. Cell Biol.* 28:577–81

168. Young SL, Fram EK, Larson EW.

1992. Three-dimensional reconstruction of tubular myelin. *Exp. Lung Res.* 18:497–504

169. Smith DS, Smith U, Ryan JW. 1972. Freeze-fractured lamellar body membranes of the rat lung great alveolar cell. *Tissue Cell* 4:457–68

170. King RJ, MacBeth MC. 1981. Interaction of the lipid and protein components of pulmonary surfactant. Role of phosphatidylglycerol and calcium. *Biochim. Biophys. Acta* 647:159–68

171. Williams MC, Hawgood S, Hamilton RL. 1991. Changes in lipid structure produced by surfactant proteins SP-A, SP-B, and SP-C. *Am. J. Respir. Cell Mol. Biol.* 5:41–50

172. Voorhout WF, Veenendaal T, Haagsman HP, Verkleij AJ, van Golde LM, Geuze HJ. 1991. Surfactant protein A is localized at the corners of the pulmonary tubular myelin lattice. *J. Histochem. Cytochem.* 39:1331–36

173. Froh D, Ballard PL, Williams MC, Gonzales J, Goerke J, et al. 1990. Lamellar bodies of cultured human fetal lung: content of surfactant protein A (SP-A), surface film formation and structural transformation in vitro. *Biochim. Biophys. Acta* 1052:78–89

174. Chi EY, Lagunoff D. 1978. Linear arrays of intramembranous particles in pulmonary tubular myelin. *Proc. Natl. Acad. Sci. USA* 75:6225–29

175. Veldhuizen RA, Yao LJ, Hearn SA, Possmayer F, Lewis JF. 1996. Surfactant-associated protein A is important for maintaining surfactant large-aggregate forms during surface-area cycling. *Biochem. J.* 313:835–40

176. Palaniyar N, Ridsdale RA, Hearn SA, Possmayer F, Harauz G. 1999. Formation of membrane lattice structures and their specific interactions with surfactant protein A. *Am. J. Physiol. Lung Cell Mol. Physiol.* 276:L642–49

177. Nakamura H, Tonosaki A, Washioka H, Takahashi K, Yasui S. 1985.

Monomolecular surface film and tubular myelin figures of the pulmonary surfactant in hamster lung. *Cell Tissue Res.* 241:523–28

178. Kashchiev D, Exerowa D. 2001. Structure and surface energy of the surfactant layer on the alveolar surface. *Eur. Biophys. J.* 30:34–41

179. Notter RH, Penney DP, Finkelstein JN, Shapiro DL. 1986. Adsorption of natural lung surfactant and phospholipid extracts related to tubular myelin formation. *Pediatr. Res.* 20:97–101

180. Benson BJ, Williams MC, Sueishi K, Goerke J, Sargeant T. 1984. Role of calcium ions in the structure and function of pulmonary surfactant. *Biochim. Biophys. Acta* 793:18–27

181. Ikegami M, Jobe AH, Whitsett J, Korfhagen T. 2000. Tolerance of SP-A-deficient mice to hyperoxia or exercise. *J. Appl. Physiol.* 89:644–48

182. Knebel D, Sieber M, Reichelt R, Galla HJ, Amrein M. 2002. Scanning force microscopy at the air-water interface of an air bubble coated with pulmonary surfactant. *Biophys. J.* 82:474–80

Annu. Rev. Physiol. 2005. 67:623–61
doi: 10.1146/annurev.physiol.67.040403.102229
First published online as a Review in Advance on October 26, 2004

LUNG VASCULAR DEVELOPMENT: Implications for the Pathogenesis of Bronchopulmonary Dysplasia

Kurt R. Stenmark and Steven H. Abman

*Developmental Lung Biology Laboratory[1], Pediatric Heart Lung Center[2], Sections of
Critical Care[1] and Pulmonary Medicine[2], Department of Pediatrics, University of
Colorado Health Sciences Center and The Children's Hospital, Denver, Colorado 80262;
email: Kurt.Stenmark@UCHSC.edu; Steven.Abman@uchsc.edu*

Key Words   angiogenesis, pulmonary circulation, pulmonary hypertension

■ **Abstract**   Past studies have primarily focused on how altered lung vascular growth
and development contribute to pulmonary hypertension. Recently, basic studies of vascular growth have led to novel insights into mechanisms underlying development of the
normal pulmonary circulation and the essential relationship of vascular growth to lung
alveolar development. These observations have led to new concepts underlying the
pathobiology of developmental lung disease, especially the inhibition of lung growth
that characterizes bronchopulmonary dysplasia (BPD). We speculate that understanding basic mechanisms that regulate and determine vascular growth will lead to new
clinical strategies to improve the long-term outcome of premature babies with BPD.

INTRODUCTION

The lung is a complex organ system whose basic physiologic function is to perform
gas exchange across a thin blood-gas interface. The lung has the largest epithelial
surface area of any mammalian organ and is capable of supporting a systemic
oxygen consumption of between 250 ml/min at rest to 5500 ml/min during exercise (1). To create such a large, diffusible interface with the circulation, the lung
epithelium must not only undergo a series of proliferative, branching, and morphogenetic steps, but also must interact continuously and in a well-coordinated
manner with mesenchymal tissue to ensure the development of functioning vascular and lymphatic systems. The lung originates as a pair of invaginations from
foregut endoderm. These endodermal buds branch and differentiate within the surrounding mesoderm, ultimately giving rise to the airways, alveoli, blood vessels,
and lymphatics of the mature lung. Lung branching morphogenesis and epithelial development and differentiation have been the subject of intense investigation
over the past 50 years, and the processes involved have been the subject of many
comprehensive recent reviews (2–9). However, studies of the mechanisms that

0066-4278/05/0315-0623$14.00   **623**

regulate lung vascular and lymphatic development and that link capillary growth with alveolarization are relatively recent and limited in scope.

Until recently, most of the information regarding pulmonary vascular development has been largely descriptive in nature and often based on model systems and methodologic techniques that may preclude accurate assessment of the diverse processes governing pulmonary vascular development. This is especially true with regard to the origin, differentiation, and maturation of the various cells types within the vascular wall of lung blood vessels. This fact is important because it has become increasingly clear that airway and vascular development are closely interactive processes and that disruption of one system may have catastrophic consequences on the development of the other and, ultimately, of lung structure and function.

Developmental abnormalities of the pulmonary circulation contribute to the pathogenesis of several neonatal cardiopulmonary disorders, including diseases associated with persistent pulmonary hypertension of the newborn (10–12). More recently, however, there has been growing recognition that the importance of understanding basic mechanisms of lung vascular growth in the context of human disease may be best highlighted in the setting of bronchopulmonary dysplasia (BPD) (13, 14). Premature birth with injury to the immature lung disrupts normal lung growth and causes severe chronic lung disease, or BPD, which remains a major cause of late morbidity and mortality of premature newborns. Histologically, BPD is characterized by arrested lung growth, with decreased alveolarization and a dysmorphic vasculature. Recent studies suggest that disruption of normal lung vascular growth may play a central role in the pathogenesis of BPD; however, little is known about basic mechanisms of pulmonary vascular injury in the immature lung, the impact of this injury on subsequent growth and development of either the pulmonary or bronchial circulations, and the contribution of abnormalities of pulmonary and/or bronchial vascular growth to either the pathogenesis of BPD or functional abnormalities that characterize the disease. The purpose of this review is to examine lung vascular development in the context of understanding how growth of the pulmonary circulation and lung structure is impaired in BPD, with an eye toward restoring the injured lung to normal.

LUNG BLOOD VESSELS: NORMAL STRUCTURE AND FUNCTION OF THE PULMONARY AND BRONCHIAL CIRCULATIONS

The vascular system of the lung is divided into the pulmonary and bronchial systems. The pulmonary arteries supply the intrapulmonary structures and ultimately regulate gas exchange; vessels branch with the airways but branch into an extensive capillary network only at the level of respiratory bronchioles and alveoli. The bronchial system is the nutrient supply to the lung and perfuses the capillary bed within the bronchial wall and the structures of the perihilar region, including lymph nodes and the adventitia of elastic and large muscular pulmonary vessels. All intrapulmonary structures drain to the pulmonary veins, whereas the hilar structures drain to the so-called true bronchial veins and then to the azygos system (15, 18).

The pulmonary artery accompanies the airways but gives off many more branches than the airway. In fact, two main types of pulmonary arterial vessels have been described: (*a*) conventional vessels, which are the long pulmonary artery branches that run within an airway, dividing as the airway divides and finally distributing to the capillary bed beyond the level of the terminal bronchioles; (*b*) supernumerary vessels, which are additional branches that arise from the pulmonary artery between the conventional branches, run a short course and supply the capillary bed of alveoli immediately around the pulmonary artery (9, 16, 17). Supernumerary arteries are thought to be a prominent component of the mature lung with the ratio of conventional to supernumerary vessels in the order of 1:2 in the prelobular region and 1:13 in the pre- and intra-acinar region (16). There appear to be distinct functional differences between supernumerary and conventional arteries. For instance, serotonin-induced vasoconstriction in supernumerary vessels is 30 times more potent than in conventional vessels (18). Others have suggested that plexiform lesions selectively develop in supernumerary vessels (19).

The pattern of veins in the lung resembles the arteries in that there are many more venous tributaries than airway branches (9, 15, 16). Similarly, at least two types of veins can be identified. Conventional veins arise from the points of division of an airway, pass to the periphery of a given lung unit, and combine to form increasingly large venous tributaries. Tributaries to the pulmonary veins also arise from the pleura and connective tissue septa. Within the lung, the anatomic distribution of the arteries and veins is characteristic. The broncho-arterial bundle includes the airway with the bronchial artery capillary bed and the adjacent pulmonary artery and lymphatic vessels, all within a single adventitial sheath. The veins always run at the periphery of any unit, whether it is the acinus, lobule, or segment. At the hilum, the veins join to form superior/inferior veins that drain to left atrium, giving them a distribution within the mediastinum different from that of the pulmonary artery.

A systemic blood supply to the lung was suggested at least 500 years ago by Galen. Its presence was confirmed by Ruysch in 1732, who designated the lung's systemic vessel as the bronchial artery (15). It is now known that the bronchial arteries may arise from the descending aorta, intercostals, subclavian, or internal mammary arteries. The bronchial arteries may be classified as extrapulmonary or intrapulmonary. The extrapulmonary artery gives off small branches to the esophagus, to mediastinal tissues, to hilar lymph nodes, and the lobar bronchus. The intrapulmonary bronchial artery distributes to the supporting tissue and structures of the intralobal bronchi (mucous membrane, muscle, perichondrium, secretory glands), to pulmonary pleura, to lymph nodes, to walls of the pulmonary artery, and to veins and nerves. Bronchial arteries on the walls of the pulmonary artery and vein are functionally vasa vasorum. They are not as dense as on the bronchial walls and cannot be easily observed on the walls of small vessels. The extrapulmonary bronchial artery blood drains into extrapulmonary bronchial veins and connect to the azygous or hemiazygous veins, which ultimately drain to the right atrium. The intrapulmonary bronchial artery drains into intrapulmonary bronchial veins and/or pulmonary veins, which then connect to the left atrium.

There are close and important relationships between the pulmonary and bronchial vascular systems, which become most evident in the diseased lung or in the setting of aberrant cardiovascular development (9, 15). For example, enlargement of the bronchial circulation is especially striking in association with congenital heart diseases, such as pulmonary atresia or transposition of the great vessels. Interestingly, ligation of the pulmonary arteries produces enlargement and dilation of the bronchial artery resulting in the increase of precapillary anastomosis and the development of various abnormal flow routes (20, 21). Similarly, severe pulmonary thromboembolic disease can result in proliferation of bronchial vessels in and around the obstructed pulmonary artery (15, 22). In bronchiectasis, a marked proliferation of bronchial vessels, which anastomosis with the pulmonary vascular system via precapillary, capillary, or postcapillary networks, is observed (23, 24). In emphysema, the normal structure of the alveolar capillary networks often disappears and enlarged branches of bronchial vessels are observed to occupy those areas, a situation obviously unfavorable to gas exchange (15). A similar proliferation of bronchial vessels is noted in the setting of chronic atelectasis (24). Thus many observations would support the idea that conditions leading to the narrowing or stenosis of blood vessels of the pulmonary vascular system are generally related to a proliferation and dilation of vessels of the bronchial vascular system. This relationship will be examined in the setting of BPD (see below).

Because the primary function of the lung is gas exchange, it is not surprising that keeping the airspaces and lung interstitium free of excess fluid is a critical component of normal lung function. The lymphatic system plays a critical role in fluid clearance, as well as in host defense, and in clearing solid particles and cells from the lung (26–28). It is also involved in the genesis of diseases such as tuberculosis and the metastasis of lung cancer. The lymph vessels form a closed circulatory system lined by endothelium. According to size, they are called lymph capillaries, lymph vessels, and collecting lymph vessels (15, 26). Lymph vessels are clearly evident at the level of the alveolar ducts (none is observed in direct relation to alveoli) and extend proximally (15, 26). Interestingly, the caliber of lymph vessels does not necessarily increase as they move toward the central lung, with lymph capillaries sometimes being larger than collecting lymph vessels. The critical nature of the lung lymphatic system is illustrated in human cases in which there is a primary developmental defect of the lung lymphatics, such as pulmonary lymphangiectasia. In the majority of neonatal cases, effective respiration is never established at birth, and the infants are stillborn or die in the first weeks of life despite aggressive support (29). In patients with bilateral pulmonary lymphangiectasia who survive, many develop chronic respiratory disease with pulmonary hypertension and cor pulmonale. Thus an understanding of the development of the lung and lung vasculature must take into account lymphatic development.

MORPHOGENESIS OF THE PULMONARY AND BRONCHIAL CIRCULATIONS

The most proximal part of the pulmonary circulation, the pulmonary trunk, is derived from the truncus arteriosus, which becomes divided into the aorta and pulmonary trunk by 8 weeks of gestation in humans by growth of the spiral aorticopulmonary septum (30). The pulmonary trunk connects to the pulmonary arch arteries, which are derived from the sixth branchial arch arteries, the most caudal of the brachial arteries, by 7-weeks gestation in humans (31). There is no consensus as to the mode of development of the sixth branchial arch artery, but recent reappraisal suggests that these pulmonary arch arteries originate from a strand of endothelial precursors that connect the ventral wall of the dorsal aortae to the pulmonary trunk (32, 33). These strands become lumenized, initially at the sites of connection with pulmonary trunk and dorsal aortae, then in between these sites of attachment. The origin of the intrapulmonary arteries has been variously described by early investigators as endothelial sprouts from either the aortic sac (31) or from the dorsal aortae (34), or as originating from a network of capillaries around the foregut (35, 36). The latter observation comes closest to the results of more recent studies suggesting that the pulmonary artery and vein develop in situ within the splanchnic mesoderm surrounding the foregut via at least two processes that likely occur concurrently: (*a*) vasculogenesis, in which new blood vessels form in situ from angioblasts and (*b*) angiogenesis, which involves sprouting of new vessels from existing ones.

The idea that some blood vessels arise de novo in the mesoderm surrounding the protruding endoderm may actually not be so new because it was initially raised some 70 years ago by Chang (37). Strong support for this idea has come from recent studies using molecular markers of endothelial progenitor cells, as well as endothelial differentiation markers (38–42). Cells expressing primitive endothelial markers appear in the mesoderm at early stages of lung development (embryonic and early pseudoglandular), which proceeds any documented connection with the established circulatory system. Lung vascular development continues throughout lung development, and requires epithelial-mesenchymal cross talk at each stage of development (8). Studies combining light and transmission electron microscopy with scanning electron microscopy of mercox vascular casts further suggest that large pulmonary arteries originate via the process of angiogenesis of central vessels, whereas distal vascular development requires vasculogenic mechanisms within the lung mesenchyme (43, 44). In addition to angiogenesis and vasculogenesis, a third process, fusion, is necessary to ultimately connect the angiogenic and vasculogenic vessels and expand the vascular network (43, 45).

At least two studies suggest similar mechanisms apply to human lung vascular development. DeMello et al. studied human embryos from the Carnegie Collection of Human Embryos and presented evidence to support the idea that both vasculogenesis and arteriogenesis operate cooperatively to form the pulmonary vessels (45). Using different techniques, Hall et al. also proposed that vasculogenesis is a primary mechanism for intrapulmonary vascular development (46). Both investigations suggested that the venous circulation develops in a fashion similar to the arterial circulation and, importantly, both showed that the venous circulation is the first to be established.

On the other hand, this concept has recently been challenged by the suggestion that the lung vasculature is formed by distal angiogenesis, a process in which the formation of new capillaries from pre-existing ones occurs at the periphery of the lung (47). In this model epithelial/endothelial interactions are decisive in inducing angiogenesis and ensure the coordinate expansion of a vascular network as branching proceeds. Newly formed vessels remodel dynamically to form the afferent (arterial) and efferent (venous) vascular systems. Therefore, questions remain regarding the process that may be related to methodologic approaches and/or semantics. Future studies using new and improved techniques will likely resolve these issues.

At the end of 16 weeks of gestation in humans, all preacinar bronchi are in place and are accompanied by the pulmonary and bronchial arteries (48, 49). However, although the conducting airways and arteries are formed by the end of this so-called pseudoglandular stage (5–17 weeks) of lung development, the development of the gas-exchanging surface of the lung has just begun. The canalicular stage (17–26 weeks) encompasses the early development of the pulmonary parenchyma during which there is a great increase in the number of lung capillaries (9, 50). The capillaries begin to arrange themselves around the air spaces and come into close apposition with the overlying cuboidal epithelium. At sites of apposition, thinning of the epithelium occurs, to form what will be the first air-blood barrier. During the saccular stage (24 weeks to birth), the airways end in clusters of thin-walled saccules and by term have formed the last generations of airways, the alveolar ducts. Capillaries form a bilayer within the relatively broad and cellular intersaccular septa at this stage (9, 50). The period of alveolarization is largely a postnatal phenomenon, with more than 90% of all alveoli being formed after birth in humans (6, 9). During the period of alveolarization the initially thick interalveolar septa are attenuated and the double capillary layer fuses to form a single layer adult form. Secondary septae initially form as low ridges that protrude into primitive airspaces to increase their surface area. During this time, the lung is undergoing marked microvascular growth and development as well (51). Multiple stimuli contribute to alveolar and vascular growth and development, and in part, involve cross talk of paracrine signals between epithelium and mesenchyme (52). For example, vascular endothelial growth factor (VEGF) is expressed in developing lung epithelium, whereas VEGF receptor-2 (VEGFR-2) is localized to angioblasts within the embryonic mesenchyme (53–57). Diffusion of VEGF to precursors of vascular endothelial cells within the mesenchyme leads to angiogenesis by stimulating endothelial proliferation, a key step in vessel development. In addition, the process of septation involves alternate upfolding and growth of capillary layers within primary septa (54). This mechanism of alveolarization suggests that the failure of capillary network formation and growth, or disruption of the infolding of the double capillary network, could potentially cause failed alveolarization. Less is known of the mechanisms contributing to development of the bronchial circulation. However, molecular marker studies suggest that intrapulmonary bronchial arteries also form in situ from the mesenchyme surrounding the epithelial buds (32, 33). Then, early in development (12 weeks in humans) connections to the systemic circulation are made. Presently, it is unclear as to what directs the proximal and distal connections between the systemic versus pulmonary vascular beds, especially when in early lung development the intrapulmonary vascular plexus are virtually indistinguishable. The question is of obvious importance because the behavior of these two circulations is vastly different. Current studies provide little insight into how the marked differences in endothelial cell and smooth muscle phenotype and function arise within the lung and its two circulations. Further, they do not adequately address how heterogeneity in phenotype and function of endothelial and smooth muscle cells arise even along the longitudinal axis of the respective circulations.

ROLE OF GROWTH AND TRANSCRIPTION FACTOR SIGNALING IN CONTROL OF LUNG VASCULAR GROWTH

Tremendous advances in our understanding of blood vessel formation within the embryo and various organs and tissues have taken place over the past 10 years. Many of the molecules involved seem to be highly conserved both across species and among various organs (Figure 1, Table 1). However, there may be subtle differences, and a complete understanding of the signaling pathways that regulate and coordinate vessel development and differentiation in a cell-specific fashion in the lung will be critical to understanding the unique functions and responses that the pulmonary and bronchial circulations exhibit in response to many pathophysiologic stimuli. Molecular programs important for the development of the lung vasculature have only recently begun to be elucidated. Among the different regulators of vessel formation, those that are specifically expressed in the endothelial lineage are best characterized during vascular development. This is the case for the ligand-receptor pairs VEGF/VEGFR, angiopoietin/TIE, ephrins/eph receptors, and notch/jagged.

Several studies have documented the appearance of VEGFR-2 (Flk-1) at very early time points of lung development in the mesenchyme (38, 58–60). VEGF also appears early in the developing epithelium and mesenchyme (58–63). It is known that VEGF-A is absolutely required for embryonic vascular development. Loss of even a single VEGF allele results in early embryonic death, before embryonic day 9.5. VEGF-A expression and function is tightly regulated during lung development. At least in the mouse, VEGF-A exists as three predominant isoforms, VEGF-A120, 164, and 188. Each isoform has differing properties, including affinity for the heparin sulfate component of the extracellular matrix, as well as for VEGFR-1 (Flt-1) and VEGFR-2 (Flk-1). VEGF-120 is highly diffusible because of lack of binding to heparin sulfate and probably serves a key early role in vascular formation

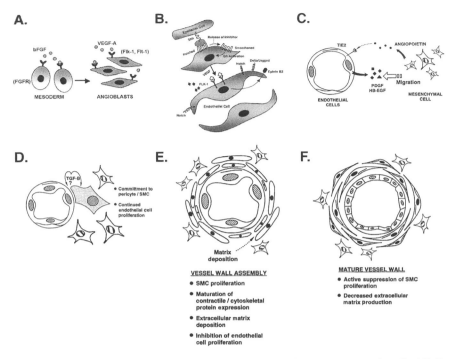

Figure 1   A proposed model for the development of a mature vessel wall. (*A*) In this model, early expansion of mesoderm, perhaps driven by bFGF, is followed by an emergence of angioblasts or endothelial precursor cells within this mesenchyme as identified by Flt-1 and Flk-1. VEGFA is critical in the expansion of this cell population. (*B*) In the lung, epithelial-mesenchymal interactions are critical for the development of the vasculature, which involves diverse molecules [adapted from Ng et al. (88)]. Recruitment of mural cells follows endothelial tube formation. Angiopoietin, produced by undifferentiated mesenchymal cells, binds to and activates the TIE2 receptor on endothelial cells (*C*), resulting in release of a recruitment/chemotactic signal (e.g., PDGF, HB-EGF) for mesenchymal cells. Migration of mesenchymal cells to the developing endothelial tube and subsequent contact with the endothelium may activate signals (*D*) (e.g., TGF-β) that are necessary for the commitment of mesenchymal cells to SMC-specific lineages. The developing vessel then undergoes structural assembly (*E*) that includes inhibition of endothelial cell proliferation, rapid SMC proliferation, and extracellular matrix deposition. Throughout this period, differentiation to a more mature cell occurs as SMC gradually accumulate contractile and cytoskeletal proteins. Importantly, the vessel remains surrounded by mesenchymal cells or fibroblasts, which may serve as a continued source for mural cells. It is possible that progenitor cells reside in even the mature adventitia. Heterogeneity in both SMC and fibroblast populations is established and, as the mature vessel morphology is achieved (*F*), SMC become quiescent, respond poorly to mitogenic stimulation, and produce minimal matrix proteins. The adventitia may provide a reservoir of cells for vessel repair after injury in postnatal life.

TABLE 1   Factors likely to be involved at different stages of blood vessel formation[a]

Molecules	Localization
Mesoderm formation (Stage A)	
VEGF-A	Endoderm
BFGF	Endoderm
VEGFR-2, VEGFR-1	Mesodermal cells
FGFR(s)	Mesodermal cells
Fibronectin	Extracellular matrix
Aggregation of angioblasts (Stage A)	
VEGFR-2	Angioblasts
VEGFR-1	Angioblasts
VE-cadherin	Angioblasts
tie-2/tek	Angioblasts
ets-1	Angioblasts
PECAM-1	Angioblasts
Fibronectin	Extracellular matrix
VEGF	Endoderm
Endothelial differentiation and formation of primary capillary plexus (Stage B)	
VEGFR-2	Endothelial cells
VEGFR-1 endothelial cells	Endothelial cells
VE-cadherin	Endothelial cells
tie-2/tek	Endothelial cells
tie-1	Endothelial cells
E-selectin	Endothelial cells
ets-1	Endothelial cells
PECAM-1	Endothelial cells
Integrins (e.g., $\alpha V\beta 3$)	Endothelial cells
Notch	Endothelial cells
delta/jagged	Endothelial cells
Ephrins	Endothelial cells
Ephs	Endothelial cells
Fibronectin	Extracellular matrix
Laminin	Extracellular matrix
Collagen IV	Extracellular matrix
VEGF-A	Endoderm
Smooth muscle cell recruitment and differentiation (Stages C, D)	
tie-2	Endothelial cells
PDGF-BB	Endothelial cells
PDGF-AA	Endothelial cells
HB-EGF	Endothelial cells
BFGF	Endothelial cells
Thromboxane A2	Endothelial cells
Angiotensin II	Endothelial cells
Endothelin	Endothelial cells

(Continued)

TABLE 1   (*Continued*)

Molecules	Localization
Leukotrienes	Endothelial cells
Substance P	Endothelial cells
Serotonin	Endothelial cells
Angiopoietin	Endothelial cells
TGF-β	Endothelial cells
IFG-II	Mesenchyme
Tissue factor	Mesenchyme
WNT	Mesenchyme
Laminin	Basement membrane
Type IV collagen	Basement membrane
Perlecan	Basement membrane
Heparan sulfate	Basement membrane

[a]Stages correspond to those illustrated in Figure 1.

through driving endothelial commitment and expansion (61). The importance of VEGF-164 and 188 isoforms was demonstrated in mice engineered to express only the VEGF-120 isoform. VEGF-120 only animals had fewer air-blood barriers and decreased airspace to parenchyma ratios compared with that of wild-type littermates. Thus as development proceeds, the pattern of VEGF isoform expression becomes more restrictive, and the VEGF isoforms serve different roles (64).

VEGF has other family members including VEGF-C and -D. These VEGFs have different affinities for specific VEGF receptors with both VEGF-C and -D demonstrating an ability to bind to VEGFR-3 (65–67). It is now recognized that VEGF-C and -D(?)/VEGFR-3 interactions are probably crucial for development of the lymphatic vascular system (65, 66). Recent studies have evaluated temporal and spatial pattern of VEGF-D expression during mouse lung development (62). The pattern of expression is distinct from that of VEGF-A, suggesting a unique function for each VEGF during lung development. In addition, the finding of VEGF-D expression in the mesenchyme by cells distinctly different from endothelial cells and smooth muscle cells SMC), i.e., fibroblasts, suggests that mesenchymal cells can influence endothelial phenotype during lung development through expression of either VEGF-A or VEGF-D (and possibly VEGF-C). Thus it is likely that subsets of cells within the mesenchyme must have specific roles pertaining to inductive events between mesenchymal and endothelial compartments and therefore may also exert influence on the generation of endothelial heterogeneity described above.

The angiopoietins (Ang) and their major receptor Tie-2 (tyrosine kinase with immunoglobulin and EGF-like domains) are critical for normal vascular development because Ang $1^{-/-}$ or Tie $2^{-/-}$ mice are embryonic lethal owing to the failure of vascular integrity (68, 69). Ang 1 is produced by lung mesenchyme and smooth muscle, whereas Tie 2, its receptor, is restricted to endothelial expression (69–71).

Ang 1 binding to Tie 2 causes receptor tyrosine phosphorylation and downstream signals for endothelial cell survival through phosphatidylinositol 3-kinase PI3K/Akt signaling (72, 73). Angiogenic actions of Ang 1 require endothelium-derived NO (74). Ang 1 promotes interactions between endothelial cells, extracellular matrix, and pericytes that are required for vessel maturation (68). Ang 1-VEGF interactions are critical for normal vascular maturation, but their interactive effects are complex and dependent upon multiple factors. In some settings, Ang1 treatment attenuates VEGF-induced angiogenesis by increasing intercellular endothelial cell junctions, but Ang 1 also makes vessels resistant to VEGF withdrawal (70, 75, 76). Ang 2 also binds Tie 2, but blocks the function of Ang 1 (72). Ang 2 is expressed only at sites of vessel remodeling where it destabilizes vessels, perhaps via Ang 1 inhibition, which may increase endothelial cell responses to VEGF. The combination of low Ang 2 and low VEGF may enhance endothelial cell death (72). Low Ang 1 to Ang 2 ratios favor decreased Tie 2 phosphorylation, which leads to weaker endothelial connections and perhaps increased responsiveness in VEGF. Little is known about Ang expression during development, especially in models of lung hypoplasia, but it is increased in nitrofen model of congenital diaphragmatic hernia (77). In addition, the role of Ang 1 in pulmonary hypertension has been extremely controversial (78). Overexpression of Ang 1 causes severe pulmonary hypertension and is increased in lungs from human patients with pulmonary hypertension (79). However, cell-based gene transfer studies with Ang 1 protected against experimental pulmonary hypertension caused by monocrotaline (80). These differences may be related to the mixed effects of Ang 1 on isolated cells: Ang 1 stimulates smooth muscle cell proliferation but inhibits endothelial cell apoptosis.

Less is known about the other ligand-receptor pairs with regard to lung vascular development. The *notched/jagged* pathways seem to lie downstream from VEGF signaling and be important during embryonic vascular development (81). Taichman et al. performed a systematic evaluation of *notch-1/jagged-1* gene and protein expression in the developing mouse lung and found that the mRNA transcripts for notch-1 and jagged-1 increased progressively from early to later lung development, accompanied by simultaneous rise in endothelial cell-specific gene expression, a pattern somewhat unique to the lung (82). Notch-1 and jagged-1 appeared initially on well-formed larger vessels within the embryonic lung and were progressively expressed on smaller developing vascular networks (82). Notch signaling may mediate its effects, at least in part, through the forkhead box (Fox) f-1 transcription factors (81). That this transcription factor family plays a critical role in development of the pulmonary vasculature was recently shown in studies demonstrating that *Fox-f1* haploinsufficiency is capable of disrupting pulmonary expression of genes in the notch-2 signaling pathway resulting in abnormal development of the lung microvasculature (83).

Another autocrine/paracrine pathway shown to be important in lung vascular development is *Wnt*. Wnt proteins are homologs of the *Drosophila wingless* gene and have been shown to play important roles in regulating cell differentiation, proliferation, and polarity (84, 85). Several *Wnt* genes have been shown to be expressed in the developing lung including *Wnt-2, Wnt-2b, Wnt-7b, Wnt-5a,* and *Wnt-11*. Of those only *Wnt-7b* is expressed at high levels exclusively in the developing airway epithelium during early lung development (86). Morrisey et al. have recently reported the generation and characterization of a *Wnt-7b* (^{lac}z) knock-in mouse (87). These mice exhibit severe pulmonary hypoplasia and lung-specific vascular defects. The vascular defects include dilation of large blood vessels in the lung with subsequent cell death and degradation of the vascular SMC layer leading to pulmonary hemorrhage. Defects in the lungs of *Wnt-7b* ($^{lac}z^{-/-}$) embryos, including early defects in mesodermal proliferation, suggest that *Wnt-7b* is required for lung mesodermal development. It has been demonstrated that the transcription factor Fox-f2, which is expressed in the developing lung mesoderm, is down-regulated in *Wnt-7b* ($^{lac}z^{-/-}$) lung tissue. Because Fox-f2 is related to Fox-f1, which, as mentioned above, has been shown through loss-of-function experiments to regulate lung vascular development (83), this suggests that Wnt signaling through forkhead box transcription factors is critical for development of vascular SMC from the lung mesoderm and for mesodermal thickening. These studies are among the first to specifically evaluate the factors involved in SMC recruitment/differentiation during lung vascular development.

Other factors have been shown to be involved in regulating the recruitment of mural cells (pericytes or SMC) to the vessel wall. However, few of these studies have been specifically performed in the developing lung. It is known that proliferating endothelial cells secrete platelet-derived growth factor-b (PDGF-b), which acts as a chemoattractant and mitogen for mural cell precursors derived from the mesenchyme surrounding the endothelial tubes (88). Other molecules thought to be involved in mural cell recruitment include angiopoietic tissue factor, notch/jagged 1, ephrin/eph, and COUP-TFII (89, 89a). Because the molecules involved in mural cell recruitment across organs appear similar, it has been assumed by many that organ-specific mesodermal cells contribute to the mural or smooth muscle layers of developing vessels, which likely results in tissue-specific functional and regulatory properties of vascular SMC and pericytes. The anatomical position of developing vessels may also contribute to mural cell heterogeneity. SMC in the pulmonary trunk appear to be largely derived from neural crest tissues, whereas mural cells in the parenchymal lung blood vessel appear to have a distinct mesenchymal origin. However, because bronchial and pulmonary vessels are apparently derived from similar (identical ?) mesenchyme and have similar anatomic locations within the lung, the cell- and tissue-specific factors that lead to the emergence of SMCs around the pulmonary and bronchial arteries with distinct functional phenotypes are unclear. Even more intriguing is how the differences in functional phenotype between SMC of conventional versus supranumery arteries arise.

Upon recruitment to the endothelial tube, newly recruited mesenchymal progenitor cells are induced toward a smooth muscle fate. This process seems to be mediated, at least in part, by the activation of transforming growth factor-β (TGF-β). Although the exact mechanisms of how TGF-β is activated and signals to cause

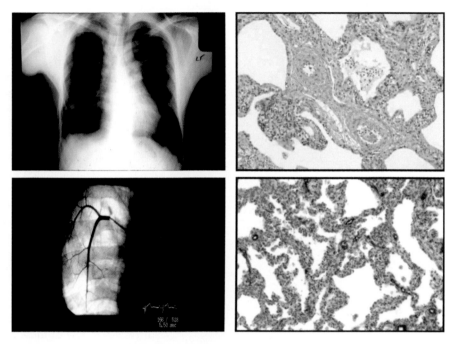

Figure 2 Radiographic and histologic features of bronchopulmonary dysplasia (BPD). This chest radiograph demonstrates late features of BPD, including marked hyperinflation (*upper left panel*). Pulmonary angiography illustrates striking pruning of the pulmonary circulation from this patient, who also had severe pulmonary hypertension (*lower left panel*). Marked fibroproliferative changes with interstitial thickening and decreased septation are shown (*right panels*). Abnormal vascular growth, as reflected by factor VIII staining of endothelium, is shown in the lower right panel.

mural cell differentiation are unknown, it is clear that this signaling system plays a critical role in vascular development (90, 91). TGF-β response elements have been identified in the promoter regions of smooth muscle (SM) genes, such as SM-α-actin, SM-22, and calponin (92). TGF-β can also induce differentiation via the up-regulation of the transcription factor serum response factor (SRF) (93). SRF binds the serum response elements in the promoter regions of mural cell–specific genes, including SM-α-actin, SM-γ-actin, SM-22α, and calponin and induces their coordinated expression (94). The acquisition of differentiated contractile SM or SM-like cells is critical not only to stabilize and maintain the integrity of the developing endothelial tubes but also to control blood flow through these tubes. This may be particularly important in the pulmonary circulation where early intense vasoconstriction capabilities must be acquired to limit blood flow to the developing lung. The mechanisms involved in this mural cell/endothelial cell interaction, which results in the acquisition of SMC with properties differing from systemic vascular smooth muscle and critical for lung function, remain to be elucidated.

Knockout studies of gap junction proteins, including CX43 and CX45, suggest that gap junction communication is required for endothelial-induced mural cell differentiation and function (95, 96). To our knowledge, developmentally regulated expression of these genes in the lung circulation has not been studied. Other requirements include cell-cell adhesion and appropriate cell-matrix interactions mediated via integrin interactions. Unfortunately, very little is known about the regulation of integrin expression, either by endothelial cells or mural cells, during pulmonary vascular development.

Several studies have suggested that homeobox genes may be involved in early vessel development and mural cell recruitment. Homeobox genes, including *Hex*, are important in early vasculogenic or angiogenic processes (97, 98). Other homeobox genes, such as the paired-related homeobox gene, *PRX-1*, may be involved in cell differentiation and stabilization of the vessel wall (99, 100). In the developing systemic vasculature, Prx-1 appears early and is highly evident in prospective connective tissues, including the endocardio-cushions, the epicardium, and the walls of great arteries and veins (99). In the chick embryo, Prx-1 mRNA expression is first evident within the primary vessel wall of coronary and pulmonary arteries. As the vessels mature and thicken, the pattern of expression becomes restricted to nonmuscle cells in the adventitial and outer medial regions (100). These data suggest that Prx-1 may control the expression of genes involved not only in early differentiation of endothelial cells but also in the assembly and segregation of different cell types within the differentiating blood vessel wall.

In addition to proangiogenic stimuli, normal vascular development may also require the expression of angiostatic molecules. Schwartz et al. have described the presence in the lung of a protein, termed endothelial monocyte-activating peptide-2 (EMAP-2), with potent angiostatic properties both in vitro and in vivo (101, 102). It appears to act by specifically causing endothelial cell apoptosis. Studies suggest that its presence may balance and therefore help shape the pulmonary vasculature in the developing mouse lung. Overexpression of EMAP-2 markedly inhibits neovascularization and also inhibits alveolar type II cell development. Other angiostatic proteins, including Ang-2, and monokine-induced by interferon-γ (MIG) and interferon-γ inducible protein (IP-10), have been identified and shown to block the angiogenic effects of angiogenic factors including VEGF. The role of these chemokines in normal lung vascular development remains to be determined.

ROLE OF OXYGEN TENSION IN THE REGULATION OF LUNG VASCULAR MORPHOGENESIS

Oxygen tension appears to be an important physiologic mediator of embryonic and fetal development, and hypoxia is known to be an important regulator of both vasculogensis and angiogenesis. As such, it is not surprising that studies of mammalian development in vitro demonstrate that proper embryonic development is dependent on low oxygen tension (3–5%) and that even short exposures to normoxic environments (20%) can be detrimental to embryonic development. The critical role of oxygen tension in lung development is strongly supported by observations regarding the development of the tracheal system of *Drosophila* (103). Initiation of tubulogenesis in *Drosophila* depends on the expression of two basic helix-loop-helix PAS transcription factors, *tracheless* and *single-minded* (*trh* and *sim*), which dimerize with the *tango* (*tgo*) gene product to direct expression of genes that govern tracheal invagination (103). These *Drosophila* genes share high sequence and functional homology with members of the mammalian HIF-1α/ARNT family where both *trh* and *sim* act as transcriptional initiators homologues to HIF-1α, which depends on an ARNT homolog, (*tgo*), for nuclear transport and the formation of the transcriptional complex (104). Experimental knockout of the complex results in complete failure of tracheal system development in *Drosophila*. Null mutation of HIF-1α in mice has no effect until embryonic day 9, a period that coincides with the initiation of lung development and vasculogenesis. These mice display apparently enlarged vascular structures, fail to initiate lung morphogenesis fully, and die by E 10.5 (105). In addition, there is a regulated loss of mesenchymal tissue suggesting a role for HIF-1α in regulating the survival and differentiation of mesenchymal progenitor cells into vascular structures (105, 106).

Recent work utilizing mammalian fetal lung explants also provides evidence that the process of lung budding and airway bifurcation is oxygen dependent (106–108). In these studies, budding and bifurcation is accelerated at PO_2s that mimic those in the fetal environment. Interestingly, the increases in branching morphogenesis appear to be confined to the periphery in explant culture, suggesting that the locus of hypoxia dependence is within the region of active airway bifurcation and vascular cell proliferation (106, 108). This effect is reversible, a finding that provides insight into the changes in gene expression and lung structure that occur at the time of birth when the lung becomes immediately exposed to alveolar (very high compared with the fetal lung) oxygen concentrations. Embryonic lung explants maintained at alveolar PO_2s show traits of alveolarization (i.e., mesenchymal thinning, epithelial flattening) and saccularization, and also express the surfactant protein C gene. Removal of these explants to fetal PO_2 for as little as 24 h results in the rapid loss of surfactant protein C expression and regrowth of airways into the saccular spaces and an increase in airway surface complexity (108). This is consistent with observations showing that neonatal mice or rats exposed to moderate hypoxia exhibit arrested postnatal lung maturation and alveologenesis, as well as surfactant protein expression and the perinatal increase in epithelial sodium transport (109–111). Thus O_2 tension may modulate both physiologic and structural characteristics of lung vascular and airway development.

Several oxygen-regulated genes appear to be involved in this process. Fibroblast growth factor (FGF)-9 expression is increased at fetal PO_2, whereas FGF-10, which is also known to be involved in branching morphogenesis, demonstrates no such O_2 dependency (108). The FGF-9 promoter has a putative binding site for C/EBP-β. Importantly, studies demonstrate that low oxygen tension results in the nuclear translocation and DNA binding of C/EBP-β. Thus O_2 regulation of FGF-9 via C/EBP-β may be a critical mediator of lung and lung vascular development. Other factors, including members of the hepatocyte nuclear factor (HNF) family, a homolog of *Drosophila forkhead* gene, may play roles in determining epithelial cell lineage fates by acting as dimerization partners for crucial developmental transcription factors such as NKX-2.1 (113, 114). The role of HNFs in hypoxic responses is demonstrated by their involvement in erythropoietin gene expression, suggesting that interactions, at least between HNF-4 and HIF-1α are necessary for regulation of erythropoietin gene expression. Similarly, Sp1 transcription factors, known to be activated by hypoxia, are dimerization partners for HNF-3. Thus HNF transcription factors may integrate the association between oxygen availability and lung development (106).

TGF family genes (TGFβ-1, -2, -3) and their receptors (TGF-βR1, 2, 3) promote proliferation of mesenchyme tissue and thereby inhibit the rate of differentiation of the lung into branched airway structures (116). TGFβ-3 is a HIF-1α−regulated gene that mediates mesenchymal proliferation under many conditions, both developmental and pathophysiologic. In embryonic lung explants, HIF-1α and TGFβ-3 appear to be up-regulated, thus supporting a role for both low oxygen concentrations and TGFβ-3 in lung development. Neutralizing antibody against TGFβ-3 results in mesenchymal thinning and arrest of airway bifurcation. On the other hand, and very importantly, exposure of neonatal rats to reduced oxygen concentrations (9.5%) results in a dramatic increase in TGFβ signaling and TGFβ-R1 receptor expression that collectively act to diminish alveolarization (117). Other groups have reported that TGFβ-1 is an effective inhibitor of stress-invoked inducible nitric oxide synthase (iNOS) activity (118). Nitric oxide generated endogenously and exogenously can increase branching morphogenesis by as much as twofold (119). All three NOS isoforms are expressed in fetal pulmonary tissues with NOS expression increasing dramatically within the final trimester (120). Thus it is possible that endogenous NO production is required for normal airway growth and development both pre- and postnatally. Although hypoxia is known to evoke and sustain NO synthesis and release, it is possible that the simultaneous augmentation of the TGFβ pathway results in inhibition of airway growth and epithelial maturation through direct repression of iNOS activity and expression.

Recent studies have begun to elucidate the signaling pathways through which hypoxia might act to modulate vascular cell proliferation and angiogenesis. With regard to hypoxia-induced angiogenesis, it is clear that PI3K activity is essential and that downstream targets, independent of Akt, are important (121). Activation of mTOR appears essential for hypoxia-mediated amplification of cell proliferation and angiogenesis and activation may occur independently of Akt activation under hypoxic conditions (122). This observation demonstrates a unique signaling aspect of hypoxia compared with that of other mitogenic stimuli. mTOR can act as an important upstream regulator of both HIF-1α and C/EBPβ and thus may represent a key point of convergence in the sensing pathways that regulate cell cycle progression under hypoxic conditions (123).

In addition, recent studies also suggest that purinergic signaling is critical in hypoxia-induced fibroblast proliferation and differentiation as well as in SMC migration and endothelial proliferation (124). This purinergic signaling loop may be autocrine in nature because hypoxia induces the release of ATP from vascular wall cells. In addition, sympathetic or sensory nerves may be an additional source of purine nucleotides and thus may exert significant trophic effects on both the nascent and the mature vasculature. Recent studies have clearly demonstrated the role of the nervous system in directing vascular development (125). However, whether this is true in the lung or just for proximal lung vessels as opposed to distal lung vessels is currently unclear.

Thus it appears unequivocal that hypoxia activates unique cellular signaling pathways critical for growth and differentiation of cells within the pulmonary vasculature. Studies in the future need to be directed at utilizing the effects of oxygen to study lung vascular development. Furthermore, we need to utilize this information to better understand the response of the premature lung, which is removed from this hypoxic environment to a hyperoxic environment. That this transition can inhibit lung development as well as vascularization often leading to significant cardiopulmonary dysfunction is clear. How the effects are mediated is not.

IMPAIRED VASCULAR GROWTH IN BRONCHOPULMONARY DYSPLASIA

Perhaps the most relevant example of how disruption of lung vascular development contributes to human disease lies in the clinical problem of BPD. Abnormalities of the pulmonary circulation, especially the development of pulmonary hypertension, have long been recognized as playing an important role in the pathophysiology and clinical outcomes of premature infants with BPD (126, 134). More recent data from animal and clinical studies suggest that impaired vascular growth may also contribute to abnormalities of lung architecture, especially decreased alveolarization, and may play a critical role in the pathogenesis of BPD. In the following section, we provide a brief overview of problems related to abnormalities of the pulmonary circulation in BPD and review recent data from animal models and clinical studies of BPD that may provide insight into how normal developmental processes are interrupted by premature delivery.

BPD–the Clinical Problem

BPD is the chronic lung disease of infancy that follows mechanical ventilation and oxygen therapy for acute respiratory distress after birth in premature newborns (14, 135, 136). As first characterized by Northway and colleagues in 1967, BPD has traditionally been defined by the presence of persistent respiratory signs and symptoms, the need for supplemental oxygen to treat hypoxemia, and an abnormal chest radiograph at 36-weeks corrected age (135) (Figure 2, see color insert). Despite improvements in perinatal care, chronic lung disease after premature birth remains a major clinical problem. With increasing survival of extremely premature newborns, BPD remains as one of the most significant sequelae of neonatal intensive care, with an estimated 10,000 affected infants in the United States each year. The medical and socio-economic impact of BPD is substantial, with many infants requiring frequent physician visits and hospitalizations due to recurrent respiratory infections, reactive airways disease, upper airway obstruction, cor pulmonale, and exercise intolerance. In many cases, signs and symptoms of chronic lung disease continue into adolescence and adulthood.

With the introduction of surfactant therapy, maternal steroids, new ventilator strategies, aggressive management of the patent ductus arteriosus, improved nutrition, and other treatments, the clinical course and outcomes of premature newborns with RDS have dramatically changed over the past 30 years. During the "presurfactant era," BPD was directly related to the severity of acute respiratory distress syndrome (RDS) and was often present in premature infants who were born at relatively large birth weights and gestational ages. For example, in the original report of Northway and coworkers, the premature infants with BPD were born at 34-weeks gestation, weighed 2200 g, and mortality was 67% (135). Since that time, mortality of preterm infants has markedly decreased, with survival increasing from less than 10% to presently over 50% in even the most extremely preterm newborn (24–26-weeks gestation). The risk of BPD rises with decreasing birth weight, with an incidence up to 85% for newborns between 500–699 g (135, 136). In extremely immature infants, even minimal exposure to oxygen and mechanical ventilation may be sufficient to contribute to BPD (137–139). A recent report demonstrated that about two thirds of infants who develop BPD have only mild respiratory distress at birth (137). These findings suggest that developmental timing of lung injury is a critical factor in the etiology of BPD.

In parallel with this changing epidemiologic and clinical pattern, key features of lung histology in BPD have also changed. Original studies of BPD described a continuous process through distinct stages of disease, progressing from acute lung injury, or an exudative phase with diffuse pulmonary edema, proteinacous debris, and inflammation, to a proliferative phase with structural features of chronic lung disease. Older reports described the gross cobblestone appearance of the lungs, representing alternating areas of atelectasis, marked scarring, and regional hyperinflation (or emphysema). Typical histologic features of BPD included marked airway changes, such as squamous metaplasia of large and small airways, increased peribronchial smooth muscle and fibrosis, chronic inflammation and airway edema, and hyperplasia of submucosal glands. Distal parenchymal lung disease was characterized by heterogenous changes, including regions of volume loss from atelectasis and septal fibrosis alternating with areas of overdistension or emphysematous regions. Mesenchymal thickening with increased cellularity and destruction of septae with alveolar hypoplasia were also noted in early autopsy studies, along with hypertensive structural remodeling of small pulmonary arteries, including smooth muscle hyperplasia and distal extension of smooth muscle growth into vessels that are normally nonmuscular.

There is now growing recognition that infants with persistent lung disease after premature birth have a different clinical course and pathology than was traditionally observed in infants dying with BPD during this presurfactant era (140, 141). The classic progressive stages that first characterized BPD are often absent owing to changes in clinical management, and BPD has clearly changed from being predominantly defined by the severity of acute lung injury to its current characterization, which is primarily defined by a disruption of distal lung growth. In contrast to classic BPD, the new BPD develops in preterm newborns who generally require minimal ventilator support and relatively low FiO_2 (fraction of inspired oxygen) during the early postnatal days. Pathologic signs of severe lung injury with striking fibroproliferative changes have become rare. At autopsy, lung histology now displays more uniform and milder regions of injury, and signs of impaired alveolar and vascular growth are more prominent. These features include a pattern of alveolar simplification with enlarged distal airspaces and reduced growth of the capillary bed, with vessels that are often described as dysmorphic because of their centralized location in the thickened mesenchyme (Figure 2).

Thus, the so-called new BPD of the postsurfactant period represents inhibition of lung development with altered lung structure, growth, and function of the distal airspaces and vasculature (see below). Physiologically, these findings suggest a marked reduction in alveolo-capillary surface area, potentially contributing to impaired gas exchange with increased risk for exercise intolerance, pulmonary hypertension, and poor tolerance of acute respiratory infections (133, 142–145, see below).

Pathogenesis of BPD

BPD represents the response of the lung to injury during a critical period of lung growth, that is, during the canalicular period (17 to 26 weeks in the human), a

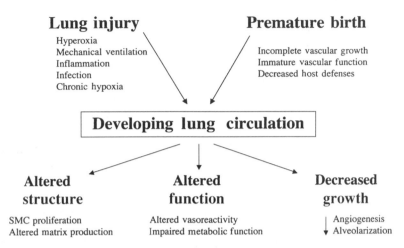

Figure 3 Abnormal pulmonary circulation in chronic lung disease of infancy (BPD) Schematic illustrating pathogenetic abnormalities of the pulmonary circulation in BPD.

time during which airspace septation and vascular development increase dramatically (9). Several factors increase the susceptibility of the premature newborn to the development of BPD, including surfactant deficiency, decreased antioxidant defenses, impaired epithelial ion and water transport function, and lung structural immaturity. Lung injury after premature birth and the subsequent arrest of lung growth results from complex interactions between multiple adverse stimuli, including inflammation, hyperoxia, mechanical ventilation, and infection, of the poorly defended developing lung (Figure 3). Most recently, prenatal exposure to proinflammatory cytokines, such as TNF-α, IL-6, IL-8, and others, due to maternal chorioamnionitis, have been shown to enhance lung maturation in utero, but increase the risk for BPD (146–148).

Hyperoxia and oxidant stress are critical factors in the development of BPD (149, 150). The transition of the premature newborn from the low-oxygen tension environment of the normal fetus to the relative hyperoxia of extrauterine life increases the risk for BPD with decreased alveolarization and a dysmorphic vasculature. This premature change in the oxygen environment is likely to impede normal epithelial-mesenchymal interactions leading to alterations in endothelial cell survival, differentiation, and organization in the microvasculature, the mechanisms of which are partly described above. The premature infant is especially susceptible to reactive oxidant species (ROS)-induced damage owing to the lack of adequate antioxidants after premature birth. Antioxidant enzymes [e.g., superoxide dismutase (SOD), catalase, and glutathione peroxidase] markedly increase during late gestation (151). In addition, the ability to increase synthesis of antioxidant enzymes in response to hyperoxia is decreased in preterm animals, suggesting that premature birth may precede the normal up-regulation of antioxidants, which persists during early postnatal life. Experimental studies have shown that hyperoxia, even in the absence of ventilation, can induce lung injury that mimics RDS and late sequelae that are similar to BPD (152, 153; see below). Endothelial and alveolar type II cells are both extremely susceptible to hyperoxia and ROS-induced injury, leading to increased edema, cellular dysfunction, and impaired cell survival and growth (152–154). The critical role of host antioxidant defenses in protecting the developing lung from hyperoxia-induced injury is further shown by studies of transgenic SOD mice and treatment with exogenous SOD (155–157).

Even in the absence of overt signs of baro- or volutrauma, treatment of premature neonates with mechanical ventilation initiates and promotes lung injury with inflammation and permeability edema, and contributes to BPD. Ventilator-associated lung injury (VALI) results from stretching distal airway epithelium and capillary endothelium, which increases permeability edema, inhibits surfactant function, and provokes a complex inflammatory cascade (158). Experimental studies have clearly shown that overdistension, and not pressure per se, is responsible for lung injury in the surfactant-deficient lung (159, 160). Even brief periods of positive-pressure ventilation, such as during resuscitation in the delivery room, can cause bronchiolar epithelial and endothelial damage in the lung, setting the stage for progressive lung inflammation and injury (161, 162).

As described above, lung inflammation, whether induced prior to birth (from chorioamnionitis) or during the early postnatal period (due to hyperoxia or VALI) plays a prominent role in the development of BPD (163–165). Numerous clinical and experimental studies have shown that the risk for BPD is associated with sustained increases in tracheal fluid neutrophil counts, activated macrophages, high concentrations of lipid products, oxidant-inactivated α-1-antitrypsin activity, and proinflammatory cytokines, including IL-6 and IL-8, and decreased IL-10 levels (166–168). Release of early response cytokines, such as TNF-α, IL-1β, IL-8, and TGF-β, by macrophages and the presence of soluble adhesion molecules (i.e., selectins) may impact other cells to release chemoattractants that recruit neutrophils and amplify the inflammatory response (169, 170). Elevated concentrations of proinflammatory cytokines in conjunction with reduced anti-inflammatory products (i.e., IL-10) usually appear in tracheal aspirates within a few hours of life in infants subsequently developing BPD. Increased elastase and collagenase release from activated neutrophils may directly destroy the elastin and collagen framework of the lung, and markers of collagen and elastin degradation have been recovered in the urine of infants with BPD (171). Infection from relatively low virulence organisms, such as airway colonization with *Ureaplasma urealyticum*, may augment the inflammatory response, further increasing to the risk for BPD (172). Finally, other factors, such as nutritional deficits and genetic factors, such as vitamin A and E deficiency or single nucleotide polymorphism variants of the surfactant proteins, respectively, are likely to increase risk for BPD in some premature newborns (173, 174). Thus multiple stimuli act on the susceptible lung at a critical stage of development after premature birth, leading to disruption of normal vascular growth and impaired alveolarization.

Pulmonary Circulation in Human BPD

In addition to adverse effects on the airway and distal airspace, acute lung injury also impairs growth, structure, and function of the developing pulmonary circulation after premature birth (13, 133, 153). Endothelial cells have been shown to be particularly susceptible to oxidant injury through hyperoxia or inflammation (152, 153, 157). The media of small pulmonary arteries may also undergo striking changes, including smooth muscle cell proliferation, precocious maturation of immature mesenchymal cells into mature smooth muscle cells, and incorporation of fibroblasts/myofibroblasts into the vessel wall (176, 177). Structural changes in the lung vasculature contribute to high pulmonary vascular resistance (PVR) through narrowing of the vessel diameter and decreased vascular compliance. In addition to these structural changes, the pulmonary circulation is further characterized by abnormal vasoreactivity, which also increases PVR (127, 130). Finally, decreased angiogenesis may limit vascular surface area, causing further elevations of PVR, especially in response to high cardiac output with exercise or stress.

Overall, early injury to the lung circulation leads to the rapid development of pulmonary hypertension, which contributes significantly to the morbidity and mortality of severe BPD. Even from the earliest reports of BPD, pulmonary hypertension and cor pulmonale were recognized as being associated with high mortality (128, 131). Walther et al. showed that elevated pulmonary artery pressure in premature newborns with acute RDS (determined from serial echocardiograms) was associated with severe disease and high mortality (131). Past studies have also shown that persistent echocardiographic evidence of pulmonary hypertension beyond the first few months of life is associated with up to 40% mortality in infants with BPD (128). High mortality rates have also been reported in infants with BPD and pulmonary hypertension who require prolonged ventilator support (132). Although pulmonary hypertension is a marker of more advanced BPD, elevated PVR also causes poor right ventricular function, impaired cardiac output, limited oxygen delivery, increased pulmonary edema and, perhaps, a higher risk for sudden death.

Physiologic abnormalities of the pulmonary circulation in BPD include elevated PVR and abnormal vasoreactivity, as evidenced by the marked vasoconstrictor response to acute hypoxia (127, 130, 132, 178). Cardiac catheterization studies have shown that even mild hypoxia causes marked elevations in pulmonary artery pressure in infants with modest basal levels of pulmonary hypertension. Treatment levels of oxygen saturations above 92–94% effectively lower pulmonary artery pressure (130). Strategies to lower pulmonary artery pressure or limit injury to the pulmonary vasculature may limit the subsequent development of pulmonary hypertension in BPD. Recent data suggest that high pulmonary vascular tone continues to elevate PVR in older patients with BPD, as demonstrated by responsiveness to altered oxygen tension and inhaled nitric oxide (178).

In addition to pulmonary hypertension, clinical studies have also shown that metabolic function of the pulmonary circulation is impaired, as reflected by the impaired clearance of circulating norepinephrine (NE) across the lung (179).

Normally, 20–40% of circulating NE is cleared during a single passage through the lung, but infants with severe BPD have a net production of NE across the pulmonary circulation. It is unknown whether impaired metabolic function of the lung contributes to the pathophysiology of BPD by increasing circulating cate-cholamine levels, or if it is simply a marker of severe pulmonary vascular disease. It has been speculated that high catecholamine levels may lead to left ventricular hypertrophy or systemic hypertension, which are known complications of BPD (179–181).

Prominent bronchial or other systemic-to-pulmonary collateral vessels were noted in early morphometric studies of infants with BPD and can be readily iden-tified in many infants during cardiac catheterization (182, 183). Although these collateral vessels are generally small, large collaterals may contribute to significant shunting of blood flow to the lung, causing edema and the need for higher FiO_2. Interestingly, this enlargement of the bronchial circulation is similar to that de-scribed in adults with emphysema, chronic atelectasis, and/or high PVR, again supporting the notion that obstruction to blood flow in the pulmonary circulation is a significant stimulus for growth of the bronchial circulation (see above). Collat-eral vessels have been associated with high mortality in some patients with severe BPD who also had severe pulmonary hypertension. Some infants have improved after embolization of large collateral vessels, as reflected by a reduced need for supplemental oxygen, ventilator support, or diuretics. However, neither the actual contribution of systemic collateral vessels to the pathophysiology of BPD nor the cellular mechanisms driving their enlargement is known.

Finally, pulmonary hypertension and right heart function remain major clinical concerns in infants with BPD. However, it is now clear that pulmonary vascular disease in BPD also includes reduced pulmonary artery density owing to impaired growth, which contributes to physiologic abnormalities of impaired gas exchange, as well as to the actual pathogenesis of BPD (184, 185). As discussed below, experimental data support the hypothesis that impaired angiogenesis can impede alveolarization (186, 187) and that strategies preserving and enhancing endothelial cell survival, growth, and function may provide new therapeutic approaches for the prevention of BPD.

Altered Signaling of Angiogenic Factors in Human BPD

As described above, multiple growth factors and signaling systems have been shown to play important roles in normal lung vascular growth (7, 13) (Table 1). Several studies have examined how premature delivery and changes in oxygen tension, inflammatory cytokines, and other signals alter normal growth factor ex-pression and signaling and thus lung/lung vascular development. The majority of studies have focused on VEGF. Impaired VEGF signaling has been associated with the pathogenesis of BPD in the clinical setting (184, 185). Bhatt and coworkers first demonstrated decreased lung expression VEGF and VEGFR-1 in the lungs of premature infants who died with BPD (184). In another study, VEGF was found to be lower in tracheal fluid samples from premature neonates who subsequently develop BPD than those who do not develop chronic lung disease (185). Experimentally, hyperoxia down-regulates lung VEGF expression, and pharmacologic inhibition of VEGF signaling in newborn rats impairs lung vascular growth and inhibits alveolarization (188–191; see below). Thus the biologic basis for impaired VEGF signaling leading to decreased vascular growth and impaired alveolarization is well established. Additionally, lung VEGF expression is impaired in primate and ovine models of BPD induced by mechanical ventilation after premature birth, further supporting the hypothesis that impaired VEGF signaling contributes to the pathogenesis of BPD.

In addition, a role for TGF-β in the pathogenesis of BPD has been raised. For example, high levels of TGF-β inhibit lung morphogenesis, and increased levels of TGF-β-1 are found in tracheal fluid samples early in the course of infants developing BPD, which may predict high risk for BPD (192). Thus while TGF-β may be necessary for certain stages in normal lung development, excessive expression, particularly at the wrong time, may induce an inhibition of lung morphogenesis and cause progressive pulmonary fibrosis (193). In fact, overexpression of TGF-β-1 inhibits alveolar growth and remodeling during the neonatal period (194). These observations make TGF-β a particularly intriguing target when considering strategies to ameliorate the structural abnormalities that characterize BPD.

Ongoing studies of the regulation and activities of diverse growth factors are likely to lead to greater understanding of the pathobiology and treatment of BPD.

Relationship of Vascular Growth and Alveolarization

As described above, close coordination of growth between airways and vessels is essential for normal lung development. Thus we and others have hypothesized that failure of pulmonary vascular growth during a critical period of lung growth (saccular or alveolar stages of development) could decrease septation and ultimately contribute to the lung hypoplasia that characterizes BPD.

To determine whether angiogenesis is necessary for alveolarization, we studied the effects of the antiangiogenesis agents, thalidomide and fumagillin, on lung growth in the newborn and infant rat (187). In comparison with vehicle-treated controls, postnatal treatment with these inhibitors of angiogenesis reduced lung vascular density, alveolarization, and lung weight. These findings suggest that angiogenesis is necessary for alveolarization during lung development and that mechanisms that injure and inhibit lung vascular growth may impede alveolar growth after premature birth. Because VEGF has potent angiogenic effects during lung development and may synchronize vascular growth with neighboring epithelium, we also studied the effects of SU5416, a selective VEGF receptor inhibitor, on alveolarization during early postnatal life (191). Treatment of neonatal rats with a single injection of SU5416 caused pulmonary hypertension, reduced lung vascular density, and reduced alveolarization in infant rats (191) (Figure 4). Thus inhibition of lung vascular growth during a critical period of postnatal lung growth

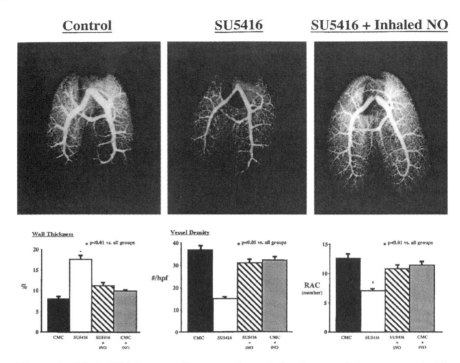

Figure 4   Inhaled NO increases lung vascular and alveolar growth in rats treated with the VEGF receptor inhibitor, SU5416. The upper panels illustrate improved vascular growth after inhaled NO therapy of SU5416-treated rats. Inhaled NO improved pulmonary artery wall thickness, vascular growth, and alveolarization (radial alveolar counts, RAC) (88).

impairs alveolarization, suggesting that endothelial-epithelial cross talk, especially via VEGF signaling, is critical for normal lung growth following birth.

The mechanisms through which impaired VEGF signaling inhibits vascular growth and alveolarization are uncertain, but, in part, may be mediated by altered NO production. Past in vitro and in vivo studies have shown that VEGF stimulates endothelial NO synthase (eNOS) expression in isolated endothelial cells from the systemic circulation (195–201), but whether VEGF is an important regulator of lung eNOS expression, especially during development, and what role NO plays on lung growth are incompletely understood. In addition to its effects on vascular tone, NO can alter angiogenesis, but data are conflicting on its effects. Although NO inhibits endothelial cell proliferation in some models, most studies have shown that NO mediates the angiogenic effects of VEGF via activation of VEGFR-2 and stimulation of the Akt-PI3K pathway (196, 198–201). Proliferating bovine aortic endothelial cells express greater eNOS mRNA and protein than confluent cells, but NOS inhibition does not apparently affect their rate of proliferation. Studies with the eNOS$^{-/-}$ fetal mouse model suggest that NO plays a critical role in vascular and alveolar growth in utero and that eNOS$^{-/-}$ newborns are more susceptible to hypoxia-induced inhibition of alveolarization (202–205). Interestingly, lung eNOS expression is down-regulated in primate and ovine models of BPD (206, 207). More recently, treatment of newborn rats with SU5416, the VEGF receptor antagonist, was shown to decrease eNOS expression and NO production during infancy, and that prolonged treatment with inhaled NO prevented the development of pulmonary hypertension, improved vascular growth, and enhanced alveolarization (208) (Figure 4). Previous studies have shown that inhaled NO attenuates hyperoxia-induced acute lung injury (209–211), which may enhance subsequent vascular and lung growth. These studies suggest that decreased VEGF signaling down-regulates lung eNOS expression, and that impaired NO production may contribute to abnormal lung growth during development. Importantly, a recent randomized single-center study has shown that inhaled NO treatment reduced the combined endpoint of BPD and death in human premature newborns with moderate RDS (212). Ongoing multicenter clinical trials are evaluating the potential efficacy of inhaled NO in reducing BPD in larger population studies.

MODELS OF BPD

For more than 20 years, exposure of newborn rats to hyperoxia has been used as a model to study mechanisms of BPD due to the effects of enriched oxygen on alveolarization and vascular growth. During the alveolar phase of lung development, the first 2 weeks of postnatal life in the rat, the lung also undergoes a striking proliferation of vascular growth. Exposure of newborn rats to either 100% or 60% FiO$_2$ markedly decreases capillary density when compared with that of air-breathing controls (153, 175). During the room air recovery period after relatively short exposures to hyperoxia, the number of capillaries and alveoli tend to increase toward normal values in infant rats. However, if the period of oxygen exposure is more prolonged, the decrease in pulmonary vascular density persists despite room air recovery.

Although multiple mechanisms are likely involved in the arrest of lung growth after hyperoxia, several studies have demonstrated marked and sustained down-regulation of lung VEGF and VEGFR-2 expression, suggesting that impaired VEGF signaling contributes to these long-lived abnormalities in lung structure. Thus hyperoxic exposure provides a useful model for the study of basic mechanisms that disrupt normal pulmonary vascular growth and development.

Recently, the gene expression profile induced by prolonged oxidative stress has been studied. Using DNA-microarray analysis, which was then largely confirmed by real-time RT-PCR, Wagenaar et al. demonstrated that prolonged oxidative stress induces changes in a complex orchestra of genes involved in inflammation, coagulation, fibrinolysis, extracellular matrix turnover, cell cycle, signal transduction, and alveolar development (213). The changes in gene expression, both up-regulated as well as down-regulated, appear consistent with the pathologic changes in immature lungs developing a BPD-like picture. One of the major effects of hyperoxia on postnatal lung development in the rat is the reduction of secondary septation and the enlargement of alveoli. This phenomenon was accompanied by a down-regulation of FGFR-4 from day 3 onward and of the Flk-1 receptor later in the course (day 10). These observations are consistent with previous findings demonstrating that the lungs of FGFR-3$^{-/-}$/FGFR-4$^{-/-}$ mice are normal at birth but then develop a complete block in alveologenesis and do not form secondary septa (214). Down-regulation of VEGFR-2 (or Flk-1) confirmed in this array analysis on day 10 coincided with the presence of enlarging alveoli and is consistent with the role of VEGF/VEGFR-2 signaling in the maintenance of alveolar structures (215). These findings provide further support for a critical role of VEGF/VEGFR-2 signaling in the pathogenesis of BPD.

Another set of genes up-regulated by hyperoxia is linked to the inflammatory response and is consistent with observations in the human as described above. The influx of leukocytes appears to be mediated by chemokines such as IL-8, CINC-1, and MIP-2 via the activation of the CXCR2 receptor. The Wagenaar study confirms the up-regulation of these molecules observed in other hyperoxic models and implicates them in the disease process. The importance of these chemokines in hyperoxia-induced lung injury is demonstrated by the fact that antichemokine treatment, which reduces neutrophil influx into the lung, preserved alveolar development in newborn rats exposed to hyperoxia (216, 217).

Several studies have demonstrated that extravascular fibrin deposits are frequently observed in the septa and alveoli of infants with BPD. Again, in the hyperoxic model described, gene array analysis demonstrates the up-regulation of the procoagulant tissue factor (TF), down-regulation of the anticoagulant thrombomodulin (TM), and up-regulation of the fibrinolytic inhibitor, PAI-1. These changes would lead to a procoagulant and antifibrolytic environment resulting in fibrin deposition. The significance of PAI-1 in hyperoxia-induced fibrin deposition has been demonstrated in PAI-1-deficient mice, which fail to develop intra-alveolar fibrin deposits and show a less severe phenotype in response to hyperoxia-induced injury (218). The accumulation of fibrin can contribute to injury in several ways. It can function as a ligand for receptors on circulating leukocytes, including the integrin Mac-1, and can promote both their migration and activation (219). Fibrin can also activate and induce proliferation and migration of fibroblasts, probably via activation of NF-κB and AP-1 (220). Therefore, unwanted activation of leukocytes and fibroblasts can contribute to the abnormalities of lung vascular growth that characterize BPD.

The combination of rat models and gene array analysis thus provides important new insights into the mechanisms through which interrupted fetal lung development can lead to significant abnormalities in postnatal pulmonary vascular and lung development.

Extensive work from the Bland laboratory has shown that prolonged mechanical ventilation of premature lambs after cesarean-section delivery at 120–130 days gestation (term 147 days) for 2–3 weeks causes a chronic lung disease (CLD) that shares many features of human BPD (221). In addition to airway abnormalities, pulmonary vascular growth, structure, and function are impaired after chronic ventilation. In comparison with controls, CLD lambs have reduced pulmonary artery density with reduced alveolarization and hypertensive pulmonary artery remodeling. Although the mechanisms of altered lung growth are unclear, lung eNOS and soluble guanylate cyclase expression are reduced in CLD lambs, suggesting that impaired NO-cGMP signaling may contribute to pulmonary vascular disease in this model.

Primate Model of BPD

An extremely important model of BPD in primates has been developed under the guidance of Coalson (141). In this model, premature baboons are delivered by cesarean section at 140 days (or 0.75 term) and treated with mechanical ventilation with high FiO_2 (222). These animals develop severe chronic lung disease, with histologic lesions that mimic human BPD (222). A newer and perhaps more relevant model for the new BPD is one in which extremely premature baboons are delivered at 125 days (or 0.67 term; which is roughly equivalent to human gestation of 26–27 weeks) and are treated with exogenous surfactant and mechanical ventilation, but are managed with lower (as needed) levels of supplemental oxygen. These animals develop structural lesions characterized by severe alveolar hypoplasia with decreased and dysmorphic vascular growth. As in the premature lamb model of BPD, lung eNOS expression is impaired (206). In addition, lung VEGF mRNA content is markedly decreased, and the localized epithelial pattern of VEGF expression is absent (223). Expression of mRNA for VEGFR-1 also decreases by 30–40% in treated animals but the VEGFR-2 mRNA expression remained unchanged. Whether interventions such as inhaled NO or VEGF treatment can reduce the abnormalities of lung vascular growth and impaired alveolarization in this model is uncertain.

CONCLUSIONS

Past studies have primarily focused on how altered lung vascular growth and development contributes to pulmonary hypertension. Recently, basic studies of vascular growth have led to novel insights into mechanisms underlying development of the normal pulmonary circulation and the essential relationship of vascular growth to lung alveolar development. These observations have led to new concepts underlying the pathobiology of developmental lung disease, especially the inhibition of lung growth that characterizes BPD. We speculate that understanding basic mechanisms that regulate and determine vascular growth will lead to new clinical strategies to improve the long-term outcome of premature babies with BPD.

The *Annual Review of Physiology* is online at
http://physiol.annualreviews.org

LITERATURE CITED

1. West JB. 2003. Thoughts on the pulmonary blood-gas barrier. *Am. J. Physiol. Lung Cell Mol. Physiol.* 285:L501–13

2. Cardoso WV. 2001. Molecular regulation of lung development. *Annu. Rev. Physiol.* 63:471–94

3. Warburton D, Schwarz M, Tefft D, Flores-Delgardo G, Anderson KD, Cardoso WV. 2000. The molecular basis of lung morphogenesis. *Mech. Dev.* 92:55–81

4. Demayo F, Minoo P, Plopper CG, Schuger L, Shannon J, Torday JS. 2002. Mesenchymal-epithelial interactions in lung development and repair: Are modeling and remodeling the same process? *Am. J. Physiol. Lung Cell Mol. Physiol.* 283:L510–17

5. Roth-Kleiner M, Post M. 2003. Genetic control of lung development. *Biol. Neonate* 8:83–88

6. Prodhan P, Kinane TB. 2002. Developmental paradigms in terminal lung development. *BioEssays* 24:1052–59

7. Kumar VH, Ryan RM. 2004. Growth factors in the fetal and neonatal lung. *Front. Biosci.* 9:464–80

8. Shannon JM, Hyatt BA. 2004. Epithelial-mesenchymal interactions in the developing lung. *Annu. Rev. Physiol.* 66:625–45

9. Burri PH. 1999. Lung development and pulmonary angiogenesis. In *Lung Development*, ed. C Gaultier, JR Bourbon, M Post, pp. 122–51. New York: Oxford Univ. Press

10. deMello DE. 2004. Pulmonary pathology. *Semin. Neonatol.* 9:311–29

11. Han RN, Babaei S, Robb M, Lee T, Ridsdale R, et al. 2004. Defective lung vascular development and fatal respiratory distress in endothelial NO synthase-deficient mice: a model of alveolar capillary dysplasia. *Circ. Res.* 94:1115–23

12. Le Cras TD, Hardie WD, Deutsch GH, Albertine KH, Ikegami M, et al. 2004. Transient induction of TGF-α disrupts lung morphogenesis causing pulmonary disease in adulthood. *Am. J. Physiol. Lung Cell Mol. Physiol.* 287:L718–29

13. D'Angio CT, Maniscalco WM. 2002. The role of vascular growth factors in hyperoxia-induced injury to the developing lung. *Front. Biosci.* 7:D1609–23

14. Jobe AJ. 1999. The new BPD: an arrest of lung development. *Pediatr. Res.* 46:641–43

15. Nagaishi C. 1972. *Functional Anatomy and Histology of the Lung.* Tokyo/Baltimore, MD: Univ. Park Press

16. deMello DE, Reid LM. 2002. Prenatal and postnatal development of the pulmonary circulation. In *Basic Mechanisms of Pediatric Respiratory Disease*, ed. CG Haddad, SH Abman, V Chernick, pp. 77–101. Hamilton, Ontario: Decker. 2nd ed.

17. Shaw AM, Bunton DC, Fisher A, McGrath JC, Montgomery I, et al. 1999. V-shaped cushion at the origin of bovine pulmonary supernumerary arteries: structure and putative function. *J. Appl. Physiol.* 87:2348–56

18. Shaw AM, Bunton DC, Brown T, Irvine J, MacDonald A. 2000. Regulation of sensitivity to 5-hydroxytryptamine in pulmonary supernumerary but not conventional arteries by a 5-HT(1D)-like receptor. *Eur. J. Pharmacol.* 408:69–82

19. Ogata T, Iijima T. 1993. Structure and pathogenesis of plexiform lesion in pulmonary hypertension. *Chin. Med. J.* 106:45–48

20. Liebow AA, Hales MR, Bloomer WE, Harrison W, Lindskog GE. 1950. Studies on the lung after ligation of the pulmonary artery. *Am. J. Pathol.* 26:177

21. Liebow AA, Hales MR, Bloomer WE. 1959. Relation of bronchial to pulmonary vascular treee. In *Pulmonary Circulation: An International Symposium*, pp. 79–98. New York/London: Grune & Stratton

22. Liebow AA, Hales MR, Lindskog GE. 1949. Enlargement of the bronchial arteries, and their anastomoses with the pulmonary arteries in bronchiectasis. *Am. J. Pathol.* 25:211

23. Cockett FB, Vass CCN. 1951. A comparison of the role of the bronchial arteries in bronchiectasis and in experimental ligation of the pulmonary artery. *Thorax* 6:268

24. Aoki S, Yoshimura M, Hisada T. 1954. Studies on the bronchus, bronchial artery and pulmonary artery using resin casts. *Nihon Ishikai Zasshi* 32:695

25. Deleted in proof

26. Lauweryns JM, Boussauw L. 1969. The ultrastructure of pulmonary lymphatic capillaries of newborn rabbits and human infants. *Lymphology* 2:108–29

27. Simer PH. 1952. Drainage of pleural lymphatics. *Anat. Rec.* 113:269–83

28. Tobin CE. 1957. Human pulmonic lymphatica: an anatomic study. *Anat. Rec.* 127:611–38

29. Noonan JA, Walters LR, Reeves JT. 1970. Congenital pulmonary lymphangiectasis. *Am. J. Dis. Child.* 120:314–19

30. Boyd JD. 1965. Development of the heart. In *Handbook of Physiology: A Critical, Comprehensive Presentation of Physiological Knowledge and Concepts: Sect. 2: Circulation*, ed. WF Hamilton, P Dow, 3:2511–44. Washington, DC: Am. Physiol. Soc.

31. Congdon ED. 1922. Transformation of the aortic arch system during the development of the human embryo. *Carnegie Inst. Contrib. Embryol.* 14:47–110

32. DeRuiter MC, Gittenberger-de-Groot AC, Poelmann RE, Van Iperen L, Mentink MM. 1993. Development of the pharyngeal arch system related to the pulmonary and bronchial vessels in the avian embryo. With a concept on systemic-pulmonary collateral artery formation. *Circulation* 87:1306–19

33. DeRuiter MC, Gittenberger-de Groot AC, Ramos S, Poelmann RE. 1989. The special status of the pulmonary arch artery in the branchial arch system of the rat. *Anat. Embryol.* 179:319–25

34. Huntingdon GS. 1919. The morphology of the pulmonary artery in the mammalia. *Anat. Rec.* 17:165–89

35. Brown AJ. 1913. The development of the pulmonary vein in the domestic cat. *Anat. Rec.* 7:299–329

36. Squier TL. 1916. On the development of the pulmonary circulation in the chick. *Anat. Rec.* 10:425–38

37. Chang C. 1931. On the origin of the pulmonary vein. *Anat. Rec.* 50:1–8

38. Gebb SA, Shannon JM. 2000. Tissue interactions mediate early events in pulmonary vasculogenesis. *Dev. Dyn.* 217:159–69

39. Schachtner SK, Wang Y, Baldwin HS. 2000. Qualitative and quantitative analysis of embryonic pulmonary vessel formation. *Am. J. Respir. Cell Mol. Biol.* 22:157–65

40. Maeda A, Suzuki S, Suzuki T, Endo M, Moriya T, et al. 2002. Analysis of intrapulmonary vessels and epithelial-endothelial interactions in the human developing lung. *Lab. Invest.* 82(3):293–301

41. Taichman DB, Loomes KM, Schachtner SK, Guttentag S, Vu C, et al. 2002. Notch1 and Jagged1 expression by the developing pulmonary vasculature. *Dev. Dyn.* 225:166–75

42. Han RNN, Post M, Tanswell AK, Lye SJ. 2003. Insulin-like growth factor-I receptor-mediated vasculogenesis/angiogenesis in human lung development. *Am. J. Respir. Cell Mol. Biol.* 28:159–69

43. deMello DE, Sawyer D, Galvin N, Reid LM. 1997. Early fetal development of lung vasculature. *Am. J. Respir. Cell Mol. Biol.* 16:568–81

44. Hislop AA. 2002. Airway and blood vessel interaction during lung development. *J. Anat.* 201:325–34

45. deMello DE, Reid LM. 2002. Embryonic and early fetal development of human lung vasculature and its functional implications. *Pediatr. Dev. Pathol.* 3:439–49

46. Hall SM, Hislop AA, Haworth SG. 2002. Origin, differentiation, and maturation of human pulmonary veins. *Am. J. Respir. Cell Mol. Biol.* 26:333–40

47. Canis Parera MC, Van Dooren M, van Kempen M, De Krijger R, Grosveld F, et al. 2004. Distal angiogenesis: a new concept for lung vascular morphogenesis. *Am. J. Physiol. Lung Cell Mol. Physiol.* In press

48. Bucher U, Reid L. 1961. Development of the intrasegmental bronchial tree: the pattern of branching and development of cartilage at various stages of intra-uterine life. *Thorax* 16:207–18

49. Hislop A, Reid L. 1972. Intrapulmonary arterial development during fetal life-branching pattern and structure. *J. Anat.* 113:35–48

50. Burri PH. 1984. Fetal and postnatal development of the lung. *Annu. Rev. Physiol.* 46:617–28

51. Burri PH. 1997. Structural aspects of prenatal and postnatal development and growth of the lung. See Ref. 224, pp. 1–35

52. Shannon JM, Deterding RR. 1997. Epithelial-mesenchymal interactions in lung development. See Ref. 224, pp. 81–118

53. Roman J. 1997. Cell-cell and cell-matrix interactions in development of the lung vasculature. See Ref. 224, pp. 365–400

54. Langston C, Kida K, Reed M, Thurlbeck WM. 1984. Human lung growth in late gestation and in the neonate. *Am. Rev. Resp. Dis.* 129:607–13

55. deMello DE, Reid LM. 1991. Pre- and postnatal development of the pulmonary circulation. In *Basic Mechanisms of Pedi-atric Respiratory Disease: Cellular and Integrative*, ed. V Chernick, RB Mellins, pp. 36–54. Philadelphia: Decker

56. Schittny JC, Djonov V, Fine A, Burri PH. 1998. Programmed cell death contributes to postnatal lung development. *Am. J. Respir. Cell Mol. Biol.* 18:786–93

57. Gerber HP, Hillan KJ, Ryan AM, Kowalski J, Keller GA, et al. 1999. VEGF is required for growth and survival in neonatal mice. *Development* 126:1149–59

58. Bhatt AJ, Amin SB, Chess PR, Watkins RH, Maniscalco WM. 2000. Expression of vascular endothelial growth factor and Flk-1 in developing and glucocorticoid-treated mouse lung. *Pediatr. Res.* 47:606–13

59. Marszalek A, Daa T, Kashima K, Nakayama I, Yokoyama S. 2001. Expression of vascular endothelial growth factor and its receptors in the developing rat lung. *Jpn. J. Physiol.* 51:313–18

60. Gebb S, Stevens T. 2004. On lung endothelial cell heterogeneity. *Microvasc. Res.* 68:1–12

61. Ng YS, Rohan R, Sunday ME, deMello DE, D'Amore PA. 2001. Differential expression of VEGF isoforms in mouse during development and in the adult. *Dev. Dyn.* 220:112–21

62. Greenberg JM, Thompson FY, Brooks SK, Shannon JM, McCormick-Shannon K, et al. 2002. Mesenchymal expression of vascular endothelial growth factors D and A defines vascular patterning in developing lung. *Dev. Dyn.* 224:144–53

63. Akeson AL, Greenberg JM, Cameron JE, Thompson FY, Brooks SK, et al. 2003. Temporal and spatial regulation of VEGF-A controls vascular patterning in the embryonic lung. *Dev. Biol.* 264:443–55

64. Galambos C, Ng YS, Ali A, Noguchi A, Lovejoy S, et al. 2002. Defective pulmonary development in the absence of heparin-binding vascular endothelial growth factor isoforms. *Am. J. Respir. Cell. Mol. Biol.* 27:194–203

65. Lohela M, Saaristo A, Veikkola T, Alitalo K. 2003. Lymphangiogenic growth factors, receptors and therapies. *Thromb. Haemost.* 90:167–84

66. Saharinen P, Petrova TV. 2004. Molecular regulation of lymphangiogenesis. *Ann. NY Acad. Sci.* 1014:76–87

67. Carmeliet P. 2003. Angiongenesis in health and disease. *Nat. Med.* 9:653–60

68. Suri C, Jones PF, Patan S, Maisonpierre PC, Davis S, et al. 1996. Requisite role of ang 1, ligand for Tie 2 receptor, during embryonic angiogenesis. *Cell* 87:1171–80

69. Sato TN, Tozawa Y, Deutsch U, Wolberg-Buchholz K, Fujiwara Y, et al. 1995. Distinct rols of the receptor tyrosine kinases Tie 1 and Tie 2 in blood vessel formation. *Nature* 376:70–74

70. Thurston G, Suri C, Smith K, McClain J, Sato TN, et al. 1999. Leakage resistant blood vessels in mice transgenically over-expressing ang 1. *Science* 286:2511–14

71. Davis S, Aldrich TH, Jones PF, Acheson A, Compton DL, et al. 1996. Isolation of ang-1 a ligand for Tie 2 receptor by secretion trap expression cloning. *Cell* 87:1161–69

72. Maisonpierre PC, Suri C, Jones PF, Bartunkova S, Wiegand SJ, et al. 1997. Ang 2 a natural antagonist for tie 2 that disrupts in vivo angiogenesis. *Science* 277:55–50

73. Kim I, Kim JH, Moon SO, Kwak HJ, Kin NG, Koh GY. 2000. Angiopoietin-2 at high concentration can enhance endothelial cell survival through the phosphatidylinositol 3′ kinsase/Akt signal transduction pathway. *Oncogene* 19:4549–52

74. Babei S, Teichert-Kuliszewska K, Zhang Q, Jones N, Dumont DJ, Stewart DJ. 2003. Angiogenic actions of Ang 1 rquires endothelium derived NO. *Am. J. Pathol.* 162:1927–36

75. Visconti RP, Richardson CD, Sato TN. 2002. Orchestration of angiogenesis and arteriovenous contribution by angiopoietins and VEGF. *Proc. Natl. Acad. Sci. USA* 99:8219–24

76. Papatrepoulos A, Garcia-Cardena G, Douglas TJ, Maisonpierre PC, Yancopoulos GD, Sessa WC. 1999. Direct actions of ang-1 on human endothelium: evidence for network stabilization, cell survival and interaction with other angiogenic growth factors. *Lab. Invest.* 79:213–23

77. Chinoy MR, Graybill MM, Miller SA, Lang CM, Kauffman GL. 2002. Ang-1 and VEGF in vascular development and angiogenesis in hypoplastic lungs. *Am. J. Physiol. Lung Cell Mol. Physiol.* 283:L60–66

78. Rudge JS, Thurston G, Yancopoulos GD. 2003. Angiopoietin 1 and pulmmonary hypertension: cause or cure? *Circ. Res.* 92:947–49

79. Du L, Sullivan CC, Chu D, Cho AJ, Kido M, et al. 2003. Signaling molecules in nonfamilial pulmonary hypertension. *N. Engl. J. Med.* 348:500–9

80. Zhao YD, Campbell AIM, Robb M, Ng D, Stewart DJ. 2003. Protective role of Ang-1 in experimental pulmonary hypertension. *Circ. Res.* 92:984–91

81. Alva JA, Iruela-Arispe ML. 2004. Notch signaling in vascular morphogenesis. *Curr. Opin. Hematol.* 11:278–83

82. Taichman DB, Loomes KM, Schachtner SK, Guttentag S, Vu C, et al. 2002. Notch1 and Jagged1 expression by the developing pulmonary vasculature. *Dev. Dyn.* 225:166–75

83. Kalinichenko VV, Gusarova GA, Kim IM, Shin B, Yoder HM, et al. 2004. Foxf1 haploinsufficiency reduces Notch-2 signaling during mouse lung development. *Am. J. Physiol. Lung Cell Mol. Physiol.* 286:L521–30

84. Cadigan KM, Nusse R. 1997. Wnt signaling: a common theme in animal development. *Genes Dev.* 11:3286–305

85. Willert K, Nusse R. 1998. Beta-catenin: a key mediator of Wnt signaling. *Curr. Opin. Genet. Dev.* 8:95–102

86. Weidenfeld J, Shu W, Zhang L, Millar SE, Morresey EE. 2002. The WNT7b promoter is regulated by TTF-1, GATA6, and

Foxa2 in lung epithelium. *J. Biol. Chem.* 277:21061–70

87. Shu W, Jiang YQ, Lu MM, Morresey EE. 2002. WNT7b regulates mesenchymal proliferation and vascular development in the lung. *Development* 129:4831–42

88. Hirschi KK, Rohovsky SA, Beck LH, Smith SR, D'Amore PA. 1999. Endothelial cells modulate the proliferation of mural cell precursors via platelet-derived growth factor-BB and heterotypic cell contact. *Cir. Res.* 84:298–305

89. D'Amore PA, Ng Y-S. 2002. Won't you be my neighbor? Local induction of arteriogenesis. *Cell* 110:289–92

89a. Hirschi KK, Skalak TC, Peirce SM, Little CD. 2002. Vascular assembly in natural and engineered tissues. *Ann. NY Acad. Sci.* 961:223–42

90. Dickson MC, Martin JS, Cousins FM, Kulkarni AB, Karlsson S, Akhurst RJ. 1995. Defective haematopoiesis and vasculogenesis in transforming growth factor-beta 1 knock out mice. *Development* 21:1845–54

91. Larsson J, Goumans MJ, Sjostrand LJ, van Rooijen MA, Ward D, et al. 2001. Abnormal angiogenesis but intact hematopoietic potential in TGF-beta type I receptor-deficient mice. *EMBO J.* 20:1663–73

92. Hautmann MB, Madsen CS, Owens GK. 1997. A transforming growth factor-beta (TGF-beta) control element drives TGF-beta-induced stimulation of smooth-muscle alpha-actin gene expression in concert with 2 CARG elements. *J. Biol. Chem.* 272:948–56

93. Hirschi KK, Lai L, Belaguli NS, Dean DA, Schwartz RJ, Zimmer WE. 2001. Transforming growth factor-beta induction of smooth muscle cell phenotype requires transcriptional and post-transcriptional control of serum response factor. *J. Biol. Chem.* 277:6287–95

94. Browning CL, Culberson DE, Aragon IV, Fillmore RA, Croissant JD. 1998. The developmentally regulated expression of serum response factor plays a key role in the control of smooth muscle-specific genes. *Dev. Biol.* 194:18–37

95. Reaume AG, de Sousa PA, Kulkarni S, Langille BL, Zhu D, et al. 1995. Cardiac malformation in neonatal mice lacking connexin43. *Science* 267:1831–34

96. Kruger O, Plum A, Kim JS, Winterhager E, Maxeiner S, et al. 2000. Defective vascular development in connexin 45-deficient mice. *Development* 127:4179–93

97. Jones PL. 2003. Homeobox genes in pulmonary vascular development and disease. *Trends Cardiovasc. Med.* 13:336–45

98. Thomas PQ, Brown A, Beddington RS. 1998. Hex: a homeobox gene revealing peri-implantation asymmetry in the mouse embryo and an early transient marker of endothelial cell precursors. *Development* 125:85–94

99. Bergwerff M, Gittenberger-de Groot AC, Wisse LJ, DeRuiter MC, Wessels A, et al. 2000. Loss of function of the *Prx1* and *Prx2* homeobox genes alters architecture of the great elastic arteries and ductus arteriosus. *Virchows Arch.* 436:12–19

100. Bergwerff M, Gittenberger-de Groot AC, DeRuiter MC, Van Iperen L, et al. 1998. Patterns of paired-related homeobox genes *Prx1* and *Prx2* suggest involvement in matrix modulation in the developing chick vascular system. *Dev. Dyn.* 213:59–70

101. Schwarz RE, Chappey O, Wautier JL, Chabot J, Lo Gerfo P, Stern D. 1999. Endothelial-monocyte activating polypeptide II, a novel antitumor cytokine that suppresses primary and metastatic tumor growth and induces apoptosis in growing endothelial cells. *J. Exp. Med.* 190:341–54

102. Schwarz MA, Zhang F, Gebb S, Starnes V, Warburton D. 2000. Endothelial monocyte activating polypeptide II inhibits lung neovascularization and airway epithelial morphogenesis. *Mech. Dev.* 95:123–32

103. Ghabrial A, Luschnig S, Metzstein MM, Krasnow MA. 2003. Branching morphogenesis of the Drosophila tracheal system. *Annu. Rev. Cell Dev. Biol.* 19:623–47

104. Crews ST. 1998. Control of cell lineage-specific development and transcription by bHLH-PAS proteins. *Genes Dev.* 12:607–20

105. Kotch LE, Iyer NV, Laughner E, Semenza GL. 1999. Defective vascularization of HIF-1α-null embryos is not associated with VEGF deficiency but with mesenchymal cell death. *Dev. Biol.* 209:254–67

106. Land SC. 2003. Oxygen-sensing pathways and the development of mammalian gas exchange. *Redox Rep.* 8:325–40

107. Gebb SA, Jones PL. 2003. Hypoxia and lung branching morphogenesis. *Adv. Exp. Med. Biol.* 543:117–25

108. Land SC, Darakhshan F. 2004. Thymulin evokes IL6-C/EBPβ regenerative repair and TNFα silencing during endotoxin exposure in fetal lung explants. *Am. J. Physiol. Lung Cell Mol. Physiol.* 286:L473–87

109. Blanco LN, Massaro D, Massaro GD. 1991. Alveolar size, number and surface area: developmentally dependent response to 13% O_2. *Am. J. Physiol. Lung Cell Mol. Physiol.* 261:L370–77

110. Baines DL, Folkesson HG, Norlin A, Bingle CD, Yaun HT, Olver RE. 2000. The influence of mode of delivery, hormonal status and postnatal O_2 environment on epithelial sodium channel (ENaC) expression in perinatal guinea-pig lung. *J. Physiol.* 522:147–57

111. Baines DL, Land SC, Olver RE, Wilson SM. 2001. O_2-evoked Na^+ transport in rat fetal distal lung epithelial cells. *J. Physiol.* 532:105–13

112. Deleted in proof

113. Galson DL, Tsuchiya T, Tendler DS, Huang LE, Ren Y, et al. 1995. The orphan receptor hepatic nuclear factor 4 functions as a transcriptional activator for tissue-specific and hypoxia-specific erythropoietin gene expression and is antag-onized by EAR3/COUP-TFI. *Mol. Cell Biol.* 15:2135–44

114. Tsuchiya T, Kominato Y, Ueda M. 2002. Human hypoxic signal transduction through a signature motif in hepatocyte nuclear factor 4. *J. Biochem.* 132:37–44

115. Deleted in proof

116. Zhao J, Sime PJ, Bringas P Jr, Gauldie J, Warburton D. 1998. Epithelium-specific adenoviral transfer of a dominant-negative mutant TGF-β type II receptor stimulates embryonic lung branching morphogenesis in culture and potentiates EGF and PDGF-AA. *Mech. Dev.* 72:89–100

117. Vicencio AG, Eickelberg O, Stankewich MC, Kashgarian M, Haddad GG. 2002. Regulation of TGF-β ligand and receptor expression in neonatal rat lungs exposed to chronic hypoxia. *J. Appl. Physiol.* 93:1123–30

118. Perrella MA, Hsieh CM, Lee WS, Shieh S, Tsai JC, et al. 1996. Arrest of endotoxin-induced hypotension by transforming growth factor β1. *Proc. Natl. Acad. Sci. USA* 93:2054–59

119. Young SL, Evans K, Eu JP. 2002. Nitric oxide modulates branching morphogenesis in fetal rat lung explants. *Am. J. Physiol. Lung Cell Mol. Physiol.* 282:L379–85

120. Shaul PW, Afshar S, Gibson LL, Sherman TS, Kerecman JD, et al. 2002. Developmental changes in nitric oxide synthase isoform expression and nitrix oxide production in fetal baboon lung. *Am. J. Physiol. Lung Cell Mol. Physiol.* 283:L1192–99

121. Brader S, Eccles SA. 2004. Phosphoinositide 3-kinase signaling pathways in tumor progression, invasion and angiogenesis. *Tumori* 90:2–8

122. Humar R, Kiefer FN, Berns H, Resink TJ, Battegay EJ. 2002. Hypoxia enhances vascular cell proliferation and angiogenesis in vitro via rapamycin (mTOR)-dependent signaling. *FASEB J.* 6:771–80

123. Hudson CC, Liu M, Chiang GG, Otterness DM, Loomis DC, et al. 2002. Regulation of hypoxia-inducible factor 1alpha expression and function by the mammalian target of rapamycin. *Mol. Cell Biol.* 22:7004–14

124. Gerasimovskaya EV, Ahmad S, White CW, Jones PL, Carpenter TC, Stenmark KR. 2002. Extracellular ATP is an autocrine/paracrine regulator of hypoxia-induced adventitial fibroblast growth. Signaling through extracellular signal-regulated kinase-1/2 and the Egr-1 transcription factor. *J. Biol. Chem.* 277: 44638–50

125. Mukouyama YS, Shin D, Britsch S, Taniguchi M, Anderson DJ. 2002. Sensory nerves determine the pattern of arterial differentiation and blood vessel branching in the skin. *Cell* 109:693–705

126. Hislop AA, Haworth SG. 1990. Pulmonary vascular damage and the development of cor pulmonale following hyaline membrane disease. *Pediatr. Pulmonol.* 9:152–61

127. Halliday HL, Dumpit FM, Brady JP. 1980. Effects of inspired oxygen on echocardiographic assessment of pulmonary vascular resistance and myocardial contractility in BPD. *Pediatrics* 65:536–40

128. Fouron JC, LeGuennec JC, Villemont D, Perreault G, Davignon A. 1980. Value of echocardiography in assessing the outcome of BPD. *Pediatrics* 65:529–35

129. Deleted in proof

130. Abman SH, Wolfe RR, Accurso FJ, Koops BL, Wiggins JW. 1985. Pulmonary vascular response to oxygen in infants with severe BPD. *Pediatrics* 75:80–84

131. Walther FJ, Bender FJ, Leighton JO. 1992. Persistent pulmonary hypertension in premature neonates with severe RDS. *Pediatrics* 90:899–904

132. Goodman G, Perkin R, Anas N. 1988. Pulmonary hypertension in infants with BPD. *J. Pediatr.* 112:67–72

133. Abman SH. 2000. Pulmonary hypertension in chronic lung disease of infancy. Pathogenesis, pathophysiology and treatment. See Ref. 225, pp. 619–68

134. Parker TA, Abman SH. 2003. The pulmonary circulation in BPD. *Semin. Neonatol.* 8:51–62

135. Northway WH Jr, Rosan RC, Porter DY. 1967. Pulmonary disease following respirator therapy of hyaline-membrane disease. *N. Engl. J. Med.* 276:357–68

136. Jobe AH, Bancalari E. 2001. Bronchopulmonary dysplasia. *Am. J. Resp. Crit. Care Med.* 163:1723–29

137. Charafeddine L, D'Angio CT, Phelps DL. 1999. Atypical chronic lung disease patterns in neonates. *Pediatrics* 103:759–60

138. Rojas MA, Gonzalez A, Bancalari E, Claure N, Poole C, Silva-Neto G. 1995. Changing trends in the epidemiology and pathogenesis of chronic lung disease. *J. Pediatr.* 126:605–10

139. Watterberg KL, Demers LM, Scott SM, Murphy S. 1996. Chorioamnionitis and early lung inflammation in infants in whom bronchopulmonary dysplasia develops. *Pediatrics* 97:210–15

140. Hussain AN, Siddiqui NH, Stocker JT. 1998. Pathology of arrested acinar development in postsurfactant BPD. *Hum. Pathol.* 29:710–17

141. Coalson JJ. 2000. Pathology of chronic lung disease of early infancy. See Ref. 225, pp. 85–124

142. Mitchell SH, Teague G. 1998. Reduced gas transfer at rest and during exercise in school age survivors of BPD. *Am. J. Resp. Crit. Care Med.* 157:1406–12

143. Slavin JD, Mathews J, Spencer RP. 1986. Pulmonary ventilation/perfusion and reverse mismatches in an infant. *Acta Radiol. Diagn.* 27:708

144. Hakulinen AL, Järvenpaa AL, Turpeinen M, Sovijarvi A. 1996. Diffusing capacity of the lung in school-age children born very preterm with and without BPD. *Pediatr. Pulmonol.* 21:353

145. Subcomm. Assem. Pediatr. Am. Thorac. Soc. 2003. Statement on the care of the

child with chronic lung disease of infancy and childhood. *Am. J. Respir. Crit. Care Med.* 168:356–96

146. Yoon BH, Romero R, Jun JK, Park KH, Park JD, et al. 1997. Amniotic fluid cytokines (interleukin-6, tumor necrosis factor-alpha, interleukin-1 beta, and interleukin-8) and the risk for the development of bronchopulmonary dysplasia. *Am. J. Obstet. Gynecol.* 177:825–30

147. Hitti J, Krohn MA, Patton DL, Tarczy-Hornoch P, Hillier SL, et al. 1997. Amniotic fluid tumor necrosis factor-alpha and the risk of respiratory distress syndrome among preterm infants. *Am. J. Obstet. Gynecol.* 177:50–56

148. Ghezzi F, Gomez R, Romero R, Yoon BH, Edwin SS, et al. 1998. Elevated interleukin-8 concentrations in amniotic fluid of mothers whose neonates subsequently develop bronchopulmonary dysplasia. *Eur. J. Obstet. Gynecol. Reprod. Biol.* 78:5–10

149. Jobe AH. 2003. Antenatal factors in the development of BPD. *Semin. Neonatol.* 8:9–17

150. Banks BA, Cnaan A, Morgan MA, Parer JT, Merrill JD, et al. 1999. Multiple courses of antenatal corticosteroids and outcome of premature neonates. North American Thyrotropin-Releasing Hormone Study Group. *Am. J. Obstet. Gynecol.* 181:709–17

151. Frank L, Groseclose EE. 1984. Preparation for birth into an O_2-rich environment: the antioxidant enzymes in the developing rabbit lung. *Pediatr. Res.* 18:240–44

152. Francis PJ, Knapp MJ, Piantadosi CA, Takeda K, Fulkerson WJ, et al. 1991. Responses of baboons to prolonged hyperoxia: physiology and qualitative pathology. *J. Appl. Physiol.* 71:2352–62

153. Roberts RJ, Weesner KM, Bucher JR. 1983. Oxygen induced alterations in lung vascular development in the newborn rat. *Pediatr. Res.* 17:368–75

154. Crapo JD. 1986. Morphologic changes in pulmonary oxygen toxicity. *Annu. Rev. Physiol.* 48:721–31

155. White CW, Avraham KB, Shanley PF, Groner Y. 1991. Transgenic mice with expression of elevated levels of copper-zinc superoxide dismutase in the lungs are resistant to pulmonary oxygen toxicity. *J. Clin. Invest.* 87:2162–68

156. Wispe JR, Warner BB, Clark JC, Dey CR, Neuman J, et al. 1992. Human Mn-superoxide dismutase in pulmonary epithelial cells of transgenic mice confers protection from oxygen injury. *J. Biol. Chem.* 267:23937–41

157. Ahmed MN, Suliman HR, Folz RJ, Nozik-Graych E, Golson ML, et al. 2003. Extracellular superoxide dismutase protects lung development in hyperoxia-exposed newborn mice. *Am. J. Respir. Crit. Care Med.* 167:400–538

158. Dreyfuss DD, Saumon G. 1998. Ventilator induced lung injury: lessons from experimental studies. *Am. J. Respir. Crit. Care Med.* 157:294–323

159. Tremblay L, Valenza F, Ribeiro SP, Li J, Slutsky AS. 1997. Injurious ventilatory strategies increase cytokines and c-fos m-RNA expression in an isolated rat lung model. *J. Clin. Invest.* 99:944–52

160. Carlton DP, Cummings JJ, Scheerer RG, Poulain FR, Bland RD. 1990. Lung overexpansion increases pulmonary microvascular protein permeability in young lambs. *J. Appl. Physiol.* 69:577–83

161. Bjorklund LJ, Ingimarsson J, Curstedt T, John J, Robertson B, et al. 1997. Manual ventilation with a few large breaths at birth compromises the therapeutic effect of subsequent surfactant replacement in immature lungs. *Pediatr. Res.* 42:348–55

162. Nilsson R, Grossman G, Robertson B. 1978. Lung surfactant and the pathogenesis of neonatal bronchiolar lesions induced by artificial ventilation. *Pediatr. Res.* 12:249–55

163. Groneck P, Gotz-Speer B, Opperman M, Eiffert H, Speer CP. 1994. Association

of pulmonary inflammation and increased microvascular permeability during the development of bronchopulmonary dysplasia: a sequential analysis of inflammatory mediators in respiratory fluids of high-risk neonates. *Pediatrics* 93:712–18

164. Pierce MR, Bancalari E. 1995. The role of inflammation in the pathogenesis of bronchopulmonary dysplasia. *Pediatr. Pulmonol.* 19:371–78

165. Groneck P, Speer CP. 1995. Inflammatory mediators and bronchopulmonary dysplasia. *Arch. Dis. Child* 73:F1–3

166. Jones CA, Cayabyab RG, Kwong KY, Stotts C, Wong B, Hamdan H, et al. 1996. Undetectable interleukin (IL)-10 and persistent IL-8 expression early in hyaline membrane disease: a possible developmental basis for the predisposition to chronic lung inflammation in preterm newborns. *Pediatr. Res.* 39:966–75

167. Merritt TA, Cochrane CG, Holcomb K, Bohl B, Hallman M, et al. 1983. Elastase and α_1-proteinase inhibitor activity in tracheal aspirates during respiratory distress syndrome. Role of inflammation in the pathogenesis of bronchopulmonary dysplasia. *J. Clin. Invest.* 72:656–66

168. Stiskal JA, Dunn MS, Shennan AT, O'Brien KKE, Kelly EN, et al. 1998. Alpha-1-proteinase inhibitor therapy for the prevention of chronic lung disease of prematurity: a randomized, controlled trial. *Pediatrics* 101:89–94

169. Ramsay PL, O'Brian SE, Hegemier S, Welty SE. 1998. Early clinical markers for the development of bronchopulmonary dysplasia: soluble E-selectin and ICAM-1. *Pediatrics* 102:927–32

170. Munshi UK, Niu JO, Siddiq MM, Parton LA. 1997. Elevation of interleukin-8 and interleukin-6 precedes the influx of neutrophils in tracheal aspirates from preterm infants who develop bronchopulmonary dysplasia. *Pediatr. Pulmonol.* 24:331–36

171. Alnahhas MH, Karathanasis P, Kriss VM, Pauly TH, Bruce MC. 1997. Elevated laminin concentrations in lung secre-

tions of preterm infants supported by mechanical ventilation are correlated with radiographic abnormalities. *J. Pediatr.* 131:555–60

172. Wang EE, Ohlsson A, Kellner JD. 1995. Association of ureaplasma urealyticum colonization with chronic lung disease of prematurity: results of a meta-analysis. *J. Pediatr.* 127:640–44

173. Tyson JE, Wright LL, Oh W, Kennedy KA, Mele L, et al. 1999. Vitamin A supplementation for extremely-low-birth-weight infants. National Institute of Child Health and Human Development Neonatal Research Network. *N. Engl. J. Med.* 340:1962–68

174. Hallman M, Haatja R. 2003. Genetic influences and neonatal lung disease. *Sem. Neonatol.* 8:19–27

175. Shaffer SG, Bradt SK, Thibeault DW. 1987. Chronic vascular pulmonary dysplasia associated with neonatal hyperoxia exposure in the rat. *Pediatr. Res.* 21:14–20

176. Jones R, Zapol WM, Reid LM. 1984. Pulmonary artery remodeling and pulmonary hypertension after exposure to hyperoxia for 7 days. *Am. J. Pathol.* 117:273–85

177. Jones R, Jacobson M, Steudel W. 1999. α-Smooth-muscle actin and microvascular precursor smooth-muscle cells in pulmonary hypertension. *Am. J. Respir. Cell Mol. Biol.* 20:582–94

178. Mourani PM, Ivy DD, Gao D, Abman SH. 2004. Pulmonary vascular effects of inhaled NO and oxygen tension in BPD. *Am. J. Respir. Crit. Care Med.* 170:1006–13

179. Abman SH, Schaffer M, Wiggins J, Washington R, Manco-Johnson M, Wolfe RR. 1987. Pulmonary vascular extraction of circulating norepinephrine in infants with bronchopulmonary dysplasia. *Pediatr. Pulmonol.* 3:386–91

180. Apkon M, Nehgme RA, Lister G. 2000. Cardiovascular abnormalities in BPD. See Ref. 225, pp. 321–56

181. Abman SH, Warady BA, Lum GM, Koops BL. 1984. Systemic hypertension in

infants with BPD. *J. Pediatr.* 104:928–31

182. Ascher DP, Rosen P, Null DM, de Lemos RA, Wheller JJ. 1985. Systemic to pulmonary collaterals mimicking patent dutus arteriosus in neonates with prolonged ventilator courses. *J. Pediatr.* 101:282–84

183. Abman SH, Sondheimer H. 1992. Pulmonary circulation and cardiovascular sequelae of bronchopulmonary dysplasia. In *Diagnosis and Treatment of Pulmonary Hypertension*, ed. EK Weir, SL Archer, JT Reeves, pp. 155–80. New York: Futura

184. Bhatt AJ, Pryhuber GS, Huyck H, Watkins RH, Metlay LA, Maniscalco WM. 2000. Disrupted pulmonary vasculature and decreased VEGF, flt-1, and Tie 2 in human infants dying with BPD. *Am. J. Resp. Crit. Care Med.* 164:1971–80

185. Lassus P, Turanlahti M, Heikkila P, Andersson LC, Nupponen I, et al. 2001. Pulmonary vascular endothelial growth factor and Flt-1 in fetuses, in acute and chronic lung disease, and in persistent pulmonary hypertension of the newborn. *Am. J. Respir. Crit. Care Med.* 164:1981–87

186. Abman SH. 2001. BPD: a vascular hypothesis. *Am. J. Respir. Crit. Care Med.* 164:1755–56

187. Jakkula M, Le Cras TD, Gebb S, Hirth KP, Tuder RM, et al. 2000. Inhibition of angiogenesis decreases alveolarization in the developing rat lung. *Am. J. Physiol. Lung Cell Mol. Physiol.* 279:L600–7

188. Klekamp JG, Jarzecka K, Perkett EA. 1999. Exposure to hyperoxia decreases the expression of vascular endothelial growth factor and its receptors in adult rat lungs. *Am. J. Pathol.* 154:823–31

189. Maniscalco WM, Watkins RH, D'Angio CT, Ryan RM. 1997. Hyperoxic injury decreases alveolar epithelial cell expression of vascular endothelial growth factor (VEGF) in neonatal rabbit lung. *Am. J. Respir. Cell Mol. Biol.* 16:557–67

190. Maniscalco WM, Watkins RH, Finkelstein JN, Campbell MH. 1995. Vascular endothelial growth factor mRNA increases in alveolar epithelial cells during recovery from oxygen injury. *Am. J. Respir. Cell Mol. Biol.* 13:377–86

191. Le Cras TD, Markham NE, Tuder RM, Voelkel NF, Abman SH. 2002. Treatment of newborn rats with a VEGF receptor inhibitor causes pulmonary hypertension and abnormal lung structure. *Am. J. Physiol. Lung Cell Mol. Physiol.* 283:L555–62

192. Le Cart C, Cayabyab R, Buckley S, Morrison J, Kwong KY, et al. 2000. Bioactive transforming growth factor-beta in the lungs of extremely low birthweight neonates predicts the need for home oxygen supplementation. *Biol. Neonat.* 77:217–23

193. Sime PJ, Xing Z, Graham FL, Csaky KG, Fauldie J. 1997. Adenovector-mediated gene transfer of active transforming growth factor-beta induces prolonged severe fibrosis in rat lung. *J. Clin. Invest.* 100:768–76

194. Gauldie J, Galt T, Bonniaud P, Robbins C, Kelly M, Warburton D. 2003. Transfer of the active form of transforming growth factor-beta 1 gene to newborn rat lung induces changes consistent with bronchopulmonary dysplasia. *Am. J. Pathol.* 163:2575–84

195. Kroll J, Waltenberger J. 1999. A novel function of VEGF receptor-2 (KDR): rapid release of nitric oxide in response to VEGF-A stimulation in endothelial cells. *Biochem. Biophys. Res. Commun.* 265:636–39

196. Shen BQ, Lee DY, Zioncheck TF. 1999. Vascular endothelial growth factor governs endothelial nitric-oxide synthase expression via a KDR/Flk-1 receptor and a protein kinase C signaling pathway. *J. Biol. Chem.* 274:33057–63

197. Hood JD, Meininger CJ, Ziche M, Granger HJ. 1998. VEGF upregulates ecNOS message, protein and NO production in human endothelial cells. *Am. J. Physiol. Heart Circ. Physiol.* 274:H1054–58

198. Ziche M, Morbidelli L, Masini E, Amerini S, Granger HJ, et al. 1994. NO mediates angiogenesis in vivo and endothelial cell growth and migration in vitro promoted by substance P. *J. Clin. Invest.* 94:2036–44

199. Morbidelli L, Chang C-H, Douglas JG, Granger HJ, Ledda F, Ziche M. 1996. NO mediates mitogenic effect of VEGF on coronary venular endothelium. *Am. J. Physiol. Heart Circ. Physiol.* 270:H411–15

200. Papapetropoulos A, Garcia-Cardena G, Madri JA, Sessa WC. 1997. NO production contributes to the angiogenic properties of vascular endothelial growth factor in human endothelial cells. *J. Clin. Invest.* 100:3131–39

201. Ziche M, Morbidelli L, Choudhuri R, Zhang HT, Donnini S, et al. 1997. NO synthase lies downstream from vascular endothelial growth factor-induced but not basic fibroblast growth factor-induced angiogenesis. *J. Clin. Invest.* 99:2625–34

202. Balasubramaniam V, Tang JR, Maxey A, Plopper CG, Abman SH. 2003. Mild hypoxia impairs alveolarization in the endothelial nitric oxide synthase-deficient mouse. *Am. J. Physiol. Lung Cell Mol. Physiol.* 284:L964–71

203. Leuwerke SM, Kaza AK, Tribble CG, Kron IL, Laubach VE. 2002. Inhibition of compensatory lung growth in endothelial nitric oxide synthase-deficient mice. *Am. J. Physiol. Lung Cell Mol. Physiol.* 282:L1272–78

204. Young SL, Evans K, Eu JP. 2002. Nitric oxide modulates branching morphogenesis in fetal rat lung explants. *Am. J. Physiol. Lung Cell Mol. Physiol.* 282:L379–85

205. Han RN, Babei S, Robb M, Lee T, Ridsdale R, et al. 2004. Defective lung vascular development and fatal respiratory distress in eNOS deficient mice: a model of alveolar capillary dysplasia. *Circ. Res.* 94:1115–23

206. Afshar S, Gibson LL, Yuhanna IS, Sherman TS, Kerecman JD, et al. 2003. Pulmonary NO synthase expression is attenuated in a fetal baboon model of chronic lung disease. *Am. J. Physiol. Lung Cell Mol. Physiol.* 284:L749–58

207. MacRitchie AN, Albertine KH, Sun J, Lei PS, Jensen SC, et al. 2001. Reduced endothelial nitric oxide synthase in lungs of chronically ventilated preterm lambs. *Am. J. Physiol. Lung Cell Mol. Physiol.* 281:L1011–20

208. Tang JR, Markham NE, Lin YJ, McMurtry IF, Maxey A, et al. 2004. Inhaled NO attenuates pulmonary hypertension and improves lung growth in infant rats after neonatal treatment with a VEGF receptor inhibitor. *Am. J. Physiol. Lung Cell Mol. Physiol.* 287:L344–51

209. Howlett CE, Hutchison JS, Veinot JP, Chiu A, Merchant P, Fliss H. 1999. Inhaled nitric oxide protects against hyperoxia-induced apoptosis in rat lungs. *Am. J. Physiol. Lung Cell Mol. Physiol.* 277:L596–605

210. McElroy MC, Wiener-Kronish JP, Miyazaki H, Sawa T, Modelska K, et al. 1997. Nitric oxide attenuates lung endothelial injury caused by sublethal hyperoxia in rats. *Am. J. Physiol. Lung Cell Mol. Physiol.* 272:L631–38

211. Nelin LD, Welty SE, Morrisey JF, Gotuaco C, Dawson CA. 1998. Nitric oxide increases the survival of rats with a high oxygen exposure. *Pediatr. Res.* 43:727–32

212. Schreiber MD, Gin-Mestan K, Marks JD, Huo D, Lee G, Srisuparp P. 2003. Inhaled NO in premature infants with respiratory distress syndrome. *N. Engl. J. Med.* 349:2099–107

213. Wagenaar GTM, ter Horst SAJ, van Gastelen MA, Leuser LM, Mauad T, et al. 2004. Gene expression profile and histopathology of experimental bronchopulmonary dysplasia induced by prolonged oxidative stress. *Free Radic. Biol. Med.* 36:782–801

214. Weinstein M, Xu X, Ohyama K, Deng C. 1998. FGFR-4 function cooperatively to direct alveogenesis in the murine lung. *Development* 125:3615–23

215. Hosford GE, Olson DM. 2003. Effects of hyperoxia on VEGF, its receptors and HIF-2α in the newborn rat lung. *Am. J. Physiol. Lung Cell Mol. Physiol.* 285:L161–68

216. Deng H, Mason SN, Auten RL Jr. 2000. Lung inflammation in hyperoxia can be prevented by antichemokine treatment in newborn rats. *Am. J. Respir. Crit. Care Med.* 162:2316–23

217. Auten RL, Richardson RM, White JR, Mason SN, Vozzell MA, Whorton MH. 2001. Nonpeptide CXCR2 antagonist prevents neutrophil accumulation in hyperoxia-exposed newborn rats. *J. Pharmacol. Exp. Ther.* 299:90–95

218. Barazzone C, Belin D, Piguet PF, Vassalli JP, Sappino AP. 1996. Plasminogen activator inhibitor-1 in acute hyperoxic mouse lung injury. *J. Clin. Invest.* 98:2666–73

219. Altieri DC, Agbanyo FR, Plescia J, Ginsberg MH, Edgington TS, Plow EF. 1990. A unique recognition site mediates the interaction of fibrinogen with the leukocyte integrin Mac-1 (CD11b/CD18). *J. Biol. Chem.* 265:12119–22

220. Sitrin RG, Pan PM, Srikanth S, Tod RF III. 1998. Fibrinogen activates NF-κB transcription factors in mononuclear phagocytes. *J. Immunol.* 161:1462–70

221. Bland RD, Albertine KH, Carlton DP, Kullama L, Davis P, et al. 2000. Chronic lung injury in preterm lambs: abnormalities of the pulmonary circulation and lung fluid balance. *Pediatr. Res.* 48:64–74

222. Coalson JJ, Winter VT, Siler-Khodr T, Yoder BA. 1992. Neonatal chronic lung disease in extremely immature baboons. *Am. J. Respir. Crit. Care Med.* 160:1333–46

223. Maniscalco WM, Watkins RH, Pryhuber GS, Bhatt A, Shea C, Huyck H. 2002. Angiogenic factors and the alveolar vasculature: development and alterations by injury in very premature baboons. *Am. J. Physiol. Lung Cell Mol. Physiol.* 282:L811–23

224. McDonald JA, ed. 1997. *Lung Growth and Development*. New York: Dekker

225. Bland RD, Coalson JJ, eds. 2000. *Chronic Lung Disease of Infancy*. New York: Dekker

Annu. Rev. Physiol. 2005. 67:663–96
doi: 10.1146/annurev.physiol.67.040403.101937
Copyright © 2005 by Annual Reviews. All rights reserved
First published online as a Review in Advance on October 19, 2004

Surfactant Protein C Biosynthesis and Its Emerging Role in Conformational Lung Disease

Michael F. Beers and Surafel Mulugeta

*Pulmonary and Critical Care Division, Department of Medicine, University of
Pennsylvania School of Medicine, Philadelphia, Pennsylvania 19104-6061;
email: mfbeers@mail.med.upenn.edu; mulugeta@mail.med.upenn.edu*

Key Words interstitial pneumonia, propeptide processing, unfolded protein
response, apoptosis, ERAD

■ **Abstract** Surfactant protein C (SP-C) is a hydrophobic 35-amino acid peptide
that co-isolates with the phospholipid fraction of lung surfactant. SP-C represents a
structurally and functionally challenging protein for the alveolar type 2 cell, which
must synthesize, traffic, and process a 191–197-amino acid precursor protein through
the regulated secretory pathway. The current understanding of SP-C biosynthesis considers the SP-C proprotein (proSP-C) as a hybrid molecule that incorporates structural
and functional features of both bitopic integral membrane proteins and more classi-
cally recognized luminal propeptide hormones, which are subject to post-translational
processing and regulated exocytosis. Adding to the importance of a detailed understanding of SP-C biosynthesis has been the recent association of mutations in the
proSP-C sequence with chronic interstitial pneumonias in children and adults. Many
of these mutations involve either missense or deletion mutations located in a region
of the proSP-C molecule that has structural homology to the BRI family of proteins
linked to inherited degenerative dementias. This review examines the current state of
SP-C biosynthesis with a focus on recent developments related to molecular and cellular
mechanisms implicated in the emerging role of SP-C mutations in the pathophysiology
of diffuse parenchymal lung disease.

INTRODUCTION

Many inherited diseases are caused by the homozygous expression of a mutant
protein isoform that induces a loss of function through interference with protein trafficking or function. In contrast to the traditional autosomal recessive diseases, the heterozygous expression of mutations of a select subset of proteins
appears sufficient to generate pathology. Such diverse disorders as Alzheimer's,
Parkinson's, and Huntington's disease, spinocerebeller ataxia, Charcot-Marie-
Tooth neuropathy, prion encephalopathies, inflammatory liver disease associated
with α-1 antitrypsin mutants, and systemic amyloidosis are part of a growing

0066-4278/05/0315-0663$14.00

list of conformational diseases that arise when a mutant protein undergoes structural rearrangement that endows an ability to elicit cellular dysfunction or toxicity (1, 2, 3). At both a recent international conference "Alpha-1 Antitrypsin Deficiency and Other Conformational Diseases"(3), as well as a National Institutes of Health Workshop on Conformational Diseases in the Lung (4), the mechanisms by which misfolded proteins cause cytotoxicity and the manner in which cells respond to production of abnormal protein conformers were highlighted as important emerging questions.

Surfactant protein C (SP-C) is a hydrophobic lung-specific protein that co-isolates with the phospholipid fraction of pulmonary surfactant. Produced exclusively by alveolar type 2 (AT2) cells, human SP-C (hSP-C) is synthesized as a 191–197-amino acid proprotein and proteolytically processed as an integral membrane protein to the 3.7 kDa mature form (hSP-C$_{3.7}$) that is subsequently transferred to the lumen of lamellar bodies for secretion into the alveolar space together with surfactant phospholipids (5). Complete biosynthesis requires four endoproteolytic cleavages of the SP-C propeptide and depends upon oligomeric sorting and targeting to subcellular processing compartments distal to the medial Golgi (6).

In addition to its role in modulating the biophysical properties of lung surfactant phospholipids in the alveolus, heterozygous expression of over 14 different mutant proSP-C forms in association with the development of diffuse parenchymal lung disease in infants and adults has underscored the importance of understanding the structure, function, and biosynthetic routing of SP-C (7–10). This review focuses on the molecular and cellular mechanisms underlying the expression of synthesized SP-C proprotein (proSP-C$_{21}$, M_r 21,000), and its targeting and subsequent post-translational processing, which lead to production of secreted mature SP-C$_{3.7}$ by the AT2 epithelial cell. The impact of recently recognized mutations in the SP-C sequence on proprotein trafficking, structure, and function, as well as on AT2 cell function, is examined. The body of evidence is summarized from studies of human, rat, and mouse SP-C isoforms in a variety of in vitro and in vivo model systems and is presented in the context of both the unique structural and functional properties of SP-C and the novel characteristics of AT2 cells. The cellular physiology of the AT2 cell is critical to understanding the normal synthetic routing, trafficking patterns, and cellular responses generated by mutant SP-C isoforms that contribute to the pathogenesis of chronic parenchymal lung disease.

STRUCTURE FUNCTION CONSIDERATIONS OF SP-C

SFTPC Genomic Structure and Polymorphisms

Figure 1 schematically depicts the domain organization for human SP-C (SFTPC) gene, mRNA, and protein. The alveolar form of human SP-C (hSP-C) represents the biosynthetic processing of a single gene product (11, 12). The SFTPC gene, with an approximate size of 3.0 kb, has been localized in humans to the short arm of chromosome 8 (8p21, distinct from SP-A, SP-B, or SP-D) (13). Multiple human SP-C cDNA clones, all ~0.9 kb in length, have been isolated and reported by a number of groups (11–15). Comparison of nucleotide sequences of these isolates with genomic SP-C clones demonstrates that the gene is organized into six exons (I through V coding, VI untranslated) and five introns. In vitro translation of human lung RNA produces SP-C primary translation products of apparent molecular mass (M_r) 21–22,000 (hSP-C$_{21}$) (14). Similar-sized products have been detected in Chinese hamster ovary (CHO) cells transfected with SFTPC cDNA (16), in cultured human fetal lung explants (17), and in primary cultures of differentiated neonatal human type 2 cells (18).

Among the deduced amino acid sequences from each of these human cDNA and genomic clones, an overall high degree of conservation exists within the proSP-C primary translation product sequence (195 of 197 amino acids) suggestive of strong evolutionary pressures exerted on the gene. Variations in residues at position 138 (Thr or Asn) and at position 186 (Ser or Asn) of the proprotein exist as the result of two single-nucleotide polymorphisms (SNPs). Four allelic variants (haplotypes) (Table 1) result from various amino acid combinations, as detected by nucleotide sequencing of isolated clones. Although formal population-based assessments have not been performed, the combination Thr138/Ser186 in the human clones appears to be the predominant sequence reported to date. Ser186 is also observed in other species (rat, mouse, and rabbit), although a high degree of discordance among species exists at position 138 (19, 20). In addition to these four potential haplotypes, further analysis of cDNA clones by several different groups also revealed the presence of two isoforms that resulted from the use of alternate splice sites at the beginning of Exon 5, which yielded an 18 bp insert encoding for proproteins varying in length by a six-amino acid stretch (Met-Glu-Cys-Ser-Leu-Gln) (12, 15). Limited data suggest that at least two of the alleles (Thr138/Ser186, Asn138/Asn186) can undergo differential splicing (12). Several other SNPs in noncoding regions of the SFTPC gene (i.e., within the 5′ flanking region, Exon 6, all introns, and the 3′ flanking region) have been reported (11, 12). In all cases the functional significance of these variations has not been defined.

The 21 kDa proSP-C protein product encoded by SFTPC (hSP-C$_{21}$) must undergo post-translational processing to liberate the biophysically active 35-amino acid peptide (hSP-C$_{3.7}$) present in the alveolus (Figure 1). For hSP-C$_{21}$ these events include directional integration into the ER membrane, covalent addition of palmitic acid residues (palmitoylation), and cleavage of -NH$_2$ and -COOH propeptide flanking regions. These post-translational events are highly relevant to the manner in which mutations of SFTPC are handled by the AT2 cell. Furthermore, because the hSP-C$_{3.7}$ sequence is contained within proSP-C, structural and functional properties of mature SP-C influence the behavior of hSP-C$_{21}$ and hSP-C propeptide intermediates. The processing events crucial to biosynthesis of hSP-C, as well as key biochemical and biophysical properties of mature SP-C relevant to understanding proSP-C in health and disease, are discussed below.

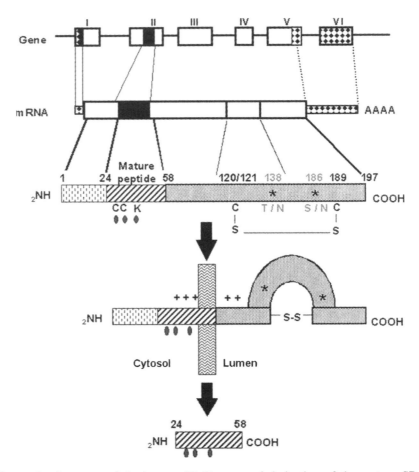

Figure 1 Structure of the human *SP-C* gene and derivation of the mature SP-C peptide. Production of alveolar hSP-C$_{3.7}$ begins with transcription and splicing of the six exons of the SP-C gene to yield a 0.9 kb mRNA that contains 5′- and 3′ untranslated regions (drawn as ◆◆◆). The primary translation product of hSP-C is a 197-amino acid integral membrane protein of M_r 21,000 in which a charge difference in the mature peptide domain (Phe24-Leu58) has been shown to mediate membrane association resulting in a bitopic, type II ($C_{lumen}/N_{cytoplasm}$) orientation in the endoplasmic reticulum (ER) membrane. Following folding of the COOH propeptide domain within the ER lumen mediated by conserved disulfide-containing cysteine residues, post-translational processing of proSP-C includes palmitoylation of two adjacent cysteine residues at positions 28 and 29 of the propeptide (C). This is followed by a sequential cleavage cascade that results in removal of first COOH-, then NH$_2$-terminal propeptide-flanking domains to yield the mature peptide. A tripalmitoylated form with addition of a palmitic acid to lysine at position 34 (K) can also occur.

TABLE 1   Common amino acid polymorphism patterns in coding region of hSP-C

Clone	Polymorphism pattern	GenBank number	References
CDNA	Thr138/Ser186	J03557	(15)
Genomic	Thr138/Ser186	U02948	(11)
Genomic	Thr138/Ser186	AY337315	Direct Genbank submission[a]
CDNA	Asn138/Asn186	J03517	(14)
CDNA	Asn138/Asn186	J03557	(15)
Genomic	Asn138/Ser186	NS[b]	M.F. Beers, unpublished observations[b]
Genomic	Thr138/Asn186	J03890.1	(12)
Genomic	Thr138/Asn186	NS[b]	M.F. Beers, unpublished observations

[a]Seattle SNPs NHLBI HL66682.
[b]Not submitted.

Mature SP-C: Primary Sequence and Higher-Order Structures

The SP-C protein product found in surfactant has been isolated and sequenced from lung lavage fractions of various species and is composed of 33–35 amino acids (15, 21–24). Between species, the majority of amino acid residues in the hSP-C$_{3.7}$ primary sequence are highly conserved. hSP-C$_{3.7}$ is a predominantly hydrophobic molecule owing to a high content of Val, Ile, and Leu ($\sim$ 60–65% of the primary sequence) (Figure 2, left). Although most of the amino acids making up SP-C are apolar, the molecule is somewhat amphipathic. The NH$_2$ terminus contains positively charged hydrophilic residues, whereas the center of the peptide is made up of a 23-amino acid polyvaline domain. In addition to its extreme hydrophobicity, SP-C is a true lipoprotein in which cysteines at positions 4 and 5 are acylated with C-16 palmitoyl chains (25, 26). Purified SP-C, when reconstituted with surfactant phospholipid components, has been shown to enhance the biophysical surface activity of these lipid mixtures in vitro and in vivo. This protein is also an active component of some clinical surfactant preparations (27–29). A complete summary of the biophysical properties of SP-C and its interactions with other surfactant proteins has been reviewed in detail elsewhere (30–33).

By SDS/PAGE, the M_r of SP-C determined under reducing conditions is reported with a range of 3000 to 6000 (15, 22). A naturally occurring dimeric form of SP-C (SP-C$_2$) has been isolated from bovine, canine, and murine lavage specimens (34–36). Although both monomeric and dimeric forms appear to exhibit biophysical activity, they exhibit differences in calcium requirements and efficiency of monolayer insertion. The relative amounts of each SP-C form

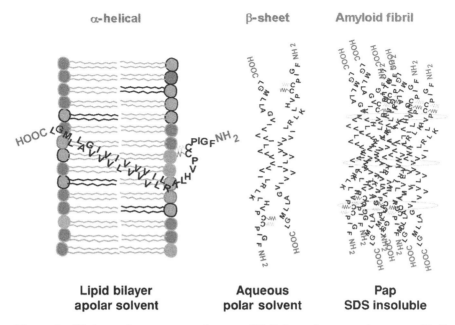

Figure 2 Higher-order structures of mature SP-C depend upon environment. (*Left*) Within a lipid membrane or in apolar solvents, a polyvaline-rich region of mature SP-C (α) imparts a high degree of α-helicity to the domain. (*Center*) Upon exposure to aqueous environment, mature SP-C forms poorly soluble aggregates and transitions to primarily β-sheet conformers. (*Right*) At higher concentrations, in the absence of adequate hydrophobic environments, such as in BAL recovered from patients with pulmonary alveolar proteinosis (PAP), detergent-insoluble aggregates form that stain positive for Congo Red birefringence, indicating the presence of amyloid fibril.

in the alveolus and the physiological significance of the SP-C$_2$ remain to be established.

Modeling of the secondary structure of mature SP-C using the translated cDNA sequence predicts that within the primary sequence, the polyvaline stretch is highly α-helical and capable of spanning phospholipid bilayers in the liquid crystalline phase. The remainder of the predicted secondary structure is mainly a random coil conformation. Studies of purified mature SP-C in apolar solvents or lipid bilayers have confirmed the existence of the valyl-rich α-helix secondary structure. Using a variety of biophysical techniques, including Fourier transform infrared spectroscopy (FT-IR), circular dichroism (CD), and ^{2}H and ^{1}H nuclear magnetic resonance (NMR) (37–39), an α-helix is observed between amino acid positions 9–34, with the long axis of this structure associated with lipid layers in an orientation near parallel to the acyl chains. (Figure 2, left). Such a stretch would encompass up to 25 of the total 35 residues and result in an α-helical length of 27–37 Å. This is sufficient to achieve a membrane-spanning orientation of most lipid bilayers and could promote the membrane association of any proSP-C forms that contain this domain.

In mixed organic solvents, monomeric SP-C$_{3.7}$ is thermodynamically unstable and readily transforms from a monomeric α-helical state into insoluble β-sheets (Figure 2, center) (40). When placed in aqueous solutions, monomeric SP-C has been shown to form amyloid fibrils. This represents an α-helical to β-sheet conversion and resembles the structural transitions that have been reported to occur in proteins and peptides associated with amyloid fibril formation in other organ systems such as the amyloid β peptide seen in Alzheimer's disease (41). By electron microscopy (EM), similar structures were also shown to occur in SDS-insoluble material from the cell-free bronchoalveolar lavage fraction of patients with pulmonary alveolar proteinosis (Figure 2, right) (42). The fibrillated hSP-C$_{3.7}$ was detectable by Congo Red birefringence and also demonstrated a loss of covalent palmitoyl residues. In vitro, removal of palmitic acid from SP-C increases the rate of fibril formation (43, 44). When compared with other amyloid-forming peptides, the mature SP-C molecule was found to contain a discordant high valine content in the α-helix (45). Taken together, the mature SP-C$_{3.7}$ domain has inherent instability imparted by native structure. It is therefore reasonable to predict that destabilizing point mutations or changes to important covalent modifications could further enhance these transitions to pathological forms.

CELLULAR MACHINERY FOR SP-C BIOSYNTHESIS

The AT2 cell plays a central role in the metabolism of all surfactant components, including SP-C. Northern and/or RT-PCR analyses of adult human, rat, mouse, and rabbit tissues have determined that SP-C mRNA expression is limited exclusively to the lung. In situ hybridization studies have further demonstrated that expression of SP-C mRNA is restricted to AT2 cells in all mammalian species examined (46–48).

When compared with other secretory cells, both morphologically and functionally, the AT2 cell is somewhat atypical. Unlike neuroendocrine cells or secretory epithelia, a dense secretory granule is not apparent. Instead, transmission EM (TEM) shows the presence of osmiophilic, lamellar organelles within the cytoplasm of AT2 cells (lamellar bodies, LBs) (49, 50). At the light microscopy level, LBs demonstrate intense fluorescence when isolated AT2 cells or lung tissue slices are incubated with phosphine 3R, which indicates the presence of lipid (51, 52). Components required for the formation of lamellar bodies and packaging of the phospholipid have not been fully elucidated. Surfactant protein B (SP-B) appears to be one of the key proteins in this process. Both newborns with hereditary SP-B deficiency and SP-B null mice have poorly formed lamellar bodies, dysfunctional surfactant, and abnormalities in SP-C biosynthesis (53, 54). Recently, loss of lamellar bodies has also been reported in two patients lacking SP-B or SP-C mutations (55, 56), suggesting that other protein factors are also involved. ATP-binding cassette (ABC) transporters are a family of membrane proteins involved in transport of macromolecular cargo across biological membranes, several of which (ABCA1, ABCA4) modulate intracompartmental movement of lipids and sterols (57). ABCA3, a 1704-amino acid family member, is highly expressed in the lung and can be localized to the limiting membrane of lamellar bodies (58, 59). Mutations in the ABCA3 gene have recently been described in neonates with respiratory failure and in abnormal electron dense bodies as seen by TEM (60). In contrast to SP-B deficiency, the effect of ABCA3 mutations and/or absence of normal lamellar bodies from other causes on biosynthesis of SP-C has not been reported.

Biochemically, lamellar bodies are similar to other lysosomal-like organelles in that they are acidic, contain lysosomal acid hydrolases, and possess the lysosomal glycoprotein marker CD63 and the integral lysosomal membrane protein LAMP-1 (61–65). Isolated LBs contain all surfactant phospholipid classes, SP-B and SP-C, as well as a fraction of total intracellular SP-A (49, 66). LBs release these contents into the alveolus via regulated exocytosis, which can be stimulated by a variety of agents acting through protein kinase C (phorbol esters, ATP) and protein kinase A pathways (terbutaline, cAMP), as well as through increases in intracellular calcium (generated by the Ca^{2+} ionophore A23187 or by mechanical stretch) (reviewed in 67). In contrast, the majority of SP-A and SP-D reaches the alveolus via constitutive pathways (68, 69).

Another important structural feature of the AT2 cell is the multivesicular body (mvb). Electron microscopy of rat and human lung has identified the presence of a circular vacuole 250–1000 nm in diameter with interior vesicles of 50–100 nm diameter (70). Similar structures have been identified in some other cells and are typically associated with sorting of internalized receptor–ligand complexes in the endocytic pathway (e.g., epidermal growth factor or transferrin receptors). The identification of proSP-B and proSP-C in the mvb of AT2 cells indicates that this organelle also participates in synthetic routing of surfactant components to LBs (64, 70–73). The mechanism of mvb fusion and LB formation is unknown, although it is believed that the process also includes an intermediate form, the composite body (or immature LB). Similar to the LB, the mvb in AT2 cells contains LAMP-1(63) and ABCA3, suggesting a close spatial and likely temporal relationship of this organelle with the LB.

In addition to containing an acidified, regulated secretory pathway with secretory lysosomal-like organelles (LB), the AT2 cell also participates in the catabolism and reutilization of its secreted products. Protein and lipid components of surfactant in the alveolus are re-endocytosed by AT2 cells via clathrin-dependent and -independent mechanisms and routed back to the central vacuolar system and LB via early endosomes [early endosome antigen-1 (EEA-1) -positive compartments] and mvbs (74, 75), directed to lysosomes for degradation (76), or quickly transported directly back to the plasma membrane for rapid resecretion (77–79). Thus the LB lies at the intersection of the secretory, endocytic, and lysosomal pathways.

In summary, the AT2 pneumocyte is a complex epithelial cell containing both a regulated secretory pathway for surfactant component biosynthesis and an endocytic pathway for surfactant degradation and/or reutilization. Thus alterations in AT2 cell function can potentially result in concomitant alterations in surfactant homeostasis.

SP-C BIOSYNTHESIS

Like many peptide hormones in exocrine and endocrine cells, the *SFTPC* gene product is initially translated as a larger proprotein that requires a series of discrete processing steps within the secretory pathway to yield the secreted biophysically active mature form of the protein (5, 6). For SP-C, these events include (*a*) translocation of apoprotein to the ER and folding, (*b*) post-translational addition of covalent palmitic acid, (*c*) sorting and exit of the lipoprotein from the Golgi, (*d*) cleavage of propeptide-flanking domains, (*e*) assembly of mature SP-$C_{3.7}$ with surfactant phospholipid and other proteins into LB, and (*f*) secretion into the alveolus. Details of this regulated series of biochemical events involved in processing of translated proSP-C_{21} have been primarily studied using a variety of in vitro models and experimental approaches (16, 17, 52, 80–86) (Figure 3, see color insert).

ER Translocation: Formation of Bitopic Type II Integral Membrane Protein

In vitro translation of hSP-C mRNA in the presence of canine pancreatic microsome vesicles results in a membrane-bound protein form resistant to salt extraction, which indicates integral association (84, 87). As mentioned above, the α-helical domain contained with the Phe24-Leu58 segment of proSP-C is predicted to span a lipid bilayer, rendering proSP-C_{21} and mature SP-C integral membrane proteins (Figure 1). Deletion of this domain abolishes membrane association, confirming that mature SP-C serves as a noncleavable signal peptide and a membrane-anchoring domain (84). Integral membrane association of proSP-C_{21} isoforms expressed in vitro has also been shown in analysis of membrane fractions of CHO cells transfected with human SP-C (16) and in ER from adult rat lung (73).

Topology studies reported by several groups have led to the further conclusion that proSP-C is positioned across the ER membrane as a type II bitopic membrane protein (COOH in ER lumen) (Figure 1) (84, 87, 88). The orientation of proSP-C_{21} is determined by the separation of charge across the transmembrane domain. Mutation of two positively charged juxtamembrane residues (Lys34 and Arg35) to neutral amino acids does not affect membrane association but results in inversion of the orientation of proSP-C in the ER membrane (COOH in lumen) (87).

In addition to the SP-C primary translation product, partially processed intermediates of proSP-C present in more distal intracellular compartments along the biosynthetic pathway are also integral membrane forms. Differential salt extraction and solubility analyses of isolated LB membranes have shown the presence of an integrally associated intermediate with M_r ~6000 (73). The post-translational processing of an integral membrane protein leading to exocytosis of the membrane-anchoring domain represents a novel mechanism for generation of a secreted mammalian protein product.

Proprotein Folding

Translocated proteins must fold correctly to facilitate anterograde trafficking. The oxidizing environment of the ER lumen contributes to disulfide-dependent folding of many proteins. Within the COOH flanking propeptide lie two conserved cysteine residues, which, based upon the type II orientation adopted by proSP-C, appear to promote folding of the COOH propeptide (Figure 1). Removal or mutagenesis of one or both of these residues results in loss of vesicular targeting and accumulation of proSP-C in juxtanuclear compartments (82, 89).

Palmitoylation

Covalent palmitic acid residues found in alveolar SP-C$_{3.7}$ are added to proSP-C in the biosynthetic pathway. Thioester linkages of palmitate to two cysteine residues near the NH$_2$ terminus of mature hSP-C domain (Cys28 and Cys29 of human proSP-C) have been shown to form in vitro, resulting in an initial increase in M_r of the SP-C primary translation product from 21,000 to 24,000 (Figure 1). This event is post-translational and Brefeldin A-insensitive, indicating that the fatty acylation of proSP-C occurs proximal to *trans*-Golgi (90). A tripalmitoylated form generated by ε-amino palmitoyl linkage at Lys34 has also been detected, although this form appears to represent less than 15% of total SP-C$_{3.7}$ that reaches the alveolus (91).

Regulated Cleavage Events

Four proteolytic cleavages of proSP-C are required to produce SP-C$_{3.7}$ (Figure 3, left). Using epitope-specific antibodies and pulse-chase labeling techniques, initial post-translational processing of palmitoylated human proSP-C$_{24}$ involves two distinct cleavages of the COOH domain producing 16 kDa and 7 kDa intermediates (17, 18). This is followed by a two-step cleavage of the NH$_2$ flanking domain to penultimately generate a 6 kDa intermediate and finally SP-C$_{3.7}$. This same sequence of events has been shown for the rat isoforms (52, 73, 81, 86).

Intracellular Sites of Cleavage

Immunoreactive proSP-C has been localized to ER, Golgi, small cytosolic vesicles, mvbs and LB of AT2 cells in human, mouse, and rat lung using ultrathin cryomicrotomy and immunogold EM (70, 72, 86, 92). Biochemically, proSP-C$_{21}$ is enriched in lung microsomes (containing rough ER), indicating that no cleavages occur in this early compartment (80). The proteolysis of proSP-C$_{21}$ can be completely blocked either by use of Brefeldin A (80, 81) or by low-temperature incubation (20°C) (86), indicating that all intracellular proteolysis of proSP-C is also occurring completely distal to the medial-Golgi. The processing scheme and spatial definition of these events are summarized in Figure 3.

Processing Enzymes

Like *SFTPC* gene expression, the complete proteolytic processing of proSP-C is AT2 cell-specific. Expression of recombinant hSP-C in CHO cells produces hSP-C_{21} and a single 16 kDa intermediate (16). Similarly, proSP-C expressed in mammary tissue via a cell-specific promoter is secreted as a partially processed proSP-C_{16} form in the transgenic milk (93). These results suggest that the initial cleavage of the COOH terminus is mediated by an enzyme that is more widespread and is not AT2-specific.

Identification of the precise constellation of processing enzymes assembled in the AT2 cell responsible for generation of mature SP-C from the 16 kDa intermediate is incomplete. Both co-localization with proSP-C and in vitro proteolysis assays using recombinant proSP-C support the notion that the cysteine protease Cathepsin H is involved in the first NH_2-terminal processing step in the mvb (94). Northern blot analysis has detected abundant expression of a novel aspartyl protease, Napsin A, in the lung (95–97). Immunogold EM has further localized Napsin A within the secretory pathway of AT2 cells (98). Using siRNA against Napsin A, the complete processing of proSP-C to mature SP-C (as well as to proSP-B to SP-B) is inhibited. Whether Napsin A cleaves proSP-C directly or whether incomplete SP-C processing occurs as a result of the loss of SP-B processing remains to be determined (99).

Processing is Dependent Upon an Intact, Functional LB Compartment

In addition to a unique AT2 cell–specific set of enzymes, intact and functional lamellar bodies are also required for biosynthesis. For many cells it is well recognized that there is a progressive decrease in intraorganellar pH along the exocytic pathway that can be an important regulatory step in the processing and trafficking of propeptide hormones. A vacuolar H^+-ATPase (v-ATPase) membrane proton pump is responsible for generation and maintenance of the internal acidic environment (100). Isolated lamellar bodies were estimated to have an internal pH ranging from 5.6 to 6.1 that is dependent upon the presence of a v-ATPase (61, 101). Complete post-translational proteolysis of proSP-C is inhibited by disruption of the *trans* LB pH gradient by the membrane-permeable weak base, methylamine, as well as by the lysosomotropic agent chloroquine, the proton ionophore monensin, and bafilomycin A_1, a specific v-ATPase inhibitor (52).

Processing of SP-C is also disrupted as a consequence of altered lamellar body morphogenesis. Patients with congenital SP-B deficiency and SP-B knockout mice have been noted to accumulate significant amounts of partially processed proSP-C intermediates (54, 92, 102, 103). Ultrastructurally, the AT2 cells found in the lungs from these patients, as well as from SP-B knockout mice, lack normal appearing

LBs (53, 92, 104). Compound transgenic mice capable of exhibiting inducible SP-B expression under control of exogenous doxycycline develop increases in misprocessed proSP-C as tissue and alveolar levels of SP-B fall (105).

Taken together, biochemical, morphological, and pharmacological data indicate that the most likely sites for proteolytic processing of proSP-C along the secretory pathway consist of acidic organellar compartments located distal to the medial Golgi, including early transport vesicles, mvbs, and lamellar bodies. The data also reinforce the conceptual notion that delivery of proSP-C from the Golgi to late compartments of the secretory pathway (mvb and LB) is essential for proper processing to mature SP-C.

Sorting and Trafficking of proSP-C

As for other secreted peptides and hormones, a regulated series of events facilitates movement of proSP-C/SP-C to the lamellar body for regulated exocytosis. Peptide motifs within proSP-C (i.e., *cis*-acting elements) appear to direct its intracellular trafficking.

Functional Signal Peptide

Unlike SP-A and SP-B, the NH_2-terminal flanking region does not contain a classic cleavable signal sequence. Rather, the mature SP-C domain (Phe24-Leu58) appears to be sufficient for ER translocation and functions as a non-cleavable signal anchor domain (83, 84).

Localization of Targeting Motifs

The bitopic type II orientation of proSP-C exposes the NH_2 propeptide to the cytosol. This arrangement is similar to other integral membrane proteins found in lysosomal-like organelles, in particular, lysosomal membrane proteins such as LAMP-1 or EGF receptor, in that all have short cytoplasmic domains (10–25 amino acids). By several lines of investigation, the intracellular trafficking of proSP-C has been shown to be dependent upon the cytosolic NH_2 propeptide (Met1-Phe23) (73, 83, 88). However, in contrast to LAMP-1 or EGF receptor, the NH_2 terminus of proSP-C does not contain classical dileucine motifs or tyrosine-sorting signals shown to direct other proteins to lysosomes. Through deletional and site-directed mutagenesis studies, the essential targeting motif within this region has been further localized to the sequence region Glu11-Thr19 (73, 88) Chimeric proteins can be directed to distal compartments by inclusion of the motif Met-Glu-Ser-Pro-ProTyr-Asp-Leu, whereas deletion of this domain results in retention of protein in ER compartments.

In contrast, the role of the COOH propeptide remains unclear. Truncations of the distal portions of COOH terminus result in retention of proSP-C in ER/Golgi (most likely owing to disruption of disulfide-mediated folding) (Figure 1) (89). However, the luminal COOH-terminal domain ($hSP-C59^-197$) of proSP-C appears to be dispensable for intracellular trafficking and secretion. Whether alone or in combination with the signal anchor (mature) SP-C domain, the COOH propeptide flanking region is insufficient to target EGFP chimeric proteins. In addition, truncation mutants of proSP-C lacking the entire COOH terminus can still be directed to distal compartments of transfected epithelial cells and PC 12 cells in vitro. When expressed in the lungs of SP-C knockout mice using adenovirus, HA-tagged proSP-C lacking the COOH propeptide can be recovered in the alveolar surfactant in vivo (88), although it is unclear if this occurs via regulated or constitutive secretion.

Sorting and Trafficking Involves Homomeric Association

Similar to other integral membrane proteins, proSP-C targeting is affected by oligomerization of proSP-C monomers. In vitro, wild-type SP-C transfected into A549 cells is capable of facilitating trafficking of mutant forms of proSP-C lacking the consensus targeting motif (SP-C24-58 or SP-C24-197), suggesting a direct interaction between the two isoforms (106). Chemical cross-linking studies of transfected A549 cell lysates expressing wild-type proSP-C with bis-malemeidohexane produced multimeric forms of wild-type SP-C. Deletion analysis localized the responsible oligomerization domain to the mature SP-C region.

Although homomeric association facilitates normal anterograde trafficking of wild-type proSP-C, and heteromeric assembly allows wild-type proSP-C to facilitate trafficking of some SP-C mutants with intact transmembrane domains but lacking targeting signals, trafficking of COOH folding mutants that form juxtanuclear aggregates blocks trafficking of coexpressed wild-type SP-C (106). Thus heterotypic oligomerization between wild-type and mutant isoforms provides a potential mechanism for development of a dominant-negative phenotype in heterozygous patients coexpressing wild-type and mutant SP-C alleles.

SP-C Mutations and Interstitial Lung Disease

A growing body of evidence links mutations in the SP-C gene to both familial and sporadic forms of interstitial lung disease (ILD). To date, over 20 separate missense, frameshift, or splice mutations in the human SP-C gene have been detected in patients with chronic parenchymal lung disease (7–10, 107, 108). The SP-C mutations are listed in Table 2, and their spatial distribution within the proSP-C primary sequence is depicted in Figure 4 (see color insert). On the basis of mutation site and phenotypic features, the known alterations in the hSP-C sequence appear to cluster in three subgroups.

Group A Mutations—The BRICHOs Domain

The index case associating a SFTPC gene mutation with ILD was reported in a full-term infant with histological features of nonspecific interstitial pneumonitis (NSIP) (8). The identified mutation, c.460 + 1 G → A, present on one allele in the

TABLE 2 Classification of published SP-C mutations associated with chronic lung disease

Mutation	Pathology	Trafficking in vitro	Other features	References
GROUP A [BRICHOS DOMAIN] (RESIDUES 94–197)				
L188Q	NSIP-Children, DIP/UIP-Adults	Aggresomes	Cytotoxicity	(109)
L188R	Pneumonitis (unspecified)			(8)
ΔExon 4	NSIP-Children, DIP/UIP-Adults	Aggresomes	Absence of mature SP-C	(10)
140delA[a]	Pneumonitis (unspecified)	ND		(8)
I126R	Pneumonitis (unspecified)	ND		(8)
P115L	Pneumonitis (unspecified)	ND		(8)
G100V	Pneumonitis (unspecified)	ND		(8)
Y104H	Pneumonitis (unspecified)	ND		(8)
GROUP B (RESIDUES 58–93)				
E66K	NSIP + PAP	Early endosome	↑ Phospho-lipid/ ↑SP-A, ↑SP-B	(108)
I73T	NSIP + PAP	Early endosome	PAS positive-staining	(8, 107)
Δ91–93	NSIP	Early endosome Non-ubiquinated aggregate	↑Phospholipid ↑SP-B	(7)
GROUP C (RESIDUES 1–35)				
P30L	NSIP/DIP	ER retention	Non-palmitoylated	(8, 90, 114)

[a]Represents a frameshift mutation associated with expression of a stable transcript.
NSIP = Nonspecific interstitial pneumonitis.
DIP = Desquamative interstitial pneumonitis.
UIP = Usual interstitial pneumonitis.
PAP = Pulmonary alveolar proteinosis.
ND = Not determined.

infant, as well as on that of the mother, who had a prior diagnosis of usual interstitial pneumonitis (UIP), produced alternate splicing of the SP-C mRNA, deletion of Exon 4, and resultant production of a defective proprotein foreshortened by 37 amino acids. Although transcript levels for the wild-type and mutant isoforms were similar, no mature SP-C was detected, suggesting that mutant protein had a dominant-negative effect on SP-C biosynthesis.

Subsequent to this observation, a separate SP-C mutation was described consisting of a heterozygous Exon 5 + 128 T → A transversion of SP-C in an extended kindred with clinical and histopathologic features of UIP in adults (age 17–57 years) and NSIP in children (4–17 months) (9). This missense substitution results in exchange of glutamine for leucine at the conserved position 188 of the COOH propeptide (L188Q), which is immediately adjacent to a conserved cysteine (Cys189) shown to be critical for proprotein folding and trafficking (89).

An additional series of 12 infants with chronic lung disease led to the discovery of other sporadic and familial missense, splice, or frameshift mutations (Table 2) (8). As for ΔExon 4 and L188Q mutations, all in this series but two (P30L and I73T) map to the distal COOH terminus of proSP-C. Of note, this region (Phe94-Ile197) has recently been linked with an emerging family of several previously unrelated proteins containing a novel BRI domain (BRICHOS) (109–113). BRICHOS is a structurally defined protein domain of approximately 100 amino acids that is found in several proteins, previously thought to be structurally and functionally unrelated, all associated with various degenerative and proliferative diseases. These proteins include the BRI family of proteins linked to familial British and Danish dementia; chondromodulin-I, which is associated with chondrosarcoma; the CA11 protein, which is associated with stomach cancer; and SP-C, which is now linked to chronic parenchymal lung diseases. In many of these proteins the conserved BRICHOS domain is located in the propeptide region that is removed during proteolytic processing. Limited data suggest that the physiological role of this domain in other proteins is related to the post-translational processing. The underlying mechanism for cellular toxicity or tissue injury from expression of BRICHOS domain mutants has not been characterized for any of the family members.

Group B Mutations: Non-BRICHO COOH Domain

Recently, several additional reports have heralded an emerging diversity of phenotypes associated with expression of mutant SP-C proteins (10, 107, 108). Although these mutations also occurred on only one allele, in contrast to Group A mutations, they cluster within a 30-amino acid span encoded on Exon 3 of the hSP-C gene, and the clinical phenotypes of each include the presence of diffuse parenchymal lung disease, abnormalities in surfactant composition and/or function, and expression of variable but detectable amounts of SP-C. A de novo heterozygous missense mutation of the SFTPC gene (g.1286T > C), resulting in substitution of threonine for isoleucine (I73T), was reported in a 13-month-old infant with a histological pattern of NSIP and PAP (107). Immunohistochemical and biochemical analyses showed an intra-alveolar accumulation of SP-A with detectable levels of SP-B, as well as mono- and dimeric mature SP-C. Interestingly, the identical mutation appeared in the recent series described by Nogee et al. (8). A second, sporadic heterozygous mutation resulting in a substitution of lysine for glutamic acid at position 66 (= E66K) was found in a full-term infant diagnosed at age 2 months with lung histology and biochemical findings consistent with NSIP. Increases in total alveolar phospholipid and SP-A were found, as well as the presence of mature SP-C. (108). A spontaneous in-frame 9-base pair deletion spanning codons 91–93 in Exon 3 of the *SP-C* gene (hSP-C$^{\Delta 91-93}$) was present on one allele of a 14-month-old infant with progressive ILD (10). The child's surfactant contained markedly decreased levels of mature SP-B and SP-C, although SP-A content was increased. Collectively, these three reports delineate a second class of spatially and phenotypically distinct SP-C mutations; the sporadic nature of their expression suggests that this region of the *SFTPC* gene represents a genetic hotspot for mutations.

Group C Mutations: Cytoplasmic Domains

A single mutation in the cytosolic NH_2 (nontransmembrane) domain of human proSP-C has been reported (8). This heterozygous proline to leucine missense substitution at codon 30 (P30L) is contained within the mature SP-C sequence (Phe24-Leu58) and was detected in an infant with a NSIP/desquamative interstitial pneumonitis (DIP) pattern on lung biopsy (114). Additional clinical details have not been reported. To date, no mutations in the NH_2-flanking propeptide (residues 1–23) have been found. Whether this is simply because of decreased statistical probabilities for this NH_2- terminal region (sixfold shorter in length than the COOH propetide) or whether mutations here represent a more lethal phenotype remains to be seen.

Mutant SP-C-Induced Lung Dysfunction: A Conformational Disease

Many diseases recessively inherited are caused by homozygous expression of a mutant gene that induces a null phenotype either through failure of protein expression or a loss of protein function. In all cases to date, the finding of an SP-C mutation on only one allele suggests either a dominant-negative effect or a toxic gain of function. The heterozygous expression of a mutant *SP-C* allele is sufficient to induce pathology and is thus similar to a seemingly diverse set of chronic degenerative disorders observed in a variety of organ systems, including brain, liver, and lung (1–3). Alzheimer's disease (115, 116), Huntington's disease (117), amyloidosis (114), spinocerebeller ataxia (118), and Charcot-Marie-Tooth disease (119) are all associated with expression or altered processing of a mutant transmembrane protein that undergoes a fatal conformational rearrangement, which unmasks an intrinsic ability either to self-associate (aggregate) or be deposited in non-native subcellular or tissue compartments. How cells respond to the production of these conformers and how misfolded proteins induce cytotoxicity and tissue inflammation remain emerging questions for cell biology and medical therapeutics.

Expression Analysis of SP-C Mutants Associated With ILD

Several lines of investigation, both in vitro and in vivo, support the concept that *hSP-C* mutations induce conformational disease through the expression and/or mistrafficking of the abnormal protein product. The effect of mutations in the proSP-C sequence on trafficking has been characterized in living cells using transfected chimeric fusion proteins containing enhanced green fluorescent protein (EGFP) analogues (10, 107, 108, 120). These studies indicate that the regional clustering of mutations described in the previous section (Table 2 and Figure 4) correlates well with distinct distribution and trafficking patterns of the EGFP fusion protein (Figure 5, see color insert).

GROUP A (BRICHOS DOMAIN)   When transfected into A549 cells, two mutations of SP-C that map to the BRICHOS domain each form perinuclear aggregates in a nonmembrane-limited compartment associated with the microtubule-organizing center (MTOC) (89, 120). In Figure 5*B*, an EGFP chimeric protein containing hSP-C$^{\Delta Exon4}$ forms juxtanuclear aggregates that colocalize with anti-ubiquitin staining. This structure has been termed an aggresome and is the product of the failure of the affected cell to degrade misfolded proteins in the standard ubiquitin proteasome system (121–123). The expression of mutant forms of either CFTR or PS-1 in HEK cells results in a similar deposition of pericentriolar membrane-free cytoplasmic inclusions containing ubiquitinated, mutant protein surrounding the MTOC. Aggresome formation and size varied with the level of mutant protein expression. Low levels of mutant protein were predominantly ER associated, whereas higher levels readily formed aggresomes. Aggresome formation could be potentiated by addition of proteasome inhibitors. Similar proSP-C aggregates have been detected in MLE15 cells and primary human AT2 cells transfected with EGFP tagged hSP-C$^{\Delta Exon4}$ (M.F. Beers, unpublished data). Additionally, a single clone of HEK 293 cells stably expressing an untagged hSP-C$^{\Delta Exon4}$ construct exhibited an ER distribution pattern (124). This suggests that at lower levels of protein expression hSP-C$^{\Delta Exon4}$ can undergo standard degradation events, as has been observed for mutant CFTR expressed in this same cell line (123). The formation of aggresomes by a mutant bitopic transmembrane protein such as proSP-C supports the emerging concept that this novel compartment is a general survival response by the cell to misfolding events when normal ER-associated degradation pathways (ERAD) are overwhelmed (discussed below).

GROUP B MUTATIONS (NON-BRICHOS)   In contrast to aggregation-prone mutants, expression of Group B mutants with missense substitutions or deletions within the proximal COOH propeptide (E66K, I73T, Δ91–93; Table 2) results in a different intracellular fate. EGFP fusion proteins containing any of these three mutants accumulate in cytosolic vesicles bearing the early endosome marker EEA-1 (Figure 5C). This finding correlates with in vivo data in which biopsies from patients with Group B mutations show abnormal cytosolic vesicular accumulation of proSP-C forms (10 107, 108).

GROUP C MUTANT   Expression of the P30L substitution in A549 cells results in retention of nonubiquitinated mutant protein in the ER (Figure 5D). A similar finding was observed when expressed untagged hSP-C^{P30L} localized to calnexin-positive (ER) compartments of CHO cells and failed to traffic to the Golgi (90).

Toxicity of SP-C Mutants Associated With ILD: Gain of Function

Several independent studies support a cause-and-effect relationship between SP-C gene mutations and chronic parenchymal lung disease. Most studies have focused on BRICHOS *hSP-C* mutants. MLE 15 cells transfected with hSP-C^{L188Q} exhibit impaired growth and evidence of increased cytotoxicity compared with cells expressing similar levels of wild-type protein. (9). Lung-specific expression of *hSP-C$^{\Delta Exon4}$* in AT2 cells of transgenic mice resulted in a profound disruption of lung morphogenesis and led to neonatal lethality in the face of two wild-type *SP-C* alleles (124). A similar phenotype was observed when mature hSP-C$_{3.7}$ alone (lacking flanking propeptide domains) was expressed (125). Given the fact that the *SP-C* knockout mouse is viable at birth (126), these results suggest that high level expression of BRICHOS mutations in the SFTPC leads to aggregation of misfolded proteins that triggers cell death during lung development. This supports the notion that *SP-C* mutations are conformational diseases, i.e., expression of a mutant phenotype has more severe consequences than the null phenotype. It is likely that lower levels of these BRICHOS SP-C isoforms under control of endogenous promoters cause a milder phenotype leading to postnatal ILD.

In vitro data show that Group B (non-BRICHOS) SP-C mutants traffic and accumulate in endosome-related compartments. In vitro toxicity assays and in vivo mouse models expressing this class of mutants have not yet been reported. However, potential mechanisms of toxicity (injury) can be extrapolated from consideration of other genetic diseases with a similar phenotype. The Hermansky-Pudlak syndrome (HPS) is a polygenetic disorder associated with defective formation of specialized lysosome-related organelles, including abnormal melanosomes, platelet granules, and lysosomes. Patients with HPS can develop pulmonary fibrosis as a complication of their disease (127). Mice with HPS1 and HPS2 gene mutations demonstrate increased lung levels of lysosomal enzymes and phospholipids in conjunction with pulmonary fibrosis, suggesting altered vesicular trafficking and abnormal accumulation of protein in these endosome/lysosome compartments (128–131). In the lung this results in the appearance of giant lamellar bodies. It has been proposed that events such as these can promote inflammation and fibrosis though transmission of proteotoxicity across cellular compartments by yet unknown mechanisms (132).

Coexpression of *SP-C* Mutants Inhibits Normal SP-C Biosynthesis: Dominant-Negative Effect

In addition to the cytotoxicity and disrupted lung morphogenesis that support induction of a toxic gain of function, characterization of surfactant of patients with SP-C-mediated ILD suggest a concomittant dominant-negative effect that appears to be mutant specific. Patients with heterozygous expression of the *hSP-C$^{\Delta Exon4}$* allele had undetectable levels of mature hSP-C$_{3.7}$ despite the presence of one wild-type allele and detectable mRNA. The in vitro studies demonstrating oligomeric trafficking of the rat proSP-C isoform suggest that hSP-C$^{\Delta Exon4}$ targeted for degradation exerts a dominant-negative effect on the trafficking and processing of coexpressed wild-type SP-C. When a hemagglutinin (HA = YPYDVPDYA) tagged hSP-C$^{1-197}$ and EGFP tagged hSP-C$^{\Delta Exon4}$ are cotransfected into A549 cells, aggregation of both constructs occurs, indicating a heterotypic oligomerization of hSP-C$^{1-197}$ and hSP-C$^{\Delta Exon4}$. This provides a molecular mechanism for the dominant-negative effect observed in vivo. Non-aggregating mutants (outside the BRICHOS region) of proSP-C have not been shown to inhibit biosynthesis of wild-type hSP-C$_{3.7}$.

How the lack of SP-C contributes to the underlying lung pathology phenotype is unclear. The SP-C null mouse is viable at birth (126), and although a recent report suggests that when SP-C$^{-/-}$ mice are back-crossed onto a different strain, they develop inflammation and emphysema (133), the BAL of multiple patients with Group B mutations (I73T, E66K) contained detectable mature SP-C$_{3.7}$. This suggests that the pathology of interstitial pneumonias (UIP, NSIP) associated with SP-C mutations is not simply through a lack of SP-C but that SP-C null mice generate an abnormal response to inflammatory stimuli.

Non-Surfactant Function of SP-C$_{3.7}$: An Immunomodulator?

Because of its hydrophobicity and tight association with surfactant phospholipid, it is also plausible to consider that interactions between the SP-C peptide and other lipid structures could take place in the alveolus. The innate or non-antibody-mediated immune system relies on the presence of pattern recognition molecules to invoke a primary interaction with inhaled pathogens and particles. Bacteria and inhaled lipopolysaccharide represent important proinflammatory stimuli that frequently impact the distal lung. In addition to the innate lung collectins SP-A and SP-D, mature SP-C was recently shown to participate in innate immunity. Isolated SP-C, but not SP-B, binds to the lipid A portion of a variety of LPS phenotypes in a specific and competitive fashion (134–137). In addition, this interaction blocks the interaction of LPS with CD14 and with macrophages, inhibiting proinflammatory cytokine and NO release by macrophages in response to LPS in vitro. Thus, apart from its physiological importance in lung biophysics, SP-C appears to have an emerging role in the local modulation of lung inflammation, and variations in alveolar levels of SP-C could serve as a phenotypic modifier in some forms of SP-C-induced lung disease.

Lost After Translation: Mechanisms and Mediators of SP-C Lung Injury

The current evidence indicates a cause-effect relationship between mutant SP-C expression, cellular toxicity, lung inflammation, and fibrosis. The cellular and molecular mechanisms linking the expression with the phenotype are currently incomplete but appear to be related to a change in protein conformation, abnormal trafficking of an integral membrane protein, and/or accumulation of translated mutant isoforms in non-native intracellular compartments.

The role of mutant SP-C in the pathogenesis of epithelial cell dysfunction and toxicity is schematically illustrated in Figure 6 (see color insert). Overall, the pathogenesis of chronic lung injury from expression of proSP-C mutants in the lung can be considered as part of a unified view of the cellular and molecular pathogenesis of a larger family of conformational diseases. The pathogenicity of mutant SP-C is due, in part, to the unique properties of SP-C and to the overlapping properties of the mutant protein (structure-function distribution) common to proteins present in other conformational diseases:

- ProSP-C is an integral membrane protein.
- Biosynthesis requires precise targeting to a unique organelle.
- The transmembrane anchor and biophysically active secreted product (mature hSP-C$_{3.7}$ domain) is inherently unstable with a tendency to undergo aggregation and amyloid fibril formation.
- SP-C mutants misfold or mistarget and, combined with the complexity of AT2 cells, this leads to the possibility of diversion of mutant isoforms to many secretory or endocytic compartments with secondary disruption of cellular function.

This approach allows for the consideration of chronic lung disease from SP-C mutations as akin to the Alzheimer's or Huntington's disease of the lung. In these cases of chronic neurodegenerative disease, it is the intracellular or extracellular accumulation of an abnormal protein isoform (such as $A\beta$—amyloid precursor protein or polyglutamine huntingtin-1) with inherent instability that results either from aberrant processing or intrinsic properties imparted by the mutation, which then leads to cellular and organ dysfunction that develops over a prolonged period of time (114, 117, 138, 139). As Figure 6 illustrates, at least five different processes are potentially perturbed in mutant SP-C expressing cells (see below).

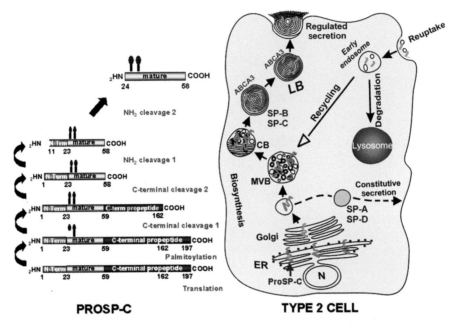

PROSP-C   **TYPE 2 CELL**

Figure 3 Schematic diagram of biochemical events, localization, and cellular machinery for SP-C biosynthesis and metabolism. Diagrammatic representation of post-translational processing of proSP-C in the adult lung. (*Left*) SP-C proteolytic processing steps: the major cleavage steps and intermediates generated from four separate cleavages of the propeptide (based on data from References 16, 17, 52, 80–86). On the basis of antibody epitopes and SDS/PAGE analysis, the approximate size of each intermediate is illustrated. (*Right*) Schematic representation of surfactant component metabolism in the AT2 cell. Biosynthesis of SP-C begins with translation of mRNA in the cytoplasm. The SP-C propeptide must be translocated to the ER, sorted in the Golgi, and then enter the regulated secretory pathway in the budding transport vesicles. Major subcellular compartments include the multivesicular body (mvb), composite body (immature LB), and LB. Proteolytic cleavage begins in post-Golgi vesicles and is completed in the LB. ABCA3 is a polytopic membrane protein shown to be critical for lamellar body genesis. SP-C is secreted into the alveolar space with SP-B and phospholipid via regulated exocytosis. In contrast, SP-A and SP-D reach the alveolar space via constitutive secretion in non-LB organelles. Both phospholipids and proteins (SP-A, SP-B, SP-C) are endocytosed and routed back either to lysosomes for degradation or to the LB for recycling via mvb.

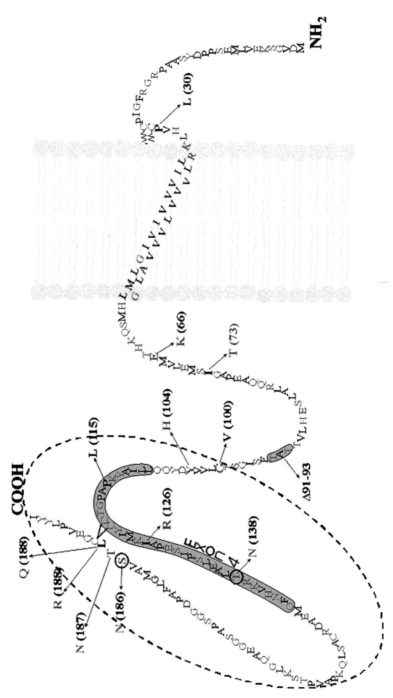

Figure 4 Spatial distribution of mutations within the human proSP-C molecule associated with interstitial lung disease. The majority of SP-C mutations (Group A) associated with chronic lung disease map to a domain of the COOH flanking propeptide (*dashed circle*), which has structural homology to a region in the BRI family of proteins associated with familial dementias (109–113). Additional mutations in the COOH propeptide (Group B) are found in the region H59-T93. To date only one mutation (hSP-CP30L) is found in the cytosolic side of proSP-C. The mature SP-C domain encompasses residues F23-L58.

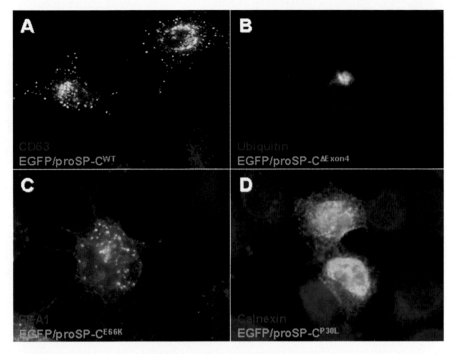

Figure 5 Distinct in vitro traffic patterns of hSP-C mutants. EGFP fusion proteins of wild-type (*A*) and mutants hSP-C$^{\Delta Exon4}$ (*B*), hSP-C^{E66K} (*C*), and hSP-C^{P30L} (*D*) were transfected into A549 cells. Double label immunocytochemistry using Texas Red conjugated CD63 (*A*), ubiquitin (*B*), EEA-1 (*C*), and calnexin (*D*) antibodies (*red fluorescence*) as indicated were colocalized (*yellow fluorescence*) with expression of corresponding EGFP fusion protein (*green fluorescence*), respectively.

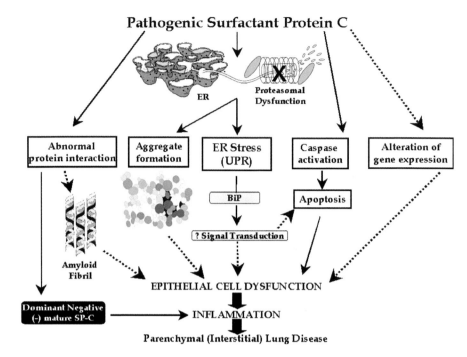

Figure 6 The role of various cellular responses in the cascade of events leading to SP-C-induced lung disease. Schematic representation of five potential pathways mediating SP-C-induced lung disease. Details are provided in the text. Solid lines represent links for which either published or preliminary data for SP-C are available. Dotted lines represent potential links based upon data in other proteins associated with conformational disease.

ER-Associated Degradation (ERAD), the Unfolded Protein Response (UPR), and ER Stress

The ER relies on an efficient quality control mechanism to prevent accumulation of unfolded or aggregated proteins (140–142), which ensures that of all the proteins translocated to the ER, only those that are properly folded transit to the Golgi compartment. Proteins misfolded in the ER are retained in an attempt to facilitate their native conformation. If unsuccessful, they are instead retrotranslocated back into the cytosol for degradation via the ubiquitin/proteasome pathway. The ER has evolved highly specific signaling pathways to ensure that its protein-folding capacity is not overwhelmed. These pathways, collectively termed the unfolded protein response (UPR), are required if the cell is to survive the ER stress that can result from expression of altered proteins or exuberant overexpression of normal proteins. Upon accumulation of abnormal proteins, the UPR is activated to reduce the amount of newly translocated proteins, upregulate the protein-folding capacity of the ER, and increase retrotranslocation and degradation of ER-localized proteins (discussed, below) (143).

One of the hallmarks of the ERAD/UPR is the upregulation of expression of the ER chaperone BiP. Expression of hSP-C$^{\Delta Exon4}$ in HEK293 cells has been associated with increases in transcription of BiP and BiP promoter-driven reporter constructs, suggesting that hSP-C$^{\Delta Exon4}$ triggers the ER stress response (124).

Cell Survival Pathways: Caspase Activation and Apoptosis

Prolonged UPR activation, owing to persistent abnormal protein expression or accumulation of trapped proteins with erroneous target signals, weakens and impairs ERAD function. Consequently, several death-promoting pathways involving apoptosis-signaling kinases and caspases are activated (144). On the basis of work in a mammalian cell system expressing polyglutamine (PolyQ), it was found that the transmission of toxicity signals from the ER to the nucleus is mediated by several proteins including Ire-1 (which senses misfolded proteins), ASK-1 (apoptosis-sensing kinase-1), TRAF2 (TNF receptor-associated factor), and stress-activated kinase cascades that include participation of JNK family members and p38 (145). Further downstream, signaling mediated UPR cell death also involves changes in ER calcium, NO production, and a cytosolic activation of a caspase cascade. The constituents triggered by UPR activation vary among cell and tissue types. In pancreatic β cells, caspase 12, caspase 9, and caspase 3 are central to this response (140, 146–148). In rat neuronal cells, polyglutamine expression resulted in caspase 8 activation and apoptosis, which was blocked by caspase inhibitors and by dominant-negative strategies directed against FADD and Bcl-2 (149). Alternative to cell death, in some cases, the ER stress response can induce NF-κB activation and production of proinflammatory cytokines while upregulating other genes by the affected cells (150).

In a preliminary report, overexpression of EGFP/hSP-C$^{\Delta Exon4}$ in vitro has been associated with activated caspase 3 expression, appearance of Annexin V staining on the plasma membrane, and cell death, suggesting that UPR and apoptosis pathways are linked in lung epithelial cells expressing mutant hSP-C (151).

Protein Accumulation and Aggregation in Foreign Compartments

In addition to ER stress, the accumulation of intracellular misfolded protein deposits outside the ER is becoming a well-accepted mechanism for proteotoxicity. There is a growing compendium of human diseases associated with aberrant trafficking of transmembrane proteins (152, 153). Cells have a limited repertoire for disposal of abnormal products. Normally, degradation of integral membrane proteins and secreted peptides unable to fold in the ER or to traffic anterograde involves one of three pathways.

RETROGRADE ROUTING TO CYTOSOLIC PROTEASOME   The ubiquitin-proteasome system functions in cellular quality control mediated by ERAD by degrading defective proteins in a well-regulated cytosolic "trashcan." When this system is overwhelmed either by protein overexpression or by inhibition of the proteasome pathway, protein aggregation develops. The cellular response is to deposit protein aggregates in a pericentriolar, membrane-free, cytoplasmic inclusion termed the aggresome (121–123, 154, 155). Intracellular deposits of proteins have been found in the brains of patients with neurodegenerative disorders, but it is unclear whether this is a cause or effect of the neuronal dysfunction (138, 156–158). Using a novel reporter cell line, transient expression of two unrelated aggregation prone proteins caused near complete inhibition of ubiquitin-dependent proteolysis (159). These data provide a link between protein aggregation and cell death. Several hSP-C mutants have been shown to accumulate in the aggresome, suggesting that proteasome inhibition by mutant aggregates can occur (89, 120).

DIVERSION TO THE CONSTITUTIVE SECRETORY PATHWAY/PLASMA MEMBRANE   Mutant transmembrane proteins can be alternately trafficked to other cellular compartments via constitutive transport. Because of their membrane anchors and lipid avidity, transmembrane proteins pose a particular problem in that secretion fails through an inability to liberate the membrane anchor. Instead proteins accumulate in membrane compartments. Overexpression of wild-type hSP-C in PC12 cells was associated with deposition of incompletely processed proprotein on the cell surface (88).

ACCUMULATION IN ENDOSOMES/LYSOSOMES   Endosomal accumulation of some SP-C mutant forms could occur by one of two pathways: Fusion of constitutive vesicles containing membrane-bound mutant to the plasma membrane and subsequent bulk membrane recycling to endosomes/lysosomes is one potential route. Alternatively, a post-Golgi pathway that is parallel to normal anterograde transport to the plasma membrane can result in the appearance of some mutant proteins in endosomal or lysosomal organelles directly from *trans* Golgi.

In either case, inundation of these organellar compartments by large amounts of proteins and/or lipid is associated with cellular dysfunction as seen in several lysosomal/endosomal storage diseases. Non-aggregating COOH SP-C mutants were shown to accumulate in EEA-1 compartments in association with alveolar accumulation of other surfactant components (107). Although the exact routing has not been defined, these findings might be explained by the accumulation of mutant SP-C protein in early endosomes resulting in an impairment of endocytosis.

Amyloidogenesis

Amyloid fibril formation occurs in a number of diseases induced by a variety of proteins and is promoted under conditions where aggregation-prone conformations present during unfolding are relatively well populated and create nuclei for precipitation (41, 139, 156, 160, 161). Given the propensity of $hSP-C_{3.7}$ to form β-amyloid conformers (42, 114), cytotoxicity from intracellular amyloid formation by proSP-C mutants containing this domain following retrotranslocation from the ER to the aqueous cytosol during quality control surveillance represents another potential mechanism for epithelial cell injury.

Alterations in Gene Expression

In lieu of inducing cell death, in some circumstances, the ER stress response can alter transcription of a secondary gene-mediated response that can amplify an inflammatory process or alter cell growth. A variety of overexpressed integral membrane proteins (including hepatitis B surface antigen, MHC class I, immunoglobulin μ-heavy chain) induce NF-κB activation (162–165). NF-κB activation and production of proinflammatory cytokines and reactive oxygen species have been reported to be induced by $CFTR^{\Delta 508}$ expression in epithelial cells (150). Thus in addition to cell death, inflammatory cascades can be amplified within the affected cell populations.

Prospects for Therapy: Molecular Strategies to Modulate SP-C Mutant-Induced Lung Disease

The realization that conformational disease results from either aberrant intermolecular aggregation or transport to non-native compartments makes possible both general and specific approaches to correction of the defects. On the basis of the mechanisms outlined in Figure 6, chemical or pharmacological treatment of conformational disease can employ strategies directed at blocking the formation of protein aggregates, disrupting amyloid accumulation, increasing turnover of protein, and redirecting protein trafficking; altering the cellular response to mutant protein expression could also be employed (2, 152).

Prevention or reversal of protein aggregation or amyloid formation represents a primary treatment strategy. The expansion of a CAG repeat coding for polyglutamine is central to neurodegenerative disorders such as Huntington's disease (117). It has been well documented that expanded polyglutamine fragments, cleaved from their respective full-length proteins, form microscopically visible aggregates in affected individuals and in transgenic mice. The azo-dye Congo red binds preferentially to β-sheets containing amyloid fibrils and can specifically inhibit oligomerization and disrupt preformed oligomers. In cell models of polyglutamine oligomerization, Congo red prevents ATP depletion and caspase activation, preserves normal cellular protein synthesis and degradation functions, and promotes the clearance of expanded polyglutamine repeats. Infusion of Congo red into a transgenic mouse model of Huntington's disease, well after the onset of symptoms, promotes the clearance of expanded PolyQ repeats in vivo, exerting marked protective effects on survival, weight loss, and motor function (117).

The modulation of optimal chaperone function in the cell or addition of exogenous chaperones represents a second therapeutic strategy. The small molecule, 4-phenylbutyric acid (4-PBA), has been shown to enhance the trafficking of CFTR$^{\Delta508}$ and α-1 antitrypsin through effects on heat shock protein expression (166–169). Oral 4-PBA has been shown to increase nasal potential in patients with a CFTR$^{\Delta508}$ variant of cystic fibrosis, suggesting enhanced plasma membrane expression of CFTR by 4-PBA (170). Addition of 4-PBA to A549 cells in vitro expressing hSP-C$^{\Delta\text{Exon4}}$ prevented aggresome formation (120). Other small molecules (chemical chaperones) that promote protein folding/trafficking include glycerol, DMSO, trimethylamine oxide (TMAO), and curcumin (171–173).

CONCLUSIONS AND UNANSWERED QUESTIONS

Since the initial isolation of SP-C two decades ago, we have come to appreciate the unique structural and functional properties that make SP-C a model for a new class of lipid-avid secretory proteins. Chronic parenchymal lung diseases [idiopathic interstitial pneumonia (IIP) and pneumonitis of infancy] have been shown to result from heterozygous expression of mutations in the proSP-C sequence, both in term infants and in kindreds of patients with histologically confirmed interstitial pneumonia (IIP/IPF). Studies spanning the past 15 years have defined the biosynthetic pathway for SP-C, including structural domains mediating proSP-C trafficking, the importance of proper protein folding, and the presence of an intact regulated secretory pathway for complete SP-C biosynthesis. Most recently, alterations in the proSP-C sequence that result in either misfolding or mistargeting of mutant protein trigger induction of intracellular aggregate formation, ER stress, and accumulation in endosomal-lysosomal compartments. From studies of chronic neurodegenerative diseases and α-1 antitrypsin deficiency, the concept of conformational diseases caused by aggregation of mutant protein in non-native conformations is being increasingly recognized. On the basis of the available body of evidence, misfolding or mistargeting of mutant forms of SP-C may actually have more dire implications than a null phenotype via toxic gain of function. Topics in need of additional investigation include a more detailed understanding of the fate of mutant, misfolded, or mistargeted proSP-C, and better understanding of signal transduction pathways linking ER stress, amyloid formation, aggregation, protein mistargeting, and other events in AT2 cells leading to the development of T cell–mediated cytokine-driven lung inflammation and myofibroblast activation. Just as for many chronic neurodegenerative diseases, improved understanding in these and other areas will potentially revolutionize molecular therapies and permit novel approaches to treatment of ILD by focusing on more upstream events in the primary dysfunctional cell (AT2 epithelial cell) in lieu of simply suppressing inflammation and cytokine production.

ACKNOWLEDGMENTS

We thank Dr. Susan Guttentag for helpful discussions and Seth T. Scanlon for graphic arts assistance.

The authors acknowledge the help and support of many mentors and collaborators over the years, including Drs. Aron B. Fisher, Philip L. Ballard, Frank Brasch, Paul N. Stevens, Larry Nogee, Aaron Hamvas, Michael Koval, Sheldon Feinstein, Linda Gonzalez, Albert Kabore, Chi Y. Kim, Amy L. Johnson, and Wen-Jing Wang. The invaluable technical assistance of Catherine A. Lomax, Scott J. Russo, and Vu Nyguyen is also appreciated. This work is supported by NIH HL-19737 (M.F.B.), HL074064 (M.F.B.) and P50-HL56401 (M.F.B.) and by a Dalsemer Research Grant in Pulmonary Fibrosis from the American Lung Association (DA 188N) (to S.M.). S.M. is a recipient of a fellowship from the Francis Families Foundation.

The *Annual Review of Physiology* is online at
http://physiol.annualreviews.org

LITERATURE CITED

1. Carrell RW, Lomas DA. 1997. Conformational disease. *Lancet* 350:134–38

2. Carrell RW, Lomas DA. 2002. Mechanisms of disease: alpha$_1$-antitrypsin deficiency—a model for conformational diseases. *N. Engl. J. Med.* 346:45–53

3. Kopito RR, Ron D. 2000. Conformational disease. *Nat. Cell Biol.* 2:E207–9

4. Zeitlin PL, Gail DB, Banks-Schlegel S. 2003. Protein processing and degradation in pulmonary health and disease. *Am. J. Respir. Cell Mol. Biol.* 29:642–45

4a. Rooney SA, ed. 1998. *Lung Surfactant: Cellular and Molecular Processing.* Austin, TX: Landes

5. Beers MF. 1998. Molecular processing and cellular metabolism of surfactant protein C. See Ref. 4a, pp. 93–124

6. Solarin KO, Wang WJ, Beers MF. 2001. Synthesis and post-translational processing of surfactant protein C. *Pediatr. Pathol. Mol. Med.* 20:471–500

7. Nogee LM, Dunbar AE, Wert SE, Askin F, Hamvas A, Whitsett JA. 2001. A mutation in the surfactant protein C gene associated with familial interstitial lung disease. *N. Engl. J. Med.* 344:573–79

8. Nogee LM, Dunbar AE III, Wert S, Askin F, Hamvas A, Whitsett JA. 2002. Mutations in the surfactant protein C gene associated with interstitial lung disease. *Chest* 121:20S–21S

9. Thomas AQ, Lane K, Phillips J III , Prince M, Markin C, et al. 2002. Heterozygosity for a surfactant protein C gene mutation associated with usual interstitial pneumonitis and cellular nonspecific interstitial pneumonitis in one kindred. *Am. J. Respir. Crit. Care Med.* 165:1322–28

10. Hamvas A, Nogee LM, White FV, Schuler P, Hackett BP, et al. 2004. Progressive lung disease and surfactant dysfunction with a deletion of surfactant protein C gene. *Am. J. Respir. Cell Mol. Biol.* 30: 771–76

11. Hatzis D, Deiter G, deMello DE, Floros J. 1994. Human surfactant protein-C— genetic homogeneity and expression in RDS—comparison with other species. *Exp. Lung Res.* 20:57–72

12. Glasser SW, Korfhagen TR, Perme CM, Pilot-Matias TJ, Kister S, Whitsett JA. 1988. Two SP-C genes encoding human pulmonary surfactant proteolipid. *J. Biol. Chem.* 263:10326–31

13. Wood S, Yaremko ML, Schertzer M, Kelemen PR, Minna J, Westbrook CA. 1994. Mapping of the pulmonary surfactant SP5 (SFTP2) locus to 8p21 and characterization of a microsatellite repeat marker that shows frequent loss of heterozygosity in human carcinomas. *Genomics* 24:597–600

14. Glasser SW, Korfhagen TR, Weaver TE, Clark JC, Pilot-Matias T, et al. 1988. cDNA, deduced polypeptide structure and chromosomal assignment of human pulmonary surfactant proteolipid, SPL (pVal). *J. Biol. Chem.* 263:9–12

15. Warr RG, Hawgood S, Buckley DI, Crisp TM, Schilling J, et al. 1987. Low molecular weight human pulmonary surfactant protein (SP5): isolation, characterization, and cDNA and amino acid sequences. *Proc. Natl. Acad. Sci. USA* 84:7915–19

16. Vorbroker DK, Dey C, Weaver TE, Whitsett JA. 1992. Surfactant protein C precursor is palmitoylated and associates with subcellular membranes. *Biochim. Biophys. Acta* 1105:161–69

17. Solarin KO, Ballard PL, Guttentag SH, Lomax CA, Beers MF. 1997. Expression and glucocorticoid regulation of surfactant protein C in human fetal lung. *Pediatr. Res.* 42:356–64

18. Gonzales LW, Angampalli S, Guttentag SH, Beers MF, Feinstein SI, et al. 2001. Maintenance of differentiated function of the surfactant system in human fetal lung type II epithelial cells cultured on plastic. *Pediatr. Pathol. Mol. Med.* 20:387–412

19. Fisher JH, Shannon JM, Hofmann T, Mason RJ. 1989. Nucleotide and deduced amino acid sequence of the hydrophobic surfactant protein SP-C from rat: expression in alveolar type II cells and homology with SP-C from other species. *Biochim. Biophys. Acta* 995:225–30

20. Boggaram V, Margana RK. 1992. Rabbit surfactant protein C: cDNA cloning and regulation of alternatively spliced surfactant protein C mRNAs. *Am. J. Physiol. Lung Cell Mol. Physiol.* 263:L634–44

21. Johansson J, Jornvall H, Eklund A, Christensen N, Robertson B, Curstedt T. 1988. Hydrophobic 3.7 kDa surfactant polypeptide: structural characterization of the human and bovine forms. *FEBS Lett.* 232:61–64

22. Johansson J, Curstedt T, Robertson B, Jornvall H. 1988. Size and structure of the hydrophobic low molecular weight surfactant-associated polypeptide. *Biochemistry (Mosc)* 27:3544–47

23. Curstedt T, Jornvall H, Robertson B, Bergman T, Berggren P. 1987. Two hydrophobic low-molecular-mass protein fractions of pulmonary surfactant; characterization and biophysical activity. *Eur. J. Biochem.* 168:255–62

24. Phelps DS, Smith LM, Taeusch HW. 1987. Characterization and partial amino acid sequence of a low molecular weight surfactant protein. *Am. Rev. Resp. Dis.* 135:1112–17

25. Curstedt T, Johansson J, Persson P, Eklund A, Robertson B, et al. 1990.

Hydrophobic surfactant-associated polypeptides: SP-C is a lipopeptide with two palmitoylated cysteine residues, whereas SP-B lacks covalently linked fatty acyl groups. *Proc. Natl. Acad. Sci. USA* 87:2985–89

26. Stults JT, Griffin PR, Lesikar DD, Naidu A, Moffat B, Benson BJ. 1991. Lung surfactant protein SP-C from human, bovine, and canine sources contains palmityl cysteine thioester linkages. *Am. J. Physiol. Lung Cell Mol. Physiol.* 261:L118–25

27. Hafner D, Germann PG, Hauschke D, Kilian U. 1999. Effects of early treatment with rSP-C surfactant on oxygenation and histology in rats with acute lung injury. *Pulm. Pharmacol. Ther.* 12:193–201

28. Hafner D, Germann PG, Hauschke D. 1998. Comparison of rSP-C surfactant with natural and synthetic surfactants after late treatment in a rat model of the acute respiratory distress syndrome. *Br. J. Pharmacol.* 124:1083–90

29. Spragg RG, Lewis JF, Wurst W, Hafner D, Baughman RP, et al. 2003. Treatment of acute respiratory distress syndrome with recombinant surfactant protein C surfactant. *Am. J. Respir. Crit. Care Med.* 167:1562–66

30. Keough KM. 1998. Surfactant composition and extracellular transformations. See Ref. 4a, pp. 1–18

31. Weaver TE, Conkright JJ. 2001. Functions of surfactant proteins B and C. *Annu. Rev. Physiol.* 63:555–78

32. Griese M. 1999. Pulmonary surfactant in health and human lung diseases: state of the art. *Eur. Respir. J.* 13:1455–76

33. Notter RH. 2000. Functional composition and component biophysicis of endogenous lung surfactant. In *Lung Surfactants: Basic Science and Clinical Applications*, pp. 171–206. New York: Marcell-Dekker

34. Creuwels LAJM, Demel RA, van Golde LMG, Haagsman HP. 1995. Characterization of a dimeric canine form of surfactant protein C (SP-C). *Biochim. Biophys. Acta* 1254:326–32

35. Baatz JE, Smyth KL, Whitsett JA, Baxter C, Absolom DR. 1992. Structure and functions of a dimeric form of surfactant protein SP-C: a Fourier transform infrared and surfactometry study. *Chem. Phys. Lipids* 63:91–104

36. Li ZY, Suzuki Y, Kurozumi M, Shen HQ, Duan CX. 1998. Removal of a dimeric form of surfactant protein C from mouse lungs: its acceleration by reduction. *J. Appl. Physiol.* 84:471–78

37. Vandenbussche G, Clercx A, Curstedt T, Johansson J, Jornvall H. 1992. Structure and orientation of the surfactant-associated protein C in a lipid bilayer. *Eur. J. Biochem.* 203:201–9

38. Johansson J, Szyperski T, Curstedt T, Wuthrich K. 1994. The NMR structure of the pulmonary surfactant-associated polypeptide SP-C in an apolar solvent contains a valyl-rich alpha-Helix. *Biochemistry (Mosc)* 33:6015–23

39. Morrow MR, Taneva S, Simatos GA, Allwood LA, Keough KMW. 1993. ^{2}H NMR studies of the effect of pulmonary surfactant SP-C on the 1,2-dipalmitoyl-*sn*-glycero-3-phosphocholine headgroup: A model for transbilayer peptides in surfactant and biological membranes. *Biochemistry (Mosc)* 32:11338–44

40. Szyperski T, Vandenbussche G, Curstedt T, Ruysschaert JM, Wuthrich K, Johansson J. 1998. Pulmonary surfactant-associated polypeptide C in a mixed organic solvent transforms from a monomeric alpha-helical state into insoluble beta-sheet aggregates. *Protein Sci.* 7:2533–40

41. Kelly JW. 1996. Alternative conformations of amyloidogenic proteins govern their behavior. *Curr. Opin. Struct. Biol.* 6:11–17

42. Gustafsson M, Thyberg J, Naslund J, Eliasson E, Johansson J. 1999. Amyloid fibril formation by pulmonary surfactant protein C. *FEBS Lett.* 464:138–42

43. Gustafsson M, Griffiths WJ, Furusjo E, Johansson J. 2001. The palmitoyl groups of lung surfactant protein C reduce unfolding into a fibrillogenic intermediate. *J. Mol. Biol.* 310:937–50

44. Dluhy RA, Shanmukh S, Leapard JB, Kruger P, Baatz JE. 2003. Deacylated pulmonary surfactant protein SP-C transforms from alpha-helical to amyloid fibril structure via a pH-dependent mechanism: an infrared structural investigation. *Biophys. J.* 85:2417–29

45. Kallberg Y, Gustafsson M, Persson B, Thyberg J, Johansson J. 2001. Prediction of amyloid fibril-forming proteins. *J. Biol. Chem.* 276:12945–50

46. Phelps DS, Floros J. 1988. Localization of surfactant protein synthesis in human lung by in situ hybridization. *Am. Rev. Resp. Dis.* 137:939–42

47. Wohlford-Lenane CL, Durham PL, Snyder JM. 1992. Localization of surfactant-associated protein C (SP-C) mRNA in fetal rabbit lung tissue by in situ hybridization. *Am. J. Respir. Cell Mol. Biol.* 6:225–34

48. Phelps DS, Floros J. 1991. Localization of pulmonary surfactant proteins using immunohistochemistry and tissue in situ hybridization. *Exp. Lung Res.* 17:985–95

49. Phizackerly PJ, Town RMH, Newman GE. 1979. Hydrophobic proteins of lamellated osmiophillic bodies isolated from pig lung. *Biochem. J.* 183:731–36

50. Gail DB, Lenfant CJM. 1983. Cells of the lung: biology and clinical implications. *Am. Rev. Resp. Dis.* 127:366–87

51. Mason RJ, Williams MC, Greenleaf RD, Clements JA. 1977. Isolation and properties of type II cells from rat lung. *Am. Rev. Resp. Dis.* 115:1015–27

52. Beers MF. 1996. Inhibition of cellular processing of surfactant protein C by drugs affecting intracellular pH gradients. *J. Biol. Chem.* 271:14361–70

53. deMello DE, Heyman S, Phelps DS, Hamvas A, Nogee L, et al. 1994. Ultrastructure of lung in surfactant protein B deficiency.

Am. J. Respir. Cell Mol. Biol. 11:230–39

54. Beers MF, Hamvas A, Moxley MA, Gonzales LW, Guttentag SH, et al. 2000. Pulmonary surfactant metabolism in infants lacking surfactant protein B. *Am. J. Respir. Cell Mol. Biol.* 22:380–91

55. Cutz E, Wert SE, Nogee LM, Moore AM. 2000. Deficiency of lamellar bodies in alveolar type II cells associated with fatal respiratory disease in a full-term infant. *Am. J. Respir. Crit. Care Med.* 161:608–14

56. Tryka AF, Wert SE, Mazursky JE, Arrington RW, Nogee LM. 2000. Absence of lamellar bodies with accumulation of dense bodies characterizes a novel form of congenital surfactant defect. *Pediatr. Dev. Pathol.* 3:335–45

57. Dean M, Hamon Y, Chimini G. 2001. The human ATP-binding cassette (ABC) transporter superfamily. *J. Lipid Res.* 42:1007–17

58. Mulugeta S, Gray JM, Notarfrancesco KL, Gonzales LW, Koval M, et al. 2002. Identification of LBM180, a lamellar body limiting membrane protein of alveolar type II cells, as the ABC transporter protein ABCA3. *J. Biol. Chem.* 277:22147–55

59. Zen K, Notarfrancesco K, Oorschot V, Slot JW, Fisher AB, Shuman H. 1998. Generation and characterization of monoclonal antibodies to alveolar type II cell lamellar body membrane. *Am. J. Physiol. Lung Cell Mol. Physiol.* 272:L172–83

60. Shulenin S, Nogee LM, Annilo T, Wert SE, Whitsett JA, Dean M. 2004. ABCA3 Gene mutations in newborns with fatal surfactant deficiency. *N. Engl. J. Med.* 350:1296–303

61. Chander A, Johnson RG, Reicherter J, Fisher AB. 1986. Lung lamellar bodies maintain an acidic internal pH. *J. Biol. Chem.* 261:6126–31

62. Gibson KF, Widnell CC. 1991. The relationship between lamellar bodies and

lysosomes in type-II pneumocytes. *Am. J. Respir. Cell Mol. Biol.* 4:504–13

63. Wasano K, Hirakawa Y. 1994. Lamellar bodies of rat alveolar type II cells have late endosomal marker proteins on their limiting membranes. *Biochemistry* 102:329–35

64. Voorhout WF, Veenendaal T, Haagsman HP, Weaver TE, Whitsett JA, et al. 1992. Intracellular processing of pulmonary surfactant protein B in an endosomal/lysosomal compartment. *Am. J. Physiol. Lung Cell Mol. Physiol.* 263:L479–86

65. Weaver TE, Na CL, Stahlman M. 2002. Biogenesis of lamellar bodies, lysosome-related organelles involved in storage and secretion of pulmonary surfactant. *Semin. Cell Dev. Biol.* 13:263–70

66. Oosterlaken-Dijksterhuis MA, van Eijk M, van Buel BLM, van Golde LMG, Haagsman HP. 1991. Surfactant protein composition of lamellar bodies isolated from rat lung. *Biochem. J.* 274:115–19

67. Rooney SA. 1998. Regulation of surfactant secretion. See Ref. 4a, pp. 139–63

68. Osanai K, Mason RJ, Voelker DR. 1998. Trafficking of newly synthesized surfactant protein A in isolated rat alveolar type II cells. *Am. J. Respir. Cell Mol. Biol.* 19:929–35

69. Gobran LI, Rooney SA. 2001. Regulation of SP-B and SP-C secretion in rat type II cells in primary culture. *Am. J. Physiol. Lung Cell Mol. Physiol.* 281:L1413–19

70. Voorhout WF, Weaver TE, Haagsman HP, Geuze HJ, Van Golde LM. 1993. Biosynthetic routing of pulmonary surfactant proteins in alveolar type II cells. *Microsc. Res. Tech.* 26:366–73

71. Brasch F, Johnen G, Winn-Brasch A, Guttentag SH, Schmiedl A, et al. 2004. Surfactant protein B in type II pneumocytes and intra-alveolar surfactant forms of human lungs. *Am. J. Respir. Cell Mol. Biol.* 30:449–58

72. Brasch F, ten Brinke A, Johnen G, Ochs M, Kapp N, et al. 2002. Involvement of cathepsin H in the processing of the hy-drophobic surfactant-associated protein C in type II pneumocytes. *Am. J. Respir. Cell Mol. Biol.* 26:659–70

73. Johnson AL, Braidotti P, Pietra GG, Russo SJ, Kabore A, et al. 2001. Post-translational processing of surfactant protein-C proprotein—targeting motifs in the NH2-terminal flanking domain are cleaved in late compartments. *Am. J. Respir. Cell Mol. Biol.* 24:253–63

74. Baritussio A, Pettenazzo A, Benevento M, Alberti A, Gamba P. 1992. Surfactant protein C is recycled from the alveoli to the lamellar bodies. *Am. J. Physiol. Lung Cell Mol. Physiol.* 263:L607–11

75. Breslin JS, Weaver TE. 1992. Binding, uptake, and localization of surfactant protein B in isolated rat alveolar type II cells. *Am. J. Physiol. Lung Cell Mol. Physiol.* 262:L699–707

76. Ikegami M, Jobe AH. 1998. Surfactant protein metabolism in vivo. *Biochim. Biophys. Acta* 1408:218–25

77. Wissel H, Lehfeldt A, Klein P, Muller T, Stevens PA. 2001. Endocytosed SP-A and surfactant lipids are sorted to different organelles in rat type II pneumocytes. *Am. J. Physiol. Lung Cell Mol. Physiol.* 281:L345–60

78. Wissel H, Zastrow S, Richter E, Stevens PA. 2000. Internalized SP-A and lipid are differentially resecreted by type II pneumocytes. *Am. J. Physiol. Lung Cell Mol. Physiol.* 278:L580–90

79. Fisher AB. 1998. Lung surfactant clearance and cellular processing. See Ref. 4a, pp. 165–90

80. Beers MF, Kim CY, Dodia C, Fisher AB. 1994. Localization, synthesis, and processing of surfactant protein SP-C in rat lung analyzed by epitope-specific antipeptide antibodies. *J. Biol. Chem.* 269:20318–28

81. Beers MF, Lomax C. 1995. Synthesis and processing of hydrophobic surfactant protein C by isolated rat type II cells. *Am. J. Physiol. Lung Cell Mol. Physiol.* 269:L744–53

82. Beers MF, Lomax CA, Russo SJ. 1998. Synthetic processing of surfactant protein C by alveolar epithelial cells. The COOH terminus of proSP-C is required for post-translational targeting and proteolysis. *J. Biol. Chem.* 273:15287–93

83. Russo SJ, Wang W-J, Lomax C, Beers MF. 1999. Structural requirements for intracellular targeting of surfactant protein C. *Am. J. Physiol. Lung Cell Mol. Physiol.* 277:L1034–44

84. Keller A, Eistetter HR, Voss T, Schafer KP. 1991. The pulmonary surfactant protein C (SP-C) precursor is a type II transmembrane protein. *Biochem. J.* 277:493–99

85. Keller A, Steinhilber W, Schafer KP, Voss T. 1992. The C-terminal domain of the pulmonary surfactant protein C precursor contains signals for intracellular targeting. *Am. J. Respir. Cell Mol. Biol.* 6:601–8

86. Vorbroker DK, Voorhout WF, Weaver TE, Whitsett JA. 1995. Posttranslational processing of surfactant protein C in rat type II cells. *Am. J. Physiol. Lung Cell Mol. Physiol.* 269:L727–33

87. Mulugeta S, Beers MF. 2003. Processing of surfactant protein C requires a type II transmembrane topology directed by juxtamembrane positively charged residues. *J. Biol. Chem.* 278:47979–86

88. Conkright JJ, Bridges JP, Na CL, Voorhout WF, Trapnell B, et al. 2001. Secretion of surfactant protein C, an integral membrane protein, requires the N-terminal propeptide. *J. Biol. Chem.* 276:14658–64

89. Kabore AF, Wang WJ, Russo SJ, Beers MF. 2001. Biosynthesis of surfactant protein C: characterization of aggresome formation by EGFP chimeras containing propeptide mutants lacking conserved cysteine residues. *J. Cell Sci.* 114:293–302

90. ten Brinke A, Batenburg JJ, Haagsman HP, van Golde LMG, Vaandrager AB. 2002. Differential effect of brefeldin A on the palmitoylation of surfactant protein C proprotein mutants. *Biochem. Biophys. Res. Commun.* 290:532–38

91. Gustafsson M, Curstedt T, Jornvall H, Johansson J. 1997. Reverse-phase HPLC of the hydrophobic pulmonary surfactant proteins: detection of a surfactant protein C isoform containing *N*epsilon-palmitoyl-lysine. *Biochem. J.* 326:799–806

92. Vorbroker DK, Profitt SA, Nogee LM, Whitsett JA. 1995. Aberrant processing of surfactant protein C in hereditary SP-B deficiency. *Am. J. Physiol. Lung Cell Mol. Physiol.* 268:L647–56

93. Wei Y, Yarus S, Greenberg NM, Whitsett J, Rosen JM. 1995. Production of human surfactant protein C in milk of transgenic mice. *Transgenic Res.* 4:232–40

94. Brasch F, ten Brinke A, Johnen G, Ochs M, Kapp N, et al. 2002. Involvement of cathepsin H in the processing of the hydrophobic surfactant-associated protein C in type II pneumocytes. *Am. J. Respir. Cell Mol. Biol.* 26:659–70

95. Chuman Y, Bergman AC, Ueno T, Saito S, Sakaguchi K, et al. 1999. Napsin A, a member of the aspartic protease family, is abundantly expressed in normal lung and kidney tissue and is expressed in lung adenocarcinomas. *FEBS Lett.* 462:129–34

96. Schauer-Vukasinovic V, Wright MB, Breu V, Giller T. 2000. Cloning, expression and functional characterization of rat napsin. *Biochim. Biophys. Acta* 1492:207–10

97. Schauer-Vukasinovic V, Bur D, Kling D, Gruninger F, Giller T. 1999. Human napsin A: expression, immunochemical detection, and tissue localization. *FEBS Lett.* 462:135–39

98. Brasch F, Johnen G, Schauer-Vukasinovic V, ten Brinke A, Kapp N, et al. 2001. Napsin A is involved in the processing of proSP-C. *Am. J. Respir. Crit. Care Med.* 163:A560 (Abstr.)

99. Ueno T, Linder S, Na CL, Rice WR, Johansson J, Weaver TE. 2004. Processing of pulmonary surfactant protein B by

napsin and cathepsin H. *J. Biol. Chem.* 279:16178–84

100. Orci L, Ravazzola M, Amherdt M, Madsen O, Perrelet A, et al. 1986. Conversion of proinsulin to insulin occurs coordinately with acidification of maturing secretory vesicles. *J. Cell Biol.* 103:2273–81

101. Chander A. 1992. Dicyclohexylcardiodimide and vanadate sensitive AtPase of lung lamellar bodies. *Biochim. Biophys. Acta* 1123:198–206

102. Ballard PL, Nogee LM, Beers MF, Ballard RA, Planer BC, et al. 1995. Partial deficiency of surfactant protein B in an infant with chronic lung disease. *Pediatrics* 96:1046–52

103. Nogee LM, de Mello DE, Dehner LP, Colten HR. 1993. Brief report: deficiency of pulmonary surfactant protein B in congenital alveolar proteinosis. *N. Engl. J. Med.* 328:406–10

104. Clark JC, Wert SE, Bachurski CJ, Stahlman MT, Stripp BR, et al. 1995. Targeted disruption of the surfactant protein B gene disrupts surfactant homeostasis, causing respiratory failure in newborn mice. *Proc. Natl. Acad. Sci. USA* 92:7794–98

105. Melton KR, Nesslein LL, Ikegami M, Tichelaar JW, Clark JC, et al. 2003. SP-B deficiency causes respiratory failure in adult mice. *Am. J. Physiol. Lung Cell Mol. Physiol.* 285:L543–49

106. Wang WJ, Russo SJ, Mulugeta S, Beers MF. 2002. Biosynthesis of surfactant protein C (SP-C). Sorting of SP-C proprotein involves homomeric association via a signal anchor domain. *J. Biol. Chem.* 277:19929–37

107. Brasch F, Griese M, Tredano M, Johnen G, Ochs M, et al. 2004. Interstitial lung disease and pulmonary alveolar proteinosis in a full-term baby with a de novo heterozygous mutation of surfactant protein C gene. *Eur. Respir. J.* 24:30–39

108. Stevens PA, Pettenazzo A, Baritussio A, Mulugeta S, Brasch F, et al. 2004. Non-specific interstitial pneumonitis, alveolar proteinosis, and abnormal proprotein trafficking in a patient with a novel spontaneous mutation in surfactant protein C gene. *Pediatr. Res.* In press

109. Sanchez-Pulido L, Devos D, Valencia A. 2002. BRICHOS: a conserved domain in proteins associated with dementia, respiratory distress and cancer. *Trends Biochem. Sci.* 27:329–32

110. Vidal R, Calero M, Revesz T, Plant G, Ghiso J, Frangione B. 2001. Sequence, genomic structure and tissue expression of human BRI3, a member of the BRI gene family. *Gene* 266:95–102

111. Vidal R, Revesz T, Rostagno A, Kim E, Holton JL, et al. 2000. A decamer duplication in the 3′ region of the BRI gene originates an amyloid peptide that is associated with dementia in a Danish kindred. *Proc. Natl. Acad. Sci. USA* 97:4920–25

112. Vidal R, Frangione B, Rostagno A, Mead S, Revesz T, Plant G, Ghiso J. 1999. A stop-codon mutation in the BRI gene associated with familial British dementia. *Nature* 399:776–81

113. Ghiso J, Vidal R, Rostagno A, Miravale L, Holton JL, et al. 2000. Amyloidogenesis in familial British dementia is associated with a genetic defect on chromosome 13. *Ann. NY Acad. Sci.* 920:84–92

114. Johansson J, Weaver TE, Tjernberg LO. 2004. Proteolytic generation and aggregation of peptides from transmembrane regions: lung surfactant protein C and amyloid beta-peptide. *Cell. Mol. Life Sci.* 61:326–35

115. Selkoe DJ. 2002. Deciphering the genesis and fate of amyloid β-protein yields novel therapies for Alzheimer disease. *J. Clin. Invest.* 110:1375–81

116. Ghiso JA, Holton J, Miravalle L, Calero M, Lashley T, et al. 2001. Systemic amyloid deposits in familial British dementia. *J. Biol Chem.* 276:43909–14

117. Sanchez I, Mahlke C, Yuan JY. 2003. Pivotal role of oligomerization in expanded

polyglutamine neurodegenerative disorders. *Nature* 421:373–79

118. Chai YH, Koppenhafer SL, Shoesmith SJ, Perez MK, Paulson HL. 1999. Evidence for proteasome involvement in polyglutamine disease: localization to nuclear inclusions in SCA3/MJD and suppression of polyglutamine aggregation in vitro. *Hum. Mol. Genet.* 8:673–82

119. Naef R, Suter U. 1999. Impaired intracellular trafficking is a common disease mechanism of PMP22 point mutations in peripheral neuropathies. *Neurobiol. Dis.* 6:1–14

120. Wang WJ, Mulugeta S, Russo SJ, Beers MF. 2003. Deletion of exon 4 from human surfactant protein C results in aggresome formation and generation of a dominant negative. *J. Cell. Sci.* 116:683–92

121. Kopito RR. 2000. Aggresomes, inclusion bodies and protein aggregation. *Trends Cell Biol.* 10:524–30

122. Kopito RR, Sitia R. 2000. Aggresomes and Russell bodies—symptoms of cellular indigestion? *EMBO Rep.* 1:225–31

123. Johnston JA, Ward CL, Kopito RR. 1998. Aggresomes: a cellular response to misfolded proteins. *J. Cell Biol.* 143:1883–98

124. Bridges JP, Wert SE, Nogee LM, Weaver TE. 2003. Expression of a human SP-C mutation associated with interstitial lung disease disrupts lung development in transgenic mice. *J. Biol. Chem.* 278:52739–46

125. Conkright JJ, Na CL, Weaver TE. 2002. Overexpression of surfactant protein-C mature peptide causes neonatal lethality in transgenic mice. *Am. J. Respir. Cell Mol. Biol.* 26:85–90

126. Glasser SW, Burhans MS, Korfhagen TR, Na CL, Sly PD, et al. 2001. Altered stability of pulmonary surfactant in SP-C-deficient mice. *Proc. Natl. Acad. Sci. USA* 98:6366–71

127. Nakatani Y, Nakamura N, Sano J, Inayama Y, Kawano N, et al. 2000. Interstitial pneumonia in Hermansky-Pudlak syndrome: significance of florid foamy swelling/degeneration (giant lamellar body degeneration) of type-2 pneumocytes. *Virchows Arch. Int. J. Pathol.* 437:304–13

128. McGarry MP, Borchers M, Novak EK, Lee NA, Ohtake PJ, et al. 2002. Pulmonary pathologies in pallid mice result from nonhematopoietic defects. *Exp. Mol. Pathol.* 72:213–20

129. Shotelersuk V, Dell'Angelica EC, Hartnell L, Bonifacino JS, Gahl WA. 2000. A new variant of Hermansky-Pudlak syndrome due to mutations in a gene responsible for vesicle formation. *Am. J. Med.* 108:423–27

130. Swank RT, Novak EK, McGarry MP, Zhang YK, Li W, et al. 2000. Abnormal vesicular trafficking in mouse models of Hermansky-Pudlak syndrome. *Pigment Cell Res.* 13:59–67

131. Swank RT, Novak EK, McGarry MP, Rusiniak ME, Feng LJ. 1998. Mouse models of Hermansky Pudlak Syndrome: a review. *Pigment Cell Res.* 11:60–80

132. Yoneda T, Urano F, Ron D. 2002. Transmission of proteotoxicity across cellular compartments. *Genes Dev.* 16:1307–13

133. Glasser SW, Detmer EA, Ikegami M, Na CL, Stahlman MT, Whitsett JA. 2003. Pneumonitis and emphysema in SP-C gene targeted mice. *J. Biol. Chem.* 278:14291–98

134. Augusto L, Le Blay K, Auger G, Blanot D, Chaby R. 2001. Interaction of bacterial lipopolysaccharide with mouse surfactant protein C inserted into lipid vesicles. *Am. J. Physiol. Lung Cell Mol. Physiol.* 281:L776–85

135. Augusto LA, Li J, Synguelakis M, Johansson J, Chaby R. 2002. Structural basis for interactions between lung surfactant protein C and bacterial lipopolysaccharide. *J. Biol. Chem.* 277:23484–92

136. Augusto LA, Synguelakis M, Espinassous Q, Lepoivre M, Johansson A, Chaby R. 2003. Cellular antiendotoxin activities of lung surfactant protein C in lipid vesicles.

Am. J. Respir. Crit. Care Med. 168:335–41

137. Augusto LA, Synguelakis M, Johansson J, Pedron T, Girard R, Chaby R. 2003. Interaction of pulmonary surfactant protein C with CD14 and lipopolysaccharide. *Infect. Immun.* 71:61–67

138. Chen F, Yang DS, Petanceska S, Yang A, Tandon A, et al. 2000. Carboxyl-terminal fragments of Alzheimer beta-amyloid precursor protein accumulate in restricted and unpredicted intracellular compartments in presenilin 1-deficient cells. *J. Biol. Chem.* 275:36794–802

139. Kelly JW. 1998. The environmental dependency of protein folding best explains prion and amyloid diseases. *Proc. Natl. Acad. Sci. USA* 95:930–32

140. Kaufman RJ. 2002. Orchestrating the unfolded protein response in health and disease. *J. Clin. Invest.* 110:1389–98

141. Sitia R, Braakman I. 2003. Quality control in the endoplasmic reticulum protein factory. *Nature* 426:891–94

142. Hampton RY. 2002. ER-associated degradation in protein quality control and cellular regulation. *Curr. Opin. Cell Biol.* 14:476–82

143. Travers KJ, Patil CK, Wodicka LF, Lockhart DJ, Weissman JS, Walter P. 2000. Functional and genomic analyses reveal an essential coordination between the unfolded protein response and ER-associated degradation. *Cell* 101:249–58

144. Patil C, Walter P. 2001. Intracellular signaling from the endoplasmic reticulum to the nucleus: the unfolded protein response in yeast and mammals. *Curr. Opin. Cell Biol.* 13:349–55

145. Nishitoh H, Matsuzawa A, Tobiume K, Saegusa K, Takeda K, et al. 2002. ASK1 is essential for endoplasmic reticulum stress-induced neuronal cell death triggered by expanded polyglutamine repeats. *Genes Dev.* 16:1345–55

146. Oyadomari S, Araki E, Mori M. 2002. Endoplasmic reticulum stress-mediated apoptosis in pancreatic beta-cells. *Apoptosis* 7:335–45

147. Oyadomari S, Takeda K, Takiguchi M, Gotoh T, Matsumoto M, et al. 2001. Nitric oxide-induced apoptosis in pancreatic beta cells is mediated by the endoplasmic reticulum stress pathway. *Proc. Natl. Acad. Sci. USA* 98:10845–50

148. Oyadomari S, Koizumi A, Takeda K, Gotoh T, Akira S, et al. 2002. Targeted disruption of the Chop gene delays endoplasmic reticulum stress-mediated diabetes. *J. Clin. Invest.* 109:525–32

149. Sanchez I, Xu CJ, Juo P, Kakizaka A, Blenis J, Yuan JY. 1999. Caspase-8 is required for cell death induced by expanded polyglutamine repeats. *Neuron* 22:623–33

150. Knorre A, Wagner M, Schaefer HE, Colledge WH, Pahl HL. 2002. DeltaF508-CFTR causes constitutive NF-kappaB activation through an ER-overload response in cystic fibrosis lungs. *Biol. Chem.* 383:271–82

151. Beers MF, Russo SJ, Ngyuen V, Mulugeta S. 2004. Mutations within the BRICHOS domain of SP-C proprotein induce aggregation, caspase 3 expression, and cell toxicity. *Am. J. Respir. Crit. Care Med.* 169: A740 (Abstr.)

152. Cobbold C, Monaco AP, Sivaprasadarao A, Ponnambalam S. 2003. Aberrant trafficking of transmembrane proteins in human disease. *Trends Cell Biol.* 13:639–47

153. Aridor M, Hannan LA. 2000. Traffic jam: a compendium of human diseases that affect intracellular transport processes. *Traffic* 1:836–51

154. Gelman MS, Kannegaard ES, Kopito RR. 2002. A principal role for the proteasome in endoplasmic reticulum-associated degradation of misfolded intracellular cystic fibrosis transmembrane conductance regulator. *J. Biol. Chem.* 277:11709–14

155. Garcia-Mata R, Bebok Z, Sorscher EJ, Sztul ES. 1999. Characterization and dynamics of aggresome formation by

a cytosolic GFP-chimera. *J. Cell Biol.* 146:1239–54

156. Kelly JW. 1997. Amyloid fibril formation and protein misassembly: a structural quest for insights into amyloid and prion diseases. *Structure* 5:595–600

157. Prusiner SB, Dearmond SJ. 1994. Prion diseases and neurodegeneration. *Annu. Rev. Neurosci.* 17:311–39

158. Holton JL, Ghiso J, Lashley T, Rostagno A, Guerin CJ, et al. 2001. Regional distribution of amyloid-Bri deposition and its association with neurofibrillary degeneration in familial British dementia. *Am. J. Pathol.* 158:515–26

159. Bence NF, Sampat RM, Kopito RR. 2001. Impairment of the ubiquitin-proteasome system by protein aggregation. *Science* 292:1552–55

160. Kelly JW. 2000. Mechanisms of amyloidogenesis. *Nat. Struct. Biol.* 7:824–26

161. Kelly JW. 2002. Towards an understanding of amyloidogenesis. *Nat. Struct. Biol.* 9:323–25

162. Pahl HL, Baeurele PA. 1995. A novel signal-transduction pathway from the endoplasmic reticulum to the nucleus is mediated by transcription factor NF-kappa B. *EMBO J.* 14:2580–88

163. Pahl HL. 1999. Signal transduction from the endoplasmic reticulum to the cell nucleus. *Physiol. Rev.* 79:683–701

164. Pahl HL, Baeuerle PA. 1997. Endoplasmic-reticulum-induced signal transduction and gene expression. *Trends Cell Biol.* 7:50–55

165. Pahl HL, Baeuerle PA. 1997. The ER-overload response: activation of NF-κB. *Trends Biochem. Sci.* 22:63–67

166. Rubenstein RC, Lyons BM. 2001. Sodium 4-phenylbutyrate downregulates HSC70 expression by facilitating mRNA degradation. *Am. J. Physiol. Lung Cell Mol. Physiol.* 281:L43–51

167. Rubenstein RC, Zeitlin PL. 2000. Sodium 4-phenylbutyrate downregulates Hsc70: implications for intracellular trafficking of DeltaF508-CFTR. *Am. J. Physiol. Cell Physiol.* 278:C259–67

168. Rubenstein RC, Egan ME, Zeitlin PL. 1997. In vitro pharmacologic restoration of CFTR-mediated chloride transport with sodium 4-phenylbutyrate in cystic fibrosis epithelial cells containing delta F508-CFTR. *J. Clin. Invest.* 100:2457–65

169. Burrows JA, Willis LK, Perlmutter DH. 2000. Chemical chaperones mediate increased secretion of mutant alpha 1-antitrypsin (alpha 1-AT) Z: a potential pharmacological strategy for prevention of liver injury and emphysema in alpha 1-AT deficiency. *Proc. Natl. Acad. Sci. USA* 97:1796–801

170. Rubenstein RC, Zeitlin PL. 1998. A pilot clinical trial of oral sodium 4-phenylbutyrate (Buphenyl) in deltaF508-homozygous cystic fibrosis patients: partial restoration of nasal epithelial CFTR function. *Am. J. Respir. Crit. Care Med.* 157:484–90

171. Brown CR, Hong-Brown LQ, Biwersi J, Verkman AS, Welch WJ. 1996. Chemical chaperones correct the mutant phenotype of the delta F508 cystic fibrosis transmembrane conductance regulator protein. *Cell Stress Chaperones* 1:117–25

172. Sato S, Ward CL, Krouse ME, Wine JJ, Kopito RR. 1996. Glycerol reverses the misfolding phenotype of the most common cystic fibrosis mutation. *J. Biol. Chem.* 271:635–38

173. Egan ME, Pearson M, Weiner SA, Rajendran V, Rubin D, et al. 2004. Curcumin, a major constituent of turmeric, corrects cystic fibrosis defects. *Science* 304:600–2

Cl⁻ CHANNELS: A Journey from Ca²⁺ Sensors to ATPases and Secondary Active Ion Transporters

Michael Pusch

Istituto di Biofisica, Consiglio Nazionale delle Ricerche
16149 Genova, Italy; email: pusch@ge.ibf.cnr.it

Cl⁻ ion channels play an increasingly recognized role in many physiological processes. Molecularly, three different classes of Cl⁻-selective ion channels have been unequivocally identified: postsynaptic GABA and glycine receptors, ClC-type Cl⁻ channels, and the cystic fibrosis transmembrane conductance regulator (CFTR). In addition, several Cl⁻ currents for which no molecular correlate has yet been identified can be measured in native cell preparations. These include Ca^{2+}-regulated Cl⁻ channels [$I_{Cl}(Ca^{2+})$] and volume-regulated (osmosensitive) Cl⁻ channels [$I_{Cl}(Vol)$]. The five reviews in this issue's special topic cover several important aspects of recent developments, without claiming to be comprehensive for the whole Cl⁻ channel field. For example GABA/glycine receptors, which mediate inhibitory postsynaptic responses, and $I_{Cl}(Vol)$, which is important for volume regulation of all cell types, are not included, despite the fact that they are of extraordinary physiological importance. An excellent review of $I_{Cl}(Vol)$ has been recently published (1). In contrast to other Cl⁻ channels, GABA and glycine receptors are traditionally well covered in the neuroscience literature and thus are not included here.

Three of the reviews are devoted to ClC proteins, the largest Cl⁻ channel family with nine distinct genes in humans. Jentsch et al. (2) provide a broad overview of the molecular cloning, the physiological functions, and the involvement of ClC proteins in human genetic diseases. Plasma membrane ClC channels are important for regulation of the membrane potential and/or Cl⁻ transport. One unexpected function of several ClC proteins is their involvement in the ion homeostasis of intracellular organelles such as endosomes, lysosomes, and synaptic vesicles. Here they are believed to be important for lumenal acidification. Several ClC-related genetic diseases are caused by a dysfunction of these still poorly characterized mechanisms. Chloride transport is of particular importance in the kidney epithelia, and Uchida & Sasaki (3) discuss in detail the role of Cl⁻ ion channel proteins (among them several members of the ClC family) in the nephron. A breakthrough for the mechanistic understanding of ClC proteins was achieved by MacKinnon et al. with the resolution of the three-dimensional structure at 2.5 Å of bacterial ClC homologues (4, 5). The structure fully confirmed the anticipated dimeric two-pore architecture of ClC proteins and opened the way for a rational structure-function analysis. Chen describes the peculiar structural features of ClC proteins

697

and reviews—in light of the structure—the extensive work on the mechanisms of muscle-type ClC channel gating and permeation (6). These mechanisms are vastly different from those of cation channels and are still poorly understood. A surprising recent finding was that the crystallized bacterial ClC homologues are actually not ion channels but secondary active H^+/Cl^- antiporters with a 1:2 stoichiometry (7). This unexpected result was not even predicted with the available crystal structure. The likely relevance of the H^+/Cl^- transporter function for mammalian ClC physiology remains to be investigated.

If ClC proteins appear to be complicated, this is nothing compared with the complexity of the regulation of the CFTR. CFTR is a chloride channel, and mutations in the gene coding for it are responsible for cystic fibrosis (8). CFTR is a member of the ABC transporter family, a broad class of primary active, ATP-consuming pumps (9). CFTR seems to be the only channel-member of the ABC family. Its activity is heavily regulated by phosphorylation/dephosphorylation by protein kinase A (PKA) and protein kinase C (PKC) at a multitude of sites and, additionally, by ATP binding and hydrolysis at its two nucleotide-binding domains. Furthermore, CFTR interacts with several cellular proteins. Despite the fact that CFTR has been extensively studied for 15 years, the precise mechanisms for how its dysfunction leads to cystic fibrosis are not entirely clear. From a therapeutic perspective, CFTR protein synthesis and maturation are of particular importance because the most common disease-causing mutation leads to a reduced functional expression in the plasma membrane. The state of the art and the recent developments in the CFTR field are comprehensively covered by Riordan (10).

Ca^{2+}-activated Cl^- play diverse roles, e.g., in olfaction and smooth muscle contraction, mostly through a depolarization of the plasma membrane stimulated by an increase of intracellular Ca^{2+}. No clear molecular correlates of these Ca^{2+}-sensing proteins have been discovered, but the situation may have changed with the relatively recent identification of bestrophins as candidates for these channels (11, 12). However, there is likely more than one class of Ca^{2+}-activated Cl^- channel, and the field is open for surprises. Hartzell et al. (13) provide a comprehensive review of these important ion channels.

Our understanding of Cl^- channels is less advanced than that of cationic channels even though there has been significant progress in the past few years. However, Cl^- channels (and transporters) are still far from being as well-understood as are K^+ channels, and scientists from many disciplines will find in them a fertile ground for further work.

LITERATURE CITED

1. Sardini A, Amey JS, Weylandt KH, Nobles M, Valverde MA, Higgins CF. 2003. Cell volume regulation and swelling-activated chloride channels. *Biochim. Biophys. Acta* 1618:153–62

2. Jentsch TJ, Poët M, Fuhrmann JC, Zdebik AA. 2005. Physiological functions of ClC Cl⁻ channels gleaned from human genetic disease and mouse models. *Annu. Rev. Physiol.* 67:779–807

3. Uchida S, Sasaki S. 2005. Function of chloride channels in the kidney. *Annu. Rev. Physiol.* 67:759–78

4. Dutzler R, Campbell EB, Cadene M, Chait BT, MacKinnon R. 2002. X-ray structure of a ClC chloride channel at 3.0 Å reveals the molecular basis of anion selectivity. *Nature* 415:287–94

5. Dutzler R, Campbell EB, MacKinnon R. 2003. Gating the selectivity filter in ClC chloride channels. *Science* 300:108–12

6. Chen T-Y. 2005. Structures and functions of ClC channels. *Annu. Rev. Physiol.* 67:809–39

7. Accardi A, Miller C. 2004. Secondary active transport mediated by a prokaryotic homologue of ClC Cl⁻ channels. *Nature* 427:803–7

8. Riordan JR, Rommens JM, Kerem B, Alon N, Rozmahel R, et al. 1989. Identification of the cystic fibrosis gene: cloning and characterization of complementary DNA. *Science* 245:1066–73

9. Schmitt L, Tampe R. 2002. Structure and mechanism of ABC transporters. *Curr. Opin. Struct. Biol.* 12:754–60

10. Riordan JR. 2005. Assembly of functional CFTR chloride channels. *Annu. Rev. Physiol.* 67:701–18

11. Sun H, Tsunenari T, Yau KW, Nathans J. 2002. The vitelliform macular dystrophy protein defines a new family of chloride channels. *Proc. Natl. Acad. Sci. USA* 99:4008–13

12. Qu Z, Fischmeister R, Hartzell C. 2004. Mouse bestrophin-2 is a bona fide Cl⁻ channel: identification of a residue important in anion binding and conduction. *J. Gen. Physiol.* 123:327–40

13. Hartzell HC, Putzier I, Arreola J. 2005. Calcium-activated chloride channels. *Annu. Rev. Physiol.* 67:719–58

Annu. Rev. Physiol. 2005. 67:701–18
doi: 10.1146/annurev.physiol.67.032003.154107
First published online as a Review in Advance on November 1, 2004

ASSEMBLY OF FUNCTIONAL CFTR CHLORIDE CHANNELS

John R. Riordan

Mayo Clinic College of Medicine, Scottsdale, Arizona, 85259;
email: riordan.john@mayo.edu

Key Words ligand-gated, hydrolyzable-ligand, monomeric channel, CFTR
domains, CFTR activation

■ **Abstract** The assembly of the cystic fibrosis transmembrane regulator (CFTR)
chloride channel is of interest from the broad perspective of understanding how ion
channels and ABC transporters are formed as well as dealing with the mis-assembly
of CFTR in cystic fibrosis. CFTR is functionally distinct from other ABC transporters
because it permits bidirectional permeation of anions rather than vectorial transport of
solutes. This adaptation of the ABC transporter structure can be rationalized by considering CFTR as a hydrolyzable-ligand-gated channel with cytoplasmic ATP as ligand.
Channel gating is initiated by ligand binding when the protein is also phosphorylated
by protein kinase A and made reversible by ligand hydrolysis. The two nucleotide-
binding sites play different roles in channel activation. CFTR self-associates, possibly
as a function of its activation, but most evidence, including the low-resolution three-
dimensional structure, indicates that the channel is monomeric. Domain assembly and
interaction within the monomer is critical in maturation, stability, and function of the
protein. Disease-associated mutations, including the most common, $\Delta F508$, interfere
with domain folding and association, which occur both co- and post-translationally.
Intermolecular interactions of mature CFTR have been detected primarily with the
N- and C-terminal tails, and these interactions have some impact not only on channel function but also on localization and processing within the cell. The biosynthetic
processing of the nascent polypeptide leading to channel assembly involves transient
interactions with numerous chaperones and enzymes on both sides of the endoplasmic
reticulum membrane.

INTRODUCTION

The cystic fibrosis transmembrane regulator (CFTR), the product of the gene mutated in patients with cystic fibrosis, has adapted the ABC transporter structural
motif to form a tightly regulated anion channel at the apical surface of many epithelia. Use of the term assembly of a functional ion channel implies the coming
together of subunits or at least smaller not-yet functional components of the active whole. In fact, on the basis of current knowledge only the CFTR polypeptide

0066-4278/05/0315-0701$14.00

701

itself is required to form an ATP- and protein kinase A-dependent low-conductance chloride channel of the type present in the apical membrane of many epithelial cells (1). The number of CFTR polypeptides that form a functional channel, however, is the subject of some debate and is considered in detail in the second section of this review after an up-to-date summary of what is known of the mechanism of action is given. The lack of high-resolution three-dimensional structural information means that a clear picture of neither the quaternary structure nor the pore is yet available. However, there have been advances toward understanding the assembly of domains within a single CFTR molecule and these are discussed in the third section. Finally, the interactions of CFTR with other proteins and their relevance to its function are considered in the fourth section.

Hypothetical Model of CFTR Channel Action

The facts that CFTR is atypical as an ABC transporter that functions as an ion channel and as an ion channel that hydrolyzes ATP, combined with its relative intractability for biochemical and structural studies, have hampered the determination of its mechanism of action. Nevertheless, a general outline of the process can be pieced together from the extensive electrophysiological data (2), a more limited set of measurements of nucleotide binding and hydrolysis (3), of phosphorylations by protein kinases A and C (PKC, PKA) and their consequences (4–6), the high-resolution structure of nucleotide-binding domain 1 (NBD1) of murine CFTR (7), and a low-resolution three-dimensional structure of the entire human protein (8). It was evident from the CFTR sequence that it was an ABC protein and therefore that it was likely regulated by ATP interactions at the NBDs (9). The presence of the unique R-domain, rich in consensus sites for phosphorylation by PKA (and PKC), also suggested that some control by phosphorylation was likely. These two levels of regulation of the CFTR channel and their integration have been extensively studied (10, 11). Under normal circumstances, phosphorylation by PKA is obligatory for activation of gating by ATP binding. The molecular mechanism of the over-riding control by phosphorylation and dephosphorylation is unclear, but the increased negative charge on the introduction of multiple phosphoryl groups is involved, as the effect can be at least partially mimicked by the introduction of carboxylate residues at these sites (12). Conformational changes at the level of the R-domain and the whole protein have been observed on PKA phosphorylation (13, 14). However, the R-domain itself is not highly structured when either phosphorylated or dephosphorylated (13, 15). On the basis of making the channel more sensitive to activation by ATP, PKA phosphorylation was proposed to enhance ATP binding or hydrolysis (11, 16). However, there is minimal evidence to support this notion at the biochemical level (16). Alternatively R-domain phosphorylation may be permissive for transmission of the conformational perturbation caused by ATP binding at the NBDs to the channel pore, but direct evidence of this has not yet been forthcoming. It should also be mentioned that nonphosphorylation-dependent activation of CFTR by ATP has also been observed in several instances, but it is not yet clear if this is physiologically relevant (17–20). Nonetheless, these observations should help elucidate the overall mechanism of gating regulation.

The relationship between ATPase activity at the NBDs of CFTR and channel gating has been interpreted in different ways. ATP hydrolysis is generally believed to be energetically coupled to the transport of solutes by other transport ATPases, but ion permeation through channels is driven only by the electrochemical gradients of the ions. Some observations of single-channel behavior were interpreted as indicating that gating probably was not at thermodynamic equilibrium and hence might be dependent of the energy of ATP hydrolysis (21). Some authors suggested there might be a direct coupling between hydrolytic events and gating transitions such that single-channel measurements would provide a direct read-out of hydrolysis (22, 23). It was also proposed that hydrolysis at one NBD might drive channel opening and that of the other channel closing (24). However, a thermodynamic analysis of a large number of opening and closing events (25) indicated that gating was at thermal equilibrium with a large E_a for opening ($\sim$100 kJ/mol) but a much lower value, similar to that of diffusion in water, for closing ($\sim$10 kJ/mol), which indicated that hydrolysis was not energetically linked to closing. Hydrolysis was found not to be essential for opening either because opening could occur through the binding of nonhydrolyzable ATP analogues or ATP in the absence of a divalent cation (26), which is absolutely essential for hydrolysis (27). Thus the large E_a for channel opening probably reflects the entropy decrease on nucleotide binding.

In this view, CFTR is similar to other ligand-gated channels in which ligand binding structurally perturbs the closed state to initiate thermally driven transitions between closed and open states. An analogy would be with cyclic nucleotide (cN) -gated channels whose ligands are cytoplasmic cyclic nucleotides (28) rather than a nucleoside triphosphate (ATP). The changes in cN concentration to which cN-gated channels respond are determined by the relative activities of cyclases and phophodiesterases, which form and degrade them, respectively (28). cN binding and dissociation and channel gating follow these concentration changes. In contrast, cellular ATP concentrations remain within a relatively narrow range under normal conditions and, hence, whereas its binding could serve to initiate channel gating, similar to ligand binding to other ligand-gated channels, a mechanism for termination of its presence at the binding site would be required to enable relaxation of the bound state. Hydrolysis of the ATP ligand by CFTR could accomplish this. In this model, CFTR would be best described as a hydrolyzable-ligand-gated channel (29). How the ABC transporter structural architecture provides for such a hydrolyzable-ligand-gated channel is not yet fully understood but can be generally rationalized by recent advances in ABC protein enzymology (30). Most significant was the proposal (31, 32) supported by substantial evidence (33–35) that rather than each forming an individual site of ATP binding and hydrolysis, the two NBDs jointly contribute ligands to the two sites (36, 37). Using the terminology of Abele & Tampé (38), site I is constituted primarily by the Walker A motif of NBD1 and the signature sequence of NBD2, and site II is constituted by the Walker A motif of NBD2 and the signature sequence of NBD1. These sites are believed to be the two sites in ABC transporters that possess identical or similar NBDs, which sequentially and alternately hydrolyze ATP during the transport cycle as originally proposed and supported experimentally with the P-glycoprotein multidrug transporter by Senior and colleagues (39). In the case of the two bacterial ABC transporters for which high-resolution three-dimensional structures have been obtained, the N- and C-terminal transmembrane domains (TMDs) associate in the membrane to form the permeation pathway at the interface, and extensive contacts are formed between the cytoplasmic loops separating helices and the NBD in the same half of the protein (40, 41). These contacts may be involved in propagating conformational movement from an NBD to the corresponding TMD. A molecular dynamics simulation of one of these transporters, BtuCD, suggests that ATP binding is the power stroke in its mechanism (30), in which case hydrolysis would essentially terminate each transport cycle allowing initiation of the next, loosely analogous to what we proposed for CFTR as a hydrolyzable-ligand-gated channel (29).

CFTR is distinctly different from ABC transporters in which the N- and C-terminal halves are identical or similar (Figure 1, see color insert). Nucleotide binding site I (NBD1 Walker A and NBD2 signature) is a stable ATP-binding site at which hydrolysis does not occur, whereas site II is hydrolytic (Figure 2) (27). A working hypothesis incorporating the possible advantage of this asymmetry to a ligand-gated channel supposes that the stably bound ATP at site I maintains a portion (TMD1) of the membrane-associated channel-forming domains in a "primed" conformation that would more easily respond to the conformational impact of ATP binding at site II where the ligand is quickly hydrolyzed to facilitate inactivation (Figure 3, see color insert). In this way, the stably-bound site I would provide the channel with a "hair-trigger" that could be rapidly pulled and released by events at site II alone. If the alternating binding and hydrolysis cycle at both sites of symmetric transporters does serve to pull the NBDs together, which in turn cause rearrangements in the TMDs, the stably occupied site I of CFTR may keep the NBDs permanently closer together. Thus at only a small further apposition would occur by binding to site II, which might be sufficient to induce the minor intramembranous shift involved in channel gating. This would be consistent with the results of a molecular dynamics simulation of BtuCD action, which showed that binding of a single MgATP altered the permeability pathway (30).

Because neither the static nucleotide interaction at site I nor the dynamic one at site II appears to be strongly influenced by PKA phosphorylation, it is presumed that R-domain phosphorylation may be essential for transmission of the power stroke to the pore. However, other unanticipated roles of the R-domain can also be imagined. For example, a regulatory domain exists C terminal of the NBDs of MalK, which may contribute to the stability of the dimer that they form (37).

Quaternary Structure of the CFTR Channel

Generation of a low-conductance chloride channel dependent on ATP and PKA by reconstitution of purified CFTR provided evidence that the CFTR polypeptide

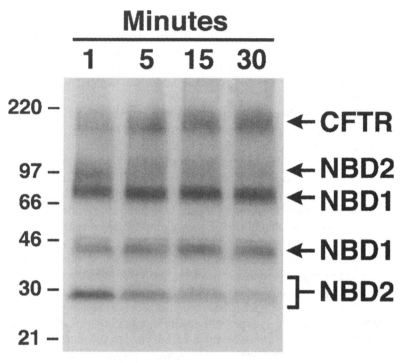

Figure 2 Photoaffinity labeling with 8-azido-ATP reveals kinetically distinct nucleotide interactions with the two NBDs of CFTR. Binding to fragments containing NBD1 is stable over the entire half hour during which time there is nearly complete turnover at NBD2.

alone was sufficient to form the regulated channel (1). However, despite extensive investigations in the intervening dozen years, neither the exact segments of the polypeptide which form the pore nor the number of CFTR molecules per pore have been determined with certainty.

The influence of charged residues within membrane-spanning segments on anion selectivity supported the reasonable assumption that these putative transmembrane helices contribute to the pore (42). Akabas and colleagues systematically analyzed the accessibility of residues in the first, third, and sixth transmembrane segments to methanethiosulfonate (MTS) reagents that bound to cysteines substituted for amino acids in the native wild-type sequence (43, 44). The influence of these MTS reagents bound at several positions in these segments on microscopic conductance in *Xenopus* oocytes and the sensitivity of the reaction rates to voltage changes suggested an involvement of these TM segments, especially TM6 in the pore (45). A major role of TM6 has been confirmed in extensive further studies employing mutagenesis and chemical modification at several positions (46, 47). These modifications strongly impacted fundamental properties of the channel, including conductance, anion selectivity and occupancy, and inhibition by blockers such as diphenylcarboxylate (DPC). This abundant and convincing evidence that residues in TM6 contribute directly to formation of the pore provides one means of testing whether single or multiple CFTR polypeptides are involved in the formation of a single pore (see below). Perturbations of permeation and gating by mutagenesis of residues in TMs 5 and 11 also implicate these two segments (48, 49), and TM12 mutations were observed to affect blocker sensitivity (50). Experiments with synthetic peptides corresponding to each of the six N-terminal TM sequences found that only TM2 and TM6 individually resulted in channel activity in planar lipid bilayers (51). Peptides modeled after TM3 and 4 were found to be α-helical (52), and a disease-associated mutation in one caused a non-native H-bond between them. None of these studies has provided insight into the relationships between any of the membrane-spanning segments in three dimensions, but a recent report demonstrates cross-linking by bifunctional MTS reagents of cysteines introduced near the cytoplasmic ends of TMs 6 and 12 (53). This is supportive of the earlier findings of the involvement of these segments in channel blocker sensitivity (50) and reflective of apparent similarity with P-glycoprotein, where TM6 and 12 interactions occur and are essential to its transport function (54).

Although the above described results support the idea that TMs from both the N- and C-terminal halves probably contribute to pore formation, there have been reports of channels formed by either half expressed separately (55, 56). However, Chan et al. (57) found that coexpression of both halves together was required to generate channel activity. Similarly, whereas Carroll et al. (58) reported that the first four TMs apparently are not essential to the generation of CFTR-like channels in *Xenopus* oocytes, these sequences do contain disease-associated mutations (42) and residues accessible to polar reagents (44). Thus not only are there insufficient data to completely identify pore-forming components, but available data and their interpretations are somewhat divergent.

Interpretation of the spatial relationships of peptide segments in CFTR that form the ion pore is, of course, absolutely dependent on knowing whether they occur intra- or intermolecularly or both. This issue has been explored experimentally at several different levels. Marshall et al. (59) concluded that CFTR did not self-associate because full-length and C-terminally truncated forms solubilized from cells in which they were coexpressed were not coimmunoprecipitated with an antibody recognizing an epitope at the C terminus. However, freeze-fracture electromicroscopy of membranes of *Xenopus* oocytes in which CFTR was expressed revealed intramembraneous particles with a cross-sectional area that might be occupied by approximately twice as many TMs as present in a CFTR monomer (60). From observations that expression of wild-type and mutant (R-domain partially deleted) CFTR sequences linked in tandem resulted in channels with intermediate gating properties, Zerhausen et al. (61) also concluded that two CFTR polypeptides may assemble to form a single pore. Subsequently, bivalent but not monovalent PDZ-domain-containing proteins were shown to enhance CFTR channel gating with a stoichiometry suggesting that promotion of dimer formation improved

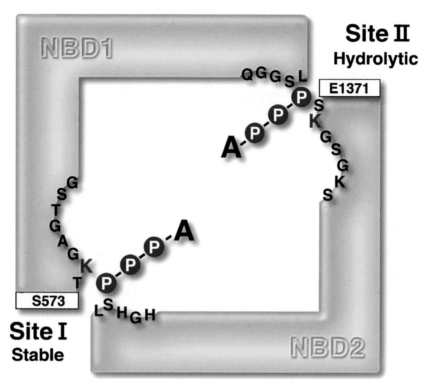

Figure 1   Cartoon depiction of asymmetric ATP-binding sites formed by the two nucleotide-binding domains (NBDs) of CFTR. Differences in the consensus Walker A and signature sequences and putative catalytic bases are indicated.

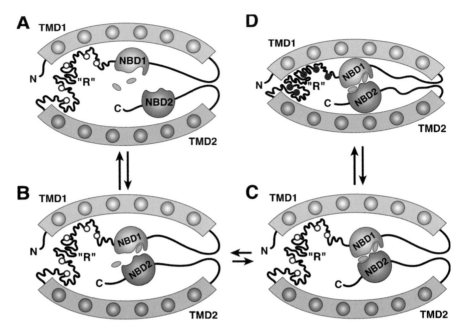

Figure 3 Schematic illustration of major events in CFTR channel regulation. In (*A*), the inactive channel is unphosphorylated and shown without ATP bound, although site I (deep cleft in NBD1) may be permanently occupied. In (*B*), NBDs are associated. In (*C*), ATP has bound at site II, but with no impact on pore-forming parts of TMDs; this occurs only with R-domain phosphorylation, which allows transmission of the structural perturbation caused by ATP binding to the pore (*D*).

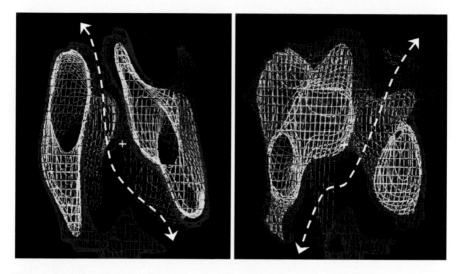

Figure 4 Side view of sections through the center of membrane-traversing regions of one CFTR crystal form. Apparent shafts through the barrel-like structure are indicated by dashed lines. The two views are rotated 90° about the vertical axis with respect to each other (taken from Reference 8).

function (62, 63). However, Ramjeesingh et al. (64), working with CFTR purified from insect Sf9 cells in which it is highly overexpressed, found that CFTR migrated on size-resolving electrophoretic and chromatographic gels with mobilities corresponding to both monomers and dimers and that both exhibited channel and ATPase activities. Chen et al. (65) concluded that CFTR behaved as a monomer both biochemically and functionally. They showed that when solubilized by a number of different detergents, from membranes of cells in which it was heterologously or endogenously expressed, the protein migrated as a monomer through a sucrose velocity gradient. When two differentially epitope-tagged forms were coexpressed and similarly solubilized, they failed to coimmuniprecipitate. Coexpressed variants with very different single-channel conductances (TM6 mutants) yielded these distinct conductances rather than intermediate values, which might have been expected if two CFTR polypeptides assembled to form a single pore. A recent in-depth study of the effects of sulfhydryl reagents on a CFTR channel in which a key charged residue in TM6 (R334) is substituted by a cysteine provided strong evidence that only one such residue was present in the pore (66, 66a). This R334C variant exhibited three subconductance states, as does the wild-type, and these states were altered simultaneously and with the same kinetics by a positively charged MTS reagent also making it seem highly unlikely that the segments from more than one CFTR polypeptide come together to form the permeation pathway.

Another recent study employing chemical cross-linking with bifunctional amino reactive reagents found that the dominant CFTR species observed in SDS-PAGE and velocity gradient centrifugation had the approximate size of a dimer (67). Most notable was the fact that these species were formed in cells under conditions that would stimulate CFTR channel activity, implying that dimerization might be part of the activation process. Importantly, however, this dimerization occurred in the absence of the binding of the C termini of monomers by PDZ-binding proteins. Thus there apparently is no correlation between this promotion of cross-linking under channel-activating conditions and the increased activity caused by bivalent PDZ proteins reported earlier (62, 63). Chemical–cross-linking experiments were also reported by Mengos et al. (68), who observed formation of multiple higher-molecular-weight species including putative dimers under channel-activating conditions but a much more prominent intramolecular cross-linking within monomers. The latter may reflect conformation changes observed owing to phosphorylation by PKA (13, 14). Such intramolecular structural changes might in turn result in increased associations between molecules. The only three-dimensional structure information yet available has come from two-dimensional crystalline arrays in which CFTR had crystallized as a monomer (Figure 4, see color insert) (8). The investigations summarized clearly do not enable a definitive conclusion to be drawn about quaternary structure of the functional CFTR anion channel. Nevertheless, an attempt at their critical integration is somewhat informative particularly in recognizing gaps in our knowledge.

Most obvious is the need for a high-resolution three-dimensional structure that would clearly define the pore. However, present evidence does not lead to the conclusion that complimentary surfaces of two or more CFTR polypeptides form the pore. In cases where activity was quantitatively enhanced by bivalent PDZ proteins, for example, there was already substantial activity before these proteins were titrated in. Of course, this may have been because some CFTR molecules were already dimerized, perhaps influenced by endogenous PDZ proteins. However, the monomeric purified CFTR of Ramjeesingh et al. (64) appeared to be as active as the fraction containing apparently dimeric CFTR. The interpretation of dimers based on freeze-fraction particle size in electron microscopy is not entirely compelling because the 9-nm diameter is similar to the 10-nm (69) and 10- to 12-nm (70) values of monomeric P-glycoprotein and the 7-nm value of crystallized CFTR monomers (8). Thus there is little support either for the notion that the CFTR ion pore is formed at contact surfaces between two or more CFTR polypeptides or that more than one pore is essential to the entire regulated activity. Nevertheless, there have been recurring suggestions of CFTR self-association. In addition to those interpretations already discussed above, Krouse & Wine (71) interpreted the heterogeneous gating modes of different CFTR channels in a patch as indicative of cooperative interactions between them. Clustering of CFTR with itself or other proteins in the membranes of activated *Xenopus* oocytes has been suggested by Schillers et al. (72, 73) from atomic force microscopy measurements. Rates of diffusion of CFTR in membranes are restricted by interactions requiring an intact C terminus, presumably to PDZ-domain proteins (74). Fairly extensive associations of CFTR with other proteins involved in its regulation have been demonstrated (75, 76). If there are specific complimentary surfaces of CFTR that favor dimerization on activation as suggested by Li et al. (67), this may be because of the apparently extensive conformational changes that occur within the monomer. Specifically, what these surfaces are is unknown, but Gupta et al. (77) have suggested that intermolecular interactions might occur between R-domains of CFTR, although in their experiments these were not influenced by phosphorylation. Just as preferred intermolecular surfaces for dimer or higher oligomer formation have not been identified, neither has the functional advantage gained by these associations. However, many interesting possibilities can be imagined and could be tested. For example, mature CFTR has a short residence time at the cell surface; it is endocytosed within several minutes but then most is recycled back into the plasma membrane (78, 79). It is conceivable that upon activation, oligomerization might prevent or diminish endocytosis and leave more active channels in the plasma membrane until dephosphorylation and inactivation have occurred. These would be dynamic events that may not have been detected in studies of the effect of PKA activation on endocytosis rates (80). In fact, activation of PKA has been observed to inhibit CFTR internalization (78, 79). In this view, self-association of CFTR would not influence the inherent functional capacity of the molecule; rather the residence time in the plasma membrane would be influenced. Other hypothetical possibilities include facilitation of interactions with other members of regulatory complexes by an oligomeric CFTR (76). It is interesting that a close relative of CFTR (ABCC7), the SUR1 sulfonylurea receptor (ABCC8), does participate in an oligomeric K_{ATP} channel (81, 82). However, this is due to the association of one SUR1 with each of the four K_{ir} 6.2 channel-forming subunits, and there is no evidence of direct interactions between individual SUR1 monomers. Also no evidence has yet been found of oligomerization of any other ABC protein in which all four principal domains (2 NBDs and 2 TMDs) are within a single peptide.

CFTR Synthesis, Conformational Maturation, and Domain Assembly

Considerable insight into the assembly of the CFTR channel can be gained from the extensive examination of its biosynthetic processing (83). This focus stems from the fact that processing and, indeed, assembly of the channel is defective in most patients with cystic fibrosis (84). Although initially observed to remain in the endoplasmic reticulum (ER) and therefore not acquire complex N-linked oligosaccharide chains, most characterization of ΔF508 CFTR processing has come from pulse-chase experiments (85, 86). Short pulses indicate that synthesis of the wild-type or mutant nascent core-glycosylated polypeptide is linear for at least 15 min, with approximately half that time being required for completion of a single chain (86, 87). This newly synthesized species then decays with a half time of approximately 30 min in both mutant and wild type, the former succumbing entirely to proteasomal degradation and the latter either partially to the same fate (50–80% depending on the host cell) or the remainder (20–50%) converted to the mature form with complex glycans (85). This rather slow and inefficient conformational maturation process has been viewed as reflecting the complexity of the sequential synthesis of hydrophilic and hydrophobic segments with folding of some domains interrupted by membrane integration of others. Overall, the immature precursor is much less compact that the mature product as reflected by the greater protease sensitivity of the former (88). The immature wild-type protein may already achieve a functional state in the ER as patch-clamp measurements of ER membrane surrounding isolated nuclei suggest (89). These active channels may be the transient population that has just matured conformationally and is ready for transit to the Golgi.

The stages at which neither the folding of each of the major domains of CFTR nor final domain assembly is completed has not yet been completely defined. However, these events are probably interdependent, as missense mutations or single-codon deletions, such as ΔF508 at different sites over the whole length of the molecule, prevent the forming of a final native structure (90). The ER quality control system, which is made up of multiple enzymes and molecular chaperones, is able to recognize the molecule in the mature structural state at which stage vesiculating segments of the ER membrane containing it are engaged by the CopI-based ER export machinery (91). Details of the domain assembly and the recognition mechanism remain to be worked out, but some clues have been obtained. A recent study of Du et al. (92) indicates that both co- and posttranslational processes probably are involved in conformational maturation and domain assembly. These investigators provided evidence that NBD1 achieves a protease-resistant conformation cotranslationally, whereas NBD2 may not. This apparently occurs even with the ΔF508 mutant, in which NBD2 permanently remains in a protease-sensitive state presumably reflecting its requirement for a wild-type NBD1. This requirement could be for a direct NBD-NBD, protein-protein interaction or an indirect effect of NBD1 through its association with TMD1. The nature and importance of domain assembly is evident from several additional studies. Expression of separate N- and C-terminal halves of CFTR simultaneously produces an active channel (65) showing that association does occur and is required, thus confirming earlier reports of TMD1/TMD2 and NBD1/NBD2 interactions (93). More recently, in what has been termed transcomplimentation, coexpression of N-terminal portions of CFTR, including the wild-type NBD1 sequence, has been shown to rescue maturation of the full-length ΔF508 molecule (94–96). This may be because the N-terminal domain of one construct can physically associate with C-terminal domain of the other. At least one study found that a misprocessing mutation in the TMD2 could be similarly rescued by coexpression of a wild-type TMD2 plus NBD2 construct (96).

Clues to the order of domain assembly have also come from determination of how much of the molecule must be formed to avoid recognition and degradation by ER quality control. Constructs with N-glycosylation sites in an extracytoplasmic loop of TMD1 were used to monitor the addition of complex oligosaccharide chains as a measure of export from ER to Golgi (96a). Both wild-type and ΔF508 versions of constructs terminating approximately at the C terminus of NBD1 (1–640) or the R-domain (1–831) were core-glycosylated, but these high mannose oligosaccharides were not trimmed and extended. Strikingly, however, when termination of the wild-type sequence was C terminal of TMD2 but did not include the NBD2, complex oligosaccharide chains were added, apparently indicating avoidance of ER quality control as effectively as that accomplished by the full-length molecule. On the basis of these observations, NBD2 does not appear to be required for CFTR to be recognized as fully assembled. Therefore, correct association of TMD1 and TMD2 may be the primary criterion of maturation recognized by ER quality control. That this can be achieved in the absence of NBD2 is evidenced by the channel activity, albeit with very low open probability, which is generated by constructs ending after TMD2 (96a). When NBD2 is present, however, it apparently relies on the influence of a prefolded NBD1 for the completion of its folding (92).

CFTR Interactions with Other Proteins: Functional Macromolecular Complexes?

Global proteomic approaches suggest that cellular proteins have highly variable numbers of interaction partners in a eukaryotic cell, with most having few but a small number having many (97). By virtue of the fact that CFTR is highly regulated by kinases and phosphatases and possibly other enzymes and is believed itself to regulate numerous membrane permeability pathways, it is likely that CFTR participates in a range of transient intermolecular interactions. Concerning activation of its own chloride channel activity by an external apical signal, adenosine receptors, Gs, adenylate cyclase, PKA, and CFTR have all been shown to be functionally coupled within the small area of airway cell apical membrane delimited within the tip of a patch-clamp pipette (75).

The C terminus of CFTR has a consensus sequence recognized by PDZ domains, and several PDZ-domain proteins have been shown to bind to it (62, 63, 98–101). These interactions may serve several functions, including localization to an apical ERM protein-containing cytoskeletal complex; AKAP-mediated association with PKA; association with the PDZ-binding β_2 adrenergic receptor, which when occupied stimulates CFTR via adenylate cyclase and PKA (76); enhancement of CFTR channel gating through dimerization on binding bivalent PDZ-domain proteins (62, 63); and mutually regulatory interactions with PDZ-binding members of the SLC26 anion exchanger family (102, 103). These and other regulatory interactions of CFTR almost certainly do not occur simultaneously and, in fact, not all have been rigorously confirmed in physiologically relevant settings. It is of interest to recall that disease-associated mutations near the C terminus precluding PDZ-domain-binding result in very mild disease, suggesting that this is not required for the most essential function of CFTR (104).

AMP-activated protein kinase binds and phosphorylates the C-terminal tail just upstream of the PDZ-binding extreme C terminus (105). Interactions with other short sequence motifs in the C-terminal extension of CFTR beyond NBD2 are involved in the endocytosis of the channel. These include the YXXΦ signal containing Y1424 recognized by the $\mu2$ subunit of the AP-2 clathrin adaptor protein (106) and the di-leucine motif further C-terminal (107).

At the N-terminal end of the molecule the SNARE protein syntaxin 1A binds to an amphiphatic helix before the first TM segments of CFTR and inhibits its channel activity (108) probably by interfering with an essential intramolecular interaction between the N-terminal tail and other cytoplasmic domains (109).

Additional indirect associations of CFTR with other proteins involved in clathrin-mediated endocytosis also occur including clathrin itself, myosin VI, and its adaptor protein Dab2, which also binds the AP2 complex (110). Participation of CFTR in this large endocytic macromolecular complex depends primarily on binding sites in the C-terminal segment, different from the PDZ-binding motif. Thus presumably these interactions would not occur simultaneously with formation of another large regulatory complex binding to the nearby PDZ-binding C terminus (76). Rather different associations are likely to form and dissociate depending on requirements for CFTR activity, changing location during its intracellular trafficking itinerary, and other factors. The possibility of these various associations in the assembly of a functional CFTR chloride channel has been discussed above.

In addition to interactions involved in channel function and endocytosis, the CFTR protein may interact directly or indirectly with other channels and transport proteins (111). An example where direct protein-protein interaction has been shown and fairly extensively characterized is the binding of the R-domain of CFTR to the sulfate transporter and anti-sigma antagonist (STAS) domain of some members of the SLC26 anion exchange family (103) that mediate bicarbonate secretion, which is defective in cystic fibrosis (112, 113). Further elucidation of the nature of this R-domain-STAS domain complex may be helpful in gaining more insight into the structure and function of the essential but somewhat enigmatic R-domain, which is apparently unique to the CFTR ion channel compared with all other known ABC transporters (114). Physical associations with other membrane-permeability proteins postulated to be influenced by CFTR have not yet been demonstrated.

A large set of cytoplasmic and ER proteins interact with nascent CFTR while it is folding at the ER, but these proteins no longer associate with nor influence the mature functional protein.

Interestingly all interactions of mature CFTR with other proteins demonstrated thus far have involved the N- and C-terminal segments and the R-domain, which reflects the fact the NBDs and the cytoplasmic loops may be primarily occupied with tight intramolecular associations of the types evident in the atomic structures of other ABC proteins (40, 41). These associations are likely mediators of the impact of nucleotide binding and hydrolysis on channel gating.

The *Annual Review of Physiology* is online at
http://physiol.annualreviews.org

LITERATURE CITED

1. Bear CE, Li C, Kartner N, Bridges RJ, Jensen TJ, et al. 1992. Purification and functional reconstitution of the cystic fibrosis transmembrane conductance regulator (CFTR). *Cell* 68:809–18

2. Dawson DC, Liu X, Zhang ZR, McCarty NA. 2003. Anion conduction by CFTR: mechanism and models. In *The CFTR Chloride Channel*, ed. KL Kirk, DC Dawson, pp. 1–34. Georgetown, TX: Lander Biosci.

3. Hanrahan JH, Gentzsch M, Riordan JR. 2003. The cystic fibrosis transmembrane conductance regulator (ABCC7). In *ABC Proteins: From Bacteria to Man*, ed. B Holland, CF Higgins, K Kuchler, SPC Cole, pp. 589–618. New York: Elsevier

4. Seibert FS, Chang XB, Aleksandrov AA, Clarke DM, Hanrahan JW, Riordan JR. 1999. Influence of phosphorylation by protein kinase A on CFTR at the cell surface and endoplasmic reticulum. *Biochim. Biophys. Acta* 1461:275–83

5. Ostedgaard LS, Baldursson O, Welsh MJ. 2001. Regulation of the cystic fibrosis transmembrane conductance regulator Cl⁻ channel by its R domain. *J. Biol. Chem.* 276:7689–92

6. Chappe V, Hinkson DA, Howell LD, Evagelidis A, Liao J, et al. 2004. Stimulatory and inhibitory protein kinase C consensus sequences regulate the cystic fibrosis transmembrane conductance regulator. *Proc. Natl. Acad. Sci. USA* 101:390–95

7. Lewis HA, Buchanan SG, Burley SK, Conners K, Dickey M, et al. 2004. Structure of nucleotide-binding domain 1 of the cystic fibrosis transmembrane conductance regulator. *EMBO. J.* 23:282–93

8. Rosenberg MF, Kamis AB, Aleksandrov LA, Ford RC, Riordan JR. 2004. Purification and crystallization of the cystic fibrosis transmembrane conductance regulator (CFTR). *J. Biol. Chem.* 279:39051–57

9. Riordan JR, Rommens JM, Kerem B, Alon N, Rozmahel R, et al. 1989. Identification of the cystic fibrosis gene: cloning and characterization of complementary DNA. *Science* 245:1066–73

10. Sheppard DN, Welsh MJ. 1999. Structure and function of the CFTR chloride channel. *Physiol. Rev.* 79:S23–45

11. Gadsby DC, Nairn AC. 1999. Control of CFTR channel gating by phosphorylation and nucleotide hydrolysis. *Physiol. Rev.* 79:S77–107

12. Rich DP, Berger HA, Cheng SH, Travis SM, Saxena M, et al. 1993. Regulation of the cystic fibrosis transmembrane conductance regulator Cl^- channel by negative charge in the R domain. *J. Biol. Chem.* 268:20259–67

13. Dulhanty AM, Riordan JR. 1994. Phosphorylation by cAMP-dependent protein kinase causes a conformational change in the R domain of the cystic fibrosis transmembrane conductance regulator. *Biochemistry* 33:4072–79

14. Grimard V, Li C, Ramjeesingh M, Bear CE, Goormaghtigh E, Ruysschaert JM. 2004. Phosphorylation-induced conformational changes of cystic fibrosis transmembrane conductance regulator monitored by attenuated total reflection-Fourier transform IR spectroscopy and fluorescence spectroscopy. *J. Biol. Chem.* 279:5528–36

15. Ostedgaard LS, Baldursson O, Vermeer DW, Welsh MJ, Robertson AD. 2000. A functional R domain from cystic fibrosis transmembrane conductance regulator is predominantly unstructured in solution. *Proc. Natl. Acad. Sci. USA* 97:5657–62

16. Li C, Ramjeesingh M, Wang W, Garami E, Hewryk M, et al. 1996. ATPase activity of the cystic fibrosis transmembrane conductance regulator. *J. Biol. Chem.* 271:28463–68

17. Reddy MM, Quinton PM. 2001. cAMP-independent phosphorylation activation of CFTR by G proteins in native human sweat duct. *Am. J. Physiol. Cell Physiol.* 280:C604–13

18. Aleksandrov AA, Aleksandrov L, Riordan JR. 2002. Nucleoside triphosphate pentose ring impact on CFTR gating and hydrolysis. *FEBS Lett.* 518:183–88

19. Reddy MM, Quinton PM. 2003. Control of dynamic CFTR selectivity by glutamate and ATP in epithelial cells. *Nature* 423:756–60

20. Himmel B, Nagel G. 2004. Protein kinase-independent activation of CFTR by phosphatidylinositol phosphates. *EMBO Rep.* 5:85–90

21. Gunderson KL, Kopito RR. 1995. Conformational states of CFTR associated with channel gating: the role of ATP binding and hydrolysis. *Cell* 82:231–39

22. Baukrowitz T, Hwang TC, Nairn AC, Gadsby DC. 1994. Coupling of CFTR Cl^- channel gating to an ATP hydrolysis cycle. *Neuron* 12:473–82

23. Bear CE, Li C, Galley K, Wang Y, Garami E, Ramjeesingh M. 1997. Coupling of ATP hydrolysis with channel gating by purified, reconstituted CFTR. *J. Bioenerg. Biomembr.* 29:465–73

24. Welsh MJ, Robertson AD, Ostedgaard LS. 1998. Structural biology. The ABC of a versatile engine. *Nature* 396:623–24

25. Aleksandrov AA, Riordan JR. 1998. Regulation of CFTR ion channel gating by MgATP. *FEBS Lett.* 431:97–101

26. Aleksandrov AA, Chang X, Aleksandrov L, Riordan JR. 2000. The non-hydrolytic pathway of cystic fibrosis transmembrane conductance regulator ion channel gating. *J. Physiol.* 528:259–65

27. Aleksandrov L, Aleksandrov AA, Chang XB, Riordan JR. 2002. The first nucleotide binding domain of cystic fibrosis transmembrane conductance regulator is a site of stable nucleotide interaction, whereas the second is a site of rapid turnover. *J. Biol. Chem.* 277:15419–25

28. Kaupp UB, Seifert R. 2002. Cyclic nucleotide-gated ion channels. *Physiol. Rev.* 82:769–824

29. Aleksandrov A, Aleksandrov L, Riordan JR. 2004. CFTR as a hydrolysable-ligand gated ion channel. *FEBS Lett.* Submitted

30. Oloo EO, Tieleman DP. 2004. Conformational transitions induced by the binding of MgATP to the vitamin B_{12} ABC-transporter BtuCD. *J. Biol. Chem.* M405084200

31. Urbatsch IL, Gimi K, Wilke-Mounts S, Senior AE. 2000. Conserved Walker A Ser residues in the catalytic sites of P-glycoprotein are critical for catalysis and involved primarily at the transition state step. *J. Biol. Chem.* 275:25031–38

32. Jones PM, George AM. 2002. Mechanism of ABC transporters: a molecular dynamics simulation of a well characterized nucleotide-binding subunit. *Proc. Natl. Acad. Sci. USA* 99:12639–44

33. Hopfner KP, Karcher A, Shin DS, Craig L, Arthur LM, et al. 2000. Structural biology of Rad50 ATPase: ATP-driven conformational control in DNA double-strand break repair and the ABC-ATPase superfamily. *Cell* 101:789–800

34. Karpowich N, Martsinkevich O, Millen L, Yuan YR, Dai PL, et al. 2001. Crystal structures of the MJ1267 ATP binding cassette reveal an induced-fit effect at the ATPase active site of an ABC transporter. *Structure* 9:571–86

35. Moody JE, Millen L, Binns D, Hunt JF, Thomas PJ. 2002. Cooperative, ATP-dependent association of the nucleotide binding cassettes during the catalytic cycle of ATP-binding cassette transporters. *J. Biol. Chem.* 277:21111–14

36. Verdon G, Albers SV, van Oosterwijk N, Dijkstra BW, Driessen AJ, Thunnissen AM. 2003. Formation of the productive ATP-Mg^{2+}-bound dimer of GlcV, an ABC-ATPase from *Sulfolobus solfataricus*. *J. Mol. Biol.* 334:255–67

37. Chen J, Lu G, Lin J, Davidson AL, Quiocho FA. 2003. A tweezers-like motion of the ATP-binding cassette dimer in an ABC transport cycle. *Mol. Cell.* 12:651–61

38. Abele R, Tampé R. 2004. The ABCs of immunology: structure and function of TAP, the transporter associated with antigen processing. *Physiology* 19:216–24

39. Senior AE, al-Shawi MK, Urbatsch IL. 1995. The catalytic cycle of P-glycoprotein. *FEBS Lett.* 377:285–89

40. Locher KP, Lee AT, Rees DC. 2002. The *E. coli* BtuCD structure: a framework for ABC transporter architecture and mechanism. *Science* 296:1091–98

41. Chang G. 2003. Structure of MsbA from *Vibrio cholera*: a multidrug resistance ABC transporter homolog in a closed conformation. *J. Mol. Biol.* 330:419–30

42. Dawson DC, Smith SS, Mansoura MK. 1999. CFTR: mechanism of anion conduction. *Physiol. Rev.* 79:S47–75

43. Cheung M, Akabas MH. 1997. Locating the anion-selectivity filter of the cystic fibrosis transmembrane conductance regulator (CFTR) chloride channel. *J. Gen. Physiol.* 109:289–99

44. Akabas MH. 2000. Cystic fibrosis transmembrane conductance regulator. Structure and function of an epithelial chloride channel. *J. Biol. Chem.* 275:3729–32

45. Cheung M, Akabas MH. 1996. Identification of cystic fibrosis transmembrane conductance regulator channel-lining residues in and flanking the M6 membrane-spanning segment. *Biophys. J.* 70:2688–95

46. Tabcharani JA, Rommens JM, Hou YX, Chang XB, Tsui LC, et al. 1993. Multi-ion pore behaviour in the CFTR chloride channel. *Nature* 366:79–82

47. Smith SS, Liu X, Zhang ZR, Sun F, Kriewall TE, et al. 2001. CFTR: covalent and noncovalent modification suggests a role for fixed charges in anion conduction. *J. Gen. Physiol.* 118:407–31

48. McCarty NA. 2000. Permeation through the CFTR chloride channel. *J. Exp. Biol.* 203:1947–62

49. Zhang ZR, McDonough SI, McCarty NA. 2000. Interaction between permeation and gating in a putative pore domain mutant in the cystic fibrosis transmembrane

conductance regulator. *Biophys. J.* 79:298–313

50. McDonough S, Davidson N, Lester HA, McCarty NA. 1994. Novel pore-lining residues in CFTR that govern permeation and open-channel block. *Neuron* 13:623–34

51. Oblatt-Montal M, Reddy GL, Iwamoto T, Tomich JM, Montal M. 1994. Identification of an ion channel-forming motif in the primary structure of CFTR, the cystic fibrosis chloride channel. *Proc. Natl. Acad. Sci. USA* 91:1495–99

52. Therien AG, Grant FE, Deber CM. 2001. Interhelical hydrogen bonds in the CFTR membrane domain. *Nat. Struct. Biol.* 8: 597–601

53. Chen EY, Bartlett MC, Loo TW, Clarke DM. 2004. The DF508 mutation disrupts packing of the transmembrane segments of the cystic fibrosis transmembrane conductance regulator. *J. Biol. Chem.* 279:39620–27

54. Loo TW, Bartlett MC, Clarke DM. 2004. Disulfide cross-linking analysis shows that transmembrane segments 5 and 8 of human P-glycoprotein are close together on the cytoplasmic side of the membrane. *J. Biol. Chem.* 279:7692–97

55. Sheppard DN, Ostedgaard LS, Rich DP, Welsh MJ. 1994. The amino-terminal portion of CFTR forms a regulated Cl⁻ channel. *Cell* 76:1091–98

56. Ramjeesingh M, Ugwu F, Li C, Dhani S, Huan LJ, et al. 2003. Stable dimeric assembly of the second membrane-spanning domain of CFTR (cystic fibrosis transmembrane conductance regulator) reconstitutes a chloride-selective pore. *Biochem. J.* 375:633–41

57. Chan KW, Csanady L, Seto-Young D, Nairn AC, Gadsby DC. 2000. Severed molecules functionally define the boundaries of the cystic fibrosis transmembrane conductance regulator's NH$_2$-terminal nucleotide binding domain. *J. Gen. Physiol.* 116:163–80

58. Carroll TP, Morales MM, Fulmer SB,

Allen SS, Flotte TR, et al. 1995. Alternate translation initiation codons can create functional forms of cystic fibrosis transmembrane conductance regulator. *J. Biol. Chem.* 270:11941–46

59. Marshall J, Fang S, Ostedgaard LS, O'Riordan CR, Ferrara D, et al. 1994. Stoichiometry of recombinant cystic fibrosis transmembrane conductance regulator in epithelial cells and its functional reconstitution into cells in vitro. *J. Biol. Chem.* 269:2987–95

60. Eskandari S, Wright EM, Kreman M, Starace DM, Zampighi GA. 1998. Structural analysis of cloned plasma membrane proteins by freeze-fracture electron microscopy. *Proc. Natl. Acad. Sci. USA* 95: 11235–40

61. Zerhusen B, Zhao J, Xie J, Davis PB, Ma J. 1999. A single conductance pore for chloride ions formed by two cystic fibrosis transmembrane conductance regulator molecules. *J. Biol. Chem.* 274:7627–30

62. Wang S, Yue H, Derin RB, Guggino WB, Li M. 2000. Accessory protein facilitated CFTR-CFTR interaction, a molecular mechanism to potentiate the chloride channel activity. *Cell* 103:169–79

63. Raghuram V, Mak DD, Foskett JK. 2001. Regulation of cystic fibrosis transmembrane conductance regulator single-channel gating by bivalent PDZ-domain-mediated interaction. *Proc. Natl. Acad. Sci. USA* 98:1300–5

64. Ramjeesingh M, Li C, Kogan I, Wang Y, Huan LJ, Bear CE. 2001. A monomer is the minimum functional unit required for channel and ATPase activity of the cystic fibrosis transmembrane conductance regulator. *Biochemistry* 40:10700–6

65. Chen JH, Chang XB, Aleksandrov AA, Riordan JR. 2002. CFTR is a monomer: biochemical and functional evidence. *J. Membr. Biol* 188:55–71

66. Zhang ZR, Cui G, Liu X, Song B, Dawson DC, McCarty NA. 2004. CFTR: one polypeptide—one pore. *J. Biol. Chem.* In press

66a. Liu X, Zhang Z-R, Fuller MD, Billingsley J, McCarty NA, Dawson DC. 2004. CFTR: a cysteine at position 338 in TM6 senses a positive electrostatic potential in the pore. *Biophys. J.* doi:10.1529/biophysj.104.050534

67. Li C, Roy K, Dandridge K, Naren AP. 2004. Molecular assembly of cystic fibrosis transmembrane conductance regulator in plasma membrane. *J. Biol. Chem.* 279:24673–84

68. Mengos A, Chang XB, Cui L, Aleksandrov L, Riordan JR. 2004. Channel activating conditions promote intramolecular cross-linking within CFTR monomers. Submitted

69. Arsenault AL, Ling V, Kartner N. 1988. Altered plasma membrane ultrastructure in multidrug-resistant cells. *Biochim. Biophys. Acta* 938:315–21

70. Rosenberg MF, Callaghan R, Ford RC, Higgins CF. 1997. Structure of the multidrug resistance P-glycoprotein to 2.5 nm resolution determined by electron microscopy and image analysis. *J. Biol. Chem.* 272:10685–94

71. Krouse ME, Wine JJ. 2001. Evidence that CFTR channels can regulate the open duration of other CFTR channels: cooperativity. *J. Membr. Biol.* 182:223–32

72. Schillers H, Danker T, Madeja M, Oberleithner H. 2001. Plasma membrane protein clusters appear in CFTR-expressing *Xenopus laevis* oocytes after cAMP stimulation. *J. Membr. Biol.* 180:205–12

73. Schillers H, Shahin V, Albermann L, Schafer C, Oberleithner H. 2004. Imaging CFTR: a tail to tail dimer with a central pore. *Cell Physiol. Biochem.* 14:1–10

74. Haggie PM, Stanton BA, Verkman AS. 2004. Increased diffusional mobility of CFTR at the plasma membrane after deletion of its C-terminal PDZ binding motif. *J. Biol. Chem.* 279:5494–500

75. Huang P, Lazarowski ER, Tarran R, Milgram SL, Boucher RC, Stutts MJ. 2001. Compartmentalized autocrine signaling to cystic fibrosis transmembrane conductance regulator at the apical membrane of airway epithelial cells. *Proc. Natl. Acad. Sci. USA* 98:14120–25

76. Naren AP, Cobb B, Li C, Roy K, Nelson D, et al. 2003. A macromolecular complex of beta 2 adrenergic receptor, CFTR, and ezrin/radixin/moesin-binding phosphoprotein 50 is regulated by PKA. *Proc. Natl. Acad. Sci. USA* 100:342–46

77. Gupta S, Xie J, Ma J, Davis PB. 2004. Intermolecular interaction between R domains of cystic fibrosis transmembrane conductance regulator. *Am. J. Respir. Cell. Mol. Biol.* 30:242–48

78. Prince LS, Workman RB Jr, Marchase RB. 1994. Rapid endocytosis of the cystic fibrosis transmembrane conductance regulator chloride channel. *Proc. Natl. Acad. Sci. USA* 91:5192–96

79. Lukacs GL, Segal G, Kartner N, Grinstein S, Zhang F. 1997. Constitutive internalization of cystic fibrosis transmembrane conductance regulator occurs via clathrin-dependent endocytosis and is regulated by protein phosphorylation. *Biochem. J.* 328:353–61

80. Bertrand CA, Frizzell RA. 2003. The role of regulated CFTR trafficking in epithelial secretion. *Am. J. Physiol. Cell Physiol.* 285:C1–18

81. Clement JPT, Kunjilwar K, Gonzalez G, Schwanstecher M, Panten U, et al. 1997. Association and stoichiometry of K(ATP) channel subunits. *Neuron* 18:827–38

82. Shyng S, Nichols CG. 1997. Octameric stoichiometry of the K_{ATP} channel complex. *J. Gen. Physiol.* 110:655–64

83. Kopito RR. 1999. Biosynthesis and degradation of CFTR. *Physiol. Rev.* 79:S167–73

84. Cheng SH, Gregory RJ, Marshall J, Paul S, Souza DW, 1990. Defective intracellular transport and processing of CFTR is the molecular basis of most cystic fibrosis. *Cell* 63:827–34

85. Lukacs GL, Mohamed A, Kartner N, Chang X-B, Riordan JR, Grinstein

S. 1994. Confirmational maturation of CFTR but not its mutant counterpart (deltaF508) occurs in the endoplasmic reticulum and requires ATP. *EMBO J.* 13:6076–86

86. Ward CL, Kopito RR. 1994. Intracellular turnover of cystic fibrosis transmembrane conductance regulator. Inefficient processing and rapid degradation of wild-type and mutant proteins. *J. Biol. Chem.* 269:25710–18

87. Loo MA, Jensen TJ, Cui L, Hou Y-X, Chang X-B, Riordan JR. 1998. Perturbation of Hsp90 interaction with nascent CFTR prevents its maturation and accelerates its degradation by the proteasome. *EMBO J.* 17:6879–87

88. Zhang F, Kartner N, Lukacs GL. 1998. Limited proteolysis as a probe for arrested conformational maturation of delta F508 CFTR. *Nat. Struct. Biol.* 5:180–83

89. Pasyk EA, Foskett JA. 1995. Mutant (delta F508) cystic fibrosis transmembrane conductance regulator Cl⁻ channel is functional when retained in endoplasmic reticulum of mammalian cells. *J. Biol. Chem.* 270:12347–50

90. Seibert FS, Loo TW, Clarke DM, Riordan JR. 1997. Cystic fibrosis: channel, catalytic, and folding properties of the CFTR protein. *J. Bioenerg. Biomembr.* 29:429–42

91. Yoo JS, Moyer BD, Bannykh S, Yoo HM, Riordan JR, Balch WE. 2002. Nonconventional trafficking of the cystic fibrosis transmembrane conductance regulator through the early secretory pathway. *J. Biol. Chem.* 277:11401–9

92. Du K, Sharma M, Lukacs GL. 2004. The deltaF508 cystic fibrosis mutation impairs domain-domain interactions and arrests posttranslational folding of CFTR. *Nat. Struct. Biol.* In press

93. Ostedgaard LS, Rich DP, DeBerg LG, Welsh MJ. 1997. Association of domains within the cystic fibrosis transmembrane conductance regulator. *Biochemistry* 36:1287–94

94. Owsianik G, Cao L, Nilius B. 2003. Rescue of functional deltaF508-CFTR channels by co-expression with truncated CFTR constructs in COS-1 cells. *FEBS Lett.* 554:173–78

95. Clarke LL, Gawenis LR, Hwang TC, Walker NM, Gruis DB, Price EM. 2004. A domain mimic increases deltaF508 CFTR trafficking and restores cAMP-stimulated anion secretion in cystic fibrosis epithelia. *Am. J. Physiol. Cell Physiol.* 287:C192–99

96. Cormet-Boyaka E, Jablonsky M, Naren AP, Jackson PL, Muccio DD, Kirk KL. 2004. Rescuing cystic fibrosis transmembrane conductance regulator (CFTR)-processing mutants by transcomplementation. *Proc. Natl. Acad. Sci. USA* 101: 8221–26

96a. Cui L, Mengos A, Hou X-Y, Gentzsch M, Aleksandrov LA, et al. 2004. Domain assembly in CFTR conformational maturation and structural stability. Submitted

97. Han JD, Bertin N, Hao T, Goldberg DS, Berriz GF, et al. 2004. Evidence for dynamically organized modularity in the yeast protein-protein interaction network. *Nature* 430:88–93

98. Short DB, Trotter KW, Reczek D, Kreda SM, Bretscher A, et al. 1998. An apical PDZ protein anchors the cystic fibrosis transmembrane conductance regulator to the cytoskeleton. *J. Biol. Chem.* 273: 19797–801

99. Wang S, Raab RW, Schatz PJ, Guggino WB, Li M. 1998. Peptide binding consensus of the NHE-RF-PDZ1 domain matches the C-terminal sequence of cystic fibrosis transmembrane conductance regulator (CFTR). *FEBS Lett.* 427:103–8

100. Hall RA, Ostedgaard LS, Premont RT, Blitzer JT, Rahman N, et al. 1998. A C-terminal motif found in the beta2-adrenergic receptor, P2Y1 receptor and cystic fibrosis transmembrane conductance regulator determines binding to the Na⁺/H⁺ exchanger regulatory factor

family of PDZ proteins. *Proc. Natl. Acad. Sci. USA* 95:8496–501

101. Gentzsch M, Cui L, Mengos A, Chang XB, Chen JH, Riordan JR. 2003. The PDZ-binding chloride channel ClC-3B localizes to the Golgi and associates with cystic fibrosis transmembrane conductance regulator-interacting PDZ proteins. *J. Biol. Chem.* 278:6440–49

102. Park M, Ko SB, Choi JY, Muallem G, Thomas PJ, et al. 2002. The cystic fibrosis transmembrane conductance regulator interacts with and regulates the activity of the HCO_3^- salvage transporter human Na^+-HCO_3^- cotransport isoform 3. *J. Biol. Chem.* 277:50503–9

103. Ko SB, Zeng W, Dorwart MR, Luo X, Kim KH, et al. 2004. Gating of CFTR by the STAS domain of SLC26 transporters. *Nat. Cell. Biol.* 6:343–50

104. Mickle JE, Macek M Jr, Fulmer-Smentek SB, Egan MM, Schwiebert E, et al. 1998. A mutation in the cystic fibrosis transmembrane conductance regulator gene associated with elevated sweat chloride concentrations in the absence of cystic fibrosis. *Hum. Mol. Genet.* 7:729–35

105. Hallows KR, McCane JE, Kemp BE, Witters LA, Foskett JK. 2003. Regulation of channel gating by AMP-activated protein kinase modulates cystic fibrosis transmembrane conductance regulator activity in lung submucosal cells. *J. Biol. Chem.* 278:998–1004

106. Weixel KM, Bradbury NA. 2001. Mu 2 binding directs the cystic fibrosis transmembrane conductance regulator to the clathrin-mediated endocytic pathway. *J. Biol. Chem.* 276:46251–59

107. Hu W, Howard M, Lukacs GL. 2001. Multiple endocytic signals in the C-terminal tail of the cystic fibrosis transmembrane conductance regulator. *Biochem. J.* 354:561–72

108. Naren AP, Nelson DJ, Xie W, Jovov B, Pevsner J, et al. 1997. Regulation of CFTR chloride channels by syntaxin and Munc18 isoforms. *Nature* 390:302–5

109. Fu J, Kirk KL. 2001. Cysteine substitutions reveal dual functions of the amino-terminal tail in cystic fibrosis transmembrane conductance regulator channel gating. *J. Biol. Chem.* 276:35660–68

110. Swiatecka-Urban A, Boyd C, Coutermarsh B, Karlson KH, Barnaby R, et al. 2004. Myosin VI regulates endocytosis of the cystic fibrosis transmembrane conductance regulator. *J. Biol. Chem.* 279:38025–31

111. Kunzelmann K. 2001. CFTR: interacting with everything? *News Physiol. Sci.* 16:167–70

112. Poulsen JH, Fischer H, Illek B, Machen TE. 1994. Bicarbonate conductance and pH regulatory capability of cystic fibrosis transmembrane conductance regulator. *Proc. Natl. Acad. Sci. USA* 91:5340–44

113. Gray MA, Plant S, Argent BE. 1993. cAMP-regulated whole cell chloride currents in pancreatic duct cells. *Am. J. Physiol. Cell Physiol.* 264:C591–602

114. Dassa E. 2003. Phylogenetic and functional classification of ABC (ATP-binding cassette) systems. In *ABC Proteins: From Bacteria to Man*, ed. B Holland, CF Higgins, K Kuchler, SP Cole, pp. 3–36. New York: Elsevier

Annu. Rev. Physiol. 2005. 67:719–58
doi: 10.1146/annurev.physiol.67.032003.154341
Copyright © 2005 by Annual Reviews. All rights reserved
First published online as a Review in Advance on October 19, 2004

CALCIUM-ACTIVATED CHLORIDE CHANNELS

Criss Hartzell,[1] Ilva Putzier,[1] and Jorge Arreola[2]

[1]*Department of Cell Biology, Emory University School of Medicine, Atlanta, Georgia 30322; email: criss.hartzell@emory.edu, iputzie@emory.edu*
[2]*Institute of Physics, University of San Luis Potosi, San Luis Potosi, SLP 78290, Mexico; email: Arreola@dec1.ifisica.uaslp.mx*

Key Words   ion channels, epithelia, bestrophin, secretion, CFTR

■ **Abstract**   Calcium-activated chloride channels (CaCCs) play important roles in cellular physiology, including epithelial secretion of electrolytes and water, sensory transduction, regulation of neuronal and cardiac excitability, and regulation of vascular tone. This review discusses the physiological roles of these channels, their mechanisms of regulation and activation, and the mechanisms of anion selectivity and conduction. Despite the fact that CaCCs are so broadly expressed in cells and play such important functions, understanding these channels has been limited by the absence of specific blockers and the fact that the molecular identities of CaCCs remains in question. Recent status of the pharmacology and molecular identification of CaCCs is evaluated.

INTRODUCTION

Many cell types express a type of Cl^- channel that is activated by cytosolic Ca^{2+} concentrations ($[Ca^{2+}]_i$) in the range of 0.2–5 μM. For example, in *Xenopus* oocytes, where these channels were first described in the early 1980s (1, 2), increases in $[Ca^{2+}]_i$ that occur upon fertilization cause Ca^{2+}-activated Cl^- channels (CaCCs) to open. This produces a depolarization of the membrane that somehow prevents the fusion of additional sperm. Similar channels have subsequently been found in many different cell types. These cell types include neurons; various epithelial cells; olfactory and photo-receptors; cardiac, smooth, and skeletal muscle; Sertoli cells; mast cells; neutrophils; lymphocytes; uterine muscle; brown fat adipocytes; hepatocytes; insulin-secreting beta cells; mammary glands; sweat glands; and *Vicia faba* guard cells. CaCCs are involved in epithelial secretion (3–5), membrane excitability in cardiac muscle and neurons (6–8), olfactory transduction (6), regulation of vascular tone (9), modulation of photoreceptor light responses, and probably other functions as well. In this review, Cl^- currents that are stimulated by Ca^{2+} (independent of the mechanism) are generically referred to as Ca^{2+}-activated Cl^- currents ($I_{Cl,Ca}$) and the channels that mediate these currents as CaCCs.

Although $I_{Cl,Ca}$ have been studied for more than 20 years, their physiological roles and mechanisms of regulation have remained somewhat cloudy. A recent

0066-4278/05/0315-0719$14.00   **719**

review on CaCCs (10) summarizes the ambivalence surrounding these channels by its subtitle "(Un)known, (Un)loved?" This ambivalence stems from the lack of specific blockers and the fact that electrophysiological studies suggest there may be several kinds of CaCCs, but their molecular identities remain in question.

The purpose of this review is to summarize the physiological functions of CaCCs, their mechanisms of activation and regulation, the mechanisms of their permeation and selectivity, and to evaluate data on their molecular identity. In the past few years, several excellent reviews and a monograph on CaCCs have been published. A collection of reviews edited by Fuller (11) provides a broad view of this family of channels. In addition, there are a number of excellent focused reviews on CaCCs in vascular smooth muscle (9), neuronal CaCCs (6), the role of CaCCs in epithelial secretion (12–14), and CaCCs in the airway (15). The molecular identity of CaCCs was recently discussed in two reviews (10, 16).

PHYSIOLOGICAL ROLE OF CaCCs

Three factors dictate the direction of Cl^- movement through CaCCs: the membrane potential, the Cl^- concentration gradient, and the $[Ca^{2+}]_i$ (Figure 1, see color insert). In most cells, the resting membrane potential is more negative than E_{Cl}. As a consequence, when $[Ca^{2+}]_i$ rises, Cl^- exits the cell, which results in a depolarization of the plasma membrane. In some cells this depolarization increases the open probability of voltage-gated Ca^{2+} channels (VGCCs), which results in additional Ca^{2+} influx and further depolarization (Figure 2). Because of osmotic forces and the requirement for charge equality, the efflux of Cl^- is accompanied by water and Na^+. If E_{Cl} is more positive than the membrane potential, opening CaCCs can lead to hyperpolarization.

Olfactory Transduction

Vertebrate olfactory receptor neurons from frog (17), newt (18), rat (19), salamander (20), mud puppy (21), and fish (22) express CaCCs that play a critical role in transduction of olfactory stimuli. Odorants bind to and activate G protein–coupled receptors in the ciliary membrane of olfactory receptor neurons (23). These receptors activate adenylyl cyclase, which produces cAMP and turns on cyclic-nucleotide-gated channels that are permeable to both Na^+ and Ca^{2+}. This leads to a membrane depolarization and an elevation of $[Ca^{2+}]_i$ in the cilium, which activates CaCCs. The Cl^- efflux (inward current) depolarizes the membrane further. Thus in olfactory receptor neurons, the Cl^- efflux through CaCCs serves as an amplification system of the odorant-activated current (19, 24). It has been estimated that the magnitude of $I_{Cl.Ca}$ can be as much as 30 times greater than the current through cyclic-nucleotide-gated channels (25). Thus the physiological role of the amplification could serve to increase the signal-to-noise ratio (24) and hence to increase sensitivity to odorants.

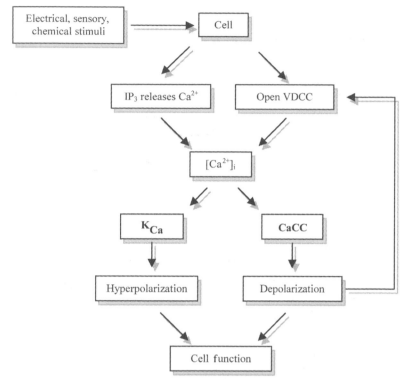

Figure 2   Regulation of membrane potential by CaCCs. Opening of CaCCs generally produces depolarization, which can feed back to open additional VGCCs. Increases in intracellular Ca can activate Ca-activated K (K_{Ca}) or Cl channels, which generally have opposite effects on membrane potential.

Pharmacological interventions support a role for $I_{Cl,Ca}$ in the odorant transducing mechanism. Niflumic acid (NFA), flufenamic acid, and DCDPC block $I_{Cl,Ca}$ in frog olfactory cilia (6) and part of the odorant-induced receptor current in olfactory receptor neurons from newt (18), rat (19), *Xenopus* (26), and salamander (20).

Taste Transduction

CaCCs are present in both mammalian and amphibian taste receptors (27, 28). Taste stimuli produce a depolarizing current in taste receptor cells that may result in a discharge of action potentials (29). In *Necturus*, action potentials in the taste receptors are followed by an outward current that is mediated by CaCCs, which open in response to Ca^{2+} influx during the action potentials (27, 30). The CaCCs produce a hyperpolarization in these cells because E_{Cl} is between -60 mV and -80 mV. This hyperpolarization is thought to play a role in taste adaptation (30).

Phototransduction

The inner segments of salamander rods and amphibian and mammalian cones express CaCCs (31–34). In addition, CaCCs are also present in the synaptic terminal of bipolar cells (35). In the dark, the vertebrate photoreceptor is depolarized by the dark current, which is mediated by cyclic-nucleotide-gated channels that conduct Na^+ and Ca^{2+} into the outer segment. The depolarization produced by the dark current opens VGCCs located at the synaptic terminal. Ca^{2+} influx through VGCCs stimulates transmitter release and activates a large conductance Cl^- channel, which favors photoreceptor depolarization. Upon illumination, the dark current turns off, the cell membrane hyperpolarizes, and transmitter release stops (36). The role of CaCCs in rods is not known, but it has been suggested that in cones they play a role in modulating lateral inhibition (31, 33, 37). Cones in the dark center of an illuminated annulus often exhibit action potentials whose repolarization is afforded by CaCCs. However, the precise role of CaCCs in photoreceptors remains ambiguous because of uncertainties about $[Cl^-]_i$. In small processes, CaCC could play a role in membrane potential stabilization by counteracting the depolarization induced by the Ca^{2+} inflow through VGCC (36).

Neuronal Excitability

CaCCs are expressed in a variety of different neurons, including dorsal root ganglion (DRG) neurons, spinal cord neurons, and autonomic neurons. In most cases, CaCCs are not expressed in all neurons of a group, but rather in a subset, suggesting that these channels perform a specific function for this subset of neurons. The functions of CaCCs in neurons remain poorly established, but it has been suggested that they are involved in action potential repolarization, generation of after-polarizations, and membrane oscillatory behavior.

About 45–90% of the somatosensory neurons from the DRG that sense skin temperature, touch, muscle tension, and pain express CaCCs (38–41). In the mouse DRG, CaCCs are expressed selectively in a subset of medium diameter (30–40 μm diameter) sensory neurons (42), suggesting that signaling in these neurons is somehow different from the rest of the population. A fraction of neurons from the quail trigeminal ganglia, which originate from the ectodermal placode, express CaCCs (38, 43) and innervate tissues different from those that do not express CaCCs. This suggests that neurons express CaCCs to process sensory information in a tissue-specific manner.

It has been proposed that CaCCs in DRG are responsible for after-depolarizations following action potentials (44, 45). The estimated $[Cl^-]_i$ in DRG neurons is 30 mM (46), which produces an E_{Cl} of -35 mV. Thus opening CaCCs by Ca^{2+} entry or Ca^{2+} release from stores would depolarize the cell membrane or produce after-depolarizations (47–49).

CaCCs are also expressed in spinal cord neurons (50, 51). As in somatosensory neurons, only a fraction of spinal cord neurons express CaCCs, which may imply some specific role in neuronal function (6). In spinal cord neurons, E_{Cl} is near

−60 mV (52–54). It has been hypothesized that opening of CaCCs would not change the membrane potential significantly but would accelerate the repolarization if channels open during the action potential (53, 54). This would tend to limit repetitive firing and trains of action potentials.

Trains of action potentials trigger long-lasting after-depolarizations in parasympathetic neurons from rabbit parasympathetic ganglia by activation of CaCCs (55, 56). Parasympathetic neurons trigger depolarizations that are extremely slow and show an oscillatory behavior possibly owing to activation of CaCCs. In bullfrog sympathetic ganglion and rat superior cervical ganglia neurons, activation of CaCCs also results in depolarization and generation of after-depolarizations (57, 58). It seems that CaCCs are located in the neuronal dendrites of the sympathetic ganglion (45) where they may regulate the response to synaptic input.

The level of expression of CaCCs in neurons is up-regulated by GTP-γ-S (39) and axotomy (42, 58, 59) and down-regulated by metabolic stress (40).

Regulation of Cardiac Excitability

In some species, CaCCs play a role in repolarization of the cardiac action potential. In many cardiac myocytes, the transient outward currents I_{to2} and I_{to1} are responsible for the initial phase of repolarization. I_{to1} is a Ca^{2+}-insensitive K^+ current blocked by 4-aminopyridine, and I_{to2} is a Cl^- current stimulated by Ca^{2+} but insensitive to 4-aminopyridine (8, 60–65). I_{to2} results from the activation of CaCCs, which may depolarize or repolarize the membrane potential depending on the membrane potential and E_{Cl}. At positive membrane potentials, $I_{Cl.Ca}$ induces Cl^- influx and drives the membrane potential toward negative values. At negative membrane potentials, $I_{Cl.Ca}$ could promote depolarization and early after-depolarizations. The state of Ca^{2+} buffering plays a key role in determining the relative contribution of CaCCs to the total transient outward current. For example, β-adrenergic stimulation of L-type channels or fast heart rates augments CaCC activation (66), which increases the contribution of $I_{Cl.Ca}$ to membrane repolarization. In contrast, early repolarization of the membrane potential limits Ca^{2+} influx through VGCCs and limits CaCC activation (67).

Certain dogs, which are genetically prone to cardiac sudden death, have an abnormal I_{to} (68), implying that CaCCs might play a role in cardiac sudden death. CaCCs also contribute to the arrhythmogenic transient inward current (I_{ti}) in many species (8, 69–71). During Ca^{2+} overload, I_{ti} can trigger oscillatory after-potentials resulting in serious cardiac arrhythmias (72–74). Cl^- channels clearly participate in arrhythmogenesis because anion substitution or pharmacologic Cl^- channel blockade protects against reperfusion and ischemia-induced arrhythmias (75–77).

Smooth Muscle Contraction

CaCCs have been extensively studied in smooth muscle cells derived from a variety of tissues and appear to be involved in both regulation of myogenic tone and contraction stimulated by agonists (9, 78). Activation of CaCCs in smooth muscle can occur by Ca^{2+} entry through VGCCs or by Ca^{2+} released from intracellular stores by inositol 1,4,5-trisphosphate (IP_3) generated through the phospholipase C (PLC) pathway (9, 78). Because E_{Cl} is positive to the resting potential in smooth muscle, opening CaCCs will produce a depolarization (79). Norepinephrine, which contracts smooth muscle by activation of G_q-coupled α-adrenergic receptors, increases membrane $^{36}Cl^-$ efflux (80), which leads to membrane depolarization (81) by activation of $I_{Cl.Ca}$ (82, 83). The depolarization is almost abolished by removing external Cl^- in pregnant guinea pig myometrium, guinea pig mesenteric vein, and the anococcygeus muscle (84, 85). This depolarization could increase the open probability of VGCCs, thereby enhancing Ca^{2+} entry and further increasing muscle contraction. Thus smooth muscle contraction is under the control of the release Ca^{2+} from intracellular stores in response to muscle activators and Ca^{2+} entry through VGCCs activated by the depolarization induced by CaCC activation.

In agreement with this idea, anthracene-9-carboxylic acid (A9C) and indanyloxyacetic acid reduce the contraction of rat portal vein strips, rat aorta, renal arteries, and arterioles induced by norepinephrine, endothelin, and angiotensin II (86–88). Further support for the role of CaCCs in smooth muscle contraction was gathered using the Cl^- channel blocker NFA (89–93). NFA blocks both rabbit portal vein $I_{Cl.Ca}$ and rat aorta contraction induced by norepinephrine by about 50%. Smooth muscle cells also express I_{KCa}. Thus an increase in $[Ca^{2+}]_i$ can open both CaCC and I_{KCa}, which will induce depolarization and hyperpolarization.

Spontaneous depolarizations, which may result from the activation of $I_{Cl.Ca}$ by Ca^{2+} sparks, have been observed in smooth muscle in the absence of agonists. Although these depolarizations could alter the smooth muscle tone, the precise physiological significance of these depolarizations are unknown (9).

Fluid Secretion by Airway and Intestinal Epithelium

Airway epithelia use ion transport mechanisms to control the level of airway surface liquid, which is important for mucous hydration and protection against infection. Secretion of fluid into the airway is accomplished by basally located transporters that accumulate Cl^- in the cell against the Cl^- electrochemical gradient and by apical Cl^- channels that permit Cl^- to flow into the extracellular space down its electrochemical gradient. Airway epithelial cells coexpress CaCCs and cystic fibrosis transmembrane regulator (CFTR) in their apical membrane (94, 95). Airway epithelial cells stimulated with ATP or UTP display a Ca^{2+}-dependent Cl^- secretion (96, 97). UTP stimulates G_q-coupled P2Y purinergic receptors to increase IP_3 production and subsequently Ca^{2+} release. This increases the short-circuit current and the airway surface liquid of murine tracheal epithelial cell line (98, 99). The control of the airway mucous layer seems to be regulated by an interplay between CFTR and CaCCs (97). The basal level of the mucous layer is controlled by CFTR, whereas CaCCs act as an acute regulator of the liquid layer. The contribution of CaCCs to airway liquid layer homeostasis in murine airway epithelium probably explains the lack of a lung phenotype in mouse models of cystic fibrosis (94, 100). Given that CFTR and CaCCs are both apical Cl^- channels, it has been proposed that the activation of CaCCs could serve as a therapy for cystic fibrosis, but this line has been hampered by the lack of specific activators of CaCCs and by uncertainty about the molecular identity of these channels.

Intestinal epithelium can secrete Cl^- transiently upon stimulation with car-bachol, histamine, and nucleotides (101–104). These agonists increase $[Ca^{2+}]_i$, which leads to activation of CaCCs and triggers a secretory response. However, it has been noted that intestinal cells from cystic fibrosis patients do not respond to Ca^{2+}-mobilizing agonists (105). Thus although the presence of CaCC in intestinal epithelium is well established, its significance remains to be resolved.

Fluid Secretion by Exocrine Glands

Acinar and duct cells from lachrymal, parotid, submandibular, and sublingual glands, as well as pancreas, express CaCCs with similar properties. See Melvin et al. (14) in this volume for a detailed discussion of the role of CaCCs in fluid and electrolyte secretion in salivary glands. These tissues secrete an isotonic, plasma-like primary fluid that is rich in NaCl. The fluid secretion is Ca^{2+} dependent and triggered by the parasympathetic neurotransmitter acetylcholine (106–109). The rise in $[Ca^{2+}]_i$ is triggered by the muscarinic receptor-induced production of IP_3, which releases Ca^{2+} from internal stores. The elevation of $[Ca^{2+}]_i$ activates CaCCs and subsequently a Cl^- efflux through the apical membrane. The exit of Cl^- drives the movement of Na^+ through the parallel pathway and drags water, resulting in salty fluid secretion. Thus CaCCs are central to the fluid secretion process because they constitute the last step in the transepithelial movement of Cl^-, which is the net driving force for the whole process.

Fast Block of Polyspermy

In *Xenopus* oocytes, $I_{Cl,Ca}$ plays a fundamental role in blocking polyspermy by generating the so-called fertilization potential (110, 111). Upon fertilization, the membrane potential changes from the resting negative value (~ -40 mV) to depo-larized ($\sim +20$ mV) for several minutes. This depolarization is from an increase in membrane conductance to Cl^- and it is preceded by an increase in $[Ca^{2+}]_i$. At fertilization, there is a Ca^{2+} rise that travels across the cytoplasm in a wave fashion, which is the result of release of Ca^{2+} from internal stores by IP_3. Thus direct injection of IP_3 or Ca^{2+} induces Cl^--dependent membrane depolarization and egg activation that are similar to those triggered by fertilization. Ca^{2+} mobiliz-ing agents such as ionomycin also induce a similar change in membrane potential (110).

IP_3 is necessary for the initiation of the fertilization potential by inducing Ca^{2+} release. However, this limited source of Ca^{2+} is not sufficient to maintain the depolarization for prolonged periods. To overcome this problem, the depleted Ca^{2+} store activates store-operated Ca^{2+} entry, which contributes to CaCC stimulation

(112, 113). Ca^{2+} entry usually is sustained and maintains $[Ca^{2+}]_i$ high enough to activate CaCCs and subsequently produce the depolarization associated with the fertilization potential. A critical test for the role of CaCC-induced depolarization in blocking polyspermy was done using the voltage-clamp technique (114). Under voltage clamp at depolarized potentials, polyspermy is blocked, thus supporting the idea that CaCC-induced depolarization is fundamental to fertilization. This mechanism is particular to amphibian eggs: Different mechanisms are responsible for fast block to polyspermy in mammals, for example.

Kidney

CaCCs are widely expressed in kidney. $I_{Cl,Ca}$ has been described in cells acutely isolated from rabbit distal convoluted tubule, proximal convoluted tubule, and cortical collecting duct (115–117); in the M1 cortical collecting duct cell line (118); in acutely isolated intramedullary collecting duct (IMCD) cells; and in the IMCD cell lines IMCD-K2 and IMCD-K3 (119–122). Although textbook models of renal ion transport suggest that the Cl^- concentration of urine is determined mainly by what remains after incomplete absorption by the proximal tubule (123), there is growing acknowledgment that fluid and salt secretion in the distal kidney may contribute to urine composition (124–129). In the IMCD, for example, there is good evidence that Cl^- secretion occurs by both CaCCs and CFTR (125, 127). Because the opening of CaCCs is hormonally controlled, one can reasonably expect that they play a role in fine-tuning the urine composition.

Endothelial Cells

Endothelial cells are involved in preventing blood clotting, immune responses, and angiogenesis and produce a variety of vasoactive substances, including NO, prostaglandins, and endothelins. Humoral substances and flow rate can also be sensed by endothelial cells. Agonists such as histamine, ATP, and thrombin increase $[Ca^{2+}]_i$-dependent Cl^- fluxes and Cl^- currents by activation of CaCCs in endothelial cells (130, 131). Yet, the physiological role of CaCC as well as other ion channels in vascular endothelial cells remains controversial. CaCCs have been implicated in the control of the membrane potential that would help to maintain the driving force for Ca^{2+} (132). They could also play a role in the control of cell volume and cell proliferation (133).

MECHANISMS OF ACTIVATION

Source of Ca^{2+}

CaCC activation requires a rise in $[Ca^{2+}]_i$. The Ca^{2+} that activates CaCCs can come from either Ca^{2+} influx or Ca^{2+} release from intracellular stores (Figure 1). In some cases, it has been documented that specific types of Ca^{2+} channels are coupled to CaCCs. In rat DRG neurons, CaCCs are activated by both Ca^{2+} influx and Ca^{2+}-induced Ca^{2+} release from internal stores. Applying caffeine to release Ca^{2+} from intracellular stores in DRG neurons can activate CaCCs (134–136). In mouse sympathetic neurons, there appears to be a selective coupling of different kinds of VGCCs to Ca^{2+}-activated Cl^- and K^+ channels: Ca^{2+} entering through L- and P-type channels activates CaCCs, whereas Ca^{2+} entering through N-type channels activates Ca^{2+}-activated K^+ channels (137). In heart, CaCCs can be activated by Ca^{2+}-induced Ca^{2+} release triggered by reverse-mode Na^+- Ca^{2+} exchange when intracellular Na^+ is elevated (74, 138). In *Xenopus* oocytes, $I_{Cl.Ca}$ has different waveform and rectification properties depending on the source of Ca^{2+} (113, 139). The current stimulated by IP_3-triggered Ca^{2+} release is outwardly rectifying and exhibits time-dependent activation and deactivation, whereas the current stimulated by Ca^{2+} influx via store-operated Ca^{2+} channels is time-independent and is not rectifying. This observation has been interpreted to mean that Ca^{2+} influx produces a greater increase in Ca^{2+} in the vicinity of the CaCCs than does Ca^{2+} release from stores, because CaCCs in excised patches switch from time-dependent and outwardly rectifying to time-independent and nonrectifying when $[Ca^{2+}]_i$ is increased from ~200 nM to ~2 μM (113, 140).

There are two possible general mechanisms for Ca^{2+} to activate CaCCs: Ca^{2+} could bind directly to the channel protein or act indirectly on the channel via Ca^{2+}-binding proteins or Ca^{2+}-dependent enzymes (Figure 3, see color insert). The distinction between directly Ca^{2+}-gated and phosphorylation-dependent currents is reflected in the observation that some CaCCs can be stably activated in excised patches by Ca^{2+} in the absence of ATP (140–146) suggesting that activation does not require phosphorylation, whereas in other preparations channel activity runs down quickly after excision, suggesting the possibility that components in addition to Ca^{2+} are required to open the channel (25, 147–151). These two mechanisms seem to operate in different cell types, but may not be exclusive.

Direct Gating by Ca^{2+}

CaCCs from salivary gland acinar cells (145), pulmonary endothelial cells (148), ventricular myocytes (141), hepatocytes (140), and glomerular mesangial cells (152) appear to be activated by direct Ca^{2+} binding to the channel protein as they do not show a rundown of channel activity in excised patches. Because little is known about the molecular identity of CaCCs, how this direct Ca^{2+}-gating mechanism works can only be speculated. One presumes that the channel protein contains Ca^{2+}-binding motifs such as EF-hands or C2 domains. The apparent affinity of CaCCs for Ca^{2+} is in the micromolar range. This affinity is generally lower than EF-hands or C2 domains. Another possible motif would be a "Ca^{2+}-bowl" motif of BK potassium channels (153). Ca^{2+} binding to CaCCs from different tissues is voltage sensitive. At high positive membrane potentials Ca^{2+} appears to bind better to the channel protein than at low negative potentials (25, 113, 154).

Evidence supporting direct gating of CaCCs by Ca^{2+} has been obtained using inside-out patches isolated from hepatocytes and from *Xenopus* oocytes exposed to increasing [Ca^{2+}] on the cytosolic face of the excised patch (113, 140). Application of Ca^{2+} to an excised patch activates both single channels and macroscopic currents even in the absence of any ATP required for phosphorylation. The quick activation of CaCCs by rapid application of Ca^{2+} to excised patches (113) or by photoreleasing Ca^{2+} in acinar cells isolated from pancreas and parotid glands (155, 156) is also compatible with the idea that CaCCs are directly gated by Ca^{2+} ions. Further evidence in favor of the direct gating mechanism has been obtained in rat parotid acinar cells treated with inhibitors of calmodulin and of CaMKII. CaCC activation by increasing [Ca^{2+}]$_i$ with ionomycin was not prevented by preexposure of cells to KN-62 or peptide inhibitors of calmodulin and of CaMKII (157).

Gating by Phosphorylation via CaMKII

Some CaCCs are stimulated by protein phosphorylation involving the calmodulin-dependent protein kinase CaMKII. Regulation of Cl^- channels by CaMKII has been shown in cells from the human colonic tumor cell line T84 (158–162), airway epithelia (3), T lymphocytes (163), human macrophages (164), biliary epithelial cells (149), and cystic fibrosis-derived pancreatic epithelial cells (165). Evidence for involvement of calmodulin is provided by experiments showing that calmidazolium and trifluoroperazine, blockers of calmodulin, decreased CaCC current (166). Although this effect is most directly explained by calmodulin blockade, which would impair channel activation, a direct effect on CaCCs was not ruled out.

Evidence supporting the participation of CaMKII in CaCC activation has been provided by the use of inhibitors of calmodulin or CaMKII or by cell dialysis with the purified enzyme in a variety of cells including colonic T84, airway epithelia, and neutrophils (3, 162, 163). The use of CaMKII inhibitors, such as KN62, or the autocamtide inhibitory peptide in neutrophils and T84 cells blocks the activation of CaCCs. In contrast, dialysis of the purified enzyme into airway epithelial cells and neutrophils activates chloride currents (3, 163).

Kinetics of Activation by Ca^{2+}

The activation kinetics of CaCCs by increasing [Ca^{2+}]$_i$ has been studied by applying constant amounts of Ca^{2+}, by photoreleasing Ca^{2+}, by inducing Ca^{2+} release from intracellular pools by IP_3, or by enhancing Ca^{2+} entry by application of Ca^{2+} ionophores (35, 112, 113, 154, 156, 166, 167). These experiments have shown that when [Ca^{2+}]$_i$ is below $\sim$1 μM, $I_{Cl.Ca}$ is both voltage and time dependent, but at higher [Ca^{2+}]$_i$, the voltage and time dependence disappears.

At [Ca]$_i$ less than $\sim$1 μM, CaCCs activate slowly reaching a steady state in $\sim$2 s. The resulting activation time constants appear voltage independent at constant [Ca^{2+}]$_i$. As [Ca^{2+}]$_i$ increases, the activation rate is accelerated. In contrast, the deactivation of CaCCs follows a time course that is described by a single exponential, with a time constant that is voltage dependent. The resulting current-voltage relationship shows outward rectification at low $[Ca^{2+}]_i$ (< 500 nM). This rectification is nearly lost by rising $[Ca^{2+}]_i$ to >1 μM (113, 154, 167).

The apparent open probability (P_o) of CaCCs is voltage dependent. As $[Ca^{2+}]_i$ is increased, the activation curve is shifted toward more negative voltages. From the analysis of the $[Ca^{2+}]$ dependence of P_o, it is estimated that more than one Ca^{2+} ion is needed to activate one channel. Hill coefficients of 2 to 5 have been estimated from dose-response curves, thus suggesting the presence of multiple Ca^{2+}-binding sites in the channel protein (113, 154). The outward rectification that is seen at low $[Ca^{2+}]_i$ may be explained by the fact that the apparent affinity of CaCCs for Ca^{2+} is voltage dependent (113, 154, 166). This implies that positive voltages induce rearrangements of the Ca^{2+}-binding sites to favor interaction with Ca^{2+} ions to increase P_o.

The activation of CaCCs in *Xenopus* oocytes and rat parotid gland has been modeled assuming that 2 or 3 Ca^{2+} ions interact with closed state(s) of the channel protein in a linear sequence. Once Ca^{2+} ions are bound, the channel reaches the open state(s). A model (Figure 4, Scheme 1) with 2 Ca^{2+} ions bound proposes 3 closed states and 1 open state (154). An extension to a more complex model (Figure 4, Scheme 2) with 3 Ca^{2+} ions bound proposes 4 closed and 3 open states (113). Both models assumed that the Ca^{2+}-binding sites were independent but have the same affinity. At low $[Ca^{2+}]_i$, both models describe reasonably well the opening of CaCCs; however, at high $[Ca^{2+}]_i$, they fail to describe the kinetics of

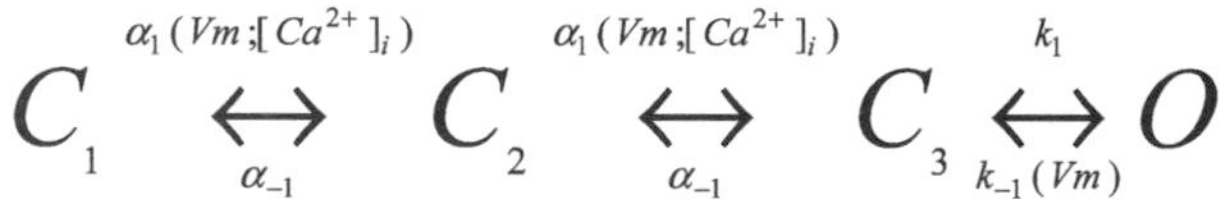

$$C_1 \underset{\alpha_{-1}}{\overset{\alpha_1(Vm;[Ca^{2+}]_i)}{\longleftrightarrow}} C_2 \underset{\alpha_{-1}}{\overset{\alpha_1(Vm;[Ca^{2+}]_i)}{\longleftrightarrow}} C_3 \underset{k_{-1}(Vm)}{\overset{k_1}{\longleftrightarrow}} O$$

Scheme 1

$$C_1 \underset{\alpha_{-1}}{\overset{\alpha_1[Ca^{2+}]_i}{\longleftrightarrow}} C_2 \underset{\alpha_{-2}}{\overset{\alpha_2[Ca^{2+}]_i}{\longleftrightarrow}} C_3 \underset{\alpha_{-3}}{\overset{\alpha_3[Ca^{2+}]_i}{\longleftrightarrow}} C_4$$

$$k_1 \updownarrow k_{-1} \qquad k_1 \updownarrow k_{-1} \qquad k_1 \updownarrow k_{-1}$$

$$O_1 \qquad\qquad O_2 \qquad\qquad O_3$$

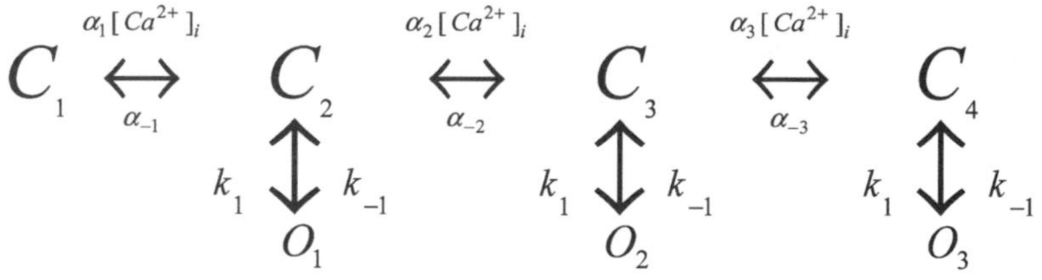

Scheme 2

Figure 4 Gating schemes for CaCCs. CaCCs are proposed to have several closed (C) and open (O) states. Rate constants are shown to be voltage (V_m) and/or Ca^{2+} sensitive. Scheme 1 was proposed by Arreola et al. (154) and Scheme 2 by Kuruma & Hartzell (113).

current deactivation. A discrepancy between these two models is the origin of the voltage dependence of CaCC gating. Scheme 1 assumes that the Ca^{2+} interaction with the channel, as well as the open-to-closed backward rate constant (k_{-1}), are voltage dependent thus conferring the V_m dependence to CaCC gating. Scheme 2, in contrast, assumes that the voltage dependence is conferred by the voltage dependence of the open-to-closed backward rate constants (k_{-1}). The voltage dependence of Ca^{2+} interaction has been assessed by rapid application of Ca^{2+} to excised patches (113). These studies show that the off-rate of Ca^{2+} binding is voltage dependent. Further insight into the mechanisms of Ca^{2+} activation would benefit from experiments involving photorelease of Ca^{2+} ions.

Single-channel data obtained from smooth muscle cells suggest that CaCCs could have 4 closed and 3 open states (168). A model similar to Scheme 2 proposed by these authors posited that the voltage dependence of CaCCs resides in the transition between C_1 and C_2. Furthermore, because the conductance of each substate decreased from 3.8 to 1.2 pS as the $[Ca^{2+}]_i$ increased, it was proposed that the occupancy of the Ca^{2+}-binding sites controlled channel conductance (168).

To make the matter more complex, the activation of CaCCs is dependent on the permeant anion (169, 170). Anions with high permeability, such as SCN^-, NO_3^-, or I^-, accelerated channel activation and slowed deactivation by a factor that nearly matched the permeability ratios. This effect was observed when permeant anions were applied on the external site of the channel and appeared to be independent of the channel affinity for Ca^{2+} (169, 170). Only a slight increase in the affinity of the *Xenopus* oocytes CaCC for Ca^{2+} has been reported when SCN^- is applied to the cytosolic site (113). Thus it appears that permeant anions favor Ca^{2+}-independent transitions that result in the channel remaining open a longer time. These data suggest that somehow the process of channel gating is coupled to the permeation mechanism (170). These observations, as well as more single-channel data, are necessary to build a model to describe the Ca^{2+} and V_m dependence of CaCC gating and the coupling of the permeation and gating mechanisms.

PERMEATION AND SELECTIVITY

CaCCs Are Relatively Non-Selective

The ability to select among various ions is a key feature of ion channels, but different kinds of channels exhibit very different kinds of selectivity. At one extreme are voltage-gated cation channels that are highly selective for one ion. For example, voltage-gated K^+ channels select for K^+ over Na^+ by a factor of >100 to 1 (Figure 5A) (171). These channels are selective for K^+ ions over other ions largely because they have a binding site in the channel pore for ions the size of K^+ (172). With this channel, and presumably others like it, the geometry of the channel and its binding site for ions is crucial for selectivity.

In contrast, most Cl^- channels including CaCCs are relatively nonselective (16, 171). Whereas voltage-gated K^+ channels exhibit >100-fold selectivity between

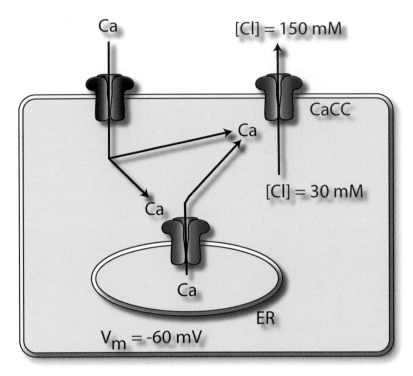

Figure 1   Factors controlling Cl flux through CaCCs. Ca^{2+} influx through plasma membrane Ca^{2+} channels (VGCCs or SOCCs) or release of Ca^{2+} from internal stores (ER) can stimulate CaCCs to open. Cl^- flux through open CaCCs depends on the membrane potential and the Cl^- concentration gradient.

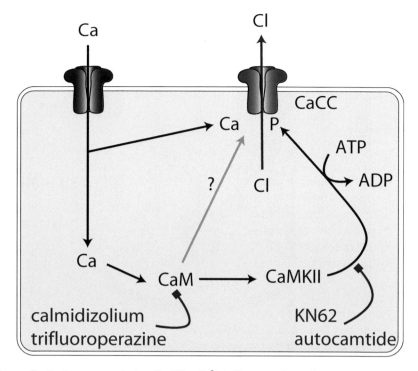

Figure 3 Pathways regulating CaCCs. Ca^{2+} influx or release from stores can stimulate CaCCs either directly of via calmodulin (CaM) -dependent pathways, including phosphorylation via CaMKII. Blockers are shown in red.

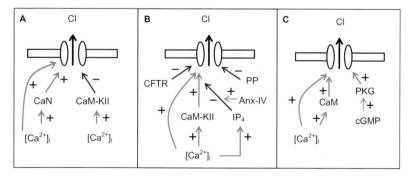

Figure 6 Alternative pathways of CaCC regulation. (*A*) Inhibition of CaCCs by CaMKII phosphorylation and calcineurin (CaN) dephosphorylation. This has been demonstrated in arterial smooth muscle. (*B*) Interaction of CaMKII, IP_4, annexin-IV, and CFTR in regulation of CaCCs. (*C*) cGMP-stimulated CaCC (270).

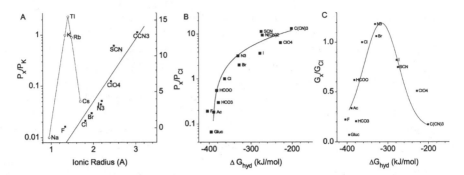

Figure 5   Selectivity and permeation through CaCCs in *Xenopus* oocytes. Data are from Qu & Hartzell (173). (*A*) Relative permeability of anions through CaCCs (*closed symbols*) as a function of anionic radius compared with the permeability of K^+ channels (*open symbols*). (*B*) Relative permeabilities of anions through CaCCs as a function of hydration energy. (*C*) Relative conductances of anions through CaCCs as a function of hydration energy.

ions having radii that differ by less than 0.5 Å, CaCCs select only $\sim$10-fold between ions that differ in radius by $\sim$1.5 Å (Figure 5*A*) (110, 173). Furthermore, CaCCs differ from K^+ channels in that there is no peak in the relationship between ionic radius and permeability. Rather, the relationship is monotonic with larger anions being more permeable than smaller ions. Most CaCCs studied, including those in *Xenopus* oocytes and rat parotid and lachrymal glands, display a selectivity sequence of $SCN^- > NO_3^- > I^- > Br^- > Cl^- > F^-$ (Figure 5*A*) (9, 148, 167, 173–175). Even the ability of CaCCs to select between anions and cations is relatively poor: Na^+ permeabilities are $\sim$10% those of Cl^- (173). These permeability features can be explained by a selectivity mechanism that is less dependent on a geometrically defined binding site for the permeant anion than K^+ channels.

Process of Ion Permeation and Conduction

Ion permeation and conduction can be viewed as consisting of several discrete steps. The ion moves from the aqueous environment and enters the channel, the ion moves through the channel, and then emerges into the aqueous environment on the other side. For the purposes of this discussion, we make a distinction between the ability of an ion to enter the channel, i.e., permeability, and the ability of an ion to pass through the channel, i.e., conductance. Permeability ratios are measured by the shift in the reversal potential when the ionic composition is changed from one in which Cl^- is the same on both sides of the membrane to another where Cl^- is replaced on one side by a substituent ion (176). Permeability ratios provide an estimate of the difference between the hydration energy in water and the solvation energy provided by the channel. Because there is no current at the reversal potential, this measurement is an indication of the ability of the ion to enter the channel. The process of moving from the aqueous environment to the channel pore involves exchanging the energy of stabilization of the ion in bulk water for the energy of stabilization of the ion by its interaction with the channel. Because ions are stabilized in bulk water by shells of water molecules surrounding the ion, stabilization of the ion in water can be characterized by its hydration energy (G_{hyd}). Stabilization of the ion by interaction with the pore can involve solvation of the ion by part of the channel protein, as in the case of the KcsA channel or other mechanisms (176, 177). The ease of ion permeation is determined by the difference between G_{hyd} in bulk water and the energy of stabilization (or solvation) by the channel. The smaller the difference, the more easily an ion enters the channel. For CaCCs, the relative permeability is related to G_{hyd} (Figure 5B) (170, 173). Generally, larger ions, which have a lower effective charge density (if the charge is uniformly distributed in the ion), have lower hydration energies. Thus larger ions are relatively more permeant than smaller ions. Permeability ratios, however, do not measure the ability of the ion to traverse the channel, which is measured by the conductance or the slope of the current-voltage relationship. To some extent, the conductance of an ion through the channel reflects how rapidly it dissociates from the ligands that stabilize the permeant ion in the channel.

The distinction between permeability and conductance becomes important because the relative permeability and conductance sequences are quite different in CaCCs (Figure 5B,C) (170, 173). The ease with which an ion enters the CaCC pore (permeability) depends on the ease with which the anion loses its bound water (Figure 5B). However, the ease with which an ion passes through the channel (conductance) exhibits a bell-shaped relationship to hydration energy (Figure 5C). These relationships suggest that anions with large hydration energies are poorly conductive because they do not enter the channel well (as shown by the P_x/P_{Cl} versus ΔG_{hyd} plots), whereas ions that have small hydration energies are poorly conductive (as shown by the G_x/G_{Cl} versus ΔG_{hyd} plot) because they become lodged in the pore, even though they enter the channel easily. In support of the idea that more hydrophobic anions stick in the pore, anions with small hydration energies, such as SCN^- and $C(CN)_3^-$, block Cl^- conductance (173). This ability of other permeant ions to block Cl^- conductance is a common feature of Cl^- channels (176, 178, 179), which suggests that the selectivity filter incorporates a hydrophobic-binding site.

From this analysis, it seems that the ionic selectivity of CaCCs can be explained by a mechanism in which ion entry into the channel is governed simply by the partitioning of anions into a tunnel with a relatively high dielectric constant (170, 173), as has been described for CFTR (170, 176–180).

Cation Permeability

The cation permeability of CaCCs is relatively large: P_{Na}/P_{Cl} is 0.1 (173). This value is comparable to values that have been reported for some other anion channels, although the values for cation permeability are often quite disperse. For example, values for CFTR range from 0.003 to 0.2 (see 181). Other anion channels differ significantly in their cation permeability. At least some $GABA_A$ receptors have undetectable K^+ permeability (182), whereas background Cl^- channels in hippocampus (183) and muscle (184) have P_{Na}/P_{Cl} ratios as high as 0.2. The explanation for this variability remains unclear, but relatively simple changes in primary sequence can modulate cation/anion selectivities significantly. Single-amino acid substitutions can increase the P_{Na}/P_{Cl} ratio in the $GABA_A$ receptor from <0.05 to >0.3 (185) and the cation-selective nicotinic ACh receptor can be made anion-selective by insertion of a proline residue (186).

Franciolini & Nonner (187, 188), studying a large-conductance Cl^- channel in hippocampal neurons, have shown that cations can permeate only in the presence of permeant anions and have proposed that anions and cations form mixed complexes while traversing the channel. It is possible that a similar mechanism may operate in CaCCs. Clearly, electrostatic interactions between permeant anions and the pore walls must play a role because of the inability of cations to permeate in the absence of permeant anions. If anion permeation commonly occurs as cation-anion complexes, it will be necessary to rethink the mechanisms of anion permeation.

Pore Dimensions

CaCCs appear to have a relatively large pore because ions as large as $C(CN)_3^-$ are highly permeable. Molecular modeling has estimated $C(CN)_3^-$ to be 0.33 × 0.75 nm (177, 189). Thus CaCCs must have a pore diameter at least that large. CaCCs are blocked in a voltage-dependent manner by A9C, which has dimensions of 0.5 × 0.94 nm, so the channel opening must lie between these values (189). From studies on voltage-dependent block by different drugs, the CaCC pore has been modeled as an elliptical cone with the larger opening facing the extracellular space (189). The outside opening is at least 0.6 × 0.94 nm because niflumic acid (NFA), which has these dimensions, enters the pore from the extracellular side.

Molecular Analysis of Permeability and Selectivity

The precise mechanisms of anion permeation in CaCCs must await the definitive molecular identification of these channels coupled with mutagenesis and structural studies. Recently, it was suggested that bestrophins may be CaCCs. Studies on anion permeation in these channels show that they have features similar to those described above for native CaCCs (190, 191). Mutational analysis has defined the putative second transmembrane domain as an important determinant of channel permeability (191–193). The conclusion from these studies is that mutations in TMD2 either destroy channel function or alter channel selectivity, consistent with the role of these residues being located in the channel pore. However, the effects of the mutations that alter selectivity are relatively modest. The effects of mutations are evident only with SCN^- as the permeant anion. P_{SCN}/P_{Cl} was changed more than fivefold, but NO_3^-, I^-, and Br^- permeabilities were not significantly affected. Furthermore, in mouse bestrophin-2 (mBest2) there was not a single amino acid residue that exhibited priority in determining anion permeability: Mutations in at least six different residues altered selectivity in similar manners. This suggests that certain details of the structure of the pore may not be crucial in determining anion permeation. Permeation may simply depend on ions partitioning into a hydrophilic channel, and as long as the channel maintains this hydrophilic pore, permeation occurs relatively normally.

Likewise, mutation of amino acids in ClC and CFTR channel pores does not produce discrete outcomes as found with K^+ channels, probably because Cl^- channels are not as highly selective as voltage-gated cation channels. In the ClC family, four stretches of noncontiguous amino acids contribute to the selectivity filter (16). Mutation of almost every amino acid in these domains changes in anion selectivity (194). In human ClC-1, for example, mutation of 17/19 amino acid residues in these regions alters channel selectivity (178, 195, 196). Moreover, the effects of these mutations are moderate. For example, one of the mutations (G233A) in the human ClC-1 channel that produces the largest changes in relative permeability increases P_{SCN}/P_{Cl} only ~8-fold and increases P_{NO3}/P_{Cl} and P_I/P_{Cl} only ~ 3-fold, compared with the 100-fold changes in Na^+/K^+ selectivity produced by mutations in the Shaker K^+ channel.

A similar situation exists with CFTR (176). The pore of CFTR has residues in transmembrane domains 6 and 11. Indeed, although numerous mutations within the transmembrane regions of CFTR alter anion binding and single-channel conductance, most mutations have rather little effect on anion selectivity. Even mutations that alter the selectivity sequence, such as F337A, do so by less than fourfold changes in relative permeabilities (197). These data have led Dawson and colleagues to propose that the detailed structure of the CFTR pore may not be a major factor determining anion selectivity (176, 177, 198). They propose that permeation is determined largely by the ease with which an anion partitions into the channel, which is a function of how easily the anion exchanges its water of hydration with residues in the channel. As long as the pore provides an adequately hydrophilic environment for the permeating anion, small perturbations in channel structure may not alter the permeability.

PHARMACOLOGY

Specific blockers are indispensable for identifying ion channels physiologically and for isolating specific currents from a mixture of currents. Blockers are also valuable tools for resolving the structure of the pore, analyzing tissue distribution, or for the affinity purification of channel proteins. Unfortunately, few specific potent anion channel blockers are available, and even fewer exist for CaCCs. Most require high concentrations to completely block Cl^- currents and may have undesirable side effects. The features of the available chloride channel blockers have been discussed in detail in several reviews (6, 10, 16). Table 1 summarizes the effects of various drugs on CaCCs in various tissues.

TABLE 1 CaCC pharmacology

	NFA	FFA	MA	DIDS	SITS	DPC	DCDPC	A9C	NPPB	FS	EA	Mib.	ArgT.	Tamox.	Glib.	Fluox.	DTT
Xenopus oocytes	17VD (286)	28 (286)	o	48VD (189)	o	111VD (189)	o	10.3VD (189)	22-68 (288)	o	o	o	o	_(189)	o	o	_(189)
IMCD	7.6[120]	o	o	+[120]	+[120]	+[120]	o	+[120]	o	o	o	o	o	_[120]	o	o	+[120]
T84	o	o	o	+[280]	o	o	o	o	o	o	o	o	o	o	o	o	o
Tracheal epithelium	20[273]	60 (273)	o	o	o	320 (273)	o	o	o	o	o	o	o	o	o	o	o
Lingual epithelium	+[287]	+[287]	o	247[287]	o	o	o	o	o					o	o	o	o
Submandib. acinar cell	o	o	o	+[277]	o	+[277]	o	+[277]	+[277]	o	o	o	o	o	o	o	o
Pulmonary arteryendothelium	+ VD (130,200)	o	o	+ VD (130,200)	o	o	+(200)	o	+nVD (130,200)	o	o	4.7 (281)	o	+nVD (130)			
Portal vein myocyte	2–3.6 (91,276)	20 (91)	70 (91)	16–210 (272,276)	640–2000 (276,285)	306 (272)	o	117–1000 (272,285)	o	500 (91)	200 (91)	o	o	o	o	o	o
Tracheal myocyte	10[275]	o	o	+[274]	o	o	o	o	o	o	o	o	o	o			
Urethral myocyte	+[278]	o	o	o	o	o	o	100[278]	o	o	o	o	o	o	o	o	o

(Continued)

TABLE 1 (*Continued*)

	NFA	FFA	MA	DIDS	SITS	DPC	DCDPC	A9C	NPPB	FS	EA	Mib.	ArgT.	Tamox.	Glib.	Fluox.	DTT
Cerebral artery myocyte	26[282]	o	o	o	o	o	o	o	o	o	o	o	o	o	o	o	o
Ventricular myocyte	+[289]	o	o	+[284,289]	o	o	o	o	o	o	o	o	o	o	61.5[290]	10.7[290]	o
DRG-neurons	+[136]	o	o	o	o	o	o	o	+[41]	o	o	o	o	o	o	o	o
Olfactory epith. neuron	44[279]	108[279]	o	o	o	o	14[17]	o	o	o	o	o	1[283]	o	o	o	o
Parasymp. neuron	o	o	o	100[271]	o	o	o	o	o	o	o	o	o	o	o	o	o
Spinal cord neuron	o	o	o	o	1000[50]	o	o	o	o	o	o	o	o	o	o	o	o

° Not determined; + blocks CaCC but IC$_{50}$ not determined; VD/nVD: block is voltage dependent/not voltage dependent.
References are in parentheses.

Blockers: (NFA) niflumic acid; (FFA) flufenamic acid; (MA) mifanamic acid; (DIDS) 4,4′-diisothiocyanato-stilbene-2,2′-disulfonic acid; (SITS) 4-acetamido-4′-isothiocyanatostilbene-2,2′-disulfonic acid; (DPC) diphenylamine-2-carboxylate; (DCDPC) 3′,5′-dichlorodiphenylamine-2-carboxylic acid; (A9C) anthracene-9-carboxylic acid; (NPPB) 5-nitro-2-(3-phenylpropylamino)-benzoic acid; (FS) furosemide; (EA) ethacrinic acid; (Mib.) mibefradil; (ArgT-636) argiotoxin-636; (Tamox.) tamoxifen; (Glib.) glibenclamide; (Fluox.) fluoxatine; (DTT) di-thiothreitol.

The most common blockers for native CaCCs are NFA and flufenamic acid (199). Both block CaCCs in *Xenopus* oocytes at concentrations in the 10 μM range (189). NFA is often considered a specific blocker and even used to identify anion currents as CaCCs in different tissues. However, NFA is far from being a perfect tool to isolate CaCCs. In addition to its blocking effect, NFA also enhances $I_{Cl.Ca}$ in smooth muscle at negative voltages (200). Undesirable effects of NFA include block of volume-regulated anion channels (VRAC)s (201) and K^+ channels (202, 203). Also, NFA can affect Ca^{2+} currents (202, 204), which complicates the interpretation of effects on $I_{Cl.Ca}$.

Other commonly used chloride channel blockers, including tamoxifen, DIDS, SITS, NPPB, A9C, and DPC, are even less effective than the flufenamates on CaCCs (6). Glycine hydrazine and acidic dacyl-ureas have recently been shown to be high-affinity blockers of CFTR (205) and VRAC (206), but they have not been tested on CaCCs. Fluoxetine and mefloquine also block Cl^- channels, but are more effective on VRAC than on CaCCs (207, 208). Chlorotoxin, a small peptide isolated from the venom of the scorpion *Leiurus quinquestriatus* (209), appears to specifically block CaCCs of rat astrocytoma cells (210). However, chlorotoxin or other related peptides were ineffective on CaCCs in T84 cells (211).

Some Cl^- channel blockers block CaCCs in a voltage-dependent manner, i.e., A9C (120, 189). A9C in the bath blocks outward current without significantly affecting inward current. Analysis of voltage-dependent block suggests that A9C binds to a site in the channel that is about 60% across the voltage field from the outside (189). Larger blocking molecules are less voltage dependent, suggesting that they lodge at sites less deep in the channel. DPC and DIDS block at a site about 30% into the voltage field, whereas NFA appears to block at the external mouth of the channel.

REGULATION

CaMKII and IP$_4$

As noted above, CaCCs can be activated by phosphorylation mediated by CaMKII. However, CaMKII activation of CaCCs is actually quite complex. For example, in arterial smooth muscle numerous data support the conclusion that CaMKII inhibits $I_{Cl.Ca}$, which is stimulated by Ca^{2+} (Figure 6A, see color insert) (212). $I_{Cl.Ca}$ in both pulmonary and coronary artery smooth muscle cells is stimulated by CaMKII inhibitors (212). Furthermore, the Ca^{2+}-dependent phosphatase calcineurin (CaN) is a positive regulator of CaCCs (213). Addition of CaN enhances the amplitude of $I_{Cl.Ca}$, whereas cyclosporin A, a CaN blocker, reduces $I_{Cl.Ca}$ amplitude. CaMKII-dependent phosphorylation also inhibits CaCCs in tracheal smooth muscle (214). In *Xenopus* A6 kidney cells, alkaline phosphatase increases $I_{Cl.Ca}$ (presumably via protein dephosphorylation) (215).

In some cell types, most notably in the human pancreatoma epithelial cell line CFPAC-1, CaCCs can be stimulated both directly by Ca^{2+} and indirectly via

CaMKII (Figure 6B) (216). In these cells, the stimulation by CaMKII is regulated by inositol 3,4,5,6-tetrakisphosphate (IP_4). IP_4, which is synthesized upon sustained PLC activation, reduces the stimulation of CaCCs by CaMKII (161, 217–220). The mechanism seems to involve inhibition of phosphorylation because IP_4 does not inhibit the channel when it is stimulated directly by Ca^{2+} (216). Furthermore, the inhibitory effect is prevented by inhibition of protein phosphatase activity, suggesting that dephosphorylation must be involved. However, IP_4 has not been shown to have a direct effect on protein phosphatase activity (161).

Similar to CFPAC-1 cells, T84 colonic carcinoma cells express CaCCs that are stimulated by CaMKII and inhibited by IP_4 (158, 159, 161, 162, 219, 221, 222). The T84 channels differ from those in CFPAC-1 cells in that they are not directly stimulated by Ca^{2+}. The inhibition of the current by IP_4 occurs under only certain conditions, suggesting that these CaCCs may have multiple phosphorylation sites such that when the channel is hyperphosphorylated, it is resistant to inhibition by IP_4 (161). A key issue in elucidating the effects of this regulatory pathway is the identification of the CaMKII-activated CaCC (see below).

The regulation of CaCCs by IP_4 may be a physiological mechanism to regulate the time course of CaCC activation (216). In response to a Ca^{2+} spike, CaCCs are turned on initially by Ca^{2+} and then in a sustained manner via CaMKII phosphorylation, which outlives the Ca^{2+} spike. In this case, IP_4 may modulate this second phase of CaCC activation.

Annexins

Annexins are phospholipid and Ca^{2+}-binding proteins that are concentrated along the apical membrane of many secretory epithelia (159, 223). Annexins can inhibit CaCCs from *Xenopus* oocytes (224) and epithelial cells (158, 159). In *Xenopus* oocytes, the IC_{50} is approximately 50 nM. The blocking effect of IP_4 is synergistic with the effect of annexin-IV: Low concentrations of annexin-IV that have no effect on CaCCs double the potency of IP_4 in blocking the current (Figure 6B) (219).

Interactions between CFTR and CaCCs

CFTR is a chloride channel that is defective in cystic fibrosis (CF). CFTR interacts with different proteins including a variety of ion channels such as ENaCs, VRACs, and CaCCs (Figure 6B) (225). In bovine pulmonary artery endothelial (CPAE) cells (226), *Xenopus* oocytes (227), and mouse parotid acinar cells (174) expression of CFTR reduces $I_{Cl.Ca}$. Airway epithelial cells from CF patients and CF-mouse models have an increased $I_{Cl.Ca}$ (174, 228–231). The regulation of CaCCs by CFTR involves the interaction of the C-terminal part of the CFTRs R-domain with CaCCs (232). The PDZ-domain of CFTR, which interacts with several other proteins (225), seems not to be involved in this interaction.

pH

A decrease in intracellular pH inhibits CaCCs in acinar cells from lachrymal and parotid glands as well as from T84 cells (233, 234). In excised patches containing CaCCs from *Xenopus* oocytes, acidification of the internal or external solution has little effect, but alkalinization of the cytoplasmic face of the patch blocks inward current (173). The mechanism of CaCC regulation by H^+ is unknown. The physiological relevance of CaCC regulation by H^+ is also not clear. It may serve as a negative feedback loop for the exit of HCO_3^- through CaCC and thereby prevent excessive cytosolic acidification of the cell (12).

cGMP-Dependent CaCC

Recently, it was shown that rat mesenteric artery smooth muscle cells have a Cl^- channel that is dependent on cGMP and is stimulated by Ca^{2+} (Figure 6*C*) (168, 235). Stimulation of this current by Ca^{2+} absolutely requires cGMP, which is thought to act via cGMP-dependent protein kinase and phosphorylation (235). Single-channel recording from inside-out patches shows that the cGMP-dependent CaCCs have substate conductances of 15, 35, and 55 pS, which are larger than those of CaCCs in rabbit pulmonary artery smooth muscle cells (168). In addition, the cGMP-dependent CaCC is potentiated by CaM but unaffected by CaMKII blockade (236). Thus smooth muscle cells appear to express various types of CaCC channels.

G Proteins

GTP-γ-S applied to the cytosolic site of cell-free, inside-out patches isolated from submandibular acinar cells induced the appearance of small-conductance CaCC (145). These data suggest that CaCCs could be up-regulated directly by activation of G proteins; however, similar data have not been reported in other preparations. On the other hand, activation of G proteins can regulate CaCCs indirectly. For example, activation of G proteins with GTP-γ-S stimulates CaCCs via the PLC-IP_3- Ca^{2+} signaling pathway in HTC hepatoma cells (237). Cl^- channel activation by GTP-γ-S occurs through an indomethacin-sensitive pathway involving sequential activation of PLC, mobilization of Ca^{2+} from IP_3-sensitive stores, and the stimulation of phospholipase A_2 ($PLCA_2$) and cyclooxygenase (COX). Surprisingly, the activation of CaCCs in these cells by Ca^{2+} was inhibited by COX inhibitors such as aspirin. Additional evidence for a role of G proteins regulation of CaCCs has been obtained with ginsenosides, the ingredients of the medical root, *Panax ginseng*, which activate $I_{Cl.Ca}$. In *Xenopus* oocytes ginsenosides seem to act specifically on the $G\alpha_{q/11}$ G protein, which is coupled to an PLC β3-like enzyme that releases Ca^{2+} from intracellular stores via IP_3 production (238, 239).

MOLECULAR STRUCTURE

How Many Types of CaCCs Exist?

CaCC currents recorded in whole-cell configuration have very similar properties in many different cell types, including *Xenopus* oocytes (139), various secretory epithelial cells (146, 154, 167, 240–244), hepatocytes (140), vascular, airway and gut smooth muscle cells (9), Jurkat T cells (163), and pulmonary artery endothelial cells (148). In general, these currents are Ca^{2+}- and voltage-sensitive; activate slowly with depolarization; exhibit a linear instantaneous IV relationship; and an outwardly rectifying steady-state IV relationship; have higher permeability to I^- than Cl^-; and are partially blocked by DIDS (100–500 μM), NPPB (100 μM), and NFA (100 μM).

Although whole-cell $I_{Cl,Ca}$ seems quite similar in different tissues, there is considerable diversity in the properties of single CaCCs. There appear to be at least four types of CaCCs in different cell types. Low-conductance CaCCs (1–3 pS) have been described in cardiac myocytes (141), arterial smooth muscle (151, 168, 245), A6 kidney cells (215), endocrine cells (246), and *Xenopus* oocytes (142). Within this class of channels, there is considerable diversity in properties. Depending on the study (or on the conditions), the channels can exhibit either linear or outwardly rectifying IV curves, K_d's for Ca^{2+} over a $\sim$500-fold range, variable voltage sensitivity, and different susceptibility to rundown after excision of the patch. Rundown after patch excision suggests that the channels are regulated by a factor that is lost upon excision. The second class of CaCCs are 8-pS with linear I–V relationships, described in endothelial cells (148) and hepatocytes (140). The 15-pS CaCCs, described in colon (150) and a biliary cell line (149) also have linear IV relationships but are blocked by CaM antagonists. The highest conductance channels (40–50 pS), described in Jurkat T cells (163), *Xenopus* spinal neurons (51), vascular smooth muscle cells (168), and airway epithelial cells (144), are outwardly rectifying. At least some of these channels are activated by CaMKII. Several other CaCCs have been described that do not fit into these four classes (144, 247, 248). In addition, there is a large conductance (310 pS) maxi-CaCC that has been described in *Xenopus* spinal neurons (51).

Whether this diversity of single-channel conductance truly reflects the variety of single channels that underlie the typical macroscopic $I_{Cl,Ca}$ remains ambiguous because rarely have investigators carefully linked single-channel measurements with macroscopic currents.

Molecular Candidates

Clearly, elucidating the molecular identity of CaCCs is an important goal in understanding the role of these channels in normal physiology, as well as in disease. The search for the molecular counterparts for CaCCs has been slow for several reasons. First, one of the favorite expression systems for expression cloning of ion channels has been the *Xenopus* oocyte, but this cell expresses huge $I_{Cl,Ca}$.

Furthermore, the tools that are available to differentiate this channel from other Cl^- channels are limited: As discussed above, there are no blockers of sufficient specificity. The absence of specific drugs that bind to CaCCs has also hampered cloning approaches that begin with purification of CaCC protein. None of the known cloned Cl^- channels—including CFTR, ligand-gated anion channels such as the $GABA_A$, and glycine receptors, and the ClC family—have properties befitting CaCCs. At present, three or possibly four molecular candidates have been proposed to be CaCCs.

ClC-3, a CaMKII-Activated Channel?

Recently it was shown that a ClC-3 homolog chloride channel, which is regulated by Ca^{2+}, might represent a Ca^{2+}-dependent chloride channel activated by CaMKII (249, 250). Experiments using $ClCn3^{-/-}$ mice show that the CaMKII-activated conductance is absent (250), but the Ca^{2+}-dependent chloride conductance is present and its properties are similar to those described in the wild-type animals (251). This suggests that the CaMKII-activated and the Ca^{2+}-activated chloride conductances are different. Additional experiments with $CaMKII^{-/-}$ mice and cloning of the gene encoding a CaCC would help to clarify if, indeed, these two conductances represent two different channels proteins.

CLCA Family

The Ca^{2+}-activated Cl^- channel family (CLCA) remains a highly contentious candidate for CaCCs (10, 16). These channels were cloned initially from a bovine tracheal cDNA expression library that was screened with an antibody generated against a purified protein that behaved as a CaCC in artificial lipid bilayers (252). Transfection of various cell types with cDNAs encoding various CLCA proteins induces Ca^{2+}-dependent currents (253–257).

Despite these findings, there has been considerable reluctance to accept the CLCA family as a candidate for CaCCs. This is partly due to the fact that the CLCA proteins have high homology to known cell adhesion proteins, and at least one family member (hCLCA3) is clearly a secreted protein (257, 258). Furthermore, despite the fact that it has been nearly 10 years since the first CLCA was cloned, structure-function analysis of this channel has not been accomplished. However, in addition, there are more fundamental questions surrounding the CLCA family, as discussed in detail by Eggermont (10). First, there are phenotypic differences between CLCA currents and those of native CaCCs. These differences include important (but perhaps only apparent) differences in Ca^{2+} sensitivity, voltage sensitivity, and pharmacology of the channel. These differences could be explained if CaCCs were heteromers with CLCA being only one of the subunits and the other subunit not yet being known. Second, CLCA proteins are proteolytically processed, but data concerning the fragment that is responsible for channel activity appear contradictory. Third, a number of cell types that express native CaCCs do not express CLCA proteins (259). One possibility to explain this is that CLCA proteins modulate endogenous Cl^- channels (260–262).

The Bestrophin Family

Recently, the Nathans group proposed that bestrophins comprise a new family of Cl^- channels (263). Mutations in human bestrophin-1 (hBest-1) produce Best vitelliform macular dystrophy, an early onset form of macular degeneration (264). It has been proposed that hBest-1 is a Cl^- channel in the basolateral membrane of retinal pigment epithelial cells (263, 265). Our laboratory and Nathans' have shown that bestrophins from several species function as Cl^- channels when expressed heterologously and that the Cl^- currents are stimulated by Ca^{2+} with a K_d of ~200 nM (190, 192, 263). Expression of dominant-negative mutants of bestrophin, notably G299E and W93C, inhibit the wild-type current. The key experiment demonstrating that bestrophin is a Cl^- channel involved mutagenesis of residues in the second transmembrane domain of mBest2. Mutation of several residues in this region (191, 193) altered the anionic selectivity and conductance of the channel. Because it is generally agreed that the selectivity of a channel is determined by the channel pore, the ability to change the selectivity by a mutation proves that bestrophin is responsible for forming the channel.

The troubled history of Cl^- channel identification (16, 266) should make one exceedingly circumspect about the conclusion that mBest2 forms the Cl^- channel pore, despite the fact that mBest2 selectivity can be changed by mutation. More detailed knowledge about the structure of the pore is required before the case is closed. Are bestrophins the molecular counterpart for classical CaCC currents? This remains to be seen (267). Both expressed bestrophin channels and classical CaCCs are gated directly by Ca^{2+}, without the involvement of kinases. Both channels exhibit the generic lyotropic anion-selectivity sequence. These similarities suggest that bestrophins could be classical CaCCs. However, as with the CLCA family, there is an important difference. Classical CaCCs exhibit voltage-dependent kinetics and outward rectification that is not seen with hBest1 or mBest2. This difference could be explained if native CaCCs have another subunit not present in the expressed homomeric channels. Furthermore, in heterologous expression systems, a large fraction of the bestrophins are intracellular, raising the possibility that bestrophins are intracellular Cl^- channels (190, 192). The conclusion that bestrophins are integral parts of CaCCs will require demonstration that disruption of bestrophin genes can knock down endogenous CaCC currents, but this has not yet been published. Also, there are no published single-channel recordings of bestrophin currents, and the pharmacology of bestrophin currents is not well described.

Tweety

Recently, two human genes (*hTTHY2* and *hTTYH3*), with homology to a gene in the flightless locus of *Drosophila* called *tweety*, have been shown to be a

Ca^{2+}-regulated maxi-Cl^- channel (260 pS) (268). This channel might correspond to the maxi-Cl^- channel found in spinal neurons (51) and skeletal muscle (269). *hTTYH3* is not expressed in salivary gland, so it is very unlikely that it plays any role in the small conductance CaCCs that are typical of acinar cells of secretory glands. A related gene, *hTTYH1*, codes for a channel that is not regulated by Ca^{2+} (268).

CONCLUDING REMARKS

CaCCs are an important family of pervasively expressed channels. Yet, our understanding of these channels is wanting for two major reasons: lack of good drugs to study them and questions about their molecular identity. On the basis of electrophysiological studies, it appears that there are a numerous types of CaCCs as seen from both their single-channel conductances and their modes of regulation by Ca^{2+} and phosphorylation. However, it has been difficult to define particular classes of CaCCs because there are no drugs that block one type or the other with adequate specificity. Indeed, this is a major problem with Cl^- channels in general. Because CaCCs are physiologically important, a concerted effort to find drugs that have specificity for these and other kinds of Cl^- channels is long overdue. High-throughput techniques are now available to make this a practical undertaking. But, another major void is the molecular identity of the channels. The two molecular candidates that have been proposed, CLCAs and bestrophins, remain unproven as subunits of classical CaCCs. Experiments investigating the effects of disruption of these genes on $I_{Cl,Ca}$ are urgently required. Other approaches to clone CaCCs, for example by expression cloning, have been tried by several investigators, but to date no success has been published. Such approaches will be important avenues to advancing our understanding of the mechanisms and physiology of these channels.

ACKNOWLEDGMENTS

This work was supported in part by National Institutes of Health grants EY104852 and GM60448 (H.C.H.); DE-09,692 and DE13539 (J.A.); and ER026 from CONA-CyT, Mexico (J.A.).

The *Annual Review of Physiology* is online at
http://physiol.annualreviews.org

LITERATURE CITED

1. Miledi R. 1982. A calcium-dependent transient outward current in *Xenopus laevis* oocytes. *Proc. Natl. Acad. Sci. USA* 215:491–97

2. Barish ME. 1983. A transient calcium-dependent chloride current in the immature *Xenopus* oocyte. *J. Physiol.* 342:309–25

3. Wagner JA, Cozens AL, Schulman H, Gruenert DC, Stryer L, Gardner P. 1991. Activation of chloride channels in normal and cystic fibrosis airway epithelial cells by multifunctional calcium/calmodulin-dependent protein kinase. *Nature* 349:793–96

4. Grubb BR, Gabriel SE. 1997. Intestinal physiology and pathology in gene-targeted mouse models of cystic fibrosis. *Am. J. Physiol. Gastrointest. Liver Physiol.* 273:G258–66

5. Boucher RC. 1994. Human airway ion transport. *Am. J. Respir. Crit. Care Med.* 150:271–81

6. Frings S, Reuter D, Kleene SJ. 2000. Neuronal Ca^{2+} -activated Cl^- channels— homing in on an elusive channel species. *Prog. Neurobiol.* 60:247–89

7. Kawano S, Hirayama Y, Hiraoka M. 1995. Activation mechanism of Ca^{2+} sensitive transient outward current in rabbit ventricular myocytes. *J. Physiol.* 486:593–604

8. Zygmunt AC. 1994. Intracellular calcium activates a chloride current in canine ventricular myocytes. *Am. J. Physiol. Heart Circ. Physiol.* 267:H1984–95

9. Large WA, Wang Q. 1996. Characteristics and physiological role of the Ca^{2+}-activate Cl^- conductance in smooth muscle. *Am. J. Physiol. Cell Physiol.* 271:C435–54

10. Eggermont J. 2004. Calcium-activated chloride channels (Un)known, (Un)loved? *Proc. Am. Thorac. Soc.* 1:22–27

11. Fuller CM, ed. 2002. *Calcium-Activated Chloride Channels*, Curr. Top. Membr. Vol. 53. San Diego: Academic

12. Begenisich T, Melvin JE. 1998. Regulation of chloride channels in secretory epithelia. *J. Membr. Biol.* 163:77–85

13. Kidd JF, Thorn P. 2000. Intracellular Ca^{2+} and Cl^- channel activation in secretory cells. *Annu. Rev. Physiol.* 62:493–513

14. Melvin JE, Yule D, Shuttleworth TJ, Begenisich T. 2005. Regulation of fluid and electrolyte secretion in salivary gland cells. *Annu. Rev. Physiol.* 67:445–69

15. Kotlikoff MI, Wang YX. 1998. Calcium release and calcium-activated chloride channels in airway smooth muscle cells. *Am. J. Respir. Crit. Care Med.* 158:S109–14

16. Jentsch TJ, Stein V, Weinreich F, Zdebik AA. 2002. Molecular structure and physiological function of chloride channels. *Physiol. Rev.* 82:503–68

17. Kleene SJ, Gesteland RC. 1991. Calcium-activated chloride conductance in frog olfactory cilia. *J. Neurosci.* 11:3624–29

18. Kurahashi T, Yau K-W. 1994. Olfactory transduction: tale of an unusual chloride current. *Curr. Biol.* 4:231–37

19. Lowe G, Gold GH. 1993. Contribution of the ciliary cyclic nucleotide-gated conductance to olfactory transduction in the salamander. *J. Physiol.* 462:175–96

20. Firestein S, Shepherd GM. 1995. Interaction of anionic and cationic currents leads to a voltage dependence in the odor response of olfactory receptor neurons. *J. Neurophysiol.* 73:562–67

21. Delay RJ, Dubin AE, Dionne VE. 1997. A cyclic nucleotide-dependent chloride conductance in olfactory receptor neurons. *J. Membr. Biol.* 159:53–60

22. Sato K, Suzuki N. 2000. The contribution of a Ca^{2+}-activated Cl^- conductance to amino-acid-induced inward current responses of ciliated olfactory neurons of the rainbow trout. *J. Exp. Biol.* 203(Pt. 2):253–62

23. Schild D, Restrepo D. 1998. Transduction mechanisms in vertebrate olfactory receptor cells. *Physiol. Rev.* 78:429–66

24. Kleene SJ. 1997. High-gain, low-noise amplification in olfactory transduction. *Biophys. J.* 73:1110–17

25. Reisert J, Bauer PJ, Yau KW, Frings S. 2003. The Ca-activated Cl channel and its control in rat olfactory receptor neurons. *J. Gen. Physiol.* 122:349–64

26. Zhainazarov AB, Ache BW. 1995. Odor-induced currents in *Xenopus* olfactory receptor cells measured with

perforated-patch recording. *J. Neurophysiol.* 74:479–83

27. McBride DW Jr, Roper SD. 1991. Ca^{2+}-dependent chloride conductance in *Necturus* taste cells. *J. Membr. Biol.* 124:85–93

28. Herness MS, Sun XD. 1999. Characterization of chloride currents and their noradrenergic modulation in rat taste receptor cells. *J. Neurophysiol.* 82:260–71

29. Lindemann B. 1996. Taste reception. *Physiol. Rev.* 76:718–66

30. Taylor R, Roper S. 1994. Ca^{2+}-dependent Cl^- conductance in taste cells from *Necturus. J. Neurophysiol.* 72:475–78

31. Barnes S, Hille B. 1989. Ionic channels of the inner segment of tiger salamander cone photoreceptors. *J. Gen. Physiol.* 94:719–43

32. Barnes S. 1994. After transduction: response shaping and control of transmission by ion channels of the photoreceptor inner segments. *Neuroscience* 58:447–59

33. Maricq AV, Korenbrot JI. 1988. Calcium and calcium-dependent chloride currents generate action potentials in solitary cone photoreceptors. *Neuron* 1:503–15

34. Bader CR, Bertrand D, Schwartz EA. 1982. Voltage-activated and calcium-activated currents studied in solitary rod inner segments from the salamander retina. *J. Physiol.* 331:253–84

35. Okada T, Horiguchi H, Tachibana M. 1995. Ca^{2+}-dependent Cl^- current at the presynaptic terminals of goldfish retinal bipolar cells. *Neurosci. Res.* 23:297–303

36. Yau KW. 1994. Phototransduction mechanism in retinal rods and cones. The Friedenwald Lecture. *Invest. Ophthalmol. Vis. Sci.* 35:9–32

37. Thoreson WB, Burkhardt DA. 1991. Ionic influences on the prolonged depolarization of turtle cones in situ. *J. Neurophysiol.* 65:96–110

38. Bader CR, Bertrand D, Schlichter R. 1987. Calcium-activated chloride current in cultured sensory and parasympathetic quail neurones. *J. Physiol.* 394:125–48

39. Scott RH, McGuirk SM, Dolphin AC. 1988. Modulation of divalent cation-activated chloride ion currents. *Br. J. Pharmacol.* 94:653–62

40. Stapleton SR, Scott RH, Bell BA. 1994. Effects of metabolic blockers on Ca^{2+}-dependent currents in cultured sensory neurones from neonatal rats. *Br. J. Pharmacol.* 111:57–64

41. Currie KP, Wootton JF, Scott RH. 1995. Activation of Ca^{2+}-dependent Cl^- currents in cultured rat sensory neurones by flash photolysis of DM-nitrophen. *J. Physiol.* 482(Pt. 2):291–307

42. Andre S, Boukhaddaoui H, Campo B, Al Jumaily M, Mayeux V, et al. 2003. Axotomy-induced expression of calcium-activated chloride current in subpopulations of mouse dorsal root ganglion neurons. *J. Neurophysiol.* 90:3764–73

43. Schlichter R, Bader CR, Bertrand D, Dubois-Dauphin M, Bernheim L. 1989. Expression of substance P and of a Ca^{2+}-activated Cl^- current in quail sensory trigeminal neurons. *Neuroscience* 30:585–94

44. Mayer ML. 1985. A calcium-activated chloride current generates the afterdepolarization of rat sensory neurones in culture. *J. Physiol.* 364:217–39

45. De Castro F, Geijo-Barrientos E, Gallego R. 1997. Calcium-activated chloride current in normal mouse sympathetic ganglion cells. *J. Physiol.* 498(Part 2):397–408

46. Kaneko H, Putzier I, Frings S, Gensch T. 2002. Determination of intracellular chloride concentration in dorsal root ganglion neurons by fluorescent lifetime imaging. See Ref. 11, pp. 167–89

47. Deschenes M, Feltz P, Lamour Y. 1976. A model for an estimate in vivo of the ionic basis of presynaptic inhibition: an intracellular analysis of the GABA-induced depolarization in rat dorsal root ganglia. *Brain Res.* 118:486–93

48. Duchen MR. 1990. Effects of metabolic inhibition on the membrane properties of

isolated mouse primary sensory neurones. *J. Physiol.* 424:387–409

49. Crain SM. 1956. Resting and action potentials of cultured chick embryo spinal ganglion cells. *J. Comp. Neurol.* 104:285–329

50. Hussy N. 1991. Developmental change in calcium-activated chloride current during the differentiation of *Xenopus* spinal neurons in culture. *Dev. Biol.* 147:225–38

51. Hussy N. 1992. Calcium-activated chloride channels in cultures embryonic *Xenopus* spinal neurons. *J. Neurophysiol.* 68:2042–50

52. Owen DG, Segal M, Barker JL. 1984. A Ca-dependent Cl⁻ conductance in cultured mouse spinal neurones. *Nature* 311:567–70

53. Barker JL, Ransom BR. 1978. Amino acid pharmacology of mammalian central neurones grown in tissue culture. *J. Physiol.* 280:331–54

54. Bixby JL, Spitzer NC. 1984. The appearance and development of neurotransmitter sensitivity in *Xenopus* embryonic spinal neurones in vitro. *J. Physiol.* 353:143–55

55. Tokimasa T, Nishimura T, Akasu T. 1988. Calcium-activated chloride conductance in parasympathetic neurons of the rabbit urinary bladder. *J. Auton. Nerv. Syst.* 24:123–31

56. Nishimura T. 1995. Activation of calcium-dependent chloride channels causes post-tetanic depolarization in rabbit parasympathetic neurons. *J. Auton. Nerv. Syst.* 51:213–22

57. Akaike N, Sadoshima J. 1989. Caffeine affects four different ionic currents in the bull-frog sympathetic neurone. *J. Physiol.* 412:221–44

58. Sanchez-Vives MV, Gallego R. 1994. Calcium-dependent chloride current induced by axotomy in rat sympathetic neurons. *J. Physiol.* 475:391–400

59. Lancaster E, Oh EJ, Gover T, Weinreich D. 2002. Calcium and calcium-activated currents in vagotomized rat primary vagal afferent neurons. *J. Physiol.* 540:543–56

60. Hiraoka M, Kawano S. 1989. Calcium-sensitive and insensitive transient outward current in rabbit ventricular myocytes. *J. Physiol.* 410:187–212

61. Zygmunt AC, Gibbons WR. 1992. Properties of the calcium-activated chloride current in heart. *J. Gen. Physiol.* 99:391–414

62. Zygmunt AC, Gibbons WR. 1991. Calcium-activated chloride current in rabbit ventricular myocytes. *Circ. Res.* 68:424–37

63. Sipido KR, Callewaert G, Carmeliet E. 1993. [Ca²⁺]ᵢ transients and [Ca²⁺]ᵢ-dependent chloride current in single Purkinje cells from rabbit heart. *J. Physiol.* 468:641–67

64. Papp Z, Sipido K, Callewaert G, Carmeliet E. 1995. Two components of [Ca²⁺]ᵢ-activated Cl⁻ current during large [Ca²⁺]ᵢ transients in single rabbit heart Purkinje cells. *J. Physiol.* 483.2:319–30

65. Zygmunt AC, Robitelle DC, Eddlestone GT. 1997. I₍to1₎ dictates behavior of I₍Cl(Ca)₎ during early repolarization of canine ventricle. *Am. J. Physiol. Heart Circ. Physiol.* 273:H1096–106

66. Nakayama T, Fozzard HA. 1988. Adrenergic modulation of the transient outward current in isolated canine Purkinje cells. *Circ. Res.* 62(1):162–72

67. Hume JR, Duan D, Collier ML, Yamazaki J, Horowitz B. 2000. Anion transport in heart. *Physiol. Rev.* 80:31–81

68. Freeman LC, Pacioretty LM, Moise NS, Kass RS, Gilmour RF Jr. 1997. Decreased density of I₍to₎ in left ventricular myocytes from German shepherd dogs with inherited arrhythmias. *J. Cardiovasc. Electrophysiol.* 8:872–83

69. Zygmunt AC, Goodrow RJ, Weigel CM. 1998. I₍NaCa₎ and I₍Cl(Ca)₎ contribute to isoporterenol-induced afterhyperpolarizations in midmyocardial cells. *Am. J. Physiol. Heart Circ. Physiol.* 275:H979–92

70. Han X, Ferrier GR. 1996. Transient inward current is conducted through two types of channels in cardiac Purkinje

fibres. *J. Mol. Cell. Cardiol.* 28:2069–84

71. Han X, Ferrier GR. 1992. Ionic mechanisms of transient inward current in the absence of Na^+-Ca^{2+} exchange in rabbit cardiac Purkinje fibres. *J. Physiol.* 456:19–38

72. January CT, Fozzard HA. 1988. Delayed after depolarizations in heart muscle: mechanisms and relevance. *Pharmacol. Rev.* 40:219–27

73. Berlin JR, Cannell MB, Lederer WJ. 1989. Cellular origins of the transient inward current in cardiac myocytes. Role of fluctuations and waves of elevated intracellular calcium. *Circ. Res.* 65:115–26

74. Hiraoka M, Kawano S, Hirano Y, Furukawa T. 1998. Role of cardiac chloride currents in action potential characteristics and arrhythmias. *Cardiovasc. Res.* 40:23–33

75. Ridley PD, Curtis MJ. 1992. Anion manipulation: a new antiarrhymic approach. Action of substitution of chloride with nitrate on ischemia- and reperfusion-induced ventricular fibrillation and contractile function. *Circ. Res.* 70:617–32

76. Curtis MJ, Garlick PB, Ridley PD. 1993. Anion manipulation, a novel antiarrhythmic approach: mechanism of action. *J. Mol. Cell. Cardiol.* 25:417–36

77. Tanaka H, Matsui S, Kawanishi T, Shigenobu K. 1996. Use of chloride blockers: a novel approach for cardioprotection against ischemia-reperfusion damage. *J. Pharmacol. Exp. Ther.* 278:854–61

78. Davis MJ, Hill MA. 1999. Signaling mechanisms underlying the vascular myogenic response. *Physiol. Rev.* 79:387–423

79. Chipperfield AR, Harper AA. 2000. Chloride in smooth muscle. *Prog. Biophys. Mol. Biol.* 74:175–221

80. Wahlstrom BA. 1973. A study on the action of noradrenaline on ionic content and sodium, potassium and chloride effluxes in the rat portal vein. *Acta Physiol. Scand.* 89:522–30

81. Bolton TB. 1979. Mechanisms of action of transmitters and other substances on smooth muscle. *Physiol. Rev.* 59:606–718

82. Byrne NG, Large WA. 1985. Evidence for two mechanisms of depolarization associated with alpha 1-adrenoceptor activation in the rat anococcygeus muscle. *Br. J. Pharmacol.* 86:711–21

83. Byrne NG, Large WA. 1988. Membrane ionic mechanisms activated by noradrenaline in cells isolated from the rabbit portal vein. *J. Physiol.* 404:557–73

84. Van Helden DF. 1988. An alpha-adrenoceptor-mediated chloride conductance in mesenteric veins of the guinea-pig. *J. Physiol.* 401:489–501

85. Large WA. 1984. The effect of chloride removal on the responses of the isolated rat anococcygeus muscle to alpha 1-adrenoceptor stimulation. *J. Physiol.* 352:17–29

86. Pacaud P, Loirand G, Baron A, Mironneau C, Mironneau J. 1991. Ca^{2+} channel activation and membrane depolarization mediated by Cl^- channels in response to noradrenaline in vascular myocytes. *Br. J. Pharmacol.* 104:1000–6

87. Takenaka T, Epstein M, Forster H, Landry DW, Iijima K, Goligorsky MS. 1992. Attenuation of endothelin effects by a chloride channel inhibitor, indanyloxyacetic acid. *Am. J. Physiol. Renal Physiol.* 262:F799–806

88. Carmines PK. 1995. Segment-specific effect of chloride channel blockade on rat renal arteriolar contractile responses to angiotensin II. *Am. J. Hypertens.* 8:90–94

89. Criddle DN, deMoura RS, Greenwood IA, Large WA. 1996. Effect of niflumic acid on noradrenaline-induced contractions of the rat aorta. *Br. J. Pharmacol.* 118:1065–71

90. Criddle DN, Meireles A, Macedo LB, Leal-Cardoso JH, Scarparo HC, Jaffar M. 2002. Comparative inhibitory effects of niflumic acid and novel synthetic derivatives on the rat isolated stomach

fundus. *J. Pharm. Pharmacol.* 54:283–88

91. Greenwood IA, Large WA. 1995. Comparison of the effects of fenamates on Ca-activated chloride and potassium currents in rabbit portal vein smooth muscle cells. *Br. J. Pharmacol.* 116:2939–48

92. Lamb FS, Barna TJ. 1998. Chloride ion currents contribute functionally to norepinephrine-induced vascular contraction. *Am. J. Physiol. Heart Circ. Physiol.* 275:H151–60

93. Yuan XJ. 1997. Role of calcium-activated chloride current in regulating pulmonary vasomotor tone. *Am. J. Physiol. Lung Cell Mol. Physiol.* 272:L959–68

94. Boucher RC, Cheng EHC, Paradiso AM, Stutts MJ, Knowles MR, Earp HS. 1989. Chloride secretory response of cystic fibrosis human airway epithelia: preservation of calcium but not protein kinase C- and A-dependent mechanisms. *J. Clin. Invest.* 84:1424–31

95. Kartner N, Hanrahan JW, Jensen TJ, Naismith AL, Sun SZ, et al. 1991. Expression of the cystic fibrosis gene in non-epithelial invertebrate cells produces a regulated anion conductance. *Cell* 64:681–91

96. Knowles MR, Clarke LL, Boucher RC. 1991. Activation by extracellular nucleotides of chloride secretion in the airway epithelia of patients with cystic fibrosis. *N. Engl. J. Med.* 325:533–38

97. Tarran R, Loewen ME, Paradiso AM, Olsen JC, Gray MA, et al. 2002. Regulation of murine airway surface liquid volume by CFTR and Ca^{2+}-activated Cl^- conductances. *J. Gen. Physiol.* 120:407–18

98. Gabriel SE, Makhlina M, Martsen E, Thomas EJ, Lethem MI, Boucher RC. 2000. Permeabilization via the P2X7 purinoreceptor reveals the presence of a Ca^{2+}-activated Cl^- conductance in the apical membrane of murine tracheal epithelial cells. *J. Biol. Chem.* 275:35028–33

99. Donaldson J, Brown AM, Hill SJ. 1989. Temporal changes in the calcium-dependence of the histamine H1-receptor-stimulation of cyclic AMP accumulation in guinea-pig cerebral cortex. *Br. J. Pharmacol.* 98:1365–75

100. Clarke LL, Grubb BR, Yankaskas JR, Cotton CU, McKenzie A, Boucher RC. 1994. Relationship of a non-cystic fibrosis transmembrane conductance regulator-mediated chloride conductance to organ-level disease in $Cftr^{-/-}$ mice. *Proc. Natl. Acad. Sci. USA* 91:479–83

101. Kachintorn U, Vajanaphanich M, Traynor-Kaplan AE, Dharmsathaphorn K, Barrett KE. 1993. Activation by calcium alone of chloride secretion in T84 epithelial cells. *Br. J. Pharmacol.* 109:510–17

102. Wasserman MA, Mukherjee A. 1988. Regional differences in the reactivity of guinea-pig airways. *Pulm. Pharmacol.* 1:125–31

103. Dho S, Stewart K, Foskett JK. 1992. Purinergic receptor activation of Cl^- secretion in T84 cells. *Am. J. Physiol. Cell Physiol.* 262:C67–74

104. Barrett KE, Keely SJ. 2000. Chloride secretion by the intestinal epithelium: molecular basis and regulatory aspects. *Annu. Rev. Physiol.* 62:535–72

105. Berschneider HM, Knowles MR, Azizkhan RG, Boucher RC, Tobey NA, et al. 1988. Altered intestinal chloride transport in cystic fibrosis. *FASEB J.* 2:2625–29

106. Botelho SY, Dartt DA. 1980. Effect of calcium antagonism or chelation on rabbit lacrimal gland secretion and membrane potentials. *J. Physiol.* 304:397–403

107. Hunter M, Smith PA, Case RM. 1983. The dependence of fluid secretion by mandibular salivary gland and pancreas on extracellular calcium. *Cell Calcium* 4:307–17

108. Melvin JE, Koek L, Zhang GH. 1991. A capacitative Ca^{2+} influx is required for sustained fluid secretion in sublingual mucous acini. *Am. J. Physiol.*

Gastrointest. Liver Physiol. 261:G1043–50

109. Douglas WW, Poisner AM. 1963. The influence of calcium on the secretory response of the submaxillary gland to acetylcholine or to noradrenaline. *J. Physiol.* 165:528–41

110. Machaca K, Qu Z, Kuruma A, Hartzell HC, McCarty N. 2002. The endogenous calcium-activated Cl channel in *Xenopus* oocytes: a physiologically and biophysically rich model system. See Ref. 11, pp. 3–39

111. Webb D, Nuccitelli R. 1985. Fertilization potential and electrical properties of the *Xenopus laevis* egg. *Dev. Biol.* 107:395–406

112. Hartzell HC. 1996. Activation of different Cl currents in *Xenopus* oocytes by Ca liberated from stores and by capacitative Ca influx. *J. Gen. Physiol.* 108:157–75

113. Kuruma A, Hartzell HC. 2000. Bimodal control of a Ca^{2+}-activated Cl^- channel by different Ca^{2+} signals. *J. Gen. Physiol.* 115:59–80

114. Jaffe LA, Cross NL. 1986. Electrical regulation of sperm-egg fusion. *Annu. Rev. Physiol.* 48:191–200

115. Barriere H, Belfodil R, Rubera I, Tauc M, Poujeol C, et al. 2003. CFTR null mutation altered cAMP-sensitive and swelling-activated Cl^- currents in primary cultures of mouse nephron. *Am. J. Physiol. Renal Physiol.* 284:F796–811

116. Rubera I, Tauc M, Bidet M, Verheecke-Mauze C, De Renzis G, et al. 2000. Extracellular ATP increases $[Ca^{2+}]_i$ in distal tubule cells. II. Activation of a Ca^{2+}-dependent Cl^- conductance. *Am. J. Physiol. Renal Physiol.* 279:F102–11

117. Bidet M, Tauc M, Rubera I, De Renzis G, Poujeol C, et al. 1996. Calcium-activated chloride currents in primary cultures of rabbit distal convoluted tubule. *Am. J. Physiol. Renal Physiol.* 271:F940–50

118. Meyer K, Korbmacher C. 1996. Cell swelling activates ATP-dependent voltage-gated chloride channels in M-1 mouse

cortical collecting duct cells. *J. Gen. Physiol.* 108:177–93

119. Boese SH, Glanville M, Aziz O, Gray MA, Simmons NL. 2000. Ca^{2+} and cAMP-activated Cl^- conductances mediate Cl^- secretion in a mouse renal inner medullary collecting duct cell line. *J. Physiol.* 523(Pt. 2):325–38

120. Qu Z, Wei RW, Hartzell HC. 2003. Characterization of Ca^{2+}-activated Cl^- currents in mouse kidney inner medullary collecting duct cells. *Am. J. Physiol. Renal Physiol.* 285:F326–35

121. Boese SH, Glanville M, Gray MA, Simmons NL. 2000. The swelling-activated anion conductance in the mouse renal inner medullary collecting duct cell line mIMCD-K2. *J. Membr. Biol.* 177:51–64

122. Boese SH, Aziz O, Simmons NL, Gray MA. 2004. Kinetics and regulation of a Ca^{2+}-activated Cl^- conductance in mouse renal inner medullary collecting duct cells. *Am. J. Physiol. Renal Physiol.* 286:F682–92

123. Weinstein AM. 2003. Mathematical models of renal fluid and electrolyte transport: acknowledging our uncertainty. *Am. J. Physiol. Renal Physiol.* 284:F871–84

124. Beyenbach KW. 1986. Secretory NaCl and volume flow in renal tubules. *Am. J. Physiol. Regul. Integr. Comp. Physiol.* 250:R753–63

125. Wallace DP, Christensen M, Reif G, Belibi F, Thrasher B, et al. 2002. Electrolyte and fluid secretion by cultured human inner medullary collecting duct cells. *Am. J. Physiol. Renal Physiol.* 283:F1337–50

126. Grantham JJ, Wallace DP. 2002. Return of the secretory kidney. *Am. J. Physiol. Renal Physiol.* 282:F1–9

127. Wallace DP, Rome LA, Sullivan LP, Grantham JJ. 2001. cAMP-dependent fluid secretion in rat inner medullary collecting ducts. *Am. J. Physiol. Renal Physiol.* 280:F1019–29

128. Grantham JJ. 1976. Fluid secretion in the nephron: relation to renal failure. *Physiol. Rev.* 56:248–58

129. Sonnenberg H. 1975. Secretion of salt and water into the medullary collecting duct of Ringer-infused rats. *Am. J. Physiol.* 228:565–68

130. Nilius B, Prenen J, Szucs G, Wei L, Tanzi F, et al. 1997. Calcium-activated chloride channels in bovine pulmonary artery endothelial cells. *J. Physiol.* 498(Pt. 2):381–96

131. Nilius B, Droogmans G. 2001. Ion channels and their functional role in vascular endothelium. *Physiol. Rev.* 81:1415–59

132. Korn SJ, Bolden A, Horn R. 1991. Control of action potentials and Ca^{2+} influx by the Ca^{2+}-dependent chloride current in mouse pituitary cells. *J. Physiol.* 439:423–37

133. Nilius B, Eggermont J, Voets T, Droogmans G. 1996. Volume-activated Cl^- channels. *Gen. Pharmacol.* 27:1131–40

134. Ivanenko A, Baring MD, Airey JA, Sutko JL, Kenyon JL. 1993. A caffeine- and ryanodine-sensitive Ca^{2+} store in avian sensory neurons. *J. Neurophysiol.* 70:710–22

135. Kenyon JL, Goff HR. 1998. Temperature dependencies of Ca^{2+} current, Ca^{2+}-activated Cl^- current and Ca^{2+} transients in sensory neurones. *Cell Calcium* 24:35–48

136. Ayar A, Scott RH. 1999. The actions of ryanodine on Ca^{2+}-activated conductances in rat cultured DRG neurones; evidence for Ca^{2+}-induced Ca^{2+} release. *Naunyn Schmiedebergs Arch. Pharmacol.* 359:81–91

137. Martinez-Pinna J, McLachlan EM, Gallego R. 2000. Distinct mechanisms for activation of Cl^- and K^+ currents by Ca^{2+} from different sources in mouse sympathetic neurones. *J. Physiol.* 527(Pt. 2):249–64

138. Kuruma A, Hiraoka M, Kawano S. 1998. Activation of Ca^{2+}-sensitive Cl^- current by reverse mode Na^+/Ca^{2+} exchange in rabbit ventricular myocytes. *Pflügers Arch.* 436:976–83

139. Kuruma A, Hartzell HC. 1999. Dynamics of calcium regulation of chloride currents in *Xenopus* oocytes. *Am. J. Physiol. Cell Physiol.* 276:C161–75

140. Koumi S, Sato R, Aramaki T. 1994. Characterization of the calcium-activated chloride channel in isolated guinea-pig hepatocytes. *J. Gen. Physiol.* 104:357–73

141. Collier ML, Levesque PC, Kenyon JL, Hume JR. 1996. Unitary Cl^- channels activated by cytoplasmic Ca^{2+} in canine ventricular myocytes. *Circ. Res.* 78:936–44

142. Takahashi T, Neher E, Sakmann B. 1987. Rat brain serotonin receptors in *Xenopus* oocytes are coupled by intracellular calcium to endogenous channels. *Proc. Natl. Acad. Sci. USA* 84:5063–67

143. Gomez-Hernandez J-M, Stühmer W, Parekh AB. 1997. Calcium dependence and distribution of calcium-activated chloride channels in *Xenopus* oocytes. *J. Physiol.* 502:569–74

144. Frizzell RA, Rechkemmer G, Shoemaker RL. 1986. Altered regulation of airway epithelial cell chloride channels in cystic fibrosis. *Science* 233:558–60

145. Martin DK. 1993. Small conductance chloride channels in acinar cells from the rat mandibular salivary gland are directly controlled by a G-protein. *Biochem. Biophys. Res. Commun.* 192:1266–73

146. Zhang G, Arreola J, Melvin JE. 1995. Inhibition by thiocyanate of muscarinic-induced cytosolic acidification and Ca^{2+} entry in rat sublingual acini. *Arch. Oral Biol.* 40:111–18

147. VanRenterghem C, Lazdunski M. 1993. Endothelin and vasopressin activate low conductance chloride channels in aortic smooth muscle. *Pflügers Arch.* 425:156–63

148. Nilius B, Prenen J, Szucs G, Wei L, Tanzi F, et al. 1997. Calcium-activated chloride channels in bovine pulmonary artery endothelial cells. *J. Physiol.* 498:381–96

149. Schlenker T, Fitz JG. 1996. Ca^{2+}-activated Cl^- channels in human biliary

cell line: regulation by Ca^{2+}/calmodulin-dependent protein kinase. *Am. J. Physiol. Gastrointest. Liver Physiol.* 271:G304–10

150. Morris AP, Frizzell RA. 1993. Ca^{2+}-dependent Cl^- channels in undifferentiated human colonic cells (HT-29). II. Regulation and rundown. *Am. J. Physiol. Cell Physiol.* 264:C977–85

151. Klockner U. 1993. Intracellular calcium ions activate a low-conductance chloride channel in smooth-muscle cells isolated from human mesenteric artery. *Pflügers Arch.* 424:231–37

152. Ling BN, Seal EE, Eaton DC. 1993. Regulation of mesangial cell ion channels by insulin and angiotensin II. Possible role in diabetic glomerular hyperfiltration. *J. Clin. Invest.* 92:2141–51

153. Bao L, Kaldany C, Holmstrand EC, Cox DH. 2004. Mapping the BK_{Ca} channel's "Ca^{2+} bowl": side-chains essential for Ca^{2+} sensing. *J. Gen. Physiol.* 123:475–89

154. Arreola J, Melvin J, Begenisich T. 1996. Activation of calcium-dependent chloride channels in rat paroid acinar cells. *J. Gen. Physiol.* 108:35–47

155. Park MK, Lomax RB, Tepikin AV, Petersen OH. 2001. Local uncaging of caged Ca^{2+}-activated Cl^- channels in pancreatic acinar cells. *Proc. Natl. Acad. Sci. USA* 98:10948–53

156. Giovannucci DR, Bruce JI, Straub SV, Arreola J, Sneyd J, et al. 2002. Cytosolic Ca^{2+} and Ca^{2+}-activated Cl^- current dynamics: insights from two functionally distinct mouse exocrine cells. *J. Physiol.* 540:469–84

157. Arreola J, Melvin JE, Begenisich T. 1998. Differences in regulation of Ca^{2+}-activated Cl^- channels in colonic and parotid secretory cells. *Am. J. Physiol. Cell Physiol.* 274:C161–66

158. Chan HC, Kaetzel MA, Gotter AL, Dedman JR, Nelson DJ. 1994. Annexin IV inhibits calmodulin-dependent protein kinase II-activated chloride conductance: a novel mechanism for ion channel regulation. *J. Biol. Chem.* 269(51):32464–68

159. Kaetzel MA, Pula G, Campos B, Uhrin P, Horseman N, Dedman JR. 1994. A role for annexin IV in epithelial cell function. Inhibition of calcium-activated chloride conductance. *J. Biol. Chem.* 269:5297–302

160. Xie W, Kaetzel MA, Bruzik KS, Dedman JR, Shears SB, Nelson DJ. 1996. Inositol 3,4,5,6-tetrakisphosphate inhibits the calmodulin-dependent protein kinase II-activated chloride conductance in T84 colonic epithelial cells. *J. Biol. Chem.* 271:14092–97

161. Xie W, Kaetzel MA, Bruzik KS, Dedman JR, Shears SB, Nelson DJ. 1998. Regulation of Ca^{2+}-dependent Cl^- conductance in a human colonic epithelial cell line (T_{84}): cross-talk between Ins(3,4,5,6)P_4 and protein phosphates. *J. Physiol.* 510.3:661–73

162. Worrell RT, Frizzell RA. 1991. CaMKII mediates stimulation of chloride conductance by calcium in T84 cells. *Am. J. Physiol. Cell Physiol.* 260:C877–82

163. Nishimoto I, Wagner J, Schulman H, Gardner P. 1991. Regulation of Cl^- channels by multifunctional CaM kinase. *Neuron* 6:547–55

164. Holevinsky KO, Jow F, Nelson DJ. 1994. Elevation in intracellular calcium activates both chloride and proton currents in human macrophages. *J. Membr. Biol.* 140:13–30

165. Chao AC, Kouyama K, Heist EK, Dong YJ, Gardner P. 1995. Calcium- and CaMKII-dependent chloride secretion induced by the microsomal Ca^{2+}-ATPase inhibitor 2,5-di-(tert-butyl)-1,4-hydroquinone in cystic fibrosis pancreatic epithelial cells. *J. Clin. Invest.* 96:1794–801

166. Nilius B, Prenen J, Voets T, Van Den Bremt K , Eggermont J, Droogmans G. 1997. Kinetic and pharmacological properties of the calcium-activated

chloride-current in microvascular endothelial cells. *Cell Calcium* 22:53–63

167. Evans MG, Marty A. 1986. Calcium-dependent chloride currents in isolated cells from rat lacrimal glands. *J. Physiol.* 378:437–60

168. Piper AS, Large WA. 2003. Multiple conductance states of single Ca^{2+}-activated Cl^- channels in rabbit pulmonary artery smooth muscle cells. *J. Physiol.* 547:181–96

169. Greenwood IA, Large WA. 1999. Modulation of the decay of Ca^{2+}-activated Cl^- currents in rabbit portal vein smooth muscle cells by external anions. *J. Physiol.* 516:365–76

170. Perez-Cornejo P, De Santiago JA, Arreola J. 2004. Permeant anions control gating of calcium-dependent chloride channels. *J. Membr. Biol.* 198:125–33

171. Hille B. 1992. *Ion Channels of Excitable Membranes*. Sunderland, MA: Sinauer

172. Doyle DA, Morais Cabral J, Pfuetzner RA, Kuo A, Gulbis JM, et al. 1998. The structure of the potassium channel: molecular basis of K^+ conduction and selectivity. *Science* 280:69–77

173. Qu Z, Hartzell HC. 2000. Anion permeation in Ca^{2+}-activated Cl^- channels. *J. Gen. Physiol.* 116:825–44

174. Perez-Cornejo P, Arreola J. 2004. Regulation of Ca^{2+}-activated chloride channels by cAMP and CFTR in parotid acinar cells. *Biochem. Biophys. Res. Commun.* 316:612–17

175. Kidd JF, Thorn P. 2000. The properties of the secretogogue-evoked chloride current in mouse pancreatic acinar cells. *Pflügers Arch.* 441:489–97

176. Dawson DC, Smith SS, Mansoura MK. 1999. CFTR: mechanism of anion conduction. *Physiol. Rev.* 79(1):S47–75

177. Smith SS, Steinle ED, Meyerhoff ME, Dawson DC. 1999. Cystic fibrosis transmembrane conductance regulator: Physical basis for lyotropic anion selectivity patterns. *J. Gen. Physiol.* 114:799–817

178. Fahlke C, Dürr C, George AL Jr. 1997.

Mechanism of ion permeation in skeletal muscle chloride channels. *J. Gen. Physiol.* 110:551–64

179. Rychkov GY, Pusch M, Roberts ML, Jentsch TJ, Bretag AH, et al. 1998. Permeation and block of the skeletal muscle chloride channel ClC-1 by foreign anions. *J. Gen. Physiol.* 111:653–65

180. Mansoura MK, Smith SS, Choi AD, Richards NW, Strong TV, et al. 1998. Cystic fibrosis transmemebrane conductance regulator (CFTR): anion binding as a probe of the pore. *Biophys. J.* 74:1320–32

181. Tabcharani JA, Linsdell P, Hanrahan JW. 1997. Halide permeation in wild-type and mutant cystic fibrosis transmembrane conductance regulator chloride channels. *J. Gen. Physiol.* 110:341–54

182. Bormann J, Hamill OP, Sakmann B. 1987. Mechanism of anion permeation through channels gated by glycine and $\bar{\gamma}$ aminobutyric acid in mouse cultured spinal neurones. *J. Physiol.* 385:243–86

183. Franciolini F, Nonner W. 1987. Anion and cation permeability of a chloride channel in rat hippocampal neurons. *J. Gen. Physiol.* 90:453–78

184. Blatz AL, Magleby KL. 1985. Single chloride-selective channels active at membrane resting potentials in culture rat skeletal muscle. *Biophys. J.* 47:119–23

185. Wang C-T, Zhang HG, Rocheleau TA, ffrench-Constant RH, Jackson MB. 1999. Cation permeability and cation-anion interactions in a mutant GABA-gated chloride channel from *Drosophila*. *Biophys. J.* 77:691–700

186. Galzi JL, Devillers-Thiery A, Hussy N, Bertrand S, Changeux JP, Bertrand D. 1992. Mutations in the channel domain of a neuronal nicotinic receptor convert ion selectivity from cationic to anionic. *Nature* 359:500–5

187. Franciolini F, Nonner W. 1994. A multi-ion permeation mechanism in neuronal background chloride channels. *J. Gen. Physiol.* 104:725–46

188. Franciolini F, Nonner W. 1994. Anion-cation interactions in the pore of neuronal background chloride channels. *J. Gen. Physiol.* 104:711–23

189. Qu Z, Hartzell HC. 2001. Functional geometry of the permeation pathway of Ca^{2+} activated Cl^- channels inferred from analysis of voltage-dependent block. *J. Biol. Chem.* 276:18423–29

190. Qu Z, Hartzell HC. 2003. Two bestrophins cloned from *Xenopus laevis* oocytes express Ca-activated Cl currents. *J. Biol. Chem.* 278:49563–72

191. Qu Z, Fischmeister R, Hartzell HC. 2004. Mouse bestrophin-2 is a bona fide Cl^- channel: identification of a residue important in anion binding and conduction. *J. Gen. Physiol.* 123:327–40

192. Tsunenari T, Sun H, Williams J, Cahill H, Smallwood P, et al. 2003. Structure-function analysis of the bestrophin family of anion channels. *J. Biol. Chem.* 278:41114–25

193. Qu Z, Hartzell HC. 2004. Determinants of anion permeation in the second transmembrane domain of the mouse bestrophin-2 chloride channel. *J. Gen. Physiol.* In press

194. Fahlke C. 2001. Ion permeation and selectivity in ClC-type chloride channels. *Am. J. Physiol. Renal Physiol.* 280:F748–57

195. Fahlke C, Beck CL, George AL Jr. 1997. A mutation in autosomal dominant myotonia congenita affects pore properties of the muscle chloride channel. *Proc. Natl. Acad. Sci. USA* 94:2729–34

196. Fahlke C, Yu HT, Beck CL, Rhodes TH, George AL Jr. 1997. Pore-forming segments in voltage-gated chloride channels. *Nature* 390:529–32

197. Linsdell P, Evagelidis A, Hanrahan JW. 2000. Molecular determinants of anion selectivity in the cystic fibrosis transmemebrane conductance regulatory chloride channel pore. *Biophys. J.* 78:2973–82

198. Smith SS, Liu X, Zhang ZR, Sun F, Kriewall TE, et al. 2001. CFTR: covalent and noncovalent modification suggests a role for fixed charges in anion conduction. *J. Gen. Physiol.* 118:407–32

199. White MM, Aylwin M. 1990. Niflumic and flufenamic acids are potent reversible blockers of Ca^{2+}-activated Cl^- channels in *Xenopus* oocytes. *Mol. Pharmacol.* 37:720–24

200. Piper AS, Greenwood IA, Large WA. 2002. Dual effect of blocking agents on Ca^{2+}-activated Cl^- currents in rabbit pulmonary artery smooth muscle cells. *J. Physiol.* 539:119–31

201. Xu WX, Kim SJ, So I, Kang TM, Rhee JC, Kim KW. 1997. Volume-sensitive chloride current activated by hyposmotic swelling in antral gastric myocytes of the guinea-pig. *Pflügers Arch.* 435:9–19

202. Doughty JM, Miller AL, Langton PD. 1998. Non-specificity of chloride channel blockers in rat cerebral arteries: block of the L-type calcium channel. *J. Physiol.* 507(Pt. 2):433–39

203. Wang HS, Dixon JE, McKinnon D. 1997. Unexpected and differential effects of Cl^- channel blockers on the Kv4.3 and Kv4.2 K^+ channels. Implications for the study of the I_{to2} current. *Circ. Res.* 81:711–18

204. Reinsprecht M, Rohn MH, Spadinger RJ, Pecht I, Schindler H, Romanin C. 1995. Blockade of capacitive Ca^{2+} influx by Cl^- channel blockers inhibits secretion from rat mucosal-type mast cells. *Mol. Pharmacol.* 47:1014–20

205. Muanprasat C, Sonawane ND, Salinas D, Taddei A, Galietta LJ, Verkman AS. 2004. Discovery of glycine hydrazide pore-occluding CFTR inhibitors: mechanism, structure-activity analysis, and in vivo efficacy. *J. Gen. Physiol.* 124:125–37

206. Helix N, Strobaek D, Dahl BH, Christophersen P. 2003. Inhibition of the endogenous volume-regulated anion channel (VRAC) in HEK293 cells by acidic di-aryl-ureas. *J. Membr. Biol.* 196:83–94

207. Maertens C, Wei L, Voets T, Droogmans G, Nilius B. 1999. Block by fluoxetine of

volume-regulated anion channels. *Br. J. Pharmacol.* 126:508–14

208. Maertens C, Wei L, Droogmans G, Nilius B. 2000. Inhibition of volume-regulated and calcium-activated chloride channels by the antimalarial mefloquine. *J. Pharmacol. Exp. Ther.* 295:29–36

209. DeBin JA, Maggio JE, Strichartz GR. 1993. Purification and characterization of chlorotoxin, a chloride channel ligand from the venom of the scorpion. *Am. J. Physiol. Cell Physiol.* 264:C361–69

210. Dalton S, Gerzanich V, Chen M, Dong Y, Shuba Y, Simard JM. 2003. Chlorotoxin-sensitive Ca^{2+}-activated Cl^- channel in type R2 reactive astrocytes from adult rat brain. *Glia* 42:325–39

211. Maertens C, Wei L, Tytgat J, Droogmans G, Nilius B. 2000. Chlorotoxin does not inhibit volume-regulated, calcium-activated and cyclic AMP-activated chloride channels. *Br. J. Pharmacol.* 129:791–801

212. Greenwood IA, Ledoux J, Leblanc N. 2001. Differential regulation of Ca^{2+}-activated Cl^- currents in rabbit arterial and portal vein smooth muscle cells by Ca^{2+}-calmodulin-dependent kinase. *J. Physiol.* 534:395–408

213. Ledoux J, Greenwood I, Villeneuve LR, Leblanc N. 2003. Modulation of Ca^{2+}-dependent Cl^- channels by calcineurin in rabbit coronary arterial myocytes. *J. Physiol.* 552:701–14

214. Wang YX, Kotlikoff MI. 1997. Inactivation of calcium-activated chloride channels in smooth muscle by calcium/calmodulin-dependent protein kinase. *Proc. Natl. Acad. Sci. USA* 94:14918–23

215. Marunaka Y, Eaton DC. 1990. Effects of insulin and phosphatase on a Ca^{2+}-dependent Cl^- channel in a distal nephron cell line (A6). *J. Gen. Physiol.* 95:773–89

216. Ho MW, Kaetzel MA, Armstrong DL, Shears SB. 2001. Regulation of a human chloride channel. a paradigm for integrating input from calcium, type II

calmodulin-dependent protein kinase, and inositol 3,4,5,6-tetrakisphosphate. *J. Biol. Chem.* 276:18673–80

217. Vajanaphanich M, Schultz C, Rudolf MT, Wasserman M, Enyedi P, et al. 1994. Long-term uncoupling of chloride secretion from intracellular calcium levels by Ins(3,4,5,6)P$_4$. *Nature* 371:711–14

218. Ismailov II, Fuller CM, Berdiev BK, Shlyonsky VG, Benos DJ, Barrett KE. 1996. A biologic function for an "orphan" messenger: D-myo-inositol 3,4,5,6-tetrakisphosphate selectively blocks epithelial calcium-activated chloride channels. *Proc. Natl. Acad. Sci. USA* 93:10505–9

219. Xie W, Kaetzel MA, Bruzik KS, Dedman JR, Shears SB, Nelson DJ. 1996. Inositol 3,4,5,6-tetrakisphosphate inhibits the calmodulin-dependent protein kinase II-activated chloride conductance in T84 colonic epithelial cells. *J. Biol. Chem.* 271:14092–97

220. Nilius B, Prenen J, Voets T, Eggermont J, Bruzik KS, et al. 1998. Inhibition by inositol tetrakisphosphates of calcium- and volume-activated Cl^- currents in macrovascular endothelial cells. *Pflügers Arch.* 435:637–44

221. Carew MA, Yang X, Schultz C, Shears SB. 2000. Myo-inositol 3,4,5,6-tetrakisphosphate inhibits an apical calcium-activated chloride conductance in polarized monolayers of a cystic fibrosis cell line. *J. Biol. Chem.* 275:26906–13

222. Renstrom E, Ivarsson R, Shears SB. 2002. Inositol 3,4,5,6-tetrakisphosphate inhibits insulin granule acidification and fusogenic potential. *J. Biol. Chem.* 277:26717–20

223. Kaetzel MA, Pula G, Campos B, Uhrin P, Horseman N, Dedman JR. 1994. Annexin VI isoforms are differentially expressed in mammalian tissues. *Biochim. Biophys. Acta* 1223:368–74

224. Jorgensen AJ, Bennekou P, Eskesen K, Kristensen BI. 1997. Annexins from

Ehrlich ascites inhibit the calcium-activated chloride current in *Xenopus laevis* oocytes. *Eur. J. Physiol.* 434:261–66

225. Kunzelmann K. 2001. CFTR: interacting with everything? *News Physiol. Sci.* 16:167–70

226. Wei L, Vankeerberghen A, Cuppens H, Eggermont J, Cassiman JJ, et al. 1999. Interaction between calcium-activated chloride channels and the cystic fibrosis transmembrane conductance regulator. *Pflügers Arch.* 438:635–41

227. Kunzelmann K, Mall M, Briel M, Hipper A, Nitschke R, et al. 1997. The cystic fibrosis transmembrane conductance regulator attenuates the endogenous Ca^{2+} activated Cl^- conductance of *Xenopus* oocytes. *Pflügers Arch.* 435:178–81

228. Clarke LL, Boucher RC. 1992. Chloride secretory response to extracellular ATP in human normal and cystic fibrosis nasal epithelia. *Am. J. Physiol. Cell Physiol.* 263:C348–56

229. Grubb BR, Vick RN, Boucher RC. 1994. Hyperabsorption of Na^+ and raised Ca^{2+}-mediated Cl^- secretion in nasal epithelia of CF mice. *Am. J. Physiol. Cell Physiol.* 266:C1478–83

230. Colledge WH, Evans MJ. 1995. Cystic fibrosis gene therapy. *Br. Med. Bull.* 51:82–90

231. Colledge WH, Abella BS, Southern KW, Ratcliff R, Jiang C, et al. 1995. Generation and characterization of a ΔF508 cystic fibrosis mouse model. *Nat. Genet.* 10:445–52

232. Wei L, Vankeerberghen A, Cuppens H, Cassiman JJ, Droogmans G, Nilius B. 2001. The C-terminal part of the R-domain, but not the PDZ binding motif, of CFTR is involved in interaction with Ca^{2+}-activated Cl^- channels. *Pflügers Arch.* 442:280–85

233. Park K, Brown PD. 1995. Intracellular pH modulates the activity of chloride channels in isolated lacrimal gland acinar cells. *Am. J. Physiol. Cell Physiol.* 268:C647–50

234. Arreola J, Melvin JE, Begenisich T. 1995. Inhibition of Ca^{2+}-dependent Cl^- channels from secretory epithelial cells by low internal pH. *J. Membr. Biol.* 147:95–104

235. Matchkov VV, Aalkjaer C, Nilsson H. 2004. A cyclic GMP-dependent calcium-activated chloride current in smooth-muscle cells from rat mesenteric resistance arteries. *J. Gen. Physiol.* 123:121–34

236. Piper AS, Large WA. 2004. Single cGMP-activated Ca^{2+}-dependent Cl^- channels in rat mesenteric artery smooth muscle cells. *J. Physiol.* 555:397–408

237. Kilic G, Fitz JG. 2002. Heterotrimeric G-proteins activate Cl^- channels through stimulation of a cyclooxygenase-dependent pathway in a model liver cell line. *J. Biol. Chem.* 277:11721–27

238. Choi S, Rho SH, Jung SY, Kim SC, Park CS, Nah SY. 2001. A novel activation of Ca^{2+}-activated Cl^- channel in *Xenopus* oocytes by ginseng saponins: evidence for the involvement of phospholipase C and intracellular Ca^{2+} mobilization. *Br. J. Pharmacol.* 132:641–48

239. Kaibara M, Nagase Y, Murasaki O, Uezono Y, Doi Y, Taniyama K. 2001. GTPγS-induced Ca^{2+} activated Cl^- currents: its stable induction by G_q alpha overexpression in *Xenopus* oocytes. *Jpn. J. Pharmacol.* 86:244–47

240. Anderson MP, Welsh MJ. 1991. Calcium and cAMP activate different chloride channels in the apical membrane of normal and cystic fibrosis epithelia. *Proc. Natl. Acad. Sci. USA* 88:6003–7

241. Cliff WH, Frizzell RA. 1990. Separate Cl conductances activated by cAMP and Ca in Cl secreting epithelial cells. *Proc. Natl. Acad. Sci. USA* 87:4956–60

242. Eguiguren AL, Sepulveda FV, Riveros N, Stutzin A. 1996. Calcium-activated chloride currents and non-selective cation channels in a novel cystic fibrosis-derived human pancreatic duct cell line. *Biochem. Biophys. Res. Commun.* 225:505–13

243. Ishikawa T, Cook DI. 1993. A Ca^{2+}-dependent Cl^- current in sheep parotid secretory cells. *J. Membr. Biol.* 135:261–71

244. Kozak JA, Logothetis DE. 1997. A calcium-dependent chloride current in insulin secreting TC-3 cells. *Pflügers Arch.* 433:679–90

245. Hirakawa Y, Gericke M, Cohen RA, Bolotina VM. 1999. Ca^{2+}-dependent Cl^- channels in mouse and rabbit aortic smooth muscle cells: regulation by intracellular Ca^{2+} and NO. *Am. J. Physiol. Heart Circ. Physiol.* 277:H1732–44

246. Taleb O, Feltz P, Bossu J-L, Felta A. 1988. Small-conductance chloride channels activated by calcium on cultured endocrine cells from mammalian pars intermedia. *Pflügers Arch.* 412:641–46

247. Groschner K, Kokovetz WR. 1992. Voltage-sensitive chloride channels of large conductance in the membrane of pig aortic endothelial cells. *Pflügers Arch.* 421:209–17

248. Nilius B, Sehrer J, Heinke S, Droogmans G. 1995. Ca^{2+} release and activation of K^+ and Cl^- currents by extracellular ATP in distal nephron epithelial cells. *Am. J. Physiol. Cell Physiol.* 269:C376–84

249. Huang P, Liu J, Di A, Robinson NC, Musch MW, et al. 2001. Regulation of human CLC-3 channels by multifunctional Ca^{2+}/calmodulin-dependent protein kinase. *J. Biol. Chem.* 276:20093–100

250. Robinson NC, Huang P, Kaetzel MA, Lamb FS, Nelson DJ. 2004. Identification of an N-terminal amino acid of the CLC-3 chloride channel critical in phosphorylation-dependent activation of a CaMKII-activated chloride current. *J. Physiol.* 556:353–68

251. Arreola J, Begenisich T, Nehrke K, Nguyen HV, Park K, et al. 2002. Secretion and cell volume regulation by salivary acinar cells from mice lacking expression of the ClC-n3 Cl^- channel gene. *J. Physiol.* 545:207–16

252. Cunningham SA, Awayda MS, Bubien JK, Ismailov II, Arrate MP, et al. 1995.

Cloning of an epithelial chloride channel from bovine trachea. *J. Biol. Chem.* 270:31016–26

253. Elble RC, Widom J, Gruber AD, Abdel-Ghany M, Levine R, et al. 1997. Cloning and characterization of lung-endothelial cell adhesion molecule-1 suggests it is an endothelial chloride channel. *J. Biol. Chem.* 272:27853–61

254. Gandhi R, Elble RC, Gruber AD, Schreur KD, Ji H-L, et al. 1998. Molecular and functional characterization of a calcium-sensitive chloride channel from mouse lung. *J. Biol. Chem.* 273:32096–101

255. Gruber AD, Elble RC, Ji H-L, Schreur KD, Fuller CM, Pauli BU. 1998. Genomic cloning, molecular characterization, and functional analysis of human CLCA1, the first human member of the family of Ca^{2+}-activated Cl^- channel proteins. *Genomics* 54:200–14

256. Gruber AD, Gandhi R, Pauli BU. 1998. The murine calcium-sensitive chloride channel (mCaCC) is widely expressed in secretory epithelia and in other select tissues. *Histochem. Cell Biol.* 110:43–49

257. Pauli BU, Abdel-Ghany M, Cheng HC, Gruber AD, Archibald HA, Elble RC. 2000. Molecular characteristics and functional diversity of CLCA family members. *Clin. Exp. Pharmacol. Physiol.* 27:901–5

258. Gruber AD, Pauli BU. 1998. Molecular cloning and biochemical characterization of a truncated secreted member of the human family of Ca^{2+}-activated Cl^- channels. *Biochim. Biophys. Acta* 1444:418–23

259. Papassotiriou J, Eggermont J, Droogmans G, Nilius B. 2001. Ca^{2+}-activated Cl^- channels in Ehrlich ascites tumor cells are distinct from mCLCA1, 2 and 3. *Pflügers Arch.* 442:273–79

260. Loewen ME, Bekar LK, Walz W, Forsyth GW, Gabriel SE. 2004. pCLCA1 lacks inherent chloride channel activity in an epithelial colon carcinoma cell line. *Am. J. Physiol. Gastrointest. Liver Physiol.* 287:G33–41

261. Loewen ME, Bekar LK, Gabriel SE, Walz W, Forsyth GW. 2002. pCLCA1 becomes a cAMP-dependent chloride conductance mediator in Caco-2 cells. *Biochem. Biophys. Res. Commun.* 298:531–36

262. Loewen ME, Smith NK, Hamilton DL, Grahn BH, Forsyth GW. 2003. CLCA protein and chloride transport in canine retinal pigment epithelium. *Am. J. Physiol. Cell Physiol.* 285:C1314–21

263. Sun H, Tsunenari T, Yau K-W, Nathans J. 2002. The vitelliform macular dystrophy protein defines a new family of chloride channels. *Proc. Natl. Acad. Sci. USA* 99:4008–13

264. Petrukhin K, Koisti MJ, Bakall B, Li W, Xie G, et al. 1998. Identification of the gene responsible for Best macular dystrophy. *Nat. Genet.* 9:241–47

265. Marmorstein AD, Mormorstein LY, Rayborn M, Wang X, Hollyfield JG, Petrukhin K. 2000. Bestrophin, the product of the Best vitelliform macular dystrophy gene (VMD2), localizes to the basolateral membrane of the retinal pigment epithelium. *Proc. Natl. Acad. Sci. USA* 97:12758–63

266. Clapham DE. 1998. The list of potential volume-sensitive chloride currents continues to swell (and shrink). *J. Gen. Physiol.* 111:623–24

267. Pusch M. 2004. Ca^{2+}-activated chloride channels go molecular. *J. Gen. Physiol.* 123:323–25

268. Suzuki M, Mizuno A. 2004. A novel human Cl^- channel family related to *Drosophila flightless* locus. *J. Biol. Chem.* 279:22461–68

269. Fahmi M, Garcia L, Taupignon A, Dufy B, Sartor P. 1995. Recording of a large-conductance chloride channel in normal rat lactotrophs. *Am. J. Physiol. Endocrinol. Metab.* 269:E969–76

270. Matchkov VV, Aalkjaer C, Nilsson H. 2004. A cyclic GMP-dependent calcium-activated chloride current in smooth-muscle cells from rat mesenteric resistance arteries. *J. Gen. Physiol.* 123:121–34

271. Akasu T, Nishimura T, Tokimasa T. 1990. Calcium-dependent chloride current in neurones of the rabbit pelvic parasympathetic ganglia. *J. Physiol.* 422:303–20

272. Baron A, Pacaud P, Loirand G, Mironneau C, Mironneau J. 1991. Pharmacological block of Ca^{2+}-activated Cl^- current in rat vascular smooth muscle cells in short-term primary culture. *Pflügers Arch.* 419:553–58

273. Chao AC, Mochizuki H. 1992. Niflumic and flufenamic acids are potent inhibitors of chloride secretion in mammalian airway. *Life Sci.* 51:1453–57

274. Hazama H, Nakajima T, Hamada E, Omata M, Kurachi Y. 1996. Neurokinin A and Ca^{2+} current induce Ca^{2+}-activated Cl^- currents in guinea-pig tracheal myocytes. *J. Physiol.* 492:377–93

275. Henmi S, Imaizumi Y, Muraki K, Watanabe M. 1996. Time course of Ca^{2+}-dependent K^+ and Cl^- currents in single smooth muscle cells of guinea-pig trachea. *Eur. J. Pharmacol.* 306:227–36

276. Hogg RC, Wang Q, Large WA. 1994. Action of niflumic acid on evoked and spontaneous calcium-activated chloride and potassium currents in smooth muscle cells from rabbit portal vein. *Br. J. Pharmacol.* 112:977–84

277. Ishikawa T. 1996. A bicarbonate- and weak acid-permeable chloride conductance controlled by cytosolic Ca^{2+} and ATP in rat submandibular acinar cells. *J. Membr. Biol.* 153:147–59

278. Jones K, Shmygol A, Kupittayanant S, Wray S. 2004. Electrophysiological characterization and functional importance of calcium-activated chloride channel in rat uterine myocytes. *Pflügers Arch.* 448:36–43

279. Kleene SJ. 1993. Origin of the chloride current in olfactory transduction. *Neuron* 11:123–32

280. McEwan GT, Hirst BH, Simmons NL. 1994. Carbachol stimulates Cl^- secretion

via activation of two distinct apical Cl⁻ pathways in cultured human T84 intestinal epithelial monolayers. *Biochim. Biophys. Acta* 1220:241–47

281. Nilius B, Prenen J, Kamouchi M, Viana F, Voets T, Droogmans G. 1997. Inhibition by mibefradil, a novel calcium channel antagonist, of Ca^{2+}- and volume-activated Cl⁻ channels in macrovascular endothelial cells. *Br. J. Pharmacol.* 121:547–55

282. Robertson BE. 1998. Inhibition of calcium-activated chloride channels by niflumic acid dilates rat cerebral arteries. *Acta Physiol. Scand.* 163:417–18

283. Sutton KG, Stapleton SR, Scott RH. 1998. Inhibitory actions of synthesised polyamine spider toxins and their analogues on Ca^{2+}-activated Cl⁻ currents recorded from cultured DRG neurones from neonatal rats. *Neurosci. Lett.* 251:117–20

284. Verkerk AO, Veldkamp MW, Bouman LN, van Ginneken AC. 2000. Calcium-activated Cl⁻ current contributes to delayed afterdepolarizations in single Purkinje and ventricular myocytes. *Circulation* 101:2639–44

285. Wang Q, Hogg RC, Large WA. 1992. Properties of spontaneous inward currents recorded in smooth muscle cells isolated from the rabbit portal vein. *J. Physiol.* 451:525–37

286. White MM, Aylwin M. 1990. Niflumic and flufenamic acids are potent reversible blockers of Ca^{2+}-activated Cl⁻ channels in *Xenopus* oocytes. *Mol. Pharmacol.* 37:720–24

287. Wladkowski SL, Lin W, McPheeters M, Kinnamon SC, Mierson S. 1998. A basolateral chloride conductance in rat lingual epithelium. *J. Membr. Biol.* 164:91–101

288. Wu G, Hamill OP. 1992. NPPB block of Ca^{2+}-activated Cl⁻ currents in *Xenopus* oocytes. *Pflügers Arch.* 420:227–29

289. Xu Y, Dong PH, Zhang Z, Ahmmed GU, Chiamvimonvat N. 2002. Presence of a calcium-activated chloride current in mouse ventricular myocytes. *Am. J. Physiol. Heart Circ. Physiol.* 283:H302–14

290. Yamazaki J, Hume JR. 1997. Inhibitory effects of glibenclamide on cystic fibrosis transmembrane regulator, swelling-activated, and Ca^{2+}-activated Cl⁻ channels in mammalian cardiac myocytes. *Circ. Res.* 81:101–9

Annu. Rev. Physiol. 2005. 67:759–78
doi: 10.1146/annurev.physiol.67.032003.153547
First published online as a Review in Advance on October 12, 2004

FUNCTION OF CHLORIDE CHANNELS IN THE KIDNEY

Shinichi Uchida and Sei Sasaki

Department of Nephrology, Graduate School of Medicine, Tokyo Medical and Dental University, Tokyo 113-8519, Japan; email: suchida.kid@tmd.ac.jp; ssasaki.kid@tmd.ac.jp

Key Words   CLC channel, Barttin, transepithelial chloride transport, nephrogenic diabetes insipidus, Bartter's syndrome

■ **Abstract**   Numerous Cl^- channels have been identified in the kidney using physiological approaches and thus are thought to be involved in a range of physiological processes, including vectorial transepithelial Cl^- transport, cell volume regulation, and vesicular acidification. In addition, expression of genes from several Cl^- channel gene families has also been observed. However, the molecular characteristics of a number of Cl^- channels within the kidney are still unknown, and the physiological roles of Cl^- channels identified by molecular means remain to be determined. A gene knockout approach using mice might shed further light on the characteristics of these various Cl^- channels. In addition, study of diseases involving Cl^- channels (channelopathies) might clarify the physiological role of specific Cl^- channels. To date, more is known about CLC Cl^- channels than any other Cl^- channels within the kidney. This review focuses on the physiological roles of CLC Cl^- channels within the kidney, particularly kidney-specific ClC-K Cl^- channels, as well as the recently identified maxi anion channel in macula densa, which is involved in tubulo-glomerular feedback.

INTRODUCTION

Three distinct Cl^- channel families, CLC, cystic fibrosis transmembrane regulator (CFTR), and ligand-gated gamma-aminobutyric acid (GABA) and glycine receptors, have been well characterized using molecular means (1). In the kidney, few people have examined the role of GABA and glycine receptors. Although CFTR is abundantly expressed in the kidney, the absence of an overt renal phenotype in cystic fibrosis (CF) patients and in mouse models of CF suggests a relatively minor role of CFTR in the kidney, compared with its physiological role in the lung, pancreas, and colon (2). Other Cl^- channels are known to be expressed in the kidney, including CLCA and p64 (CLIC) Cl^- channels, which have been described as Ca^{2+}-activated Cl^- channels and intracellular Cl^- channels, respectively (1). However, the physiological role of these Cl^- channels in the kidney remains unknown, primarily because of a lack of information with regard to the intrarenal and intracellular localization of these channels, as well as a lack of experimentation in knockout mice. In contrast, recent human and mouse genetic studies have identified clear roles for ClC-K1 (3), ClC-K2 (4), and ClC-5 (5). After a short overview of CLC Cl⁻ channel expression, this review focuses on the physiological roles of CLC Cl⁻ channels in the kidney on the basis of information obtained using knockout mice. In addition, human diseases caused by mutations in CLC Cl⁻ channel genes are discussed in order to provide further insight into the physiological function of Cl⁻ channels.

CLC Cl⁻ CHANNELS EXPRESSED IN THE KIDNEY

Nine CLC Cl⁻ channels have been identified in mammals (6–16). With the exception of skeletal muscle-specific ClC-1, all the CLC Cl⁻ channels identified to date are expressed in the kidney. Although ClC-K1 and ClC-K2 were initially thought to be kidney-specific CLC Cl⁻ channels, they have since been identified in the inner ear (17–19). ClC-5 was also first reported as a kidney-specific channel (20); however, broad expression of ClC-5 has now been observed throughout the body, including the colon, brain, liver, lung, and testis (10, 15). ClC-3 and ClC-4, which form a sub-branch with ClC-5, showed relatively broad expression patterns in the body compared with ClC-5. An in situ hybridization study of ClC-3 has demonstrated its presence in the type B intercalated cells of collecting ducts (21). However, owing to the lack of an appropriate ClC-3-specific antibody for use in immunohistochemical staining, intrarenal and intracellular localization of ClC-3 cannot be confirmed. A renal phenotype has not been observed in ClC-3 knockout mice (22, 23). Recently, Mohammad-Panah et al. have established a ClC-4 specific antibody and shown colocalization of ClC-4 with ClC-5 in the endosomes of proximal tubules (24). However, they have not specified whether ClC-4 is present elsewhere in the nephron. Ubiquitous expression of ClC-2 has been observed (7), although the precise intrarenal and cellular localization of ClC-2 in the kidney has not been demonstrated by immunohistochemical means. Obermuller et al. have reported localization of ClC-2 mRNA in the S3 segments of proximal tubules (21). However, a renal phenotype has not been observed in ClC-2 knockout mice (25) or humans (26). Expression of ClC-6 and ClC-7 in the mouse kidney has been demonstrated by Brandt & Jentsch (11). Kida et al. (27) have further localized expression of ClC-6 and ClC-7 to the proximal tubules following in situ hybridization studies in mice. Nonetheless, an overt renal phenotype has not been observed in ClC-7 knockout mice (28). These data suggest that some cells in the kidney express more than one CLC Cl⁻ channel. Proximal tubular cells probably express ClC-4, -6, and -7, in addition to ClC-5. However, deletion of ClC-5 alone is sufficient to cause Dent's disease, suggesting that a loss of ClC-5 is not compensated by expression of other ClCs in proximal tubules. A similar lack of compensation has been observed in the hippocampus of the ClC-3 knockout mouse. Although coexpression of ClC-3 and ClC-4 has been observed in hippocampal neurons (22), knockout of ClC-3 alone induces hippocampal degeneration (22, 23). These observations suggest that the various ClCs have different roles within cells, despite the fact that they function as Cl⁻ channels. With regard to intracellular localization, ClC-7 (28) and ClC-5 (29, 30) have shown different localization, whereas differences in intracellular localization between ClC-3, ClC-4, and ClC-5 remain to be established, as does the exact intracellular localization of ClC-6.

ClC-K Cl⁻ CHANNELS

Cloning and Intrarenal Localization

The first member of CLC Cl⁻ channel, ClC-0, was identified by the expression cloning strategy (31) from the electric organ of the *Torpedo marmorata*. Following identification of ClC-0, two mammalian CLC Cl⁻ channels (ClC-1 and ClC-2) were isolated by low stringency hybridization (6, 7). Sequencing of the ClC-1 and ClC-2 Cl⁻ channels has enabled utilization of reverse-transcription PCR using degenerate primers in order to identify other CLC Cl⁻ channels. Uchida et al. (12) and Adachi et al. (13) have isolated two ClC Cl⁻ channels from the rat kidney, called ClC-K (kidney)1 and ClC-K2, owing to their predominance of expression in the kidney. ClC-K1 and K2 are composed of 687 amino acids and possess the same signature sequences as other CLC channels. There is 81% homology between the amino acid sequences of ClC-K1 and ClC-K2 in rats. Subsequently, two human ClC-K channels have been cloned (16) and called Ka and Kb because the high degree of amino acid identity between these channels (91% homology between Ka and Kb) makes it difficult to determine species orthologs by simple sequence comparisons (Figure 1). Takeuchi et al. have also isolated a human ClC-K channel, and the intrarenal expression of this channel, as determined by RT-PCR, suggests that it represents ClC-Kb (14).

Although ClC-K1 and ClC-K2 are highly homologous, different intrarenal localizations of these channels have been observed. Yoshikawa et al. (32) have shown

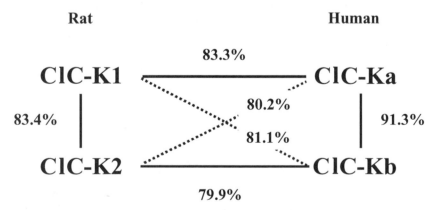

Figure 1   Amino acid identity between rat and human ClC-K Cl⁻ channels.

a predominance of K1 expression in the inner medulla, and K2 expression in the outer medulla and outer cortex alone, by in situ hybridization using $3''$-untranslated regions of K1 and K2 as probes, which share less than 20% nucleotide identity. This pattern of ClC-K1 mRNA expression is consistent with that observed in rats following immunohistochemical staining of rat ClC-K1 using an antipeptide antibody (33). Staining for rat ClC-K1 has been observed in both the apical and basolateral plasma membranes of the thin ascending limb of Henle's loop (tAL). The exact intrarenal localization of ClC-K2 could not be established prior to the generation of ClC-K1 knockout mice owing to the lack of a ClC-K2-specific antibody. However, Kobayashi et al. have since determined the intrarenal and cellular localization of mouse ClC-K2 in ClC-K1 knockout mice using an antibody that reacts with both mouse K1 and K2 (34). ClC-K2 has thus been localized to the basolateral surface of the thick ascending limb of Henle's loop (TAL), as well as to the distal tubules and connecting tubules. In addition, it has been found in the intercalated cells of collecting ducts. This pattern of localization matches quite nicely with the results of an in situ hybridization study of rat ClC-K2 (32). Vandevalle et al. have reported on ClC-K immuno-localization within the rat kidney using an antibody specific for both K1 and K2 (35). In addition to basolateral staining of the tAL, as well as staining of distal nephron segments, they have shown ClC-K immunostaining in proximal tubules (S3 segments) and inner medullary collecting ducts. Unlike Uchida et al. who observed apical localization of ClC-K1 in the tAL, apical ClC-K immunostaining was not observed by Vandevalle et al. This discrepancy might arise from differences in the specificity and sensitivity of the antibodies used in these studies. Recently, Kobayashi et al. generated transgenic mice harboring an enhanced green fluorescence protein (EGFP) gene driven by a human *CLCNKB* gene promoter (18). The EGFP expression was observed in the TAL, distal tubules, connecting tubules, and intercalated cells in the collecting ducts, which confirms that human ClC-Kb corresponds to rat and mouse ClC-K2. This pattern of localization of ClC-Kb fits with the phenotype observed in Bartter's syndrome, which is caused by mutations in the *CLCNKB* gene (4). Andreoli's group has identified a rabbit ClC-K channel, which they call rbClC-Ka (36, 37). Expression of rbClC-Ka has been observed in the TAL, suggesting that rbClC-Ka might be a rabbit homologue of rat and mouse ClC-K2 and human ClC-Kb. They have also described two mouse ClC-K channels, mcClC-K and mmClC-K (38). The nucleotide sequences of mcClC-K and mmClC-K appear to correspond with mouse ClC-K1 and ClC-K2, respectively. However, although mcClC-K was isolated from cortical TAL cells, localization of mouse ClC-K1 to cortical TAL has not been observed in vivo; rather, ClC-K1 has been localized to the tAL in vivo (3).

Functional Characterization of Rat ClC-K1

Uchida et al. described the functional expression of rat ClC-K1 in *Xenopus oocytes* (33). Earlier controversy with regard to the reproducibility of these results has

TABLE 1 Comparison of functional characteristics of ClC-K1 and a native chloride transport in the thin ascending limb of Henle's loop

ClC-K1 expressed in *Xenopus* oocytes[a]	In vitro perfusion of tAL
Anion selectivity	
$Br^- > Cl^- > I^-$	$Br^- > Cl^- > I^-$ [b]
pH sensitivity	
Inhibited by acidic pH	Inhibited by acidic pH
$K_d = \sim 6.9$	$K_d = 6.3$ [c]
Extracellular Ca^{2+} sensitivity	
Ca^{2+} free	
Cl^- current reduced to $\sim$40% of control	Decreased V_{Cl} and $^{36}Cl^-$ flux to $\sim$40% of control[d]
Drug sensitivity	
DIDS: $K_i = \sim 30\ \mu M$	80% reduction of V_{Cl} by 1 mM DIDS[e]
Furosemide: $K_i = \sim 100\ \mu M$	10% reduction of V_{Cl} by 1 mM furosemide[e]
DPC: $K_i = \sim 0.5$ mM	10% reduction of V_{Cl} by 1 mM DPC[f]

[a]Reference 33, [b]Reference 84, [c]Reference 85, [d]Reference 86, [e]Reference 87, [f]Reference 88. V_{Cl}: diffusion voltage of Cl^- ions, DIDS: 4,4'-diisothiocyanostilbene-2,2'-disulfonic acid, DPC: diphenylamine carboxylate.

subsided following confirmation by Waldegger & Jentsch (39). Recently, Barttin, a small membrane protein mutated in Bartter's syndrome type IV (51), was identified as necessary for the fully functional expression of ClC-K channels (17). It is unclear why only rat ClC-K1 (not mouse ClC-K1 or human ClC-Ka) can be functionally expressed in oocytes. The functional characteristics of oocyte-expressed rat ClC-K1 (33), compared with those of Cl^- transport in the tAL previously characterized in in vitro perfusion studies, are summarized in Table 1. Cl^- permeability in the tAL was extraordinarily high among nephron segments (Figure 2). Although some differences in the characteristics of rat ClC-K1 and Cl^- transport in the tAL exist, for the most part they show striking similarities. These results, combined with those pertaining to the immuno-localization of rat ClC-K1 within the tAL (33), strongly suggest that ClC-K1 is responsible for the high degree of Cl^- permeability noted in the tAL. This has been confirmed by studies of ClC-K1 knockout mice (3). Recently, Lourdel et al. succeeded in recording Cl^- channels on the basolateral membrane of microdissected distal-convoluted tubules using the patch-clamp technique (33a). This may be the first characterization of native ClC-K at the single-channel level.

A recent structural analysis of a bacterial CLC Cl^- channel has revealed an important role for a glutamate (E) residue contained within a conserved CLC signature (GK_EGP) sequence in gating of the CLC channel (40). Interestingly, all ClC-K channels lack E in this position, instead having valine or threonine (40). In ClC-0, when E166 was mutated to valine, the fast gate appears to stay open, and the

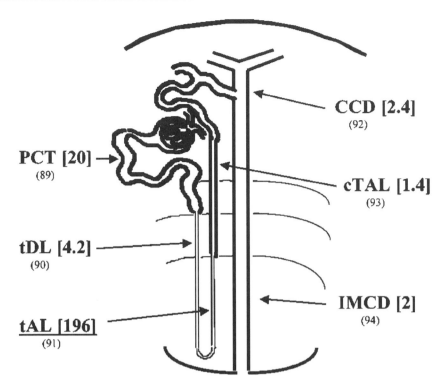

Figure 2 Transepithelial Cl⁻ permeability (P_{Cl}) along the nephron. The numbers in brackets represent P_{Cl} ($\times 10^{-5}$ cm/s). PCT, proximal tubules; tDL, thin descending limb of Henle's loop; tAL, thin ascending limb of Henle's loop; cTAL, cortical thick ascending limb of Henle's loop; CCD, cortical collecting ducts; IMCD, inner medullary collecting duct. The numbers in parentheses are references.

effect is most obvious at negative test voltages where gradual reduction of inward current observed in the wild-type channel is nearly absent (40). Conversely, a V166E mutation of ClC-K1 has been observed to result in strong voltage-dependent gating (39), and macroscopic currents of V166E mutant of ClC-Ka expressed in oocytes with Barttin displayed outward rectification (S. Uchida, unpublished observation). Thus it appears that insertion of a neutral amino acid at this position disrupts the voltage-dependent gating of ClC-K channels. The lack of voltage dependence might be necessary for an efficient transepithelial Cl⁻ transport in the plasma membranes.

ClC-K1 Knockout Mice

Matsumura et al. have reported on nephrogenic diabetes insipidus (NDI) in *Clcnk1⁻ᐟ⁻* mice (3). Differences in the physical appearance of *Clcnk1⁻ᐟ⁻* and *Clcnk1⁺ᐟ⁺* mice, or in their organ morphology, have not been observed.

Alterations in plasma creatinine, Na^+, K^+, and HCO_3^- levels have not been observed in $Clcnk1^{-/-}$ mice. The presence of normal $HCO3^-$ and K^+ values excludes the possibility that the loss of ClC-K1 leads to Bartter's syndrome, unlike the loss of function of ClC-Kb in patients with Bartter's syndrome (4). An approximately fourfold increase in the volume of excreted urine has been observed in $Clcnk1^{-/-}$ mice, compared with that in $Clcnk1^{+/+}$ mice allowed free access to water. Water deprivation has been associated with a 27% reduction in body weight in $Clcnk1^{-/-}$ mice, compared with ~10% in $Clcnk1^{+/+}$ mice. A decrease in excessive urinary volume has not been observed following intraperitoneal injection of dDAVP, a V2R agonist, in $Clcnk1^{-/-}$ mice, confirming the presence of NDI. A lack of pH- and NPPB sensitivity has been observed in the tAL of $Clcnk1^{-/-}$ mice upon measuring the transepithelial diffusion voltage of Cl^- ions in the tAL. Liu et al. have also directly measured the transepithelial flux of $^{22}Na^+$ and $^{36}Cl^-$ at the tAL, thereby confirming an absence of Cl^- permeability but no alteration in Na^+ permeability in $Clcnk1^{-/-}$ mice (41). They have also confirmed that Na^+ transport in the tAL occurs via a paracellular pathway. These data clearly establish that the high degree of Cl^- permeability noted in the tAL is mediated by ClC-K1. Akizuki et al. have further investigated the mechanism by which NDI occurs in $Clcnk1^{-/-}$ mice (42). Increased urinary excretion of AVP has been observed in $Clcnk1^{-/-}$ mice, even under normal conditions, indicating that the hypothalamic response to high plasma osmolarity is not impaired in these mice. Akizuki et al. have also shown similar total excretion of osmotic substances (urea, Na^+, and Cl^-) in $Clcnk1^{-/-}$ and $Clcnk1^{+/+}$ mice, confirming that the polyuria observed in these animals is because of water diuresis, not osmotic diuresis. It is thought that Cl^- transport in the tAL contributes to the countercurrent system of the inner medulla, in which a hypertonic environment is established without the requirement for active transport. In order to determine whether generation and maintenance of interstitial hyperosmolarity is impaired in the knockout mice, Akizuki et al. measured the total osmolarity of the inner medulla. As shown in Figure 3, a reduction in tissue osmolarity at the tip of papilla has been observed in $Clcnk1^{-/-}$ mice. Furthermore, impairment of Na^+ and urea, as well as Cl^-, accumulation has been observed in the inner medulla of $Clcnk1^{-/-}$ mice (42). This clearly suggests that a loss of one component of the countercurrent system (i.e., Cl^- transport within the tAL) is enough to disrupt the entire system. Thus the data obtained from $Clcnk1^{-/-}$ mice have clarified the in vivo role of ClC-K1 in the kidney. To date, there have been no reports of the *CLCNKA* gene mutations in patients with NDI. This situation might parallel that of aquaporin 1 (AQP1) knockout mice (43) and AQP1-deficient patients (44). AQP1 knockout mice have demonstrated an impaired urine concentrating ability, probably the result of an impaired countercurrent system because AQP1 is a major water channel within the thin descending limb of Henle's loop. Similar daily urine output values have been observed in $Aqp1^{-/-}$ and $Clcnk1^{-/-}$ mice (4–5 ml/day); however, much greater values have been observed in $Aqp3/4$ double knockout mice (over 10 ml/day). This indicates that direct perturbation of water permeability in the collecting ducts results in more severe polyuria than does

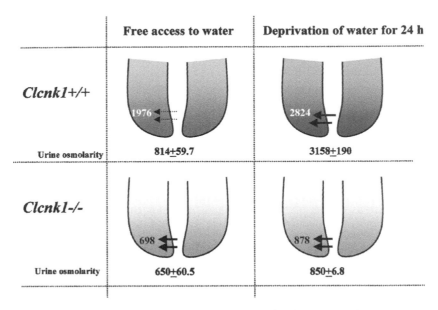

Figure 3   Tissue and urine osmolarity in $Clcnk1^{-/-}$ and $Clcnk1^{+/+}$ mice. Tissue osmolarity at the tip of the papilla was estimated from the total measured osmolarity of the inner medulla on the basis of an equation detailed in a previous report (42). Water permeability of the collecting ducts is indicated by the arrows.

impairment of the countercurrent system. This may hold true in humans because AQP1-deficient humans have shown only mild impairments in urinary concentrating ability (44). AQP1-deficient patients can be identified in so far as they are Colton blood group-negative; however, they have never been identified as having NDI. In humans, dysfunction of the countercurrent system may not lead to overt NDI.

Postnatal Development of ClC-K1

Apart from studies in knockout mice, other evidence exists in support of the roles of ClC-K1 and the countercurrent system in concentrating urine, i.e., postnatal development of urine concentrating ability in mammals (45, 46). In mice and rats, maximum urine concentrating ability has been observed 2–3 weeks after birth. A gradual increase in ClC-K1 protein expression within the ascending limb of Henle's loop in the inner medulla (there is no thin ascending limb in the neonate) has also been observed to occur over this time. Conversely, NKCC2, Na/K/2Cl cotransporter, present in the ascending limb of Henle's loop at birth, gradually disappeared in the mature tAL. Liu et al. have observed marked similarities between inner medullary organization in the neonatal period in mammals and in the adult period in avian species (46). Increased urinary concentrating ability appears to correlate with maturation of the inner medulla, i.e., ClC-K1 appearance and NKCC2 disappearance in the tAL, suggesting that the countercurrent system of the inner medulla is required for maximal urine concentrating ability in mammals. However, birds may not require ClC-K1 and may have only one ClC-K, as do fish (see below).

ClC-K in Fish

ClC-K channels from *Xenopus laevis* (47) and *Oeochromis mossambicus* (a fish called tilapia that can adapt to both sea and fresh water) have been cloned (48). One ClC-K channel has been cloned from each species. OmClC-K, an *Oeochromis mossambicus* ClC-K channel, has been localized to the basolateral plasma membrane of distal tubules, and colocalization with Na/K-ATPase channels has been observed. The tilapia nephron lacks a loop of Henle; however, tilapia distal tubules share similar properties with the TAL in mammals, such as active Na^+ and Cl^- transport through the apical Na/K/2Cl cotransporter, basolateral Na/K-ATPase, and a lack of water permeability, suggesting that this is a nephron segment to generate free water. Tilapia can adapt to sea water and fresh water. Thus if OmClC-K has a role in generating free water, it should be active when tilapia are in fresh water because they must conserve NaCl and secrete water in fresh water conditions. In fact, Miyazaki et al. have shown that OmClC-K is expressed only when tilapia are in fresh water (48), which suggests that the ClC-K channel evolved to enable tilapia to tolerate fresh water. ClC-K2 in mammals might be akin to ClC-K in tilapia. It is possible that gene duplication resulted in ClC-K1 in mammals, thus enabling even greater urine concentrating ability. Intrarenal localization of *Xenopus* ClC-K has been reported by Maulet et al. (47) following immunohistochemical staining of ClC-K at the basolateral surface of proximal tubules and the apical surface of distal nephrons. However, this observation of intrarenal and cellular localization of *Xenopus* ClC-K does not fit with the above observations regarding the role of ClC-K channels in mammals and fish. Recently, however, Vize has determined that *Xenopus* ClC-K is present in pronephric distal tubule and duct (49) but not in the proximal tubules. Further characterizarion of *Xenopus* ClC-K may be necessary.

ClC-K2 Cl^- Channel and Bartter's Syndrome

The physiological role of mammalian ClC-K2 in nephron segments distal to the tAL is similar to that of OmClC-K. This has been verified by identification of mutations in the *CLCNKB* gene (encoding ClC-Kb) in patients with Bartter's syndrome (4). Thus far, 22 mutations have been reported in the Human Gene Mutation Database. The mechanism by which some of these missense ClC-Kb mutations resulted in a loss of function was not entirely clear until Barttin was identified as a β-subunit of ClC-K channels (17) since ClC-K2 cannot be heterologously expressed alone, unlike rat ClC-K1. The currents induced in oocytes by the injection of rat ClC-K2 cRNA only (13) might be activated endogenous Cl^- currents. Thus disruption of Cl^- transport in the basolateral membrane of the TAL and distal tubules likely impairs transepithelial NaCl reabsorption in distal nephron segments, which results in Bartter's syndrome. Differences in the clinical manifestations of the various types of Bartter's syndrome have been reported. For example, in Bartter's syndrome type III (caused by *CLCNKB* mutations), nephrocalcinosis is not observed and hypercalciuria is not a consistent finding, both of which are common in Bartter's syndrome types I and II (4). Loss of ClC-Kb function leads to a loss of transepithelial NaCl transport in the TAL and in the distal tubules. Decreased Ca^{2+} reabsorption in the TAL, as seen in types I and II Bartter's syndrome, might be offset by increased Ca^{2+} reabsorption in the distal tubules, as occurs in Gitelman's syndrome (50).

Barttin is a β-Subunit of ClC-K Channels

The *BSND* gene encoding Barttin has been identified as a fourth gene responsible for Bartter's syndrome (51). Because consistent colocalization of Barttin with ClC-K1 and K2 has been observed, Estevez et al. examined whether Barttin might be a functional subunit of ClC-K channels in *Xenopus* oocytes (17). Co-injection of ClC-K and Barttin cRNA has been observed to generate a pH-sensitive Cl^- current. This induction was specific for ClC-K and was attained by the translocation of ClC-K to the plasma membrane. Disease-causing mutations other than G10S failed to induce ClC-K currents in oocytes, thus confirming the role of Barttin mutations in Bartter's syndrome. Hayama et al. recently examined the pathogenesis of Bartter's syndrome caused by Barttin mutations using stable cell lines expressing wild-type and some mutant Barttin proteins (52). They have shown that wild-type Barttin reaches the plasma membrane but that a disease-causing Barttin mutant (*R8L*) is confined to the endoplasmic reticulum (ER). ClC-K, which is predominately expressed in the Golgi apparatus without Barttin coexpression, might be recruited to the plasma membrane by Barttin. It is not clear whether Barttin binds and recruits ClC-K to the appropriate location or whether Barttin masks Golgi-localizing signals inherent in ClC-K when it binds to ClC-K, thus releasing it from the Golgi apparatus. In addition, the Barttin mutant (*R8L*) has been observed to bind to CLC ClC-K2 (52). As a result, ClC-K2 becomes trapped in the ER with Barttin and does not reach the plasma membrane. In addition to regulation of intracellular trafficking, marked enhancement of ClC-K2 protein stability has been observed with Barttin coexpression. The mechanism by which this occurs is unknown. On the other hand, mutation of the PY motif of Barttin has been associated with altered regulation of Barttin protein itself. Alteration of a putative PY motif enhances the induction of ClC-K current in oocytes by Barttin (17). In Madin-Darby canine kidney (MDCK) cells, enhanced localization of a *Y98A* mutant in the plasma membrane has been observed (52). These results suggest that isolation and characterization of proteins (probably containing the WW domains of ubiquitin protein ligases) that bind to the PY motif might reveal novel regulatory mechanisms of Barttin/ClC-K channels. Apart from Barttin's role in regulating the cellular localization of ClC-K channels, the effect of Barttin coexpression on the channel function of ClC-K is unknown. Waldegger et al. (53) have not noted any change in the electrophysiological properties of ClC-K1 channels when coexpressed with Barttin, except for a change in sensitivity to extracellular Ca^{2+}. Increasing extracellular Ca^{2+} from 1.8 mM to 6.8 mM has been observed to increase the amplitude of current in oocytes expressing rat ClC-K1 alone, but not rat ClC-K1 coexpressed with Barttin. More detailed studies are necessary to clarify this issue. Andreoli's group (53a) has reported regulation of Cl^- channels in the TAL, likely ClC-K2 channels, by protein kinase A (PKA). Also, Takahashi et al. (53b) have reported increased Cl^- permeability in the tAL by PKA. However, phosphorylation of ClC-K by PKA has not been described. In fact, in vitro phosphorylation of ClC-K1 protein by PKA has not been observed (S. Uchida, unpublished observation). It is possible that regulation of ClC-K in vivo by PKA, or by some of the other known regulatory mechanisms, occurs indirectly via regulation of Barttin.

The presence of Barttin in the inner ear explains why deafness has been noted in some patients with Barttin mutations. On the other hand, the absence of deafness in patients with Bartter's syndrome type III (*CLCNKB* mutation) suggests that, in addition to ClC-Kb, ClC-Ka must be present in the inner ear and that expression of normal ClC-Ka in the ear allows for adequate Cl^- transport. Estevez et al. have confirmed the presence of ClC-K1 and ClC-K2 in the inner ear in mice (17), although direct evidence of human ClC-Ka in the inner ear has not been reported. Recently, combined ClC-Ka and ClC-Kb mutations have been observed in deaf patients with Bartter's syndrome. As expected from studies in mice, this evidence supports the role of ClC-Ka in the inner ear (54).

ClC-5 Cl^- Channel

Mutations in the *CLCN5* gene located on chromosome Xp11.22 have been identified in patients with Dent's disease (5, 55), which is a renal tubular disorder characterized by low-molecular-weight proteinuria (LMWP), hypercalciuria, nephrocalcinosis, nephrolithiasis, and progressive renal failure. The *CLCN5* gene encodes a human ClC-5 Cl^- channel, which is classified as a sub-branch of the CLC family, along with ClC-3 (8, 56) and ClC-4 (24, 57–60). At present, 46 mutations have been listed in the Human Gene Mutation Database. Wu et al. recently analyzed 15 mutations (one in-frame insertion and 14 missense mutations) thought to form a complete architecture of the channel by aligning the human ClC-5 sequence with bacterial ClCs, the X-ray crystal structures of which have recently been established (61). They found that 12 of the 15 mutations are clustered at the interface between two homodimeric ClC-5 channels, suggesting the dimer formation might be crucial for proper ClC channel function. Several reports have described the intrarenal and cellular localization of ClC-5. Rat and mouse ClC-5 have been observed in the early endosomes in the subapical regions of proximal tubules and α-intercalated cells and colocalized with H^+-ATPase (29, 30, 62). In addition, Devuyst et al. reported ClC-5 localization in the TAL in the human kidney (62). Luyckx et al.

also reported localization of rat ClC-5 in the TAL (63). However, their antibody recognized the proximal tubules in only S3 segments. The phenotype of ClC-5 knockout mice (64, 65) strongly suggests the presence of ClC-5 in early proximal tubules. Accordingly, the existence of ClC-5 in TAL must be re-evaluated. Endosomal localization of ClC-5 has also been confirmed in cultured cells (29).

Disease-causing mutations of ClC-5 have been noted to abolish Cl^- currents induced by heterologous expression systems, suggesting that LMWP, as a result of impairment of endocytosis in proximal tubules, might somehow be related to a failure of endosomal ClC-5 Cl^- channels (5). It has been postulated that the Cl^--conducting pathway must be present in acidic organelles as a counterion transport system with H^+-ATPase. However, the biophysical properties of heterologously expressed ClC-5 are not consistent with this possibility because voltages more positive than $+20$ mV are required to induce Cl^- currents, and the channel is inhibited by acidic pH (10, 57). To examine this further, two groups generated and characterized ClC-5 knockout mice (64, 65). Luyckx et al. have reported decreased ClC-5 expression in rebozyme-expressing mice (66). These mice have not been observed to demonstrate any obvious phenotype, apart from a slight increase in serum Ca^{2+}, which is controlled by a low-Ca^{2+} diet. In addition, no dysfunction of proximal tubules (manifesting as LMWP) has been observed. It is possible that rebozyme does not completely inhibit the expression of ClC-5. Piwon et al. have described the phenotypes of $Clcn5^{-/-}$ mice (64), which include LMWP and elevated urinary phosphate concentrations in the absence of hypercalciuria. A reduction in endocytosis (to less than 30% of control values) has also been observed. An approximately twofold reduction in megalin protein, which plays a role in the receptor-mediated endocytosis of a number of proteins, has been observed in $Clcn5^{-/-}$ cells, perhaps explaining the observation of LMWP. Christensen et al. have also reported a reduction in cubilin, a transferrin receptor, in $Clcn5^{-/-}$ mice (67). However, they have reported that the overall reduction in megalin is not sufficient to explain LMWP since $megalin^{-/+}$ mice have not demonstrated LMWP. Thus Christensen et al. have theorized that redistribution of megalin and cubilin might result in failure of apical protein endocytosis. They demonstrated selective loss of two receptors at the brush border of proximal tubules by subcellular fractionation and immunogold labeling and concluded that redistribution resulting in translocation of these receptors from the brush border to intracellular endocytotic organelles might result in LMWP. Recently, Moulin et al. examined renal biopsy specimens from patients with Dent's disease (68). They have consistently noted inversion of H^+-ATPase polarity in the proximal tubules to a basolateral distribution, in contrast to its apical localization in the normal kidney. Further studies are required to elucidate the mechanism by which translocation of various membrane proteins occurs in the absence of ClC-5.

Phosphaturia has been explained by the observation that defective endocytosis in the absence of ClC-5 results in an elevated luminal PTH concentration, thereby inducing the endocytosis of NaPi-2. This is supported by the fact that (*a*) increased urinary PTH excretion has been observed in $Clcn5^{-/-}$ mice and that (*b*) apical localization of NaPi-2 has been observed in the early segments of proximal tubules where a lack of endocytosis would be expected to have a negligible effect, compared with its subapical vesicular localization in most other parts of the proximal tubules (64). With regard to hypercalciuria (one of the symptoms observed in patients with Dent's disease), examination of different strains of knockout mice has yielded conflicting reports. The knockout mice generated by Piwon et al. (64) do not develop hypercalciuria, whereas those in the model produced by Wang et al. have shown a twofold increase in urinary Ca^{2+} excretion (65). It is possible that disruption of the $Clcn5$ gene has opposing effects on the production of $1,25(OH)_2$ VitD3. Increased luminal PTH normally increases α-hydroxylase levels, resulting in conversion of $25(OH)$VitD3 to $1,25(OH)_2$ VitD3 (the active form of vitamin D). However, the impaired endocytosis in the $Clcn5^{-/-}$ mice also leads to a decreased availability of $25(OH)$VitD3 in the proximal tubules. Hypercalciuria may be determined by the level of serum $1,25(OH)_2$ VitD3, which depends on the balance between these factors. This might also explain why a range of severity of hypercalciuria has been noted among patients with Dent's disease, as well as in mouse models.

The results of studies of mouse models have clearly demonstrated the requirement for ClC-5 for efficient endocytosis within proximal tubules, as well as Ca^{2+} homeostasis. In addition, these studies have shown that hyperphosphaturia in Dent's disease is an indirect result of defective apical endocytosis of PTH and $25(OH)$VitD3. Further studies are necessary to elucidate the mechanism by which loss of ClC-5 results in impaired endocytosis. Slower endosomal acidification has been demonstrated in $Clcn5^{-/-}$ mice, and this is thought to contribute to defective endocytosis. However, other evidence has surfaced to suggest that impaired acidification might not be sufficient to impair endocytosis alone (69, 70). In addition, direct evidence suggesting that ClC-5 is a Cl^- shunt pathway is lacking, a reality further complicated by the electrophysiological properties of ClC-5.

OTHER CLC Cl^- CHANNELS IN THE KIDNEY

ClC-3 and ClC-4 are classified along with ClC-5 as a subcategory of CLC Cl^- channel. Unlike ClC-5, for which a definite role in the kidney has been established, the functions of these channels within the kidney remain largely unknown. In the past, a lack of ClC-3- and ClC-4-specific antibodies made it difficult to identify their role within the kidney. Recently, a ClC-4-specific antibody was established by Mohammad-Panah et al. with which they have observed colocalization of ClC-4 and ClC-5 in the form of a heterodimer in the early endosomes in the proximal tubules (24). Friedrich et al. have characterized the function of ClC-4 in *Xenopus* oocytes (57). ClC-4 has been observed to elicit an outwardly rectifying Cl^- current, similar to that generated by ClC-5. In light of the fact that there is colocalization of ClC-4 and ClC-5 within the endosomes of proximal tubules and that ClC-4 generates a current similar to that of ClC-5, it is not clear why ClC-4 does not compensate for a lack of ClC-5 in $Clcn5^{-/-}$ mice or in patients with Dent's disease.

Further studies using *Clcn4* knockout mice and ClC4/ClC-5 double knockout mice are necessary to clarify this issue. Several groups have generated *Clcn3*$^{-/-}$ mice. Stobrawa et al. reported a loss of hippocampus and retinal degeneration in *Clcn3*$^{-/-}$ mice (22). They determined the cellular localization of ClC-3 by Western blot of fractionated tissue homogenate exposed to a sucrose gradient and observed ClC-3 in an endosomal fraction of the liver, as well as in synaptic vesicles within the brain. They have also demonstrated impairment of acidification within synaptic vesicles in *Clcn3*$^{-/-}$ mice. Yoshikawa et al. also generated *Clcn3*$^{-/-}$ mice and report similar phenotypes (23). In addition to the report by Stobrawa et al., Yoshikawa et al. further observed degeneration of ilial mucosa. These phenotypes are similar to those observed in cathepsin D knockout mice. Interestingly, cathepsin D knockout mice are used to model a human neuro-degenerative lysosomal storage disease, known as neuronal ceroid lipofuscinosis (NCL). In fact, electron dense bodies have been observed in the lysosomes of degenerating hippocampal cells in *Clcn3*$^{-/-}$ mice. F1F0ATPase subunit c, a highly hydrophobic protein known to accumulate in NCL lysosomes, has been observed to accumulate in the degenerating hippocampal cell lysosomes. This finding suggests that *Clcn3*$^{-/-}$ mice might also be used to study NCL. Yoshikawa et al. have clearly demonstrated an increased intravesicular pH in *Clcn3*$^{-/-}$ mice. It seems that impairment of endosomal and/or lysosomal acidification somehow damages protein degradation within organelles. These two studies using *Clcn3*$^{-/-}$ mice clearly support a role for ClC-3 as an intracellular channel. There has been controversy surrounding the potential role of ClC-3 as a volume-regulated Cl$^-$ channel (71). The generation of *Clcn3*$^{-/-}$ mice has not resolved this issue because conflicting evidence remains. Stobrawa et al. reported that swelling-activated Cl$^-$ currents are unchanged in the mice (22). However, Yamamoto-Mizuma et al. recently suggested that in response to deletion of the *Clcn3* gene compensatory changes in the expression of other anion channels may occur and produce currents similar to that of native ClC-3 by a different mechanism (72). Thus they have cautioned against negating ClC-3 as a volume-regulated Cl$^-$ channel on the basis of data obtained from *Clcn3* knockout mice. RNAi technology in a cell culture system may eventually be used to answer this question. Whether ClC-3 is a volume-regulated Cl$^-$ channel, the physiological role of ClC-3 in the kidney remains unclear. At present, there is a lack of definitive information regarding the intrarenal and cellular localization of ClC-3 in the kidney, and a renal phenotype has not been identified in *Clcn3*$^{-/-}$ mice. Further studies using double and triple knockout mice of ClC-3, ClC-4, and/or ClC-5 may answer this question.

ADDITIONAL NON-CLC Cl$^-$ CHANNELS IN THE KIDNEY

In addition to CLC Cl$^-$ channels, CLCA and CLIC (putative) Cl$^-$ channels are also expressed in the kidney (1, 73–75). As reviewed previously (1), the physiological roles of these channels within cells and organs, including the kidney, remain unclear. Further studies of knockout mice are required to elucidate their function.

There has also been a growing interest in ATP conductance through epithelial cell anion channels. Although ATP conduction through the CFTR has been described (76, 77), these findings could not be reproduced by all laboratories, and it is still uncertain whether this actually occurs or is an artifact. Bell et al. recently reported (78) a large conductance voltage-dependent anion channel (maxi anion channel) in macula densa cells, which releases ATP upon entry of NaCl into cells. After being metabolized to adenosine, the released ATP may directly or indirectly trigger tubulo-glomerular feedback signaling at the afferent arteriolar smooth muscle cells. Complete inhibition of this macula dense anion channel has been observed with 50 μM Gd^{3+}, which is known to inhibit ATP release from other cell types. However, sensitivity of the maxi anion channel to DPC and NPPB has not been observed, nor has it shown sensitivity to Ca^{2+}. Dutta et al. have identified a similar channel in mouse mammary C127 cells, which shows a large unitary conductance ($\sim$400 pS), as well as bell-shaped, voltage-dependent ATP permeability (79). Maxi anion channels are broadly similar to the voltage-dependent anion channel (VDAC) of the outer mitochondrial membrane (79, 80). The suggestion that VDACs might exist in extramitochondrial locations has received support from several recent studies (81–83). An isoform of mitochondrial VDAC (pl-VDAC) contains a leader sequence targeting it to the plasma membrane (81). Colocalization of VDAC with markers of plasma membrane lipid rafts (83) and caveolin (82) has also been shown. Therefore, it is possible that VDAC itself, or VDAC-like channels, comprise the ATP-release channel in the macula densa. This anion channel might have other important roles in kidney, such as transepithelial anion transport and cell volume regulation. Molecular characterization of maxi anion channels within the kidney would contribute significantly to future Cl^- channel research.

Although studies on CLC channels have greatly clarified the physiological roles of Cl^- channels in the kidney, entire families of these channels postulated to be important in kidney physiology have not been identified at the molecular level. Undoubtedly, the recent advances in systematic genomic and proteomic approaches will facilitate this process.

ACKNOWLEDGMENTS

Work in our laboratory is supported in part by grants-in-aid from the Ministry of Education, Culture, Sports, Science, and Technology of Japan.

The *Annual Review of Physiology* is online at
http://physiol.annualreviews.org

LITERATURE CITED

1. Jentsch TJ, Stein V, Weinreich F, Zdebik AA. 2002. Molecular structure and physiological function of chloride channels. *Physiol. Rev.* 82:503–68

2. Devuyst O, Guggino WB. 2002. Chloride channels in the kidney: lessons learned from knockout animals. *Am. J. Physiol. Renal Physiol.* 283:1176–91

3. Matsumura Y, Uchida S, Kondo Y, Miyazaki H, Ko SB, et al. 1999. Overt nephrogenic diabetes insipidus in mice lacking the CLC-K1 chloride channel. *Nat. Genet.* 21:95–98

4. Simon DB, Bindra RS, Mansfield TA, Nelson-Williams C, Mendonca E, et al. 1997. Mutations in the chloride channel gene, CLCNKB, cause Bartter's syndrome type III. *Nat. Genet.* 17:171–78

5. Lloyd SE, Pearce SH, Fisher SE, Steinmeyer K, Schwappach B, et al. 1996. A common molecular basis for three inherited kidney stone diseases. *Nature* 379:445–49

6. Steinmeyer K, Ortland C, Jentsch TJ. 1991. Primary structure and functional expression of a developmentally regulated skeletal muscle chloride channel. *Nature* 354:301–4

7. Thiemann A, Grunder S, Pusch M, Jentsch TJ. 1992. A chloride channel widely expressed in epithelial and non-epithelial cells. *Nature* 356:57–60

8. Kawasaki M, Uchida S, Monkawa T, Miyawaki A, Mikoshiba K, et al. 1994. Cloning and expression of a protein kinase C-regulated chloride channel abundantly expressed in rat brain neuronal cells. *Neuron* 12:597–604

9. Jentsch TJ, Gunther W, Pusch M, Schwappach B. 1995. Properties of voltage-gated chloride channels of the ClC gene family. *J. Physiol.* 482:19S–25

10. Steinmeyer K, Schwappach B, Bens M, Vandewalle A, Jentsch TJ. 1995. Cloning and functional expression of rat CLC-5, a chloride channel related to kidney disease. *J. Biol. Chem.* 270:31172–77

11. Brandt S, Jentsch TJ. 1995. ClC-6 and ClC-7 are two novel broadly expressed members of the CLC chloride channel family. *FEBS Lett.* 377:15–20

12. Uchida S, Sasaki S, Furukawa T, Hiraoka M, Imai T, et al. 1993. Molecular cloning of a chloride channel that is regulated by dehydration and expressed predominantly in kidney medulla. *J. Biol. Chem.* 268:3821–24

13. Adachi S, Uchida S, Ito H, Hata M, Hiroe M, et al. 1994. Two isoforms of a chloride channel predominantly expressed in thick ascending limb of Henle's loop and collecting ducts of rat kidney. *J. Biol. Chem.* 269:17677–83

14. Takeuchi Y, Uchida S, Marumo F, Sasaki S. 1995. Cloning, tissue distribution, and intrarenal localization of ClC chloride channels in human kidney. *Kidney Int.* 48:1497–503

15. Sakamoto H, Kawasaki M, Uchida S, Sasaki S, Marumo F. 1996. Identification of a new outwardly rectifying Cl^- channel that belongs to a subfamily of the ClC Cl^- channels. *J. Biol. Chem.* 271:10210–16

16. Kieferle S, Fong P, Bens M, Vandewalle A, Jentsch TJ. 1994. Two highly homologous members of the ClC chloride channel family in both rat and human kidney. *Proc. Natl. Acad. Sci. USA* 91:6943–47

17. Estevez R, Boettger T, Stein V, Birkenhager R, Otto E, et al. 2001. Barttin is a Cl^- channel beta-subunit crucial for renal Cl^- reabsorption and inner ear K^+ secretion. *Nature* 414:558–61

18. Kobayashi K, Uchida S, Okamura HO, Marumo F, Sasaki S. 2002. Human CLC-KB gene promoter drives the EGFP expression in the specific distal nephron segments and inner ear. *J. Am. Soc. Nephrol.* 13:1992–98

19. Maehara H, Okamura HO, Kobayashi K, Uchida S, Sasaki S, Kitamura K. 2003. Expression of CLC-KB gene promoter in the mouse cochlea. *Neuroreport* 14:1571–73

20. Fisher SE, Black GC, Lloyd SE, Hatchwell E, Wrong O, et al. 1994. Isolation and partial characterization of a chloride channel gene which is expressed in kidney and is a candidate for Dent's disease (an X-linked hereditary nephrolithiasis). *Hum. Mol. Genet.* 3:2053–59

21. Obermuller N, Gretz N, Kriz W, Reilly RF, Witzgall R. 1998. The swelling-activated chloride channel ClC-2, the chloride channel ClC-3, and ClC-5, a chloride channel mutated in kidney stone disease, are expressed in distinct subpopulations of renal epithelial cells. *J. Clin. Invest.* 101:635–42

22. Stobrawa SM, Breiderhoff T, Takamori S, Engel D, Schweizer M, et al. 2001. Disruption of ClC-3, a chloride channel expressed on synaptic vesicles, leads to a loss of the hippocampus. *Neuron* 29:185–96

23. Yoshikawa M, Uchida S, Ezaki J, Rai T, Hayama A, et al. 2002. CLC-3 deficiency leads to phenotypes similar to human neuronal ceroid lipofuscinosis. *Genes Cells* 7:597–605

24. Mohammad-Panah R, Harrison R, Dhani S, Ackerley C, Huan LJ, et al. 2003. The chloride channel ClC-4 contributes to endosomal acidification and trafficking. *J. Biol. Chem.* 278:29267–77

25. Bosl MR, Stein V, Hubner C, Zdebik AA, Jordt SE, et al. 2001. Male germ cells and photoreceptors, both dependent on close cell-cell interactions, degenerate upon ClC-2 Cl⁻ channel disruption. *EMBO J.* 20:1289–99

26. Haug K, Warnstedt M, Alekov AK, Sander T, Ramirez A, et al. 2003. Mutations in CLCN2 encoding a voltage-gated chloride channel are associated with idiopathic generalized epilepsies. *Nat. Genet.* 33:527–32

27. Kida Y, Uchida S, Miyazaki H, Sasaki S, Marumo F. 2001. Localization of mouse CLC-6 and CLC-7 mRNA and their functional complementation of yeast CLC gene mutant. *Histochem. Cell Biol.* 115:189–94

28. Kornak U, Kasper D, Bosl MR, Kaiser E, Schweizer M, et al. 2001. Loss of the ClC-7 chloride channel leads to osteopetrosis in mice and man. *Cell* 104:205–15

29. Gunther W, Luchow A, Cluzeaud F, Vandewalle A, Jentsch TJ. 1998. ClC-5, the chloride channel mutated in Dent's disease, colocalizes with the proton pump in endocytotically active kidney cells. *Proc. Natl. Acad. Sci. USA* 95:8075–80

30. Sakamoto H, Sado Y, Naito I, Kwon TH, Inoue S, et al. 1999. Cellular and subcellular immunolocalization of ClC-5 channel in mouse kidney: colocalization with H⁺-ATPase. *Am. J. Physiol. Renal Physiol.* 277:F957–65

31. Jentsch TJ, Steinmeyer K, Schwarz G. 1990. Primary structure of *Torpedo marmorata* chloride channel isolated by expression cloning in *Xenopus* oocytes. *Nature* 348:510–14

32. Yoshikawa M, Uchida S, Yamauchi A, Miyai A, Tanaka Y, et al. 1999. Localization of rat CLC-K2 chloride channel mRNA in the kidney. *Am. J. Physiol. Renal Physiol.* 276:F552–58

33. Uchida S, Sasaki S, Nitta K, Uchida K, Horita S, et al. 1995. Localization and functional characterization of rat kidney-specific chloride channel, ClC-K1. *J. Clin. Invest.* 95:104–13

33a. Lourdel S, Paulais M, Marvao P, Nissant A, Teulon J. 2003. A chloride channel at the basolateral membrane of the distal-covoluted tubules: a candidate ClC-K channel. *J. Gen. Physiol.* 121:287–300

34. Kobayashi K, Uchida S, Mizutani S, Sasaki S, Marumo F. 2001. Intrarenal and cellular localization of CLC-K2 protein in the mouse kidney. *J. Am. Soc. Nephrol.* 12:1327–34

35. Vandewalle A, Cluzeaud F, Bens M, Kieferle S, Steinmeyer K, Jentsch TJ. 1997. Localization and induction by dehydration of ClC-K chloride channels in the rat kidney. *Am. J. Physiol. Renal Physiol.* 272:678–88

36. Zimniak L, Winters CJ, Reeves WB, Andreoli TE. 1995. Cl⁻ channels in basolateral renal medullary vesicles. X. Cloning of a Cl⁻ channel from rabbit outer medulla. *Kidney Int.* 48:1828–36

37. Zimniak L, Winters CJ, Reeves WB, Andreoli TE. 1996. Cl⁻ channels in

basolateral renal medullary vesicles XI. rbClC-Ka cDNA encodes basolateral MTAL Cl⁻ channels. *Am. J. Physiol. Renal Physiol.* 270:F1066–72

38. Mikhailova MV, Winters CJ, Andreoli TE. 2002. Cl⁻ channels in basolateral TAL membranes. XVI. MTAL and CTAL cells each contain the mRNAs encoding mmClC-Ka and mcClC-Ka. *Kidney Int.* 61:1003–10

39. Waldegger S, Jentsch TJ. 2000. Functional and structural analysis of ClC-K chloride channels involved in renal disease. *J. Biol. Chem.* 275:24527–33

40. Dutzler R, Campbell EB, MacKinnon R. 2003. Gating the selectivity filter in ClC chloride channels. *Science* 300:108–12

41. Liu W, Morimoto T, Kondo Y, Iinuma K, Uchida S, et al. 2002. Analysis of NaCl transport in thin ascending limb of Henle's loop in CLC-K1 null mice. *Am. J. Physiol. Renal Physiol.* 282:F451–57

42. Akizuki N, Uchida S, Sasaki S, Marumo F. 2001. Impaired solute accumulation in inner medulla of *Clcnk1⁻/⁻* mice kidney. *Am. J. Physiol. Renal Physiol.* 280:F79–87

43. Ma T, Yang B, Gillespie A, Carlson EJ, Epstein CJ, Verkman AS. 1998. Severely impaired urinary concentrating ability in transgenic mice lacking aquaporin-1 water channels. *J. Biol. Chem.* 273:4296–99

44. King LS, Choi M, Fernandez PC, Cartron JP, Agre P. 2001. Defective urinary-concentrating ability due to a complete deficiency of aquaporin-1. *N. Engl. J. Med.* 345:175–79

45. Kobayashi K, Uchida S, Mizutani S, Sasaki S, Marumo F. 2001. Developmental expression of CLC-K1 in the postnatal rat kidney. *Histochem. Cell Biol.* 116:49–56

46. Liu W, Morimoto T, Kondo Y, Iinuma K, Uchida S, Imai M. 2001. "Avian-type" renal medullary tubule organization causes immaturity of urine-concentrating ability in neonates. *Kidney Int.* 60:680–93

47. Maulet Y, Lambert RC, Mykita S, Mou-

ton J, Partisani M, et al. 1999. Expression and targeting to the plasma membrane of xClC-K, a chloride channel specifically expressed in distinct tubule segments of *Xenopus laevis* kidney. *Biochem. J.* 340:737–43

48. Miyazaki H, Kaneko T, Uchida S, Sasaki S, Takei Y. 2002. Kidney-specific chloride channel, OmClC-K, predominantly expressed in the diluting segment of freshwater-adapted tilapia kidney. *Proc. Natl. Acad. Sci. USA* 99:15782–87

49. Vize PD. 2003. The chloride conductance channel ClC-K is a specific marker for the *Xenopus pronephric* distal tubule and duct. *Gene Expr. Patterns* 3:347–50

50. Simon DB, Nelson-Williams C, Bia MJ, Ellison D, Karet FE, et al. 1996. Gitelman's variant of Bartter's syndrome, inherited hypokalaemic alkalosis, is caused by mutations in the thiazide-sensitive Na-Cl cotransporter. *Nat. Genet.* 12:24–30

51. Birkenhager R, Otto E, Schurmann MJ, Vollmer M, Ruf EM, et al. 2001. Mutation of BSND causes Bartter syndrome with sensorineural deafness and kidney failure. *Nat. Genet.* 29:310–14

52. Hayama A, Rai T, Sasaki S, Uchida S. 2003. Molecular mechanisms of Bartter syndrome caused by mutations in the BSND gene. *Histochem. Cell Biol.* 119:485–93

53. Waldegger S, Jeck N, Barth P, Peters M, Vitzthum H, et al. 2002. Barttin increases surface expression and changes current properties of ClC-K channels. *Pflügers Arch.* 444:411–18

53a. Winters CJ, Reeves WB, Andreoli TE. 1991. Cl⁻ channels in basolateral renal medullary membrane vesicles IV. Analogous channel activation by Cl⁻ or cAMP-dependent protein kinase. *J. Membr. Biol.* 122:89–95

53b. Takahashi N, Kondo Y, Ito O, Igarashi Y, Omata K, et al. 1995. Vasopressin stimulates Cl⁻ transport in ascending thin limb of Henle's loop in hamster. *J. Clin. Invest.* 95:1623–27

54. Schlingmann KP, Konrad M, Jeck N, Waldegger P, Reinalter SC, et al. 2004. Salt wasting and deafness resulting from mutations in two chloride channels. *N. Engl. J. Med.* 350:1314–19

55. Thakker RV. 1998. The role of renal chloride channel mutations in kidney stone disease and nephrocalcinosis. *Curr. Opin. Nephrol. Hypertens.* 7:385–88

56. Kawasaki M, Suzuki M, Uchida S, Sasaki S, Marumo F. 1995. Stable and functional expression of the ClC-3 chloride channel in somatic cell lines. *Neuron* 14:1285–91

57. Friedrich T, Breiderhoff T, Jentsch TJ. 1999. Mutational analysis demonstrates that ClC-4 and ClC-5 directly mediate plasma membrane currents. *J. Biol. Chem.* 274:896–902

58. Hebeisen S, Heidtmann H, Cosmelli D, Gonzalez C, Poser B, et al. 2003. Anion permeation in human ClC-4 channels. *Biophys. J.* 84:2306–18

59. Mohammad-Panah R, Ackerley C, Rommens J, Choudhury M, Wang Y, Bear CE. 2002. The chloride channel ClC-4 co-localizes with cystic fibrosis transmembrane conductance regulator and may mediate chloride flux across the apical membrane of intestinal epithelia. *J. Biol. Chem.* 277:566–74

60. Vanoye CG, George AL Jr. 2002. Functional characterization of recombinant human ClC-4 chloride channels in cultured mammalian cells. *J. Physiol.* 539:373–83

61. Wu F, Roche P, Christie PT, Loh NY, Reed AA, et al. 2003. Modeling study of human renal chloride channel (hCLC-5) mutations suggests a structural-functional relationship. *Kidney Int.* 63:1426–32

62. Devuyst O, Christie PT, Courtoy PJ, Beauwens R, Thakker RV. 1999. Intrarenal and subcellular distribution of the human chloride channel, CLC-5, reveals a pathophysiological basis for Dent's disease. *Hum. Mol. Genet.* 8:247–57

63. Luyckx VA, Goda FO, Mount DB, Nishio T, Hall A, et al. 1998. Intrarenal and sub-cellular localization of rat CLC5. *Am. J. Physiol. Renal Physiol.* 275:F761–69

64. Piwon N, Gunther W, Schwake M, Bosl MR, Jentsch TJ. 2000. ClC-5 Cl$^-$-channel disruption impairs endocytosis in a mouse model for Dent's disease. *Nature* 408:369–73

65. Wang SS, Devuyst O, Courtoy PJ, Wang XT, Wang H, et al. 2000. Mice lacking renal chloride channel, CLC-5, are a model for Dent's disease, a nephrolithiasis disorder associated with defective receptor-mediated endocytosis. *Hum. Mol. Genet.* 9:2937–45

66. Luyckx VA, Leclercq B, Dowland LK, Yu AS. 1999. Diet-dependent hypercalciuria in transgenic mice with reduced CLC5 chloride channel expression. *Proc. Natl. Acad. Sci. USA* 96:12174–79

67. Christensen EI, Devuyst O, Dom G, Nielsen R, Van der Smissen P, et al. 2003. Loss of chloride channel ClC-5 impairs endocytosis by defective trafficking of megalin and cubilin in kidney proximal tubules. *Proc. Natl. Acad. Sci. USA* 100:8472–77

68. Moulin P, Igarashi T, Van der Smissen P, Cosyns JP, Verroust P, et al. 2003. Altered polarity and expression of H$^+$-ATPase without ultrastructural changes in kidneys of Dent's disease patients. *Kidney Int.* 63:1285–95

69. van Weert AW, Dunn KW, Gueze HJ, Maxfield FR, Stoorvogel W. 1995. Transport from late endosomes to lysosomes, but not sorting of integral membrane proteins in endosomes, depends on the vacuolar proton pump. *J. Cell Biol.* 130:821–34

70. Mellman I, Fuchs R, Helenius A. 1986. Acidification of the endocytic and exocytic pathways. *Annu. Rev. Biochem.* 55:663–700

71. Duan D, Winter C, Cowley S, Hume JR, Horowitz B. 1997. Molecular identification of a volume-regulated chloride channel. *Nature* 390:417–21

72. Yamamoto-Mizuma S, Wang GX, Liu LL, Schegg K, Hatton WJ, et al. 2004. Altered

properties of volume-sensitive osmolyte and anion channels (VSOACs) and membrane protein expression in cardiac and smooth muscle myocytes from *Clcn3*$^{-/-}$ mice. *J. Physiol.* 557:439–56

73. Pauli BU, Abdel-Ghany M, Cheng HC, Gruber AD, Archibald HA, Elble RC. 2000. Molecular characteristics and functional diversity of CLCA family members. *Clin. Exp. Pharmacol. Physiol.* 27:901–5

74. Landry D, Sullivan S, Nicolaides M, Redhead C, Edelman A, et al. 1993. Molecular cloning and characterization of p64, a chloride channel protein from kidney microsomes. *J. Biol. Chem.* 268:14948–55

75. Redhead CR, Edelman AE, Brown D, Landry DW, al-Awqati Q. 1992. A ubiquitous 64-kDa protein is a component of a chloride channel of plasma and intracellular membranes. *Proc. Natl. Acad. Sci. USA* 89:3716–20

76. Schwiebert EM, Egan ME, Hwang TH, Fulmer SB, Allen SS, et al. 1995. CFTR regulates outwardly rectifying chloride channels through an autocrine mechanism involving ATP. *Cell* 81:1063–73

77. Reisin IL, Prat AG, Abraham EH, Amara JF, Gregory RJ, et al. 1994. The cystic fibrosis transmembrane conductance regulator is a dual ATP and chloride channel. *J. Biol. Chem.* 269:20584–91

78. Bell PD, Lapointe JY, Sabirov R, Hayashi S, Peti-Peterdi J, et al. 2003. Macula densa cell signaling involves ATP release through a maxi anion channel. *Proc. Natl. Acad. Sci. USA* 100:4322–27

79. Dutta AK, Okada Y, Sabirov RZ. 2002. Regulation of an ATP-conductive large-conductance anion channel and swelling-induced ATP release by arachidonic acid. *J. Physiol.* 542:803–16

80. Sabirov RZ, Dutta AK, Okada Y. 2001. Volume-dependent ATP-conductive large-conductance anion channel as a pathway for swelling-induced ATP release. *J. Gen. Physiol.* 118:251–66

81. Buettner R, Papoutsoglou G, Scemes E, Spray DC, Dermietzel R. 2000. Evidence for secretory pathway localization of a voltage-dependent anion channel isoform. *Proc. Natl. Acad. Sci. USA* 97:3201–6

82. Bathori G, Parolini I, Tombola F, Szabo I, Messina A, et al. 1999. Porin is present in the plasma membrane where it is concentrated in caveolae and caveolae-related domains. *J. Biol. Chem.* 274:29607–12

83. Bahamonde MI, Fernandez-Fernandez JM, Guix FX, Vazquez E, Valverde MA. 2003. Plasma membrane voltage-dependent anion channel mediates antiestrogen-activated maxi Cl$^-$ currents in C1300 neuroblastoma cells. *J. Biol. Chem.* 278:33284–89

84. Isozaki T, Yoshitomi K, Imai M. 1991. Selectivity of ion permeability across ascending thin limb of Henle's loop: interaction of Cl$^-$ and other halogens with anion transport system. *Kidney Int. Suppl.* 33:S113–18

85. Kondo Y, Yoshitomi K, Imai M. 1987. Effect of pH on Cl$^-$ transport in TAL of Henle's loop. *Am. J. Physiol. Renal Physiol.* 253:1216–22

86. Kondo Y, Yoshitomi K, Imai M. 1988. Effect of Ca^{2+} on Cl$^-$ transport in thin ascending limb of Henle's loop. *Am. J. Physiol. Renal Physiol.* 254:F232–39

Annu. Rev. Physiol. 2005. 67:779–807
doi: 10.1146/annurev.physiol.67.032003.153245

Physiological Functions of CLC Cl⁻ Channels Gleaned from Human Genetic Disease and Mouse Models

Thomas J. Jentsch, Mallorie Poët, Jens C. Fuhrmann, and Anselm A. Zdebik

Zentrum für Molekulare Neurobiologie Hamburg (ZMNH), Universität Hamburg, Falkenried 94, D-20251 Hamburg, Germany; email: Jentsch@zmnh.uni-hamburg.de, Mallorie.Poet@zmnh.uni-hamburg.de, Fuhrmann@zmnh.uni-hamburg.de, azdebik@zmnh.uni-hamburg.de

Key Words   anion channels, endocytosis, transepithelial transport, volume regulation, cystic fibrosis

■ **Abstract**   The CLC gene family encodes nine different Cl⁻ channels in mammals. These channels perform their functions in the plasma membrane or in intracellular organelles such as vesicles of the endosomal/lysosomal pathway or in synaptic vesicles. The elucidation of their cellular roles and their importance for the organism were greatly facilitated by mouse models and by human diseases caused by mutations in their respective genes. Human mutations in CLC channels are known to cause diseases as diverse as myotonia (muscle stiffness), Bartter syndrome (renal salt loss) with or without deafness, Dent's disease (proteinuria and kidney stones), osteopetrosis and neurodegeneration, and possibly epilepsy. Mouse models revealed blindness and infertility as further consequences of CLC gene disruptions. These phenotypes firmly established the roles CLC channels play in stabilizing the plasma membrane voltage in muscle and possibly in neurons, in the transport of salt and fluid across epithelia, in the acidification of endosomes and synaptic vesicles, and in the degradation of bone by osteoclasts.

INTRODUCTION

The molecular diversity of anion channels may not rival that of cation channels, but these channels belong to several structurally unrelated classes (1). The known Cl⁻ channels can be grouped into (*a*) CLC chloride channels, which are often gated by voltage; (*b*) ligand-gated $GABA_A$ and glycine receptors, which are related to ligand-gated cation channels such as the nicotinic acetylcholine receptor; (*c*) the cystic fibrosis transmembrane conductance regulator (CFTR), the only member of the ABC transporter family known to function as a chloride channel; and, probably,

0066-4278/05/0315-0779$14.00

(d) bestrophins, which may function as Ca^{2+}-activated Cl^- channels (2). The case for bestrophins directly mediating anion currents has been significantly strengthened by structure-function studies (3). This list is most likely incomplete, as some Cl^- channels (e.g., a class of swelling-activated channels) have not yet been identified at the molecular level. For a recent in-depth review on Cl^- channels, see (1).

Chloride channels are present in the plasma membrane and in membranes of intracellular organelles. They are involved in a broad range of functions, including the stabilization of membrane potential, synaptic inhibition, cell volume regulation, transepithelial transport, extracellular and vesicular acidification, and endocytotic trafficking. Many of these functions were discovered through the phenotypes resulting from their inactivation in human inherited disease or in mouse models. In this review, we focus on the surprisingly diverse functions of mammalian CLC chloride channels that were unraveled since the discovery of the CLC gene family in 1990 (4). The role of other classes of chloride channels in health and disease is discussed in recent reviews, such as on the involvement of glycine receptors in myoclonus and startle syndromes (5) and GABA receptor mutations in certain forms of epilepsy (6). Bestrophin (7) is mutated in Best macular dystrophy and is not discussed here.

GENERAL PROPERTIES OF CLC CHANNELS

To understand the pathogenic effects of mutations in CLC genes, it is useful to recall some general properties of these channels. For example, they function as dimers in which each monomer has its own pore (double-barreled channels). The two-pore architecture was first postulated based on an analysis of single channels reconstituted from *Torpedo* electric organ (8). After cloning (4) of this channel (later named ClC-0 (9)), protein purification (10), site-directed mutagenesis, and concatemers (11–13) strongly supported this notion and additionally suggested that each pore is completely contained within one subunit (11, 13). For bacterial CLC proteins, this has now been confirmed by crystallography (14), suggesting that all CLC proteins display the same basic architecture. Interestingly, bacterial CLC proteins may function as cotransporters rather than channels (15), but a mutational analysis of ClC-1 showed that the structure of the bacterial exchanger is highly conserved in the mammalian channel (16).

Each pore of the dimer retains its individual properties such as ion selectivity and single-channel conductance even when forced together in an artificial heterodimer, as shown, e.g., for ClC-0 and ClC-1 or ClC-2 (13). At least in ClC-0, ClC-1 (17), and ClC-2 (13), each pore (protopore) can be opened and closed by an individual gate. In ClC-0 (the best studied channel in this respect), protopore gating is independent of the state of the other gate (8, 18). In addition to the protopore gate (also called the fast gate in ClC-0), there is also (at least in ClC-0 and ClC-1) a common gate that closes both pores in parallel. In ClC-0, its kinetics led to the name slow gate, but it is much faster in ClC-1 (17). A glutamate side chain that obstructs the pore may play a role in protopore gating (14, 19). The structural basis for the common gate is still obscure.

This architecture has important consequences for the impact of CLC mutations. In contrast to tetrameric K^+ channels, mutations that reduce single-channel conductance or protopore gating will generally not have dominant-negative effects on coexpressed wild-type subunits. Dominant effects can be obtained by those mutations that affect the common gate, which closes both subunits, or by mutations resulting in proteins that, while retaining their ability to associate with wild-type subunits, cause the missorting or degradation of the resulting abnormal dimer. No dimerization signals have been identified in CLC channels. However, genetic data and in vitro studies indicate that most, if not all, truncations within the transmembrane part lack dominant-negative effects, suggesting that these proteins are unable to associate to dimers. Crystal structures of bacterial CLC proteins (14) revealed a broad interface between the subunits, which involves helices H, I, P, and Q.

These considerations suggest that dominant-negative mutations occur less frequently, for example, with CLC channels than with K^+ channels. The dimeric channel structure implies that even the strongest dominant-negative mutations are unlikely to decrease currents to less than 25% in the heterozygous state. This compares with the strong reduction (down to 6%) with dominant mutations in tetrameric K^+ channels. The moderate dominant-negative effects possible with CLC channels explain that dominant myotonia congenita (mutations in ClC-1) (20–22) and osteopetrosis (ClC-7) (23, 24) are generally less severe than the recessive variants, in which both alleles are mutated and thus may be associated with a total loss of function.

Whereas CLC channels can function as homodimers (and this may be the most common situation in vivo), in vitro studies have shown that heterodimerization is possible within the branch of plasma membrane channels (ClC-0, -1, -2) (13, 25). Co-immunoprecipitation experiments suggested that ClC-4 and ClC-5 might interact (26). Whether this occurs to a sizeable degree in vivo is still unknown.

So far, the highly related channels ClC-Ka and ClC-Kb (ClC-K1 and ClC-K2 in rodents) are the only CLC channels known to require a β-subunit (barttin) for functional expression (27). The auxiliary subunit barttin (28) is crucial for an efficient transport of ClC-K proteins to the plasma membrane (27).

The currents through CLC channels can be modulated by voltage, extra- and intracellular anions (29–32), pH (27, 31, 33), extracellular Ca^{2+} (27, 34), cell swelling (33, 35), and phosphorylation (36). CLC channels lack a charged voltage sensor of the type seen in voltage-dependent cation channels (S4-segment). The voltage-dependence of protopore gating is thought to result from the movement of the permeant anion within the pore (30, 37), with the anion acting as gating charge (30). This simple model renders gating dependent on both Cl^- concentration and voltage. Crystal structures of bacterial CLC proteins revealed a glutamate whose side chain obstructs the pore a short distance to the extracellular side of the central Cl^--binding site (14, 19). Indeed, mutations at the equivalent position in

ClC-0 (19), ClC-1 (38), ClC-2 (32), ClC-K1 (39), ClC-4 and -5 (40), and ClC-3 (41) strongly influenced or abolished gating. The pH-dependence of gating was proposed to be due to a protonation of the gating glutamate, and the pH-dependence of gating was indeed abolished when this glutamate was mutated in ClC-0 (19). However, other parts of the protein may also contribute to pH-sensitivity, as ClC-K channels are modulated by pH but lack a glutamate at this position (27, 34, 42).

CLC channels have functions in the plasma membrane (ClC-1, -2, -Ka, -Kb) or in intracellular membranes of the endocytotic-lysosomal pathway (ClC-3 through ClC-7) (Figure 1, see color insert). The roles of plasma membrane CLC channels include the stabilization of membrane potential, transepithelial transport, and cell volume regulation, whereas endosomal/lysosomal CLC channels are thought to provide an electric shunt for the efficient pumping of the H^+-ATPase (1). Some vesicular channels may also be inserted into specialized domains of the plasma membrane. For instance, the late endosomal/lysosomal ClC-7 has an important role in the ruffled border of osteoclasts (23).

CLC-1 AND MYOTONIA

ClC-1 (9) is the closest homologue of the *Torpedo* electric organ Cl^- channel, ClC-0 (4). ClC-1 is nearly exclusively expressed in skeletal muscle. In rodents, ClC-1 transcripts are upregulated after birth (9). ClC-1 expression depends on muscle electrical activity (43). Immunocytochemistry located ClC-1 primarily to the outer, sarcolemmal membrane of skeletal muscle (44, 45), although physiological investigations revealed that muscle Cl^- conductance is mainly found in t-tubules (46).

ClC-1 has a small single-channel conductance of about 1–1.5 pS (13, 17, 47) and is blocked by 9-anthracene-carboxylic acid and 4-chloro-phenoxy-acetic acid derivates: Their binding sites have been mapped by mutagenesis and molecular modeling (16). As is true for other CLC channels, ClC-1 is partially blocked by I^-. Both the protopore gate and the common gate, which is much faster than that of ClC-0, are activated by depolarization (9, 17).

In an exceptional situation, the resting conductance of skeletal muscle is dominated by Cl^- rather than K^+. This equilibrates the electrochemical potential of Cl^- with the resting potential, which is ultimately generated by the K^+ gradient as in other cells. The large Cl^- conductance stabilizes the resting potential and helps to repolarize action potentials. In skeletal muscle, t-tubules propagate the electrical excitation deep into the muscle fibers, where the voltage-dependent activation of L-type Ca^{2+} channels eventually leads to intracellular Ca^{2+} release and muscle contraction. If the repolarization of t-tubular membranes occurred primarily through K^+ channels, $[K^+]$ would increase in the small space inside these tubules during prolonged muscle activity, thereby leading to a long-lasting moderate depolarization. When Cl^- channels are used for repolarization, the same absolute change in t-tubular $[Cl^-]$ (which is in the 100 mM range as is $[Cl^-]_o$ in general) will lead to a much smaller relative change in $[Cl^-]$, which will not appreciably change the t-tubular voltage. For this reason, evolution has chosen Cl⁻ channels to electrically stabilize and repolarize skeletal muscle membranes.

Bryant and colleagues (48, 49) demonstrated a severely reduced Cl⁻ conductance in the skeletal muscle of human patients with myotonia congenita and of a myotonic goat strain. Myotonia, a symptom found in several genetic diseases, is an impairment of skeletal muscle relaxation after voluntary contraction. It results from an increase in muscle excitability that can be detected in electromyograms in the form of myotonic runs, i.e., long trains of action potentials. In humans, there are two forms of pure nonsyndromic myotonia: autosomal recessive Becker-type myotonia congenita, and autosomal-dominant myotonia or Thomsen disease.

Soon after ClC-1was cloned (9), it was shown (50) that the open reading frame of ClC-1 was destroyed by the insertion of a transposon in the myotonic mouse strain *adr*. This demonstrated that ClC-1 is the major skeletal muscle Cl⁻ channel essential for maintaining normal muscle excitability. Soon afterward, it was found that ClC-1 also accounts for human myotonia congenita (51). Mutations were identified in dominant myotonia (Thomsen's disease) (20, 52), including mutations in family members of Dr. Thomsen (20), who was also affected. The Thomsen mutation (P480L) exerted a strong dominant-negative effect on wild-type channels coexpressed in *Xenopus* oocytes (20).

More than 80 different ClC-1 mutations have been identified in human myotonia (for a recent review, see 53). Most mutations result in recessive myotonia. This includes all truncations in the membrane portion of the channel. Although these mutations could lead to nonsense-mediated decay of RNA or to protein instability, a likely reason for a lack of a dominant-negative effect is the inability of truncated proteins to associate with wild-type subunits. As discussed above, the broad interface between the two subunits of the dimer also suggests that truncations before helix Q may lead to an inability to associate to dimers. Some recessive mutations, e.g., M485V, drastically reduce single-channel conductance (54). Satisfyingly, the crystal structure of the bacterial CLC protein (14) revealed that the highly conserved phenylalanine directly neighboring this methionine participates in coordinating a Cl⁻ ion in the narrowest part of the permeation pathway. As the pores of CLC channels are entirely contained within each subunit of the dimer (13, 14), pore mutations are unlikely to affect the conductance of the second subunit in wild-type/mutant heteromeric channels and therefore will generally lack dominant-negative effects. Several recessive mutations, including M485V, also strongly changed the voltage-dependence of ClC-1 gating (54–56).

With the exception of truncations very close to the carboxy terminus of ClC-1, all dominant mutations are missense mutations. Almost all these mutations exert dominant-negative effects by shifting the voltage-dependence of gating of the dimeric channel towards positive voltages where the channel has no impact on membrane repolarization (22). The shift is because of an effect on the common gate that acts on both pores in parallel (17). Indeed, many but not all mutations in dominant myotonia change residues close to the subunit interface, in particular in helices H and I. Site-directed mutagenesis of residues in helices I, G, H, P, and Q affect the common gate of ClC-1 (57–59).

Myotonic mutations in ClC-1 that change the voltage-dependence of gating of homomeric mutant channels have different effects on the voltage-dependence of wild-type/mutant heterodimers, which partly explains the variable penetrance of some of these mutations (60). This variability might be caused by differential effects on the common versus the protopore gate, or by differences in subunit affinities that may lead to preferential assembly of homodimers. Although myotonia has traditionally been classified into recessive (Becker) and dominant (Thomsen) forms, it is now clear that the "border" between these inheritance patterns is blurred. There are indeed mutations that are associated with recessive myotonia in some, and with dominant myotonia in other families (21, 60, 61). On the basis of studies in a small number of patients, it was recently proposed that differences in allelic expression may determine the penetrance of some mutations, thereby influencing the apparent pattern of inheritance (62).

As both alleles are mutated in patients with recessive myotonia, a total loss of ClC-1 channel function may ensue. In contrast, at least 25% of wild-type currents will remain in heterozygous patients carrying dominant-negative mutations, as expected from the dimeric channel architecture. Accordingly, recessive myotonia is clinically more severe than the dominant Thomsen form. As discussed below, even more dramatic differences in disease severity are observed with mutations in ClC-7 that underlie recessive and dominant osteopetrosis (23, 24).

Myotonia is a symptom of myotonic dystrophy (DM), a more severe disease that also affects several other tissues, e.g., the heart and the eye. Skeletal muscle displays the hyperexcitability that is typical for myotonia. In contrast to myotonia congenita, DM is associated with muscle dystrophy that develops with age. Electrophysiological studies on muscle biopsies from patients displayed several abnormalities, including a variable decrease in Cl^- conductance (63). Myotonic dystrophy is caused by CTG or CCTG (i.e., DNA base) expansions in the 3' end of the DM protein kinase (*DMPK*) gene (in DM1) or in an intron of the zinc finger 9 (*ZNF9*) gene (in DM2), respectively. The aberrant RNAs accumulate in the nuclei. Recent work has shown that this results in mis-splicing of the ClC-1 RNA (45, 64), possibly by recruiting the CUG-binding muscleblind protein, whose knockout in mice also leads to mis-splicing of ClC-1 and myotonia (65). The overall levels of ClC-1 RNA may then decrease by nonsense-mediated decay. However, the repeat-containing DMPK RNA may also sequester transcription factors, which results in an additional reduction of ClC-1 transcription (66). Although several aspects, including the relative specificity for ClC-1, are not yet fully understood, these studies highlight again the importance of ClC-1 in muscle physiology.

CLC-2: A UBIQUITOUSLY EXPRESSED CHANNEL WITH MANY CANDIDATE FUNCTIONS

ClC-2, an almost ubiquitously expressed Cl^- channel (67), is activated by hyperpolarization, cell swelling, and weakly acidic extracellular pH (33, 35, 67). At unphysiologically strong acidic pH values (<pH6), however, its open probability

decreases (68). Similar to ClC-0 and ClC-1, ClC-2 probably has a common gate and protopore gates (69), although the common gate was not detected in single-channel recordings (13). ClC-2 has a single-channel conductance of about 2–3 pS (13). Single-channel currents with similar properties were observed in astrocytes (70), which express ClC-2 (71), as confirmed by the absence of these currents in astrocytes isolated from ClC-2 knockout mice (72). Similar to gating in other CLC channels, the gating of ClC-2 is influenced by anions. However, in contrast to ClC-0 (30), ClC-2 is mainly affected by intracellular Cl⁻, with extracellular anions having minor effects (32, 73, 74). In addition, ClC-2 has a Cl > I selectivity (similar to other CLC channels) that was reported to be modulated by cyclin-dependent protein kinases (36, 75) and can be phosphorylated by protein kinase A. However, this does not change its channel activity (76, 77). The knockout of ClC-2 unexpectedly led to testicular and retinal degeneration (78).

The ubiquitous expression of ClC-2 and the various possibilities to modulate its channel activity have invited many speculations on its function. The activation by cell swelling (35) suggests a role in cell volume regulation. However, the swelling-activated Cl⁻ channel (called VRAC or VSOAC) that is thought to dominate cell volume regulation has quite different properties, most prominently an outward rectification and an I > Cl selectivity (1, 79). When heterologously expressed in *Xenopus* ooyctes (77) or Sf 9 cells (80), ClC-2 accelerated their regulatory volume decrease (RVD) after hypotonic swelling. However, parotid acinar cells from ClC-2 knockout mice recovered their volume as fast as did wild-type cells (81). It is an open question whether RVD of other cells depends on ClC-2. This issue may be best studied using $Clcn2^{-/-}$ mice.

The expression of ClC-2 in neurons that display an inhibitory response to GABA suggested that it may play a role in establishing a low chloride concentration in neurons (82). Because GABA$_A$- and glycine receptors are ligand-gated Cl⁻-channels, their opening results in de- or hyperpolarization when $[Cl^-]_i$ is above or below its electrochemical equilibrium, respectively. Early in development, GABA and glycine are excitatory in most neurons. The excitation later gives way to inhibition as $[Cl^-]_i$ decreases (83). It is now known that the main process lowering $[Cl^-]_i$ is the K-Cl cotransporter, KCC2 (84, 85). Other transport proteins, such as KCC3 (86) or ClC-2, may play additional roles. Indeed, when ClC-2 was transfected into dorsal root ganglion cells, their normally excitatory response to GABA was converted to inhibition (87). This is expected from an equilibration of the electrochemical Cl⁻ potential with the membrane potential. Opening of GABA$_A$ receptors will then yield neither hyperpolarization nor depolarization, but will stabilize the voltage close to its resting value. In discussing roles of ClC-2 in the central nervous system (CNS), one should be aware that ClC-2 is expressed not only in neurons, but also, prominently, in glia (71, 72, 88) where it may serve in the homeostasis of extracellular ion composition.

According to the proposed role of ClC-2 in lowering neuronal $[Cl^-]_i$, one might expect that its lack gives rise to epilepsy. This, however, was not observed in $Clcn2^{-/-}$ mice (78). On the other hand, a locus for multigenic idiopathic generalized epilepsy was mapped to 3q26 close to the ClC-2 locus (89). Screening a large cohort of patients revealed three sequence abnormalities that cosegregated in an apparently autosomal-dominant fashion with epileptic symptoms in three pedigrees (90). One mutation truncates ClC-2 in helix F, directly predicting a loss of function. The mutant reportedly also exerted a dominant-negative effect on wild-type ClC-2 (90). However, our laboratory and others (91) did not observe such a dominant effect. Indeed, similar truncations in other CLC channels lack dominant-negative effects, as discussed above for ClC-1. The second mutation deletes 11bp of an intron (90). This was reported to alter splicing, thereby increasing the abundance of a protein that lacks 44 amino acids in the intramembrane portion. Expression of a corresponding cRNA was reported to exert a strong dominant-negative effect that exceeded the 75% reduction that is possible with a dimer. Using minigenes, another study (91) failed to detect an effect of the deletion on splicing. Finally, Haug et al. (90) identified a missense mutation (G715E) in a family with three affected siblings. G715 lies between CBS1 and CBS2 in the cytoplasmic tail. G715E was reported to shift the voltage-dependence in a $[Cl^-]_i$-dependent manner to more positive voltages. This is equivalent to a gain of function, contrasting with the loss of function of the truncated channel (90). However, no such gain of function was observed in our laboratory (T.J. Jentsch, unpublished results), nor by others (91). CBS domains were recently shown to bind ATP and other nucleotides (92). Although G715 is located between CBS domains, the G715E mutation decreased the affinity of the ClC-2 carboxy terminus for AMP in vitro (92). In an electrophysiological study (91), the G715E mutant was indistinguishable from wild-type ClC-2 in the presence of 1 mM cytoplasmic ATP, but showed different gating kinetics when ATP was replaced by 2 mM AMP. Whether these conditions occur during development of epileptic seizures is unclear.

Hence, the strongest case for a causative role of ClC-2 in epilepsy is a single family whose affected members are heterozygous for a truncating mutation. In the likely absence of dominant-negative effects, the mutation may act via haploinsufficiency. This contrasts with the lack of epilepsy in mice totally lacking ClC-2 (78). In order to firmly establish a causative role of ClC-2 in epilepsy, it seems desirable to find further epilepsy-associated ClC-2 mutations that compromise channel function. In a recent study, no *CLCN2* mutations were found in 55 families with idiopathic generalized epilepsy (93), but clearly screening of more patients may be required to settle this question.

ClC-2 was also thought to be important for gastric acid secretion (94), a proposal which could not be confirmed in the ClC-2 knockout mouse (78). Because the lung needs to secrete salt and water during development, and because—in contrast to CFTR—ClC-2 is expressed in the lung early on, ClC-2 was proposed to be important for lung development (95–97). The expansion of fetal lung cysts in vitro could be inhibited with an antisense-oligonucleotide directed against ClC-2 (98). This study, however, is inconclusive as it employed the same antisense oligonucleotide that was used previously to knock down an inwardly rectifying Cl^- channel in choroid plexus cells (99). The knocked-down channel differed in several biophysical properties (most notably ion selectivity) from ClC-2, and experiments on $Clcn2^{-/-}$ mice revealed that it does not correspond to ClC-2 (100).

Obviously, this oligonucleotide (98, 99) lacks specificity. Convincing evidence against an essential role of ClC-2 in lung development comes from the ClC-2 knockout mouse (78). Lung morphology appeared normal even when both ClC-2 and CFTR were disrupted (101).

The fact that ClC-2 is also found in epithelia fueled speculations that it might modulate the phenotype of cystic fibrosis (CF), a severe disease that is caused by mutations of the cAMP-activated Cl⁻ channel CFTR. The latter apical channel mediates Cl⁻ secretion in many epithelia. An optimistic hypothesis holds that pharmacological activation of ClC-2 might create an alternative pathway for apical Cl⁻ secretion that could be of benefit for CF patients (102, 103). Of course, this requires an apical localization of ClC-2. Depending on the antibody used, however, divergent results were obtained. In lung epithelia, ClC-2 was described in apical membranes (97). It was also reported to localize to apical junctional complexes in an intestinal cell line and to contribute to their anion secretion (103). However, using two different antibodies, other groups described ClC-2 as being in basolateral membranes of intestinal epithelia (74, 104, 105).

If ClC-2 provides a pathway for Cl⁻ secretion in parallel to CFTR, it would be expected that a disruption of both channels would yield a more severe CF phenotype than the knockout of only CFTR. In particular, there might be symptoms in tissues affected in humans but spared in mice (as lung and pancreas), and the intestinal phenotype of CFTR mouse models could get worse. However, this was not the case (101). Surprisingly, colon from $Clcn2^{-/-}$ mice displayed larger cAMP-stimulated Cl⁻ secretion than wild-type colon. This would be compatible with a basolateral rather than apical localization of ClC-2. Equally surprising, $Cftr^{\Delta F508/\Delta F508}/Clcn2^{-/-}$ mice survived better than $Cftr^{\Delta F508/\Delta F508}$ mice (101). A deletion of phenylalanine 508 (ΔF508) is the most common CFTR mutation in Caucasians and leads to a trafficking defect of an otherwise functional channel to the plasma membrane. The better survival of $Cftr^{\Delta F508/\Delta F508}/Clcn2^{-/-}$ mice was hypothesized to be from a slight enhancement of the residual colonic Cl⁻ secretion by the disruption of basolateral ClC-2 channels (101). The apical exit of chloride may occur through ΔF508 CFTR or another unidentified apical Cl⁻ channel. The disruption of ClC-2 unexpectedly led to selective male infertility and blindness (78). The observed degeneration of germ cells and photoreceptors may be from a defect in transepithelial transport across Sertoli cells and the retinal pigment epithelium, respectively, which are important to provide an appropriate environment for these cells. Indeed, Ussing chamber experiments revealed a reduction of transepithelial current and resistance across the retinal pigment epithelium (78).

CLC-K/BARTTIN: SALT TRANSPORT IN THE KIDNEY AND THE INNER EAR

Two highly homologous CLC channels are predominantly expressed in the kidney, hence the name ClC-K. The sequences of ClC-Ka and ClC-Kb (ClC-K1 and ClC-K2 in rodents) are about 90% identical (42, 106, 107). Their genes are located very close to each other on human chromosome 1p36 (108, 109), suggesting a recent duplication. Both channels need the small accessory β-subunit barttin for functional expression (27). The distribution of ClC-K channels has been studied by RT-PCR (27, 42, 106, 107, 110), in situ hybridization (111), immunocytochemistry (27, 34, 112–115) in conjunction with ClC-K1 knockout mice (113, 116), and by the transgenic expression of a reporter gene driven by the ClC-Kb promoter (117). In the kidney, ClC-K1 (ClC-Ka) is expressed in the ascending thin limb of the loop of Henle, where it was found in both apical and basolateral membranes (34). However, in another study, it was found only in basolateral membranes (112). In contrast, ClC-K2 (ClC-Kb) is expressed in basolateral membranes of the thick ascending limb, of the distal convoluted tubule, and of intercalated cells of the collecting duct. In the inner ear, both channels are expressed in basolateral membranes of secretory epithelia, i.e., in marginal cells of the stria vascularis and in dark cells of the vestibular organ (27, 110). Both in the kidney and in the inner ear, ClC-K proteins always colocalize with their β-subunit barttin, which in turn always colocalizes with ClC-K proteins (27). Given the high sequence identity between ClC-Ka and ClC-Kb, these isoforms could not be distinguished by the antibody (27).

Rodent ClC-K1 is the only ClC-K channel known to yield currents by itself (34, 39, 107). In retrospect, the currents published for ClC-K2 (106) were probably endogenous to *Xenopus* oocytes used for expression. Surprisingly, even the human ortholog ClC-Ka did not yield currents when expressed alone (42). The failure to yield currents was puzzling, in particular because immunocytochemistry showed the presence of ClC-K proteins in renal plasma membranes (34, 112), a localization strongly supported by the renal transport defect in Bartter syndrome type III, which is caused by ClC-Kb mutations (109). The solution to this problem came from human genetics: Hildebrandt and colleagues identified a small integral membrane protein, named barttin (see also above), as being mutated in Bartter syndrome type IV (28). Shortly afterward, it became clear that barttin was necessary for the functional expression of ClC-Ka and ClC-Kb and works as a β-subunit for those channels. Barttin drastically increased currents of ClC-K1, both in oocytes and transfected mammalian cells (27). This was due to a large increase in surface expression (27). Currents had an anion selectivity sequence of $Cl \geq Br > NO_3 > I$ for ClC-Ka/barttin and $Cl > Br = NO_3 > I$ for ClC-Kb/barttin. They show only little voltage-dependent gating, consistent with the fact that the gating glutamate that obstructs the pore (14, 19) is replaced by valine in both ClC-Ka and ClC-Kb. Indeed, introducing a glutamate at that position generated voltage-dependent gating (39). Currents through either channel were increased by raising $[Ca^{2+}]_o$ or pH_o (27). Whereas ClC-K1 currents were strongly increased by $[Ca^{2+}]_o$, this effect was blunted upon coexpression with barttin (118). Hence, this β-subunit changes current properties in addition to increasing surface expression. As ClC-K1 channels are normally present in complexes with barttin, this observation also cautions against attributing an important physiological regulatory role to $[Ca^{2+}]_o$.

Barttin, a small integral membrane protein, has two predicted amino-terminal transmembrane domains. Both amino and carboxy termini are inside the cell (27, 28). Barttin does not belong to a larger gene family. ClC-K channels are apparently the only CLC proteins that interact with barttin. Barttin carries a putative PY-motif in its carboxy-terminal tail. Similar motifs have been identified in ClC-5 (119) and in the epithelial Na^+ channel ENaC (120, 121), where they were shown to bind to WW-domain (a protein domain characterized by typical tryptophane residues) containing ubiquitin ligases that down-regulate channel activity by enhancing their endocytosis. Compatible with a similar role in barttin, mutations in its putative PY motif increased currents and surface expression of ClC-K/barttin (27).

As with several other CLC channels, the physiological importance of ClC-K/barttin channels is highlighted by human inherited disease (28, 109) and a mouse model (116). ClC-Kb is mutated in Bartter syndrome type III (109). This syndrome is associated with severe renal salt loss, predominantly through a loss of NaCl reabsorption in the thick ascending limb of Henle's loop. In this nephron segment, NaCl is taken up in a secondary active process by the apical Na-K-2Cl cotransporter (NKCC2). K^+ ions are recycled over the apical membrane by the ROMK (Kir1.1) K^+ channel, whereas Cl⁻ leaves the cell basolaterally through ClC-Kb/barttin Cl⁻ channels. This transport model is now strongly supported by genetic evidence: Mutations in NKCC2 lead to Bartter syndrome type I (122), those in ROMK to Bartter syndrome II (123), those in ClC-Kb to Bartter syndrome III (109), and finally, those in barttin to Bartter syndrome IV (28).

ClC-K1 (the rodent ortholog of ClC-Ka) has been disrupted in mice (116), leading to renal water loss reminiscent of diabetes insipidus. A high Cl⁻ permeability in the ascending thin limb, the site of ClC-K1 expression, is essential for establishing the high osmolarity of the renal medulla in a countercurrent system. Accordingly, the solute accumulation in the inner medulla of ClC-K1 knockout mice was severely impaired (124). No corresponding human disease is known so far, but mutations in both ClC-Ka and ClC-Kb were found in one family (125). As expected, the patients presented with symptoms resembling Bartter syndrome IV, in which the loss of barttin eliminates the function of both Cl⁻ channels as well.

The renal symptoms in Bartter syndrome III are slightly different (e.g., concerning Ca^{2+} handling) from Bartter syndromes type I and II (126), as might be expected from the additional presence of ClC-Kb in the distal convoluted and collecting tubules. Likewise, renal symptoms in Bartter syndrome IV are more severe than in the other forms as the lack of functional barttin disrupts the function of both ClC-K isoforms. Interestingly, a common polymorphism in the human *CLCN-KB* gene led to dramatically increased ClC-Kb/barttin currents in heterologous expression (127). No pathological consequences have been reported as yet.

Barttin mutations in Bartter syndrome type IV additionally lead to congenital deafness (28). This may be explained by a secretory defect of the stria vascularis. This epithelium secretes K^+ into the scala media and generates a positive voltage of this unique extracellular compartment. Its high K^+ concentration ($\sim$150 mM) and potential (+90 mV) are necessary to drive K^+ through apical mechanosensitive cation channels into sensory hair cells. Marginal cells of the stria, which face the lumen of the scala media, take up K^+ through basolateral Na/K-ATPase pumps and NKCC1 Na-K-2Cl cotransporters. K^+ ions leave the cells apically through KCNQ1/KCNE1 K^+ channels. Similar to ROMK, which recycles K^+ for the NKCC2 cotransporter across apical membranes of the thick ascending limb, ClC-K/barttin channels are needed to recycle Cl^- for the NKCC1 cotransporter over the basolateral membrane of marginal cells (27). RT-PCR indicated the presence of both ClC-K1 and ClC-K2 in the inner ear (27), readily explaining that the disruption of ClC-Kb in Bartter syndrome III does not lead to deafness. The lack of barttin, however, which compromises both channels, reduces Cl^- recycling to such a degree that deafness ensues. This model (27) has been confirmed recently by the finding that the lack of both ClC-Ka and ClC-Kb also results in deafness (125).

CLC-3, AN ENDOSOMAL Cl^- CHANNEL THAT IS ALSO EXPRESSED ON SYNAPTIC VESICLES

ClC-3, -4, and -5 form their own branch of the CLC gene family. They share about 80% sequence identity. These channels are located mainly in membranes of intracellular vesicles, mostly in the endocytotic pathway. Whereas the role of ClC-5 in endocytosis is well established, less is known about the physiological functions of ClC-3 and in particular of ClC-4.

Cell fractionation and transfection studies localized ClC-3 to an endosomal compartment, where it partially colocalized with rab4 and lamp-1 and, additionally, to synaptic vesicles (128). The localization in endosomal/lysosomal compartments and in synaptic vesicles has now been confirmed in several studies (41, 129–131). AP3-deficient mice and cell lines were used to show that the AP3 adaptor complex plays a role in targeting ClC-3 to synaptic vesicles (131). The overexpression in mammalian cells often results in artificial, large intracellular vesicles that are acidic and stain for ClC-3 and lysosomal markers (41; T. Breiderhoff & T.J. Jentsch, unpublished data). However, some ClC-3 protein was also detected at the plasma membrane of ClC-3-overexpressing Chinese hamster ovary (CHO) cells (129).

ClC-3 knockout mice show a severe degeneration of the retina and the brain (128). After a few months, the hippocampus had totally degenerated and was replaced by fluid-filled spaces (128). The neurodegeneration, a consistent feature of three independent mouse models (128, 130, 132), was not restricted to the hippocampus, but was also seen in other brain regions. It was associated with a moderate storage of the subunit c of mitochondrial ATPase in lysosomes (130), a feature considered typical for human neuronal lipofuscinosis (133). The neurodegeneration was associated with an activation of microglia and astrogliosis (128, 132). In spite of the severe neurodegeneration, mice survived more than a year

(128), although there was increased lethality (132). ClC-3 knockout mice show several behavioral abnormalities, including hyperactivity (128, 132). The mechanism by which a loss of ClC-3 leads to neurodegeneration remains unclear and may be related to cellular trafficking defects (128).

Vesicular Cl⁻ channels are thought to provide an electric shunt for the electrogenic H⁺-ATPase, thereby facilitating the acidification of synaptic vesicles and compartments in the endosomal/lysosomal pathway. Indeed, the disruption of ClC-3 partially inhibited the acidification of synaptic vesicles in vitro (128), and the luminal pH of a vesicle fraction mainly representing endosomes was elevated (130). ClC-3 was also reported to be expressed on insulin-containing granules of pancreatic β-cells, but this result hinges on the quality of the antibody (134). It was also proposed to participate in the acidification of insulin-containing granules of pancreatic β-cells and to play an important role in insulin secretion (134). However, ClC-3 knockout mice did not exhibit hyperglycemia, neither under resting conditions, nor following a glucose load (132; A.A. Zdebik & T.J. Jentsch, unpublished data).

As expected from a mainly intracellular localization, many laboratories, including our own, have been unable to obtain plasma membrane currents of ClC-3 upon heterologous expression (129, 135). There are numerous contradictory reports on putative ClC-3 currents. ClC-3 was variably reported to yield protein kinase C-inhibitable, moderately outwardly rectifying Cl⁻ currents in *Xenopus* oocytes (136), strongly rectifying, $[Ca^{2+}]_i$-inhibitable currents in CHO cells (137), a moderately outwardly rectifying current dramatically activated by CaM-kinase II in tsA cells (138, 139), and moderately activated outward currents that were further activated by cell swelling (140, 141). In spite of these conflicting properties, all these currents (136–141) share an I⁻ > Cl⁻ selectivity. This contrasts with the preference of Cl⁻ over I⁻ of ClC-0, ClC-1, ClC-2, ClC-K/barttin, ClC-4, and ClC-5 (40, 142, 143) (the latter two being structurally highly related to ClC-3), and even of a bacterial CLC protein (144). On the other hand, Weinman and coworkers reported currents that share the consensus Cl⁻ > I⁻ selectivity of CLC channels in CHO cells overexpressing ClC-3 (41, 145). The extremely strong outward rectification of these currents closely resembled that of the highly homologous ClC-4 and ClC-5 channels (40, 142). Furthermore, a neutralizing mutation of the gating glutamate drastically changed the I/V-curve to an inwardly rectifying behavior (41), closely resembling the effect of similar mutations in ClC-4 and ClC-5 (40). Thus these currents (41, 145) almost certainly represent the real ClC-3 currents, whereas the other currents might be endogenous to the expression systems. It seems highly unlikely that the ion selectivity of CLC channels might be significantly changed by accessory proteins potentially present in other expression systems, as their pores are contained within a single protein subunit (13, 14).

The notion that ClC-3 might represent a swelling-activated Cl⁻ channel ($I_{cl,swell}$) (140) was also invalidated by ClC-3 knockout mouse models (128, 146, 147). Swelling-activated Cl⁻ currents were not changed in pancreatic and hepatic cells (128), salivary gland cells (146), or in cardiac myocytes (147). The argument

(147) that another swelling-activated Cl^- channel with exactly the same properties is upregulated to compensate for the loss of ClC-3 is not plausible, as the only reasonable candidates are ClC-4 and ClC-5. Their transcription, however, was unchanged in $Clcn3^{-/-}$ mice (128, 147), and their channel properties differ drastically from $I_{cl,swell}$ (as, almost certainly, do ClC-3 currents) (41, 145). The different regulation of $I_{cl,swell}$ reported for ClC-3 knockout cells (147) may be from indirect effects, as similar findings were previously observed with multidrug resistance P-glycoprotein (mdr) or $I_{cl,n}$, proteins that are no longer regarded as Cl^- channels (1, 148). Ca^{2+}-activated Cl^- currents were also found unchanged in two independent ClC-3 knockout mouse models (1, 128, 146). However, CamKII-activated Cl^- currents were reported to be strongly reduced in $Clcn3^{-/-}$ mice (139), possibly suggesting an effect of ClC-3 on regulatory pathways.

CLC-5, A VESICULAR CHANNEL IMPORTANT FOR RENAL ENDOCYTOSIS

ClC-4 and ClC-5 are about 80% identical to ClC-3. Both reside in endosomal membranes but can reach the plasma membrane to some degree in heterologous expression. Both channels are extremely outwardly rectifying and mediate measurable currents at voltages only more positive than $\sim+20$ mV. These have a selectivity sequence of $NO_3 > Cl > Br > I$ and are inhibited by extracellular acidic pH (40, 142, 143). The physiological significance of the extremely strong outward rectification of these channels is unclear, because voltages more positive than $+20$ mV are unlikely to be reached under physiological conditions in the plasma membrane or intracellular vesicles. Mutagenesis, particularly a mutation that neutralized the gating glutamate, confirmed that the observed currents are directly mediated by these channels (40). The single-channel properties of ClC-5 and ClC-4 are not yet established. Two groups reported very different values for the single-channel conductance of ClC-4 (143, 149). Overall, the biophysical properties of ClC-4 and ClC-5 are very similar to those reported later for ClC-3 (41, 145). ClC-5, but not ClC-4 or ClC-3, carries a PY motif between the two CBS domains in its carboxy-terminal cytoplasmic tail (119). Destroying this motif by mutagenesis increased surface expression and currents about twofold. This motif probably interacts with the WW-domains of the HECT-ubiquitin ligase WWPII, which is present in the kidney (119, 150). Although ClC-4 and ClC-5 are expressed in most tissues to some degree, ClC-4 is more abundant in brain, muscle, and liver (151, 152), whereas ClC-5 is predominantly found in kidney and intestine (142, 153).

There is still little known about the functional roles of ClC-4. Overexpression of ClC-4 in cell lines led to an increase of copper incorporation into ceruloplasmin (154), which resembles the function of the yeast homologue *Gef1p* in iron metabolism (155–157). However, it is currently unknown whether ClC-4 is necessary for this process, a question that could be answered by analyzing ClC-4 knockout mice.

In contrast, we have a fairly good understanding of the roles of ClC-5. This channel is expressed predominantly in endosomes, where it colocalized with rab5 and endocytosed markers (158). In the proximal tubule of the kidney, it is mainly expressed in apical endosomes below the brush border, where it colocalizes with the H$^+$-ATPase (158–160). It is also highly expressed in acid-transporting intercalated cells of the collecting duct, where it colocalized with the proton pump in acid-secreting α-, but not acid-reabsorbing β-cells (158). In the intestine, ClC-5 is expressed in apical endosomes as well (153). Although we do not yet understand the role of ClC-5 in intercalated cells, there is now ample evidence that it is essential for apical endocytosis in the proximal tubule. A similar function is likely for ClC-5 in the intestine.

The physiological importance of ClC-5 is illustrated by Dent's disease (161), the pathophysiological mechanism of which has been elucidated by knockout mouse models (162–164). Dent's disease is an X-linked disorder associated with low-molecular weight proteinuria, hyperphosphaturia, and hypercalciuria, which ultimately leads to kidney stones in the majority of patients (165). In addition, there is a variable presence of other symptoms of proximal tubular dysfunction such as glucosuria and aminoaciduria. By now, many different ClC-5 mutations have been identified in patients with Dent's disease. These include early truncations, but also missense mutations. For reasons that are still unclear, the latter mutations cluster at the interface between the two subunits (166). Many missense mutations have been studied in the *Xenopus* oocyte expression system and were shown to abolish or drastically reduce ClC-5 currents (167, 168).

The knockout of ClC-5 led to low-molecular weight proteinuria (162, 164), and, depending on the mouse model, also to hyperphosphaturia (162) or hypercalciuria (164). The proteinuria is the result of a cell-autonomous defect in endocytosis, which extends to fluid-phase endocytosis, receptor-mediated endocytosis, and the endocytosis of integral plasma membrane proteins such as the Na-PO$_4$ cotransporter NaPi-IIa or the Na/H exchanger NHE3 (162). Endocytosis, however, is not totally abolished, but strongly reduced. The amount of the endocytotic receptor megalin, which mediates the uptake of a wide variety of proteins and other substrates, was significantly reduced, and its presence in the brush border appeared to be reduced (162). This observation probably indicates a role of ClC-5 in recycling endosomes. A strong reduction of megalin in the brush border was also revealed by immunoelectron microscopy (169). A reduction of megalin plasma membrane expression is expected to reduce receptor-mediated endocytosis even further. Renal cortical endosomes, which are predominantly derived from proximal tubules, had a lower rate and extent of acidification than wild-type endosomes in vitro (162, 163). This strongly supports the hypothesis that the Cl$^-$ conductance provided by ClC-5 is needed to dissipate the voltage created by the electrogenic H$^+$-ATPase, thereby enabling efficient acidification in the endosomal pathway.

How does a defect in endosomal acidification, which in turn leads to a defect in endocytosis, eventually cause kidney stones? Several hormones, including parathyroid hormone (PTH) and vitamin D$_3$, are filtered into the primary urine.

After binding to megalin, these hormones are normally endocytosed by proximal tubular cells. In the absence of ClC-5, the reduced endocytosis of PTH is expected to result in a progressive increase of luminal PTH concentration along the length of the proximal tubule, while serum concentrations of the hormone remain unchanged (162). The increased luminal levels of PTH will stimulate apical PTH receptors, which in turn enhance the endocytosis of the apical Na-PO$_4$ cotransporter NaPi-IIa. Indeed, immunocytochemistry revealed that the majority of NaPi-IIa had shifted to intracellular vesicles in knockout mice, whereas it resided in the brush border of wild-type proximal tubules (162). This PTH-dependent decrease of NaPi-IIa in the plasma membrane readily explains the hyperphosphaturia that was observed in the knockout (162) and was found in patients with Dent's disease (165).

PTH is also known to stimulate the transcription of the enzyme α-hydroxylase, the enzyme that converts the inactive precursor 25(OH)-VitD$_3$ to the active hormone 1,25(OH)$_2$-VitD$_3$ in proximal tubular cells. As expected from the increased luminal concentration of PTH, mRNA levels of α-hydroxylase and its enzymatic activity were increased in ClC-5 knockout mice (162, 163). These findings suggest increased levels of the active hormone 1,25(OH)$_2$-VitD$_3$. Indeed, many patients with Dent's disease have slightly elevated serum concentrations of active VitD$_3$ (165, 170). These may lead to an increased intestinal reabsorption of Ca^{2+}, which in turn could cause the hypercalciuria and kidney stones in Dent's disease. However, there is a major complication: A large part of the precursor 25(OH)-VitD$_3$ is taken up into proximal tubular cells through megalin-dependent apical endocytosis. The amounts of VitD-binding protein and VitD$_3$ were drastically increased in the urine of *Clcn5$^-$* mice (162, 163). Therefore, there are two opposing effects (upregulation of the activating enzyme and loss of substrate) that may lead to an increase or decrease of active VitD$_3$. The outcome will depend on many factors (including genetic and dietary ones), and may explain the clinical variability of Dent's disease. Such a variability was even observed between the two ClC-5 knockout mouse models: Whereas the knockout mouse generated in our laboratory has decreased serum levels of 1,25(OH)$_2$-VitD$_3$ and no hypercalciuria (162, 163), the model from Guggino's laboratory has slightly elevated levels of the active hormone and displays hypercalciuria (164, 171). Thus the unifying hypothesis described above explains the complex and variable symptoms of Dent's disease through changes in calciotropic hormones that stem from defects in proximal tubular endocytosis, which in turn result from a defective acidification of endosomes (162, 163).

CLC-7, A LATE ENDOSOMAL/LYSOSOMAL Cl$^-$ CHANNEL IMPORTANT FOR NEURONS AND OSTEOCLASTS

ClC-6 and ClC-7 belong to a third branch of the CLC family and share about 45% sequence identity (108). Both channels are nearly ubiquitously expressed. Like ClC-3, -4, and -5, they are localized in membranes of intracellular compartments (23, 172). This subcellular expression readily explains why our (108) laboratory

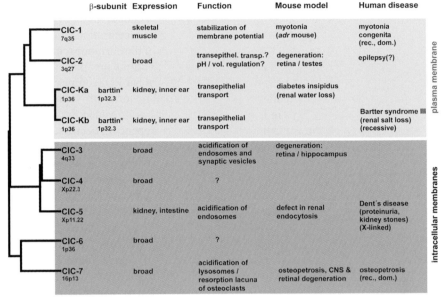

β-subunit	Expression	Function	Mouse model	Human disease
ClC-1 7q35		skeletal muscle	stabilization of membrane potential	myotonia (*adr* mouse)	myotonia congenita (rec., dom.)
ClC-2 3q27		broad	transepithel. transp.? pH / vol. regulation?	degeneration: retina / testes	epilepsy(?)
ClC-Ka 1p36	barttin* 1p32.3	kidney, inner ear	transepithelial transport	diabetes insipidus (renal water loss)	
ClC-Kb 1p36	barttin* 1p32.3	kidney, inner ear	transepithelial transport		Bartter syndrome III (renal salt loss) (recessive)
ClC-3 4q33		broad	acidification of endosomes and synaptic vesicles	degeneration: retina / hippocampus	
ClC-4 Xp22.3		broad	?		
ClC-5 Xp11.22		kidney, intestine	acidification of endosomes	defect in renal endocytosis	Dent's disease (proteinuria, kidney stones) (X-linked)
ClC-6 1p36		broad	?		
ClC-7 16p13		broad	acidification of lysosomes / resorption lacuna of osteoclasts	osteopetrosis, CNS & retinal degeneration	osteopetrosis (rec., dom.)

mutated in Bartter syndrome IV (renal salt loss and deafness)

Figure 1 A phylogenetic tree of all mammalian CLC channels, their chromosomal localization in humans, their known subunits, and an overview of diseases associated with loss of these channels in mice and humans. Channels in the gray box mainly function in the plasma membrane, whereas those in the blue box are primarily localized in intracellular membranes. Mouse genes are abbreviated *Clcn*X; the respective genes in humans are written in upper case letters. For proteins, the systematic nomenclature is ClC-X.

and others (173) were unable to measure plasma membrane currents from these channels. The currents reported for ClC-7 (174) may be endogenous to *Xenopus* oocytes in particular, because similar currents were observed by one group (175) upon expression of ClC-4 —but these currents differ in many respects from the established properties of ClC-4 (40, 143). Whereas little is known about the functions of ClC-6, a ClC-7 knockout mouse that resulted in osteopetrosis and neurodegeneration has yielded considerable insights into the roles of ClC-7 (23).

ClC-7 resides in late endosomes and lysosomes (23). Whereas ClC-3 to ClC-7 are all present in the endocytotic pathway, ClC-7 is the only one of these channels that localizes to a large degree to lysosomes (23). It is also present in the ruffled border of osteoclasts, a specialized domain of the plasma membrane that faces the resorption lacuna of these bone-degrading cells. The ruffled border is generated by the exocytotic insertion of late endosomal/lysosomal membranes (176). Hence it contains lamp-1, a marker for late endosomes/lysosomes, and significant amounts of a V-type H^+-ATPase in addition to ClC-7. The exocytotic buildup of the ruffled border is paralleled by a secretion of lytic enzymes such as cathepsins into the resorption lacuna. These lysosomal enzymes, which degrade the organic bone matrix, require an acidic pH for optimal activity. The acidification of the resorption lacuna, which is also needed for the chemical dissolution of inorganic bone material, is mediated by the electrogenic H^+-ATPase of the ruffled border. Similar to the acidification of renal endosomes or synaptic vesicles, which relies on ClC-5 (162, 163) or ClC-3 (128), respectively, it is thought that the neutralizing current through ClC-7 is required for an efficient acidification of the resorption lacuna. ClC-7 might also play a role in the exocytotic trafficking of late endosomal/lysosomal vesicles because the ruffled border was less developed in ClC-7 knockout osteoclasts (23).

The disruption of ClC-7 in mice caused blindness and severe osteopetrosis (23), a disorder characterized by dense, fragile bones devoid of bone marrow and hence associated with extramedullary hematopoiesis (177). In the knockout mice, osteoclasts were present in normal numbers, but could not acidify the resorption lacuna and failed to resorb bone (23). These findings quickly led to the identification of human *CLCN7* mutations underlying infantile malignant osteopetrosis (23). Subsequently, more ClC-7 mutations were identified in this condition as well as in autosomal-dominant osteopetrosis type II (ADOII, Albers-Schönberg disease) and in intermediate forms (24, 178–180). The dimeric structure of CLC channels predicts that the reduction of ClC-7 currents is unlikely to exceed 75% with heterozygous dominant mutations. Indeed, ADOII patients are generally not blind and develop osteopetrosis (whose phenotype is less severe than in malignant infantile osteopetrosis) only during adult life.

Mutations in the a3 subunit of the V-type H^+-ATPase also cause osteopetrosis in mice (181, 182) and humans (183, 184), underscoring the need to acidify the resorption lacuna. Mutations in this gene lead to a recessive, severe infantile form of the disease. As in $Clcn7^{-/-}$ mice, the ruffled border of osteoclasts from mice lacking the a3 subunit was poorly developed.

In addition to its function in bone resorption, ClC-7 plays other essential roles in the organism. ClC-7 transcripts were detected in all tissues analyzed, including brain, eye, kidney, liver, and testis, and ClC-7 protein was detected in cultured mouse fibroblasts (23, 108). In brain, expression is strongest in neurons (D. Kasper, R. Planells-Cases, J.C. Fuhrmann, T.J. Jentsch, in preparation). Consistent with the lysosomal localization of the channel, ClC-7-deficient mice develop a neuronal pathology with clear symptoms of lysosomal dysfunction (D. Kasper, R. Planells-Cases, J.C. Fuhrmann, T.J. Jentsch, in preparation). Cortical and hippocampal neurons accumulate osmiophilic storage material positive for periodic acid-Schiff stain. In the brain, the levels of various lysosomal enzymes are increased, as is the amount of subunit c of the mitochondrial ATP synthase. The latter increase, a hallmark of certain lysosomal storage diseases (133), is much more pronounced than in $Clcn3^{-/-}$ mice (130; D. Kasper, R. Planells-Cases, J.C. Fuhrmann, T.J. Jentsch, in preparation). Taken together, these changes are reminiscent of the symptoms observed in a group of human pathologies classified as lysosomal storage disease or neuronal ceroid lipofuscinosis. The underlying genetic causes are heterogeneous. In most cases there are mutations in genes encoding lysosomal enzymes, or mutations in other proteins that impair the function of lysosomes (133, 185).

In addition to central neurodegeneration, the disruption of ClC-7 led to a massive loss of photoreceptors and retinal neurons, causing blindness before the age of four weeks (23). Interestingly, osteopetrosis caused by mutations in the a3-subunit of the H^+-ATPase, as in osteosclerotic (oc/oc) mice (181), is not accompanied by a fast, primary retinal degeneration and lysosomal storage in central neurons (D. Kasper, R. Planells-Cases, J.C. Fuhrmann, T.J. Jentsch, in preparation). This finding is of clinical relevance because it suggests that molecular diagnosis might influence the choice of therapy.

SUMMARY AND OUTLOOK

Almost immediately after the cloning of the first mammalian voltage-gated Cl^- channel, ClC-1 (9), it became clear that its inactivation causes myotonia (50), an inherited muscle stiffness. During the past ten years, many other CLC channelopathies have been discovered. Progress in this area came from human genetics, as with the roles of ClC-Kb (109) and barttin (28) in two different forms of Bartter syndrome (renal salt loss) or with ClC-5 in Dent's disease (kidney stones) (161), or from mouse models, which led to the identification of human disease genes, as with ClC-7 and osteopetrosis (23). The pathologies associated with CLC defects (muscle stiffness, renal salt loss, deafness, blindness, neurodegeneration, male infertility, osteopetrosis, proteinuria, and kidney stones) revealed a hitherto unsuspected range of physiological functions of chloride channels and stressed their obvious medical importance. The greatest surprise was probably the discovery that the majority of CLC Cl^- channels function in intracellular vesicles and that their disruption leads to specific and highly diverse diseases. The discoveries that ClC-5 has an important role in renal endocytosis (162) or that ClC-7 is crucial for osteoclast function (23) would have been much more difficult or impossible without the clues provided by mouse models and human disease. As the vesicular CLC channels may have overlapping subcellular expression patterns, which in most cases remain to be rigorously defined, one may expect that additional functions will emerge upon the elimination of redundancies by double knockouts. Furthermore, the pathologies observed with the elimination of ubiquitously expressed channels such as ClC-2, ClC-3, or ClC-7 may represent only the tip of an iceberg, with many interesting discoveries still to be made. Increasingly sophisticated mouse models, as well as a broad spectrum of physiological, biophysical, and cell biological techniques, will be needed to address these issues.

ACKNOWLEDGMENTS

Work in this laboratory is supported by the Deutsche Forschungsgemeinschaft, the European Union, and the BMBF. M.P. is a recipient of a Marie-Curie-fellowship from the European Union.

The *Annual Review of Physiology* is online at http://physiol.annualreviews.org

LITERATURE CITED

1. Jentsch TJ, Stein V, Weinreich F, Zdebik AA. 2002. Molecular structure and physiological function of chloride channels. *Physiol. Rev.* 82:503–68

2. Sun H, Tsunenari T, Yau KW, Nathans J. 2002. The vitelliform macular dystrophy protein defines a new family of chloride channels. *Proc. Natl. Acad. Sci. USA* 99:4008–13

3. Qu Z, Fischmeister R, Hartzell C. 2004. Mouse bestrophin-2 is a bona fide Cl⁻ channel: identification of a residue important in anion binding and conduction. *J. Gen. Physiol.* 123:327–40

4. Jentsch TJ, Steinmeyer K, Schwarz G. 1990. Primary structure of *Torpedo marmorata* chloride channel isolated by expression cloning in *Xenopus* oocytes. *Nature* 348:510–14

5. Schofield PR. 2002. The role of glycine and glycine receptors in myoclonus and startle syndromes. *Adv. Neurol.* 89:263–74

6. Noebels JL. 2003. The biology of epilepsy genes. *Annu. Rev. Neurosci.* 26:599–625

7. Petrukhin K, Koisti MJ, Bakall B, Li W, Xie G, et al. 1998. Identification of the gene responsible for Best macular dystrophy. *Nat. Genet.* 19:241–7

8. Miller C. 1982. Open-state substructure of single chloride channels from *Torpedo* electroplax. *Philos. Trans. R. Soc. London Ser. B* 299:401–11

9. Steinmeyer K, Ortland C, Jentsch TJ. 1991. Primary structure and functional expression of a developmentally regulated skeletal muscle chloride channel. *Nature* 354:301–4

10. Middleton RE, Pheasant DJ, Miller C. 1994. Purification, reconstitution, and subunit composition of a voltage-gated chloride channel from *Torpedo* electroplax. *Biochemistry* 33:13189–98

11. Ludewig U, Pusch M, Jentsch TJ. 1996. Two physically distinct pores in the

dimeric ClC-0 chloride channel. *Nature* 383:340–43

12. Middleton RE, Pheasant DJ, Miller C. 1996. Homodimeric architecture of a ClC-type chloride ion channel. *Nature* 383:337–40

13. Weinreich F, Jentsch TJ. 2001. Pores formed by single subunits in mixed dimers of different CLC chloride channels. *J. Biol. Chem.* 276:2347–53

14. Dutzler R, Campbell EB, Cadene M, Chait BT, MacKinnon R. 2002. X-ray structure of a ClC chloride channel at 3.0 Å reveals the molecular basis of anion selectivity. *Nature* 415:287–94

15. Accardi A, Miller C. 2004. Secondary active transport mediated by a prokaryotic homologue of ClC Cl^- channels. *Nature* 427:803–7

16. Estévez R, Schroeder BC, Accardi A, Jentsch TJ, Pusch M. 2003. Conservation of chloride channel structure revealed by an inhibitor binding site in ClC-1. *Neuron* 38:47–59

17. Saviane C, Conti F, Pusch M. 1999. The muscle chloride channel ClC-1 has a double-barreled appearance that is differentially affected in dominant and recessive myotonia. *J. Gen. Physiol.* 113:457–68

18. Ludewig U, Pusch M, Jentsch TJ. 1997. Independent gating of single pores in CLC-0 chloride channels. *Biophys. J.* 73:789–97

19. Dutzler R, Campbell EB, MacKinnon R. 2003. Gating the selectivity filter in ClC chloride channels. *Science* 300:108–12

20. Steinmeyer K, Lorenz C, Pusch M, Koch MC, Jentsch TJ. 1994. Multimeric structure of ClC-1 chloride channel revealed by mutations in dominant myotonia congenita (Thomsen). *EMBO J.* 13:737–43

21. Meyer-Kleine C, Steinmeyer K, Ricker K, Jentsch TJ, Koch MC. 1995. Spectrum of mutations in the major human skeletal muscle chloride channel gene (*CLCN1*) leading to myotonia. *Am. J. Hum. Genet.* 57:1325–34

22. Pusch M, Steinmeyer K, Koch MC, Jentsch TJ. 1995. Mutations in dominant human myotonia congenita drastically alter the voltage dependence of the ClC-1 chloride channel. *Neuron* 15:1455–63

23. Kornak U, Kasper D, Bösl MR, Kaiser E, Schweizer M, et al. 2001. Loss of the ClC-7 chloride channel leads to osteopetrosis in mice and man. *Cell* 104:205–15

24. Cleiren E, Benichou O, Van Hul E, Gram J, Bollerslev J, et al. 2001. Albers-Schönberg disease (autosomal dominant osteopetrosis, type II) results from mutations in the ClCN7 chloride channel gene. *Hum. Mol. Genet.* 10:2861–67

25. Lorenz C, Pusch M, Jentsch TJ. 1996. Heteromultimeric CLC chloride channels with novel properties. *Proc. Natl. Acad. Sci. USA* 93:13362–66

26. Mohammad-Panah R, Harrison R, Dhani S, Ackerley C, Huan LJ, et al. 2003. The chloride channel ClC-4 contributes to endosomal acidification and trafficking. *J. Biol. Chem.* 278:29267–77

27. Estévez R, Boettger T, Stein V, Birkenhäger R, Otto M, et al. 2001. Barttin is a Cl^--channel β-subunit crucial for renal Cl^--reabsorption and inner ear K^+-secretion. *Nature* 414:558–61

28. Birkenhäger R, Otto E, Schürmann MJ, Vollmer M, Ruf EM, et al. 2001. Mutation of *BSND* causes Bartter syndrome with sensorineural deafness and kidney failure. *Nat. Genet.* 29:310–14

29. Richard EA, Miller C. 1990. Steady-state coupling of ion-channel conformations to a transmembrane ion gradient. *Science* 247:1208–10

30. Pusch M, Ludewig U, Rehfeldt A, Jentsch TJ. 1995. Gating of the voltage-dependent chloride channel ClC-0 by the permeant anion. *Nature* 373:527–31

31. Rychkov GY, Pusch M, Astill DS, Roberts ML, Jentsch TJ, Bretag AH. 1996. Concentration and pH dependence of skeletal muscle chloride channel ClC-1. *J. Physiol.* 497:423–35

32. Niemeyer MI, Cid LP, Zuniga L, Catalan M, Sepulveda FV. 2003. A conserved pore-lining glutamate as a voltage- and chloride-dependent gate in the ClC-2 chloride channel. *J. Physiol.* 553:873–79

33. Jordt SE, Jentsch TJ. 1997. Molecular dissection of gating in the ClC-2 chloride channel. *EMBO J.* 16:1582–92

34. Uchida S, Sasaki S, Nitta K, Uchida K, Horita S, et al. 1995. Localization and functional characterization of rat kidney-specific chloride channel, ClC-K1. *J. Clin. Invest.* 95:104–13

35. Gründer S, Thiemann A, Pusch M, Jentsch TJ. 1992. Regions involved in the opening of ClC-2 chloride channel by voltage and cell volume. *Nature* 360:759–62

36. Furukawa T, Ogura T, Zheng YJ, Tsuchiya H, Nakaya H, et al. 2002. Phosphorylation and functional regulation of ClC-2 chloride channels expressed in *Xenopus* oocytes by M cyclin-dependent protein kinase. *J. Physiol.* 540:883–93

37. Chen TY, Miller C. 1996. Nonequilibrium gating and voltage dependence of the ClC-0 Cl⁻ channel. *J. Gen. Physiol.* 108:237–50

38. Fahlke C, Yu HT, Beck CL, Rhodes TH, George AL Jr. 1997. Pore-forming segments in voltage-gated chloride channels. *Nature* 390:529–32

39. Waldegger S, Jentsch TJ. 2000. Functional and structural analysis of ClC-K chloride channels involved in renal disease. *J. Biol. Chem.* 275:24527–33

40. Friedrich T, Breiderhoff T, Jentsch TJ. 1999. Mutational analysis demonstrates that ClC-4 and ClC-5 directly mediate plasma membrane currents. *J. Biol. Chem.* 274:896–902

41. Li X, Wang T, Zhao Z, Weinman SA. 2002. The ClC-3 chloride channel promotes acidification of lysosomes in CHO-K1 and Huh-7 cells. *Am. J. Physiol. Cell Physiol.* 282:C1483–91

42. Kieferle S, Fong P, Bens M, Vandewalle A, Jentsch TJ. 1994. Two highly homologous members of the ClC chloride channel family in both rat and human kidney. *Proc. Natl. Acad. Sci. USA* 91:6943–47

43. Klocke R, Steinmeyer K, Jentsch TJ, Jockusch H. 1994. Role of innervation, excitability, and myogenic factors in the expression of the muscular chloride channel ClC-1. A study on normal and myotonic muscle. *J. Biol. Chem.* 269:27635–39

44. Gurnett CA, Kahl SD, Anderson RD, Campbell KP. 1995. Absence of the skeletal muscle sarcolemma chloride channel ClC-1 in myotonic mice. *J. Biol. Chem.* 270:9035–38

45. Mankodi A, Takahashi MP, Jiang H, Beck CL, Bowers WJ, et al. 2002. Expanded CUG repeats trigger aberrant splicing of ClC-1 chloride channel pre-mRNA and hyperexcitability of skeletal muscle in myotonic dystrophy. *Mol. Cell* 10:35–44

46. Palade PT, Barchi RL. 1977. Characteristics of the chloride conductance in muscle fibers of the rat diaphragm. *J. Gen. Physiol.* 69:325–42

47. Pusch M, Steinmeyer K, Jentsch TJ. 1994. Low single channel conductance of the major skeletal muscle chloride channel, ClC-1. *Biophys. J.* 66:149–52

48. Lipicky RJ, Bryant SH. 1966. Sodium, potassium, and chloride fluxes in intercostal muscle from normal goats and goats with hereditary myotonia. *J. Gen. Physiol.* 50:89–111

49. Lipicky RJ, Bryant SH, Salmon JH. 1971. Cable parameters, sodium, potassium, chloride, and water content, and potassium efflux in isolated external intercostal muscle of normal volunteers and patients with myotonia congenita. *J. Clin. Invest.* 50:2091–103

50. Steinmeyer K, Klocke R, Ortland C, Gronemeier M, Jockusch H, et al. 1991. Inactivation of muscle chloride channel by transposon insertion in myotonic mice. *Nature* 354:304–8

51. Koch MC, Steinmeyer K, Lorenz C, Ricker K, Wolf F, et al. 1992. The skeletal muscle chloride channel in dominant

and recessive human myotonia. *Science* 257:797–800

52. George AL, Jr., Crackower MA, Abdalla JA, Hudson AJ, Ebers GC. 1993. Molecular basis of Thomsen's disease (autosomal dominant myotonia congenita). *Nat. Genet.* 3:305–10

53. Pusch M. 2002. Myotonia caused by mutations in the muscle chloride channel gene CLCN1. *Hum. Mutat.* 19:423–34

54. Wollnik B, Kubisch C, Steinmeyer K, Pusch M. 1997. Identification of functionally important regions of the muscular chloride channel ClC-1 by analysis of recessive and dominant myotonic mutations. *Hum. Mol. Genet.* 6:805–11

55. Fahlke C, Rüdel R, Mitrovic N, Zhou M, George AL Jr. 1995. An aspartic acid residue important for voltage-dependent gating of human muscle chloride channels. *Neuron* 15:463–72

56. Zhang J, Sanguinetti MC, Kwiecinski H, Ptacek LJ. 2000. Mechanism of inverted activation of ClC-1 channels caused by a novel myotonia congenita mutation. *J. Biol. Chem.* 275:2999–3005

57. Duffield M, Rychkov G, Bretag A, Roberts M. 2003. Involvement of helices at the dimer interface in ClC-1 common gating. *J. Gen. Physiol.* 121:149–61

58. Lin YW, Lin CW, Chen TY. 1999. Elimination of the slow gating of ClC-0 chloride channel by a point mutation. *J. Gen. Physiol.* 114:1–12

59. Accardi A, Ferrera L, Pusch M. 2001. Drastic reduction of the slow gate of human muscle chloride channel (ClC-1) by mutation C277S. *J. Physiol.* 534:745–52

60. Kubisch C, Schmidt-Rose T, Fontaine B, Bretag AH, Jentsch TJ. 1998. ClC-1 chloride channel mutations in myotonia congenita: variable penetrance of mutations shifting the voltage dependence. *Hum. Mol. Genet.* 7:1753–60

61. George AL Jr, Sloan-Brown K, Fenichel GM, Mitchell GA, Spiegel R, Pascuzzi RM. 1994. Nonsense and missense mutations of the muscle chloride channel gene

in patients with myotonia congenita. *Hum. Mol. Genet.* 3:2071–72

62. Duno M, Colding-Jorgensen E, Grunnet M, Jespersen T, Vissing J, Schwartz M. 2004. Difference in allelic expression of the *CLCN1* gene and the possible influence on the myotonia congenita phenotype. *Eur. J. Hum. Genet.* 10.1038/sj.ejhg.5201218

63. Franke C, Hatt H, Iaizzo PA, Lehmann-Horn F. 1990. Characteristics of Na$^+$ channels and Cl$^-$ conductance in resealed muscle fibre segments from patients with myotonic dystrophy. *J. Physiol.* 425:391–405

64. Charlet BN, Savkur RS, Singh G, Philips AV, Grice EA, Cooper TA. 2002. Loss of the muscle-specific chloride channel in type 1 myotonic dystrophy due to misregulated alternative splicing. *Mol. Cell* 10:45–53

65. Kanadia RN, Johnstone KA, Mankodi A, Lungu C, Thornton CA, et al. 2003. A muscleblind knockout model for myotonic dystrophy. *Science* 302:1978–80

66. Ebralidze A, Wang Y, Petkova V, Ebralidse K, Junghans RP. 2004. RNA leaching of transcription factors disrupts transcription in myotonic dystrophy. *Science* 303:383–87

67. Thiemann A, Gründer S, Pusch M, Jentsch TJ. 1992. A chloride channel widely expressed in epithelial and non-epithelial cells. *Nature* 356:57–60

68. Arreola J, Begenisich T, Melvin JE. 2002. Conformation-dependent regulation of inward rectifier chloride channel gating by extracellular protons. *J. Physiol.* 541:103–12

69. Zuniga L, Niemeyer MI, Varela D, Catalán M, Cid LP, Sepúlveda FV. 2004. The voltage-dependent ClC-2 chloride channel has a dual gating mechanism. *J. Physiol.* 555:671–82

70. Nobile M, Pusch M, Rapisarda C, Ferroni S. 2000. Single-channel analysis of a ClC-2-like chloride conductance in cultured rat cortical astrocytes. *FEBS Lett.* 479:10–14

71. Ferroni S, Marchini C, Nobile M, Rapisarda C. 1997. Characterization of an inwardly rectifying chloride conductance expressed by cultured rat cortical astrocytes. *Glia* 21:217–27

72. Makara JK, Rappert A, Matthias K, Steinhäuser C, Spat A, Kettenmann H. 2003. Astrocytes from mouse brain slices express ClC-2-mediated Cl⁻ currents regulated during development and after injury. *Mol. Cell Neurosci.* 23:521–30

73. Pusch M, Jordt SE, Stein V, Jentsch TJ. 1999. Chloride dependence of hyperpolarization-activated chloride channel gates. *J. Physiol.* 515:341–53

74. Catalán M, Niemeyer MI, Cid LP, Sepúlveda FV. 2004. Basolateral ClC-2 chloride channels in surface colon epithelium: Regulation by a direct effect of intracellular chloride. *Gastroenterology* 126:1104–14

75. Zheng YJ, Furukawa T, Ogura T, Tajimi K, Inagaki N. 2002. M phase-specific expression and phosphorylation-dependent ubiquitination of the ClC-2 channel. *J. Biol. Chem.* 277:32268–73

76. Park K, Begenisich T, Melvin JE. 2001. Protein kinase A activation phosphorylates the rat ClC-2 Cl⁻ channel but does not change activity. *J. Membr. Biol.* 182:31–37

77. Furukawa T, Ogura T, Katayama Y, Hiraoka M. 1998. Characteristics of rabbit ClC-2 current expressed in *Xenopus* oocytes and its contribution to volume regulation. *Am. J. Physiol. Cell Physiol.* 274:C500–12

78. Bösl MR, Stein V, Hübner C, Zdebik AA, Jordt SE, et al. 2001. Male germ cells and photoreceptors, both depending on close cell-cell interactions, degenerate upon ClC-2 Cl⁻-channel disruption. *EMBO J.* 20:1289–99

79. Okada Y. 1997. Volume expansion-sensing outward-rectifier Cl⁻ channel: fresh start to the molecular identity and volume sensor. *Am. J. Physiol. Cell Physiol.* 273:C755–89

80. Xiong H, Li C, Garami E, Wang Y, Ramjeesingh M, et al. 1999. ClC-2 activation modulates regulatory volume decrease. *J. Membr. Biol.* 167:215–21

81. Nehrke K, Arreola J, Nguyen HV, Pilato J, Richardson L, et al. 2002. Loss of hyperpolarization-activated Cl⁻ current in salivary acinar cells from *Clcn2* knockout mice. *J. Biol. Chem.* 26:23604–11

82. Smith RL, Clayton GH, Wilcox CL, Escudero KW, Staley KJ. 1995. Differential expression of an inwardly rectifying chloride conductance in rat brain neurons: a potential mechanism for cell-specific modulation of postsynaptic inhibition. *J. Neurosci.* 15:4057–67

83. Ben-Ari Y. 2002. Excitatory actions of Gaba during development: the nature of the nurture. *Nat. Rev. Neurosci.* 3:728–39

84. Rivera C, Voipio J, Payne JA, Ruusuvuori E, Lahtinen H, et al. 1999. The K⁺/Cl⁻ co-transporter KCC2 renders GABA hyperpolarizing during neuronal maturation. *Nature* 397:251–55

85. Hübner C, Stein V, Hermanns-Borgmeyer I, Meyer T, Ballanyi K, Jentsch TJ. 2001. Disruption of KCC2 reveals an essential role of K-Cl-Cotransport already in early synaptic inhibition. *Neuron* 30:515–24

86. Boettger T, Rust MB, Maier H, Seidenbecher T, Schweizer M, et al. 2003. Loss of K-Cl co-transporter KCC3 causes deafness, neurodegeneration and reduced seizure threshold. *EMBO J.* 22:5422–34

87. Staley K, Smith R, Schaack J, Wilcox C, Jentsch TJ. 1996. Alteration of GABA_A receptor function following gene transfer of the CLC-2 chloride channel. *Neuron* 17:543–51

88. Sik A, Smith RL, Freund TF. 2000. Distribution of chloride channel-2-immunoreactive neuronal and astrocytic processes in the hippocampus. *Neuroscience* 101:51–65

89. Sander T, Schulz H, Saar K, Gennaro E, Riggio MC, et al. 2000. Genome search for susceptibility loci of common

idiopathic generalised epilepsies. *Hum. Mol. Genet.* 9:1465–72

90. Haug K, Warnstedt M, Alekov AK, Sander T, Ramirez A, et al. 2003. Mutations in CLCN2 encoding a voltage-gated chloride channel are associated with idiopathic generalized epilepsies. *Nat. Genet.* 33:527–32

91. Niemeyer MI, Yusef YR, Cornejo I, Flores C, Sepúlveda FV, Cid P. 2004. Functional evaluation of human ClC-2 chloride channel mutations associated with idiopathic generalized epilepsies. *Physiol. Genom.* 10.1152/physiolgenomics.00070.2004

92. Scott JW, Hawley SA, Green KA, Anis M, Stewart G, et al. 2004. CBS domains form energy-sensing modules whose binding of adenosine ligands is disrupted by disease mutations. *J. Clin. Invest.* 113:274–84

93. Marini C, Scheffer IE, Crossland KM, Grinton BE, Phillips FL, et al. 2004. Genetic architecture of idiopathic generalized epilepsy: clinical genetic analysis of 55 multiplex families. *Epilepsia* 45:467–78

94. Malinowska DH, Kupert EY, Bahinski A, Sherry AM, Cuppoletti J. 1995. Cloning, functional expression, and characterization of a PKA-activated gastric Cl- channel. *Am. J. Physiol. Cell Physiol.* 268:C191–200

95. Blaisdell CJ, Edmonds RD, Wang XT, Guggino S, Zeitlin PL. 2000. pH-regulated chloride secretion in fetal lung epithelia. *Am. J. Physiol. Lung Cell Mol. Physiol.* 278:L1248–55

96. Murray CB, Morales MM, Flotte TR, McGrath-Morrow SA, Guggino WB, Zeitlin PL. 1995. ClC-2: a developmentally dependent chloride channel expressed in the fetal lung and downregulated after birth. *Am. J. Respir. Cell Mol. Biol.* 12:597–604

97. Murray CB, Chu S, Zeitlin PL. 1996. Gestational and tissue-specific regulation of ClC-2 chloride channel expression. *Am. J. Physiol. Lung Cell Mol. Physiol.* 271:L829–37

98. Blaisdell CJ, Morales MM, Andrade AC, Bamford P, Wasicko M, Welling P. 2004. Inhibition of CLC-2 chloride channel expression interrupts expansion of fetal lung cysts. *Am. J. Physiol. Lung Cell Mol. Physiol.* 286:L420–26

99. Kajita H, Omori K, Matsuda H. 2000. The chloride channel ClC-2 contributes to the inwardly rectifying Cl$^-$ conductance in cultured porcine choroid plexus epithelial cells. *J. Physiol.* 523:313–24

100. Speake T, Kajita H, Smith CP, Brown PD. 2002. Inward-rectifying anion channels are expressed in the epithelial cells of choroid plexus isolated from ClC-2 'knock-out' mice. *J. Physiol.* 539:385–90

101. Zdebik AA, Cuffe J, Bertog M, Korbmacher C, Jentsch TJ. 2004. Additional disruption of the ClC-2 Cl$^-$ channel does not exacerbate the cystic fibrosis phenotype of CFTR mouse models. *J. Biol. Chem.* 279:22276–83

102. Schwiebert EM, Cid-Soto LP, Stafford D, Carter M, Blaisdell CJ, et al. 1998. Analysis of ClC-2 channels as an alternative pathway for chloride conduction in cystic fibrosis airway cells. *Proc. Natl. Acad. Sci. USA* 95:3879–84

103. Mohammad-Panah R, Gyomorey K, Rommens J, Choudhury M, Li C, et al. 2001. ClC-2 contributes to native chloride secretion by a human intestinal cell line, Caco-2. *J. Biol. Chem.* 276:8306–13

104. Lipecka J, Bali M, Thomas A, Fanen P, Edelman A, Fritsch J. 2002. Distribution of ClC-2 chloride channel in rat and human epithelial tissues. *Am. J. Physiol. Cell Physiol.* 282:C805–16

105. Catalán M, Cornejo I, Figueroa CD, Niemeyer MI, Sepúlveda FV, Cid LP. 2002. ClC-2 in guinea pig colon: mRNA, immunolabeling, and functional evidence for surface epithelium localization. *Am. J. Physiol. Gastrointest. Liver Physiol.* 283:G1004–13

106. Adachi S, Uchida S, Ito H, Hata M, Hiroe M, et al. 1994. Two isoforms of a chloride

channel predominantly expressed in thick ascending limb of Henle's loop and collecting ducts of rat kidney. *J. Biol. Chem.* 269:17677–83

107. Uchida S, Sasaki S, Furukawa T, Hiraoka M, Imai T, et al. 1993. Molecular cloning of a chloride channel that is regulated by dehydration and expressed predominantly in kidney medulla. *J. Biol. Chem.* 268:3821–24. Erratum. 1994. *J. Biol. Chem.* 269:19192

108. Brandt S, Jentsch TJ. 1995. ClC-6 and ClC-7 are two novel broadly expressed members of the CLC chloride channel family. *FEBS Lett.* 377:15–20

109. Simon DB, Bindra RS, Mansfield TA, Nelson-Williams C, Mendonca E, et al. 1997. Mutations in the chloride channel gene, *CLCNKB*, cause Bartter's syndrome type III. *Nat. Genet.* 17:171–78

110. Ando M, Takeuchi S. 2000. mRNA encoding 'ClC-K1, a kidney Cl⁻ channel' is expressed in marginal cells of the stria vascularis of rat cochlea: its possible contribution to Cl⁻ currents. *Neurosci. Lett.* 284:171–74

111. Yoshikawa M, Uchida S, Yamauchi A, Miyai A, Tanaka Y, et al. 1999. Localization of rat CLC-K2 chloride channel mRNA in the kidney. *Am. J. Physiol. Renal Physiol.* 276:F552–58

112. Vandewalle A, Cluzeaud F, Bens M, Kieferle S, Steinmeyer K, Jentsch TJ. 1997. Localization and induction by dehydration of ClC-K chloride channels in the rat kidney. *Am. J. Physiol. Renal Physiol.* 272:F678–88

113. Kobayashi K, Uchida S, Mizutani S, Sasaki S, Marumo F. 2001. Intrarenal and cellular localization of CLC-K2 protein in the mouse kidney. *J. Am. Soc. Nephrol.* 12:1327–34

114. Sage CL, Marcus DC. 2001. Immunolocalization of ClC-K chloride channel in strial marginal cells and vestibular dark cells. *Hear. Res.* 160:1–9

115. Kobayashi K, Uchida S, Mizutani S, Sasaki S, Marumo F. 2001. Developmental expression of CLC-K1 in the postnatal rat kidney. *Histochem. Cell Biol.* 116:49–56

116. Matsumura Y, Uchida S, Kondo Y, Miyazaki H, Ko SB, et al. 1999. Overt nephrogenic diabetes insipidus in mice lacking the CLC-K1 chloride channel. *Nat. Genet.* 21:95–98

117. Kobayashi K, Uchida S, Okamura HO, Marumo F, Sasaki S. 2002. Human CLC-KB gene promoter drives the EGFP expression in the specific distal nephron segments and inner ear. *J. Am. Soc. Nephrol.* 13:1992–98

118. Waldegger S, Jeck N, Barth P, Peters M, Vitzthum H, et al. 2002. Barttin increases surface expression and changes current properties of ClC-K channels. *Pflügers Arch.* 444:411–18

119. Schwake M, Friedrich T, Jentsch TJ. 2001. An internalization signal in ClC-5, an endosomal Cl⁻-channel mutated in Dent's disease. *J. Biol. Chem.* 276:12049–54

120. Staub O, Dho S, Henry P, Correa J, Ishikawa T, et al. 1996. WW domains of Nedd4 bind to the proline-rich PY motifs in the epithelial Na⁺ channel deleted in Liddle's syndrome. *EMBO J.* 15:2371–80

121. Schild L, Canessa CM, Shimkets RA, Gautschi I, Lifton RP, Rossier BC. 1995. A mutation in the epithelial sodium channel causing Liddle disease increases channel activity in the *Xenopus laevis* oocyte expression system. *Proc. Natl. Acad. Sci. USA* 92:5699–703

122. Simon DB, Karet FE, Hamdan JM, DiPietro A, Sanjad SA, Lifton RP. 1996. Bartter's syndrome, hypokalaemic alkalosis with hypercalciuria, is caused by mutations in the Na-K-2Cl cotransporter NKCC2. *Nat. Genet.* 13:183–88

123. Simon DB, Karet FE, Rodriguez-Soriano J, Hamdan JH, DiPietro A, et al. 1996. Genetic heterogeneity of Bartter's syndrome revealed by mutations in the K⁺ channel, ROMK. *Nat. Genet.* 14:152–56

124. Akizuki N, Uchida S, Sasaki S, Marumo F. 2001. Impaired solute accumulation in

inner medully of *Clcnk1⁻/⁻* mice kidney. *Am. J. Physiol. Renal Physiol.* 280:F79–87

125. Schlingmann KP, Konrad M, Jeck N, Waldegger P, Reinalter SC, et al. 2004. Salt wasting and deafness resulting from mutations in two chloride channels. *N. Engl. J. Med.* 350:1314–19

126. Jeck N, Konrad M, Peters M, Weber S, Bonzel KE, Seyberth HW. 2000. Mutations in the chloride channel gene, *CLC-NKB*, leading to a mixed Bartter-Gitelman phenotype. *Pediatr. Res.* 48:754–58

127. Jeck N, Waldegger P, Doroszewicz J, Seyberth H, Waldegger S. 2004. A common sequence variation of the *CLCNKB* gene strongly activates ClC-Kb chloride channel activity. *Kidney Int.* 65:190–97

128. Stobrawa SM, Breiderhoff T, Takamori S, Engel D, Schweizer M, et al. 2001. Disruption of ClC-3, a chloride channel expressed on synaptic vesicles, leads to a loss of the hippocampus. *Neuron* 29:185–96

129. Weylandt KH, Valverde MA, Nobles M, Raguz S, Amey JS, et al. 2001. Human ClC-3 is not the swelling-activated chloride channel involved in cell volume regulation. *J. Biol. Chem.* 276:17461–67

130. Yoshikawa M, Uchida S, Ezaki J, Rai T, Hayama A, et al. 2002. CLC-3 deficiency leads to phenotypes similar to human neuronal ceroid lipofuscinosis. *Genes Cells* 7:597–605

131. Salazar G, Love R, Styers ML, Werner E, Peden A, et al. 2004. AP-3-dependent mechanisms control the targeting of a chloride channel (ClC-3) in neuronal and non-neuronal cells. *J. Biol. Chem.* 279:25430–39

132. Dickerson LW, Bonthius DJ, Schutte BC, Yang B, Barna TJ, et al. 2002. Altered GABAergic function accompanies hippocampal degeneration in mice lacking ClC-3 voltage-gated chloride channels. *Brain Res.* 958:227–50

133. Ezaki J, Kominami E. 2004. The intracellular location and function of proteins of neuronal ceroid lipofuscinoses. *Brain Pathol.* 14:77–85

134. Barg S, Huang P, Eliasson L, Nelson DJ, Obermuller S, et al. 2001. Priming of insulin granules for exocytosis by granular Cl⁻ uptake and acidification. *J. Cell Sci.* 114:2145–54

135. Borsani G, Rugarli EI, Taglialatela M, Wong C, Ballabio A. 1995. Characterization of a human and murine gene (*CLCN3*) sharing similarities to voltage-gated chloride channels and to a yeast integral membrane protein. *Genomics* 27:131–41

136. Kawasaki M, Uchida S, Monkawa T, Miyawaki A, Mikoshiba K, et al. 1994. Cloning and expression of a protein kinase C-regulated chloride channel abundantly expressed in rat brain neuronal cells. *Neuron* 12:597–604

137. Kawasaki M, Suzuki M, Uchida S, Sasaki S, Marumo F. 1995. Stable and functional expression of the ClC-3 chloride channel in somatic cell lines. *Neuron* 14:1285–91

138. Huang P, Liu J, Robinson NC, Musch MW, Kaetzel MA, Nelson DJ. 2001. Regulation of human ClC-3 channels by multifunctional Ca²⁺/calmodulin dependent protein kinase. *J. Biol. Chem.* 276:20093–100

139. Robinson NC, Huang P, Kaetzel MA, Lamb FS, Nelson DJ. 2004. Identification of an N-terminal amino acid of ClC-3 critical in phosphorylation-dependent activation of a CaMKII-activated chloride current. *J. Physiol.* 556.2:353–68

140. Duan D, Winter C, Cowley S, Hume JR, Horowitz B. 1997. Molecular identification of a volume-regulated chloride channel. *Nature* 390:417–21

141. Duan D, Cowley S, Horowitz B, Hume JR. 1999. A serine residue in ClC-3 links phosphorylation-dephosphorylation to chloride channel regulation by cell volume. *J. Gen. Physiol.* 113:57–70

142. Steinmeyer K, Schwappach B, Bens M, Vandewalle A, Jentsch TJ. 1995. Cloning and functional expression of rat ClC-5, a

chloride channel related to kidney disease. *J. Biol. Chem.* 270:31172–77

143. Vanoye CG, George AG Jr. 2002. Functional characterization of recombinant human ClC-4 chloride channels in cultured mammalian cells. *J. Physiol.* 539:373–83

144. Maduke M, Pheasant DJ, Miller C. 1999. High-level expression, functional reconstitution, and quaternary structure of a prokaryotic ClC-type chloride channel. *J. Gen. Physiol.* 114:713–22

145. Li X, Shimada K, Showalter LA, Weinman SA. 2000. Biophysical properties of ClC-3 differentiate it from swelling-activated chloride channels in Chinese hamster ovary-K1 cells. *J. Biol. Chem.* 275:35994–98

146. Arreola J, Begenisch T, Nehrke K, Nguyen HV, Park K, et al. 2002. Secretion and cell volume regulation by salivary acinar cells from mice lacking expression of the *Clcn3* Cl⁻ channel gene. *J. Physiol.* 545.1:207–16

147. Yamamoto-Mizuma S, Wang GX, Liu LL, Schegg K, Hatton WJ, et al. 2004. Altered properties of volume-sensitive osmolyte and anion channels (VSOACs) and membrane protein expression in cardiac and smooth muscle myocytes from *Clcn3⁻/⁻* mice. *J. Physiol.* 557:439–56

148. Clapham D. 2001. How to lose your hippocampus by working on chloride channels. *Neuron* 29:1–3

149. Hebeisen S, Heidtmann H, Cosmelli D, Gonzalez C, Poser B, et al. 2003. Anion permeation in human ClC-4 channels. *Biophys. J.* 84:2306–18

150. Pirozzi G, McConnell SJ, Uveges AJ, Carter JM, Sparks AB, et al. 1997. Identification of novel human WW domain-containing proteins by cloning of ligand targets. *J. Biol. Chem.* 272:14611–16

151. Jentsch TJ, Günther W, Pusch M, Schwappach B. 1995. Properties of voltage-gated chloride channels of the ClC gene family. *J. Physiol.* 482:19S–25S

152. van Slegtenhorst MA, Bassi MT, Borsani G, Wapenaar MC, Ferrero GB, et al. 1994. A gene from the Xp22.3 region shares homology with voltage-gated chloride channels. *Hum. Mol. Genet.* 3:547–52

153. Vandewalle A, Cluzeaud F, Peng KC, Bens M, Lüchow A, et al. 2001. Tissue distribution and subcellular localization of the ClC-5 chloride channel in rat intestinal cells. *Am. J. Physiol. Cell Physiol.* 280:C373–81

154. Wang T, Weinman SA. 2004. Involvement of chloride channels in hepatic copper metabolism: ClC-4 promotes copper incorporation into ceruloplasmin. *Gastroenterology* 126:1157–66

155. Schwappach B, Stobrawa S, Hechenberger M, Steinmeyer K, Jentsch TJ. 1998. Golgi localization and functionally important domains in the NH₂ and COOH terminus of the yeast CLC putative chloride channel Gef1p. *J. Biol. Chem.* 273:15110–18

156. Greene JR, Brown NH, DiDomenico BJ, Kaplan J, Eide DJ. 1993. The *GEF1* gene of *Saccharomyces cerevisiae* encodes an integral membrane protein; mutations in which have effects on respiration and iron-limited growth. *Mol. Gen. Genet.* 241:542–53

157. Davis-Kaplan SR, Askwith CC, Bengtzen AC, Radisky D, Kaplan J. 1998. Chloride is an allosteric effector of copper assembly for the yeast multicopper oxidase Fet3p: an unexpected role for intracellular chloride channels. *Proc. Natl. Acad. Sci. USA* 95:13641–45

158. Günther W, Lüchow A, Cluzeaud F, Vandewalle A, Jentsch TJ. 1998. ClC-5, the chloride channel mutated in Dent's disease, colocalizes with the proton pump in endocytotically active kidney cells. *Proc. Natl. Acad. Sci. USA* 95:8075–80

159. Devuyst O, Christie PT, Courtoy PJ, Beauwens R, Thakker RV. 1999. Intrarenal and subcellular distribution of the human chloride channel, CLC-5, reveals a pathophysiological basis for Dent's disease. *Hum. Mol. Genet.* 8:247–57

160. Sakamoto H, Sado Y, Naito I, Kwon TH, Inoue S, et al. 1999. Cellular and subcellular immunolocalization of ClC-5 channel in mouse kidney: colocalization with H⁺-ATPase. *Am. J. Physiol. Renal Physiol.* 277:F957–65

161. Lloyd SE, Pearce SH, Fisher SE, Steinmeyer K, Schwappach B, et al. 1996. A common molecular basis for three inherited kidney stone diseases. *Nature* 379:445–49

162. Piwon N, Günther W, Schwake R, Bösl MR, Jentsch TJ. 2000. ClC-5 Cl⁻-channel disruption impairs endocytosis in a mouse model for Dent's disease. *Nature* 408:369–73

163. Günther W, Piwon N, Jentsch TJ. 2003. The ClC-5 chloride channel knock-out mouse—an animal model for Dent's disease. *Pflügers Arch.* 445:456–62

164. Wang SS, Devuyst O, Courtoy PJ, Wang XT, Wang H, et al. 2000. Mice lacking renal chloride channel, CLC-5, are a model for Dent's disease, a nephrolithiasis disorder associated with defective receptor-mediated endocytosis. *Hum. Mol. Genet.* 9:2937–45

165. Wrong OM, Norden AG, Feest TG. 1994. Dent's disease; a familial proximal renal tubular syndrome with low-molecular-weight proteinuria, hypercalciuria, nephrocalcinosis, metabolic bone disease, progressive renal failure and a marked male predominance. *Q. J. Med.* 87:473–93

166. Wu F, Roche P, Christie PT, Loh NY, Reed AA, et al. 2003. Modeling study of human renal chloride channel (hCLC-5) mutations suggests a structural-functional relationship. *Kidney Int.* 63:1426–32

167. Lloyd SE, Günther W, Pearce SH, Thomson A, Bianchi ML, et al. 1997. Characterisation of renal chloride channel, CLCN5, mutations in hypercalciuric nephrolithiasis (kidney stones) disorders. *Hum. Mol. Genet.* 6:1233–39

168. Igarashi T, Günther W, Sekine T, Inatomi J, Shiraga H, et al. 1998. Functional characterization of renal chloride channel, CLCN5, mutations associated with Dent's Japan disease. *Kidney Int.* 54:1850–56

169. Christensen EI, Devuyst O, Dom G, Nielsen R, Van der Smissen P, et al. 2003. Loss of chloride channel ClC-5 impairs endocytosis by defective trafficking of megalin and cubilin in kidney proximal tubules. *Proc. Natl. Acad. Sci. USA* 100:8472–77

170. Scheinman SJ. 1998. X-linked hypercalciuric nephrolithiasis: clinical syndromes and chloride channel mutations. *Kidney Int.* 53:3–17

171. Silva IV, Cebotaru V, Wang H, Wang XT, Wang SS, et al. 2003. The ClC-5 knockout mouse model of Dent's disease has renal hypercalciuria and increased bone turnover. *J. Bone Miner. Res.* 18:615–23

172. Buyse G, Trouet D, Voets T, Missiaen L, Droogmans G, et al. 1998. Evidence for the intracellular location of chloride channel (ClC)-type proteins: colocalization of ClC-6a and ClC-6c with the sarco/endoplasmic-reticulum Ca²⁺ pump SERCA2b. *Biochem. J.* 330:1015–21

173. Buyse G, Voets T, Tytgat J, De Greef C, Droogmans G, et al. 1997. Expression of human pICln and ClC-6 in *Xenopus* oocytes induces an identical endogenous chloride conductance. *J. Biol. Chem.* 272:3615–21

174. Diewald L, Rupp J, Dreger M, Hucho F, Gillen C, Nawrath H. 2002. Activation by acidic pH of CLC-7 expressed in oocytes from *Xenopus laevis. Biochem. Biophys. Res. Commun.* 291:421–24

175. Kawasaki M, Fukuma T, Yamauchi K, Sakamoto H, Marumo F, Sasaki S. 1999. Identification of an acid-activated Cl⁻ channel from human skeletal muscles. *Am. J. Physiol. Cell Physiol.* 277:C948–54

176. Boyle WJ, Simonet WS, Lacey DL. 2003. Osteoclast differentiation and activation. *Nature* 423:337–42

177. Kornak U, Mundlos S. 2003. Genetic disorders of the skeleton: a developmental approach. *Am. J. Hum. Genet.* 73:447–74

178. Campos-Xavier AB, Saraiva JM, Ribeiro LM, Munnich A, Cormier-Daire V. 2003. Chloride channel 7 (CLCN7) gene mutations in intermediate autosomal recessive osteopetrosis. *Hum. Genet.* 112:186–89

179. Waguespack SG, Koller DL, White KE, Fishburn T, Carn G, et al. 2003. Chloride channel 7 (ClCN7) gene mutations and autosomal dominant osteopetrosis, type II. *J. Bone Miner. Res.* 18:1513–18

180. Frattini A, Pangrazio A, Susani L, Sobacchi C, Mirolo M, et al. 2003. Chloride channel *ClCN7* mutations are responsible for severe recessive, dominant, and intermediate osteopetrosis. *J. Bone Miner. Res.* 18:1740–47

181. Scimeca JC, Franchi A, Trojani C, Parrinello H, Grosgeorge J, et al. 2000. The gene encoding the mouse homologue of the human osteoclast-specific 116-kDa V-ATPase subunit bears a deletion in osteosclerotic (*oc/oc*) mutants. *Bone* 26:207–13

182. Li YP, Chen W, Liang Y, Li E, Stashenko P. 1999. *Atp6i*-deficient mice exhibit severe osteopetrosis due to loss of osteoclast-mediated extracellular acidification. *Nat. Genet.* 23:447–51

183. Frattini A, Orchard PJ, Sobacchi C, Giliani S, Abinun M, et al. 2000. Defects in TCIRG1 subunit of the vacuolar proton pump are responsible for a subset of human autosomal recessive osteopetrosis. *Nat. Genet.* 25:343–46

184. Kornak U, Schulz A, Friedrich W, Uhlhaas S, Kremens B, et al. 2000. Mutations in the a3 subunit of the vacuolar H⁺-ATPase cause infantile malignant osteopetrosis. *Hum. Mol. Genet.* 9:2059–63

185. Holopainen JM, Saarikoski J, Kinnunen PK, Jarvela I. 2001. Elevated lysosomal pH in neuronal ceroid lipofuscinoses (NCLs). *Eur. J. Biochem.* 268:5851–56

Annu. Rev. Physiol. 2005. 67:809–39
doi: 10.1146/annurev.physiol.67.032003.153012
First published online as a Review in Advance on October 12, 2004

STRUCTURE AND FUNCTION OF CLC CHANNELS

Tsung-Yu Chen

*Center for Neuroscience and Department of Neurology, University of California, Davis,
California 95616; email: tycchen@ucdavis.edu*

Key Words   CLC family, chloride channel, gating, permeation, transporter

■ **Abstract**   The CLC family comprises a group of integral membrane proteins
whose major action is to translocate chloride (Cl^-) ions across the cell membranes.
Recently, the structures of CLC orthologues from two bacterial species, *Salmonella
typhimurium* and *Escherichia coli*, were solved, providing the first framework for
understanding the operating mechanisms of these molecules. However, most of the
previous mechanistic understanding of CLC channels came from electrophysiological
studies of a branch of the channel family, the muscle-type CLC channels in vertebrate
species. These vertebrate CLC channels were predicted to contain two identical but
independent pores, and this hypothesis was confirmed by the solved bacterial CLC
structures. The opening and closing of the vertebrate CLC channels are also known
to couple to the permeant ions via their binding sites in the ion-permeation pathway.
The bacterial CLC structures can probably serve as a structural model to explain the
gating-permeation coupling mechanism. However, the CLC-ec1 protein in *E. coli* was
most recently shown to be a Cl^--H^+ antiporter, but not an ion channel. The molecular basis to explain the difference between vertebrate and bacterial CLCs, especially
the distinction between an ion channel and a transporter, remains a challenge in the
structure/function studies for the CLC family.

INTRODUCTION

CLC Cl^- channels are a well-established channel family, and they play many functional roles, including maintenance of membrane potential, regulation of transepithelial Cl^- transport, and control of intravesicular pH. The physiological importance of these channels is best illustrated by the hereditary diseases caused by
defects of these channels, such as myotonia congenita (1–5), Dent's disease (6–8),
Bartter's syndrome (9–11), osteopetrosis (12), and idiopathic epilepsy (13). Before
the discovery of this channel family, CLC channels were poorly understood, and the
current through these anion channels was thought to be nonspecific leak current on
the membrane. Molecular cloning of various CLC members revealed the protein sequences, making it possible to study the structure/function relationship of the channel molecules. Meanwhile, the identification of these channels also facilitated the
clarification of their physiological roles by genetic analysis of hereditary diseases
(1, 3, 9, 12, 13) and by gene knock-out approaches (12, 14, 15) that revealed the

0066-4278/05/0315-0809$14.00

mutational consequences of several CLC channels in a variety of tissues. This review emphasizes the mechanistic side of CLC channel operations. [For discussion of the physiological roles of CLC channels and the genetic basis of hereditary diseases caused by CLC channel mutations, see several excellent review articles (16–20).]

CLC CHANNEL FAMILY

The members of the CLC family are widely distributed in species ranging from bacteria to human beings. The identification and cloning of the first member, CLC-0, from the electric organ of the *Torpedo* ray set the foundation for establishing the CLC channel family (21). Through homology cloning in the early 1990s, nine members in mammalian species (CLC-1 to CLC-7, CLC-Ka, and CLC-Kb) have been molecularly cloned. In addition, several members in invertebrate species, such as bacteria (22–24), yeast (25–27), and *Caenorhabditis elegans* (28–31), have also been identified by searching the genomic data bases. Comparisons of gene sequences and classifications of functional properties divide vertebrate CLC channels into three subfamilies (16, 18). CLC-0 and its mammalian homologues, CLC-1, CLC-2, CLC-Ka, and CLC-Kb, share ~50–60% sequence identity among each other. These channels are all present on the cell membrane to control the Cl^- flux and the membrane potential, and therefore are grouped together in the subfamily called muscle-type CLC channels. The second branch of the family includes CLC-3, CLC-4, and CLC-5. These channels were recently found to be located on the membrane of intracellular vesicles and are thought to be important in maintaining the pH of these vesicles (14, 32, 33). A functional role of CLC-3 in regulating cell volume was also suggested (34–36), although this issue has been controversial (37, 38). The third subfamily comprises CLC-6 and CLC-7, which are broadly expressed in various tissues. These two proteins cannot be functionally expressed in heterologous expression systems (39). They may be present also in the intracellular organelles to regulate the vesicular pH (12). Although genetic analysis of human hereditary diseases and recent gene knock-out approaches have provided directions for understanding the physiological roles of the nonmuscle-type CLC channels, the functional operations of these channel molecules are poorly understood. Therefore, I focus mostly on the muscle-type CLC channels in the discussion that follows.

Although CLC channels have been studied for more than two decades since the first discovery of CLC-0, previous functional approaches, which mainly used electrophysiological recordings, provided only a limited structural picture—the channels were dimers and contained two pores. The recent breakthrough in obtaining high-resolution crystal structures from bacterial CLC homologues has been of great benefit, although the structures of eukaryotic CLC channels are not yet available. On the other hand, functional studies of the bacterial CLCs are currently at a primitive stage (40, 41). This review thus correlates the functional properties of the muscle-type CLC channels with the bacterial CLC structures.

STRUCTURES OF BACTERIAL CLC MOLECULES

The discovery of CLC members in bacterial species (23) made it possible to purify a large quantity of CLC proteins suitable for crystallographic work. The first experiments to determine the structure of CLC channels took advantage of the electronmicrographic image analyses of two-dimensional crystals of a CLC member from *E. coli* (42). The projection map that integrated the electron density of the protein along the direction normal to the cell membrane was generated at $\sim$6.5-Å resolution, and this two-dimensional structure revealed an overall rhombus-shaped molecule. This picture confirmed a dimeric architecture implied from functional studies, but the exact locations of the expected two pores and the molecular details of the protein had to wait for three-dimensional crystals, a breakthrough achieved shortly thereafter by the MacKinnon group (43, 44).

The bacterial CLC molecules consist of only $\sim$400 amino acids (43). In comparison, the muscle-type CLC channels contain $\sim$800–1000 amino acids (21, 45–47). The proteins encoded by bacterial CLCs correspond to the N-terminal half of their vertebrate counterparts, a critical portion of the protein that is embedded in the lipid bilayer and forms the ion-permeation pathway (43). The bacterial CLC structure shows two identical subunits, each containing a potential anion-permeation pathway with respect to a twofold symmetry axis perpendicular to the membrane plane (see Figure 1, see color insert). This double-barreled picture was predicted long before the bacterial CLC structure was available based on the functional experiments from CLC-0 (48–56).

The protein structure of bacterial CLC molecules is made up wholly by α-helices but not β-sheets (Figure 1a,b). There are 18 α-helices in one subunit, named helix A to R from the N to C terminus of the protein (43, 44). These helices wrap around the ion-permeation pathway, in which two and three ions have been identified in the high-resolution X-ray structures of the wild-type protein and the point mutant of residue E148, respectively (Figure 2a,b, see color insert). From the extracellular entrance toward the intracellular end of the pore, the ion-binding sites are named S_{ext}, S_{cen}, and S_{int}, respectively, to reflect their relative locations in the pore. These binding sites are formed by amino acid residues that are located on loop regions on the N-terminal end of helices D, R, F, and N. S_{cen} is located in the middle of the ion-permeation pathway, and the Cl^- that is bound at this location is coordinated by the side-chain hydroxyl of S107 (corresponding to S123 of CLC-0) in helix D and that of Y445 (corresponding to Y512 in CLC-0) in helix R. The bound Cl^- ion was also shown to interact with the main-chain nitrogen atoms from I356 and F357 of helix N (43).

The ion-permeation pathway shown in the bacterial CLC molecules is constricted on the extracellular end by the N termini of helix F and helix N. Just 4 Å external to the S_{cen} site, a glutamate residue (E148 in bacterial CLCs or E166 in CLC-0) projects its side chain into the pore, and the Cl^- exit pathway is occluded by this negatively charged side chain (Figure 2a). When this glutamate is mutated to a noncharged amino acid (Figure 2b), a Cl^- ion is seen at the position normally occupied by the negatively charged side chain of this glutamate (44). This Cl^--binding site, which is occupied by Cl^- only in the absence of the wild-type glutamate residue, is named S_{ext} because it is located on the external side of S_{cen}. This glutamate residue is conserved throughout the vertebrate CLC channels, except CLC-K. It has been suggested that the side chain of this critical glutamate residue serves as the gate of the channel (43, 44) and that the competition between Cl^- and the glutamate side chain provides a potential way for coupling the channel gating to ion permeation (Figure 3, see color insert)

In addition to S_{cen} and S_{ext}, a third Cl^--binding site is found ~ 6 Å internal to S_{cen}, and thus is named S_{int} (Figures 2 and 3). This site is located at the junction where the intracellular aqueous solution meets the pore. The Cl^- ion at this position is coordinated by the backbone amide group from G106 and S107 (Figure 3). On the intracellular side of this binding site, there may be room for water molecules to hydrate the bound Cl^-. Thus the Cl^- ion at the S_{int} site may be partially hydrated. However, the intracellular pore entrance is restricted so that the bound Cl^- ion at S_{cen} is hardly visible when the pore is viewed from the intracellular pore entrance.

The bacterial CLC structure reveals that Cl^- ions at these three binding sites do not make direct contact with positively charged amino acids. Rather, they are stabilized by the positive end (N terminus) of helix dipoles. This arrangement is thought to be important because the partial charge from helix dipoles can stabilize a bound Cl^- ion while still allowing rapid ion diffusion to occur (43). Binding of the permeant ions at these binding sites should presumably be affected by the orientations of α-helices, which determine the orientation of the macrodipoles. However, a recent theoretical calculation showed that, in contrast to K^+ channels, the contribution of helix macrodipoles to chloride binding in bacterial CLCs was only marginal. The stabilizing forces of protein on the bound Cl^- ions were thought to come from hydrogen bonding as well as from favorable electrostatic interactions (57). At what is still an early stage of the postcrystallographic era, such a conclusion derived from theoretical calculations awaits more supporting experimental evidence.

Although the bound Cl^- ions do not make direct contact with charged amino acid side chains, there are charged residues at the pore entrances that were thought to be important to draw Cl^- ions into the pore (43). Experimentally, these charged residues were indeed shown to be critical in controlling the channel properties (49, 53). With very few exceptions, the charged residues in the pore are conserved in the muscle-type CLC channels. Also conserved are the serine (S107) and tyrosine residue (Y445) that were found to coordinate the Cl^- ion at S_{cen}. Thus, even though the bacterial CLCs and the muscle-type CLC channels share only $\sim 18\%$–20% amino acid sequence identity, all the important residues are there. Thus, the overall architectures of the mammalian CLC channels are very likely to be similar to that of their bacterial counterparts. Recent attempts using two different approaches to directly compare the structures between bacterial and vertebrate CLCs support this conclusion (58, 59). These comparisons lend confidence to correlation of the structures of bacterial CLCs with the functions of vertebrate

CLCs. In the following discussion, I use the basic architecture of the bacterial CLC structures as a framework (60) through which to view the functional properties of the muscle-type CLC channels, even though bacterial CLC proteins have recently been shown to be transporters (41).

FUNCTIONAL PROPERTIES OF VERTEBRATE CLC CHANNELS

Functional properties are available in the muscle-type CLC channels but not in other CLC members because of the technical feasibility of electrophysiological recording of channels expressed on the cell membrane but not on the membrane of intracellular organelles. Over a decade of functional studies have indicated that only CLC-0, CLC-1, and CLC-2 have a robust functional expression. Although CLC-Ks also belong to the subfamily of muscle-type CLC channels, their expression appears to require an auxiliary protein, bartin, which provides for an efficient targeting/trafficking of the channel protein to the cell membrane (61, 62). For the nonmuscle-type CLCs, either the molecules were not convincingly demonstrated to generate functional Cl^- current (for example, CLC-6 and CLC-7), or the generated Cl^- current was described only as a phenomenon without a mechanistic interpretation (for example, CLC-4 and CLC-5) (39, 63, 64). Thus, only the structure-function relations of the muscle-type CLC channels, especially CLC-0, CLC-1, and CLC-2, are discussed. [For a review of the CLC-K channels, see Uchida & Sasaki (65) in this volume.]

Gating Mechanism of the Muscle-Type CLC Channels

The gating of CLC-0, CLC-1, and CLC-2 is controlled by multiple mechanisms. A good example is the single-channel recording trace of CLC-0 shown in Figure 4, from a channel expressed in *Xenopus* oocyte. This single-channel current was recorded at -60 mV by the excised, inside-out patch configuration with symmetrical 120 mM Cl^-. Similar recording traces can be obtained from channel proteins directly purified from *Torpedo* electric organs or from a mammalian cell line expressing the CLC-0 channel gene (52, 53). The recording trace shows long, nonconducting states (indicated by "I" in Figure 4) between bursts of channel activities. Within each burst of activity are three current levels: the nonconducting level (L_0), the middle level (L_1), and the fully open level (L_2). Thus, direct examination of the recording trace already suggests two kinds of gating processes: one within a burst, which has a gating kinetics in the millisecond range, is called fast gating; and the other, which corresponds to the appearance and disappearance of bursts, has a transition time in the range of seconds. The latter is called slow gating, or inactivation gating because the long, nonconducting events are called inactivation events. The characteristics for the channel activity within the burst— three equally spaced current levels and the binomial distribution of the probabilities

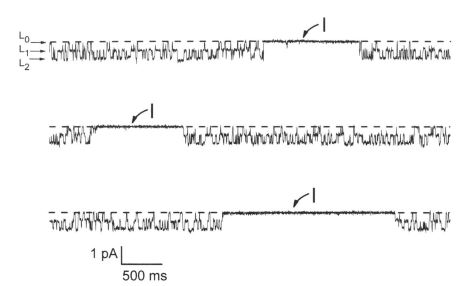

Figure 4 Single-channel recording trace of the wild-type CLC-0. Continuous 15-s recording in symmetrical 120 mM Cl⁻ solutions under the excised inside-out configuration at −60 mV. L_0, L_1, and L_2 represent the three current levels within a burst of channel activities. They correspond to the current levels when 0, 1, and 2 pores are open, respectively. The letter "I" indicates the inactivation event, which is due to the closure of the slow gate.

of these three current states—argue that a single-channel molecule contains two ion-permeation pores (54, 55, 66). The three equidistant current levels with binomial distribution are also found in CLC-1 (67). In addition, when heterodimeric channels were generated by coexpressing genes for CLC-0 and CLC-1, CLC-0, and CLC-2, or CLC-1 and CLC-2, single channels with two nonidentical but independent pores were observed (56). Thus, it is reasonable to conclude that the double-barreled model can be applied at least to CLC-0, CLC-1, and CLC-2. Although bacterial CLC structures are also consistent with such a double-barreled model, a more general extension of this picture to all members of the CLC family requires direct experimental evidence from the remaining CLC members.

Thus, the muscle-type CLC channels contain two subunits, each forming its own independent pore. According to the kinetics of the gating transition, there appear to be two mechanisms, the fast and slow gating. However, the gating of these channels may be more complicated than just two mechanisms, at least based on the different types of gating modulations, as discussed below.

SLOW GATING   Several factors, such as membrane potential, temperature, chloride ion, and pH, affect the slow gating. Early studies on the native CLC-0 channel from

Torpedo electric organ indicated that the slow gating favors a nonconducting, inactivated channel state upon membrane depolarization and a low external pH (pH_o) favors the opening of the slow gate (68, 69). Moreover, the operation of the slow gate is affected by the Cl^- gradient across the channel pore (70), and a reduction of the extracellular chloride concentration ($[Cl^-]_o$) favors the inactivated state (71). The effect of Cl^- flux on a nonequilibrium gating cycle of the slow gating provided the first piece of evidence that the ClC-0 gating is coupled to ion permeation (70).

Recently, the slow gating of ClC-0 was found to be extremely temperature dependent, with a Q_{10} of ∼40 (72, 73), which suggests that the slow gating may involve a large conformational change of the channel protein. Extracellular zinc ions (Zn^{2+}) facilitate this inactivation process, thus inhibiting the macroscopic current of the wild-type ClC-0 channel (73). The inactivating effect of the extracellular Zn^{2+} ion on the slow-gating process is supported by the observation that a mutant C212S, whose slow gate is locked in the open state, loses its sensitivity to Zn^{2+} inhibition (74). Except for an absence of inactivation, the fast gating and the pore properties of the C212S mutant of CLC-0 are nearly identical to those of the wild-type channel (74).

Although the slow gating of CLC-0 can be affected by numerous point mutations, including C213G, C480S (74), P522G, and L524I (72), little is known about the molecular nature of the slow gate and how these mutations affect the slow gating. Deleting the C-terminal intracellular domain was also shown to affect the slow-gating kinetics (75), suggesting that, in addition to the membrane spanning part of the protein, the C-terminal half of the channel is also important. The structural and functional characterization of the C terminus of CLC channels was rarely performed, although a sequence comparison revealed a so-called CBS (cystathionine-L-synthase) domain. This intracellular domain was thought to play an important role for CLC channel function, because deleting the C-terminal intracellular domain, or making mutations at the specific residues in this domain, either resulted in poor expression or led to nonfunctional channels (76–78). In addition, split-channel approaches showed that coexpressions of the membrane portion (corresponding to the bacterial CLC structure) and the remaining C-terminal domain could generate functional channels in both CLC-0 and CLC-1 (79, 80), suggesting that the C terminus of CLC channels should play a complementary function to the membrane-embedded portion of the channel.

The slow gating in CLC-1 is different from that of CLC-0 in its kinetics. The slow gating of CLC-1 is much faster, with transition time on the order of tens of milliseconds, comparable to that of the CLC-1 fast gating. Therefore, when membrane potential is altered, the relaxation kinetics at the millisecond time range is usually composed of two exponential processes. Analysis of the open probability directly from the measurement of the tail currents of macroscopic recordings, although useful for CLC-0 to obtain the fast-gate open probability, completely loses a mechanistic meaning in CLC-1 because such an open probability is a combination of the open probabilities of the fast and the slow gates. Dissecting the fast gating from the slow gating using two-exponential curve fitting is thus necessary. The fast gating was also separated from the slow gating by using so-called envelope protocols that employed a tail pulse to -140 mV after stepping the voltage to a certain test potential for increasing durations (81, 82). Using these different approaches, the slow-gating was shown to be $\sim$ threefold slower than the fast gating, and its open probability increased upon membrane depolarization. Analysis of the dwell-time histograms of single-channel recording traces of CLC-1 showed two populations of nonconducting events, with durations differing from each other by $\sim$ threefold (67). Examination of the temperature dependence of the CLC-1 slow gate showed a Q_{10} value of $\sim$6 to 7 (83), again quite a large number. Mutation of C277S, which is the corresponding mutant for C212S in CLC-0, results in a relaxation of macroscopic current composed of only a single-exponential process (84), suggesting that this mutation also locks open the slow gate of CLC-1.

CLC-1 is also inhibited by extracellular zinc (Zn^{2+}) or cadmium (Cd^{2+}) ions (85–87). Mutations of several residues in CLC-1 reduced this Zn^{2+} inhibition (86). However, there is no direct evidence to conclude whether the mutated residues that affect the Zn^{2+} inhibition, including C212S of CLC-0, or C242, C254, and C546 of CLC-1, participate in forming the Zn^{2+}-binding site. This is always the problem of using functional results to infer structural information. Take Zn^{2+} inhibition of CLC-0 as an example. Since the C212S mutation has changed the slow-gating transition to favor the open state of the slow gate, it is reasonable that the C212S mutation will reduce Zn^{2+} inhibition even if C212 has nothing to do with the Zn^{2+}-binding site, a classic example of coupling ligand binding to channel gating. Indeed, the residue in the bacterial CLC protein that corresponds to the C212 of CLC-0 is located at a buried position, making it less likely that this residue is involved in forming a Zn^{2+}-binding site. So far, how a point mutation of C212S in CLC-0 locks the slow gate in the open state is still unknown. The slow-gating mechanism of CLC channels was thought to be a process involving the rearrangement of the two subunits (55), and several residues at the interface of the two subunits were shown to affect the slow gating of CLC-1 (88). Perhaps C212 of CLC-0 (or C277 of CLC-1) is critical to translate the energy from dimer interactions to the gate in the ion-permeation pathway.

The gating mechanism of CLC-2 is less defined than those of CLC-0 and CLC-1. This channel gating is dominated by hyperpolarization-induced channel activities. Early mutational studies discovered that deletion of an N-terminal region (aa. 16–61) in CLC-2 rendered the channel constitutively open, and that this domain could be transplanted to different regions of the channel without loss of function (89). A follow-up study further identified a region from $\sim$ residue 350–365 that, when deleted, conferred the same phenotype as that caused by the deletion of the N-terminal region (90). A "ball-and-chain"-type mechanism similar to the N-type inactivation of Shaker K^+ channel (91, 92) was proposed as an explanation for the gating mechanism of CLC-2, with the N-terminal region being the ball and the region corresponding to residues 350–365 being the ball receptor (89, 90).

Although interesting, this proposal was perhaps premature because there was no other key supporting experimental evidence, such as knock-off effects by permeant ions or competitions of the ball domain with pore blockers, as has been shown in Shaker K^+ channels (93, 94). A recent study of rat CLC-2 expressed in HEK293 cells argued that the ball-and-chain mechanism to explain the CLC-2 gating may be false (95). The effects of deleting the N-terminal domain on CLC-2 gating was attributed to intracellular events, including potential intracellular diffusible components or the osmotic state of cytoskeleton structure (95). From the now-available CLC structure (43, 44), the ball-and-chain mechanism is also dubious because both the proposed ball and receptor regions are located far away from the ion-permeation pathway.

Since the cloning and expression of CLC-2, it was not known if the hyper-polarization-induced gating corresponds to the fast gating or to the slow gating of CLC-0, or to both. A recent study (96) examined the slow-gate locked-open effect of a mutation at the residue C256, which corresponds to C212 of CLC-0 or C278 of CLC-1. The mutation C256A indeed led to a significant fraction of constitutively open channels at all potentials. Extracellular Cd^{2+}, which inhibits wild-type CLC-2 almost completely, inhibited the C256A mutant only by ~50%. These results collectively suggested that CLC-2 possesses a common gate that is also sensitive to Cd^{2+}, and that C256A mutation may also lock this common gate in an open position—a phenomenon similar to that found in CLC-0 and CLC-1 slow-gating mechanisms. Thus, it appears that CLC-0, CLC-1, and CLC-2 are all gated by at least two different gating mechanisms (96).

FAST GATING   The fast gating of ClC-0 (and ClC-1) is also controlled by the membrane potential—the more depolarized the membrane potential, the higher the open probability. Since Cl^- current has a reversal potential at ~ -80 mV in muscle cells, this gating mechanism in CLC-1 is important to regulate the contraction and relaxation of skeletal muscle because it is fast enough to counteract the depolarization generated by the opening of Na^+ channels in an action potential. Mutations that reverse the voltage dependence of ClC-1 result in certain forms of myotonia because these mutant channels are unable to contribute hyperpolarizing forces after membrane depolarization. Although the membrane potential plays a critical role in the fast gating, this gating mechanism is also regulated by Cl^- and H^+ in at least three different ways, two in the opening process and one in the closing process of the fast gate. The fast-gating mechanisms of CLC-0 and CLC-1 are phenomenologically similar to each other. However, the apparent voltage dependence of CLC-2 gating is opposite to that of CLC-0 and CLC-1. Among the gating mechanisms of these three channels, the fast gating of CLC-0 is by far the best-studied mechanism. In the following sections, different aspects of the CLC-0 fast gating are discussed, followed by a general description of other muscle-type CLC channels.

Depolarization-favored fast-gate opening is tightly coupled to external Cl^-   The "prominent" voltage dependence of the CLC-0 fast gating is similar to that found

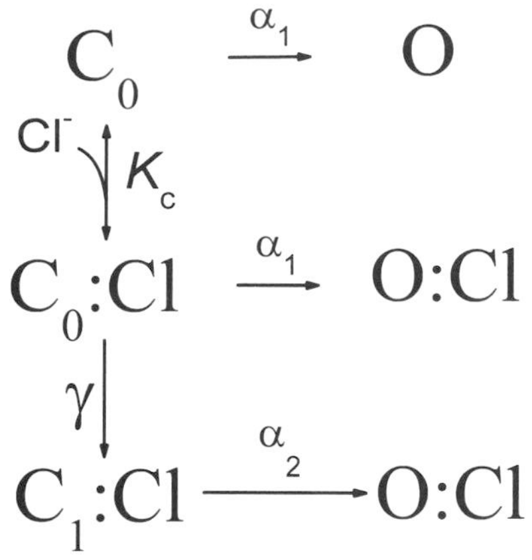

Figure 5 A six-state scheme describing the fast-gate opening mechanism of CLC-0. In this scheme, C and O represent the closed and open states of the fast gate, respectively. Cl represents a Cl^- ion, and K_c is the binding constant for external Cl^- to bind to the initial binding site S_{out}. C_0 and C_1 are used to indicate two different closed states. See Chen & Miller (71) for detailed descriptions.

in the voltage-gated cation channels: The open probability (P_o) is higher at more depolarized membrane potentials (52, 54, 55, 66, 71, 74, 97). This depolarization-induced activation of ClC-0 (and also its mammalian homologue, ClC-1) is thought to arise from the linkage of extracellular Cl^- to the gating (71, 97, 98). The first demonstration of such a linkage was a shift of the fast-gate open probability-voltage (P_o-V) curve toward a more depolarized membrane potential upon reducing external Cl^- concentration ([Cl^-]$_o$) (97). Subsequent studies at the single-channel level found that external Cl^- significantly increases the opening rate but only has a very small effect on the closing rate (49, 71, 99). Based on the fast-gate opening kinetics, a kinetic scheme like that shown in Figure 5 was proposed. This scheme, which only describes the opening process of the fast gate, suggests that the channel can open in two different modes. One opening mode depends on the presence of external Cl^-. In this external Cl^--dependent fast-gate opening mechanism, Cl^- first binds (with a binding constant K_c), in a voltage-independent fashion, to a site located at the external end of the conduction pathway. Subsequently, the Cl^--bound channel then undergoes a conformational change, which translocates the bound Cl^- to an inner site within the membrane electric field, with a translocation rate γ. Once the channel reaches the state C_1:Cl (see Figure 5), the channel opens with a very high opening rate α_2. Because Cl^- carries a negative charge, an inward movement of the bound Cl^- (which is the rate-limiting γ step in this Cl^--dependent process) in the membrane electric field favors a depolarization-induced channel opening.

Thus, when the channel is bound with Cl^-, it opens in response to membrane depolarization. On the other hand, the opening of the channel is favored by membrane hyperpolarization in the absence of external Cl^-. The opposite voltage dependence in the presence or absence of a bound Cl^- thus suggests that the voltage dependence of this opening mechanism indeed arises from Cl^- (71). This mechanism is completely different from that of the voltage-gated cation channels. In S4-type cation channels, the voltage dependence arises from the translocation of the fixed

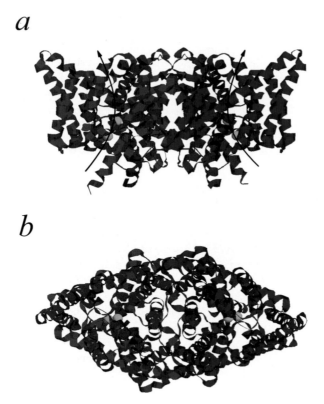

Figure 1 Architecture of bacterial CLC protein molecules. (*a*) Side view of the CLC-ec1 molecule in *E. coli*. The two identical subunits are shown by blue and red colors, respectively. Each subunit contains 18 α-helices. Curved arrows roughly represent the two ion-permeation pathways. The two green spheres in each subunit represent the Cl⁻ ions bound in the ion-permeation pathway of the wild-type molecule. (*b*) Extracellular view of the same molecule. The pictures are based on the coordinate, 1OTS, from Protein Data Bank (PDB), with the cocrystallized antibodies being removed.

a   *b*

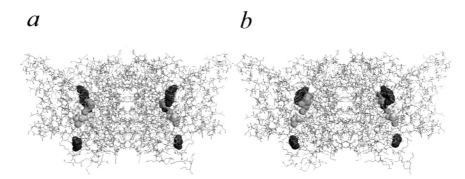

Figure 2 Comparison of the high-resolution structures between (*a*) wild-type CLC-ec1 from *E. coli* (PDB code 1OTS) versus (*b*) E148Q mutant (PDB code 1OTU). Colors for the space-filled residues are: purple, R147; yellow, S107; blue, T452. They correspond to K165, S123, and K519 of CLC-0, respectively. The critical glutamate E148 (corresponding to E166 of CLC-0) and the replaced glutamine are shown in CPK color. Most of the protein structures are nearly identical except for slight changes in the side-chain orientations of R147 and E148. In the wild-type molecule (*a*), the negatively charged E148 side chain projects more horizontally. When a neutral residue glutamine is placed at this position (*b*), the side chain is oriented toward external pore entrance, making room for Cl$^-$ to stay at S$_{ext}$.

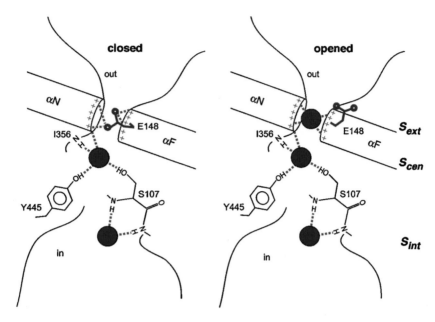

Figure 3 Competition between Cl⁻ and glutamate side chain explains the gating-permeation coupling mechanism of CLC channels. Cl⁻ ions at the binding sites (labeled as S_{ext}, S_{cen}, and S_{int}) in the ion-permeation pathway of the CLC-ec1 molecule in *E. coli* are depicted as red spheres. Interactions of the bound ions with neighboring stabilizing forces are shown. The ion at S_{ext} does not exist in the closed state of the wild-type molecule (*left*). However, when the side chain of E148 swings away from the ion-permeation pathway (*right*), an anion can be stabilized at the S_{ext} site by the charge from the positive end of helix dipoles. Reprinted figure with permission from Dutzler et al. (44). Copyright 2004 AAAS.

charge in the channel protein, the S3b-S4 paddle, across the membrane electric field (100, 101). In ClC-0, however, it is the movement of the permeant ion in the electric field that renders the channel voltage dependent.

For CLC-1 channel, the external Cl^- appears to have an effect on the fast gating similar to that of CLC-0. Although a voltage-dependent gating mechanism based on intrinsic voltage sensors similar to those in the voltage-gated cation channels had been proposed for CLC-1 gating (102, 103), the evidence that led to this argument was not compelling. One argument for the voltage-sensor type gating was based on the altered voltage dependence of channel opening upon D136G mutation in CLC-1 (103). The corresponding mutation D70G in CLC-0 conferred the same phenotypical change (104). However, a charge-conserved mutation in CLC-0, D70E, also exhibited the same alteration of the voltage dependence (104). Thus, D70 of CLC-0 or D136 of CLC-1 is important for the normal voltage dependence of the fast gating, not because of its side-chain charge but perhaps because of a complex allosteric control. Another argument for the presence of an intrinsic voltage sensor in CLC-1 was based on the left shift of the P_o-V curve in CLC-1 when external Cl^- ions were replaced with methane sulfonate ($CH_3SO_3^-$) ions (103). Because the fast gate became easier to open when the permeant ion Cl^- was replaced with the nonpermeant ion $CH_3SO_3^-$, it was thought that the external Cl^--activated opening mechanism cannot possibly be correct for CLC-1 fast gating. However, although CLC-0 or rat CLC-1 showed very low permeability to $CH_3SO_3^-$, this bulky anion did permeate through human CLC-1 channel, with a permeability ratio $P_{CH_3SO_3^-}/P_{Cl^-}$ of ~0.2 (105). The anion-binding sites responsible for the external Cl^--dependent gating in CLC-0 and CLC-1 may not be completely identical, resulting in a discrepancy for $CH_3SO_3^-$ to gate these homologous channels.

Hyperpolarization-activated fast-gate opening is favored by external H^+, but not by external Cl^-   Another mode of the fast-gate opening process is not related to external Cl^-, the channel can open with a small rate α_1 (see Figure 5) whether the channel is bound with Cl^- or not. In single-channel recordings of CLC-0, the durations of nonconducting events within a burst of channel opening (the fast-gating transition) are described by a single-exponential distribution (66, 71, 74, 99). The average duration of these nonconducting events at L_0 can be used to calculate the fast-gate opening rate, which is a function of membrane voltage—the more hyperpolarized the membrane potential, the smaller the fast-gate opening rate. However, when the membrane potential is made very negative (more negative than -140 or -150 mV), the opening rate of the fast gate becomes paradoxically faster (71). Thus, the overall opening rate as a function of the membrane voltage does not follow a single-exponential function, suggesting two different processes for the fast gate to open. Besides a depolarization-activated opening mechanism modulated by extracellular Cl^-, there is a hyperpolarization-activated opening mechanism for the fast gate of CLC-0, the α_1 step shown in Figure 5 (71). This hyperpolarization-activated fast-gate opening mechanism in CLC-0 is not regulated by $[Cl^-]_o$ but is highly sensitive to extracellular pH (pH_o)—when pH_o is reduced, the fast-gate P_o-V curve is changed because of an increase of minimal P_o (98, 99). The same phenotype also occurs in CLC-1 (98). The effect of pH_o on the fast gating (an increase of minimal P_o) is in contrast to the shift of the P_o-V curve along the voltage axis in response to the change of $[Cl^-]_o$. Thus, these two opening processes not only have different voltage dependence but also show distinct regulations by external Cl^- and H^+, suggesting that the fast gating could contain two opening mechanisms. The molecular natures of these two opening mechanisms, the $[Cl^-]_o$-dependent, depolarization-favored fast-gate opening and the pH_o-modulated, hyperpolarization-favored opening, have been subject to broad speculation (106), and will likely be the focus of future studies.

The hyperpolarization-favored fast gating is not a totally unexpected gating mechanism. First, the prominent gating in ClC-2, which has the same degree of sequence homology with ClC-0 as that with ClC-1, is an opening mechanism favored by membrane hyperpolarization (47, 89, 95, 96, 107, 108). Second, various manipulations on the channel proteins, including single-point mutations in ClC-0 and ClC-1, render these two channels susceptible to activation by slight membrane hyperpolarization (79, 80, 103, 104, 109–111). The hyperpolarization-activated opening mechanism of CLC-2 is sensitive to both external pH and internal Cl^- (107, 108). However, the slow gating of CLC-0 is also sensitive to pH and Cl^- (68, 70, 71, 99). Still undetermined is the relation between the normal hyperpolarization-activated mechanism in the wild-type CLC-0 with that of CLC-2 or with the abnormal hyperpolarization-activated gating of the CLC-0 mutants.

Fast-gate closing rate is reduced by internal Cl⁻   The scheme in Figure 5 describes only the coupling of external Cl^- to the opening process of ClC-0 fast gating. For fast-gate closing, a profound effect of Cl^- on the closing rate was also observed (71, 112). In this case, an increase of internal Cl^- concentration ($[Cl^-]_i$) reduces the closing rate (Figure 3), a phenomenon similar to the foot-in-the-door effect first described in voltage-gated cation channels (113). Intracellular Cl^- ions, however, have little effect on the opening rate of the fast gate (71, 112). Thus, although both internal and external Cl^- ions regulate CLC-0 gating, the kinetic aspects of these two regulations appear to be quite different. Furthermore, the respective Cl^--binding sites that control the opening and the closing mechanisms may also be different because the apparent Cl^- affinities in modulating these two processes differ by more than 20-fold (71, 112).

More evidence to support the disparity between Cl^--dependent fast-gate opening and closing mechanisms can be found in mutational studies. For mutations at the external pore entrance of CLC-0, for example, K165, the effect is on the opening rate but not the closing rate (49). On the other hand, for a mutation at the intracellular pore entrance of CLC-0, K519, the effect is more on the closing but not the opening rate (112). Taken together, these results strongly suggest that the Cl^--binding site responsible for the $[Cl^-]_o$-dependent fast-gate opening and that for the $[Cl^-]_i$-dependent fast-gate closing may be different, and that these two aspects of the fast-gating mechanisms may not be reversible processes of each other (112). If a gating mechanism is coupled to Cl⁻ permeation, the mechanism must be a nonequilibrium process, because a Cl⁻ ion, whether coming from the extracellular or the intracellular side of the membrane, binds first and then must pass through the pore (18, 70). The exact mechanisms whereby the fast gating functions as a nonequilibrium cycle and how fast gating is coupled to the slow gating are still elusive, even with an available CLC structure.

Channel Conductance and Ion Permeation

All vertebrate ClC channels studied so far conduct Cl⁻, Br⁻, NO_3^-, or larger anions such as I⁻ and SCN⁻ (18). Unlike cation channels, which can be much more selective for a specific cation, the permeability ratios between these anions in CLC channels usually are within a factor of 10. These permeability ratios were mostly obtained from whole-cell recording experiments (105, 114–117). Without rigorous, single-channel studies using permeant blockers such as those performed in voltage-gated cation channels (118–122), it is difficult to infer ion permeation mechanisms from these experiments. Nevertheless, previous work has demonstrated at the phenomenological level the relative permeability sequence: SCN⁻ ~ NO_3^- > Cl⁻ > Br⁻ > I⁻. However, caution is needed regarding the permeation studies in CLC channels because permeant ions also affect the gating of these channels, and the P_o of the channel never reaches zero, even at extremely negative potentials. Thus, if the experiment is carried out without a good solution control (for example, recording whole oocyte current), it is always a concern whether substitution of extracellular permeant ions will change the intracellular ion concentration.

The ion-permeation mechanism of CLC channels was understood only at a very superficial level because of the warning described above and because of the small conductance of most of CLC channels (18). Consequently, structural inference from functional data was limited. However, since the bacterial CLC structure was solved, the ion-permeation pathway has been directly suggested from the bound Cl⁻ ions seen in the X-ray structure together with the conserved residues previously identified by mutagenesis. For example, on both external and internal ends of the CLC-0 pore, two charged residues, K165 and K519, respectively, have been identified as critical (49, 53). At the center of the permeation pathway, S123T mutation is also known to reduce the single-channel conductance (50). These residues thus serve as landmarks to identify the ion-permeation pathway (see Figure 3). However, even with identification of the permeation pathway, the mechanism of ion permeation is still not clear. For example, the single-channel current-voltage (I-V) relation of CLC-0 is linear (52, 53), whereas that of CLC-1 is inwardly rectified, as deduced from the instantaneous I-V curve (83, 84, 87, 102, 105, 115). In addition, the single-channel conductance for each member within the muscle-type subfamily varies significantly from one member to another. At symmetrical 150 mM Cl⁻, the single-channel conductance of CLC-0, CLC-1, and CLC-2 is ~10 pS, ~1 pS, and ~3 pS, respectively (54, 56, 67, 123).

Before the high-resolution structures of these channels become available, it will be extremely challenging to tackle the molecular basis for the variation in single-channel conductance.

Nevertheless, since the bacterial CLC structure became available, there have been several attempts to understand the ion permeation mechanisms of CLC channels using theoretical calculation methods. These have included Monte Carlo computation of anionic electrostatic potential energy profile along the ion-permeation pathway, Poisson-Boltzmann calculations of ion-binding energy, all atoms molecular dynamic simulations, and Brownian dynamics simulations (57, 124–126). One major hurdle for these theoretical approaches is that the apparent ion-permeation pathway in bacterial CLCs is blocked by the side chain of pore residues, such as that of E148. In addition, the intracellular pore entrance to the S_{cen} site is also significantly narrowed by the S107 side chain. Therefore, some studies have tried to "open" the pore by removing these obstructions. These theoretical approaches confirmed the importance of the experimentally identified charged residues. They also made predictions on previously unnoticed amino acids. For example, one study for the first time pointed out the potential important role of residue 318 of bacterial CLCs. The corresponding residues in CLC-0 and CLC-1 are glycine and glutamate, respectively. By using Brownian dynamic simulations, this study showed that the charge at this position may determine Cl^- exit rate toward extracellular solution, which explains the differences between CLC-0 and CLC-1 in single-channel conductance as well as in the rectification property of I–V curves (126). By calculating the source of the stabilizing energy of the bound Cl^- ion, another study predicted that the side chain of K131 and R340 (corresponding to residues 149 and 401 of CLC-0) contribute significantly to stabilize the bound Cl^- in the ion-permeation pathway (57). R340 is particularly interesting as the charge of this residue was suggested to contribute an important stabilizing force for Cl^- to stay at the most external Cl^--binding site, which was not seen in the crystal structure of bacterial CLCs. It would be very interesting to examine the corresponding mutations in vertebrate CLC channels to experimentally verify these predictions.

CLC FAST-GATING MECHANISMS: STRUCTURE AND FUNCTION CORRELATION

Before the high-resolution structure of the bacterial CLC molecule was available, functional studies using site-directed mutagenesis and electrophysiological recordings in the muscle-type CLC channels had only minimal success in revealing the structural/functional properties of CLC channels. The pioneering work can be classified in two directions. First, the mutations were made following the suggestions of "myotonia" sites including D136G and G230E in CLC-1 (2, 103, 127, 128). The other approach was based on sequence comparisons that identified conserved regions among members of the CLC family (74, 97, 115, 116). Systemic mutations were then made, and some of the mutants showing significant functional consequences were studied in more details. These early studies more or less identified several important residues. The critical roles of some of these residues immediately become apparent after the bacterial CLC structure is available (for example, S123, K165, E166, and K519 of CLC-0; see Figure 2). It is, however, unclear how other residues, for example, C212S of CLC-0, or several mytonia mutations such as P480L, Q522 of CLC-1, control CLC channel function.

The high-resolution structure of bacterial CLC molecules has been valuable in interpreting mutagenesis studies. For example, the residues corresponding to the critical glutamate in several vertebrate CLC channels were mutated before the bacterial CLC structure was available (64, 115, 129). Although the mutation caused a dramatic effect on the channel gating, the molecular basis for these results was unclear until the crystallographic structure of bacterial CLC revealed the pore-lining position of this critical glutamate. The corresponding residues in CLC-0, CLC-1, and CLC-2 for this residue are E166, E232, and E217, respectively. The phenotype of the mutation of this critical glutamate has been rigorously examined following the suggestion from the bacterial CLC structure. The mutant channels all lost a major part of gating transition—they were in the open state most of the time (44, 108). On the other hand, CLC-K represents the only exception of the presence of a glutamate gate. This kidney channel does not have a glutamate in the conserved GKEGP motif. Not surprisingly, the gating of the wild-type CLC-Ks did not show the same gating relaxation as those observed in the other three muscle-type CLC channels. When the corresponding residue valine in CLC-K was mutated into glutamate, the channel revealed an inward rectification, very similar to the CLC-2 gating, or the hyperpolarization-activated channel opening (129). Thus, the negative charge on these glutamate residues in CLC-0, CLC-1, and CLC-2 must play important roles in generating the gating relaxation. The most straightforward hypothesis is that the side chain of this critical glutamate acts as the gate for the channels. There are two E148 residues, each seen in one pore. Thus, if the side chain of E166 in CLC-0 is indeed the gate, it must be at least part of the fast gate. The X-ray crystal structures of the E148A and E148Q mutants of bacterial CLCs showed that when the negative charge on this critical glutamate is removed, a Cl^- ion is found to occupy the location of the removed negative charge (see Figure 2), suggesting that the coupling of ion permeation with the fast gating may be due to a competition mechanism between the Cl^- ion and the negatively charged gate (Figure 3).

How does the competition of Cl^- with this key glutamate side chain explain the CLC fast-gating mechanisms? When the importance of this glutamate side chain was first determined in the bacterial CLC structures, it was suggested that Cl^- ions from the extracellular solution enter the pore to compete with the carboxylate group of the key glutamate residue at S_{ext}—a mechanistic interpretation for the $[Cl^-]_o$-dependent fast-gate opening mechanism of CLC-0 (43). This proposal was puzzling in light of several functional observations (130). First, the mechanisms of fast-gate opening and closing are controlled by Cl^- ions in very different ways (71). If the glutamate side chain is the gate opened by the external Cl^-, what is the mechanism responsible for the internal Cl^- control of the fast-gate closure? Second, in the original functional recordings, the external Cl^--binding site for the fast-gate opening was proposed to be located outside the membrane electric field (71). However, the location of S_{ext} appears to be deep in the ion-permeation pathway. In fact, recent theoretical work using Poisson-Boltzmann calculations for the transmembrane potential suggests that S_{ext} is located at the center of the membrane electric field—the electric distance from external bulk solution to S_{ext} is ~0.5 (57). Third, the apparent affinity for external Cl^- to control fast-gate opening is ~50 mM. When $[Cl^-]_o > 150$–300 mM, this gating effect was nearly saturated (71, 97). It is doubtful that S_{ext} would have such a high affinity for Cl^- given that Cl^- competes with a fixed negative charge in the pore. Finally, previous mutational studies also provide puzzling data. The mutations of K165 and K519 in CLC-0 both affect the fast gating (49, 53, 97), even though these two residues are separated by more than 25 Å. In fact, they are facing the extracellular and intracellular solutions, respectively. How can such two distant residues regulate the same gating mechanism? These functional observations together raise concerns about the original proposal: Is S_{ext} indeed the gating-permeation coupling site? Does the movement of the side chain of the key glutamate explain all fast gating? Or even more fundamentally, is the side chain of E166 the fast gate of CLC-0?

Several approaches described below have been used to examine the structural and functional correlations of CLC channels. These experiments were performed mostly in CLC-0 because of the technical advantages in studying this *Torpedo* channel. Most of these studies agree that the side chain of E166 represents at least part of the fast gate of CLC-0, although modifications from a simple swinging of the glutamate side chain as the gating mechanism have been proposed.

Substituted Cysteine Accessibility Approaches

Introduction of cysteines at various positions in the channel, followed by modifications of the introduced cysteines with methane thiosulfonate (MTS) reagents, has been widely used to study ion channels. Pioneering studies in nicotinic acetylcholine receptors and voltage-gated cation channels have provided very informative results regarding the structural and functional correlations of these channels (131–136). In particular, state-dependent modifications of the introduced cysteine—the cysteine is modified much faster in the open state than in the closed state—were used to infer the presence of a moving structure that gates the access of the MTS reagents (132). This idea was applied in exploring the pore of CLC-0 by mutating residues in helix R, a helix lining the pore from S_{cen} to the intracellular pore entrance (58), as suggested by the X-ray structures of bacterial CLCs. The accessibility of the introduced cysteine to two charged MTS compounds was then compared to examine not only the accessibility of the introduced cysteine but also the intrinsic electrostatic potential in the pore. Three conclusions were derived from this study. First, the overall modification pattern of these residues was consistent with a helix arrangement—the residues with higher MTS accessibility all faced the permeant ions seen in the bacterial CLC structure. Second, there was no observation of a physical structure that gates the access of internally applied MTS reagents. Thus, the idea that the side chain of the key glutamate serves as the fast gate is applicable in CLC-0. Third, the experiments reveal a prominent intrinsic electrostatic potential that is important not only in determining the MTS modification rates but also in controlling the channel functions (58).

A series of MTS modification experiments have also been carried out on CLC-1 (109, 115, 137). Since the experiments were performed before the available structural guide, some of the interpretations may need to be modified in light of the bacterial CLC structures solved later. This series of MTS modification experiments concluded that the CLC-1 consists of a narrow constriction pore region between K231 on the external end and H237 on the internal end. The narrow region was neighbored by wide vestibules on both sides of the membrane. The narrow region was estimated to have a diameter between 4–6 Å, and the vestibules were thought to be wider than 10 Å (115, 137). Although the suggested location for residue K231 was correct (but was wrong in terms of pore stoichiometry), the conclusion on the internal pore region was not consistent with the bacterial CLC structure because a wide internal pore vestibule at the level of G190 of CLC-1 (corresponding to G108 of bacterial CLC) was not seen in the bacterial CLC structures. Unless the architecture of CLC-1 is very different from that of bacterial CLC, it is difficult to correlate the conclusion derived from functional experiments in CLC-1 with the bacterial CLC structure. The conclusions in CLC-1 were based on the accessibility (or the modification rate) of the introduced cysteine to bulky MTS reagents. The observation that the electrostatic potential could affect the MTS modification rate (58) introduces a variable into the interpretations of the experiments. However, since the structure of CLC-1 is not available, the suggested difference between CLC-1 and bacterial CLC structures may well be real. It would be interesting to further examine this issue using both structural and functional approaches to see if CLC-1 indeed contains a wide internal pore vestibule that cannot be explained by the current picture suggested from bacterial CLC structures.

Intrinsic Pore Potentials and Channel Properties

One structurally insightful precrystallographic study was the discovery of the electrostatic control of CLC-0 conductance by the charge from the side chain of K519 (53). The wild-type CLC-0 channel (containing lysine at position 519) has a conductance of $\sim$10 pS in physiological Cl^- concentrations. When mutations of this residue were made, the conductance of the channel pore was reduced according to the charge placed at this position. On the other hand, the size or shape of the side chain of the introduced residues had only a minimal effect. Such a pure electrostatic control of the single-channel conductance was used to imply that K519 is located close to the inner pore entrance (53).

To examine the mechanism underlying this charge effect on the pore conductance, a series of charge mutations at residue K519 was rigorously studied at various $[Cl^-]_i$ (138). The charge effect from the side chain of K519 diminished in the presence of high $[Cl^-]_i$. However, the application of high concentrations of nonpermeant ions did not change the channel conductance. Thus, a pure surface charge mechanism is not a satisfactory explanation for the charge regulation of the pore conductance. Instead, the results suggest a specific binding site near K519 that might be responsible for mediating the effect. Under the guidance of bacterial CLC structure, a negatively charged residue E127, which is only ∼4 Å away from K519, was also found to play a critical role in controlling the pore conductance. In the E127Q mutation background, the mutational effects of K519 on the pore conductance were nearly eliminated, as if the regulation by the charge at position 519 had to go through the negative charge at position 127. The alterations of single-channel conductance by K519 and/or E127 mutations showed a functional behavior very different from that caused by S123T mutation. K519E and S123T mutations both reduce the wild-type pore conductance by ∼five- to sevenfold in physiological Cl^- concentrations. However, the reduced channel conductance in K519E can be reversed by high concentrations of $[Cl^-]_i$, whereas the conductance of S123T mutant cannot be significantly increased by the same maneuver. Consequently, a dichotomy exists in the pattern of the conductance-concentration curve for these two mutants (138). These results therefore suggest at least two conductance determinants in the pore that are controlled by K519 and S123, respectively. As crystal structures of the wild-type bacterial CLCs show two bound ions (44), these two Cl^--binding sites are thought to correspond to the two conductance determinants (138). Detailed mutagenesis experiments that map the influences of residues at nearby regions of the Cl^--binding sites should be helpful in further examining this hypothesis.

Besides the effect on pore conductance, the charge from residue 519 is also critical in regulating the fast gating of the channel (112). This fast-gate regulation is more significant in the closing rate than the opening rate of the channel. The effects of these charge mutations on the fast-gate closing rate mirror the effects on the pore conductance. When residue 519 was made more negative, the channel conductance was smaller while the fast-gate closing rate was slower. Meanwhile, the E127Q mutation greatly eliminated the effects on both the conductance and the fast-gate closing rate by K519 mutations. The effects of K519 mutations are mostly on the closing but not on the opening rate of the fast gate, an effect similar to that which occurs by increasing $[Cl^-]_i$. Because bacterial CLC structures have suggested the competition of Cl^- with the negatively charged glutamate side chain at the S_{ext} site, the internal Cl^- effects on the fast-gate closing rate appear to illustrate this competition quite nicely—a foot-in-the-door mechanism (Figure 3). The electrostatic control on the fast-gate closing rate adds another layer of color in this picture. The side-chain charge of K519, through altering the intrinsic pore potential, may change the locations of Cl^- occupancy in the pore and therefore affect the fast-gate closing. Meanwhile, because Cl^- ions carry negative charges, the binding of Cl^- to the pore would presumably also change the intrinsic potential of the pore. Thus, alterations of the fast-gate closing rate could come from a combination of two effects: a competition of Cl^- with the glutamate side chain and the more negative potential generated either by increasing the Cl^- occupancy in the pore (by increasing $[Cl^-]_i$) or by a direct introduction of negative charges in the inner pore region (112).

If the competition of Cl^- with the negatively charged side chain of E166 explains the internal Cl^- effect on the fast-gate closing, what is the mechanism for the external Cl^--dependent fast-gate opening? Where is the ion-binding site for the external Cl^--dependent fast gating? Before the bacterial CLC structure was solved, the location of Lys165 of CLC-0 was hypothesized in the external pore region, and this residue was thought to be involved in both Cl^- permeation and channel gating (49). In CLC-0, when Lys165 was mutated into cysteine, the mutant channel was not functional. The channel function, however, could be recovered when the introduced cysteine was modified with 2-aminoethyl methane thiosulfonate (MTSEA). However, the function of the MTSEA-modified K165C mutant is not exactly the same as that of the wild-type channel. The side chain of the MTSEA-modified K165C is ~1.5 Å longer than the side chain of lysine residue in the wild-type channel. Yet the fast-gate P_o-V curve of the mutant is shifted by ~45 mV in the depolarization direction. The shift in the P_o-V curve is mostly due to a reduction of the opening rate (49), an effect similar to that which occurs by reducing $[Cl^-]_o$ (71). These results thus suggest that K165 is likely to play an important role in the external Cl^--dependent fast-gate opening process. In CLC-1, the corresponding residue is K231. Mutation of this residue into cysteine also alters the channel gating—the mutant channel is activated by hyperpolarization (109). The introduced K231C residue can be modified by MTS reagents, and the modification rate is reduced in the presence of external SCN^-, suggesting that SCN^- may bind to the nearby region of K231 and protect the cysteine from MTS modification (137).

Consistent with the experimental data, theoretical calculations also suggest that the positive charge of the corresponding residue Arg147 of the bacterial CLC is important in stabilizing the Cl^- ions in the pore (124). One study suggested that a region external to this conserved R/K site could be the initial Cl^--binding site (called S_{out}) responsible for the $[Cl^-]_o$-dependent fast-gate opening (57). From the calculation of bacterial CLC structure, the Cl^- ion at this very external pore entrance is coordinated by the N-H dipoles of Gly315 and Gly316. A nearby positively charged residue Arg340 fixes the orientation of the carbonyls of Gly314 and Gly315 and contributes to the stabilization of Cl^- ion at this position. In addition to Arg340 and the backbone amide dipoles of Gly315 and Gly316, the stabilization of Cl^- at the S_{out} site also relies on other interactions, including the conserved Arg147. These theoretical predictions, albeit interesting, await experimental verification.

Probing the Channel by Inhibitors and Blockers

In the history of functional studies of ion channels, inhibitors were widely used to explore the mechanistic operations of channel molecules. Good examples include blockades of K^+ channels by small venom toxins (139–142) and by organic compounds such as quarternary ammoniums (143, 144). No effective toxin molecules have been found to block CLC channels. Although an early study showed that a chemical compound 4,4′-diisothiocyano-2,2′-stilbenedisulfonate (DIDS) is an irreversible inhibitor of CLC-0 (55), the mechanism underlying this inhibition has not been examined. Thus, probing the channels with inhibitors or blockers was not even suggested for the structure-function studies of CLC channels for nearly two decades.

Recently, two groups of organic compounds, clofibric acid derivatives and anthracene carboxylic acids, were demonstrated to be useful inhibitors for muscle-type CLC channels (145). The clofibric acid derivatives include 2-(p-chlorophenoxy)butyric acid (CPB) (84, 146), 2-(p-chlorophenoxy)propionic acid (CPP) (147–151), and 2-(p-chlorophenoxy)acetic acid (CPA) (152, 153), all of which contain a p-chlorophenoxy group attached with various lengths of aliphatic carboxylic acid. For CPB and CPP, the stereoisomers exist, but only the S(-) forms are active inhibitors (147, 151). These compounds block CLC-0 and CLC-1 with affinities that range from several micromolar to millimolar, depending on which compound was used and which channel was studied. For example, S(-)-CPP significantly inhibited CLC-1 at a concentration of 50–300 μM, whereas it had negligible effect on CLC-2. On the other hand, CPA, a shorter compound, has a lower apparent affinity in blocking CLC-0 and CLC-1. Depending on the membrane voltages, the half blocking concentration of CPA in CLC-0 ranged from 1 mM to 100 mM (152). In contrast to the simple pore-blocking mechanism for numerous blockers of potassium channels, the mechanisms of these drug actions on various CLC channels may be complicated. For example, S(-)-CPP shifts the fast-gate P_o-V curves of CLC-0 and CLC-1 to the right, but CPA appears to act on CLC-0 by plugging into the pore because the compound reduces the single-channel conductance of CLC-0. More experiments are required to determine whether these apparently different actions on gating and conductance are through the same or separate mechanisms.

The anthracene carboxylic acid derivatives have long been known to block the muscle chloride current, which is contributed predominantly by CLC-1 (154–156). The particular compound 9-anthracene carboxylic acid (9-AC) is especially useful, but its blocking kinetics on CLC channels is slower than that of clofibric acid derivatives (145). These two types of compounds also show different blocking capabilities on various CLC channels. Taking advantage of the difference in inhibition by 9-AC between CLC-0, CLC-1, and CLC-2, the binding site of 9-AC in the CLC channel pore was recently mapped (59). In particular, a serine residue (S537) between helices O and P of CLC-1 was found to be crucial for high 9-AC blocking affinity. Standing on this basis, several residues were further identified through mutagenesis studies under the guidance of the bacterial CLC structure, and the mutational effects of these high-impact residues on the blocking affinities of 9-AC and CPA were compared. The results suggest that a partially hydrophobic pocket close to the Cl$^-$-binding site S$_{cen}$ may be responsible for both 9-AC and CPA blocks. Although the exact binding sites for these two compounds were not identical, they likely overlap with each other (59). Because the binding sites are accessible from the intracellular pore entrance, the blockers likely access the blocking sites through intracellular pore entrance.

CPA has also been used to probe the gating mechanism of CLC-0 (152, 153). It was found that the apparent blocking affinity of CPA on CLC-0 is voltage dependent. At a very negative potential, for example -140 mV, the affinity is ~ 1 mM. At $+60$ mV, the affinity decreases to ~ 65 mM. Such a large difference of CPA blocking affinity is attributed to a state-dependent binding of CPA to the channel— the blocker binds to the closed-state channel much better than to the open-state channel (152). The large difference in CPA affinity between the open and closed states of the channel was used to argue that the fast-gate opening involves not just a local swinging of the side chain of E166. Instead, a large conformational change at the inner-pore region during CLC-0 fast gating was suggested (152). In accordance with the large conformation rearrangement at the inner pore region during fast-gate opening, the CPA blocking affinities for E166A and E166S mutants, which are constantly open because their fast gate has been removed, are >200-fold higher than that of the wild-type channel (153). At -140 mV, the apparent K_D for CPA in these two mutants is ~ 4 μM. Thus, if fast gating consists of only the movement of the glutamate side chain, why does the E166A mutant, which is nearly open all the time, show a higher affinity to CPA block than the wild-type channel (106)?

The implication of a large conformational change upon fast gating based on studies of blockers is not without flaws. The MTS modification experiments have shown a prominent role for electrostatic potential in anion binding in the pore (58). CPA is a negatively charged molecule. The binding of CPA in the pore could be affected by the intrinsic electrostatic potential, which is likely influenced by the charge status at the position 166. More experiments are necessary to address this unsettled issue.

FUNCTIONS OF BACTERIAL CLC PROTEINS: TRANSPORTERS OR ION CHANNELS?

Very little was known about the functions of bacterial CLCs at the time when their molecular structures were solved. The first and the only functional study prior to the crystallographic era was performed using a flux measurement (23), a crude assay for studying the ion channel function. This functional study furnished two pieces of information: the homodimeric structure of the bacterial CLC molecule, and the permeability sequence for anions, which is roughly similar to those of CLC-0 and CLC-1: $Cl^-, Br^- > NO_3^- > I^-, F^- \gg H_2PO_4^-$. These two conclusions did not provide information beyond what had been known in vertebrate CLCs. However, the similarity in overall subunit stoichiometry and in ion permeation properties provided some confidence in using bacterial CLCs as a model molecule for the CLC channel family. This is especially useful in light of the difficulty in crystallizing vertebrate or mammalian CLC channels. However, regarding the functions of bacterial CLCs, the flux assay may not have given a satisfactory description of how bacterial CLC molecules work. Most recently, the functional assay using a higher time-resolution method, the lipid bilayer recording, was performed on

CLC molecules from *E. coli* with surprising findings (40, 41): CLC-ec1 allowed not only Cl^- but also H^+ to permeate through the protein. Furthermore, the reversal potential predicted from the Cl^- and H^+ concentration gradients by using the diffusion model (i.e., the conventional Goldman-Hodgkin-Katz equation) did not corroborate with the experimentally determined value. On the other hand, the calculation based on a $Cl^- $-$H^+$ antiporter model with a coupling transport ratio of 2 Cl^-:1 H^+ closely predicted the experimental results. The bacterial CLC was thus proposed to be a transporter but not an ion channel (41). The key glutamate residue, E148, again plays a critical role in the antiporter function. When the wild-type glutamate residue was mutated to alanine, the coupled transporter behavior described above disappeared: H^+ had negligible permeation ability through the mutant protein, and the reversal potential closely followed the equilibrium potential of Cl^- calculated from the Nernst equation—the mutant protein became a Cl^- channel or a Cl^- uniporter.

The surprising discovery raises an interesting question: Do vertebrate or mammalian CLCs also function as transporters? Clearly, the observed currents from muscle-type CLC channels are Cl^- fluxes through ion-diffusion pores because the reversal potentials for these channels follow faithfully the imposed Cl^- gradients and are not altered by external or internal pH (66, 85, 98, 99). In addition, recordings for the muscle-type CLC channels showed sizable single-channel conductance, which would be too large for the slow turnover rate of a transporter. It is unclear, however, whether members of the other two subbranches of CLC family may function as transporters. Mammalian CLC-3 to CLC-7 are now thought to reside in the membrane of intracellular organelles, and their functions may be related to the regulation of the intravesicular pH (157), a physiological role similar to that played by bacterial CLC proteins in maintaining the normal pH within *E. coli* (24). By overexpressing these proteins in heterologous expression systems, these other vertebrate CLC proteins can appear on the plasma membrane for electrophysiological recordings. However, no convincing single-channel recordings have been obtained to date from these two other branches of mammalian CLC members. It would be interesting to examine if Cl^- fluxes through these mammalian CLCs are also coupled to H^+ fluxes.

Another even more fundamental question arises with the discovery of the transporter function of bacterial CLC molecules: What is the fine line between ion channels and transporters, two types of molecules whose actions have been thought to be very different? If the overall molecular designs of bacterial CLCs and vertebrate CLCs are similar, as we believe, the distinction between ion channels and transporters must be very subtle. Indeed, recent studies have shown many examples of a close relation between transporters and ion channels. For example, the structural architecture of CFTR belongs to that of the widespread ABC transporters, yet this membrane protein exhibits a bona fide ion channel function (158, 159). The P-type ATPase is known to transform into an ion channel under the influence of toxins (160, 161). Several members of the glutamate transporter family also contain pore-like ion-permeation pathways to conduct Cl^- ions (162–166). In this last example, it was shown that the glutamate transport activity and the Cl^- conduction were mediated through two distinct conformational states of the protein (162). One hypothesis was that the monomer of homopentameric glutamate transporter could function as a transporter, whereas the assembly of monomers into a pentameric complex forms a central pore that is used to conduct Cl⁻ fluxes (163, 166). For CLC members, however, the key glutamate side chain appears to be critical both for the channel-gating function in vertebrate CLC channels and for the antiporter activity in bacterial CLC molecules. It would be important to examine if the Cl⁻ transport pathway for the antiporter functions in the bacterial CLC indeed follows the path shown in the high-resolution crystal structure, and if this pathway is the ion-conducting pathway in vertebrate CLC channels. Until these critical issues are solved, it may not be easy to define the separation line between ion channels and transporters.

CONCLUSION

The molecular functions of CLC members have been full of surprises. The unexpected findings of the double-barreled pore structure in the early 1980s, the discovery of gating-permeation coupling properties in the 1990s, and the surprising antiporter behavior recently found in bacterial CLC molecules have expanded the scope of research in the field. However, even though many functional properties as well as the structural features of CLC channels have been unveiled, the majority of the members in the family are still poorly understood. Studying their structural and functional properties remains a great challenge, especially for the nonmuscle-type CLC channels. However, the framework laid out by the function of the muscle-type CLC channels and by the structure of bacterial CLCs should serve as a compass by which to explore the uncharted CLC territory.

ACKNOWLEDGMENTS

I thank Drs. Bob Fairclough, Tzyh-Chang Hwang, Chris Miller, and Jim Trimmer for their comments on this review. I also thank Mary Edwards for stylistic advice in the preparation of the manuscript. The work of my laboratory is supported by a grant from the National Institutes of Health (GM-65447).

The *Annual Review of Physiology* is online at
http://physiol.annualreviews.org

LITERATURE CITED

1. Koch MC, Steinmeyer K, Lorenz C, Ricker K, Wolf F, et al. 1992. The skeletal muscle chloride channel in dominant and recessive human myotonia. *Science* 257:797–800

2. Jentsch TJ, Lorenz C, Pusch M, Steinmeyer K. 1995. Myotonias due to CLC-1 chloride channel mutations. *Soc. Gen. Physiol. Ser.* 50:149–59

3. Kubisch C, Schmidt-Rose T, Fontaine B,

Bretag AH, Jentsch TJ. 1998. ClC-1 chloride channel mutations in myotonia congenita: variable penetrance of mutations shifting the voltage dependence. *Hum. Mol. Genet.* 7:1753–60

4. Zhang J, Bendahhou S, Sanguinetti MC, Ptacek LJ. 2000. Functional consequences of chloride channel gene (CLCN1) mutations causing myotonia congenita. *Neurology* 54:937–42

5. George AL Jr, Crackower MA, Abdalla JA, Hudson AJ, Ebers GC. 1993. Molecular basis of Thomsen's disease (autosomal dominant myotonia congenita). *Nat. Genet.* 3:305–10

6. Lloyd SE, Gunther W, Pearce SH, Thomson A, Bianchi ML, et al. 1997. Characterisation of renal chloride channel, CLCN5, mutations in hypercalciuric nephrolithiasis (kidney stones) disorders. *Hum. Mol. Genet.* 6:1233–39

7. Lloyd SE, Pearce SH, Fisher SE, Steinmeyer K, Schwappach B, et al. 1996. A common molecular basis for three inherited kidney stone diseases. *Nature* 379:445–49

8. George AL Jr. 1998. Chloride channels and endocytosis: ClC-5 makes a dent. *Proc. Natl. Acad. Sci. USA* 95:7843–45

9. Konrad M, Vollmer M, Lemmink HH, van den Heuvel LP, Jeck N, et al. 2000. Mutations in the chloride channel gene CLC-NKB as a cause of classic Bartter syndrome. *J. Am. Soc. Nephrol.* 11:1449–59

10. Thakker RV. 2000. Molecular pathology of renal chloride channels in Dent's disease and Bartter's syndrome. *Exp. Nephrol.* 8:351–60

11. Hebert SC. 2003. Bartter syndrome. *Curr. Opin. Nephrol. Hypertens.* 12:527–32

12. Kornak U, Kasper D, Bosl MR, Kaiser E, Schweizer M, et al. 2001. Loss of the ClC-7 chloride channel leads to osteopetrosis in mice and man. *Cell* 104:205–15

13. Haug K, Warnstedt M, Alekov AK, Sander T, Ramirez A, et al. 2003. Mutations in CLCN2 encoding a voltage-gated chloride channel are associated with idiopathic generalized epilepsies. *Nat. Genet.* 33:527–32

14. Stobrawa SM, Breiderhoff T, Takamori S, Engel D, Schweizer M, et al. 2001. Disruption of ClC-3, a chloride channel expressed on synaptic vesicles, leads to a loss of the hippocampus. *Neuron* 29:185–96

15. Bosl MR, Stein V, Hubner C, Zdebik AA, Jordt SE, et al. 2001. Male germ cells and photoreceptors, both dependent on close cell-cell interactions, degenerate upon ClC-2 Cl⁻ channel disruption. *EMBO J.* 20:1289–99

16. Jentsch TJ, Friedrich T, Schriever A, Yamada H. 1999. The CLC chloride channel family. *Pflügers Arch.* 437:783–95

17. Jentsch TJ, Stein V, Weinreich F, Zdebik AA. 2002. Molecular structure and physiological function of chloride channels. *Physiol. Rev.* 82:503–68

18. Maduke M, Miller C, Mindell JA. 2000. A decade of CLC chloride channels: structure, mechanism, and many unsettled questions. *Annu. Rev. Biophys. Biomol. Struct.* 29:411–38

19. Waldegger S, Jentsch TJ. 2000. From tonus to tonicity: physiology of CLC chloride channels. *J. Am. Soc. Nephrol.* 11:1331–39

20. Jentsch TJ, Poët M. Fuhrmann JC, Zdebik AA. 2005. Physiological functions of ClC Cl⁻ channels gleaned from human genetic disease and mouse models. *Annu. Rev. Physiol.* 67:779–807

21. Jentsch TJ, Steinmeyer K, Schwarz G. 1990. Primary structure of *Torpedo marmorata* chloride channel isolated by expression cloning in *Xenopus* oocytes. *Nature* 348:510–14

22. Purdy MD, Wiener MC. 2000. Expression, purification, and initial structural characterization of YadQ, a bacterial homolog of mammalian ClC chloride channel proteins. *FEBS Lett.* 466:26–28

23. Maduke M, Pheasant DJ, Miller C. 1999. High-level expression, functional reconstitution, and quaternary structure of a

prokaryotic ClC-type chloride channel. *J. Gen. Physiol.* 114:713–22

24. Iyer R, Iverson TM, Accardi A, Miller C. 2002. A biological role for prokaryotic ClC chloride channels. *Nature* 419:715–18

25. Schwappach B, Stobrawa S, Hechenberger M, Steinmeyer K, Jentsch TJ. 1998. Golgi localization and functionally important domains in the NH2 and COOH terminus of the yeast CLC putative chloride channel Gef1p. *J. Biol. Chem.* 273:15110–18

26. Gaxiola RA, Yuan DS, Klausner RD, Fink GR. 1998. The yeast CLC chloride channel functions in cation homeostasis. *Proc. Natl. Acad. Sci. USA* 95:4046–50

27. Hechenberger M, Schwappach B, Fischer WN, Frommer WB, Jentsch TJ, Steinmeyer K. 1996. A family of putative chloride channels from Arabidopsis and functional complementation of a yeast strain with a CLC gene disruption. *J. Biol. Chem.* 271:33632–38

28. Schriever AM, Friedrich T, Pusch M, Jentsch TJ. 1999. CLC chloride channels in *Caenorhabditis elegans*. *J. Biol. Chem.* 274:34238–44

29. Strange K. 2002. Of mice and worms: novel insights into ClC-2 anion channel physiology. *News Physiol. Sci.* 17:11–16

30. Bianchi L, Miller DM 3rd, George AL Jr. 2001. Expression of a ClC chloride channel in *Caenorhabditis elegans* gamma-aminobutyric acid-ergic neurons. *Neurosci. Lett.* 299:177–80

31. Nehrke K, Begenisich T, Pilato J, Melvin JE. 2000. Into ion channel and transporter function. *Caenorhabditis elegans* ClC-type chloride channels: novel variants and functional expression. *Am. J. Physiol. Cell Physiol.* 279:C2052–66

32. Gunther W, Piwon N, Jentsch TJ. 2003. The ClC-5 chloride channel knock-out mouse—an animal model for Dent's disease. *Pflügers Arch.* 445:456–62

33. Mohammad-Panah R, Harrison R, Dhani S, Ackerley C, Huan LJ, et al. 2003. The chloride channel ClC-4 contributes to endosomal acidification and trafficking. *J. Biol. Chem.* 278:29267–77

34. Duan D, Winter C, Cowley S, Hume JR, Horowitz B. 1997. Molecular identification of a volume-regulated chloride channel. *Nature* 390:417–21

35. Duan D, Zhong J, Hermoso M, Satterwhite CM, Rossow CF, et al. 2001. Functional inhibition of native volume-sensitive outwardly rectifying anion channels in muscle cells and *Xenopus* oocytes by anti-ClC-3 antibody. *J. Physiol.* 531:437–44

36. Yamamoto-Mizuma S, Wang GX, Liu LL, Schegg K, Hatton WJ, et al. 2004. Altered properties of volume-sensitive osmolyte and anion channels (VSOACs) and membrane protein expression in cardiac and smooth muscle myocytes from Clcn3$^{-/-}$ mice. *J. Physiol.* 557:439–56

37. Majid A, Brown PD, Best L, Park K. 2001. Expression of volume-sensitive Cl$^-$ channels and ClC-3 in acinar cells isolated from the rat lacrimal gland and submandibular salivary gland. *J. Physiol.* 534:409–21

38. Weylandt KH, Valverde MA, Nobles M, Raguz S, Amey JS, et al. 2001. Human ClC-3 is not the swelling-activated chloride channel involved in cell volume regulation. *J. Biol. Chem.* 276:17461–67

39. Brandt S, Jentsch TJ. 1995. ClC-6 and ClC-7 are two novel broadly expressed members of the CLC chloride channel family. *FEBS Lett.* 377:15–20

40. Accardi A, Kolmakova-Partensky L, Williams C, Miller C. 2004. Ionic currents mediated by a prokaryotic homologue of CLC Cl$^-$ channels. *J. Gen. Physiol.* 123:109–19

41. Accardi A, Miller C. 2004. Secondary active transport mediated by a prokaryotic homologue of CLC Cl$^-$ channels. *Nature* 427:803–7

42. Mindell JA, Maduke M, Miller C, Grigorieff N. 2001. Projection structure of a

ClC-type chloride channel at 6.5 Å resolution. *Nature* 409:219–23

43. Dutzler R, Campbell EB, Cadene M, Chait BT, MacKinnon R. 2002. X-ray structure of a ClC chloride channel at 3.0 Å reveals the molecular basis of anion selectivity. *Nature* 415:287–94

44. Dutzler R, Campbell EB, MacKinnon R. 2003. Gating the selectivity filter in ClC chloride channels. *Science* 300:108–12

45. O'Neill GP, Grygorczyk R, Adam M, Ford-Hutchinson AW. 1991. The nucleotide sequence of a voltage-gated chloride channel from the electric organ of *Torpedo californica*. *Biochim. Biophys. Acta* 1129:131–34

46. Steinmeyer K, Ortland C, Jentsch TJ. 1991. Primary structure and functional expression of a developmentally regulated skeletal muscle chloride channel. *Nature* 354:301–4

47. Thiemann A, Grunder S, Pusch M, Jentsch TJ. 1992. A chloride channel widely expressed in epithelial and non-epithelial cells. *Nature* 356:57–60

48. Bauer CK, Steinmeyer K, Schwarz JR, Jentsch TJ. 1991. Completely functional double-barreled chloride channel expressed from a single *Torpedo* cDNA. *Proc. Natl. Acad. Sci. USA* 88:11052–56

49. Lin CW, Chen TY. 2000. Cysteine modification of a putative pore residue in ClC-0: implication for the pore stoichiometry of ClC chloride channels. *J. Gen. Physiol.* 116:535–46

50. Ludewig U, Pusch M, Jentsch TJ. 1996. Two physically distinct pores in the dimeric ClC-0 chloride channel. *Nature* 383:340–43

51. Ludewig U, Pusch M, Jentsch TJ. 1997. Independent gating of single pores in CLC-0 chloride channels. *Biophys. J.* 73:789–97

52. Middleton RE, Pheasant DJ, Miller C. 1994. Purification, reconstitution, and subunit composition of a voltage-gated chloride channel from *Torpedo* electroplax. *Biochemistry* 33:13189–98

53. Middleton RE, Pheasant DJ, Miller C. 1996. Homodimeric architecture of a ClC-type chloride ion channel. *Nature* 383:337–40

54. Miller C. 1982. Open-state substructure of single chloride channels from *Torpedo electroplax*. *Philos. Trans. R. Soc. London Ser. B* 299:401–11

55. Miller C, White MM. 1984. Dimeric structure of single chloride channels from *Torpedo electroplax*. *Proc. Natl. Acad. Sci. USA* 81:2772–75

56. Weinreich F, Jentsch TJ. 2001. Pores formed by single subunits in mixed dimers of different CLC chloride channels. *J. Biol. Chem.* 276:2347–53

57. Faraldo-Gomez JD, Roux B. 2004. Electrostatics of ion stabilization in a ClC chloride channel homologue from *Escherichia coli*. *J. Mol. Biol.* 339:981–1000

58. Lin CW, Chen TY. 2003. Probing the pore of ClC-0 by substituted cysteine accessibility method using methane thiosulfonate reagents. *J. Gen. Physiol.* 122:147–59

59. Estevez R, Schroeder BC, Accardi A, Jentsch TJ, Pusch M. 2003. Conservation of chloride channel structure revealed by an inhibitor binding site in ClC-1. *Neuron* 38:47–59

60. Miller C. 2003. ClC channels: reading eukaryotic function through prokaryotic spectacles. *J. Gen. Physiol.* 122:129–31

61. Estevez R, Boettger T, Stein V, Birkenhager R, Otto E, et al. 2001. Barttin is a Cl^- channel beta-subunit crucial for renal Cl^- reabsorption and inner ear K^+ secretion. *Nature* 414:558–61

62. Waldegger S, Jeck N, Barth P, Peters M, Vitzthum H, et al. 2002. Barttin increases surface expression and changes current properties of ClC-K channels. *Pflügers Arch.* 444:411–18

63. Vanoye CG, George AL Jr. 2002. Functional characterization of recombinant human ClC-4 chloride channels in cultured mammalian cells. *J. Physiol.* 539:373–83

64. Friedrich T, Breiderhoff T, Jentsch TJ.

1999. Mutational analysis demonstrates that ClC-4 and ClC-5 directly mediate plasma membrane currents. *J. Biol. Chem.* 274:896–902

65. Uchida S, Sasaki S. 2005. Function of chloride channels in the kidney. *Annu. Rev. Physiol.* 67:759–78

66. Hanke W, Miller C. 1983. Single chloride channels from *Torpedo electroplax*. Activation by protons. *J. Gen. Physiol.* 82:25–45

67. Saviane C, Conti F, Pusch M. 1999. The muscle chloride channel ClC-1 has a double-barreled appearance that is differentially affected in dominant and recessive myotonia. *J. Gen. Physiol.* 113:457–68

68. Miller C, White MM. 1980. A voltage-dependent chloride conductance channel from *Torpedo electroplax* membrane. *Ann. NY Acad. Sci.* 341:534–51

69. White MM, Miller C. 1981. Chloride permeability of membrane vesicles isolated from *Torpedo californica electroplax*. *Biophys. J.* 35:455–62

70. Richard EA, Miller C. 1990. Steady-state coupling of ion-channel conformations to a transmembrane ion gradient. *Science* 247:1208–10

71. Chen TY, Miller C. 1996. Nonequilibrium gating and voltage dependence of the ClC-0 Cl⁻ channel. *J. Gen. Physiol.* 108:237–50

72. Pusch M, Ludewig U, Jentsch TJ. 1997. Temperature dependence of fast and slow gating relaxations of ClC-0 chloride channels. *J. Gen. Physiol.* 109:105–16

73. Chen TY. 1998. Extracellular zinc ion inhibits ClC-0 chloride channels by facilitating slow gating. *J. Gen. Physiol.* 112:715–26

74. Lin YW, Lin CW, Chen TY. 1999. Elimination of the slow gating of ClC-0 chloride channel by a point mutation. *J. Gen. Physiol.* 114:1–12

75. Fong P, Rehfeldt A, Jentsch TJ. 1998. Determinants of slow gating in ClC-0, the voltage-gated chloride channel of *Tor-*

pedo marmorata. *Am. J. Physiol. Cell Physiol.* 274:C966–73

76. Hryciw DH, Rychkov GY, Hughes BP, Bretag AH. 1998. Relevance of the D13 region to the function of the skeletal muscle chloride channel, ClC-1. *J. Biol. Chem.* 273:4304–7

77. Hebeisen S, Biela A, Giese B, Muller-Newen G, Hidalgo P, Fahlke C. 2004. The role of the carboxy-terminus in ClC chloride channel function. *J. Biol. Chem.* 279:13140–47

78. Estevez R, Pusch M, Ferrer-Costa C, Orozco M, Jentsch TJ. 2004. Functional and structural conservation of CBS domains from CLC channels. *J. Physiol.* 557:363–78

79. Maduke M, Williams C, Miller C. 1998. Formation of CLC-0 chloride channels from separated transmembrane and cytoplasmic domains. *Biochemistry* 37:1315–21

80. Schmidt-Rose T, Jentsch TJ. 1997. Reconstitution of functional voltage-gated chloride channels from complementary fragments of CLC-1. *J. Biol. Chem.* 272:20515–21

81. Accardi A, Pusch M. 2000. Fast and slow gating relaxations in the muscle chloride channel CLC-1. *J. Gen. Physiol.* 116:433–44

82. Aromataris EC, Rychkov GY, Bennetts B, Hughes BP, Bretag AH, Roberts ML. 2001. Fast and slow gating of CLC-1: differential effects of 2-(4-chlorophenoxy) propionic acid and dominant negative mutations. *Mol. Pharmacol.* 60:200–8

83. Bennetts B, Roberts ML, Bretag AH, Rychkov GY. 2001. Temperature dependence of human muscle CLC-1 chloride channel. *J. Physiol.* 535:83–93

84. Accardi A, Ferrera L, Pusch M. 2001. Drastic reduction of the slow gate of human muscle chloride channel (ClC-1) by mutation C277S. *J. Physiol.* 534:745–52

85. Rychkov GY, Astill DS, Bennetts B, Hughes BP, Bretag AH, Roberts ML. 1997. pH-dependent interactions of Cd²⁺

and a carboxylate blocker with the rat ClC-1 chloride channel and its R304E mutant in the Sf-9 insect cell line. *J. Physiol.* 501:355–62

86. Kurz LL, Klink H, Jakob I, Kuchenbecker M, Benz S, et al. 1999. Identification of three cysteines as targets for the Zn^{2+} blockade of the human skeletal muscle chloride channel. *J. Biol. Chem.* 274:11687–92

87. Kurz L, Wagner S, George AL Jr, Rudel R. 1997. Probing the major skeletal muscle chloride channel with Zn^{2+} and other sulfhydryl-reactive compounds. *Pflügers Arch.* 433:357–63

88. Duffield M, Rychkov G, Bretag A, Roberts M. 2003. Involvement of helices at the dimer interface in ClC-1 common gating. *J. Gen. Physiol.* 121:149–61

89. Grunder S, Thiemann A, Pusch M, Jentsch TJ. 1992. Regions involved in the opening of ClC-2 chloride channel by voltage and cell volume. *Nature* 360:759–62

90. Jordt SE, Jentsch TJ. 1997. Molecular dissection of gating in the ClC-2 chloride channel. *EMBO J.* 16:1582–92

91. Hoshi T, Zagotta WN, Aldrich RW. 1990. Biophysical and molecular mechanisms of Shaker potassium channel inactivation. *Science* 250:533–38

92. Zagotta WN, Hoshi T, Aldrich RW. 1990. Restoration of inactivation in mutants of Shaker potassium channels by a peptide derived from ShB. *Science* 250:568–71

93. Choi KL, Aldrich RW, Yellen G. 1991. Tetraethylammonium blockade distinguishes two inactivation mechanisms in voltage-activated K^+ channels. *Proc. Natl. Acad. Sci. USA* 88:5092–95

94. Demo SD, Yellen G. 1991. The inactivation gate of the Shaker K^+ channel behaves like an open-channel blocker. *Neuron* 7:743–53

95. Varela D, Niemeyer MI, Cid LP, Sepulveda FV. 2002. Effect of an N-terminus deletion on voltage-dependent gating of the ClC-2 chloride channel. *J. Physiol.* 544:363–72

96. Zuniga L, Niemeyer MI, Varela D, Catalan M, Cid LP, Sepulveda FV. 2004. The voltage-dependent ClC-2 chloride channel has a dual gating mechanism. *J. Physiol.* 555:671–82

97. Pusch M, Ludewig U, Rehfeldt A, Jentsch TJ. 1995. Gating of the voltage-dependent chloride channel ClC-0 by the permeant anion. *Nature* 373:527–31

98. Rychkov GY, Pusch M, Astill DS, Roberts ML, Jentsch TJ, Bretag AH. 1996. Concentration and pH dependence of skeletal muscle chloride channel ClC-1. *J. Physiol.* 497:423–35

99. Chen MF, Chen TY. 2001. Different fast-gate regulation by external Cl^- and H^+ of the muscle-type ClC chloride channels. *J. Gen. Physiol.* 118:23–32

100. Jiang Y, Ruta V, Chen J, Lee A, MacKinnon R. 2003. The principle of gating charge movement in a voltage-dependent K^+ channel. *Nature* 423:42–48

101. Jiang Y, Lee A, Chen J, Ruta V, Cadene M, et al. 2003. X-ray structure of a voltage-dependent K^+ channel. *Nature* 423:33–41

102. Fahlke C, Rosenbohm A, Mitrovic N, George AL Jr, Rudel R. 1996. Mechanism of voltage-dependent gating in skeletal muscle chloride channels. *Biophys. J.* 71:695–706

103. Fahlke C, Rudel R, Mitrovic N, Zhou M, George AL Jr. 1995. An aspartic acid residue important for voltage-dependent gating of human muscle chloride channels. *Neuron* 15:463–72

104. Ludewig U, Jentsch TJ, Pusch M. 1997. Inward rectification in ClC-0 chloride channels caused by mutations in several protein regions. *J. Gen. Physiol.* 110:165–71

105. Rychkov GY, Pusch M, Roberts ML, Jentsch TJ, Bretag AH. 1998. Permeation and block of the skeletal muscle chloride channel, ClC-1, by foreign anions. *J. Gen. Physiol.* 111:653–65

106. Pusch M. 2004. Structural insights into chloride and proton-mediated gating of

CLC chloride channels. *Biochemistry* 43:1135–44

107. Pusch M, Jordt SE, Stein V, Jentsch TJ. 1999. Chloride dependence of hyperpolarization-activated chloride channel gates. *J. Physiol.* 515:341–53

108. Niemeyer MI, Cid LP, Zuniga L, Catalan M, Sepulveda FV. 2003. A conserved pore-lining glutamate as a voltage- and chloride-dependent gate in the ClC-2 chloride channel. *J. Physiol.* 553:873–79

109. Fahlke C, Rhodes TH, Desai RR, George AL Jr. 1998. Pore stoichiometry of a voltage-gated chloride channel. *Nature* 394:687–90

110. Warnstedt M, Sun C, Poser B, Escriva MJ, Tranebjaerg L, et al. 2002. The myotonia congenita mutation A331T confers a novel hyperpolarization-activated gate to the muscle chloride channel ClC-1. *J. Neurosci.* 22:7462–70

111. Zhang J, Sanguinetti MC, Kwiecinski H, Ptacek LJ. 2000. Mechanism of inverted activation of ClC-1 channels caused by a novel myotonia congenita mutation. *J. Biol. Chem.* 275:2999–3005

112. Chen TY, Chen MF, Lin CW. 2003. Electrostatic control and chloride regulation of the fast gating of ClC-0 chloride channels. *J. Gen. Physiol.* 122:641–51

113. Swenson RP Jr, Armstrong CM. 1981. K^+ channels close more slowly in the presence of external K^+ and Rb^+. *Nature* 291:427–29

114. Fahlke C, Durr C, George AL Jr. 1997. Mechanism of ion permeation in skeletal muscle chloride channels. *J. Gen. Physiol.* 110:551–64

115. Fahlke C, Yu HT, Beck CL, Rhodes TH, George AL Jr. 1997. Pore-forming segments in voltage-gated chloride channels. *Nature* 390:529–32

116. Ludewig U, Jentsch TJ, Pusch M. 1997. Analysis of a protein region involved in permeation and gating of the voltage-gated *Torpedo* chloride channel ClC-0. *J. Physiol.* 498:691–702

117. Hebeisen S, Heidtmann H, Cosmelli D,

Gonzalez C, Poser B, et al. 2003. Anion permeation in human ClC-4 channels. *Biophys. J.* 84:2306–18

118. Neyton J, Miller C. 1988. Discrete Ba^{2+} block as a probe of ion occupancy and pore structure in the high-conductance Ca^{2+}-activated K^+ channel. *J. Gen. Physiol.* 92:569–86

119. Neyton J, Miller C. 1988. Potassium blocks barium permeation through a calcium-activated potassium channel. *J. Gen. Physiol.* 92:549–67

120. Kuo CC, Hess P. 1993. Block of the L-type Ca^{2+} channel pore by external and internal Mg^{2+} in rat phaeochromocytoma cells. *J. Physiol.* 466:683–706

121. Kuo CC, Hess P. 1993. Characterization of the high-affinity Ca^{2+} binding sites in the L-type Ca^{2+} channel pore in rat phaeochromocytoma cells. *J. Physiol.* 466:657–82

122. Kuo CC, Hess P. 1993. Ion permeation through the L-type Ca^{2+} channel in rat phaeochromocytoma cells: two sets of ion binding sites in the pore. *J. Physiol.* 466:629–55

123. Pusch M, Steinmeyer K, Jentsch TJ. 1994. Low single channel conductance of the major skeletal muscle chloride channel, ClC-1. *Biophys. J.* 66:149–52

124. Miloshevsky GV, Jordan PC. 2004. Anion pathway and potential energy profiles along curvilinear bacterial ClC Cl^- pores: electrostatic effects of charged residues. *Biophys. J.* 86:825–35

125. Cohen J, Schulten K. 2004. Mechanism of anionic conduction across ClC. *Biophys. J.* 86:836–45

126. Corry B, O'Mara M, Chung SH. 2004. Conduction mechanisms of chloride ions in ClC-type channels. *Biophys. J.* 86:846–60

127. Steinmeyer K, Lorenz C, Pusch M, Koch MC, Jentsch TJ. 1994. Multimeric structure of ClC-1 chloride channel revealed by mutations in dominant myotonia congenita (Thomsen). *EMBO J.* 13:737–43

128. Pusch M, Steinmeyer K, Koch MC,

Jentsch TJ. 1995. Mutations in dominant human myotonia congenita drastically alter the voltage dependence of the ClC-1 chloride channel. *Neuron* 15:1455–63

129. Waldegger S, Jentsch TJ. 2000. Functional and structural analysis of ClC-K chloride channels involved in renal disease. *J. Biol. Chem.* 275:24527–33

130. Chen TY. 2003. Coupling gating with ion permeation in ClC channels. *SciSTKE* 2003:pe23

131. Akabas MH, Stauffer DA, Xu M, Karlin A. 1992. Acetylcholine receptor channel structure probed in cysteine-substitution mutants. *Science* 258:307–10

132. Liu Y, Holmgren M, Jurman ME, Yellen G. 1997. Gated access to the pore of a voltage-dependent K^+ channel. *Neuron* 19:175–84

133. Yang N, George AL Jr, Horn R. 1996. Molecular basis of charge movement in voltage-gated sodium channels. *Neuron* 16:113–22

134. Larsson HP, Baker OS, Dhillon DS, Isacoff EY. 1996. Transmembrane movement of the shaker K^+ channel S4. *Neuron* 16:387–97

135. Wilson GG, Pascual JM, Brooijmans N, Murray D, Karlin A. 2000. The intrinsic electrostatic potential and the intermediate ring of charge in the acetylcholine receptor channel. *J. Gen. Physiol.* 115:93–106

136. Pascual JM, Karlin A. 1998. State-dependent accessibility and electrostatic potential in the channel of the acetylcholine receptor. Inferences from rates of reaction of thiosulfonates with substituted cysteines in the M2 segment of the alpha subunit. *J. Gen. Physiol.* 111:717–39

137. Fahlke C, Desai RR, Gillani N, George AL Jr. 2001. Residues lining the inner pore vestibule of human muscle chloride channels. *J. Biol. Chem.* 276:1759–65

138. Chen MF, Chen TY. 2003. Side-chain charge effects and conductance determinants in the pore of ClC-0 chloride channels. *J. Gen. Physiol.* 122:133–45

139. MacKinnon R, Latorre R, Miller C. 1989. Role of surface electrostatics in the operation of a high-conductance Ca^{2+}-activated K^+ channel. *Biochemistry* 28:8092–99

140. MacKinnon R, Miller C. 1989. Functional modification of a Ca^{2+}-activated K^+ channel by trimethyloxonium. *Biochemistry* 28:8087–92

141. MacKinnon R, Miller C. 1989. Mutant potassium channels with altered binding of charybdotoxin, a pore-blocking peptide inhibitor. *Science* 245:1382–85

142. MacKinnon R, Miller C. 1988. Mechanism of charybdotoxin block of the high-conductance, Ca^{2+}-activated K^+ channel. *J. Gen. Physiol.* 91:335–49

143. Yellen G, Jurman ME, Abramson T, MacKinnon R. 1991. Mutations affecting internal TEA blockade identify the probable pore-forming region of a K^+ channel. *Science* 251:939–42

144. MacKinnon R, Yellen G. 1990. Mutations affecting TEA blockade and ion permeation in voltage-activated K^+ channels. *Science* 250:276–79

145. Pusch M, Accardi A, Liantonio A, Guida P, Traverso S, et al. 2002. Mechanisms of block of muscle type CLC chloride channels. *Mol. Membr. Biol.* 19:285–92

146. Pusch M, Accardi A, Liantonio A, Ferrera L, De Luca A, et al. 2001. Mechanism of block of single protopores of the *Torpedo* chloride channel ClC-0 by 2-(p-chlorophenoxy)butyric acid (CPB). *J. Gen. Physiol.* 118:45–62

147. De Luca A, Tricarico D, Wagner R, Bryant SH, Tortorella V, Conte Camerino D. 1992. Opposite effects of enantiomers of clofibric acid derivative on rat skeletal muscle chloride conductance: antagonism studies and theoretical modeling of two different receptor site interactions. *J. Pharmacol. Exp. Ther.* 260:364–68

148. Pusch M, Liantonio A, Bertorello L, Accardi A, De Luca A, et al. 2000. Pharmacological characterization of chloride channels belonging to the ClC family by

the use of chiral clofibric acid derivatives. *Mol. Pharmacol.* 58:498–507

149. Liantonio A, De Luca A, Pierno S, Didonna MP, Loiodice F, et al. 2003. Structural requisites of 2-(*p*-chlorophenoxy) propionic acid analogues for activity on native rat skeletal muscle chloride conductance and on heterologously expressed CLC-1. *Br. J. Pharmacol.* 139:1255–64

150. Liantonio A, Accardi A, Carbonara G, Fracchiolla G, Loiodice F, et al. 2002. Molecular requisites for drug binding to muscle CLC-1 and renal CLC-K channel revealed by the use of phenoxyalkyl derivatives of 2-(*p*-chlorophenoxy) propionic acid. *Mol. Pharmacol.* 62:265–71

151. Aromataris EC, Astill DS, Rychkov GY, Bryant SH, Bretag AH, Roberts ML. 1999. Modulation of the gating of ClC-1 by S-(-) 2-(4-chlorophenoxy) propionic acid. *Br. J. Pharmacol.* 126:1375–82

152. Accardi A, Pusch M. 2003. Conformational changes in the pore of CLC-0. *J. Gen. Physiol.* 122:277–93

153. Traverso S, Elia L, Pusch M. 2003. Gating competence of constitutively open CLC-0 mutants revealed by the interaction with a small organic inhibitor. *J. Gen. Physiol.* 122:295–306

154. Harris GL, Betz WJ. 1987. Evidence for active chloride accumulation in normal and denervated rat lumbrical muscle. *J. Gen. Physiol.* 90:127–44

155. Betz WJ, Caldwell JH, Harris GL. 1986. Effect of denervation on a steady electric current generated at the end-plate region of rat skeletal muscle. *J. Physiol.* 373:97–114

156. Betz WJ, Caldwell JH, Kinnamon SC. 1984. Physiological basis of a steady endogenous current in rat lumbrical muscle. *J. Gen. Physiol.* 83:175–92

157. Li X, Wang T, Zhao Z, Weinman SA. 2002. The ClC-3 chloride channel promotes acidification of lysosomes in CHO-K1 and Huh-7 cells. *Am. J. Physiol. Cell Physiol.* 282:C1483–91

158. Zhou Z, Hu S, Hwang TC. 2002. Probing an open CFTR pore with organic anion blockers. *J. Gen. Physiol.* 120:647–62

159. Zou XQ, Hwang TC. 2001. ATP hydrolysis-coupled gating of CFTR chloride channels: structure and function. *Biochemistry* 40:5579–86

160. Artigas P, Gadsby DC. 2003. Na^+/K^+-pump ligands modulate gating of palytoxin-induced ion channels. *Proc. Natl. Acad. Sci. USA* 100:501–5

161. Artigas P, Gadsby DC. 2002. Ion channel-like properties of the Na^+/K^+ pump. *Ann. NY Acad. Sci.* 976:31–40

162. Ryan RM, Vandenberg RJ. 2002. Distinct conformational states mediate the transport and anion channel properties of the glutamate transporter EAAT-1. *J. Biol. Chem.* 277:13494–500

163. Slotboom DJ, Konings WN, Lolkema JS. 2001. Glutamate transporters combine transporter- and channel-like features. *Trends Biochem. Sci.* 26:534–39

164. Fairman WA, Vandenberg RJ, Arriza JL, Kavanaugh MP, Amara SG. 1995. An excitatory amino-acid transporter with properties of a ligand-gated chloride channel. *Nature* 375:599–603

165. Wadiche JI, Amara SG, Kavanaugh MP. 1995. Ion fluxes associated with excitatory amino acid transport. *Neuron* 15:721–28

166. Eskandari S, Kreman M, Kavanaugh MP, Wright EM, Zampighi GA. 2000. Pentameric assembly of a neuronal glutamate transporter. *Proc. Natl. Acad. Sci. USA* 97:8641–46

Subject Index

and, 636
Escherichia coli
  cellular stress response
    and, 226, 235, 242,
    244–45
  CLC channels and, 809,
    811, 829
  colonic crypt transport and,
    477, 482
Esophagus
  intestinal Na^+/H^+
    exchange and, 416
Estrogen
  endocrinology of stress
    response and, 271
  ligand control of
    coregulator recruitment to
    nuclear receptors and,
    310–16, 319–24
Estrogen and progesterone
    signaling
  allosteric modulators,
    346–47
  bone, 345–46
  cardiovascular system,
    338–41
  cell membrane, 354–57
  CNS, 341–44
  coupling with signal
    pathways, 357–62
  endocrine pancreas, 344
  estrogen target tissues,
    344–46
  $GABA_A$, 346–47
  gene transcription, 336–37
  germ cell maturation,
    347–54
  introduction, 335–36
  macrophages, 344–45
  oocyte maturation, 350–53
  ovarian granulosa cells,
    354
  oxytocin receptors, 346–47
  progesterone/progesterone
    metabolites, 341, 346–47
  rapid extranuclear and
    nuclear signaling, 362–64

signal transduction, 341–44
sperm cell acrosome
  reaction, 347–50
steroid hormones, 336–38
steroid receptor
  localization, 354–57
subcellular targeting, 357
summary, 364–66
triggering responses,
  357–62
Estrogen biology
  knockout and transgenic
    animals
  AP-1 tethered mode of
    ER signaling, 295
  ArKO, 294–95
  breast cancer models,
    301–3
  converging signals,
    299–301
  ER, 301–3
  $ER\beta$, 293–94
  ER-growth factor cross
    talk, 298
  ER-null environment,
    294–95
  ERKO model, 288–94,
    298
  estrogen-free
    environment, 294–95
  estrogen receptor
    molecule, 285–87
  implantation-associated
    signals, 299
  male fertility, 293
  mechanisms, 285–87,
    295–97
  mouse models, 287–88,
    295–97
  NERK1, 295
  ovary, 292–93
  PRKO model, 288–92
  structures, 285–87
  summary, 303
N-Ethylmaleimide
  bicarbonate secretion in
    pancreatic duct and, 382

"Evo-devo" paradigm
  comparative developmental
    physiology and, 210–11
Evolution
  cellular stress response
    and, 225–47
  comparative developmental
    physiology and, 203–17
Excitability
  Ca^{2+}-activated Cl^-
    channels and, 719,
    722–23
Excitation-contraction (EC)
    coupling
  intracellular calcium
    release channels and
    cardiac disease, 69,
    78–82, 86
Exercise
  endocrinology of stress
    response and, 267, 273
Exocrine glands
  Ca^{2+}-activated Cl^-
    channels and, 725
Exocytosis
  CLC Cl^- channels and,
    795
  lung surfactant and
    ALF, 605–8
    Ca^{2+}, 598–603
    Ca^{2+}-independent
      pathways, 602–3
    compound exocytosis,
      604–5
    convertase, 609
    cycling, 608–9
    fusion pore expansion,
      603–4, 606–7
    introduction, 595–612
    lamellar body fusion,
      598, 601–5
    mechanistic model,
      601–2
    modes of stimulation,
      597–98
    post-fusion phase,
      603–4

CUMULATIVE INDEXES

CONTRIBUTING AUTHORS, VOLUMES 63–67

CHAPTER TITLES, VOLUMES 63–67

Comparative Physiology

Endocrinology

Gastrointestinal Physiology

Neurophysiology

Respiratory Physiology

Muscle Physiology

Special Topics

Chloride Channels

ANNUAL REVIEWS

Intelligent Synthesis of the Scientific Literature

Annual Reviews – Your Starting Point for Research Online
http://arjournals.annualreviews.org

- Over 900 Annual Reviews volumes—more than 25,000 critical, authoritative review articles in 31 disciplines spanning the Biomedical, Physical, and Social sciences—available online, including all Annual Reviews back volumes, dating to 1932

- Current individual subscriptions include seamless online access to full-text articles, PDFs, Reviews in Advance (as much as 6 months ahead of print publication), bibliographies, and other supplementary material in the current volume and the prior 4 years' volumes

- All articles are fully supplemented, searchable, and downloadable—see http://physiol.annualreviews.org

- Access links to the reviewed references (when available online)

- Site features include customized alerting services, citation tracking, and saved searches

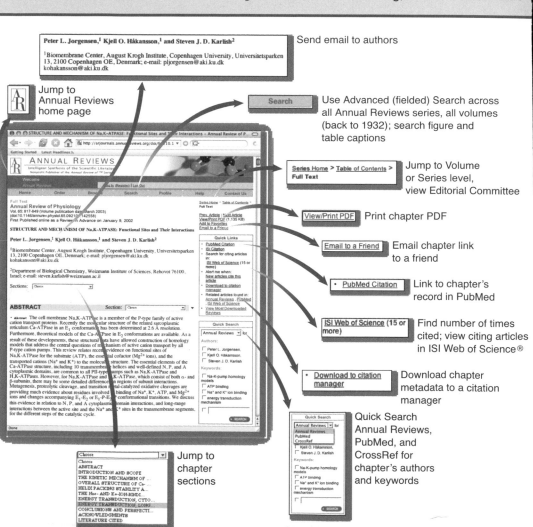